D0948134

WORLDMARK
CHRONOLOGY
of the Nations

WORLDMARK
CHRONOLOGY
of the Nations

Volume 4 – Europe

Timothy L. Gall and Susan B. Gall, Editors

GALE GROUP

Detroit
San Francisco
London
Boston
Woodbridge, CT

The Gale Group

Editorial
Shelly Dickey, Project Editor
Matthew May, Assistant Editor
With contributions from the Cultures and Customs Team
Rita Runchock, Managing Editor
Production
Mary Beth Trimper, Production Director
Evi Seoud, Production Manager
Wendy Blurton, Senior Buyer
Product Design
Cynthia Baldwin, Production Design Manager
Michelle DiMercurio, Senior Art Director
Graphic Services
Barbara J. Yarrow, Graphic Services Director
Randy Bassett, Image Database Supervisor
Pamela R. Reed, Imaging Coordinator
Permissions
Maria Franklin, Permissions Manager
Margaret Chamberlain, Permissions Specialist

Library of Congress Cataloging-in-Publication Data

Worldmark chronology of the nations / Timothy L. Gall and Susan Bevan Gall, editors.
    p. cm.
Includes bibliographical references and index.
v. 1. Africa -- v. 2. Americas -- v. 3. Asia -- v. 4. Europe.
ISBN 0-7876-0521-2 (set) -- ISBN 0-7876-0522-0 (v. 1) -- ISBN 0-7876-0523-9 (v. 2)
-- ISBN 0-7876-0524-7 (v. 3) -- ISBN 0-7876-0525-5 (v. 4)
1. Geography. I. Gall, Timothy L. II. Gall, Susan B.
G133.W93   1999
910 21--dc21

    99-044217
    CIP

While every effort has been made to ensure the reliability of the information presented in this publication, Gale does not guarantee the accuracy of the data contained herein. Gale accepts no payment for listing; and inclusion in the publication of any organization, agency, institution, publication, service, or individual does not imply endorsement of the editors or publisher. Errors brought to the attention of the publisher and verified to the satisfaction of the publisher will be corrected in future editions.

The paper used in this publication meets the minimum requirements of American National Standard for Information Sciences—Permanence Paper for Printed Library Materials, ANSI Z39,478-1984.

This publication is a creative work fully protected by all applicable copyright laws, as well as by misappropriation, trade secret, unfair competition, and other applicable laws. The authors and editors of this work have added value to the under lying factual material herein through one or more of the following: unique and original selection, coordination, expression, arrangement, and classification of the information.

All rights to this publication will be vigorously defended.
Copyright © 1999 by
The Gale Group
27500 Drake Rd.
Farmington Hills, MI 48331–3535

Contents

Contributors

Editors: Timothy L. Gall and Susan Bevan Gall
Associate Editors: Daniel M. Lucas, Eleftherios Netos
Editor, Photo Research: Michael Cikraji
Typesetting and Design: Bridgette Nadzam
Graphics: Rebecca Kimble, Hannah Lissauer
Editorial Assistants: Jill Coppola, Rebecca Kimble, Susan Stern, Jennifer Wallace

ADVISORS

JANE L. THOMAS. Librarian, McNeil High School, Round Rock Independent School District, Austin, Texas
WENDI GRANT. Teacher, Geography, McNeil High School Library, Round Rock Independent School District, Austin, Texas
FLO RANKIN. Library Media Specialist, Hoover High School, North Canton, Ohio
MARK KURT. Teacher, Lake Forest High School, Lake Forest, Illinois
KEITH PLATTE. Teacher, Social Studies, Kalamazoo Central High School, Kalamazoo, Michigan

CONTRIBUTORS

OLUFEMI A. AKINOLA, Ph.D. W.E.B. DuBois Institute, Harvard University
LESLIE ASHBAUGH, Ph.D. Department of Sociology, Seattle University
VICTORIA J. BAKER. Department of Anthropology, Eckerd College
IRAJ BASHIRI. Professor of Central Asian Studies, Department of Slavic and Central Asian Languages and Literature, University of Minnesota
THEA BECKER. Researcher/Writer, Cleveland, Ohio
HEATHER BOWEN. Researcher/Writer, Washington, D.C.
GABOR BRACHNA. Researcher/Writer, Cleveland, Ohio
SALVADOR GARCIA CASTANEDA. Department of Spanish and Portuguese, The Ohio State University
ERIK CHING. History Department, Furman University
FRANCESCA COLECCHIA. Modern Languages Department, Duquesne University
LEAH ERMARTH. Worldspace Foundation, Washington, D.C.

JENNIFER FORSTER. Researcher/Writer, Kent, Ohio
ALLEN J. FRANK, Ph.D., Researcher/Writer
DIDIER GONDOLA. History Department, Macalester College, St. Paul, Minnesota
ROBERT GROELSEMA. Ph.D. African Bureau, U.S. Agency for International Development (USAID)
HIMANEE GUPTA. Researcher/Writer, Honolulu, Hawaii
BRUCE HEILMAN. University of Dar es Salaam in Tanzania
ROSE M. KADENDE-KAISER. Director, Women's Studies Program, Mississippi State University
EZEKIEL KALIPENI. Department of Geography, University of Illinois at Urbana-Champaign
RICHARD A. LOBBAN, Jr., Ph.D. Department of Anthropology, Rhode Island College
IGNACIO LOBOS. Journalist, Honolulu, Hawaii
DERYCK O. LODRICK. Visiting Scholar, Center for South Asian Studies, University of California, Berkeley
DANIEL M. LUCAS. Researcher/Writer, Cleveland, Ohio
PATRIZIA C. MCBRIDE. Department of German, Scandinavian, and Dutch, University of Minnesota
MEGAN MENTREK. Researcher/Writer, Cleveland, Ohio
WILLIAM MILES. Department of Political Science, Northeastern University
EDITH T. MIRANTE. Project Maje, Portland, Oregon
CAROL MORTLAND. Crate's Point, The Dalles, Oregon
NYAGA MWANIKI. Department of Anthropology and Sociology, Western Carolina University
ELEFTHERIOS NETOS. Ph.D. Candidate, Kent State University
BRUCE D. ROBERTS, Department of Anthropology and Sociology, University of Southern Mississippi
GAIL ROSEWATER. Researcher/Writer, Cleveland, Ohio
JENNIFER SPENCER. Researcher/Writer, Columbus, Ohio
JEANNE-MARIE STUMPF. Researcher/Writer, Cleveland, Ohio.
CARMEN URDANETA. M.A. Researcher/Writer, Boston, Massachusetts
KIMBERLY VARGO. Researcher/Writer, Cleveland, Ohio
JEFF WASICK. Kent State University.
GERHARD H. WEISS. Department of German, Scandinavian, and Dutch, University of Minnesota
ROSALIE WIEDER. Researcher/Writer, Cleveland, Ohio
JEFFREY WILLIAMS, Ph.D. Cleveland State University

Preface

Worldmark Chronology of the Nations contains entries on 192 countries of the world. Arranged in four volumes—Africa, Americas, Asia, and Europe—*Worldmark Chronology of the Nations* follows the organizations of its sister sets, *Worldmark Encyclopedia of the Nations* and *Worldmark Encyclopedia of Cultures and Daily Life*. Within each volume, entries are arranged alphabetically.

Each volume begins with a general timeline of world history. This timeline provides a history of the world from prehistoric times to the present, lending a context to the entries on individual nations.

The nations profiled are those that exist at the end of the twentieth century. Profiles cover the territory within the modern-day geographic borders. History of earlier nation-states and empires is included in entries for each modern-day nation that has evolved within their region. Emphasis is on social and cultural history, helping the user of *Worldmark Chronology of the Nations* gain an understanding of a country's history beyond its succession of political leaders and military conflicts.

Each entry begins with an overview essay. This essay provides a general introduction to the country, its history, and its people from the beginning of time to the present. Trends in all aspects of a nation's development—politics, the military, literature, art, industry, religion, society, and relations with its neighbors and the other nations of the world—are included.

Following the introduction is a timeline of history, with dated entries describing the key people and events that shaped the nation. Notable people whose achievements helped shape the society and who contributed to world history are profiled. Entries may cover an era, a decade, a range of years, a specific year, a period of months, a specific month, or a specific date. A heading summarizes the event's significance, while the paragraphs that follow provide further details.

A current map accompanies each entry; in addition, over 100 historic maps illustrate the location of earlier empires, kingdoms, and depict transfers of territory between nation-states. Sidebars provide context for events, empires, people, and organizations of significance, such as the Cold War, Ottoman Empire, and League of Nations.

A comprehensive bibliography following the organization of the work (by continent and country), glossary of terms, and comprehensive index appear at the back of each volume.

Over fifty university professors, professional writers, journalists, and expert reviewers contributed to the preparation of this *Worldmark Chronology of the Nations*. Many have carried out historical research in the country about which they wrote. All are skilled researchers with expertise in their chosen area of study.

Acknowledgments

The editors express appreciation to staff of The Gale Group who were involved at various stages of the project: Linda Irvin and Kelle Sisung, who assisted with development of the initial concept of the work; Shelly Dickey, Project Editor, Matthew May, Assistant Editor, who guided the editorial development of the entries with contributions from the Cultures and Customs Team, Rita Runchock, Managing Editor; Cynthia Baldwin and Michelle DiMercurio, who were responsible for the design of the volumes' covers; and Mary Beth Trimper, Evi Seoud, and Wendy Blurton, who supervised the printing and binding process. In addition, the editors express appreciation to William Becker, archivist of the Cleveland State University Photo Archive, for his assistance with photo research. Also helping with images were members of the Gale Group Graphic Services department, Barbara J. Yarrow, Randy Bassett, and Pamela R. Reed; and Permissions department, Maria Franklin and Margaret Chamberlain.

SUGGESTIONS ARE WELCOME: The first edition of a work of this size and scope is an ambitious undertaking. We look forward to receiving suggestions from users on ways to enhance and improve future editions. Please send comments to:

Editors
Worldmark Chronology of the Nations
The Gale Group
27500 Drake Road
Farmington Hills, MI 48331–3535
(248) 699-4253

Timeline of World History

c. 2,600,000 B.C. *Homo Australopithecus* **walks the earth**

Homo Australopithecus, the earliest ancestor of *Homo Sapiens Sapiens* (present-day human beings) lives in sub-Saharan Africa. *Homo australopithecans* are hunter-gatherer nomads who fashion stone tools.

c. 100,000 B.C. **Emergence of *Homo Sapiens Sapiens* in Africa**

The earliest ancestors of the modern human species, *Homo Sapiens Sapiens,* emerges in Africa. They have larger brains than their predecessors and are hunter-gatherers who communicate by means of sophisticated language. By 30,000 B.C. they are the only human-like hominid (two-legged primate) left.

c. 8000 B.C. **Agricultural revolution**

Humans begin raising crops and domesticating animals. This results in the greatest change in human lifestyle to this point; humans start settling in fertile river valleys. This change in life patterns leads to the development of the first civilizations.

c. 3200 B.C. **Sumerians establish first civilization**

The first civilization emerges in Sumer, in Mesopotamia, along the Euphrates River. The establishment of civilization brings with it government, religion, urbanization, and a specialized economy (including trade between cities). Significantly, the Sumerians establish a pictographic system of writing called Cuneiform. A writing system allows for record keeping as well as communication, particularly for trading purposes.

c. 2850 B.C. **Old Kingdom established in Egypt**

During the period of the Old Kingdom, the lands along the Nile River in northern Africa are the second region of the world to form a civilization. Arguably the most impressive achievement of the Old Kingdom are the pyramids which are built as the tombs of pharaohs (kings). The Egyptians of the Old Kingdom also establish a calendar based on the annual flooding of the Nile and develop a form of writing known as

hieroglyphics which they often write on papyrus scrolls (early paper).

2200 B.C. **Emergence of Chinese civilization**

Chinese civilization emerges along the Yellow River with the establishment of the Xia (Hsia) dynasty. This is the first step in the formation of a Chinese empire. To this day, Chinese civilization is the oldest in existence.

1728–1686 B.C. **Code of Hammurabi**

King Hammurabi of Babylonia writes a law code for his subjects. The laws, which seek to maintain stability in the kingdom, pertain to topics as diverse as marriage, taxes, and business contracts. The Code of Hammurabi is the oldest law code in the world.

c. 1500–1000 B.C. **Emergence of Indian civilization, basis of Hinduism**

Indian civilization emerges along the Indus River valley. An early achievement of the Indus civilization is the compilation of the first *Vedas*, spiritual texts which form the basis of Hinduism.

c. 1200 B.C. **Emergence of Hebrew civilization, Judaism**

The emergence of the Hebrew civilization gives rise to Judaism, a monotheistic religion. Although the Hebrews subsequently succumb to internal division and foreign invasion, their religion is one of the most influential in the world. Both Christianity and Islam are offshoots of Judaism.

c. 1000 B.C. **Emergence of the Kingdom of Kush**

The Kingdom of Kush emerges in present-day Sudan. Centered around the city of Meroe, the Kingdom of Kush is the first sub-Saharan African civilization.

c. 1000 B.C. **Emergence of first American civilizations in Mexico and Peru**

Civilization emerges in the Americas in the areas of present-day Mexico and Peru.

8th century B.C.   Emergence of Classical Greek civilization, *Iliad* and the *Odyssey* compiled

Classical Greek civilization arises out of a dark age. The Greeks organize themselves into city-states both on the Greek mainland as well as on hundreds of the islands that make up the Aegean and Ionian seas. Great seafarers, the Greeks establish colonies and trade throughout the Mediterranean and Black Seas.

Greek civilization is also well-known for its many cultural achievements. Homer's compilation of the *Iliad* and the *Odyssey* are among the oldest works of Western literature. Greek civilization reaches its height during the Athenian Golden Age (see 5th century B.C.).

563–483 B.C.   Life of Buddha

Siddharta Gautama, an Indian prince, seeks eternal happiness as a way of escaping suffering in the world. After a personal odyssey, he adopts the name Buddha (enlightened one) and forms a belief system that becomes known as Buddhism. In subsequent centuries the Buddhist faith spreads throughout much of Asia including Japan, Korea, and China (where it becomes the dominant form of religious belief).

551–479 B.C.   Life of Confucius

Chinese philosopher Confucius advocates human behavior based upon morality and justice. His writings, compiled in the *Analects,* are part of a program of political reform.

509 B.C.   Roman Republic founded

The Roman monarchy is overthrown and replaced by a republican system of government. Under this system, citizens with voting rights (landholding males) elect representatives to legislate for them. The foundation of the republic leads to stronger and more effective government which eventually unites all of the Italian peninsula. The republican system of government breaks down when Rome subsequently expands to encompass all of the Mediterranean world as well as much of southern and western Europe. The Roman Republic becomes the Roman Empire.

480–431 B.C.   Athenian Golden Age

During the Athenian Golden Age, Athens becomes the center of Greek politics and culture. The city's democratic government allows all of its citizens (roughly a third of its population) to participate in government at some point in their lives. Art, philosophy, and theater flourish. The Parthenon is erected, Socrates (469–399) debates pupils and populace, and Euripides (485?–406?) writes plays. The Golden Age ends with the outbreak of the Peloponnesian War (431–404) against Sparta, which ends in an Athenian defeat.

c. 4th century B.C.   Emergence of Mayan civilization

Mayan civilization emerges in the area of present-day Mexico. The Maya are the first civilization to emerge in the Americas and are best-known for their sophisticated mathematics which they apply to the study of astronomy and which enables them to devise a highly accurate calendar.

356–323 B.C.   Life of Alexander the Great

In only thirteen years King Alexander III (the Great) of Macedon creates an empire that extends from Greece to the Indus River. His conquests spread Greek civilization throughout the Middle East and into Asia.

4th–2nd centuries B.C.   Hellenistic Era

In the wake of Alexander's death, his empire fragments as his successors quarrel for the spoils. Nevertheless, a new cosmopolitan culture emerges that combines Greek and Eastern influences.

3rd century B.C.   Erection of the Great Wall of China

In an effort to keep foreign invaders at bay, the Chinese emperor constructs a wall stretching across the northern frontier of China. Known as the Great Wall, this fortification stretches for over a thousand miles (1,620 kilometers) and becomes one of the Seven Wonders of the World.

A.D. 1st century   Axumite Kingdom established

The Axium Kingdom emerges in Ethiopia. This state soon adopts Christianity and becomes a Christian stronghold in sub-Saharan Africa as it has remained to the present.

5 B.C.–A.D. 29   Life of Jesus, Rise of Christianity

In Roman Judea, Jesus of Nazareth claims to be the Jewish messiah (savior). Although he gains a following among those who embrace his message, the Roman authorities view him as an obstacle to their rule and he is sentenced to death. After his crucifixion, his followers claim that he is resurrected. His life is the basis of a new religion, Christianity, that comes to dominate the Roman Empire by the fourth century. It subsequently spreads throughout Europe into Asia, Africa, and, ultimately, the Americas and Australia. It is presently the largest faith in the world with over one billion followers.

5th century   Middle Ages, rise of feudalism

The Middle Ages follow the fall of the Roman Empire in Western Europe. This period is characterized by the establishment of small kingdoms based upon the economic system of feudalism. Under this regime, peasant serfs work for lords who provide protection from invaders. In turn, the lords provide services for a king in return for land.

440–61   Reign of Pope Leo the Great

Under Pope Leo the Great, Rome increases its power over the other four patriarchates (the five original seats of church government: Rome, Constantinople, Antioch, Jerusalem, Alexandria) of Christianity. Subsequently, the pope becomes the leader of the Roman Catholic branch of Christianity.

528   Justinian's Code

The Byzantine (Eastern Roman) emperor Justinian, (483–565) establishes a code of laws for his empire called the *Corpus Juris Civilis*, which forms the foundation of much of European law today.

7th–8th centuries   Rise and spread of Islam

Under the direction of Mohammed (570–632), the Islamic religion emerges among the Arabs. An offshoot of Judaism and Christianity, Mohammed claims that he is the last prophet of God. After Mohammed's death, the Muslims (followers of Islam) create an Islamic empire that stretches from the North African coast and Spain as far east as Persia. Islam later spreads throughout much of sub-Saharan Africa, and central and southeast Asia. It is presently the second-largest religion in the world.

8th century   Emergence of Ghana

The African kingdom of Ghana emerges in the northwest part of Africa and is the first sub-Saharan African civilization. The kingdom establishes trade across the Sahara with Morocco. Although Ghana is an advanced civilization it falls victim to invasion and disintegrates in the eleventh century.

8th–9th centuries   Carolingian Renaissance

The rise of the Carolingian Empire in western Europe, results in a renewed stress on learning. Newly-established monasteries become centers of education and preservation where monks study Greek and Roman classics (as well as the scriptures) while also copying and translating the old texts.

10th century   Rise of Russia

The Varangians establish the first Russian state centered in Novgorod. This later becomes the foundation of the Russian Empire that stretches across Siberia.

1054   East-West split in Christianity

Christianity splits into two branches: Roman Catholicism, centered in Rome; and Eastern Orthodoxy, centered in Constantinople, the capital of the Byzantine Empire. The split is of vital importance over time as the two halves of Europe undergo significantly different historical development.

11th–13th centuries   Crusades foster greater East-West contact

In an effort to oust the Muslims from the Holy Land, Christian warriors from western Europe wage a series of holy wars (crusades) against the Muslim Turks (who control the Middle East). All but the first crusade are military defeats for the Europeans. (Indeed, the Fourth Crusade, rather than attacking the Holy Land, captures Constantinople in 1204.) Nonetheless, they help bring western Europe out of its isolation during the Middle Ages.

1192   Feudal Japan emerges

Feudalism emerges in Japan. Under this system, the emperor becomes a figurehead and political power rests in the hands of the *Shogun*, the chief warlord. Japan remains under this system until the Meiji Restoration in 1868.

c. 1200   Height of Inca civilization

The Inca establish an empire in the Quechua mountains along the Pacific coast in South America. The Inca government is a theocracy in which the emperor holds absolute authority. Among the Inca achievements are the creation of thousands of suspension bridges across rivers and numerous religious sculptures.

12th century   Rise of the Mongol Empire

Under the leadership of Genghis Khan (1162–1227), the Mongols establish an empire that stretches from China, through central Asia, into Europe.

c. 1300   Aztec civilization

The Aztec civilization in present-day Mexico reaches its height in the fourteenth century. The Aztecs have an advanced civilization that includes sophisticated cities and government. Along with the great pyramids and temples they build, they are known for their bloody ritual human sacrifices meant to please their gods.

1347–50   Black Death sweeps through Europe

Brought in from Asia, Bubonic Plague (known as the Black Death) spreads across Europe and kills approximately half its population.

14th–16th centuries   Era of the Renaissance

A period of intellectual and cultural rebirth, the Renaissance, begins in northern Italy and later takes root in northern Europe, particularly Holland. The central characteristic of the Italian Renaissance is the emphasis on classical (ancient Greek and Roman) civilization. Renaissance artists include Leonardo da Vinci (1452–1519), Michaelangelo (1475–1564), and Raphael (1483–1520). Much of their work graces churches and cathedrals in Rome, most notably St. Peter's.

1445   Gutenberg invents printing press

Johann Gutenberg (1400–67) invents the printing press which allows for the widespread publication of books. Mass production of books leads to widespread access to ideas and is instrumental in bringing people closer together.

1453   Constantinople falls to the Ottoman Turks

Constantinople, the capital of the Byzantine Empire, falls to the Ottoman Turks. The Ottomans, a Muslim people, establish an empire that, at its height, includes all of Asia Minor, all of the Balkans and central Europe as far as Vienna, the North African coast, the Fertile Crescent, and the Caucasus. The empire lasts until 1923.

Late 15th century–19th century   African slave trade

The African slave trade begins with the Portuguese. Over the course of four centuries approximately twenty-two million black Africans are uprooted forcibly. The half who survive the voyage are sent primarily to the Americas where they primarily work in agriculture.

1492   Columbus makes first voyage to the Americas

While searching for a shorter route to Asia by sailing west, Christopher Columbus (an Italian in the service of the Spanish) instead reaches the West Indies and the South American coast. However, not recognizing his mistake, he embarks on three return journeys over the next decade. His voyages promote further European exploration that results in the eventual colonization of North and South America.

Early 16th century   Protestant Reformation

A revolt against certain practices of the Catholic Church led by German monk Martin Luther (1483–1546) results in a full-scale revolt against the Papacy in western Europe. Religious wars ensue between the Catholics and Protestants (those who split from Rome) and leave much of western Europe politically and economically devastated.

1500–1700   European exploration and colonization of the Americas

European explorers make a series of voyages to the Americas. Although their early journeys focus on discovery, they soon turn to expansion. Spain and Portugal conquer most of the Americas (although England, France, and the Netherlands stake claims as well) and establish vast empires in the "New World." The Europeans soon begin exploiting their new territories for mineral and agricultural wealth (often through slave labor). Most of the native population dies of disease or warfare while the Europeans begin establishing their territories.

1519–21   Magellan circumnavigates the globe

An expedition led by Portuguese explorer Ferdinand Magellan (1480–1521) sets sail to circumnavigate the globe. Although Magellan and most of his crew die, the survivors return to Portugal in 1521 and prove that the world is round.

1519–25   Spanish conquest of the Aztecs

The Spanish, under Hernan Cortez (1485–1547) conquer the Aztec Empire in Mexico. The rapid conquest of the Aztecs illustrates the advantages held by Europeans over non-Europeans in weapons technology.

1564–1616   Life of William Shakespeare

English playwright and poet William Shakespeare creates his artistic and literary masterpieces. Not only do his works become widely-acclaimed as the best writings in the English language, they receive international recognition, are translated into dozens of languages, and his plays are performed throughout the world.

1581   Beginning of Russian expansion into Siberia

Russian traders begin expanding their activities eastward into Siberia, incorporating the land into the Russian Empire as they travel. This begins one of the greatest expansions in history as the Russian Empire eventually grows to encompass one-sixth of the entire land area of the world.

17th–18th centuries   Age of Reason, Enlightenment

The principles of liberal political thought emerge during the Enlightenment. Writers such as John Locke (1632–1704), Voltaire (1694–1778), and Jean-Jacques Rousseau (1712–78) stress the importance of individual liberty based on private property, religious toleration, and the social contract between a government and its people. These ideas become the key ideology behind the American and French Revolutions and remain an important part of Western political thought through the end of the twentieth century.

1618–48   Thirty Years War

When Catholic Habsburg prince Ferdinand II becomes king of Protestant Bohemia, Protestant and Catholic states in Germany, and eventually throughout central Europe, fight one another for political and religious dominance.

c. 1750   Industrial Revolution

Industrialization begins to develop in England. Characterized by steam-powered manufacturing and mass production, the industrial revolution sets the stage for the modern age. The development of industry soon spreads to western Europe and North America and propels them to the forefront of world economic, political, and military power. With the West dominant throughout the globe, the rest of the world attempts to industrialize as well.

1770   Cook charts the east coast of Australia

English Captain James Cook (1728–79) charts the east coast of Australia while on a South Sea expedition. His discovery paves the way for the eventual European colonization of the island-continent.

1776   Adam Smith writes *The Wealth of Nations*

British economist Adam Smith writes *The Wealth of Nations*. This work stresses that the key to expanding a nation's wealth lies in the establishment of a capitalist economic system characterized by free enterprise, private property, and an absence of government interference in the economy.

1776–83   American War of Independence

In the name of Enlightenment principles of political freedom, Great Britain's thirteen American colonies declare their independence and, by 1783, emerge victorious as the United States of America. The new country comes to dominate the North American continent and by the twentieth century becomes the most powerful nation in the world.

1789   French Revolution

With its goals of *Liberté*, *Egalité*, and *Fraternité* (Liberty, Equality, and Fraternity), revolution breaks out in France. By 1792 it overthrows the monarchy and institutes a republic. Although the revolution ultimately degenerates into the imperial dictatorship of Napoleon Bonaparte (1769–1821) it serves as a model for subsequent liberal nationalist revolutions throughout the world.

Early 1800s   Latin American revolts against Spain

Spain's colonies in Latin America rise in revolt against Spain. By the mid-1820s, the once vast Spanish Empire in South America is reduced to minor island possessions.

1839–42   Opium War

The Chinese forbid the importation of opium into their ports by the the British, who ship it from India. The British wage war on China forcing them to reopen their ports. The Chinese defeat is a severe setback to the forces that oppose Western imperialism.

1848   Revolutions in Europe

Liberal nationalist revolutions sweep across the European continent. In the short-term, the revolutions fail, although their long-term impact is profound as they form the basis of future liberal nationalist political programs that play an important role in European politics in the late nineteenth century.

1848   Karl Marx publishes the *Communist Manifesto*

In the wake of revolutions that sweep across Europe, Karl Marx publishes his *Communist Manifesto*, which advocates the overthrow of capitalism in favor of socialist societies in which all property is communal and workers rule. Although his work has no immediate impact, his ideas form the basis of future Communist revolutions in the twentieth century.

1868   Meiji Restoration

In an effort to combat Western imperialism, Japanese patriots stage a revolution that "restores" the emperor's authority. As a result of this successful revolt, Japan embarks upon a policy of rapid modernization that makes it a leading power by the first decade of the twentieth century. Its defeat of Russia in the Russo-Japanese War (1904–05) is the first time that a non-European power defeats a European power in modern times.

1885   Karl Benz designs first automobile

The German Karl Benz designs the first automobile. Within decades, his invention becomes the primary means of ground transportation in the industrialized world.

1903   Wright brothers make first powered flight

The American Wright brothers make the first powered flight in a light airplane in Kitty Hawk, North Carolina. By the late twentieth century, air travel becomes one of the safest and primary means of long-distance transportation.

1911–49   Chinese Revolution

In 1911, the Manchu dynasty falls and is replaced by a republic led by Sun Zhongshan (Sun Yat-Sen, 1866–1925). The revolution is aimed at modernizing China and throwing off Western imperialism. Sun is succeeded by Jiang Jieshi (Chiang Kai-Shek, 1887–1975). By the late 1920s, however, a Communist movement led by Mao Zedong (Mao Tse-Tung, 1893–1976) challenges Jiang for control and a bloody civil war ensues. By 1949 the Communists are victorious and drive Jiang's forces off the mainland to the island of Formosa (Taiwan).

1914–18   World War I

World War I pits two great alliances against each other: the Triple Alliance (Germany, Austria-Hungary, the Ottoman Empire, and, later, Bulgaria) versus the Triple Entente (Great Britain, France, Russia, later joined by the United States) and their associates. The war results in an Entente victory but only after most of Europe is destroyed. The war kills around nine million people. Political instability follows the war and leads to World War II in 1939.

1917 Russian Revolution

The monarchical tsarist regime falls in Russia. After a brief period in which a provisional government takes over, Russia is governed by the Bolsheviks, who establish the first Communist government in the world. The Communists seek to create a socialist state run by the working class. In practice, however, their one-party regime relies upon force and widespread suppression of human rights to remain in power.

1939–45 World War II

World War II pits the expansionist Axis (Germany, Italy, and Japan) and its associates against the Allies (Great Britain, France, China, the United States, and the Soviet Union) and its partners. By 1945 the Allies are victorious. Their victory, however, comes at great cost as most of Europe and Asia are destroyed. The war kills around sixty million people, including twenty-seven million Soviet citizens and six million European Jews. Jews are singled out for extermination by Nazi Germany in a policy known as the Holocaust.

1947–91 Cold War

The wartime alliance between the United States and the Soviet Union (known as the superpowers) breaks down into a relationship of mutual distrust and hostility. The Cold War is the commonly used name for the prolonged rivalry and tension between the United States and the Soviet Union which lasts from the end of World War II to the break-up of the U.S.S.R. in 1991. The Cold War encompasses the predominantly democratic and capitalist nations of the West, which are allied with the U.S., and the Soviet-dominated nations of eastern Europe, where Communist regimes are imposed by the U.S.S.R. in the late 1940s. Although the United States and the Soviet Union never go to war with each other, the Cold War results in conflicts elsewhere in the world, such as Korea, Vietnam, and Afghanistan, where the superpowers fight wars either by proxy or with their own troops against allies of their superpower rival.

1950s–60s Decolonization

Throughout the world European colonial empires collapse and previously subjugated peoples receive their independence. Freedom from colonial rule presents many challenges to the newly-independent states. Economic underdevelopment, widespread poverty, and tenuous political stability all plague these new countries.

1969 Man walks on the moon

On July 20, 1969, U.S. astronauts Neil Armstrong (b. 1930) and Edwin "Buzz" Aldrin (b. 1930) become the first men to walk on the moon. This marks the first time human beings have ever set foot upon another celestial body. It is only sixty-six years since the Wright brothers' first flight and only eight years since Soviet cosmonaut Yuri Gagarin made the first manned space flight.

1980–Present Information revolution

The increasing importance of computers in daily life—particularly for the transmission of information results in the so-called "Information Revolution."

1989 Collapse of Communism in Eastern Europe

As part of the liberalizing policies of Soviet leader Mikhail Gorbachev (b. 1930), the U.S.S.R. allows its East European satellites to go their own way. As a result, Communist regimes throughout Eastern Europe collapse.

1991 Collapse of the Soviet Union

Gorbachev's liberalization policies lead to a rise in nationalism among the peoples of the Soviet Union. When his reforms prove unable to rescue Communism, the Communist Soviet Union disintegrates into its constituent republics. Russia, the chief republic of the old U.S.S.R. is a mere shadow of its predecessor and is plagued by political and economic instability for the rest of the decade.

1992 European Union formed

The nations of the European Community (Belgium, Denmark, France, Germany, Greece, Ireland, Italy, Luxembourg, Netherlands, Portugal, Spain, and the United Kingdom) form the European Union. More than an economic union (as its predecessor was), the European Union seeks to establish a common currency, defense and foreign policy.

1997: Summer Asian economic crisis

Beginning in Thailand, a financial crisis strikes East Asia. Caused by over-investment and currency devaluation, the crisis brings economic growth in East Asia to a halt and threatens global prosperity. Prior to this economic reverse East Asian economies are the fastest-growing in the world and comprise one-third of the world economy.

Albania

Introduction

Albania is a tiny country on the Balkan Peninsula with an area of about 11,000 square miles. Bounded on the southwest by the Adriatic Sea, its neighbor to the south is Greece, and to the north is the former Yugoslavia. Macedonia is its neighbor to the east. Albania is also known by the ancient Albanian name *Republika e Shqiperise*, which means "Land of the Eagle." The people known today as Albanians can trace their heritage to a civilization of people dating from about 2000 B.C. who referred to themselves as Illyrians. This population was a Bronze Age (3000 B.C.–1000 B.C.) culture that developed in the mountainous regions of the Balkan Peninsula. These descendants of the Indo-European or Caucasian community split into two distinct populations, one to the north and one to the south. The northern people were known as Ghegs, and the southern people were called Tosks.

Albania has a Mediterranean climate, dry, hot summers and mild, wet winters. The terrain is mainly mountainous. Many forests have been lost due to clearing and overgrazing. Three rivers separate the north, central, and southern regions: the Shkumb to the north, the Viossa in the central region, and the Drin in the south. Natural mineral resources include iron, coal, gold, silver, and copper. Agriculture consists mostly of grains cultivation; in fact, Albania is the main supplier of corn for the region. Tobacco, wine, and olive oil are also exported.

In 1998, the population of Albania was estimated at 3,293,252 people, with most Albanians living in the capital city, Tirane. Other heavily populated cities are Elbasin and Durres. Ethnic Albanians comprise 95 percent of the population and Greeks 3 percent. Most of the Albanians are Muslim (70 percent), while 20 percent are Albanian Orthodox, and another 10 percent are Roman Catholic. As of the late 1990s the government of Albania is a republic with a history of serious turmoil dating back two thousand years.

History

The earliest evidence of Albanian culture dates from about 2000 B.C. A Bronze Age culture, called Illyria, occupied the area then. Tribes occupied the western part of the Balkans, Slovenia, and Greece. (In the 1990s, many of these countries changed names and borders, but the kingdom of the Illyrians included Dalmatia, Croatia, Bosnia, Herzegovina, and Montenegro. The capital was Shkodra (Scutari) in present-day northern Albania.) The pagan Illyrians believed in an afterlife and buried their dead with cultural artifacts. Their distinct language separates them from later-day Greek and Roman cultures. Albanian influence on the names of Greek gods indicates the language's ancient origins. Modern Albanian has only had a written alphabet since 1927, a combination of the Gheg and Tosk dialects. Although the Illyrians had warlike and peaceful contacts with the Greeks, theirs was a distinct culture. The earliest known king, Hyllus (the Star), died in 1225 B.C. In the fourth century B.C. various tribes were united under the rule of Bardhylus (White Star), whose reign included the kingdoms of Illyria, Epirus, and much of Macedonia. His rule fell apart under the attacks of Philip of Macedon (382–336 B.C.), father of Alexander the Great.

The Illyrians had continuing battles with the Greeks, yet in 312 B.C., King Glaucius of Illyria succeeded in expelling the Greeks. In the year 232 B.C., Queen Teuta occupied the throne of Illyria. She was a successful ruler and is often referred to as the Catherine the Great of Illyria.

After the Greek occupation, the Romans pushed into Illyria. They forced Queen Teuta to accept their peace terms. The last Illyrian king, Gentius, was defeated by the Romans in 165 B.C., and Illyria became a Roman dependency. During the Roman occupation, the Illyrians resisted assimilation, but their language acquired many Latin words. The Illyrians had great military skills which influenced the Roman army.

In the fourth century A.D. the Roman Empire was divided into east and west. Illyria became part of the eastern or Byzantine Empire but remained under the jurisdiction religiously of the pope in Rome. One Byzantine emperor, Justinian I (491–565) was Illyrian. Also in the fourth century, the barbarian invasion of Europe by the Goths, Huns, Visigoths, and Ostogoths extended to the Balkans. The tribes of northern Illyria were assimilated by these invaders, but the southern provinces (including modern-day Albania) resisted, thus preserving local language and customs.

In 1054 the Christian Church was divided into two groups: one that supported the pope in Rome and the other

ALBANIA

0   25   50   75 Miles

0   25   50   75 Kilometers

During this time, roughly from the ninth to the fourteenth century, Albania suffered invasions by the Normans (England), the Angevins (Italy), and the Serbs. This last invasion by the Serbs led many Albanians to seek refuge in other countries such as Greece and Turkey. In the fourteenth century the Ottoman Turks invaded Albania and succeeded in controlling it for about four hundred years with one exception—the twenty-five-year rebellion led by Skanderbeg. Born Gjergj Kastrioto in 1405, Skandebeg was sent to pay homage to the Ottoman sultan. There he studied at military school and became one of the best officers. He recaptured his father's land in Kruja and maintained the emblem of the black double-headed eagle on a red background which is present-day Albania's flag. He managed to keep the Ottoman Empire at bay and was a fierce defender of the Roman Catholic Church. Skanderbeg, who died in 1468, became a national hero. The Albanians continued their resistance, but no one else was able to contain the Ottomans. The Ottoman Empire was so repressive that Albania had no connection with Renaissance Europe. Under the Turks, the Albanians were forced to convert to Islam and required to pay tribute to the Byzantine capital, Constantinople. More than two-thirds of the Albanians abandoned the Roman Catholic Church in order to escape torture and death.

In 1878, influential Albanians met in Prizren to form the Prizren League for establishing Albanian independence and to re-unify the Albanian people through the use of a written Albanian language. The awareness of their heritage helped Albanians finally declare their independence in November, 1912.

During World War I (1914–18), the Serbs and Greeks battled to control Albania. Although the Albanians were outnumbered, they refused to admit defeat. A conference in Paris in 1920 planned to divide parts of Albania among Yugoslavia, Italy and Greece. The Albanians formed the Lushnje National Assembly and vowed to resist at all costs. The Assembly formed a parliamentary government with the capital to be Tirane. Woodrow Wilson, then president of the United States (1913–21), supported Albania by recognizing its official representative in Washington and by admitting it to the League of Nations (predecessor of the United Nations). However, Albania's borders remained in doubt. During this time, Albania continued to fight the occupation of Italy and Yugoslavia who finally surrendered claims to Albania after the League of Nations sent in troops to protect the newly formed nation.

Even though Albania was now a new nation, it continued to be torn apart by its feudal past. Tribal landowners were unwilling to submit to the new government made up of liberals and intellectuals who wished to modernize Albania and liberate it from its Ottoman past. The parties were led by Ahmet Bey Zogu, also known later as King Zog I (1895–1961), who supported the feudal state, and Fan S. Noli, who wanted a western form of government. The power shifted

that supported the Byzantine emperor Leo III. With the split of the Church, Albania was also divided with the northern region forced to pledge allegiance to Rome while the southern half was forced to support the Byzantine or Eastern Church. With this internal division, Albania entered the Middle Ages, developing a feudal society in which peasants were forced to serve mighty warlords. The leading Albanian families of this feudal society were the Thopias, Balshas, Muzakas, Dukagjinis, and the Kastriotis. The Kastriotis gave rise to Albania's greatest leader, Skanderbeg (1403–68).

between the two men with Zogu finally succeeding in overthrowing Noli in December, 1924.

Between World War I and World War II, Albania was wooed by emerging communist leaders. However, communism, which appealed mainly to middle-class workers, was rejected in Albania especially by northern Tosk intellectuals. The southern peasants were mired in a medieval feudal state. Then, during World War II (1939–45), Albania was invaded by Italy. King Zog was forced to abdicate, and he fled the country. As fragmented as the political and social life of Albania was during this period, the people managed one of the most humane acts of kindness and decency to the Jews and Gypsies of eastern Europe. While these peoples were being systematically murdered by the Nazis under Adolf Hitler, Albania rescued and hid many refugees. They also refused to surrender any persecuted people to Nazi Germany. Near the end of World War II, two communist leaders, Enver Hoxha (1908–85) and Mehmet Shehu, emerged to take control of Albania, ousting the Italians and getting help from Yugoslavian communist leader, Josip Tito. Yugoslavia recognized Albanian independence in April, 1944.

Following the war, under Hoxha's regime, Albania severed ties with Yugoslavia and also with previous western allies, such as the United States and Britain. Hoxha systematically murdered or imprisoned opponents of his regime while maintaining Albania's isolation. His regime was primarily governed by the Sigurimi, or secret police. Citizens who complained were systematically tortured, sent to labor camps, or even executed. The United States tried to intervene by granting post-war economic aid, known as the Marshall Plan, but Albania refused. The communist doctrine of peasant equality was clearly at odds with old tribal alliances favored by the Albanians who still prized privileges previously enjoyed by old tribal chifs protected by the laws of heredity. Women benefited some by having their status upgraded to that of men. However, they have continued to remain underrepresented in the government and in skilled professions. Under the communists, Albania then became economically dependent on the Soviet regime headed by Joseph Stalin (1879–1953). Hoxha, following the party line of atheism, closed churches and mosques throughout Albania. Pro-Yugoslavian communists were purged, and many other atrocities committed.

Following the communist takeover, Albania remained even more isolated from the rest of the world. Finally, Hoxha, disillusioned with the Soviet (Russian) communists, became an ally of China and supported Mao Tse-tung (1873–1976) the communist head of China. This support was short-lived though, and after Mao's death, Hoxha became angry at the new regime which seemed to be forming diplomatic ties with the United States.

In 1976, Hoxha declared Albania to be a people's socialist republic. He tapped Ramiz Alia to be his successor, rebuking his ally, Mehmet Shehu. Shortly after, Shehu was found dead, said to be a suicide, but most likely he was murdered by Hoxha's supporters. After Hoxha's death, Alia attempted to end Albania's isolation. His regime signed a peace accord with Greece, and Albania made overtures to re-enter the free world. It announced intentions to re-establish relations with the United States and Great Britain. The People's Assembly established laws allowing the freedom of religion, and they opened some mosques and churches. They also allowed freedom to travel abroad, which had previously been denied.

At the close of the twentieth century, Albania is still in turmoil. The government, despite favoring popular elections, has yet to show signs of much stability. In April, 1992, Sali Berisha became the first democratically elected president. However, there have been many threats to his rule. There are many local rebels and armed gangs that control rural areas and are opposed to Berisha's regime. Berisha himself is accused of having his own gangs who threaten the people. Without order, the rest of the world will not invest in Albania's economy.

The second problem facing Albania is connected to the disputed territory of Kosovo located in previously-controlled Yugoslavia. Almost ninety percent of the Kosovo population is ethnic Albanian and wants to be reunited with Albania. The Serbs, who control a small minority of the population, refuse to relinquish the territory to Albania. There have been mass murders of ethnic Albanians and other atrocities. The United Nations and other governments have tried to end the violence, but as a result of the dissolution of the Soviet Union, the region known as Yugoslavia is now comprised of only Serbia and Montenegro, while Bosnia-Herzegovina, Croatia, Macedonia, and Slovenia are now independent countries. While Kosovo is trying to assert its independence, the Serbian government is using force to stop the separation.

Several famous Albanians have influenced the world in terms of culture and philosophy. Perhaps the most well-known is Mother Teresa who was born Gonxhe Agnes Bojaxhiu in 1910 in Skopje in present-day Macedonia. She was known for her dedication to the poor of India and took the name of Teresa after entering the Catholic order of the Sisters of Loreto. Winner of the Noble Peace Prize, she died in 1997 after a lifetime of dedication to the poor. Second to Mother Teresa is Ismail Kadare, a famous Albanian author who was born in southern Albania and studied at the University of Tirane. He is especially well-known in France, and in 1991, his novel *The Concert* received a French literary prize. Third, the actor, Aleksander Moisiu, appeared in many stage and film productions from about 1910 to 1935.

Timeline

2000 B.C.   Evidence of Illyrian culture in Albania

Artifacts indicate a Bronze Age (3500 B.C.) civilization exists in present-day Albania. Evidence shows a culture which developed independently of Greece, but which had cultural and trade contacts with the Greeks. Some linguists think that the Greek names for gods and goddesses come from the ancient Illyrian language. Bronze age people inhabit an area that ranges from the Danube River in the north across much of the Balkan Peninsula. Two distinct people develop: the Ghegs in the north and the Tosks in the south.

1225 B.C.   King Hyllus dies

The earliest known king of the Illyrians, Hyllus (The Star), dies.

400 B.C.   King Bardhylus (White Star)

King Bardhylus of Illyria unites the kingdoms of Illyria, Molossia (Epirus), and Macedonia. The unification of previously warring tribes creates a more forceful nation.

358 B.C.   Philip II of Macedonia defeats Illyrians

Philip, father of Alexander the Great (382–336 B.C.), defeats the Illyrians. Frequent wars force the Illyrians to form tribal federations for protection.

312 B.C.   King Glaucius expels the Greeks from Durres

The Illyrian king is victorious in forcing the Greeks from the city of Durres on the Adriatic Sea.

232 B.C.   Queen Teuta of Illyria rules

Queen Teuta, often referred to as the "Catherine the Great" of Illyria, makes significant inroads on Roman commercial enterprises. Her strong navy worries the Romans, and in 227 the Roman Senate, after years of warring against Teuta, forces her to come to peaceful terms.

165 B.C.   Illyria becomes a Roman dependency

King Gentius, the last Illyrian king, is captured and brought to Rome as a prisoner. Henceforth, Illyria is under Roman rule. The kingdom of Illyria is divided into three republics known as Shkoder, Durres, and Ulqin in Montenegro. The Illyrians have a well-developed civilization but are regarded as pagans by the Romans. Women are the equals of men, and some are even tribal leaders. Illyrians are skillful metal workers and boat builders. Romans build roads and aqueducts and connect Illyria to other Roman cities, even as far away as Constantinople.

1st Century A.D.   Christianity comes to Illyria

Under Roman rule, Illyrians are introduced to Christianity.

A.D. 2   Illyria becomes Albanoi

The first mention of Illyria as Albania (Albanoi) is documented by the geographer Ptolemy of Alexandria.

A.D. 9   Romans divide Albania

Albania is now under complete control of the Romans who divide the Albanian territory into three parts: Dalmatia, Epirus, and Macedonia.

395   Roman Empire is divided

The Roman Empire is in two parts. The eastern part is known as the Byzantine Empire, and Albania is located in this jurisdiction. Both parts are still controlled by the Church in Rome. The Illyrians resist Roman rule and are not assimilated into Roman culture. However, some Illyrians, known for their expertise in war, come to occupy important Roman posts. The emperor Justinian I (491–565) is of Illyrian background.

500   Decline of the Roman Empire

The land known as the Roman Empire is systematically invaded by barbarians: the Goths, the Visigoths, the Huns, and the Bulgars. Each successive invasion weakens the Roman Empire, and by the fifth century it is no longer a political entity. The Albanians in the south are not assimilated and keep their original language. They are the only Illyrian tribe to keep their heritage and language. Now the nation formerly known as Illyria is Albania.

732   Albania is controlled by the Byzantine Empire

Leo III (680–741), called the Isaurian, is head of the Byzantine Church. He is at odds with the Roman Catholic Church over idol worship. He controls the ancient Illyrian people in ecclesiastic matters.

1042   Split in Catholic Church

Albania is now divided by the ideologic split in the Roman Catholic Church. Northern Albania is ruled by Rome; the southern portion is ruled by Leo III and Constantinople.

1081   Feudal System is established

Historical Illyria is now referred to as the Principality of Arberi (Albania). Kruja is its capital, and the Albanians form a feudal system with serfs farming the land for their masters. Most of these feudal territories are in the isolated mountainous regions. The lords of these systems continuously fight the Byzantine Empire. These wars last into the twelfth and thirteenth centuries.

1271–1368   Sicilian Kings rule

The Sicilian kings (House of Anjou) rule what they call the "Kingdom of Albania." They declare war on the Byzantine Empire.

1331–58   Stefan Dushan, Monarch of Albania

Stefan Dushan (1308–55), a Serbian, manages to control a large part of the Balkans including Albania, Macedonia,

Greece, and Serbia. He calls himself *Imperator Romaniae Slavoniae et Albaniae*. He draws up a legal code known as Code of Dushan. He conquers Bosnia and Macedonia. As he sets out for Constantinople he dies. After his death, the kingdom again splits up.

1385 Ottoman Empire now controls Albania

The Ottoman Empire with a huge standing army fights for the religion of Islam. The Ottoman Empire (Turks) occupies Asia Minor and up until the twentieth century occupies parts of Europe, northern Africa, and southwestern Asia.

1389 Ottomans establish their authority

In the battle on the plain of Kosovo, the Albanians under Prince Lazar are overpowered by the Ottomans. They establish their system of feudalism. They divide the land into fiefdoms called *timars* that are awarded to local rulers in return for loyalty and service to the Ottoman Empire. In the beginning, little attempt is made to convert the Christians to Islam. However, the Turks do fill their armies with the Christian children taken as "taxes" from their families and indoctrinate them in Islam. The Ottomans train them to be military leaders. These conscripted soldiers are called *janissaries*.

1403 Gjergj Kastrioti is born

Later known as Skanderbeg, the Albanian national hero Gjergj Kastrioti (George Castriota, 1403–68) is born. He is taken from his family and educated in the military by the Ottomans who give him the Turkish name Iskander Bey (Iskender Bey), which means Lord Alexandre. The Kastriotis, whose symbol is the double-headed eagle, are one of four leading families in Albania. (The others are the Balshajs, whose symbol is the wolf; the Topiajs, whose symbol is the lion with a crown; and the Dukagjinis, whose symbol is the eagle.) The Kastriotis's symbol of the double-headed eagle eventually becomes part of the flag of modern-day Albania; Albania means "land of the eagle."

1419–32 Skanderbeg continues resistance to Ottoman Empire

Even though he is educated in the Ottoman tradition and converts to Islam, Skanderbeg returns to Albania to fight for independence.

1421–51 Albania under Murad II

Albania is now controlled by the Ottoman state and is ruled by the sultan Murad II (1403–51).

1444 Skanderbeg forms League of Lezha

Skanderbeg recaptures his father's land at Kruja. He establishes a league of Albanian princes (League of Lezha) and commands them to do battle against the Turks. He embraces the Albanian flag, red with a black double-headed eagle. The Ottoman Empire tries many times to oust Skanderbeg, but he prevails. He converts to Christianity and receives support from the King of Naples, the Pope, and Venice. For the next two decades, Skanderbeg is successful in fending off the Turks. Skanderbeg is well-known and respected throughout his career. Later, poets and composers honor him in verse and opera, giving tribute to him as a great leader.

1453 Fall of Constantinople

Constantinople, the capital of the Byzantine Empire, finally falls to Turkish rule, and Islam, not Eastern Orthodoxy, is the main religion of the region.

1461 Skanderbeg and Turks agree to cease-fire

Skanderbeg and the Turks agree to stop battling for control in the region; Turks control Constantinople, and Skanderbeg controls the region that is modern-day Albania.

1468 Skanderbeg dies

Skanderbeg, leader of the Albanian fight for independence, loses the support of his Christian allies but decides to lead an attack against the Turks anyway. Albanians continue resisting Ottoman domination even after his death, and the Ottomans do not gain full control of Albania until the sixteenth century.

1480 Albania falls to Ottomans

Twelve years after Skanderbeg's death, Albania finally falls to the Ottoman Empire. Many Albanians seek refuge in Italy, Greece, and Egypt. Those remaining are forced to accept Islam. During the five hundred years of Ottoman occupation, two-thirds of the Albanians accept Islam under threat of torture and death. Also Christians are forced to pay huge taxes if they still refuse to convert. Catholics are most severely persecuted. The Orthodox Christians, still under the protection of the Eastern Byzantine Church, fare a little better. Some Christians go underground and pretend to be Moslem, taking Moslem names for public use, but practicing Christianity in private.

1520–66 Reign of Suleyman (Suleiman) the Magnificent

One of the most famous sultans of the Ottoman Empire, Suleyman the Magnificent, expands his reign to include the Balkans.

17th century Under Ottoman rule

The Ottoman Empire continues to hold a firm grasp on the Balkans. The Albanians who choose to convert to Islam find opportunities under the Ottoman rule.

The Ottoman Period

The Ottoman Empire began in western Turkey (Anatolia) in the thirteenth century under the leadership of Osman (also spelled Othman) I. Born in Bithynia in 1259, Osman began the conquest of neighboring countries and began one of the most politically influential reigns that lasted into the twentieth century. The rise of the Ottoman Empire was directly connected to the rise of Islam. Many of the battles fought were for religious reasons as well as for territory.

The Ottoman Empire, soon after its inception, became a great threat to the crumbling Byzantine Empire. Constantinople, the jewel in the Byzantine crown, resisted conquest many times. Finally, under the leadership of Sultan Mehmed (1451–81), Constantinople fell and became the capital of the Ottoman Empire.

Religious and political life under the Ottoman were one and the same. The Sultan was the supreme ruler. He was also the head of Islam. The crown passed from father to son. However, the firstborn son was not automatically entitled to be the next leader. With the death of the Sultan, and often before, there was wholesale bloodshed to eliminate all rivals, including brothers and nephews. This ensured that there would be no attempts at a coup d'état. This system was revised at times to the simple imprisonment of rivals.

The main military units of the Ottomans were the Janissaries. These fighting men were taken from their families as young children. Often these were Christian children who were now educated in the ways of Islam. At times the Janissaries became too powerful and had to be put down by the ruling Sultan.

The greatest ruler of the Ottoman Empire was Suleyman the Magnificent who ruled from 1520–66. Under his reign the Ottoman Empire extended into the present day Balkan countries and as far north as Vienna, Austria. The Europeans were horrified by this expansion and declared war on the Ottomans and defeated them in the naval battle of Lepanto in 1571. Finally the Austrian Habsburg rulers were able to contain the Ottomans and expand their empire into the Balkans.

At the end of the nineteenth century, the Greeks and the Serbs had obtained virtual independence from the Ottomans, and the end of a once great empire was in sight. While Europe had undergone the Renaissance, the Enlightenment, and the Industrial Revolution, the Ottoman Empire rejected these influences as being too radical for their people. They restricted the flow of information and chose to maintain strict religious and governmental control. The end of Ottoman control in Europe came in the First Balkan War (1912–13) in which Greece, Montenegro, Serbia, and Bulgaria joined forces to defeat the Ottomans.

During the First World War, (1914–18), the Ottoman Empire allied with the Central Powers and suffered a humiliating defeat. In the Treaty of Sévres, the Ottoman Empire lost all of its territory in the Middle East, and much of its territory in Asia Minor. The disaster of the First World War signaled the end of the Ottoman Empire and the beginning of modern-day Turkey. Although the Ottoman sultan remained in Constantinople (renamed in 1930, "Istanbul"), Turkish nationalists under the leadership of Mustafa Kemal (Ataturk) (1881–1938) challenged his authority and, in 1922, repulsed Greek forces that had occupied parts of Asia Minor under the Treaty of Sévres, overthrew the sultan, declared the Ottoman Empire dissolved, and proclaimed a new Republic of Turkey. That following year, Kemal succeeded in overturning the 1919 peace settlement with the signing of a new treaty in Lausanne, Switzerland. Under the terms of this new treaty, Turkey reacquired much of the territory—particularly, in Asia Minor—that it had lost at Sévres. Kemal is better known by his adopted name "Ataturk", which means "father of the Turks". Ataturk, who ruled from 1923 to 1938, outlawed the existence of a religious state and brought Turkey a more western type of government. He changed from the Arabic alphabet to Roman letters and established new civil and penal codes.

1718: June 22 Vivaldi opera, *Scanderbeg,* premiers in Italy

Vivaldi chooses his opera, *Scanderbeg,* to be performed at the reopening of the Teatro de la Pergola in Florence, Italy. The library in Turin, Italy, has *arias* from Vivaldi's operatic version of the life of the folk hero, Skanderbeg, in its collection. A copy of the *libretto* is among the holdings of a library in Bologna, Italy.

1757–1831 Bushati family and Ali Pasha Tepelena gain control

With the decline of the Ottoman Empire, two warring families—Bushati and Ali Pasha Tepelena—continue to resist Turkish rule. However, these groups never get beyond feuding and are unsuccessful in combining their efforts to overturn the Ottoman Empire.

1785 Kara Mahmud Bushati named governor of Shkoder territory

By virtue of the family power, Bushati comes to control some of Albania.

1822 Ali Pasha Tepelena is overthrown

The Ottomans are displeased with the power and defiance of the Bushati family. They assassinate Ali Pasha. With the end of the powerful Pashas at hand, other secondary leaders continue fighting against the Ottomans.

1830: August 26 Turkish leaders massacre Albanians

In an effort to deal with Albanian resistance to Ottoman rule, Turkish leaders invite 1000 Albanian leaders for a meeting and massacre 500 of them. This effectively ends the local control that Albanian leaders hold and stymies any national movement for much of the nineteenth century.

1836 Ottomans introduce political reforms

In response to continued resistance, the Ottomans introduce the *Tanzimat* which contains political reforms. They promise to eliminate corruption and to guarantee rights to all people. They also introduce forced military service which angers Albanians.

1856 New reforms organize Albanians

The Turkish leaders open schools, but the schools do not use the Albanian language. The purpose of the schools is to promote allegiance to the Ottoman Empire and to convert people to Islam. Many Albanians choose not to attend.

1878 Formation of Prizren League

Turkey is defeated by Russia and territory is to be divided. Albania is arbitrarily divided, and Albanians form the Prizren League to indicate that they do not wish to be part of any other country. This new nationalist movement is not regarded with favor by the Turks who seek to crush it by military force. They succeed in disbanding Albanian leaders. However, this movement serves to identify Albanians as a distinct people by virtue of the fact that they all speak the Albanian language. They are divided by religion and territory, but they do possess a sense of heritage. Now national schools wish to promote learning the Albanian language, but the process is complicated because it does not yet have an alphabet.

1879 Birth of actor Aleksander Moisiu

Actor Aleksander Moisiu (1879–1935) is born in Durrës. At age twenty, Aleksander, his two sisters, and his mother settle in Vienna, Austria. There he meets Austrian actor Josef Kainz (1858–1910), who encourages Moisiu to pursue an acting career. He performs in Czechoslovakia and Germany, where he joins the acting ensemble of Max Reinhardt (1873–1943). Moisiu's best-known roles include Oedipus, Hamlet, Faust, Fedya (in Leo Tolstoy's *The Living Body),* and Dubedat in George Bernard Shaw's *The Doctor's Dilemma.* Moisiu also appears in film productions—seven silent and three "talking." Both the High College of Drama in Tirana and the Professional Theater of Durrës bear his name. (See also 1995.)

1880s Frasheri Brothers and Albanian awareness

Political and literary leaders Abdyl (1839–94), Naim (1846–1900), and Sami (1850–1903) Frasheri bring to the forefront the cause of Albanian independence. Abdyl is a skillful political leader who organizes pro-Albanian propaganda. Sami is a scholar who publishes in both Turkish and Albanian. All are well-educated and are aware of the ideals of the French Revolution (1789) which continue to influence nationalist movements throughout Europe.

Naim is a poet whose works deal with subjects ranging from the national hero, Skanderbeg, to the Albanian language and religious tolerance. His nationalistic work serves as emotional inspiration to the Albanian people. Among his best-known works is the poem, *Bagëti e Bujqësi* (Herds and Pastures), which describes in romantic language the landscape of Albania.

1912–13 First Balkan War

Albanians declare an independent state in the wake of the Ottoman defeat in the First Balkan War. However, Greek, Serbian, and Montenegrin armies try to divide the region for themselves.

1912: November 28 Albania declares independence

At a conference in Vlore, Albania declares its independence and sets up a provisional government with Ismail Kemal as its leader.

Men in traditional costume from the alpine region of Albania join the resistance forces fighting the invasion by Italy. (EPD Photos/ CSU Archives)

1913 Albania's borders defined

At a conference in London, Austria, Hungary and Russia define Albania's borders. However, the province of Kosovo is given to Serbia. Even though Albania is now a nation, many ethnic Albanians continue to live in Kosovo. The problems of this region at the close of the twentieth century result from this decision.

1914 Wilhelm von Wied appointed as Albanian prince

The powers of Europe responsible for Albania's independence now choose a ruler, Prince Wilhelm von Wied. With little knowledge of the area, he arrives in March, 1914 and proclaims Durres as the capital city. He leaves six months later, unable to govern.

1918 Albania after World War I

After the defeat of the Austro-Hungarian Empire at the close of World War I (1914–18), Albania is once again at the mercy of its neighbors. This time, as the result of a secret treaty,

Italy declares its sovereignty over Albania. U.S. President Woodrow Wilson lends his support to the idea of independence for Albania, and his position prevails.

1922 Zogu becomes leader

Ahmet Bey Zogu (1895–1961) becomes the leader of the government, which is immediately divided into two factions—those who support Zogu and his tribal chiefs (a leftover from Ottoman rule) and those who support the more liberal leader Fan Noli (1882–1965) who wants to abolish the last vestiges of feudalism and establish a true democratic government. A Western-educated bishop of the Russian Orthodox church, Noli is a poet and writer of distinction.

1922 Zogu overthrown

Within six months, Zogu's government is overthrown and replaced by Fan Noli.

1924: December Zogu overthrows Noli

Zogu assembles an army while in exile in Yugoslavia and returns to overthrow Noli's attempt at democratic rule. Zogu establishes authoritarian rule.

1928: September 1 Zogu proclaims himself King Zog I

Zog establishes himself as ruler and forges an alliance with Italy. Even though the country is governed by autocratic rule, this is a peaceful period for Albania. Zog starts an educational system but does little to improve the life of Albanian peasants who are in the majority.

1928–39 Mussolini tries to control Albania

Benito Mussolini (1883–1945), dictator of Italy during World War II, exerts his influence over Albania. With its gifts of money and soldiers Italy largely governs the Albanian army. Thus, Italy has control over Zog. Under Mussolini's influence Zog changes the political system to a monarchy which does not increase his popularity with his countrymen.

1936: January 28 Birth of Ismail Kadare

Novelist Ismail Kadare (b. 1936) is born in Gjirokastra in southern Albania. He studies history and philology (study of the written word) at the University of Tiranë. His books and plays, dealing with historic Albanian themes and events, become popular throughout Europe. Among his works are *The General of the Dead Army, The Wedding,* and *Chronicle in Stone,* the story of his childhood. His novel *The Citadel* (or *The Castle*), describes the defeat of the Turks by Scanderbeg. He is frequently nominated for the Nobel Prize for Literature.

1939: April 7 Mussolini invades Albania

Mussolini tries to extend his territory by invading Albania on Good Friday. There is resistance, but soon Zog is forced into exile. Albania remains under Italian rule through World War II.

1939–45 Albania during World War II

During World War II, Albania is occupied by the Italian army which uses it as a springboard for its invasion of Greece in October, 1940. Within weeks the invasion stalls, and Greek forces counter-attack occupying a quarter of Albania. The Italian debacle forces Germany's Adolf Hitler (1889–1945) to bail out his beleaguered ally in April 1941 by invading Yugoslavia as well as Greece. The quick German conquests of Albania's neighbors result in territorial changes for Albania. Mussolini annexes Kosovo (a Yugoslav territory) to Albania as well as Epirus in Greece. Meanwhile, a communist opposition movement springs up in Albania dedicated to overthrowing foreign control and creating a socialist state.

1943 Hoxha works for women's rights

Nexhmije Hoxha, wife of the future leader of Albania, organizes a women's group to expand women's rights.

1943: July 6 Massacre at Borova

The village of Borova in southern Albania is invaded by the Nazi army. All of the approximately 450 buildings are destroyed by fire or in the attack by the Germans. Over 100 people are killed, including five entire families. The German attack is to retaliate against the people of Borova because they supported the Albanian fighters who previously attacked German troops who had occupied their village. This massacre is commemorated by Albanian composer Thoma Gaqi in his orchestral work, *Poem Simfonik Borova*.

1944 Communist partisans in Albania

At the close of World War II, a new era in government begins in Albania. Communist leaders, espousing the ideology of Karl Marx (1818–83), organize the National Liberation Movement. This group resists the occupation of Albania. There are also non-communist resistance organizations.

1944: May Communists occupy Southern Albania

Communists begin to control most of Albania. Enver Hoxha (1908–85) emerges as the head of the communists in Albania. He is its executive chairman and leader of the National Liberation Army. Son of a Muslim merchant, Hoxha studies for a time in France. He is a propagandist for the liberation of Albania and an excellent politician. The Allied Forces (United States, England, France, and Russia) are anxious to defeat Germany and support Hoxha in his quest for Albanian freedom.

1945 Elections held

The first "democratic" elections take place in Albania, but only Communist Party (Democratic Front) members are

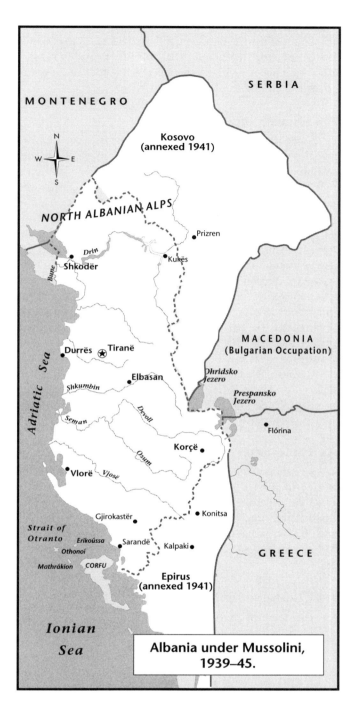

Albania under Mussolini, 1939–45.

allowed to stand for election. Enver Hoxha is the leader of the Albanians as commander-in-chief, prime minister, defense minister, and foreign minister. He effectively sets up a "quasi-dictatorship."

The issue of the Kosovo province is still unsettled with both Yugoslavia and Albania declaring sovereignty.

1948 Albania establishes ties with Russia (Soviet Union)

Albania breaks ties with Josip Broz-Tito (leader of Yugoslavia) and establishes economic relations with the USSR. The

United States tries to influence a democratic government in Albania by offering economic aid (Marshall Plan) to Albania. Albania refuses. Hoxha purges all former Yugoslav sympathizers from the government and becomes a staunch supporter of Joseph Stalin, head of the Soviet Union.

1948 Hoxha's agenda

Hoxha becomes a leader in the style of his hero, Russian leader Joseph Stalin (1879–1953). He rids himself of his enemies either by exiling them or by executing them after conducting secret trials. He creates forced labor camps and creates the *Sigurimi* (secret police) to control opposition to his regime. Hoxha uses the Soviet-styled method of the "five-year" plan to incease education and industrial goals in order to bring Albania up to the level of other industrial nations. He changes the old semi-feudal system to one of collective land ownership. Women are now considered the equals of men, but traditional gender roles continue much as before. Under his regime, the dialects of Gheg and Tosk are combined to form one language. He bans the outward practice of religion and destroys many churches and mosques declaring atheism to be the religion of the Albanians. Literature and the arts are strictly controlled. Hoxha also bans foreign travel which further isolates Albania. Stalin uses Albania to spread communist propaganda and to infiltrate Greece. Hoxha supports Stalin in these efforts.

Late 1940s–1990 The Sigurimi

The secret police, or Sigurimi, help Hoxha control all people opposed to Hoxha's rule. Those who dare to disagree with him publicly are often sent to forced labor camps. Hoxha executes others. One of the documents coming out of these camps is the journal of imprisoned Albanian poet Arshi Pipa, *The Prison Notebook,* which details the horrors of the prison camps.

1954 Hoxha shares power

After the death of Stalin, the Soviet Union goes from one-man rule to a committee, and Albania follows suit. Hoxha shares limited power with Mehmet Shehu.

1955 Khrushchev forms ties with Yugoslavia

Nikita Khrushchev (1894–1971), the new Communist leader of the USSR, re-establishes ties with Yugoslavia which worries Albanians because they have political disagreements with Yugoslavians.

1959 Communist nations court Albania

Communist countries such as the Soviet Union and China begin giving economic aid to Albania in an attempt to cement relations. Albania supports China against the Soviet Union in matters of ideology. Ramiz Alia, Hoxha's advisor, supports him in these decisions.

1960 Albania formally supports China

In response to Albania's support for China, Moscow tries unsuccessfully to unseat Hoxha and Shehu. Reprisals are carried out against Soviet supporters.

1961 More rhetoric against the Soviets

Hoxha and Shehu disagree with Khrushchev's desire for Albania to concentrate on agriculture. Hoxha's regime wants instead to improve the industrial economy. Soviets deny economic aid to Albania, but Albania receives support from the Chinese.

1964 Khrushchev falls from power

The new administration still refuses to deal with Albania. Albania has China for support.

1966: February Hoxha begins cultural and ideological revolution

In response to unrest following a repressive period, Hoxha, as head of the Albanian Party of Labor (APL), promises to institute reforms such as equal wages for all workers. For the first time women are mentioned as equal partners and encouraged to take jobs outside the home. Child care is provided to make it easier for women to work.

1968 Albania withdraws from Warsaw Pact

Soviets lead the invasion of Czechoslovakia, and Albania resigns from the Warsaw Pact, a military alliance of Soviet countries, in protest.

1973 Women in work force

Now at least forty percent of Albanian women are in the work force.

1974 Kosovo remains part of Yugoslavia

The province of Kosovo, comprised mainly of ethnic Albanians, remains an autonomous region in Yugoslavia.

1976 Albania adopts a constitution

Albania changes its name officially to the People's Socialist Republic of Albania but remains under the same communist rule as before. The only political party remains the Albanian Party of Labor (APL).

1976: September Hoxha becomes disillusioned with China

After Chairman Mao's death, Hoxha begins to criticize the new Chinese regime.

1978   China and Albania rupture ties

Albania ends all diplomatic and economic ties with China. Albania tries to become self-sufficient, but in reality, it is isolated and in economic ruin.

1980   Hoxha does not nominate Shehu as successor

Hoxha, perhaps fearing Shehu's power, refuses to name him as his successor. He instead names Ramiz Alia (b. 1925).

1981   Shehu's alleged suicide

Shehu is found dead. Hoxha's agents allege that he committed suicide. Most likely he was murdered.

1981   Demonstrations in Kosovo

People in the Yugoslav Republic of Kosovo demand autonomy. Yugoslavia quells the rebellion.

1982: November   Alia becomes head of government

Ramiz Alia is one of Hoxha's strong supporters and is also favored by Hoxha's wife, Nexhmije. He manages to survive all the political turmoil. Alia realizes the deplorable state of Albania's economic situation and makes some reforms. He also notices the unrest and political criticism directed at the government. He is aware of human rights violations and releases some political prisoners. He also ends a period of isolation by re-establishing ties with Greece, Turkey, Italy and Yugoslavia. He relaxes some bans on tourism in order to increase trade.

1985   Enver Hoxha dies

Enver Hoxha, the last of the original East European Communist heads of state, dies. His death results in a gradual relaxation of some restrictions at home and a slow opening toward the rest of the world. Nevertheless, Albania remains the poorest country in Europe and extremely isolated in international affairs.

1986–89   Government offers reforms

The government attempts to quell opposition by releasing some long-term political prisoners. In an effort to modernize the country, Alia makes reforms in the courts and the legal system and lifts bans on open religious worship. Some churches and mosques are re-opened.

1987   Economic crisis

Alia realizes there is a huge economic crisis due to mismanagement of the system. He introduces some relaxation of foreign trade and tries to decentralize the economy. There are also demonstrations protesting human rights and calling for an end to political repression. Alia's attempts to improve these situations are meager and are opposed by his government officials who fear they will be overthrown if they give in to popular demands.

1987   Greece and Albania end state of war

Albania officially ends the state of war with Greece existing since World War II.

1990   Albania seeks to restore ties with Soviet Union and United States

Albania, which has severed ties with the Soviets since the death of Stalin, indicates that it wishes to re-establish diplomatic relations.

1990: December   People call for Alia's resignation

The first opposition party, the Albanian Democratic Party (ADP), is recognized. This is the first time people can have a government of more than one party and enjoy some representation.

1991: January   Opposition newspaper begins publishing

The newspaper *Rilindja Demokratike* is openly critical of the government but is allowed to publish.

1991: March   United States and Albania resume diplomatic relations

Albania's break with the United States since the end of World War II is over.

1991: April   Alia re-elected

Some constitutional reforms take place. A committee is appointed to draft new constitution.

1991: June   Cabinet resigns

Prime Minister Fatos Nano and the rest of Alia's cabinet resign after general strikes and protests in the city of Shkoder. A coalition government begins rule. Some of the reforms passed allow private ownership of property and foreign investments. The Sigurimi is abolished. The United States opens an embassy in Tiranë, and Alia declares new elections for 1992.

1992: March   Elections

Two major parties, the Socialist Party of Albania and the Albanian Democratic Party, vie for power. The Albanian Democratic Party wins and Sali Berisha, the leader of the ADP, becomes the first democratically elected president. He is also the first non-communist president since before World War II.

1992: July   Communist Party is banned

The Albanian government officially passes a law banning the Communist Party.

1992: September–December   Anti-government officials arrested

Many government officials, including Alia and Hoxha's wife, are arrested. Even though the government is moving toward the democratic process, the old methods of arresting dissidents is still in effect.

1993   Fatos Nano jailed for corruption

Fatos Nano, the Socialist Party president, uses force to control his enemies. He is jailed on charges of corruption and misappropriating funds. A mob frees him from prison to begin political life anew.

1994   Berisha tries for more power

Sali Berisha tries to maintain power by changing the constitution to grant more power to the president.

1995   Anniversary of the death of Alexander Moisiu

The year is dedicated to the actor Aleksander Moisiu (see 1879), in honor of the sixtieth anniversary of his death. The dedication is sponsored by the Aleksander Moisiu Foundation based in Durrës, Albania.

1995: June   Ilir Hoxha indicted

Ilir Hoxha, son of former leader Enver Hoxha, is arrested for inciting hate crimes. Soon the govenment passes a genocide law which prohibits former communists from holding public office until the year 2002.

1995: November   Discovery of mass graves

The discovery of mass graves near Shkoder leads the government to press charges against former president Alia. Alia and other officials receive death sentences in May, 1996.

1996: May   National Elections Monitored

Elections to the Albanian government are monitored by an international committee. However, the demonstrations that follow invalidate the results. There are riots and voter intimidation.

1997: March   Berisha elected to five-year term

In another disputed election, Berisha is the new president, but immediately thereafter, riots erupt and Berisha must impose a national state of emergency. He finally signs an agreement for an interim government which is called the Government of National Reconciliation. Included in this agreement is the release of former political prisoners, including former president Alia and Prime Minister Fatos Nano.

1998   Violence continues

Albania is again ruled by violence as governments are determined more by coups than by elections. Sali Berisha, a supporter of the Kosovo Liberation Army, manages to control the government until he, too, is blamed for supporting many illegal pyramid schemes which force people to lose all their savings. He is at odds with Fatos Nano, the Socialist Party President. In 1998, the murder of Azem Hajdari, a Kosovo sympathizer, forces Nano (now prime minister under President Rexhep Meidani) to resign. Rioting in the streets results from Berisha's efforts. Berisha threatens to return to political life, but there is now a warrant for his arrest.

The resolution of the Kosovo problem (whether it should be part of Serbia or part of Albania), cannot solve the problems of poverty and violence in Albania. Poverty is still the lot of most Albanians and continuing violence discourages other countries from economic ventures other than drug smuggling. Many people are still without running water and electricity.

Bibliography

Albania: A Country Study. Diane Publishing Company, n.d.

Albania: General Information. Tirana: 8 Nentori Publishing House, 1984.

Battiata, Mary. "Albania's Post-Communist Anarchy," *Washington Post,* March 21, 1998 A1–A18.

Biberaj, Ekiz. *Albania: A Sicuakust Maverick.* Boulder, CO: Westview Press, 1990.

Durham, M.E. *High Albania.* Boston: Beacon Press, 1985.

"Hold Your Breath," *The Economist.* July 5, 1997.

"In a Land Adrift, the Albanian People Drift Too," *The New York Times.* July 16, 1997.

Jelavich, Barbara. *History of the Balkans.* Cambridge: Cambridge University Press, 1983.

Logoreci, Anton. *The Albanians: Europe's Forgotten Survivors.* Boulder, CO: Westview Press, 1977.

Pipa, Arshi. *The Politics of Language in Socialist Albania.* New York: Columbia University Press for Eastern European Monographs, 1989.

Post, Susan. *Women in Modern Albania.* McFarland and Company.

Victers, Miranda. *The Albanians: A Modern History.* St. Martin's Press.

Andorra

Introduction

Andorra is a micro-state of 181 square miles (468 square kilometers) located along the Pyrenees mountains between France and Spain. As of 1998, its predominantly Roman Catholic population is roughly 64,000, most of whom speak Catalan (although French and Spanish are also spoken by many Andorrans). However, of these residents, only about 13,000 hold Andorran citizenship. Modern-day Andorrans are the descendants of barbarian (mostly Visigoth) invaders who swept through Europe in the fifth century A.D. and helped bring about the collapse of Roman rule in Western Europe.

The largest city and capital of Andorra is Andorra la Vella, which has a population of about 25,000. Other cities include Encamp, La Massána, Les Escaldes, Ordino, and San Juliá de Lòria. The terrain of the principality is largely mountainous, and the altitude of the country ranges from 3,000 feet (900 meters) to 9,655 feet (2,946 meters). A significant part of the country is still forested and provides scenic views for tourists. Winters are quite cold, and the temperature varies significantly between day and night. Andorra's economy is primarily commercial and service-oriented with a strong emphasis on tourism.

Like the other European micro-states, Andorra is a historical anomaly—a vestige of a bygone era of small duchies and principalities. Indeed, Andorra is still governed officially by two co-princes. Archaeological records indicate human settlement back to the Neolithic era. Evidence for the period prior to the Middle Ages is sketchy, even though Andorra was a part of the Roman Empire. After the fall of Rome, Andorra became a crossroads for barbarian invaders who moved south into the Iberian peninsula. Andorra's history as a distinct political entity can be traced back to the era of the Frankish conqueror Charlemagne, who wrested control of the region from the Moors in the eighth century A.D. In the wake of the Carolingian Empire's division after Charlemagne's death, control of the Valleys of Andorra (as the region was then called) passed to his son, Louis I. He, in turn, granted the region to the Count of Urgel in 819. In 988 control passed to the Bishop of Urgel. To this day, the Bishop remains one of the principality's co-princes.

The second co-prince reflects French influence upon the region beginning with the Count of Foix's inheritance of Andorra through marriage in 1208. The Count of Foix's claim to the land placed him in dispute with the Bishop of Urgel. Not until 1278 was the matter resolved through a compromise that provided that the Bishop of Urgel and the Count of Foix rule over Andorra as co-princes in perpetuity. The Count of Foix ascended to the French throne in 1590 as King Henry IV but remained a co-prince. Although the abolition of the monarchy in 1792 resulted in a brief period of disrupted ties with France, relations were restored in 1806 following an Andorran petition of French emperor Napoleon Bonaparte (1769–1821). Since then, the French co-prince's position has been filled by the French head of state, be it emperor, king, or president.

Andorra in the Twentieth Century

The last century has brought the greatest amount of transformation in the principality's history. The two most significant developments have been the establishment of democratic self-government and economic expansion. For much of the nineteenth and twentieth centuries, government under the co-princes proved both a benefit and a hindrance to Andorrans. As inhabitants of a tiny state with no significant source of wealth, Andorrans viewed co-princely rule as a way to preserve a measure of their own freedom from foreign domination. At times, however, the disputes of the co-princes hampered Andorran efforts to achieve international recognition and partake in international affairs. France proved particularly troublesome when Andorrans entered into postal agreements with Spain and when Andorra attempted to enter international regimes such as the International Opium Convention in 1925, the Universal Copyright Convention in 1952, and the Hague Convention for the Protection of Cultural Property in the Event of Armed Conflict in 1954. In each case, France argued that only it could handle Andorra's foreign relations because Andorra was not a state in international law. Spain, meanwhile, supported the Bishop of Urgel, who

ANDORRA

FRANCE

PYRENEES

El Serrat

Llorts

Pic de Coma Pedrosa 9,652 ft. 2942 m.

Arinsal

Valira del Nord

Soldeu

Ordino

Valira d'Orient

Pal

La Massána

Anyos

Encamp

Pas de la Casa

Andorra la Vella

Estany d'Engolasters

Les Escaldes

Santa Coloma

Valira

Sant Juliá de Lòria

Farga de Moles

Arcabell

SPAIN

Segre

Valira

Andorra

countered these French attempts to claim supremacy in Andorran affairs.

The French, and at times the Spanish, also intervened in Andorra's domestic politics. French troops intervened in the 1930s during a period of political unrest. In that same decade, the co-princes voided certain General Council legislation and issued decrees in their place.

Not until the 1970s did the co-princes take seriously an Andorran request that the principality's government be revamped. Even then, it took nearly two decades for the reforms to take effect. When they finally did in 1993, they marked the most significant governmental change in the principality's history. The new constitution retained the Bishop of Urgel and the President of the French Republic as co-princes. Although they still sanctioned legislation and appointed the head of government (who was elected by the General Council and six ministers) in other areas their rights were severely circumscribed. The constitution also provided for a government that featured separation of powers between the government and judiciary. No longer hampered by French objections, Andorra embarked upon a policy of international recognition

and participation. In 1993 Andorra joined the United Nations and the Council of Europe, and in 1995 the principality gained full membership in the World Trade Organization.

Concurrent with these developments, Andorra has also undergone a radical economic transformation. Beginning after World War II, Andorrans took advantage of their country's scenic location and began promoting tourism as a way to acquire much needed foreign currency exchange. As a result, Andorra takes in about 12 million tourists per year. The general population of the principality has not gone unaffected by this influx of visitors. As of 1995 World Bank estimates put the country's per capita GNP over $10,000 per year. Andorra's liberal business laws make it favorable to foreign investment, and, indeed, many of the principality's leading hoteliers and restauranteurs are foreigners.

Timeline

c. 6000-3000 B.C.   Neolithic settlement

Human settlers in the valleys of Andorra leave the earliest artifacts from the region.

c. 2000 B.C.   Bronze Age settlements

During the Bronze Age, human settlement in Andorra occurs near the modern villages of Cedre and Sa Serra d'Enclar. Some scholars argue that these inhabitants are related to the Basques or Celts.

154 B.C.   Roman conquest

The Romans begin their conquest of the Iberian peninsula. Roman rule brings Andorra under the control of the most powerful Western state of its time. Roman rule over Andorra lasts for over five centuries.

5th century A.D.   Visigoths invade

The Visigoths, a barbarian Germanic tribe, overrun much of the Western Roman Empire. In 410 they sack Rome. Eventually, they settle in the Iberian peninsula and southern Gaul (France). The Visigoths are the ancestors of today's Andorrans. Eventually, the barbarian conquerors convert to Christianity, thus forging a continuity in religious belief among Andorrans that lasts to the present day.

Early 8th century   Arab conquest

Muslim Arabs conquer the Iberian peninsula and threaten to overrun all of Western Europe. They are finally defeated by the Frankish king Charles Martel at the Battle of Poitiers in Gaul in 732. Not until later in the century are they driven from Andorra. Nonetheless, the Arab impact on Andorra is minimal. Unlike most of the lands south of Pyrenees where

Muslims rule until the fifteenth century, Andorrans remain Christian and have little interaction with the invaders.

Late 8th century Charlemagne conquers modern-day Andorra

The Great Frankish conqueror Charlemagne (742–814) wrests the region of modern-day Andorra from the Moors. Charlemagne eventually expands the territory of his Frankish kingdom to include all of Gaul (modern-day France), parts of the Iberian peninsula, most of Germany, and the northern half of the Italian peninsula. His empire is the greatest Western European state since the fall of Rome. As if to underscore Charlemagne's achievement, the pope crowns him Roman Emperor in 800 and the conqueror's kingdom becomes known as the Holy Roman Empire.

819 Count of Urgel receives Andorra

Charlemagne's son, Louis I, grants the Valleys of Andorra to the Count of Urgel.

988 Possession of Andorra passes to Bishop of Urgel

A descendant of the Count of Urgel exchanges the Valleys of Andorra with the Bishop of Urgel in exchange for other parishes.

11th century Caboet family provides protection

The Bishop of Urgel grants Andorra in fief to the Caboet family in exchange for military protection. Such an arrangement is typical of the feudal system then in place in western Europe. Under the feudal regime, lords (such as the Bishop of Urgel, in this instance) grant land (called a *fief*) to a warrior vassal (such as Caboet) in return for protection.

1208 Rights to Andorra pass to Count of Foix

The Count of Foix gains the right to the lands of Andorra through marriage to Ermessenda de Castellbó, daughter of Arnalda de Caboet who had inherited the Valleys of Andorra. This event places the Count of Foix in conflict with the Bishop of Urgel.

1278: September 6 Mediated agreement between the Bishop of Urgel and the Count of Foix

Following mediation by the Bishop of Valencia, the Bishop of Urgel and the Count of Foix sign the *Pariatge*, an agreement which defines their rights and guarantees their rule over Andorra in perpetuity. This treaty gives the Count the right to send a permanent representative to Andorra, the *veguer*. In addition, both the Count and the Bishop of Urgel are to receive a biennial contribution, the *qüèstia*, from the Andorran population.

1288: December 6 Second *pariatge* is signed

A second *pariatge* is signed to resolve some minor disputes that remained after the first agreement.

1419 Council of the Land created

The two co-princes accede to a request by Andorrans that a council be created to deal with local issues. This body, known initially as the Council of the Land, eventually evolves into the General Council, the parliamentary body still in existence in 1999.

1590 Count of Foix becomes King Henry IV

Henry of Navarre, the Count of Foix, ascends to the French throne as King Henry IV (d. 1610) beginning a twenty-year reign. A Protestant on the French throne during a time of interminable religious strife between French Protestants and Catholics, Henry pursues a policy of conciliation that culminates in his returning to the Catholic fold in 1588 and issuing the Edict of Nantes, a proclamation of religious tolerance. Despite his efforts for reconciliation, however, Henry IV is assassinated in 1610.

1607 Henry IV issues the Edict of Reunion

Henry IV issues the *Edit de Réunion* (Edict of Reunion) which unites all of his lands under the French Crown. Significant for Andorra, this event makes the ruler of France one of the co-princes of the tiny country (with the exception of the years 1793–1806). The presence of a French ruler as co-prince also undercuts the power of successive Bishops of Urgel, men who have nowhere near the resources of France with which to wield influence.

1748 Andorra declared neutral

The *Manuel Digest* declares that Andorra is to be a neutral state in the event of war between France and Spain, or, indeed, any other war.

1793–1806 Period of no French co-prince of Andorra

In the aftermath of the French Revolution that begins in 1789, feudalism is outlawed in France. As a result, the French revolutionary government refuses to accept the *qüèstia* from Andorra. The Andorrans oppose this French refusal to accept payment because they fear that by paying only the Bishop, they may become subject to his rule.

1806: March 27 Napoleon restores the *qüèstia*

Following an Andorran petition, French emperor Napoleon Bonaparte agrees to restore the status quo, accepts the Andorrans' *qüèstia*, and appoints a *veguer*.

1866: April 22   New Reform introduced

The New Reform provides for the election of the General Council by male suffrage limited to the heads of households.

1878   Overtures made for introduction of postal service

The *Union Universelle Postale* (UPU) approaches Spain about the organization of a postal service in Andorra.

Twentieth Century

1906   Spain organizes postal service

The UPU entrusts Spain with the organization of the Andorran postal service. This Spanish run post replaces the joint French-Spanish courier service.

1925   League of Nations requests leads to diplomatic squabble

The Secretary General of the League of Nations invites the Syndic of the General Council to adhere to the International Opium Convention of February 19, 1925. The Syndic agrees, but the French nullify the decision, saying that only it has the right to handle Andorra's international relations. Spain, in turn, rejects the French claim. This dispute causes a series of disputes between the two co-princes over the extent of their rights and Andorra's position in international law.

1929   France protests Spanish postal rights

France protests Spanish postal rights in Andorra. The UPU recognizes the French rights and urges negotiations.

1930: June 30   New postal agreement

France and Spain reach a new postal agreement. The new arrangement establishes a joint French-Spanish postal administration. As a result of this decision, however, Andorra is excluded from the UPU.

1931–33   Unrest in Spain causes unrest in Andorra

In 1931, the Spanish monarchy is overthrown and a republic is established. The Spanish government asks the Bishop of Urgel to cede his rights in Andorra to the Spanish government in Madrid, but the bishop refuses. France, led by the French co-prince (the president of the French Republic), opposes the Spanish government's request and supports the bishop.

1931–33   Clashes between General Council and co-princes

In addition to the battle between the Spanish government and the Bishop of Urgel, the Andorran General Council (the country's parliament) engages in a struggle with the two co-princes over the establishment of a gambling house in the Andorra. After the General Council approves the house's opening, the co-princes announce their opposition to the mea-sure. As a consequence, the co-princes cancel the General Council's agreement with Spain to install a telephone network in Andorra.

When the General Council attempts to institute universal male suffrage, the co-princes retaliate by voiding the law, dissolving the legislative body, and replacing it by a provisional council. Only after new elections are called do the co-princes issue a decree establishing universal male suffrage.

1933   Political unrest hits Andorra, French intervention

The attempt of the Russian Boris de Skossyreff to declare himself King Boris I of Andorra sparks political unrest in Andorra. De Skossyreff is supported by certain politicians as well as many citizens. Forces at the disposal of the Bishop of Urgel ultimately remove de Skossyreff but continuing unrest causes the Tribunal de Corts (Andorra's judiciary) to unseat the General Council. In response to the continued chaos fifty French policemen are sent to Andorra to quell the unrest. The General Council sends the League of Nations a telegram protesting the incursion. However, the League rejects the request on the basis that only the co-princes can handle Andorra's external affairs. This action appears to support the French claim that Andorra is not a state in international law.

Once order is reestablished, new elections are held and the voting age is lowered to twenty-four. Males over thirty years of age can hold office.

1936–40   French troops on Andorran border

France places troops along the Andorran border to prevent a possible spillover of fighting from the Spanish Civil War. Although the civil war that rages in Spain pits the forces of the government (the Republicans) against the rebel forces under General Francisco Franco (Nationalists), the conflict takes on international dimensions as the Republicans gain aid from left-wing supporters (including the communist Soviet Union) while the Nationalists are supported by Nazi Germany and Fascist Italy. In 1939, the Nationalists prevail.

1944   French and Spanish troops in Andorra

French troops enter Andorran territory, and the bishop requests that the Spanish *Guardia Civil* (Civil Guard) do like-wise. The Spanish forces are quickly withdrawn, however, while the French forces remain a while longer under the authority of the co-princes.

1952, 1954   Disputes between France and Spain over Andorra's international status

Attempts by Andorra to enter into the Universal Copyright Convention in 1952 and the Hague Convention for the Protection of Cultural Property in the Event of Armed Conflict in 1954 bring renewed disagreement between France and Spain over the status of Andorra. In the 1954 dispute, France, which opposes Andorran support of the convention, invokes the colonial clause of the agreement, while Spain signs in the

Andorran citizens gather to discuss the political situation and share views on the presence of French peacekeepers.
(EPD Photos/CSU Archives)

name of the bishop of Urgel (although the bishop, himself, opposes the convention).

1967   Automatic telephone service

The installation of an automatic service allows for easier telephone use.

1970: April 14   Women's suffrage

Andorran women finally gain suffrage through a decree of the co-princes.

1970   Telex system introduced

A telex system is installed in the principality.

1971: January 31   Juan Martí Alanis becomes new co-prince

Juan Martí Alanis, Bishop of Urgel, becomes the new Spanish Episcopal co-prince.

1971   Voting age lowered

Andorra lowers the voting age to twenty-one years of age.

1974   National Library founded

Andorra founds its National Library in Andorra la Vella, the first of a series of public centers that Andorra establishes in the 1970s and 1980s.

1975   National Archives founded

Andorra founds its National Archives in Andorra la Vella. This institution serves as a repository of documents relating to the history of the principality.

1975   Official call for change in structure of Andorran government

The General Syndicate (the permanent commission of the General Council) sends a letter to the co-princes calling for a change in the structure of Andorra's government and institutions. The co-princes agree, but it is nearly two decades before the reforms take effect.

1978 Seventh parish created, Andorra applies for WTO membership

A seventh parish (administrative region), Les Escaldes, is created in Andorra. Andorra applies for membership in the World Trade Organization (WTO).

1981 New government radio stations established

Following the expiration of the contracts for Spanish *Radio Andorra* and French *Sud-Radio*, the Andorran government establishes two new radio stations, *ORTA* and *Radio Valira*.

1981: January 15 Executive Council created

A new governmental body, the Executive Council, is formed. This body consists of the Head of Government (elected by the General Council and approved by the co-princes) and a cabinet of four to six ministers with portfolios in defense, education, finance, foreign affairs.

1983 Andorra Scientific Society founded

The Andorra Scientific Society is founded in Andorra la Vella.

1985 Tax on imports

Andorra levies an indirect tax on all imports.

1987 Museums founded

A general interest museum in Excaldes-Engordany and a decorative arts museum are founded.

1988 National Motor Car Museum founded

Andorra founds the National Motor Car museum in Encamp. Its collection features old automobiles, bicycles, and motorbikes.

1990: June 28 Customs union with EC

Andorra signs a customs union with the European Community to take effect the following year.

1990 Call for a draft of a constitution

The head of the Executive Council calls for a draft of a constitution based on a separation of powers to be ratified by referendum. The Tripartite Commission, composed of representatives to the General Council and the two co-princes, is assigned this task. The Tripartite Commission begins its work in December, 1990.

1991: July 1 Customs union takes effect

The customs union with the EC takes effect which ends the customs regime with France that has existed since 1967.

1993: February 2 General Council approves the constitution

The General Council approves unanimously the proposed Andorran constitution.

1993: March 14 Voters approve constitution

A majority of 74.2 percent of Andorran voters approve the constitution by referendum.

1993: May 4 Constitution takes effect

The constitution becomes the law of the land. Governmental powers are transferred from the co-princes to the General Council and the Government. Sovereignty rests with the Andorran people, while the two co-princes are defined as the Bishop of Urgel and the President of the French Republic. Foreign affairs remain largely the domain of the co-princes. The *veguers* and permanent representatives of the co-princes are abolished and replaced by one representative of each co-prince. These representatives reside in the principality.

A chief feature of the Andorran constitution is the provision of a separation of powers among the different branches of government. The General Council forms the parliamentary body and consists of twenty-eight members. Fourteen of these representatives are elected on a parish basis (two from each parish) while fourteen representatives are elected by national constituency. All representatives are elected to four-year terms by Andorrans with political rights. Andorran voters also have the opportunity to express their desires through referenda. Indeed, a referendum is mandatory for ratification of any amendment to the constitution. The General Council has the power to draft laws and propositions for draft laws. In addition, the co-princes must sanction legislation, as must the Head of Government.

The Head of Government is appointed by the co-princes after having been elected by the General Council and six ministers. The Head of Government plays a leading role in national and international policy and holds administrative and regulative power. The government is responsible to the General Council which has the right to dismiss the former in certain circumstances. Members of the government cannot be members of the General Council. The judiciary remains a separate branch of the Andorran government while the constitution establishes a Constitutional tribunal to examine the legality of laws and procedures.

1993: June 3 Treaty with Spain

Andorra signs the Treaty of vicinage, friendship, and cooperation with Spain. This signifies a renewal of Andorra's historic ties with Spain following the establishment of constitutional government in Andorra.

1993: June 9   Application for UN membership

Andorra files an application with the Secretary General for membership in the United Nations (UN).

1993: July 28   Andorra admitted to the UN

Andorra becomes a member of the United Nations by acclamation.

1993: December   First elections under new constitution

Andorra holds its first elections since the ratification of its constitution. Five political parties participate. The distribution of seats in the General Council are as follows: National Democratic Grouping (AND), eight seats; and five seats each for the Liberal Union (UL), New Democracy (ND), National Andorran Coalition (CAN), and the National Democratic Initiative (IDN). The AND forms a coalition government.

1994   Government falls

The Andorran government led by the AND falls and is replaced by a new coalition government led by the UL.

1994: November 10   Andorra joins the Council of Europe

In keeping with its post-1993 efforts to play a greater role in world affairs, Andorra joins the Council of Europe with two representatives in the organization. Andorra has sought entry earlier, but the council has recommended greater democratization and the protection of rights in a constitution.

1995   Andorra has diplomatic relations with forty-eight states, full membership in WTO

Only two years after establishment of its constitution, Andorra enjoys diplomatic relations with forty-eight states. This change reflects the success of the new effort of the Andorran government to establish wide-ranging international political and economic ties. After a seventeen-year wait, Andorra becomes a full member of the WTO.

1995: May 17   Jacques Chirac becomes new co-prince

Newly-elected French President Jacques Chirac (b. 1932) becomes Andorra's new French co-prince.

1997: February   Liberal Union wins elections

In the second elections held under Andorra's 1993 constitution, the UL prevails with sixteen seats. Other parties' representation in the General Council is as follows: AND, six seats; and two seats each for the IDN, ND, and CAN. Marc Forné Molné of the UL is elected president of the General Council. In this election, 8,842 voters (81.6 percent of those eligible) go to the polls.

Bibliography

Carter, Youngman. *On to Andorra.* New York: W. W. Norton, 1964.
Deane, Shirley. *The Road to Andorra.* New York: William Morrow, 1961.
Duursma, Jorri. *Fragmentation and the International Relations of Micro-States: Self-determination and Statehood.* Cambridge: Cambridge University Press, 1996.
Johnson, Virginia W. *Two Quaint Republics: Andorra and San Marino.* Boston: Dana Estes, 1913.
Leary, Lewis Gaston. *Andorra: The Hidden Republic.* New York: McBride, Nast, 1912.

Armenia

Introduction

Armenia occupies approximately 11,506 square miles (29,800 square kilometers) in the region southwest of Russia between the Black and Caspian Seas. Bordered on the north by Georgia, on the east by Azerbaijan, on the south by Iran, on the southwest by the Nalchichevan Autonomous Republic of Azerbaijan, and on the west by Turkey, modern Armenia is small and landlocked. Historical Armenia sprawled across territory now held by Turkey. The crumbling remains of ancient Armenian kingdoms can be found in the valley of the Aras River and around Lake Van. The medieval kingdom of Cilicia was a historical curiosity, a powerful Armenian state located on the Mediterranean Sea in southern Turkey, hundreds of miles away from the center of the culture in southern Transcaucasia.

The Lesser Caucasus mountain range dominates the topography of northern Armenia, stretching between Lake Sevan and Azerbaijan. Southwest of the range lies the Armenian Plateau, encircled by small mountains and extinct volcanoes. At 4430 meters, Mt. Aragat is the highest point in Armenia. The valley of the Aras River, with an elevation of only 380 meters is Armenia's lowest point.

As of the year 1999, most of Armenia's 3.8 million people lived in urban areas, especially the capital city of Yerevan. Armenians are the dominant ethnic group. There are also small numbers of Russians and Yezidi Kurds. In the 1989 census, the Azerbaijani population was estimated at 2.6 percent of the total. However, by 1993 virtually all Azerbaijainis had fled the region. The majority of Armenians belong to the Armenian Apostolic (Orthodox) Church. Minority religions include Islam, Roman Catholicism, and a variety of Protestant denominations. Armenian is the official language of the country, but Russian is still spoken in cities.

History

Armenia was settled as early as the 6000 B.C. By the ninth century B.C., the kingdom of Urartu was a formidable power in the area around Lake Van. Greek historical works of the sixth century B.C. recorded a new group of people living in the area who called themselves the Hai, after a legendary folk hero. The Greeks named them "Armenians." By the end of the century, a distinct ethnic group had emerged from the mingling of the Hai and Urartu peoples.

After centuries of Persian domination, Armenia reached its greatest size and influence in 95 B.C., when Tigran the Great unified all the regions inhabited by Armenians. His realm stretched from the Caspian Sea to the Mediterranean Sea. The kingdom became a tributary state of the Roman Empire in 72 B.C.

In A.D. 301 Armenia became the first state to adopt Christianity after the conversion of King Tiridates III by St. Gregory the Illuminator. The invention of the Armenian alphabet in the fifth century A.D. allowed for the translation of religious writings. However, a fifth-century schism further isolated the Armenian Church.

Politically, Armenia had become a pawn in the rivalries between Rome and Persia. The two empires divided the territory in A.D. 387, with Persia receiving the larger portion. Persian Armenia was lost to Arab conquerors in A.D. 640. Thirteen years later, Byzantine (Roman) Armenia met the same fate. Many Armenians fled to Byzantium, settling in Cilicia, on the Mediterranean Sea in modern southern Turkey. An Armenian kingdom arose there in 1080, serving as a point-of-departure for the Armenian Diaspora, as thousands of Armenians fleeing massacre and invasion immigrated to Cairo, Cyprus, Holland, India, Marseilles, and Venice. Cilicia fell to the Mamluk Turks in 1375.

For the next few centuries, the empires of Ottoman Turkey, Safavid Persia, and later tsarist Russia fought over Armenia. The Ottomans and the Persians divided the territory in 1639, with the Ottomans retaining control over most of the region. Persian Armenians appealed for Russian aid as early as 1725. The 1813 Treaty of Gulistan awarded the Persian province of Karabakh to Russia. In 1828, the Russians defeated the Persians and were awarded the province of Yerevan by the Treaty of Turkmenchai. Although tsarist rule brought a number of modernizing reforms to Armenia, the imperial government's policy of "Russification" in the late-nineteenth and early twentieth centuries led to the formation of a number of nationalist movements in the area.

ARMENIA

after an invasion by the Soviet Red Army. Briefly grouped with Azerbaijan and Georgia in the Transcaucasian Soviet Federated Republic, Armenia became a single republic within the Soviet Union in 1936. In 1988, a strong nationalist movement formed in response to the endemic corruption of the communist-dominated government and to the status of Nagorno-Karabakh, an autonomous region populated mostly by Armenians but controlled by Azerbaijan. As the conflict in Nagorno-Karabakh escalated, calls for national sovereignty grew. The Armenian Supreme Soviet declared its independence from Moscow in September, 1991, and in October Levon Ter-Petrosian was elected president. By 1993, the economy of Armenia was in a state of collapse and fuel supplies were scarce. Robert Kocharian was elected president of the self-proclaimed Republic of Nagorno-Karabakh in 1994. In 1996, Ter-Petrosian and Kocharian were re-elected to their posts. That same year, Ter-Petrosian invited Kocharian to form a government. In February, 1998, Ter-Petrosian resigned over the issue of Nagorno-Karabakh. Kocharian was elected president of Armenia in March, 1998. Armenia's many economic and ecological problems remain unsolved.

Timeline

6000 B.C.   First human settlements

Human settlements appear in the regions of historical Armenia.

c. 1200 B.C.   Urartu kingdom

King Aramu founds the kingdom of Urartu in the region surrounding Lake Van in modern Eastern Turkey. The Urartu kingdom becomes a formidable agricultural and commercial power, known for its highly advanced metalwork and stonemasonry.

714 B.C.   Sargon II invades

Sargon II of Assyria invades Urartu, weakening but not destroying the kingdom.

c. 614 B.C.   Medes invade

The Medes of northern Persia conquer Urartu after seizing power in Assyria.

c. 550 B.C.   Armenians mentioned

Greek historical works mention a new group of people whom they call "Armenians" living among the people of Urartu. The Armenians refer to themselves as Hai and their land as Haiasta or Haiastan, after their heroic ancestor Haik.

A reformist movement called the *Tanzimat* swept over the Ottoman Empire in the mid-nineteenth century. In the reaction that followed, the position of the Armenians began to deteriorate. In 1878, Ottoman Armenian delegates to the Congress of Berlin appealed without success to the European powers for help. The "Armenian Question" continued to be a source of diplomatic contention.

Amidst accusations of disloyalty, the Ottomans massacred 300,000 Armenians in 1895. Thirteen years later, the sultan was overthrown by the Young Turks, a revolutionary group who saw the Armenians as an impediment to their nationalist goals. Invasions by the British and the Russians during the early years of World War I provided the pretext for genocide. In 1915, the government killed between 600,000 and 2 million Armenians by execution and forced death marches.

During World War I, the Russian Empire collapsed and Armenia emerged as an independent state. In 1920, Armenia was forced to accept a communist-dominated government

520 B.C.   Armenian tribute

Herodotus lists the lands occupied by the Urartu people and the Armenians among the different *satrapies* or autonomous tributary regions of Persia.

c. 404 B.C.   Orontes seizes power

Orontes, a local magnate, seizes power in the Armenian satrapy and declares his rule independent of the Persian Empire

331 B.C.   Alexander the Great

Alexander the Great conquers Persia and brings Armenia into contact with Hellenistic culture.

323 B.C.   Death of Alexander the Great

At the death of Alexander the Great, the Seleucid dynasty takes control of Persia.

The power of the Orontid dynasty grows in the wake of the ineffectual rule of the Seleucids.

c. 190 B.C.   Roman victory at Magnesia

The Romans defeat the Seleucids at Magnesia, and in the ensuing struggle for power in the area, a local ruler named Artaxias overthrows the Orontid Dynasty.

95–55 B.C.   Reign of Tigran the Great

Through a series of military victories, Tigran the Great unifies all of the regions inhabited by Armenians. He builds a new capital called Tigranakert on the site now occupied by the modern town of Silba and takes the Persian title "King of Kings." Armenia reaches its greatest size under his rule, extending all the way from the Caspian Sea to the Mediterranean Sea.

69 B.C.   Fall of Tigranakert

A Roman army under the command of the Roman General Lucullus (110 B.C. –56 B.C.) captures Tigranakert.

66 B.C.   Tigran surrenders

Tigran surrenders to the Roman General Pompey (106 B.C.– 48 B.C.) and is forced to pay tribute to the Roman Empire.

30 B.C.   Roman conquest of Armenia

The Roman Empire conquers Armenia.

A.D. 53   Arsacid dynasty founded

King Tiridates founds the Arsacid dynasty in Armenia with the help of his brother Vologeses I, king of Parthia.

59   Romans defeat Parthians

The Romans defeat the Parthians and drive Tiridates from the throne of Armenia. He is later restored by the Roman Emperor Nero.

62   Romans vanquished

Tiridates I defeats the Romans in battle and again assumes the throne of Armenia.

66   Nero crowns Tiridates

The Roman emperor Nero crowns Tiridates king of Armenia.

226   Sasanid dynasty

The Sasanid dynasty of Persia attempts to introduce Zoroastrianism, a Persian religion, into Armenia.

301   Christianity

St. Gregory the Illuminator converts the Armenian ruler Tiridates III (238–314) to Christianity. Armenia becomes the first country to adopt Christianity, and the Armenian Apostolic (Orthodox) Church develops independently of Rome and Constantinople. A fifth-century schism places Armenia even further outside the Byzantine fold.

387   Division of Armenia

The Roman and Persian Empires divide Armenia, with Persia receiving most of the territory.

404   Mesrop-Mashtots invents Armenian alphabet

The Armenian scribe Mesrop-Mashtots invents the Armenian alphabet, written from left to right. The alphabet becomes an important source of Armenian cultural identity in the centuries that follow.

428   Turmoil in Persian Armenia

The local nobility petitions the Persian ruler to turn the country into a satrapy.

451   Battle of Avarayr

Persian attempts to introduce Zoroastrianism into Armenia spark violent resistance from Armenian nobles at Avarayr. Despite their defeat, Armenians continue to resist Persian efforts to displace Christianity.

485   Liberty of worship

The Sasanid Dynasty of Persia allows Armenians to practice Christianity freely after attempts to introduce Zoroastrianism fail.

640   Arab conquest

The Arabs conquer the Persian Armenia.

653 **Byzantines cede Armenia**

The Byzantine Empire turns over control of western Armenia to the Arabs.

703–704 **Armenian dispersal**

The massacre of a large number of Armenian nobles by Arab forces prompts a mass exodus of Armenians to the Byzantine Empire.

700s–900s **Kingdom of Ani**

Rise of the Armenian kingdom of Ani in the lands of modern-day Turkey. Ruled by the Armenian-Georgian Bagratid Dynasty, the short-lived kingdom exchanges overlords many times, falling in 1054 to the Byzantines, in 1064 to the Seljuk Turks, and in 1236 to the Mongols.

806 **Bagratid Dynasty**

Favored by the Caliph, the Bagratid Dynasty comes to power in Armenia.

900s **Armenian dispersal**

The Byzantine Empire settles immigrant Armenians in Cilicia, located on the Mediterranean Sea in southern Turkey. The Muslim population of the area flees. The initial settlements eventually result in the emergence of an Armenian kingdom hundreds of miles from the historic lands of Armenia.

1000s **Armenian Diaspora begins**

Amid a centuries-long process of invasion, migration, conversion, deportation, and massacre that will eventually reduce the Armenian population to a minority in their historic lands, Armenians immigrate to India, the Middle East, Poland, Russia, and Western Europe.

1054 **Byzantine takeover in Ani**

The Bagratid king of Ani abdicates. His kingdom becomes part of the Byzantine Empire. Drained of money and armaments, the former kingdom cannot adequately defend itself against invaders.

1064 **Seljuk Turks invade**

Seljuk Turks invade and destroy the Byzantine outpost of Ani.

1071 **Battle of Manzikert**

The Turkish defeat of the Byzantine Empire at the Battle of Manzikert brings all of Armenia under Seljuk domination.

1080 **Cilician independence**

Prince Rupen of Cilicia declares his independence from the Byzantine Empire. The principality remains an independent

state until 1375, when it falls to the forces of the Mamluk Turks. A great trading center for silk and spices, Cilicia is also a stopping point for Armenian immigrants on their way to Cairo, Cyprus, Holland, Marseilles, and Venice.

1100s–1300s **Turmoil in Cilicia**

Caught up in the European Crusades, Cilicia is weakened over time by internal struggles between the Catholic and Armenian churches and by external assaults by Seljuk and Mamluk Turks.

1236 **Mongol invasion**

Mongol forces conquer the historic lands of Armenia.

1342 **Cilicia under Frankish control**

The Armenian kingdom of Cilicia passes through the female line to the Frankish family of de Lusignan.

1386–94 **Tamerlane invades**

The Central Asian warlord Tamerlane invades the lands of historic Armenia, wreaking havoc and massacring a large part of the population, but never ruling the territory. Instead, the region is dominated by the rival Turkoman dynasties of the White Rams and the Black Rams.

1375: April **Mamluk Turks**

Cilician Armenia falls to the armies of the Mamluk Turks.

1400s **Ottoman domination**

Most of Western Armenia falls under Ottoman domination.

1469 **Black Rams vanquished**

In Eastern Armenia, Urzun Hasan of the White Rams crushes the Black Rams and brings the region under his control.

1502 **Safavid conquest**

Shah Ismail, founder of the Safavid Dynasty of Persia, defeats the White Rams and annexes Eastern Armenia.

1605 **Armenian dispersal**

Thousands of Armenians from Eastern Armenia are uprooted by Shah Abbas the Great of the Persian Empire for strategic reasons and resettled in New Julfa, outside Isfahan.

1639 **Partition of Armenia**

The Ottoman and Safavid Empires partition Armenia, with the majority of the land going to Turkey.

The Ottoman Period

The Ottoman Empire began in western Turkey (Anatolia) in the thirteenth century under the leadership of Osman (also spelled Othman) I. Born in Bithynia in 1259, Osman began the conquest of neighboring countries and began one of the most politically influential reigns that lasted into the twentieth century. The rise of the Ottoman Empire was directly connected to the rise of Islam. Many of the battles fought were for religious reasons as well as for territory.

The Ottoman Empire, soon after its inception, became a great threat to the crumbling Byzantine Empire. Constantinople, the jewel in the Byzantine crown, resisted conquest many times. Finally, under the leadership of Sultan Mehmed (1451–81), Constantinople fell and became the capital of the Ottoman Empire.

Religious and political life under the Ottoman were one and the same. The Sultan was the supreme ruler. He was also the head of Islam. The crown passed from father to son. However, the firstborn son was not automatically entitled to be the next leader. With the death of the Sultan, and often before, there was wholesale bloodshed to eliminate all rivals, including brothers and nephews. This ensured that there would be no attempts at a coup d'état. This system was revised at times to the simple imprisonment of rivals.

The main military units of the Ottomans were the Janissaries. These fighting men were taken from their families as young children. Often these were Christian children who were now educated in the ways of Islam. At times the Janissaries became too powerful and had to be put down by the ruling Sultan.

The greatest ruler of the Ottoman Empire was Suleyman the Magnificent who ruled from 1520–66. Under his reign the Ottoman Empire extended into the present day Balkan countries and as far north as Vienna, Austria. The Europeans were horrified by this expansion and declared war on the Ottomans and defeated them in the naval battle of Lepanto in 1571. Finally the Austrian Habsburg rulers were able to contain the Ottomans and expand their empire into the Balkans.

At the end of the nineteenth century, the Greeks and the Serbs had obtained virtual independence from the Ottomans, and the end of a once great empire was in sight. While Europe had undergone the Renaissance, the Enlightenment, and the Industrial Revolution, the Ottoman Empire rejected these influences as being too radical for their people. They restricted the flow of information and chose to maintain strict religious and governmental control. The end of Ottoman control in Europe came in the First Balkan War (1912–13) in which Greece, Montenegro, Serbia, and Bulgaria joined forces to defeat the Ottomans.

During the First World War, (1914–18), the Ottoman Empire allied with the Central Powers and suffered a humiliating defeat. In the Treaty of Sévres, the Ottoman Empire lost all of its territory in the Middle East, and much of its territory in Asia Minor. The disaster of the First World War signaled the end of the Ottoman Empire and the beginning of modern-day Turkey. Although the Ottoman sultan remained in Constantinople (renamed in 1930, "Istanbul"), Turkish nationalists under the leadership of Mustafa Kemal (Ataturk) (1881–1938) challenged his authority and, in 1922, repulsed Greek forces that had occupied parts of Asia Minor under the Treaty of Sévres, overthrew the sultan, declared the Ottoman Empire dissolved, and proclaimed a new Republic of Turkey. That following year, Kemal succeeded in overturning the 1919 peace settlement with the signing of a new treaty in Lausanne, Switzerland. Under the terms of this new treaty, Turkey reacquired much of the territory—particularly, in Asia Minor—that it had lost at Sévres. Kemal is better known by his adopted name "Ataturk", which means "father of the Turks". Ataturk, who ruled from 1923 to 1938, outlawed the existence of a religious state and brought Turkey a more western type of government. He changed from the Arabic alphabet to Roman letters and established new civil and penal codes.

1701   Russian aid sought

The Armenian adventurer Israel Ori (d. 1711) is received by Peter the Great of Russia and appeals to him in vain for help against the Persians.

1772   First Armenian book of political philosophy

Shahamir Shahamirian, a Madras Armenian, publishes the first work of Armenian political philosophy. Written by Movses Baghramian, a Madras Armenian living in Karabakh, and entitled *Nor tetrak, Vor Kochi Hordorak* (A New Tract, Entitled Admonishment), the work encourages Armenians to restore their lost homeland.

1813   Treaty of Gulistan

The Treaty of Gulistan awards the Persian province of Karabakh to the Russians. Armenians are encouraged to settle in the area.

1828: February 10–22   Treaty of Turkmenchai

The Treaty of Turkmenchai awards Russia the Yerevan province, an Armenian land historically linked to Persia. Armenians are encouraged to settle in the area.

1829   Birth of Mikayel Nalbandian

Mikayel Nalbandian is born. A champion of Armenian independence from Russia, Nalbandian composes the song *Hiusisapayl* (Aurora Borealis), later adopted as the national anthem of Armenia. He dies in jail in Russia in 1866.

1835   Birth of Hakob Melik-Hakobian

Hakob Maliq-Hakobian (or Melik-Hakobian, 1835–88) is born. Writing under the pen name Raffi, Maliq-Hakobian becomes the most important Armenian novelist of the period through the publication of such works as *Jelaleddin* and *Khente.*

1840s   First Armenian novel written

The first Armenian novel, *Verk Haiastani* (The Wounds of Armenia) is written by Khachatur Abovian.

1850s   Reform in Turkey

The reform movement called the *Tanzimat* is launched in the Ottoman Empire in an effort to modernize the country.

1863   Armenians granted constitution

A special constitution is granted the Ottoman Armenians following the introduction of Tanzimat reforms in the Ottoman Empire.

1870s   Reactionary violence

The Armenian position in the Ottoman Empire deteriorates as reactionary violence breaks out against Tanzimat reforms.

1878   Armenian Question

At the Congress of Berlin, Ottoman Armenian delegates plead for help from Europe, marking the emergence of the "Armenian Question" as an international diplomatic issue.

1895   Armenian massacre

The Ottomans massacre 300,000 Armenian subjects on charges of disloyalty.

1903   Russification

In Russian Armenia, officials close Armenian-language churches and schools and confiscate church property following Tsar Nicholas II's effort to "Russianize" his empire.

1915–22   Armenian genocide

The Ottoman Empire institutes a policy of genocide towards its Armenian population. Through execution and forced death marches, between 600,000 and 2 million Armenians are killed out of a population of 3 million. Survivors flee western Armenia, shifting the heartland of historical Armenia to the eastern region held by Russia, the site of modern-day Armenia.

1915–17   Russian occupation

Russia occupies almost all of Ottoman Armenia.

1917   Transcaucasian Federation formed

Armenia, Azerbaijan, and Georgia form the independent Transcaucasian Federation.

1918: May–1920: December   Independent Armenia

Following the defeat of the Ottoman Empire in World War I, Armenia emerges as an independent state.

Soviet Era, 1920–91

1920–91   Soviet dominance

Armenia falls under the dominance of the Union of Soviet Socialist Republics.

1920   Treaty of Sevres

The Treaty of Sevres signed by Turkey and the Allies enlarges independent Armenia to include most of historic Armenia.

1920: November   Red Army invades

Armenia is forced to accept a communist-dominated government following an invasion by the Soviet Red Army. Armenia signs the Treaty of Aleksandropol with Turkey, ceding the northern Kars region.

1922   Transcaucasian Soviet Federalist Socialist Republic

The Soviet Union groups Armenia, Azerbaijan, and Georgia into a single republic called the Transcaucasian Soviet Federalist Socialist Republic.

1924   Nalchichevan Autonomous Republic of Azerbaijan

The Soviet Union creates the Nalchichevan Autonomous Republic of Azerbaijan and places the Nagorno-Karabakh region under Azerbaijani control. Ninety-four percent of the population of Nagorno-Karabakh is Armenian. From 1923 to 1979, the Armenian population of the region drops to 76 percent while the Azerbaijani population climbs to 24 percent.

1936   Change of status for Armenia

The Soviet Union divides Armenia, Azerbaijan, and Georgia into separate republics. Armenia becomes known as the Armenian Soviet Socialist Republic.

1970s   New emigrations

The temporary relaxation of Soviet emigration controls leads to a new wave of Armenian immigration to America, increasing the size of the community there to 600,000.

1974   Karen Demirchian

In an attempt to clean up corruption, the Soviet Union places control of Armenia in the hands of Karen Demirchian, who later is removed on charges of corruption.

1977   Referendum demanded

The Nationalist Union Party (NUP), a dissident nationalist party, calls for a referendum on the establishment of a new state in the region of historic Armenia. Karabakh and Nalchichevan are to be included in the proposed republic. The NUP declaration is the first to link the issue of Karabakh with Armenian independence.

1988   Unification declared

The Armenian Supreme Soviet demands the unification of Nagorno-Karabakh with Armenia. In response, Moscow enacts direct rule over the enclave.

1988: February 20   Unification vote

The Armenian-dominated Nagorno-Karabakh government votes to unify with Armenia. Armenians make up the majority of the population in this autonomous region of Azerbaijan.

1988: February 26–28   Anti-Armenian pogroms

Anti-Armenian pogroms (organized persecutions) break out in Sumgait, a city in Azerbaijan, over the issue of Nagorno-Karabakh. The Armenian death toll is placed at thirty-two men, women, and children.

1988: December 7   Earthquake devastates Armenia

A massive earthquake destroys the northern Armenian cities of Leninakan and Spitak, killing 25,000. The economy is adversely affected, since much of the productive land is devastated by the earthquake.

1989: Spring   Karabakh Committee released

Mass demonstrations lead to the release of members of the Karabakh Committee, arrested by the Soviets as members of an illegal nationalist movement. The Karabakh Committee will later merge with other nationalist parties to form the Armenian National Movement.

1989: September   Fuel blockade

Azerbaijan institutes a blockade of Armenian fuel and supply lines over the issue of Nagorno-Karabakh.

1989: November   Unification declared

The Nagorno-Karabakh National Council declares the unification of Nagorno-Karabakh with Armenia.

1989: December   Unification vote

The Armenian parliament votes to unite with Nagorno-Karabakh.

1990: January 13   Mob violence in Baku

The explosive issue of Nagorno-Karabakh leads to mob violence against Armenians living in Baku.

1990: August 23   Secession

Armenia declares its intention to leave the Soviet Union. The Armenian independence charter calls for the creation of armed forces, Turkish recognition of the 1915 genocide, and the unification of Karabakh with Armenia

1990: August 4   Levon Ter-Petrosian

Levon Ter-Petrosian, a member of the Armenian Pan-National Movement, is chosen chairman of the Armenian Supreme Soviet.

1991: February   Referendum decided

The Armenian parliament decides to hold a referendum on the question of independence from Russia.

1991: September 21   Independence approved

Armenian voters approve national independence from Russia in a referendum.

Independent Armenia

1991: September 23   Independence declared

Armenia formally declares its independence from Russia.

1991: October 16   Ter-Petrosian elected president

Levon Ter-Petrosian is elected president of Armenia.

1991: November   Azerbaijani embargo

Azerbaijan imposes an energy embargo on Armenia over the issue of Nagorno-Karabakh. The need to find alternative sources of fuel in Armenia leads to massive deforestation across most of the country.

1991: December   Independent state of Nagorno-Karabakh

Armenians in Nagorno-Karabakh declare their independence from Azerbaijan.

1991: December–1992: February   Shelling of Stepanakert

Azerbaijani forces shell the Armenian city of Stepanakert with multi-rocket launchers.

1992: February 25–27   Azerbaijanis massacred

Armenian snipers massacre over one hundred Azerbaijanis in Nagorno-Karabakh, as the men, women, and children try to cross an exposed valley leading into the city of Khojali.

1992: Spring   Lahin occupied

Armenian armed forces occupy the Lachin corridor, a strip of land linking Nagorno-Karabakh to Armenia.

1992: April   Agreement signed

Armenia becomes the first former Soviet republic to sign a comprehensive bilateral trade agreement with the United States and the first to receive Most Favored Nation trading status.

1993: February 2   Prime minister dismissed

President Levon Ter-Petrosian dismisses Prime Minister Khosrov Harutunian for failing to develop a sound economic policy. One week later he appoints Grant Batagrian, an avocate of radical economic reform, to the post.

1993: June   Peace accord signed

Armenia, Azerbaijan, and Karabakh Armenians sign a peace accord.

1993: July 23   Agdam falls

Armenian forces capture the Azerbaijani town of Agdam, a key site in the war over Nagorno-Karabakh.

1993: August   Jebrail and Fizuli captured

Armenian forces capture the Azerbaijani towns of Jebrail and Fizuli. Tens of thousands of Azerbaijanis flee to Iran.

1993: Fall   No end to Nagorno-Karabakh conflict

Despite international negotiations, the Nagorno-Karabakh conflict continues into 1994, with officials of the United Nations High Commissioner for Refugees estimating that one million Azerbaijanis have been displaced from Karabakh and surrounding regions.

1994: May   Cease-Fire in Nagorno-Karabakh

A cease-fire is declared in Nagorno-Karabakh. The Azerbaijani population of the region has been completely displaced. The area is now inhabited by about 160,000 Armenians. 600,000 more Azerbaijani refugees are created when the Armenian army declares the area around the Nagorno-Karabakh province a buffer zone to protect the Armenian population.

1994: December   Elections

Robert Kocharian is elected president of Nagorno-Karabakh. He holds the position until 1996 when he is invited by Armenian president Levon Ter-Petrosian to form a government in Armenia.

1995: July   Constitution adopted

A new constitution is adopted by Armenia. It is modeled after the constitution of the Fifth French Republic and provides for strong presidential powers.

1996: April   Agreement signed

Armenia signs a partnership and co-operation agreement (PCA) with the European Union.

1996: September 22   Ter-Petrosian re-elected

Levon Ter-Petrosian is re-elected president of Armenia.

1996: October–November   Armenia and Azerbaijan discuss Nagorno-Karabakh

Advisors for the presidents of Armenia and Azerbaijan meet to discuss the status of Nagorno-Karabakh. Armenia rejects an Azerbaijani offer of full autonomy for the region, supporting full independence instead.

1996: November 24   Elections

Robert Kocharian is re-elected president of the self-proclaimed Republic of Nagorno-Karabakh.

1997   Economic downturn

Armenia's economy takes a downturn as inflation continues to rise.

1998: January   Governmental crisis over Nagorno-Karabakh

President Levon Ter-Petrosian is on the defensive with his prime minister, Robert Kocharian, and his cabinet after he adopts the Organization for Security and Co-operation in Europe's (OSCE) Minsk Group's compromise plan on Nagorno-Karabakh.

1998: February 3   Ter-Petrosian resigns

An isolated Levon Ter-Petrosian resigns the presidency.

1998: March 30   Kocharian elected

Prime Minister Robert Kocharian is elected president of Armenia.

Bibliography

Batalden, Stephen K. and Sandra L. Batalden. *The Newly Independent States of Eurasia: Handbook of Former Soviet Republics.* Phoenix, Ariz.: The Oryx Press, 1993

Croissant, Michael P. *The Armenia-Azerbaijani Conflict: Causes and Implications.* Westport, Conn.: Praeger, 1998.

Curtis, Glenn E., ed. *Armenia, Azerbaijan, and Georgia: Country Studies.* Washington, D.C.: Federal Research Division, Library of Congress, 1995.

Economist Intelligence Unit, The. *Country Profile: Georgia, Armenia, 1998–1999.* London: The Economist Intelligence Unit, 1999.

Economist Intelligence Unit, The. *Country Report: Georgia, Armenia, First Quarter 1999.* London: The Economist Intelligence Unit, 1999.

Goldberg, Suzanne. *Pride of Small Nations: The Caucasus and Post-Soviet Disorder.* London, Atlantic Heights, NJ: Zed, 1994.

Kaeger, Walter Emil. *Byzantium and the Early Islamic Conquests.* Cambridge: University Press, 1992.

Lang, David Marshall. *Armenia: Cradle of Civilization.* London, Boston: Allen and Unwin, 1978.

McEcedy, Colin. *The New Penguin Atlas of Medieval History.* London: Penguin, 1992.

Walker, Christopher J. *Armenia: The Survival of a Nation.* New York: St. Martin's Press, 1980.

Austria

Introduction

The Federal Republic of Austria is a small state situated in southern-central Europe. It consists of nine provinces (Burgenland, Carinthia, Lower Austria, Salzburg, Styria, Tirol, Upper Austria, Vorarlberg, and the capital Vienna) and shares borders with eight countries (Germany, the Czech Republic, Slovakia, Hungary, Slovenia, Italy, Switzerland, and Liechtenstein). Austria's territory has a total area of 32,368 square miles (83,858 square kilometers). A section of the eastern Alps stretches from the western panhandle into the center, forming the country's mountainous backbone. The river Danube cuts through the northern hills, continuing its course into the Vienna Basin and the Pannonian Lowlands (in the east). Almost half (forty-six percent) of the republic's territory is covered by forests, which makes the country one of the most heavily wooded in Europe. In 1996, Austria had an estimated population of 8.06 million. The great majority of its inhabitants (ninety-eight percent) are German-speaking. Other officially recognized ethnic groups include Croats and Hungarians (many of whom live in Burgenland), Slovenes (in southern Carinthia and southern Styria), Czechs and Slovaks (in Lower Austria), and Romany and Sinti (in Burgenland). Seventy-eight percent of Austrians are Roman Catholic, while five percent are Protestant and four-and-one-half percent belong to other religious groups.

Although the first mention of a territory called "Austria" is found in a document from the tenth century, the present-day state is a relatively recent creation. It originated at the end of World War I from the collapse of a multinational empire stretching from the Balkans to today's Czech and Slovak Republics and from Northern Italy to Hungary. The fate of the territories that form today's Austria is intimately connected with the fortunes and misfortunes of this empire, which was ruled for centuries by the powerful Habsburg dynasty. The origins of the Habsburg empire go back to the thirteenth century, when Rudolf of Habsburg, king of the Holy Roman Empire of the German Nation, extended his control over the lands held by the extinct dynasty of the Babenbergs. The primarily German-speaking territories, which coincide approximately with present-day Austria, became Habsburg hereditary possessions in 1282. As the cradle of a powerful central-European empire, they remained under Habsburg rule until 1918.

The Habsburg Empire

Between the fourteenth and the sixteenth century, the Habsburgs succeeded in enlarging and consolidating their territorial possessions by pursuing a shrewd policy of military conquest, political alliance, and intermarriage with some of the most powerful dynasties in Europe. When Charles of Habsburg became emperor of the Holy Roman Empire (as Charles V, in 1519), he found himself ruling a multi-ethnic empire extending from Spain and the Netherlands in the west to the Austrian hereditary lands in the east. The unmanageable size of the empire prompted Charles to relinquish the Austrian lands to his brother Ferdinand. This step split the Habsburg dynasty into a Spanish line (under Charles) and a properly Austrian line (under Ferdinand). As a result of Ferdinand's marriage, the kingdoms of Bohemia and Hungary were also added to the Austrian hereditary lands. A decisive factor in establishing the political preeminence of the Habsburgs in central Europe was their hold on the crown of the Holy Roman Empire, the loose political confederation composed of weak and divided German states. Although the German crown was offered to a sovereign chosen by the German princes and never became hereditary, it was held by the Habsburgs almost without interruption after the sixteenth century.

Two powerful menaces confronted the Austrian Habsburgs in the sixteenth and seventeenth centuries. On the domestic front, the mid-1500s saw the rapid spread of Luther's Reformation in their territories, the majority of which converted to Protestantism. Remaining faithful to Catholicism, the Habsburg sovereigns sought to hold on to political power by supporting the Counter-Reformation, the ideological counteroffensive of the Catholic Church. Through campaigns of forcible conversion and the expulsion of large Protestant groups from their lands, the Habsburgs succeeded in re-imposing Catholicism on the great majority of their subjects. Their political and territorial power was further strengthened by the outcome of the Thirty Years War (1618–48), the conflict which finally settled the political struggle

between Protestants and Catholics in central Europe. The house of Habsburg presented itself as well as a bulwark of Catholicism and Western civilization in its fight against its other major enemy of this period, the Turks of the Ottoman empire. After conquering most of the Balkan peninsula, the Turks had been making incursions into the southern and eastern territories of the Habsburg empire since the 1530s. They advanced as far as the capital Vienna, which was besieged twice (in 1529 and 1683). The second siege of Vienna prompted an Austrian, German, and Polish coalition to react forcefully against the Ottoman aggressors, who were soundly defeated in 1697 at Zenta (in the Balkan province of Vojvodina). The treaty of Carlowitz (1699) definitively settled the dispute between the Habsburg and the Ottoman empires, setting a lasting boundary between the two states.

The history of the Habsburg empire in the eighteenth century was profoundly shaped by the far-reaching reforms introduced by Empress Maria Theresa (r. 1740–80) and her son Joseph II (1780–90). After fighting wars against Prussia and France, which had challenged the legitimacy of her succession to the throne, Maria Theresa implemented policies aimed at strengthening the central authority of the Habsburgs

over the disparate lands of the empire (which at this point encompassed the Austrian hereditary lands and the kingdoms of Bohemia, Hungary, and Croatia). The measures introduced by the empress aimed at establishing a secular, modern state with a powerful central administration. They included reforming the bureaucracy and the army, curtailing the power of local nobility and of the Catholic Church, and promoting education and scientific development. Her son Joseph II pursued her policy of reforms, though not always with lasting success. His major achievements were the abolition of serfdom, which since the Middle Ages had given the nobility great power over the peasants inhabiting their lands, and proclamation of the Toleration Edict, which introduced religious freedom in the empire and further weakened the power of the Catholic Church.

The order established at the end of the eighteenth century suffered a blow during the Napoleonic Wars (1803–14). French General Napoleon Bonaparte pushed well into the Habsburg territories in the wake of his expansionistic campaigns. Foreseeing the near collapse of the Holy Roman Empire of the German Nation under Napoleon's onslaught, the then German King Francis of Habsburg assumed the title of Austrian emperor (1806). Following Napoleon's defeat, an international congress (1814–15) was held at Vienna under the presidency of Austrian Chancellor Clemens von Metternich. Thanks to Metternich's political skills, the Austrian empire retained its political preeminence over the German Confederation, the successor to the dissolved German empire. Nevertheless, in the two decades following the Congress of Vienna, Prussia came to pose an increasingly powerful challenge to Austrian authority in central Europe. Through a network of customs agreements, the emerging Prussian superpower managed to establish its hold on the economies of the German states. This development led to the foundation of the German Customs Union in 1834. Austria's exclusion from the economic union resulted in a considerable erosion of its political and economic control over the German states and greatly weakened its international standing.

The Congress of Vienna inaugurated a period of internal peace and stability, which lasted until the revolutions of 1848. The Austrian empire experienced unprecedented economic growth in the wake of the Industrial Revolution. Economic development was also accompanied by increasing demands for political democratization and economic liberalization. In particular, liberal politicians rallied for the abolition of censorship and other repressive measures put in place by Metternich's regime. These demands were at the heart of the revolutions that swept through the Austrian lands between March, 1848, and the early part of 1849. Though the uprisings were successfully suppressed by the central government, Emperor Ferdinand was forced to resign. The long rule of his successor, Francis Joseph, coincided with the final phase in the history of the Habsburg empire.

The second half of the nineteenth century witnessed the unstoppable decline of the multi-ethnic empire, which was increasingly weakened by economic and political backwardness and by a series of disastrous military conflicts. On the domestic front, the push for independence by non-German nationalities led first to the loss of the northern Italian territories (1859), which soon thereafter were incorporated into the newly established Italian monarchy. After long negotiations with the Hungarians, Emperor Francis Joseph agreed to the political reorganization of the Austrian empire into a dual monarchy in 1867. The new political unit comprised two distinct monarchies—Austria and Hungary—that were symbolically united in the person of the emperor. This step did not resolve the grievances of other non-German nationalities, particularly the Czechs. Their continued fight for independence greatly impaired Austria's domestic and foreign policy. Prussia in particular was able to take advantage of Austria's internal weaknesses. Its rivalry with Austria over control of the German states in central Europe was definitively settled in 1866, after a brief military confrontation. Prussia's swift victory in the conflict resulted in Austria's complete exclusion from all German affairs and confirmed the waning international status of the dual monarchy.

The Twentieth Century

In 1914 the assassination in Sarajevo of the Austrian heir to the throne by a Bosnian nationalist triggered a series of events that led to the outbreak of World War I. The conflict dealt the crumbling monarchy a final blow. Following the demise of the military coalition of which Austria was part, Emperor Charles I abdicated (November, 1918). This marked the end of centuries of Habsburg rule in Austria and sealed the empire's dismemberment. In the same month, a republic was proclaimed in these German-speaking territories. Enormous economic difficulties and a chaotic political climate marred the brief life of the first Austrian Republic. The burden of reconstruction and taxing war reparations imposed by the victorious Allies was compounded, at the end of the 1920s, by a world-wide economic crisis. The democratic experiment of the first republic ended in 1933 with the establishment of an authoritarian regime under Chancellor Dollfuss. After 1936 Austria became increasingly unable to oppose Adolf Hitler's pressures for a merger with Germany. Hitler's march into Austria (in 1938) finally opened the way for the forcible unification of the two states (the so-called *Anschluss*). Austria ceased to exist as an independent unit and became part of Germany's Third Reich.

At the end of World War II, the reconstituted political parties proclaimed the second Republic under the leadership of the esteemed Social Democrat Karl Renner. The first democratic elections were held in 1945. Although the winning Allies (Britain, France, Russia, and the United States) regarded Austria as the first victim of Nazi Germany, rather than as Germany's collaborator, they nevertheless divided the new republic into four sectors of occupation. The unfolding of the Cold War in the following decade further hampered Austria's negotiations to recover full sovereignty. A breakthrough came in 1955, with the signing of the State Treaty. The document restored Austria's sovereignty and opened the way for the departure of the occupational armies. In return, the new republic committed itself to remaining permanently neutral. Also in 1955, Austria joined the United Nations (UN), becoming a full member of the international community.

The postwar period has been a time of great political stability and economic prosperity for Austria. Political life has been shaped by the principle of social partnership, which fosters cooperation and consensus among political and economic forces. Moreover, Austria's involvement with the international community has grown considerably since the end of World War II. As a member of the UN, Austria has taken part in various peace-keeping operations (Vienna has been one of the permanent UN seats since 1979). If in the postwar period the republic has tended to play a relatively marginal role as a neutral state between the two Cold War blocks, the end of the Cold War deeply transformed Austria's international standing and role. Since the 1990s, Austria has found itself at the heart of an increasingly united Europe. In 1995 Austria was invited to participate in the Partnership for Peace of the North Atlantic Treaty Organization (NATO), which resulted in its involvement in NATO peace operations in the Balkans. In a referendum held in 1995, the majority of the population supported Austria's membership in the European Union. This step sealed Austria's full political and economic integration within Europe.

Timeline

800–400 B.C.   Illyrian settlements and kingdom of Noricum

The Illyrians, an Indo-European population, settle in the territories that constitute present-day Austria. (Archaeological findings document that these territories are a land of transit and migration since the middle paleolithic era, c. 80,000 B.C.) Their civilization is often named after Hallstatt, the most important site of their culture; they are particularly active in trading (salt and ore). Around 400 B.C. Celtic tribes begin conquering the lands inhabited by the Illyrians. They found the kingdom of Noricum (in the eastern Alps), which is the first known "state" established on Austrian territory.

c. 15 B.C.   Romans invade kingdom of Noricum

The region's iron riches, as well as its strategic geographical position, attract Roman invaders. They occupy territories roughly coinciding with present-day Austria. which become the Roman provinces of Noricum, Pannonia, and Rhaetia. The Romans build an extensive network of roads and found several settlements, including Vindobona (Vienna) and Iuvavum (Salzburg). Christianity starts spreading around the second century A.D.

The Romans' hold on the provinces begins to crumble in the third century A.D., as invasions of Germanic and Slavic tribes become increasingly frequent. By the fifth century several Germanic tribes (Goths, Heruli, Langobardi) settle on Austrian territory.

6th–9th century   Migrations and Charlemagne

The era following the demise of the Romans is a time of mass migrations and devastations. Various Germanic and Slavic tribes vie for control over the former Roman provinces and the Roman cultural legacy declines rapidly. In the 790s the territories come under the control of Charlemagne, king of the Franks (r. 771–814). They become marches, that is, border provinces of his expanding empire (which in the early 800s encompasses most of western and central Europe).

976   Otto I grants Babenberg family territory in Danube valley

Leopold of Babenberg, descendent of a noble family from Bavaria, receives from German King Otto I a fief in the valley of the Danube river, between the Enns and Traiser rivers, in what is today Lower Austria (*fief* is the term for lands that medieval rulers entrust to subordinate lords in exchange for allegiance). The new territory represents the historical cradle of Austria as a political and territorial unit.

996   Oldest record of Old German word for Austria

Ostarrîchi, the Old German word for Austria, appears for the first time in a church document from 996. It means "kingdom of the East" (*ostar*-eastern; *rîchi*-kingdom, realm).

1156   Duchy of Austria founded

Following a petition to German Emperor Frederick Barbarossa, the march (border province) granted to the Babenbergs becomes a duchy and achieves greater independence from imperial authority. In the same year Duke Henry II moves his definitive residence to Vienna. By the end of the twelfth century the Babenbergs' holdings extend from the Czech and Slovak border in the north to the province of Styria in the south and to the March and Leitha rivers in the east.

Cultural life flourishes at the Vienna court in the second half of the thirteenth century. The court attracts some of the major German poets of the time, most notably Walther von der Vogelweide (c. 1170–1230). Vogelweide is associated with the new poetic genre of the *Minnesang*, which praises the medieval virtue of chivalrous love as well as the bonds of trust and faithfulness between knight and master. Monastic culture, centered around the copying and preservation of manuscripts, blossoms at the monasteries and abbeys founded by the Babenbergs. This era also sees the development of Romanesque architecture.

1246: June 15   Frederick II of Babenberg dies in battle over border dispute

The Austrian duke leaves no legitimate heirs. His death marks the end of the Babenberg dynasty and provokes a bitter struggle for succession to the throne. King Ottokar II of Bohemia (in today's Czech Republic), who had married a Babenberg princess, incorporates the Austrian duchy into his territories.

1276   Rudolf of Habsburg invades Austrian territories

Rudolf (r. 1273–91), king of the Holy Roman Empire of the German Nation (a loose confederation of German territories), challenges the rule of Bohemian King Ottokar II in Austria. The conflict comes to a head in 1278, when Rudolf's army defeats the Bohemians on a field northeast of Vienna. Ottokar himself dies in the battle. The duchies of Austria and Styria become Habsburg hereditary possessions in 1282.

The Habsburgs are a noble family from Swabia, in southwestern Germany. Beginning with Rudolf's reign, they rule Austria for the next six centuries, almost without interruption.

1300s–1500s   Habsburg empire grows

Habsburg sovereigns enlarge and consolidate their territorial possessions in central Europe. They achieve this through both military conquest and a policy of intermarriage with some of the most powerful European dynasties. A case in point is Maximilian I (r. 1493–1519) and his immediate descendants. Through his marriage to the heiress of Burgundy (1477), the Habsburgs gain a realm extending from eastern France to the Flemish provinces (in the Netherlands). In 1496 Maximilian's son Philip marries the Spanish heiress, thereby adding the Spanish lands of Castile and Aragon to the family holdings. When Maximilian's grandson Charles is elected emperor of the Holy Roman Empire of the German Nation (as Charles V, r. 1519–58), he finds himself ruling the Habsburg hereditary lands (corresponding to today's Austria) as well as the three most important political entities of his time. These are the German empire (comprising an array of German states in central Europe), Spain, and the Crowns of Bohemia, Hungary, and Croatia. The Habsburg policy of intermarriage and family alliances is immortalized in the famous saying, "Bella gerant alii, tu felix Austria nube" (Let others fight wars; you, lucky Austria, marry).

Because of the unmanageable size of the empire, Charles V agrees to cede the Austrian territories to his brother Ferdinand. The Habsburg house thus splits into two main lines: on the one hand, the Spanish Habsburgs under Charles V; on the other, the properly Austrian Habsburgs under Ferdinand. Ferdinand, who has married the daughter of the king of Bohemia and Hungary, becomes the king of these territories when the last Bohemian king dies without male heirs (1526). The union of the Archduchy of Austria and the kingdom of Bohemia and Hungary under Ferdinand lays the foundations for a large Austrian empire in central Europe.

14th–15th century   Social and economic conditions in the Austrian lands

The structure of late medieval society is based on a hierarchy of estates (social groups whose status is defined by birth), with the emperor and the royal family at the top. The high clergy and the aristocracy form the next tier; the wealthy burghers in the towns follow. At the bottom of the pyramid are the peasantry, who far and away represent the largest class. Under the system of feudalism, peasants owe personal services to their lords (either labor or payment of rent for the lands they cultivate). Local lords (whether secular or ecclesiastic) are also administrators of justice. Generally, the aristocracy and the clergy enjoy great autonomy and special privileges (for instance, tax exemptions). They elect representatives to councils (*diets*), which counsel the sovereign on taxes and other administrative matters.

More than four-fifths of the population lives outside towns, although a large number of towns is founded in the Alpine lands at this time. The backbone of the towns' economy is artisans and craftsmen, who are organized in professional associations known as guilds. Commerce and industry flourish in the Alpine lands due to their position at the intersection of transit routes connecting eastern and western Europe, as well as Italy and Germany. The river Danube develops as a main route of commerce between Germany, Bohemia, and Hungary (traded wares include cloth, salt, wine, and cattle). Iron foundries in Styria and the glass industry in Tirol are among the growing industries.

c. 1450s–1520s   Renaissance culture blossoms at Habsburg court

The court of the Habsburgs becomes a center for the blossoming movements of Humanism and the Renaissance (cultural and intellectual currents that stem from the study of the texts of classical antiquity; they oppose ignorance and superstition and promote the ideal of human self-fulfillment). Among the personalities that embody the new culture is Eneas Silvius Piccolomini (1405–64), private secretary and adviser of Austrian sovereign Frederick III (r. 1440–93) and author of didactic tracts and travelogues on Bohemia, the Rhineland, Austria, and Vienna. Beyond the Habsburg dynasty, other forces driving the cultural and artistic renaissance are the aristocracy and the Church.

1529: September   Turks lay siege to Vienna

The siege is the result of the expansionistic politics of the Turks of the Ottoman Empire, who have conquered most of the Balkan peninsula in the course of the fifteenth century. Though the siege is broken in October, the Turks become a permanent threat to the Habsburg empire. They repeatedly raid eastern and southern Habsburg territories and besiege Vienna again in 1683. Because of its strategic position, the Austrian empire comes to be a political and military, as well as cultural, bulwark against Turkish expansion in southeastern Europe. The Habsburgs present themselves as champions of the Christian faith and Western civilization which are assailed by Islam and the Orient.

c. 1540s–1650s   Habsburgs lead Counter-Reformation in southern Europe

The new religious doctrine of Protestantism, founded by the German theologian Martin Luther as a reform movement against the corruption of the Catholic Church, spreads quickly through the Austrian lands. The majority of Habsburg subjects converts to Protestantism by the 1550s. Remaining faithful to Catholicism, the ultra pious Habsburgs assume a leading role in the Counter-Reformation, the ideological counteroffensive of the Catholic Church. Through often violent campaigns of forced conversion, Catholicism is reinstated as the majority religion in the Austrian lands. This objective is also achieved through mass migrations of Protestants (who resettle in the still Protestant German states and in the imperial cities of southern Germany). The campaign of religious reconquest finds its pillar in the Society of Jesus, a religious order of educators and teachers founded in 1534 by the Spaniard Ignatius Loyola. By the end of the sixteenth century, the Jesuits impose their influence on various universities, including those at Vienna and Graz. They also found gymnasiums (elite secondary schools) in the major cities of the Austrian lands, including Graz, Innsbruck, Linz, and Salzburg. If today approximately eighty percent of all Austrians are Catholic, then this is due to the long-term effects of the Counter-Reformation.

1618   Ferdinand of Habsburg unleashes devastating war over religious issues

In the spirit of the Counter-Reformation, Archduke of Austria Ferdinand II takes the occasion of his election to the Bohemian throne to begin a campaign to curtail the influence of the Protestant nobility in Bohemia (which is part of the Habsburg lands). This action provokes widespread revolts in Bohemia, which in turn unleash a conflict between Catholic and Protestant German rulers (the first of the armed confrontations that come to be known as the Thirty Years War). Ferdinand (German emperor since 1619) heads the Catholic league. He hopes to reestablish Catholicism in all German states and to strengthen the waning authority of the Holy Roman Empire. In the course of the 1630s, the conflict spreads beyond the borders of the German states and involves several European countries (including Sweden, Spain, and France). After three decades of devastation, the warring parties sign the Peace of Westphalia (1648), marking a further consolidation of Catholicism in the Habsburg territories and the strengthening of Vienna's central control.

The effects of the prolonged war on the economy are disastrous. The conditions of the weaker social group, the peasants, worsen considerably, while nobility and church often acquire new lands in the territorial reorganization.

c. 1600s–1700s   Baroque culture flourishes in Counter-Reformation Austria

Deeply influenced by the religious fervor of the Counter-Reformation and by the devastation of the Thirty Years War, the culture of the Baroque era in the Austrian lands develops themes of the transitoriness of life, the powerlessness of humans in the face of fate, and an obsession with death and the afterlife. The highest achievements of Austrian Baroque are in painting and architecture. In these arts the style of the Baroque distinguishes itself for its exuberant forms and strong visual contrasts, ornamentation characterized by curved rather than straight lines, and tendency toward over-decoration. The architects Johann Bernhard Fischer von Erlach the Elder (1656–1723) and Johann Lukas von Hilde-

brandt (1660–1750) are among the most celebrated creators of Baroque art in Austria. The former designed the Schönbrunn palace, the summer residence of the Habsburgs in Vienna, and the church of St. Charles Borromäus, Vienna's monumental Baroque church, along with numerous buildings in Graz and Prague. Hildebrandt is renowned for his elegant palaces, which include Vienna's Belvedere Palace. Built as the residence of Prince Eugene of Savoy, the Belvedere represents one of the most notable examples of Baroque architecture in the German-speaking countries.

1683   Second Turkish siege of Vienna

An Austrian, German, and Polish coalition forms to push back the Ottoman aggressors. The coalition army attains a decisive victory near Zenta (in the Balkan peninsula) in 1697, under the command of Prince Eugene of Savoy, whose military feats become legendary. Peace negotiations between the Habsburg and Ottoman governments result in the Treaty of Carlowitz (1699), which establishes a common boundary between the two empires. With the exception of the Habsburg annexations of the Balkan lands of Banat (1718) and Bukovina (1775), this boundary remains intact until 1878.

1700s   The Habsburg empire in the early 18th century: a political and ethnic mosaic

At the beginning of the century, the possessions of the Austrian Habsburgs include territories with different ethnic compositions and historic traditions. There are three major divisions: 1) the so-called hereditary lands, which include the provinces of Vorarlberg, Styria, Upper and Lower Austria, Carniola, and Carinthia (and some scattered German lands); 2) the kingdom of Bohemia, which consists of Bohemia, Moravia, and Silesia (the lands of the Crown of St. Wenceslas); and 3) the kingdom of Hungary, comprising Hungary, Transylvania, and the kingdom of Croatia (the lands of the Crown of St. Stephens). Because there is no uniform administrative system, the person of the ruler constitutes the primary link uniting the disparate Habsburg holdings. Nonetheless, the emperor enjoys only limited powers. Although he has full discretion over military matters and questions of foreign policy (treaties and alliances), his lands are under the direct administration of the nobility, who exercise unlimited authority over the peasants on their estates, administer justice, and provide some social services.

The peasantry continues to form the largest social group. They represent the backbone of the largely agrarian economies of the Habsburg lands. With the exception of some free peasant estates in the Alpine lands (particularly in Tirol), most peasants remain tied to the land by the feudal bond of serfdom and have virtually no contact with the central administration of the empire. A class of merchants, artisans, and professionals shapes the life of towns and cities, which often enjoy extended rights of self-government. However, the shift of major trade routes to the Atlantic (following the colonization of the American continent) and the devastations of the Thirty Year War greatly hamper the economic growth of cities.

1713   Emperor Charles VI proclaims Pragmatic Sanction

With this decree, Charles VI (r. 1711–40) establishes that the Austrian lands are indivisible and that they can be inherited by both female and male rulers of the Habsburg house. In this way, Charles seeks to prevent the Austrian Habsburgs from sharing the fate of the Spanish line of the family. The last Spanish Habsburg died without male heirs in 1700 and as a result the dynasty became extinct.

1740   Maria Theresa becomes first female ruler of Habsburg dynasty

At the death of Emperor Charles VI without male heirs, his daughter Maria Theresa accedes to the Habsburg throne. Soon she must wage wars against numerous enemies, including Prussia, Bavaria, and France. These hostile nations either attempt to carve out a portion of the Habsburg lands for themselves or challenge the legitimacy of the empress's succession (Silesian War, 1740–1748; Seven Years War, 1756–1763). Maria Theresa succeeds at retaining the Habsburg lands, with the exception of the rich province of Silesia, which comes under Prussian rule. Prussia emerges as the power that challenges the influence of the Habsburgs within the German states of the Holy Roman Empire.

Maria Theresa proves to be a politically gifted ruler, capable of selecting a team of talented statesmen and counselors. She implements a wide range of reforms aimed at unifying political, economic, and social structures in the diverse Habsburg territories. They include a reorganization of the administration and of the army, the development of a reliable bureaucracy, the promotion of education and scientific research, and the extension of state control over religious matters. These measures help reduce the authority of the local nobility and of the Church. They lay the foundations for the absolutist rule of the sovereign and strengthen the territorial power of the Habsburg dynasty. Maria Theresa's government is shaped by the ideals of the Enlightenment, the intellectual movement that exalts human rationality as a guide in promoting the equality, freedom, and personal fulfillment of all human beings. The empress perceives it as her duty to devote her life to the betterment of the conditions of her subjects, many of whom regard her as a benevolent mother.

1781   Wolfgang Amadeus Mozart moves to Vienna

A native of Salzburg, the gifted composer (b. 1756) achieves renown as a child wonder under the vigilant care of his father Leopold, also a musician. Following his early success at Maria Theresa's court, Mozart undertakes a concert tour of Europe and subsequently enters the service of the archbishop of Salzburg (1769). A falling out with his powerful patron causes him to move to the Viennese court, where he composes some of his most celebrated works (among them are the operas *The Marriage of Figaro, Don Giovanni,* and *The Magic Flute*). In spite of his genius, Mozart fails to secure a stable income during his Viennese period. He dies in Vienna in 1791, plagued by poverty and illness.

1781   Emperor Joseph II issues Toleration Edict

The Edict represents one of the most significant reforms introduced by Joseph II, son and immediate successor of Maria Theresa (r. 1780–90). It grants most non-Catholics the right to exercise their religion. Further measures (introduced in individual provinces) allow for the partial emancipation of the Jews, who have been the target of persecution and discrimination since the Middle Ages, because of their different appearance, life style, and religious practices. Joseph's policies continue the reforms initiated by his mother, often taking them in an even more radical direction. He curtails the power of the Church by bringing many monasteries and church properties under state control, introduces agrarian reforms, abolishes barriers to domestic trade, and further strengthens the central administration of the state. His greatest achievement is the abolition of serfdom (Serfdom Patent, 1781), which frees peasants to choose their profession, marry, or even leave the land without the lord's permission. In spite of the emperor's good intentions, however, his radical reforms often lack the necessary support measures and encounter great resistance in many territories. They are frequently implemented only in part and are sometimes even repealed.

1750s–1800s   Vienna emerges as Europe's musical capital

In the course of the eighteenth century, opera, a new musical genre imported from Italy, which combines theatrical and musical performance, flourishes at the court of Vienna. It undergoes a renaissance under Christoph Willibald Gluck (1714–87), who promotes and reforms the genre during his term as composer at the imperial court. To counterbalance the spread of Italian opera, Emperor Joseph II takes steps to encourage the composition of works in the German language (one of the results is Mozart's celebrated opera *Die Entführung aus dem Serail*). In addition to Mozart, two other great composers cross paths in Vienna in the last three decades of the eighteenth century. Joseph Haydn (1732–1809), a friend and mentor of Mozart, finds employment as music director under the tutelage of Prince Esterházy (in Eisenstadt, near Vienna). Here he composes church music along with numerous symphonies, operas, and string quartets. Among Haydn's pupils is German-born composer Ludwig van Beethoven (1770–1827), who has moved to Vienna to study with Haydn. After establishing his residence in Vienna, Beethoven earns his living as one of the first free-lance composers (i.e., not under the direct patronage of the court or of members of the aristocracy).

This golden period of musical renaissance becomes known as Vienna classicism. It has an extraordinary impact on the development of modern music, particularly on the new Romantic movement. Romanticism in music reaches its peak in the works of Viennese composer Franz Schubert (1797–1828). He is the creator of the *Kunstlied*, an innovative genre based on setting poems to music.

1806 Holy Roman Empire of the German Nation dissolved

The collapse of the Holy Roman Empire results from the upheavals caused by the French Revolution (1789). After establishing a dictatorial government in France (1799), General Napoleon Bonaparte starts the systematic conquest of Europe and invades most of the Habsburg territories. German Emperor (and Habsburg sovereign) Francis II (r. 1792–1835) foresees the near collapse of the Holy Roman Empire and assumes the new title of emperor of Austria (as Francis I). The dissolution of the Holy Roman Empire at Napoleon's hands marks the beginning of a decline in the influence exerted by the Habsburgs on the German states of central Europe. The imperial crown has been in Habsburg hands almost without interruption since the fifteenth century.

1814: September   Historic congress opens at Vienna

After over twenty years of turmoil caused by the Napoleonic Wars (1803–14), the major European powers (Britain, Prussia, Russia, and Austria) convene a congress in Vienna, which meets through June, 1815. It reestablishes the boundaries of several European states and creates a territorial order that will endure in Europe until the outbreak of World War I (1914). Count Clemens von Metternich, the Austrian chancellor and foreign minister, presides over the political event (g. 1809–48). Through his exceptional diplomatic skills, Metternich secures for Austria a position of preeminence within the German Confederation (which is the successor to the dissolved Holy Roman Empire of the German nation). In the newly established European order, the Austrian empire comes to occupy a key position between the expanding Prussian empire in the northwest, the Russian empire in the east, and the crumbling Ottoman empire in the Balkan peninsula. The Habsburgs retain the territories they had held before the war, with some exceptions: they lose the Austrian Netherlands and scattered German lands but gain permanent rights over coastal territories on the Adriatic Sea, including Dalmatia, Venetia, and Istria; they give up some Polish territory but retain the major portion of their previous Galician holdings.

Both in domestic and international affairs, Metternich's policies center on suppression of the revolutionary ideals of freedom, equality, and fraternity, which have spread in the Napoleonic period. His government inaugurates a period of political repression that will last for the next three decades.

This choir performs on the 130th anniversary of the first performance of "Silent Night." (EPD Photos/CSU Archives)

1819   "Silent Night" is written

The Christmas carol, "Silent Night," is written in Oberndorf, near Austria, for the choir to perform during a period when the church organ was out of order.

1819   Carlsbad decrees introduce censorship

Using the political murder of German journalist Kotzebue as a pretext, Chancellor Metternich introduces legislation that imposes censorship measures and strict state supervision over education in schools and universities (the decrees are named after the German city of Carlsbad). Metternich's repressive policies involve dismissing allegedly subversive teachers, dissolving student organizations (the so-called *Burschenschaften*), suppressing some newspapers, and censoring all publications of fewer than twenty *Bogen* (roughly corresponding to 300 pages).

1820s–40s   The economy booms in the Austrian lands

Following the economic stagnation caused by the Napoleonic Wars, the economy picks up again in the wake of the Industrial Revolution. Industrial development is triggered by the introduction of more effective machinery in many manufacturing branches. Another decisive factor is the extraordinary improvement of the transportation network, for example, the introduction of steamboats and railroads, as well as the building or improvement of roads and canals. The textile industry flourishes in Bohemia, Moravia, and the region around

Vienna. The Alpine regions register a significant increase of iron and steal production. The government (led by Emperor Francis and his advisers) generally does not intervene to promote this development for fear that any change may upset the social and political order of the empire.

An outcome of industrialization is the rapid growth of the working classes in urban centers. The concentration of industrial workers causes Vienna to double its population (from around 200,000 in 1780 to over 400,000 by the 1850s). Working conditions are generally very poor: long hours, low pay, and insufficient housing prevail. In spite of the swift expansion of industries, agriculture remains the predominant branch of Austrian economy. Around the middle of the century, farming and related occupations represent most of the economy. They employ roughly three quarters of the population, while only seventeen percent find employment in industry and mining.

1825   Johann Strauss (the father) becomes court dance director at Vienna

The prominent Vienna-born composer (1804–49) achieves renown for his dance music, particularly the waltz, an innovative whirling dance that becomes the rage of Viennese high society. His children Johann (1825–99), Eduard (1835–1916), and Josef (1827–70) follow in his footsteps. They distinguish themselves in the composition of dance music and operettas (a comedic genre that combines singing, dance, and drama).

1832   Inauguration of first railroad speeds up development of transportation system

The new railroad connects the cities of Budweis (in the modern-day Czech Republic) and Linz. The government quickly realizes the importance of the new means of transportation for military and economic purposes and assumes a leadership role in developing the railroad system. The center of the railroad network is the capital, Vienna. This period also sees the rapid expansion of steam navigation on the Austrian rivers. The Danube Steamship Company (*Donaudampfschiffahrtsgesellschaft or DDSG*), founded in 1829, contributes greatly to developing the commercial life of the river Danube.

1832: June, July   Six articles added to Carlsbad decrees

The decrees, promoted by Metternich, intensify censorship and introduce repressive measures aimed at foiling any type of protest or disturbance. Yet they are effective only in part, as they prove powerless to halt the spread of political materials introduced from abroad. Despite the climate of ideological repression, liberalism, a political doctrine which conceives of the state as the promoter of the freedom and welfare of all citizens, spreads in the Austrian lands (particularly in Lower Austria and in Vienna). It finds a fertile terrain among the Austrian middle class, composed of merchants and manufacturers, bureaucrats, lawyers, doctors, and teachers. Liberal reformers agitate for legislative and tax reforms, for basic civil liberties (e.g., right to free speech and an uncensored press), for more effective representation of citizens in elected assemblies, and for the abolition of barriers to trade and commerce. Their ideals do not necessarily include the establishment of a democracy open to all social classes. Rather, they generally favor restricting suffrage (the right to vote) to middle and high-income citizens.

1820s–60s   Theater flourishes in Vienna

Theater becomes the entertainment of choice for many Viennese, as it is one of the few public expressions tolerated by the censor. Vienna has two main theaters within the city walls and three in the suburbs. Whereas the theaters in the center play serious dramas, the suburban theaters specialize in comedies of the *Volkstheater* (popular theater). This dramatic genre combines elements of fairy tales, parodies of antique mythology, and allusions to social and political life in contemporary Vienna. It reaches its peak in the works of Vienna-born Johann Nestroy (1801–62), an actor and playwright who becomes a celebrity in the Viennese dramatic scene. Because of the caustic social critique and crude language that characterize his plays, Nestroy is repeatedly targeted by Austrian censors. The tradition of serious drama and classical tragedy reaches its highpoint in the works of Franz Grillparzer (1791–1872). His dramas weave together history and myth to recount the Austrian past and the great deeds of the centuries-old Habsburg dynasty.

1834   Austria excluded from German Customs Union

Under pressure from Prussia, the states of the German Confederation agree to unify tariffs and customs and to abolish trade barriers. Austria decides not to join the newly established customs union. Its exclusion causes a further erosion of its influence among the German states. At the same time it strengthens the position of Prussia, Austria's main antagonist in German and central-European politics.

1848: March 3   Revolution sweeps through the Habsburg lands

As revolutionary uprisings initiate in Paris (February, 1848) and spread throughout Europe, revolts break out in Vienna and Budapest. Peasants' insurrections and workers' revolts soon follow in all parts of the empire. Causes for the discontent include the well-to-do middle-classes' restricted political power, the hardships suffered by the growing class of industrial workers in the economic crisis of 1846–47, and the aspirations to national independence of the many ethnic groups in the Austrian empire. In Vienna, insurgents demand the establishment of a more liberal government. Count Metternich, symbol of the repressive regime established after the Congress of Vienna (1814–15), is forced to resign and flees to Great Britain.

Although Emperor Ferdinand grants many of the requests of the insurgents (including the enactment of a constitution and the abolition of censorship), the revolutionary movement continues to grow. Hungarians, Croats, Czechs, and Italians make demands for autonomy and even national independence. Nevertheless, the Austrian army manages to regain control of the rebellious territories in a series of military campaigns. It crushes the Czech revolutionary movement and establishes a military dictatorship in Bohemia; under the command of General Radetzky, it regains control of the Italian province of Lombardy after defeating the Piedmontese-Italian army. Tsar Nicholas of Russia comes to the aid of Austria to suppress the Hungarian insurrection. Vienna, which had become a bulwark of the revolutionaries, is bombarded. By August, 1849, all uprisings are put down. Yet the constitutional reforms granted by Emperor Ferdinand remain in place. Liberals obtain the right to municipal self-government in Vienna, which is thereby removed from direct imperial rule.

1848: December   Francis Joseph ascends to Austrian throne

Eighteen-year old Francis Joseph (b. 1830) becomes Austrian emperor after Ferdinand I abdicates (following the upheavals of the 1848 revolution). His reign is one of the longest in Austrian history, extending until his death in 1916. The young emperor inherits an ethnically diverse empire. German speakers enjoy a privileged status everywhere in the empire, though they represent only twenty percent of the some fifty million subjects. The other groups include Czechs, Slovaks, Poles, Ukrainians, Hungarians, Rumanians, Serbs, Croats, Slovenes, and Italians. In the second half of the nineteenth century, the Austrian empire is increasingly weakened by demands for autonomy and independence by non-German nationalities.

As a ruler, Francis Joseph is known for his piety, mediocre intellect, and extremely conservative views. Because of his inability to understand the changes brought by industrialization, urbanization, the rise of the working class, and mass politics, the emperor resists the introduction of much-needed political and economic reforms. He also fails to take adequate measures to meet the challenge posed by the independence movements of non-German nationalities. As a result, Austria becomes increasingly mired in difficulties and loses terrain in the international arena. In spite of his shortcomings, the monarch becomes a quasi-legendary figure, toward the end of his reign. For the German-speaking population in particular he comes to embody the cultural and political heritage of the venerable Habsburg empire.

1851   Constitution suspended; Bach system introduced

Under the influence of conservative politicians, the Austrian government suspends constitutional guarantees, revoking many of the liberties granted to various nationalities during the 1848 revolution. Alexander Bach, minister of the interior, promotes a policy of Germanization of the non-German lands and repression of ethnic diversity, which comes to be known as the Bach system. A German-speaking bureaucracy takes over the local governments of the various provinces.

At the same time, the government realizes the importance of stimulating and modernizing the economy. It abolishes tariffs that hamper trade and helps peasants become fully emancipated. It also continues to build up the railroad network, including the opening, in 1854, of the first mountain railroad, connecting the Semmering Pass (near Vienna) with Trieste (in Northern Italy).

1855   Austrian government seals concordat with Catholic Church

The new agreement aims at weakening the independence movements in the various Habsburg lands by strengthening the position of the Catholic Church. It grants the Church extensive power, particularly over education. It breaks with the tradition of containing Church influence on social and cultural matters (established in the 1780s by Emperor Joseph II).

1856   Austria involved in devastating war in Crimea

The conflict breaks out when France and Britain attempt to prevent Russia from extending its influence in the Balkans. The Western allies threaten to support the cause of Italian and Hungarian separatism if Austria refuses it to take a stance against the Russian emperor. Austria ultimately sides with the Western powers, but the prolonged mobilization and overall costs of the war leave the state finances in a shambles. This military confrontation exposes the internal weaknesses of the Austrian empire and undermines Austria's international position in Europe. On the domestic front, it opens the way for the secession of the Italian territories and the establishment of Hungarian autonomy in the following decade.

1857: December   Decree permitting the redevelopment of area around Vienna's inner city

Emperor Francis Joseph's decree involves demolishing the military fortifications and *glacis* (embankment in front of a fortification) that separate the inner city from the suburbs. Work begins on the construction of the *Ringstrasse*, a grand boulevard planned to encircle and beautify the inner city. By 1900 the new artery displays a number of monumental buildings, including the new town hall, the parliament, the university, and a magnificent opera house, as well as elegant apartment houses and mansions. They imitate great styles of the past (neo-gothic, neo-Renaissance, neo-Baroque) and are intended to communicate a sense of the stability, continuity, and undiminished importance of the Habsburg monarchy. Only the aristocracy and wealthy middle classes can afford to live in the residential buildings that line the majestic boulevard.

While pursuing the monumental *Ringstrasse* project, city officials do little to ease the abysmal living conditions of the lower classes. Vienna's population increases dramatically in the wake of the industrial expansion; since the 1860s thousands of immigrants have been flocking to the capital from all territories of the empire. To accommodate them, countless tenement houses spring up in the suburbs outside the boulevard. Due to the lack of laws regulating the rental market, wealthy landlords charge rents that are disproportionate to the workers' wages. This forces families with many children to live in apartments consisting of one or two rooms and lacking basic sanitary facilities. The growing practice of bedhiring exemplifies the dire living conditions of the lower classes. To help pay rent, poor people resort to renting out their sleeping quarters while they are not used during the day.

1859: April   Austria declares war on Piedmont

Piedmont's Savoy dynasty has caused agitations in the Austrian lands of Lombardy-Venetia and promoted the cause of Italian unification. A few weeks into the conflict, however, Francis Joseph begins peace negotiations with the Italians. Several reasons prompt him to interrupt the war. They include the poor fighting conditions of the Austrian army, apprehensions about the possibility of concomitant insurrections in Hungary, and the intervention of France on the side of the Italians. In a compromise agreement signed in the Italian town of Villafranca, the emperor agrees to cede Lombardy to the Italians. As a concession to other nationalities in the empire, the government repeals the Bach system. These conciliatory policies are interpreted as a sign of weakness, however, and insurrections soon break out all over Italy. In 1861, the Savoy dynasty heads the unification of all Italian states (with the exception of Venetia, still in Austrian hands, and Rome, under French protection). The creation of a unified Italy represents an international humiliation for Austria and gives further ammunition to separatist efforts in other parts of the empire, particularly in Hungary.

1861: February   February Patent establishes new constitution

Under pressure from both liberal forces and the non-German nationalities of the empire, Emperor Francis Joseph issues a constitution that establishes parliamentary rule (that is, a system in which representatives of the people administer the power of creating and amending laws in freely elected assemblies). However, the new constitution grants the German-speaking middle class a political influence disproportionate to their numbers. At the same time, it fails to provide the autonomy and rights invoked by the other national groups, especially the Hungarians.

1866: June–August   Seven-Weeks war between Austria and Prussia

The war represents the final show-down in the decade-long rivalry between Prussia and Austria for control over the German states. Hoping to exploit Austria's domestic troubles with the non-German minorities (particularly the Hungarians), Prussian Chancellor Otto von Bismarck provokes the monarchy into risking a military confrontation. The ill-organized Austrian army is soundly defeated at Königgrätz (today's Hradec Krállovéc, in the Czech Republic). Following this defeat, Austria signs a peace treaty in which it relinquishes Venetia to Italy. The conflict enables Prussia to achieve its main objective, namely, the complete exclusion of Austria from German affairs. This opens the way for Germany's unification under Prussian rule (in 1871).

1867: October   Austrian government grants Hungary autonomy

The agreement, known as the *Ausgleich* or compromise, is the outcome of negotiations begun in 1862 with the leaders of the Hungarian independence movement. It lays the foundations for a reorganization of the empire based on a dual system, that is, on the coexistence of two distinct states: the kingdom of Hungary (which encompasses the lands of the Crown of Saint Stephens) and Austria (comprising the remaining Habsburg holdings). The empire thus becomes a dual monarchy (often referred to as Austria-Hungary). The two governments agree to share a common defense and foreign policy; they otherwise enjoy complete autonomy in internal matters. The person of the emperor functions as the link between the two states. This is the last of the major territorial and political adjustments undertaken by Austria in the second half of the nineteenth century. It establishes a territorial order that will last until 1918. The constitution granted in 1861 remains the foundational document for the Austrian lands.

1871   Carl Menger publishes path-breaking economic treatise

The work, *Grundsätze der Volkswirtschaftslehre* (Principles of National Economy) earns Austrian economist Carl Menger (1840–1921) a professorship at Vienna's university. Menger is one of the founders of the Austrian School of Economics. His greatest achievement consists in the so-called theory of value. According to this theory, the value of a good does not depend on any quality proper to it but rather on its ability to satisfy human needs.

1888:   Social Democratic Workers Party founded

The party, which unifies the working class movement in Austria, vows to defend the cause of the lower classes by following the communist doctrine of Karl Marx and Friedrich Engels. Its leader is Victor Adler (1852–1918), a physician from a well-to-do Jewish family.

The latter two decades of the nineteenth century also see the rise of the other two political forces that will shape Austrian politics well into the twenty century. These are the Christian Socials and the German Nationals. The former party pursues a program of social reform inspired by the principles of Catholicism and appeals primarily to the common man of the lower middle class. Among its most prominent representatives is Karl Lueger (1844–1921), the powerful mayor of Vienna between 1896 and 1910. Lueger's political success relies in part on his shrewd manipulation of existing prejudices against the Jewish population. His opportunistic statements against the Jews foster the growth of antisemitism (the hatred of Jews) in Vienna. Antisemitism is also a primary ideological component in German Nationalism. Under the leadership of Georg von Schönerer (1842–1921), the German Nationals advocate restoring the cultural and racial purity of ethnic Germans. Along with the Jews, a main target of their attacks are the Slavic populations in the multinational empire.

1890s–1920s   Vienna emerges as one of Europe's cultural capitals

At the turn of the century, the hallmark of Vienna's cultural life are countless coffeehouses, which become the meeting places of choice for intellectuals and artists (many of whom are Jewish). Coffeehouses still play a central role in present-day Vienna's social and cultural life.

Among the themes that dominate Viennese culture at this time are a fascination with death and the decadence of traditional culture, a literary style based on the communication of the writer's immediate impressions (impressionism), and the search for a new poetic language, more appropriate for conveying the meaning of modern experience. Lyricist and dramatist Hugo von Hofmannsthal (1874–1929) questions in his works the boundaries and function of poetic language, while mourning the alleged decline of Austria's cultural tradition. More critical tones are found in Arthur Schnitzler's (1862–1931) dramas and novellas, which contain a merciless critique of hypocritical sexual behavior, lingering antisemitism and xenophobia (fear of foreigners), and the petty morality of middle-class Vienna. In a similar vein, essayist Karl Kraus (1874–1936) denounces the corruption of every-day language and moral values in the journalistic practice of the rising mass press.

Continuing a tradition dating back to the eighteenth century, the capital of Austria-Hungary remains at the forefront of musical innovation. Jewish composer Gustav Mahler (1860–1911), director of Vienna's prestigious Court Opera between 1897 and 1907, forges a musical style that weaves traditional and innovative elements. A more radical turn takes place in the works of composer Arnold Schönberg (1874–1951), also of Jewish extraction. His twelve-tone technique, a revolutionary method of composition based on the systematic use of all twelve tones of the chromatic scale, marks a departure from traditional harmonic music and lays the foundations for modern atonal music. At the forefront of popular music, the developing genre of the operetta celebrates Vienna's happy and supposedly carefree lifestyle. It contains a mixture of dramatic and song performance and represents a precursor of the musical. An immensely successful form of popular entertainment, it reaches new heights in the works of Franz Léhar and Johann Strauss.

1897: April 5   Promulgation of Badeni language ordinances

The decrees promoted by Austrian Minister Badeni establish that the Czech language is to be treated on a par with German in all public offices of Bohemia and Moravia. This move acknowledges Czech demands for equal treatment. However, the Germans, who are afraid of losing their privileged status, stage violent protests to resist the alleged "Slavification" imposed by the government. This forces Badeni to resign on November 28. The failure of Badeni's linguistic policies is a symptom of the unsolved tensions among the monarchy's ethnic groups.

1897: May 25   Viennese Secession founded

Displeased with the outdated models promoted in painting and architecture by the academic establishment (and represented in the monumental buildings of the *Ringstrasse*), nineteen artists leave the *Künstlerhaus*, Vienna's powerful artists' association, and found their own association, the Secession. At their head is painter Gustav Klimt (1862–1918) whose works best represent the new style of the Secession movement, known as *Jugendstil* (Viennese Art Nouveau). It is characterized by light, ornamentally stylized figures as well as floral and geometrical designs. In architecture, the new style is exemplified in the works of Otto Wagner (1841–1918). His functional and elegant buildings (among which are the Postal Savings Office, the stations for the metropolitan railway, the Steinhof Church, as well as various apartment houses in downtown Vienna) reshape the capital's urban landscape and are at the forefront of modernist architecture.

1899: November   Sigmund Freud publishes *The Interpretation of Dreams*

The work marks the beginning of psychoanalysis, Freud's innovative approach to human psychology. In particular, it lays the foundation for the scientific investigation of the non-rational dimension of the human psyche. At the heart of the new discipline is a theory of sexuality based on the systematic investigation of the realms outside consciousness, which Freud calls the unconscious. According to Freud (1856–1939), the unconscious manifests itself in the work of dreams and in mental illnesses, like neuroses and psychoses. Freud's psychoanalytic theories, which he continues to develop in the following four decades, have a large impact on the fields of

Austro-Hungarian Empire, 1914.

psychology and the disciplines of anthropology and cultural and literary studies.

1907: January 26   Law introduces universal suffrage for parliamentary elections

The introduction of the universal right to vote is part of the socialists' agenda and aims at a democratization of Austrian politics. Since 1909 the government increasingly rules by decree, however, which deprives the parliament of any effective power.

1908   Austria annexes Bosnia and Herzegovina

The annexation seals the Austrian occupation of Bosnia and Herzegovina, two Balkan provinces that had been under Austrian protectorate since 1878. The move violates several international treaties and worsens the already tense situation in the Balkans. In particular, it shatters the dream of a state uniting all southern Slavs and provokes the hostility of Serbia.

1914: June 28   Heir to the Austrian throne assassinated in Sarajevo, Bosnia

During a visit to the Bosnian capital, Archduke Francis Ferdinand and his wife Sophie are the victims of a terrorist attack by a nineteen-year-old Bosnian-Serb nationalist. The Austrian government uses the incident as an excuse for challenging Serbia with a humiliating ultimatum. The goal is to provoke a regional military confrontation and curtail Serbia's growing influence in the Balkan peninsula. However, with the intervention of Germany on the side of Austria-Hungary, and Russia, France, and Britain in support of the Serbs, what has begun as a localized conflict turns into the First World War.

1914–18   World War I brings end to Hapsburg empire

Following Austria-Hungary's declaration of war against Serbia (on July 28, 1914), France, Britain, and Russia quickly enter the conflict against Austria-Hungary and its German ally. The latter's hope in a quick victory soon reveals itself to be a miscalculation. Direct military confrontations on the eastern front with Russia and on the western front with France soon give way to an enervating trench war. Germany soon takes the lead in the conflict, relegating its Austrian partner to a subordinate position. The conflict broadens when Italy and Rumania join the Serb-Russian-French coalition. It takes a further turn in 1917, with the intervention of the United States on France's side and the outbreak of the Russian Revolution, which forces the Russians out of the war.

In 1916, Emperor Francis Joseph dies after sixty-eight years of reign. With him goes the last symbol of unity and legitimation of the dual monarchy. His successor, his grand-nephew Charles I, recognizes the impossibility of Austria-Hungary winning the war but hesitates to abandon the alliance with Germany. Meanwhile the situation in the non-German parts of the monarchy escalates, with the Czechs and the southern Slavs in the process of founding separate states. In summer 1918, the German and Austrian armies begin to crumble. Despite the concessions made by Emperor Charles I to the nationalities in the Austrian half of the empire, the dissolution of the empire appears unavoidable. On November 11, 1918, Charles I declares his withdrawal from state affairs. The creation of the sovereign states of Czechoslovakia, Hungary, Rumania, and Yugoslavia seals the dissolution of the Habsburg empire.

1915 Prague-born novelist Franz Kafka publishes *The Metamorphosis*

The short story is one of most widely read works by the Jewish prosaist (1883–1924), who belongs to Prague's German-speaking middle class. After receiving a degree in jurisprudence, Kafka leads an uneventful life as the employee of an insurance company in Prague. His early death, from tuberculosis, near Vienna, prevents him from completing any of his three major novels (*The Castle*, *The Trial*, and *America*). Although his works find appreciation only in restricted literary circles during his life-time, they later are recognized as some of the most influential prose of twentieth-century literature. While lacking direct references to the Habsburg monarchy, Kafka's prose contains a clear critique of the malfunctioning bureaucracy, the corrupt judicial system, as well as the hypocrisy, narrow-mindedness, and greed of the middle class in Austrian-dominated Prague.

1918: November 12 Republic of German-Austria proclaimed

The new republic encompasses the German-speaking countries of the dissolved dual monarchy (roughly corresponding to present-day Austria). Karl Renner (a Social Democrat) becomes chancellor of the provisional government. Renner (1870–1950) begins peace negotiations with the victorious allies, who regard Austria as the heir to the defeated Habsburg monarchy. The allies' main concern is to prevent Austria from merging with the new German Republic (also

founded in November, 1918). As they fear, this would again lay the foundations for a strong German state in central Europe. The outcome of the negotiations is the Treaty of Saint Germain (near Paris), which forbids German-Austria from uniting with Germany, changes the name of the republic to Austria, and imposes severe war reparations. In the new republic, the harsh conditions of the treaty cause great resentment toward the West.

The newly established Austrian republic lacks a definite national identity and must immediately confront enormous economic problems. The loss of the non-German territories causes acute shortages in raw materials (Czech and Polish coal, food stuffs from Hungary) as well as the loss of vital markets for Austrian goods. Skyrocketing unemployment, the frequent standstill of the transportation network, and the collapse of the administrative and security systems characterize the republic's chaotic first years.

1919: November Election of Constitutional National Assembly

The assembly is the result of the first free elections in the Republic. It has the task of drafting a constitution and defining the form of the new state. The new constitution, ratified in October 1920, provides for a president with only representative functions and a strong parliament, which comprises a powerful national chamber and a weaker provincial chamber. Vienna becomes the capital of the republic.

1922 Ignaz Seipel becomes Austrian chancellor

Seipel (1876–1932), a Catholic priest and professor of moral theology, is the leading figure of the Christian Socialists. He shapes the course of Austria's politics until his death. Seipel succeeds in normalizing relations between Austria and the Western powers; he obtains international loans that help stabilize the economy. As a measure to fight inflation, in 1924 his government introduces the present-day Austrian currency, the *schilling*. Seipel's government also attempts to pacify the riotous political climate in Austria, which since the early 1920s is dominated by skirmishes between two paramilitary groups. These are the *Heimwehr* (which means home defense), a German nationalist group that supports the conservative Christian Socials, and the *Republikanischer Schutzbund* (that is, republican protection league), which is allied with the Social Democrats.

While conservative coalitions (dominated by the Christian Socials) govern Austria between 1920 and 1933, Vienna remains firmly in the hands of the Social Democrats. The Socialist government implements ambitious programs aimed at improving the living conditions of the working classes. These include the building of affordable housing (often in enormous apartment complexes like the still extant Karl-Marx-Hof), the implementation of school reforms and adult education programs, and the introduction of health services aimed at combating widespread illnesses like tuberculosis.

1927: July 14   Palace of Justice set ablaze during riots

In the course of demonstrations against a controversial trial verdict (involving a clash between the nationalist group of the *Heimwehr* and the leftist *Schutzbund*), a mob sets Vienna's Palace of Justice on fire. The episode exemplifies the tense social and political situation in the Republic. It generates fear of more widespread unrest and introduces an authoritarian turn in Austrian politics.

1931: May 11   Credit Anstalt collapses

The immediate culprit for the devastating collapse of Austria's leading financial institution is France, who withholds a sorely needed loan from the Austrian bank. France's move aims at undermining Austria's plans for a customs union with Germany. However, the collapse also indicates the worsening of Austria's economy in the late 1920s in the wake of a world-wide recession.

1933: March 4   Parliamentary government suspended

Exploiting a moment of confusion at the end of a parliamentary session, Chancellor Engelbert Dollfuss shuts down the parliament. Dollfuss has been the head of a conservative coalition of Christian Socials and Agrarians since 1932. After outlawing all political parties (including the Nazi party), the chancellor proceeds to reorganize the state according to a corporative structure. That is, representatives of various professional organizations (agriculture, industry, public service, commerce, etc.) take the place of freely elected party officials. This effectively establishes an authoritarian regime under Dollfuss' leadership. One unresolved issue concerns relations with Nazi Germany, where Hitler has just established a dictatorship. Hitler's success in Germany galvanizes the Austrian Nazis, who agitate for a union of Austria and Germany.

1934: July 25   Chancellor Dollfuss killed in Nazi coup

During a failed attempt to overthrow the government, Austrian Nazis shoot and kill Dollfuss. His successor, Kurt Schuschnigg, vows to resist Hitler's increasing pressures for a merger of the two states. After 1936, however, Austria loses the protection of its strongest ally against Hitler, namely, Mussolini's Italy. Nazi riots and demonstrations make the country increasingly unmanageable.

1938: March 13   Austria's union with Nazi Germany proclaimed

The confrontation between Austria and Germany escalates when Chancellor Schuschnigg announces a popular referendum on the issue of unification with Germany. Hitler demands the postponement of the vote and forces Schuschnigg to resign. On March 12, German troops begin invading Austria, encountering no resistance. When the new chancellor, Seyss-Inquart, formally declares the merger with

Germany, Austria effectively ceases to exist as a distinct territorial and political unit. Its history during World War II coincides with that of Hitler's Third Reich.

1945: April   Soviet troops occupy Vienna

Vienna's fall marks the collapse of the Nazi regime in Austria. In their policy toward the vanquished enemy, the Western allies decide to regard Austria not as a collaborator but rather as a victim of Hitler's regime. They permit the foundation of political parties and entrust the esteemed Socialist leader Karl Renner with the task of forming a provisional government.

1945: April 27   Second Republic founded

The three newly founded political parties, the Communist Party of Austria (*Kommunistische Partei Österreichs*, KPÖ), the Socialist Party of Austria (*Sozialistische Partei Österreichs*, SPÖ), and the Austrian People's Party (*Österreichische Volkspartei*, ÖVP) proclaim the Second Republic. Karl Renner becomes its first chancellor.

1945: July 9   Austria divided in four occupation sectors

In spite of the proclamation of the republic, the victorious Allies (France, Great Britain, United States, and the Soviet Union) sign an agreement dividing Austria into four zones of occupation (each controlled by one victorious power). The declared aim is to monitor the democratic development of the new country. Occupied Austria soon becomes a pawn in the Cold War confrontation that develops in the immediate postwar period between Russia and the Western allies, however. Both blocks struggle to draw the new state into their sphere of influence. This hampers Austria's attempt at ending the occupation. Negotiations for reinstating the republic's full sovereignty drag on for the next decade.

In the elections of November, 1945, the first free electoral consultation since 1930, the Socialist Party and the Austrian People's Party receive the great majority of votes. In subsequent decades they confirm themselves as the dominant political players of the postwar period. Among the problems faced by the new government is the prosecution of former Nazis, many of whom lose their right to vote and other civil rights, have been removed from public service, and must pay penalties or special taxes. The government tries 23,000 Austrians and convicts 13,600 of Nazi crimes; it executes 30 war criminals. A law passed in 1947 seeks to address the difficulty of dealing with the large number of Austrians who have been implicated with Nazism, whether forcibly or not. It formulates guidelines for distinguishing between militant Nazi believers and those who joined the party out of fear or opportunism and allows for the reintegration into society of over half a million Austrians. It is followed by a general amnesty for less incriminated Nazis (1948). None of these measures solves the predicament entailed in evaluating Austria's involvement with Nazi Germany. In this regard, two extreme

Adolf Hitler addresses a crowd of Austrians in Vienna after annexing Austria. Hitler says, "I proclaim now for this land its new mission. The oldest eastern province of the German people shall be from now on the youngest bulwark of the German nation." (EPD Photos/CSU Archives)

positions confront each other. On the one hand, there are those who regard the first Austrian republic as the first victim of the Third Reich. On the other, Austrians are seen as all too willing supporters of Hitler's Germany.

1945–52 Economic reconstruction

With the economy in shambles, the new government implements policies aimed at bringing heavy industry and banking under state control (nationalization). Financial aid from the United Nations and the American Marshall Plan enables Austria to begin rebuilding the industries and infrastructures destroyed during the war.

1955: May 15 State Treaty signed

The treaty, negotiated between the Austrian government and the four occupation powers, restores Austria's full territorial sovereignty (within the borders of the first Austrian Republic) and paves the way for the departure of the occupying armies.

It also grants special rights for the Slovene and Croatian minorities in Carinthia, Styria, and Burgenland, and forbids a reunification with Germany. In December, 1955, Austria joins the United Nations and becomes a full member of the international community.

1955: October 26 Parliament passes law establishing Austria's neutrality

The constitutional law establishing permanent neutrality primarily entails Austria's commitment to abstain from international, armed conflicts. It aims at reassuring the countries engaged in the Cold War that a sovereign Austria will not contribute to altering the balance of power in central Europe.

1950s–90s Austrian politics and social partnership

After World War II, the principle of social partnership becomes a special feature of Austrian political and social exchange. It entails a system in which social partners—representatives of particular interest groups such as the chambers of commerce, industry, trade, agriculture, and the labor unions—agree to discuss common concerns (e.g., wages) independently of political parties. This practice builds compromise and consensus and avoids bitter antagonisms. It is generally beneficial for the economy (for instance, it reduces the incidence of strikes) and ensures political stability. It is one of the factors that account for Austria's post-war prosperity.

1970–83   Kreisky era

Bruno Kreisky (1911–90), one of the most distinguished postwar politicians, becomes the head of the first all-Socialist government in 1970. Under his leadership, Austria experiences a period of economic prosperity and political stability. He introduces a series of reforms, including the reorganization of the legal code (spearheaded by Minister of Justice Christian Broda). In the late 1970s and early 1980s, his government is marred by difficulties with his plans of privatization of national industries and by a series of scandals involving ministers in his cabinet. He resigns when the Socialist Party loses its absolute majority in the elections of 1983.

1978   Greens rally against nuclear power plant near Vienna

Due to growing environmental and security concerns, the Kreisky government calls a popular referendum to decide the fate of a new nuclear reactor in Zwentendorf (twenty-five miles from Vienna). By a narrow margin, Austrians vote against opening the power plant, and Zwentendorf becomes synonymous with the debacle of atomic energy in Austria. It represents a high-profile victory for the Greens, a growing political movement that places top priority on the protection of the environment. The steady growth of the Green Party in the following decade reflects the increasing attention Austrians pay to environmental issues, which include the reduction of pollution, the conservation of natural resources like water, and the development of environmentally safe energies. The Greens enter the Austrian parliament in 1986.

1986   Presidential candidate Kurt Waldheim involved in scandal over alleged Nazi crimes

During the presidential campaign of 1986, the press publishes allegations that Kurt Waldheim, former Secretary General of the United Nations, has participated in Nazi crimes while stationed in the Balkans. Though a panel of experts finds no evidence that corroborates the allegations, Waldheim's failure to disclose his involvement as a low-ranking officer in Nazi Austria generates widespread criticism. He nonetheless wins the presidential elections. The scandal exposes the difficulties Austria faces in its confrontation with the Nazi past and deals a blow to its international image.

1995: January 1   Austria joins the European Union (EU)

In the early 1990s, Austrians appear deeply split on the issue of membership in the EU, an organization which pursues the political and economic integration of the European nations. Concerns include possible economic drawbacks and a loss of political independence. However, in a popular referendum held in June, 1997, sixty-seven percent of the voters express themselves in favor of Austria's participation in the EU. The EU membership greatly strengthens Austria's political and economic ties to Europe.

Bibliography

Janik, Allan, and Stephen Toulmin. *Wittgenstein's Vienna.* New York: Simon and Schuster, 1973.

Jelavich, Barbara. *Modern Austria. Empire and Republic 1815–1986.* Cambridge: Cambridge University Press, 1987.

Johnson, Lonnie. *Introducing Austria: A Short History.* Riverside, Calif.: Ariadne, 1989.

Johnston, William M. *The Austrian Mind: An Intellectual and Social History 1848–1938.* Berkeley: University of California Press, 1972.

Kann, Robert A. *A History of the Habsburg Empire 1526–1918.* Berkeley: University of California Press, 1974.

Morton, Frederic. *A Nervous Splendor: Vienna, 1888–89.* Boston: Little Brown, 1979.

———. *Thunder at Twilight: Vienna 1913–14.* New York : Scribner, 1989.

Pauley, Bruce F. *The Habsburg Legacy 1867–1939.* Malabar, Fla.: Robert E. Krieger Publishing Company, 1972.

Schorske, Carl E. *Fin-de-Siècle Vienna: Politics and Culture.* New York: Vintage, 1961.

Steiner, Kurt, ed. *Modern Austria.* Palo Alto, Calif.: Sposs Inc, 1981.

Steininger, Rolf, and Michael Gehler, eds. *Österreich im 20. Jahrhundert.* 2 vol. Vienna: Böhlau, 1997.

Waissenberger, Robert, ed. *Vienna 1890–1920.* Secaucus, N.J.: Wellfleet, 1984.

Weissensteiner, Friedrich. *Der ungeliebte Staat. Österreich zwischen 1918 und 1938.* Vienna: Österreichischer Bundesverlag, 1990.

William, Cedric E. *The Broken Eagle: The Politics of Austrian Literature from Empire to Anschluss.* London: Paul Elek, 1974.

Belarus

Introduction

Belarus, one of the westernmost republics of the former Soviet Union, has been an independent nation since 1991. Like Ukraine to its south, Belarus has been dominated for most of its history by two neighboring powers—Poland to the west and Russia to the east. Lithuania, with which it shares a border, has also played a major role in Belarussian history. Belarus's geographic location at the northern border of Ukraine brought disaster in 1986, when the republic absorbed around seventy percent of the radioactive fallout from the explosion at the Chernobyl nuclear power plant. Since independence Belarus has retained close ties with Russia, and the leadership of both nations was moving toward political and economic union as the 1990s came to a close.

Belarus is a landlocked country with an area of 80,154 square miles (207,600 square kilometers). Part of the Great Plain of Eastern Europe, Belarus consists mostly of low-lying terrain, with rolling hills to the north and marshland to the south in the Pripet and Dnieper river basins. Minsk, the capital of Belarus, is also the capital of the Commonwealth of Independent States, a loose federation of former Soviet republics. As of the late 1990s, Belarus has an estimated population of 10.5 million people.

History

Slavs first settled in present-day Belarus in the first millennium B.C. By the eighth century A.D., several principalities had been formed in the region. In the following century, Belarus became part of Kievan Rus, the forerunner of present-day Ukraine and, at the time, the largest independent state in Europe. Mongol invasions in the thirteenth century led to the decline of Kievan Rus, and Belarus came under the control of the Grand Duchy of Lithuania, which was expanding its power. Culturally, however, the Belarussians remained relatively autonomous, free to practice Eastern Orthodoxy and use their native language.

With the political union of Lithuania and Poland in 1569, Belarus came under the control of the Polish monarchy, which intervened much more closely in its affairs, splitting its communal farmland into individual estates, establishing a system of serfdom under which the Polish peasants were oppressed for generations, and spreading its own language and culture in Belarus, where the upper classes began speaking Polish and also adopted Poland's religion, Roman Catholicism. In the following century, a new force gained power in the region when the Ukrainian Cossacks seeking freedom from Poland formed an alliance with Russia, and Russia and Poland waged a thirteen-year war over Ukraine, much of it fought on Belarussian soil. Belarus itself remained under Polish control until the late eighteenth century, when Poland was partitioned by Russia, Austria, and Prussia. At that time it was merged with Russia, which undertook an extensive Russification program to incorporate Belarus culturally as well as politically. Once again, the transmission of Belarussian language and culture fell largely to the peasantry, while the upper classes adopted the Russian language. By 1840, the Russian legal code was also imposed on Belarus, which was now called the Northwest Region.

The unsuccessful 1863–64 Polish uprising against Russia, in which many Belarussians took part and the renewed Russian repression that followed it led to a renewal of Belarussian nationalism. However, no political changes occurred until after World War I, when an independent Belarussian republic was declared following the overthrow of the Russian tsar. However, Russia's new rulers, the Bolsheviks, were not willing to let go of Belarus. In January, 1919, they formed the Belarussian Soviet Socialist Republic, which became one of the first member republics of the Soviet Union in 1922. By 1921, Belarus was once again divided between Russia and Poland, when Russia ceded 38,600 square miles of Belarus to its traditional adversary after a year of warfare over both Ukraine and Belarus.

The early years of Soviet rule were relatively liberal, and Belarussian nationalism was accepted and allowed to flourish. In the late 1920s and 1930s, however, Belarus was subjected to severe repression under the rule of Soviet leader Josef Stalin (1879–1953). Many Belarussians, accused of nationalism or other offenses, were murdered or sent to labor camps, and farmers were required to give up their individual farms and work on state-run agricultural collectives. During the same period, widespread urbanization began in Belarus,

BELARUS

0   50   100   150 Miles

0   50   100   150 Kilometers

RUSSIA

LATVIA
Daugavpils
Velikye Luki

LITHUANIA
Polatsk
Zach.
Dvina
Vitsyebsk
Pastayy
Smolensk
Vilnius
Maladzyechna
Orsha
Dzerzhinskaya Gora
1,135 ft.
346 m.
Mahilyow
Lida
Nyoman
Minsk
Hrodna
Krychaw
Vawkavysk
Asipovichy
POLAND
Baranavichy
Babruysk
Bryansk
Salihorsk
Zhlobin
Dynaprowska-
Buhski Kanal
Rechitsa
Homyel
Pinsk
Bug
Mazyr
Brest
Pripyats
Pinsk Marsh
Chernihiv
UKRAINE
Kiev

Belarus

N
W   E
S

as industrialization led many Belarussians to move from rural to urban areas.

World War II brought death and destruction to Belarus, which was occupied at different points in the war by German and Soviet troops. An estimated 2.5 million Belarussians died, and the region's cities suffered extensive damage. After liberation by the Soviet Army in 1944, all of Belarus became part of the Soviet Union. In the postwar period, the war-torn republic's agricultural and industrial resources were restored, and Belarus became one of the USSR's major industrial centers. However, Belarussian nationalism continued to be suppressed. By the late 1970s, three out of four Belarussians considered Russian their first language.

The Chernobyl nuclear accident in the spring of 1986 was a major disaster for Belarus, as the nuclear plant was located in northern Ukraine, just south of the shared border between the two republics. Anger over attempts by the Soviet government to cover up the full extent of the disaster contributed to a growing nationalist movement, which was also fueled by the liberalization set in motion by Communist party leader Mikhail Gorbachev. By 1990 Belarus's parliament had passed a declaration asserting its sovereignty, and that sover-

eignty became a political reality a year later, when the Soviet Union dissolved and Belarus became an independent republic and a member of the newly formed Commonwealth of Independent States. Since then Belarus has faced the economic problems common to the former Soviet republics, worsened by drought and flooding in 1993–94. Its first president, Aleksandr Lukashenko, elected in 1994, has dismayed observers by the authoritarian nature of his administration. In 1996 the Belarussian electorate approved a revised constitution giving Lukashenko sweeping new powers, including the right to appoint members of parliament and the right to rule by decree.

Lukashenko has also supported close ties with Russia. Since his election in 1994, he and Russian president Boris Yeltsin have gradually moved the two republics toward political and economic union.

Timeline

1st millennium B.C.   Slavs settle in Belarus

Slavic groups referred to as East Slavs settle in present-day Belarus. These include the Milograd and Zarubinsky civilizations.

6th century A.D.   Slavic tribes are linguistically linked

The Slavs in Belarus are divided into tribes that share a common language.

8th century   Principalities are formed

The Slavic groups in Belarus form political structures governed by princes (principalities) in Minsk, Pinsk, Slutsk, Polotski, and Turov.

9th century   Belarus becomes part of Kievan Rus

Belarus becomes part of Kievan Rus, the first organized Slavic state in the region and Europe's largest independent state.

1240   Mongols invade Kievan Rus

With the invasion of Mongols from the east, the power of Kievan Rus declines, and the Grand Duchy of Lithuania begins expanding its power in the region. However, Belarus has considerable cultural autonomy, retaining its own language and Eastern Orthodoxy.

1386   Polish and Lithuanian monarchies merge

The ruling houses of Poland and Lithuania are united when the Lithuanian grand duke Ladislaw Jagiello marries Queen Jadwiga of Poland. He becomes king of Poland and rules as Ladislaw II, inaugurating the long-lived Jagiellonian dynasty

The Polish-Lithuanian State, 15th century.

(1386–1572). The union of the two monarchies is the first step in a political union that eventually places all of the former Kievan Rus, including Belarus, under Polish rule.

1517   Bible is translated into Belarussian

Frantsyk Skaryna produces the first Belarussian translation of the Bible. He also publishes a series of religious pamphlets in Belarussian.

Polish Rule

1569   Union of Lublin

Poland and Lithuania agree to full political union, placing Belarus under Polish control and resulting in major social, economic, and cultural changes. Communal farms are broken up into individual estates distributed to the Polish nobility, while Ukrainian peasants become their serfs. The Polish language is adopted by Belarussians, especially those of the wealthier class, many of whom also convert to the Polish religion, Roman Catholicism.

1654–67   Russo-Polish War

Ukrainian Cossacks agree to unite with Russia against Poland, and a thirteen-year war follows for control of Ukraine, much of its fought on Belarussian soil. Belarus again becomes a battleground in the Great Northern War between Russia and Sweden (1700–21). These two wars together kill almost half its population.

1697   Belarussian is banned

Poland bans the use of the Belarussian language for official purposes.

Russian Rule

1773–95   Partition of Poland

Russia, Austria, and Prussia partition Poland in separate stages, dividing its territories among them. Belarus is incorporated into Russia, which undertakes a campaign of linguistic and cultural assimilation in the region. Russian becomes the language of the upper classes, while the peasantry continues to speak Belarussian, which is regarded as a local dialect.

1830–31   Uprising fails

Poles and Belarussians rise up against their Russian rulers but are defeated. After this revolt, Russia steps up its campaign to culturally assimilate Belarus.

1840   Russian legal code is imposed

Russia imposes its own legal code on Belarus and bans the use of the name *Belarus,* replacing it with the term "Northwest Region." Many Belarussians emigrate to maintain their cultural identity and seek economic opportunities.

1863–64   Polish uprising sparks nationalist revival

Belarussians take part in the "January Insurrection," a Polish rebellion against Russia.

After two years of guerrilla warfare, the rebels are defeated, and Russian rule becomes more repressive than before. In spite of Russification policies, a clandestine political and cultural nationalist revival emerges in Belarus. Interest in Belarussian history and the Belarussian language is renewed and underground publications appear.

1866 Art school opens in Vilnius

A technical and applied drawing academy opened in Vilnius spurs a revival of art education in Lithuania and Belarus.

1880s Railroad construction begins

Railroad construction begins in Belarus.

1882 Birth of author Yakub Kolas

Kolas, one of Belarus's two great modern writers, is born Kanstancin Mickievic. He is raised in a village near Minsk. Although born into a modest peasant family, he learns to read at a young age and is introduced to Belarussian literature early. He attends teacher training college but is expelled for participating in revolutionary activities aimed at Belarussian independence from Russia. Kolas eventually settles in Vilnius (now part of Lithuania), where he is associated with the journal *Nasa Niva*. Kolas becomes well known for the sympathy with which he portrays the situation of everyday people in his novels, poems, and short stories. Kolas dies in 1956.

1882 Birth of author Janka Kupala

Janka Kupala is the pseudonym of Ivan Lucevic, who is born in central Belarus to a poor family and suffers hardship and tragedies in his youth. He becomes the breadwinner for his family at a young age. Kupala receives very little formal education but is self-educated. After initial literary efforts in Polish, Kupala begins writing in Belarussian and becomes one of the main writers for the journal *Nasa Niva*. The author of highly regarded poetry collections and four plays, Kupala dies in 1942. He is regarded as one of Belarus's greatest authors.

1887: July 7 Birth of artist Marc Chagall

Marc Chagall is born in Vitebsk and receives a traditional Hasidic upbringing and education. Between 1907 and 1910 he studies art in St. Petersburg and continues his career in Paris where he maintains a studio in the La Ruche artists' colony. His first paintings exhibited in Paris include *The Fiddler* (1912–13), *Pregnant Woman* (1913), and *Self-portrait with Seven Fingers* (1912–13). In 1914 Chagall travels to Vitebsk for a short vacation but political events—World War I and the Russian Revolution—oblige him to stay, and he remains there for eight years. Following the Bolshevik Revolution of 1917, Chagall is appointed Commissar for the Arts in Vitebsk. However, political developments interfere with his career, and in 1922 he and his family leave for Berlin where he takes up etching, which he subsequently uses for book illustrations.

A year later he moves to Paris, where he settles permanently except for a sojourn in the United States during and immediately after World War II. In 1937 Chagall becomes a French citizen, but because he is a Jew he is forced into exile by World War II, when France is occupied by Nazi Germany. Between 1941 and 1948, Chagall lives in New York. He later returns to France, settling near Nice in 1950. In the last thirty years of his life, Chagall becomes famed for his brilliantly colored stained-glass windows, produced for French cathedrals, the United Nations, and the Hadassah Medical Center in Jerusalem. In spite of his international career, Chagall's Belarussian background continues to play an important role in his art. Chagall dies at his home in France in 1985.

1898 First Marxist party is founded

Russia's first Marxist political party, the Russian Social Democratic Labor Party, is founded in Minsk, Belarus.

1906–15 Nasa Niva literary journal is published

The weekly journal *Nasa Niva* becomes the focal point for Belarussian nationalist literature. Belarus's most prominent writers, including Yakub Kolas (see 1882) and Janka Kupala (see 1882), are associated with it. In addition to poetry and prose, the journal publishes social and political commentary.

1918: March 25 Independent democratic republic is founded

As the Russian Revolution brings the collapse of Tsarist Russia, Belarus becomes one of many regions to set up its own state. The All-Belarussian Congress proclaims the Belarussian Democratic Republic. Russia's Bolshevik (Communist) leaders refuse to recognize the new republic, and it is overthrown by the Red Army.

1919–22 Revolution brings artistic renaissance

Belarus becomes a center for *avant-garde* (unconventional) art in the period immediately following the Russian Revolution, which is sometimes referred to as the Vitebsk Renaissance.

1919 State art exhibition is held

The first State Exhibition of Paintings by local and Moscow artists is held in Vitebsk.

Soviet Republic

1919: January 1 Soviet republic is formed

Communists establish the Belarussian Soviet Socialist Republic.

1921 Treaty of Riga

Following a year of hostilities between Russia and Poland, Russia cedes western areas of Belarus totaling 38,600 square miles (100,000 square kilometers) to Poland. The Belarussian Soviet Socialist Republic retains control of 27,000 square miles (70,000 square kilometers) with a population of five million.

1922–28 Nationalist revival under Soviet rule

In the early years of Soviet rule, Belarussian nationalism is encouraged. Belarussians wield the political power within their own republic, and a cultural revival takes place. A university and science academy are founded, and universal education is instituted.

1922 Belarus becomes part of the Soviet Union

The area of Belarus not under Polish control becomes one of the first four constituent republics of the USSR (Union of Soviet Socialist Republics).

1922: February 18 Birth of painter Mikhail Savitsky

Savitsky (b. 1922), born in Vitebsk, is one of the most prominent artists of Soviet-era Belarus. He fights in the Soviet army in World War II and is imprisoned in the German concentration camps of Dachau and Buchenwald. In line with the tendencies of Soviet art, many of his paintings have socially relevant themes, either involving a political event or focusing on the theme of labor. The concentration camps of World War II figure in his cycle of paintings *Figures on the Heart* (1974–80), while paintings such as *Loaf* (1968) celebrate rural labor. In 1987 Savitsky paints a series of pictures inspired by the disaster at the Chernobyl nuclear plants.

1922: July 22 Birth of composer Heinrich Wagner

Heinrich Wagner (Genrikh Vagner), although born in Poland, is forced to move to Belarus as a teenager at the start of World War II and builds his career there as a Belarussian composer. He studies piano at the Belarussian Conservatory, graduating in 1948. After graduation, he teaches piano and begins building a reputation as a composer, returning to school for a composition degree in the early 1950s. His works include the ballet *Belorusskaya syuita* (The False Bride; 1958); the symphonic poem *Vachno zhviye* (Those Who Live Forever; 1959), dedicated to those killed in World War II; the ballet *Svet I teni* (Light and Shadows; 1962); the Piano Concerto (1964), and the Second Symphony (1971). Wagner's earlier works are largely traditional; those composed after 1960 are more influenced by twentieth-century musical developments.

1928–39 Belarus suffers Stalinist repression

After enjoying relative cultural autonomy in the 1920s, Belarus, together with many other regions of the USSR, suffers under the totalitarian regime of Soviet leader Josef Stalin (1879–1953). Farmers are thrown off their land and sent to work on collective farms, intellectuals are persecuted, and many Belarussians are either killed or sent to labor camps. As many as a million or more Belarussians die during this period as a result of government repression.

1930s Industrialization brings urbanization

The introduction of widespread industrialization leads many Belarussians to migrate from rural to urban areas. These become one of the most highly urbanized populations in the USSR.

1939: August Polish areas annexed to Belarus

Under the Molotov-Ribbentrop nonaggression pact between the USSR and Germany, large areas of eastern Poland are annexed to Belarus.

World War II

1939: September 1 Germany invades Poland

In violation of the non-aggression pact between the two countries, Germany invades Poland, launching World War II. As many as 2.5 million people—a quarter of its population—are killed in the war, and its cities and factories suffer extensive wartime damage. Over a million buildings are destroyed, including entire villages.

1939: September 17 Soviets occupy Belarus

To counter the German invasion of Poland, Soviet troops move into the western parts of Belarus and Ukraine, which have been under Polish rule. At first regarded as liberators, the Soviets impose harsh conditions on these areas. In the next year and a half, some 300,000 Belorussians are deported to labor camps in the Soviet Union.

1941: June Germans invade Belarus

Launching their invasion of Russia, German troops invade and occupy Belarus. Initially they are welcomed for freeing the Belarussians from Communist occupation, but their occupation is worse than that of the Soviets. The Germans kill or deport three-fourths of the population and attempt to assimilate the remaining one-fourth.

1942 Death of author Janka Kupala

Poet and playwright Janka Kupala, considered one of Belarus's two greatest modern writers, dies. (See 1882.)

1944   Belarus is liberated

Belarus is liberated by the Soviet Army in summer, 1944. Many Belorussians flee the country, settling in Germany, other parts of Western Europe, and the United States.

The Postwar Period

1950s–70s   Economy recovery and suppression of nationalism

The Soviets launch an economic recovery program in Belarus, rebuilding the region's industry and agriculture. Belarus becomes one of the Soviet Union's major industrial centers. However, it also becomes one of its most politically repressive republics, and the Belarussian national identity is severely suppressed. In urban areas, Russian replaces the Belarussian native language. Russian, rather than Belarussian, is the language of instruction in the schools, and Russian-language publications outnumber those in Belarussian. By the late 1970s, three out of four Belarussians regard Russian as their native language.

1956   Death of author Yakub Kolas

Yakub Kolas, one of the founders of modern Belarussian literature, dies. (See 1882.)

1985   Liberalization takes place under Gorbachev

When Mikhail Gorbachev (b. 1931) becomes the leader of the Soviet Union, his policies of greater openness (*glasnost*) lead to a revival of Belarussian nationalism. The Byelarussian Popular Front (BPF) is formed and advocates the establishment of an independent Belarussian state. Other parts of its nationalist platform include the teaching of Belarussian in the schools.

1985: March 28   Death of artist Marc Chagall

Marc Chagall, the most famous Belarussian-born artist, dies in France. (See 1887.)

1986: April 26   Belarus hit hard by the Chernobyl nuclear accident

As much as seventy percent of the radioactive fallout from the explosion at the Chernobyl nuclear plant in Ukraine near the Belarussian border hits Belarus. The worst nuclear accident in history, the Chernobyl explosion releases approximately nine tons of radioactive matter into the world's air, soil, and water. An estimated 2.5 million Belarussians are directly affected by the accident, and the damage is increased by the Soviet cover-up of the accident. Health problems related to the accident are widespread over the coming decade. It is estimated that 640,500 acres (260,200 hectares) of Belarus's farmland have been rendered permanently unusable. As the full extent of the disaster is discovered, it becomes another catalyst in the movement for Belarussian independence.

1986: December   Petition voices cultural demands

Belarussian intellectuals send a petition to Mikhail Gorbachev protesting suppression of Belarussian culture and demanding the reinstatement of the Belarussian language in government, education, and the media.

1988: June   Mass graves of political victims uncovered

Graves of as many as a quarter million Belarussians murdered during the Stalinist era are found near Minsk and reinforce calls for independence from the Soviet Union.

1988: October   Demonstration dispersed by police

Sparked by the mass graves uncovered in June, mass demonstrations supporting the Belarussian Popular Front are held by some 10,000 Belarussians. They are broken up by riot police.

1990: June 27   Government declares its sovereignty

Following the passage of a similar measure by Russia, the supreme soviet (government assembly) of Belarus passes a declaration of its sovereignty. However, Belarus still supports the preservation of the Soviet Union and the continuation of its association with it.

1991: April   Strikers express growing discontent

Striking Belarussians demand higher wages and protest new sales taxes. Some demand the resignation of the government.

Independence

1991: August 25   Independence is declared

Following the failed coup he supported that exposes the weakness of Mikhail Gorbachev's government, the head of Belarus's government, Nikolai Dementei, is forced to resign. Following the lead of Estonia, Latvia, and Ukraine, Belarus declares its independence and changes its name from the Belarussian Soviet Socialist Republic to the Republic of Belarus (or Byelarus).

1991: December 8   The CIS is formed

As the Soviet Union dissolves, Belarussian leader Stanislaw Shushkyevich signs an agreement forming the Commonwealth of Independent States (CIS) with Russia and Ukraine. Eleven other former Soviet republics join the CIS later in the month.

1993–94   Drought and flooding imperil harvests

Severe flooding followed by drought reduces the nation's harvests, further reduces already declining agricultural production, and puts additional stress on an already struggling economy.

1994: March 30   New constitution is adopted

Belarus adopts a new constitution providing for separation of powers and freedom of religion and creating the office of president.

1994: July 10   First president is elected

Belarus elects former Soviet official Aleksandr Lukashenka as its first president in a runoff election with over eighty percent of the vote. Lukashenka, who advocates closer ties with Russia, imposes authoritarian rule on Belarus, censoring the media and silencing political opposition.

1995: May   Parliamentary elections are held

Belarus holds its first parliamentary elections since proclaiming independence. A majority of eligible voters turn out in only 119 out of 260 voting districts, meaning that only 119 members can be elected to parliament because the results in the remaining districts are considered invalid. With only 119 members, the parliament cannot meet, since a quorum of 179 is required under the new constitution. Thus Belarus effectively has no parliament, and President Lukashenko rules by decree until new elections can be held.

In the same polling, over eighty percent of Belarussian voters approve a referendum calling for reestablishing Russian as the country's official language and moving toward an economic union with Russia. The new flag adopted by Belarus is also replaced by the old Soviet-era flag.

1995: May 26   Customs union formed with Russia

Belarus and Russia agree to form a customs union that will abolish border checkpoints between the two nations.

1995: July   Belarus gets IMF loan

The International Monetary Fund (IMF) extends a $300 million standby loan to Belarus based on promises of economic reform. However, Belarus continues to delay reforms and has yet to implement privatization of government-owned entities.

1995: November–December   New parliamentary elections

Belarus successfully elects its first post-independence parliament in its second scheduled election. The Communist and Agrarian parties control about eighty seats, further reducing chances of major economic reforms.

1996: November 9–24   New constitution is approved

In a popular referendum, seventy percent of Belarussian voters approve a new constitution that significantly reduces their government's separation of powers and greatly increases the power of President Aleksandr Lukashenko, who is authorized to appoint one-third of the members of a newly created upper house of parliament and half the members of the highest court. Lukashenko is also given the power to rule by decree, and his term is extended an additional two years to 2001. The new constitution is denounced by the Council of Europe and the U. S. State Department.

1996: November 17   Demonstrators protest new constitution

Between 5,000 and 10,000 protesters in Minsk express their disapproval of the new constitution giving President Aleksandr Lukashenko increased powers.

1996: November 24   Last nuclear arms are shipped to Russia

President Lukashenko announces that the last of the Soviet nuclear arms stored in Belarus—fourteen intercontinental ballistic missiles—have been shipped to Russia to be dismantled.

1997: April 2   Belarus and Russia sign treaty of union

The presidents of Russia (Boris Yeltsin) and Belarus (Aleksandr Lukashenko) sign a treaty providing for greater political and economic integration between the two countries, whose citizens can now own property in each other's countries and vote in each other's local elections, as well as move freely across their common border. An earlier version of the agreement, which was scrapped due to Russian objections, created a joint supreme council made up of the presidents, prime ministers, and parliamentary speakers from both countries.

Bibliography

Gross, Jan Tomasz. *Revolution from Abroad: The Soviet Conquest of Poland's Western Ukraine and Western Belorussia.* Princeton, N.J: Princeton University Press, c1988.

Kipel, Vitaut and Zora Kipel, eds. *Byelorussian Statehood: Reader and Bibliography.* New York : Byelorussian Institute of Arts and Sciences, 1988.

Loftus, John. *The Belarus Secret.* New York: Knopf, 1982.

Lubachko, Ivan S. *Belorussia under Soviet Rule, 1917–57.* Lexington: University Press of Kentucky, 1972.

Marples, David R. *Belarus: From Soviet Rule to Nuclear Catastrophe.* London: Macmillan, 1996.

Sword, Keith, ed. *The Soviet Takeover of the Polish Eastern Provinces, 1939–41.* New York: St. Martin's Press, 1991.

Urban, Michael E. *An Algebra of Soviet Power: Elite Circulation in the Belorussian Republic, 1966–86.* New York: Cambridge University Press, 1989.

Zaprudnik, I.A. *Belarus: At a Crossroads in History.* Boulder, Colo.: Westview Press, 1993.

Belgium

Introduction

As one of Europe's smallest countries, Belgium, with its strategic location bordering the great European powers of France and Germany and the North Sea coast, has been vulnerable to attack and occupation over hundreds of years. However, this same location has also made the country a crossroads for commerce since the Middle Ages and brought Belgium prosperity and a rich cultural history. In recent decades, the nation's ethnic and linguistic divisions—between the Dutch-speaking Flemish in the north and the French-speaking Walloons in the south—have moved to the forefront of national concerns, and the major regions of Belgium have gradually moved toward internal autonomy, or self-rule. Today, Belgium is officially a federation of three regions (Flanders, Wallonia, and Brussels) and also comprises three language communities (French, Dutch, and German).

With a triangular shape that resembles a bunch of grapes, Belgium consists of three major geographical regions. The coastal plain in Flanders is made up of dunes, pasture, and *polders* (land reclaimed from the sea). The central region of the country is a low plateau containing rich farmland. To the southeast lie the wooded highlands of the Ardennes, which stretch into France. Smaller regions include the Kempenland Plateau, a center of mining and industry bordering the Netherlands to the northeast, and Belgian Lorraine at the southern tip of the country. With an estimated population estimated at roughly ten million people in the 1990s, Belgium is one of the world's most densely populated countries.

The Low Countries

Belgium is part of a region historically referred to as the Low Countries (or Lowlands) because much of it is close to, or even below, sea level. Today this region comprises the nations of Belgium, Luxembourg, and the Netherlands. When the Romans, led by Julius Caesar, conquered the southern Lowlands (present-day Belgium) in around 50 B.C., they found resistance by the local Celtic people (whom they called *Belgae*) stronger than that in any of the neighboring territories. In addition to unifying the area politically, the Romans brought with them their laws and language. They were followed by the Franks, who invaded in the fourth and fifth cen-

turies A.D., introducing their Germanic tongue, and with it the beginnings of a language division that has lasted to the present day. Inhabitants in the north—the direction from which the Franks came—who adopted the new language were the forebears of today's Dutch-speaking Flemish. Those in the south retained the Latin dialect that later became French, which is still spoken by the Walloons, who live in that part of the country. The empire of the Franks reached its high point during the reign of Charlemagne (768–814), after which it was divided into three parts that eventually dissolved into a number of independent principalities under the feudal system of the Middle Ages.

In the thirteenth and fourteenth centuries, the southern Lowlands prospered with the growth of commerce, and coastal lands were reclaimed from the sea for agriculture by a system of dikes and walls to help feed an ever-expanding population. Between the fourteenth and sixteenth centuries, the feudal states of the Low Countries came under the control of the French dukes of Burgundy. This was a period of great artistic achievement. In the visual arts, Flemish painters from Jan Van Eyck to Peter Paul Rubens refined the art of oil painting, creating great religious canvases as well as portraits and other secular works for wealthy private patrons. The great Flemish composers, including Jean d'Ockeghem and Josquin des Prez, brought the art of Renaissance polyphony to its highest level.

In the sixteenth century the Low Countries (which were also collectively referred to as the Netherlands at this time) came under Spanish rule as a result of marriages between the royal houses of different countries. During the reign of Philip II in the latter part of the century, the largely Protestant population of the northern Netherlands, which suffered severe religious persecution by the Catholic monarch, successfully rebelled against Spanish rule and formed the United Provinces. The south, however, which was heavily Catholic, remained loyal to Spain and became known as the Spanish Netherlands. This name was changed to the Austrian Netherlands in 1714, when the region was ceded to the Austrian Habsburg Empire at the close of the War of the Spanish Succession. For most of the eighteenth century, Habsburg rule was indirect and allowed for a great degree of autonomy. However, Emperor Joseph II, who succeeded Maria Theresa, imposed unpopular religious and administrative reforms that

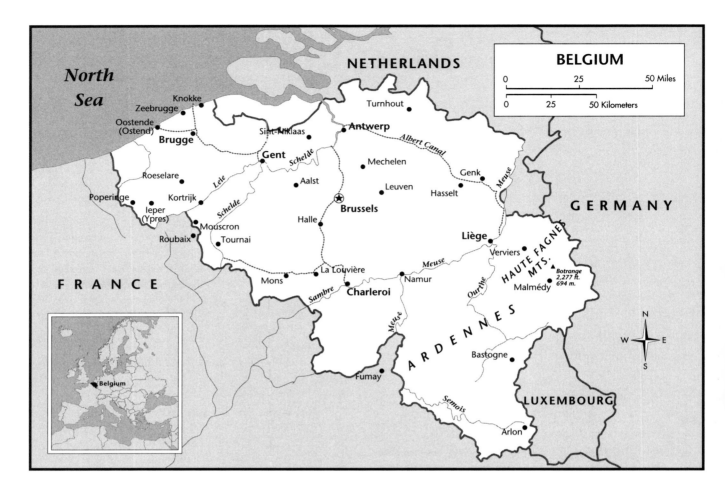

led to an unsuccessful revolt by the Belgians in 1789. This was soon followed by French invasions leading to a twenty-year occupation.

At the Congress of Vienna following the downfall of Napoleon in 1815, the European powers reunited the northern and southern Netherlands, in hopes of creating a strong nation that could halt any French aggression that might occur in the future. William I of Orange was named as ruler of the newly formed Kingdom of the Netherlands, whose formation soon led to discontent among the Belgians, who felt that they were treated as second-class citizens in a country largely run by the Dutch. Dutch was declared the official language, in spite of the fact that a majority of Belgians spoke French. In addition, the Belgians, who had the larger population, received no more parliamentary representation than the Dutch, who also were given most of the important appointed posts. In 1830 the Belgians staged armed rebellions and unilaterally seceded from the Kingdom of the Netherlands, declaring their independence. They drafted a constitution, formed their own kingdom, and chose Prince Leopold of Saxe-Coburg-Gotha (Leopold) to be its ruler. The Dutch attempted to suppress the Belgian rebellion, but the other European powers intervened, and a truce was declared. In 1839 the Netherlands officially recognized the kingdom of Belgium, which in return ceded some of the territory it had claimed, including Luxembourg.

To help protect the new nation from future invasions, it was officially recognized as a neutral nation. Belgium's European neighbors respected its neutrality for over seventy years, although relations with the Netherlands remained tense for some time.

Independent Belgium

During the reign of Leopold I, Belgium underwent rapid industrialization, benefiting from its extensive coal reserves to become one of Europe's leading industrial powers. The Belgians, who already enjoyed the natural advantage of excellent harbors, further expanded their trade capacity with the construction of a railroad and a canal system. Under Leopold's successor, Leopold II, who ruled from 1865 to 1909, Belgium joined the ranks of the European colonial powers, exploring and annexing the Congo River Basin in Africa.

With the growth of industrialization came the formation of groups to protect against the modern state's worst excesses. By the 1880s, Belgium's socialist party was powerful enough to effect the passage of important reform legislation to improve housing and labor conditions in the nation's cities. In 1893 all Belgian males won the right to vote without restrictions based on property or income. Five years later a law was passed giving equal status to Flemish and French. Unofficially, however, French continued to be regarded as the language of the elite, reflecting the greater clout enjoyed by Wallonia, the more populous southern part of Belgium and its industrial center. A knowledge of French remained the passport to a good education and professional advancement.

After more than seventy years, Belgian neutrality was breached at the beginning of World War I, when Germany invaded the country, launching a four-year occupation that ravaged the nation's resources and economy. Located between France and Germany, Belgium was the site of many of the worst battles of the war, including two engagements at Ypres that cost the lives of approximately 400,000 Allied soldiers. Casting off its prewar neutrality, Belgium signed a defense pact with France after the war and participated in the French occupation of the Ruhr district in Germany in 1923. During World War II, Belgium once again underwent German occupation. The prime minister and cabinet fled to London, where they established a government-in-exile. King Leopold III, however, elected to remain in Belgium under the occupation, a decision that later led to his abdication. Following the war, Belgium played an active role in European economic cooperation, joining the European Coal and Steel Community in 1952 and becoming a founding member of the European Economic Community in 1957. In 1960 the colonial era ended for Belgium with the granting of independence to the Congo, (renamed Zaire in 1971–96). Since the 1960s, a major economic shift has occurred within Belgium, as the traditional heavy industries, which were centered in Wallonia, have given way to high-tech enterprises, which have increasingly clustered in Flanders, shifting the nation's economic base northward.

In the postwar decades, autonomy for Belgium's Flemish and Walloon communities became a prominent political concern. By 1963 the government had recognized separate linguistic zones for Flemish, French, and German, designating the capital city of Brussels a bilingual area. In the early 1970s a constitutional amendment provided for the establishment of regional autonomy contingent on the vote of a two-thirds parliamentary majority. However, the required majority was not achieved, and implementation of the plan stalled. In 1980 representatives of Flanders and Wallonia agreed on a new plan providing for regional assemblies and granting each regional jurisdiction over its own economic, environmental, and other internal matters. A new framework providing for even more decentralization of political power was voted into law in 1993.

Timeline

c. 50 B.C.   Roman conquest

The Roman emperor Julius Caesar conquers the Low Countries, whose southern part comprises present-day Belgium.

The resistance of the Celtic people living there, whom the Romans call *Belgae* (the origin of the name *Belgium*), is so fierce that Caesar calls them the bravest inhabitants of the region, which he calls Gallia Belgica. The Romans unify the area politically and introduce Roman laws and the Latin language.

4th–5th centuries A.D.   Conquest by the Franks

The Franks, a Germanic people, invade from the north. The linguistic division of present-day Belgium begins when the inhabitants in the north of the region begin speaking a Germanic language that later evolves into Dutch, while those in the south continue speaking a Latin dialect, leading them eventually to become French speakers.

768–814   Reign of Charlemagne

During the reign of Charlemagne, the greatest of the Frankish rulers, Belgium achieves its highest degree of unity.

843   Treaty of Verdun divides Charlemagne's empire

The empire of Charlemagne is divided into three parts by the Treaty of Verdun. Belgium and the rest of the Lowlands are assigned to the duchy of Lower Lorraine, governed by King Lothair.

Feudalism

10th–14th centuries   Feudal rule

Following the breakup of Charlemagne's empire, the political unity of the southern lowlands gradually dissolves, and the region is divided into autonomous feudal states whose survival is often threatened by their vulnerable location between France and Germany.

13th–14th centuries   Southern Lowlands flourish

With the growth of foreign trade, the feudal states of the southern Lowlands, including Flanders, Brabant, Hainaut, Namur, and Liège, enjoy political stability and economic prosperity, based on the wool trade and other forms of commerce. The new commercial classes thrive and challenge the feudal dominance of the nobility. Due to the food needs of a growing population, low-lying land is reclaimed and intensive agriculture is developed.

1242   Rebuilding begins on Tournai cathedral

The rebuilding of the Cathedral of Notre Dame at Tournai, with its buttresses and vaulted roofs, becomes the first example of Gothic architecture in Belgium.

The Burgundian Era

14th–16th centuries   Burgundian rule begins

The semi-autonomous states of the Low Countries—present-day Belgium as well as the Netherlands and Luxembourg—come under the control of the French dukes of Burgundy.

1384   Philip becomes the first Burgundian ruler

Philip the Bold of Burgundy, the brother of the French king, becomes the ruler of Flanders and Artois, beginning the Burgundian rule of the Low Countries, which is later expanded to include Holland, Zeeland, Hainaut, Namur, Limburg, and Luxembourg.

c. 1395   Birth of painter Jan Van Eyck

Jan Van Eyck is the most famous member of a family of painters that includes his brothers Hubert and Lambert and his sister, Margaret. The family is thought to have lived in Flanders. Jan receives his training from Hubert, the eldest brother. By 1422 he is established in The Hague as a master painter employed by the Count of Holland, a position he retains until the count's death three years later. In 1425 Van Eyck moves to Bruges to assume a position with Philip the Good, Duke of Burgundy. His duties include voyaging to Lisbon to paint a portrait of Princess Isabella, the daughter of John I of Portugal, during the negotiations preceding the Duke's engagement to her.

By 1431 Van Eyck, still employed at court, is living on his own in Bruges. He continues to produce works commissioned by the Duke and serves as his general advisor in artistic matters. In addition, he also receives commissions from other patrons, many of them members of Bruges' Italian community. Van Eyck is renowned for his detailed depiction of his subjects and his skill in integrating them with fully realized settings, either interior or exterior. His mastery of oil painting is so outstanding that he has been mistakenly credited with inventing the technique. However, he is its most skilled practitioner during his era. Among his many paintings are *The Arnolfini Wedding, St. Jerome in His Study, Man in a Red Chaperon,* and *Adoration of the Mystic Lamb,* a painting he completed for his brother Hubert after the latter's death. Van Eyck dies in Bruges in 1441.

c. 1410   Birth of composer Jean d'Ockeghem

Ockeghem, one of the greatest Flemish composers of the Renaissance, is employed as a court composer by three French kings in a row: Charles VII, Louis XI, and Charles VIII. He has a strong influence on the succeeding generation of composers through both his works and his teaching. His works are known for their unbroken flow of sound, achieved by having cadences (a series of notes that indicate a partial or complete conclusion) occur at different times for each melodic voice, and the rich sound produced by their choral harmony. Ockeghem also extends the bass range lower than previous composers have. The known works attributed to Ockeghem include ten motets (a polyphonic song of a sacred nature, often unaccompanied), fourteen masses, and twenty chansons (songs). Among the texts set to music in Ockeghem's motets are *Ave Maria, Salve regina,* and *Alma redemptoris mater.* Ockeghem dies in 1497.

1423–32   The Ghent Altarpiece is produced

Hubert and Jan Van Eyck create the Ghent Altarpiece for the cathedral of St. Bavo. The twenty-four-panel altarpiece is widely considered the greatest painting of the period, as well as the pinnacle of fifteenth-century altarpiece decoration. Modeled on the Van Eycks' example, elaborate altarpiece paintings are created for many churches throughout the Netherlands, funded by private patrons (whose likenesses are often included in the altarpieces). The altarpiece's *Adoration of the Lamb* is particularly majestic, with its expansive landscape and multitude of saints. The Ghent Altarpiece has remained at St. Bavo continuously except during World War II, when it was taken to Germany by Nazi leader Hermann Goering.

c. 1430   Birth of painter Hans Memling

Flemish painter Hans Memling, known for his religious paintings, is born near Frankfurt-am-Main, Germany. He studies art in Cologne and Brussels, settling in Bruges in 1465 and entering the painters' guild there two years later. Memling's best-known works are painted for the Hospital of St. John in Bruges, although he also receives commissions from patrons elsewhere in Europe. Among Memling's many paintings are the *Chatsworth Triptych,* the *Passion of Christ,* and *St. Sebastian.* He also gains renown for painting altarpieces, including the *Adoration of the Magi,* the *Mystic Marriage of St. Catherine,* and the *Floriens Altar.* Memling dies in 1494.

c. 1440   Birth of composer Josquin des Prez

The major musical innovator—and most famous composer—of the early Renaissance, Josquin is thought to have been born in the province of Hainaut. Details of his life are sketchy. From 1474 to 1479 he works for the Duke of Milan, and from 1486 to 1494 for the Church in Rome. In the early 1500s he is employed in Ferrara. Toward the end of his life (c. 1521) he holds positions in Brussels and Condé. Among his many accomplishments, Josquin is known for his mastery of imitative counterpoint and his development of the *cantus firmus* (variations on an existing melody), as well as his rhythmic innovations. He writes twenty masses, more than one hundred motets, and over fifty secular pieces, various psalms, hymns, and other works. Some of his best-known compositions are the "De Beata Virgine" and "Pange lingua" Masses, the *Stabat Mater,* and the "Miserere." Josquin dies in about 1521.

1441: June   Death of painter Jan Van Eyck

Renowned painter Jan Van Eyck dies in Bruges. (See 1395.)

1477   Habsburgs gain control of the Low Countries

Mary of Burgundy marries Holy Roman Emperor Maximilian I, bringing the Low Countries, now consisting of seventeen provinces (the area that currently comprises the Benelux countries), under the control of the Habsburg dynasty.

1497: February 6   Death of composer Jean d'Ockeghem

Jean d'Ockeghem, one of the greatest Flemish Renaissance composers, dies in France. (See 1410.)

1514: January 1   Birth of anatomist Andreas Vesalius (1514–64)

Vesalius, a pioneer in the study of human anatomy, is born in Brussels. He receives his medical training there and in Louvain, Paris, and Padua, where he receives his medical degree and accepts a position teaching anatomy and surgery. Dissatisfied with the model of human anatomy current at the time, based on writings of the Roman physician Galen that are over a thousand years old, Vesalius transforms the field by carrying out studies on human corpses (as opposed to previous research, which has been conducted largely on animals). The result is a landmark seven-volume study entitled *De humani corporis fabrica* (On the Structure of the Human Body), which is published in 1543. It becomes famous for the high quality and accuracy of its woodcut illustrations.

Vesalius's fame grows, and he accepts a position as court physician to Holy Roman Emperor Charles V, ruler of the Netherlands (of which Belgium is a part). As part of this job, he works as a military doctor for the royal army and as personal physician to the king and is also able to establish an extensive private practice, becoming the most renowned doctor in Europe. Following the abdication of Charles V, Vesalius is required to move to Spain and serve Charles's son, Philip II. Desiring to resume his onetime teaching position in Padua, he takes a leave of absence from the court and travels to Italy and the Holy Land. However, Vesalius's ship encounters rough weather on the return voyage from Jerusalem, and he does not survive the trip, dying in 1564.

The Spanish Netherlands

1516   Low Countries come under Spanish control

Charles, the grandson of Mary of Burgundy and son of Johanna of Castile and Aragon, inherits the Spanish throne, becoming King Charles I of Spain and Holy Roman Emperor, and the Low Countries come under Spanish rule.

c. 1521   Death of composer Josquin des Prez

Josquin des Prez, the most renowned of the early Flemish composers, dies. (See 1440.)

c. 1525   Birth of artist Pieter Brueghel the Elder (c. 1525–69)

Flemish painter Pieter Brueghel the Elder is among the most renowned artists of the Low Countries. Little is known of his early life. As a young man, he travels in Italy and later settles first in Antwerp and then in Brussels, where he lives after 1563. He is professionally successful, enjoying commissions from wealthy patrons as well as the respect of his peers. His sons Pieter the Younger (c. 1564–1638) and Jan (1568–1625) both become well-known artists in their own right. Brueghel is known for the essentially medieval moral dimension of his paintings, which subordinate the individual to larger forces, and for transforming landscape painting into an art form. Well-known paintings by Brueghel include *The Fall of Icarus* (c. 1558), *Fight Between Carnival and Lent* (c. 1559), *Procession to Calvary* (1564), and the series of five landscapes called *Seasons,* or *Months* (1565). Brueghel dies in 1569.

1531   Financial exchange opens in Antwerp

The Antwerp bourse, one of the world's first financial exchanges, opens it doors.

1532   Birth of composer Orlandus Lassus

Orlandus Lassus (also known as Orlando di Lasso) is born at Mons, in Hainaut, eventually making his way to Naples and Rome, where he is appointed choirmaster at the basilica of St. John. The music of Italy—especially the madrigal—becomes an important influence on Lassus's work, where it merges with the musical traditions of the Low Countries, whose composers have flourished for over a hundred years. After a period of travel and some time spent in Antwerp, Lassus becomes part of the court of Duke Albert, the prince of Bavaria, at Munich, where he remains for the rest of his life, composing and taking charge of all aspects of music at court. He and the duke enjoy a warm personal bond that strengthens their professional relationship, and Lassus is accepted as a social equal within the royal circle. Lassus's reputation spreads throughout Europe, and he is offered other positions but signs a lifetime contract with Duke Albert. He dies in 1594 and is buried in Munich.

Lassus composes a wide variety of both sacred and secular works, including 1,500 motets, 100 magnificats (a song, poem, or hymn of praise), 53 masses (many using melodies taken from secular songs), 150 madrigals, and many short songs in both French and German. His most renowned religious works are his motets, of which a highly regarded example is the *Seven Penitential Psalms.* Wit and satire are prominent in many of Lassus's secular music, such as his widely performed madrigals.

1556   Rule of Philip II begins

The Low Countries (also collectively referred to as the Netherlands during this period) come directly under Spanish rule when Charles I abdicates and turns the region over to his son, who becomes King Philip II. Philip tries to force the Protestants of the northern provinces to abandon their religious beliefs and embrace Catholicism.

1564: October 15   Death of physician Andreas Vesalius

Andreas Vesalius, renowned for revolutionizing the field of anatomy, dies in a shipwreck on a voyage to Jerusalem. (See 1514.)

1568   Northern provinces revolt

The Protestants of the northern provinces, persecuted for their religion by King Philip II, revolt against Spain under the leadership of William of Orange. The struggle with Spain will last through various stages (including a twelve-year truce) until 1648, becoming known as the Eighty Years' War. The southern provinces, which are primarily Catholic, remain loyal to Spain.

1577: June 28   Birth of painter Peter Paul Rubens

Peter Paul Rubens, one of the greatest Flemish painters, is born in Westphalia, the son of a lawyer. He receives an extensive education in academic subjects as well as painting and masters six languages. After studying with three different artists, Rubens is admitted to the guild of master painters in 1598. In 1600 he leaves for Italy to study and paint, returning eight years later when his mother dies. He establishes a successful studio in Antwerp, and his paintings are in great demand. Their subject matter includes battle and hunt scenes, religious themes, and mythological topics. Rubens's works are commissioned by wealthy private patrons, both in Belgium and abroad, and churches. In the 1620s he undertakes extensive projects for the French monarchy. In addition to his painting, Rubens also pursues a career as a diplomat, taking part in major peace negotiations. He dies in 1640.

1579   Union of Arras is formed

The largely Catholic southern provinces of the Low Countries form a confederation that recognizes the sovereignty of Philip II. These provinces become known as the Spanish Netherlands. The division between the Protestant north (which will become the present-day Netherlands) and the Catholic south (which will become Belgium) is confirmed when the northern provinces declare their independence from Spain in the same year and form the United Provinces. Although the Spanish are effectively driven from the northern provinces within the first years of the war, Spain continues its efforts to retake them. Full Spanish recognition of Dutch independence does not come for eighty years. Thus the rebellion is known as the Eighty Years' War.

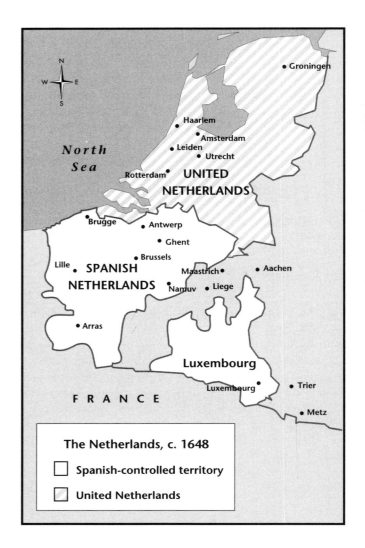

The Netherlands, c. 1648

☐ Spanish-controlled territory

☐ United Netherlands

1594   Death of composer Orlandus Lassus

Flemish composer Orlandus Lassus, famed for both sacred and secular works, dies in Munich. (See 1532.)

1598   The Spanish Netherlands becomes a state

Under the rule of Archduke Albert and his wife, Isabella, the Spanish Netherlands becomes a separate state under Spanish rule. However, when Albert and Isabella die without leaving an heir, the region returns to direct Spanish rule.

1599: March 22   Birth of painter Anthony Van Dyck

Renowned portrait painter Anthony Van Dyck (1599–1641) is born in Antwerp and enters the painters' guild at the age of nineteen, working as an assistant to Peter Paul Rubens (see 1577) for three years. He travels to England and then Italy, where he stays for six years. Eventually Van Dyck settles in London. His portraits of the British aristocracy achieve such success that he is knighted. In the final years before his early death in 1641, the painter divides his time between Antwerp,

London, and Paris. His use of color influences later artists including Gainsborough, Watteau, and Murillo.

1640: May 30   Death of painter Peter Paul Rubens

One of Belgium's greater painters, Peter Paul Rubens, dies. (See 1577.)

1641: December 9   Death of painter Anthony Van Dyck

Belgian painter Sir Anthony Van Dyck dies in London. (See 1599.)

1648   Treaty of Westphalia

The northern provinces of the Netherlands win formal recognition of independence from Spain, and become known as United Netherlands, while the southern provinces remain under Spanish rule.

1663: July   Art academy is authorized

Prince Philip IV of Spain authorizes the establishment of an art education facility—the Academie voor Schone Kunsten—in Antwerp modeled on French and Italian academies. David Teniers the younger is put in charge of the project, which is to teach painting, sculpture, and book production.

1695   Brussels is besieged

As part of an attack by the French under King Louis XIV, Brussels is bombarded by cannons, which destroy over two thousand buildings and its central square, the Grand Place. A new square is later built.

Austrian and French Rule

1714   Southern provinces come under Habsburg rule

At the close of the War of the Spanish Succession, the Spanish Netherlands are ceded to Austria and become known as the Austrian Netherlands. The Belgians retain a high degree of autonomy. Neither Charles VI, the Austrian monarch at the time of the takeover, nor Maria Theresa, the next ruler, ever sets foot in the provinces, which are governed through Austrian-appointed governors in Brussels.

1744–48   French occupation

Belgium is occupied by France during the War of the Austrian Succession. The occupation is ended by the Treaty of Aix-la-Chapelle.

1780s   Austrian-imposed reforms arouse discontent

Religious and administrative reforms imposed by the Austrian monarch Joseph II, who succeeds Empress Maria Theresa, arouse resentment among the Belgians, who feel that their institutions and way of life are being threatened.

1788   Belgians withhold tax payments

To protest reforms by Austrian emperor Joseph II, the Belgian provinces of Brabant and Hainaut withhold their tax payments.

1789   Belgians revolt against Austria

A full-scale revolt against Austria, referred to as the Brabanconne Revolution, takes place, with the Belgian legislative body proclaiming the United States of Belgium.

1790: February   Revolt against Austria is crushed

Austrian emperor Leopold II crushes the Belgian rebellion and overthrows the fledgling Belgian state.

1792   Invasion by France

France invades Belgium following the French Revolution, disillusioning Belgian supporters of the Revolution. Another invasion occurs two years later, followed by annexation.

1795   Fine arts museum is founded

The Royal Museum of Fine Arts is established, divided into two main collections. The older collection includes works by Flemish masters of the fourteenth through seventeenth centuries. The modern collection features nineteenth and twentieth-century works by artists from both Belgium and other countries.

1797   French takeover

France claims sovereignty over Belgium in the Treaty of Campo Formio. Belgium remains under French occupation until the downfall of Napoleon Bonaparte in 1815.

1812: December 3   Birth of author Hendrik Conscience (1812–83)

Hendrik Conscience, the author of the first Flemish novel, plays such a large role in the establishment of a Flemish literary tradition that it is said, "He taught his people to read." Born in Antwerp, he works as a teacher and then enlists in the army. In 1837 he publishes *In 't wonderjaar* (In the Year of Miracles), his first work in Flemish, which has largely been regarded as an inferior language not suited for literature. The following year, *De leeuw van Vlaanderen* (The Lion of Flanders) gains fame both for its overall quality as a historical novel. It is the first historical novel to be published in Flemish. After 1840 Conscience turns from his early interest in French Romanticism toward more realistic works describing urban and rural Belgian life. He also writes more historical novels and in 1862 publishes *Het Goudland* (The Land of Gold), the first adventure novel in Flemish.

By the mid-nineteenth century, Conscience achieves a reputation beyond Belgium, and his works are translated into other languages. The publication of the author's one hundredth book in 1881 occasions public celebration in Brussels. Conscience dies two years later, and the city of Antwerp erects a statue in his honor.

A Reunified Kingdom

1815   Northern and southern Netherlands are reunited

Hoping to create a strong buffer state to prevent future French aggression, the Congress of Vienna creates a new Kingdom of the Netherlands uniting the northern and southern provinces of the former Low Countries. A prince of the House of Orange, crowned William I, is chosen to rule over both the Dutch and Belgians as head of the new kingdom.

Although union with the Netherlands provides some economic advantages, the Dutch maintain the upper hand in governing the new kingdom, and the Belgians feel that they are treated like second-class citizens. Dutch is the official language, in spite of the fact that many Belgians speak French. Belgian representation in parliament does not reflect its greater population, and most top appointed posts are given to the Dutch.

1822: December 10   Birth of composer César Franck

Belgian-born composer César Auguste Franck, who becomes one of the giants of late-nineteenth-century European music, is born in Liège, where he studies music at the Ecole Royale de Musique and later at the Paris Conservatory. His father attempts to prepare him for the career of a concert pianist, but Franck shows an early interest in composition, writing piano pieces that display his various abilities. In 1835 Franck moves to Paris with his family and studies with composer Antonin Reicha, later entering the Paris Conservatory as a composition student. He publishes his first serious compositions in 1842 and 1843 and begins an active career as a teacher. He is also appointed organist at Notre Dame de Lorette.

In 1872 Franck is appointed to teach organ at the Paris Conservatory, introducing his students not only to performance techniques but also to his own compositions, which bypass currently popular French styles and draw on traditions established by Bach and Beethoven. Franck's teachings influence many notable composers, including Vincent d'Indy, Ernest Chausson, and Henri Duparc. The composer's great mature works include the Piano Quintet (1878–79), the Prelude, Chorale, and Fugue for piano (1884), the Symphonic Variations for piano (1885), the Violin Sonata in A major, composed for the great violinist Eugen Ysäye (1886), the Symphony in D minor (1886–88), the String Quartet in D major (1889), and the Three Chorales for organ (1890). Franck dies in Paris from injuries suffered in a street accident in 1890.

1828   Belgian political union is created

The two main rivals for Belgian political power, the Liberals and the Catholics, are brought together by their opposition to Dutch rule and form a powerful political union, called *unionisme,* to help Belgium regain its independence.

1830: May 1   Birth of poet Guido Gezelle

Writing in Flemish, Gezelle produces some of the finest European lyric poems of the nineteenth century. Gezelle is born in Bruges and later lives in Roulers, where he becomes a teacher and is ordained as a priest in 1854. In 1860 Gezelle returns to Bruges. In addition to his religious and literary commitments, Gezelle is also an ardent Flemish nationalist and political journalist. He founds a cultural periodical, *Rond den Heerd*, in 1865 and later begins publishing a scholarly journal. He is also known for his translation of the poem *Hiawatha* by the American poet Henry Wadsworth Longfellow. Gezelle dies in 1899.

1830: August 25   Uprisings begin

Inspired by the July Revolution in France, Belgians in Brussels and Liège stage revolts, which spread throughout the country. Dutch forces soon enter Belgium to retake the provinces and are met with resistance.

Independence

1830: October 4   Belgian independence is declared

Following a number of uprisings against Dutch rule, Belgium declares itself independent, and a provisional government is formed.

1831–65   King is chosen to rule Belgium

The Belgian Parliament chooses Prince Leopold of Saxe-Coburg-Gotha as ruler of the new kingdom (Leopold I). Rapid industrial growth occurs during his reign.

1831: February   Constitution is drafted

The Belgians draft a progressive constitution that provides for a limited monarchy governed by a king, a parliament, and a prime minister who require parliamentary approval. Secret ballot elections are set up, with voting restrictions based on income, and linguistic equality is guaranteed.

1831–33   Dutch attacks on Belgium

King William of the Netherlands refuses to accept Belgian independence and invades, but his forces are driven back. The European powers intervene to ease the conflict, and a treaty is drawn up in London, but the Dutch king rejects it and invades Belgium again. His forces are rebuffed with help from the French.

1833   Dutch-Belgian truce

The Netherlands and Belgium sign a truce agreement.

1835   Museum of art and history is founded

The Royal Museum of Art and History is established in Brussels, housing artifacts from the ancient civilizations of Egypt, Greece, Rome, and the Middle East. Its collection includes Chinese ceramics, Flemish tapestries, and a variety of glass and jewelry collections, as well as exhibits on Belgian folklore.

1838   Publication of *The Lion of Flanders*

This book by Hendrik Conscience is a landmark in the campaign to gain legitimacy for the Flemish language, which has a lower status, both officially and unofficially, than French.

1839: April   The Netherlands recognizes Belgium

The Netherlands accepts Belgian independence through the Treaty of the XXIV Articles, an agreement settling boundary and debt disputes between the two nations. Belgium cedes Maastricht, plus parts of Limburg and Luxembourg to the Dutch. Belgian independence and neutrality are recognized both by the Netherlands and the other European powers, including Great Britain, Austria, France, Russia, and Prussia. Relations between Belgium and the Netherlands remain tense for decades.

1847–70   Period of liberal power

A liberal government, favoring secularism in education, is in power.

1847   Voting requirements are lowered

Property qualifications for voting are relaxed, helping to avert revolutionary activity such as that occurring in France and other countries during this period.

1862: August 29   Birth of author Maurice Maeterlinck (1862–1949)

Acclaimed poet and playwright Maeterlinck is a leading figure in the turn-of-the-century Symbolist movement in literature. He is associated with the Jeune Belgique (Young Belgium) poetry movement (see 1881). His first major published work is the poetry collection *Serres chaudes* (Hot House Blooms; 1889). Maeterlinck is best known for his drama *Pelléas et Mélisande* (1892), which is set to music ten years later as an opera by French composer Claude Debussy (1862–1918). Other plays by Maeterlinck include *L'Oiseau bleu* (The Bluebird; 1908) and *Le Bourgmestre de Stilmonde* (The Burgermaster of Stilmonde; 1918). Maeterlinck is also a widely read author of prose works including *La Vie des abeilles* (The Life of the Bee; 1901) and *L'Intelligence des fleurs* (The Intelligence of Flowers; 1907). Maeterlinck is

awarded the Nobel Prize for Literature in 1911. He dies in 1949.

1865–1909   Leopold II succeeds to the throne

Leopold II succeeds Leopold I as ruler of Belgium. During his reign, Belgium begins building its colonial empire through exploration and settlement of the Congo River Basin.

1879: January 2   Lemmens Institute opens

The Lemmens Institute, a school for training in religious music, is founded at Mechelen by Jaak Nikolaas Lemmens. Church organists and choirmasters are educated at the facility, and future clergy study music there as well.

1880s   Socialist party gains power

A Socialist party, the *Parti ouvrier belge* (POB), under the leadership of Emile Vandervelde, is instrumental in the passage of reform legislation in the areas of housing and labor.

1880   Commemorative park is built

The *Parc du Cinquantenaire* (Fiftieth Anniversary Park) is created in Brussels to commemorate Belgium's fiftieth anniversary as an independent nation. The Royal Museum of Art and History (see 1835) is later situated in this park.

1881   *La Jeune Belgique* is founded

La Jeune Belgique, an influential literary review, is founded by poet Max Waller. It is associated with a movement of the same name whose aim is to create a distinctly Belgian poetry, freed of the worn out conventions of Romanticism. Aside from Waller, prominent poets involved in the group include Maurice Maeterlinck, Emile Verhaeren, and Max Elskamp.

1883: September 10   Death of novelist Hendrik Conscience

Hendrik Conscience, author of the first Flemish novel (*The Lion of Flanders*; 1838), dies in Brussels. (See 1812.)

1884–85   Leopold wins personal control of the Congo

The Berlin Conference dividing the African continent among the European powers agrees to place the Belgian Congo under the personal jurisdiction of Leopold II, rather than entrusting it to the government as a whole. Leopold subsequently draws international censure for allowing the maltreatment of the Congolese by European commercial interests.

1890s   Group of authors promotes Flemish literature

The *Van nu straks* (Of Now and Later) literary circle, associated with the periodical of the same name, flourishes, advancing the recognition accorded to literature in Flemish. Authors include August Vermeylen, Prosper van Langendonck, Alfred Hegenscheidt, and Emmanuel Karel de Bom.

1890: November 8   Death of composer César Franck

Belgian-born composer César Franck dies in Paris. (See 1822.)

1893   Universal manhood suffrage is enacted

All males, regardless of income level, win the right to vote. This measure is widely seen as a victory for Catholics, who enjoy widespread support among the lower classes.

1898   Language law is passed

Legislation makes French and Flemish equal in the eyes of the law. Informally, however, French continues to be accorded a higher status, and a knowledge of French is generally necessary for professional advancement.

1898: November 21   Birth of artist René Magritte

Magritte, a major Surrealist and probably the best-known twentieth-century Belgium painter, is born in Hainaut province and studies at the *Académie des Beaux Arts* (Academy of Fine Arts) in Brussels. By 1919 he is exhibiting posters he has painted. Early influences include Futurism and abstract art. The artist produces his first Surrealist works in the 1920s and plays an important role in the formation of a Belgian Surrealist group, although most of its members are writers. From 1927 to 1930, Magritte lives in Paris, becoming acquainted with the French Surrealists.

The Belgian Surrealists, unlike the French, challenge conventional concepts of reality by skewing ordinary representational conventions rather than by an emphasis on the subconscious. Magritte accomplishes this by creating new objects; altering the usual form and order of objects (as in *Empire of Light,* which shows a nighttime street scene set against an afternoon sky); and manipulating the relationship between words and images (as in the *Treachery of Images,* which features a picture of a pipe with the inscription "Ceci n'est pas une pipe" [This is not a pipe.]). Magritte is able to paint full-time after signing contracts with two galleries in 1926. He mounts his first one-man show in March 1927 at Galerie Le Centaure, thus making his debut as a Surrealist. To weather the economic depression of the 1930s, Magritte produces commercial art together with his brother, Paul. Magritte's first one-man show in Paris takes place in 1948. His reputation continues to grow in the postwar period, and he influences later artistic movements, including Pop art and conceptual art. Magritte dies in Brussels in 1967.

1899   Proportional representation is instituted

The passage of legislation mandating proportional representation in the legislature eases discontent in both rural and urban constituencies.

1899: May 24   Birth of author Henri Michaux

Surrealist poet and painter Henri Michaux is born in Namur and settles in Paris in 1922. Both his poems and paintings are known for their richly imagined depictions of an inner life that defies the rationality of external existence. His poetry collections include *L'Espace du dedans* (The Space Within; 1944), *Ailluers* (Elsewhere; 1948), and *La vie dan les plis* (Life Within the Folds; 1950). Michaux dies in 1984.

1899: November 27   Death of poet Guido Gezelle

Lyric poet Guide Gezelle dies in Bruges. (See 1830.)

1903: February 13   Birth of author Georges Simenon

Detective fiction author Georges Simenon is born in Liège and begins working as a journalist while in his teens. Moving to Paris by the age of nineteen, he publishes over two hundred books of popular fiction under various pseudonyms over a period of ten years. In his 1931 novel *Pietr-le-Letton* (the first to be published under his own name), Simenon introduces the fictional detective Inspector Maigret, who will be featured in about eighty more novels and become one of the most popular of fictional characters. Simenon also authors short story collections and memoirs. He dies in Lausanne, Switzerland, in 1989.

1903: June 8   Birth of author Marguerite Yourcenar (1903–87)

Author Marguerite Yourcenar is born Marguerite de Crayencour in Brussels. When she is a young woman, her father dies, leaving her independently wealthy, and she travels throughout Europe, living and writing in various places until the beginning of World War II, when she emigrates to the United States. Yourcenar writes historical novels, short stories, essays, and other works as well as producing translations. Her most famous work is the 1951 novel *Mémoires d'Hadrien* (Memoirs of Hadrian), which reconstructs the life of a second-century Roman emperor. Marguerite Yourcenar is the first woman elected to the distinguished French intellectual body, the *Académie Francaise* (French Academy). She dies in 1987.

1908   Belgian government is put in charge of the Congo

The Congo is transferred from the individual management of King Leopold II to the jurisdiction of the Belgian government.

1909   Albert I becomes king

Albert I, nephew of Leopold II, becomes king of Belgium.

1911 Author Maeterlinck wins Nobel Prize

The Nobel Prize for Literature is awarded to Maurice Maeterlinck, author of the drama *Pelléas et mélisande* and other works. (See 1862.)

The World Wars

1914–18 World War I

Belgium's seventy-five years of political neutrality come to an end with World War I. Belgium is invaded by Germany and undergoes a four-year occupation that destroys its industrial base and plunders its resources. The Belgian economy is further decimated by the disruption of normal foreign trade. The economic problems generated by the war result in severe inflation. Belgium is also the site of some of the heaviest fighting of the war, including two battles at Ypres, at which as many as 400,000 men die.

1914: August 3–4 Germany invades Belgium

After refusing to allow German troops to pass through on their way to attack France, Belgium itself is invaded by the Germans.

1919 German territory ceded to Belgium

Under the Treaty of Versailles, Belgium gains control of the Eupen, Malmédy, St. Vith, and Moresnet districts.

1920 Military pact is signed with France

Belgium throws off its prewar policy of neutrality by signing a defense pact with France, which is later brought into play when the Ruhr is occupied (1923).

1921–33 Pro Arte concert series promotes modern music

The Pro Arte series of concerts, organized by Paul Collaer, introduces the Belgian public to the music of Schoenberg, Berg, Webern, Stravinsky, Milhaud, and other modern composers.

1923 Occupation of the Ruhr

Following disagreements over German payment of wartime reparations, Belgium and France occupy the Ruhr region.

1925 Belgian borders are guaranteed by pact

Belgium's boundaries are affirmed by the Locarno Pact, which also establishes the borders of France, Germany, and the Rhineland.

1925 Synthétiste composers' group is formed

Seven students of composer Paul Gilson, influenced by composers including Ravel, Stravinsky, Hindemith, and Honegger, and by neoclassicism in general, form a musical circle known as the Synthétistes.

1926 Progressive art institute is founded

The Hoger Instituut voor Sierkunsten is established in Brussels. It becomes Belgium's first art education institution to embrace avant-garde schools, such as the Bauhaus, a school of architecture characterized by functionality in design.

1929 Depression begins

Belgium begins to feel the effects of the global depression that will last into the 1930s.

Unemployment grows rapidly, banks fail, and economic growth stalls.

1929: April 8 Birth of singer-songwriter Jacques Brel

Popular composer and entertainer Jacques Brel is born in Brussels and learns to play guitar as a teenager. After working in his family's business for four years and serving in the military, he moves to Paris and embarks on a performing career with the support of theatrical agent Jacques Canetti. In 1953, Canetti presents Brel, accompanying himself on guitar and backed by a small combo, in concerts at the Theatre des Trois Baudets in Pigalle. The following year Brel performs at the Paris Olympia. By the early 1960s, Brel's material begins to change from love songs to songs of social commentary and protest. In addition to the content, language in his songs alienates some listeners, but the singer still retains a wide following.

Brel makes his U.S. debut in Carnegie Hall in 1965 and records selections from a 1967 tour for Reprise Records. Poet Eric Blau and songwriter Mort Shuman expand Brel's audience even more with their 1968 Broadway musical *Jacques Brel is Alive and Well and Living in Paris,* which includes twenty-five of Brel's songs in translation. A number of Brel's songs also become widely known in versions recorded by other artists, including Dusty Springfield, Neil Diamond, Ray Charles, Frank Sinatra, and the Kingston Trio. Brel retires from performing in 1966, later moving to Polynesia. The singer battles cancer for several years in the 1970s, returning to France for treatment in 1977. He dies the following year.

1930 Radio broadcast institute is created

The Belgian government forms the Institut National de Radiodiffusion to broadcast radio programming.

1930 Administrative division of Belgium into regions

Belgium is divided into Flemish and Walloon regions for administrative purposes.

1935 Economic program is launched

The Belgian government introduces an economic recovery program comparable to the United States New Deal.

1936 National orchestra is founded

The state-supported Belgian National Orchestra, formerly the Brussels Symphony Orchestra, is established.

1940–44 Second German occupation

In World War II, Belgium is once again invaded and occupied by Germany. Prime Minister Hubert Pierlot and the Belgian cabinet flee to London and set up a government-in-exile, while an underground resistance is carried out in the occupied country itself. King Leopold III remains in the country as a prisoner of war, an act that creates controversy afterwards.

1940: May 10 German invasion

Germany stages a surprise invasion of Belgium, bombing transport and communications centers.

1940: May 28 King Leopold surrenders

King Leopold III surrenders unconditionally and becomes a prisoner of the Germans in occupied Belgium for the remainder of the war.

1940: October Government-in-exile is established

The Belgian government-in-exile is set up in London.

1944: September Allies liberate Belgium

Belgium is liberated by Allied forces. The Germans keep King Leopold III a prisoner, taking him back to Germany as they flee Belgium. The returning government-in-exile installs Leopold's brother, Prince Charles, on the throne as regent in his absence.

The Postwar Period

1947 Benelux customs union is formed

The bill creating a customs union among the three Benelux countries—Belgium, the Netherlands, and Luxembourg—is ratified.

1948 Full female suffrage is granted

Women win the right to vote on the same basis as men.

1949: April 4 NATO membership

Belgium becomes a member of the North Atlantic Treaty Organization (NATO).

An estimated 50,000 demonstrators carry portraits of the late King Albert and the late Astrid in a parade to protest King Leopold's decision to remain in Belgium during World War II. (EPD Photos/ CSU Archives)

1949: May 6 Death of author Maurice Maeterlinck

Nobel Prize-winning poet and playwright Maurice Maeterlinck dies in Nice, France. (See 1862.)

1950 Belgian vote favors return of King Leopold

In a special referendum, slightly over 50 percent of voters favor the return to Belgium of King Leopold, a controversial figure who voluntarily remained in Belgium during the German occupation and has lived in Switzerland since the end of the war. However, Belgium's two major population groups, the Flemings and Walloons, are divided on the issue, and civil unrest ensues.

1950: July 22 King Leopold returns to Belgium

King Leopold III returns to Belgium for the first time since the end of World War II. His return is marked by strikes and other protests, primarily by liberals and socialists.

1951: July 16   Abdication of King Leopold

King Leopold III, abdicates under pressure from his detractors, turning the throne over to his son, Prince Baudouin.

1952   Belgium joins European steel producers' group

Belgium becomes a founding member of the European Coal and Steel Community, formed to promote international trade by cutting taxes on imports of coal and iron ore and other measures.

1955   Law cuts aid to private schools

Legislation reducing state support for private religious schools is passed, spurring demonstrations by Belgians on both sides of this incendiary issue.

1957: March 25   European Community membership

Belgium joins the newly formed European Economic Community (EEC).

1958   Compromise education bill is passed

A new government headed by the Christian Socialist Party attempts to resolve the controversy over school funding with compromise legislation limiting the number of parochial schools that can receive government aid.

1958   Electronic music studio is formed

Internationally recognized composer Henri Pousseur founds Belgium's first electronic music studio, known as APELAC, in Brussels. It later becomes part of the Centre de Recherches Musicales de Wallonie in Liège.

1960   Composers' association is formed

The *Union des Compositeurs Belges* (Union of Belgian Composers) is formed by Belgium's foremost composers to promote Belgian musically both at home and internationally.

1960: June 30   Congo wins independence

Belgium grants independence to the Belgian Congo, which is renamed Zaire.

1962: November 20   An income tax is instituted

As part of a more inclusive tax law, an income tax is adopted, covering personal income from all sources.

1963   Legislation creates linguistic regions

Laws enacted by parliament divide Belgium into four linguistic regions, including Brussels as a bilingual zone.

1967: May 22   Department store fire kills hundreds in Brussels

Brussels becomes the scene of the most deadly store fire in history when a blaze breaks out at L'Innovation department store at a peak shopping time with a total of four thousand people, both clerks and shoppers, on the premises. The building has no sprinkler system, and firefighter response is too slow to control the fire until hundreds of people have died, many from being trampled in the panic to reach windows and exits. The explosion of butane gas cannisters stored on the roof for summertime sale to campers makes the disaster even worse. Altogether, 322 people die in the blaze, many of them reduced to ashes.

Regional Autonomy

1970   Constitutional revision creates language zones

A constitutional revision incorporating aspects of previous legislation (see 1963) gives each part of Belgium one of two official languages. The provinces of Antwerp, Limburg, East Flanders, and West Flanders in the north are the Flemish (Dutch) zone. The French region comprises the provinces of Hainaut, Liège, Luxembourg, and Namur in Wallonia. The province of Brabant, located in the center of the country, is divided into three zones, one of which is bilingual.

1971   Constitution is revised to provide for regional autonomy

In the wake of growing discord between Flanders and Wallonia, the Belgian constitution is reformed to pave the way for economic and cultural autonomy for the two regions.

1974: July   Conditions for autonomy are set by legislature

A law is enacted providing for regional autonomy for the regions of Flanders and Wallonia and the Brussels area upon the attainment of a two-thirds majority in parliament. However, proponents of autonomy are not able to gather the required majority.

1977   Federalist autonomy plan offered

Following the failure of a previous plan for regional autonomy, a Christian Socialist coalition draws up a new plan for federal representation of Belgium's regions, but parliament fails to pass it.

1978   Equal rights laws for women are passed

New laws guarantee equal pay and employment opportunities for women and men. (Into the 1990s, however, women earn nearly one-third less than men, and their unemployment rate is almost twice as high.)

1978: October 10   Singer Jacques Brel dies

Popular 1960s singer-songwriter Jacques Brel dies of cancer near Paris. (See 1929.)

1979   Martens become prime minister

Wilfried Martens, leader of the Christian Social Party (CVP) begins thirteen years as prime minister at the head of thirteen different cabinets. During his first two years in office, three governments are toppled by the nation's serious economic problems (especially a high rate of unemployment) and the unpopular austerity measures framed to relieve them.

1980: August 5   Terms of autonomy are agreed on

Flanders and Wallonia agree on terms for limited regional autonomy, and new constitutional amendments are introduced to meet them. Regional assemblies are created, with posts to be filled by the members of the current Belgian parliament who represent each existing constituency. The regions have responsibility for economic, environmental, energy, and housing matters.

1982–86   Martens is granted special powers for economic reform

By a parliamentary vote, Prime Minister Martens is granted special powers to deal with the economy. He devalues the franc, freezes wages and prices, reduces hours worked by persons who are employed, and implements other austerity measures.

1982   Economic austerity program is imposed

A recession, rising oil prices, and Belgium's large national debt lead the government to impose an economic austerity program.

1984–85   Bombings protest NATO missile deployment

Communist extremists bomb twenty-seven economic targets, including banks, stores, and fuel supply lines, to protest the proposed deployment of NATO cruise missiles in Wallonia. Due to opposition, the plan to deploy forty-eight NATO cruise missiles is modified, and only sixteen missiles are installed.

1984: October 18   Death of Surrealist Henri Michaux

Surrealist poet and painter Henri Michaux, born in Namur, dies in Paris. (See 1899.)

1985: May 29   Soccer riot kills thirty-nine and brings down government

Thirty-nine spectators are killed in rioting, mostly by British fans, following the European Cup finals soccer match between England and Italy. Most of the deaths are caused by the collapse of a stadium wall. The Belgian police come under fire for failing to provide adequate security, and many in parliament call for the resignation of the minister of the interior. His refusal to resign ultimately leads to the downfall of the ruling coalition.

1985: July 16   Prime minister tenders resignation

Having lost support in parliament, Prime Minister Wilfried Martens tenders his resignation to King Baudouin, who refuses to accept it. However, early elections are called in October.

1986   Government tightens economic controls

The government's economic austerity program is expanded to include wage controls, tax reform, and public spending cuts. Trade unions strongly protest the new policies.

1987   Linguistic squabble brings down the government

The ruling coalition is toppled by an incident involving language rights. The mayor of Voeren, a Fleming-administered district located in French-speaking Wallonia, refuses to learn Dutch as a protest against existing language laws, touching off a controversy that ultimately brings down the current government.

1987: December 17   Death of author Marguerite Yourcenar

Belgian-born author Marguerite Yourcenar dies in Northeast Harbor, Maine. (See 1903.)

1988: May   New coalition assumes power

A new center-left coalition takes office and grants greater autonomy to Belgium's regions and linguistic communities, also providing a regional council for Brussels. Martens remains prime minister.

1989   Regional assembly elected in Brussels

Brussels elects its first regional assembly, with the same status as assemblies in Flanders and Wallonia.

1989: September 4   Death of author Georges Simenon

Detective fiction author Georges Simenon, creator of the popular detective Inspector Maigret, dies in Lausanne, Switzerland. (See 1903.)

1990: April   King avoids ruling on abortion law

King Baudouin requests to abdicate the throne for one day to avoid having to sign an abortion bill. The bill is passed, and abortion is legalized.

1991: June   Women are allowed to take over the monarchy

A constitutional amendment makes it legal for a woman to ascend the throne of Belgium.

1991: November Center-left coalition forms government

The CVP, under the leadership of Jean-Luc Dehaene, forms a center-left government.

1992 Agusta probe begins

The government begins investigating allegations that the Belgian Socialist party accepted bribes from the Agusta helicopter company in 1988 in return for influencing the awarding of government contracts.

1992: March Dehaene becomes prime minister

Christian Democrat Jean-Luc Dehaene takes office as prime minister, bringing an end to the thirteen-year period for most of which Wilfried Martens has held the post.

1992: July Maastricht treaty is ratified

The Belgian legislature approves the Maastricht Treaty providing for European economic union.

1993: February Constitutional reforms enacted

Constitutional changes create a federal system, shifting substantial political power from the central government to the regional assemblies of Flanders, Wallonia, Brussels, and a small German-speaking enclave recognized as a separate region. The assemblies of these regions receive authority over employment, housing, agriculture, transportation, energy, and environmental concerns.

1993: July 14 Federal plan goes into effect

Belgium's new federal system, mandated by constitutional changes (see February 1993), goes into effect. Four regions and three language jurisdictions (French, German, and Flemish) are recognized.

1993: July 31 King Boudouin dies

King Baudouin dies suddenly while in Spain on a vacation. In the absence of heirs, he is succeeded by his brother, Prince Albert.

1994 Conscription is abolished

The military draft operating in conjunction with voluntary enlistment is terminated.

1994 Agusta probe results in top-level resignations

Three ministers belonging to the Socialist Party resign as a result of the ongoing bribery investigation related to the Agusta helicopter company.

1995: February Agusta probe touches international officials

The Agusta bribery investigation is expanded to include two international officials who were active in Socialist Party politics at the time of the alleged bribes: Willy Claes, former economic affairs minister and current secretary general of the North Atlantic Treaty Organization (NATO), and Karel Van Miert, a European Union commissioner who formerly served as president of the Socialist Party.

1995: May 12 Claes is questioned by highest court

NATO secretary general Willy Claes, coming under increasing pressure to resign over alleged involvement in the Agusta scandal, is questioned by Belgium's highest court.

1995: May 21 Elections return ruling coalition to power

The center-left coalition led by Prime Minister Jean-Luc Dehaene is returned to power in the first general elections since the 1993 constitutional reforms creating a federal system (see August 1993). The coalition retains its majority in the Chamber of Representatives, which has been reduced from 212 seats to 150 under the revised constitution.

1995: December 19 Strikes cripple transportation system

Strikes against Belgium's state-owned railroad (Belgian Railways) and airline (Sabena World Airways) nearly bring all transportation in the country to a halt. Strikers protest the proposed layoff of over eight thousand transport employees over a ten-year period.

1997: December 30 The Antwerp Bourse closes

The Antwerp Bourse, one of the world's oldest financial exchanges, closes permanently.

Bibliography

Files, Yvonne. *The Quest for Freedom : The Life of a Belgian Resistance Fighter.* Santa Barbara, CA: Fithian Press, 1991.

Fitzmaurice, John. *The Politics of Belgium: A Unique Feudalism.* London: Hurst, 1996.

Fox, Renie C. *In the Belgian Chateau: The Spirit and Culture of a European Society in an Age of Change.* Chicago: I.R. Dee, 1994.

Gutman, Myron P. *War and Rural Life in the Early Modern Low Countries.* Princeton, NJ: Princeton University Press, 1980.

Hilden, Patricia. *Women, Work, and Politics: Belgium 1830–1914.* Oxford: Clarendon Press, 1993.

Hooghe, Liesbet. *A Leap in the Dark: Nationalist Conflict ad Federal Reform in Belgium.* Ithaca: Cornell University Press, 1991.

Lijphart, Arend, ed. *Conflict and Coexistence in Belgium: The Dynamics of a Culturally Divided Society.* Berkeley, CA: University of California at Berkeley, Institute of International Studies, 1981.

Warmbrunn, Werner. *The German Occupation of Belgium: 1940–1944.* New York: P. Lang, 1993.

Wee, Herman van der. *The Low Countries in Early Modern Times.* Brookfield, VT: Variorum, 1993.

Wilenski, Reginald H. *Flemish Painters, 1430–1830.* New York: Viking Press, 1960.

Bosnia and Herzegovina

Introduction

Bosnia and Herzegovina was one of the six republics that made up the country of Yugoslavia until 1991. By the end of the twentieth century, it was an independent nation struggling to recover from a devastating civil war among its three major ethnic groups: the Bosnian Serbs, the Croats, and the Bosnian Muslims, the inheritors of a unique Slavic-speaking Islamic culture that arose during the centuries when Bosnia was ruled by the Ottoman Empire. Although hostilities were formally ended by the Dayton Peace Accords of 1995, interethnic tensions in Bosnia and among its Balkan neighbors continue to threaten political stability in the region.

Bosnia and Herzegovina is located in the heart of the Balkans, between Serbia, Montenegro, and Croatia. Its mountainous terrain has historically provided the region a certain amount of isolation, especially in medieval and early modern eras when its unusual religious traditions were being forged. Much of the land is forested, and plains cover the northern third of the country. As of 1999, Bosnia and Herzegovina has an estimated population of over 2.5 million people. The capital city, Sarajevo, located in the central part of the country, has a population of about half a million.

The first known inhabitants of present-day Bosnia and Herzegovina were Illyrians who settled there in the second and first millennia B.C. From about 400 B.C. the region was successively occupied by Greeks, Celts, Romans, and Goths. The first Slavs migrated to the Balkans in the sixth century A.D. The South Slavs, who today account for most of the region's population, began arriving in the seventh century. (The word *Yugoslavia* is derived from the root words *yugo*—south and *slav.*) The two major south Slavic groups that settled in the region—the Serbs and Croats—formed different religious affiliations.

The Serbs, settling in the east, affiliated themselves with the Eastern Orthodox Church; the Croats, to the west, became Roman Catholics. This religious division, which also entailed linguistic and cultural differences, was a major factor in the estrangement and rivalry between the two groups. Bosnia, which was situated between Serbia and Croatia, was exposed to both religions; however, due to its geographical isolation and other factors, neither religion took hold as strongly there, and the Bosnians developed their own religious practices using elements of both traditions.

The Bosnians enjoyed a period of independence, but by the eleventh century, control of the region shifted between Hungary, Serbia, and the Byzantine Empire. In the twelfth century, Hungary occupied most of Bosnia, administering it through local rulers called *bans*. Under one of these *bans* (Ban Kulin, r. 1180–1204), Bosnia achieved considerable power and autonomy and expanded its territory. However, the Bosnians' religious practices were drawing the attention of their neighbors, and in 1203 the pope sent a delegation to Bosnia to investigate the church there. Although Kulin and the Bosnians affirmed their belief in the church, Hungary condemned the Bosnian church for heresy, and in the thirteenth century, it instituted a religious purge that served as an occasion to regain political control of Bosnia. At the time, Bosnia became a center of a dualistic *Bogomil* heresy that worshipped both God and the devil. In the fourteenth century, Bosnia won back some of its lost autonomy, especially under the rule of Stefan Tvrtko (1353–91), who broke away from Hungary altogether and named himself king of Bosnia in 1377. Tvrtko expanded Bosnia's territory and influence, giving Bosnia an outlet to the Adriatic coast and also acquiring the region of Hum (present-day Herzegovina).

Rule by the Ottomans

By the end of the fourteenth century, however, the Ottoman Empire had begun its conquest of the Balkans. Bosnia, whose power was already on the wane, fell to the Ottomans in the latter half of the fifteenth century and, like the rest of the Balkans, remained under Turkish rule for some four hundred years. This period saw the emergence of a cultural phenomenon that was to have great significance for the future politics of the region: a large number of Bosnians converted to Islam (largely for tax purposes), and a new cultural group—which also adopted Turkish dress, manners, and cuisine—was created, clearly distinct from its Christian neighbors, the Serbs and Croats. The importance of these divisions stems from the Ottoman system of administration. The Ottoman government divided its population into *millets* (religious communities)

each under the jurisdiction of its religious leader who was answerable to the sultan. Under this system, Muslims, Catholics, Orthodox Christians, and Jews each belonged to their own *millet*; according to this hierarchy, the Muslim *millet* took precedence over the other three groups. Having adopted the religion and culture of the Ottomans, the Bosnian Muslims rose to the top of the social order, while the Christians mostly filled the lower ranks of society. Many worked as serfs

for Muslim nobles. Thus the religious divide in Bosnia also took on aspects of a class conflict.

By the seventeenth century, the Ottoman Empire was weakening, and Russia and Austria were competing for control of its lands, including Bosnia. At the end of the century, Austrian forces led by the Duke of Savoy burned down the city of Sarajevo, the Bosnian capital. In the eighteenth century, Austria temporarily won control of parts of the region.

Nationalist sentiment among the Slavs increased sharply in the nineteenth century. In addition, there were growing tensions between the local Muslim elite and the Ottoman government, as well as rising unrest among the peasants. Regional instability came to a head in 1875, when a peasant rebellion in a region of Herzegovina spread throughout the Balkans, evolving into the wider Serbo-Turkish War, which Russia entered in 1877. Under the 1878 Treaty of Berlin, Austria-Hungary was granted a mandate to administer Bosnia and Herzegovina and it occupied the region.

Austrian Rule

The Austrians took a much more active role in the administration of Bosnia than the Turks had, improving its infrastructure and developing agriculture and industry. However, nationalist feeling in the region was too strong for foreign rule to be passively accepted. In 1908 Austria-Hungary formally annexed Bosnia and Herzegovina, largely to head off a threatened Serb takeover. Following the Balkan wars of 1912 and 1913, World War I was touched off by the assassination of Austrian archduke Franz Ferdinand by a Serbian radical nationalist in Sarajevo in July, 1914. During the war, Bosnian Serbs and Croats under Austro-Hungarian rule were obliged to fight against their brethren, because Serbia and Montenegro were allied with the opposing side (known as the Triple Entente). In addition, Serbian guerrilla fighters, called *Chetniks,* carried out attacks on Bosnian Muslims. With the defeat of the Austrians, the dream of an independent South Slav nation became a reality, and the Kingdom of Serbs, Croats, and Slovenes was proclaimed on December 4, 1918.

The new country inherited serious social, economic, and political problems, including the long-standing antagonism between its diverse ethnic groups. Chetnik activity against Bosnian Muslims continued, and the latter formed their own political party, the Yugoslav Muslim Association. The country was renamed Yugoslavia in 1929 and came under the dictatorial rule of King Alexander I, who partitioned Bosnia between the Serbs and Croats. During World War II, Bosnia became part of a Nazi-allied Croatian puppet state. Bosnian Muslims pursued different courses during the war—some fought with the fascist Croatian Ustashe. Indeed, Heinrich Himmler, (1900–45) the leader of the Waffen SS (the military wing of the Nazi storm troopers) authorized the creation of a Bosnian Muslim division known as the *Handzar* (scimitar, a type of sword) division. However, more joined the anti-Nazi Partisans, led by Josip Broz Tito (1892–1980). Once again, the Muslims suffered at the hands of Serbian Chetniks, who made up a third fighting force in the region.

At the close of World War II, the Yugoslav monarchy was dissolved, and the country became a Soviet-style federation of six republics led by Tito, who was now celebrated as a war hero for leading the Partisans. After 1948 Yugoslavia pursued its own communist path separately from the Soviet Union due to a rift between Tito and Soviet leader Josef Sta-

lin (1879–1953). The hallmark of communist rule under Tito was political and economic decentralization, with state-owned enterprises managed by local workers' councils. Tito also positioned Yugoslavia as a non-aligned nation carving out its own path between East and West in the Cold War. With Tito's death in 1980, an important restraint on the nation's ethnic rivalries was gone. In addition, the nation faced widespread economic problems triggered by problems in Russia and Eastern Europe.

Independence

A constitutional amendment allowed for multiparty elections in 1990. In Bosnia and Herzegovina, the Communists were voted out of power in favor of nationalist groups representing Bosnia's three major constituencies. The following year, Yugoslavia began to unravel as Croatia and Slovenia declared independence. Bosnia followed suit in 1992, but Bosnian-Serb forces (covertly supported by Serbian President Slobodan Milosevic) soon took over much of the country, launching a three-year civil war that killed as many as 250,000 people, produced an estimated three million refugees, and inflicted massive physical damage on the country. Bosnian-Serbs opposed the breakup of Yugoslavia. In the event of a secession of Bosnia, they sought the right to secede from Bosnia and join with Serbia and Montenegro in order to create a Greater Serbia. To that end, the Bosnian-Serb strategy called for the creation of ethnically homogenous areas contiguous to the Republic of Serbia. In order to achieve this aim, the Bosnian-Serb forces engaged in a policy of the forced removal of non-Serbs (so-called ethnic cleansing) from a given region. Opposing the Bosnian-Serbs were the forces of the internationally recognized government of Bosnia and Herzegovina led by President Alija Izetbegovic, who fought to preserve a multi-ethnic state. Croatia, under President Franjo Tudjman, generally supported the Izetbegovic government although Bosnian-Croats engaged in fighting against forces loyal to Izetbegovic several times during the war. Like the Bosnian-Serbs who favored unification with Serbia, many Bosnian-Croats sought to unite their areas of Bosnia to Croatia.

The war gained worldwide attention for its brutality, which included repeated rape and mass executions, and charges of genocide were leveled at Serbs. In 1994 an international war crimes tribunal for Yugoslavia was established in The Hague, and UN peacekeeping forces were stationed in the country. The following year NATO launched air strikes against the Bosnian Serbs following mass executions in Srebrenica and Zepa. In late 1995 the presidents of Bosnia, Serbia, and Croatia agreed to a peace plan that divided Bosnia into two parts: a Serbian area (called Republica Srpska) and a Bosnian-Croatian area (the Federation of Bosnia and Herzegovina). The Dayton Peace Accord, as the agreement was called, was signed in Paris in December, 1995. Early in 1996, military forces in Bosnia withdrew to the new borders specified by the accord, and the country began a rebuilding effort that the World Bank estimated would cost $1.8 billion in the first year alone. By the end of 1995, the international tribunal in The Hague had charged forty-three Serbs and six Croats—including Bosnian Serb leader Radovan Karadzic—with war crimes, but only one person had been taken into custody. The court heard its first case in May, 1996. As of 1998 Karadzic was still at large.

Internationally monitored elections were held in Bosnia in September, 1996. Both halves of the country elected their own officials, and the agreed-on three-member presidency for the country as a whole was also elected.

Timeline

2nd–1st millennia B.C.   Illyrians settle in region

Illyrian tribes settle in present-day Bosnia and Herzegovina.

c. 400 B.C.   Greeks arrive

Greeks colonize Bosnia and Herzegovina.

c. 300 B.C.   Celtic migration

Celts migrate to the region.

2nd–1st centuries B.C.   Roman conquest

Over a period of about 150 years, the Romans take control of the area, which they assimilate culturally and incorporate into the province of Illyricum.

A.D. 395   Roman Empire is divided

When Rome is divided into two parts by the heirs of Roman Emperor Theodosius, Bosnia is in the region of the dividing line, a fact that will carry great significance for its ultimate ethnic makeup.

4th–5th centuries   Goth invasions

Goth forces invade Bosnia and Herzegovina, inflicting defeat on its Roman occupiers, only to be vanquished in turn by the Byzantine Empire.

Slavic Migration and Occupation

6th century   Goths are expelled; Slavic migration begins

Byzantine Emperor Justinian I drives the Goths out of Bosnia and Herzegovina, and the first Slavs settle in the area.

7th century   South Slavs migrate to the Balkans

South Slavs, mainly Croats and Serbs, begin migrating to the Balkans, where they pursue communal agriculture, living in clans ruled by a council of chieftains. One branch settles in the western part of the area, the other branch in the east. The western group eventually becomes the Croats, while those who settle in the east become the Serbs. Bosnia, in the center of the region, is not dominated by any one group. Unlike Serbia and Croatia, its name is derived from a geographical feature—the Bosna River—rather than a tribal or ethnic group.

In the first part of the seventh century, the South Slavs ally themselves with the Byzantine emperor against two Turkic tribes, the Avars and the Bulgars, and become the dominant cultural force in the region. Politically and militarily, however, they vie with Bulgaria and Byzantium intermittently for centuries.

7th century   Religious divisions arise

The cultural division between Serbs and Croats is strengthened as they form affiliations with two competing churches, both of which are well established by the end of the century. Settling along the Adriatic Sea, the Croats look westward for their religious guidance, accepting the authority of Holy Roman Emperor Charlemagne and the Catholic Church. The Serbs retain their allegiance to Eastern Orthodoxy.

Of the entire Balkan region, this division is especially disruptive in Bosnia, which lies at the center, and is thus divided between the two groups. Because of this competition and due to the isolation resulting from Bosnia's rough terrain, the influence of the two competing churches is weakest in Bosnia, whose church develops independently and is later charged with heresy.

11th–12th centuries   Rival powers vie for control

Control of Bosnia and Herzegovina shifts between the Byzantine Empire, Hungary, and Serbia.

12th century   Hungarian rule

Following its royal union with Croatia, Hungary occupies most of present-day Bosnia, administering it through local rulers called *bans*.

1180–1204   Ban Kulin in power

Ban Kulin, a Hungarian viceroy (local ruler), gains control of most of the area along the source of the Bosna River, naming it Bosnia and greatly increasing the region's autonomy. Under his administration, mining is developed, immigration by skilled workers is encouraged, and trade is expanded. The individual regions of Bosnia retain much of their local political and cultural autonomy as well.

1203   Ban Kulin affirms faith at special gathering

Growing in isolation from other Roman Catholic nations but in close proximity to the Eastern Orthodox religion, Bosnian Christianity develops unique characteristics that are considered heretical by the Catholic hierarchy in Rome. (For example, the Bosnians have adopted a Slavic liturgy rather than the Latin one, and use the form of the Cyrillic alphabet used by Russians and other eastern Slavs.) When the Vatican sends a papal legate to Bosnia to investigate Bosnian religious practices, Kulin convenes a special council to reaffirm his loyalty to the Church and pledge religious reforms. The medieval Bosnians have been widely described as belonging to a heretical sect called *Bogomilism*, a dualistic belief that worships both God and the devil, which originated in Bulgaria. However, modern scholarship suggests that the Bosnians had evolved their own variation on Catholicism rather than subscribing to a preexisting religious philosophy.

13th century   Religious purge increases Hungarian influence

Hungary's Catholic monarchy condemns the spread of religious heresy in Bosnia and tries strenuously to stamp it out, sending papal emissaries and later missionaries to Bosnia. In the process, Hungary also extends its political control of the region as the century progresses.

1299   Croatian administers Bosnia

The Hungarian monarch appoints a Croatian noble, Paul Subic, to take charge of Bosnia. Subic brings Franciscan missionaries with him to help eradicate the heretical Bogomil religious sect. Rulership of Bosnia is later passed down to his son, Mladen.

1322   Croatian ruler is overthrown

Local Bosnian nobles overthrow their Hungarian-appointed Croatian ruler, Mladen, son of Paul Subic. A native Bosnian, Stefan Kotromanic, replaces him.

1353–91   Reign of Stefan Tvrtko

Stefan Tvrtko succeeds his uncle, Stefan Kotromanic, as ban of Bosnia. Under his rule, Bosnia becomes more autonomous and powerful. The landlocked territory gains control of part of Dalmatia.

1377   Stefan declares himself king

Bosnia's ruler, Stefan Tvrtko, abandons his allegiance to Hungary and declares himself king of Bosnia, whose territorial gains under Tvrtko include the region called Hum (present-day Herzegovina), which gives Bosnia access to the Adriatic Sea. For a period, Bosnia eclipses Serbia as the most powerful kingdom in the region.

Ottoman Conquest and Occupation

1386   Turkish conquest of the Balkans begins

The Ottoman Turks begin their conquest of the Balkan region, which is accomplished over a period of nearly a century. Bosnia is conquered in the mid-fifteenth century.

1389: June 28   Battle of Kosovo

Turks led by Sultan Murad enter the Balkans from the Ottoman Empire in the east and defeat a Serbian force. This battle is the turning point in the Ottoman conquest of the Balkans.

15th century   Bosnian power wanes

Following the reign of Stefan Tvrtko, Bosnian power declines as Hungary retakes some of its former territory and the monarch loses his power over the feudal lords.

1449   Herzegovina gets its present name

The region of Hum becomes known as Herzegovina when its ruler, Stefan Vukcic, adopts the title of *herceg*.

1463–1528   Ottomans conquer Bosnia

The Ottoman Turks conquer Bosnia and later Herzegovina, taking control of the region gradually, although most of it is under Ottoman rule within the fifteenth century.

1463–1878   Ottoman rule

Together with the entire Balkan region, Bosnia is ruled by the Turkish Ottoman empire for over four hundred years. The Turks merge all the Balkan states into a single administrative entity. Few Turks actually occupy the land, and the inhabitants are generally allowed to retain the Christian religion as long as they do not resist Turkish rule. Bosnia is unique in that large numbers of inhabitants, whose ties to the church were relatively weak to begin with, willingly convert to Islam over a period of several generations, giving Bosnia privileged status within the Ottoman Empire. The Bosnian Muslims adopt Turkish dress, manners, and titles as well as its language, cuisine, and architecture. A unique Slavic Muslim culture emerges. The region retains its territorial integrity and even incorporates parts of Croatia.

Although Turkish governors are installed at the capitals of Bosnia and Herzegovina, the ruling elite is mostly made up of Bosnian Muslims, while the serfs are largely Christian. Thus the religious tensions in the region also take on aspects of a class conflict. Life for the ordinary rural dweller becomes harder as individual small farms give way to large estates worked by serfs who have to pay high taxes to the Ottoman sultan (emperor).

The Ottoman Period

The Ottoman Empire began in western Turkey (Anatolia) in the thirteenth century under the leadership of Osman (also spelled Othman) I. Born in Bithynia in 1259, Osman began the conquest of neighboring countries and began one of the most politically influential reigns that lasted into the twentieth century. The rise of the Ottoman Empire was directly connected to the rise of Islam. Many of the battles fought were for religious reasons as well as for territory.

The Ottoman Empire, soon after its inception, became a great threat to the crumbling Byzantine Empire. Constantinople, the jewel in the Byzantine crown, resisted conquest many times. Finally, under the leadership of Sultan Mehmed (1451–1481), Constantinople fell and became the capital of the Ottoman Empire.

Religious and political life under the Ottoman were one and the same. The Sultan was the supreme ruler. He was also the head of Islam. The crown passed from father to son. However, the firstborn son was not automatically entitled to be the next leader. With the death of the Sultan, and often before, there was wholesale bloodshed to eliminate all rivals, including brothers and nephews. This ensured that there would be no attempts at a coup d'état. This system was revised at times to the simple imprisonment of rivals.

The main military units of the Ottomans were the Janissaries. These fighting men were taken from their families as young children. Often these were Christian children who were now educated in the ways of Islam. At times the Janissaries became too powerful and had to be put down by the ruling Sultan.

The greatest ruler of the Ottoman Empire was Suleyman the Magnificent who ruled from 1520–1566. Under his reign the Ottoman Empire extended into the present day Balkan countries and as far north as Vienna, Austria. The Europeans were horrified by this expansion and declared war on the Ottomans and defeated them in the naval battle of Lepanto in 1571. Finally the Austrian Habsburg rulers were able to contain the Ottomans and expand their empire into the Balkans.

At the end of the nineteenth century, the Greeks and the Serbs had obtained virtual independence from the Ottomans, and the end of a once great empire was in sight. While Europe had undergone the Renaissance, the Enlightenment, and the Industrial Revolution, the Ottoman Empire rejected these influences as being too radical for their people. They restricted the flow of information and chose to maintain strict religious and governmental control. The end of Ottoman control in Europe came in the First Balkan War (1912–13) in which Greece, Montenegro, Serbia, and Bulgaria joined forces to defeat the Ottomans.

During the First World War, (1914–18), the Ottoman Empire allied with the Central Powers and suffered a humiliating defeat. In the Treaty of Sévres, the Ottoman Empire lost all of its territory in the Middle East, and much of its territory in Asia Minor. The disaster of the First World War signaled the end of the Ottoman Empire and the beginning of modern-day Turkey. Although the Ottoman sultan remained in Constantinople (renamed in 1930, "Istanbul"), Turkish nationalists under the leadership of Mustafa Kemal (Ataturk) (1881–1938) challenged his authority and, in 1922, repulsed Greek forces that had occupied parts of Asia Minor under the Treaty of Sévres, overthrew the sultan, declared the Ottoman Empire dissolved, and proclaimed a new Republic of Turkey. That following year, Kemal succeeded in overturning the 1919 peace settlement with the signing of a new treaty in Lausanne, Switzerland. Under the terms of this new treaty, Turkey reacquired much of the territory—particularly, in Asia Minor—that it had lost at Sévres. Kemal is better known by his adopted name "Ataturk", which means "father of the Turks". Ataturk, who ruled from 1923 to 1938, outlawed the existence of a religious state and brought Turkey a more western type of government. He changed from the Arabic alphabet to Roman letters and established new civil and penal codes.

1567   *Qur 'an* is transcribed

A copy of the *Qur 'an*, the Muslim holy book, is produced in Sarajevo. It is an example of the art of manuscript illumination that flourishes in Bosnia during this period.

17th century   Ottoman empire begins to decline

As the Ottoman empire weakens, rival powers Austria and Russia begin to vie for control of Bosnia. This rivalry is a part of what, in subsequent centuries, becomes known as the Eastern Question, the question of what will become of the Ottoman Empire as it declines. Both Russia and Austria seek southward expansion and their designs clash. Russia also has cultural ties to the region through the Orthodox Christian faith. As Great Britain becomes a Mediterranean power in the eighteenth and nineteenth centuries, it becomes the biggest Russian rival as its seeks to impede Russian access to a warm water port (which Russia lacks).

1683–99   Habsburg-Ottoman war sends refugees to Bosnia

Following Austrian conquest of Hungary and Slavonia from the Ottomans, large numbers of Muslim refugees flee to Bosnia, increasing its already large Muslim population (estimated at forty percent).

1697   Sarajevo burns

During a war between Austria and the Ottoman Empire, the Duke of Savoy leads Austrian forces into Bosnia and burns down the capital city.

1718–39   Austria controls part of Bosnia

Following the Treaty of Passarowitz (1718) Austria rules parts of Bosnia and Herzegovina, but the territory reverts to the Ottomans in 1739 under the Treaty of Belgrade. These territorial revisions stem from a series of wars fought between the Austrian Habsburg and Ottoman Empires.

19th century   Writers espouse nationalist causes

Bosnian Croatian and Serb writers take up the nationalist causes of those regions. Prominent among these writers is Ivan Franko Jukic, a Croat and a Franciscan monk who edits the first literary journal in Bosnia, *Sosanski prijatelj* (Bosnian Friend). Bosnian Serb writers include Sima Milutinovic Sarajlija and Nicifor Ducic.

Early 19th century   Political turmoil grows

Turmoil grows with tensions between Bosnia's Muslim upper class and the Ottoman government, as well as increasing unrest among the peasantry due to increased taxes. Fervent nationalist movements in Serbia and Croatia add to the confusion: each side claims the Bosnian Muslims as its own, while Bosnian Muslims feel no allegiance to either.

1839   Feudal nobility is abolished

In the midst of tensions between the central Ottoman government and Bosnia's Muslim nobility, the Ottoman empire abolishes the feudal nobility and declares all Bosnians equal before the law. This reform is part of a general Ottoman reform movement that sweeps the empire in the 1830s and 1840s.

1848   Forced labor system is abolished

The Ottoman government ends the right of landowners to receive unpaid labor from serfs.

1875: July   Rebellion breaks out

Famine caused by poor harvests spurs a peasant rebellion in the Nevesinje region of Herzegovina. Unrest then spreads throughout the Balkans and launches a wider European war.

1876: July   War between the Balkans and the Ottomans

Serbia and Montenegro declare war on the Ottoman Empire.

1877: July   Russia enters Balkan war

Russia becomes involved in the war between the Ottomans and Serbia and Montenegro (also known as the Serbo-Turkish War). Russian intervention tips the scales against the Ottomans who sue for peace the following year.

Austrian Control

1878   Treaty of Berlin

Under the Treaty of Berlin, Austria-Hungary is charged with restoring order to Bosnia and Herzegovina and occupies the region. Technically, it remains the province of Turkey, but the Austrians are the de facto rulers.

1883–1903   Kállay administration

Bosnian and Herzegovina undergoes rapid economic development during the period when it is administered by Benjamin von Kállay, a Hungarian official. Agriculture is modernized; tobacco and lumber production expands. A public works program is inaugurated, resulting in the construction of a new road network and railroads. However, the introduction of Austrian bureaucrats into the administration of the region fuels Slavic nationalism among the Serbs and other South Slavic nationalists.

1888   National Museum is established

A National Museum is founded in Sarajevo to preserve and exhibit the country's archaeological, ethnographic, natural history, and artistic treasures.

1892: October 10   Birth of novelist Ivo Andric

Andric, a Bosnian-Serb, who becomes Yugoslavia's foremost twentieth-century writer, is born near Travnick, Bosnia. As a youth during World War I he takes part in nationalist activities and is arrested and imprisoned. When the state of Yugoslavia is formed following World War I, he serves in its diplomatic corps. He begins publishing short stories in 1920, with collections appearing periodically. He writes three novels during World War II, of which the best known is *The Bridge over the Drina*, 1945. Other novels by Andric that have been translated into English include *Bosnian Story*, 1945, *The Damned Yard*, and *Conversation with Goya*. Andric is awarded the Nobel Prize for Literature in 1961. He dies in Belgrade on March 13, 1975.

Early 20th century   Europeanization of Bosnian art

With annexation by the Austro-Hungarian empire, Bosnian art becomes Europeanized, as its leading artists study in Vienna. These artists, who become active early in the twentieth century, include the following: Impressionist painter Lazar Drljaca (1881–1970), *plein air* (outdoor landscape) painter Atanasije Popovic (1881–1948), naive artist Milenko Atanackovic (1875–1955), and Expressionist painter Djoko Mazalic (1888–1965).

1908   Austrian annexation

Concerned about the threat posed by Serbian nationalism, Austria-Hungary annexes Bosnia and Herzegovina, making its de facto rule official. At the close of Turkish rule, deep divisions remain among the area's three major population groups: Muslims, Croats, and Serbs.

1912–13   First Balkan War

Serbia, Montenegro, Greece, and Bulgaria join together to expel the Ottoman Turks from the Balkans. Serbia gains control of much of the remaining Ottoman lands in Europe.

1913   Museum buildings are completed

The four buildings in Sarajevo that are to house the country's National Museum are completed. The institution becomes one of southern Europe's leading museums. (The facility and its contents sustain heavy damage in the civil war of 1992–95.)

1913   Second Balkan War

After Bulgaria attacks its former South Slavic allies, they respond and the Bulgarians are defeated.

1914: June 28   Bosnian-Serb assassinates Austrian archduke, triggering World War

Austrian archduke Franz Ferdinand and his wife, the archduchess, are assassinated by a Bosnian-Serb student, Gavrilo Princip, during a state visit to Sarajevo. Princip is associated with the nationalist secret society, the Black Hand. Secret societies formed by Serbian radicals have already made several assassination attempts on Austrian officials. It is an ultimatum designed to be rejected.

1914: July 23   Ultimatum by Austria-Hungary

Austria Hungary threatens to attack Serbia unless it can participate in bringing the archduke's killer to justice, and Serbia agrees to ban secret societies.

1914: July 28   War is declared

Although the Serbian demand is conciliatory, Austria-Hungary declares war on Serbia. Together, Austria and its allies—Germany and the Ottoman Empire—are known as the Central Powers. Serbia allies itself with France, Britain, and Russia, whose alliance is known as the Triple Entente. The war pits ethnic Serbs, Croats, and Slovenians in territories ruled by Austria Hungary—including Bosnia— against their brethren in Serbia and Montenegro. In Bosnia, Serbian guerrilla fighters called *Chetniks* attack Muslim landowners. In spite of ethnic and religious division, the idea of an independent state uniting the South Slavs gains force during this period.

1917   Art exhibit in Sarajevo

A major exhibition by Bosnian artists is held in Sarajevo.

1918: November 1918   End of World War I

By the time World War I ends in victory for the Triple Entente, the South Slavic lands have been liberated from Austro-Hungarian control.

Independence

1918: December 1   South Slavs win independence

The inhabitants of the South Slavic lands, including Bosnia, proclaim an independent Kingdom of Serbs, Croats and Slovenes. Alexander Karadjordjevic, King of Serbia, becomes the country's ruler as Alexander I. (Ten years later he renames the country Yugoslavia.) The new nation is troubled politically and economically from the outset. It inherits a legacy of ethnic rivalry and centuries of religious, linguistic, and cultural difference. In addition, diplomatic tensions arise immediately because much of the new nation is assigned land that has previously been promised under the secret 1915 Treaty of London that gives much of the Dalmatian coast to Italy. Serbs and Croats also clash over the issue of government centralization versus federalism. Economically, the kingdom is faced with wartime damage, heavy debt, labor shortages, and a pressing need for land reform created by centuries of feudalism.

Serb antagonism toward the Bosnian Muslims is undiminished, and some Chetnik terrorist activity continues. In the face of Serb and Croat nationalism, the Bosnians Muslims form their own political party—the Yugoslav Muslim Organization (YMO)—to advance their interests within the new federation.

1921–22   Little Entente is formed

At the instigation of France, the South Slavic kingdom forms a military alliance with Romania, Czechoslovakia, and France to prevent attempts at treaty revision by the defeated Austrians, Bulgarians, Germans, and Hungarians. The alliance is known as the Little Entente.

1929   Royal dictatorship is established

King Alexander imposes a royal dictatorship and renames the country Yugoslavia (Land of the South Slavs). He dashes Bosnian hopes of autonomy by dividing the region among several administrative districts. Throughout Yugoslavia, non-Serbs resent continued domination at the hands of the administration in Belgrade that is dominated by Serbs.

1930   First art gallery opens

The country's first art gallery, exhibiting more than 600 artworks, is launched in Sarajevo.

1931   Constitution is adopted

The autocratic rule of King Alexander is modified by adoption of a constitution, and political parties are legalized. However, a number of other freedoms are still restricted.

1931   Global economic slump hits Yugoslavia

Bosnia and the rest of Yugoslavia feel the effects of the worldwide economic depression, which leads to bankruptcies and unemployment. The crisis is worsened by weather conditions that produce famine in rural areas.

1934   Balkan Entente is formed

Yugoslavia joins Turkey, Greece, and Romania in a defense alliance known as the Balkan Entente. This treaty is aimed at stifling Bulgarian attempts at revising the World War I peace settlement.

1934: October   King Alexander is assassinated

King Alexander is murdered in Marseilles, France by an assassin working for *Ustashe,* a Croatian underground terrorist organization. His death brings fears that Yugoslavia will collapse. Alexander's son, Petar II, becomes the country's regent. Three officials are appointed to rule for him while he is still a minor.

1939   Part of Bosnia ceded to Croatia

In an agreement (*Sporazum*) negotiated between Serb and Croat leaders, parts of Bosnia and Herzegovina with Croatian majorities are ceded to the newly formed autonomous Croatian territory, leading Bosnian Serbs to consider the rest of the region theirs to control. This reorganization of Bosnia is part of the effort to create regional autonomy for non-Serbs. Bosnia's Muslims are not considered a separate group until after the Second World War.

1941: March 25   Pact is signed with Germany

Under military pressure and surrounded by pro-Nazi countries, Yugoslavia's government agrees to join the Tripartite Pact (the Axis alliance concluded among Germany, Italy, and Japan) in return for German guarantees of nonaggression.

1941: March 27   Yugoslavian government is overthrown

Because of its cooperation with Nazi Germany, the Yugoslavian government is overthrown in a coup led by Yugoslav Air Force officers, and sixteen-year-old Petar II, the regent and son of the slain king Alexander, is declared king. In spontaneous demonstrations, the populace expresses its hostility toward the Nazis and their allies.

1941: April 6   Germany bombs and invades Yugoslavia

The Yugoslavian capital of Belgrade is bombed by the German *Luftwaffe* (air force), and ground forces invade the country. The government flees and the military surrenders unconditionally. Yugoslavia is divided among the Axis powers, and Bosnia is absorbed into a Croatian puppet state controlled by the Nazis and led by Ante Pavelic, the head of a Croatian fascist group called the Ustashe. In accord with the Nazis' theories of racial purity, the Ustashe declares Croatians to be racially superior Aryans (despite their Slavic roots), while the Serbs are singled out for elimination through deportation or extermination, together with Jews and gypsies. The atrocities committed by the Ustashe during the war add yet more fuel to the hostility between Serbs and Croats and will be repeatedly invoked by the Serbs during the Serb-Croat war and the Bosnian civil war of the 1990s.

1941: July   Partisans stage anti-German revolts

Partisans (anti-Nazi freedom fighters) organized by long-time communist leader Josip Broz Tito (1892–1980) carry out rebellions in the countryside and remain active throughout the war in spite of German retaliation. Although most Partisans are Serbs, all ethnic groups are represented (Tito himself is Croatian-born). Bosnian Muslims come down on both sides in the conflict—some fight with the pro-Nazi Ustashe forces, but more join the Partisans. Royalist Serbian Chetnik forces also organize to fight the Ustashe and Nazi forces, but their primary motive is ethnic hostility, and they also murder thousands of Bosnian Muslims. Indeed, they subsequently collaborate with the Nazi occupation forces against Tito's partisans.

1943   Newspaper *Oslobodjenje* is founded

The daily newspaper *Oslobodjenje* (Liberation) is established in Sarajevo as an anti-Nazi resistance paper. It later achieves distinction as the only newspaper to maintain publication through the siege of Sarajevo in the 1990s. Two of its editors are honored by the *World Press Review.*

1943: September   Italy surrenders

The surrender of Italy bolsters the Partisan effort by providing access to the Adriatic coast as well as a supply of arms and a supply route.

1944: October 20   Red army liberates Belgrade

The Soviet army, aided by Partisan forces, marches into Belgrade and liberates Yugoslavia. Some 1.7 million Yugoslavians have died since the war began—more than half at the hand of other Yugoslavs. Cities are left in ruins, and the countryside is also devastated.

1945   Art school and artists' association are formed

The State School of Painting and the Association of Artists of Bosnia and Herzegovina are established.

Yugoslavia under Tito

1945: March 7   Provisional government is installed

A communist-dominated provisional government, led by Tito, takes office.

1945: November 29   Yugoslavia's monarchy is dissolved

The new Yugoslav parliament meets for the first time. The monarchy is officially dissolved, and the country is named the Federal People's Republic of Yugoslavia (later renamed the Socialist Federal Republic of Yugoslavia). Bosnia becomes one of six republics in a Soviet-style federation with a strong centralized government. In an effort to dilute the power of Serbia, Macedonia and Montenegro are recognized as republics, Kosovo is established as an autonomous Albanian province, and Vojvodina becomes an ethnically mixed autonomous province.

1945–66   Serbs dominate Bosnia

In the postwar years Serbs control Bosnian politics locally.

1948   Yugoslav-Soviet split

Following World War II, Yugoslavia allies itself closely with the Soviet Union. By 1948, however, growing tensions, begun with disagreement over the management of joint enterprise, create a rift between the two countries. Unlike the heads of other Eastern European satellites of the USSR, Tito does not owe his influence or position to the Soviets and will not allow them to dictate policy for Yugoslavia. He becomes the only Eastern European communist leader to break with Stalin and remain in power. Trade with the Soviet Union and other Communist bloc nations is reduced, and Tito is forced to turn to the West for new trade partners.

1949   University of Sarajevo is founded

Bosnia's oldest institution of higher learning is established. It offers programs in humanities, sciences, law, medicine, engineering, and the social sciences.

1950s   Self-management economic system develops

Yugoslavia develops its communist economy through a system in which enterprises are locally managed by workers' councils, and economic goals are first formulated locally and coordinated centrally. As in other communist countries, these goals are organized into five-year plans. In spite of rapid industrialization throughout the country, Bosnia remains primarily rural.

1953   Constitution incorporates workers' councils

The constitution is modified to authorize the creation of workers' councils to run state-owned enterprises.

1955   Birth of film director Emir Kusturica

After attending film school in Prague, Kusturica works as a director for Yugoslavian television before making his film debut in 1981 with the hit movie *Do You Remember Dolly Bell?,* which wins the Golden Lion awards at the Venice Film Festival. His subsequent films also win international recognition: *When Father Was Away on Business* (1985) wins the Palme d'Or at the Cannes Film Festival and is nominated for an Academy Award in the Best Foreign Film category. Kusturica wins the Best Director award at Cannes for *Time of the Gypsies* (1989). In the 1990s, Kusturica teaches film directing in the United States at Columbia University. His most recent films are *Arizona Dream* (1993), *Once Upon a Time There Was a Country* (1994), and *Underground* (1996).

1960s   Croatian nationalism grows

Resentment of Serbian political, economic, and cultural domination fuels the growth of Croatian nationalism. One of its expressions is a decade-long controversy over the status of the Croatian language. Although Serbs and Croats (as well as Bosnian Muslims) speak the same language, Serbo-Croatian, the Orthodox Serbs use the Slavic Cyrillic alphabet while the Catholic Croats (and Bosnian Muslims) use the Latin alphabet. In addition to the differences in alphabet, regional dialects differ. Nationalists seize upon these differences to argue that theirs is a separate language. By the 1990s, references are made to the "Bosnian" language, distinct from Serbo-Croatian.

1961 Ivo Andric wins Nobel Prize

Bosnian-born novelist Ivo Andric is awarded the Nobel Prize for Literature.

1961 Nuclear technology institute is founded

The Institute for Thermal and Nuclear Technology is founded in Sarajevo.

1961: September Conference of nonaligned nations

A conference of nonaligned nations held in Belgrade confirms Yugoslavia's international leadership position in the group of nations that have proclaimed their neutrality in the Cold War, including India and Egypt.

1965 Market socialism is introduced

Yugoslavian industry grows rapidly in the 1950s, but imports still exceed exports by a wide margin, and some industrial inefficiency remains. In response to an economic crisis of the early 1960s, the government modifies its economic policy, eliminating price controls and export subsidies while cutting import duties.

1967–68 Yugoslavian republics gain more power

Constitutional amendments increase the power of the Yugoslavian republics at the expense of the central government. In Bosnia, hard-line local communists keep Croat and Serb nationalism under control, while supporting the aspirations of the Bosnian Muslims.

1968 Tito condemns Soviet invasion of Czechoslovakia

Yugoslav leader Tito condemns the USSR's forcible halt to Czech liberalization under Alexander Dubcek.

1969–71 Croat and Slovene springs

Paralleling developments in Czechoslovakia, liberals assume control in Croatia and Slovenia, leading to demands for greater independence from the central government, which sets in motion a purge of liberal leaders in both republics.

1970s Popular music flourishes in Sarajevo

Pop and rock music thrives in Sarajevo. The most popular band of the era is *Bijelo Dugme* (White Button).

1970s New universities are founded

Universities are founded in Banja Luka and Tusla. Both offer programs in science, engineering, and other fields.

1971 Bosnian Muslims recognized as a nation

The Yugoslav census grants the Bosnian Muslims equal status with the other national groups of the country when it lists them in their own category.

1972 Academy of Arts is launched

The Academy of Arts opens in Sarajevo. Among the new generation of artists trained there are painters Radoslav Tadic and Nusret Pasic; sculptor Kemal Selakovic; and graphic artist Petar Waldegg.

1974 New constitution is adopted

A new constitution gives the Yugoslav republics considerably greater political and economic autonomy. It lays out an interlocking framework of local, republic, and national representation and a seven-member presidency, with two members from each of the country's main ethnic constituencies and one additional member.

1975: March 13 Death of novelist Ivo Andric

Nobel-Prize winning novelist Ivo Andric dies in Belgrade.

1980s Politics and the economy pull Yugoslavia apart

Following the death of Marshal Tito, the stability achieved over decades is threatened by age-old ethnic rivalries and affected by widespread economic problems in Russia and Eastern Europe. Yugoslavia is beset by rising prices, labor strikes, food shortages, and financial scandals, as tensions grow between the different republics. Serbian nationalism, which has been kept in check in Bosnia during the Tito years, gains strength in spite of government crackdowns.

1980: May 4 Death of Marshal Tito

Josip Broz Tito, the leader who has held Yugoslavia together since World War II, dies. He is widely mourned at home, and forty-nine foreign nations send representatives to his funeral.

1981 Yugoslav film is honored at Cannes

Sjecas li se Dolly Bell? (Do You Remember Dolly Bell?), a film by prominent Yugoslavian director Emir Kusturica, receives the Golden Lion award at the Cannes Film Festival in France.

1983: April Muslim activists tried

As Serb activism grows, the Bosnian Muslims band together as well. An activist group goes on trial for advocating the establishment of a separate Bosnian Muslim republic. One member of this group is Bosnia's future President Alija Izetbegovic (b. 1925).

1984 Winter Olympics are held in Sarajevo

The well-organized and successful Winter Olympic Games held in Sarajevo are a testament to peace and cooperation among the city's diverse ethnic groups. Ten years later, during the civil war that rocks the country, thousands of coffins are built of wood taken from structures built for the Olympics.

1990 Multiparty elections are allowed

The constitution is amended to allow for multiparty elections.

1990: November 18 Multiparty elections are held

After approving multiparty legislative elections, the Communist Party is removed from power by new national parties representing the country's three major ethnic groups: the Muslim party wins eighty-six seats, the Serbs win seventy, and the Croats win forty-five.

1990: December Izetbegovic is elected president

Alija Izetbegovic, president of the Party of Democratic Action, a Muslim party, is elected president by a coalition of non-communist parties.

The Collapse of Yugoslavia

1991: June 25 Croatia and Slovenia declare independence

The Croats and Slovenes proclaim independent republics, beginning the dissolution of Yugoslavia.

1991: October Bosnia prepares for independence

After Serbia and Montenegro take control of Yugoslavia's central government, Bosnia's legislature passes a measure laying the framework for an independent Bosnian state. However, the Serbian delegation refuses to vote and walks out. Bosnian Serbs oppose the creation of an independent Bosnia in which they would be dominated by Muslims. Instead, they reason that in the event of Bosnian independence, they should have the right to secede from Bosnia and join their fellow Serbs in the Republic of Serbia. Bosnian-Serbs begin organizing militarily in order to fight for the creation of a contiguous Bosnian-Serb area that could be united to Serbia.

1991: December 20 Bosnia seeks EU recognition

Bosnia's government declares its sovereignty and seeks EU recognition as an independent state.

1992: February 29 Referendum is held on independence

Bosnians hold a referendum on independence from Yugoslavia. Since independence would sever Bosnia's political ties with Serbia, Bosnian Serbs (over a third of Bosnia's popula-

tion) abstain from voting. The remainder of the electorate chooses independence by an overwhelmingly majority (99.7 percent).

1992: March 18 Bosnian leaders agree framework for independence

The elected leaders of Bosnia's three ethnic communities agree on a framework for sharing power in an independent republic. Meanwhile, though, Serbian activists are forming their own Serbian Republic of Bosnia and Herzegovina.

1992: April Recognition of Bosnian independence

The European Union and the United States recognize Bosnia as an independent republic. This is in stark contrast to their reluctance to extend diplomatic recognition to Croatia and Slovenia for nearly half a year.

Civil War

1992: April Hostilities break out

Bosnian Serbs declare a separate state. With military equipment provided by the Yugoslavian army (Yugoslavia at this point consists of Serbia and Montenegro), Bosnian Serb forces start a civil war, taking control of seventy percent of Bosnia's territory. Over the next three years, the country is shattered by the conflict, as cities are bombed and cease-fires repeatedly fail. Serb atrocities against Bosnian Muslims and Croats (so-called ethnic cleansing) are reported, including repeated rape and mass execution. Serbian forces lay siege to Sarajevo for over a year and a half, until they are turned back by the threat of NATO air strikes.

1992–95 Theater productions continue despite war

Of Sarajevo's five theater companies, three continue producing plays during the war. Room 55, one of the companies, runs as many as three productions at a time and offers free admission to its plays, which continue despite shelling of the theater and casualties among its actors and theater staff. The plays are well attended. The Academy of Theater Arts remains open as well.

1993: October Film festival is defeated by forces of war

Theater director Haris Pasovic organizes an international film festival to give war-weary Sarajevans a break from their troubles and remind the world of the city's prewar cultural reputation. More than two dozen videotapes are acquired, and actors Jeremy Irons and Vanessa Redgrave schedule personal guest appearances at the event. However, the forces of war interfere with festival plans—the city suffers a power outage that affects the theater, and the British government, fearing that Irons and Redgrave will speak out against its policies, interferes with their travel plans. Film-goers are left frustrated and disappointed.

1994 War crimes tribunal is established

The United Nations sets up the International War Crimes Tribunal in the Hague for war crimes in the former Yugoslavia.

1994: February NATO ultimatum

Following a heavy shelling of Sarajevo that kills sixty-six people, NATO issues an ultimatum to the Serbs, and they agree to the presence of UN peacekeeping forces, and the pressure on Sarajevo subsides. However, the Serbs now focus their offensive on the cities of Tusla and Gorazde, continuing to attack in spite of UN protests.

1994: July Milosevic endorses new UN peace plan

Serbian President Slobodan Milosevic (b. 1941) angers the Bosnian Serbs by endorsing a new UN peace plan that proposes to divide the country and give the Serbs forty-nine percent. Milosevic agrees to end assistance to the Bosnian Serbs until they endorse the new peace plan, and he closes the border between Serbia and Bosnia. The Bosnian Serbs begin attacking UN peacekeepers.

1994: December Carter mediates truce

Former U.S. president Jimmy Carter (b. 1924) mediates a truce, which lasts four months, after which the siege of Sarajevo is renewed, and the Bosnian Serbs take 350 UN peacekeepers hostage.

1995: July Top Serbs indicted for war crimes

Serb political leader Radovan Karadzic and military leader Ratko Mladic are indicted for war crimes. Karadzic is charged with ordering the siege of Sarajevo and the holding of UN personnel as hostages.

1995: July Massacres take place in Srebrenica and Zepa

The UN-protected areas of Srebrenica and Zepa are invaded by Serb forces, who carry out mass killings of Bosnian men. Several thousand are shot and buried in mass graves. Children are killed in front of their parents, and hundreds of people are buried alive or undergo other forms of torture.

1995: August NATO launches air strikes against Bosnian Serbs

In response to international pressure, NATO forces carry out air strikes against strategic Serb military targets, including command posts and radar installations.

1995: September Serbs suffer military setbacks

Bosnian and Croatian troops rout Serb forces in the Bihac and Krajina regions. Bihac is a part of Bosnia along the Croatian border, while the Krajina is a region of Croatia populated by Serbs who seceded from Croatia in 1991 and created the Serb Republic of the Krajina. The Croatian forces that lead the drive against the Serbs are modernized and well-trained unlike their war-weary, ill-equipped, and poorly led Serb opponents.

1995: September Breakthrough agreement is reached

For the first time, the Bosnian government agrees to the creation of an autonomous Serbian entity in exchange for Serbian recognition of Bosnia-Herzegovina, paving the way for a peace accord.

1995: November Karadzic receives second war crimes indictment

Bosnian Serb leader Radovan Karadzic receives his second war crimes tribunal indictment, this time for killing thousands of Muslims at Srebrenica in July, 1995. However, the Serbs refuse to turn Karadzic or indicted general Ratko Mladic over to the tribunal. As of 1998 both men are still at large.

By the end of 1995, the war crimes tribunal has seen 300 hours of videotape and reviewed some 65,000 pages of documents.

1995: November 21 Dayton peace accord is drawn up

After a month of talks at Wright-Patterson Air Force Base in Dayton, Ohio, a peace agreement is framed by the presidents of Bosnia, Serbia, and Croatia. It maintains Bosnia as a single nation, but divided into two parts: one governed by the Bosnian Serbs (Republica Srpska) with forty-nine percent, and another by Bosnians and Croats (the Federation of Bosnia and Herzegovina) with fifty-one percent. Each is allotted roughly half of Bosnia's territory. Both entities are to have their own legislatures and presidents, although there will also be a central government. NATO will maintain peacekeeping forces totaling 60,000 troops in the country. Elections are scheduled for September, 1996.

1995: December Dayton Accords are signed

The Dayton peace accords ending the Bosnian civil war are signed in Paris. Estimates of the number of deaths in the three-and-a-half-year war range from 25,000 to 250,000, and the number of refugees is estimated at three million.

The Postwar Period

1996 Rebuilding begins

Reconstruction of the war-torn country begins. The World Bank estimates that it will spend $1.8 billion in the first year alone on repairing wartime damage and restoring normal life to Bosnia. Seventy percent of the nation's electrical supply has been lost due to power plant damage. Half the nation's houses have been damaged, creating severe housing shortages: twenty thousand housing units are scheduled for repair or replacement within a year. In addition to the massive physical damage, refugees returning to Bosnia find widespread unemployment and corruption fostered by the growth of organized crime.

1996: February–March   Forces withdraw to new borders

Both sides in the Bosnian civil war move their troops to the new agreed-on borders. However, the return of refugees specified in the Dayton peace accord does not take place.

1996: May   First war crimes trial opens

The international tribunal for Bosnia opens its first trial for war crimes.

1996: September 14   Elections are held

Internally monitored elections are held in both halves of the newly divided country. A new three-member presidency is also elected to govern the entire country. It will be headed by former president Alija Izetbegovic.

1997: February   Croat police fire on Muslim crowd

Bosnia's worst violence occurs when Bosnian Croat police open fire on a crowd of Muslims at a cemetery in Mostar, wounding twenty persons and killing one. The Croats then evict some 100 Muslims from the western part of the city.

1997: April   Visit by Pope John Paul II

Pope John Paul II visits Sarajevo.

1997: July 10   NATO troops catch war criminals in raid

NATO troops from Britain carry out a raid in northwestern Bosnia-Herzegovina to capture Bosnian Serbs indicted for war crimes. One Serb is killed and another is arrested.

1998: April   Plan to capture renegade Serb leader is scrapped

A plan by NATO forces to capture former Bosnian Serb leader and indicted war criminal Radovan Karadzic is abandoned following an intelligence leak.

1998: September 12–13   Nationwide elections are held

Nikola Poplasen, a political hard-liner, is elected president of the Republika Srpska, replacing the more moderate Biljana Plavsic. The international community perceives his election as a setback for peace efforts in Bosnia-Herzegovina. A more moderate Bosnian Serb, socialist Zivko Radisic, is elected to the country's three-member presidency.

1999: March 5   European monitor fires Bosnian-Serb president

Under powers granted by the Dayton Peace Accords, an international observer, Carlos Westendorp, fires Nikola Poplasen, the hard-line president of Republica Srpska, the Serbian part of Bosnia-Herzegovina. This action and an international arbitrator's decision to take the disputed town of Brcko out of Serb control are both rejected by the Bosnian-Serb parliament and are followed by weeks of street protests.

Bibliography

Burg, Steven L., and Paul S. Shoup. *The War in Bosnia-Herzegovina: Ethnic Conflict and International Intervention.* Armonk, NY: M.E. Sharpe, 1999.

Drakulic, Slavenka. *How We Survived Communism and Even Laughed.* 1st American ed. New York: W.W. Norton, 1992.

———. *The Balkan Express: Fragments from the Other Side of War.* 1st American ed. New York: W.W. Norton, 1993.

Filipovic, Zlata. *Zlata's Diary: A Child's Life in Sarajevo.* New York: Viking, 1994.

Gapinski, James H. *The Economic Structure and Failure of Yugoslavia.* Westport, Conn.: Praeger, 1993.

Gjelten, Tom. *Sarajevo Daily: A City and Its Newspaper Under Siege.* New York: HarperCollins Publishers, 1995.

Glenny, Misha. *The Fall of Yugoslavia: The Third Balkan War.* New York: Penguin, 1996.

Gutman, Roy. *A Witness to Genocide: The 1993 Pulitzer Prize-winning Dispatches on the "Ethnic Cleansing" of Bosnia.* New York: Macmillan, 1993.

Lampe, John R. *Yugoslavia as History: Twice there was a Country.* Cambridge: Cambridge University Press, 1996.

Mojzes, Paul. *Yugoslavian Inferno: Ethnoreligious Warfare in the Balkans.* New York: Continuum, 1994.

Owen, David. *Balkan Odyssey.* New York: Harcourt Brace, 1995.

Pinson, Mark, ed. *The Muslims of Bosnia-Herzegovina: Their Historic Development from the Middle Ages to the Dissolution of Yugoslavia.* 2nd ed. Cambridge, Mass.: Harvard University Press, 1996.

Prstojevic, Miroslav. *Sarajevo Survival Guide.* Trans. Aleksandra Wagner with Ellen Elias-Bursac. New York: Workman Publishing, 1993.

Rogel, Carole. *The Breakup of Yugoslavia and the War in Bosnia.* Westport, Conn.: Greenwood, 1998.

Stankovic, Slobodan. *The End of the Tito Era.* Stanford, Calif.: Hoover Institution Press, 1981.

West, Rebecca. *Black Lamb and Grey Falcon.* Reprint. New York: Penguin Books, 1982.

Bulgaria

Introduction

The first known people in the region of modern Bulgaria were the Thracians, at least as early as 3500 B.C. Until final defeat by the Roman Empire about the turn of the millennium, these skilled horse-riders dominated southeastern Europe and beyond. Their archaeological remnants include masterful works in gold and stone burial sites. As the Roman Empire fell into its decline, Central Asian tribes began to occupy the hinterlands of Roman control, including the Balkans. In this early period of Asiatic invasion, the Bulgars arrived from the steppes of Ukraine.

The Bulgars originally spoke a Turkic language, but by the ninth century, stone inscriptions show that they tended to assimilate the Slavic language of other area inhabitants. In 681, a Bulgar state was recognized by Byzantium (the heir to the Roman Empire of antiquity). Modern Bulgaria recognizes this date as the beginning of Bulgarian national history. This period, known as the First Bulgarian Empire, encompassed the spread of Christianity to the region as well as the spread of literacy.

The followers of Cyril and Methodius led the initial missions, which the Bulgarian ruler Boris patronized. The missions established centers for the translation of Christian texts in monasteries. Slavic literary and religious culture originated with these centers of translation and transcription. Before the printing press, all records were reproduced for preservation and dissemination by painstaking hand-copying. The tenth century capital of Bulgaria, Preslav, was one of these centers.

Territorial expansion and political power in the First Empire peaked in the reign of Simeon, before his armies were defeated by the Byzantines in 924. For one hundred years, Bulgars fought the Byzantines, until Emperor Basil II of Byzantine vanquished the Bulgars. Until 1185, Bulgaria remained under Byzantine control. The religious patriarch of Bulgaria was also reduced to a subordinate position, that of an archbishop.

The Second Bulgarian Empire began with a revolt against Byzantium in 1185. Byzantium was weakened by its war with western Europeans, who captured Constantinople in 1204. Until 1261, the Bulgars had to contend with the Latin Empire, established by the Crusaders who defeated the Byzantines. Two major internal phenomena rose to prevalence in Bulgaria: *Bogomilism* (a dualistic belief in God and the devil) and *Hesychasm* (a form af contemplative mysticism). Considered heresy, these social movements established alternative religious modes of living and worldviews that came to influence Bulgarian literature and culture for centuries.

The Second Empire contained the literary flourishing known as the Silver Age. The central figure of the Silver Age was the monk Evtimii. Also the last patriarch of the Bulgarian Orthodox Church before the Ottoman victory, his original work included biographies of saints and a treatise on living as an ascetic. Evtimii was sent into exile following the Ottoman occupation of Bulgaria in 1393. Ottoman rule lasted until late in the nineteenth century; the Bulgarian literary tradition survived in Serbia, Romania, and isolated monasteries until its revival in the 1800s.

The nineteenth century witnessed the greatest decline of Ottoman power. As a result of this decline, Balkan states such as Bulgaria began to regain cultural identity and eventually political identity. Bulgarian-language schools were established as early as 1835 which revived national identity with the help of nationalistic writings from the eighteenth century such as the *Stematografia* of Hristofor Zhefarovich, a cultural history. In 1862, Georgi Rakovski led the first modern group to take up arms against the Ottoman overlords. The Bulgarian Orthodox Church was made autonomous in 1870, and political autonomy followed soon after, in 1878. The first government of the modern Bulgaria was based on the Tirnovo Constitution of 1879 that established civil rights and parliamentary government but left the highest authority in the hands of a monarch. This arrangement persisted through two regional wars and came to its end in the second of two world wars, when post-World War II Bulgaria was re-formed as a communist state.

In the 1890s, political parties developed to represent different elements of society and related outlooks. The Bulgarian Communist Party, which existed into the 1980s, was founded as the Social Democratic Party, in 1894. The Agrarian Party was founded in 1899 to represent peasant interests. Bulgaria's royal leader, Ferdinand (1861–1948), declared the state independent of the Ottoman Empire in 1908. Indepen-

BULGARIA

0   25   50 Miles

0   25   50 Kilometers

ROMANIA

Bucharest

YUGOSLAVIA

MACEDONIA

GREECE

TURKEY

Black Sea

Aegean Sea

Zaječar • Vidin
Craiova
Constanța
Dunav (Danube)
Lom Kozloduy
Silistra
Alexandria
Ruse
Ogosta
Mikhaylovgrad
Beleňe
Razgrad
Dobrich
Iskur
Pleven
Vratsa
Veliko Tŭrnovo
Tŭrgovishte
Shumen
Varna
BALKAN
Osum
Kremikovtsi
Gabrovo
MOUNTAINS
Kamchiya
Sofia
Stryama
Kazanlŭk
Sliven
Pernik
Panagyurishte
Tundzha
Burgas
Priboj
Struma
Stara Zagora
Yambol
Musala 9,596 ft. 2925 m.
Pazardzhik
Plovdiv
Maritsa
Elkhovo
Blagoevgrad
RHODOPE
Asenovgrad
Svilengrad
Mesta
Strumica
Smolyan
Arda
Kŭrdzhali
Rudozem
Madan
MTS.
Alexandroupolis
Thessaloníki

Bulgaria

dence lasted peacefully for only four years. In 1912, Bulgaria, Serbia, and Greece fought the Ottomans (Turkey) with substantial military skill on the Bulgarians' part. While they were fighting the Turks, the Serbs and the Greeks captured Ottoman lands and annexed them, leading to the Second Balkan War of 1913. Bulgarians felt wronged by their allies and proceeded to attack them without even a declaration of war. The ensuing defeat led to territorial diminishment in the Treaty of Bucharest and abiding resentment among the Bulgarians against their Balkan neighbors. Consequently, the Bulgarian alliance with the Central Powers in World War I (Bulgaria was involved from 1915–18) was heavily influenced by the potential territorial gains in victory.

The Central Powers and Bulgaria faced defeat in the war, and the resultant chaos was only resolved when a popular political organization came to power. From 1919 to 1923, the Agrarian Party ruled Bulgaria under the leadership of Alexander Stamboliiski. The government by the peasant party ended when reactionaries assassinated Stamboliiski. The period from 1923 to 1934 was one of popular rule, although very conservative, even fascistic. The Tsankov and then the Liapchev governments presided over a period of economic distress across Europe. The transition from the fascist period of government to the royal dictatorship which lasted until 1943 was bridged by the Zveno government in 1934. Zveno achieved power in a coup and intended to reduce party politics to rule on an individual level in parliament. Instead of leading to a new period of democracy, the ban on political organizations paved the way for King Boris III to rule Bulgaria autocratically until his death in 1943.

Bulgaria sided with the Axis Powers in World War II. The orientation towards Germany and Italy was rooted in the ties of Boris' administration to Germany and his marriage to an Italian princess.

War effort was mostly limited to the Greek and Yugoslav frontiers by negotiations of the king with Hitler. Most of Bulgaria's Jewish population was spared the horrific treatment of many other European Jews, although the Jewish minorities in Macedonian and Greek territories under Bulgarian control were deported. As the tide of the war turned against the Axis Powers, once again Bulgaria lost a war. On September 5, 1944, Soviet troops crossed the Danube River. The post-war government in Bulgaria was established by 1946 with the Bulgarian Communist Party ruling, with strong Soviet backing.

By the end of the 1950s, the state appropriated all of Bulgarian private property and ninety-two percent of the land. More than forty years of communist rule was conducted as the economy was centrally planned by the political establishment. Large industrial concerns employed the majority of Bulgarians, and farms were run as collectives. Health care, education, and pensions were granted to all. Bulgarians were made equal, theoretically. The ethnic Turks of Bulgaria were not granted the benefits of the communist state, however. From 1950 to 1951, 155,000 were expelled. Periodically, until the 1990s, ethnic Turks were expelled from Bulgaria or forced to assimilate by changing surnames. Bulgarian language schools were closed, and the issue was unapproachable under the absence of freedom of speech. In 1962, Nikita Khrushchev approved Todor Zhivkov as prime minister, a post he held until 1989. Bulgaria became the Soviet Union's closest ally, making every political move in concert.

In the late 1980s, Mikhail Gorbachev's policy of *glasnost* (openness) in the government of the Soviet Union made waves throughout the Warsaw Pact nations, including Bulgaria. Popular informal organizations of scholars, workers, and ecologists began to create opposition to official policy in Bulgaria. In 1988 an ecological demonstration protesting pollution at the Danube River port city of Ruse led to violent suppression by the government. The government allowed an international meeting in Sofia addressing environmental concerns, ostensibly to allay fears, but in the wake of the Chernobyl nuclear disaster of 1986 tensions were heightened throughout Europe. The meeting went very poorly for the Bulgarians and exposed strict limits on civil liberties. Combined with the final forced exodus of 310,000 Turks, international criticism mounted, even from the Soviet Union, and in response the Bulgarian Communist Party forced Premier Zhivkov from power.

After Zhivkov's departure from the government, the newly renamed Bulgarian Socialist Party (dropping the discredited term "communist") unsuccessfully attempted to reestablish firm control. Popular dissent became powerful enough that the BSP opened elections to multiple parties. The elec-

tions in June, 1990, returned a majority for the BSP, but strikes and other forms of protest against the government continued until economic reforms were effected. The Arable Land Law of 1991 began to return appropriated landholdings to the original owners, and private property began to be returned in April, 1992. After a short-lived opposition government of the Union of Democratic Forces (UDF) from October, 1991, until October, 1992, the BSP returned to power for five years. In this span, ex-government officials were prosecuted for mishandling public funds, the ecological degradation of forty years of communism was revealed, and economic reforms did little but destroy confidence of Bulgarians and foreign investors alike.

In the April 23, 1997 general elections, the UDF gained a 137 seat share of the 240 seat National Assembly (compared to fifty-eight for the BSP). The new prime minister, Ivan Kostov, in tandem with the newly elected and widely popular president, Petar Stoyanov, made economic policies which brought inflation under control. The events of 1997 and 1998 brought Bulgaria to the negotiating table for accession (joining) the European Union and membership in NATO.

Timeline

3500 B.C.   Thracians appear

Seminomadic pastoralists, the Thracians are the first group known to have inhabited the area of modern Bulgaria.

c. 500 B.C.   First government forms

The Thracians develop a state government, which functions for several centuries.

c. 300 B.C.   Thracians repel Macedonians

The Thracians successfully defend an aggressive attack on their government by the Macedonian.

c. 1st century A.D.   Thracians absorbed into Roman Empire

After a 150 year struggle against the Romans, the Thracians are absorbed into the Roman Empire about the end of the millennium.

c. 370 A.D.   Central Asian tribes invade Balkans

In several waves of invasions, Bulgars, Huns, Goths, and Avars overrun the region known as the Balkans in southeastern Europe. The Bulgars probably originate as a Turkic-speaking tribe.

c. 630   First federation of Bulgar tribes

The Bulgar tribes rise up in unison against the previously dominant Avars. Crossing the frontier of the Danube River, the Bulgar tribes occupy the area of modern Bulgaria for the first time, subjugating the Slavic-speaking people of the area, only to eventually adopt Slavic folkways and language.

681   Byzantium recognizes the existence of independent Bulgaria

In the seventh and eighth centuries, the Bulgar state in the Balkans is established largely due to Byzantine distraction with the Arabs in and around territories in Asia Minor. Byzantium is the most powerful city-state in the eastern Mediterranean region, tracing its history directly to the Roman Empire. In exchange for a cessation of attacks, the Byzantines must pay an annual tribute to the ruler of the Bulgarian state.

716   Treaty between Bulgaria and Byzantium

A treaty between Bulgaria and Byzantium moves the frontier closer to Constantinople and establishes tribute (a type of tax) at thirty pounds of gold per year, to be paid to the Bulgar *khan* (local ruler).

750–900   First recorded writing of Bulgars

Bulgars produce stone inscriptions in Greek letters, nearly 100 of which survive to the present day in the northeastern Bulgarian city of Pliska. Some employ the Turkic language of the original Bulgar conquerors, while others contain Slavic terminology. All use the Greek alphabet. The inscriptions are essentially official in nature: chronicles of events, names of fortresses and of battles, records of titles and buildings, and military regulations and peace treaties.

816   New treaty between Bulgaria and Byzantium

A century of intermittent conflict results in a new treaty line similar to that of 716. It leaves Byzantium with a strip of land forty miles wide on the northern shore of the Aegean Sea.

c. 820   Construction of barriers along borders

On the new Bulgar-Byzantine frontier (see 816), the Bulgars build the Great Ditch of Thrace, a high rampart and trench, that merchants are permitted to cross only with letters of safe-conduct. In the north, the Bulgars build a system of triple earthworks for protection against the horsemen of the *steppe* (treeless plain) to the north of the Danube River.

863–64   Boris exerts power of Bulgar state

Khan Boris (r.852–89) asks the East Franks, rather than the Byzantines, to send Christian missionaries to Bulgaria. He hopes to limit Byzantine interference in the internal affairs of the Bulgarian state. Upon hearing of Boris's request, the government at Constantinople declares war on Bulgaria and forces the Bulgars to renounce the Frankish connection. Boris, dissatisfied with the degree of control that Constantinople seeks to exert over the new Bulgarian Church, attempts to persuade the Greek patriarch to grant Bulgaria its own archbishopric. The attempt fails. Boris then offers his allegiance to the Roman pope, in exchange for an autonomous church. The pope is likewise reluctant to grant this authority. Boris's tactics gain him better status in relation to the Orthodox hierarchy, and Bulgaria is granted a limited version of ecclesiastical independence (Bulgaria's patriarch must still be consecrated in Constantinople). In symbolic retaliation, the pope excommunicates all Greek bishops and clergy serving in Bulgaria.

885   Arrival of Methodius's followers in Bulgaria

Some of the followers of Methodius arrive in Bulgaria as refugees from Greek territories. (Cyril (827–69)and Methodius (826–85), brothers who are eventually known as saints in the Eastern Orthodox faith, make it their mission to teach the Christian Gospels to the relative newcomers to Europe, the Bulgars.) Boris and his successors support the continuation of their work translating religious texts from Greek to Slavic.

890–930   Golden Age of First Bulgaria Empire

Bulgarian Tsar Simeon (r. 893–927), the most powerful monarch in eastern Europe, is described as "Emperor and Autocrat of all the Bulgars and Greeks," a title acknowledged by the Roman Catholic pope.

893   Liturgical and secular literature translated

The centers of Slavic translation in Europe found at Preslav, capital of Bulgaria, and Ohrid in Macedonia are patronized by the Tsar Simeon. (*Tsar* is derived from the Roman Empire's "caesar", or emperor.) The translations include: the Gospels, the Psalms, liturgical books, the New and most of the Old Testaments, biographies of the saints, and writings of the Fathers of the Church. Secular works include: books of dreams, prophecies, and natural disasters; fables and wisdom literature; and pseudo-histories like that of the Trojan War or the romance of Alexander the Great. Most original Old Slavic literature is of Bulgarian origin.

Clement of Ohrid produces many sermons for special occasions, while Bishop Constantine of Preslav writes the Didactic Gospel. Constantine (the first organizer of the Bulgarian Church) writes this text, beginning each line with a different letter to form an acrostic (a verse in which certain letters on each line spell out a word) of the alphabet as a whole. The text, intended to be a teaching tool, is also the first known original poem in Old Slavic. The Bulgarian monk Hrabr argues in favor of the Slavic language as a medium for religious purposes in his 85-line poem, "On Letters/Apology." John the Exarch (of the Bulgarian Church) writes an account of the creation of the world, known as *The Six Days*.

This text is a type of encyclopedia of natural history, praising God for the perfection of His world and includes criticism of pagans and heretics.

893   Cyrillic made official alphabet

The Bulgarian *Sabor* (assembly) proclaims the Cyrillic alphabet, which is based on the Greek capital (*uncial*) letters, as official for secular and religious purposes in Bulgaria. At the Bulgar capital of Preslav, all Slavic texts previously written in the Old Slavic, or Glagolitic, alphabet, are transcribed into Cyrillic.

913   Tsar Simeon advances to the walls of Constantinople

The first Bulgarian Empire reaches its greatest expansion when Simeon leads the Bulgarian armies to the nearly impenetrable walls of Constantinople, capital of Byzantium.

917 or 919   Simeon makes Bulgarian archbishop a patriarch

Tsar Simeon unilaterally raises the archbishop of Bulgaria to the status of a patriarch, in an attempt to support his imperial ambitions.

924   Simeon defeated by Byzantines

The First Empire begins to decline after Tsar Simeon's forces are defeated by Byzantine armies.

Mid-10th century   Bulgarian priest writes tract criticizing Bogomilism

Kozma, a priest in the Bulgarian Orthodox Church, criticizes members of a dualist sect called Bogomilism for what he considers a hypocritical stance in their opposition to social injustices. (See 1211.)

967–72   Prince of Kiev invades Bulgarian territory

Prince Sviatoslav of Kiev invades Bulgaria at Byzantine invitation and captures various towns on the lower Danube.

971   Regent John Tzimiskes restores Byzantine rule in eastern Bulgaria

A force of Bulgars, Rus, Pechenegs, and Magyars is defeated by the Byzantines under Regent (later Emperor) Tzimiskes, establishing the Byzantine border at the Danube River.

991–1014   Western Bulgaria is conquered by Basil

Emperor Basil II of Byzantium (r. 976-1025) gains territory in western Bulgaria. In a series of campaigns beginning in 998, his forces invade Bulgaria every year. He establishes base camp at Plovdiv and systematically advances into Bulgar territory.

1014   Emperor Basil II definitively defeats Tsar Samuel

In a decisive battle near Salonika Emperor Basil surprises the Bulgarian army and captures 14,000 soldiers. He orders the blinding of 99 of each 100, leaving the 100th with one eye with which to lead the others home.

1018   Bulgaria reverts to Byzantine rule

After a long period of war, defeated Bulgaria remains under Byzantine control until 1186. The patriarchate of Bulgaria, centered at Ohrid in Macedonia, is reduced to the rank of archbishopric.

1185–97   Reign of John and Peter Asen

The Asen brothers in Bulgaria successfully revolt against Byzantium and reestablish an independent Bulgar state with capital at Tirnovo. Bulgaria is to remain an independent state until 1393. This period is known as the Second Bulgarian Empire.

1202   Peace with Byzantium

Tsar Kaloyan (r. 1197-1207) makes peace with the Byzantine Empire. Bulgaria again achieves full independence.

1204   Western European states expand eastward

The Fourth Crusade captures Constantinople, establishing the Latin Empire. Recognizing the new regional power, Bulgaria signs a treaty with Rome, acknowledging the pope as the highest religious authority and thus consolidating the western border of the Bulgarian Empire

1204-64   Byzantine government moves to Nicaea

The Latin Empire (Constantinople and surrounding territories) is ruled by Western Christians, while Byzantine emperors rule from Nicaea.

1205   Bulgars defeat Latins

Bulgar Tsar Kaloyan defeats the Latins at The Battle at Adrianople and occupies western Macedonia and Thrace.

1211   Bogomilism declared heretical

The Bulgarian church synod at Tirnovo summoned by Tsar Boril (r. 1207–18) proclaims the Bogomil doctrine to be heretical. Bogomilism is a dualist creed which flourishes in Bulgaria from the tenth to the fourteenth centuries. Believers hold that God and Satan have both existed eternally: whereas God is the author of the invisible, spiritual world, Satan has created the visible, material world. On the basis of this central tenet, physical sacraments such as water for baptism or bread and wine in the Eucharist are considered to be within the realm of Satan. Thus, salvation occurs for Bogomils not through physical sacrament but through fasting and prayer. Bogomils reject Christian sacraments, images, the cross, and the Mass because they belong to the material, not the spiritual world. The most esteemed members of the sect, called the Perfect (monks), renounce all earthly goods and lead strictly ascetic lives. They reject violence, marriage, and meat-eating, and spend much time praying. All of the Perfect, men and women, are allowed to preach. The sect has no ecclesiastical hierarchy, but a bishop heads the community. In Bulgaria, the sects splits into two branches: the Dragovitsian, which is strictly dualist, and the Bulgarian, which regards Satan as a fallen angel, not eternal. Bogomilism is possibly an offshoot of Manichaeism, a dualist religion of Persian origin that was widely popular in the late Roman Empire. The thirteenth century is the high point of the sect; the church synod called by Boris is an element of the Tsar's active persecution of the sect during his reign.

1230   Theodore of Epirus defeated

Theodore of Epirus attacks Bulgaria but is defeated at Klokotnic.

1235   John Asen II exerts power over Byzantines

The Byzantine government at Nicaea recognizes the title tsar of John Asen II (r. 1218–41) and establishes the second Bulgarian patriarchate at Tirnovo, in exchange for an alliance against the Latin Empire.

1241   Tatar raids

Tatar raids and feudal factionalism begin, causing social and political disorder.

1246   Nicaeans occupy Bulgarian territory

Bulgaria gets a child ruler who is weak. This enables the Nicaeans to begin occupying Bulgar territory.

1261   Byzantium reestablished

Nicaeans recapture Constantinople, overthrowing the Latin Empire.

1277   Peasant tsar Ivailo

In a peasant revolt, the swineherd Ivailo leads Bulgarians against the Tatars and becomes tsar himself.

1331–71   Reign of John Alexander

John Alexander is the major patron of the literary revival in medieval Bulgarian literature known as the Silver Age. He encourages original writing in Slavic, as well as translating, copying, and illuminating manuscripts. Compositions from this era are overwhelmingly religious and contemplative in nature, reflecting the influence of Hesychasm and the ever-present Ottoman Turkish threat.

1355   Synod curses Bogomils and Hesychasts

Bulgarian Church Synod at Tirnovo pronounces a curse on Bogomils and Hesychasts and condemns Jews and heretics. Hesychasm, like Bogomilism, is an alternative religious practice that develops in the Bulgarian realm. It originates in the Orthodox monasteries of Mount Athos. The practice encourages contemplative silence as a means of attaining personal inner tranquillity and knowledge of God. Adherents devote themselves to solitary meditations causing them to see a supernatural light, which leads to a feeling of ecstasy. Despite the pronouncement of the Synod, Hesychasm survives and inspires an important Bulgarian literary movement. Hesychasts regard language as not just a vehicle for expression but as an intrinsic component of divine truth—thus they are eager to correct existing translations of Old Slavic texts on the grounds that textual errors could lead to heresy.

1365   Bulgaria divided

Tsar John Alexander divides his realm between his two sons.

1365–69   Vidin region ruled by Hungarians

Hungarians rule in the Vidin region of western Bulgaria and force many Orthodox Christians to be baptized as Catholics.

Mid-14th century   Formerly Bulgarian territory of Dobrudja secedes

Dobrudja is named for its second ruler, Dobrotitsa. Eventually, the region becomes a part of modern Romania.

1369   Ottoman Turks capture Adrianople (Edirne)

Adrianople is the first major urban acquisition in Europe by the expanding Ottoman state.

1371   Emergence of two Bulgarian kingdoms

After the death of Tsar John Alexander, two separate Bulgarian kingdoms emerge: Vidin in the west and Tirnovo in the east. Two Macedonian states appear around the cities of Velbuzd (on the upper course of the Struma River) and Prilep.

1371   Tirnovo made vassal of Ottoman Empire

The Bulgarian king of Tirnovo is forced to accept the status of an Ottoman vassal after the defeat of the armies of Macedonian princes at Chirmen, a small village on the lower Maritsa River.

1375   Central figure of Silver Age of Bulgarian literature is monk Evtimii

Evtimii (also known as Euthymius) becomes the last patriarch of the pre-Ottoman Bulgarian Church. He authors several biographies of Bulgarian saints, including St. John of Rila, and a treatise on ascetic life. He introduces spelling reform, updating Cyrillic to match spoken Bulgarian. Most importantly he founds the Tirnovo literary school: a group of dedicated monks whose principal activity is to compare Old Slavic liturgical texts to Greek originals to eliminate errors and incorporate a more ornate literary style. This literary practice originates in the Hesychast movement (see 1355).

c. 1385   Sofia and Nis captured

The Ottoman Turks conquer major Bulgarian cities, Sofia and Nis.

1388   Alliance of Balkan states

John Stratismir of Vidin unites with Lazar I of Serbia and Tvrto I of Bosnia to battle Sultan Murad I of the Ottoman Empire. This alliance gains a victory over the Ottoman Turks at Plocnik, a small village west of Nis. Subsequently, the Sultan invades Vidin, forcing the state to accept Ottoman overlordship.

1389   Fall of Serbia

Ottoman Turks with help of Christian vassal forces defeat Serbs and Bosnians at Kosovo Polje, exposing remaining Bulgarian territory to Ottoman occupation and beginning five hundred years of Ottoman rule over the Balkans.

c. 1393   Vidin Bulgaria overrun

Hungarians invade the Vidin Kingdom, allied with Wallachians who conquer Tirnovo. Wallachian forces are led by Prince Mircea cel Batrin (r.1386-1418). The Wallachian occupy Dobrudja and the city of Silistra on the Danube River.

1393   Ottomans occupy Bulgaria

Sultan Bayezid I (r. 1389-1402) expels the Wallachians from Silistra and Dobrudja. He declares Tirnovo (Danubian) Bulgaria to be an Ottoman province. The last ruler of medieval Bulgaria, John Shishman, is accused of collaboration with the enemy and is executed on the sultan's orders.

1393   Last Patriarch of Bulgarian Church is deposed

Evtimii is deposed and sent into exile following the Ottoman conquest of Tirnovo.

1396   Vidin Kingdom becomes an Ottoman province

King Sigismund of Luxembourg, emperor of the Holy Roman Empire and King of Hungary (r. 1387-1437), with the assistance of French knights and the Venetians, leads an army to defeat at Nikopolis, on the Danube River, on September 25. Because Vidin, a vassal state of the Ottoman Empire, allows the army onto its territory, it is subordinated to the position of a province of the Empire.

1410   Flight of Constantine Kostenecki

Kostenecki, also known as "the Philosopher," flees Bulgaria and settles in Serbia. He is an adherent of the Evtimiian policy of literary orthodoxy and a major author. His presence in Serbia contributes to the Bulgarian influence on Serbian literature. In the fifteenth century, literary life in Bulgaria is mainly confined to monasteries in the western part of the country.

1444: November 10   Christian army defeated at Varna

A Hungarian-Wallachian army, encouraged by the pope and the Byzantines, crosses the Danube and marches through Bulgaria to Edirne. Former sultan Murad II comes out of retirement to command the Ottoman forces to victory at Varna on the Black Sea.

1453   Fall of Constantinople

Ottoman Turks under Sultan Mohammed II capture Constantinople, bringing the Byzantine Empire to its end.

1492   Emigration of Spanish Jews

Many Jews settle in Ottoman Bulgaria after being expelled from Spain.

Late 14th century   Turkish settlement in Balkan lands

This period sees a strong influx of Turkoman settlers in eastern areas of Ottoman Europe.

1508   First printed work in Bulgarian language

The first printed Bulgarian work, a liturgy, appears in Romania.

1520–30   Ottoman census

A census of taxable hearths (households) in the Empire reveals high numbers of Muslim households in Bulgarian districts. Two of the twenty-eight European kazas (Ottoman administrative regions), Silistra and Chirmen, show a Muslim majority in households. The modern-day Bulgarian capital of Sofia registers a 66.4 percent Muslim population.

c. 1600   Height of Ottoman Empire

Ottoman Empire reaches zenith of power and territorial control.

1651   Modern Bulgarian language in print

The Abagar, primarily a prayer book, is printed in Rome. It comes to be known as the first publication to contain elements of modern Bulgarian language.

c. 1666 Forced conversions

A forced drive for conversion to Islam is conducted in the Rhodope Mountains region.

1688 End of Catholic influence

The suppression of a Bulgarian revolt against Ottomans at Chiprovets ends Catholic influence in Bulgaria.

c. 18th century Bulgarian elite forms

The Bulgarian *khorbachis*, a quasi-middle class between the Turkish overlords and the Bulgarian peasants, acquires economic privileges and minor administrative posts in the Ottoman Empire.

1722–98 Life of Father Paisii

Paisii is born in 1722 in Bansko, in northern Macedonia. At age twenty-three, he becomes a monk at the Hilendar Monastery on Mount Athos. In the libraries of monasteries on Mt. Athos, he collects material relating to Bulgarians. In 1762 he publishes the *Slavo-Bulgarian History*, written in old Church Slavonic. This work refers to the glory of Bulgaria's past and includes discussion of the origins of Christianity in Bulgaria and the achievements of the Bulgarian tsars.

1739–1815 Life of Stoiko Vladislavov

Vladislavov is one of the few Bulgarians to rise to the position of bishop in the Eastern Orthodox Church. He is born in Kotel and meets Paisii in 1765. Subsequently, he becomes known as *Sophronii* and is made the Bishop of Vratsa. In 1806, his book of sermons known as the *Sunday Book* is the first book published in Bulgaria. His autobiography, *Life and Sufferings of Sinful Sophronius*, popularizes Paisii's ideas and nationalist goals. It is also among the earliest works written in good modern literary Bulgarian.

1741 *Stematografia*

Hristofor Zhefarovich completes the *Stematografia*, the seminal work on Bulgarian cultural history.

1804 Serbian revolt

Serbia is the first Slavic land to take arms against Ottoman empire.

1815 Bulgarians support Serbs

Bulgarian volunteers join Serbian independence fighters.

c. 1820 End of *kurdzhaliishtvo*

This anarchic period, known as the *kurdzhaliishtvo*, is precipitated by a breakdown of Ottoman authority in Bulgarian territory.

1835 First Bulgarian-language school

Neofit Rilski opens the first school where teaching occurs in Bulgarian, using Petur Beron's secular educational system.

1840 Girls' school

First girls' school where teaching occurs in Bulgarian language opens.

1844 First periodical

In 1844, the first periodical is printed in Bulgaria.

1856 First public reading room

First *chitalishte* (public reading room) opens in Sofia.

1860 Bulgarian diocese independent

Bishop Ilarion Makariopolski declares the Bulgarian diocese of Istanbul independent of the Greek Orthodox patriarchate.

1862 Armed group for independence

Georgi Rakovksi forms the first armed group for Bulgarian independence.

1870 Orthodox Church granted autonomy

The Bulgarian Orthodox Church is declared a separate *exarchate* (autonomous ecclesiastical jurisdiction) by the Ottoman Empire. Religious autonomy is thus restored to Bulgarians. Contested areas of Ottoman rule are defined through *plebiscites* (binding votes by the general population). The Macedonian region is claimed hereafter by both Greeks and Bulgarians. The dispute includes Serbia after 1878. The region contains an ethnic amalgam: mainly Turks, Greeks, and Bulgarians, but also Albanians, Vlachs, Roma (gypsies), Armenians, Jews, and some Serbs.

1875 September Uprising

September Uprising, the first general Bulgarian revolt against Ottoman rule, is crushed.

1876 April Uprising

The April Uprising spurs massacres of Bulgarians by Ottomans and a European conference on autonomy for Christian subjects of the Ottoman Empire.

1878 San Stefano Bulgaria

The Russo-Turkish War of 1877–78 ends in Treaty of the San Stefano, creating an autonomous Bulgaria stretching from the Aegean Sea to the Danube River.

Autonomous Bulgaria, 1878.

tsar over the union of Eastern Rumelia with Bulgaria in 1885, he finds himself without domestic political support. Liberals and *Russophobe* nationalists led by Stefan Stambolov advance to take charge.

1887 Stambolov serves as prime minister

Stefan Stambolov begins seven years as prime minister, a period which exhibits accelerating economic development. Despite a climate of hostility originating with the Russian tsar, Catholic Ferdinand of Saxe-Coburg-Gotha accepts the Bulgarian throne. Agitation by the Orthodox hierarchy against Ferdinand is kept in check by Stambolov's authoritarian measures.

1888 Establishment of University of Sofia

The University is intended to promote the domestic training of teachers, civil servants, and jurists with national and patriotic elements in its curriculum that foreign colleges would not provide. Its three faculties cover history and philology, physics and mathematics, and law. Students in other disciplines continue to enroll in more western institutions.

1891 Origins of Bulgarian Communist Party

The Social Democratic Party, antecedent to the Bulgarian Communist Party, is founded.

1894 Diplomatic relations strengthened with Russia

The new Russian tsar, Nicholas II, and Ferdinand establish good diplomatic relations between their two states. By baptizing his son, Boris, in the Orthodox faith, Ferdinand gains the favor and political backing of Russia, which enables him to oust Stambolov as prime minister and to replace him with Konstantin Stoilov, the founder of the Conservative People's Party. Stoilov is a more pliant politician and serves until 1899. Ferdinand establishes a network of loyal subordinates within the government, particularly through control of the ministers of war.

1894 IMRO founded

The Internal Macedonian Revolutionary Organization (IMRO) is founded by Bulgarian Macedonians. The IMRO seeks either autonomy within the Ottoman Empire or annexation by Bulgaria.

1899 Origins of Bulgarian Agrarian Party

The Bulgarian Agrarian Union is founded to represent peasant interests.

1900 Bulgarian economy

Bulgaria is a poor agricultural country. In a population of 3,774,000, over eighty percent are peasants. Only 6.8 percent of households use metal plows. Handicrafts are the dominant

1878 Autonomous Bulgaria

In the Treaty of Berlin, western Europe forces a revision of San Stefano Treaty, returning the area south of the Balkan Mountains to the Ottoman Empire; a smaller Bulgaria retains autonomy within the Empire.

1879 Tirnovo constitution

The Tirnovo constitution is written as the foundation of modern Bulgarian state. Alexander of Battenburg is elected prince of the Bulgarian constitutional monarchy. The constitution complies with the Treaty of Berlin stipulation that there be freedom of religion for all citizens and foreign residents in Bulgaria; no discrimination on the basis of religion in the exercise of civil, political, and economic rights; and no impediments to the internal organization of the religious communities and their relations with superiors abroad. It provides for the franchise of all male citizens over twenty-one; eligibility for elective office for those over thirty and literate; a unicameral legislature; freedom of speech, press, and assembly; broad local self-government; and free elementary education. It also invests the ruler with authority as head of state and control of the executive branch, without tying responsibility for executive decisions to his office.

1886 Prince abdicates

The first prince of modern Bulgaria, Alexander von Battenberg, abdicates. He is a favorite nephew of the Russian tsar Alexander II, whose assassination leads to the conservative reaction of his successor, Tsar Alexander III, and enables the prince to suspend the constitution and rule Bulgaria under a "personal regime." In so doing, he loses liberal and conservative political support. After a confrontation with the Russian mode of non-agricultural production, in terms of employment: 129,000 are employed in handicraft workshops as opposed to fewer than 8,000 workers in industrial establishments. The late 1890s witnesses the completion of the Bulgarian gap in the Vienna-Constantinople railroad (the Orient Express) and the construction of trunk lines to ports on the Danube and the Black Sea. Peasant interests are neglected by the established political parties, so rural teachers and intellectuals form a Peasant Union.

1901   U.S. diplomatic contacts in Bulgaria

The U.S. establishes diplomatic relations with Bulgaria. The first American diplomatic agent, Charles M. Dickinson, tours the country. His report notes the general openness of Bulgarians toward progressive ideas.

1903   Illinden-Preobrazhensko Uprising

The revolt sends large numbers of Macedonian refugees into Bulgaria and inflames Macedonian issues. The IMRO instigates this ill-planned and ill-prepared uprising in the Bulgarian population centers of Turkey.

1908   Unilateral declaration of independence

Ferdinand declares Bulgaria fully independent of Ottoman Empire and himself tsar.

1912   Prelude to war

After the Treaty of Berlin has returned several territories to Turkish rule (which have been awarded to Bulgaria under the Treaty of San Stefano), war with Turkey is anticipated. By 1912, the nation's foreign debt is 1.3 billion *leva*. Handling the foreign debt is necessary for the state to remain creditworthy but requires periodic tax increases and the establishment of special state monopolies over common goods to cover the interest.

1912   Increased population of cities creates miserable living conditions

Sofia, which in 1878 has a populace of 20,000, now contains 120,000. New citizens come from Bulgarian population centers outside of the Bulgarian state, such as Turkish Macedonia, Thrace, and Romanian Dobrudja; many also are peasants seeking work. Socialists attempt to form a political base in this *proletariat* (working class) with little electoral success. By 1912, only one deputy in the National Assembly, Ianko Sakuzov, is a socialist.

1912   The First Balkan War pushes the Ottoman Empire completely out of Europe

In alliance with Serbia, Greece, and Montenegro, Bulgarian units soundly defeat Turkish forces, advancing to but not taking Constantinople. Meanwhile, Serbian and Greek forces defeat Turkish garrisons in Macedonia. By virtue of their occupation of this territory, Greece and Serbia fail to honor treaties established with Bulgaria for the award of Macedonia. As a result of the war, Bulgaria gains territory in Turkish Thrace including the city of Adrianople (Edirne). Bulgarian leaders are disappointed by the territorial gains they have made.

1913   Second Balkan War

The Second Balkan War ends in shameful defeat for Bulgaria. Ferdinand and his commander in chief, General Mikhail Savov, order an attack on Serbian and Greek positions without declaring war. This attempt to take additional territory by force ends when Romania invades from the north, and Turkey retakes Adrianople. The Peace Treaty of Bucharest on August 10 divides Macedonia between Greece and Serbia and awards southern Dobrudja to Romania. Bulgaria gains only a small part of Macedonia, known as the Pirin region, and is allowed to annex a small part of Aegean Thrace.

1915–18   Bulgaria and World War I

Bulgaria fights in World War I on the side of the Central Powers. The Entente (Britain, France, and Russia) offer territorial gains to Bulgaria only in Turkey, a Central Power. In contrast, the Central Powers offer all of Serbian-held Macedonia and territories of Greece and Romania if those two countries join the Allied side. Ferdinand and his German-educated prime minister, Vasil Radoslavov, commit Bulgaria to the Central Powers on September 6, agreeing to attack Serbia within thirty-five days. Stamboliiski, Agrarian Union leader, publishes an exchange between himself and Ferdinand in which he challenges the king's divergence from the policy of strict neutrality and is consequently imprisoned.

1916: August   Romania joins the Allies and Bulgaria advances

Bulgarian troops converge with those of Austria and Germany and reach Bucharest in December.

1917   The United States enters on the war on the Entente side

The American entrance into the war signals the impending defeat of the Central Powers. Despite pressure from its allies, Bulgaria refuses to declare war on the U.S. Numerous contacts made through missionary and educational work, especially that involving Robert College in Constantinople preclude an American declaration of war on Bulgaria, one of the Teutonic allies.

1918   Ferdinand replaces Radoslavov and attempts to reconcile with the Entente

In an attempt to negotiate a separate peace with the Entente, Ferdinand installs Alexander Malinov as both prime and foreign minister. Before Bulgaria can argue its case at an international peace conference, its military collapses under rejuvenated Allied offensives on the Macedonian front. At Dobro Pole on September 15, Bulgarian troops are forced to retreat. Some troops march to Sofia, while others seize Kiustendil, the military headquarters. On September 25, Stamboliiski and his protégé Raiko Daskalov are released from prison in hopes that they will gain control of the mutinous peasant soldiers. Stamboliiski first attempts to ally with the leader of the Narrow (radical) Socialists. Blagoev refuses to unite with the Peasant Party because Stamboliiski asserts that peasants must be allowed to retain the right of private property. Blagoev maintains that on this basis, peasants are part of the property-holding *bourgeoisie* (middle class), and therefore class enemies. Daskalov attempts to create a new state called the Radomir Republic, formed of about 8,000 mutinous soldiers. The Radomir Republic is defeated by forces under General Alexander Protogerov, a leader of the IMRO, who is commanding elements of the Sofia garrison, IMRO volunteers, and some German troops. On October 3, Ferdinand abdicates, in fear for his life. He leaves his young son Boris as prince and flees to Germany by train. On December 29, with the assistance of the American Consul-general Murphy, the Entente agrees to an armistice with harsh terms. Bulgaria's part in the World War I is thus ended.

1919: August Parliamentary elections show strong support for radical parties

After Ferdinand abdicates, a government of national unity takes power. Led by Teodor Teodorov of the People's Party and including Stamboliiski and other Agrarians, it calls for parliamentary elections in August. The Agrarians win eighty-five seats (180,648 votes), the newly renamed Communists (formerly Narrow Socialists until their union with Lenin's Communist International in March) win forty-seven seats (118,671 votes), and the Broad Socialists win thirty-six seats (82,826). The Democrats have the best showing of the so-called bourgeois parties, gaining only twenty-eight seats (65,267 votes). Stamboliiski and the Agrarians control eighty-five out of 233 seats in the National Assembly, too few to form a government. The Communists refuse to form a coalition, on the hope that disorder will allow them to orchestrate a revolutionary process like the Bolshevik Revolution in Russia (October 1917) that will put them in control. The Broad Socialists demand control of several important ministries, including the army and police. Stamboliiski considers their demand too high and turns to parties on the right.

1919: November Stamboliiski signs the Treaty of Neuilly

In the pattern of the Treaty of Versailles (which ended Germany's engagement in the war), the Entente presents a punitive treaty to Bulgaria. Territorial losses to its neighbors include western Thrace and areas along the Serbian border.

Huge reparations in the amount of 2.25 billion gold francs are mandated, with a payment over thirty-seven years at five percent interest. Thousands of head of livestock and large amounts of coal are to be delivered to Serbia, Romania, and Greece. The Bulgarian army is limited to 20,000 volunteers.

1919 Life of Alexander Stamboliiski

In 1919, at age forty, Stamboliiski finally leads the government. He uses this opportunity to institute reforms based on the ideology of Agrarianism which he has been formulating since 1909. This political ideology is established in political writings such as *Politicheski partii ili suslovni organizatsii?* (Political parties or estatist organizations?). His views on the communist attitude towards the peasantry in Bulgaria and in Russia are sharply critical. His foreign policy aims to align Bulgaria with European democracies and to establish a Balkan Federation with Yugoslavia.

1923: June 9 Stamboliiski is murdered

A coalition of IMRO elements, army units, and armed civilians capture Sofia. Stamboliiski is killed in his home village. King Boris appears to gain a favorable outcome as the Agrarian Party loses its ruling status. Members of the new government led by Alexander Tsankov as well as leaders of the Military League and IMRO joined in the plot to assassinate Stamboliiski. The Tsankov government calls itself the Democratic Alliance. Tsankov is a professor of political economy; before World War I he is a socialist, but his politics evolve toward fascism and nationalism. He and his foreign ministers ally with Mussolini's Italy. The minister of war, General Ivan Vulkov, is an associate of King Boris and of IMRO leaders.

1923: September September uprising of Communists and Agrarians

The ill-planned uprising fails as the urban centers, supposedly the bases of communist power, yield very little resistance. The exiled leaders of the communists and Agrarians negotiate on the idea of uniting against Tsankov's "Fascist" regime, but the Agrarians refuse to allow the communists the level of control they demand. The negotiations (which are held in Vienna and Moscow) and the idea of collaboration are henceforth abandoned until the conclusion of World War II (1945).

1926: January Tsankov resigns over loan negotiations

After failing to gain financial support from British bankers, Tsankov resigns. The loan, necessary to maintain the functions of the government, is refused because of social and political instability in Bulgaria. A new government is formed from within the governing coalition. Liapchev, the new prime minister, relaxes some of the government's repressive measures. His policy allows amnesty for persons convicted under the Law on the Defense of the State and provides communist legal access to politics as the Worker's Party. The Liapchev

On the fifteenth anniversary of the signing of the Treaty of Neuilly, armed mounted police are called out to control huge crowds of demonstrators, unhappy with conditions in Bulgaria in the years after the treaty. (EPD Photos/CSU Archives)

regime oversees free elections, held on June 31, 1926, but also downward trends in the economy worsened by the world economic crisis after 1929.

1926: June 31   Liapchev government administers free elections

In the elections, the Liapchev coalition loses massively. The biggest winner is the People's Bloc. The Bloc is composed of the Democratic Party (led by Alexander Malinov and Nikola Mushanov), the centrist Agrarians (Gichev, Muraviev, and Dimov), the Radicals (Kosturkov), the Liberals and a few other minor parties.

1930   King Boris marries Princess Giovanna of Italy

The royal wedding of the ruling families of Bulgaria and Italy draws the two nation-states closer diplomatically and politically.

1934   Balkan Entente

In Balkan Entente, Greece, Romania, Turkey and Yugoslavia reaffirm existing Balkan borders. Bulgaria refuses participation and remains isolated.

1934: May   *Zveno* government is installed in bloodless coup

Zveno (meaning Link) is an attempt to govern on a level above the political parties and yet to involve individuals from the political, professional, and intellectual elites of a variety of backgrounds. The Military League carries out the coup on May 19, installing a government led by Kimon Georgiev. Damian Velchev is the intellectual source of Zveno's ideas but remains in the background to reduce the appearance of grasping for power. His policies urge austerity and self-sacrifice at all levels of the new administration. Although short-lived, the Zveno government establishes several important policies. Authoritarian rule, based on Article 47 of the Tirnovo Constitution, enables the king to rule by decree during emergencies. It dissolves the National Assembly and bans all political parties and activities. Zveno also manages to eliminate the IMRO.

1935: January   End of Zveno; beginning of royal dictatorship

Royalist officers in the Military League, many of whom favor alignment with Germany and Italy, urge Georgiev, the head of the Zveno coalition, to resign. Thus, King Boris is awarded a political system through which he can govern personally and directly. The royal dictatorship comes to an end in 1943.

1938   Royally directed elections held

The parliamentary elections of 1938 are held to lend an appearance of democracy to Boris's dictatorial rule. A dependable majority of pro-fascist members is maintained. In the election, no one from any of the banned parties is allowed to run, candidates are screened by the courts and faced with tight censorship, and the overall number of seats is reduced from 274 to 160. Interestingly, certain (presumably conservative) segments of the female population are granted the right to vote in this election.

1939: July   Kioseivanov visits Berlin

Boris's foreign minister Georgi Kioseivanov meets with Hitler and his foreign minister Ribbentrop to strengthen diplomatic ties between Bulgaria and Germany.

1940: February   New Cabinet is formed, headed by Professor Bogdan Filov

Filov, a renowned archaeologist and art historian, educated in Germany, is chosen as prime minister to cement Bulgaria's diplomatic orientation in favor of Germany. Filov presides over Bulgaria's government until September 9, 1944.

1940: Summer   Bulgaria gains territory in southern Dobrudja

Acting under the terms of the secret protocol the Nazi-Soviet Non-Aggression Pact (the agreement between the two governments to jointly invade Poland but not to attack each other), the Soviet Union urges Romania to cede Bessarabia and other areas. In the ensuing collapse of Romania, its borders are redrawn to grant southern Dobrudja to Bulgaria.

1941: January   Law on the Protection of the Nation curtails civil rights of Bulgarian Jews

To please Hitler, measures against the Jews based on the German Nuremberg laws have been under official consideration since July, 1940. In January, 1941, the parliament passes, and Boris executes the Law on the Protection of the Nation which is subsequently expanded and intensified to impose on Bulgarian Jews many repressive measures. The law arouses wide protests; among other bodies in opposition, the Synod of the Bulgarian Orthodox Church is noted for its declaration, "We are all sons of a heavenly Father." The Synod's official position is that in the case of a threat to the nation, measures should be taken against individuals and not against ethnic and religious groups.

1941: Spring   The United States sends an envoy to meet with King Boris

William J. Donovan visits Sofia to notify Boris that the U.S. intends to see Great Britain victorious and warns him of the consequences of collaboration with the Nazis.

1941: March 1   German troops cross the Danube River, heading for Greece

Bulgarian-German agreements allow Nazi army units to traverse Bulgarian territory from occupied Romania to invade British-allied Greece.

1941: April 19   Boris meets Hitler in Vienna

This meeting in Vienna establishes that Bulgaria will receive Yugoslav Macedonia and parts of Greek Thrace in return for military and diplomatic allegiance to the Axis (Germany, Italy, and Japan).

1941   Bulgaria signs treaty with Nazis

Bulgaria signs the Tripartite Pact, allying with Nazi Germany in World War II. Bulgaria refrains from action against the Soviet Union for the duration of the war.

1941: November 25   Bulgaria joins the anti-Comintern pact

Comintern (Communist International) is a political organization headed by the Soviet Union intended to foster the expansion of communism. By joining the anti-Comintern pact, Bulgaria gains the favor of the Axis but the enmity of the Soviet Union.

1941: December 13   Bulgaria declares war on the United States

Shortly after Japan attacks Pearl Harbor, Bulgaria joins Germany and Italy in a declaration of war.

1943: March   Deportation of Jews

Jews from Macedonian and Greek territories under Bulgarian control are deported.

1943: August   Death of King Boris III

After meeting with Hitler on August 14 and 15, Boris III dies, leaving a three-man regency to rule for his underage son Simeon II. Suspicions of assassination by poison are widespread but never proven.

1944: January to March   Allied air raids damage Sofia heavily

Following major Allied air raids on Sofia, the activity of anti-war factions in Bulgaria increases.

1944: June 1   Imminent defeat

The Bulgarian leadership attempts to limit damage of imminent defeat. On June 1, regents of Simeon II, the young heir to the throne, form a new cabinet. While rebuilding the government, the regents authorize contacts with the American diplomats in Istanbul.

1944: September 2   More politically-leftist cabinet is formed

In order to gain more popular support, the three-man regency forms a new cabinet more representative of the politically active members of Bulgarian society.

1944: September 5   Soviet troops invade

As the Bulgaria government seeks peace with the Allies, the Red Army invades. A temporary Bulgarian government is formed and subsequently overthrown by a communist-led coalition.

1944: September 9   Bulgarian regency is overthrown in coup

The Bulgarian monarchy comes to its end in an overthrow.

1944: October 9   Percentage Agreement between Allies

Secret negotiations between Winston Churchill of Great Britain and Joseph Stalin of the Soviet Union establish a system to divide political influence in the post-war Bulgarian political arena. Under this system, the Soviet Union is granted an eighty percent share of political control, with twenty percent reserved for British and American influence. The coalition government formed after the coup metes out 2,138 death sentences and 10,000 prison terms.

1945: February   Yalta Agreement gives hope of freedom to European masses

The diplomatic agreement between the Allied powers, victors over Germany and its allies (like Bulgaria), includes language regarding "the right of peoples to freely choose the form of government under which to live." Like several other Eastern European states, the Yalta Agreement soon rings hollow in Bulgaria as Soviet-sponsored factions succeed in dominating the political arena.

1946   Dimitrov made prime minister

Georgi Dimitrov of the Bulgarian Communist Party (BCP) becomes prime minister of the new People's Republic of Bulgaria.

1947   The Dimitrov constitution goes into effect

Remaining opposition parties to the BCP are silenced. State confiscation of private property is completed.

1948-49   Religious freedoms are eliminated

The Muslim, Orthodox, Protestant, and Roman Catholic religious organizations are restrained or banned.

1949: July   Death of Dimitrov

Georgi Dimitrov, first prime minister of communist Bulgaria, dies.

1949   Joseph V. Stalin chooses Vulko Chervenkov to succeed Dimitrov

A period of Stalinist popularity, purges of the Bulgarian CP and strict cultural and political orthodoxy begin. Chervenkov is the brother-in-law of Bulgaria's first communist premier, Georgi Dimitrov. His Stalinist political alignment may be attributed in part to his training at the Marx-Engels Institute in Moscow. Before returning to Bulgaria, he spends several years working in the Communist International (Comintern), an organization guided by Soviet policy and intended to foster the establishment of worldwide communist rule.

1950   Large-scale collectivization of agriculture

Collective agriculture, in which formerly privately held land and tools are consolidated into large centrally-directed farms, is a major element of Soviet-style communism. By 1952, sixty-one percent of Bulgarian farmland is organized under the collective system; by 1956, seventy-seven percent. In 1958, ninety-two percent of Bulgarian farmland is managed collectively.

1950   United States severs diplomatic relations with Bulgaria

In response to Bulgaria's turn away from the west and severely curtailed civil rights, the United States withdraws its diplomatic mission.

1950: August   Deportation of ethnic Turks

Chervenkov announces that 250,000 Turks will be deported. The communist leadership of Bulgaria considers the Turkish minority too resistant to land-use policies and atheist political indoctrination. Turkey closes its border in November, 1951, by which time 155,000 Bulgarian Turks have been expelled.

1953   Death of Stalin

The death of Stalin leads to loosening of Chervenkov's control and easing of party discipline.

1956   Life of Todor Zhivkov

Zhivkov is born in Pravets, a village northeast of Sofia, in 1911. He migrates to Sofia, where he works as a printer then joins the Bulgarian Communist Party (BCP) in 1932. During the war, he is the liaison between the party leadership in Sofia and the *Chardar* guerrilla brigade. In 1945, Zhivkov becomes a candidate member (non-voting) of the BCP Central Committee. His rise to the leadership of the party continues as he is promoted to voting status in the Central Committee in 1948. In 1950, Zhivkov is appointed a candidate member of the *Politburo* (main governing council of the Communist Party) and made the Secretary of the Central Committee.

1957–58   Response to invasion of Hungary

After the Soviet invasion of Hungary, Bulgaria cracks down on non-conformism to party line in culture and politics.

1962   Purge of Yugov

Chervenkov removes the last obstacle to the highest authority in Bulgaria by purging (removing) Ivan Yugov. He then becomes general secretary and premier, thus controlling both the government and the BCP.

1962   Rise of Zhivkov

Nikita S. Khrushchev anoints Zhivkov as the successor to Chervenkov. Zhivkov becomes prime minister and remains the unchallenged leader of the BCP for the next twenty-seven years.

1968   Repercussions of Soviet invasion of Czechoslovakia

In response to a native version of communism called "socialism with a human face," Soviet troops allied with those of the Soviet satellites (including Bulgaria) occupy Czechoslovakia.

Domestic Bulgarian politics take a more repressive turn, ostensibly to prevent Soviet overdetermination.

Early 1970s   Repression of Pomak ethnic group

Pomaks (Bulgarian-speaking Muslims) are forced to change their Turkish or Arabic names to Slavic equivalents. The term *Pomak* itself is banned. From 1968 to 1978, 130,000 of the Turkish ethnic minority in Bulgaria emigrate under a diplomatic agreement with Turkey.

1971   Zhivkov constitution

A new constitution specifies the role of the BCP in Bulgarian society and politics. The new constitution establishes the State Council as the supreme policy-making body. As president of the State Council, Zhivkov is head of state; he sets general policies but removes himself from everyday responsibilities. Zhivkov's policies are subsequently executed by the Council of Ministers, headed by a series of pliant prime ministers.

1978   Dissident Georgi Markov is assassinated in London, England

On a London sidewalk, Markov is assassinated for his political views by a member of the Bulgarian secret police. He is stabbed with a poisoned umbrella tip.

1981   New Economic Model

Restructuring under the New Economic Model (NEM) brings a temporary economic upswing, but no long-term improvement.

Late 1988 –early 1989   Dissident groups begin to form around environmental and human rights issues

The rise of informal organizations is a manifestation of political and social resistance. The term "informal" provides these groups a measure of reprieve from government scrutiny. Writers' groups form in Sofia and Plovdiv. An ecological association called *Ekoglasnost* based in Sofia has as its most prominent members Petur Slabakov, Petur Beron, and Alexander Karakachanov. Khristof Subev figures prominently in a committee for religious rights, freedom of conscience, and spiritual values in Veliko Turnovo. *Podkrepa* (Support) is organized as an independent federation of labor, similar to Poland's Solidarity. The Club for Glasnost and Democracy is chaired by, among others, Dr. Zhelin Zhelev, future leader of post-communist Bulgaria.

1988   Ecological demonstration in Ruse

Ruse, Bulgaria's main port on the Danube river, has air heavily polluted by Romanian manufacturing plants across the river at Giurgiu. Government forces violently suppress the demonstration and a support meeting held in Sofia.

1989   Forced exodus of 310,000 Turks

In the summer, the second Turkish assimilation program brings massive Turkish emigration, increased dissident activity, and international criticism. Disapproval is handed down by the Soviet Union.

1989: October   EcoForum

Internationally-attended ecological summit held in Sofia exposes internal suppression of freedoms to outside world.

1990   Zhivkov is forced out

Three BCP-dominated governments are formed and dissolved. Round-table discussions between the BCP and opposition parties begin to formulate reform legislation.

1990: June   Multiparty elections

First multiparty election since World War II yields majority in National Assembly to the Bulgarian Socialist Party, the renamed BCP. A large opposition block is gained by the Union of Democratic Forces (UDF), which refuses to participate in the government.

1990: July   Sofia civil disobedience

A tent-city demonstration begins in Sofia, continuing through the summer.

1990: August   President chosen from opposition party

UDF leader Zhelim Zhelev is chosen to be president.

1990: September   International contacts made by Zhelev

Zhelev meets with French and American leaders and receives pledges of economic support.

1990: November to December   General strike

A general strike forces the resignation of the government of Prime Minister Andrei Lukanov. An interim coalition government is formed under Dimitur Popov.

1991: January   Move to freer markets

The initial phase of economic reforms goes into effect. It includes elimination of price controls on some commodities.

1991: Spring   Land reform

The Arable Land Law begins a redistribution of land to private farmers.

1991: July   Democratic Constitution

The National Assembly approves a new constitution. National elections are set for October.

1991: August 4   BSP reveals reform program

The economic reform program, intended to improve relations with both the Soviet Union and the United States, also sets elections using proportional representation.

1991: August 22   End of miners' strike

The nine-day strike involves 27,000 of Bulgaria's 75,000 miners. The miners want salary increases and improved conditions but end the strike without immediate concessions from the government.

1991: October 14   National Assembly election

In this election, the Union of Democratic Forces wins a plurality (thirty-six percent), but not a majority, of seats. In order to form a government, the UDF must join with another party; in this case, the UDF unites with the Movement for Rights and Liberty, which is known to represent the interests of ethnic Turks. The Bulgarian Socialist Party wins about thirty-two percent of the seats.

1991: November 8   National Assembly approves new Cabinet

For the first time since World War II, the Bulgarian government excludes communists. The new prime minister is Filip Dimitrov of the UDF.

1991: December 7   Nationwide strikes

Strikes across the country and in many different trades coalesce under the banner of *Podkrepa*, the Support labor organization. Demands are for wage increases, protection against layoffs, and changes in management. Massive power shortages result from the striking of miners and transporters of coal.

1992: January 19   Zhelev elected president

In run-off elections, Dr. Zheliu Zhelev is Bulgaria's first directly elected president. Running on the UDF party, Zhelev takes fifty-four percent of the vote; BSP candidate Velko Valkanov retains forty-five percent. Nationally, seventy-five percent of eligible voters participate.

1992: February 23   Electrical supply reduced by half

Problems in the Kozloduy nuclear plant require that it be removed from the national power grid.

1992: March 9   Church leader removed

The Bulgarian Orthodox Church removes its highest authority, Patriarch Maxim, from authority on grounds of collaboration with the communist regime.

1992: April 23   Privatization Law

The National Assembly passes a law to privatize state-owned enterprises, which constitute ninety-five percent of the nation's property.

1992: May 19   Former president indicted

Government prosecutors charge Petar Mladenov with inciting discrimination and national hatred. The charges stem from his support for the Bulgarianization of the names of members of the Turkish minority members.

1992: June 25   Bulgaria joins regional economic union

The regional union is based on the Black Sea Economic Cooperation Treaty and includes Bulgaria, Russia, Albania, Ukraine, Georgia, Romania, Greece, Turkey, Moldova, Armenia, and Azerbaijan.

1992: July 9   Former prime minister arrested

Andrei Lukanov is charged with mismanaging the Bulgarian economy and misappropriating up to $500 million in state funds.

1992: September 4   Zhivkov sentenced

Former premier Todor Zhivkov is sentenced to seven years for embezzlement. He is later acquitted of the same charges by a higher court, after serving about four years under house arrest in his grandfather's luxury villa.

1992: September 24   National Assembly President Savov resigns

Stefan Savov resigns because of political disagreements between the two members of the ruling coalition, the UDF and the MRL.

1992: October 28   Movement for Rights and Liberty votes no-confidence

The MRL votes against the coalition with the UDF. The MRL and the BSP combine for a 120 to 111 decision that will lead to dissolution of the current government.

1993: January 28   G-7 reactor fund targets Bulgaria

The G-7 (the world's leading industrialized countries) designate the Bulgarian nuclear reactor system as Europe's least safe. The reactor fund is intended to improve safety of Soviet-designed reactors.

1994: January 29   Bulgaria and Romania sign joint agreement

The joint agreement establishes guidelines for military operations conducted on a joint basis between the two countries.

1994: January 29   Prosecutions for illegal arms sales

Prosecutor Ivan Tatarchev, never a member of the Bulgarian Communist Party, is challenged by BSP leadership after he prosecutes twenty-two former officials in the communist regime.

1994: February 8   Land reform

As of this date, 47.7 percent of appropriated land has been returned to its original owners. A 1991 statute has begun the process.

1994: March 28   Military agreement with Germany

In the first military ties since World War II alliance, Bulgaria and Germany agree to regularly exchange information and staff between their armies.

1994: May 19   General strike

More than 100,000 go on strike to protest the government's social policy.

1994: May 31   UDF boycotts National Assembly

The UDF legislators boycott the proceedings of the Assembly to protest the failure of the BSP to gain adequate support for its proposed Cabinet.

1994: September 2   Government resigns

After failing to obtain support from the National Assembly for the proposed cabinet, the BSP government resigns. The resignation is accepted by the Assembly, 219 to four on September 8.

1994: September 14   Radioactive materials seized

Nineteen containers of radioactive materials, including plutonium, strontium, and cesium, are removed from two Sofia cellars.

1994: October 17   Zhelev calls new elections

President Zhelev dissolves the Assembly and sets new elections for December 18. Zhelev appoints Bulgaria's first female in the office, economist Reneta Indjova, as interim prime minister.

1994: November 18   Anti-communist protest

In Sofia, over 20,000 participate in marches protesting the continued influence of communists in Bulgarian politics.

1994: December 18   New general elections

In the elections, the BSP retains power in government with a 43.5 percent share (125 seats). The UDF fares poorly, gaining only 23.2 percent (69 seats). The MRL gains eighteen seats and a newcomer, the Business Bloc, wins thirteen seats.

1995: January 26   New Cabinet approved by National Assembly

The new government is led by BSP Prime Minister Zhan Videnov. The opposition criticizes the inclusion in the Cabinet of Ilcho Dimitrov, who is partly responsible for the program in the 1980s to force the assimilation of the ethnic Turkish minority.

1995: April 1   New ferry line revives the Silk Road

Ferries begin to transport trucks from the Bulgarian port of Varna to the Georgian port of Poti across the Black Sea. The route revives the historic Silk Road, the overland route for Chinese silk trade with Europe.

1995: July 27   U.S. report declares Kozloduy reactor most dangerous

The U.S. Energy Department report declares that several East European reactors are "accidents waiting to happen," with Kozloduy at the top of the list.

1995: August 29   Rise in cancer rate

The head of the hematology and oncology clinic in Sofia announces that incidence of thyroid cancer among young Bulgarians has risen since the 1986 Chernobyl nuclear disaster in the (then) Soviet Union.

1996: January 12   Wheat and flour shortages

Commerce Minister Kiril Tsotchev and Agriculture Minister Vassil Tchitchibaba resign under fire from the political opposition, which blames them for a nationwide food shortage.

1996: April 5   Manufacturer Rover leaves Bulgaria

After seven months, the British automaker Rover decides to close its factory in Varna because of the predominance of former communists in political positions.

1996: May 12   National Assembly approves financial measures

In a move designed to gain the favor of the International Monetary Fund (IMF), the government announces plans to close several loss-making government enterprises. Two days later the Assembly passes a law which forces poorly managed banks into bankruptcy.

1996: May   Kozloduy reactor closed

The Bulgarian government closes the Kozloduy reactor to upgrade and modernize four of the six reactors in the complex.

1996: May 26   King Simeon II returns to Bulgaria

Former Bulgarian ruler Simeon II has lived since 1946 as an exile in Spain. His visit is greeted by nearly 500,000.

1996: June 4   Zhelev defeated in primary

In the UDF party primary election, Zhelev is defeated by Petar Stoyanov.

1996: August 11   Famous clairvoyant dies

Vangelis Gushtenova (b. 1912), also known as Aunt Vanga, was visited for advice and predictions by many, including Leonid Brezhnev of the Soviet Union and Indira Gandhi, former prime minister of India.

1996: November 4   Stoyanov elected president

UDF candidate Petar Stoyanov wins approximately sixty percent of votes cast. His stated views favor Bulgarian integration into the European Union (EU) and membership in NATO (North Atlantic Treaty Organization).

1996: November 19   Bank panic

The State Savings Bank announces that it has paid its debts with depositors' assets. Customers are able to retrieve up to $738 each.

1996: December 21   BSP prime minister resigns

Zhan Videnov resigns, and BSP officials replace him with Georgi Parvanov.

1997: January 11   Demonstrations escalate to violence

Police clash with proUDF protesters who have blocked the exit of government offices belonging to 100 BSP legislators. The protesters demand new elections for the National Assembly. Injuries are in the dozens on both sides.

1997: January 10   UDF boycotts National Assembly

The UDF move is intended as protest of the BSP's formation of a new cabinet under Nikolai Dobrev.

1997: January 19   Stoyanov takes office

In response to public demands, Stoyanov announces that elections will be held early.

1997: February 7   BSP boycotts National Assembly

After a meeting between President Stoyanov and Prime Minister Dobrev on February 4, the BSP leadership agrees to hold new elections in April. In response, the rank-and-file of the BSP stages a protest boycott.

1997: April 23   UDF victorious in general elections

The UDF wins 137 seats in the 240 member National Assembly. The BSP seats total only fifty-eight. President Stoyanov announces the new prime minister will be UDF party head Ivan Kostov.

1997: May 21   Kostov and Cabinet overwhelmingly approved

In a 179-54 vote, the National Assembly approves the UDF government.

1997: February 17   Plan for NATO membership

The BSP Foreign Ministry announces that Bulgaria will apply for membership in NATO.

1997: June 11   Bulgarian currency board established

With IMF guidance, the Bulgarian currency, the *lev*, has its value tied to the Deutsche Mark of Germany. This move reduces government interference in monetary policy and stabilizes the economy.

1997: November 4   Balkan cooperation meeting

Leaders of Balkan region countries meet in Crete for two days. The meeting produces pledges to cooperate in efforts against drug trafficking and terrorism.

1997: November 25   Crime agreement

In a follow-up to the November 4 Balkan cooperation meeting, Bulgaria, Romania and Turkey sign a joint agreement in regards to international crime.

1997: October 13   Bulgaria and Russia sign natural gas agreement

Unlike typical state to state deals under communism, negotiations are conducted directly between commercial entities in this agreement. Bulgargas and Russia's Gazexport sign a ten-fifteen-year deal to supply Bulgaria with Russian natural gas. The deal is tied to another agreement which allows Russia to transship its natural gas to western Turkey via a Bulgarian-owned pipeline.

1998: May 22   Multinational force in Balkans

A two- to three-thousand member peacekeeping force for the Balkans is established. The agreement includes Bulgaria, Macedonia, Albania, Turkey, Romania, Greece, Slovenia, and Italy. The U.S. endorses the force, which will be deployed in peace support and humanitarian missions.

1999: February 10   Dispute over Macedonian language resolved

Tensions between Bulgaria and Macedonia arise from the Bulgarian contention that Macedonian is just a dialect of Bulgarian. The agreement of February 10 asserts that all official communication between the two will be recorded in the official languages specified by each state's constitutions.

Bibliography

Bar-Zohar, Michael. *Beyond Hitler's Grasp: The Heroic Rescue of Bulgaria's Jews*. Holbrook, Mass.: Adams Media Corp., 1998.

Crampton, R.J. *A Concise History of Bulgaria*. Cambridge, New York: Cambridge University Press, 1997.

Curtis, Glenn E., ed. *Bulgaria, a Country Study*. 2nd ed. Federal Research Division, Library of Congress. Washington, D.C., 1993.

Jelavich, Barbara. *The Establishment of the Balkan National States, 1804–1920. A History of East Central Europe*, 8. Seattle and London: University of Washington Press, 1977.

MacDermott, Mercia. *Bulgarian Folk Customs*. Philadelphia, London: Jessica Kingsley, 1998.

Melone, Albert P. *Creating Parliamentary Government: The Transition to Democracy in Bulgaria*. Columbus: Ohio State University Press, 1998.

Minaeva, Oksana. *From Paganism to Christianity: Formation of Medieval Bulgarian Art (681–972)*. Frankfurt am Main, New York: P. Lang, 1996.

Paskaleva, Krassira, ed. *Bulgaria in Transition: Environmental Consequences of Political and Economic Transformation*. Brookfield, VT: Ashgate Publications, 1998.

Perry, Duncan M. *Stefan Stambolov and the Emergence of Modern Bulgaria, 1870–95*. Durham: Duke University Press, 1993.

Pundeff, Marin. "Bulgarian Nationalism." *Nationalism in Eastern Europe*. Ed. Peter F. Sugar and Ivo J. Lederer. Seattle, London: University of Washington Press, 1969. 93–165.

Rothschild, Joseph. *East Central Europe between the Two World Wars. A History of East Central Europe*, 9. Seattle, London: University of Washington Press, 1974.

Sedlar, Jean W. *East Central Europe in the Middle Ages, 1000–1500. A History of East Central Europe*, 3. Seattle, London: University of Washington Press, 1994.

Sugar, Peter F. *Southeastern Europe under Ottoman Rule, 1354–1804. A History of East Central Europe*, 5. Seattle, London: University of Washington Press, 1977.

Croatia

Introduction

Croatia is one of the six republics that constituted Yugoslavia until the 1990s. Situated at the northwestern corner of the Balkans, Croatia has historically found itself at a dividing line—between Eastern and Western Europe, between the Ottoman and Habsburg empires, and between the Eastern Orthodox and Roman Catholic churches. The Adriatic coastline that long made Croatia attractive to foreign invaders enabled it to build a booming tourist industry that became a mainstay of the Yugoslav economy during the post-World War II communist era. Croatian independence in 1991, with the dissolution of the old Yugoslavia, came at a high price—war with the Serb-dominated Yugoslav Federal Army—but it fulfilled the thousand-year-long dream of a people who had been ruled by a succession of foreign powers.

Croatia comprises three major regions: Slavonia to the east, Istria to the west, and Dalmatia to the south along the Adriatic Sea. With an area of 21,829 square miles (56, 538 square kilometers), the country has two major geographical regions: the Dinaric Alps in the south of the country and the lower-lying lands of the Sava and Drava river valleys to the north. Croatia also includes over one thousand islands in the Adriatic Sea, which geologically are extensions of the Dinaric Alps. Almost all are uninhabited. As of 1999, Croatia has an estimated population of slightly over five million. Its capital, Zagreb, is located in the northern part of the country; metropolitan Zagreb accounts for roughly one-fifth of the nation's population.

History

The South Slavs (*Jugoslavs*, or Yugoslavs) first migrated to the Balkans in the seventh and eighth centuries A.D. Among the regions they settled in were the former Roman seacoast provinces of Dalmatia and Pannonia. One of the South Slavic clans, the *Hrvats*, were the ancestors of today's Croats. In the ninth century came occupation by Frankish conqueror and Holy Roman Emperor Charlemagne (742–814), under whose rule the inhabitants of the region converted to Roman Catholicism. The Croats reasserted their independence in the follow-

ing century under their first native ruler, Tomislav, who declared himself king in 925 and expanded Croatian territory and power. Before long, however, Croatia's powerful neighbors were asserting claims to the region. By the turn of the eleventh century, Venice had established its first foothold in the coastal region. One hundred years later, a Hungarian noble was crowned king, and in 1102 a political union was formed between Hungary and Croatia. Croatia maintained a degree of internal autonomy, retaining its traditional assembly, but its nobles pledged their allegiance to the Hungarian crown.

Over the next three centuries, Hungary and Venice fought twenty-one separate wars over Dalmatia, which finally was established as Venetian territory by the fifteenth century. By then, the Ottomans had begun their conquest of the Balkans. In 1493 many of the Croats' military leaders lost their lives to Ottoman invaders in the Battle of Krbava. After the Ottomans inflicted a bitter defeat on Hungary in the Battle of Mohács in 1526, they occupied large areas of both Hungary and Croatia. Following this defeat, Hungary and Croatia placed themselves under Habsburg rule for protection from further Ottoman attacks. However, these attacks continued throughout much of the century, killing thousands of Croats and destroying their villages. Austria created a buffer zone, called the military frontier, or *krajina*, along the Croatian border and settled it with Serbs and other non-Croatians, effectively creating a minority ethnic enclave whose existence was to have serious political repercussions for Croatia and its neighbors through the twentieth century. Also during this period, the Protestant Reformation reached Croatia and was embraced by many Croatian nobles. The first Croatian-language Bible was printed in 1562.

The Reformation was followed by the Counter Reformation in the seventeenth century. The Croatian assembly outlawed all religions except Catholicism. Jesuit efforts to spread the Catholic religion aided the cultural advancement of the region as they published books and opened new schools. The written version of Croatian that had been developed for liturgical use—called *Glagolitic*—made advances as a literary language.

Beginning in the latter part of the seventeenth century, the Habsburgs' grip on Croatia tightened with the inaugura-

tion of absolutist rule by Leopold I (1790–1865). By then, Ottoman power in the area was waning, and with the Treaty of Karlowitz in 1699 the Ottomans renounced their claims to Hungary and Croatia. Habsburg rule continued unchallenged through the eighteenth century, and the Austrian Habsburgs made increasing efforts to bring the German culture and language to the region. Germanization decrees in the Austrian half of the empire toward the end of the century led to a *Magyarization* campaign by the Hungarians in their half of the empire, who attempted to make Hungarian the official language of Croatia. At the end of the eighteenth century, how-

ever, another power was added to the growing list of Croatia's rulers: revolutionary France. At first the French left Croatia under Austrian rule, but then took it over from 1805 to the downfall of Napoleon Bonaparte (1769–1821) in 1814, when it reverted to the Austrians once again.

The nineteenth century saw the growth of Croatian nationalism, most notably in the Illyrianist movement started by linguist and author Ludevit Gaj (1809–72) in the 1830s. The cornerstone of the movement was Gaj's promotion of a South Slavic dialect used by both Croats and Serbs as a national language. Another important feature of Illyrianism was pan-Slavism: Gaj envisioned a movement that included Croats, Serbs, and other South Slavs. Although the Illyrianists failed to attract Serbs to their cause in significant numbers, they did inspire both literary and musical revivals within Croatia. Gaj founded a literary journal, *Danica* (Morning Star), and the Croatian language was used in the Croat national assembly for the first time in hundreds of years.

During the latter part of the nineteenth century, Croatia shifted between Austrian and Hungarian rule. The policies of both countries further fueled Croatian nationalist sentiment, especially during the administration of the Hungarian viceroy appointed at the end of the century. Early in the twentieth century, Croat nationalists formed a political party, and moderate Croats formed a coalition with Serb nationalists to further the South Slavic cause. During World War I, Serb, Croat, and Slovene leaders joined together to work for the creation of a South Slavic state in anticipation of the defeat of Austria-Hungary. On December 1, 1918, after this defeat had occurred, an independent Kingdom of Serbs, Croats and Slovenes was proclaimed with Prince Alexander Karadjordjevic (1888–1934) of Serbia as its king.

In spite of the cooperation shown by the South Slavic nationalist leaders in forming the new state, it inherited a troubling legacy of ethnic and religious rivalry, and the Croats and Slovenes found themselves eclipsed by the political clout of the Serbs, who dominated the country's government in the interwar years. Following the assassination of a Croat political leader, Stjepan Radic in 1928, King Alexander suspended the country's parliamentary government and instituted a dictatorial regime in 1929. Five years later, he was assassinated, and his cousin Paul took over as regent for Alexander's son, Peter. Paul was deposed after signing an agreement with Nazi Germany on the eve of World War II.

The Nazis invaded the country in 1941 and established a puppet government run by the head of the *Ustashe,* a Croatian fascist organization. Eventually the situation became a three-way struggle between the Ustashe, the Serbian *Chetnik* forces (whose motivation was primarily anti-Croat), and the Partisan forces, headed by communist leader Josip Broz Tito (1892–1980), which encompassed both Serbs and Croats whose primary goal was the defeat of fascism.

As the head of the Partisan forces, Tito gained sufficient credibility to install a communist government when the war ended in 1945. The Yugoslav monarchy was dissolved, and the country became the Federal People's Republic of Yugoslavia, with Croatia forming one of six federated republics. In spite of early support from the Soviet Union, Tito soon clashed with Soviet leader Josef Stalin (1879–1953), and Yugoslavia moved out of the Soviet sphere of influence to carve its own political and economic path. By maintaining a strong central government, Tito was able to keep the region's traditional ethnic rivalries from raging out of control, and his economic policies brought the country a high standard of living, of which Croatia was one of the major beneficiaries.

However, by the late 1960s and 1970s, nationalism and demands for liberal reforms were beginning to grow, and Tito's hard-won political stability unraveled following his death in 1980, the process hastened by a growing economic crisis.

In Croatia, a nationalist party, the Croat Democratic Union defeated the Communists in free elections in the spring of 1990, and its leader, Franjo Tudjman (b.1922), became the president of Croatia. After a vote the following year, the Croats withdrew from Yugoslavia and formed an independent state—their first in nearly a thousand years. The status of Croatia's Serb-populated Krajina region created a conflict in which the Yugoslavian federal government (now consisting solely of Serbia and Montenegro) became involved almost immediately. The result was months of continuous warfare until a UN-sponsored cease-fire at the beginning of 1992 and intermittent fighting until the 1995 Dayton Peace Accords, at a cost of thousands of lives and widespread property destruction. Serbia and Croatia renewed diplomatic relations in 1996, although Croatia was still receiving criticism from the international community for the treatment of its Serb minority. Franjo Tudjman, who had been president of the nation since independence, was reelected in 1997.

Timeline

4th century B.C.   Celts occupy Croatia

Celts occupy present-day Croatia.

3rd–1st centuries B.C.   Roman conquest

Over a two-hundred-year period, the Romans conquer Croatia, turning the region into three provinces: Illyricum, Dalmatia, and Pannonia. They found cities, build roads and bridges, introduce improved farming methods, and develop mining.

4th–6th centuries A.D.   Barbarian invasions

As the Roman Empire declines, Croatia is invaded by a variety of barbarian (i.e. non-Roman) groups, including the Lombards, Huns, Avars, and Slavs.

7th–8th centuries   South Slavs enter Croatia

The ancestors of today's Croatians migrate to the Balkans from the Carpathian Mountain region. They settle along the Adriatic coast in the former Roman provinces of Dalmatia and Pannonia, living in tribal groups. Together with the Serbs and other South Slavs, they become embroiled in the conflicts between the Byzantine Empire and the Avars, a Turkic tribe. They are organized by clans, one of which is called *Hrvat,* the name from which the modern word *Croat* is derived.

A.D. 800   Conquest by Charlemagne

The Frankish ruler Charlemagne, now crowned Holy Roman Emperor, conquers Dalmatia and Pannonia, and the inhabitants are converted to Roman Catholicism. The Franks rule through Croatian princes called *knezes.*

9th century   Croatia under native and Frankish rule

A succession of native rulers unify the Croatian tribes of Dalmatia and Pannonia. For parts of the century, they are under Frankish rule; some rebel successfully against the Franks for periods of time, but control of the region shifts repeatedly.

910   Tomislav unites the Croatian provinces

Tomislav becomes the first native Croatian ruler to unite the provinces of Pannonia and Dalmatia, ending Frankish rule of the region permanently.

925   Tomislav declares himself king

Tomislav declares himself king, becoming the first Croatian monarch. He is recognized by the pope and also establishes relations with the Byzantine church. Under his rule, Croatia gains power and expands to include parts of present-day Bosnia and Montenegro. The Croatian kingdom survives until the beginning of the eleventh century, when Venice assumes power in Dalmatia.

Rule by Venice and Hungary

1000   Venice invades

At the turn of the eleventh century, Venice invades Dalmatia, establishing its first foothold on the Adriatic coast. As the power of the Byzantine Empire declines, Croatia and Venice struggle for control of Dalmatia.

1058–74   Reign of Kresimir IV

With the accession of King Kresimir IV, who gains the throne with the help of the pope, Croatia regains control of Dalmatia.

1074–89   Reign of Zvonimir

Zvonimir, a noble from Pannonia, is crowned by a papal legate. His rule is weakened by the closeness of his relationship with the papacy.

1091   Hungarian noble is crowned king

After Zvonimir's death, a group of Croatian nobles choose their next king: Laszlo, a Hungarian and King Zvonimir's brother-in-law.

1094   Bishopric of Zagreb is founded

King Laszlo founds the bishopric of Zagreb, which later becomes the religious center of the Croatian church.

1102   Union of Hungary and Croatia

Through the *Pacta Conventa,* King Kalman of Hungary effects the personal union of Hungary and Croatia. In exchange for allegiance to Hungary, the Croatian nobles win privileges including exemption from taxes, and they are promised internal self-government as well. Croatia has its own governor, or *ban,* and its own assembly, called the *Sabor.* The Hungarians introduce coinage and new agricultural methods.

1115–1420   Prolonged struggle for control of Dalmatia

Upon the death of King Kalman, Venice launches new assaults on Dalmatia. Hungary and Venice continue to fight over Dalmatia and access to the Adriatic Sea for three hundred years, waging twenty-one separate wars over this period. In addition, Bosnia and Serbia make incursions into Dalmatian territory.

1222   Nobles' privileges are limited

Following a rebellion by town dwellers and lesser nobles, or gentry, King Endre II is forced to restrict some of the feudal privileges of the wealthiest landowners and expand the rights of lower classes.

1241–42   Mongol invasions

Dalmatia and Pannonia are devastated by Mongol invasions. During the fifteenth century, Croatia is also torn apart by warfare between its cities. Venice gains additional territory during this period.

1301–82   House of Anjou wins control

The House of Anjou wins control of Croatia and establishes an elective monarchy (i.e. each new king is elected by the nobility). This system brings Croatia a succession of kings from different royal houses in the course of the century. During this period, the Croatian nobles remain largely autonomous and are ruled primarily by their local *ban,* or viceroy, and their own assembly.

1342–82   King Ludovik wins back Dalmatia

King Ludovik I of Hungary reclaims Dalmatia from Venice.

15th–16th centuries   Literary renaissance in Dubrovnik

The free city of Dubrovnik, enjoying wealth and independence, is the site of a literary flowering inspired by the Italian Renaissance. Among its earliest well-known authors are poets Sisko Mencetic (1457–1527) and Djore Drzi (1461–1501), both of whom are influenced by the Italian poet and humanist Petrarch (1304–74). Dubrovnik's great comic playwright Marin Drzic (1510–67) flourishes in the sixteenth century, and the seventeenth century encompasses the career of Croatia's most celebrated author, Ivan Gundulic (see 1589).

15th century   Venice's power increases

By early in the century, Venice has won control of Dalmatia, which it retains until the Napoleonic invasion of 1797. Under Venetian rule, Dalmatia remains impoverished and backward. Venice uses up the region's natural resources, including timber, olive oil, figs, fish, and wine, but does nothing to develop roads or education. The autonomy of its inhabitants is restricted, a trade monopoly is imposed, and the region is prevented from developing its own industries so it will remain dependent on Venice. The beleaguered population, also subjected to a series of epidemics, is near starvation by the end of the eighteenth century. The only Dalmatian city to escape Venetian rule is Dubrovnik, which becomes powerful and wealthy by playing off European interests against the growing power of the Ottomans.

1493   Battle of Krbava

The Croats suffer a devastating defeat at the hands of the Ottomans at Krbava. Many of the Croat leaders are killed in battle, and the Croats' appeals to their European neighbors for aid in fighting the Ottomans go unheeded.

Habsburg Rule

16th century   Turkish conquest and Habsburg rule

Following the Battle of Mohács (see 1526), Hungarian and Croatian nobles elect a Habsburg prince, Ferdinand I (1503–64), as King of Hungary and Croatia with the understanding that he will fend off further Ottoman conquests. The Habsburgs restrict the power of the native Croat assembly and create a buffer zone in the south of Croatia along the Bosnian and Serb borders called the *Vojna Krajine* (military frontier, or border). They settle it with non-Croatians whom they bring in with promises of independence, free land, and religious freedom (most are Orthodox Christians).

Nevertheless, Ottoman attacks continue through the century, inflicting death and destruction on the region. Villages are burned, churches destroyed, and thousands of people killed. Thousands of Croats flee their homes, are deported, or are sold into slavery by the Turks. Turkish victories continue until the 1560s.

1526   Battle of Mohács

Ottoman forces inflict a devastating defeat on Hungary at the Battle of Mohács and occupy large portions of Hungary and Croatia.

1562   First Croatian Bible is printed

Many Croatian nobles embrace the Protestant Reformation in the mid-sixteenth century, and the first Croatian-language Bible is published.

1571   Turks are turned back at Lepanto

The Battle of Lepanto signals the first major Turkish defeat.

1573   Peasant rebellion

Croatian peasants stage an organized rebellion to protest price increases.

1589: January 8   Birth of author Ivan Gundulic

Ivan Gundulic, widely considered Dubrovnik's greatest author, is a poet and playwright best known for the epic poem *Osman,* which describes a Slavic victory over the Ottoman Turks. In addition to his literary career, Gundulic is also a government official in the free city of Dubrovnik. He dies in 1638.

17th century   Literary flowering in Dubrovnik

Influenced by the literary Renaissance in Italy, Dubrovnik develops its own provincial literature.

1606   University library is established

The National University Library is founded in Zagreb. In the 1990s it has 2.3 million volumes.

1609   Assembly issues anti-Protestant decree

The Counter Reformation, the effort of the Catholic Church to fight the spread of Protestantism, makes inroads into Croatia in the sixteenth century. The Sabor, the Croatian assembly, issues a decree outlawing all religions except Roman Catholicism. As the Jesuits, a Catholic religious order created as part of this Catholic movement, organize schools and publish books, the Counter Reformation leads to the growth of educational and literary activities and aids in the development of Croatian as a literary language.

1626   Epic poem *Osman* is composed

The epic poem *Osman* is written by Ivan Gundulic (see 1589). It describes a Polish victory over the Ottoman sultan Osman II at the Battle of Khotin in 1621.

1638: December 8   Death of Ivan Gundulic

Poet and dramatist Ivan Gundulic dies in Dubrovnik.

Late 17th century   Habsburgs institute absolutist rule

Leopold I establishes an absolutist monarchy that eclipses the power of the nobility. Under his rule, Austria wrests Croatia up to the Sava River from the Ottomans.

1669   University of Zagreb is founded

The University of Zagreb, Croatia's first university, is established.

1683   Siege of Vienna

The Ottoman Turks are routed at the Siege of Vienna, which begins a series of Turkish defeats. This marks the second time the Ottomans are turned back at the gates of Vienna. Their previous defeat takes place in 1529.

1693   First private museum is opened

Author and philosopher Pavao Ritter Vitezovic opens Croatia's first private art museum in his home in Zagreb.

1699   Treaty of Karlowitz

With the Treaty of Karlowitz (also called the Treaty of Sremski Karlovci), the Ottoman Turks renounce nearly all territorial claims in Hungary and Croatia.

1715–90   Habsburg rule unchallenged

The Austrian Habsburgs retain their hold on Hungary and Croatia. They bring German military officers into the military frontier (Krajina) area and become actively involved in the economy of the region.

1717   Habsburgs try to gain control of the coast

The Habsburg monarchy challenges Venetian control of the Adriatic coast.

1780–90   Decrees implement Germanization

The Austrian monarchy attempts to strengthen its presence in Hungary and Croatia by expanding German culture and language into these areas. Emperor Joseph II issues decrees making German the official language.

1790   Hungarians respond with Magyarization

The Hungarians reject the Habsburgs' attempt to spread German culture to their regions and institute a Magyarization policy that makes Hungarian the official language of Hungary and territories it controls, including Croatia, whose official language has previously been Latin.

1796   Napoleonic invasion

French military leader Napoleon Bonaparte leads French revolutionary forces into the Italian states, ending the power of the Venetian Republic and turning Dalmatia over to Austria.

1805   French take over Dalmatia

Following the Battle of Austerlitz, Austria cedes Dalmatia to France, which reconstitutes Dalmatia, western Croatia, and the free city of Dubrovnik as the Illyrian Provinces and undertakes extensive development of the area.

1814–15   Congress of Vienna

Following the downfall of Napoleon Bonaparte (1775–1821), Croatia reverts to Austrian rule.

1816   Kingdom of Illyria is formed

Hungary renames France's former Illyrian Provinces, including Croatia, the Kingdom of Illyria, but the name is used for administrative purposes only, and no real independence comes with it.

The Growth of Nationalism

1830s   Language-based nationalist movement influences politics and the arts

Croatians, led by nationalist writer Ludevit Gaj (1809–72), develop their own nationalist movement to counteract Hungary's Magyarization efforts. To give their movement greater credibility they attempt to involve other South Slavs, especially Serbs and Slovenes, and they promote the adoption of the Zagreb dialect, used by both Serbs and Croats within the Habsburg Empire, as a national language. Gaj devises a written version of the language that uses a Latin-based script. A member of the Croatian assembly presents an address in Croatian for the first time in centuries. The official language of Croatia has long been Latin. For everyday activities, northern Croatians mostly speak German or Hungarian, while upper-class Croats in Dalmatia speak Italian. Politically, the movement in unsuccessful in attracting other Slavs, but it has a linguistic impact.

Gaj's language reforms also inspire a literary renaissance in Croatia centering on the periodical *Danica* (Morning Star), which is founded by Gaj. Important writers of this period include poets Ivan Mazuranic (1814–90) and Petar Preradovic (1818–72). Musical composition of the period also centers on Illyrianism, which is associated with Romantic nationalism and the use of folk material. Opera is the favored genre for the promotion of a national consciousness. The musical styles developed during this period have a resurgence in the early twentieth century.

1834   Croatian-language newspaper is established

In connection with the Illyrian movement, a Croatian-language newspaper is established.

1846   Natural history museum is founded

The Croatian Natural History Museum is inaugurated in Zagreb.

1847   Assembly votes on national language

The Sabor, the Croatian assembly, votes to declare the Zagreb dialect, under the name Illyrian, the national language in spite of Hungarian disapproval.

1848 Croats support Austria against Hungary

During the revolutionary activities that sweep Europe in 1848, Hungary revolts against Austrian control. The Croats, fearing that they will be subject to Hungarian rule, ally themselves with Austria, hoping that the South Slavic provinces will later be united under Austrian rule. However, except for the abolition of serfdom, conditions only become more repressive under Austria, which imposes absolutist rule and Germanization and suspends the Croatian constitution. Nevertheless, bourgeois Croats gain power in relation to the aristocracy, and for the first time some rise to the position of *ban*. As the economic power of the aristocrats wanes, many of their landholdings are divided up and sold to members of the middle class and even the peasantry.

1855: August 4 Birth of painter Vlaho Bukovac (1855–1922)

Bukovac, the founder of an influential school of painting in Zagreb, studies in Paris at the Ecole des Beaux-Arts and exhibits his paintings at the Paris Salon. He remains in Paris until 1893, when he returns to Croatia, where he is at the center of a group of young painters called the colorful school, who gain international recognition for their use of bright colors and unorthodox techniques. He is among the founders of the Society of Croatian Artists and is instrumental in the construction of Zagreb's Art Pavilion in 1898. Among his best-known compositions is the large painting *Gundulic's Dream* (1894), whose title refers to celebrated Croatian author Ivan Gundulic (see 1589). Bukovac is appointed professor at the Prague Academy of Fine Arts in 1903. He dies in Prague in 1922.

1867 Formation of Austria-Hungary

With the formation of the dual-monarchy Austro-Hungarian Empire, Croatia comes under Hungarian jurisdiction, while Dalmatia is still administered by Austria. The Croatians achieve limited autonomy but need approval from Austria and Hungary on financial and other matters.

1867 South Slav Academy of Arts and Sciences is founded

Bishop Josip Strossmayer founds the South Slav Academy of Arts and Sciences to further the cause of South Slavic nationalism.

1879 Arts and crafts museum is founded

The Museum of Arts and Crafts is established in Zagreb.

1881 The Krajina is merged into Croatia

Austria incorporates the Krajina (the southern military border region) into Croatia. Since the region is populated by Serbs, this action raises the total percentage of Serbs in Croatia to twenty-five percent.

1883–1903 Appointed official imposes Magyarization

When a troubled economy fuels peasant uprisings and other forms of civil unrest, the Hungarian government appoints Karoly Khuen-Hedervary of Slavonia as *ban* (viceroy) of Croatia to restore order. Khuen-Hedervary attempts to impose Hungarian culture on the Croatians by exploiting tension between the Serbs and Croats. His policies provoke opposition and spur the spread of Croatian nationalism.

The Croat nationalists split into two camps. One, under the leadership of Bishop Josip Juraf Strossmayer, favors unity among the South Slavic peoples (*Jugoslvenstvo*) within the Austro-Hungarian Empire. The other, led by Ante Starcevic and the Party of Rights, favors a completely independent, separate, and expanded Greater Croatia.

1884 Old Masters' gallery is founded

Bishop Josip Juraj Strossmayer founds a gallery for the exhibition of paintings by the Old Masters (later renamed the Strossmayer Gallery of Old Masters).

1892: May 25 Birth Josip Broz Tito

Josip Broz, who later adopts the name "Tito," is born in Kumrovec, Croatia, to a family with fifteen children. He becomes a socialist by the age of eighteen. He fights in the Austro-Hungarian army in World War I and is taken prisoner by the Russians but escapes. During the Bolshevik Revolution he joins the Communist Party and serves in the Red Army. During the 1920s and 1930s, he becomes a revolutionary and rises through the ranks of the Communist Party. By 1937 he is the secretary general of the Yugoslavian party.

During World War II, Tito organizes the Partisan guerrilla resistance to the pro-Nazi Croatian Ustashe government. By the end of the war, he is the country's foremost leader. In 1945 he organizes a communist government and is elected president of the country. Although at first allied to the Soviet Union, Tito refuses to follow all of its orders which causes a split with Soviet leader Josef Stalin (1879–1953) and denunciation by the Soviets by 1948. After this, Yugoslavia pursues its own communist path, which includes decentralization of government and greater personal freedoms than are found elsewhere in the communist bloc. Tito is elected president for life in 1963. In the 1970s he paves the way for his successors by instituting a three-member "collective presidency" and increasing the autonomy of the individual republics. In spite of his precautions, however, Yugoslav stability unravels following his death in 1980.

1893: July 7 Birth of author Miroslav Krleza (1893–1981)

Playwright and novelist Miroslav Krleza, born in Zagreb, becomes one of Croatia's foremost twentieth-century literary figures. He is also the founder of a left-wing literary review following World War I, and later serves as president of the

Yugoslav Writers' Union. Among his works are the novel *Povratak Filipa Latinovicza* (The Return of Philip Latinovicz; 1932); the dramatic trilogy *Glembajevi* (The Glembaj Family; 1932); and the short story collection *Hrvatski bog Mars* (The Croatian God Mars; 1936). Krleza dies in 1981.

1903   Crackdown by Hungary

Croat demonstrations and civil unrest bring down the administration of Khuen-Hedervary and lead to a crackdown on Croatian freedoms. Hungary restricts the freedom of the Croatian press and denies Croatia's requests for financial autonomy.

1904   Croat Peasant Party is formed

Ante and Stjepan Radic form the Croat Peasant Party, which is influential during the early years of Yugoslavian statehood.

1905   Serbs and Croats form a coalition

Political leaders from Dalmatia ally themselves with moderate Serbs and Hungarians to form a new political coalition.

1905   Modern art gallery opens

The Gallery of Modern Art in Zagreb is launched, with more than 2,000 eighteenth- and nineteenth-century works.

1908   Serbs and moderate Croats unite

The coalition of Serbs and moderate Croats wins a majority of seats in the Sabor. The new government denounces Austria's annexation of Bosnia and Herzegovina, which has been placed under Austrian administration by the Treaty of Berlin in 1878. A new Austrian governor brings trumped-up charges of treason against Croatian Serb leaders.

World War I

1914: June 28   Bosnian assassinates Austrian archduke, triggering World War

Austrian archduke Franz Ferdinand and his wife, the archduchess, are assassinated by a Bosnian student, Gavrilo Princip, during a state visit to Sarajevo. Princip is associated with the nationalist secret society, the Black Hand. Secret societies formed by Serbian radicals have already made several assassination attempts on Austrian officials.

1914: July 23   Ultimatum by Austria-Hungary

Austria Hungary threatens to attack Serbia unless it can participate in bringing the archduke's killer to justice, and Serbia agrees to ban secret societies.

1914: July 28   War is declared

Austria-Hungary declares war on Serbia. Together, Austria and its allies—Germany and the Ottoman Empire (and Bulgaria after 1915)—are known as the Central Powers. Serbia allies itself with France, Britain, and Russia, whose alliance is known as the Triple Entente. The war pits ethnic Serbs, Croats, and Slovenians in territories ruled by Austria Hungary—including Bosnia— against their brethren in Serbia and Montenegro. In Bosnia, Serbian guerrilla fighters called *Chetniks* attack Muslim landowners. In spite of ethnic and religious division, the idea of an independent state uniting the South Slavs gains force during this period.

1915   Treaty of London

As an inducement to join the Triple Entente, the Entente powers secretly promise Italy territory on the eastern Adriatic coast.

1915   Yugoslav committee-in-exile is formed

Serb, Croat, and Slovene political leaders in exile form a Yugoslav committee to work toward the creation of an independent state for the South Slavs.

1917   Serbian government supports Yugoslav committee

The Serbian government-in-exile agrees to support the creation of a Yugoslav state in the form of a constitutional monarchy ruled by the Karadjordjevic dynasty.

1918: October   Austro-Hungarian Empire collapses

As Austria-Hungary collapses, the Croatian assembly (the *Sabor*) votes to end all political ties to Austria and Hungary. In their place, it recognizes the authority of a joint Serbian-Slovenian-Croatian council.

1918: November   End of World War I

By the time World War I ends in victory for the Triple Entente, the South Slavic lands have been liberated from Austro-Hungarian control.

The Kingdom of Serbs, Croats, and Slovenes

1918: December 1   South Slavs win independence

Representatives of the South Slavic peoples meet with Regent Alexander Karadjordjevic, prince of Serbia, and proclaim an independent Kingdom of Serbs, Croats, and Slovenes. Alexander becomes the country's ruler, ascending the throne as Alexander I. (Ten years later he renames the country Yugoslavia.)

The new nation is troubled politically and economically from the outset. It inherits a legacy of ethnic rivalry, and the balance of power within the new government favors the Serbs. The capital, Belgrade, is located in Serbia, and Croatia is under-represented in the new country's balance of power. Croatia's assembly, the Sabor, is disbanded, and the Croats are under-represented in the legislature of the new state. In addition, currency reforms favor the Serbs at the expense of the Croats. Serbs and Croats also clash over the issue of government centralization versus federalism. Economically, the kingdom is faced with wartime damage, heavy debt, labor shortages, and a pressing need for land reform created by centuries of feudalism.

1921–23 Art journal is published

The avant-garde journal *Zenit* debuts in Zagreb. (It is later moved to Belgrade.)

1921–22 Little Entente is formed

At the instigation of France, the South Slavic kingdom forms a military alliance with Romania, Czechoslovakia, and France to prevent attempts at treaty revision by the defeated Austrians, Bulgarians, Germans, and Hungarians. The alliance is known as the Little Entente.

1921 Arts academy is founded

Croatia's Academy of Arts in Zagreb becomes the first such institution in the former Yugoslavia.

1921 Constitution is adopted

A constituent assembly adopts a constitution for the country. The Serb position prevails, and a strong centralized government is created.

1922 Birth of Franjo Tudjman

Franjo Tudjman, first president of the modern independent state of Croatia, is born in Zagorje. He begins a military career during World War II, serving until 1961, when he becomes a historian. In the 1970s and 1980s he is active in the Croat nationalist cause and is jailed twice. In the 1980s he founds the Croatian Democratic Union, which becomes the nation's ruling party in the next decade.

After Croatia votes to withdraw from Yugoslavia in 1990, its parliament elects Tudjman as president. Two years later, he is elected by popular vote. In 1995 Tudjman, along with the presidents of Bosnia and Serbia, signs the Dayton Peace Accords, although he has since been criticized for not abiding by its terms, especially in the area of refugee repatriation. At the close of his five-year term, Tudjman is reelected although he is suffering from cancer.

1922: April 23 Death of Ivan Bukovac

Painter Ivan Bukovac, leader of the turn-of-the-century Zagreb colorful school of painting dies in Prague. (See 1855.)

1924 Natural science museum is founded

The Museum of Natural Sciences is established in Split.

1925 Croat Peasant Party is outlawed

The government outlaws the Croat Peasant Party, which opposes provisions of the constitution enacted in 1921. The party's leader, Stjepan Radic, is imprisoned and forced to agree to a coalition with the ruling Serbian party. However, he later resumes his dissident stance.

1928 Opposition leader Radic is assassinated

Croat Peasant Party founder and leader Stjepan Radic and several other Croats are assassinated by a Serb representative at a session of the Yugoslavian parliament.

1929 Progressive arts group is founded

The Earth Group, composed of progressive artists and architects, is established in Zagreb and continues until 1935, when it is banned by the government.

1929: January 6 Royal dictatorship is established

Using the assassination of Croat leader Stjepan Radic as justification for tighter government control, King Alexander imposes a royal dictatorship, abolishing the country's parliamentary government and outlawing political parties. He also renames the country Yugoslavia (Land of the South Slavs).

Ante Pavelic forms the Croatian Liberation Organization, or *Ustashe,* to oppose the dictatorship of King Alexander. He later flees to Italy, where his organization becomes allied with fascist leader Benito Mussolini (1883–1945).

1931 Constitution is adopted

The autocratic rule of King Alexander is modified by adoption of a constitution, and political parties are legalized. However, a number of other freedoms are still restricted.

1931 Global economic slump hits Yugoslavia

Yugoslavia feel the effects of the worldwide economic depression, which leads to bankruptcies and unemployment. The crisis is worsened by weather conditions that produce famine in rural areas.

1934 Balkan Entente is formed

Yugoslavia joins Turkey, Greece, and Romania in a defense alliance known as the Balkan Entente. This treaty is aimed at stifling Bulgarian attempts at revising the World War I peace settlement.

1934: October King Alexander is assassinated

King Alexander is murdered in Marseilles, France, by a Bulgarian assassin working for Ustashe, the Croatian pro-fascist liberation organization. His death brings on fears that Yugoslavia will collapse. Alexander's son, Petar II, becomes the country's regent. Three officials are appointed to rule for him while he is still a minor.

1939   Part of Bosnia ceded to Croatia

In an agreement (*Sporazum*) negotiated between Serb and Croat leaders, parts of Bosnia and Herzegovina with Croatian majorities are ceded to the newly formed autonomous Croatian territory, leading Bosnian Serbs to consider the rest of the region theirs to control. This reorganization of Bosnia is part of the effort to create regional autonomy for non-Serbs. Bosnia's Muslims are not to be considered a separate group until after the Second World War.

World War II

1941: March 25   Pact is signed with Germany

Under military pressure and surrounded by pro-Nazi countries, Yugoslavia's government agrees to join the Tripartite Pact (the Axis alliance concluded among Germany, Italy, and Japan) in return for German guarantees of nonaggression.

1941: March 27   Yugoslavian government is overthrown

Because of its cooperation with Nazi Germany, the Yugoslavian government is overthrown in a coup led by Yugoslav Air Force officers, and sixteen-year-old Petar II, the regent and son of the slain king Alexander, is declared king. In spontaneous demonstrations, the populace expresses its hostility toward the Nazis and their allies.

1941: April 6   Germany bombs and invades Yugoslavia

The Yugoslavian capital of Belgrade is bombed by the German Luftwaffe (air force), and ground forces invade the country. The government flees and the military surrenders unconditionally. Yugoslavia is divided among the Axis powers, and Croatia is placed under the rule of the Independent State of Croatia (NDH), a puppet state controlled by the Nazis and led by Ante Pavelic, head of the Ustashe.

In accord with the Nazis' theories of racial purity, the Ustashe declares Croatians to be racially superior Aryans, while the Serbs, being Slavs, are singled out for elimination through deportation or extermination, together with Jews and gypsies. The atrocities committed by the Ustashe during the war add yet more fuel to the hostility between Serbs and Croats and will be repeatedly invoked by the Serbs during the Serb-Croat war and the Bosnian civil war of the 1990s.

1941: July   Partisans stage anti-German revolts

Partisans (anti-Nazi freedom fighters) organized by long-time communist leader Josip Broz (1892–1980), popularly known as Tito, carry out rebellions in the countryside and remain active throughout the war in spite of German retaliation. Although most Partisans are Serbs, all ethnic groups are represented (Tito himself is Croatian-born). The Ustashe also faces armed opposition by Serbian Chetnik forces, whose primary motive is ethnic antagonism and the opportunity to massacre Croats. They also murder thousands of Bosnian Muslims.

1943: September   Italy surrenders

The surrender of Italy bolsters the Partisan effort by providing access to the coast as well as a supply of arms and a supply route. The Partisans win control of much of the country.

1944: October 20   Red army liberates Belgrade

The Soviet army, aided by Partisan forces, marches into Belgrade and liberates Yugoslavia. Some 1.7 million Yugoslavians have died since the war began—more than half at the hand of other Yugoslavs. Cities are left in ruins, and the countryside is also devastated.

Yugoslavia under Tito

1945: March 7   Provisional government is installed

A communist-dominated provisional government, led by Tito, takes office.

1945: November 29   Yugoslavia's monarchy is dissolved

The new Yugoslav parliament meets for the first time. The monarchy is officially dissolved, and the country is named the Federal People's Republic of Yugoslavia (later renamed the Socialist Federal Republic of Yugoslavia). Croatia becomes one of six republics in a Soviet-style federation with a strong centralized government. In an effort to dilute the power of Serbia, Macedonia and Montenegro are recognized as republics, Kosovo is established as an autonomous Albanian province, and Vojvodina becomes an ethnically mixed autonomous province. Tito also keeps ethnic rivalries in check by setting up a centralized, single-party communist government and suppressing political opposition and organized religion.

1945–66   Serbs dominate Bosnia

In spite of the measures taken by Tito to neutralize the power of Yugoslavia's ethnic constituencies, Serbs come to dominate Yugoslavian politics in the postwar years.

1948   Yugoslav-Soviet split

Following World War II, Yugoslavia allies itself closely with the Soviet Union. By 1948, however, growing tensions, begun with disagreement over the management of joint enterprise, create a rift between the two countries. Unlike the heads of other Eastern European satellites of the USSR, Tito does not owe his influence or position to the Soviets and will not allow them to dictate policy for Yugoslavia. He becomes the only Eastern European communist leader to break with Stalin and remain in power. Trade with the Soviet Union and other communist bloc nations is reduced, and Tito is forced to turn to the West for new trade partners.

1948   Folk art institute is founded

The Institute for Folk Arts is established in Zagreb to promote folk music research and publication and collect manuscripts and recordings.

1950s   "Self-management" economic system develops

Yugoslavia develops its communist economy through a system in which enterprises are locally managed by workers' councils, and economic goals are first formulated at the local level and coordinated centrally. As in other communist countries, these goals are organized into five-year plans.

1953   Constitution incorporates workers' councils

The constitution is modified to authorize the creation of workers' councils to run state-owned enterprises.

1960s   Croatian nationalism grows

Resentment of Serbian political, economic, and cultural domination fuels the growth of Croatian nationalism. One of its expressions is a decade-long controversy over the status of the Croatian language. Although Serbs and Croats (as well as Bosnian Muslims) speak the same language, Serbo-Croatian, the Orthodox Serbs use the Slavic Cyrillic alphabet while the Catholic Croats (and Bosnian Muslims) use the Latin alphabet. In addition to the differences in alphabet, regional dialects differ. Nationalists seize upon these differences to argue that theirs is a separate language. By the 1990s, references are made to the "Bosnian" language, distinct from Serbo-Croatian.

1960s   Zagreb School animators gain international acclaim

The Zagreb Film Studio, popularly known as the Zagreb School, wins international recognition for its animated films. One of its co-founders, Dusan Vukotic, becomes the first director to receive an Academy Award in the animation category for a film produced outside the United States. The Zagreb Studio's development of reduced animation processes that create more radical, less lifelike forms, has an influence on animators around the world. Vukotic also does groundbreaking work combining animation with live action film.

1961   Contemporary music festival is established

The prestigious Zagreb Biennale of Contemporary Music is inaugurated and becomes an important event in the Yugoslavian musical scene, introducing the Yugoslav public to European avant garde music and influencing a new generation of composers, who move away from folk-inspired music and toward serialism and other newer compositional techniques. The Biennale is instrumental in resurrecting Zagreb as a leading musical center.

1961: September   Conference of nonaligned nations

A conference of nonaligned nations held in Belgrade confirms Yugoslavia's international leadership position in the group of nations that have proclaimed their neutrality in the Cold War, including India and Egypt.

1965   Market socialism is introduced

Yugoslavian industry grows rapidly in the 1950s, but imports still exceed exports by a wide margin, and some industrial inefficiency remains. In response to an economic crisis in the early 1960s, the government modifies its economic policy, eliminating price controls and export subsidies while cutting import duties.

1966   Tito purges top Serbian leader

Because the Serbs have once again risen to the position of political dominance they enjoyed following World War I, Tito dismisses Serbian leader Aleksandar Rankovic and disbands Rankovic's powerful secret police.

1967–71   Period of liberalization and growing nationalism in Croatia

Liberals dominate the Croat leadership and institute political reforms similar to those passed during Czechoslovakia's Prague Spring, before the Soviet invasion of that country.

1967–68   Yugoslavian republics gain more power

Constitutional reforms expand the power of the individual Yugoslavian republics relative to the central government.

1968   Tito condemns Soviet invasion of Czechoslovakia

Yugoslav leader Tito condemns the USSR's forcible halt to Czech liberalization under Alexander Dubcek, straining Yugoslav relations with the Soviets. Tito prepares the Yugoslavian military to resist a possible Soviet invasion.

1971   Croatian nationalism antagonizes Serbia

Those in the vanguard of the Croatian leadership begin demanding even greater economic and political autonomy, including the establishment of a Croatian army, the restoration of the traditional national assembly, and even complete independence. The Serbs begin urging Tito to restrain the growing nationalist fervor that grows in Croatia.

1971: November   University students strike

The Croatian government does nothing to restrain mass student demonstrations.

1971: December   Tito imposes crackdown on liberal Croats

Tito purges Croatia's liberal communist leadership. Thousands of officials are arrested and either jailed or expelled from the party. Included among them is future Croatian president Franjo Tudjman (b. 1922). The purges continue for over a year and spread to the other Yugoslavian republics also. They expand into a wider crackdown that includes press censorship, dissident arrests, and pressure on university professors to conform with the party line.

1974   New constitution is adopted

A new constitution gives the individual republics and autonomous provinces a greater role in governing themselves. The role of the federal government is limited to foreign policy, defense matters, and certain economic powers. The constitution also lays out the structure of the collective presidency that Tito has created to assure the continuity of the Yugoslav federation after his death. It is to have two members from each of the country's three main ethnic constituencies and one additional member.

1980s   Politics and the economy pull Yugoslavia apart

Following the death of Marshal Tito, the stability achieved over decades is threatened by age-old ethnic rivalries and affected by widespread economic problems in the Soviet Union and Eastern Europe, as well as Yugoslavia's inability to service its excessive foreign debt. The country is beset by inflation, labor strikes, food shortages, and financial scandals, as tensions grow between the individual republics. Serbian nationalism, which had been kept in check in Bosnia during the Tito years, gains strength in spite of government crackdowns.

1980: May 4   Death of Marshal Tito

Josip Broz Tito, who had greater success in unifying the South Slavs than any other leader, dies. He is widely mourned at home, and forty-nine foreign dignitaries attend his funeral.

1981: December 29   Death of author Miroslav Krleza

Krleza, one of the dominant figures of twentieth-century Croatian literature, dies in Zagreb.

1984   Winter Olympics are held in Sarajevo

The Winter Olympic Games are held in Sarajevo, the capital of Bosnia and Herzegovina.

1990: March–April   Communists defeated in free elections

The nationalist Croat Democratic Union (HDZ) edges out the Communist party in free elections, forty percent to thirty percent, and its leader, Franjo Tudjman, becomes president of Croatia.

1990: December   New constitution is enacted

The Croats enact a new constitution weakening the rights of Croatia's Serb minority of over half a million. Many Serbs are fired from government jobs.

1991   Stock exchange is launched

The Zagreb Stock Exchange is inaugurated.

Croatian Independence

1991: May   Croats choose independence

An overwhelming ninety-three percent majority of Croats vote for independence from Yugoslavia.

1991: June 26   Independence is proclaimed

Croatia officially proclaims its independence. The Serb-populated Krajina region declares its independence from Croatia and hostilities begin, with the Serb militia supported by the Serb-dominated Yugoslav Federal Army of 180,000.

1991: August   Yugoslav forces attack Vukovar

When Croatia sets up a blockade of Yugoslav military installations, Yugoslav forces besiege the city of Vukovar.

1991: September   Arms embargo imposed by UN

Yugoslav forces control at least one-fourth of Croatia. Due to the increasingly volatile situation in the Balkans, the United Nations declares an embargo on selling arms to all the former Yugoslav republics.

1991: October   Dubrovnik comes under attack

Dubrovnik is attacked by the Yugoslav army as well as forces from Montenegro. An attempt to assassinate Croatian President Franjo Tudjman by shelling the presidential palace in Zagreb fails.

1991: November 19   Vukovar falls

The city of Vukovar falls after a three-month siege. It is stormed by 30,000 Yugoslav troops with 600 tanks.

1991: December   UN protection force is organized

An estimated 10,000 Croatians have died since June. Following negotiations with UN special envoy Cyrus Vance, Serb leaders agree to the stationing of a UN Protection Force (UNPROFOR) of 14,000 peacekeeping troops in Croatia.

1992   National bank is established

The National Bank of Croatia is established with the authority to issue currency and regulate the banking industry.

1992: January   Yugoslav forces withdraw from Croatia

Under the terms of a UN-sponsored cease-fire, Yugoslavia withdraws its federal forces from Croatia, and the cease-fire mostly holds. By this time, however, the city of Vukovar has been almost totally destroyed. The UN peace plan calls for local Serb militias to disarm, refugees to be repatriated, and the return of the Krajina to Croatia.

1992: May   Croatia joins the United Nations

Croatia becomes a member of the United Nations.

1992: August   Tudjman reelected

Franjo Tudjman is reelected to the presidency in Croatia's first direct presidential election.

1993: January   Croatian offensive in the Krajina

Croatia launches an offensive in the Serb-occupied Krajina region to protect its position in Dalmatia. The offensive is successful, but the Serbs proclaim their own Republic of Serbian Krajina.

1993: June   Krajina Serbs join the Bosnian Serbs

The Serbs of the Krajina vote to ally themselves with the Bosnian Serbs in the Bosnian civil war and incorporate the region into Greater Serbia.

1994: March   Cease-fire agreement calms the Krajina

Although the terms of the original Vance peace plan have not been met, the Krajina Serbs agree to a cease-fire that brings about a temporary cessation of hostilities between Serbia and Croatia.

1995: May   Croatia recaptures Slavonian territory

Croatians recapture territory in western Slavonia under the rule of the Krajina Serbs. The Serbs respond by shelling Croatian cities, including Zagreb. However, their allies in Serbia itself are now no longer actively assisting them.

1995: August 4   Croats attack Knin

Croatian forces attack the capital of the Krajina Serbs, Knin, routing Serb military forces and thousands of civilians.

Attacks on Serbian villages, designed to drive all Serbs out of the region, continue for months and lead to accusations of human rights abuses by the international community.

1995: December   Signing of the Dayton Peace Accords

The Dayton Peace Accords are signed in Paris. The agreement recognizes Croatia's traditional borders and mandates the return of eastern Slavonia.

1996: May 30   Council of Europe postpones admitting Croatia

The council of Europe delays its planned admission of Croatia because of the country's treatment of its minority Serb population.

1996: August 7   Tudjman and Milosevic meet in Athens

Croatian president Franjo Tudjman and Serbian president Slobodan Milosevic (b. 1941) meet in Athens, Greece, and agree to establish diplomatic relations for the first time since the beginning of hostilities between their countries in the spring, 1991.

1996: September 9   Croatia and Serbia renew diplomatic ties

Croatia renews diplomatic ties with Serbia following an agreement between the presidents of the two countries.

1996: September 20   UN criticizes Croat treatment of Serb minority

The United Nations Security Council criticizes Croatia for its unequal treatment of Croat and Serb refugees, claiming it has discouraged Serbs from returning to their homes, while Croats receive preferential treatment.

1997: June 19   Tudjman is reelected

Croatian President Franjo Tudjman is reelected for the second time with over sixty percent of the vote. However, international monitors declare the vote marred by biased news coverage in which opposition candidates are called enemies of the state.

1998: January   East Slavonia is returned to Croatia

East Slavonia reverts to Croatian rule.

1998: July 11   Croatia finishes third in World Cup soccer tournament

In its World Cup debut at the games in Paris, Croatia finishes third, defeating the Netherlands 2–1.

1998: October 3   Pope beatifies World War II-era archbishop

During a three-day visit to Croatia, Pope John Paul II beatifies Alojzije Stepinac, who is archbishop of Zagreb during World War II. (Beatification is a preliminary step toward sainthood.) Stepinac is celebrated for his resistance to communism, but the beatification is controversial because of his support of the fascist Ustashe organization that persecuted Serbs, Jews, and other minorities. Stepinac has died in 1961 under house arrest by the Communists for his role in the war.

Bibliography

Cuvalo, Ante. *The Croatian National Movement, 1966–72.* New York: Columbia University Press, 1990.

Despalatovic, E.M. *Ljudevit Gaj and the Illyrian Movement.* New York: Columbia University Press, 1975.

Glenny, Michael. *The Fall of Yugoslavia: The Third Balkan War.* New York: Penguin, 1992.

Irvine, Jill A. *The Croat Question: Partisan Politics in the Formation of the Yugoslav Socialist State.* Boulder, Colo.: Westview Press, 1993.

Kadic, Ante. *From Croatian Renaissance to Yugoslav Socialism.* The Hague: Mouton, 1969.

Omrcanin, Ivo. *Diplomatic and Political History of Croatia.* Philadelphia: Dorrance, 1972.

———. *Croatian Spring.* Philadelphia: Dorrance, 1976.

Tanner, Marcus. *Croatia: A Nation Forged in War.* New Haven, CT: Yale University Press, 1997.

Tudjman, Franjo. *Horrors of War: Historical Reality and Philosophy.* Trans. Katarina Mijatovic. New York: M. Evans, 1996.

Vladovich, Simon. *Croatia: The Making of a Nation.* Oklahoma City, Okla.: Vladovich International Pub., 1995.

Czech Republic

Introduction

The Czech Republic has only been in existence since 1993, when the former Czechoslovakia split into two countries following the demise of communism in Eastern Europe. One of the world's most heavily industrialized nations, it has pursued a policy of rapid conversion to a free-market economy. Although their country is young, the Czechs have a venerable cultural history: Charles University in Prague, founded in 1348, is one of the oldest universities in Europe, and the Czech Republic has rich literary and musical traditions. The importance of learning and the arts in Czech life is reflected in the background of its political heroes. Tomas Masaryk (1850–1937), a co-founder of Czechoslovakia and its first president, was a respected author and a professor of philosophy, while the Czech Republic's current president, Vaclav Havel (b. 1936), is an internationally acclaimed playwright.

Bordered by Austria, Germany, Poland, and the Slovak Republic, the Czech Republic has an area of 30,387 square miles (78,703 square kilometers) and consists of two major regions: the mountain-ringed hills and plains of Bohemia to the west, and to the east the Moravian lowlands, broken up by a series of deep ridges running southwest to northeast. As of the late 1990s, the population of the Czech republic was estimated at 10.3 million, and its capital is the historic city of Prague.

Slavic settlers arrived in the present-day Czech Republic in the fifth century A.D., settling in agricultural villages. The first unified state in the region was founded in 625, but it was short-lived. In the ninth century the Great Moravian Empire was founded in the region and became the first political entity to include both Czechs and Slovaks, as well as Poles and Hungarians. It was broken up in the tenth century by Magyar invasions, which once again divided the Czechs and Slovaks, thus playing an important role in the two groups' development of separate ethnic identities. The Czechs came under the the Bohemian Kingdom, ruled by the Premyslid dynasty, that emerged to the west and eventually conquered Moravia as well (thus occupying roughly the same territory as the present-day Czech Republic). The Bohemian kingdom lasted as an autonomous entity for hundreds of years, reaching its peak in the fourteenth century, by which time it was ruled by Charles IV of Luxembourg. This period, known as the Czech "Golden Age," was one of political strength and cultural achievement. The capital city of Prague was renovated and expanded, and, in 1348, Charles University—one of the oldest universities in Europe—was founded.

In the fifteenth century, the Protestant Reformation came to the region in the form of the Hussite movement, led by clergyman Jan Hus (c. 1369–1415), whose calls for religious reform drew many followers but brought condemnation by the church when they became too radical for it to tolerate. Hus was burned at the stake in 1415 for refusing to recant his beliefs and has been a revered national figure ever since.

The Austrian house of Habsburg gained control of the Czech lands in 1526 when Archduke Ferdinand became the king of Bohemia (Habsburg territory was later expanded to include Moravia and Slovakia as well). Habsburg rule brought with it religious tensions between the Catholic rulers and their largely Protestant subjects. These tensions erupted into warfare in the seventeenth century following the notorious 1618 "Defenestration of Prague," in which two royal officials were thrown out of a castle window by Protestant leaders. The Czech revolt that followed was crushed two years later in the Battle of White Mountain, and thousands of Czech Protestants fled to other countries, fearing reprisals.

The enlightened rule of Austrian monarchs, Maria-Theresa and Joseph II, in the eighteenth century lifted much of the repressiveness of Habsburg rule, in terms of political and religious rights and economic opportunities. Partly as a result of expanded educational opportunities, nationalist sentiment took root among the Czech population and grew throughout the nineteenth century, reflected by cultural institutions such as the museum of the Bohemian Kingdom and in the music of Bedrich Smetana (1824–84) and Antonin Dvorak (1841–1904), which incorporated themes from Czech folk music. Czech nationalists were caught up in the wave of revolutionary activity that swept Europe in 1848, holding mass demonstrations and holding the first Slavic Congress to further unity among the Slavic peoples of Central and Eastern Europe. The response of the Habsburgs was to crack down even further on their Czech subjects, but a series of military defeats in the following years weakened their position. In

1861 the Czechs were granted a constitution, and the Dual Monarchy with Hungary was formed in 1867.

From the turn of the twentieth century, Czech nationalists began to pursue a new goal: political union with neighboring Slovakia in an independent state. During World War I, this course was actively pursued by Czech leader Tomas Masaryk, who founded the Czechoslovak National Council from his base in London. With the dissolution of the Habsburg empire in the war, the creation of the Czechoslovak state was approved by the Allied powers. Czechoslovakian independence was declared on October 28, 1918, and Masaryk became its first president, holding this office until he resigned in 1935. The new nation had a strong industrial base but struggled to create unity among its diverse ethnic population. Its most serious challenge was the problem of the German-speaking Sudetenland to the south, where a movement toward reunification with Germany grew up as the Nazis, under

Adolf Hitler, came to power in Germany. By the late 1930s Hitler was demanding outright that the Sudetenland be ceded to Germany, and, hoping to forestall a military confrontation, the European powers agreed to his demand in the 1938 Munich Pact. In fact, this agreement led to the dismemberment of Czechoslovakia, which lost other lands as well. By the following spring, Bohemia and Moravia were declared German protectorates, and a six-year Nazi occupation of the region began.

During the war, Czech leaders established closer ties with the Soviet Union, and it was Soviet troops that liberated the Czech lands from the Nazis. In 1945 the sovereignty of Czechoslovakia was restored, and a new government was formed that was closely allied with the Soviet Union but still a democracy. Three years later, however, the Communist Party took effective control of the country, forming a repressive government modeled on that of the Soviet Union and instituting a series of purges that climaxed with the execution of top government leaders, notably Rudolf Slansky, the party secretary. Even when liberalization began in the Soviet Union after the death of Josef Stalin (1879–1953), rule by the Czech Communist government remained harsh under Antonin Novotny (1904–75), who served as president from 1957 to 1968.

In 1968 Alexander Dubcek (1921–92), the Communist leader of the Slovaks, was named as the country's top Communist leader, and he instituted a series of reforms aimed at creating "Socialism with a human face." The period that followed was characterized by unprecedented freedom of expression, as media censorship was lifted and plans were made for economic reforms also. However, the government leaders of the Soviet Union and Czechoslovakia's other Communist neighbors felt threatened by the growing freedom so near their borders and criticized the new Czech policies. On August 20, 1968, the brief period of freedom known as the "Prague Spring" ended when Soviet tanks rolled into Czechoslovakia. Dubcek and other top leaders were taken into Soviet custody. A new government was formed under the hard-line Communist Gustav Husak (1913–91), under whose direction nearly one-third of the Communist Party's members were dismissed between 1970 and 1975 because they were considered too liberal. In an important gesture of protest, over two hundred leading Czech intellectuals signed a petition in 1977 criticizing the government for human rights violations. The main spokesperson for the signers of Charter 77, as it was called, was playwright and future president Vaclav Havel.

In the 1980s a major liberalization took place in the Soviet Union under Prime Minister Mikhail Gorbachev (b. 1931), but Czech president Gustav Husak did not declare support for Gorbachev's policies until 1987, and anti-government feeling grew in Czechoslovakia throughout the 1980s. In November 1989, the brutal government response to a large student demonstration in Wenceslas Square triggered the "Velvet Revolution," a ten-day period of protests, rallies, and strikes that led to the resignation of the country's Communist leaders in December and the parliament's election of dissident playwright Vaclav Havel to the presidency.

The country was renamed the Czech and Slovak Federal Republic, and the first free elections since 1946 were held in June, 1990. However, tensions between the Czechs and Slovaks grew, and the end of 1992 saw the "Velvet Divorce," by which the political union of the two groups was dissolved and separate Czech and Slovak republics were formed. Vaclav Havel was elected president of the Czech Republic at the beginning of 1993 and reelected five years later. During this time, the country underwent a rapid transformation to a free-market economy under a coalition government led by Prime Minister Vaclav Klaus (b. 1941). In 1997 NATO (the North Atlantic Treaty Alliance) announced that Czechoslovakia was among three former Eastern Bloc nations that would be invited to join the organization in 1999.

Timeline

c. 50 B.C.   Celts occupy Bohemia and Moravia

The present-day Czech Republic, comprising the territories of Bohemia and Moravia, is occupied by Celtic tribes, who are its earliest recorded inhabitants.

5th century A.D.   Slavs settle in the region

Slavic settlers arrive from the east and develop an agricultural economy, building traditional circular villages called *okrouhlice.*

6th century   Avar invasions

A foreign people known as the Avars invade and conquer some of the Slavic territories between the Elbe and Dnieper rivers, establishing a loosely linked empire.

625–658   Samo unifies the Slavs

The first unified Slavic state in the region is founded by Samo, a member of a Germanic tribe called the Franks. Centered in Bohemia, the state dissolves upon the death of its founder.

796   Avar empire is destroyed

The Frankish emperor Charlemagne destroys the Avar empire, aided by the Czechs of Moravia, who are rewarded with limited land rights.

9th century   The Great Moravian Empire

A kingdom is established in Moravia by a Slavic leader named Mojmir. Later expanded to include parts of Bohemia, Slovakia, Poland, and Hungary, it becomes known as the Great Moravian Empire. It unites the Czechs and Slovaks politically for the first time.

9th century   Cyril and Methodius convert Czechs to Christianity

Located between the Frankish Roman Catholic lands to the west and the Orthodox Christian lands of Byzantium to the east, the Moravian Empire is visited by missionaries from both churches. The famous Orthodox monks Cyril and Methodius convert many Czechs to the eastern church, also introducing a new script (the Cyrillic alphabet) to render religious texts in Slavic. Ultimately, however, Catholicism becomes the predominant religion in the region.

The Bohemian Kingdom

10th century   Magyar invasion ends the Moravian Empire

Invasion by the Magyars spells the end of the Moravian Empire, as the Czechs and Slovaks become politically separated. The Czechs ally themselves with the Franks, and their political influence becomes restricted to Bohemia, where a new kingdom emerges. The Slovaks come under the rule of the Magyars and their Kingdom of Hungary. This political division is a major contributing factor in the cultural separation of the Czechs and Slovaks and the development of distinct national identities by the two ethnic groups.

10th century   The Bohemian kingdom is founded

Under the Premyslid dynasty, the Bohemian Kingdom is founded and allies itself with the German Franks to the west to receive protection from the Magyars to the east. The Bohemians continue to choose their own kings, but they are ultimately under the control of the Holy Roman Emperor.

973   Bishopric founded at Prague

The bishopric (seat of a Roman Catholic bishop) of Prague is founded.

1029   Moravia is conquered by Bohemia

The Kingdom of Bohemia wins political control of Moravia following warfare with Hungary and Poland, but Moravia retains separate internal rule, and the Czech national identity is centered primarily in Bohemia.

13th century   Period of increased Premyslid power

With the greater imperial powers occupied with problems elsewhere in Europe, Bohemia enjoys a period of relative autonomy that leads to political expansion, especially under Otakar II (r. 1253–78), who marries into the Habsburg family and gains control of parts of Austria.

13th century   Germans immigrate to Bohemia

The thirteenth century is a period of increased immigration by Germans, who settle near the borders and in some parts of the interior. The Germanic law they bring with them is an important influence on the development of laws governing commerce.

The "Golden Age"

14th century   The Czech "Golden Age"

Bohemia enjoys a period of political and cultural advancement under the Luxembourg dynasty that succeeds the Pre-

Kingdom of Bohemia, 14th century.

myslids, especially during the reign of Charles IV (1342–78). Charles strengthens the Bohemian crown and the kingdom's religious authority and oversees the rebuilding and expansion of Prague, its capital city.

1348   Charles University is founded

Charles IV of Luxembourg, the king of Bohemia, founds Charles University in Prague as part of his plan to make the city an international center of learning.

1355   Charles IV becomes Holy Roman Emperor

Charles IV, king of Bohemia, is crowned Holy Roman Emperor, which places him in a position to increase Bohemia's political standing and influence.

c. 1369   Birth of Jan Hus

Jan Hus, the foremost Czech religious reformer and a national hero, is born in Bohemia and serves as both a priest and a lecturer at the University of Prague, where he is named rector in 1409. Early in his career, Hus becomes familiar with the principles espoused by previous religious reformers and begins advocating church reform. At first his ideas are tolerated by the church, but when he begins to preach the radical philosophy of English reformer John Wycliffe (1330–84), he is excommunicated. After a period of self-imposed exile, Hus goes before a general council in Constance to defend his views but is imprisoned and burned at the stake on July 6, 1415, when he refuses to retract his statements.

15th century   Hussite movement

The Hussite movement brings Bohemia into the forefront of opposition to the Catholic Church. Clergyman Jan Hus (see 1369) embraces the reformist teachings of the Englishman John Wycliffe (1330–84), whose objections to church practices foreshadow the Protestant Reformation. Hus criticizes the wealth and corruption of the church and advocates changes in the communion ritual and clerical vows of poverty. After initially enjoying the support of King Wenceslas IV (r. 1378–1419), Hus is removed from his university post and exiled from Prague. He is recalled to answer charges of heresy and burned at the stake, becoming a national hero.

Armed religious conflict lasting for decades breaks out after Hus's death. Riots break out in Prague, and Hussite extremists massacre German Roman Catholics, leading many Germans to flee the country. The Taborites, one faction of the Hussites, led by Jan Zizka, roam the countryside storming churches and monasteries and appropriating church-owned lands. Hussite doctrines are also spread among the Slovaks when Hussite rebels and refugees enter Slovakia.

1415: July 6   Jan Hus is burned at the stake

Religious reformer Jan Hus is charged with heresy at the Council of Constance and burned at the stake. (See 1369.)

1468   Printing is introduced

The first Czech book is printed.

Habsburg Rule

1526–1867   The Czechs come under Habsburg rule

Following victories by the Ottoman Turks and the Habsburgs, the Bohemian nobles elect a Habsburg Austrian ruler, Archduke Ferdinand, as their king, inaugurating three centuries of Habsburg rule. Austrian rule brings religious tensions, as the Habsburgs are Catholics and by this time most Czechs are Protestants. During this period, neighboring Moravia and Slovakia are also part of the Habsburg empire.

1618–48   Thirty Years' War

Religious warfare between German Protestants and the Holy Roman Emperor begins at the same time that Czech Protestants revolt against their Habsburg rulers (see 1618) but continues beyond the Czech revolt, which is crushed after two years. Czechs continue to fight in the larger war, mostly on the Protestant side, and many battles take place on the soil of their homeland, causing widespread destruction.

1618   Prague "defenestration"

When Bohemia's Habsburg rulers close two Protestant churches, Protestant leaders kill two royal governors by throwing them out of the window of a Prague castle, an event that becomes famous as the "Defenestration of Prague." Czech leaders overthrow their Catholic king, replace him with a Protestant, and raise an army to resist the Habsburgs. A two-year religious war between the Czechs and their Austrian rulers ensues.

1620: November 8   Battle of White Mountain

Rebellious Czech Protestants are decisively defeated by their Catholic rulers, who regain control of the region. The leaders of the revolt are executed, and thousands of Czech Protestants (including the vast majority of the Czech nobility) flee to other countries. After their departure, their lands are appropriated. Much of this land is given to Catholic immigrants who flock to Bohemia from southern Germany.

1622   Charles University is merged with religious college

Historic Charles University in Prague is merged with the Jesuit Academy, and Bohemia's entire education system is placed under the supervision of the Catholic Church.

1648   Peace of Westphalia

The Peace of Westphalia, which ends the Thirty Years' War, confirms Habsburg rule over Bohemia and authorizes large-scale immigration of Germans to Czech lands, which results in the spread of German culture in the region.

18th century   Enlightened rule

There are major changes in Habsburg rule of Bohemia in the eighteenth century thanks to the enlightened policies of Empress Maria-Theresa (r. 1740–80) and Joseph II (r. 1780–90). These policies eliminate the most repressive aspects of government and offer new economic opportunities to the Czechs. In line with the general eighteenth-century Enlightenment trend toward secularization, the power of the Catholic Church is curbed, which allows the Czechs greater religious freedom. Serfs are granted more rights, and many move to cities, drawn by the dawning of the industrial age.

Together with greater religious freedom and economic and educational opportunities comes some loss of national identity as Bohemia is officially merged with Austria for greater bureaucratic centralization and efficiency. The Czechs are stripped of whatever localized political power they had preserved, and German becomes the official language of all Habsburg lands.

1741   Prussia invades Bohemia

The Prussian Hohenzollern dynasty, allied with Bavaria and Saxony, invades Bohemia under the leadership of King Frederick II (1712–86). The Habsburgs counter-attack and gain back most of the captured territory.

Czech Nationalism Grows

1781   Edict of Toleration

The Edict of Toleration extends religious freedom to Protestants.

1791   Czech language chair is established at university

A chair in Czech language and literature is created at the Charles-Ferdinand University. The revival of Czech nationalism, which is in its early stages, centers on the restoration of the Czech language as a literary medium and a means of unifying the Czechs, who speak a variety of dialects. Formal instruction in Czech in the schools is also stressed.

1818   Bohemian museum is founded

The Museum of the Bohemian Kingdom is established to further Czech scholarship.

It becomes the venue for the publication of a Czech journal.

1824: March 2   Birth of composer Bedrich Smetana

Smetana is the first composer to incorporate Czech themes in classical music. A gifted pianist and conductor as well as composer, Smetana begins his own music school in Prague in 1848. Beginning in 1856 he spends two periods of time working as an orchestra conductor in Goteberg, Sweden, returning to Bohemia in 1863, where he establishes a reputation as a conductor, composer, and music critic. Smetana shares the affliction of deafness with his great predecessor Beethoven (1770–1827), going totally deaf by the age of fifty. By 1882 the composer becomes mentally ill, and he dies in 1884. Smetana's most famous works are the opera *The Bartered Bride*; the cycle of symphonic tone poems *Ma Vlast* (My Country), whose second part, *The Moldau* (a depiction of Bohemia's major river), is Smetana's single most famous piece; and the String Quartet in E minor, called *From My Life*.

1841: September 8   Birth of composer Antonin Dvorak

The most famous Czech composer, Dvorak, is the son of an innkeeper. He leaves home to acquire a musical education in Prague, where he earns money playing the violin and viola. With the support of Johannes Brahms (1833–97), Dvorak quickly establishes his reputation as a composer. In 1892 he receives an appointment to head the National Conservatory of Music in New York and moves to the United States. Because he misses his homeland, he remains in the U.S. for only three years but writes some of his most acclaimed music while there, including the Symphony No. 9 in E minor (*From the New World*) and the Cello Concerto in B minor. Extending his practice of using Czech folk themes in his music, Dvorak uses American folk material—including both black and Indian folksongs—in the music he writers in America, most notably in the New World Symphony and the "American" String Quartet in F major.

Following his sojourn in the U.S., Dvorak returns to Prague, where he becomes director of the conservatory in 1901. He dies in 1904. Other famous works by Dvorak include the Slavonic Dances, the Symphony No. 5, the Piano Quintet in A major, the "Dumky" Trio for Piano and Strings; and the Stabat Mater for chorus, orchestra, and soloists.

1848: March 11   Czechs demonstrate in Prague

Czechs hold a demonstration to demand language rights, freedom of the press, the abolition of serfdom, and their own parliament.

1848: June   First Slavic Congress is convened

In the wake of rebellions throughout Europe, the Czechs hold the first Pan-Slav Congress under the leadership of Czech nationalist Frantisek Palacky as a means of uniting the various Slavic peoples under Habsburg rule, including Poles, Slovaks, Serbs, Croats, Slovenes, and Ruthenians, as well as Czechs. The Austrians' initial response to the aspirations of Slavic nationalists is the imposition of a repressive absolute monarchy. However, military reverses of the 1850s and 1860s eventually force a compromise.

1850: March 7   Birth of statesman Tomas Masaryk

Tomas Masaryk, a founder of Czechoslovakia and its first president, is born in Moravia in modest circumstances and works hard to receive an education, eventually pursuing postgraduate studies abroad and becoming a lecturer in philosophy at the Czech University in Prague. In 1890 Masaryk becomes active in politics and serves briefly in parliament. For most of the decade he edits a liberal journal and publishes books on social and political topics.

During World War I, Masaryk joins with the Czech Eduard Benes (1884–1948) and the Slovak Milan Stefanik (1880–1919) in planning the establishment of a Czechoslovak state when the war ends. The three men found the Czechoslovak National Council and gain international support for their cause. When the new republic is formed in 1918, Masaryk becomes its first president. A popular, effective, and respected leader, he is reelected in 1920, 1927, and 1934, resigning in 1935 due to advanced age and the need for younger leaders to confront the growing Nazi threat in Europe. Presented with a country estate by a grateful nation, Masaryk retires. He dies two years later.

1854: July 3   Birth of composer Leos Janacek

Janacek is a composer of highly regarded instrumental works and considered one of the nineteenth century's foremost composers of opera. Born into a family of musicians and teachers, he continues the family tradition by teaching and working as a choir director in Brno. He is an early champion of the works of fellow Czech composer Antonin Dvorak (see 1841), who also becomes a friend. Janacek also founds and serves as director of an organ school in Brno. Czech nationalism and pan-Slavicism (the desire for unity of the Slavic peoples) find a voice in Janacek's operas, all of which are based on Bohemian, Moravian, or Russian themes. Janacek's operas include *Jenufa, From the House of the Dead,* and *The Cunning Little Vixen.* Other works include the Sinfonietta and the Glagolithic Mass for chorus and orchestra. Janacek dies in 1928.

1861 Constitutional rule is granted

Austria agrees to implement a constitutional monarchy in Bohemia.

1867 Compromise of 1867

Under pressure after a series of military disasters, the Habsburg Empire agrees to share power with Hungary, forming the Dual Monarchy of Austria-Hungary, which have a common ruler but are independent and equal in most respects. The Czechs attempt unsuccessfully to gain an equal share in this partnership.

1868–81 Construction of the National Theater

The decoration of the interior of the newly built National Theater in Prague triggers a revival in Czech painting. Artwork for the interior, which emphasizes the Czech landscape and themes from the country's history, is provided by a group of artists who later become known as the National Theater Generation. Included are Miklos Ales, Vaclav Brozik, Julius Marak, and Vojtech Hynais.

1880 Language law makes Czech and German equal

A new law gives Czech and German equal legal and administrative status in Bohemia.

1882 University is divided between Czechs and Germans

Charles-Ferdinand University is divided into two separate entities, one German and one Czech.

1883: April 30 Birth of writer Jaroslav Hasek

Satirical writer Jaroslav Hasek is born in Prague and begins his writing career by writing newspaper pieces. He serves in World War I and is taken prisoner by the Russians. After the war, he works on his renowned four-volume work *The Good Soldier Schweik,* which remains incomplete at his death in 1923.

1883: July 3 Birth of writer Franz Kafka (1883–1924)

Franz Kafka, esteemed as one of the great literary figures of the twentieth century, is born in Prague and spends most of his life there. As a Jew, he considers himself outside the mainstream of Czech society, although he is an atheist. From 1907 he works for an insurance company while writing (in German) stories and novels that express an alienation and sense of absurdity considered by many to be emblematic of the human condition in the twentieth century. In 1923 Kafka moves to Berlin, intending to devote himself to writing full time but dies of tuberculosis the following year. During his lifetime, Kafka publishes some shorter works including *The Judgment* (1916), *The Metamorphosis* (1915), and *In the Penal Colony* (1919). He instructs his friend and literary executor Max Brod to burn his longer works after his death, but Brod disregards his request. Between 1925 and 1927 Brod publishes Kafka's posthumous novels, which will become masterpieces of twentieth-century literature: *The Trial, The Castle,* and *Amerika.*

1884: May 12 Death of composer Bedrich Smetana

Nationalist composer Bedrich Smetana dies in Prague. (See 1824.)

1890s Nationalist groups clash in Bohemia

Czech and German militant nationalist groups are established in Bohemia and come into conflict with each other. Czech and Slovak nationalists begin discussing cooperation and the possible formation of a "Czechoslovak" state.

1890: December 8 Birth of composer Bohuslav Martinu

Martinu, one of the foremost twentieth-century Czech composers, has a varied career full of unexpected twists and turns, some due to political circumstances. As a student, he is expelled from the Prague Conservatory for lack of interest in his studies. He goes on to teach and compose music during World War I and plays violin with the Czech Philharmonic Orchestra. Moving to Paris in 1923, he studies with the composer Albert Roussel (1869–1937) and begins to establish a professional reputation. By the early 1930s, his works are achieving international recognition, and he begins integrating Czech folk material into works including the ballet *Spalicek* and the opera *Hry o Marii.* The advent of World War II forces Martinu to leave France, and he emigrates to New York, where he continues his prolific compositional output, writing symphonies, chamber works, and other pieces. The ascendancy of Communism in his homeland prevents him from returning there after the war. In the 1950s he teaches composition at Princeton University and the Curtis Institute and spends time in Nice and Rome. He dies in 1959.

1900 Tomas Masaryk founds nationalist party

Tomas Masaryk, a university professor and future president of the Czech Republic, founds the Czech Progressive Party, which supports Czech autonomy in conjunction with liberal parliamentary politics and universal suffrage.

1901: September 23 Birth of poet Jaroslav Seifert

Seifert, the first Czech to win the Nobel Prize for Literature, writes poetry throughout his life but also works as a journalist until 1950. His earliest work reflects his support for communism and the Russian revolution, but his poetry soon becomes more personal and lyrical. By 1929 he is disillusioned with communism. Much of Seifert's poetry is concerned with events in his homeland, such as *Switch Off the Lights* (1938), written in the wake of the Munich agreement that preceded the German takeover of Czechoslovakia in World War II. Altogether, Seifert publishes some thirty volumes of poetry, as well as children's books, journal articles, and memoirs. He dies in 1986.

1904: March 1 Death of composer Antonin Dvorak

Renowned composer Antonin Dvorak dies in Prague. (See 1841.)

1907: April 18 Avant-garde artists' group opens first exhibit

A group of Czech painters known as The Eight holds its first exhibit in a Prague storefront. Influenced by avant-garde trends elsewhere in Europe, the group advocates innovations in color and expression. Among its members are Emil Filla, Bohumil Kubista, Antonin Prochazka, and Otokar Kubin. A second exhibit is held the following year.

1912 Prague art workshops are established

The Prague Art Workshops (PUD), a design center for arts, crafts, and furniture, are established by architects Josef Gocar and Pavel Janak. Its goal is to produce household furnishings distinguished by functionality and good taste that also rise to the level of art. Some furniture with Cubist designs is produced.

1914–18 World War I

Many Czech soldiers defect who feel a closer bond to their fellow Slavs, the Russians, than to their Austrian rulers. Defectors form the Czech Legion, which fights on the Russian side.

1916 Czechoslovak National Council is formed

Czech nationalist leader Tomas Masaryk (see 1850), together with fellow Czech Eduard Benes and Slovak leader Milan Stefanik (1880–1919), founds the Czechoslovak National Council, headquartered in France. The three leaders lobby for

recognition by the Western powers while their fellow Czech leaders at home form an underground resistance group.

1918 The Allies recognize the Czechoslovak council

The Allied powers recognize the Czechoslovak National Council and approve the formation of a Czechoslovak state.

Czechoslovakia is Founded

1918: October 28 Masaryk declares Czech independence

A provisional Czech government is set up, with Tomas Masaryk (see 1850) named as president. Still in exile in the United States, Masaryk issues a declaration of Czechoslovak independence. The Slovaks issue the Declaration of Turciansky Svaty Martin, officially approving the Czech-Slovak union. Masaryk proclaims that the new state will occupy the same territory as the former Bohemian Kingdom, which includes the German-speaking Sudetenland.

With a population of over 13.5 million, the new nation will encompass a variety of ethnic groups, including Czechs, Slovaks, Moravians, Ruthenians, and Sudeten Germans. It will have a large part of the industrial base that formerly belonged to the Austro-Hungarian empire, including china, glass, automobile, and beer manufacturing. However, far more of the industrialization is in the Czech part of the country than in Slovakia.

1918: November 11 Czechs occupy the Sudetenland

Czech troops occupy the Sudetenland.

1919: September 10 Ruthenia becomes part of Czechoslovakia

The Treaty of Saint-Germain includes provisions for Subcarpathian Ruthenia (a rural area in the Carpathian mountains) to become part of the new nation of Czechoslovakia with internal autonomy. The region had previously been under the control of the Ottoman empire and then the Hungarians. Most of its population belongs to the Uniate Church, which combines elements of the Catholic and Orthodox faiths.

1920–21 Formation of the Little Entente

Czechoslovakia, Yugoslavia, and Rumania form the Little Entente for mutual protection against a possible restoration of the Habsburg monarchy.

1920s Czechoslovakia struggles to accommodate diverse ethnic groups

Through the 1920s, Czechoslovakia maintains a stable central government with a vigorous party system, and its strong industrial base brings prosperity to the new nation. However, the conflicting interests and expectations of its diverse ethnic constituency create serious problems for the country, especially since much of that constituency, including the Slovaks and Ruthenians, expect more autonomy than it is ultimately granted.

The most difficult problem arises in the populous, German-speaking Sudetenland. In addition to the traditional hostility between the region's Czechs and Germans, tensions are aroused over a variety of specific controversies, including the absence of Sudeten representation in the drafting of the 1920 constitution. Opposition to the central government decreases during the 1920s but then undergoes a resurgence with the advent of the Depression and the rise to power of Adolf Hitler (1889–1945) in Germany. There is a growing movement favoring union of the Sudetenland with Germany.

1920   Constitution is adopted

Czechoslovakia adopts its first permanent constitution, which provides for a parliamentary democracy with a bicameral national assembly and a president elected to a seven-year term who shares executive power with an appointed cabinet.

1921: June 26–29   First Worker Olympiad

The Czechoslovak Worker Gymnastics Association sponsors the first unofficial Worker Olympics. Leftists oppose the regular Olympics, which they regard as an elitist event for the wealthy and privileged that emphasizes competition over cooperation and encourages rivalry between nations. The Prague Olympics, which attracts participants from twelve European countries and the United States, also features cultural events, including musical and theatrical performances and art exhibits.

1923: January 3   Death of author Jaroslav Hasek

Jaroslav Hasek, author of *The Good Soldier Schweik,* dies. (See 1883.)

1924   Defense pact is signed with France

Czechoslovakia and France sign a mutual defense treaty aimed at protection against possible German aggression in the future.

1924: June 3   Franz Kafka dies

Author Franz Kafka dies of tuberculosis near Vienna. (See 1883.)

1928: August 12   Death of composer Leos Janacek

Famed opera and instrumental composer Leos Janacek dies. (See 1854.)

1929: April 1   Birth of author Milan Kundera

Milan Kundera, writer of fiction, plays, and poetry, is born in Brno. He studies at the Academy of Music and Dramatic Arts in Prague and is appointed to the faculty of its film department. From the beginning, his writing draws criticism from the communist government. His early poetry is censured for its erotic content and subversive tone. In the repressive atmosphere following the Soviet invasion of 1968, the government bans publication of his second novel, *Life is Elsewhere* (1969). He is also fired from his teaching jobs and thrown out of the Communist Party. In 1975 Kundera emigrates with his wife and settles in France. Among his best-known novels are *The Joke* (1967), *The Book of Laughter and Forgetting* (1979), and *The Unbearable Lightness of Being* (1984), which is made into a motion picture. Following the dissolution of Czechoslovakia's Communist regime in 1989, Kundera's books, which had been banned, are once again allowed to be published in his homeland.

1932: February 18   Birth of film director Milos Forman

Born to a Jewish father and a Protestant mother, Forman loses both parents in World War II and is raised by relatives. Graduating from the Academy of Music and Dramatic Art in Prague, he begins his film career by writing screenplays. In the 1960s he becomes the leading figure in the Czech New Wave of filmmaking, gaining acclaim for innovations made possible by the liberalization of the 1960s that leads up to the "Prague Spring." His best-known movies of this period are *Black Peter, Loves of a Blonde,* and *The Firemen's Ball.* These films are characterized by their casting of non-actors together with professionals, use of improvisation, skillful choice of music, and the humor with which they portray the ironies inherent in the everyday lives of ordinary people.

Forman is in Paris when the Russians invade Czechoslovakia in 1968, and he elects to remain in the West, emigrating to the United States (he becomes a U.S. citizen in 1975). With the 1971 film *Taking Off,* Forman launches a successful U.S. career that includes two multiple-Oscar-winning smash hits, *One Flew Over the Cuckoo's Nest* (1975) and *Amadeus* (1984), as well as the films *Hair* (1971), *Ragtime* (1981), *Valmont* (1989), and *The People vs. Larry Flynt* (1996). Together with Czech emigre writer Jan Novak, Forman authors the 1994 memoir *Turnaround.*

1935   Masaryk resigns

Tomas Masaryk (see 1850), president of Czechoslovakia since 1918, resigns due to age and fragile health, and is succeeded by Eduard Benes.

1935   Sudetendeutsche Partei is formed

The *Sudetendeutsche Partei* (Sudeten German Party) is formed by Konrad Henlein and gains increasing popularity among Sudeten Germans. Henlein has ties with the Nazi party in Germany and his party imitates Nazi slogans and uniforms.

School girls perform national folk dances during an exhibition for President Benes at Masaryk Stadium. (EPD Photos/CSU Archives)

1936: October 5   Birth of Vaclav Havel

Playwright, dissident, and political leader Vaclav Havel serves, first, as president of Czechoslovakia when Communist rule ends, and then as president of the newly formed Czech Republic after the separation from Slovakia. He is born to an upper-middle-class family in Prague and begins his association with the theater by holding low-level theatrical jobs while attending the Prague Academy of Film Arts. His first play, *The Garden Party,* is premiered in 1963. It is followed by *The Memorandum* in 1965. Because of his critical stance toward his country's communist regime, Havel's plays are banned in Czechoslovakia after 1968, but they gain an international following, and he wins a number of prestigious literary awards abroad.

Havel gains a higher political profile when he serves as spokesperson for the dissident Charter 77 group that signs a historic protest petition in 1977. As a result of his activities, he is sentenced to the first of four jail terms. In 1989, with the communist regime on the brink of collapse. Havel forms the Civil Forum, an alliance of pro-democracy groups that becomes the ruling party when the Velvet Revolution brings Communist rule to an end in Czechoslovakia. On December 29 he is named president of the country, and his position is confirmed in the elections of July, 1990. Following the break-up of Czechoslovakia into separate republics, Havel is elected president of the Czech Republic in January 1993.

1937: September 14   Death of Tomas Masaryk

Czechoslovakia's founder and longtime president dies two years after retirement at his country estate in Lany. (See 1850.)

Nazi Occupation

1938: July   State athletic association demostration

Several thousand members of the state athletic association present an exhibition at the Masaryk Stadium in Prague. President and Mrs. Benes are among the honored guests.

1938: September 29–30   Munich Pact

In response to threats by Nazi leader Adolf Hitler (1889–1945), the major Western powers pressure Czechoslovakia to cede the Sudetenland to Germany. Representatives of Britain, France, Italy, and Germany meet in Munich to sign an agreement with Germany, with no representation by Czechoslovakia. Hoping to avert a war, they sign the Munich Pact, which deprives the Czechs of the Sudetenland and other territories. Czechoslovakia loses its most industrialized and prosperous region, as well as a large percentage of its population.

1938: October 15   **Benes resigns and leaves the country**

Czech president Eduard Benes resigns his post and goes into exile in London.

1938: November   **New government is formed**

A new government is formed under Emil Hacha and named the Federated Republic of Czechoslovakia (also known as the Second Republic). Hacha's right-wing government attempts to appease the Germans by cooperating with their demands, adopting a new constitution that weakens the nation further by granting substantial autonomy to Slovakia and Subcarpathian Ruthenia.

1939: March 15   **Germany completes takeover of Czechoslovakia**

Bohemia and Moravia become German protectorates, effectively ending the existence of Czechoslovakia as a sovereign nation. Czech lands are occupied by German troops. Hungary annexes Subcarpathian Ruthenia.

1940   **Government in exile is formed in London**

Following the outbreak of World War II in 1939, Czech president Eduard Benes forms a government in exile in London. It is recognized by both Britain and the Soviet Union.

1943   **New Czech-Soviet pact is signed**

Feeling that they have been betrayed by the West and anticipating that it is the Soviets who will liberate Czech lands from their German occupiers, Czech leaders strengthen their ties to the Soviet Union throughout the war years. This includes the signing of a 20-year mutual defense pact.

1945: April   **New government is formed**

A new Czech government is formed with Eduard Benes as president. The government is controlled by parties friendly to the Soviet Union and communists hold important posts, but Czechoslovakia remains a democracy, and the Soviet Union pledges to respect its independence.

Communist Rule

1948: February   **Communists take control of government**

Twelve non-communist cabinet ministers resign to protest the actions of the ministry of the interior, which include alleged police protection of communist terrorists. Instead of raising support for their position, the move consolidates the power of the communists. The twelve ministers are replaced by communists and communist sympathizers.

1948: March 10   **Jan Masaryk dies in fall from window**

Foreign Minister Jan Masaryk (1886–1948), son of Czechoslovakia's first president and the only independent minister left in the cabinet, dies by falling from a window in an alleged suicide widely thought to be a political assassination.

1948: May 9   **New constitution is adopted**

Czechoslovakia adopts a new constitution setting up a people's republic based on the Soviet model.

1948: June 8   **Benes resigns**

Eduard Benes resigns the presidency. He is succeeded by leading Communist Klement Gottwald (1896–1953).

1950–52   **Political purges**

A series of political purges is implemented to strengthen communist control of Czechoslovakia by removing known or suspected dissidents from the government and the Communist Party. Party Secretary Rudolf Slansky and Foreign Minister Vladimir Clementis are among top officials tried and hanged as traitors. Other institutions, including the press and the universities, are also purged of non-communists, repression of church activities is implemented, and forced labor camps are set up for the punishment and "reeducation" of dissidents.

1953   **Academy of sciences is founded**

Czechoslovakia's Academy of Sciences, modeled after that of the Soviet Union, is established.

1953: March   **Gottwald dies**

Klement Gottwald (1896–1953) dies and is succeeded as president by Antonin Zapotocky (1884–1957).

1953: June   **Mass unrest spurs liberalization**

Elimination of rationing and new currency policies lead to a workers' revolt in Plzen and demonstrations elsewhere. In response the government institutes limited reforms.

1956   Writers union criticizes Czech government

Czech writers protest government censorship at their national conference. Social realist author Jan Drda is removed from his post as chairman of the Writers' Union.

1956: May   Students demonstrate in Prague

Students organize a protest calling for greater cultural and intellectual freedom. The suspected leaders of the protest are punished, but there is no major repression by the government.

1957–68   Novotny presidency

Upon the death of Antonin Zapotocky, Antonin Novotny (1904–75) becomes president, remaining in office for over a decade. He institutes repressive measures that are at odds with the liberalization taking place in the Soviet Union under premier Nikita Krushchev (1894–1971) following the death of Josef Stalin (1879–1953).

1959: August 28   Death of composer Bohuslav Martinu

Martinu, one of the most acclaimed twentieth-century Czech composers, dies in Switzerland. (See 1890.)

1962: December   Destalinization begins

The Twelfth Congress of the Czechoslovak Communist Party institutes a program of destalinization (repudiation of former repressive measures inspired by the policies of Soviet leader Josef Stalin). Hanged leaders Vladimir Clementis and Rudolf Slansky are exonerated from the charges against them.

Prague Spring and Soviet Invasion

1968: January   Dubcek is named head of the Communist Party

Alexander Dubcek (1921–92), first secretary of the Slovak Communist Party, replaces Antonin Novotny as leader of Czechoslovakia's Communist Party.

1968   "Prague Spring"

Under the leadership of Alexander Dubcek (1921–92), wide-ranging liberal reforms are implemented with the goal of creating "Socialism with a human face." Media censorship is lifted, and freedom of expression is allowed to an unprecedented degree. Plans for economic liberalization are announced as well. The reforms are strongly criticized by the Soviet Union and Czechoslovakia's Eastern European neighbors.

1968: August 20–21   Soviets invade Czechoslovakia

The Soviet Union, aided by troops from other Warsaw Pact nations, invades and occupies Czechoslovakia. Top government officials, including Alexander Dubcek (1921–92), are seized and taken to the Soviet Union. However, there is strong opposition to the invasion among the Czech people, and the remaining Czech leaders refuse to cooperate by forming a puppet government whose policies will be dictated by the Soviets.

1969   Czech student sets himself on fire to protest Soviet actions

Czech resistance to the Soviets culminates in the self-immolation of Jan Palach, a Czech student.

1969: April   Dubcek steps down

Following anti-Soviet riots after a soccer game, Czech leader Alexander Dubcek (1921–92) is forced to resign his post and is replaced by Gustav Husak (1913–91).

1970–75   Liberals are purged from party

The Czech Communist Party loses nearly one-third of its members as the purging of liberals continues under Gustav Husak (1913–91).

1977: January   Dissidents publish manifesto

A group of over 200 Czech intellectuals call for government compliance with the human rights provisions of the Helsinki Agreement in a manifesto (Charter 77), which is published in the West. It is not taken seriously as a threat by the Czech government. Czech police arrest many of those who sign the document. Prominent among the signers is future president Vaclav Havel (see 1936).

1980s   Widespread environmental damage

Government economic policies lead to excessive burning of lignite, which harms the environment through acid rain and serious deforestation. Documented adverse effects on the nation's health include shorter life expectancy and higher infant mortality.

1983: October   Deployment of Soviet missiles is announced

The deployment of Soviet tactical nuclear missiles in Czechoslovakia is announced and triggers antinuclear demonstrations in 1983–84.

1984   Poet Jaroslav Seifert receives the Nobel Prize

To the embarrassment of the Czech government, poet Jaroslav Seifert (see 1901), a signer of the protest manifesto Charter 77, is awarded the Nobel Prize for literature, becoming the first Czech to receive this award.

1985: July   Mass gathering for religious commemoration

A crowd of 150,000 gathers to celebrate the 1,100th anniversary of the death of Saint Methodius, one of two priests who are credited with bringing Orthodox Christianity to the Czechs. The occasion is a sign of a growing religious revival in defiance of government repression of church activity.

1986: January 10   Poet Jaroslav Seifert dies

Nobel Prize-winning poet Jaroslav Seifert dies in Prague. (See 1901.)

1986: April   Fallout from Chernobyl affects Czechoslovakia

The Chernobyl nuclear accident occurs in the Soviet Union, causing radioactive nuclear fallout over many parts of Czechoslovakia.

1987   Husak belatedly endorses Soviet liberalization

After initial resistance, Gustav Husak (1913–91) reluctantly declares his support of Soviet leader Mikhail Gorbachev's (b. 1931) liberal policies of *perestroika* (restructuring) and *glasnost* (openness).

The Velvet Revolution

1989: January–October   Growing anti-government protests

Increasingly larger anti-government protests take place in Prague throughout the year. The government reacts by suppressing the demonstrations and arresting their leaders.

1989: November 17   Youths stage mass protest in Wenceslas Square

Some 3,000 young people demonstrate in Wenceslas Square, calling for free elections. Violent government suppression of the protest sparks meetings, demonstrations, and a general strike—the "velvet revolution"—that lead to the resignation of the Jakes government and the retirement of Gustav Husak (1913–91).

1989: November 20   The Civic Forum is formed

Led by Vaclav Havel, the Czech dissident groups join together to form the Civic Forum (comparable Slovak groups form the Public Against Violence, or VPN).

1989: November 25   750,000 rally in Prague

The "Velvet Revolution" culminates in a huge demonstration in Prague and a short general strike two days later.

1989: December   Communist leaders resign

Czechoslovakia's communist leaders apologize to the country and resign, and the constitution is amended to abolish the power of the Communist Party.

1989: December 28–29   Havel is named as president

Dissident playwright Vaclav Havel becomes president by acclamation. Alexander Dubcek (1921–92), who was the Czech leader during the "Prague spring" of 1968, is named chairman of the national assembly.

1990: February   Soviets pledge to withdraw troops

The Soviet Union apologizes to the Czechs for the 1968 invasion and occupation of their country and promises to withdraw its troops from Czechoslovakia by July 1991.

1990: June 20   Free elections are held

Czechoslovakia (now renamed the Czech and Slovak Federal Republic) holds its first free elections since 1946, with a turnout of 96 percent of eligible voters. The Civic Forum and its Slovak counterpart win a decisive victory, and the government begins a program of economic reform.

1990: July 5   Assembly confirms Havel as president

The newly elected assembly reelects Vaclav Havel (see 1936) to a two-year term as president.

1991–92   Slovak separatist sentiment grows

The historic tension between the Czechs and Slovaks becomes increasingly evident, and pressure for separation builds. The Czechs want a strong central government and a rapid shift to a free-market economy, while the Slovaks desire greater autonomy and a slower pace of reform.

1992: June   Czechs and Slovaks agree to form separate republics

The Czech and Slovak leaders agree to the "Velvet Divorce" dividing their territories into two separate republics and ending the political union begun in 1918. It is agreed that the two republics will go their separate ways as of December 31, 1992.

1992: December 31   Czechoslovakia is dissolved

Czechoslovakia is dissolved as the Czechs and Slovaks form separate political entities.

The Czech Republic

1993: January 26   Havel is confirmed as president of new republic

Vaclav Havel (see 1936) is elected to a five-year term as president of the newly formed Czech Republic by its parliament.

1994: March   Government announces war reparations

The Czech government announces that it will use the proceeds from selling off property formerly owned by the Czech Communist Party to pay a total of $33.7 million in reparations to persons imprisoned or sent to concentration camps during the German occupation of Czechoslovakia during World War II.

1996   Czech students excel in international science test

Czech middle-school-level students have the second highest test scores in science in a global test administered by the U.S. Department of Education to gauge the effectiveness of math and science education in forty-one countries. Among the nations included in the survey, the Czech republic is second only to Singapore in the science portion of the test.

1996: May 31–June 1   First parliamentary elections are held

Czechs hold their first parliamentary elections since the separation of the Czech and Slovak Republics. The new republic's acting parliament since 1993 has been the former regional parliament elected when Czechoslovakia was in existence. The ruling coalition led by Prime Minister Vaclav Klaus (b. 1941) narrowly misses winning a majority, and President Vaclav Havel (see 1936) asks Klaus to form a new government. The Social Democrats score big gains, winning an additional sixteen seats, for a total of sixty-one.

1997   Prospective Czech NATO membership announced

The North Atlantic Treaty Alliance (NATO) announces that the Czech Republic, along with Poland and Hungary, will formally be invited to join the organization in 1999. This will be the first expansion of NATO since Spain joined in 1982, as well as the first admission of nations formerly belonging to the Warsaw Pact, the alliance of Soviet Bloc nations that was NATO's adversary during the Cold War.

1997: November 30   Klaus resigns as prime minister

Vaclav Klaus, Czech prime minister since 1992, resigns amid allegations that his party accepted huge campaign donations from former tennis star Milan Srejber in exchange for government favors to Srejber's company.

1998: January 20   Havel is reelected

Vaclav Havel is elected to a second five-year term as president by the Czech parliament.

1998: June 19–20   Social Democrats triumph in elections

The Czech Social Democratic Party wins 32.3 percent of the vote in parliamentary elections and announces it will form a minority government rather than forming a coalition with other parties.

Bibliography

Benes, Eduard. *Memoirs: From Munich to New War and New Victory.* Boston: Houghton Mifflin, 1954.

Bradley, J. F. N. *Czechoslovakia's Velvet Revolution: A Political Analysis.* New York: Columbia University Press, 1992.

Kalvoda, Josef. *The Genesis of Czechoslovakia.* New York: Columbia University Press, 1986.

Kaplan, Karel. *The Short March: The Communist Takeover in Czechoslovakia, 1945–1948.* New York: St. Martin's, 1987.

Korbel, Josef. *Twentieth-Century Czechoslovakia: The Meanings of Its History.* New York: Columbia University Press, 1977.

Kriseova, Eda. *Vaclav Havel: The Authorized Biography.* New York: St. Martin's Press, 1978.

Leff, Carol Skalnik. *The Czech and Slovak Republics: Nation Versus State.* Boulder, Colo.: Westview Press, 1997.

Novak, Jan. *Commies, Crooks, Gypsies, Spooks, and Poets: Thirteen Books of Prague in the Year of the Great Lice Epidemic.* South Royalton, Vt.: Steerforth Press, 1995.

Schwartz, Harry. *Prague's 200 Days: The Struggle for Democracy in Czechoslovakia.* New York: Praeger, 1969.

Unterberger, Betty Miller. *The United States, Revolutionary Russia, and the Rise of Czechoslovakia.* Chapel Hill: University of North Carolina Press, 1989.

Valenta, Jiri. *Soviet Intervention in Czechoslovakia, 1968.* Baltimore: Johns Hopkins University Press, 1991.

Denmark

Introduction

Linking Scandinavia to continental Europe and surrounded by the sea, Denmark has traditionally been a peaceful and tolerant country and a crossroads for new ideas. At the same time, their political dominance of Scandinavia in the centuries leading up to the modern era made the Danes instrumental in the formation of a distinctive Nordic culture. Denmark has a long tradition of looking after its people, from the model adult education system begun in the nineteenth century to the comprehensive welfare state established in the twentieth. The country's popular twentieth-century monarchs have successfully brought one of Europe's oldest kingdoms to the edge of a new millennium.

The peninsula of Jutland, bordering Germany to the south, accounts for about seventy percent of Denmark's 16,629 square miles (43,069 square kilometers). The rest consists of 406 islands to the east of Jutland, of which the largest are Sjoelland (Zealand), Fyn (Funen), Lolland, and Falster. (The capital city of Copenhagen is located on the east coast of Sjoelland.) Coastal lands occupy much of Denmark, and most of the land is flat or gently rolling. In addition to Denmark itself, the Kingdom of Denmark also comprises two possessions: Greenland, at 840,000 square miles (2,175,600 square kilometers) the world's largest island, and the Faroe Islands, with an area totaling 540 square miles (1,399 square kilometers). Denmark has an estimated population of approximately 5.3 million, about a quarter of whom live in metropolitan Copenhagen, the largest city.

Vikings Expand Denmark

The Viking raids of the ninth to eleventh centuries began a period of expansion for Denmark, which, in addition to controlling other parts of Scandinavia, also invaded England in the latter part of the ninth century and conquered an area that came to be called the Danelaw. Christianity was introduced in the same period in the tenth century, and Harald Bluetooth became Denmark's first Christian monarch. In 1282 the Danes adopted their first constitution, which limited the powers of the monarchy and created the Danish parliament, or Rigsrad. By the late fourteenth century, Denmark had assumed effective control of Norway when Queen Margrethe I (r.1375–87), the widow of Norway's King Haakon VI, began acting as regent for her infant son, Olaf.

In 1397 the Kalmar Union, politically uniting Denmark, Norway, and Sweden, was formed. It also included the Danish possessions of Greenland, Iceland, and the Faroe Islands. The union lasted over a hundred years and was then disrupted when Sweden revolted and withdrew in 1523, beginning over two centuries of hostilities between the two countries, during which Denmark gradually lost its dominant political position. (Denmark remained united with Norway until 1814 and with Iceland until 1943.) The sixteenth century also brought the Protestant Reformation to Denmark, which formally became a Lutheran country during the reign of King Christian III (1535–59).

Absolute Monarchy and the Enlightenment

With Denmark weakened by the Thirty Years' War (1618–48), which bankrupted the country and left Sweden the most powerful nation in Scandinavia, King Frederick III (r.1648–70) established an absolute monarchy in 1660, a move that tightened the king's control over the nobility and strengthened the merchant class. By the end of the eighteenth century, however, the ideals of the Enlightenment (reason, rationalism) came to influence Denmark's rulers, especially Christian VII (r. 1766–1808), who gave the country a free press, abolished serfdom, and banned slavery in all land under Danish rule. Soon afterward, Denmark's fortunes declined when it sided with France in the Napoleonic wars. Great Britain attacked the Copenhagen harbor, nearly leveling the city, and imposed a naval blockade that cut off Danish trade. By 1813 the country was bankrupt; the following year, it lost two of its territories—Norway and Pomerania—in the Treaty of Kiel.

Constitutional Monarchy

The nineteenth century saw a rise in liberalism and the spread of democratic ideals. In the 1840s Denmark's system of folk high schools was inaugurated, providing general adult educa-

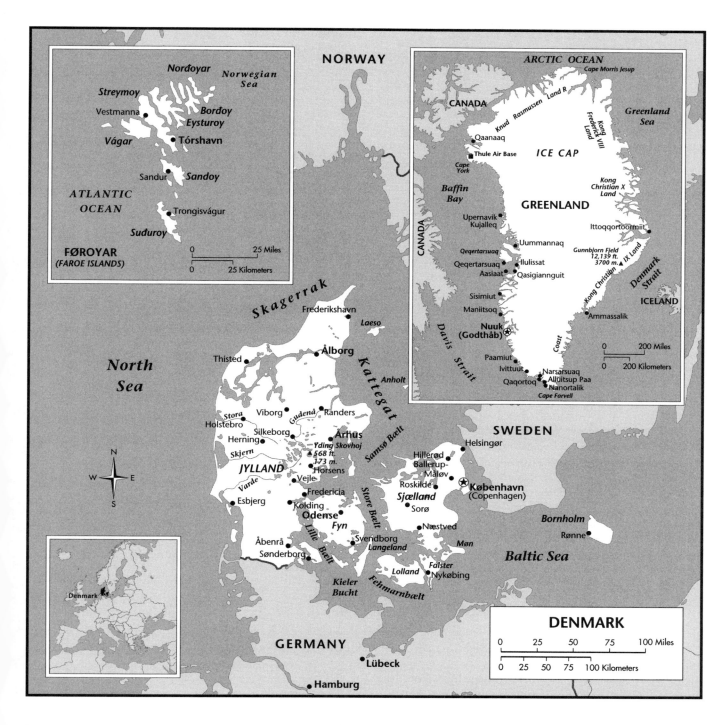

tion to the rural population and serving as a model for similar programs in other countries. King Frederick VII (r. 1848–63) put an end to Denmark's absolute monarchy in 1849, signing a new constitution that provided for the formation of a bicameral (having two legislative chambers) legislature.

The nationalism of the period provoked a revolt in Schleswig and Holstein, the southern provinces of Jutland bordering Germany. Unresolved tensions in this region brought on a war between Denmark and Prussia in 1864, ending with the Treaty of Vienna, which ceded the two provinces—accounting for nearly one-third of Denmark's total territory—to Austria.

In 1866 conservative interests controlled the Danish government and adopted a new constitution that increased the power of the upper house of parliament (the Landsting) over the lower house (the Folketing).

The Two World Wars

Denmark remained neutral in World War I (1914–18), although its foreign trade was disrupted by German mines. At the end of the war, Iceland was granted internal self-rule (complete independence came in 1943). In a plebiscite (a direct vote by the people), the residents of North Schleswig voted to make that territory part of Denmark once again. In the 1930s, the Danish government put into effect a comprehensive plan to battle the effects of the global economic depression and succeeded in rescuing the nation's economy through trade controls, debt measures, guaranteed farm prices, and social reforms that established today's comprehensive welfare state.

In spite of a 1939 nonaggression pact with Nazi Germany, Denmark, along with Norway, was invaded by the Germans in April 1940 and occupied throughout World War II (1939–45). By 1943 a vigorous underground resistance movement encompassed many facets of daily life, while Danish citizens resigned en masse (as a whole) from organizations and activities sponsored by the official Nazi-approved government. In addition, the Danish merchant marine fled the country, and the sailors offered their services to the Allied navies. Denmark was liberated by British troops in May 1945.

After World War II the Social Democrats became Denmark's leading political party. The country's postwar economic recovery was hastened by aid from the United States's Marshall Plan, and Denmark enjoyed an economic boom between the late 1950s and early 1970s. Renouncing its former policy of neutrality, Denmark joined the North Atlantic Treaty Organization (NATO) in 1949. In 1953 a new constitution abolished the upper house of parliament, making it a unicameral (with one chamber) body. The rules governing royal succession were also changed to permit women to succeed to the throne.

Denmark Heads into the Future

In 1973 Denmark strengthened its ties to southern Europe by joining the European Community. The global oil crisis in the 1970s ended the nation's long postwar economic upsurge, as rising inflation and other problems threatened the continuation of its extensive social welfare programs. In 1983 a conservative-liberal coalition government put into effect a series of social and economic reforms that included freezing government spending and giving the government the right to impose settlements in deadlocked labor disputes.

In the 1990s Denmark was once again confronted with a decision about its economic connections to southern Europe—the decision whether or not to ratify the Maastricht Treaty providing for greater European economic integration. After rejecting the treaty in a 1992 referendum, Danish voters approved it in 1993 with a special provision, giving Denmark the option of remaining outside the monetary union that adopted the new unified currency, the euro, in 1999.

Timeline

c. 10,000 B.C.   First human habitation

Danes are known to have inhabited Jutland.

c. 4,000–3,000 B.C.   Agriculture is introduced

Early forms of agriculture are introduced to present-day Denmark. Using wooden plows, farmers cultivate cereal grains. They live in circular huts or longhouses and use flint tools.

1500–400 B.C.   Bronze Age

Bronze-Age Danes have a relatively well-developed culture and have established trade relations with other parts of Europe.

9th–11th centuries A.D.   Viking era

Danes are in the forefront of Viking raids on Western Europe.

The Kingdom of Denmark

9th century   Christianity is introduced

Benedictine missionaries Willibrord and Anschar begin converting the Danes to Christianity, and the first churches and monasteries are founded. The two men are later recognized as saints.

Late 9th century   Danish Vikings invade England

Danish Vikings invade England during the reign of Alfred the Great (871–899), winning control of an area in the east and southeast which they call the Danelaw.

c. 910–85   First Christian king reigns

Harald Bluetooth becomes the first Christian king of Denmark when he is converted during his reign.

986–1014   Reign of Sweyn Forkbeard

Sweyn, the son of Harald Bluetooth, succeeds him on the throne and conquers England near the end of his reign. He also introduces a form of tribute in conquered territories called the Danegeld.

1017–35   Reign of Canute II

Under King Canute II (Canute the Great), Denmark is united with Norway and England. After Canute's death, however, Norway withdraws, and the union dissolves.

1157–82   Waldemar I reigns

Triumphing over his rivals, Waldemar I is crowned king and brings peace to Denmark after a period of civil unrest. He takes military action against the Wends (Slavs) to prevent them from overrunning Jutland.

1202–41   Reign of Waldemar II

Waldemar II ("the Victorious") carries out internal administrative reforms, including the establishment of a code of law and the implementation of Denmark's first census. Waldemar also pursues territorial expansion abroad, waging successful campaigns throughout northern Europe. However, he is captured and held for ransom during his attempt to take Estonia, and a defeat at Bornhöved substantially reduces the size of Danish holdings.

1231   First census is carried out

Under order of King Waldemar II, Denmark has its first census count.

1259–86   Reign of Eric V

Under the reign of Eric V, the Danes wrest control of their land from the counts of Holstein, in an ongoing rivalry over control of Jutland. Under Eric, Denmark gets its first constitution (the "Great Charter"), and a governing assembly, the Rigsrad, is formed.

1282   Eric signs the Great Charter

Eager to limit the monarch's ability to use their taxes for warfare, Danish landowners force Eric V to sign the *Handfaestning* (the "Great Charter"), creating the Rigsrad (parliament).

14th century   Nobles clash with Denmark's kings

Denmark's kings are forced to accede to the growing power of feudal lords.

1340–75   Reign of Waldemar IV

Waldemar IV regains much of the power lost to the nobility by previous monarchs. He regains Danish possessions lost by other kings as well as control over the entrance to the Baltic Sea, enabling Denmark to impose a shipping toll in the area. However, Waldemar is eventually forced to recognize the growing power of the German Hanseatic merchants.

1370   Treaty of Stralsund

King Waldemar IV is forced to make commercial concessions to the Hanseatic League.

1375–87   Margrethe rules as regent

When Waldemar IV dies with no male heirs, his daughter, Margrethe, rules as regent for her own infant son, Olaf, also becoming regent of Norway, as she is the widow of the Norwegian monarch Haakon VI.

1387   Olaf dies

Olaf Haakonsson, heir to both the Danish and Norwegian thrones, dies at the age of seventeen, leaving his mother, Margrethe, to rule both countries.

The Kalmar Union

1397   The Kalmar Union is formed

Through the Kalmar Union, Sweden, Denmark, and Norway are united politically under the Danish crown. Uniform systems of coinage and taxation are introduced. Besides the three Scandinavian countries, the union also includes their possessions, the Faroe Islands, Greenland, and Iceland. The Kalmar Union's first ruler is Eric VII, duke of Pomerania, although Queen Margrethe exercises control until her death in 1412.

1448–81   Reign of Christian I

Christian I succeeds Eric VII as ruler of the Kalmar Union and founds the Oldenburg dynasty. During his reign, Schleswig and Holstein, the two northernmost regions of Germany, come under Danish rule.

1479   University of Copenhagen is founded

Denmark's oldest university, the University of Copenhagen, is founded.

1513–23   Reign of Christian II

The short but eventful reign of Christian II is marked by uprisings that ultimately lead to Swedish independence and to his own overthrow.

1520–23   Swedes revolt

Following the "Stockholm Bloodbath," in which many members of the Swedish nobility are murdered at the order of Christian II, the Swedes revolt and win their independence under the leadership of Gustavus Ericsson Vasa.

1521   Christian II launches reform program

Christian II antagonizes the nobility by implementing laws favoring the common people. This leads to a revolt by the

Danish nobles and eventual civil war, and Christian is forced into exile. When he returns he is captured and imprisoned.

1523   Kalmar Union ends

In the wake of a successful Swedish revolt, Sweden withdraws from the Kalmar Union, ending it. The Swedes elect revolutionary leader Gustavus Ericsson Vasa as their king.

1535–59   Christian III brings Protestantism to Denmark

Under Christian III, Lutheranism becomes the state religion of Denmark.

1546: December 14   Birth of astronomer Tycho Brahe

Danish astronomer Tycho Brahe is acclaimed for the most accurate observations of the heavens before the invention of the telescope. He studies at the universities of Copenhagen, Denmark and Leipzig, Germany. His discovery of a supernova ("Tycho's Star") in 1572 poses a serious challenge to the Ptolemaic system, which postulated the immutability of stars (and also placed the earth at the center of the universe). Tycho sets up a complex observatory on the island of Hveen, making major discoveries about the sun and moon, the other planets, and comets. After moving to Prague, Czechoslovakia, Tycho becomes acquainted with Johannes Kepler, with whom he shares his theories and to whom he leaves his research. Tycho Brahe dies in 1601.

Rivalry with Sweden

1559–88   Reign of Frederick II

War breaks out with Sweden during the reign of Frederick II. A peace agreement is reached (see 1570), but the antagonism between the two countries is not resolved.

1570   Peace of Stettin

A treaty ends the warfare between Denmark and Sweden, but a lasting peace is not achieved for two more centuries.

1588–1648   Reign of Christian IV

Christian IV presides over an architectural renaissance that transforms Copenhagen and the Norwegian capital of Christiania (Oslo) and founds international trading companies to pursue opportunities in the New World and the Far East. He also involves Denmark in the Thirty Years' War (1618–48) and in renewed hostilities with Sweden (see 1611).

1601: October 24   Death of astronomer Tycho Brahe

Tycho Brahe, Danish astronomer whose work poses an important challenge to the geocentric (with the Earth as the center) Ptolemaic system, dies in Prague. (See 1546: December 14)

1611–13   Kalmar War

Denmark defeats Sweden in a renewal of conflict between the two countries.

1618–48   Thirty Years' War

Under Christian IV, Denmark enters the Thirty Years' War, in which it loses territory to Sweden, which in turn emerges as the dominant country in Scandinavia. The war bankrupts Denmark and lays waste to the land, leaving the country in a vulnerable position.

1629   Treaty of Lübeck

The Treaty of Lübeck, signed by Christian IV, acknowledges the successful invasion of Jutland by the armies of the Catholic League, an anti-Protestant alliance under Maximilian, duke of Bavaria.

1648–70   Reign of Frederick III

The reign of Frederick III coincides with the continued military ascendancy of Sweden over Denmark. During Frederick's reign, Denmark loses additional territory as well as toll collection rights.

1655   Frederick builds royal art gallery

King Frederick III has a special extension built onto his castle to house his art collection, resulting in Denmark's first museum. It also houses the royal library.

1658–60   Wars with Sweden

Frederick III fights two wars with Sweden. Denmark is defeated in both, losing land both at home and in Norway, as well as toll rights to the Swedes. The boundaries that emerge at the end of these wars become the modern boundaries of Denmark, Sweden, and Norway.

Absolute Monarchy

1660   Frederick III establishes an absolute monarchy

With Denmark weakening, the Rigsrad (parliament) grants Frederick absolute power, giving the country an inherited rather than an elected monarchy. The king's new power favors the merchant class and puts the provincial nobles at a disadvantage. Frederick institutes a number of reforms, overhauling Denmark's legal system, standardizing its currency and its weights and measures, extending its road network, and building new schools.

1670–99   Reign of Christian V

Christian V, a son of Frederick III, continues the absolute monarchy begun by his father and attempts to restore the

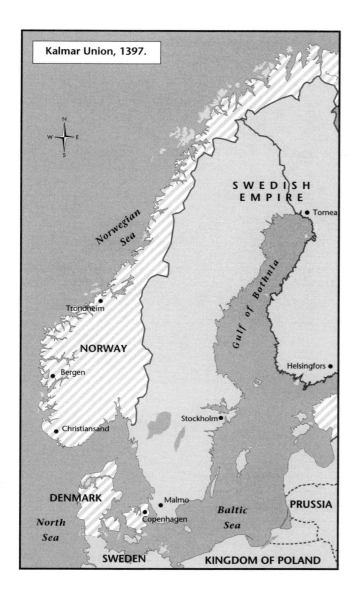

Kalmar Union, 1397.

SWEDISH EMPIRE

• Tornea

Norwegian Sea

Gulf of Bothnia

Trondheim •

NORWAY

• Bergen

Helsingfors •

• Christiansand

Stockholm•

DENMARK

• Malmo

PRUSSIA

North Sea

Copenhagen •

Baltic Sea

SWEDEN   KINGDOM OF POLAND

---

former preeminence of Denmark, forming alliances with Prussia and the Netherlands. During his reign, a powerful Danish navy inflicts military defeats on Sweden but still cannot decisively win the upper hand in the rivalry.

1676: June 1   Battle of Oland

Admiral Niels Juel wins one of Denmark's great naval battles when he triumphs over Sweden in the Battle of Oland, which is followed by a subsequent victory in Koge Bay.

1684: December 3   Birth of writer Ludvig Holberg

Norwegian-born Ludvig Holberg, considered both a Norwegian and a Dane, is the foremost Scandinavian writer of the eighteenth century. Holberg studies at the University of Copenhagen and in England. After a period of travel, he settles in Denmark in 1716 and is appointed to the faculty of the University of Copenhagen the following year. The first of the

humorous works for which Holberg is renowned is a parody of Virgil's *Aeneid* entitled *Peder Paars,* which is published in 1719. It is considered a milestone in Danish literature. In 1722 Holberg begins producing a string of theatrical comedies that establishes his reputation as "the Molière of the North." These plays, which satirize both contemporary Denmark and universal human nature, include *Jean de France* (1723), *Den stundeslose* (The Fussy Man; 1731), and *Erasmus Montanus* (1731). Among Holberg's other works is a satirical novel, *Nicolai Limii iter subterraneum* (The Journey of Niels Klim to the World Underground; 1731). Denmark recognizes Holberg's achievement by making him a baron in 1747. Ludvig Holberg dies in 1754.

1699–1730   Frederick IV reigns

Under Frederick IV Denmark forms an alliance with Prussia, Poland, and Russia and once again becomes involved in warfare against Sweden, this time under the rule of Charles XII. After twenty years of intermittent warfare (the Great Northern War), Frederick turns his attention to reforming the Danish economy and educational system.

1700–21   The Great Northern War

Denmark joins forces with Russia and Poland to counter Swedish dominance in the Baltic region.

1754   Art school is founded in Copenhagen

The Kongelige Danske Kunstakademi, originally launched by King Christian VI, receives its official charter from Frederick V.

1754: January 28   Death of author Ludvig Holberg

Acclaimed playwright and satirist Ludvig Holberg dies in Copenhagen. (See 1684: December 3)

1766–1808   Reign of Christian VII

The reign of Christian VII is the high point of the Enlightenment (a movement based on the belief in reason) in Denmark. Christian's adviser and court physician Count Johann Struensee rises to a position of great influence and is responsible for major administrative reforms, a new legal code, health-care improvements, and the granting of freedom of the press. Many of his achievements, however, are later reversed.

1768: November   Birth of sculptor Bertel Thorvaldsen

Thorvaldsen, Denmark's most famous traditional sculptor, is also known as a collector of paintings; today his collection is housed in the Thorvaldsens Museum in Copenhagen. Thorvaldsen studies at the Academy of Fine Arts in Copenhagen and begins his career making woodcarvings and portrait busts. In 1797 he arrives in Rome with a scholarship to travel and study in Italy. A major statue commission allows Thorvaldsen to remain in Rome after the scholarship expires, and he spends most of his career there. He also receives a number of commissions from patrons in Germany and Poland. Other than a brief stay in 1819, Thorvaldsen does not return to Denmark until 1838, when he begins to plan the creation of a museum to house his art collection. Bertel Thorvaldsen dies in Copenhagen in 1844.

1788 Serfdom is abolished

Serfdom (condition where the serf is bound to the master's land and transfers with the land to a new owner) is ended in Denmark, and the peasants become owners of their own land.

1792 Denmark bans slavery in its territories

Denmark becomes the first European power to abolish slavery in its territories overseas.

The Napoleonic Era

1801 British attack Copenhagen

Antagonized by Denmark's participation in the League of Armed Neutrality formed to resist British searches of vessels at sea, Britain, under the command of Lord Horatio Nelson (1758–1805), attacks Copenhagen's harbor, destroying much of the Danish naval fleet. The Danish capital, which is nearly destroyed, sustains massive damage.

1805: April 2 Birth of Hans Christian Andersen

World-famous fairy tale author Hans Christian Andersen studies at the University of Copenhagen. His first significant published work, a fantasy modeled on the tales of the German author E.T.A. Hoffmann, establishes his reputation as a writer. His next success is with a play (*Mulatten*; 1840) that critiques the institution of slavery. Andersen then writes a series of largely autobiographical novels, including *Improvisatoren* (The Improviser; 1835) and *O.T.* (1836).

In the 1830s Andersen begins publishing the fairy tale collections that will make him a famous and beloved author around the world. His first collection, called *Eventyr, fortalte for born* (Tales, Told for Children; 1835) includes "The Tinderbox" and "The Princess and the Pea." More volumes follow in 1837, 1840, 1843, 1847, and 1852. Among the famous stories included in these works are: "The Snow Queen," "The Ugly Duckling," "The Little Mermaid," and "The Red Shoes." Andersen spends the years from 1840 to 1857 traveling extensively in Europe, Asia, and Africa, and publishes a number of travel books. Hans Christian Andersen dies in 1875.

1803–15 Napoleonic wars

As the armies of the French Napoleon Bonaparte (1769–1821) spread across Europe, Denmark, forced to choose

This house in Odense where Hans Christian Andersen, author of fairy tales, was born is now a museum. (EPD Photos/CSU Archives).

between France and Britain, chooses to side with France, a decision for which it will later pay a steep price. By the end of the conflict, Denmark has lost two of its territories and faces economic bankruptcy.

1813: May 5 Birth of philosopher Soren Kierkegaard

Kierkegaard, a figure whose influence extends beyond his own era to twentieth-century philosophy and theology, receives much of his intellectual training from his father. Kierkegaard is best known for his opposition to Hegelian objective philosophy and stesses faith instead. He graduates from the University of Copenhagen in 1830 and completes an advanced degree in theology in 1841. He writes a series of books dealing with issues of Christian faith, aesthetics (the study of art and beauty), and the nature of human morality. His thinking influences such well-known figures as Henrik Ibsen (1828–1906), Anton Chekhov (1860–1904), Martin Heidegger (1889–1976), and Karl Jaspers (1883–1969). The earliest version of the twentieth-century philosophy of existentialism, developed by Heidegger and Jaspers, reflected the ideas of Kierkegaard, although the better-known postwar variety of existentialism does not. Works by Kierkegaard include *Either/Or, Fear and Trembling,* and *Edifying Discourses* (all published in 1843); *The Present Age* (1846); and *The Sickness Unto Death* (1846–48). Kierkegaard dies in Copenhagen in 1855.

1814: January 14   Treaty of Kiel

Denmark cedes Norway to Sweden to make reparations for having sided with Napoleonic France. The Danish are also forced to cede Pomerania to Prussia.

The Rise of Liberalism

1817: February 22   Birth of composer Niels Gade

Gade, the outstanding musical figure of nineteenth-century Denmark, is born into a musical family and makes his debut as a violinist at the age of sixteen. His early orchestral overture *Efterklange of Ossian* (1840), which wins a prize from the Copenhagen Musical Society, reflects the influence of the Romantic movement in literature and music. Famed composer Felix Mendelssohn (1809–47) becomes a supporter of Gade's music and arranges for the performance of his Symphony No. 1 in C Minor in Germany. He also arranges for Gade a position teaching music at the Leipzig Academy. With the outbreak of war over Schleswig and Holstein in 1848, Gade returns to Copenhagen, establishing a permanent orchestra under the sponsorship of the Musical Society and leading the first Danish performance of the *St. Matthew Passion* by Johann Sebastian Bach (1685–1750). In addition to his influence as a teacher and conductor, Gade has an important influence on Danish music by being the first composer to combine the German Romantic style of composers such as Schumann and Mendelssohn with traditional Scandinavian elements, although this combination is more evident in his early works than in his later ones. Gade dies in 1890.

1836   Music society is founded

The Musikforeningen (Music Society) is founded in Copenhagen and begins to sponsor orchestral and choral concerts. For a century, it is at the center of musical life in Denmark. Preeminent composer and conductor Niels Gade (see 1817: February 22) serves as its music director from 1849 until his death in 1890.

Mid-19th century   System of folk high schools is begun

Denmark institutes its tradition of adult education with a system of folk high schools, special educational facilities planned mostly for rural farmers, who live at the schools for forty weeks and take general adult education courses. There are no examinations or degrees, and individualized study is favored over lecture classes. Receiving support from both the government and private sources, the schools serve as models for comparable programs in other countries.

1844: March 24   Death of sculptor Bertel Thorvaldsen

Sculptor and art collector Bertel Thorvaldsen dies in Copenhagen. (See 1768: November.)

1848   Thorvaldsens Museum opens

The museum planned by sculptor Bertel Thorvaldsen to house the extensive art collection he amassed over forty years in Rome opens four years after the artist's death.

1848–63   Reign of Frederick VII

The rise of liberalism (a political philosophy based upon economic and political freedom and property rights) in Denmark culminates in Frederick VII renouncing absolutism and agreeing to a constitutional monarchy.

1848–50   War over Schleswig and Holstein

The ethnic Germans living in the Danish possessions of Schleswig and Holstein oppose nationalist provisions of the new constitution that make Danish the official language, and they carry out a three-year revolt, aided by Prussia.

Constitutional Monarchy

1849: June 5   Frederick approves a new constitution

Frederick VII signs a new constitution creating a bicameral (having two legislative chambers) Rigsdag (legislature).

1852   Protocol ends Schleswig-Holstein revolt

The revolt of Germans in Schleswig and Holstein ends with an agreement that maintains Danish sovereignty over both duchies but reaffirms the connection between them. Peace is achieved, but major issues underlying the conflict are not resolved.

1855: May 5   Death of Soren Kierkegaard

Philosopher Soren Kierkegaard dies in Copenhagen. (See 1813: May 5.)

1857: June 2   Birth of author Karl Gjellerup

Religion is a central focus for Nobel Prize-winning poet and novelist Karl Gjellerup. Gjellerup, whose father is a parson, studies theology in his youth but then abandons traditional religion in favor of Darwinism and other contemporary theories. Later in his life, however, he becomes interested in religion once again, eventually turning to Buddhism and other Eastern religions. He spends the last years of his life in Germany. In 1917 he receives the Nobel Prize for Literature jointly with fellow Danish author Henrik Pontopoppidan. Gjellerup's works include *En idealist shildring af Epigonus* (An Idealist, A Description of Epigonus; 1878) and *Germanernes lærling* (The Teutons' Apprentice; 1882), both of which express the evolution of Gjellerup's ideas about religion; *Minna* (1889), a contemporary novel set in Germany; and *Pilgrimen Kamanita* (The Pilgrim Kamanita; 1906), which is set in India and deals with reincarnation. Gjellerup dies in 1919.

1857: July 24　Birth of author Henrik Pontoppidan

Nobel Prize-winning author Henrik Pontoppidan is a prolific novelist whose works present a complex, austere, and epic picture of Denmark in his time. Pontoppidan's early works *Landsbybilleder* (Village Pictures; 1883) and *Skyer* (Clouds; 1890) portray rural Danish life. Works from the 1890s, centering mostly on moral and aesthetic issues, include *Nattevagt* (Night Guard; 1894) and *Hoisang* (High Song; 1896). At the turn of the century, Pontoppidan writes the novel *Lykke-Per* (Lucky Per; 1898–1904), much of which is based on events in his own life. Subsequent works include several novel cycles and a four-volume memoir, *Undervejs til mig selv* (En Route to Myself), published between 1933 and 1940. In 1917 Pontoppidan is awarded the Nobel Prize for Literature jointly with Karl Gjellerup (see 1857: June 2), another Danish writer. Pontoppidan dies in 1943.

1863–1906　Reign of Christian IX

The conservatives gain power during the reign of Christian IX, who tries to impose new restrictions on Schleswig and Holstein, thus bringing on a new war with Prussia over the two duchies. On the whole, however, Denmark prospers due to rapid industrialization.

1864　Danish-German War

When Christian IX tries to impose tighter controls over Schleswig and Holstein, Denmark and Prussia go to war, and Prussia enjoys a rapid victory.

1864: October 30　Treaty of Vienna

Denmark ends its war with Prussia by signing the Treaty of Vienna, ceding Schleswig to Prussia and Holstein to Austria. Together, the two territories account for nearly one-third of Denmark's total area.

1865: June 9　Birth of composer Carl Nielsen

Denmark's most famous composer is the seventh of twelve children born to a poor family. He demonstrates musical gifts early in life, composing his first pieces in childhood. He receives his earliest training from his father, who is a painter and musician, and is able to attend the Copenhagen Conservatory with help from a wealthy patron. His first successful composition, the *Little Suite op. 1 for Strings*, is published in 1888, and Nielsen goes on to refine and develop his unique personality as a composer, rejecting the influence of Romanticism and turning back to Classicism as his major model. His First Symphony, premiered in 1894, gets a warm reception.

By 1901 Nielsen receives an annual stipend from the Danish government, allowing him to devote himself to composing without any teaching responsibilities. In 1903 he signs a long-term contract with a publishing house. His 1908 appointment as *kapelmeister* of Denmark's Royal Theatre places him at the center of the capital's musical life, and his compositions receive increasing acclaim. He makes frequent trips abroad to conduct his works. Between 1915 and 1927 he serves as conductor of Copenhagen's Musical Society (*Musikforeningen*). He also serves on the faculty and board of directors of the Copenhagen Conservatory. Nielsen dies of angina on October 3, 1931. Among Carl Nielsen's most famous works are his six symphonies; the opera *Saul og David* (1898–1901); his songs, which had a major influence on the development of Danish vocal music; the Violin Concerto (1911); the Wind Quintet (1922); and the Flute Concerto (1926).

1866　New constitution is adopted

A new conservative constitution makes the upper house of parliament (the Landsting) dominant, expanding its power beyond that of the lower house (the Folketing).

1868　New seaport expands trade with Britain

A new port, Esbjerg, on Jutland's western coast, bolsters trade with Great Britain.

1868　Chamber music society is formed

The Kammermusikforeningen (Chamber Music Society) is formed in Copenhagen to promote the performance of chamber music.

1869: June 1　Birth of working-class novelist Martin Andersen Nexo

Martin Andersen Nexo, raised in poverty, works for social justice through his novels. Nexo is educated with the aid of a wealthy patron and works as a teacher at one of Denmark's folk high schools, which are adult-education facilities, until 1901, when he quits to devote himself to writing full time. He writes two long multivolume novels that illustrate the plight of the disadvantaged through the experience of a working-class protagonist. Part of the first novel, *Pelle the Conqueror* (1906–10), is later made into an award-winning movie directed by Bille August (see 1948). The second novel is *Ditte, mennskebarn* (Ditte, Daughter of Man; 1917–21). Following the Russian Revolution, Nexo joins the Communist Party and makes several visits to the Soviet Union. He writes his memoirs in the 1930s and publishes a sequel to *Pelle the Conqueror* (Morten the Red) in 1945. In 1949 he emigrates to East Germany, where he dies in 1954.

1873: January 20　Birth of author Johannes Jensen

Nobel Prize-winning author Johannes Jensen writes tales, poems, novels, and essays. Among his early works is a three-volume fictionalized biography of King Christian II, *Kongens Fald* (1900–1901). Jensen's first volume of poetry, *Digte*, is published in 1906, and he does not publish more poetry until 1943. Jensen's best-known work is a series of novels that portray the evolution of human beings from prehistoric times to the colonization of the New World. Published between 1908 and 1922, these works are acclaimed both for their literary qualities and their anthropological insights. Jensen is awarded the Nobel Prize for Literature in 1944. He dies in 1950.

1875: August 4   Death of Hans Christian Andersen

Renowned fairy tale author Hans Christian Andersen dies in Copenhagen. (See 1805: April 2.)

1885: April 17   Birth of author Isak Dinesen

Isak Dinesen (pseudonym for Karen Dinesen, Baroness Blixen-Finecke) is a twentieth-century writer known for her Gothic short stories and memoir about her sojourn in Africa. Dinesen is educated in Copenhagen and marries her cousin, Baron Bror Blixen. Together they emigrate to Africa, where they operate a coffee plantation in Kenya. Both Dinesen and her husband also take up big-game hunting. The couple is divorced in 1921, but Dinesen continues running their coffee plantation for ten more years and then returns to Denmark, where she pursues a writing career. Dinesen's first book is her acclaimed memoir *Out of Africa* (1937) about her years in Kenya. During World War II (1939–45), Dinesen writes a novel satirizing conditions under the German occupation. The chief works on which Dinesen's literary reputation rests are her books of imaginative tales, including *Seven Gothic Tales* (1934), *Winter's Tales* (1942), and the posthumously published *Last Tales* (1957). Dinesen dies in 1962.

1885: October 7   Birth of physicist Niels Bohr

Niels Bohr, one of the greatest modern scientists, is born to an academically distinguished family in Copenhagen and educated at the University of Copenhagen, where he earns a doctorate in 1911. His doctoral thesis on the electron theory of metals is still studied today. Between 1912 and 1916 Bohr works on his atomic particle theories with Ernest Rutherford in England, publishing important papers on the subject in 1913 and 1914. In 1916 Bohr accepts a professorship at the University of Copenhagen. In 1920 the University creates the Institute of Theoretical Physics so that Bohr can serve as its director, which he does for the remainder of his career. By 1922 Bohr has been awarded the Nobel Prize for Physics.

In 1939 Niels Bohr visits the United States and works with John A. Wheeler at Princeton on a theory of nuclear fission. After returning to Denmark, Bohr is once again obliged to leave in 1943 to escape the Nazi occupation and returns to the United States, where he advises the scientists working on the first atomic bomb. However, because of moral objections, Bohr never works directly on the project himself. After the war, he returns to Copenhagen and resumes his career there. In addition to the Nobel Prize, Bohr receives many other honors, including formal recognition by the Danish government and appointment to major positions within the scientific community. In 1957 he wins the first Atoms for Peace award. Bohr dies in Copenhagen in 1962.

1888   Women's art school opens

The Kvindelige Kunstskole (Women's Art School) opens as part of Denmark's major art academy, the Kongelige Danske Kunstakademi (Royal Danish Art Academy).

1890: December 21   Death of composer Niels Gade

Composer and conductor Niels Gade dies in Copenhagen. (See 1817: February 22.)

1896   State art museum is formed

The Statens Museum for Kunst, or state art museum, is formed from a merger of other facilities. The collection includes Danish art from the eighteenth century forward, as well as the art of other Scandinavian countries and modern art from France.

1915   Women's suffrage is enacted

Danish women win the right to vote as part of a newly adopted constitution that also lowers the voting age for men and eliminates special voting privileges for the wealthy.

The World Wars

1914–18   World War I

Denmark remains neutral in World War I but still suffers disruption of trade from German mines planted in its waters.

1917   Nobel Prize is jointly awarded to two Danish authors

The Nobel Prize for Literature is awarded jointly to two Danish authors both born in the same year (1857): novelist and short fiction writer Henrik Pontoppidan (see 1857: July 24) and Karl Gjellerup (see 1857: June 2), a poet and novelist, several of whose works reflect his search for religious faith in the modern world.

1918   Iceland becomes autonomous

Denmark grants Iceland self-rule. However, the Danish king remains Iceland's official head of state.

1919: October 11   Death of author Karl Gjellerup

Nobel Prize-winning author Karl Gjellerup dies in Germany. (See 1857: June 2.)

1920   Part of Schleswig rejoins Denmark

Following a plebiscite (a direct vote of the peopole on a polit-ical issue), North Schleswig once again becomes part of Den-mark.

1922   Niels Bohr receives the Nobel Prize

Scientist Niels Bohr is awarded the Nobel Prize for Physics in recognition of his work on the quantum theory of matter and other achievements. (See 1885: October 7.)

1930s   Denmark weathers the Depression

To counter the effects of global depression, Denmark's gov-ernment implements an economic rescue plans that includes trade controls, guaranteed farm prices, and measures to relieve the nation's debt.

1930   New music society is founded

Det Unge Tonekunstnerselskab (DuT), a society for new music, is created from the merger of two existing groups. Publishing contemporary musical scores and sponsoring per-formances of new music, it becomes Denmark's major pro-moter of contemporary compositions.

1932   Social Democrats come to power

The Social Democratic Party gains power, leading to the inauguration of Denmark's welfare state (a system whereby the government provides a series of welfare benefits for all of its citizens during the course of a lifetime).

1933   International court rules on the status of Greenland

Denmark's claim to Greenland is upheld by the International Court of Justice in The Hague.

1939: May 31   Denmark and Germany sign pact

Denmark and Germany sign a ten-year nonaggression pact which the Germans violate within a year.

1940: April 9   Germany invades Denmark

Germany launches a surprise invasion of both Denmark and Norway. Most of Denmark's merchant fleet escapes and cooperates with the Allies. Throughout the war, the Germans occupy Denmark. The Danish populace becomes known for its efforts to protect Denmark's Jewish population. When the Nazis order all Jews to wear yellow arm bands to make them easily identifiable, many non-Jewish Danes—including the king—wear the arm bands also to defeat their purpose.

1943   Danish resistance escalates

In response to repression by the German occupation, the Dan-ish government ends all collaboration with the occupiers, and the entire cabinet resigns. The Freedom Council, the Danish underground government, expands its opposition activities.

1943: August 21   Death of author Henrik Pontoppidan

Nobel Prize-winning novelist Henrik Pontoppidan dies. (See 1857: July 24.)

1944   Author Johannes Jensen receives the Nobel Prize

Novelist, poet, and essayist Johannes Jensen is awarded the Nobel Prize for Literature. (See 1873: January 20.)

1945: May   Denmark is liberated by the British

British troops liberate Denmark from its German occupiers. A temporary coalition government is formed until elections can be held.

The Postwar Era

1945: October   Elections are held

Denmark's first postwar elections are held, and the Social Democrats begin their tenure as Denmark's leading political party.

1947–72   Reign of Frederick IX

Frederick IX reigns as king of Denmark.

1948–53   Denmark receives aid through the Marshall Plan

Denmark's postwar economy recovery is hastened by $350 million in aid from the Marshall Plan run by the United States.

1948   Birth of film director Bille August

Bille August, Denmark's foremost film director, trains as a commercial photographer and then studies cinematography at the Danish Film School, graduating in 1971. He works on films including *Karleken* (1980) and *The Grass Is Singing* (1981). By the early 1980s he turns to directing. His first major success is *Zappa* (1983), followed by an even more popular sequel, *Twist and Shout* (1986). The following year August gains international recognition for *Pelle the Con-queror,* based on an early-twentieth-century novel by Nobel Prize-winning Danish author Martin Andersen Nexo (see 1869: June 1). August's film wins both the Golden Palme at the Cannes Film Festival and the Best Foreign Film Academy Award in the United States. Other films directed by August include *The Best Intentions* (1992), based on a screenplay by Swedish director Ingmar Bergman; *The House of the Spirits* (1993), based on the novel by Chilean author Isabel Allende; and most recently, *Smilla's Sense of Snow* (1997), based on an acclaimed novel by Peter Hoeg.

1948: March 23   The Faroe Islands win home rule

The Faroe Islands, located in the Atlantic Ocean midway between Iceland and Norway, are granted home rule under Danish sovereignty. The islands have their own flag, currency, and postage stamps. Involvement by the central Danish government is limited to foreign affairs and international trade matters. The Faroese use their own language and teach Danish in the schools as a foreign language.

1949: April 4   Denmark joins NATO

Abandoning its former policy of neutrality, Denmark becomes a founding member of the North Atlantic Treaty Organization (NATO).

1950: November 25   Author Johannes Jensen dies

Nobel Prize-winning novelist Johannes Jensen dies in Copenhagen. (See 1873: January 20.)

1952   Denmark is a founding member of the Nordic Council

Denmark helps found the Nordic Council to promote cooperation among the Scandinavian countries.

1953   New constitution is adopted

Denmark enacts a new constitution that makes major changes in the organization of its government. The *Folkesting*, formerly the lower chamber in a bicameral (with two legislative chambers) parliament, becomes the only chamber, as the upper house, or *Landsting*, is abolished. The line of succession to the throne is also altered to permit the succession of women. Local governments are given greater power. The island of Greenland also gets a new constitution, giving it the same status as Denmark's other territories.

1954: June 1   Death of author Martin Nexo

Working-class novelist Martin Nexo dies in East Germany. (See 1869: June 1.)

1955   Kiel Declaration

West Germany guarantees the rights of the Danish minority in South Schleswig, which was formerly part of Denmark.

1958–73   Economic "golden years"

Denmark enjoys a long period of economic prosperity fueled by a strong European economy and expanding world trade.

1958   Modern art museum opens

The Louisiana Museum opens in Humlebæk. Built to house the collection of its director, Knud W. Jensen, it becomes Denmark's foremost modern art museum.

1960   Denmark joins the EFTA

Denmark becomes a member of the newly formed seven-member European Free Trade Association (EFTA), whose formation spurs an increase in trade among Scandinavian countries. The other original members include: Austria, Great Britain, Norway, Portugal, Sweden, and Switzerland.

1962: September 7   Death of author Isak Dinesen

Isak Dinesen, author of short stories and the memoir *Out of Africa,* dies in Rungsted. (See 1885: April 17.)

1962: November 18   Death of physicist Niels Bohr

Nobel Prize-winning physicist Niels Bohr dies in Copenhagen at the age of seventy-seven. (See 1885: October 7.)

1969: May 30   Boycott imposed on trade with South Africa

Denmark imposes a government boycott on all trade with South Africa to protest that nation's apartheid policy of racial separation and segregation.

1972   Reign of Queen Margrethe II

Margrethe II succeeds her father, Frederick IX, on the throne of Denmark. (Since 1953 women have been in the royal line of succession [see 1953].)

1972   Danish voters endorse Common Market membership

Danish voters approve Denmark's entry into the European Economic Community (EEC), also known as the Common Market.

1973–74   World oil crisis dampens Denmark's economy

The global oil crisis sends Denmark's economy into a downturn, raising inflation, unemployment, and trade deficits, and threatening the continuation of the nation's comprehensive social welfare program.

1973   Economic crisis unseats Social Democrats

The Social Democrats and other parties that favor increased social spending suffer an electoral setback as the economic underpinnings of Denmark's social welfare programs are threatened by the global oil crisis. The Social Democrats lose more than a quarter of their support, and a period of shaky coalition governments follows. In the absence of a strong nonsocialist coalition, the Social Democrats return to power after a hiatus (a gap) of two years.

1973: January 1   Denmark joins the EEC

Denmark is formally admitted to the European Economic Community (EEC). This is the leading Western European economic organization that has as its goals the creation of a common market throughout the Continent. The organization is later known as the European Community (EC), and, in 1993, the European Union (EU)

1979 Greenland gets home rule

Greenland receives home rule while remaining a Danish territory. It is represented in the Danish parliament by two legislators.

1983 Conservative-Liberal coalition comes to power

A strong Conservative-Liberal coalition with Poul Schlüter as prime minister comes to power and carries out social and economic reforms. Schlüter freezes government spending and assumes the right to impose wage settlements in cases of deadlocked labor negotiations. In order to lower inflation, wage levels are no longer indexed to the cost of living.

1985 Greenland withdraws from the EC

Due to international tension over fishing rights and other matters, the Danish possession of Greenland withdraws from the European Community (EC).

1985 Queen designs commemorative stamp

Denmark issues a stamp commemorating the fortieth anniversary of liberation from Germany. The stamp is designed by Queen Margrethe, an artist who has illustrated several books.

1985: March–April Widespread strikes cripple Denmark

Day-to-day economic operations are paralyzed by massive strikes. Resorting to powers conferred under recent economic reforms, Denmark's parliament imposes a settlement.

1989 Marriage laws are liberalized

Denmark's marriage laws are reformed to confer legal marital status on all couples in a "stable partnership," whether they are heterosexual or homosexual, making them eligible for all rights enjoyed by traditionally married couples, except the right of adoption.

1990–91 Denmark sends troops to the Persian Gulf

Denmark participates in the Allied coalition during the Persian Gulf War.

1992: June Voters reject Maastricht Treaty

Danish approval of the Maastricht Treaty providing for European economic integration is narrowly rejected by voters in a referendum.

1993: January Immigration scandal topples Schlüter government

The government of Poul Schlüter resigns in the wake of a scandal involving illegal efforts to prevent the immigration of Sri Lankan Tamil refugees with family members already in Denmark and a subsequent cover-up. The affair highlights the controversy over Denmark's growing non-European immigrant population, which will account for 2.2 percent of the population by 1994.

1993: January Center-left government coalition is formed

Following the resignation of Poul Schlüter's government because of the "Tamil Affair" (see above 1993: January), the Social Democrats, led by Nyrup Rasmussen, form a coalition government.

1993: May Maastricht Treaty wins approval in second vote

After defeating the Maastricht Treaty in a previous referendum, Danish voters approve the measure following the addition of a special provision giving Denmark the option of staying out of the monetary union to be formed in 1999 if Danish voters are against it at that time.

1995: July 17 Sweden-Denmark tunnel contract is awarded

The Danish and Swedish governments hire a consortium (partnership or alliance) of European companies to build the proposed Oresund tunnel which will connect Copenhagen with Malmo, Sweden. The 2.3 mile (3.7 kilometer) tunnel is expected to cost 3.8 billion Danish krona (U.S. $680 million).

1996: May 20 Film by Danish director takes second place at Cannes

The film *Breaking the Waves* by Danish director Lars von Triers wins the second-highest prize at the Cannes Film Festival, the Grand Prix.

1996: July 21 Dane wins the Tour de France

Bjarne Riis becomes the first Dane to win the 2,423 mile (3,900 kilometer) Tour de France bicycle race, winning a prize of 2.2 million French francs (U.S. $400,000).

1998: March 11 Social Democrats make strong showing in elections

The Social Democrats' showing in parliamentary elections is stronger than anticipated, and the ruling center-left coalition remains in power under Prime Minister Poul Nyrup Rasmussen. The Social Democrats win sixty-five seats (up from their previous total of sixty-two), while the Liberal Party wins forty-three. The right-wing Danish People's Party, campaigning on an anti-immigration platform, sees its representation rise from four seats to thirteen.

1998: April 27   Labor stages national walk-out

A nationwide strike idles roughly twenty percent of Denmark's work force, as the Confederation of Danish Trade Unions rejects a pay proposal because of inadequate vacation time. Factory operations and transportation are disrupted, as are distribution of food and fuel, and schools are forced to close.

1998: May 5   Nonstriking workers locked out to pressure strikers

About 60,000 nonstriking workers are locked out of their jobs to place pressure on employees who have been on strike since April 27 over a contract dispute.

1998: May 7   Parliament forces strikers to return to work

The Danish parliament passes legislation forcing the Confederation of Danish Trade Unions (LO) to end a nationwide strike and attempting to find a compromise between the labor and management positions on vacation pay. Workers receive two extra vacation days per year.

1998: June 14   World's second-longest suspension bridge opens

A suspension bridge connecting the islands of Sjoelland and Funen is officially opened by Queen Margrethe II. At 5,320 feet (1,620 meter), it is the world's second-longest suspension bridge, and Europe's longest. The bridge will make it possible to travel from Copenhagen to the European mainland by land.

Bibliography

Borish, Steven M. *The Land of the Living: The Danish Folk High Schools and Denmark's Non-violent Path to Modernization*. Nevada City, Calif.: Blue Dolphin Publishers, 1991.

Goldberger, Leo, ed. *The Rescue of the Danish Jews: Moral Courage Under Stress*. New York: New York University Press, 1987.

Holbraad, Carsten. *Danish Neutrality: A Study in the Foreign Policy of a Small State*. New York: Oxford University Press, 1991.

Jensen, Jorgen. *The Prehistory of Denmark*. New York: Methuen, 1983.

Johansen, Hans Christian. *The Danish Economy in the Twentieth Century*. New York: St. Martin's Press, 1987.

Jones, W. Glyn. *Denmark: A Modern History*. Wolfeboro, N.H.: Longwood, 1986.

Kjrgaard, Thorkild. *The Danish Revolution, 1500–1800: An Ecohistorical Interpretation*. Cambridge: Cambridge University Press, 1994.

MacHaffie, Ingeborg S., and Margaret A. Nielson. *Of Danish Ways*. New York: Harper & Row, 1984.

Miller, Kenneth E. *Denmark, a Troubled Welfare State*. Boulder, Colo.: Westview Press, 1991.

———. *Friends and Rivals: Coalition Politics in Denmark, 1901–95*. Lanham, Md.: University Press of America, 1996.

Monrad, Kasper. *The Golden Age of Danish Painting*. New York: Hudson Hills Press, 1993.

Pundik, Herbert. *In Denmark It Could Not Happen: The Flight of the Jews to Sweden in 1943*. New York: Gefen Publishing House, 1998.

Estonia

Introduction

Until recently, Estonia was part of the Union of Soviet Socialist Republics or USSR and was known as the Estonian Soviet Socialist Republic. The collapse of the Soviet Union in August, 1991, meant freedom for the Baltic states and other Eastern and Central European countries. Estonia reestablished the independence it had experienced once before during the interwar years, from 1918 to 1940, and officially became a parliamentary democracy known as the Republic of Estonia *(Eesti Vabariik)*.

Estonia has a total area (land and water) of 17,413 square miles (45,100 square kilometes), which is just a little smaller in size than the combined area of Vermont and New Hampshire. Estonia is bordered by the Gulf of Finland to the north, Latvia to the south, Russia to the east and the Baltic Sea to the west. Finland lies north of the Gulf of Finland, and Sweden lies west of the Baltic Sea. Geographically Estonia is mainly flat, with approximately forty-eight percent of the country covered by forest land and over 1,400 natural and manmade lakes. Its territory includes 1,521 islands in the Baltic Sea, and its coastline measures 3,800 kilometers.

Tallinn, the capital of Estonia and its largest city, is located on the north coast and has a population of over 415,200. As of 1999, the total population of Estonia is an estimated 1,445,100. Estonian is the official language, and Lutheranism the leading religion. Ethnic Estonians account for roughly sixty-five percent of the total population. The second largest ethnic group are Russians who follow the Orthodox religion and account for 28.1 percent of the population.

History

The Baltic coastal region attracted inhabitants as early as the eighth century B.C. Among those tribes settling there sometime between 2,500 and 1,500 B.C. were the Finno-Ugric and proto-Baltic, who, together with those peoples already in the region, became the ancestors of the Estonians. Influences at the beginning of the Christian era included the Roman Empire and Germanic tribes. By the twelfth century the Baltic peoples still lacked a centralized form of government. Numbering about l00,000 in population, they were active in trade and maintained their independence, resisting the military advances of Vikings and Slavic tribes.

In the thirteenth century, the Baltic peoples were conquered by crusading German knights of the Teutonic Order, who imposed Christianity on the region. In the mid-thirteenth century the Danes conquered Northern Estonia, which included Tallinn, the present-day capital. For about 100 years the Baltic peasants resisted control of their land by oppressive landlords. From 1343–45 Estonian peasants staged a series of unsuccessful revolts against the Germans and Danes. Ultimately, Denmark ceded Northern Estonia to the Germans, and more than two centuries of peace followed. Major Estonian cities developed, including Tallinn and Tartu, as trade flourished.

By the sixteenth century, Russia posed a threat to German control of the Baltic states. In 1558 Ivan IV (the Terrible), Czar of Russia from 1533 to 1584, successfully invaded Estonia. His victory, however, was short-lived. In 1561 Sweden seized control of northern Estonia while Poland took southern Estonia. The Russian armies fought to maintain their positions but were forced out of Estonia in 1583. With Russia removed, Sweden and Poland competed for control of the land, and by 1629 Sweden had triumphed.

The Swedes imposed serfdom on the Estonian peasants while allowing the German nobility to retain its privileges in return for its support of Swedish rule. In spite of the hard life lived by most Estonians during this period, economic progress was made, and educational advances were achieved. Under the Swedes Tartu University was established in 1632. Swedish control also brought the Protestant Reformation to Estonia with the introduction of Lutheranism as the official religion.

At the beginning of the eighteenth century, Peter I (the Great) (1682–1725) drove the Swedes out of Estonia and imposed Russian imperial rule, which lasted for over two centuries. As was the case under Sweden, the privileged minority of German nobles retained their privileges, while oppression continued for the Estonian peasants. By the early nineteenth century, the revolutionary ideals of the Enlightenment had led to the abolition of serfdom. By mid-century, a nationalist movement was gaining strength, encouraged by a cultural

revival that encompassed the composition of the national epic, the *Kalevipoeg*. Estonian nationalism survived a concerted effort to impose Russian culture on the populace (Russification) and led to a declaration of independence following World War I.

Between the two world wars, Estonia was an independent republic governed as a parliamentary democracy. However, in 1934 the growing threat from fascists within the country prompted the administration of elected president Konstantin Paets to suspend ordinary political processes, making it possible for Paets to remain in power as a moderate dictator for the remainder of the decade. Estonia was nevertheless unable to withstand authoritarian rule; it came from the Soviet Union, which gained control of the country under its 1939 nonaggression pact with Germany, which placed the Baltic states within the Soviet sphere. The Soviet army invaded Estonia in September, 1939, and annexed the country as a Soviet republic. The Soviet Army was driven out by Germany in 1941 but retook control in 1944.

During the postwar period, the USSR launched an intensive effort to assimilate Estonians and erase their spirit of national and cultural identity, implementing mass immigration of non-Estonians into the republic and deporting large numbers of ethnic Estonians. Nevertheless, Estonian nationalism remained strong, and, along with its Baltic neighbors, Estonia was one of the first Soviet republics to declare independence, although its 1990 declaration was rejected by

Soviet leader Mikhail Gorbachev (b.1931). Following the aborted coup that led to the demise of the Soviet Union, Estonia became an independent republic in August, 1991. Since that time it has been a parliamentary democracy, with free national elections held in 1992, 1995, and 1999. On September 17, 1991 Estonia was granted full membership in the United Nations. Negotiations over membership in the European Union were initiated in 1998.

Timeline

2500–1500 B.C.   Arrival of Finno-Ugric and proto-Baltic tribes

Finno-Ugric and proto-Baltic tribes settle in the region sometime between 2,500 and 1,500 B.C. They mix with the existing peoples to create the forerunners to the Estonians. During this era the runic folk song, the oldest form of Estonian folk culture, begins. Religion is practiced in the form of totem worship, shamanism, and pagan rituals.

8th century B.C.   Early inhabitants arrive in the Baltic

Inhabitants are attracted to the Baltic coastal region due to its maritime (seafaring) characteristics. The early inhabitants of this region are visited by many different tribal groups migrating to the east.

1st–6th centuries A.D.   Spread of Roman and Germanic culture

The Roman Empire and numerous Germanic tribes provide the Estonians and their Baltic neighbors tremendous opportunities for social and cultural development and increased trade. Estonians gradually come together in small clusters called parishes which are then grouped together in larger areas called counties.

8th–12th centuries   Baltic peoples remain independent

The Estonians and their Baltic neighbors live freely as peasants. Although they are organized into parishes and counties, they still lack a centralized form of government. Numbering about 100,000, they are commercially active and wage war with the Scandinavian Vikings (mainly from Norway, Sweden, and Finland) and Slavic tribes in order to maintain their independence.

11th century   Christianity is introduced

Christianity appears in the Baltic territories.

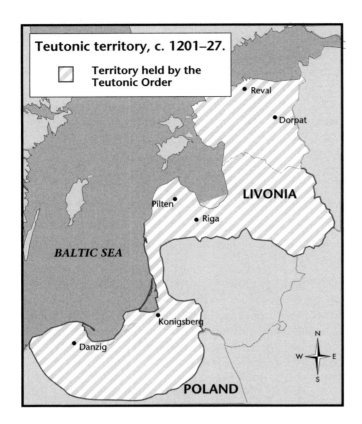

Teutonic territory, c. 1201–27.

☐ Territory held by the Teutonic Order

Reval

Dorpat

LIVONIA

Pilten

Riga

BALTIC SEA

Konigsberg

Danzig

N
W E
S

POLAND

German and Danish Rule

1201–27   The Baltic region is conquered by the Teutonic Order

The Baltic peoples are conquered by crusading German knights, known as the knights of the Teutonic Order, who undertake to convert the populace to Christianity. They seize control of southern Estonia, Latvia, and Lithuania, merging southern Estonia and northern Latvia into a region they call Livonia.

Mid-13th century   Northern Estonia is conquered by Denmark

The Danes, allied with the Teutonic knights, conquer Northern Estonia, including Tallinn. Estonia's land is divided into small feudal states governed by the German and Danish nobility. The first examples of written Estonian are traced to this period.

1343–45   Peasant revolts

Baltic peasants, increasingly enslaved to the land, have sought to regain control of it from their oppressive landlords for a period of about 100 years. Between 1343 and 1345, Estonian peasants lead a series of revolts, known as the St. George's Night Uprising, against the Germans and Danes.

14th–16th centuries   German rule

Although the nobility defeats peasant attempts at independence, Denmark decides to turn northern Estonia over to the Germans. As a result, peace settles on the Baltic states for over 200 years, allowing for the growth of towns, commerce and industry, increased east-west trade, and the development of major Estonian cities such as Tallinn and Tartu.

1525   Publication of the first Estonian book

The first book in the Estonian language is printed.

Estonia Becomes a Battleground

1558   Ivan the Terrible invades Estonia

Due to the strategic location of Estonia as a trading center, the Germans face increasing threats to their control. Soon the German nobility is unable to repel rivals from the north and east. Ivan IV, the Terrible, Czar of Russia from 1533–84, successfully invades Estonia, beginning a twenty-five-year struggle for control of the region. He is challenged by both Sweden and Poland.

1561   Estonia is invaded by Sweden and Poland

Sweden seizes control of northern Estonia while Poland takes southern Estonia.

1583   Russia is expelled from Estonia

After over two decades of fighting, the Russian armies are forced out of Estonia by Sweden and Poland, who then compete against each other for control of the land.

Control by Sweden

1629   Sweden wins control of Estonia

Sweden succeeds in forcing Poland out of Estonia. Under Swedish rule, life in Estonia is often harsh for Estonian peasants, who have lost much land and many lives during the war years. A system of serfdom is imposed under which the peasants, or serfs, can be bought or sold with the land on which they live and work, and they are taxed regularly. The German nobility retains its landed privileges and manors in exchange for their support of Sweden. However, as difficult as times are for Estonians, the Swedes are instrumental in helping Estonia advance in several key areas, including manufacturing and education. It is during the Swedish period that Tartu University, Estonia's largest, is founded. Later in the century, several parish schools open. The population grows to a total of almost 400,000 during this period, and Lutheranism becomes the official state religion.

1632   Tartu University is founded

Estonia's largest university is established. It later becomes a leading scientific and cultural center in Imperialist Russia.

Czarist Rule

1710   Russia reconquers Estonia

Peter I (the Great) (r. 1682–1725) takes control of Estonia from Sweden, inaugurating two centuries of Russian Imperialist rule ending with the last czar, Nicholas II (1868–1918). The German nobility, also called the Baltic Germans, who represent a small percentage of the population, thrive under Russian rule, while the peasants have to organize and fight for reform. Russian Orthodox Christianity replaces Lutheranism as the leading religious doctrine.

1739   Bible is translated into Estonian

The Bible is published into Estonian.

1803: December 26   Birth of national poet F. R. Kreutzwald

Poet and folklorist Friedrich Reinhold Kreutzwald is the author of Estonia's national epic, the *Kalevipoeg* (see 1857). A physician, Kreutzwald serves as medical administrator in Voru for most of his career. In 1838, already planning the composition of a folk epic, he founds the Estonian Learned Society to promote the collection of ethnic folk material. The poem, which combines folk elements with original poetry in the German Romantic tradition, is published between 1857 and 1861. Kreutzwald dies in 1882.

1816–19   Reforms advance peasants' rights

The nineteenth century opens with much promise for Estonia. The intellectual revolution of the previous century has created an increased awareness of the need for reforms to improve peasants' lives. Between 1816 and 1819 peasants win increased rights to their personal property and rights to farm their land in perpetuity or without interruption. These reforms lead to the formal abolition of serfdom in Estonia by the end of this period. Peasants become landowners and open schools for their children. They even attend Tartu University.

1830s   The Industrial Revolution begins

The Industrial Revolution gives peasants the opportunity to move off the land and become urban dwellers and skilled laborers. The growth and development of metal and machinery industries and cotton and wood processing industries improve Estonia's economy and, in turn, improve living conditions for many. Peasants who prefer farm life form agricultural cooperatives and farmers' associations.

1843   Birth of poet Lydia Koidula

Lydia Koidula (1843–86) is Estonia's first woman poet and is often called the poet of national awakening. She plays a central role in Estonia's cultural awareness because she uses poetry to express her deep love for her country. She helps motivate Estonians, especially women, to raise their political consciousness.

1857–61   Publication of the *Kalevipoeg,* Estonia's national epic

Poet F. Reinhold Kreutzwald composes the *Kalevipoeg*, a nationalist text based on folk material gathered by F.R. Faehlmann and others, and deliberately presented in the guise of a historic text. (One English translation is titled *Kalevipoeg: An Ancient Estonian Tale.*) The work, in which original poetry by Kreutzwald is combined with folktales and folk songs, recounts the adventures of the title character.

1869   First song festival is organized

The first Estonian song festival takes place in Tartu. One thousand singers participate, and virtually the entire population of Tartu (15,000) attends the event.

1870   Theaters are founded

The Estonia Theatre in Tallinn and the Vanemuine Theatre in Tartu are established.

1880s   Estonian nationalism survives Russification policies

A new spirit of nationalism (pride in one's people and culture) and the desire for independence emerge in mid-nineteenth century Estonia. Everywhere Estonians begin to redefine their history and society in ways meant to clearly distinguish them from Imperial Russia. Peasants start their own language schools, the first Estonian national song festival is held, and national theaters are established. A national epic, the *Kalevipoeg*, is composed from collected folk material (see 1857).

Estonian demands for more sweeping reforms force Imperial Russia to tighten its hold on the Baltic state by strengthening its own nationalism through a policy of Russification. This policy calls for Russianizing ethnic Estonians using local authorities operating under central Russian control. Social, cultural, and political organizations supporting the national Estonian movement are censored and, if necessary, eliminated. But the Estonian national spirit remains strong and outlasts Russification, which ends with the century.

1882: August 25   Death of poet F. R. Kreutzwald

F. Reinold Kreutzwald, author of Estonia's nineteenth-century epic, the *Kalevipoeg*, dies in Tartu. (See 1803.)

c. 1900   Professional musicians flourish in Estonia

As professionally educated composers and performers begin to multiply in Estonia, orchestra and choral concerts become common.

c. 1902–17   Young Estonia movement

A widely based cultural movement, encompassing theater, the visual arts, and literature, is spurred by the growth of Estonian nationalism in the early twentieth century. Politically it is linked to socialism, but it also advocates the importance of the individual. The foremost figure of the movement is poet Gustav Suits (1883–1956), who originates many of its philosophies. These philosophies are further developed in the periodical *Noor-Eesti,* published in 1910–11. Many of the artists associated with Young Estonia undertake an integration of European trends, such as Postimpressionism and Symbolism, with native Estonian themes. Estonian landscapes are an especially popular subject among these artists, who include Konrad Magi (1878–1925), Aleksandr Tassa (1882–1957), Nikolai Triik (1884–1940), and a number of others.

1905   Russian revolution sparks nationalist activity

The failed revolution against the Russian czar spurs Estonian demands for political and cultural autonomy. Once the czarist regime is restored, it refuses to meet any of these demands.

1905: June 18   Birth of composer Eduard Tubin

Tubin, the foremost Estonian composer of his generation and also a well-known conductor, studies at the Tartu Academy (1924–30). He conducts the Vanemuine Theatre Orchestra (1930–44). Between 1934 and 1973 he composes ten symphonies, in addition to concertos, ballets, violin sonatas, solo songs, and two operas. Early influences on his compositions include Igor Stravinsky (1882–1971) and the Hungarian composers Bela Bartok (1881–1945) and Zoltan Kodaly (1882–1967), whom he meets in Budapest. In 1944 Tubin moves to Sweden. He dies in Stockholm in 1982.

Interwar Independence

1918: February 24   Estonia declares independence

The collapse of Imperialist Russia in 1917 and Germany's defeat in World War I in 1918 give Estonia its long-awaited opportunity to become an independent nation, and an Estonian Provisional Government assumes power. This date is celebrated as a national holiday, Independence Day.

1918–39   The interwar period

During the interwar years (time between the end of World War I and the beginning of World War II) the new Estonian government emphasizes the development and growth of agriculture and industry, especially textiles and machinery. A series of major land reforms divide large estates belonging to the German nobility and parcel them out to small farmers and war veterans. As a result the number of independently owned small farms grows to over 125,000. In order to further reduce German influence and increase Estonian self-awareness, numerous language reforms are introduced during the 1920s and 1930s. With Soviet economic markets closed to it, Estonia has to seek new markets in Western Europe. By the end of the 1930s Estonian industry is flourishing.

1919: October   Major land reforms are passed

Land reform laws break up most of Estonia's landed estates and distribute the land to rural laborers.

1920   Tartu Music Academy is founded

The Tartu Music Academy is established.

1920   Women's suffrage is passed

Estonian women win the right to vote.

1920: December 21   Estonian constitution is adopted

A new Estonian constitution, approved by the national assembly, goes into effect. It provides for a unicameral one hundred-member legislature called the *Riigikogu.*

1920: February 20   Tartu Peace Treaty is signed

Within less than a year after declaring independence, Estonia faces a new threat from Soviet Russia, which attempts to permanently annex the region. By 1919, two-thirds of Estonia is under Soviet control. However, the Estonians fight back and regain control of their land in a war of independence. With the signing of the Tartu Peace Treaty, the Republic of Estonia is born.

1921   Estonia joins the League of Nations

Estonia becomes a member of the League of Nations.

1923   Music school is founded in Tallinn

The Tallinn Conservatory is established.

1924: December   Communist revolt is foiled

A Soviet-sponsored Communist revolt in Tallinn fails, and the Communist Party is forced to go underground.

1926   Radio broadcasts begin

Estonian radio begins offering regular broadcasts.

1929: March 29   Birth of Lennart Meri

Lennart Meri, Estonia's first post-communist-era president, is born in Tallinn. In the wake of the Soviet takeover of Estonia during World War II, Meri's family is exiled to Siberia

(1941–46). Graduating from Tartu University in 1953, Meri embarks on a career as a writer and filmmaker. He is appointed foreign minister of Estonia in 1990 and participates in the creation of the Baltic Council, formed by the newly independent Baltic republics. In 1992 he is named ambassador to Finland. Meri is elected president of Estonia in 1992.

1933: October   Voters approve new right-wing constitution

The right-wing Association of Estonian Freedom Fighters, which has gained strength over several years as an anti-communist group, proposes the adoption of a new constitution with a strong chief executive. In a referendum, Estonian voters approve the constitution, which provides for a popularly elected president elected to a five-year term. Executive power is transferred from the legislature to the president.

1934: January   Paets seizes power

In order to head off an electoral victory by the fascist Freedom Fighters candidate, Prime Minister Konstantin Paets seizes control of the government and sets up a dictatorial but politically moderate regime. Paets rules by decree, postponing presidential elections indefinitely. He bans the Freedom Fighters and places a ban on all political assemblies.

1935: February   Political parties are banned

The Paets government abolishes all political parties.

1935: September 11   Birth of composer Arvo Pärt

One of Estonia's leading twentieth-century composers, Arvo Pärt studies at the Tallinn Conservatory. Between 1957 and 1967, he works as sound director for Estonian radio. His early musical influences include Russian composers Serge Prokofiev (1891–1953) and Dmitri Shostakovich (1906–75). His later works employ serialism and other *avant garde* (unconventional) compositional techniques, including Charles Ives-style *collages* (an art form in which unrelated objects are pasted together) incorporating well-known melodies in unusual and dissonant contexts. Well-known works by Pärt include the *First Symphony,* the *Perpetuum Mobile, Pro et Contra,* and *Laul Armastatule.* Pärt emigrates to Berlin in 1982.

1937   Tallest radio tower is built in Turi

The tallest radio tower in Europe is built in Turi.

1937   Birth of conductor Nëeme Jarvi

Internationally known orchestra conductor, Nëeme Jarvi is born in Tallinn. He is educated in Tallinn and at the Leningrad Conservatory. He serves as the music director of the Estonian State Symphony Orchestra and the director of the Estonian Opera, making his Metropolitcan Opera debut in the United States in 1979, when he conducts Tchaikovsky's opera *Eugene Onegin.* Jarvi emigrates to the U.S. the following year. His subsequent appointments include principal guest conductor of the City of Birmingham Symphony Orchestra (1981–84), principal conductor of the Scottish National Orchestra (1984–88), and music director of the Detroit Symphony Orchestra beginning in 1990.

1937: February   New constitution is adopted

A new constitution provides some return to parliamentary government but retains a strong presidency. The election of opposition members of parliament is allowed once again.

1938   Academy of Sciences is founded

The Estonian Academy of Sciences, located in Tallinn, is founded. It consists of fifty-six different institutes representing various scientific, social science, and humanities disciplines.

World War II (1939–45)

1939: August   Soviet-German pact places Estonia under Soviet control

The Molotov-Ribbentrop nonaggression pact between Germany and the Soviet Union places Estonia and the other Baltic states within the Soviet sphere of influence.

1939: September 28   Soviet troops enter Estonia

With Estonia having been forced to sign an alliance allowing the stationing of Soviet troops in the country, the first troops enter Estonia.

1940   All property is nationalized

The Soviet Union takes government control of all Estonian property. Many people are forced to surrender their single-family homes and live in multistory dwellings.

1940: July   A Soviet republic is declared in Estonia

A new legislature consisting solely of Communists declares a Soviet republic and votes to join the Soviet Union. All political parties other than the Communist Party are banned from Estonia, and many Estonians are arrested.

1941–44   Germans occupy Estonia

After Germany launches its invasion of the Soviet Union in June, 1941, German troops overrun and occupy all the Baltic states.

1942   Birth of printmaker Tonis Vint

Vint, who becomes one of Estonia's premier postwar artists, is born and educated in Tallinn. He becomes the leader of a

It was reported in early September 1939 that the Estonian authorities allowed a Polish submarine to "hide" in the harbor of Tallinn, angering Russian officials. Russian troops enter Estonia later that month. (EPD Photos/CSU Archives)

group of young artists, called ANK '64, who rebel against artistic restrictions imposed by Soviet rule and gain wide influence through his position as editor of Estonia's main art journal, *Kunst*. However, he is harassed by the Soviet authorities, who interfere with his career. In one incident, Vint is forced to tear by hand pages with a nude painting of his out of 36,000 copies of a magazine. In addition to his other achievements, Vint, who is also influential among the artistic communities of the other Baltic nations, is known as an expert on Baltic folklore.

1944: Autumn   Soviets retake Estonia

As the Germans retreat westward, the Soviet Union resumes its control of Estonia.

Postwar Soviet Rule

1944–91   "Sovietization" of Estonia

The USSR attempts to thoroughly assimilate the Baltic States through its collectivization and immigration policies. The Sovietization of Estonia deprives the people of their national identity. In addition, anti-Christian legislation is introduced to severely restrict church worship and activities and close down churches and confiscate church property. Official anti-religious bias and repression by the Soviet government continue until such legislation is repealed in 1990. The immigration to Estonia of nearly a quarter-million ethnic Russians and the deportation of over 20,000 ethnic Estonians to Siberia help weaken ethnic Estonian dominance. More than 80,000 Estonians flee their native land, either by way of the Baltic Sea or Gulf of Finland. By 1989 ethnic Estonians make up only about sixty percent of the republic's population. Forced industrialization and agricultural collectivization do, however, help stabilize the economy and enable Estonia to have a higher standard of living.

1945–53   Anti-Soviet resistance

Estonian rebels stage guerrilla warfare against the Soviets but do not succeed in driving them from the country.

1947: December 29   Birth of artist Leonhard Lapin

Lapin, a painter and graphic artist, is educated as an architect. He joins ANK '64, the movement started by Tonis Vint (see

1942), and generally resists attempts by the Soviet authorities to dictate the content of his art. He is a leader in introducing pop art to Estonia, founding a group called SOUP '69 in honor of U.S. artist Andy Warhol's (1927–87) famous painting of a Campbell's soup can. The subject matter and spirit of Lapin's art change with the achievement of Estonian independence in the 1990s.

1949 Tallinn Jazz Festival is launched

The prestigious week-long Tallinn Jazz Festival is founded. It becomes famous for attracting jazz musicians from the Soviet Union and Western countries.

1960s New wave of immigration

Immigration to Estonia expands once again as the Soviet government begins development of military facilities in the Narva region. Non-Estonians eventually form a large majority of the area's inhabitants.

1961 Chamber orchestra is formed

The Tallinn Chamber Orchestra is founded under the direction of well-known Estonian conductor Neeme Jarvi (see 1937). It is particularly acclaimed for its performances of modern Estonian music.

1982 Death of composer Eduard Tubin

Estonian composer Eduard Tubin dies in Stockholm, Sweden. (See 1905.)

Mid-1980s Gorbachev's policies spur Estonian nationalism

Estonia takes advantage of Mikhail Gorbachev's policies of *glasnost* (openness) to renew their demands for greater political and cultural autonomy.

1988 Estonian Popular Front is established

The Popular Front of Estonia is founded and presses for economic autonomy. The Estonian Supreme Soviet breaks with the central Soviet authorities, declaring Estonia has the right to political sovereignty.

1989: October Popular Front calls for sovereignty

The Popular Front of Estonia officially declares Estonian sovereignty and demands independence from the Soviet Union.

1990–93 Property owners are compensated for nationalization

After fifty years, property reforms are introduced either to restore the rights of property owners in order for them to reclaim their property or to compensate them monetarily for their loss. This policy of reform is called *privatization*.

1990: February Popular Front scores electoral victory

The Popular Front of Estonia wins a majority of seats in the Estonian parliament. Popular Front leader Edgar Savisaar is elected prime minister, becoming Estonia's first non-communist head of government since World War II.

1990: May 8 Estonia declares independence

Following the other two Baltic republics, Estonia declares independence, but Soviet leader Mikhail Gorbachev rejects the declaration and decrees it invalid.

The Republic of Estonia

1991: August 20 Estonia wins independence

While the coup against Mikhail Gorbachev is in progress, the Supreme Council of the Republic of Estonia issues a decision on the re-establishment of independence based on the historical continuity of statehood.

1992: January 12 Treaty on military withdrawal by Russia signed

The Treaty on Interstate Relations Between the Republic of Estonia and the Russian Federation is signed, the purpose of which is stated as establishing a time for the complete withdrawal of Russian armed forces from Estonia by August 31, 1994. However, border disputes continue between Estonia and Russia.

1992: February 26 Controversial citizenship law is approved

Estonia's legislature passes a citizenship law aimed at redressing the Sovietization of the preceding decades. Citizenship is now based on one of two criteria, first, that one birth parent be an ethnic Estonian or, two, that Estonian be the individual's primary language. Under the new law, residents who were not citizens of Estonia before 1940—or those who are not descendants of pre-1940 citizens—are declared "resident aliens" and must undergo a naturalization process to regain their citizenship. The new system invalidates the citizenship of 600,000 persons, or approximately thirty-eight percent of Estonia's population. Anyone who does not qualify for citizenship cannot vote in national elections or have any political influence.

Ethnic Russians decry the policy as a violation of human rights. It attracts international attention and an examination of Estonian minority policies. The human rights issue becomes linked to the pullout of Russian troops from Estonia. Russia will not remove all troops unless citizenship is restored to ethnic Russians, and Estonia refuses to change its policy unless the troops are removed. Estonia's position is strengthened by the strong international support it receives and the failure of any human rights groups to find clear violations.

1992: April   Traditional Estonian parliament is reinstated

The traditional single-chamber Estonian parliament—the *Riigikogu*—replaces the Supreme Council of the Soviet era. The office of chairman of the Supreme Council is abolished and replaced by that of the president.

1992: June 28   New constitution is adopted

Estonia's new constitution defines government responsibility as follows: to implement domestic and foreign policies; to direct and coordinate government institutions; to create and implement legislation, the resolutions of Parliament, and edicts of the president; to submit draft laws and foreign treaties to Parliament; to manage the state budget; and to organize international relations.

1992: August   New currency is adopted

Estonian economic reform begins with the introduction of the *kroon* as the official currency. One Estonian kroon (EEK) = 100 cents.

1992: October   Lennart Meri is elected president

Former author, filmmaker, and foreign minister Lennart Meri is elected to the largely ceremonial post of president in Estonia's first national elections following independence. (See 1929.)

1993: June   Privatization law is passed

A privatization law is passed by the parliament which, in effect, enables Estonians to qualify for state funding toward the purchase of an apartment or house.

1994: August 31   Last Soviet troops leave Estonia

In accordance with a deadline negotiated by the Estonian and Latvian leaders and Russian president Boris Yeltsin (b. 1931), the Soviet military presence in Estonia and Latvia comes to an end, fifty-four years after the Soviet Union first occupies the region during World War II.

1995: March 5   National elections

In Estonia's second national elections since independence, voters overwhelmingly reject the reformist Fatherland Party that has led the country since 1992 in favor of politicians with closer ties to the communist-era government. The upset is attributed to voter displeasure with the hardships imposed by the rapid economic reform pushed by the reform party, several of whose young leaders are only in their twenties and early thirties.

1996: August 23   Orthodox church jurisdiction is divided

The Orthodox Church of Moscow and the Eastern Orthodox Church, based in Constantinople, agree to share jurisdiction over Estonia's eighty-four Orthodox Christian congregations. The agreement supersedes an earlier unilateral declaration in which the *patriarchate* (ecclesiastical jurisdiction) in Constantinople had claimed jurisdiction over all the congregations in the country, creating tensions with Moscow.

1996: September 20   President Meri is reelected

President Lennart Meri is elected to a second five-year term as president of Estonia, winning 196 votes from an electoral college made up of Estonian legislators. Meri's closest rival, former communist-era president Arnold F. Ruutel, wins 126 votes.

1998   Women's political round table is formed

The Round Table of Political Women's Organizations is founded. The group issues a demand that political parties and women voters support more women candidates for top offices.

1998: March 30   EU entry talks are begun

Negotiations aimed at the eventual entry of Estonia, together with four other Eastern and Central European nations, into the European Union are launched in Brussels, Belgium.

1998: December 8   Controversial citizenship laws are modified

In a move aimed at easing the country's eventual entry into the European Union (EU), Estonia's parliament amends the controversial citizen laws passed in 1992 by granting automatic citizenship to children born to ethnic Russians after 1991.

1999: February 25   Baltic military college is founded

The Baltic states officially open a joint military college which they hope will increase their chances of eventually being accepted into NATO. Located in Tartu, Estonia, the college will be staffed by NATO officers. Classes at the new Baltic defense college, informally called BALTDEFCOL, will begin in August 1999. English will be the main language of instruction for Baltic cadets. The college is financed by Nordic and NATO members including the United States, France, and Germany.

1999: March 7   Center-right coalition leads in national elections

A center-right coalition formed by three political parties all favoring free-market reforms and pro-Western policies—the Pro Patria Union, the Estonian Reform Party, and the Moderates—wins fifty-three of Estonia's 101 parliamentary seats and forms a majority government.

There are 508 women candidates in the March 7, 1999 general elections, up from 222 women candidates in 1995. The party with the largest number of women candidates is the Russian Party in Estonia with 44.6 percent. The smallest number of women candidates belong to the Estonian Country People's Party with 17.4 percent. Overall, women account for 26.9 percent of the total number of candidates. Three women run as independents.

Bibliography

Gerner, Kristian, and Stefan Hedlund. *The Baltic States and the End of the Soviet Empire.* London and New York: Routledge, 1993.

Hiden, John and Patrick Salmon. *The Baltic Nations and Europe.* London and New York: Longman, 1994.

Iwaskiw, Walter R. *Estonia, Latvia, and Lithuania: Country Studies.* Washington, DC: Federal Research Division, Library of Congress, 1996.

Laar, Mart. *War in the Woods: Estonia's Struggle For Survival, 1944–56.* Washington, DC: Compass Press, 1992.

Lieven, Anatol. *The Baltic Revolution.* New Haven and London: Yale University Press, 1993.

Rank, Gustav. *Old Estonia: the People and Culture.* Bloomington: Indiana University, Research Center for Language and Semiotic Studies, 1976.

Raun, Toivo U. *Estonia and the Estonians.* 2nd ed. Stanford, Calif.: Hoover Institution Press, 1991.

Taagepera, Rein. *Estonia: Return to Independence.* Boulder, Col.: Westview Press, 1993.

Von Rauch, George. *The Baltic States: The Years of Independence.* Berkeley and Los Angeles: University of California Press, 1970.

Finland

Introduction

Throughout its history, Finland has fought to achieve or maintain independence. Sandwiched between Sweden to the west and Russia to the east, Finnish territory has been controlled at different times in history by one or the other of these two countries. Finnish culture bears influences of Swedish and Russian culture, and many Finns speak Swedish as well as Finnish. Even after achieving independence in 1917, Finland struggled to avoid being absorbed by its powerful neighbors. Modern Finland maintains a position of neutrality in international relations. This posture was especially challenging to maintain, due to Finland's proximity to Russia, during the Cold War years, when diplomatic tension existed between the United States (and its Western allies) and the Soviet Union (and its communist allies).

With an area of 130,128 square miles (337,030 square kilometers), Finland is slightly smaller than Montana. The country is densely forested, and has thousands of small islands along its coast, and thousands of lakes dotting its interior.

Battle Ax to Viking Age

Archaeologists speculate that the earliest inhabitants of Finland, unlike those of their Scandinavian neighbors, traveled overland from northeastern Europe, perhaps from as far away as Siberia. Around 1800 B.C., people of the Battle Ax culture migrated to Finland from the Baltic region. ("Battle Ax" refers to the axes used as weapons by this culture group during this period.) The Battle Ax people left extensive evidence of their skill in fashioning implements from stone. Five hundred years later metalworking with bronze developed, and by this time the people living in Finland had turned their vision away from the European continent toward the sea. Finland's national epic, *The Kalevela,* recounts legends of metalworking, fishing, and trapping by the earliest Finns.

Swedish, Danish, and Norwegian sailors, known as Vikings, passed around Finland on their trading route to Constantinople, a city in modern-day Turkey. Finns traded furs with the Vikings. By the end of the Viking era, there were three principal areas of settlement in Finland—the southwest; the lake district in the western part of the interior, accessible to the sea by rivers; and the eastern region around Lake Ladoga. Because of its vast natural resources—metal ores, forests, and wild game and fish—Finnish territory was coveted by its neighbors.

Swedish Rule, c. 1145–1809

Some portion of Finland was part of the Swedish kingdom for almost seven centuries. In the middle of the twelfth century, the Swedes began the conquest of Finland, part of a general expansion across the Baltic Sea. King Erik of Sweden and Bishop Henry of Uppsala led a religious crusade into Finland, and by the end of the thirteenth century, Swedish control of most of southwestern Finland and the region around the Gulf of Bothnia was firmly established. Stockholm, the capital of Sweden, functioned as the capital for the Finnish provinces until 1809. Swedish was the official language in Finland until late in the 1800s, when it shared dual language status with Finnish.

From 1388–1488, the monarchs of countries comprising modern-day Scandinavia—Norway, Sweden, Finland, and Denmark—cooperated to rule the four states as one state, initially under Denmark's regent, Margaret (or Margareta, 1353–1412). This is known as the Three Crowns era, or the era of the Kalmar Union. However, Danish rule was challenged repeatedly, and in the early 1500s, Swedish leader Gustavus Vasa (1523–60) drove the Danish monarchy out of Sweden, establishing himself as king. In 1550, as King Gustavus I Vasa, he created Helsinki, one of the world's first planned cities, to compete with Tallin in neighboring Estonia. In 1556, he declared the Finnish territory a duchy (territory controlled by a duke) under the control of his son, Johan (John). Duke Johan and his wife held court at Turku from 1556–63. Johan himself became king of Sweden in 1568. During this era, the first books were printed in Finnish by Michael Agricola, a bishop who championed preservation of the Finnish language.

In the seventeenth century, under King Gustavus II (Gustavus Adolphus), Sweden won additional Finnish territory from Russia in a war that ended in 1617. King Gustavus

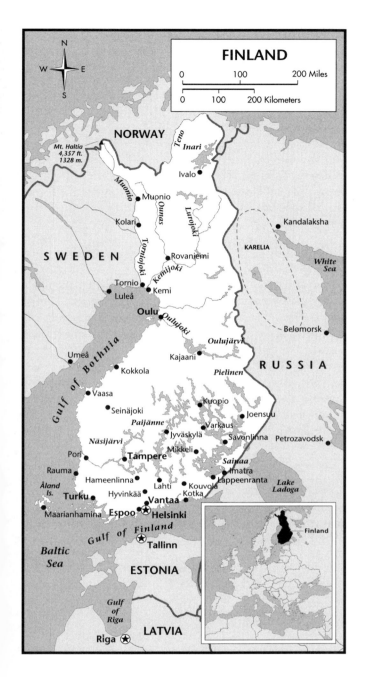

FINLAND

Finland as Grand Duchy of Russia, 1809–1917

In 1805 Sweden became involved in the Napoleonic Wars. In 1809, the control of Finland transferred from Sweden to Russia. Russian Tsar Alexander I gave Finland a special autonomous status, in accordance with the grand duchy's constitutional tradition. Three men, all of whom would play a role in the development of Finnish national culture, were born during the first decade of the nineteenth century. Elias Lönnrot, born April 9, 1802, compiled the first printed version of the *Kalewala* (or *Kalevala*), the Finnish national epic, in 1835. Johan Ludwig Runeberg, born in 1804, became one of the most important writers of the era of Swedish rule. His poem, *Maamme* (*Vårt Land* in Swedish; Our Land) is the national anthem, and his wife, Fredrika Tengström Runeberg was a early activist for women's rights. Johan Wilhelm Snellman, born in 1806, was a leader of the movement for Finnish nationalism. In 1865, Finland's best-known composer, Jean Julius Christian Sibelius was born. Many of his works brought Finnish myths to an international audience, His *Finlandia* became a rallying point for Finnish independence.

By the first decade of the twentieth century, women had won the right to vote in elections, the first in Europe to do so.

Independent Finland, 1917–present

Despite the influences of Sweden and Russia, feelings of nationalism among the Finnish people were strong. But when Finland declared independence on December 6, 1917, all factions were not satisfied. Radicals in the Social Democratic Party (known as the Reds) agitated for a real revolution, and in January they seized control of the southern part of the country. The resulting civil war saw the government (known as the Whites) garner support from Germany; the Reds were backed by Russia. Led by Gustaf Mannerheim (1867–1951), the Whites won victory in May 1918. The street signs in Finland were still in three languages—Finnish, Swedish, and Russian—at the time of independence in 1917.

Over the next two decades, the government gradually shifted its alliance from Germany to Scandinavia, and Finland fought the Soviet Union, militarily and politically, to maintain territory and independence. During World War II, Finland's Baltic neighbors were forced to ally with Russia while Sweden remained neutral. Following World War II, Finland established a peaceful coexistence with neighboring Soviet Union.

During the 1950s Finland reinforced its position of neutrality in world affairs. Artists and craftsmen thrived during this decade, establishing Finnish glassware and silversmithing on the international markets. A great novelist of this era, Väinö Linna (1920–92), published the best-seller, *Unknown Soldier*, and other works dealing with Finnish history in the 1800s and 1900s. He was typical of Finnish writers who examined nationalism and character in light of the country's

II then entered the Thirty Years' War on the side of the Protestants fighting against the Habsburg Holy Roman Empire (1618–48).

In the early 1700s, Finland and Russia fought a twenty-year war known in Finland as the Great Wrath. When the war ended, Finland relinquished territory to Russia. As is the case through much of history, both Sweden and Russia sought to expand their own territory. At the end of the eighteenth century, the two countries engaged in war. Finns, until now fairly loyal to Sweden, began to question whether Sweden would protect Finnish land and interests. Sweden regained control of Finland for the time being.

Sauna

Almost every household in Finland has a sauna, a special type of bath that has been adopted by health clubs, spas, hotels, and some private citizens around the world. The sauna may be a special room in a house, or a separate structure. There is a stove in the sauna to heat rocks to temperatures ranging from 175°–212°F (80°–100°C). In some cases, water is poured over the rocks to intensify the heat, but the basic principle of sauna is dry heat.

The sauna is lined with wooden benches, usually on two levels. Bathers—as those using or "taking" a sauna are called–sit or lie on the benches, sometimes beating themselves gently with birch twigs, until they perspire freely. Bathers leave the sauna to plunge into a lake, pool, or cold shower. The process is repeated.

Finns consider the sauna a social setting. It is not uncommon for whole families to entertain guests in the sauna. In winter, Finns whose sauna is near a lake may cut a hole in the ice to jump through as part of the process.

constant struggle against oppression at the hands of its neighbors.

Modern Finland society has embraced technology. Statistics show that Finland has the highest percentage of households connected to the Internet, and a Finnish cellular phone company (Nokia) is among the leaders in cellular communication.

Timeline

8000 B.C. Human culture established in northern coastal regions

Archaeologists have discovered relics that indicate the existence of human culture comprised of primitive fishers and hunters in the newly emerged coastal lands around the Arctic Ocean. These humans inhabit a region that lies north of the modern-day Finnish border.

7200 B.C. Human culture exists in southern regions

Ancestors of the Finns—hunters and fishers—come to the region that is modern Finland by crossing through the Baltic region surround the Baltic Sea (present-day Latvia, Lithuania, and Estonia). They enter the region from the south, and settle to the east and north. These early settlers are make simple tools and weapons from stone and bone.

3000 B.C. Comb-Ceramic period features pottery

Pottery, introduced into Finland from southern cultures, is employed by people living in the south. The inhabitants of southern Finland survive by fishing and hunting and, therefore, live in regions near the fishing waters. It is believed that these early inhabitants of Finland are members of a long-skulled race of people, short in stature, who migrated from the plains regions of eastern Europe.

1800 B.C. Battle Ax culture

People of the Battle Ax culture, known for their stone implements and weapons, migrate to Finland from the Baltic region. The Battle Ax people engage in agriculture, especially in tending herds of cattle. They decorate special axes with animal likenesses and often bury these weapons with the dead. Battle Ax people learn to sail and turn their interests to their Swedish neighbors, away from the European continent. Swedish, Danish, and Norwegian sailors, known as Vikings, pass around Finland on their trade route to Constantinople, a city in modern-day Turkey. Finns trade furs with the Vikings.

1300 B.C. Bronze metalworking develops

Bronze implements are introduced, probably from western neighbors. It is likely that most of the metal comes into Finland through trading, since the country's metal reserves are not yet discovered.

The Viking Age

c. A.D. 793 Vikings trade with Finns

The Vikings pass through Finland on their trading route from Norway and Sweden to Constantinople (modern-day Istanbul, Turkey). They trade furs with the Finns. By the end of the Viking era, there were three principal areas of settlement in Finland—the southwest; the lake district in the western part of the interior, accessible to the sea by rivers; and the eastern region around Lake Ladoga. Because of its vast natural resources—metal ores, forests, and wild game and fish—Finnish territory is coveted by its neighbors.

c. 1100 References to bog iron and copper mining in Finland

The *Kalevala*, Finland's national epic poem, suggests that iron and copper are being mined in Finland.

Swedish Rule, c. 1145–1809

c. 1145   Swedish religious crusaders are believed to exert control over region

Traditional history suggests that Swedish religious crusaders carry out a number of missions near the middle of the twelfth century. The first of these occurs around 1145, when King Erik of Sweden joins the English missionary, Bishop Henry of Uppsala, in undertaking a crusade into Finnish territory. According to some accounts, the two were instrumental in the conversion of the Finns to Christianity. Other historians believe that the Finns are largely Christian before the Swedish crusades. Nevertheless, Bishop Henry later becomes patron saint of Finland.

1150–1809   Finland is part of the Swedish Kingdom

At least some portion of Finland is a province of the Swedish kingdom for almost seven centuries.

1172   Pope Alexander II reports about Finns

The first written evidence of Christianity in Finland appears in a bull (an official document) issued by the Pope. Pope Alexander III (c.1105–c.81) issues a Papal Bull, *Gravis admodum*, that refers to his concern that the Finns are not strong enough in their faith or obedient enough to their priests. This suggests that Christianity is well-established in Finland at this point.

1293   Swedish control of Finnish territory is established

Swedes gain control of most of the southwestern region of present-day Finland, and of the lands along the Gulf of Bothnia, the body of water separating Finland and Sweden. The geographical region that will become modern-day Finland is a group of individual provinces. Stockholm, the capital of Sweden, functions as the capital for the Finnish provinces as well until 1809.

1300s   Folklore develops around Finnish heroes

A Finnish hero, Knut Porse, enjoys spectacular military success in battles against the Russians. In Finnish folklore, Porse's success is attributed to magic: he defeats the Russians in battle by using a bomb made of frogs, chalk, and mercury. In a battle defending a castle, folklore recounts that Porse climbs to the top of the castle and shakes out a feather pillow. The feathers transform into armed men as they land and quickly defeat the Russians.

1300   Finns accept Swedish rule

Most Finns accept Swedish rule and influence on their culture, and the Swedes, in turn, make little effort to overrun Finland. Most Finns acknowledge that the first key to success under Swedish rule is to learn the Swedish language.

1323   Peace treaty between Sweden and Novgorod (Russia)

Although Novgorod (Russia) is interested in Finnish territory, a peace treaty negotiated with Sweden assigns most Finnish lands to Sweden. Only Karelia, a province in eastern Finland, becomes part of the Russian empire. The prominent religion there is Eastern Orthodox.

1362   Finns participate in election of Swedish king

The Swedish government gives Finns the right to send representatives to participate in the election of the king.

1388–1488   Norway, Sweden, Finland, and Denmark ruled as one state

Norway, Sweden, Finland, and Denmark are ruled as one united state. This era, known as the Three Crowns, evolves through intermarriage between monarchs of the region.

When Haakon VI (1339–80), king of Norway, dies, he leaves his kingdom to his wife and heir, Margaret of Denmark. When Margaret's father (the king of Denmark) dies, Margaret also inherits his throne. Sweden elects her queen, too, to escape the oppressive threat of the German monarch, Albert of Mecklenberg. Seven years of war with Albert follow, and Finland supplies one-third of the soldiers in the Swedish army.

1493   First map to name Finland is printed

A book of maps, *Liber Chronicarum*, is published by German historian Hartman Schedel (1440–1514). For the first time, the word Finland is used on a map. It identifies the wrong territory, however. Finland is placed on the western (Swedish) coast of the Gulf of Bothnia. Around the same time, German geographer Martin Behaim creates the first globe.

1506   Birth of Michael Agricola, first writer in Finnish

Michael Agricola (1506?–57) is born. He is believed to be the first writer to publish in the Finnish language. He publishes *Abckiria* (ABC Book) in 1543 or 1544. During his life, he serves as bishop of Abo (later Turku). While holding this office, he publishes a number of religious works, icluding a version of the New Testament in 1548.

1527   King Vasa establishes Lutheran Church

The Lutheran Church is established in Finland (which is under Swedish control), when King Gustavus Vasa (1523–60) breaks Sweden's link with Roman Catholic Church and establishes the Lutheran Church in his domain.

1550   City of Helsinki is planned

Helsinki is one of the first planned cities in the world. King Gustavus Vasa of Sweden wants to establish a city to compete with Tallinn in neighboring Estonia. He issues a decree that the people in four towns should pack up all their belongings and businesses and relocate to a spot near the Vantaa River. This is the precursor to the modern city of Helsinki (see 1640).

1551–52 Selections from the Old Testament published in Finnish

After publishing a Finnish language version of the New Testament (see 1548), Michael Agricola (1506–57), bishop of Abo (later, the city of Turku), publishes three volumes containing selected books of the Old Testament of the Bible in Finnish. A complete Bible in Finnish is published in Stockholm in 1642.

1556 King Vasa gives Finnish territory to his son

King Gustavus I Vasa (1523–60) of Sweden donates Finland as a duchy (territory controlled by a duke) to his son Johan (John) (1537–92). Duke Johan and his wife Catharina (Katarina) Jagellonica hold court at Turku from 1556–63. In 1563, Johan is captured and imprisoned by his brother Erik XIV at Turku Castle until 1567. With the help of their younger brother Karl, Johan escapes and deposes Erik, taking the Swedish throne for himself and becoming King Johan III.

1564 Pori is chartered

Pori, capital of the modern-day province of Turku-and-Pori in southwest Finland, is chartered. It is located near the mouth of the Kokemaenjoki River and is an export center for timber and metals. (See 1840.)

1595 Peace treaty between Sweden and Russia

When Johan is crowned king of Sweden, he makes Finland a Grand Duchy. The eastern border begins by running through Lake Ladoga (modern-day Ladozhskoye Ozero in Russia) in the south. The path of the border as it runs northward follows almost the same boundary as exists in modern Finland.

Early 1600s Finnish soldiers recruited into Swedish army

The army of the Finnish Grand Duchy (controlled by Sweden) has its troops depleted by the needs of the Swedish Army fighting to defend Swedish interests in Germany and Russia.

1616 The city of Uusikaupunki chartered

Uusikaupunki, a city in southwest Finland on the Gulf of Bothnia, is chartered. As a port city, it is a center for trade. Uusikaupunki's small business ventures include sawmills for processing lumber and machine shops for serving the shipyard. In the surrounding region, granite is quarried.

1621 Tornio in northwest Finland chartered

Tornio, a city in Lapland in northwest Finland, is chartered. Forest products are shipped from its harbor at the mouth of the Torne River where it empties into the Gulf of Bothnia. In 1684, a wooden church is built in Tornio. The church remains standing into the late twentieth century, making it Finland's oldest wooden church structure.

1639 The city of Savonlinna is chartered

Savonlinna in the Saimaa lake region of southwest Finland is chartered. The city is centered around the fifteenth-century fortress of Olavinlinna (see 1475). It becomes the site of the world-renowned Savonlinna Music Festival held each summer.

1640 Helsinki moves

Helsinki (see 1550), a planned city on the banks of the Vantaa River, moves to its present location on the Gulf of Finland, to gain access to a harbor.

1640 Turku Academy, the first university in Finland, is founded

Turku Academy, a university, is founded at Turku, where Michael Agricola is bishop. The academy offers training for civil servants. In 1828, it will move to Helsinki, the capital of Finland. Swedish Count Per Brahe is instrumental in its founding.

1642 Finnish language Bible published in Stockholm, Sweden

A complete version of the Bible is published in the Finnish language in Stockholm, Sweden. Parts of the Bible (New Testament and selected books of the Old Testament) are first published in Finnish (see 1550) by Michael Agricola (see 1506). Because Finland is ruled by Sweden from c.1150–1809, most writers choose to write in Swedish, to reach a larger audience.

1651 City of Kajaani is chartered

Kajaani, a city in Oulu Province in central Finland, is chartered. Kajaani lies on the Kajaaninjoki River, where paper goods and other lumber products are produced in the 1900s. The city, developed around the Kajaneborg fortress, is also a transportation center.

1700 The Great Wrath, a war with Russia, begins

A war, known in Finland as the Great Wrath and elsewhere as the Great Northern War, begins. It lasts over two decades, and Russia occupies all of Finland for several years during the period. (See 1721)

1710 Plague decimates Helsinki

A plague hits the town of Helsinki. Two-thirds of the population—1,200 of the 1,800 inhabitants—die.

1721 The Great Wrath ends with loss of Finnish territory

For over two decades, Finland and Russia fight over territory (see 1700). Uusikaupunki, a city on the Gulf of Bothnia in southwest Finland, is the site for the signing of the Treaty of Nystad ending the conflict. When the conflict ends, Finland cedes the province of Vyborg to Russia.

1739 Birth of Henrik Gabriel Porthan, historian

Henrik Gabriel Porthan (1739–1804) is born. In the late 1700s, he becomes a prominent historian in Finland, introducing the notion of distinct national history for the country.

1779 City of Tammerfors is founded

The Tammerkoski River is chosen as the location for a new settlement, Tammerfors, since expeditions to northern Finland and Lapland are often launched from here. Tammerfors is later known as Tampere.

1788–90 Sweden and Russia at war over Finland

Finnish territory is sought by both Sweden and Russia, and war breaks out between the two countries. During the war, which lasts until 1790, Finns begin to question whether Sweden will protect Finnish land and interests. Some citizens push for an independent Finland within Russia, but this proposal does not win sufficient support to succeed. Sweden retains control of Finnish territory.

1802: April 9 Birth of Elias Lönnrot

Elias Lönnrot (April 9, 1802–March 19, 1884) is born in southern Finland in the town of Sammatti. Lönnrot enters the University of Turku in October 1822, and successfully completes the examination in philosophy in spring 1827. From 1828–44, he takes numerous trips to eastern Finland, especially Karelia, to collect poems. In 1833, he becomes a district doctor, requiring him to travel extensively in rural areas; during these trips, he continues his poetry studies. Until 1853, he continues to practice medicine. Lönnrot compiles the first printed version of the *Kalewala* (or *Kalevala*), the Finnish national epic (see 1835). In 1853, Lönnrot leaves medicine to accept the position as professor of Finnish language. He dies on March 19, 1884 in southern Finland.

1804 Birth of Johan Ludwig Runeberg

Johan Ludwig (Ludvig) Runeberg (1804–77) is born in Pietarsaari, a Swedish-speaking city in western Finland on the Gulf of Bothnia. He is considered the most important writer of the era of Swedish rule and is Finland's national poet. His writing, mainly in Swedish, includes *Elgskyttaråe*

(Elk Hunters; 1832) and the two-volume *Fänrik Ståls Sägner* (Tales of Ensign Stål; 1848–60). The latter deals with the action of Finnish troops in the war with Russia (see 1808–09). His wife, Fredrika Tengström Runeberg (see 1807) is an early activist for women's rights in Finland.

1806 Birth of J. W. Snellman, advocate for Finnish nationalism

Johan Wilhelm Snellman (1806–81) is born. During his time as a student leader at the University of Helsinki, he works to promote the recognition of Finnish as an official language, along with Swedish. At first, the University is reluctant to offer him a position on the faculty because of his controversial ideas, so he becomes the principal of a secondary school in Kuopio, a small, remote town (see 1844). He eventually becomes a professor at the University of Helsinki (see 1855). The Russian government under Alexander II (1818–81) issues the Language Decree (see 1863) as a result of the campaign led by Snellman.

1807 Birth of Fredrika Runeberg, author and feminist pioneer

Frederika Tengström Runeberg (1807–79), author and feminist, is born in Pietarsaari. She shows early literary talent, writing small poems and stories as a small child. At the age of twenty-three, she marries Johan Ludwig Runeberg (see 1804), Finland's national poet. In 1861, she publishes a collection of her articles in a book entitled *Dreams and Pictures* in Swedish.

1808–09 Russia and Sweden at war

Russia invades Finland, which is under Swedish control. Russia hopes to pressure Sweden into joining a blockade of Britain. Sweden's position is weak, and Russia succeeds in winning control of the Finnish territory controlled by Sweden.

1808 Helisinki destroyed by fire

During the war between Russia and Sweden, the city of Helsinki is destroyed by fire. The Finns rebuild the city following a design created by German architect Johan Carl Ludwig Engel (1778–1840).

Finland as Grand Duchy of Russia, 1809–1917

1809: March Russia annexes Finland

Finland becomes an autonomous grand duchy of Russia. Alexander I (1777–1825) emperor of Russia from 1809–25, gives Finland freedom to control its own affairs, employing the government systems established under Swedish rule.

Thus, the Lutheran Church is retained as the national church, and Swedish remains the official language. In addition, Russia allows Vyborg (see 1721) to be returned to Finland.

1812 Helsinki replaces Turku as capital of Finland

Until this year, Turku is the capital of Finland. The Russians, fearing that the Swedes control too much of Turku's business and culture, move the capital to Helsinki.

1820s Nationalism sparks demand for Finnish language instruction

As the spirit of nationalism grows in Finland, citizens demonstrate their interest in having Finnish become the language of instruction in schools. Until now, the predominate language among the educated classes is Swedish. The University of Turku establishes a teaching position in the Finnish language to satisfy the groundswell of interest. Thus, the university becomes the center of nationalist thinking. Student organizations here are among the first in Finland to adopt Finnish as their official language (the one used to record by-laws, minutes, motions, and other official organizational records). This nationalistic fervor is referred to by observers as "Fennomania."

1820 Textile factory established

Scots industrialist James Finlayson (1772–1852), establishes a spinning mill at Tammerfors (modern-day Tampere). The city grows to be a major weaving center and hub of other industrial development. Finlayson's company, Oy Finlayson Ab, continues to operate and is Scandinavia's largest textile manufacturer as of the end of the twentieth century.

1820 Hydrotherapy spa established on Lake Saimaa

Lappeenranta, a city on Lake Saimaa in southeast Finland, is the site of a hydrotherapy center. Lappeenranta, established in 1649, has lumber mills and cement factories in addition to the therapy center.

1827 Turku is destroyed by fire

Turku, Finland's third-largest city and former capital (see 1812), is destroyed by a fire started by candles in a butcher shop. Only the stone buildings survive. Following the fire, the University of Turku is relocated to Helsinki.

1831 Finnish Literary Society founded

Suomalaisen Kirjallisuuden Seura (Finnish Literary Society) is founded to promote research in the field of Finnish studies. The society publishes works on Finnish history and language, including Elias Lönnrot's (see 1835) Finnish-Swedish dictionary. (See 1841.)

Kantele

When he compiles Finland's national epic, *Kalevala* (see 1835), Elias Lönnrot describes a musical instrument that provides power to the hero of the epic, Läinämöinen. This instrument, the kantele, is described as being made of the jawbone of a pike (a large fish). The kantele, the oldest known Finnish instrument, is made of wood, usually pine, spruce, or alder. The most basic style of instrument has five strings, fixed at one end and stretched to a tuning peg at the other. To play the kantele, the performer holds the instrument in his or her lap (or places it on a table). The strings are plucked and strummed to create melody and accompanying chords. By the mid-nineteenth century, kanteles were made with as many as thirty strings. In the 1920s, a thirty-six-string chromatic kantele was produced, and a mechanism is added to change the key easily.

1834 Aleksis Stenwall Kivi is born

Aleksis (Alexis) Stenwall Kivi (or "Kiwi") (1834–72), considered the founder of modern Finnish-language literature, is born. His work, *The Seven Brothers,* published in 1870, is the first Finnish novel. The son of a tailor, Kivi also writes both poetry and comic and dramatic works for the theater. There is a statue of Kivi in front of the Finnish National Theater in Helsinki. His works continue to be performed there through the twentieth century.

1835 *Kalevala* is published

Elias Lönnrot (1802–84) compiles the first printed version of the *Kalevala* (or *Kalewala*), the Finnish national epic.

1838 City of Mikkeli founded

Mikkeli, in south central Finland, is founded. It is capital of Mikkeli Province and seat of a Lutheran bishopric (district of the Lutheran church headed by a bishop). Mikkeli's location in the Saimaa lake region makes it an important lake port, transportation hub, and resort area.

1841 Finnish Literary Society begins publication of *Suomi*

Suomalaisen Kirjallisuuden Seura (Finnish Literary Society; see 1831) begins publication of *Suomi* (Finland), an annual devoted to Finnish studies.

1844 Minna Canth, noted playwright, is born

Minna Canth (1844–97) is born Ulrika Vilhelmina Johnsson in the city of Tampere. She enters the teachers' seminary at Jyväskylä, the first such institution in Finland, marries a member of the faculty, J. F. Canth, and has eight children. Her husband dies, and she supports her family by returning to Kuopia and managing a business, while producing the powerful plays for which she is best known. These include *Työmiehen vaimo* (A Working-Class Wife; 1885) and *Kovan onnen lapsia* (Children of Misfortune; 1888). Her later work includes *Anna Liisa,* written in 1895, a psychological drama about a woman's plight and her own ability to change.

1854 The Murole canal opens

Finland's first canal, the Murole canal, connecting the resort of Ruovesi to Näsijärvi, opens. Most of Finland's rivers are short and are not used for shipping. The exception are those rivers that flow between lakes. These are used in a limited way for navigation, most notably by the logging industry to float logs from one point to another.

The second canal to open, the Saimaa canal, connects the Saimaa River to the port city of Viipuri. This canal provided an important navigational link that was lost for a time following World War II (1939–45) when Viipuri was ceded to the Soviet Union. Under an agreement made later (see 1962), Finland regained the right to use the Saimaa canal.

1858 First Finnish-language secondary school established

The first secondary (high) school to offer instruction in Finnish opens at Jyvaskyla, a city in south-central Finland. Built on the shores on Lake Paijanne, the city is an important port.

1861 Tsar Alexander II founds city of Maarianhamina

Maarianhamina is founded on Ahvenanmaa Island in southwest Finland by Tsar Alexander II. It is the capital of Ahvenanmaa Province and a popular summer resort.

1861 First train operates through the Finnish countryside

The first train track opens in Finland. Modern Finland has 3,600 miles of railroad track in a network serving all parts of the country.

1861 Political leader P. E. Svinhufvus and author Juhani Aho are born

P. E. Svinhufvus (1861–1944), political leader at time of independence, is born. Svinhufvus is the leader of the Senate in 1917 (see 1917: December 6) when it approves the declaration of independence.

The same year, Juhani Aho (1861–1921), master of Finnish prose, is also born.

1863 Language Decree launches Finnish as an official language of government

The government of Russian ruler Alexander II issues the Language Decree, beginning the process whereby Finnish replaces Swedish as the official language of government. Swedish remains a dominant language in government in Finland for another forty years.

1864 Law increases unmarried women's rights

A law is passed that allows unmarried women over age twenty-five to handle their own affairs. Prior to passage of this law, an unmarried woman is considered a ward of her father or other male relative. Married women are legally wards of their husbands. (See 1929.)

1865 Birth of composer Jean Sibelius

Finnish composer Jean Julius Christian Sibelius (1865–1957) is born at Tavastehus. He studies at Helsinki, Berlin (Germany), and Vienna (Austria) and joins the faculty of the music conservatory in Helsinki, eventually becoming the principal of the institution. His *Kullervo Symphony* (1892) and symphonic poems, *Finlandia* and *Oceanides,* are based on Finnish myths, largely drawn from *Kalevala* (see 1835 and 1849). *Finlandia* becomes a rallying point for Finnish independence. Other works performed by orchestras worldwide include *Valse Triste* and incidental music for *As You Like It, The Tempest* (both works by Shakepeare), and *Scaramouche.*

1865 Zachris Topelius begins collecting Finnish folktales

Zachris Sakari Topelius (1818–98), whose father has the same name, begins collecting Finnish folktales. Topelius is professor of Finnish history at University of Helsinki (1854–78). He publishes several volumes of folktales and is known as the "Finnish Hans Christian Anderson" after the Danish folktale collector.

1867 Birth of military leader and president Gustaf Mannerheim

Carl Gustaf (Gustav) Emil Mannerheim (1867–1951) is born. He fights in the Russian army during the Russia's war with Japan (1904–05) and in World War I (1913–18). When Finland wins its independence, Mannerheim becomes supreme commander of the Finnish military. He leads the Finnish troops in their successful battle against the Russians (see 1939: November 30). He serves as president of Finland (1944–46).

1870 The first novel in the Finnish language is published

The Seven Brothers by Aleksis (Alexis) Stenwall Kivi (or Kiwi) is the first novel to be published in Finnish (see 1834). It is the story of a family of brothers who prefer the life of the forest over education and the city.

1878   Eino Leino, Finnish lyric poet, is born

Eino Leino (1878–1926), pseudonym of Armas Eino Leopole Lönnbohn, regarded as Finland's greatest lyric poet, uses many folk themes drawn from *Kalevala* in his work. Leino translates many works, including Dante's *Divine Comedy*, into Finnish.

1878   Selim Palmgren, Finnish composer of works for the piano, is born

Selim Palmgren (1878–1951), Finnish composer of works for the piano, is born. During his lifetime, he becomes known as the "Chopin of the North," referring to the prolific Polish composer, Frederic Chopin (1810–1849). (See Poland.)

1887   Composer Leevi Madetoja is born

Composer Leevi Madetoja (1886–1947) is born. His best-known work, an opera, is entitled *Pohjalaisia* (The Ostro-bothnians). It is set during the Russian occupation of Finland and evokes the atmosphere of despair that permeates Finland during this period.

1898   Finnish architect Alvar Aalto is born

Hugo Henrik Alvar Aalto (1898–1976) is born in Kuortane. He becomes one of Finland's most renowned citizens and is considered the father of Modernism in Scandinavian architecture and design. At Helsinki Polytechnic, he begins to shape his unique style, bases his architectural style on asymmetric forms, and makes extensive use of natural building materials such as wood and rough stone. In the 1930s, Aalto creates a style of molded plywood furniture, using birchwood, that becomes a signature of Finnish furniture design for decades. His designs include the library at Viipuri, a wood pulp mill in Kotka, and the Finlandia Concert Hall in Helsinki.

1899   Movement to absorb Finland into Russia gains momentum

From the early 1800s when Finland fall under Russian control, a succession of Russian leaders allow Finland to function independently as a state within a state. Finland issues its own money (the mark), postage stamps, and has its own legislature and army. During the reign of Alexander III (1881–94) and his successor Nicholas II (1894–1917), a movement grows to wipe out Finnish separatism and to impose a process of "Russification" to make the Finns more a part of Russia. The 1905 Revolution in Russia causes the government to focus on other problems, and Finland takes advantage of the opportunity to reorganize its government. It creates a one-house (unicameral) parliament and awards universal suffrage to all men over age twenty-four.

1899: February 15   "February Manifesto" issued by the Russian government

The "February Manifesto" under Tsar Nicholas II essentially neutralizes the power of the Finnish diet (the semi-autonomous legislature). In addition, Nicholas appoints a Russian governor to the post of dictator over Finland and implements an order to merge the Russian and Finnish military forces. Russian officials attempt to force the Finnish people to use the Russian language for dealings with the government and military. Finnish resistance to this dramatic change in self-rule is strong.

1900   Finnish craftsmen work for Fabergé

The majority of the master-craftsmen who execute the design of Russian enamelist and jeweler Carl Faberge (1845–1920) are Finnish. These Finnish jewelers return to Helsinki after the Russian Revolution (1917) and establish workshops.

1902   Finnish and Swedish are dual official languages

The Finnish language is used equally with Swedish for official business and is spoken and understood by a majority of the Finnish population. The Language Decree (see 1863) began the process of giving Finnish the status of an official language.

1904–14   Eliel Saarinen designs railroad station

Eliel Saarinen (1873–1950) designs a new railroad station for Helsinki. The structure establishes his international reputation as an architect. Saarinen is born in Rantasalmi in 1873 and emigrates to the United States in 1923. His designs in the United States include the Cranbrook Academy of Art near Detroit, Michigan, where Saarinen serves as president from 1932–48. He opposes the skyscraper building form, preferring designs more closely tied to their site. His son, Eero, becomes his partner in architecture in 1937. The two design such buildings as the Jefferson Memorial Arch in St. Louis, Missouri; Columbia Broadcasting System (CBS) headquarters in New York; American embassies in London, England, and Oslo, Norway; and the TWA terminal at Kennedy Airport in New York.

1905: November   National strike against Russian government

For six days, the Finnish resistance to Russian intervention in their ability to self-govern takes the form of a national strike. As a result, Tsar Nicholas II capitulates to their pressure and reinstates Finnish self-government.

1906: March 7   Finnish citizens elect their first parliament; Finnish women participate in the election

Finnish citizens vote in a general election to establish a parliament, known as *Eduskunta* (or *Riksdag* in Swedish). All women and men in Finland age twenty-four and older are eligible to vote. This is the first time Finnish women exercise the right to vote; they are the first European women to receive the right to vote on the same basis as men. (The only citizens prohibited from voting in this election are those who are supported by the Finnish government.)

1906: March   Oskari Tokoi is sworn in as prime minister of Finland

Oskari Tokoi is sworn in as the first democratically elected socialist prime minister in the world.

1907–17   Russia continues to pressure Finland

Tsar Nicholas II continues to pressure Finland to become "Russianized." Russian troops remain a constant presence in Finland, and the presence of Russian business and government interests is unsettling to the Finnish citizenry at times.

1912   Discovery of copper is made at Outokumpu

Copper is discovered at Outokumpu in Kuopio province. The mine is the most important source of metal ore in Finland, but serious exploitation of the ore field does not begin until the government takes over the mine in the 1920s. Besides copper, the ore field contains large deposits of iron, zinc, cobalt, nickel, tin, gold, silver, and sulfur. The largest nickel deposit, in the Petsamo area, was ceded to the Soviet Union in 1920 under the terms of a promise made in 1864 by the Russian tsar controlling Finland at that time.

1914–17   World War I rages in Europe

Finland remains neutral in World War I but suffers adverse effects from it. The Gulf of Bothnia is blockaded, preventing Finnish merchant ships from transporting food and other critical supplies to Finland. Finnish people suffer food shortages during the war.

Independent Finland, 1917–present

1917: December 6   Parliament declares independence

The Finnish parliament approves the declaration of independence drawn up under the leadership of P. E. Svinhufvus (see 1861). The declaration follows the Russian Revolution (also known as the Bolshevik Revolution), overthrowing the tsar. The new communist government in Russia recognizes Finland but leaves some of its own troops stationed there.

1918   Finland adopts national flag and coat of arms

The national flag consisting of a blue cross on a white background is adopted. The design of the coat of arms officially adopted by the parliament of the new Republic of Finland dates back to the 1500s.

1918: January–May   Civil war rages in newly independent Finland

In the first six months following independence, a civil war erupts. Left wing socialists, (known as the Red Guard) battle nonsocialists (known as the White Guard). The Red Guard is supported by Russia, and the White Guard receives aid from Germany. In May, White Guard forces win victory under the leadership of General Gustaf Mannerheim (1867–1951).

1919: Summer   Finland officially becomes a republic

The Finnish parliament adopts a republican constitution which establishes a single-chamber parliament whose members will be elected. Kaarlo Juho Ståhlberg (1865–1952) is elected Finland's first president, a post he holds until 1925. The Finnish government's foreign policy is based on establishing cooperative relations with neighboring nations across the Baltic Sea: Latvia, Lithuania, Estonia, and Poland. Relations with neighboring Sweden and Russia remain tense. Finland asks that Russia cede control of eastern Karelia to Finland (like the rest of the province of Karelia), When Russia does not agree to this, the Finns demand that eastern Karelia be given its independence. This dispute fuels tensions between the two nations for years.

1920s   Gallen-Kallela develops a style of Finnish Romantic painting

Akseli Valdemar Gallen-Kallela (1865–1931), born in Pori, trains in Helsinki, and develops his style in Paris, France, where he studies with French master William Bouguereau (1825–1905) from 1884–90. In 1900, he returns to Paris to create frescoes for the Finnish pavilion at the Paris World's Fair. When he returns to Finland, he pioneers a nationalistic Romantic style of painting and chooses themes from Finnish myths, especially *Kalevala*. From 1909–10, Gallen-Kallela and his family lived in British East Africa (modern-day Kenya); he creates150 paintings while there.

1921   League of Nations settles dispute with Sweden over Aland Islands

Sweden and Finland disagree over the right to the Aland Islands in the Gulf of Bothnia, twenty-five miles from Sweden and fifteen miles from Finland. Unable to forge a settlement themselves, the two countries turn the to the League of Nations. The League's decision assigns the islands to Finland.

1923   Freedom of religion is introduced in Finland

Finns enjoy complete freedom of religion for the first time, and the Evangelical Lutheran Church of Finland and the Orthodox Church of Finland, as national churches, have special legal status and the right to collect church tax.

1924: July 10   Paavo Nurmi wins two Olympic Gold Medals

Paavo Nurmi (1897–1973) competes in the Olympic Games. He wins both the 1,500 meter and the 5,000 meter races on July 10, with only one hour rest between the two events.

1926–27   Social Democrats form a minority government

The Social Democrats, who have played a role in the government since its beginning, form a government of their own.

1929–32   The Lapua Movement demands an end to communist activities

The Lapua Movement, inspired by the fascists of Italy (see Italy), demand that the government ban communist activities in Finland. In 1932, the Lapua Movement stages a failed armed revolt against the government.

1929   Law grants married women new independence

The 1929 Marriage Act gives women new independence. Until now, married women are considered wards of their husbands. (See 1864.)

1939   Sillanpää win the Nobel Prize for literature

Frans Eemil Sillanpää (1888–1964) wins the Nobel Prize for literature. His works draw from his own life experiences living in Finland during Russia rule and after independence. *Hurskas kurjuus* (Meek Heritage; 1919) describes rural life during the final decades of Russian rule in Finland and the trauma of civil war.

1939–45   Russia demands ships from Finnish shipbuilders

The Russian navy demands that Finland supply shipping vessels equaling 365,000 tons to meet Russia's needs for World War II. This production level taxes the Finland's capacity; during the 1930s, Finnish shipbuilders produce a total of 23,000 tons of shipping vessels. Finnish producers somehow meet the quota, and the result is a much stronger shipbuilding industry in Finland for the post-World War II era. Finland never officially allies with either side during World War II.

1939: November 30   Winter War is launched with attack on Finland

The army of the Soviet Union invades Finland. Despite being vastly outnumbered (the population of Finland is only 3.5 million), Finland resists the attack. Under the leadership of General Gustaf Mannerheim (see 1867), Finnish troops fight fiercely, sometimes on skis.

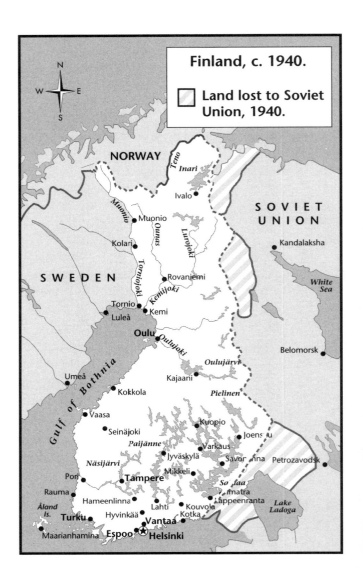

Finland, c. 1940.

☐ Land lost to Soviet Union, 1940.

1939: December 3   Foreign minister speaks to American audience via radio

The NBC radio network broadcasts a live speech by Foreign Minister Väinö Tanner to listeners in the United States three days after the Soviet Union attacks Finland. He affirmed Finland's commitment to independence: "the Finnish government will not refuse to take part in negotiations for the restoration of peace. Nevertheless, anyone who believes that the Finnish people can be brought by the threat of force, and the terror already launched, to make concessions that would denote in reality the loss of their independence is mistaken." (Excerpted from Virtual Finland, available at http://virtual.finland.fi/finfo/english/tanner.html)

1940: March 13   Finland surrenders to the Soviet Union

Finland surrenders to the Soviet Union, ending the Winter War. A peace treaty is prepared in Moscow, ending the conflict and requiring that territory in southeastern Finland (southern Karelia) be given to the Soviet Union. The area given up makes a significant impact on Finland: It represents ten percent of Finland's land, twelve percent of its population, and includes Lake Ladoga and Vyborg (Viipur), the second largest city.

1939–45 Sweden provides support during World War II

Sweden provides volunteer personnel, donations of food, clothing, and other necessities, weapons, aircraft, ammunition, and other war material. In addition, Sweden accepts nearly 40,000 Finnish children as refugees from the fighting during this period.

1939: December 14 The League of Nations expels the Soviet Union

The League of Nations, citing violations of the Treaty of Nonaggression, condemns the Soviet Union for its attack on Finland. The League votes unanimously to expel the Soviet Union from the organization. This is the last official act taken by the League of Nations before it dissolves.

1941 Finland allows Germany to station troops on its territory

Finland, after losing territory to the Soviet Union in late 1939 and 1940, decides to allow Germany to position troops in northern Finland as a base from which to attack Russia. Finland's goal is the recapture of territory lost to Russia. In retaliation, Russia bombs Finland in what becomes known as the Continuation War. Finnish troops regain control of southern Karelia. In December, the United States naval base at Pearl Harbor, Hawaii, is bombed by Japan, drawing the U.S. into the war.

1944 Russia continues its attack on Finland

Russian troops penetrate deeper and deeper into Finnish territory, until Finland decides to abandon further attempts to resist.

1944 Gustaf Mannerheim becomes president

Gustaf Mannerheim (1867–1951) becomes president of Finland. He is seventy-seven years old at the time he takes office and resigns two years later due to failing health.

1944: September 19 Finland withdraws totally from World War II

Finland gives up on its resistance to Russian troops, withdraws from the fighting of World War II, and drives all German forces out of the northern part of the country. As they leave, the Germans burn forests, towns, and villages. Peaceful relations are established with Russia, although Russia retains control of territory in southeastern Finland. In addition, Finland gives up Petsamo on the Arctic Ocean to Russia and is forced to pay over $200 million in reparations (money paid to make up for damages, usually from war). Russia leases a naval base at Porkkala near Helsinki. Finland, devastated by its war losses, maintains a position of neutrality for the remainder of the World War.

1945 Biochemist Artturi Virtanen wins Nobel Prize

Artturi Ilmari Virtanen (1895–1973) wins the Nobel Prize for chemistry. His research involves the processes by which nitrogen is absorbed and released by certain plant roots, the nutritional requirements of plants, and the processing of carotene and vitamin A. He is professor of biochemistry at the Finland Institute of Technology (1931–39) and at the University of Helsinki (1939–48).

1945 Tove Jansson publishes first work

Tove Jansson (b. 1914), author and artist, publishes *Smaatrollen och den stora oversvamningen* (The Small Trolls and the Large Flood). This is her only title never to be published in English. Many of her works are translated from Finnish into other languages, including English. Her children's books feature the world of a fantasy species, the roly-poly Moomins. Jansson eventually writes and illustrates a total of twelve Moomintroll books. The daughter of an artist mother and sculptor father, Jansson is one of Scandinavia's best-known authors and illustrators.

1945 Mika Waltari publishes *Sinhue, the Egyptian*

Mika Toimi Waltari (1908–79) publishes *Sinuhe, egyptiläinen* (Sinhue, the Egyptian), a work of historical fiction set in ancient Egypt. The work becomes an international bestseller, finding readers in the United States, Britain, and Germany.

1946 Juho Paasikivi assumes duties of president

Juho (or Juo) Kusti Paasikivi (1870–1956) takes over the office of president when Gustaf Mannerheim (1867–1951) resigns for health reasons. Paasikivi is later elected in his own right to a full presidential term (see 1950).

1947: June Finnish Festival Games in Helsinki

Over 70,000 athletes converge in Helsinki to compete in the Finnish Festival Games.

1948 Family allowance paid by the government

The government begins a program whereby families are paid an annual allowance for every child under age sixteen.

1948 Finland and the Soviet Union sign a treaty

The Republic of Finland and the Union of Soviet Socialist Republics (known as the Soviet Union) sign an Agreement of Friendship, Cooperation, and Mutual Assistance that outlines the terms for the relationship between the two nations.

Precision gymnasts perform in the exhibition at Helsinki's Olympic Stadium during the 1947 Finnish Festival Games.
(EPD Photos/CSU Archives)

Known as the YYA Treaty (an acronym for its Finnish language title), it defines ways of resolving conflict and establishes a framework for Finnish-Soviet relations.

1950   Juho Paasikivi, elected president, affirms position of neutrality

Juho Paasikivi is elected president for the first time, although he has held the office of president since the resignation of Gustaf Mannerheim (see 1946). Paasikivi sets a clear course of neutrality on the international front. He forges close ties with Finland's neighbors, Sweden, Denmark, Norway, and Russia.

1950   Population reaches four million

In the first half of the twentieth century, the population of Finland grows rapidly, from three million in 1914 to four million in 1950.

1952   Olympic Games held in Helsinki

Helsinki hosts the summer Olympic Games. Paavo Nurmi (1897–1973), legendary Finnish runner, carries the torch into the arena at the opening ceremony.

1954   Väinö Linna publishes *Unknown Soldier*

Väinö Linna (1920–92) publishes *Tuntematen sotilas*, (Unknown Soldier). It recounts the story of a patrol of Finnish troops fighting the Russians. The book achieves some renown outside Finland when it is made into a film that is distributed internationally. *Unknown Soldier*, with its antiwar message, has also been adapted for the stage.

1955   Finland regains Porkkala and renews agreement with Russia

President Juho Paasikivi succeeds in negotiating with Russia for the return of the naval base, Porkkala, near Helsinki (see 1944: September 14), to Finnish control. The 1948 Agreement of Friendship is also renewed.

1955: December 14   Finland joins the United Nations and the Nordic Council

Finland becomes a member of both the United Nations and the Nordic Council with Norway, Sweden, Denmark, and Iceland. Citizens of these countries may work, travel, and receive social benefits such as health care in any other member country without special permits, passports, or visas. This leads to a significant migration of Finns to Sweden.

1956 Urho Kekkonen elected president

Urho Kaleva Kekkonen (1900–86), upon his election as president, establishes a strong commitment to Finland's position of neutrality in international relations. He maintains a position of cautious but friendly relations with the Soviet Union and fosters openness with Finland's neighbors. In 1980, he receives a Lenin Peace Prize for his leadership.

1958 Conductor and composer Esa-Pekka Salonen is born

Esa-Pekka Salonen, internationally renowned conductor and composer, is born. His academic training includes study of the French horn at the Sibelius Academy and study of composition and conducting in Finland and Italy. In 1983, he establishes his international reputation when he conducts Gustav Mahler's Third Symphony in London to notable critical acclaim. He holds permanent posts with the Los Angeles Philharmonic in California (since 1992) and the Swedish Radio Symphony Orchestra (since 1985).

1962 Agreement gives Finland access to the Saimaa canal

An agreement between Finland and Soviet Union gives Finland access to the Saimaa canal.

1965 Gross national product is double the 1945 level

In the twenty years following World War II, Finland doubles its gross national product. This is possible in part due to the increased capacity in shipbuilding and related industries that developed to meet Russian military demands during World War II. (See 1939–45.)

1969–70 Population of Finland declines

Large-scale emigration to neighboring Sweden combines with a slower birth rate to result in a decrease in population in Finland between 1969 and 1970.

1973 Composer Joonas Kokkonen wins the International Sibelius Prize

Joonas Kokkonen (b. 1921), composer and member of the Academy of Finland, wins the International Sibelius Prize for composition. Kokkonen teaches at the Sibelius Academy from the 1940s until the 1960s. His works include operas, symphonies, and chamber works. His opera, *The Last Temptations*, has been performed at the Metropolitan Opera in New York.

1973: January Finnish parliament passes a bill to extend president's term

The Finnish parliament, in a move to reinforce the position of neutrality forged by president Urho Kekkonen and to acknowledge Kekkonen's popularity, passes a bill to extend

his term of office. Kekkonen then holds office until his resignation in 1981.

1975: Summer Peace conference held in Helsinki

Finland, a state recognized for its neutrality, is the site for the Conference on Security and Cooperation in Europe. The conference, attended by representatives from thirty-five nations, focuses on ways the countries of Europe, including Scandinavia, can share resources to foster greater security and economic and cultural exchange.

1976 Pertti Karpinnen wins Olympic gold medal

Oarsman Pertti Karpinnen (b. 1953) wins the gold medal in the Olympic single skulls event. (See also 1980 and 1984.)

1976 Finnish architect Alvar Aalto dies

Alvar Aalto, perhaps Finland's most famous architect, dies. (See 1898.)

1979 Aulis Sallinen writes symphony for Turku anniversary

Aulis Sallinen (b. 1935) writes his Symphony Number 4 for the 750th anniversary of the Finnish city of Turku.

1980 Women work outside the home in greater numbers

About seventy percent of women are employed outside the home, compared to only ten percent in 1920. Women hold sixty percent of all public sector positions, although only about ten percent of management-level positions are held by women. Women's salaries are estimated to be about seventy-five percent of men's salaries.

1980 Pertti Karpinnen wins Olympic gold medal

Oarsman Pertti Karpinnen wins the Olympic gold medal in single skulls for the second time. (See also 1976 and 1984.)

1981: September Mauno Koivisto becomes president

Mauno Koivisto (b. 1923) assumes the duties of president of Finland when Urho Kekkonen, president since 1956, requests a leave of absence due to ill health.

1981: October Kekkonen resigns

Urho Kekkonen resigns as president due to poor health.

1982: January Koivisto elected president

Mauno Koivisto is elected president and reelected in 1988.

1984   Matti Nykanen wins gold medal at Winter Olympics

Finnish ski jumper Matti Nykanen (b. 1964) wins the gold medal in the ninety-meter ski jump at the Winter Olympics in Sarajevo, Yugoslavia. (See 1992.)

1984   Pertti Karpinnen wins Olympic gold medal

Oarsman Pertti Karpinnen wins his third gold medal in single skulls. He and a Russian oarsman (V. Ivanov) share the record of three gold medals in single skulls. (See 1976, 1980.)

1985   Deep-water port opens at Pori

A deep-water harbor complex opens to handle oil and coal imports.

1986   Finland joins EFTA

Finland becomes a full member of the European Free Trade Agreement (EFTA).

1987   *The King Goes Forth to France* is performed at Covent Garden

Finnish composer Aulis Sallinen's opera, *The King Goes Forth to France*, is performed at Covent Garden in London, England. The work is commissioned by the British Broadcasting Corporation (BBC), Covent Garden (a famous performing arts hall in London, England), and the Savonlinna Festival (an international music festival held in Finland).

1987   Women win new rights

Parliament passes the Equality Act to prevent discrimination because of gender in employment. The Finnish Evangelical Lutheran Church agrees to allow women to study for and be ordained as priests.

1990s   Finns embrace technology

Statistics show that Finland has the highest percentage of households connected to the Internet among the countries of the world..

1991   Population growth

The population of Finland grows steadily during the twentieth century (except for a brief period in the late 1960, see 1969), and reaches 5 million this year.

1991: April 26   Elections result in number of women in parliament

The parliament is nearly forty percent female, with women candidates winning seventy-seven of the total two hundred parliamentary seats. Esko Aho (b. 1954) takes office as prime minister following general elections

1992   Finnish athletes win gold medals at the Winter Olympics

Ski jumper Matti Nykanen wins a record three gold medals in ski jumping at the Winter Olympics in Calgary, Alberta, Canada. He is the first man to win both the seventy-meter and ninety-meter titles. His third medal was in the ninety-meter team event. (See 1984.)

1992: January   Finland recognizes Russia as successor to the Soviet Union

Finland and Russia conclude a treaty defining good relations between the two countries. No military articles are included in the treaty, and the two countries declare that the Treaty on Friendship, Cooperation, and Mutual Assistance between Finland and Soviet Union is null and void (see 1948).

1992: March   Finland applies for membership in the European Community

With neighboring Sweden applying for membership in the European Community in 1991 and the Soviet Union breaking up the same year, Finland decides to apply to join the European Community. (See 1994: May)

1992: May   European Economic Area (EEA) treaty is signed

EFTA and the European Community sign the European Economic Area (EEA) treaty is signed. To Finland, the European Economic Area treaty fulfills the country's goal to participate in economic growth opportunities with its European neighbors.

1994: March 1   Martii Ahtisaari takes office as president

Martii Ahtisaari takes office as the tenth president of Finland, having received 54 percent of the vote in the election.

1994: May   Finland wins acceptance into the European Union

Finland's application to join the European Union (formerly the European Community) is approved. Yet it takes a year for Finland to formally join the European Union.

1994: October   Referendum held on membership in the European Union.

In a referendum on Finnish membership in the European Union, 57 percent of the voters support membership. In November, the Finnish parliament approves membership beginning in 1995 by a vote of 152–45.

1995: March   Center Party government topples

The Center Party government topples as a result of recession and high unemployment. The Social Democrats then emerge as the largest party in Finland, with Paavo Lipponen as prime minister.

1997 Lake Tuusula Centenary

The famous artists' colony of the Lake Tuusula area, the home of the celebrated composer Jean Sibelius, celebrates its centenary.

1999 Finland joins the EMU (European Monetary union)

Bibliography

Derry, T. K. *A History of Scandinavia: Norway, Sweden, Denmark, Finland, and Iceland.* Minneapolis: University of Minnesota Press, 1979.

Jutikkala, Eino, and Kauko Pirinen. rev. ed. Paul Sjöblom, trans. *A History of Finland.* London: Heinemann, 1979.

Lander, Patricia Slade. *The Land and People of Finland.* New York: HarperCollins, 1990.

Maude, George. *Historical Dictionary of Finland.* Metuchen, N.J.: Scarecrow Press, 1994.

Poutasuo, Tuula. *Finnish Silver: From the 2nd World War to Post-Modernism.* Kirjayhtyma, Helsinki, Finland: 1989.

Rajanen, Aini. *Of Finnish Ways.* New York: Barnes & Noble Books, 1981.

Schoolfield, George C. *Helsinki of the Czars: Finland's Capital, 1808–1918.* Columbia, S.C.: Camden House, 1996.

Singleton, Fred. *A Short History of Finland,* 2nd ed. Cambridge, Eng.: Cambridge University Press, 1998.

France

Introduction

France is a country with a complex and eventful history and a rich cultural tradition. As the birthplace of the Enlightenment, Romanticism, Modernism, and poststructuralist theory, France has historically been at the forefront of Western cultural and intellectual developments. Its tumultuous political history has encompassed extremes ranging from the absolute monarchy of Louis XIV (1638–1715) to the radical revolutionary ideals of the Jacobins (a society of radical democrats that met in a Jacobin monastary). Following the devastation of two world wars, France has recovered and prospered in the postwar era, achieving political stability under the Fifth Republic, established in 1958, and becoming a leader in the expanded European integration that will include monetary union.

Located at the extreme western edge of Europe, France is the second-largest country on the continent (and the largest in Western Europe), with an area of approximately 211,000 square miles (547,000 kilometers). Its varied geography ranges from the fertile Paris Basin (the main lowland area) to the Massif Central (a region of granite plateaus that occupies one-sixth of the country) to the French Alps, the highest of several mountain ranges that ring the country and the site of Mont Blanc, at 15,771 feet (4,807 meters) the highest point in Europe. France has a current estimated population of 58 million people, of whom about one-sixth live in the metropolitan area that includes the capital city of Paris.

Roman and Frankish invasions

The first great recorded period in French history was the Roman occupation that took place between about 51 B.C. and the fifth century A.D. The Romans who conquered present-day France under the leadership of Julius Caesar (c. 100–44 B.C.) called its Celtic inhabitants Gauls (and referred to the region itself as Gaul). The foundation of French culture was laid during this period with the introduction of the Latin language, from which French later evolved, and of Roman laws and institutions. Christianity was introduced during this period as well. The Romans were followed by Germanic invaders, notably the Franks, whose greatest king, Charlemagne (747–814), made France part of a vast empire encompassing most of present-day Western Europe. After Charlemagne's death, the Carolingian empire he inherited and strengthened so greatly was split into three parts. It was further weakened by centuries of raids by the Scandinavian Vikings.

Meanwhile, in 1066 William II (c. 1028–87), Duke of Normandy, invaded and conquered England, founding a new dynasty whose claims to French land would eventually lead to a century of war between France and England. The Hundred Years' War was actually a series of conflicts between a variety of French and English monarchs between 1337 and 1453, a period during which France was ravaged not only by warfare but also by the bubonic plague, which killed more than one-fourth of its population. Although the English had the upper hand for more than one period during this long conflict, it ended with France recapturing almost all of the land claimed by England following the historic victory of Charles VII (1403–61), the dauphin (heir to the throne) of France, at Orléans, where French troops rallied under the leadership of Joan of Arc (1412–31).

Influence of the Renaissance and Protestant Reformation

By the end of the fifteenth century, the French were battling a new foe: the Habsburg dynasty of Spain (1516–1700). A series of wars in Italy was significant culturally as well as militarily, for they exposed the French to the scholarship and arts of the Italian Renaissance. An early exponent of the humanist ideals of the Renaissance was the satirist François Rabelais (1494?–1553?). The sixteenth century was also the era of the Protestant Reformation, whose French adherents were called Huguenots. Decades of religious warfare between Huguenots and Catholics ended with the coronation of King Henry IV (1553–1610) and the Edict of Nantes (1598), which protected the religious freedom of the Huguenots.

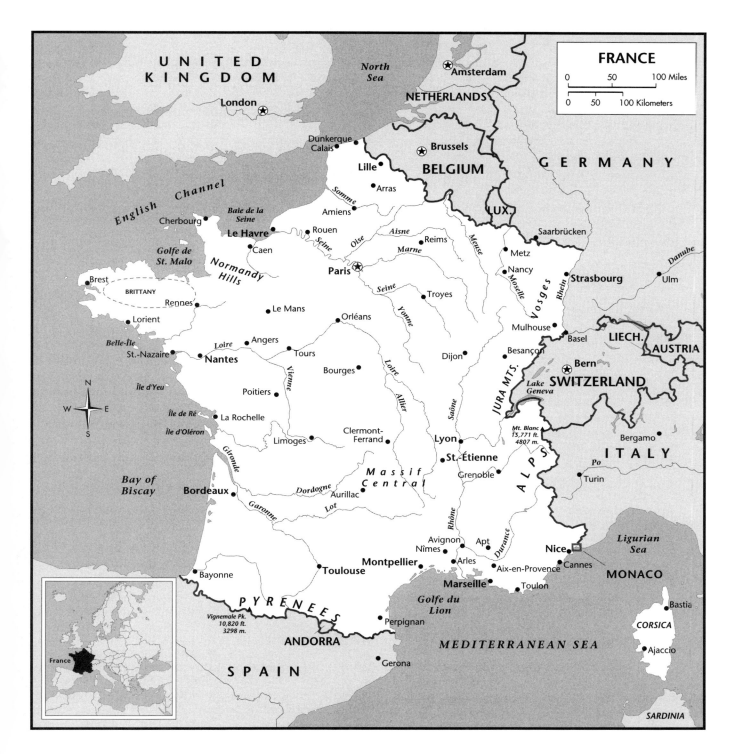

Reign of Louis XIV

In the seventeenth century, the French monarchy gained increasing power, first through the policies of Louis XIII's (1601–43) powerful adviser, Cardinal Richelieu (1585–1642) and then under the rule of Louis XIV (the Sun King; 1638–1715), whose reign was famous for its lavish displays of wealth and fashion as well as its cultural achievements. These included the plays of Jean Racine (1639–99), Thomas Cor-

neille (1625–1709), and Molière (1622–73) and the founding of the famous Comédie Française. The composers Francois Couperin (1668–1733) and Jean Philippe Rameau (1683–1764) made their musical contributions during this time of rich culture. Unfortunately, Louis XIV involved France in four costly wars, and by the end of the War of the Spanish Succession (1701–14), France had been seriously weakened and lost most of its colonial possessions.

For the remainder of the eighteenth century, inept monarchs allowed the nation's economic situation to deteriorate even further, exacerbated by poor harvests and food shortages in the countryside, while they continued to embroil the nation in unsuccessful warfare with England and impose new taxes to deal with France's mounting debt. Popular discontent with the French kings was fueled yet further by the writings of the eighteenth-century French philosophers, including François Voltaire (1694–1778) and Jean-Jacques Rousseau (1712–78), who criticized the monarchy and espoused democratic ideals. The stage was set for the French Revolution, which would not only have far-reaching effects in France but also influence events in other nations.

The French Revolution and the Rise of Napoleon

Initially the revolutionaries took a moderate approach, and King Louis XVI (1754–93) in 1790 agreed to a new constitution providing for a limited monarchy. However, events culminated in the extremes foreshadowed by the storming of the Bastille (a state prison in Paris) in 1789. Between 1793 and 1795, thousands of French men and women lost their lives during the Reign of Terror and the period of reprisals that followed it. The Directoire that emerged in 1795 as the governing body was overthrown four years later by military hero Napoleon Bonaparte (1769–1821), who became First Consul and then Emperor of France. Napoleon instituted wide-ranging internal reforms, bringing new administrative efficiency to domestic affairs. Abroad, he embarked on a series of military campaigns in an attempt to bring all of Western Europe under French domination. By 1810 he had nearly succeeded in this goal, but then his military fortunes declined, and he was permanently removed from power in 1815.

Monarchy Restored; Second Republic Established

Following the ouster of Napoleon, the Bourbon monarchy was restored, but then it, too, was overthrown in the July Revolution of 1830, and Louis-Philippe (1773–1850) of Orléans was installed as king. Known as the "citizen king," he favored the interests of the rising class of French industrialists. Along with the growth of industrialization in the nineteenth century came the railroad. French rail construction began in the 1840s and with it an unfortunate milestone—the first recorded railroad accident, which ironically occurred en route from the glittering seventeenth-century palace of Versailles, as revelers were returning to Paris from a celebration of Louis-Philippe's birthday. The mid-nineteenth century was also the era of Realism in French fiction by authors including Honoré de Balzac (1799–1850), Gustave Flaubert (1821–80), and Emile Zola (1840–1902).

In 1848 France was caught up in a new wave of revolutionary fervor that swept through Europe on the heels of a severe crop failure. King Louis-Philippe was overthrown, and the Second Republic was established. Louis Napoleon (1808–73), the nephew of Napoleon Bonaparte, was elected president. Near the end of his term, he engineered a coup and proclaimed the Second Empire, with himself as Emperor Napoleon III (the number "II" was skipped in honor of Napoleon's son, who had died). Napoleon III presided over a period of economic prosperity and colonial expansion. Like the tenure of his namesake, however, the reign of Napoleon III was ended by his involvement in warfare with a foreign power, in this case Prussia. Following the disastrous Franco-Prussian War of 1870–71, Napoleon was ousted, and the Third Republic was proclaimed. Basic civil liberties were guaranteed under a new constitution, as was the separation of church and state.

In the arts, the late nineteenth century was an era of innovation and experimentation. Impressionist painters, including Claude Monet (1840–1926), Camille Pissarro (1830–1903), Auguste Renoir (1841–1919), and Édouard Manet (1832–83), produced works in which the solid outlines and realistic representations of traditional painting gave way to shimmering bursts of color that emphasized the fleeting effects of light, movement, and weather. A similar focus on the immediate experience of a given moment was evident in the works of Symbolist poets, including Stéphane Mallarmé (1842–98), Paul Verlaine (1844–96), and Arthur Rimbaud (1854–91). In music, composers Claude Debussy (1862–1918) and Maurice Ravel (1875–1937) pioneered the use of new harmonies and sonorities.

Both socially and politically, the struggle between progressives and reactionaries coalesced, or united, around the Dreyfus Affair at the end of the nineteenth century. The treason conviction of Captain Alfred Dreyfus (c. 1859–1935), a prominent Jewish army officer, and the long campaign that led to his retrial and vindication, divided the nation for twelve years and became a leading cause among French intellectuals.

World War I and II

As a member of the Triple Entente with Russia and Britain, France was plunged into World War I against Germany, Austria-Hungary, and Italy between 1914 and 1918 and was the scene of some of the war's worst fighting. Under the Treaty of Versailles following the war, France regained the provinces of Alsace and Lorraine. However, the nation was devastated by the loss of around one and a half million men in four years of bitter trench warfare. The war was followed by a succession of weak governments and the financial hardship of the Depression.

In September 1939 France declared war on Germany following the invasion of Poland. The following May the Germans breached France's allegedly impenetrable Maginot Line fortifications and occupied the country. The north and west were occupied directly by Germany; the remainder was administered by the collaborationist (one who cooperates with an enemy invader) Vichy government until 1942, when Germany completed its occupation of the entire country. Having fled to England, a young general named Charles de Gaulle (1890–1970) organized a government-in-exile and a resistance force, known as the Free French, that fought the Germans from within France and also joined in the Allied invasion that liberated the country in 1944.

de Gaulle and the Fifth Republic

Following the war, de Gaulle formed a provisional, or temporary, government recognized by the Allies. The constitution approved by French voters for the new Fourth Republic, however, did not meet with his approval, and he resigned from office. Like other nations in Western Europe, France was greatly helped in its recovery from the war by the Marshall Plan implemented by the United States, and French cooperation with the U.S. was expanded further when it became a founding member of the North Atlantic Treaty Alliance (NATO). In addition, conflicts over its colonial possessions played a prominent role in French foreign policy in the postwar period, first in Indochina (1946–54), and then in North Africa (1954–62). France's last major colony, Algeria, won its independence in 1962.

France's political leadership was seriously weakened during the conflict in Algeria. Asked to take over the faltering government, Charles de Gaulle agreed on condition that the executive branch be strengthened. French voters agreed, approving the constitution of the Fifth Republic in 1958 and making de Gaulle its first president. The short-lived stability that followed was shattered by the student protests and general civil unrest of 1968, which resulted in labor and educational reforms. Economic difficulties following the worldwide energy crisis of the 1970s led to the election in 1981 of France's first Socialist president, François Mitterrand (1916–96), who served two seven-year terms. His initial policies, including nationalization of major banks and corporations and measures favoring labor and social welfare programs, were well received. However, his government was eventually obliged to adopt Conservative-style austerity measures to deal with high inflation and the declining value of the franc. Conservative gains in the 1986 legislative elections forced Mitterrand to appoint and work with a conservative prime minister, Jacques Chirac (b. 1932), for two years in a political arrangement referred to in French politics as "cohabitation."

The 1990s have seen the most serious student protests since 1968; the appointment of France's first female prime minister; the deployment of French troops in the Gulf War; the opening of the Channel Tunnel making rail service between England and France a reality; and French approval of the Maastricht Treaty providing for European monetary and economic union. In 1995 former prime minister Jacques Chirac was elected president. Two years later Chirac, like his predecessor, found himself in a "cohabitation" situation with a prime minister of the opposing party when the Socialists made an unexpectedly strong showing in legislative elections.

Frenzied celebrations took place in 1998 when France, which had hosted the World Cup soccer tournament, won its first championship title.

Timeline

c. 30,000 B.C.   First humans populate France

France's first human inhabitants are thought to have lived about 30,000 years ago, leaving behind cave paintings including those at Lascaux.

600 B.C.   Greeks and Phoenicians settle in the south

Greek and Phoenician settlers found cities on the southern coast.

58–51 B.C.   Roman conquest

The Romans, under Julius Caesar, conquer present-day France, which is inhabited by the Celtic peoples whom they call "Gauls." (They refer to France itself as "Gaul".)

51 B.C.–c. A.D. 500   Roman occupation

The Romans occupy present-day France from its conquest by Julius Caesar (c. 100–44 B.C.) to the decline of their empire in the fifth century A.D. As in other parts of their empire, they bring with them their language and laws, and build cities and roads. Christianity is introduced early in this period.

Early 5th century A.D.   Germanic tribes invade France

Germanic tribes, including the Franks and the Visigoths, invade France.

Merovingian Dynasty

A.D. 486   Clovis I ousts the Romans

The last Roman governor of France (then called Gaul) is overthrown by the Frankish king Clovis (466–511) at the Battle of Soissons. Clovis's rule inaugurates the Merovingian kingdom.

496   Clovis converts to Christianity

King Clovis accepts Christianity and is baptized, giving France its first Christian ruler.

511 Death of Clovis begins Merovingian decline

Following the death of Clovis, the Merovingian kingdom enters a two-hundred-year period of civil warfare and is divided into two kingdoms, Austrasia and Neustria. Both kingdoms have weak monarchs, and by the early seventh century, actual power is wielded by each king's chief minister, called the mayor of the palace.

732 Battle of Poitiers

Charles Martel (c. 688–741) inflicts a historic defeat on Muslim invaders from Spain.

Carolingian Kings

751 Carolingian dynasty begins

Pepin III (the Short; c.715–768), son of Charles Martel, overthrows the Merovingian king and inaugurates the Carolingian dynasty.

768–814 Reign of Charlemagne

Frankish rule reaches its peak during the reign of Charlemagne (747–814) who expands its territories, administers them efficiently, and promotes education and the arts. Under Charlemagne, the Carolingian empire reaches its greatest size, encompassing the lands that today make up France, Germany, the Netherlands, Belgium, Luxembourg, and northern Italy.

800 Charlemagne is crowned emperor of the West

Pope Leo III (c. 750–816) crowns Charlemagne emperor of the Western Roman Empire, further strengthening the status of Christianity in his kingdom.

843 The Carolingian empire is divided

Following the death of Charlemagne, the Carolingian empire is divided into three parts by the Treaty of Verdun. The most powerful two parts correspond roughly to present-day France and Germany.

911 Vikings conquer Normandy

Scandinavian invaders known as Vikings, who have attacked and plundered France since the late 700s, occupy the northern region between the Somme River and Brittany that later becomes known as Normandy. It is ceded to them by Charles III (the Simple; 879–929).

The Capetian Line

987 Capetian dynasty is founded

The feudal lords of France elect Hughe Capet (c. 938–996) as their king, inaugurating the Capetian line that will govern France for over three hundred years.

1066 Norman Conquest strengthens Normandy

William II, duke of Normandy, conquers England, becoming William I (the Conqueror; c. 1028–87). He not only gains control over new territory, he also strengthens the wealth and power of his duchy in France, whose holdings on the continent are extensive.

1108–37 Louis VI consolidates royal power

Louis VI (the Fat; 1081–1137) is the first Capetian king to solidify the dynasty's control of its own fiefdom (its heritable land), consisting of Paris and the surrounding area, issuing royal charters to the area's growing towns. He also arranges the marriage of his son Louis (later Louis VII; c. 1120–80) to Eleanor of Aquitaine (c. 1122–1204), heiress to extensive lands.

1130 Birth of troubadour Bernart de Ventadorn

Troubadour musician Bernart de Ventadorn (1130–95) is born. He becomes one of the most famous troubadours (poet-musicians) of France. Eleanor of Aquitaine invites him to England after her marriage to Henry II (see 1154).

1152 Marriage of Louis VII and Eleanor is annulled

The marriage of Louis VII, the Capetian king, to Eleanor of Aquitaine is annulled. She later marries the French Henry of Anjou (the future king of England; 1133–89), giving him claim to Aquitaine and Normandy.

1154 Henry II is crowned king of England

Henry of Anjou becomes Henry II of England (1133–89). In addition to ruling England, he controls extensive territories in France, some of them gained through his recent marriage to Eleanor of Aquitaine (see 1152). He also wins control of Brittany through the marriage of his son Geoffrey. Henry's extensive holdings threaten the power of Louis VII.

1180–1223 Reign of Philip II

Philip Augustus (Philip II; 1165–1223), son of Louis VII, succeeds to the throne of France. Capturing much of the Angevin land (land in Anjou) controlled by the kings of England (first from Henry II and then from his sons, Richard the Lion-Hearted [1157–99] and John [1167–1216]), he greatly increases the power of the French monarchy.

1200 University of Paris is founded

The University of Paris, a prominent center of medieval learning, is established. The great scholars Saint Albertus Magnus (c. 1200–80) and his pupil, Saint Thomas Aquinas (1225–74), pursue their theological studies here.

1202–04 Philip II gains most Angevin territories in France

Warring with the forces of the English king John, the son of Henry II, Philip strips the English Angevin kings of all their land in France except Gascony.

1226–70 Reign of Louis IX

Louis IX (1214–70) maintains and consolidates the royal power won by his predecessor, Philip Augustus (see 1180–1223), through effective administrative reforms.

1285–1314 Reign of Philip the Fair

Philip IV (Philip the Fair; 1268–1314), grandson of Louis VII, reigns over France. He increases the already considerable power of the king, creating the model for an absolute monarchy. He challenges the power of the Church, resisting its interference in matters of state and eroding the powers of the clergy, including its immunity from taxation. Ultimately, he has the papacy itself moved to France, where French kings control it for nearly a century.

1302 Flemish rebels defeat French troops

A French attempt to take over Flanders fails when the French are driven out in a successful revolt. A detachment of troops is massacred in Bruges, and reinforcements are driven back at the Battle of Courtrai.

1303 Philip has Pope Boniface arrested

In the wake of growing disagreements between Philip IV and Pope Boniface VIII (c. 1235–1303), men employed by the king kidnap the pope at Anagni and detain him. He dies shortly afterward.

1305 A French pope is elected

King Philip uses his influence to have a French prelate (a high ranking church official and supporter of the king), Pope Clement V (c. 1260–1314), elected to succeed Pope Boniface VIII.

1309–77 The papacy is moved to Avignon

Pope Clement V moves the seat of the papacy to Avignon in Provence. Here it is dominated by French kings for most of the century in what is often referred to as the "Babylonian captivity."

1328 Philip VI inaugurates the Valois dynasty

Philip VI (1293–1350), the nephew of Philip the Fair, succeeds to the French throne in accord with Salic Law because Philip's sons leave no direct heirs. With Philip's accession, the Capetian dynasty ends, and the line of Valois kings begins.

The Hundred Years' War

1337–1453 The Hundred Years' War

The Hundred Years' War is launched when the English king Edward III claims the French throne, challenging the rule of Philip VI. Warfare is not continuous throughout the entire period but rather consists of a series of shorter conflicts and skirmishes. England gains the upper hand in the first decades of the war, a difficult period when France is also ravaged by the Plague (see 1348–50). In the later stages, however, France regains almost all the territory it has previously lost, and its monarchy emerges strengthened, although the country itself has been decimated.

1337 Edward III claims the French throne

On the grounds of his descent from Philip the Fair on his mother's side, the English king Edward III challenges Philip VI's claim to the French throne.

1346 Battle of Crécy

In the first major battle of the Hundred Years' War (see 1337–1453), the French are badly beaten by the English in a battle that pits armored knights (notably superior in number) using outmoded battle tactics against English longbowmen.

1348–50 The Black Death strikes France

France falls victim to the Black Death (bubonic plague) ravaging Europe. It is one of a series of epidemics that reduces France's population by over one-fourth in the latter half of the fourteenth century.

1356 Battle of Poitiers

In another major defeat of the Hundred Years' War, King John II of France (1319–64) is taken prisoner by English forces led by the son of Edward III, known as the Black Prince.

1358 Peasant rebellion is subdued

Charles, the dauphin (eldest son of France's king) and future king of France, restores the reputation of the monarchy by putting down the Jacquerie, a powerful peasant rebellion. The Jacquerie is provoked by the sour economic conditions due to the Hundred Years War and by overly-burdensome demands from the nobles. In their revolt, the peasants decimate a large area, demolish fortifications, loot provisions, and cause violent havoc. The leader of the rebellion, Guillaume Karle, is captured and beheaded by Charles, and in retaliation, the nobles butcher thousands of peasants.

1360 Treaty of Brétigny

Following major military defeats, France is forced to cede the southwest part of its kingdom to England.

1364–80 Charles V restores French lands

King Charles V (1337–80), with the aid of famed military commander Bertrand du Guesclin, wins back most of the territory captured by England in the Hundred Years' War.

1415 Battle of Agincourt

Henry V of England, taking advantage of a weakened French monarchy under the mentally unstable Charles VI (1368–1422) and of factional rivalry between the dukes of Burgundy and Orléans, invades France, decisively defeating the French forces at Agincourt.

1420 Treaty of Troyes

Under the Treaty of Troyes, Catherine, the daughter of Charles VI, is forced to marry Henry V, who will become king of France when Charles dies.

1422 Death of Henry V and Charles VI

Charles VI dies, and Henry V, who is supposed to succeed Charles VI as king of France, dies in the same year. Henry VI, the infant son of Henry and Catherine, Charles's daughter, is crowned king of France, but his claim to the throne is challenged by Charles's son, the dauphin of France.

1429 Jeanne d'Arc leads Charles's forces to victory at Orléans

When English forces fighting for Henry VI drive southward to conquer the lands held by Charles VII (1403–61), dauphin of France, the French forces are rallied by a peasant girl named Jeanne d'Arc (Joan of Arc; 1412–31). They repulse the English attack and restore the morale of the French. Charles is crowned Charles VII at Reims.

1431 Jeanne d'Arc is burned at the stake

Jeanne d'Arc, who has become a hero to the French, falls into enemy hands and is burned at the stake as a heretic.

1453 French recapture Bordeaux

French troops recapture Bordeaux, ending the Hundred Years' War. Only Calais remains in English hands.

Consolidation of Royal Power

1461–83 Reign of Louis XI

King Louis XI (1423–83) consolidates the absolute power of the French monarchy following the Hundred Years' War. Louis eliminates all threat to the crown by the dukes of Burgundy and other feudal lords. He guides the country's economic recovery from the devastation of the war, building roads and canals and supporting textile production. He also supports trade by manipulating tariffs and signing agreements with foreign powers. Louis's policies are also significant in terms of cooperation between the king and the rising commercial classes.

1470–98 Reign of Charles VIII

Charles VIII ascends the throne at the age of thirteen, inheriting absolute rule over a powerful and united country from his father, Louis XI (see 1461–83). By his marriage to the duchess Anne of Brittany, he regains control of that province for France. He also inaugurates the French wars with Italy by his invasion of that country, which is significant, even though it is soon thwarted.

1477 Defeat of the Duke of Burgundy at Nancy

Charles the Bold, duke of Burgundy, is killed in battle at Nancy, ending the Burgundian military threat to the supremacy of the French monarchy.

1494–1559 Italian wars

A series of French attacks on Italy, inaugurated by Charles VIII and pursued by successive monarchs, inaugurates a long-lived rivalry between France and the Spanish Habsburg dynasty. Aside from Charles, other kings who make war on Italy include Francis I (r. 1515–47) and Henry II (r. 1547–59). The Habsburgs achieve a decisive victory over the Valois dynasty of France at the Battle of Gravelines in northern France. Although the campaign against Italy is a military failure for France, it results in a cultural gain through the exposure to the flowering of Renaissance Italy.

c. 1494 Birth of Rabelais

The great satirist François Rabelais is born in Poitou. He takes Franciscan orders in 1521, later leaving and joining the Benedictines. Eventually he begins studying medicine, lecturing and publishing his own editions of the works of the ancient Greek physicians. From 1532 he practices medicine in Lyons. Between 1532 and 1564 he publishes (under the pseudonym Alcofribas Nasier) a series of comic novels based on the exploits of the giant Gargantua and his son, Pantagruel. Of these the most notable is *Gargantua and Pantagrue*. In his works he satirizes a variety of targets, from outmoded medieval ideas to the social foibles of his day and promotes the ideals of Renaissance humanism (the emphasis on the ideals, forms and studies of classical thought). Rabelais dies in Paris in 1553.

1515–47   Reign of Francis I

Francis I (1494–1547), the former Count of Angoulême and cousin to Louis XII, succeeds to the throne. Through his continuation of the wars against Italy begun by his immediate predecessors, the humanistic scholarship and artistic achievements of the Italian Renaissance become important cultural influences in France. It is also during this period that the Protestant Reformation takes hold among the French. The Reformation, a religious revolution against the oppressive medieval Catholic church, not only severally erodes church power, but firmly establishes the Protestant churches (hence the name) and leads to the freedom of dissent.

1530   The Collège de France is founded

The Collège de France is established. In the twentieth century, it is still a popular venue for lectures by eminent scholars from around the world.

1558   Calais is taken by France

Calais, the last English-controlled territory in France, is reclaimed by the French.

1562–98   Religious differences lead to war

France is torn by conflict between Catholics and Protestants (called Huguenots in France) that escalates into a series of civil wars punctuated by unstable truces based on the granting of limited rights of worship to the Huguenots. The most notorious incident of these wars is the St. Bartholomew's Eve Massacre (see 1572: August 23–24).

1572: August 23–24   St. Bartholomew's Eve Massacre

The St. Bartholomew's Eve Massacre ends a period of truce in the series of religious civil wars between France's Protestants (Huguenots) and Catholics. King Charles IX (1550–74), encouraged by his mother Catherine de' Medici, orders an attack on Huguenots gathered in Paris to celebrate St. Bartholomew's Day and the wedding of Henry of Navarre (later King Henry IV) to Margaret of Valois. Two thousand Huguenots are killed within a day, and the massacre spreads from Paris to the provinces. It ultimately results in as many as fifty thousand deaths and the resumption of the Wars of Religion.

1585: September 9   Birth of Cardinal Richelieu

Cardinal Richelieu is born Armand Jean du Plessis. He enters the Church to fill the bishopric of Luçon, left vacant by his elder brother, thus keeping it in the du Plessis family. Richelieu's political career begins with his participation as a representative of his province in the Estates General of 1614, which attracts the notice of Marie de' Medici, the Queen Mother. He is named to be her chaplain and becomes an adviser as well. By 1624 he is appointed first minister to King Louis XIII (1601–43), a position he retains until his death nearly twenty years later. During this time, he works to secure the power of the king and curb that of the nobles, becoming the single most influential figure in the development of an absolute monarchy in France. He also reduces the power of the Huguenots (French Protestants) by removing their armed strongholds and involves France in the Thirty Years' War against Spain.

Throughout his long tenure as the chief architect of France's domestic and foreign policy, Richelieu succeeds in outmaneuvering his political opponents, among them Marie de' Medici, who had been his initial patron. He is named a cardinal in 1622 and becomes a duke in 1631. In addition to his other accomplishments, Richelieu founds the Académie Française, the institution assigned to maintain language standards in France. Richelieu dies in 1642, on the eve of France's decisive victory over Spain at Rocroi (see 1643).

1589   Last Valois monarch is assassinated

Henry III (1551–89), the last king in the Valois line, is assassinated by a Dominican monk for choosing a Protestant successor, Henry of Navarre, who succeeds to the throne as Henry IV, beginning the Bourbon line of monarchs.

The Bourbon Dynasty

1593   Henry IV converts to Catholicism

In order to gain control of the strongly Catholic city of Paris and assure the viability of his kingship, Henry IV (1553–1610) converts to Catholicism. He is crowned at Chartres the following year and becomes a popular and successful monarch, ending thirty-five years of religious warfare and restoring order to the country.

1598   Edict of Nantes

The religious wars between France's Protestants and Catholics are ended by the Edict of Nantes issued by Henry IV, guaranteeing limited religious and civil rights to Huguenots (French Protestants) and allowing them to retain control of fortified strongholds, mostly in southern and southwestern France.

1610   Henry IV is assassinated; Louis XIII succeeds to the throne

King Henry IV, a Protestant who has accepted Catholicism, is assassinated by a Catholic extremist, and his nine-year-old son, Louis, ascends the throne. The Queen Mother, Marie de Médicis, acts as regent during his minority.

1622: January   Birth of Molière

Molière, France's greatest comic playwright, is born Jean-Baptiste Poquelin in Paris. Destined for the law by his family, he instead chooses a career in the theater by the age of nineteen, joining a theater troupe. For twelve years, the young actor and mime tours the provinces of France, honing the observation of human nature that will later serve him as a satirist and dramatist. In 1658 his troupe returns to Paris, where it wins the patronage of King Louis XIV, performing at the royal theater that later becomes the famed Comédie Française.

Molière soon launches his career as a playwright with *Les Précieuses Ridicules*, achieving great popular success but also arousing controversy and censure on religious grounds throughout his career. His greatest plays include *L'École des femmes* (The School for Wives), *Tartuffe, Dom Juan, Le Misanthrope* (The Hater of Mankind), *L'Avare* (The Miser), *Le Malade imaginaire* (The Imaginary Invalid), and *Le Bourgeois gentilhomme* (The Bourgeois Gentleman). In February 1673, Molière, in ill health for years, collapses at the end of a performance and dies shortly afterward.

1624–42   Cardinal Richelieu in power

Cardinal Richelieu (Armand Jean du Plessis) becomes the chief minister to King Louis XIII (1601–43). He is the power behind the throne for nearly twenty years and helps France achieve political supremacy in Europe. He subdues the rebellious French nobles and reduces the power of the Huguenots, eliminating any threat of armed religious rebellion. In foreign affairs, he resumes France's struggle against the Habsburgs.

1629   Peace of Alais

Richelieu reaches a settlement with the Huguenots after their key stronghold of La Rochelle is seized by government troops. They retain all the religious freedoms guaranteed under the Edict of Nantes (see 1598) but surrender their fortified towns.

1635   France enters the Thirty Years' War

Under the guidance of Cardinal Richelieu, France takes part in the Thirty Years' War (1618–48) against the Habsburgs of Austria and Spain. Though the war concerns most of the large states of Europe, most of the fighting takes place in what is now Germany. The war symbolizes the struggle against institutional Catholicism, and the outcome is devastating for Germany in particular. A horrific annihilation of its population (by some accounts one-third of the population is killed), and a decimating affect on the economic structure of the area, leaves the Holy Roman Empire a mere shell in the succeeding centuries.

1635   Jardin du Roi is established

The Jardin du Roi (Royal Garden), a botanical garden and research center that includes laboratories and an amphitheater for lectures, is established. Also housing facilities for research in astronomy and chemistry, it is France's first government-supported scientific institution.

1635   Académie Française is founded

The Académie Française is established (see 1585: September 9). Membership is limited to forty prominent figures in the intellectual life of the nation. The dictionary and grammar guideline published by the Académie set the official standard for use of the French language.

1639: December   Birth of dramatist Jean-Baptiste Racine

Jean-Baptiste Racine, the most revered author of classical French tragedy, defies a strict Protestant upbringing to choose a career in the theater. His first plays are staged by the company of comic dramatist Molière (see 1622: January), but he switches to the Hôtel de Bourgogne as the venue for most of his works. His greatest tragedies include *Britannicus* (1670), *Bérénice* (1671), *Bajazet* (1672), and *Phèdre* (1677). The first two are set in ancient Rome, the third is based on recent Turkish history, and the last is based on a work by the Greek dramatist Euripedes (480–406 B.C.). Racine is admitted to the Académie Française in 1674. He and the noted critic Nicolas Boileau are appointed to write the official history of the reign of King Louis XIV. Racine dies in 1699.

The Sun King

1643   Battle of Rocroi

A year after Richelieu's (see 1585: September 9) death, France, following the policies he set, enjoys the definitive military success that ends Spain's European dominance, as forces commanded by the Duc de Condé inflict a devastating defeat on the Spanish infantry.

1643   Accession of Louis XIV

Louis XIV (1638–1715) succeeds to the throne at the age of five following the death of his father, Louis XIII. His mother, Anne of Austria, acts as regent, but Cardinal Mazarin, a protégé of Richelieu, has the major influence over policy decisions.

1648–53   French nobles stage their last revolt (Le Fronde)

The French nobles revolt against the Crown one last time in the uprising that comes to be known as the Fronde. It is successfully defeated under the leadership of Mazarin. At one point, the rebels plot to seize the young Louis XIV, who is whisked out of the palace by his retainers in the middle of the night.

1648   Treaty of Westphalia

The Thirty Years' War ends with the Treaty of Westphalia, whose terms are highly favorable for France, and it gains control of Alsace.

1659   Treaty of the Pyrenees

Under the Treaty of the Pyrenees, Habsburg dominance over Europe ends, and France becomes the major power.

1660s   Royal Academy of Science is founded

Louis XIV establishes the Royal Academy of Science, charging its members with calculating the exact length of one degree of longitude, a task they complete successfully.

1661–1715   The mature reign of Louis XIV

With the death of his advisor, Mazarin, Louis XIV takes over primary responsibility for governing the country. The mature reign of Louis XIV, popularly known as the "Sun King," (so-called because of the lavish brilliance of the court he had built under the direction of Jules Mansart . . . Versailles) is an age of sumptuous fashions and cultural brilliance, all due to his great support of the arts, which he patronizes, finances and personally encourages. His policies are aimed at enhancing the absolute power of the monarchy, both domestically and abroad. At home, he curbs the nobles by building the palace at Versailles and requiring them to live there supported by royal pensions, thus making them dependent on him. He develops an efficient method of collecting taxes to fund the massive army with which he aspires to control all of Europe. Under Louis's rule, France takes part in four expansionist wars that ultimately weaken the country. The most devastating is the War of the Spanish Succession (see 1701–14). During Louis's reign, France also begins building its colonial empire in the New World and Asia.

1667   War of Devolution

In its first expansionist war under Louis XIV, France attempts to seize control of the Spanish Netherlands. The Netherlands joins with England and Sweden in forming the Triple Alliance to drive back the French invasion.

1668: November 10   Birth of composer François Couperin

Couperin, famed for his keyboard and other instrumental pieces, is born to a prominent musical family. In 1685 he takes over the position of organist at the church of St. Gervais in Paris and remains in this post until his death. He becomes a favorite of the royal family both as an organist and performer on the clavecin (a forerunner of the piano comparable to the harpsichord), and serves as music teacher to the young Louis XV. Among Couperin's best-known pieces are a series of suites for harpsichord called *Ordres,* published between 1713 and 1730; the set of trio sonatas called *Les Nations* (1726), and the *Concerts royaux* (1714–15), written for Louis XIV. Couperin dies in Paris in 1733. The twentieth-century composer Maurice Ravel (see 1875: March 7) wrote a suite of keyboard pieces, *Le Tombeau de Couperin* (later arranged for orchestra as well), in homage to Couperin.

1672–78   Dutch War

In a secret alliance between Louis XIV and Charles II, France joins England in its third war against the Dutch. They invade the United Provinces (the northern provinces of the Netherlands), but their forces are driven back by flooding when the dikes open.

1680   Comédie Française is established

Louis XIV founds the Comédie Française, France's oldest and most venerable theater, as a venue (scene or locale) for the performance of plays by Molière, Racine, and others.

1683: September 24   Birth of composer Jean-Philippe Rameau

Rameau is famed as a composer and also makes a significant contribution to the development of music theory. He holds a variety of church posts, including organist at Clermont cathedral in Lyons. From the age of forty he lives in Paris. Rameau's best-known works are his four volumes of harpsichord pieces. However, his operas, which he begins writing at the age of fifty, are also highly regarded and include *Hippolyte et Aricie* (1733), *Castor et Pollux* (1737), and *Dardanus* (1739). In addition to his achievements as a composer, Rameau writes a number of articles and books on music theory and is the first theoretician to explain harmonic rules scientifically on the basis of acoustics. Rameau dies in Paris in 1764.

1685   Louis revokes the Edict of Nantes

King Louis XIV revokes the Edict of Nantes (see 1598) granting religious freedom to the Huguenots (French Protestants) and outlaws the practice of all religions except Catholicism. Huguenots are also forbidden to emigrate, but some 200,000 Huguenots flee France, taking their training and work skills to other countries.

1688–97   War of the League of Augsburg

Louis XIV embroils France in its third expansionist war, generated by the rivalry between the Bourbon and Habsburg monarchies and uncertainty over the succession to the Spanish throne due to the inability of Charles II to produce heirs. France opposes the Grand Alliance, comprising England, the United Provinces of the Netherlands (the northern, Protestant provinces not under Spanish rule), and the Habsburgs of Austria. A long, grueling struggle that stretches over nine years and expands to the overseas possessions of the embattled countries, the war ends in a stalemate that does little to resolve the underlying conflict, which again surfaces in the War of the Spanish Succession (see 1701–14).

1699: April 21   Death of Racine

France's greatest tragic dramatist dies in Paris. (See 1639: December.)

1701–14   War of the Spanish Succession

France suffers devastating losses in the final war carried out during the reign of Louis XIV. It is touched off by conflict over the succession to the Spanish throne, a conflict that also sparked the nine-year War of the League of Augsburg (see 1688–97) but remains unresolved. England and Austria effectively counter French military power, and France loses most of its colonial possessions and exhausts its treasury.

Economic and Military Decline

1715   Louis XIV dies

Louis XIV dies and is succeeded by his five-year-old great-grandson, Louis XV (1710–74). The Duke of Orleans serves as regent for the new king. Under Louis XV's financial mismanagement, the debt created by his predecessor's expensive wars continues to grow and helps lead to the French Revolution by contributing to popular discontent with the monarchy.

1720   Scandal over Mississippi Scheme

John Law's Mississippi Scheme, a risky speculative venture, collapses, and government involvement in it is revealed. The Scheme involves the selling of large amounts of land to settlers that is at best wildly exaggerated in its description, if it even exists at all. Though Law's "Mississippi Bubble" does burst, and in general the venture is a financial calamity, the Scheme is responsible for the largest settling of Louisiana up to this time.

1733: September 12   Death of composer François Couperin

Composer François Couperin, known for his keyboard pieces and trio sonatas, dies in Paris. (See 1668: November 10.)

1740   War of the Austrian Succession

The War of the Austrian Succession involves France in warfare with England, mostly in the colonial territories of the two nations.

1744–88   Buffon records his research in *Natural History*

Appointed to head the Jardin du Roi (Royal Garden; see 1635), the Comte de Buffon records his research in the seven-volume work, *Natural History,* which begins appearing in 1749 (notably, the work on the project does not halt with his death and is continued until 1804). Offering theories in areas including geology, anthropology, and zoology, it makes eighteenth-century scientific thought accessible to the general public and is one of the most widely read and influential books of its period.

1751–80   Publication of the *Encyclopédie*

The thirty-six-volume *Encyclopédie,* a compilation of scholarly writing in a broad range of fields, is published (note: the original *Encyclopédie's* volumes were completed in 1772, but a five-volume supplement and two-volume index were added in 1780). Edited by Denis Diderot, it focuses on the new intellectual attitudes and developments that characterize the liberal spirit of the Enlightenment and exerts a significant influence on political and philosophical thought in France.

1756–63   Seven Years' War

France opposes England in a second colonial war and is once again defeated.

1758: May 6   Birth of Maximilien Robespierre

Robespierre, the foremost Jacobin leader of the French Revolution, is born in Arras, where he returns to begin a law practice after studying law in Paris. An idealist inspired by the writings of Jean-Jacques Rousseau, Robespierre dedicates himself to bringing down the old social and political order and establishing an ideal democratic state. A member of the radical Jacobins, Robespierre is an active participant in the States-General convened in 1789 (see 1789: May 5) and in its successors, the National Assembly (see 1789: June 17), the Paris Commune (see 1792: July), and the National Convention (see 1792: September). He becomes the major guiding force behind the Terror of 1793–94 (see 1793–94), in which thousands of aristocrats and moderate supporters of the revolution are killed. Forcibly toppled by his opponents, Robespierre is sent to the guillotine on July 28, 1794.

1759   Publication of *Candide*

The famous philosophical novel by Voltaire (1694–1778) is published and becomes his most popular work. Narrating the adventures of the naive young man of the title, it satirizes the metaphysical optimism of the German philosoper Leibniz in its famous phrase, "All is for the best in the best possible of all worlds."

1762 *Le Contrat sociale* **appears**

Philosopher Jean-Jacques Rousseau (1712–78) publishes *Le Contrat social* (The Social Contract), his profoundly influential tract about the fundamental relationship between the individual and society, and the tension between authority and freedom. *Le Contrat social* becomes a major influence on the participants in the French Revolution and on framers of democracies in other parts of the world.

1763   The Peace of Paris

In the treaty ending the Seven Years' War, France loses most of its colonies, including possessions in India and Canada.

1764: September 12   Death of composer Jean-Philippe Rameau

Famed composer and music theoretician Jean-Philippe Rameau dies in Paris. (See 1683: September 24.)

1769: August 15   Birth of Napoleon Bonaparte

Napoleon Bonaparte becomes dictator of France and, for a brief time, ruler of Europe following the French Revolution. He is born in the French possession of Corsica and educated at military academies in France. After an erratic early career, his prospects rise when he successfully thwarts a Royalist uprising against the Directory (one of the governments set up after the Revolution) in 1795 and is given command of French forces in Italy, where he defeats the armies of Austria and Sardinia by 1797. In 1798 he leads an invasion of Egypt. Napoleon returns home the following year and takes part in a coup that overthrows the Directory and replaces it with the Consulate. Napoleon is named First Consul, becoming the leader of France. Following more French military victories, both Austria and Britain sign peace treaties with France.

At home, Napoleon sets about making wide-ranging administrative reforms. He centralizes local governments by creating the system of departments, draws up new laws (including the Napoleonic Code), and reorganizes the nation's financial institutions and educational system. In 1802 he is named consul for life. Two years later, he is crowned Emperor of France. By 1805 the other European powers and Russia have formed a new coalition, and Napoleon leads the French armies to a series of victories over the following years. By 1810 his empire is at its height. However, after his failed invasion of Russia two years later, foreign resistance to French domination grows. Napoleon's army suffers a crushing defeat in October 1813, Paris is taken the following March, and Napoleon abdicates in April. He is exiled to the island of Elba, from which he escapes and is briefly restored to power in 1815. He suffers his final defeat at the Battle of Waterloo in central Belgium on June 18, 1815 and spends the remaining six years of his life on the British island of St. Helena.

1777–81   Necker serves as finance minister

Swiss-born banker Jacques Necker (1732–1804) is brought in to rescue the country from its dire financial straits. Following his dismissal, in 1781 he publishes a pamphlet *Compte rendu* (Account Rendered), which for the first time discloses the state of the royal treasury.

1778–83   France aids American revolutionaries

Louis XVI (1754–93) aids the rebellious colonists in the American Revolution to exact revenge on its old enemy, England. General Count Rochambeau (1725–1807) is dispatched to the colonies with about six thousand troops. France's participation in the war only exacerbates its existing debt crisis. In addition, Louis inadvertently publicizes the ideals of the revolution by supporting them militarily.

1783–87   Calonne serves as finance minister

Charles Calonne (1734–1802), appointed as finance minister, promotes a land tax reform plan as the only way to save the faltering economy, but it is hotly opposed by the French nobles.

1783: January 23   Birth of Stendhal

The novelist and essayist Stendhal (1783–1842) is known primarily for his two masterworks, *Le Rouge et le noir* (Scarlet and Black) and *La Chartreuse de Parme* (The Monastery of Parma). Stendhal (the literary pseudonym of Marie-Henri Beyle) is born in Grenoble. After filling a bureaucratic position with the French army under Napoleon, he settles in Milan, Italy, between 1814 and 1821. Returning to Paris, he establishes himself in the literary salons of the capital, where he is known for his wit.

Le Rouge et le noir, published in 1830 (English translation 1938), recounts the adventures of Julien Sorel, an ambitious young man who becomes embroiled in two seductions that lead to his downfall. It paints a vivid, satirical portrait of French society in the post-Napoleonic period of 1815–1830. Like *Le Rouge et le noir, La Chartreuse de Parme* also tells the story of a young, idealistic admirer of Napoleon (this time in Parma, Italy) who attempts to gain power and prestige by rising through the ranks of the Church but complicates his life through his romantic adventures.

In the latter part of his career, Stendhal is appointed to a diplomatic post in an obscure location outside France, returning to Paris only on leaves of absence. He dies of a stroke in 1842.

The French Revolution

1789: May 5   The Estates-General is convened

Unable to resolve the nation's dire economic situation, (which most historians agree has to do with the antiquated social and economic institution of feudalism) King Louis XVI convenes the national legislative body, Estates-General, at Versailles. It includes representatives of the three estates: the nobility, the clergy, and the third estate, or Commons.

1789: June 17   National Assembly is formed by third estate

Dissatisfied with its representation in the Estates-General, the third estate proclaims a separate body, the National Assembly, and invites participation by delegates from the other estates. Its members vow to create a constitution for France.

1789: July 14   Storming of the Bastille

In response to a royal troop buildup near Versailles, a mob in Paris storms the royal prison, the Bastille. Many members of the nobility flee the country. This date is celebrated thenceforth as a national holiday in France.

1789: August 4   Peasant rents are abolished

Following a series of peasant uprisings, the National Assembly votes to end peasant rents and other obligations.

1789: August 17   Declaration of the Rights of Man

The National Assembly adopts a statement of political principles called the "Declaration of the Rights of Man." Inspired by the American Declaration of Independence and the Enlightenment, it officially lists the "inalienable rights" of the individual, and the rights to "liberty, property, security, and resistence to oppression." The French Declaration asserts the basic equality of men and soverignty of the people, and the responsibilty due to officials to uphold these values. Directed at assailing the institutions of the ancien régime and feudalism, it has a massive effect on liberal thought for the entire next century.

1789: October   Women march on Versailles

Spurred by severe food shortages in Paris, thousands of women march on the palace at Versailles to demand food and threaten the royal family.

1790: July   Constitution is adopted

The National Assembly presents Louis XVI with a constitution that provides for a limited monarchy with separation of church and state.

1790: July 12   Civil Constitution of the Clergy is enacted

The National Assembly enacts legislation enforcing the separation of church and state. The state seizes church lands and takes over responsibility for education and charity, as well as support for members of the clergy. The posts of priest and bishop are turned into elective offices.

1791: June 20   Louis XVI attempts to flee

King Louis XVI and his family attempt to flee from the revolutionary government, but they are caught and returned to Paris. After a brief interval, Louis is reinstated to his former position.

1792: April 20   Legislative Assembly is convened

France declares war on Prussia and Austria, who are preparing to intervene and save the French monarchy in order to protect their own. By autumn, French revolutionary troops occupy Savoy, Nice, part of the Rhineland, and the Austrian Netherlands, and are offering to help revolutionaries of other countries.

1792: July   The Commune is established

The moderate Girondin government falls and is replaced by the revolutionary Commune, under the influence of the radical Jacobins.

1792: August   Storming of the Tuileries

The Tuileries, the royal palace in Paris, is stormed. King Louis XVI's Swiss guards are murdered, and the king is imprisoned.

1792: September   Republic is proclaimed

The National Convention (the successor to the National Assembly) convenes and proclaims the First French Republic.

1792: December   Louis XVI is tried

King Louis XVI is tried on charges of conspiring with foreign powers and sentenced to death. Calling on Austria and Prussia (then at war with France) to rescue him from confinement by the revolutionaries, along with his attempted escape of France in disguise, confirms to many his guilt of conspiring with enemy powers.

1793–94   Reign of Terror

The radical Jacobins, under the leadership of Maximilien Robespierre; (1758–94; see 1758: May 6) with the support of Georges Danton (1759–94) and Jean Paul Marat (1743–93) and inspired by the ideals of the French *philosophes* (philosophers), seize control of the government. Marat is murdered in the bath when a Girondin-supporter, Charlotte Corday (1768–93), poses as a messenger and stabs him to death. This inaugurates the Reign of Terror against anyone perceived as a counterrevolutionary, including both royalists and moderate republicans. Thousands are executed between May 1793 and June 1794, most by the guillotine or in mass drownings called *noyades*. Show trials are carried out by revolutionary tribunals, and the only sentence for those found guilty is death.

1793: January 21   Execution of Louis XVI

Convicted by the Convention by an absolute majority of one, King Louis XVI is executed.

1793: April   The Committee of Public Safety is formed

The Committee of Public Safety, guided by Maximilien Robespierre, is created to deal with the nation's economic plight, safeguard the Revolution by rooting out counterrevolutionaries, and respond to the military threat from Austria and Prussia. It inaugurates the Reign of Terror.

1793: September   Hungry mobs riot in Paris

Thousands riot in the streets of Paris, spurred by hunger.

1793: October   Marie Antoinette is executed

Queen Marie Antoinette (1755–93), known derisively as the "Widow Capet," is guillotined by sentencing of the Revolutionary Tribunal for essentially the same charges that were claimed against her husband, Louis XVI. The very reason she was married to Louis XVI was for better Austrian-French relations, and this combined with constant drive for Austrian intervention on the side of the monarchy in France during the Revolution insured her death on the guillotine.

1794–95   The White Terror

With the downfall of Robespierre and the end of the Reign of Terror, the enemies of the Jacobins exact revenge in what becomes known as the White Terror, or the Thermidorian Reaction (due to 9 Thermidor, (or July 27, 1794) when the Reaction took place, according to the French Revolutionary Calendar). In a counterpart to the earlier lawlessness, mobs break into prisons and kill incarcerated Jacobins, sometimes committing brutal atrocities in the process. By the summer of 1795, a general amnesty is extended to both sides in the conflict, in which an estimated 23,000 to 40,000 have lost their lives.

1794: July 27   Execution of Robespierre

Jacobin leader Robespierre is removed from power and executed. The date, according to the new calendar devised by the Jacobins, is the ninth of Thermidor.

1795   The Directory is formed

With the Reign of Terror over, the National Convention is replaced by the five-member Directory as France's governing body, and a new, moderate constitution is enacted.

The Napoleonic Era

1799: November 9   Coup of 18 Brumaire

Military leader Napoleon Bonaparte (see 1769: August 15) leads a coup d'état that ousts the Directory and forms a new government known as the Consulate (Brumaire was a month on the Revolutionary, or Jacobin Calendar). Wielding dictatorial power as First Consul, Napoleon restores order domestically through effective administrative reforms and pursues foreign wars against France's European neighbors.

1800   Successful military campaigns

France launches successful attacks on Austria, Italy, and southern Germany. Under the Treaty of San Ildefonso, Spain cedes its New World possession, Louisiana, to France.

1801   Relations with the Church are resumed

The Catholic Church is reestablished in France through the Concordat, an agreement with Pope Pius VII initiated by Napoleon Bonaparte.

1801: February   Treaty of Lunéville

Napoleon imposes territorial and policy changes on other countries, notably Germany, following French military victories the previous year.

1802   Treaty is signed with England

A peace treaty, the Treaty of Amiens, between France and England is signed at Amiens, bringing (though arguably) the French Revolutionary Wars to an end.

1802   Legion of Honor is established

Napoleon establishes the Legion of Honor, which rewards exemplary civil and military deeds.

1803: December 11   Birth of composer Hector Berlioz

Hector Berlioz, the greatest French composer of the Romantic era, is born to a small-town doctor in the region of Grenoble. His father is a music lover, but Berlioz's parents are against music as a career for their son and send him to Paris to study medicine. There he enrolls in the Conservatoire, but his medical studies are short-lived. By 1830 he has won the school's top composition prize and already composed what is probably his best-known masterpiece, the *Symphony Fantastique*.

After a two-year residence in Rome required by his conservatory prize, he returns to Paris, where he continues to compose. Starting in 1842, Berlioz undertakes a series of concert tours to England, Germany, and Russia, conducting the works of Beethoven and other well-known composers and also introduces his own compositions to a wider public. The

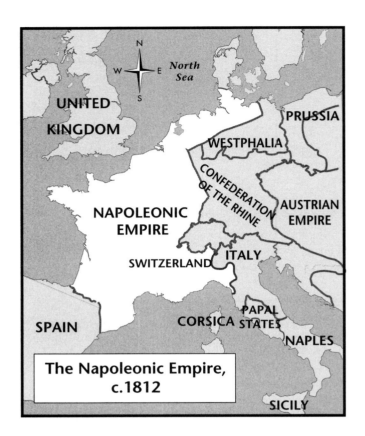

The Napoleonic Empire, c.1812

composer also serves as the acclaimed music critic of the *Journal des Débats*. In addition to the *Symphony Fantastique*, Berlioz's best-known works include *Harold in Italy* (1834), the *Requiem Mass* (1837), *Romeo and Juliet* (1839), *Nuits d'Eté* (Summer Nights, 1841). Berlioz dies in Paris in 1869.

1804   Napoleon is elected emperor

Following the discovery of an assassination plot against Napoleon, the French Senate votes to have him establish a hereditary dynasty by becoming emperor of France.

1804: December 2   Napoleon is crowned

Napoleon Bonaparte is crowned Napoleon I in a ceremony at the cathedral of Notre Dame in Paris. He appoints relatives and allies as rulers of territories that France conquers, creating a new set of royalty.

1805: December 2   Battle of Austerlitz

In one of his greatest military victories, Napoleon defeats the Russian and Austrian armies at Austerlitz.

1806   Blockade on British trade is established

Napoleon implements the Continental System, an embargo on trade with Britain designed to weaken Britain by sabotaging its economy. The blockade leads France into further con-

flicts when it retaliates against nations that refuse to observe it.

1808: April 20   Birth of Napoleon III

Charles Louis Napoleon Bonaparte is born in Paris to Louis Bonaparte, the younger brother of Napoleon I and king of Holland. When the son of Napoleon I dies in 1832, Louis Napoleon, considering himself the heir apparent, begins trying to gain control of the throne of France. He makes unsuccessful coup attempts in 1836 and 1840 and is jailed after the second episode but eventually escapes.

The revolution of 1848 that unseats King Louis Philippe brings Louis Napoleon to power as the first president of the republic (see 1848: December 10). When his term of office is on the eve of expiring, he stages a coup d'état and becomes a dictator, first as president and then as Emperor Napoleon III. His regime brings financial prosperity to France but at the cost of individual liberty. Napoleon does liberalize his rule significantly in 1860, but defeat in the Franco-Prussian War (1870–71) topples his regime, and he lives the remaining two-and-a-half years of his life in exile near London, dying in 1873.

1809   Treaty of Schoenbrun

Following yet another victory over Austria and the signing of the Treaty of Schoenbrun, Napoleonic France reaches its high point. Napoleon's empire includes all of Europe west of Russia and east of Portugal.

1812   Failed invasion of Russia

Napoleon's fortunes begin to turn when he attempts to invade Russia and is forced to retreat from Moscow in the harsh Russian winter. He finds the city deserted and burning, and Czar Alexander refuses to surrender. Out of an army of 600,000 men, about 500,000 are captured, die, or desert in the course of the disastrous Russian campaign. Russia gathers a powerful new coalition to oppose the French ruler, including Prussia, Sweden, and Britain.

1813: October 16–19   Defeat at Leipzig

Napoleon suffers his worst defeat (in a single battle) to date in the Battle of the Nations at Leipzig, at the hands of combined enemy forces. Some 30,000 of his men are killed, and he is forced to retreat.

1814: April 11   Napoleon abdicates

After Paris is captured, Napoleon abdicates (formally gives up the throne) and goes into exile on the island of Elba, off the coast of Italy. The French monarchy is restored with Louis XVIII (1755–1824), a brother of the executed Louis XVI, installed as king.

1814: May 30   Treaty of Paris

The European Allies set terms for peace following Napoleon's abdication and the restoration of the Bourbon monarchy. France is returned to its 1792 boundaries but not required to pay an indemnity (compensation for damages).

1815: March 20   Napoleon retakes Paris

Napoleon escapes from the island of Elba with about one thousand followers, gathering additional men on his triumphant march from Cannes to Paris and routing the newly installed monarch, Louis XVIII, who flees the city.

1815: June 18   Final defeat for Napoleon at Waterloo

Napoleon is defeated for the last time at Waterloo by the combined armies of Britain (commanded by the Duke of Wellington) and Prussia (led by Marshal Blücher). He abdicates a second time and is held until his death as a British prisoner on the island of St. Helena, a remote English-occupied island in the South Atlantic, 1,200 miles west of Africa. He is buried on the island, but his remains are later moved to the Hôtel des Invalides in Paris, where he had wished to be interred.

The Monarchy is Restored

1815: November 20   Second Treaty of Paris

The European Allies sign a second treaty with France, imposing harsher conditions, including a three-year period of military occupation and payment of an indemnity.

1818   Military occupation ends

The occupation of France by the European Allies (Great Britain, Austria, Prussia and Russia) ends following the Congress of Aix-la-Chapelle.

1821: May 5   Death of Napoleon

Napoleon Bonaparte dies on the island of St. Helena. (See 1769: August 15.)

1821: December 12   Birth of novelist Gustave Flaubert

Flaubert is widely regarded as the master of Realism in French fiction; his masterpiece *Madame Bovary* is one of the greatest nineteenth-century novels. For much of his life, Flaubert lives a secluded existence at Croisset, near Rouen, pursuing his literary work. *Madame Bovary,* published in serial form in the fall of 1856, tells the story of a bourgeois (middle-class) woman whose boredom and discontent lead her into an adulterous relationship that eventually destroys her. The novel's subject matter leads to the prosecution of its author, who is brought to trial and nearly convicted. Flaubert's other major works include *L'Éducation sentimentale* (A Sentimental Education; 1869), *La Tentation de Saint Antoine*

(The Temptation of Saint Anthony; 1874), and *Trois contes* (Three Tales; 1877). Flaubert dies in 1880.

1824–30   Reign of Charles X

The moderate Louis XVIII is succeeded by Charles X (1757–1836), an ultraroyalist who attempts to restore the conditions of absolute monarchy that existed prior to the revolution, as well as the dominance of the Catholic Church.

1826–37   Braille reading system for the blind is developed

After learning of a military technique for communicating at night without using lights, Louis Braille (1809–52), blind since early childhood, develops the touch system of raised dots used by the blind for reading and writing.

1830s   Social impact of industrialization is felt

The introduction of the factory system in France evokes the first expressions of labor unrest, which are brutally suppressed.

The July Revolution

1830: July–August   July Revolution

A political alliance of bourgeois (middle-class) and leftist forces overthrows Charles X and installs Louis Philippe (1773–1850) of the house of Orléans as king. (He is known as the "citizen king.") His regime is favorable to the interests of the rising class of industrialists, an alliance from which he personally benefits financially.

1835–37   Daguerre develops photographic process

Painter and inventor Louis Daguerre (1789–1851) develops a process for capturing images with mercury vapor on an iodized silver plate and fixing them with a salt solution. The picture that results is called a daguerreotype in honor of its inventor.

1838   Birth of composer Georges Bizet

Composer Georges Bizet (1838–75) is born. He composes *Symphony in C* when he is just seventeen. His other works include opera (*Carmen, The Pearl Fishers,* and *The Fair Maid of Perth*) and incidental music for the theater (*L'Arlésienne,* or The Girl from Arles, and *Jeux d'enfants,* or Children's Games).

1839: January 19   Birth of painter Paul Cézanne

Painter Paul Cézanne is a central figure in the transition from nineteenth-century Impressionism to early-twentieth-century painting styles, including Fauvism and Cubism. Cézanne is born in Aix-en-Provence and studies in Paris in the 1860s, when he comes under the influence of Camille Pissarro, Edouard Manet, and others experimenting with new techniques of landscape painting. Works by Cézanne are included in the major Impressionist exhibitions of the 1870s. After 1877, however, he distances himself from his colleagues and retires to Aix, where he spends his mature years developing his painting in new directions, giving his works a more solid structure than those of the Impressionists. He attemps to "re-create nature" by using more simplified forms, his process is to utilize color and break down the forms in his art to basic geometiric equivalents to allow the landscape to express the true essence of how it is. Most of Cézanne's paintings are still lifes or landscapes, but he also produces portraits and scenes that include figures. Well-known paintings by Cézanne include *Still-life with Basket of Apples* (1890–94), *Boy with the Red Vest* (1895), and *Card Players* (1892–95). Close to the end of Cézanne's life, his works are discovered by a new generation of artists when they are shown at the Salon d'Automne in 1904. They subsequently influence the development of twentieth-century painting. Cézanne dies at Aix in 1906.

1840s Railroad construction begins

The first railroads are built, touching off a major investment boom.

1840: November 14 Birth of painter Claude Monet

Monet, the painter most closely associated with the Impressionist school of French painting, is born in Paris and grows up in Le Havre. Monet starts out painting *plein air* (out-of-doors) canvases that include lifelike figures in a realistic style. However, he soon develops his distinctive style of creating landscapes whose emphasis is on representations of shifting light and weather. His colleagues include Auguste Renoir, Camille Pissarro, and Eduard Manet. Two famous sets of paintings by Monet are the "Haystack" series done between 1889 and 1891 and the huge lily pond paintings, done late in life and based on a Japanese water garden that Monet engineers himself on a country estate. For the last twenty-five years of his life, Monet suffers impaired vision from cataracts but continues to paint. He dies in 1926, a successful artist revered by his countrymen.

1842: March 23 Novelist Stendhal dies

One of France's great novelists, Stendhal (Marie-Henri Beyle), dies in Paris. (See 1783: January 23.)

1842: May 8 First recorded railroad disaster occurs at Versailles

An overcrowded train carrying revelers from Versailles back to Paris following a celebration of King Louis Philippe's birthday catches fire when its cars collide due to a broken axle. The official death toll in the world's first recorded railroad disaster is 54, but the actual number is believed to be over 100.

1847 Poor harvests produce famine

A general crop failure throughout Europe causes famine conditions in many parts of France and has dire effects on already impoverished urban laborers.

Revolution and Republic

1848: February Revolution erupts

A second revolution shakes France, ending the reign of King Louis Philippe (see 1830: July–August), who is forced to abdicate, and resulting in the establishment of the Second Republic.

1848: November New constitution is adopted

The Second Republic adopts a new constitution providing for universal suffrage, a unicameral (having a single legislative chamber) representative assembly, and a strong president elected by direct popular vote.

1848: December 10 Louis Napoleon is elected president

Louis Napoleon (see 1808: April 20), a nephew of Napoleon Bonaparte, is elected as France's first president.

The Second Empire

1851: December 2 Second Republic is overthrown in coup d'état

Louis Napoleon, president of the Second Republic, leads a coup that overthrows the republic and leads to the establishment of the Second Empire.

1852–71 Second Empire

Following a coup d'état, Louis Napoleon proclaims the Second Empire with himself as Emperor Napoleon III. France prospers economically during this period as it expands its colonial empire. Napoleon introduces new credit institutions and launches major public works projects, including the modernization of Paris under the guidance of the municipal administrator Georges Haussmann (1809–91). Influenced by the ideas of utopian thinker St. Simon, he launches model farms and other agricultural projects. He also involves France in three foreign wars: the Crimean War (see 1854–56), the war for Italian unification (1859), and the disastrous Franco-Prussian War (see 1870–71).

1854–56   Crimean War

France allies itself with Britain in combating Russian expansion into territory belonging to the Ottoman Turks.

1859   War for unification of Italy

France supports the Italian nationalists of Piedmont in achieving Italian independence and unification by driving the Austrians out of Italy.

1860s–1880s   Impressionist painting flourishes in France

The late 1860s to the mid-1880s is the heyday of the school of painting commonly referred to as Impressionism. Its primary exponents include Claude Monet, Auguste Renoir, Eduard Manet, Camille Pissarro, Berthe Morisot, Frédéric Bazille, and Alfred Sisley. (The name *Impressionism,* which is first coined by an art critic [see 1874], is not used by the artists themselves.) Impressionist painting is characterized by an emphasis on fleeting effects of light, weather, or movement, captured through a technique in which individual brushstrokes are readily apparent and often produce a shimmering effect. The paintings are also known for their use of intense colors, sometimes including the juxtaposition (side-by-side placement) of contrasting colors. Impressionism is also closely associated with *plein-air* painting—works actually painted out of doors from direct observation of nature rather than from sketches. Most Impressionist paintings are either landscapes or scenes from daily life. The Impressionist painters present eight exhibitions in Paris between 1874 and 1886.

1861   Unsuccessful intervention in Mexico

The French intervene unsuccessfully in Mexico, establishing a short-lived empire under the archduke of Austria Maximilian I.

1862: August 22   Birth of composer Claude Debussy

The works of Claude Debussy, who is often referred to as an Impressionistic composer, form an important transition point between Western music of the nineteenth and twentieth centuries. In his youth Debussy is recognized as a gifted pianist and enrolls in the Paris Conservatoire at the age of eleven. In 1884 he wins the Conservatoire's Prix de Rome and spends two years studying in Italy. The diverse influences on Debussy's music include the Symbolist poets, Javanese music, the music of Wagner, Gregorian chant, and Impressionist painting. His harmonic innovations, which are influential in the breakdown of tonality in twentieth-century music, include the use of pentatonic and whole-tone scales.

Debussy's most famous orchestral pieces include *Prélude à L'Après-midi d'un faune* (1894), *La Mer* (1905), the *Nocturnes* (1899), and *Images* (1912). His *Pelléas et Mélisande* (1902) is regarded as a milestone in the develop-

ment of opera after Wagner. Debussy is also renowned for his piano pieces, including *Estampes* (1903), *Children's Corner* (1906–1908), the two-volume *Préludes* (1910–13), and the *Études* (1915). Debussy's works also include chamber music and a ballet, *Jeux,* written for the great Russian choreographer Diaghilev. Debussy dies in Paris in 1918.

1869: May 8   Death of composer Hector Berlioz

Famed composer Hector Berlioz dies in Paris. (See 1803: December 11.)

1869: December 31   Birth of painter Henri Matisse

Matisse, known for his intense and expressive use of color, is born in northern France. Arriving in Paris in 1892, he studies at the Académie Julien and with Gustave Moreau, moving away from a representative and toward an expressive aesthetic of painting. His works shown in the "fauvist" exhibition at the Salon d'Automne in 1905 demonstrate the characteristic two-dimensional perspective, "primitive" drawing style, and vivid colors that are associated with his mature style. Fauvism, which becomes integral to art in the twentieth century, is largely expressionist in style, and is characterized by vivacious color and a daring distortion of forms. Among Matisse's best-known works are *Luxe, Calm, et Volupté* (1904–05), *The Joy of Life* (1895–96), *Bathers with a Turtle* (1908), *Bathers by a River* (1916), and *The Music Lesson* 1917). Matisse dies in Nice in 1954.

1870–71   Franco-Prussian War

France is forced to face the powerful Prussian army when Otto Von Bismarck deliberately provokes the French to declare war. The conflict results in a decisive defeat for France, the downfall of the Second Empire, and the end of French predominance in Europe. By the Treaty of Frankfurt, France loses the provinces of Alsace and Lorraine in northeastern France and is forced to pay massive reparations.

1870: September 2   France surrenders to Prussia

Napoleon III surrenders to the Prussians at Sedan.

1870: September 4   The Third Republic is proclaimed

Following defeat by Prussia, Napoleon III is overthrown, and a provisional government (the Government of the National Defense) proclaims the Third Republic.

1871: February   National Assembly is elected

France elects a National Assembly.

1871: March   The Paris Commune is reinstituted

Radicals, protesting France's surrender in the Franco-Prussian War, seize control of Paris and reestablish the Paris Commune, a revolutionary government, which foments (incites) civil war for two months.

1871: May 21–28   "Bloody Week"

The rebel Commune government in Paris is brutally crushed by government forces, which execute or deport its supporters.

1871: July 10   Birth of author Marcel Proust

Proust is the author of the greatest French novel of the early twentieth century, the seven-part *À la recherche du temps perdu* (Remembrance of Things Past). Proust, in delicate health throughout his life, nevertheless is an active participant in the upper-class social life of Paris in the 1890s when he begins publishing short fiction in periodicals. Proust's involvement in the crusade to free the wrongly convicted army officer Alfred Dreyfus (Proust, like Dreyfus, is of Jewish origin) and his own declining health lead to the author's gradual withdrawal from the Parisian social circles he had frequented. Once he begins work on his great novel in 1909, his retirement from society becomes total.

Published between 1913 and 1927 (five years after Proust's death), *À la recherche du temps perdu* is based on Proust's own life and incorporates realistic and satirical observations of the social milieu he has frequented. Its concentration on the inner psychological life of the narrator, which includes stream-of-consciousness elements, makes Proust an important figure in the emergence of early-twentieth-century literary modernism. Proust dies in 1922.

1873: January 9   Death of Napoleon III

Napoleon III (Louis Napoleon Bonaparte), former emperor of France, dies in England. (See 1808: April 20.)

1874   First Impressionist exhibition

Thirty artists, including most of those associated with the Impressionist school of painting, show their works in an independent exhibition in Paris. The name "Impressionism" is coined as a result of this exhibit when a critic uses the term pejoratively, or derogatorily, in describing the painting *Impression: Sunrise* by Claude Monet.

The Third Republic

1875   New constitution is adopted

The Third Republic adopts a new constitution following the failure of conservative political factions to reestablish the monarchy. The new constitution guarantees freedoms of speech, press, and association and the separation of church and state.

1875: March 7   Birth of composer Maurice Ravel

Maurice Ravel, one of the great twentieth-century composers, is known for his experimentation with harmony and sonority. His name is often paired with that of his near-contemporary Claude Debussy (see 1862: August 22) as an "Impressionistic" composer. Ravel studies piano and composition at the Paris Conservatoire, where his professors include the composer Gabriel Fauré. Between 1901 and 1903 the young composer is widely expected to win the Conservatoire's Prix de Rome. However, it is denied him for three years in a row, although he has already written his famous *Pavane pour une infante défunte,* the brilliant piano piece *Jeux d'eau,* and the String Quartet in F major. Except for service as an ambulance driver in World War I (1914–18) and a tour of the United States, Ravel remains in Paris for the rest of his life and continues composing. He dies in 1937 of brain injuries suffered in a car accident four years earlier.

Ravel's orchestral works include *La Valse,* the popular *Bolero,* and the often-performed orchestration of Mussorgsky's *Pictures from an Exhibition.* Ravel is particularly noted for his piano works, which include two piano concertos (one for left hand alone); the *Tombeau de Couperin,* a suite of pieces in homage to the great eighteenth-century French composer (see 1668: November 10); and the *Sonatine* for piano. Ravel's ballet *Daphnis et Chloé* is another of the composer's most notable works.

1877: May 16   *Seize Mai* crisis reduces the power of the president

Executive power in the Third Republic is dramatically weakened as the result of a showdown between provisional president Marshal Marie MacMahon and the National Assembly over the retention of Prime Minister Jules Simon. The legislature's ultimate victory over MacMahon results in the president's resignation and, more importantly, sets a precedent of ministerial accountability for the Assembly.

1880s   Education is made compulsory

Free, compulsory education is inaugurated.

1880: May 8   Death of novelist Flaubert

Novelist Gustave Flaubert, the author of *Madame Bovary,* dies at Croisset. (See 1821: December 12.)

1883   Birth of fashion designer Coco Chanel

Gabrielle "Coco" Chanel (1883–1971) is born in France, but both her parents die when she is quite young. She works with her sister as a milliner (hat maker), but opens her own shop in 1912. She works as a nurse during World War I (1914–18) and opens a couture (specialty fashion) shop in Paris in 1924. In this period, she designs her famous chemise dress that becomes known later as "the little black dress," considered essential in every woman's wardrobe. She launches the cardigan jacket without collar or lapels in 1924 and introduces the concept of accessorizing clothes with many necklaces, pins, and other costume jewelry in the same period. She launches her famous fragrance, Chanel No. 5, one of the most enduring perfumes ever created.

1884   Trade unions are legalized

With the labor and socialist movements gaining support, trade unions are legalized.

1886–1905   Postimpressionists react to Impressionism

In the decade following the last exhibition of Impressionist paintings, several important French painters adopt a new style that rejects the Impressionists' emphasis on lifelike representation of light and color, and moves toward a more abstract and symbolic mode of painting. The artists commonly identified as Postimpressionists include Paul Gauguin (1848–1903), Paul Cézanne (see 1839: January 19), Georges Seurat (1859–91), and Vincent van Gogh (1853–90). The term *Postimpressionism* is coined in 1910 by the English art critic Roger Fry as the title of his exhibition of French art at London's Grafton Galleries.

1889: January   Boulanger affair

A popular general, Georges Boulanger, attempts to undermine the ruling government. However, his efforts fall short of an outright coup, and he flees the country.

1890–92   The Third Republic gains papal support

The Catholic Church, accepting the end of the monarchy and the loss of its former privileged status, actively endorses the legitimacy of the Third Republic through the *Ralliement* (Rallying) movement instituted by the pope.

1890: November 22   Birth of Charles de Gaulle

Charles de Gaulle (1890–1970), France's foremost twentieth-century statesman, is born in Lille. He attends Saint Cyr military school and enters the army in 1911 and serves with distinction in World War I (1914–18). After the war de Gaulle serves as an aide to Marshal Pétain (World War I commander and leader of the Vichy government in World War II [1939–45]). He also teaches military history and authors several books on warfare. When the Germans invade France during World War II, de Gaulle flees to London, where he sets up a French government-in-exile and a resistance force, popularly known as the Free French. When the Allies invade Europe, de Gaulle's forces aid them in defeating the Nazi occupiers. De Gaulle heads a provisional government after the war but resigns when French voters adopt a constitution with executive powers he considers too limited. He forms his own political party in 1947 but retires from political life six years later to concentrate on writing several volumes of memoirs.

A governmental crisis precipitated by nationalist revolts in Algeria brings de Gaulle back onto the political stage in 1958. With the French government on the verge of collapse, the former leader is asked to take over and agrees on condition that he be granted emergency powers to deal with the situation. In September, voters approve a new constitution providing for a stronger executive, and de Gaulle is sworn in as the first president of the Fifth Republic, a position he holds for the next eleven years. He presides over the granting of Algerian independence, distances France from the North Atlantic Treaty Organization (NATO) alliance, forges new ties with West Germany and the Communist bloc, and commits France to a nuclear buildup. Following the student unrest of 1968 and voter rejection of proposed constitutional changes, de Gaulle resigns from the presidency in April 1969. He dies the following year.

1892   Panama Canal scandal

Allegations of corruption in connection with a French company that has contracted to build a canal across Panama extend to members of the French government.

1894–1902   The Dreyfus Affair

Captain Alfred Dreyfus (1859–1935), falsely accused of treason, is arrested, convicted, and imprisoned. The case raises an outcry worldwide, largely based on allegations that Dreyfus, a Jew, is a victim of antisemitism. A campaign for retrial is undertaken and succeeds when new evidence in Dreyfus's support surfaces, including admissions of forgery. Dreyfus is retried and found innocent and receives a pardon from the French government.

1894   Lumière brothers invent the *cinématograph*

Inspired by a Paris demonstration of Thomas Edison's Kinetoscope, Auguste (1862–1954) and Louis (1865–1948) Lumière invent the first film projector used commercially. Called the cinématograph, it projects images at a rate of sixteen frames per second. Their film of workers leaving the family factory (*La Sortie des ouvriers de l'usine Lumière*; 1895) is regarded as the world's first motion picture. The first showing of the cinématograph in Paris evokes great public enthusiasm and launches the French film industry. The Lumières go on to produce over forty short films showing scenes from everyday life.

1903   The Curies are awarded the Nobel Prize

Marie Sklodowska Curie (1867–1934) and her husband, Pierre Curie (1859–1906), receive the Nobel Prize for physics for discovering the principles of radioactivity.

1903   Tour de France bicycle race is launched

Journalist and cyclist Henri Desgrange establishes the Tour de France, which becomes the world's most prestigious and

Bicyclists appear as dots along the winding road during the Tour de France. The road winds toward the summit of Col du Galibier, an 8,386-foot peak, the highest point in the course of the race. (EPD Photos/CSU Archives)

well-known bicycle race. Since its founding, the race has expanded to 2,500 miles (4,000 kilometers), covered in twenty-one stages run over a three-week period, with the prize going to the contestant who has the highest aggregate (total) score for all the stages. The course includes both flat terrain and mountainous stretches in the Alps and Pyrenees. In addition to France, it extends into Belgium and also briefly to Spain, Germany, Switzerland, and Italy. With participants from as far away as North and South America and Australia, the Tour de France is an international sporting event surpassed in popularity only by the Olympics and the World Cup soccer tournament.

1904   Entente Cordiale

Britain and France enter into an informal pact of friendship, motivated by the growing threat of a unified, industrialized Germany being the master of Europe.

1905   Church and state are declared separate

France's religious minorities finally attain full religious freedom with the formal declaration of separation between church and state.

1905   Autumn Salon launches Postimpressionism

The Autumn Salon, an exhibition in Paris, provides an early public exposure of works by Henri Matisse (see 1869: December 31) and other Postimpressionist painters.

1905: June 21   Birth of author Jean-Paul Sartre

Existentialist author Jean-Paul Sartre (1905–80) is born in Paris, where he attends the École Normal Supérieure. Between his graduation in 1929 and World War II (1939–45), he teaches at various schools and writes philosophical works influenced by phenomenology (the study of perceptual experience that is purely subjective). Sartre's first novel, *La Nausée* (Nausea), is published in 1938. It is followed by a short story collection, *Le Mur* (The Wall; 1939). With the outbreak of World War II, Sartre is drafted. He is captured by the Germans, imprisoned for under a year, and then released. He becomes active in the French Resistance and continues his writing, publishing the major philosophical essay *L'Être et le néant* (Being and Nothingness; 1943) and producing two plays, *Les Mouches* (The Flies; 1943) and *Huis Clos* (No Exit; 1945).

As the leading proponent of existentialism, a philosophical school centered on the freedom of the individual, Sartre is a prominent intellectual figure in postwar France. Together with Simone de Beauvoir (1908–86) and Maurice Merleau-Ponty (1908–61), he founds the literary and political periodical *Les Temps modernes*. Although Sartre is a political leftist, he has no formal affiliation with the Communist Party and is critical of its military intervention in Eastern Europe. In 1964 Sartre is awarded the Nobel Prize for Literature, but he refuses to accept it on principle. He remains active as a writer until 1970, completing an exhaustive psychological and Marxist-oriented biography of novelist Gustave Flaubert (see 1821: December 12). In his last years, he loses his eyesight, and his overall health deteriorates. Sartre dies in Paris in 1980.

1906: March 10   Coal mine explosions kill over 1,000 miners

In the worst mine disaster to occur in Europe up to this time, a coal mine pit in the Courrières Colliery in northern France explodes, killing 1,060 miners. The explosion is the result of gases seeping into a closed-off pit where a fire had started the day before. When the explosion occurs, nearly 2,000 men and boys are at work in the mine. Of those who are killed, many are trapped by collapsed tunnels and cannot be reached by rescuers. A rescue party of 40 becomes trapped below ground as well. Rescue efforts continue for a week, and soldiers are brought in to maintain order at the scene of the grisly tragedy.

1906: May 6   General strike

A general strike launched by France's unions is a sign that faith in government reform has given way to a preference for direct action by labor. The strike establishes May 1 as a day dedicated to workers.

1906: October 22   Death of painter Paul Cézanne

Postimpresionist painter Paul Cézanne dies at Aix. (See 1839: January 19.)

1910   Railroad strike

The proliferation of union activity in the preceding decade climaxes in a massive rail strike.

1913: November 7   Birth of author Albert Camus

Novelist, playwright, and essayist Albert Camus (1913–60), winner of the 1957 Nobel Prize for Literature, is one of the most influential modern French literary figures. Coming of age between the world wars, Camus writes of modern alienation and the moral dilemma involved in choosing to act in the face of meaninglessness and absurdity. Camus is born and raised in Algeria, where he experiences poverty early in life. He studies philosophy at the University of Algiers and later becomes involved in the theater, producing, writing, adapting, and acting in plays. Shortly before the beginning of World War II (1939–45), he begins working as a journalist in Algeria. During the war he edits a resistance newspaper, *Combat*. After the war he resumes his involvement in the theater and follows his first novel, *L'Étrangère* (The Stranger; 1942) with two more: *La Peste* (The Plague; 1947) and *La Chute* (The Fall; 1956)*. Camus is best known for his novels and essays, especially *The Myth of Sisyphus,* first published in 1942. In addition, two of his plays (*Le Malentendu* [The Misunderstanding] and *Caligula*) are considered important contributions to the Theater of the Absurd. Camus dies in an automobile accident in 1960.

World War I (1914–18)

France becomes the site of much of the worst fighting of the war, as the French and Germans dig in for four years of costly trench warfare in a stalemate after the German offensive against France is halted at the Marne. Under the leadership of Georges Clemenceau, France rightly shares in the Allied victory, for France pays a severe price in loss of life. Nearly all the warfare takes place on French soil and France's war dead number almost one and a half million, this number only topped by the losses of Russia.

1916: October 26   Birth of François Mitterrand

François Mitterrand (1916–96) is one of France's most important post-World War II (1939–45) leaders. After reviving the French Socialist party in the 1970s, he serves two seven-year terms as president of France, from 1981 to 1995, when he is succeeded by his former conservative prime minister Jacques Chirac (b. 1932). Mitterrand dies in 1996.

1918: March 25   Death of composer Claude Debussy

Composer Claude Debussy dies in Paris. (See 1862: August 22.)

1918: November 11   Armistice is declared

The armistice ending World War I is proclaimed.

1919   Treaty of Versailles

France regains the provinces of Alsace and Lorraine under the Treaty of Versailles following the end of World War I (1914–18), as well as mandates to administer territories in Syria and Africa. Germany is ordered to pay heavy war reparations.

Between the Wars

1919   Abortion is ruled illegal

To compensate for the massive number of French lives lost in World War I, the government declares abortion illegal, with fines and prison sentences mandated for anyone performing or receiving an abortion. The law is widely broken, with thousands of abortions performed every year. By 1970 an estimated five hundred lives are lost to illegal abortions annually.

1919: May 9   Women are able to vote in regional elections

A bill is passed enabling French women to vote for communal and departmental representatives. They do not win unrestricted voting rights until 1944.

1922: November 18   Death of Marcel Proust

The greatest French novelist of the early twentieth century dies in Paris. (See 1871: July 10.)

1923–25   Occupation of the Ruhr

France (and Belgium) occupy Germany's Ruhr district when German reparation payments are halted. The Ruhr district (often referred to as the Ruhr Basin) is located in central-western Germany, is rich in natural resources and is the industrial heartland of Germany as well.

1925   Locarno Pact

The Locarno Pact restores relations between France and Germany and establishes the boundaries between the two nations.

1926: December 5   Painter Claude Monet dies

Impressionist painter Clause Monet dies at Giverny. (See 1840: November 14.)

1931   Lifar takes post with Paris Opera

Russian-born dancer and choreographer Serge Lifar (1905–86) is named ballet master and lead dancer at the Paris Opera, beginning a period of revitalization for ballet in France.

1933   Financial scandal rocks government

The Stavisky Affair, a financial scandal involving numerous government officials, shakes the public's faith in the government and results in mass demonstrations.

Serge Alexandre Stavisky (c. 1886–1934) moves to Paris in 1900 and becomes a citizen in 1914. He engages in various fraudulent business practices and in 1922 is discovered handling fraudulent bond transactions. His dealings involve a number of government officials, and when the authorities pursue him he either commits suicide or is murdered by the police. Many allege that Stavisky is unlawfully executed to cover-up his dealings with officials, and it enrages both the left and right. The scandal becomes so troubled that violent riots rock Paris, and the political parties allegedly involved, and even parlimentary democracy itself, become discredited as a consequence.

1934   Depression brings unemployment

France begins to feel serious effects of the Great Depression (begins in 1929 and extends through most of the 1930s), including unemployment.

1934: February 6–7   Mass protests in Paris

Demonstrators throng Paris, protesting a financial scandal with widespread government involvement. It is feared they will storm the Chamber of Deputies.

1936   Popular Front government is installed

Economic difficulties due to the Depression and public loss of faith in government due to the Stavisky Affair (see 1933 and 1934: February 6–7) lead to the electoral victory of a left-wing coalition, the Popular Front, led by a Socialist prime minister, Léon Blum (1872–1950). The Popular Front government implements labor and agricultural reforms and nationalizes some of France's industries.

1937: December 27   Death of composer Maurice Ravel

Composer Maurice Ravel dies in Paris. (See 1875: March 7.)

1938   Conservative government is elected

The conservative government that follows the Popular Front is led by Edouard Daladier (1884–1970), who concurs with the British policy of appeasement toward Nazi Germany.

World War II (1939–45)

1939: September   France declares war on Germany

Following the German invasion of Poland, France, together with the United Kingdom, declares war on Germany.

1940–42   German and Vichy rule

Germany places nearly half of France under direct occupation. The rest is administered by a government, formed by World War I hero Philippe Pétain (1856–1951) at Vichy in central France, until the German occupation of the entire country is completed.

1940: May 10   Germany attacks France and the Low Countries

Germany launches a surprise attack on France and the Low Countries, overcoming France's Maginot Line fortifications.

1940: June   De Gaulle launches resistance movement

A young general, Charles de Gaulle (see 1890: November 22), forms a French resistance movement (the Free French) from a base in London. Increasing numbers of French join him as the Nazis impose harsh conditions on occupied France, attempting to send Frenchmen to Germany to perform forced labor. The Free French take part in combat in French territories in North Africa, later establishing a provisional government in Algiers.

1944–46   Provisional government is established

De Gaulle establishes a provisional government in Paris following the liberation of France by the Allies in the summer of 1944.

1944   Female suffrage is enacted

French women, able since 1919 to vote in regional elections, gain the right to vote on the same basis as men.

1944: October   Free French government is recognized by the Allies

The provisional government formed by General Charles de Gaulle is officially recognized by the United States, Britain, and the Soviet Union.

1945   France joins the UN

France becomes a member of the United Nations (UN).

The Fourth Republic

1946–54   Indochina War

France becomes involved in costly warfare as it attempts to reestablish control over its possessions in Southeast Asia (present-day Laos, Cambodia, and Vietnam).

1946   The Fourth Republic is formed

A national referendum approves a new constitution creating a republic with a weak executive branch and a bicameral (two legislative chambers) legislature whose lower house, the National Assembly, holds most of the power. The administration of France's colonial territories is reorganized under the Constitution of the French Union. Through this Constitution France attempts to thwart growing anti-colonialism by creating a Union of French territories. This Union is composed of two categories: individual departments as overseas colonies, and separate protectorates with their members having representation in the French parliament.

1947   U.S. forms the Marshall Plan

The United States announces the Marshall Plan to aid European economic recovery.

1949   France joins NATO

France becomes a founding member of the North Atlantic Treaty Organization (NATO).

1949   Publication of feminist classic *The Second Sex*

The Second Sex by author Simone de Beauvoir (1908–86) is published. It becomes a landmark work in the new wave of feminism that begins in the 1960s in France, the United States, and other countries.

1951   *Cahiers du Cinèma* is founded

The influential film journal *Cahiers du Cinèma* is established. In the late 1950s and early 1960s, the young film critics and directors associated with the periodical develop the school of filmmaking known as the New Wave. Among these directors are François Truffaut (1932–84), whose works include *The Four Hundred Blows* (1959) and *Jules and Jim* (1961); Jean-Luc Goddard (b. 1930), whose first feature film is *Le Bout de souffle* (Breathless; 1959); and Alain Resnais (b. 1922) *Hiroshima, My Love;* 1960; and *Last Year at Marienbad*; 1962.

1954   Indochina War ends

Newly installed prime minister Pierre Mendès-France (1907–82) negotiates a peace agreement to end the fighting in Indochina.

1954   Rebellions take place in North Africa

Following the end of the Indochina War (see 1946–54), France's North African territories, Morocco, Tunisia, and Algeria, are shaken by agitation for independence. Morocco and Tunisia are granted independence within three years, but France sends troops to Algeria in its attempt to put down a nationalist rebellion staged by the Muslim Federation for National Liberation (FLN). Eventually some half a million troops are stationed there in the attempt to maintain control of the region. As the French government is weakened by the Suez crisis (see 1956), the military takes advantage of the situation to commit excesses against the Algerian rebels, with the support of extremists among the French colonists. The controversy raised by their actions helps bring down the Fourth Republic.

1954: November 3   Death of painter Henri Matisse

Painter Henri Matisse, one of France's foremost early twentieth-century painters, dies in Nice. (See 1869: December 31.)

1956   Suez crisis

To counter Egypt's support for nationalist rebels in Algeria, France joins the ill-fated British attack on the Suez Canal, further weakening support for the French government.

1957   France helps found the EEC

France becomes a founding member of the European Economic Community (EEC).

1957   Author Albert Camus wins the Nobel Prize for Literature

Novelist, essayist, and playwright Albert Camus is awarded the Nobel Prize for Literature. (See 1913: November 7.)

1958: May   Right-wing extremists threaten to topple government

Right-wing extremists, aided by the French military, seize power in Algiers. Unrest spreads to the French territory of Corsica and even to Paris, threatening the government's control over the country.

1958: June 1   De Gaulle takes control

The leaders of the Fourth Republic call on general and war hero Charles de Gaulle (see 1890: November 22) to take control of the government and pull it back from the brink of chaos. De Gaulle agrees on condition that a new constitution be drafted providing stronger powers for the executive branch.

1958: September   De Gaulle is endorsed in popular referendum

French voters overwhelmingly endorse Charles de Gaulle's (see 1890: November 22) leadership and his proposed constitution, which increases the power of the executive and limits that of the legislature. The president is authorized to rule by decree in an emergency, appoint ministers without approval by the legislature, and dissolve the legislature at will. In addition, the length of the annual legislative session is reduced to five months.

1958: November   Gaullist candidates sweep elections

In elections for a new legislature, Gaullists (candidates who support de Gaulle) win by a wide margin.

The Fifth Republic

1958: December   The Fifth Republic is inaugurated

The Fifth Republic is officially inaugurated, and Charles de Gaulle (see 1890: November 22) takes office as its first president. During his tenure in office, he presides over the transition to Algerian independence. He also pursues a strongly independent foreign policy that distances France from the United States and the United Kingdom, withdrawing French troops from the NATO-unified command and getting the NATO headquarters moved out of France. In addition, France recognizes the People's Republic of China and refuses to sign the major international agreements curbing the testing and proliferation of nuclear weapons. France also opposes the admission of Britain to the European Economic Community.

1960: January   New revolt in Algeria

New unrest occurs among French settlers in Algeria, where hostilities have continued.

De Gaulle, using special emergency powers conferred on him by the new constitution, replaces Algeria's top-ranking colonial officers with others who are his supporters.

1960: January 4   Death of author Albert Camus

Nobel Prize-winning author Albert Camus dies. (See 1913: November 7.)

1962: March 18   Algerian cease-fire is declared

The French government and Algerian rebels sign a cease-fire and peace agreement providing for the establishment of an independent nation closely linked to France and receiving French aid.

1962: July 1   Algerian independence

Algeria ends it colonial relationship with France, becoming one of the last French colonies in Africa to gain independence.

1964   France recognizes Communist China

France establishes diplomatic relations with the People's Republic of China, further distancing itself from the political position of the United States, which still recognizes the government on Taiwan as the legitimate Chinese government.

1965   De Gaulle is reelected

President de Gaulle (see 1890: November 22) is elected to a second seven-year term as president following a run-off election.

1966   France withdraws from NATO

France withdraws its forces from the North Atlantic Treaty Organization (NATO) and requires that NATO military personnel and headquarters be withdrawn from French soil.

1967: March   Gaullists lose parliamentary majority

The Gaullists (supporters of Charles de Gaulle) suffer a significant setback in parliamentary elections.

1968   Winter Olympics held at Grenoble

French ski racer Jean-Claude Killy (b. 1944) wins three gold medals for slalom, giant slalom, and downhill when the Winter Olympic Games are held at Grenoble.

1968: May   Student protests shake French stability

Student demonstrations sparked by criticism of France's educational system and government and discontent with the affluence and materialism of French society as a whole escalate into riots, sparking workers' strikes and nearly bringing down the government. De Gaulle calls for new legislative elections in June. Ultimately, the strikes result in major educational reforms that dramatically decentralize the university system. The University of Paris is divided into thirteen independent institutions.

1968: June   Gaullists retain power in special elections

Frightened by the violence and disorder unleashed by the recent student protests and strikes, the French electorate returns the Gaullists to power with a large legislative majority.

1968: November   Currency crisis

The stability of France's economy is threatened by a currency crisis stemming from depletion of the nation's gold reserves by inflation and speculative excesses. De Gaulle resists pressure to devalue the franc, instead implementing unpopular economic austerity measures to bring the situation under control.

The De Gaulle Era Ends

1969: April 28   De Gaulle resigns as president

Charles de Gaulle (see 1890: November 22) ends his eleven-year tenure as president, resigning when constitutional reform measures he supports to strengthen the executive branch are defeated in a constitutional referendum.

1969: June 15   Georges Pompidou becomes president

Georges Pompidou (1911–74), a Gaullist and former prime minister, becomes the second president of the Fifth Republic. Over the five years he is in office, the Gaullists gradually lose political power.

1969: August   Pompidou devalues the franc

Pompidou responds to economic pressure by devaluing the franc, which lowers workers' buying power.

1970: April 28   Death of Charles de Gaulle

Military leader and statesman Charles de Gaulle dies at his home in Colombey. (See 1890: November 22.)

1974: March 3   DC-10 crashes near Paris in worst air disaster to date

The most deadly aviation disaster to date occurs when a Turkish Airlines DC-10 crashes in a field near Paris shortly after takeoff from Orly Airport. The accident occurs following decompression caused by a faulty cargo door blowing off the plane. Following a similar cargo door failure two years earlier over Canada, McDonnell Douglas, the manufacturer of the DC-10, is ordered to make safety modifications in the door's latching mechanism, but these orders are not uniformly enforced. In addition, the inspection sequence designed to detect improper latching fails due to a variety of human errors. At 11,500 feet, the door blows open, and the cabin decompresses. Two rows of passengers are blown out of the plane still strapped to their seats. The rest perish when the plane plummets, hitting the ground at 497 miles per hour. All 346 passengers and crew aboard the crowded flight die, and most of the bodies are battered beyond identification.

1974: April 2   Pompidou dies

President Georges Pompidou (see 1969: June 15) dies in office.

1974: May 19   Pompidou is replaced by Valéry Giscard d'Estaing

Independent Republican candidate Valéry Giscard d'Estaing (b. 1926) defeats Socialist François Mitterrand in a runoff election to become France's president following the death in office of Georges Pompidou. Giscard d'Estaing, a conservative, adopts policies to reduce the role of government in the economy, paring the rolls of public employees and removing many price controls. Giscard d'Estaing continues the independent, active foreign policy of de Gaulle (see 1890: November 22), sending French troops to serve in a peacekeeping force in Lebanon and offering assistance to the government of Chad against Libyan-backed rebels.

1974: July   Giroud is named to cabinet-level women's post

Publisher Françoise Gourdji Giroud (b. 1916) is appointed as France's first secretary of state for the Condition of Women in France. Areas falling under the domain of this newly created post include maternity, child care, and retirement benefits, sex discrimination, tax law, and sexism in the media.

Giroud is born in Geneva, Switzerland, in 1916. She begins her career as a typist, but she eventually applies her talents to editing. She cofounds *l'Express,* a news magazine. Her writings include *Ce que je crois* (What I Believe), *La Comédie du pouvoir* (The Comedy of Power), and *Les hommes et les femmes* (Men and Women), which becomes a bestseller.

1977   The Beaubourg opens

The Pompidou Cultural Center, informally known as the Beaubourg, opens in Paris. With its pipes and ductwork painted in bright colors and visible outside the building, the museum's design sparks heated controversy. The facility, however, which features mostly contemporary art, gains wide popularity, becoming the most-visited tourist site in France.

1979   Ariane rocket is launched

The Ariane rocket, developed by French scientists at the *Centre National d'Études Spatiales* (National Center for Space Research), is fired for the first time. It will be used to launch telecommunications and meteorological satellites.

1980   Arianespace is formed

Arianespace, the world's first commercial space transportation firm, is set up, with sixty percent of its funding coming from France and the rest from a variety of European companies and banks.

1980: April 15   Death of Jean-Paul Sartre

Existentialist author Jean-Paul Sartre dies in Paris. (See 1905: June 21.)

1981   First woman is named to the Académie Française

Belgian-born novelist Marguerite Yourcenar (pseudonym for Marguerite de Crayencour; 1903–87) becomes the first female member of the Académie Française, which has been in existence since 1635. Her works include *Les Memoires d'Hadrien* (Memoirs of Hadrian), *L'Oeuvre au noir* (Work in Black), and a prose poem, "Feux" (Fires). Her autobiography, *Souvenirs Pieux* (Pious Memories) is published in 1977.

1981   High-speed train service begins

The Train á Grande Vitesse (TGV) begins service between Paris and Lyons. With an average speed of 155 miles (250 kilometers) per hour, it is the world's fastest train. Service is later inaugurated between Paris and Lausanne, Switzerland.

The Mitterrand Years

1981: May   Mitterrand is elected president

François Mitterrand (1916–96; see 1916: October 26) is elected France's first Socialist president and forms a left-wing government that includes four Communist cabinet members. His government nationalizes some major banks and corporations, expands social welfare programs, raises the minimum wage, reduces the role of the central government on the local level, and outlaws capital punishment. His economic policies initially are successful, but France's economy later stalls due to recession.

1982   Corsica is granted greater autonomy

The French territory of Corsica, an island off the western coast of Italy, wins a greater say in electing its own assembly. However, this move does not appease militant separatists, who engage in terrorist activities throughout the decade.

1983–84   Basque separatists resort to violence

Basque extremists carry out a series of terrorist attacks in southwestern France. France and Spain cooperate in antiterrorist efforts.

1983   French forces fight in Chad

French troops aid in resisting Libyan-backed insurgents.

1983: March   Economic austerity program announced

The government of François Mitterrand is forced to cut public spending and implement an austerity program because of high inflation and the falling value of the French franc.

1985   Outcry caused by sinking of environmentalists' ship

The Mitterrand government suffers embarrassment and international censure when French agents are found to have had a hand in the New Zealand sinking of the Greenpeace flagship *Rainbow Warrior* carrying environmentalists protesting French nuclear testing in the South Pacific. One person is killed in the incident. France reduces its tests in the region from eight to six per year.

1986   Terrorist attacks in Paris bring crackdown

Terrorist bombings in the capital result in the passage of legislation to step up security measures.

1986   Musée d'Orsay opens in Paris

The Musée d'Orsay, featuring nineteenth- and twentieth-century French paintings, opens in a former railroad station. Its collection includes many famous Impressionist and Postimpressionist works.

1986: March   Conservatives triumph in general elections

Legislative elections result in a major setback for the Socialist party. A conservative, Jacques Chirac of the Gaullist RPR (Rassemblement pour la République), becomes prime minister, heading a center-right cabinet, and Socialist president Mitterrand pursues a course of "cohabitation" as he attempts to work with the right-wing opposition. Backed by the legislature, Chirac pushes through the privatization of sixty-five state-owned enterprises.

1988: May   Mitterrand is elected to a second term

Challenged by Prime Minister Jacques Chirac, Socialist François Mitterrand is elected to a second seven-year term as president with fifty-four percent of the vote, and his party wins a comfortable majority in legislative elections one month later. He appoints Michel Rocard (b. 1930) prime minister.

1989: November 9   Collapse of Berlin Wall brings foreign policy changes

With the end of the Cold War in sight, heralded by the demise of the Berlin Wall, France supports greater European integration to counter any future German aggression.

1990–91   French troops fight in Gulf War

When Mitterrand fails to persuade France's erstwhile ally, Saddam Hussein, to withdraw from Kuwait peacefully, France sends 12,600 air, naval, and ground troops to fight in the Gulf War, creating extensive controversy at home. In addition to the two nations' political ties, Iraq is the largest Middle East customer for French exports. In addition, France has a large Arab population. As part of the multinational alliance opposed to Saddam, Mitterrand also takes the unprecedented and controversial step of allowing French troops to fight under an American operational command.

1990   Political finance reform law is passed

The National Assembly enacts laws reforming the financing of political parties.

1990   Student demonstrations and riots

Secondary school students hold mass demonstrations calling for improvements in the educational system. The protests climax in rioting, the worst since the student unrest of 1968. The government agrees to increase the country's education budget.

1991   Cresson becomes France's first woman prime minister

Following the dissolution (breaking up) of the government of Prime Minister Michel Rocard (see 1988: May), Socialist Edith Cresson (b. 1934) is named prime minister, becoming the first woman in French history to hold that post.

1991: June–July   Wave of urban violence spurs job program creation

Urban areas with large numbers of immigrants suffer an increase in violence. Prime Minister Cresson sets up an emergency youth job training program in response.

1992   Cresson's government is dissolved

After only a year, the government of Edith Cresson, France's first female prime minister, is dissolved as Socialists suffer large losses in local and regional elections. Former finance minister Pierre Bérégovoy is named to replace Cresson.

Maastricht and the Euro

1992: June   Maastricht Treaty is narrowly approved

The Maastricht Treaty, signed in Maastricht, Netherlands, and providing for the creation of a European monetary and economic union, is ratified by a narrow margin of 51.05 percent to 48.95 percent.

1992: November 1   Smoking is banned in public places

The French government bans smoking in public places, but the ban is widely ignored.

1993   Former premier charged in HIV transfusion scandal

Former prime minister and Socialist party chairman Laurent Fabius (b. 1946) is charged in a case involving government negligence that allowed HIV-infected blood to be administered to hemophiliacs, killing 300.

1993: February   Prime minister implicated in financial wrongdoing

Allegations link Prime Minister Pierre Bérégovoy to a financier involved in insider trading.

1993: March   Conservatives gain in general election

The French electorate is so disillusioned with the Socialist government that one-third stay away from the polls in legislative elections, and those who do vote hand the Socialists a devastating defeat. Conservative candidates win eighty percent of the seats in the National Assembly, and President Mitterrand is once again paired with a conservative prime minister, this time Edouard Balladur, in a "cohabitation."

1993: May   Bérégovoy commits suicide

Former Socialist prime minister Pierre Bérégovoy kills himself following allegations of involvement in an insider trading scandal.

1994   Mitterrand's former extramarital affair exposed

The widely read weekly *Paris Match* publishes photographs of President François Mitterrand with his daughter from an extramarital affair two decades earlier. There is widespread criticism of the newspaper's action, and neither Mitterrand nor the Socialist party is affected politically by disclosure of the matter, which Mitterrand has already admitted to in the past.

1994: November 14   Channel Tunnel opens

Nearly seven years after it is begun, the thirty-one mile (fifty kilometer) Channel Tunnel between England and France opens to the public. With a reported cost of between $15 and $16 billion, the tunnel (nicknamed the "Chunnel") carries passengers between terminals at Folkestone, England, and Calais, France, on high-speed Eurostar trains that make the run in thirty-five minutes, making it possible to travel from London to Paris via ground transport in three hours.

1995–96   French nuclear testing comes under fire

Nuclear testing in the South Pacific and seizure of the Greenpeace ship *Rainbow Warrior II* (see 1985) arouse international protest. President Chirac pledges that France will sign the Comprehensive Test Ban Treaty in 1996.

1995: May   Chirac is elected president

Former prime minister Jacques Chirac is elected president, defeating Socialist Lionel Jospin in the second round of voting. He appoints former foreign minister Alain Juppé as prime minister and begins preparing France for admission to the new European monetary union in 1999. He also pledges to alleviate France's unemployment rate of around twelve percent, which is among the highest in Western Europe.

1995: June–August   Bombings rock Paris

A series of terrorist bombings frighten both Parisians and tourists and prompt tougher government security measures.

1995: December   Wave of strikes protests Chirac's policies

Hundreds of thousands of French people from all walks of life participate in the most widespread wave of strikes since 1968. The strikes protest President Chirac's reversal from the policies outlined in his campaign promises to an economic austerity program designed to pave the way for admission to the European monetary union. The government makes some concessions but still suffers from a general loss of confidence.

1996: January 8   Death of François Mitterrand

Former two-term president François Mitterrand dies in Paris. (See 1916: October 26.)

1996: September 24   France signs test ban pact

Together with the world's other nuclear powers—the United States, Britain, Russia, and China—France signs the Comprehensive Test Ban Treaty (CTBT) in a ceremony at United Nations (UN) headquarters in New York. The treaty outlaws all nuclear weapons tests of any kind, regardless of weapon size or test location.

1996: November 18   Fire halts Channel Tunnel service

In the first major accident since the opening of the Channel Tunnel between France and Britain (see 1994: November 14), a train transporting trucks through the "Chunnel" catches fire, damaging about 1,980 feet (600 meters) of the tunnel. Three of the train's crew members and a number of the truck drivers aboard the train are injured, but there are no fatalities, and rescuers are praised for their efficiency. It is expected that normal service through the tunnel will be disrupted for several weeks.

1997: May–June   Chirac calls parliamentary elections

President Jacques Chirac calls parliamentary elections a year early to demonstrate support for his economic austerity measures, but the strategy backfires, and the left wing wins a handsome victory, discrediting the president's policies. The center-right loses the large majority it had won in 1993, and Chirac is forced to appoint Socialist leader Lionel Jospin, whose support for monetary union is only lukewarm, as prime minister. With the Socialist victory, the number of women in the National Assembly, the lower house of Parliament, jumps from thirty-seven to sixty-two.

1998: February 6   Corsican leader is slain by assassin

Claude Erignac, the highest-ranking government official in the French possession of Corsica, is assassinated in a street attack by two assailants. Most suspicions center on Corsican separatists, who have waged a twenty-year terrorist campaign to win autonomy for the island. However, possible links are mentioned to Erignac's recent probes of organized crime on Corsica.

1998: July 12   France wins World Cup victory over Brazil

France, the host nation for the 1998 World Cup championship games in soccer, wins its first ever World Cup victory, defeating Brazil 3–0 in the final round. The win, before a live crowd of 80,000 people and a television audience estimated at 1.7 billion, sets off a frenzy of celebration in France. A victory parade attended by over half a million people is held the following day along the Champs-Elysées in Paris to fete the winning team.

Bibliography

Agulhon, Maurice. *The French Republic, 1879–1992.* Cambridge, Mass.: B. Blackwell, 1993.

Cerny, Philip G., ed. *Social Movements and Protest in France.* New York: St. Martin's Press, 1982.

Cook, Malcolm, ed. *French Culture Since 1945.* New York: G.P. Putnam's, 1983.

Corbett, James. *Through French Windows: An Introduction to France in the Nineties.* Ann Arbor, Mich.: University of Michigan Press, 1994.

Duby, Georges and Robert Mandrou. *A History of French Civilization from the Year 1,000 to the Present.* Trans. James Blakely Atkinson. New York: Random House, 1964.

Gildea, Robert. *France Since 1945.* Oxford: Oxford University Press, 1996.

Gough, Hugh, and John Horne. *De Gaulle and Twentieth-Century France.* New York: Edward Arnold, 1994.

Greene, Nathanael. *From Versailles to Vichy: The Third French Republic, 1919–40.* Arlington Heights, Ill.: Harlan Davidson, 1970.

Hirsch, Arthur. *The French New Left: An Intellectual History from Sartre to Gorz.* Boston: South End, 1981.

Hollifield, James F., and George Ross, eds. *Searching for the New France.* New York: Routledge, 1991.

Lefebre, Georges. *The French Revolution.* 2 vols. New York: Columbia University Press, 1962–64.

Noiriel, Gérard. *The French Melting Pot: Immigration, Citizenship, and National Identity.* Minneapolis, Minn.: University of Minnesota Press, 1996.

Northcutt, Wayne. *Mitterrand: A Political Biography.* New York: Holmes & Meier, 1992.

———. *The Regions of France: A Reference Guide to History and Culture.* Westport, Conn.: Greenwood Press, 1996.

Winchester, Hilary. *Contemporary France.* New York: J. Wiley & Sons, 1993.

Young, Robert J. *France and the Origins of the Second World War.* Basingstoke, England: Macmillan, 1996.

Georgia

Introduction

With the commercial advantages of its location along the Black Sea coast, Georgia has been invaded and dominated by powerful neighbors for most of its history. Although nominally ruled by a single dynasty, the Bagratids, for nearly one thousand years, it was divided and controlled by foreigners from the time of the Mongol invasions in the thirteenth century. Russia, the last of these ruling powers, gradually annexed Georgia in the course of the nineteenth century. For most of the twentieth, Georgia was a Soviet republic. Between the 1930s and the early 1950s, it suffered the worst excesses of Soviet repression under the authoritarian rule of Josef Stalin (1879–1953), even though Stalin himself was a Georgian. With the demise, or death, of the Soviet Union in 1991, Georgia gained its independence, but its new government confronted multiple challenges, including separatist rebellions, revolts by supporters of ousted president Zviad Gamsakhurdia, and a faltering economy.

Most of Georgia is located in the Caucasus Mountains, so over two-thirds of the country's 26,911 square miles (69,700 square kilometers) is mountainous. To the west lie the Colchis Lowlands, with the Kakhetian and Kartlian plains to the east. To the south, the Lesser Caucasus Mountains and Southern Georgian Highlands form the border with Turkey and Armenia. Georgia has an estimated population of over 5.2 million people, of whom over half live in the capital city of Tbilisi.

Bagratid Empire

Georgia was first unified into one kingdom, called Iberia, in the fourth century A.D., when Christianity was adopted by the populace. Their continuing loyalty to the Orthodox Christian faith has been one of the main distinguishing features of Georgian culture. After several centuries of invasions by Persia and the Byzantine Empire, a new dynasty—the Bagratids—arose to unite the Georgians. Between the ninth and thirteenth centuries, they expanded Georgia's territory until it grew into a powerful medieval state whose boundaries reached from the Caspian Sea to the Black Sea and from Armenia to the Caucasus Mountains. The high point of

Bagratid rule came with the rule of Queen Tamar in the twelfth century, a period known as Georgia's Golden Age. Queen Tamar's rule saw a renaissance of Georgian literature that included the composition of the national epic, *The Knight in a Panther's Skin* by Shota Rustaveli (c.1172–c.1216), Georgia's most honored poet.

Mongol invasions in the 1220s brought an end to the glory of the Bagratid empire. The dynasty survived until the nineteenth century but largely under the control of foreign powers. In the mid-sixteenth century, Persia and Turkey, which had been warring over Georgia for centuries, divided the region into eastern and western spheres of influence. The areas under Persian rule were permitted increasing degrees of autonomy over the following centuries, and their Bagratid rulers asserted a real measure of control over their lands. Nevertheless, the Georgians periodically looked to other Christian nations for help in countering the power of the Muslim Turks and Persians.

Russian Rule

By the late eighteenth century, the rulers of eastern Georgia were appealing to Russia for aid in combating aggression by the Ottoman Turks. In 1783 the Georgian king Herekle II (1720–98) and Russian empress Catherine the Great (1729–96) signed the Treaty of Georgievsk, guaranteeing Russian protection to eastern Georgia in return for political control of the region. Nonetheless, the Russians did not save Tbilisi from being sacked by the Persians in 1795.

In 1801 Russia annexed eastern Georgia and gradually took control of the rest of Georgia over the next sixty years. In the course of military victories over Persia and Turkey, the Russians undertook large-scale modernization efforts in Georgia, building roads, mines, factories, and eventually, railroads. However, Russian efforts to assimilate Georgia culturally spurred periodic revolts and eventually led to the formation of nationalist groups. At the end of the nineteenth century, the foremost nationalist group was the socialist "Third Group," which counted among its members a young Georgian named Josef Dzhugashvili (1879–1953). One of his two nicknames was "Stalin" ("man of steel"), and it was under this name that he later became the most feared dictator

of the twentieth century, wielding an iron fist over his homeland as well as the rest of the Union of Soviet Socialist Republics (USSR).

Georgia's socialists, like those in Russia, split into two rival factions in the early twentieth century: the radical Bolsheviks and the more moderate Mensheviks. When Georgia declared independence after the Russian Revolution (1917), the Mensheviks were in control of the newly formed Georgian Social Democratic Republic. By 1921, however, the Bolsheviks had seized power and formed a Soviet republic, which became part of the Soviet Union the following year. Although Stalin and his most powerful associate, Lavrenty Beria, were both Georgians, Georgia was not spared the worst excesses of the Stalinist period, and thousands lost their lives to government repression, while many more were jailed or exiled, as the Communists attempted to root out all opposition to Stalin's regime.

Government repression was relaxed following the death of Stalin in 1953, but the increased freedom also became an opportunity for corruption to flourish. Georgia developed an extensive network of "underground business" and became known as one of the most corrupt republics in the USSR.

Rise of Shevardnadze and Georgian Independence

In 1972 Eduard Shevardnadze (b. 1928), a Communist official known for opposing corruption, was appointed to head the Georgian Communist Party and launched an extensive anticorruption campaign. Georgian nationalism once again came to the forefront in the 1970s and 1980s, as Zviad Gamsakhurdia, son of a prominent Georgian author, began to speak out against the Communist government and to attract followers to the nationalist cause. With the Soviet liberalization that accompanied the tenure of Mikhail Gorbachev as Soviet leader in the 1980s, Georgians were further encouraged to work for independence.

Georgia declared its independence in the spring of 1991, before the failed coup that led to the demise of the Soviet Union. Gamsakhurdia was elected president, but his authoritarian leadership style aroused such great opposition that he was ousted by the beginning of 1992, and Eduard Shevardnadze was invited to take over leadership of the government. Elected to a five-year term as president in 1995, Shevardnadze, who has survived two assassination attempts, has contended with a faltering economy, widespread crime and corruption, and separatist movements in the autonomous regions of Abkhazia and South Ossetia.

Timeline

5000–4000 B.C.   Early inhabitants

Neolithic hunter-gatherers inhabit present-day Georgia.

12th century B.C.   First recorded civilization

Migrants from Asia Minor and the Hittite Empire occupy western Georgia, leaving the first records of civilization in the region.

8th century B.C.   Greeks visit Georgia

Greek traders with bases in the Black Sea region visit Georgia.

730 B.C.   Foreign invasions

Cimmerians and Scythians invade Georgia and occupy much of the Caucasus, dispersing the tribes they find there.

65 B.C.   Rome invades

The Roman Empire, under Pompey (106–48 B.C.), conquers Georgia. They establish good relations with the culture already there, building new roads and expanding trade in the area.

4th century A.D.   Iberian kingdom is formed

Georgia's first independent kingdom is founded by the merger of two different tribes. Its capital is at Mtskheta. For the first time, the inhabitants of Georgia are united by a common language.

A.D. 330   Christianity is adopted

Georgia, under the rule of King Marian III, accepts Christianity, linking it more closely to the Byzantine Empire.

5th–7th centuries   Warfare between foreign powers

Iberia becomes a battleground in the power struggle between Persia to the east and the Byzantine Empire to the west.

645   Arabs conquer Iberia

Arab invaders capture the city of Tbilisi, setting up a capital there.

The Bagratid Kingdom

9th century   The Bagratid kingdom is founded

The Bagratids, an Armenian noble family, come to dominate Georgia, uniting it under a single kingdom and inaugurating a "Golden Age" in the region (see 12th century). Even after its most glorious period of rule is ended, the Bagratids continue to rule over at least a portion of Georgia for almost a millennium.

813–30   Rule of the first Bagratid monarch

Ashot I gains control over southern Georgia, becoming the first Bagratid ruler.

975–1014   Bagratid III unites Georgia

Under Bagratid III, the eastern and western areas of Georgia are united for the first time.

The Golden Age

12th century   "Golden Age" under Bagratid rule

Georgia experiences a "Golden Age" under the rule of the Bagratids, who unite and expand the region and drive out foreign occupiers. Along with political power comes a cultural renewal that includes the founding of renowned academies and the composition of Georgia's national epic, *The Knight in the Panther's Skin* by court poet Shota Rustaveli (see 1172). Cultural achievements of the period also include church building and the creation of masterpieces by Georgian goldsmiths.

1089–1125   Reign of David IV

The "Golden Age" of the Bagratid kingdom begins under David IV (known as "The Builder"), who takes over the throne from his father, King Giorgi II, at the age of sixteen. Driving the Seljuk Turks from Georgia, David begins expanding the power and influence of the Bagratids further eastward and southward. Trade and culture flourish during his reign.

c. 1172   Birth of poet Shota Rustaveli

Shota Rustaveli, Georgia's most honored writer, authors its national epic, *The Knight in the Panther's Skin*. Few biographical details are known about Rustaveli. He is active during the reign of Queen Tamar, which marks the high point of the literary flowering that takes place during medieval Georgia's "Golden Age." In addition to his epic poem, Rustaveli writes odes in honor of the queen and is rewarded by an appointment as court treasurer. Scholars have found Greek, Persian, and Chinese influences in Rustaveli's poetry. Rustaveli is thought to have died around 1216.

1184–1213   Reign of Queen Tamar

The Bagratid kingdom reaches its high point under the rule of Queen Tamar, who rules over an area stretching from the Caspian Sea to the Black Sea, and from present-day Armenia to the Caucasus. Georgian literature flourishes during this period (see c.1172).

Mongol, Persian, and Turkish Invasions

1220s   Mongol invasion

The Bagratid kingdom falls to a brutal Mongol invasion. The invaders kill the Bagratid king, Giorgi IV Lasha, in battle and seize control of Georgia for nearly a century.

1314–46   Bagratid power is restored

Under King Giorgi V (The Brilliant), Georgian independence is regained from a weakened Mongol empire. However, it is soon lost again to another Mongol conqueror, Tamerlane. (See 1386.)

1386–1404   Invasions by Tamerlane

After a brief respite from Mongol dominance, Georgia is again subdued by Mongols under the leadership of the warrior Tamerlane (1336–1405), who destroys Tbilisi and inflicts widespread destruction on the Georgian kingdom in eight separate attacks.

1553   Peace of Amasia

After conquering Constantinople in 1453, the Ottoman Turks enter a period of dominance in Europe, vying with the Safavid Persians for control of Georgia. Eventually the two powers agree to share control of Georgia, dividing it into eastern and western sectors, with the Turks controlling the west and the Persians dominating the east. Persia permits local Georgian rulers to remain in power, but under strict Persian control. With Islamic powers dominating their homeland, many Georgians are persecuted for their allegiance to Christianity, and the Georgians periodically seek the help of other Christian nations.

18th century   Georgian autonomy increases

By the eighteenth century, Persian control of Georgia is weakening, and the Bagratid monarchs Taimuraz II (r. 1744–62) and Herekle II (1720–98) are able to reassert a greater measure of control. However, raids by Muslim attackers subdue the Georgians once again.

1762   King Herekle reunites Georgia

King Herekle II strengthens the Georgian monarchy, unites the eastern parts of Georgia, and expands trade.

1773   Herekle seeks Russian help

King Herekle seeks Russian protection from the Ottoman Turks, involving Georgia in the bitter political and military rivalry between Russia and Turkey. Georgia undergoes periods of occupation by Russian troops.

1783   Treaty of Georgievsk

Herekle II and Russian empress Catherine the Great (1729–96) sign the Treaty of Georgievsk, which guarantees Russian protection to the eastern Georgian kingdom of Kartli-Kakheti in return for authority over the region.

1795   Persians invade Tbilisi

Persian forces invade and sack the Georgian capital of Tbilisi. Despite its promise of protection under the Treaty of Georgievsk, Russia does not come to the aid of Georgia.

Russian Rule

1801–64   Russian annexation of Georgia

Under Czar Alexander I (1777–1825), Russia takes control of eastern Georgia in 1801, gradually annexing the rest over the next sixty years. Georgia's existing kingdom is dissolved. The Bagratid monarch is forced to abdicate and is replaced by Russian governors. Economic conditions improve under Russian modernization, as factories, mines, highways, and railroads are built, and there is a large-scale migration to the cities to find work in the area's emerging industries. Tbilisi becomes an important government and trade center. Russia alienates many Georgians, however, by embarking on a campaign to assimilate Georgia politically and culturally, imposing its own system of social stratification and education on the region. The Georgians periodically revolt against their Russian rulers, but all their rebellions are quickly suppressed.

1851   Theater is built in Tbilisi

The Italian opera *Lucia di Lammermoor* by Italian composer Gaetano Donizetti (1797–1848) is performed in a newly built theater in Tbilisi. It is followed by many other opera performances.

1862: May   Birth of painter Niko Pirosmanashvili

Georgia's most acclaimed painter is born in modest circumstances and orphaned at a young age. He makes a living painting signs, murals, and other types of public art while evolving a unique style of painting based on elements of Georgian folk art. Pirosmanashvili gains professional acclaim in 1912 when a circle of artists discovers and begins promoting his work, which is displayed in an exhibition in Moscow. In spite of his new-found public acclaim, Pirosmanashvili remains impoverished, and his situation worsens during World War I (1914–18). He dies in 1918 and is buried in a pauper's grave. Today his works are highly regarded and are among the most prized items displayed by Georgia's State Museum of Art. About 200 of his paintings are owned by the museum and private collectors.

1864   Serfdom is abolished

Three years after a general decree abolishing serfdom in Russia, it is also terminated in Georgia.

1877–78   Russia annexes ports of Poti and Batumi

The Black Sea ports of Poti and Batumi are added to Russian Georgia during the Russo-Turkish War (1877–78).

1879: December   Birth of Josef Stalin

Josef Stalin, the most brutal and powerful authoritarian leader of the twentieth century, is born Josef Dzhugashvili in the Georgian town of Gori. (He later adopts the name Stalin, which means "man of steel.") He is born into an impoverished family, with a pious mother and an alcoholic father. As a youth, Stalin enrolls in a seminary in Tbilisi but is expelled and turns to revolutionary activity while still in his teens. In the early years of the twentieth century, he preaches socialism to workers in the Georgian countryside and becomes active in the Bolshevik branch of Russia's socialist party, led by the future founder of the Soviet Union, Vladimir Lenin (1870–1924). Stalin continues his political activities through the World War I (1914–18) period in spite of numerous arrests.

Following the Russian Revolution in 1917, Stalin assumes posts of increasing power within the Communist Party and is appointed as general party secretary in 1922. He uses this position to consolidate his influence, and by 1929 he is the most powerful leader in the Soviet Union. Stalin institutes programs of mass agricultural collectivization and large-scale industrialization. He also shows himself to be a ruthless dictator who presides over the exile, incarceration, or murder of millions of people, especially intellectuals and professionals. His political purges of the late 1930s and the post-World War II years (World War II: 1939–45) create mass terror and suffering and disillusion many Communist sympathizers throughout the world. Following his death from a stroke in 1953, the Soviet government officially repudiates many of his policies and actions.

1880   First Russian opera is performed in Tbilisi

A visiting company gives the first Georgian performances of Russian operas, including works by Mikhail Glinka (1804–57).

1883   Musical society supports concerts

A Georgian branch of the Russian Musical Society is formed. It organizes orchestra and chamber music concerts, including some by internationally known musicians including the Russian composer Peter Ilyich Tchaikovsky (1840–93).

1890s   Nationalist groups are formed

Georgians form illegal nationalist groups working for independence from Russia. The most prominent one, called *Mesame-dasi* (the Third Group), is a socialist organization affiliated with the Social Democratic Party. One of its two leaders is a young Georgian named Josef Dzhugashvili (see 1879: December). Later, under the name Josef Stalin (see 1879: December), he becomes the most powerful and feared leader of the Soviet Union.

1903   Socialists split into factions

Organized Georgian socialists split into two major factions. The Mensheviks, led by Noe Zhordania (see 1918: May 26), favor gradual change, while the Bolsheviks, led by Josef Stalin, seek radical revolutionary action.

1905–06   Failed revolution sparks further unrest

The failed revolutionary effort within Russia leads to general industrial strikes, which are harshly suppressed, and a peasant revolt in western Georgia.

1905   Georgian Philharmonic Society promotes Georgian music

The Georgian Philharmonic Society is established to promote the collection and performance of Georgian folksongs and to promote operas based on Georgian texts.

1914–18   World War I

Russian forces invading Turkey, together with the rest of the Caucasus, turn Georgia into a battleground, and economic conditions in the region deteriorate, helping to lead to the demise (death) of the Russian empire in the coming Russian Revolution (1917).

1918: April 7   Death of artist Niko Pirosmanashvili

Acclaimed painter Niko Pirosmanashvili dies in Tbilisi. (See 1862: May.)

Transcaucasian Soviet Federative Socialist Republic

- - - - - Transcaucasian border

———— Modern borders

1918: May 26   Georgia declares independence

Czarist rule over Georgia is ended when the Russian czar Nicholas II (1868–1918) is overthrown in 1917 and the Bolsheviks, led by Vladimir Ilyich Lenin (1870–1924), seize control of Russia. After a brief union with Armenia and Azerbaijan in a transcaucasian republic, Georgia declares independence and forms the Georgian Social Democratic Republic, under the control of a moderate government led by Noe Zhordania, a Menshevik revolutionary. It is recognized by over twenty foreign powers, including the Soviet Union under Vladimir Lenin. In spite of large-scale land redistribution, unfavorable economic conditions continue.

1919   Georgian operas are performed

The first operas by Georgian composers are staged. They include *Legend of Shota Rustaveli* by Arakishvili, *Abesalom da Eteri* by Paliashvili, and *Keto and Kote* by Viktor Dolidze.

1920s   Economic reconstruction

Georgia's industrial facilities are rapidly expanded, and its economy reaches prewar levels by the mid-twenties. In addition, transportation networks are rebuilt, illiteracy is reduced, and institutions of higher education are established.

1921: February 25   Soviet republic is formed in Georgia

Bolshevik leaders, aided by the Russian army, stage an uprising against Georgia's Menshevik leaders, seize control of Georgia, and form a Soviet government.

Soviet Republic

1922: March   Georgia becomes part of Transcaucasian republic

Georgia is incorporated into the Transcaucasian Soviet Federated Socialist Republic, which becomes part of the Soviet Union later in the year.

1923   National art museum is founded

The State Museum of Arts is founded in Tbilisi, housing the country's major art collection, from medieval icons to nineteenth- and twentieth-century Russian paintings. Georgian artists represented in its collection include Lado Gudiashvili, Yelena Akhlediana, David Kakabadze, and Niko Pirosmanashvili (see 1862: May).

1923–29   State Museum of Georgia is created

The State Museum of Georgia is established to house archaeological and ethnographic material relating to Georgian history. Its collections include objects dating back to the Paleolithic era, as well as the Middle Ages. It also displays many items from the era of Czarist rule and the early years of Georgian independence following World War I (1914–18).

1924   Mass political executions take place

Some five thousand Georgians are executed in retaliation for a Menshevik-led revolt against the Soviet government.

1928: January 25   Birth of political leader Eduard Shevardnadze

Shevardnadze, Georgia's foremost political leader following independence from the Soviet Union, is the son of a teacher. He becomes active in the Communist Party as a youth, rising through the ranks to become first secretary of the Georgian party by 1972. In the 1970s, he becomes active in the Soviet government on the national level, joining the Politburo in 1978. In 1985 Mikhail Gorbachev names Shevardnadze to succeed Andrei Gromyko as the nation's foreign minister, a post he holds until 1990, negotiating arms deals with the United States and the Soviet withdrawal from Afghanistan.

In early 1992, following the collapse of the Soviet Union, Shevardnadze returns to Georgia, whose government is in disarray following the ouster of its first elected president, Georgian nationalist leader Zviad Gamsakhurdia. Shevardnadze is named chairman of Georgia's parliament, an appointment confirmed by his election to that post in the fall of that year. Winning election to a five-year presidential term in 1995, Shevardnadze remains in power as Georgia's top political leader throughout the decade. During this time, he confronts separatist movements in Abkhazia and South Ossetia, the growing power of paramilitary groups and organized crime, and a faltering economy. He also survives two assassination attempts, one in August 1995 and a second in February 1998.

1930s–50s   Rapid industrialization

Under centralized Soviet economic planning, Georgia undergoes rapid industrialization and the resulting urbanization of its population.

1931–38   Repression under Beria

Stalin appoints Lavrenty Beria, a fellow Georgian, as first secretary of Georgia's Communist Party. Beria carries out the collectivization of Georgian agriculture and institutes a campaign of terror aimed at eliminating all political opposition to Stalin's regime.

1936–37   Georgians suffer in Soviet purge

At the behest (command) of Communist leaders Stalin and Beria, two-thirds of Georgia's elected Communist leaders are exiled or executed in major political purges. Artists and intellectuals are persecuted as well.

1936: December   Georgian republic is created

A separate Georgian republic is formed as part of the Soviet Union following the dissolution of the Transcaucasian republic to which Georgia previously belonged.

1939–45   World War II

Wartime patriotism strengthens Georgian cooperation with the Soviet government and defuses nationalism in the region, even though Georgia itself is never invaded by Germany. Half a million Georgian men fight in the war, and Georgia supplies vital armaments and textiles to the war effort. About 300,000 Georgians die serving in the Soviet military.

1950s–70s   Decentralization leads to corruption

With the greater regional freedoms that follow the death of Josef Stalin (see 1953), Georgia regains greater control over its economic and cultural affairs but also develops a thriving underground economy that is not accountable to the government. An extensive network of corruption flourishes during the tenure of Vasili Mzhavanadze as Georgian Communist Party leader. Another adverse effect of decentralization is discrimination against minority ethnic groups within Georgia.

1953   Death of Josef Stalin

Georgian-born Soviet leader Josef Stalin dies (see 1879: December) and is succeeded by Nikita Khrushchev (1894–1971). On one hand, Georgia welcomes the relaxation of authoritarian control by the central Soviet government. However, because Stalin was from Georgia, Georgians take offense at Khrushchev's denunciations of the former leader, whose memory remains honorably preserved in the republic even while it is virtually erased elsewhere in the Soviet Union.

1956: March   Georgians are killed in de-Stalinization protest

Hundred of Georgians lose their lives when Soviet troops are called in to break up demonstrations against the "de-Stalinization" policy implemented by the Soviet government under Nikita Khrushchev.

1970s   Dissident movement is launched

Antigovernment nationalists form an organized opposition group under the leadership of Zviad Gamsakhurdia, son of famed Georgian author Konstantine Gamsakhurdia.

1972   Shevardnadze assumes leadership post

Reform politician Eduard Shevardnadze (see 1928: January 25) is appointed first secretary of the Georgian Communist Party and works to eradicate the widespread corruption and inefficiency of the Georgian government. His anticorruption efforts are only partially successful.

1978   Soviets order change in language status

The central Soviet government in Moscow attempts to abolish the official status of the Georgian language but backs down in response to mass demonstrations.

1978   Abkhaz leaders threaten to secede

Bringing to the forefront an interethnic conflict that will remain a problem in the coming decades, leaders of the Abkhazian Autonomous Republic threaten to secede from Georgia. They claim that the Abkhazians are the victims of cultural, economic, and political discrimination. The crisis is defused by Georgian leader Eduard Shevardnadze (see 1928: January 25).

1978   Gamsakhurdia receives Nobel nomination

Georgian dissident and nationalist leader Zviad Gamsakhurdia is nominated for the Nobel Peace Prize for championing ethnic rights in Georgia.

1985   Nationalist movement is formed

Georgians form an organized nationalist movement spurred by the liberalization of the central Soviet government in Moscow under Mikhail Gorbachev (b.1931).

1989: April 9   Georgian demonstrators are killed in Tbilisi

The movement toward Georgian independence receives added impetus from the "April Tragedy," in which twenty Georgians are killed by Soviet troops accused of using sharpened shovels and poisonous gas to disperse demonstrators in Tbilisi. Nationalist spokesperson Zviad Gamsakhurdia is jailed for protesting the government's actions.

1990: October   Nationalists lead in free elections

When Mikhail Gorbachev allows free elections in the Soviet republics, the Round Table-Free Georgia nationalist coalition wins sixty-two percent of the vote in Georgia. With nationalists in control, the Georgian parliament elects nationalist hero Zviad Gamsakhurdia as the republic's leader.

Independence

1991: April   Georgians declare independence

Following a referendum on independence approved by ninety percent of Georgian voters, Georgia declares independence from the Soviet Union several months before the dissolution of the USSR.

1991: May 26   Gamsakhurdia is elected president

Georgian nationalist leader Zviad Gamsakhurdia wins the republic's first presidential election with eighty-six percent of the vote. His supporters are disappointed, however, when he turns out to be an autocratic leader, censoring the media and suppressing dissent. Gamsakhurdia's failure to publicly condemn the aborted coup against the Gorbachev government in

August causes opposition to mount, and his regime becomes increasingly brutal.

1991: September–December   Opposition to Gamsakhurdia grows

Opposition to Georgian president Gamsakhurdia causes increasing civil unrest, during which the leader increasingly takes refuge inside the parliament building.

1992: January 6   Gamsakhurdia flees the country

President Gamsakhurdia flees to Armenia, later taking up residence in western Georgia, where he attempts to rally support. He meets with some success but not enough to regain power. A state council forms a provisional government until elections can be held and invites Eduard Shevardnadze (see 1928: January 25), the former Georgian leader who had risen to a position of national prominence under Mikhail Gorbachev, to head the new government.

1992: August   Abkhazia declares independence

The autonomous republic of Abkhazia votes to return to its 1925 constitution, making it independent of Georgia. Georgian leader Shevardnadze sends in troops, which meet with resistance. The dispute is settled through a cease-fire mediated by Russian president Boris Yeltsin (b. 1931).

1992: October   Parliamentary elections are held

New parliamentary elections are held. Former Soviet foreign minister Eduard Shevardnadze is elected parliamentary speaker.

1993: September   Gamsakhurdia launches a rebellion

Former Georgian leader Zviad Gamsakhurdia attempts to regain power, starting a six-week rebellion in western Georgia. Russia aids the Georgian government in putting down the revolt, and in return, Georgia agrees to join the Commonwealth of Independent States (CIS), composed of twelve former Soviet republics.

1994: January   Gamsakhurdia's death is reported

The death of former Georgian president Zviad Gamsakhurdia is reported from Grozny, the capital of Chechnya, a breakaway area of Russia. Conflicting accounts are given of his death. His wife claims he committed suicide, while Georgian authorities say he was murdered.

1994: February   Georgia joins the CIS

Georgia joins the Commonwealth of Independent States (CIS). Other members of the CIS include: Armenia, Azerbaijan, Belarus, Kazakhstan, Kyrgyzstan, Moldova, Russia, Tajikistan, Turkmenistan, and Ukraine. It also agrees to the establishment of Russian military bases in return for military aid and trade concessions. The defense agreement between Georgia and Russia is opposed by political factions in both countries.

1995: February 10   Abkhazia modifies its demands

Hopes for an end to the Abkhazian conflict are raised by an announcement revoking Abkhazia's demand for complete independence.

1995: August 24   Parliament adopts new constitution

The Georgian parliament adopts a new constitution, the republic's first since gaining independence in 1991. The document provides for a president who is both head of state and head of the executive branch of government and limits his tenure to two five-year terms.

1995: August 29   Shevardnadze survives assassination attempt

Georgian leader Eduard Shevardnadze receives only minor wounds when a bomb explodes near a motorcade in which he is traveling en route to sign a new constitution approved by parliament. Shevardnadze's security minister, Igor Georgadze, is later implicated in the incident and escapes to Russia.

1995: November 5   Shevardnadze is elected president

Eduard Shevardnadze, Georgia's parliamentary speaker and acting head of state, is elected to a five-year term as president with seventy-five percent of the vote. Shevardnadze's party, the Citizens' Union, leads the balloting in the parliamentary elections held on the same day, winning twenty-four percent of the vote.

1996: November 10   President is elected in South Ossetia

South Ossetia, one of two breakaway regions of Georgia, elects Lyudvig Chibirov—its current separatist leader—president. The Georgian government condemns the election as illegal and voices fears that it will lead to further conflict in the region.

1996: November 23   Ardzinba wins Abkhazian elections

Vladislav Ardzinba wins the presidential elections held in Abkhazia and goes on to rule the region as an autonomous state, although Georgia rejects the validity of the elections because ethnic Georgians in Abkhazia are barred from voting.

1998: February 9   Shevardnadze survives second assassination attempt

A second murder attempt is made on the life of President Eduard Shevardnadze. Like the previous attack (see 1995:

August 29), this one is made on a presidential motorcade in Tbilisi. Two dozen attackers fire on the motorcade with automatic weapons and launch grenades, killing three people, including two bodyguards, and injuring four more. Shevardnadze escapes injury when his driver manages to divert his car from the line of fire. Supporters of former president Zviad Gamsakhurdia are implicated in this attack, as in the previous one.

1998: February 19–25   UN observers are taken hostage

Four United Nations (UN) observers and six other persons are abducted from a UN office and taken hostage by gunmen, said to be linked to former Georgian president Zviad Gamsakhurdia. The hostages are released unharmed following negotiations with Georgian and Russian government personnel. Two of the kidnappers escape, and two surrender to authorities.

1998: October 19   One-day revolt in western Georgia

Military personnel stage a one-day revolt in western Georgia, the stronghold of former president Gamsakhurdia. The rebels surrender following negotiations with the government. Their leaders are later charged with treason.

Bibliography

Braund, David. *Georgia in Antiquity: A History of Colchis and Transcaucasian Iberia, 550 BC–AD 562*. New York: Oxford University Press, 1994.

Brook, Stephen. *Claws of the Crab: Georgia and Armenia in Crisis*. London: Sinclair-Stevenson, 1992.

Gachechiladze, R. G. *The New Georgia: Space, Society, Politics*. College Station, Tex.: Texas A&M University Press, 1995.

Goldstein, Darra. *The Georgian Feast: The Vibrant Culture and Savory Food of the Republic of Georgia*. New York : HarperCollins, 1993.

Lang, David Marshall. *The Georgians*. New York, Praeger, 1966.

———. *A Modern History of Soviet Georgia*. New York: Grove Press, 1962.

Schwartz, Donald V., and Razmik Panossian. *Nationalism and History: The Politics of Nation Building in Post-Soviet Armenia, Azerbaijan and Georgia*. Toronto, Canada : University of Toronto Centre for Russian and East European Studies, 1994.

Suny, Ronald Grigor. *The Making of the Georgian Nation*. 2nd ed. Bloomington, Ind.: Indiana University Press, 1994

———, ed. *Transcaucasia, Nationalism, and Social Change: Essays in the History of Armenia, Azerbaijan, and Georgia*. Ann Arbor, Mich.: University of Michigan Press, 1996.

Germany

Introduction

Germany has a long and checkered history. While it has been shaped by forces reaching to the times of ancient Rome, Germany's status as a united country in central Europe dates back no farther than 1871. Before that, the German area was a loose confederation of largely independent states, principalities, and cities, with a weak and often totally ineffective central governing body. In contrast to other major European countries, prior to 1871 Germany had no real capital or cultural focal point, resulting in great regional variations which are still noticeable today. The division of Germany after the country's defeat in World War II (1939–45) into a Western (democratic) state called Federal Republic of Germany and an Eastern (communist) state named German Democratic Republic lasted from 1949 to 1990 and also left its mark on today's reunited country.

Germany is located in the heart of Europe and since time immemorial its territory has been the transit route between east and west, north and south. This fact explains the country's exposure to innumerable cultural influences as well as its frequent involvement in European conflicts. The earliest ancestors of today's Germans arrived in Central Europe around 1000 B.C. By the first century B.C. the Romans had come in contact with them along the Rhine River. Indeed, in A.D. 9, Germanic forces halted Roman expansion across the Rhine and repulsed a Roman force on its way to the Elbe River. This Germanic victory halted Roman expansion in central Europe.

Germanic tribes swept south and inundated the Roman Empire from the fourth to the seventh centuries. This movement, one of the greatest mass migrations in history, led to the collapse of Roman power in Western Europe and the beginning of the Middle Ages. The roots of modern German civilization emerged during this era. As Germanic tribes settled, they soon established their own kingdoms and absorbed many of their neighbors' cultural traits, particularly the Christian faith.

The Franks formed the most significant Germanic kingdom of the early Middle Ages. Located in the regions of present-day France and western Germany, the Franks formed the most powerful state in Western Europe. At its height in 800, the pope crowned the Frankish king, Charles, Roman Emperor. History remembers Emperor Charles (r. 768–814) as Charles the Great (Ger: *Karl der Große*, or French: Charlemagne) and his state as the Carolingian Empire.

Following Charlemagne's death, his empire disintegrated in disputes among his heirs. A semblance of order returned through the Treaty of Verdun (843) which divided the empire in three kingdoms. The eastern kingdom formed the basis of modern Germany, the western kingdom was the foundation of modern France, while the central kingdom remained a contested region. However, not until the reign of Otto of Saxony (Otto the Great; 912–973) did a strong leader emerge in the eastern kingdom. While in the service of the pope, Otto took the title Holy Roman Emperor and referred to his kingdom as the Holy Roman Empire. Under Otto's successors his state weakened, the empire became a confederation of states, and the emperor's title became merely symbolic. Nevertheless, under the Habsburg dynasty (which took over the imperial title in 1273) the Holy Roman Empire remained a political entity until 1806 and formed the basis of German cultural identity, if not actual political unity. In many respects the Middle Ages were a formative period in German history. As trade grew, Germany's geographic location made it a center of commerce. As a result, towns grew in importance and an urban culture developed.

A distinct turning point in German history that seriously affected its development was the Protestant Reformation, a revolt against Roman Catholicism begun in the sixteenth century by a German monk, Martin Luther. The political consequences of this split in the Church proved ruinous for Germany as intermittent warfare continued for over a century ending only with the end of the Thirty Years' War in 1648. Afterward, Germany remained as fragmented as ever, divided into over 300 states, some no larger than a town. Generally, the north German states (including Hohenzollern-led Brandenburg-Prussia) adopted Protestantism, while the southern states, particularly the Habsburg Empire, adhered to Roman Catholicism.

In the two centuries that followed the end of the Thirty Years' War, the most significant developments in Germany were the rise of Brandenburg-Prussia and the emergence of German nationalism. Under a succession of able rulers, the Hohenzollerns managed to turn their small kingdom into the

preeminent military power in central Europe. As a result, Prussia (as Brandenburg-Prussia became known by the eighteenth century) became the dominant state in northern Germany and posed a challenge to the power of the Habsburgs.

The French Revolution and the Napoleonic Wars (1792–1815) brought about a rise in German national consciousness. Influenced by the ideas of the French Revolution, young Ger-

man intellectuals began advocating the formation of a German national state. But what form would such a state take and who would lead the way? The Napoleonic reorganization of Germany reduced the number of German states to thirty-nine, a decision that was upheld at the Congress of Vienna (1815) after Napoleon's defeat. In addition, Napoleon had dissolved the Holy Roman Empire in 1806. Yet both Prussia and Austria (as the Habsburg Empire was now formally called) remained as the two leading German states—rivals which did not see German unity as their ultimate objective. Liberal nationalist movements continued to grow but remained suppressed; the abortive all-German Frankfurt Parliament in 1848 symbolized the failure of German liberal nationalism.

When it occurred, German unification was led by conservative Prussia under Otto von Bismarck (1815–98) and did not unite all Germans into a national state (Austria retained its own multi-ethnic empire). Bismarck, more a Prussian patriot than a German nationalist, forged a German Empire under Hohenzollern rule through a series of short, successful wars against Denmark (1864), Austria (1866), and France (1870–71). Bismarck's creation became the dominant military power in Europe. Yet instead of bringing instability, Bismarck's adroit diplomacy resulted in a myriad of alliances that secured a balance of power throughout his days as Imperial Chancellor (1870–90).

By contrast, during the reign of *Kaiser* (Emperor) Wilhelm II (r. 1888–1918) instability plagued Europe and culminated in the formation of rival alliances and the outbreak of World War I (1914–18). German forces, in alliance with Austria-Hungary, Bulgaria, and the Ottoman Empire, fought Britain, France, Russia, and their allies (including, eventually, the United States). Although Germany enjoyed tremendous successes well into the war, it was eventually defeated and forced to sign the humiliating Treaty of Versailles in 1919.

The defeat left Germany in a much weaker position. It lost all of its overseas colonies as well as European territory to Belgium, France, Denmark, and Poland. By signing the Versailles treaty, Germany also admitted its guilt in having started the war and agreed to pay reparations for damages inflicted on its enemies. Forced to deal with this situation were the leaders of the new Weimar republic (Wilhelm II abdicated at the time of the armistice in 1918; the republic took its name from the city of Weimar where its constitution was drafted). Democratic forces faced attacks from both the far left (which agitated for a worldwide communist revolution) and the far right (which did not accept Germany's defeat and called for a forced revision of the treaty). Reparation obligations strained the economy to the breaking point and by 1923 hyperinflation set in as the German government resorted to printing ever more banknotes in order to meet its financial needs.

World War II

Although stability returned with the economic prosperity of the mid–late 1920s, the onset of the worldwide Great Depression in 1929–30 brought a resurgence of radical political movements, chief among them, the *Nationalsozialistiche Deutsche Arbeiterpartei* (National Socialist German Workers' Party, or Nazi), an extreme right-wing racist, hypernationalist party led by Adolf Hitler (1889–1945). By 1932 they became the largest party in the *Reichstag*. Yet they remained

out of power until 1933 when President Paul von Hindenburg appointed Hitler chancellor of a right-wing coalition government. Through coercion and violation of the Weimar constitution (although the Nazis always made it appear that their actions were legal) Hitler established a dictatorship. Once he consolidated his power inside Germany, he proceeded upon a program of military expansion that ultimately resulted in World War II (1939–1945).

Without even firing a shot, Hitler succeeded in abrogating key provisions of the Treaty of Versailles. He annexed Austria in March 1938, the Sudetenland (the ethnically-German region of Czechoslovakia) in October 1938, and occupied the rest of Czechoslovakia in March 1939. In addition, he concluded an alliance with Italy in May 1939 (to which Japan acceded in 1940) and a non-aggression pact with the Union of Soviet Socialist Republics (USSR). War began with the German invasion of Poland on September 1, 1939. Two days later, Britain and France declared war against Germany. War continued for the next five and one half years. Although Germany succeeded in defeating France, conquering the Low Countries, much of Scandinavia, North Africa, the Balkans, and European Russia, Germany went down to total defeat in May 1945. Major turning points included the failure to subdue Britain in 1940, the failure of the invasion of the Soviet Union in 1941 (Hitler abrogated his non-aggression treaty), and the decision to declare war against the United States on December 11, 1941 after the Japanese attack against Pearl Harbor, Hawaii. By 1943 the tide had turned against Germany and its allies (Italy, Bulgaria, Finland, Hungary, Romania, and Japan). With Soviet forces only hundreds of yards from his Berlin bunker, Hitler committed suicide on April 30, 1945; Germany surrendered eight days later on May 7. Unlike the defeat in 1918, Germany in 1945 was destroyed. Its cities lay in ruins from round-the-clock Allied bombing attacks and its economy was disrupted. To this day, Germans refer to 1945 as *Stunde Null* (Zero Hour).

Worldwide over fifty-five million people died during the Second World War, including approximately twenty-seven million Soviet citizens and about six million Jews. The Nazis singled out the Jews along with Gypsies, homosexuals, the physically and mentally handicapped, and other "undesirables" for systematic extermination. This episode of mass annihilation, known as the Holocaust, resulted in the deaths of an estimated eleven million people.

Post-war Germany

In 1945 Germany ceased to exist as a national entity. It was occupied by the four victorious powers: the Soviets in the east, the British in the northwest, the Americans in the west and southeast, and the French in the southwest. There was no German government. All directives were supposed to come from a joint Allied Control Council, established on June 5, 1945, in the old capital, Berlin. Berlin itself, though located within the Soviet zone, was—like the country as a whole—

divided into four Sectors (American, British, French, and Soviet). With the advent of the Cold War (c. 1947–89), tensions between the Soviet Union and the Western allies (US, Britain, and France) led to the breakdown of any common administration and, in 1949, to the establishment of a democratic Germany in the west (Federal Republic of Germany, capital: Bonn) and a communist state in the east (German Democratic Republic [GDR or, in German, DDR], capital: East Berlin).

The divided city of Berlin became the focal point of many confrontations. From June 24, 1948, to May 12, 1949, West Berlin, an enclave in the Soviet Zone, was cut off by Soviet forces from all ground transportation, necessitating a massive airlift by the Western Allies to keep its population alive. As conditions improved in West Germany and stagnated or worsened in the East, Berlin offered an escape route for disgruntled Easterners (officially, leaving the GDR was not allowed). To stop this population drain, on August 13, 1961, GDR authorities sealed off their part of the city with a concrete wall and established a barbed wire and death strip zone at the borders between their country and West Germany (Federal Republic). With the fall of the Communist regime in East Germany in 1989, the wall was opened again (November 9, 1989), and on October 3, 1990, after complex international negotiations, the German Democratic Republic acceded to the Federal Republic. The event is now celebrated as a national holiday. Most Germans were pleased with the reunification. Those in the east hoped to enjoy not only greater personal freedom but also major improvements in their standard of living. After fifty years of separation, the welding together of the two sectors turned out to be more difficult than expected, with many misunderstandings, with mistrust, and continuing economic problems.

The original Federal Republic consisted of ten states (plus West Berlin, which had special legal status). After unification, this number was now sixteen. Berlin regained its pre-World War II position of national capital, replacing the "provisional capital" of Bonn. The constitution (called *Grundgesetz* or Basic Law) guarantees a free, democratic order for all German citizens. It was adopted for West Germany on May 23, 1949, and, since October 3, 1990, is the law of the land for the entire reunited country. After Germany's traumatic experience with the Nazi dictatorship (1933–45), the Basic Law is particularly strong in protecting civil rights. As a federated state, considerable power is left in the hands of local governments. Each *Land* (State) has its own constitution which, of course, must conform to the national document.

Political parties play a major role in Germany. The two largest are the *Christlich Demokratische Union* (CDU, or Christian Democratic Union), which—together with its Bavarian sister party *Christlich Soziale Union* (CSU, or Christian Social Union)—represents liberal-conservative points of view and the *Sozialdemokratische Partei Deutschlands* (SPD, or Social Democratic Party of Germany) which has its historical roots in the nineteenth century labor movement and now is a left-of-center party. There are also a number of smaller parties. The most important among them are the *Freie Demokratische Partei* (FDP, or Free Democratic Party) which emphasizes individual initiative and less government involvement and *Die Grünen* (The Green Party), which is the voice of environmentalists. In the states of former East Germany, the *Partei des Demokratischen Sozialismus* (PDS, or Party of Democratic Socialism) has gained votes as successor to the previous communist/socialist ruling party. To be represented in the Bundestag or in one of the state parliaments, a party must attain at least five percent of the vote. Since the two major parties rarely obtain a clear electoral majority, they usually form a coalition with one of the smaller groups. For example, the CDU/CSU, the governing party from 1982 to 1998, depended on the FDP as its coalition partner.

The Federal Republic of Germany, since its inception in 1949, has been a loyal partner of the western democracies. It is a member of the North Atlantic Treaty Organization (NATO) and is the most outspoken proponent of a United Europe. It has worked hard on its rehabilitation from a pariah state at the end of World War II, guilty of genocide and crimes of aggression, and it has established a particularly close relationship with the United States. Germans appreciate the help they received through the Marshall Plan (U.S. assistance to war-torn Europe) and U.S. protection during the days of the Cold War.

Modern Germany

Modern Germany is surrounded by nine neighbors: The Netherlands, Belgium, Luxembourg, and France in the west; Switzerland and Austria in the south; the Czech Republic and Poland in the east; and Denmark in the north. With an area of 137,826 square miles (356,970 square kilometers), present-day Germany is a little smaller than Montana. Its distance from the northernmost to the southernmost point measures about 550 miles (880 kilometers), from the westernmost to the easternmost point approximately 400 miles (640 kilometers). In the United States, such distances may mean little; in Germany they encompass considerable topographical, cultural, and even speech variations (dialects).

Moving from north to south, present-day Germany can be divided into three geographically distinct parts. The northernmost third of the country (known as the North German Plains) is low-lying and flat. It reaches from the coastal areas of the North Sea and Baltic Sea to the central German plateau.

Since the country rises from the flatlands in the north to the alpine regions in the south, most major rivers flow from south to north. Among these are the Rhine, the Weser, the Elbe, and the Oder (shared with Poland). Only the Danube flows from the South German plateau in an easterly direction toward Austria. These rivers, now connected by man-made canals, have been ancient trade routes and to this day form an important transportation net not only for Germany, but also for much of Europe. They make it possible, for example, for goods from the Czech Republic or Poland to be shipped by barge to ports in Germany, the Netherlands, or France, or goods from France to be transported to eastern Europe, to Austria, Hungary, and even to the Black Sea.

Until the early nineteenth century, the German area was predominantly agricultural. While there is still a viable agricultural base, German industrialization and the common market of Europe have greatly reduced agricultural production.

As of the end of the twentieth century, Germany was primarily an urban, industrialized country. Its cities (the biggest among them are Berlin with 3,477,900; Hamburg with 1,703,800; Munich with 1,251,100; and Cologne with 963,300 inhabitants) have become centers of manufacturing and commerce. Though many of them were badly damaged by bombing and fighting during World War II, they have been rebuilt as modern metropolitan centers, connected by an excellent net of railroads, super-highways (the Autobahn), airports, and telecommunication. Germany has become the hub of Europe. Because its industrial production, while suffering a slump in the 1990s, ranks among the highest internationally, Germany has been called the "locomotive" of European economics. Since much of the raw material and the sources of energy have to be imported, however, Germany is heavily dependent on the fluctuations of the world market.

The language spoken is German, which is related to Dutch, English, and the Scandinavian languages. In addition to the standard official German (*Hochdeutsch*) used in writing, by the media, and in public discourse, there also exist regional dialects which sometimes may be difficult to understand for outsiders. The dialects spoken in the north, the low lying districts of Germany, are commonly referred to as Low German, (*Plattdeutsch*); those in the south are Upper German (*Oberdeutsch*).

Until recently, most Germans belonged at least pro forma to one of the two major Christian denominations. The north was largely Protestant, the south Catholic. This regional division was the result of historical developments growing out of the Protestant Reformation in the sixteenth century, which led to the establishment of State Churches While constitutionally church and state have been totally separated since 1919, the state still collects church taxes (about ten percent of the income tax). Since many Germans have become indifferent to their church and resent the tax, membership has dwindled. In 1993, about thirty-five percent of all Germans were Protestant and thirty-four percent Roman Catholic. There is a tiny Jewish community, consisting mainly of Jews from Russia who immigrated after World War II. The once thriving German Jewish community (which in 1933 had numbered 560,000, less than one percent of the country's population at that time) largely fell victim to the Holocaust. With the influx of Turk-ish and other Muslim workers and their families, the Islamic presence in Germany is growing.

In the fields of literature, art, and music, Germany has a long and rich tradition. The poet Johann Wolfgang von Goethe (1749–1832; especially well known for his drama *Faust*); the painter Albrecht Dürer (1471–1528); the composers Johann Sebastian Bach (1685–1750; most famous for his organ preludes, chorales, cantatas, oratorios and concerti), Georg Friedrich Händel (1685–1759; best known for his oratorio *Messiah*), and Johannes Brahms (1833–97; composer of the *German Requiem*) are classic representatives of German art. Many recent artists can be added to these names, among them the writers Thomas Mann (1875–1955), Nelly Sachs (1891–1970), and Heinrich Böll (1917–85), all recipients of the Nobel Prize in Literature.

Particularly notable are the German contributions to science and technology. Here are just a few examples: Wilhelm Conrad Röntgen (1845–1923), discovery of the x-rays (1895); Max Planck (1858–1947), formulation of the quantum theory (1900); Albert Einstein (1879–1955), theory of relativity (1905); Johannes Gutenberg (c. 1397–1468), invention of the printing press with movable type (c. 1450); Rudolf Diesel (1858–1913), invention of the diesel engine (1892); Gottlieb Daimler (1834–1900) and Carl Benz (1834–1900), the inventors of the automobile (1884–90); Wernher von Braun (1912–77) and his team, perfection of the long-range liquid fuel rocket (1944).

Timeline

Premodern Regional Chronology

c. 1000 B.C. The Germanic people arrive in Central Europe

The date of the arrival of the Germanic people in Central Europe is largely conjecture. Individual groups seem to have moved into the area from the northeast at various times, replacing or mixing with indigenous Celtic people. They are independent tribes with no sense of common nationhood.

c. 330 B.C. Phytheas of Massilia mentions Germanic people

The Greek geographer Pytheas (356–323 B.C.) travels from Marseilles (Massilia) to Britain and the coast of the North Sea. His reports offer the first written record of the existence of Germanic people in northern Europe. While his own works do not survive, we know about them through citations in the works of some of his contemporaries.

51 B.C.   Julius Caesar's armies reach the Rhine

Between 58 and 51 B.C., the Roman emperor Julius Caesar (100–44 B.C.) conquers Gaul (modern France) and moves his armies east to the Rhine River, subduing the Germanic tribes living there. In his book *De bello Gallico* (Gallic War) he describes them as fierce and a threat to the Roman empire. The territories west of the Rhine and south of the Danube become Roman colonies, developing a flourishing Roman-Germanic culture.

c. 50 B.C.–c. A.D. 50   Roman camps become thriving cities

In their new colonies between the Rhine and Danube, the Romans establish army posts and trading centers which soon become important towns in which Romans and natives freely mingle. One of the most prominent is the city of Cologne (Ger. Köln, Lat.Colonia Agrippinensis; the birthplace of Agrippina, wife of Roman emperor Claudius). Here remains of Roman walls, streets, mosaics, aqueducts and all kinds of artifacts have been excavated. Today, a museum in the center of the city offers a rich impression of the developing Roman-Germanic culture. Other cities are Koblenz (Lat.Confluentis) at the confluence of Rhine and Moselle, Bonn (Lat.Bonna), Trier (Lat.*Augusta Treverorum*), Regensburg (Lat.*Castra Regina*), Vienna (Ger.Wien, Lat.Vindobona). In these colonies west of the Rhine and south of the Danube, the Romans introduce civil government and the rule of law. The construction of stone buildings, paved roads, and bridges create a civilized environment. Romans also establish the first vineyards and orchards along the Rhine and Moselle.

A.D. 9   Germanic tribe defeats Roman legions

Arminius (Hermann) (c.16 B.C.–A.D 21), a Germanic tribal prince trained by the Roman army, ambushes and annihilates three Roman legions who have crossed the Rhine on their way to the Elbe River. The battle takes place in a wilderness area not far from what is today the city of Osnabrück. Legend later situates the battle in the Teutoburg Forest nearby and elevates Arminius (Hermann the German) to the rank of national hero and liberator. While it is true that there is no further Roman penetration into Germanic lands east of the Rhine, Arminius thinks primarily in tribal and not in national terms. In 1875, German nationalists erect a huge Hermann's Monument in the Teutoburg Forest. German immigrants to the United States build a replica in New Ulm, Minnesota (1888).

98   Tacitus writes book on Germanic tribes

The Roman historian Publius Cornelius Tacitus (c. 45–c. 116) publishes a monograph *De origine et situ Germanorum* (About the Origin and the Location of the Germanic People). While it is not always reliable (since Tacitus has never visited the area, much is based on hearsay or is polemic, contrasting Germanic simplicity with Roman decadence), the book offers a glimpse of life among non-Roman occupied Germanic tribes. Tacitus discusses their social patterns, their customs and mores; he praises their courage in battle, their marital faithfulness, their hospitality. Women are responsible for work in the house and field, while men hunt, fight, and gamble. They are often addicted to alcohol, and Tacitus suggests that the Germanic tribes can be more easily defeated with wine than with weapons.

c. 300–c. 600   Germanic tribes are on the move

At the end of the second century A.D., the Roman empire begins to decline and eventually disintegrates, creating a power vacuum. Germanic tribes from the north and east penetrate the Roman borders on the Rhine and Danube, some moving deep into Italy and as far as Spain and North Africa. One tribe, the Vandals, sack the city of Rome in 455—a deed still remembered in the word "vandalism." The causes for this Great Migration are varied and some are still debated. The obvious ones are: 1) the attraction of Roman wealth and luxury for people who still live in a very primitive environment; 2) the need for food and land during periods of bad harvest and famine; and 3) escape from hostile invasions from the east (especially the Huns). What makes the Great Migration particularly complex is the fact that in some instances whole tribes begin to move, and in others, only the surplus population.

486   Beginning of the Frankish Kingdom

The Germanic tribe of the Franks, having spread from the middle and lower Rhine into Roman Gaul, becomes the first to form an organized state, combining Roman and Germanic traditions. Its king, Clovis, crowned in 486, rules over a realm that reaches from the Pyrenees Mountains at the Spanish border to what is now central Germany. The country, with its mixed Germanic-Celtic-Roman population, becomes the nucleus for the development of modern Europe. The country name "France (Ger.Frankreich) and that of the German area of Franconia (Ger.Franken) are reminders of their Frankish origin.

496: December 25   Clovis is baptized

On Christmas Day, King Clovis, together with 3000 of his followers, is baptized by the Roman Catholic bishop in the cathedral of Rheims. With this, the Frankish kingdom becomes linked to the Roman church, beginning a close relationship that will dominate European culture and history up to modern times.

716   Boniface brings Christianity to pagan tribes

Bonifatius (Boniface; 673–754), a missionary from Wessex in England, arrives to convert the Germanic tribes in the Frankish heartlands, Hesse, Thuringia, and Frisia. Known as the

Apostle of the Germans, Boniface establishes monasteries and bishoprics, many at places that before have been centers of pagan cults. These monasteries, and those founded by other missionaries, become important centers of culture and education in the German area. Boniface himself is slain in 754 by pagan Frisians. He is buried in the cathedral of Fulda.

786 First official mention of a German language

According to Hermann Paul (*Deutsches Wörterbuch*), Latin sources describe the language of the eastern Frankish kingdom as *lingua theodisca*, i. e. "language of the people." *Theod* (people) is the root for modern German, *Deutsch*.

800: December 25 Frankish king Charles crowned Emperor

While attending mass in Rome on Christmas day 800, allegedly much to his surprise, Charles (Ger.Karl 747–814), king of the Franks, is crowned Roman Emperor by the pope. He becomes the first ruler of a Christian western and central Europe and is honored in history as Charles the Great (Ger:*Karl der Große*; French:Charlemagne).

768–814 The rule of Charlemagne

King Charles of the Franks (since 800 Emperor Charles, ruler of most of Christian Europe) governs a multi-cultural and multi-lingual realm, an early "United States of Europe." He secures the country's borders against threats from the east and west, establishes a rule of law, controls the collection of taxes, establishes schools, and supports the arts (in hope of reviving the glory of the Roman empire). He brings important advisers to his court in Aachen (Aix-la-Chapelle), among them the Anglo-Saxon theologian and educator Alcuin (730–804). Given its cultural impact, his reign is often referred to as the Carolingian Renaissance. During the Middle Ages, he becomes the source of many legends and is immortalized in poetry, stories, pictures and sculptures as the ideal Christian ruler. Since 1950, the German city of Aachen offers the coveted Karlspreis (Charlemagne-Prize) to recipients noted for their special contributions to the unification of Europe.

843 The Frankish Empire is divided

After the death of Charlemagne, the empire disintegrates, leading to numerous internal wars. At the treaty of Verdun in 843, the country is finally divided into an eastern kingdom (which will become modern Germany), a western kingdom (France), and an often contested area in the middle (Lorraine—named after its first ruler, Lothar [795–855]).

936 Otto I King of the East Frankish lands

Following in the tradition of Charlemagne, Otto of Saxony (912–73) is crowned King in the cathedral of Aachen. Because of his strong and successful rule, bringing stability to

the country and securing its eastern borders, historians often refer to him as Otto the Great.

962 Founding of the Holy Roman Empire of the German Nation

While in Rome protecting papal interests, Otto I is crowned Roman Emperor. Otto's realm, considered a Christian continuation of the Roman empire, is initially referred to as the Holy Roman Empire. Since a large portion of its subjects are German, it eventually becomes known as the Holy Roman Empire of the German Nation and as such exists until 1806. Through much of its history, it retains close links to the church and the papacy. Many church officials also serve the state, and the pope in Rome often has great influence on the selection of the Emperor and on his secular policies.

c. 1000–c. 1400 Medieval Germany

After the death of Otto I (973), successive emperors are often quite weak, leading to power struggles and the rise of regional rulers who act as semi-sovereigns in their lands. This divisiveness inhibits any development of a nation-state, making the Holy Roman Empire of the German Nation a loose confederation without a real national identity.

c. 1180–c. 1220 Courtly culture flourishes

During the twelfth century the courts of the nobles and the castles of the knights become centers of culture, strongly influenced by developments in France and Britain (Arthurian legends). A code of ethics requires courtly behavior (English "courteous"), in contrast to "boorish" (related to Ger."Bauer"-peasant) manners. German minstrels (often low-level knights) compose stylized poetry of unrequited love (Minnesang), addressed to beautiful aristocratic ladies, or write extensive epics, telling the tales of heroes such as Charlemagne and his warriors in the Rolandslied (Song of Roland), courageous Siegfried in the Nibelungenlied (Song of the Nibelungen), or tragic lovers, like *Tristan und Isolde*. This period marks the first golden age of what can be called German literature.

1000s Beginning of German urban culture

The Middle Ages are dominated by the major and minor courts in Germany, the extensive land holdings of the church, and an economy based on a system in which dependent peasants maintain the fields for their lords (who in turn are supposed to protect them)—a system known as feudalism. Towns develop relatively late. The earliest are mainly of Roman or Celtic origin. From the tenth century on, however, small towns evolve around market places, river crossings, castles, and monasteries. Some become flourishing trading centers with their own laws and armies, offering relative safety and freedom behind their walls. The town cathedral and later the town hall become symbols of a city's wealth and importance.

Cities thrive on commerce (much of it from pilgrims visiting cathedrals) and the highly developed skills of their craftsmen and artisans. A prime example for the quality of medieval German urban culture is the city of Nürnberg.

1100s German eastward expansion

Beginning in the twelfth century, but especially in the thirteenth, Germans move eastward into the open spaces beyond the Elbe River. This *Drang nach Osten* (Urge to move east) is motivated by a need for land for a growing population, by the desire to stabilize the eastern frontiers, and—in the crusading spirit of the time—by a missionary zeal to force the Christian faith on the indigenous pagan Slavs. The areas eventually colonized include what are today the *Länder* of Brandenburg and Mecklenburg-Vorpommern, portions of today's Poland, and, in the northeast, the territory then occupied by the Prussians, a tribe belonging to the Baltic people.

1273 Rudolf of Habsburg becomes German king

After years of anarchy, the German princes elect Rudolf of the house of Habsburg (1218–91) German king. With this begins the reign of the Habsburg dynasty, which rules the Holy Roman Empire until its end in 1806 and Austria until 1918.

1348–50 The plague is blamed on Jews

Since the days of Charlemagne, Jewish communities exist in the major trading centers of the empire, cities such as Worms, Speyer, Cologne, Mainz, Trier and Regensburg. Jews are considered outsiders, tolerated by the king and local authorities. They are not permitted to own land or belong to the guilds of craftsmen. Their livelihood is restricted to trade and banking. Living in special sections of town, the *Ghetto*, and forced to wear special, identifying clothes, they are often made the scapegoats for outside calamities, leading to looting, abuse, burning of synagogues, forced baptism, and even killing. When in 1348 the bubonic plague (also known as Black Death) spreads over much of Europe, killing large sections of the population, Jews are accused of having caused the disease by poisoning the wells which supply the Christians with drinking water. Rioting mobs destroy many Jewish communities and murder their inhabitants. Other Jews escape to Poland, where they are well received. This is the beginning of a rich Eastern European Jewish culture, lasting until the Holocaust during World War II (1939–45). The language spoken by these Jews is Yiddish, which preserves much of medieval German.

1348 First university established by German emperor

With increased interest in learning and the need to have well-trained people in law, medicine and theology, universities are organized in Italy, France and England. The first university on German soil is established by Emperor Charles IV in the city of Prague (capital of Bohemia [today: Czech Republic]), the emperor's residence. Subsequent universities are established at Vienna (1365), Heidelberg (1385), Cologne (1388), Erfurt (1392) and, in the fifteenth century and thereafter, at many additional places.

1358 Lübeck leads the Hanseatic League

To meet the threats of growing insecurity on the highways and at sea, and in order to strengthen their market positions, a number of towns along the North Sea and Baltic Sea coasts band together in an alliance of mutual support. This alliance is known as Hansa, or Hanseatic League. The city of Lübeck on the Baltic, famous for its successful trade with Scandinavia, assumes leadership. Its laws, language, and currency dominate the entire Baltic area, reaching as far as Russia, Finland and Sweden. For many years, the League controls much of the North German and North European trade, with posts in England, Norway, and many places on the continent. In addition to Lübeck, the cities of Hamburg and Bremen play major roles. As Länder (states) of the Federal Republic of Germany today, they reflect their proud heritage through their official names: "Hansestadt Hamburg" and "Hansestadt Bremen." The Hanseatic League loses its effectiveness in the fifteenth century and afterwards ceases to exist as a functioning organization.

1415 Friedrich of Hohenzollern becomes Elector

To establish law and order in Brandenburg, an eastern district of Germany, Friedrich of Hohenzollern is appointed herditary prince of the area. His position, and that of his heirs, is that of Kurfürst (Elector). Electors form the electoral college, which selects the German emperors. Friedrich's appointment marks the beginning of the Hohenzollern rule in Brandenburg, then Prussia, and finally Germany, ending in 1918.

c.1455 Johannes Gutenberg completes printing of Bible

To meet the growing interest in and need for books, Johannes Gutenberg (c.1397–1468) of Mainz invents a printing method, using movable metal types. This invention makes the dissemination of books and knowledge much faster and cheaper than the previous system of hand-copying (manuscripts). Gutenberg's bible (in Latin, with forty-two lines per page) is the first book printed with the new types. Forty-seven copies of this book are still in existence.

c.1500 Nürnberg becomes a center of art and culture

At the end of the Middle Ages, the city of Nürnberg, a flourishing commercial and crafts center, attracts many artists and scholars. Among them are Albrecht Dürer (1471–1528), known for his portraits of important personages, his oil paintings of religious scenes, and his magnificent etchings; the trader and geographer Martin Behaim (c.1459–1507), who designes the first globe (1491); and Hans Sachs (1494–1576), a most prolific shoemaker-turned poet, whose songs and short plays represent the height of the Meistersängerschool. Meistersänger, or master singers are townspeople who follow a rigidly prescribed set of poetic rules—the ultimate mastery of which makes them masters of the craft of poetry.

Modern History

1517: October 31   Martin Luther protests church abuse

Protesting the sale of indulgences (writs claiming to release the holder from divine punishment for committed sins), Martin Luther (1483–1546), an Augustinian monk and professor of theology at the University of Wittenberg, posts or distributes ninety-five theses—points of debate for the reform of the Catholic church. This event marks the beginning of the Protestant Reformation.

1519   Charles V elected German emperor

Charles (1500–58) of the Habsburg dynasty becomes the most powerful ruler in Europe. His realm includes not only the Holy Roman Empire of the German Nation but also Spain and the Spanish conquests in Central and South America.

1521   Luther refuses to recant at the Diet of Worms

During the imperial *diet* (parliament) meeting in the city of Worms, Martin Luther is asked to retract his remarks and publications criticizing the existing church and its teachings. He refuses, is declared a heretic, and on his return journey to Wittenberg is given refuge at Wartburg Castle.

1521–22   Luther translates the New Testament

While in hiding at the Wartburg, Luther, with the help of several scholar friends, translates the New Testament from the original Greek into German, using a standardized form of the language, easily understandable in all regions of the land. This translation marks the beginning of a written language that eventually becomes the standard German used in literature and educated speech.

1524–26   Peasants revolt in Germany

Groups of peasants in southern and central Germany rise up against their feudal lords, demanding greater economic, religious, and personal freedoms. The rebellion is brutally crushed by the authorities.

1555: September 25   Religious factions sign compromise

After years of sometimes bloody struggle between Catholics and Protestants, a peace treaty is signed in the city of Augsburg, specifying that the local ruler of each region may determine the denomination to which his territory should belong.

The result of this agreement is that to this day most areas of north and central Germany (and many of the independent city states) are Protestant, while much of the south and the west (along the Rhine) is Catholic. The emperor himself remains faithful to the Catholic church. The peace between the factions is a precarious one with each watching the other with great suspicion.

c. 1555–c. 1650   Counter Reformation strengthens Catholics

Under the leadership of the Jesuit order, the Catholic church rebounds in many areas of Germany. The Jesuits contribute much to the revival of Catholic education, stressing the humanities, mathematics, and the natural sciences. They also have considerable political influence at some of the princely courts.

1618–48   Germany embroiled in thirty years of war

The political tensions fostered by the religious division in the Empire come to an explosion when the Protestant nobles in Bohemia rebel against the emperor's appointment of a Catholic ruler for their land. Other Protestant princes come to the aid of the Bohemians, while Catholic princes rally around the emperor. A bloody war ensues, in which the Catholics initially have the upper hand. In 1630, Gustavus Adolphus (1611–32), King of Sweden, comes to the aid of the Protestants, and the emperor's forces retreat. Eventually, the war loses its religious aspect and becomes a political struggle for the hegemony in Europe. Protestant Sweden and Catholic France fight together against the Austrian-Spanish forces of the emperor. Germany becomes the battle ground for the first European war. Marauding bands of mercenaries devastate large areas of the country, torturing peasants, and pillaging towns. War and accompanying disease may have claimed approximately forty-five percent to fifty-nine percent of the rural, and twenty-five to thirty percent of the urban population.

1648: October 24   Peace of Westphalia

The peace treaty, signed in Münster and Osnabrück, with France and Sweden as co-signatories, confirms the confessional agreements reached in Augsburg in 1555. In addition, it guarantees almost complete sovereignty to the approximately 300 territorial rulers of Germany, including the right to form alliances and conduct their own foreign affairs. The Netherlands and Switzerland secede from the German Empire and become fully independent. The treaty remains the basis for the political structure of the Empire until 1806.

1669   Grimmelshausen publishes *Simplicissimus*

Even though the war has devastated much of Germany, the period is not without important cultural achievements. Some first-rate poets (Martin Opitz [1597–1636]; Paul Fleming

[1609–1640]; Andreas Gryphius [1616–1664], and others) produce poetry and plays, often alluding to the horrors of war. One of the most important works is Hans Jacob Christoph von Grimmelshausen's (1621–76) novel *Simplicissimus* (in German: *Der Abentheuerliche Simplicissimus Teutsch*), which tells the story of a young man growing up in the midst of the Thirty Years War.

1683   Turks lay siege to Vienna

A Turkish army, approaching from the Balkans, lays siege to Vienna and seriously threatens the city. Armies from Germany and Poland liberate the city and end the Turkish danger once and for all. Legend has it that fleeing Turks leave sacks of coffee behind, which give rise to the famous Vienna coffee houses.

1685   French refugees settle in Brandenburg

Frederick William (1620–88), the Great Elector, ruler of Brandenburg, transforms his poor and war-devastated country into a modern, well-run state, respected in all of Europe. He reforms the tax system, drains swamps to improve agriculture, builds canals and roads with the help of Dutch engineers. In 1685, he invites French protestants (Huguenots), persecuted in their homeland because of their faith, to settle in Brandenburg, and especially in Berlin, its capital city. These refugees represent a social elite and soon assume positions of leadership in the territory's commerce and government.

1700   Leibniz proposes organizing Academy of Science

While serving at the court of the Duke of Brunswick in Hanover, Gottfried Wilhelm von Leibniz (1646–1716), one of the great philosophers and mathematicians of the time, proposes the organization of a *Sozietät der Wissenschaften* (Academy of Sciences), to serve as a forum for scientific discourse. In addition to major contributions to philosophy and to mathematics (calculus), Leibniz also develops an early version of an adding machine and designs an international language based on algebraic formulas.

1701: January 18   Elector of Brandenburg becomes King in Prussia

The Great Elector's son, Elector Frederick III (1657–1713), aspires to greater honors than just being a regional prince. Ruling a flourishing land and loving luxury and pomp, he crowns himself King Frederick I in Prussia. The coronation takes place in the city of Königsberg (since 1945 part of Russia and called "Kaliningrad"). Prussia is the easternmost domain of the Hohenzollern family. It later becomes known as East Prussia (Ger.Ostpreussen). This event marks the beginning of the Prussian state, which exists until 1945.

1713–40   King Frederick William I—soldier king

Frederick William I (1688–1740) establishes Prussia's efficient army and governs his country as an absolute monarch, austere and often ruthless but also totally devoted to serving the state. At the time of his death, Prussia and Austria are the two leading powers in the German Empire.

1723   Bach becomes organist and choirmaster in Leipzig

Johann Sebastian Bach (1685–1750), one of Germany's greatest composers, is appointed to the prestigious post of organist and choirmaster at St. Thomas Church in Leipzig, a position he holds until his death. Prior to this appointment he serves as court musician at several small princely courts (Weimar, Köthen). Bach is especially known for his organ works, his approximately 200 church cantatas and oratorios (among them the monumental *St. Matthew Passion* [1729]), and his six *Brandenburg Concertos* (1721).

1740   New rulers in Austria and Prussia.

In 1740, two young rulers ascend to the throne: Frederick II (1712–86) of Prussia and Maria Theresa (1717–80) of Austria (after 1745 also Empress of the Holy Roman Empire). Both are highly competent, enlightened monarchs who introduce many reforms in their countries (revision of the judicial and penal codes, educational reforms, fairer taxation, reduction of privileges, limited religious tolerance, and others). Frederick II, later known as Frederick the Great, considers himself the "first servant of the state," subject to, and not above, the laws of the land.

1740–1763   Beginning of Austro-Prussian struggle

Historians often speak of German Dualism. By this they mean the struggle between the two strongest powers in the Empire, Austria and Prussia. Catholic Austria, home territory of the powerful Habsburgs, the traditional German emperors, and largely Protestant Prussia, an aggressive "upstart" under Hohenzollern leadership, vie for dominance within the Empire. The struggle begins in 1740, when Frederick II invades the Austrian province of Silesia and annexes it. There are altogether three Silesian Wars (1740–42; 1744–45; 1756–63), in which Prussia prevails, often just marginally.

1740–86   Life in Prussia under Frederick II

Even though the costly and destructive wars are a major burden for the population, the standard of living improves during the reign of Frederick II. The king introduces the potato as a major crop, thereby reducing frequent famines; he establishes state factories (porcelain; textiles); and he encourages the manufacture of silk (which fails, because silk worms do not survive in Prussia's climate). His goal, which he does not reach, is to make Prussia independent from foreign imports. As philosopher king, Frederick surrounds himself with some of the great thinkers of the time, among them the Frenchman

Voltaire (1694–1778). Berlin, the capital, and Potsdam, the king's preferred residence, attract intellectuals and artists from all over Germany. The educated urban middle class gains in prominence. Jews, until now excluded from Prussian cultural life, suffer fewer restrictions. The Jewish philosopher Moses Mendelssohn (1728–86) becomes an important part of the Berlin Enlightenment and contributes substantially to Jewish emancipation.

1770   Kant becomes Professor of Philosophy in Königsberg

Immanuel Kant (1724–1804), the greatest philosopher of the German Enlightenment, is well known for his works such as the *Kritik der reinen Vernunft* (Critique of Pure Reason; 1781), in which he discusses the possibilities and limits of human cognition. He also formulates the "categorical imperative," a moral guideline proposing that one should always behave in such a way that one's own actions could also be taken as "universal law," governing everybody's action. Kant defines Enlightenment as man's ability to shed his "immaturity" and his willingness to think and act as an independent, mature, and critical being (1784).

1774   Goethe publishes his *Werther*

Johann Wolfgang Goethe (1749–1832), a young lawyer from Frankfurt, composes a highly emotional epistolary novel (a story told in letters), in which he explores the wounded soul of a young man who suffers from unrequited love and finds his final peace in suicide. *Die Leiden des jungen Werthers* (The Sorrows of Young Werther) becomes an immediate success wordwide, makes the poet famous, and marks the beginning of a highly emotional literature (Storm and Stress, Romanticism).

1775   Goethe moves to Weimar

Goethe is called to the court of Karl August, Duke of Weimar, initially as educator and companion, but soon assuming important offices of state (in charge of the finance office, director of mines, director the the theater). It is also in Weimar, where Goethe lives until his death in 1832, that he writes his major works, his poems, his dramas (among them *Faust*), his scientific treatisies and his novels. Goethe attracts many other poets, philosophers, scholars, and artists to Weimar (among them the poet and history professor Friedrich Schiller [1759–1805]), making the little princely court world-famous as a center of German culture.

1779   Lessing completes his dramatic poem *Nathan der Weise*

Gotthold Ephraim Lessing (1729–1781), a leading German literary critic, poet, and philosopher, who has spent many years in Frederick the Great's Berlin and has become a friend of Moses Mendelssohn, writes the drama *Nathan der Weise* (Nathan the Wise) as a proclamation of humanism and religious tolerance. The protagonists are a Jew, a Christian, and a Muslim who in the end recognize that they have more in common than separates them.

1785   Prussia signs trade agreement with United States

The trade agreement between Prussia and the United States (though in the end not much comes of it), marks the first treaty between a European country and the newly independent U.S.

1789: July 14   Beginning of the French Revolution

The storming of the Bastille, a prison in Paris, by angry crowds marks the beginning of the French Revolution, whose slogan "Freedom, Equality, Brotherhood" soon resounds through all of Europe. There are small-scale uprisings along the German Rhineland. The old order of Europe is beginning to disintegrate.

1791: December 5   Mozart dies

Wolfgang Amadeus Mozart (1756–1791), child prodigy musician at the Salzburg court and thereafter highly prolific composer in Vienna, is the creator of forty-one symphonies, many concertos and some of the greatest operas ever to be composed. One of the world's greatest musical geniuses, he dies a pauper on December 5, 1791.

1792–1805   Coalition Wars against France

Austria, Prussia, and other German states become involved in a series of wars against the French Revolutionary Army. Their mercenary troops are thoroughly trounced by the highly motivated French citizen soldiers and their brilliant leader Napoleon Bonaparte (1769–1821, Emperor of France from 1804–1815). France annexes the left bank of the Rhine.

1792   Beethoven arrives in Vienna

Ludwig van Beethoven (1770–1827), who is born in Bonn, establishes his permanent residence in Vienna, the music capital of the world. A student of the composer Franz Joseph Haydn (1732–1809), he himself achieves world-renown as composer of nine symphonies, many piano sonatas as well as other orchestral works, and an opera (*Fidelio*).

c. 1800   Literary salons flourish in Berlin

The homes of several highly educated Jewish women become gathering places for philosophers, poets, academics, statesmen and other middle or upper class intellectual leaders. They come together to discuss literary, philosophical, and aesthetic issues. The most important "hostesses" of these salons are Henriette Hertz (1764–1847); Dorothea Schlegel (1763–1839), who is the oldest daughter of Moses Mendelssohn; and Rahel (Levin) Varnhagen (1771–1833).

1803: February 25   Territorial reorganizations

At a meeting of German princes in Regensburg, the territorial boundaries of many German principalities are revised. To compensate for the losses on the left bank of the Rhine, small independent states (city states, church-owned territories) are divided among the major principalities which substantially reduces the number of semi-sovereign units in the Empire.

1806: July   Federation of the Rhine

Napoleon encourages the formation of the Rheinbund, a federation of sixteen small German states along the Rhine. They become allies of France and place themselves under French protection. They also adopt many of the reforms brought by the French Revolution, including the adoption of civil rights and the abolition of serfdom.

1806: August 6   Holy Roman Empire is abolished

Emperor Francis II (*Kaiser Franz II*, 1768–1835), facing the reality of a disintegrating German Empire, resigns from the imperial throne and thereby officially dissolves the Holy Roman Empire of the German Nation. He continues to rule as Francis I, Emperor of Austria.

1806: October 9   Prussia declares war on France

Prussia, neutral since 1795, is finally drawn again into the conflict. Its army, which has not been reformed since the days of Frederick the Great, is quickly beaten by the French, and King Frederick William III (1770–1840) has to sue for peace.

1807: July 7–9   Peace of Tilsit

In the Peace of Tilsit between France and Prussia, Prussia loses all its land west of the Elbe River, has to pay extensive war reparations, and is obligated to support Napoleon's future campaigns.

1812   Grimm Brothers publish fairy tales

The brothers Jacob (1785–1863) and Wilhelm (1786–1859) Grimm gather fairy tales told by villagers and publish them with scholarly annotations under the title *Kinder- und Hausmärchen* (Fairy Tales for Children and the House). They have a tremendous influence on literature and the development of folklore.

1807–13   Prussian Reforms

After its crushing defeat in 1807, Prussia initiates a number of major reforms. The Prussian ministers Karl vom Stein (1757–1831 [in office: 1807–1808]) and Karl August von Hardenberg (1750–1822 [in office: 1810–22]) abolish the serfdom of peasants, give cities limited rights of self-governance, proclaim equality before the law, abolish the inhibiting powers of the guilds, and give Jews the right to citizenship. The military reforms are largely the work of Gerhard von Scharnhorst (1755–1813). He follows the French example in creating a People's Army with conscripted citizen soldiers. They replace the mercenaries of the old professional army. To serve in the military becomes the patriotic duty of every young Prussian male.

The reformers also recognize that a modern state requires an educated citizenry. To this end, schools and universities need to be reorganized. Wilhelm von Humboldt (1767–1835) restructures the Prussian school system and establishes the Gymnasium, an elite secondary school with heavy emphasis on the languages and ideas of ancient Greece and Rome. Humboldt's most important accomplishment is the founding of the Friedrich-Wilhelm-Universität in Berlin (1810) (since 1945: *Humboldt Universität*). This university, based on the concept of the integration of teaching, learning, and research, becomes the model for modern research universities all over the world.

1813–15   The Wars of Liberation

After Napoleon's defeat in Russia in 1812, Prussia, Austria, and other German states rise up against the French (1813). In alliance with England, they achieve final victory over Napoleon in the battle of Waterloo (June 18, 1815). The war generates much patriotic enthusiasm on the German side, among the civilian population as well as among the soldiers. It is "the people's war," fought not for a king, but for the liberation of the country, which many hope will be a united national state with constitutional rights for its citizens.

1814–15   Congress of Vienna

In October, 1814, the statesmen and rulers of Europe convene in Vienna to debate the reorganization of the continent in the post-Napoleonic period. The sometimes stormy deliberations conclude on June 9, 1815. The dream of many citizens, a united Germany, is not realized. While a German Confederation *Deutscher Bund* is organized to take the place of the former Holy Roman Empire, it is a loose association of some thirty-eight principalities, whose ambassadors hold occasional meetings in Frankfurt. There is no national government, no common citizenship, no common currency or law. Each member state remains autonomous and sovereign. The hope for a constitution is also not fulfilled. Basically, the major powers at the Congress are eager to return to the period before the Napoleonic Wars, and they see their main task in snuffing out subversive ideas left over from the French Revolution. Prussia and Austria remain the two most powerful states within the confederation.

1817: October 31   Jena university students rally for Germany

Celebrating the tricentennial of the beginning of the Reformation, nationalistic student groups from Jena rally at the Wartburg. They cheer the idea of a united Germany and burn the books of those they consider hostile.

1818   Hegel becomes Professor in Berlin

Georg Friedrich Wilhelm Hegel (1770–1831), one of the most influential German philosophers, accepts a chair at the university, where he teaches and shapes the intellectual environment until his death in 1831. Hegel's ideas of the dialectic development of history and his concept of the importance of the state influence political thought to the present time.

1819: March 23   August von Kotzebue assassinated

The prolific playwright, political observer, and sometime Russian diplomat August von Kotzebue (1761–1819) is an enemy of the German student movement and rumored to be a reactionary Russian spy. He is stabbed to death at his home in Mannheim by Karl Sand, a student from Jena.

1819: August 6–31   German governments clamp down

The governments of Prussia, Austria and several other German states use the assassination of Kotzebue as a pretext for rigorous measures against liberals (and those agitating for German unity). A series of decrees against the "demagogues" call for rigorous supervision of the universities and strict censorship of the press (Karlsbad Decrees). Many liberals and nationalists flee to France or England; others face lengthy incarceration.

1832: May 27–30   Rally for freedom and unity

In spite of the Karlsbad Decrees, between 20,000 to 30,000 German burghers rally at the Maxburg near Hambach to celebrate the idea of German unity and freedom.

1834: January 1   Beginning of a German common market

With German economy slowly shifting from agriculture to industry, the cumbersome customs regulations of the various principalities become a major stumbling block for expansion. The *Deutscher Zollverein* (German Customs Union) abolishes the collection of duties among its member states. Initiated by Prussia, the union at first has fewer than twenty members. Eventually, almost all German states join, except Austria. The Customs Union signifies the beginning of German economic unification.

1835: December 7   Beginning of rail traffic in Germany

The first steam-powered railway train links the neighboring towns of Nürnberg and Fürth. Two years later, a second line is opened between Dresden and Leipzig. By 1850, more than 4,000 miles of track are in operation, and by the end of the century, almost all parts of Germany are connected by rail.

1841   "Deutschland, Deutschland, über alles . . ."

August Heinrich Hoffmann von Fallersleben (1798–1874), a strong proponent of a free and united Germany, admonishes his countrymen to put Germany above their narrow territorial allegiances. This message occurs in the opening lines of his poem: "Germany, Germany, above everything . . .," composed in 1841. In 1919, his poem becomes the German national anthem. Today, only the third stanza is sung, which emphazises unity, justice, and freedom.

1844: June   Weaver rebellion

The import of cheap textiles from England and the introduction of milling machines seriously threaten the livelihood of the weavers in the Silesian mountains. Once engaged in a flourishing cottage industry, the weavers now have become destitute. In their despair, they attack local factories and destroy machines. The rebellion is crushed with brutal force.

1846–47   Social tensions

The potato blight and repeated poor harvests create serious problems among the rural population, leading to riots and unrest. Whole villages emigrate to the United States.

1847   Krupp works become armament factory

Under the leadership of Alfred Krupp (1812–87), the family-owned Krupp works in Essen, famous for their cast steel, expand to include the manufacture of military weapons. The quality of their products soon makes them the leaders in the armament industry—a position Krupp maintains to the end of World War II.

1848: March   Revolts shake German states

After the February revolution in Paris, revolutionary fever also reaches Germany. Revolts in Vienna, Berlin, and other major cities lead to a temporary victory of the revolutionary movement. Among of the main demands are the establishment of a constitutional monarchy in a united Germany and the guarantee of a free press. The "freedom fighters" come primarily from the ranks of the educated middle class.

1848: May 18   German National Assembly meets in Frankfurt

As a result of the initial success of the revolution, an elected assembly gathers at St. Paul's Church in Frankfurt to draft a German constitution and debate the forming of a united country. Much time is devoted to the question whether a united Germany should include Austria (which itself is a commonwealth of many nationalities, most of them not German). Deliberations continue for almost a year.

1848   Karl Marx publishes the *Communist Manifesto*

Karl Marx (1818–83), journalist, political philosopher, and economic theorist, together with his friend Friedrich Engels (1820–95), issues the *Manifest der kommunistischen Partei* (at a time when such a party does not really exist). The *Manifesto*, which subsequently becomes one of the basic documents of Marxism, offers a critical analysis of capitalism and proclaims the need for the proletariat to have control of the means of production.

1849: March 28   King of Prussia refuses German crown

By a narrow vote in Frankfurt, the assembly opts for the "small Germany" solution (Kleindeutschland), i.e. a German state without Austria. King William IV (1795–1861) of Prussia is offered the hereditary imperial crown, which he refuses. This event, for all practical purposes, marks the end of the Frankfurt Parliament, and shortly thereafter, of the revolution. Many participants are jailed; others flee to Switzerland, France, England, and to the United States. In Germany, the old order is largely restored.

1858   Rudolf Virchow publishes book on cellular pathology

Rudolf Virchow (1821–92), director of the Pathological Institute in Berlin, is the discoverer of cell pathology. His book, *Die Cellularpathologie*, represents a major breakthrough in the field of medicine. Virchow also plays a leading role in reforming the Berlin sanitation system.

1862   Johannes Brahms moves to Vienna

The composer and pianist Johannes Brahms (1833–97), after successful years in various German cities, arrives in Vienna, where he is soon recognized as one the great musicians of the century. His works (symphonies, concertos, choral pieces—among them the *Deutsches Requiem* [1868]) contribute to Vienna's claim as "musical capital of the world."

1862: October 8   Bismarck becomes prime minister of Prussia

King William I of Prussia appoints Otto von Bismarck (1815–98), a Prussian *Junker* (a member of the landed gentry) with a conservative family background, to serve as prime minister. Since parliament is unwilling to grant an increase in the military budget, Bismarck rules without parliamentary consent, creating a major constitutional crisis.

1863: May 23   Founding of German Workers Association

With the transition from an agricultural to an industrial society, serious social problems face the country. Industry attracts workers from the rural areas, uprooting them, and at the same time making them dependent upon the whims of the industrial marketplace. Long working hours (up to sixteen hours per day), exploitation, unemployment, child labor, and dismal living conditions are some of the issues confronting the factory workers. Their answer is the establishment of a Workers Association (*Allgemeiner Deutscher Arbeiterverein*) in Leipzig, the beginning of an organized German labor movement.

1864   War with Denmark

Largely at the instigation of Bismarck, Prussia and Austria join forces to annex the area of Schleswig-Holstein, claimed by the King of Denmark but inhabited by a largely German population.

1866   War between Prussia and Austria

Bismarck's goal is the elimination of Austria as a possible contender for leadership in a future united Germany. In a seven-week war the Prussian army, using the railroads for transportation and employing techniques learned in the American Civil War, decisively defeats the Austrians. A non-vindictive peace treaty excludes Austria from further involvement in German affairs but lays the basis for a friendly alliance between two sovereign states. This agreement also means the end of the German Confederation of 1815.

1869   Social Democratic Party founded

Under the leadership of August Bebel (1840–1913) and Wilhelm Liebknecht (1826–1900), German workers form a political party, the *Sozialdemokratische Arbeiterpartei* (Social Democratic Workers' Party), at a rally in Eisenach.

1869   Construction of Neuschwanstein begins

King Ludwig II of Bavaria (1845–86), a Romantic dreamer infatuated with the Middle Ages, begins construction of a magnificent castle in medieval style. It is one of several such castles constructed by him which bankrupt the state and lead to the king's removal from government on grounds of mental incompetence.

1870: July 19   France declares war on Prussia

The French government under Napoleon III feels threatened by the possibility that a distant relative of the Prussian king might ascend to the throne in Spain. They ask for guarantees from the king that this will not happen, and, with tensions mounting, France finally declares war on Prussia—much to Bismarck's delight. Most German states rally around Prussia, and together they quickly move against the enemy.

1870: September 2   Battle of Sedan

The rapidly advancing German armies win a decisive victory at Sedan. Emperor Napoleon III is taken prisoner. Until 1914, this day is celebrated as a national holiday in Germany. The

The national legislature meets in the Reichstag building in Berlin. (EPD Photos/CSU Archives)

German armies move on to Paris, which is encircled and forced to surrender on January 28, 1871.

1871: January 18   King William of Prussia becomes German Emperor

In the Hall of Mirrors of the Palace of Versailles, King William I of Prussia is proclaimed German Emperor. With this, the German states (with the exclusion of Austria) are finally united under one government. Prussia plays a dominant role in the new German Empire. Bismarck is appointed German chancellor (prime minister), and Berlin becomes the capital.

1871: May 10   Peace treaty is signed

The peace treaty, signed in Frankfurt, stipulates the annexation of Alsace and part of Lorraine by Germany. It also compels France to pay a war indemnity of five billion francs. This harsh treaty poisons German-French relations for years to come.

1871–1918   Government in the new Empire

The new Germany is a limited constitutional monarchy. The parliament (Ger:) is elected by universal male suffrage. The chancellor, appointed by the emperor, is the head of government and is responsible only to the emperor. The individual states within the empire are represented in the upper chamber, called Federal Council, (Ger. Bundesrat), in which Prussia has the decisive majority.

1871–73   Economic boom and bust

The sudden infusion of money from the French war indemnity and the new opportunities of a united Germany create an overheated economy, which provides considerable wealth for the entrepreneurial class but often impoverishes laborers, small craftsmen, and family businesses. German historians call this period Gründerjahre, i.e. Founder Years.

1875 German socialists unite

At a congress in Gotha, the *Allgemeine deutsche Arbeiterverein* and the *Sozialdemokratische Arbeiterpartei* join forces as the Socialist Workers Party of Germany (Sozialistische Arbeiterpartei Deutschlands, or SAPD).

1876 First Bayreuth Festival

Richard Wagner (1813–83), famous composer of many operas, is greatly admired by King Ludwig II of Bavaria. Wagner's themes of Germanic sagas (Götterdämmerung) and medieval legends (Parsifal) strike a responsive cord in the king's heart. They are also enthusiastically received by many middle-class Germans to whom these operas are an escape from the crass materialism of modern times. To offer Wagner a center for his operatic productions, the king supports the construction of a vast festival hall in Bayreuth. After 1876, the Bayreuth Festivals are annual events of international importance.

1878–90 Laws against socialists

Bismarck sees the socialist movement as a threat to the established order of the state. After two assassination attempts on the emperor (actually not by socialists), he passes a series of decrees, drasticly limiting the activities of the Socialist Party. These laws remain in force until 1890.

1881 First electric streetcars in Berlin

Werner von Siemens (1816–92), leader of the German electro-industry, inventor of the dynamo machine (1866), and much involved in establishing an international telegraph net, constructs the first working electric streetcar for service in a Berlin suburb. Soon electric streetcars replace horsedrawn carriages.

1882 Robert Koch discovers tubercule bacillus

Tuberculosis is one of the major killers in nineteenth century Germany. Malnutrition and inadequate housing contribute to the prevalence of the disease. When Robert Koch isolates the tubercule bacillus, a major step is taken toward the treatment of tuberculosis.

1883–89 Beginning of the social safety net

To meet the acute needs of the working people and to woo them away from the promises of the socialists, the Bismarck government institutes a number of important social security measures: Universal Health Insurance (1883); Accident Insurance (1884); and Old Age and Disability Insurance (1889).

1883–85 Nietzsche writes Zarathustra

The philosopher Friedrich Nietzsche (1844–1900) is one of the major critics of his time. He questions traditional morality and observes cultural hypocrisy as well as the mediocrity of his contemporaries. He espouses nihilistic ideas ("God is dead"). In his treatise *Also sprach Zarathustra* (Thus Spake Zarathustra) he proclaims the coming of a new man, the Übermensch, a human being beyond good and evil.

1884 Germany becomes a colonial power

Bismarck is not interested in colonies, but the pressure of German business and of German nationalists who want to see their country an equal to Britain and other colonial powers, forces him to agree to the acquisition of territory in Africa. Later additional territories are added in the South Sea Islands and even in China (Kiaochow), which serves as a German naval base. German colonial domain yields little and is terminated at the end of World War I (1919).

1885–86 Invention of the automobile

In 1885, Carl Benz (1844–1929) constructs the first (three-wheel) gasoline powered motorcar. In 1886, Gottlieb Daimler (1834–1900) introduces the first four-wheel automobile.

1888 The year of the three emperors

Emperor William I dies on March 9, 1888. His son, Emperor Frederick III (1831–88),who in 1858 has married a daughter of Queen Victoria of Britain, supports a more liberal government, based on the British system. He dies on June 15 of throat cancer. He is succeeded by his son, Emperor William II (1859–1941), who serves as German Emperor from 1888–1918, i.e. to the end of the Empire.

1889 First university preparatory school for women

Helene Lange (1848–1930), a schoolteacher and leader of the German women's movement, establishes in Berlin the first "Gymnasium" that prepares women for entry to the university.

1889 The Berlin stage *Freie Bühne* creates theater scandals

Young German dramatists, influenced by French and Scandinavian writers, produce plays that are starkly naturalistic, depicting social problems (alcoholism, poverty) and disease (tuberculosis, mental illness), up to then taboo subjects for the stage. Their playhouse is the *Freie Bühne* (Free Stage). Some of their productions result in riots. The most talented among the playwrights is Gerhart Hauptmann (1862–1946), who wins the Nobel Prize in literature in 1912.

A dragon is maneuvered from storage to be positioned for a 1951 performance of the Wagnerian opera, *Siegfried*, at the Bayreuth Festival, founded by Richard Wagner in 1876. (EPD Photos/CSU Archives)

1890: March 20   Bismarck dismissed

Young Emperor William II, for personal and political reasons, dismisses Bismarck as chancellor of the German Empire. His departure has dire consequences for German foreign policy.

1891   Socialist party changes name

At a rally in Erfurt, the Socialist Workers Party of Germany changes its name to Social Democratic Party of Germany, (*Sozialdemokratische Partei Deutschlands* or SPD) and remains a major player in German politics to the present (making it the oldest continuing political party in Germany).

1891   Worker's Protection Laws

The Reichstag passes legislation guaranteeing workers more free time on Sundays and greater wage protection.

1895   Röntgen discovers x-rays

Wilhelm Conrad Röntgen (1845–1923), a physicist, discovers penetrating rays, which he calls "x-rays." They soon find broad application, especially in medicine.

1898   Beginning of German naval expansion

To protect its colonies and to show the flag as a super-power, Germany embarks on an ambitious naval building program. Under the leadership of Admiral Alfred von Tirpitz (1849–1930), and with the support of heavy industry (Krupp) and nationalistic circles, Germany invests large sums in the construction of a modern battle fleet, thereby straining its relations with Britain.

1900: January 1   Common Civil Law for all of Germany

The common Civil Law code (*Bürgerliches Gesetzbuch*) becomes the law of the land and, with revisions, is still in force today.

1900   Women admitted to German universities

While women are admitted to most German universities, in Prussia they wait until 1908. Many German women, seeking a university degree, study in Switzerland, where women have been enrolled since 1840.

1900: July 2   Flight of the first Zeppelin

Count (Ger.:*Graf*) Ferdinand von Zeppelin (1838–1917), a retired army officer (who in 1863 has also served with the

Union Army during the American Civil War), designs a dirigible, rigid airship, which, after its successful test flights, becomes the prototype for 129 future "Zeppelins."

1905 German painters break with tradition

A group of German painters (among them Ernst Ludwig Kirchner [1880–1938] and Karl Schmidt-Rottluff [1884–1976]) organize in Dresden *Die Brücke* (The Bridge), an association of artists who are exponents of a new style of art, Expressionism.

1913 William II celebrates his twenty-fifth anniversary

After twenty-five years as Emperor of Germany, William II is pleased with the state of affairs. Though he has committed many blunders and is considered unstable by the international community, his country has become a European super-power, with a strong military and a booming industry. The standard of living has greatly improved, and a social safety net protects the working class. However, major social problems still exist, and the class structure continues to create friction. The privileged class of the nobility—though largely impoverished—still has considerable political clout.

1914: June 28 Austrian Archduke assassinated

While visiting the city of Sarajevo in Bosnia-Herzegovina, Francis Ferdinand (1863–1914), Archduke of Austria (crown prince), and his wife Sophie (1868–1914), are shot to death by a Serbian nationalist, Gavrilo Princip (1894–1918).

1914: August 1–1918: November 11 World War I

The tensions created by the assassination of the archduke and a general power play among the major European countries lead to the outbreak of hostilities, which soon engulf not only Europe, but much of the world. Germany, Austria-Hungary, Bulgaria and Turkey fight against Russia, Britain, France, Italy, Japan, and—after 1917—also the United States.

Germany has considerable initial success, with its army reaching the vicinity of Paris. Strong resistance brings their advance to a halt, stabilizing the front, and leading to years of fierce trench warfare and battles of attrition, causing mass slaughter on both sides without much territorial gain. The introduction of modern weaponry (poison gas, heavy artillery, airplanes, submarines) contribute to the horror of this war (total killed, about 10 million; Germans about 1.8 million).

The blockade of Germany by allied naval forces cuts the country off from all imports, leading by 1917 to serious food shortages. Malnutrition among the civilian population becomes a serious health problem. With men drafted into the army, many women enter the workforce in fields previously reserved for males.

1917: December 15 Russians sign armistice

Russia, wracked by revolution and badly mauled by German and Austrian forces, requests the cessation of hostilities at its fronts.

1918: March 3 Peace of Brest-Litovsk

Russia signs a devastating peace treaty, dictated by Germany and Austria-Hungary.

1918: November 9 Germany becomes Republic

With starvation and unrest at home and with the discipline in the armed forces disintegrating, Germany sues for an armistice. Emperor William II abdicates and flees to neutral Holland. Friedrich Ebert (1871–1925), a Social Democrat, is appointed chancellor. Germany becomes a republic.

1918: November 11 Germany signs armistice

Germany signs an armistice agreement with the allied forces, ending four years of fighting. The allied food blockade continues until the signing of the peace treaty.

1919: January Unrest in Germany

There is widespread unrest in major cities of Germany, with street-fighting among supporters of the new government and radicals of the extreme right and left. Two prominent leaders of the left, Rosa Luxemburg (1870–1919) and Karl Liebknecht (1871–1919) are brutally murdered by a right-wing group.

1919: January 18 Germans vote for national assembly

By universal suffrage, including women, Germans elect an assembly, whose deputies draft a constitution for the new republic.

1919: February 6 Constitutional assembly begins its work

The constitutional assembly meets in Weimar, since the days of Goethe a place identified with the best in German culture. The constitution offers guarantees of civil rights and justice and outlines a truly democratic system of government. Historians refer to it as the Weimar Constitution and to the republic as the Weimar Republic. Friedrich Ebert becomes Germany's first president.

1919: June 28 Peace treaty of Versailles

The German government is forced to sign a harsh peace treaty, later referred to by demagogues as the "Versailles Dictation." Germany loses Alsace and Lorraine, portions of Schleswig-Holstein, its province of West Prussia (leaving East Prussia as territory separated from Germany by the "Polish Corridor"), parts of Upper Silesia, and all its colonies. The German armed forces are reduced to 100,000 men, and heavy

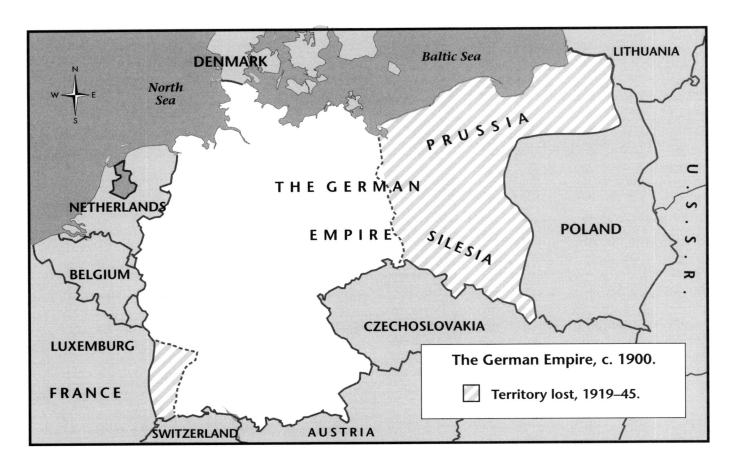

The German Empire, c. 1900.

▨ Territory lost, 1919–45.

weapons are not permitted. In addition, Germany has to pay substantial reparations. What angers many Germans most, however, is a clause in which Germany must admit having been the guilty party in starting the war.

1920–21 Unrest continues in Germany

After the signing of the peace treaty, the young Republic faces a number of serious challenges from right-wing and Communist radicals. The latter want to make Germany a Soviet Republic; the former hope to restore the old order. Many Germans become sceptical of the new democracy and its ability to cope with Germany's problems. A tendency develops to blame the Republic for conditions caused by predecessor governments and by the lost war.

1922: June 24 German foreign minister assassinated

Foreign Minister Walter Rathenau (1867–1922), a leading German industrialist, writer and politician of Jewish descent, helps shape the immediate post-war German foreign policy and is especially successful in establishing contacts with Russia. He becomes the target of anti-Semitic and anti-government propaganda and is murdered on his way to the Foreign Office.

1923: January 11 Belgian and French troops occupy the Ruhr area

To enforce reparation payments, French and Belgian units occupy the industrial center of Germany in the Ruhr area, creating a great deal of resentment and leading to a passive resistance movement.

1923: October Hyperinflation

The German economy, weakened by war debts, reparation payments, and large amounts of paper money in circulation, suffers from run-away inflation. Personal savings are all but wiped out, causing much resentment. On October 13, a currency reform stabilizes the currency and ends the hyperinflation.

1923: November 9 Right-wing riot in Munich

Adolf Hitler (1889–1945), leader of the fledgling National Socialist Party (*Nationalsozialistische Deutsche Arbeiterpartei* [NSDAP] [Nazi Party]), an extremist right-wing, anti-Semitic organization, attempts unsuccessfully to topple the Bavarian, and with it, the German government. A number of people are killed; Hitler himself is briefly jailed.

1924: August 16   German reparations regularized

An international agreement, the Dawes Plan, regularizes German reparation payments and thereby further stabilizes the German economy.

1925   Bauhauscampus established in Dessau

The Bauhaus, founded in 1919 in Weimar by Walter Gropius (1883–1969), is an avant-garde school combining the study and production of architecture, interior design, arts and crafts, and other art forms connected with modern functionalism. It achieves international acclaim and has influenced modern architecture and interior design in Europe and the United States.

1925: April 26   Hindenburg becomes president

Though democracy is now functioning, many Germans look with nostalgia to the past. They elect Paul von Hindenburg (1847–1934), a former field-marshall and war hero, as president of the republic, because he symbolizes the old order.

1926: September 8   Germany joins League of Nations

Gustav Stresemann (1878–1929; foreign minister from 1923–1929) pursues a policy of reconciliation. He succeeds in restoring international confidence in Germany, leading to its admission to the League of Nations.

1927: June 16   Unemployment compensation introduced

With the passage of a law covering unemployment compensation and job placement, the social safety net is further improved.

1928   Bertolt Brecht produces Dreigroschenoper

Germany, and especially Berlin, become a center of avant-garde theater, literature, and art. The Berlin Cabarets, with their jazz music and political satires, are world famous. Bertolt Brecht (1898–1956), together with the composer Kurt Weill (1900–50), shock the theater world with their Dreigroschenoper (Threepenny Opera), an irreverent parody of grand operas, full of social criticism.

1929   Thomas Mann receives Nobel Prize

The author Thomas Mann (1875–1955), one of the leading German novelists, is honored with the Nobel Prize in Literature. His novels, such as *Die Buddenbrooks* (1900)—tracing the history of a family through several generations, or *Der Zauberberg* (The Magic Mountain; 1924), offering a critique of the time just before World War I, are part of the canon of world literature and have been translated into many languages.

1930: June 30   French troops leave Rhineland

The last contingents of the French occupation forces are withdrawn. The Rhine area remains a demilitarized zone.

1931: July 13   German banks become insolvent

As a result of the world economic crisis (Black Friday on Wall Street, October 24, 1929), the German economy suffers serious setbacks, leading to the closing of many financial institutions and panic among the population. Workers are laid off, and whole factories shut down.

1932: February   Massive unemployment

About 6.1 million Germans are unemployed, almost ten percent of the total population.

1932–33   Political turmoil

With severe unemployment and state finances in ruin, Germans lose all confidence in their government. Chancellors and their cabinets change every few months, and no political party has a clear majority. The German political system suffers from too many splinter parties, making governing practically impossible. As the political process deteriorates, the extremists gain in power. Among them are on the right the National Socialists under Hitler, and on the left the communists. Hitler's party promotes a strong, authoritarian, and highly nationalistic state that will break "the fetters of the treaty of Versailles." He blames "international Jewery" for Germany's troubles and vows to eliminate the Jews as an "inferior race." He sees in capitalism and communism a "Jewish plot." His ideology is based on a crude social Darwinism that considers the Nordic race superior to all others. The communists, on the other hand, want to establish a kind of "Soviet" Germany. They spend much of their time fighting the Social Democrats rather than the Nazis, their most serious opponents. All radical parties have in common the desire to bring down the existing government and with it the Weimar Republic.

1933: January 30   Hitler becomes chancellor

After yet another political crisis, President Hindenburg asks Adolf Hitler to form a coalition government of conservatives and National Socialists.

1933: March 24   Weimar Constitution suspended

Under pressure from Hitler, the German parliament passes a law "for the alleviation of the suffering of people and country," suspending the Constitution and the rights it guarantees. Political parties are dissolved, except for Hitler's own. Germany becomes a totalitarian one-party state, where any opposition is prosecuted as a criminal act.

With this begins a regime of terror, in which real or imagined dissidents (communists, socialists, members of the clergy, etc.) are placed in concentration camps (Dachau,

Oranienburg, later many more), where they are tortured, beaten, and often killed.

1933: October 14   Germany leaves League of Nations

As an act of defiance toward the international community, Germany withdraws from the disarmament conference and from the League of Nations.

1934: August 2   Hindenburg dead

With the death of Germany's president von Hindenburg, Hitler unites the offices of president and chancellor in his own person as *Führer und Reichskanzler* (Leader and Chancellor). He is commonly referred to as *der Führer*.

1935: January 13   Saar Territory returns to Germany

In 1919, the Saar area is placed under League of Nations trusteeship, with economic rights going to France. A plebiscite is to determine the area's ultimate fate. On January 13, the people of the Saar overwhelmingly vote for a return to Germany.

1935: March 16   Reinstatement of military draft

In clear defiance of the Versailles Treaty, Hitler orders the reinstatement of universal military service. German industry begins massive rearmament of Germany (thereby ending unemployment).

1935: September 15   Nürnberg Laws

A series of laws, "for the protection of German blood and honor," exclude Jews from German citizenship, prohibit the marriage between Jews and members of the "German race," define what constitutes a "racial Jew," and go into detail about the degree of mongrelization among offsprings of existing mixed marriages.

1936   Olympic Games in Germany

The Winter Olympics in the Bavarian Alps and the Summer Olympics in Berlin are a major publicity coup for the Nazi regime. Foreign visitors are shown a youthful, exhuberant, spruced-up country that seems to belie the horror stories circulating around the world.

1936: March 7   German army re-occupies Rhineland

Again in defiance of the Versailles Treaty, German military forces cross the Rhine to re-occupy the demilitarized zone.

1938: March 13   Annexation of Austria

With support of the Austrian Nazi Party, Hitler destabilizes the Austrian government. To "restore order," he then sends in his troops—who receive a hero's welcome. With the annexation of Austria, Germany is now Grossdeutschland (Greater Germany).

1938: October 29   Appeasement in Munich

Hitler proclaims that he has to liberate the "much oppressed" ethnic German population in the Sudetenland, the mountain area at the border between Germany and Czechoslovakia. To prevent a possible war, the leaders of Britain, France, and Italy come to Munich to find a compromise. In the end, they give in to his demands, and the German army occupies the Czech border areas. Ironically, Hitler calls it his "last territorial demand," and British Prime Minister Chamberlain speaks of having secured "peace in our time."

1938: November 9   Violent action against Jews

On November 7, a distraught young Polish Jew kills a German embassy official in Paris. The Nazi regime uses this as pretext for the organized destruction of Jewish property. Nazi gangs destroy some 280 synagogues, more than 7,000 Jewish business establishments, herd about 30,000 Jews into concentration camps, and murder at least ninety-one. In addition, Jews collectively have to pay 1 billion marks for the damages incurred. After this, the remaining Jews desperately try to emigrate, but the immigration laws of most countries make this a slow, cumbersome, and usually futile endeavor.

World War II, 1939–45

1939–45   Nazi ideology of the "Pure German Race"

Hitler's belief in the superiority of the German race leads to the harsh treatment of Poles and Russians, who are considered inferior stock. Many are shipped to Germany to work as slave laborers. Particularly cruel is the persecution of Jews and Gypsies. The Jewish communities of Poland and of occupied Russia are destroyed. Jews in Germany and all occupied areas have to wear a distinguishing yellow star (beginning September, 1941), and at a special conference in 1942 (Wannsee-Konferenz), the total elimination of all Jews in German-occupied Europe is planned (Hitler's "final solution" of the "Jewish problem"). In extermination camps like Auschwitz, approximately 6 million European Jews die in the gas chambers (Holocaust).

1939–41   Elimination of the mentally retarded

Following a theory that deformed, retarded, and mentally ill individuals are "unfit to live," Hitler initiates a Euthenasia Program, through which about 5,000 children and 100,000 adults are killed. Through church intervention, the program is stopped in 1941.

1939: March 15   Germany occupies the rest of Czechoslovakia

Although Hitler has guaranteed Czech independence at the Munich compromise, he now orders the army to occupy the rest of the country.

Adolf Hitler, followed by his chief ministers, arrives at the Kroll Opera House in Berlin to deliver a speech. (EPD Photos/CSU Archives).

1939: September 1   German army invades Poland

Ostensibly to help oppressed German ethnic groups in Poland and to punish Poland for alleged incursions on German soil, Hitler orders the army to attack. The Poles fight back, and, on September 3, Britain and France also declare war on Germany, thus beginning World War II, which lasts to May 8, 1945. Italy and Japan become Germany's allies—most other major countries eventually join the side of Germany's enemies.

1939   World War II—A summary of military events

The well-trained and well-equipped German army is initially successful. Poland is conquered in eighteen days. In 1940, Denmark, Norway, Holland, Belgium and France are defeated. A sustained bombing raid on Britain in preparation for a possible landing fails (Battle of Britain). In 1941, Yugoslavia and Greece fall to Germany, and German forces attack the Soviet Union, making big inroads in that vast country. In December, after Pearl Harbor, Germany declares war on the United States, in fulfillment of its treaty commitments with Japan. In 1942, German losses in the Soviet Union mount. In addition, German forces are needed in Africa to help the

weakening Italian army. In 1943, the German army suffers a crushing defeat at Stalingrad and begins its gradual retreat from the Soviet Union. The Nazi regime proclaims the "Total War," calling on everyone to participate in the war effort. In Africa, the remaining German forces surrender to the advancing British-American armies. The Allies invade Sicily and Italy. The Italians surrender, but the battle continues against German forces occupying the country. British and American bombers begin systematic raids on German cities and military installations. In 1944, the Soviet Army continues to advance. On June 6, allied forces land in Normandy and, by the end of the year, have liberated France and Belgium and crossed into German territory. In 1945, Soviet armies advance from the east and allied forces from the west. Berlin falls to the Red Army on May 2, after Hitler has committed suicide in his bunker on April 30. On May 8, the German army surrenders unconditionally. It is estimated that about 5.25 million Germans are killed during the war.

1944   Attempts to kill Hitler

Several attempts to remove Hitler end with the execution of those involved. Most noteworthy are: the *Weisse Rose* (White

Rose) movement of a group of Munich students—among them Hans and Sophie Scholl (1942–43) and the assassination attempt of July 20, 1944, involving leading military officers and other prominent persons.

1945: June 5   Victorious powers govern Germany

The Allied Control Council (U.S., Britain, Soviet Union, and subsequently France) assumes all government functions in Germany. The country is dividied into four occupation zones. Berlin, divided into four sectors, is the seat of the Control Council.

1945: July 17–August 2   Potsdam Conference

Allied leaders deliberate about the future of Germany. Pending a final peace settlement, the territories east of the rivers Oder and Neisse are ceded to the Soviet Union and Poland. Approximately 6.75 million German are expelled, under often inhuman conditions.

1945: Summer and fall   Political parties reappear

With permission of the occupation authorities, several political parties emerge. Among them are the Christian Democratic Union (CDU) and the Social Democrats (SPD—the only party with a tradition going back to 1863). In the Soviet Zone, all parties unite in an Anti-Fascist Block under the leadership of the communists.

1945: Winter–spring , 1946   Starvation in Europe and Germany

The economic conditions in devastated Europe rapidly deteriorate. In Germany, an extreme shortage of fuel and food causes people to barter their valuables to obtain the essentials to survive (Black Market). German money is practically worthless and is often replaced by American cigarettes as trading currency.

1946: September 30–October 1   Judgment at Nurenberg

An international war crimes tribunal, established by the victorious powers, sentences twelve top political and military leaders of the Nazi regime to death and others to various prison terms as punishment for their crimes against humanity.

1948: March 20   Soviets leave Allied Control Council

With increased tensions between the Soviet Union and the Western allies (Iron Curtain; Beginning of the Cold War), the Soviet representative leaves the Allied Control Council, thereby terminating any further collaboration between East and West.

1948: April   Congress passes Economic Cooperation Act

A plan for European recovery through U.S. help, proposed by U.S. Secretary of State George Marshall, also includes the Western zones of Germany. The Marshall Plan (1948–52), eventually contributes 1.4 billion dollars to support the reconstruction of West Germany.

1948: May 31   Western zones united

After the economic and administrative fusion of the British and American zones, the Western allies now also add the French zone.

1948: June 20   Currency reform

The introduction of a new currency, the Deutschmark (German Mark or DM), in the western zones, signals the beginning of German economic recovery. Since the new currency is not accepted in the Soviet Zone, it contributes to the final split between East and West.

1948: June 24–May 12 , 1949   Berlin Blockade

The Soviets use the introduction of a new currency in the west as basis for the demand to integrate West Berlin in the Soviet Zone. When the Western allies refuse, Soviet forces block all land access to the city in an attempt to starve it into submission. An enormous airlift by American, British, and French planes assures the survival of West Berlin until the blockade is lifted.

1949: May 8   The Federal Republic of Germany established

Representatives from the various regions in western Germany draft a provisional constitution, called *Grundgesetz* (Basic Law), a document guaranteeing a free, democratic order for a newly constituted federal Germany. The new Germany is decentralized, with extensive powers vested in the ten constituent states. The legislative structure of the Federal Republic of Germany consists of a lower house (Bundestag), elected by universal suffrage, and an upper house (Bundesrat), whose deputies are delegated by the state governments. The major legislative power rests with the Bundestag. The federal chancellor (Bundeskanzler), is the political head of the government, and the federal president (Bundespräsident) serves as official representative of the Republic and performs a number of ceremonial duties. He is not a policy maker but often has considerable moral clout. Konrad Adenauer (1876–1967) becomes the first chancellor, Theodor Heuss (1884–1963) the first president. Bonn is selected as the country's provisional capital.

1949: October 7   The German Democratic Republic established

The German Democratic Republic (GDR) (*Deutsche Demokratische Republik* or DDR) evolves as a centralized, socialist state within the boundaries of the Soviet Zone. Under Communist leadership (officially SED, i.e. *Sozialistische Einheitspartei Deutschlands* (Socialist Unity Party of Germany), the country becomes part of the Soviet block.

Industry, agriculture and many businesses are nationalized. East Berlin becomes the capital.

1949–end of 1950s   Economic miracle

Under the leadership of Ludwig Erhard (1897–1977), West Germany's economics minister (1949–63), and with substantial help from the Marshall Plan and a hard-working population, economic conditions in the Federal Republic rapidly improve. The German mark becomes one of the most stable currencies in the world. Erhard's "social marketpolicy" (capitalism with a social conscience) is the economic guideline governing German politics to the present. A generous benefits package and extensive vacations assure a high quality of life for every citizen.

1953: June 17   Political protests in the GDR

Workers in the GDR, unhappy about working conditions and discontent with the regime, openly riot in East Berlin and other cities. The protest is quelled with the help of Soviet tanks.

1955: May 5   West Germany joins NATO

Adenauer pursues a policy of reconciliation with France and of making Germany a partner of the western security alliances. Following the request of the United States, the Federal Republic establishes its own defense force which, however, is part of NATO (North Atlantic Treaty Organization).

1959   Günter Grass publishes *Die Blechtrommel*

With the publication of *Die Blechtrommel* (The Tin Drum) the author Günter Grass (b. 1927) offers a satirical-critical perspective of the Nazi era as seen through the eyes of a hero who deliberately stunts his growth and communicates his commentary by beating his tin drum.

1961: August 13   Construction of the Berlin Wall

Unhappy with conditions in East Germany and attracted by the economic boom in the west, many people escape from the GDR via the divided city of Berlin. (In July, 1961, alone, more than 30,000 flee to the west.) The East German government attempts to stop this manpower drain by constructing a concrete wall, separating East and West Berlin.

1966   Nelly Sachs receives Nobel Prize

Nelly Sachs, a German-Jewish poet who has fled to Sweden to escape the Nazis, is awarded the Nobel Prize in Literature.

1967   European Community established

Through a fusion of several international groups, West European countries, including the Federal Republic, form the European Community, as first step toward a broader European union.

1967–68   Student unrest in West Germany

Students, critical of West German materialism and consumerism and upset over the tendency of their elders to cover up their Nazi past, take to the streets and riot. Their rallies are often combined with angry protests against the U.S. and the war in Vietnam.

1970   Easing of tension

West Germany's chancellor Willy Brandt (1913–92; chancellor from 1969–74) pursues a policy of détente with the East. He visits Warsaw and apologizes for German atrocities during World War II. He also meets with the government leaders of East Germany to negotiate an improvement of relations. In 1971, he is awarded the Nobel Prize for Peace.

1972   Heinrich Böll receives Nobel Prize

The writer Heinrich Böll (1917–85), well-known for his novels and short stories describing life in post-war Germany, is honored with the Nobel Prize for Literature.

1972   Olympic Games in Munich

By hosting the Summer Olympics in Munich, the Federal Republic hopes to present to the world a new, democratic Germany, quite different from the Germany of 1936, when the Olympic Games are held in Berlin. The games are marred when Palestinian terrorists take Israeli athletes hostage and kill them.

1973   West and East Germany become members of the United Nations

1973   2.6 million foreign workers in Germany

The economic miracle creates a labor shortage in West Germany. Since the early 1960s, German industry recruits workers from Italy, Spain, Yugoslavia, Greece, Turkey, and other countries who settle in the major industrial areas. The number peaks at 2.6 million. As the economic opportunities diminish, the number declines. With German unemployment rising, the presence of foreign workers at times creates friction.

1982: October 1   Helmut Kohl becomes federal chancellor

Helmut Kohl (b. 1930), leader of the Christian Democratic Union (CDU) and former prime minister (governor) of the state of Rheinland-Pfalz, succeeds Social Democrat Helmut Schmidt (b. 1918) as chancellor of the republic.

1985   Changes in Soviet policy

With the election of Mikhail Gorbachev (b.1931) to the post of general secretary of the Soviet Communist Party (de facto head of government), a major shift toward liberalization takes place in the Soviet Union.

1988   Demands for greater freedom in East Germany

The liberalization in the Soviet Union and in other East Block countries, especially Poland and Hungary, encourages East German citizens to engage in peace rallies and protest marches (often originating in a church). They are broken up by police and State Security (Staatssicherheitsdienst or Stasi).

1988: November   A woman becomes president of Bundestag

Rita Süssmuth (b. 1937), member of the CDU and Federal Minister for Youth, Family, Women, and Health (1985–88), is elected president of the federal parliament (Speaker of the House).

1989: Summer   East Germans flee the country

Taking advantage of the liberalizations in Hungary and Czechoslovakia, many East Germans travel to these countries, ostensibly on vacation, but then cross over to West Germany or Austria. Others seek refuge in the West German embassies in Budapest, Prague, Warsaw, and even in East Berlin.

1989: October 7   East Germany celebrates fortieth anniversary

The official celebration of the fortieth anniversary of the German Democratic Republic is marred by protesters, demonstrating against the dictatorship of the Socialist Unity Party (SED).

1989: October 9   "We are the people!"

On Monday, October 9, about 100,000 people march peacefully through the East German city of Leipzig, chanting "We are the people" (Wir sind das Volk). There is no police interference.

1989: October 18   Erich Honecker resigns

After eighteen years as head of state of the GDR, Erich Honecker (1912–94) is removed from office.

1989: November 4   Massive protests in Berlin

About 1 million people in East Berlin demonstrate against the regime.

1989: November 9   Berlin Wall opens

When East German officials announce a liberalization of travel regulations for citizens of the GDR, this is interpreted to include the movement between East and West Berlin. When the wall is opened, large numbers of East Berliners stream into West Berlin, where they are greeted with great enthusiasm.

1990: February 14   Talks about German unification

The foreign ministers of the four World War II victors and of East and West Germany begin discussion about possible German unification.

1990: March 18   First free elections in East Germany

1990: April 19   East Germany ready for unification

Lothar de Maizière, newly elected prime minister of the GDR, declares his government's commitment to German unification.

1990: July 1   Economic and monetary unification

The German mark (DM) becomes the currency of both states, and their economies are integrated.

1990: October 3   The GDR joins the Federal Republic.

With great enthusiasm, the people of East and West Germany celebrate the unification of their country. The five Länder (states) of the former GDR become part of the Federal Republic of Germany as free and democratic members of the federation. German unity and full independence is restored. Chancellor Kohl is at the height of his popularity as the chancellor of German unification.

1991   Problems of economic integration

The economic conditions of East Germany are much worse than anticipated. Much of the industry does not meet modern western standards and cannot compete on the world market. The infrastructure (roads, telecommunication, etc.) needs to be overhauled. Pollution of water, soil, and air is widespread. The privatization of state-owned facilities proceeds slowly and often encounters complex legal hurdles concerning original ownership. West German taxpayers have to contribute much more than expected to support the East which creates hard feelings on both sides. The problems continue over the following years. With the reorganization of industry, unemployment in some areas of the former GDR rises above twenty percent.

1991: June 20   Parliament votes to restore Berlin as capital

After a heated debate on whether to retain the capital in Bonn or move it to Germany's pre-war seat of government, Berlin, the vote is cast in favor of Berlin. This decision requires massive new construction at huge government expense.

1993   Treaty of Maastricht

The member countries of the European Community agree on even closer integration, including a common currency (the euro, to be introduced by 1999). The Community changes its name to European Union EU). Chancellor Kohl is one of the strongest supporters of a unified Europe.

1994: July   Supreme Court allows use of German army abroad

The Basic Law (Constitution) of the Federal Republic severely restricts the use of German military forces except in case of defense. A ruling of the Constitutional Court authorizes the deployment of German soldiers abroad, subject to parliamentary approval. This ruling permits German participation in peacekeeping efforts in the Balkans.

1998   Election Year

There are many issues facing the electorate in 1998. Unemployment is high; the economy is stagnant. The introduction of the euro is imminent and worries many Germans who believe in the sanctity of the German mark. The main contenders are Helmut Kohl (CDU), chancellor since 1982, and Gerhard Schröder (b. 1944) SPD. The major political parties are seen as unresponsive and unable to cope with the country's problems.

Bibliography

Cantor, Norman. *Medieval History.* New York: Macmillan, 1969.

Craig, Gordon A. *The Germans.* New York: Putnam's Sons, 1982.

Davies, Norman. *A History of Europe.* New York: Oxford University Press, 1996.

Deutsche Geschichte. Vol. 1: Mittelalter; Vol. II: Frühe Neuzeit; Vol. III: 19. und 20. Jahrhundert. Göttingen: Vandenhoeck & Ruprecht, 1985.

Dülffer, Jost. *Nazi Germany 1933–1945: Faith and Annihilation.* London: Arnold, 1996.

Eley, Geoff, ed. *Society, Culture, and the State in Germany: 1870–1930.* Ann Arbor: The University of Michigan Press, 1996.

Ellis, William S. "The Morning After: Germany Reunited," *National Geographic Magazine,* Vol. 180, No.3 (September 1991), 2-41.

Facts About Germany. Frankfurt/Main: Societäts-Verlag, 1997. (Regularly updated).

Friedländer, Saul. *Nazi Germany and the Jews.* Volume I - "The Years of Persecution, 1933–1939." New York: HarperCollins, 1997.

Fulbrook, Mary. *A Concise History of Germany.* Updated edition. New York: Cambridge University Press, 1994.

Gies, Frances and Joseph. *Cathedral, Forge, and Waterwheel: Technology and Invention in the Middle Ages.* New York: HarperCollins, 1994.

Hoffmeister, Gerhart and Frederic C. Tubach. *Germany: 2000 Years.* Volume III: "From the Nazi Era to the Present." New York: Ungar Publishing Co., 1986.

Kramer, Jane. *The Politics of Memory: Looking for Germany in the New Germany.* New York: Random House, 1996.

Questions on German History. 19th edition. Bonn: Deutscher Bundestag, 1996.

Reinhardt, Kurt. *Germany: 2000 Years.* Volume I: "The Rise and Fall of the 'Holy Empire.'" Volume II: "The Second Empire and the Weimar Republic." New York: Frederick Ungar, 1961.

Greece

Introduction

Culturally and historically, Greece is a land of contrast and paradox. Home to the civilization that inspired the Renaissance in Western Europe, Greece itself has a culture that is both Mediterranean and Balkan as well as European, with the additional uniqueness bestowed by its allegiance to the Greek Orthodox Church rather than the major Catholic or Protestant faiths of the rest of Western Europe. Although Greece was the birthplace of democracy during the Golden Age of Athens, its own path to democracy has been a long and tortuous one. It has included four hundred years of Ottoman rule, the imposition of foreign-dominated monarchies in the nineteenth century, and, in the twentieth, a long and bloody civil war and a period of military rule. Since 1974, however, Greece has been both a democracy and an active participant in European affairs as a full member of the European Economic Community.

Greece is the southernmost country in Europe. About one-fifth of its area of 50,942 square miles (131,940 square kilometers) consists of some 1,400 islands in the Aegean and Ionian seas. Four-fifths of Greece's terrain is mountainous, with the major plains area located in Thessaly, to the east of the Pindus Mountains that dominate mainland Greece, running from north to south. To the south lies the hand-shaped Peloponnesus, separated from the main part of the Greek peninsula by the Gulf of Corinth. At the end of the twentieth century Greece has an estimated population of 10.5 million, of whom nearly one-third live in the metropolitan area of Athens, its capital.

History

The foundation for ancient Greece's contribution to Western culture was laid with the creation of the Greek city-state, or *polis,* between 1,000 and 500 B.C. These political entities were governed separately, but a common language and religion united them culturally. After defeating Persia in two wars, Athens became the most powerful and influential city-state and, under the political leadership of Pericles, entered a golden age that lasted from roughly 450 to 400 B.C. This period witnessed an unprecedented series of artistic and intellectual achievements, including the works of the great tragic playwrights Aeschylus (c. 525–456 B.C.), Sophocles (c. 496–406 B.C.), and Euripedes (c. 485–406 B.C.) as well as the comic dramatist Aristophanes (c. 450–385). The philosopher Socrates (c. 470–399 B.C.) lived and taught during these years, the Parthenon was built, and great sculptural masterpieces were produced.

The Peloponnesian Wars that lasted intermittently from 431 to 404 B.C. ended the dominance of Athens and marked the ascendancy of the militaristic state of Sparta. In the second half of the fourth century Philip II of Macedon gained control of Greece. Upon his death in 336, it passed to his son Alexander (later known as Alexander the Great), who in the next thirteen years used it as a base from which to build the greatest empire the world had known, stretching as far as Afghanistan and India. After Alexander's death, the Greek states succumbed to the growing power of Rome, first as protectorates and then directly under Roman rule. Although they lost their autonomy under Rome, they prospered during the two-hundred-year *Pax Romana,* (Roman Peace) a long period of peace, and their culture survived and was transmitted by the Romans throughout their world.

As the power of Rome declined after 180 A.D., its empire was divided in two. The western part remained centered in Rome, while in the eastern part, the Byzantine Empire, with its capital in Constantinople, grew in power and influence. It was to last a thousand years, surviving the cultural dislocations caused by massive Slavic migration in the sixth century, as well as Islamic invasions in the seventh and eighth centuries. Greek culture remained a powerful force in the Byzantine Empire, while the Eastern Orthodox religion became an important part of Greek life and remained so, even after Constantinople fell to the Ottoman Turks in 1453. In fact, the Ottomans actually strengthened the role of the church in Greek life by their practice of organizing ethnic minorities within their empire into communities (called *millets*) based on religion and headed by religious leaders.

In spite of the religious autonomy allowed, life under Turkish rule was harsh, and large numbers of Greeks emigrated during the period of Ottoman rule, resulting in a phe-

nomenon that remains important to this day—the Greek *diaspora* (Greeks living in other countries).

It was the Greeks of the diaspora who took the lead in the development of eighteenth-century Greek nationalism, which, together with the ideals of the Enlightenment and the example of the French Revolution, provided the impetus for the Greek struggle for independence in the nineteenth cen-

tury. The long and bloody war for independence began in 1821 and lasted for most of the decade. The great European powers of France, Great Britain, and Russia turned the tide for the Greeks in 1827 but exacted a price for their help, essentially turning the Greek state into a protectorate with a Catholic, European monarch. In addition, the territory of the new nation was smaller than the Greeks had expected. In the

Balkan Wars of 1912–13, however, Greece nearly doubled its territory, acquiring the island of Crete, islands in the Aegean Sea, southern Macedonia, and other areas.

Twentieth-century Greece has seen its share of political turmoil. Greek allegiance in World War I was such a divisive issue that the country's king, who had backed Germany, was forced to abdicate. A series of further abdications followed the war, and a republic was declared in 1924. However, economic problems and political divisions resulted in the restoration of the monarchy in 1935, after which Greece fell under the sway of prime-minister-turned-dictator Ioannis Metaxas. World War II brought great suffering to Greece, which fought off Italian invaders only to be occupied by Germany for three years. These events were followed immediately by a long and bitter civil war between royalist and communist forces that lasted until 1949.

For much of the postwar period, foreign affairs were dominated by the Cold War and by the issue of *enosis* (union) with the island of Cyprus, a majority of whose residents are ethnic Greeks. Following a right-wing coup in 1967, Greece was ruled by a military *junta* (group of officers) until 1974, when democratic civilian government was restored. Greece elected its first socialist government in 1981, when the Panhellenic Socialist Movement, led by Andreas Papandreou, came to power. After falling out of favor in the late 1980s following a financial scandal, the party staged a comeback and returned in power in the early 1990s.

In 1997 it was announced that Greece, home to the original Olympiad and birthplace of the modern Olympic Games, would host the summer games in the year 2004, carrying an ancient cultural legacy forward to a new millennium.

Timeline

12,000–3000 B.C.   Neolithic era

Stone-age populations live in farming villages in present-day Greece.

Early Civilizations

c. 3000–1100 B.C.   Minoan civilization

The Minoan civilization emerges on the island of Crete. Its capital is at Knossos. Named for the legendary King Minos, it is known for its sculpture, pottery, and palaces. The Minoans also become the first Greek civilization to employ writing (although not the Greek language).

1600 B.C.   Beginning of Mycenaean civilization

The earliest civilization on mainland Greece, known as the Mycenaean or Helladic, is inaugurated by Indo-European tribes who have migrated to the region, including the Dorians, Aeolians, Ionians, and Achaeans. They introduce the Greek language.

The Polis

1000 B.C.–c. 500 B.C.   Formation of city-states

The Greek city-state, or *polis,* develops. Initially the city-states are ruled by kings. Although governed separately, the Greeks are unified culturally by their language and religion. Toward the end of this period, most of the kings are overthrown by dictators.

c. 9th century B.C.   Homer writes the *Iliad* and the *Odyssey*

The poet Homer authors Greece's two great epic poems. Although nothing can be documented about Homer's life, it is thought probable that he was an Ionian, and he is traditionally thought to have been blind. Rather than being written down by Homer, both poems were probably passed on at first through the oral poetry tradition. The *Iliad,* consisting of twenty-four books, recounts the actions of the hero Achilles and his followers during the Trojan war. The *Odyssey* narrates the adventures of Odysseus, the king of Ithaca, as he journeys home from the same war. Both poems contribute to a sense of ethnic and cultural unity and shared values among the ancient Greeks and helped lay the foundation for Western literature.

776 B.C.   The first Olympiad

The traditional date for the founding of the Greek Olympiad, the model for the modern Olympic games, is 776 B.C. During the games, the competing city-states declare a truce from all political conflict. The Olympiad is not only a great athletic event but also the occasion for the composition of literary works and the creation of sculpture.

750–500   Era of colonization

Overpopulation and the search for new trade outlets spur a wave of Greek colonization in the Mediterranean region and the Black Sea. Among the colonized regions are southern Italy and Sicily, known collectively as *Magna Graecia* (Greater Greece). Greek influence during this period extends as far as Spain and Egypt.

6th century B.C.   Athens-Sparta rivalry

Athens and Sparta emerge as the two most powerful Greek city-states. Athens is a democracy governed by the middle class, while Sparta is a strongly militarized state dominated by an aristocracy.

c. 525–456   Life of Aeschylus

The first of classical Greece's three great tragic dramatists, Aeschylus introduces into traditional Greek drama innovations that will characterize the works of all three: He makes the solo actors more important relative to the chorus, raises their number from one to two, and heightens the role of dialogue. Out of approximately ninety plays composed by Aeschylus, only seven survive: *The Persians, Seven Against Thebes, The Suppliants, Prometheus Bound,* and the three plays of the *Oresteia: Agamemnon, The Libation-Bearers,* and *The Eumenides.* The *Oresteia,* commonly regarded as the dramatist's greatest achievement, centers on the tragic fate of the hero Agamemnon, who leads the Greeks in the Trojan War.

c. 496–406   Life of Sophocles

Sophocles, the author of *Oedipus the King, Antigone,* and other great plays, builds on and extends Aeschylus's modifications of the Greek tragedy, increasing the number of actors to three and further diminishing the role of chorus. Compared with the plays of Aeschylus, those of Sophocles focus less on the workings of fate and more on the human nature of those subjected to the will of the gods. As with Aeschylus, only seven of Sophocles's plays survive, including (in addition to those named above) *Ajax, Trachiniae, Electra, Philoctetes,* and *Oedipus at Colonus.*

490   First Persian War

The Athenians, led by Miltiades, strengthen their power and prestige by defeating the Persians, who invade Attica under the command of King Darius. Following this defeat, Persia withdraws from the Aegean region and retreats to Asia, and the development of Athenian democracy continues unimpeded.

c. 485–406   Life of Euripides

Euripides, the third great Greek dramatist of the fifth century B.C., continues to move the classic Greek drama toward an emphasis on human motives and psychology. Among his eighteen surviving plays are *Medea, Trojan Women, Orestes, Helen, Iphigenia at Aulis.* The plays of Euripides are especially notable for their portrayals of the extremes of human passion, such as that found in the story of Media, who revenges herself on her husband, Jason, for his unfaithfulness by killing not only the woman for whom he abandons her but also her own two sons because they are Jason's children.

481–479   Second Persian War

The Persians, led by King Xerxes, launch an invasion of Greece. On land, the Greeks suffer defeat at Thermopilai, but they inflict a decisive naval defeat on the Persians at the Bay of Salamis and force them to retreat. The Persian Wars create a breach in the previous unity between Europe and Asia.

c. 470–399   Life of Socrates

The first of Greece's great philosophers leaves his legacy through the oral teachings passed on by his pupils, foremost among them Plato (see 427 B.C.). (A satirical portrait is provided in Aristophanes's comic play, *The Clouds.*) Tradition depicts him as living in poverty and devoting himself to teaching his students. He is also famous for the "Socratic method," consisting of teaching by asking his pupils a series of questions. Socrates is tried in 399 for corrupting youth by his teachings and sentenced to death. He refuses to compromise by requesting a sentence of exile instead, and hemlock (a type of poison) is administered to him.

464   Earthquake rips Sparta apart

Sparta suffers massive damage from an earthquake that rocks a large portion of Greece. Sparta's deaths are estimated at 20,000. In addition, some connect the disaster with the Sparta/Athens rivalry that eventually results in the Peloponnesian War—a connection stemming from Sparta's ingratitude for Athenian help in quelling the slaves who break free and terrorize the city following the earthquake.

The Golden Age of Athens

c. 450–400   Golden Age of Athens

Following the defeat of Persia, Athens comes to dominate the Aegean region, assuming control of the other Greek city-states and collecting tribute from them. The wealth generated from this tribute lays the foundation for a great period of cultural and intellectual development that encompasses literature, philosophy, art, and architecture. These advances accompany innovations in democratic government under the Athenian leader Pericles (c. 495–429).

c. 450–385   Life of Aristophanes

Aristophanes, the best-known Greek author of comedies, writes plays that are both topical and timeless, in that they satirize specific contemporary targets but also human nature. The subjects of Aristophanes's plays range from the judicial system of Athens (*Wasps*) to the tragic playwrights Aeschylus and Euripides (*Frogs*), and even include the philosopher Socrates (*Clouds*). These three are among a total of eleven surviving plays by Aristophanes.

438   The Parthenon is dedicated

Built between 447 and 432 B.C., the Parthenon, a temple atop the Acropolis (the sacred hill of Athens) is regarded as the pinnacle of classical Greek architecture. It is dedicated to Athena, the patron goddess of the city.

431–421   First phase of the Peloponnesian War

Athens's enemies join together under the leadership of Sparta to challenge Athenian naval power. Neither side wins a decisive victory, and a peace accord ends the conflict.

c. 427–347   Life of Plato

The great philosopher and teacher Plato grows up in an environment of strong dedication to public affairs and is closely involved in politics early in his career. However, various events, including the trial and execution of his teacher, Socrates, disillusion him, and he retires to a life of teaching and philosophy, founding the Academy, a forerunner of the modern university. Plato's works, referred to as "Dialogues" because they are organized as series of questions and answers, include the *Sophist,* the *Politicus,* the *Timaeus,* the *Laws,* the *Apology,* and the *Crito,* among others. Perhaps Plato's most famous dialogue is the *Republic,* which outlines the organization of the ideal state.

414–404   Conclusion of the Peloponnesian War

After thwarting an attempted Athenian invasion of Sicily, Sparta overcomes the Athenian navy at Aigispotamoi, an event that spells the end for Athenian dominance of the Aegean, which passes to Sparta. Sparta, however, is eventually weakened by internal discord and wars with Persia.

338   Philip of Macedon annexes Greece

After seizing control of the Balkans, Philip II of Macedon conquers Greece's territories. It passes to his twenty-year-old son, Alexander, when Philip is assassinated two years later.

336–323   The empire of Alexander the Great

Alexander (356–323), son of Philip II of Macedon, comes to power following the death of his father and creates an empire greater than any that has existed before. From its base in Greece, Alexander's empire eventually reaches to Asia Minor, Persia, Egypt, Mesopotamia, Afghanistan, and India. Alexander's rule spreads the Greek language and culture throughout much of the Near East. After only thirteen years in power, Alexander dies of malaria at the age of thirty-three, and his empire soon crumbles.

c. 280   Greeks battle the Romans in southern Italy

Greeks under the leadership of King Pyrrhus of Epirus confront the rising power of Rome in the struggle for control of Magna Graetia, Greece's territories in Southern Italy. The Greeks also become involved in the Punic Wars between Rome and its main adversary, Carthage.

200–197 B.C.   Second Macedonian War

At the conclusion of Rome's first major invasion of Greece, the reigning Macedonian king, Philip V, loses all his territo-ries outside Macedonia, and the Greek city-states become protectorates of Rome, although they retain their autonomy.

The Roman Period

146   Greece comes under Roman control

After Rome's victory in the Fourth Macedonian War, Macedonia becomes a province of Rome, which gradually takes control of much of the territory that had belonged to the Macedonian Empire of Alexander the Great.

86   Rome conquers Athens

The city-state of Athens, the center of Greek culture, is conquered by Rome.

31 B.C.–A.D. 180   *Pax Romana*

Following the Battle of Actium, Greece is incorporated into the Roman Empire, playing an important role in the ascent to power of Octavian (Augustus Caesar) (27 B.C.–A.D. 14), who bases his armies there. Greek learning and culture continue to flourish and are spread to other parts of the Empire, as the Greeks benefit from the *Pax Romana* (Roman Peace), the long period of peace that lasts until the death of the emperor Marcus Aurelius (A.D. 121–180). The Greek cities prosper, and a Graeco-Roman culture emerges as the Greek and Roman cultures interact.

The Byzantine Empire

A.D. 395   The Roman empire is divided

Following the death of Roman emperor Theodosius (347–395), the empire is divided into eastern and western parts. The eastern part, which includes Greece, has its capital at Constantinople and is known as the Byzantine Empire. The western part retains its capital at Rome but becomes less and less important as it is weakened by the invasions of Germanic tribes. Greek culture remains a powerful force in the Byzantine empire, although the cultural center moves from Athens to Constantinople.

6th century   Slavic migration begins

Slavs from the region north of the Danube River migrate to the Balkan Peninsula, dislodging the Greeks and their culture from the region.

A.D. 527–65   Reign of Justinian

The emperor Justinian (483–565) greatly expands the power and territory of the Byzantine Empire by conquest and administrative reform, including the unification of the Roman legal code and the creation of effective centralized government administration.

530s   Hagia Sophia church is built

Byzantine culture is famed for its architecture, of which the most famous example is the Hagia Sophia church, located in present-day Istanbul, Turkey.

7th–8th centuries   Muslim invasions

Islamic invaders from the Arabian Peninsula invade the Near East, seizing control of Persia, Iraq, and Syria, and nearly toppling the Byzantine Empire.

9th century   Islamic invaders are driven back

Under the control of a new dynasty of rulers, the Byzantines drive back Muslim invaders, regaining control of Crete, the Aegean Sea, and nearby regions including Georgia and Armenia. The Byzantine empire flourishes under its new rulers. Agriculture and trade flourish, and the population grows.

1054   The Great Schism

The eastern (Eastern Orthodox) and western (Roman Catholic) branches of the Christian church undergo an official rupture. This split has profound historical impact for it engenders deep animosity between the two groups that persists (to some less extent) to this day.

1071   Capture of Romanus IV

Byzantine emperor Romanus IV is captured by the Seljuk Turks during the Battle of Manzikurt, an action seen as an important turning point in the decline of the Byzantine Empire.

1204   Crusaders plunder Constantinople

The Byzantine capital of Constantinople is sacked by Catholic participants in the Fourth Crusade, who remove many valuable treasures from the city. This act reinforces the suspicions harbored by Orthodoxy toward Catholicism.

Ottoman Rule

1453   Fall of Constantinople

After winning Asia Minor and the Balkans, the Ottoman Turks, who have risen to power following the Mongol invasions, lay siege to Constantinople and conquer the city, winning control of most of Greece by 1460. They continue conquering Greek lands in the next century until they rule all of Greece except the Ionian Islands, which are ruled by Venice. The Ottoman conquest marks the end of the Byzantine Empire.

1453–1821   The Greeks come under Ottoman rule

The Greeks remain under Ottoman rule during this period. They are allowed to maintain the Christian Orthodox religion and, like all the empire's religious minorities, are organized into communities called *millets*, thus putting the Greek Orthodox Church at the center of Greek life during this period. The religious leader of Constantinople becomes the head, or *ethnarch,* of the Greek community. Although they allow religious autonomy, the Ottomans impose taxes on the non-Muslim taxes (see sidebar).

Many Greeks emigrate during this period, going to Romania, the Black Sea region, Russia, and territories controlled by the Habsburg Empire. Greek culture is carried throughout the lands included in this scattering, or *diaspora.* The Greeks also rise to the forefront of trade in the region, forming merchant communities throughout the diaspora and playing a dominant role in the commercial dealings of the Ottomans.

c. 1541   Birth of painter El Greco

Greece's most famous painter is born Domenikos Theotocopulos, on the island of Crete, where Byzantine art serves as a formative influence. In about 1565 he moves to Venice, where he studies with Titian and is influenced by other great painters of the Italian Renaissance, including Michelangelo (1475–1564) and Tintoretto (1518–94). In 1570 he moves to Rome and six years later arrives in Spain, attracted by the splendor of the Spanish court of Philip II. Failing to find favor with Philip, the artist retires to Toledo in 1577 and remains there until his death in 1614. It is in Spain that he becomes known as *El Greco,* meaning "the Greek." El Greco is known for the highly personal, subjective, and expressive style of his paintings and for their depth of religious feeling. His use of elongated, twisting forms links his work to that of the Mannerists. Famous works by El Greco include *The Dream of Philip II; Crucifixion, Resurrection,* and *Assumption; St. Martin with Beggar; View of Toledo;* and the altar of Santo Domingo el Antiguo in Toledo. El Greco dies in Toledo in 1614.

1778–79   Orlov rebellion

During the eighteenth century Greek nationalism grows, led by the Greeks of the diaspora and influenced by the ideals of the Enlightenment and the American and French Revolutions. An unsuccessful Russian-led rebellion against the Ottomans is mounted in the Peloponnesian region. The Ottomans respond with repressive measures that increase Greek discontent with Ottoman rule.

1798: April 8   Birth of poet Dhionisios Solomos

Solomos, considered the national poet of Greece, is closely associated with the Greek struggle for independence. He is also the first major poet to use Demotic Greek—the spoken form of the language—in his poetry. Well-known poems by Solomos include "Hymn to Liberty" (1823); a famous poem about the death of the English poet Lord Byron (1788–1824);

The Ottoman Period

The Ottoman Empire began in western Turkey (Anatolia) in the thirteenth century under the leadership of Osman (also spelled Othman) I. Born in Bithynia in 1259, Osman began the conquest of neighboring countries and began one of the most politically influential reigns that lasted into the twentieth century. The rise of the Ottoman Empire was directly connected to the rise of Islam. Many of the battles fought were for religious reasons as well as for territory.

The Ottoman Empire, soon after its inception, became a great threat to the crumbling Byzantine Empire. Constantinople, the jewel in the Byzantine crown, resisted conquest many times. Finally, under the leadership of Sultan Mehmed (1451–81), Constantinople fell and became the capital of the Ottoman Empire.

Religious and political life under the Ottoman were one and the same. The Sultan was the supreme ruler. He was also the head of Islam. The crown passed from father to son. However, the firstborn son was not automatically entitled to be the next leader. With the death of the Sultan, and often before, there was wholesale bloodshed to eliminate all rivals, including brothers and nephews. This ensured that there would be no attempts at a coup d'état. This system was revised at times to the simple imprisonment of rivals.

The main military units of the Ottomans were the Janissaries. These fighting men were taken from their families as young children. Often these were Christian children who were now educated in the ways of Islam. At times the Janissaries became too powerful and had to be put down by the ruling Sultan.

The greatest ruler of the Ottoman Empire was Suleyman the Magnificent who ruled from 1520–66. Under his reign the Ottoman Empire extended into the present day Balkan countries and as far north as Vienna, Austria. The Europeans were horrified by this expansion and declared war on the Ottomans and defeated them in the naval battle of Lepanto in 1571. Finally the Austrian Habsburg rulers were able to contain the Ottomans and expand their empire into the Balkans.

At the end of the nineteenth century, the Greeks and the Serbs had obtained virtual independence from the Ottomans, and the end of a once great empire was in sight. While Europe had undergone the Renaissance, the Enlightenment, and the Industrial Revolution, the Ottoman Empire rejected these influences as being too radical for their people. They restricted the flow of information and chose to maintain strict religious and governmental control. The end of Ottoman control in Europe came in the First Balkan War (1912–13) in which Greece, Montenegro, Serbia, and Bulgaria joined forces to defeat the Ottomans.

During the First World War, (1914–18), the Ottoman Empire allied with the Central Powers and suffered a humiliating defeat. In the Treaty of Sévres, the Ottoman Empire lost all of its territory in the Middle East, and much of its territory in Asia Minor. The disaster of the First World War signaled the end of the Ottoman Empire and the beginning of modern-day Turkey. Although the Ottoman sultan remained in Constantinople (renamed in 1930, "Istanbul"), Turkish nationalists under the leadership of Mustafa Kemal (Ataturk) (1881–1938) challenged his authority and, in 1922, repulsed Greek forces that had occupied parts of Asia Minor under the Treaty of Sévres, overthrew the sultan, declared the Ottoman Empire dissolved, and proclaimed a new Republic of Turkey. That following year, Kemal succeeded in overturning the 1919 peace settlement with the signing of a new treaty in Lausanne, Switzerland. Under the terms of this new treaty, Turkey reacquired much of the territory—particularly, in Asia Minor—that it had lost at Sévres. Kemal is better known by his adopted name "Ataturk", which means "father of the Turks". Ataturk, who ruled from 1923 to 1938, outlawed the existence of a religious state and brought Turkey a more western type of government. He changed from the Arabic alphabet to Roman letters and established new civil and penal codes.

the unfinished revolutionary poem, *Lambros* (begun in 1826); and "The Woman of Zante." Solomos dies in 1857.

The Struggle for Independence

1814   Society of Friends is formed

The *Filiki Etaireia* (Society of Friends), the most influential of the Greek revolutionary organizations, is formed by expatriates in Odessa, Russia, and attracts a wide membership.

1821: March 25   The Greek revolution begins

Greek patriots led by Archbishop Germanos of Patras claim independence for Greece, and a war of liberation is begun. The Greeks marshal a small but effective force, including a small navy, and engage the Turks in numerous skirmishes. A pattern of massacre and retaliation develops and includes the murder of the Greek Orthodox Patriarch by the Turks.

Although the Ottoman Empire has been weakened by economic conditions following the fall of Napoleon (1769–1821), the struggle for Greek independence—the first successful rebellion against the Ottomans—is long and bloody and eventually involves intervention by several other countries. It is also accompanied by serious internal struggles among the Greeks themselves. Liberals throughout Europe espouse the Greek cause. Many volunteer to aid the Greeks, including the English poet Lord Byron, who dies in the struggle. However, the Greek rebels do not receive assistance from any foreign governments until 1827.

1825   Egypt comes to the aid of Turkey

Egypt, which is part of the Ottoman Empire, comes to Turkey's aid in the war against Greece, inflicting major defeats on the Greek rebels.

1827: July 6   Treaty of London

Great Britain, France, and Russia sign the Treaty of London and threaten to establish diplomatic relations with the Greek revolutionaries unless an armistice is declared between the Greeks and the Turks. The three powers send naval forces to the region to monitor activity there but remain neutral unless attacked.

1827: October 20   Foreign alliance comes to the aid of Greece

Following the murder of a British officer, the allied European powers engage the Turks in a naval battle that cripples the Turkish and Egyptian navies.

1828   War between Russia and Turkey

Russia attacks the Ottomans on land, making them divert their forces from Greece.

1829: September   The Ottomans recognize Greek independence

By the Treaty of Adianople, which ends the Russian-Turkish war, the Turkish sultan recognizes Greek independence and agrees on the boundaries of the new state, which extends only as far north as Thessaly. It is also agreed that Greece will be a hereditary monarchy and that, although under the protection of the allied powers, its prince will not belong to any of their ruling dynasties.

1830   The London Protocol

An agreement between the major international powers establishes a Greek state smaller than is hoped for by Greek nationalists. The new nation is composed of the Peloponnesus, the Cyclades, and a region north of the Gulf of Corinth. Although formally independent, it essentially becomes a joint protectorate of France, Britain, and Russia.

The Kingdom of Greece

1832   Otto becomes king of Greece

Following the assassination of Greek nationalist leader Count John Kapodistrias, the major powers arrange for Otto (1815–67), the German king of Bavaria, to become Greece's first monarch. Upon his selection, Otto adopts the Greek version of his name, Othon. Othon, a Roman Catholic king in a Greek Orthodox country, establishes an authoritarian, highly centralized government dominated by Germans.

1836   School of Fine Arts opens

The School of Fine Arts (later the Higher School of Fine Arts) opens in Athens, becoming the first arts education institution in modern Greece. Its curriculum includes architecture, engineering, painting, and sculpture.

1844   Constitutional monarchy is created

A peaceful coup d'état forces King Othon to accept a constitution providing a bicameral national assembly and a prime minister and guaranteeing basic individual freedoms, including freedom of assembly and freedom of the press.

1857: November 21   Death of poet Dhionisios Solomos

The man regarded as Greece's national poet dies in Corfu. (See 1798.)

1862   Othon is deposed in a coup

Othon, the German monarch, is overthrown.

1863–1913   Reign of George I

Othon is replaced by a Danish prince who adopts the title George I (1845–1913).

1864   Ionian Islands are ceded to Greece

Britain cedes the Ionian Islands to Greece.

1864   New constitution is adopted

Under King George I, Greece adopts a new constitution that provides for a unicameral legislature.

1864: August 23   Birth of Eleftherios Venizelos

Greece's most important statesman of the early twentieth century is born on Crete to a family with a history of revolutionary activity. He leads an unsuccessful revolt against the Turks during the Greco-Turkish War of 1897 and later becomes the island's minister of justice when it is granted internal autonomy after the war. In the early 1900s, Venizelos is asked to lead a new Greek revolutionary group, the Military League. From 1910 to 1915, he serves as Greece's prime minister, implementing reforms at home and doubling Greece's territory through concessions won during the Balkan Wars.

During World War I, Venizelos supports the Allied cause, coming into conflict with King Constantine, who supports Germany. Ultimately, Venizelos organizes the resistance that forces the king to abdicate in 1917, when Greece declares war on the Germans and their allies (the Central Powers). Venizelos wins further territorial gains for Greece at the peace talks in Paris that follow the war. When the Greek people vote to restore Constantine to the throne in 1920, Venizelos goes into exile in Paris. When the second Hellenic Republic is formed in 1924, he returns to Greece to head the Liberal Party and serves another four-year term as prime minister from 1928 to 1932. When the monarchy is restored yet another time in 1935, Venizelos leaves the country for a final time, dying in Paris the following year.

1874   Acropolis Museum opens

The Acropolis Museum, built to house sculpture from the Acropolis, is completed. It is designed in a Classical style in keeping with the artifacts of ancient Greece.

1881   Thessaly is annexed

Greece seizes Thessaly from the Turks.

1885: December 2   Birth of author Nikos Kazantzakis

Greece's best-known twentieth-century novelist is born on Crete. He studies law in Athens and philosophy (with Henri Bergson) in Paris. After a period of travel, he returns to Greece before World War II. After the war he holds government posts and also works for the United Nations agency UNESCO. Kazantzakis is well known for his best-selling novels, which have been widely translated into other languages and include *Vios kai politia tou Alexi Zormpa* (Zorba the Greek; 1946), *O Kapetan Mikhalis* (Freedom or Death; 1950), and *O televtaios pirasmos* (The Last Temptation of Christ; 1955). This last work gets him excommunicated from the Greek Orthodox Church. However, he also writes a highly regarded sequel to the *Odyssey* (Odissa; 1938) and translates classics including *The Divine Comedy*. Kazantzakis dies in Germany in 1957.

1894   Women are admitted to art school

For the first time, women are allowed to enroll in the Higher School of Fine Arts.

1896   First modern Olympics are held

The first modern Olympic games, modeled on the Olympiad of ancient Greece, are held at Athens.

1897: April–May   Greco-Turkish War

Greece and Turkey go to war over the predominantly Greek island of Crete, which is still held by the Turks. The Greek forces are overwhelmingly defeated and agree to an armistice. Following intervention by the major European powers, Crete is granted internal autonomy.

1900   National Gallery opens

The National Gallery and the Alexandros Soutzos Museum, planned as repositories for the country's major public and private art collections, open in Athens. Holdings include modern sculpture, paintings, and the decorative arts.

1900: March 13   Birth of poet George Seferis

Nobel Prize-winning poet George Seferis is hailed as the preeminent poet of the generation that introduced Symbolism to Greek poetry. Trained as a lawyer, Seferis serves with the Greek diplomatic service after World War II in the Middle East and is Greece's ambassador to London from 1957 to 1962. He wins immediate recognition with the publication of his first volume of poetry, *I strofi* (The Turning Point) in 1931. Other volumes include *Imeroloyion katastromatos I* (Log Book I; 1940), *Kikhli* (Thrush; 1947), and *Imeroloyion katastromatos III* (Log Book III; 1955). In 1947 Seferis's achievement is recognized by the Academy of Athens. In 1963 he wins the Nobel Prize for Literature. His works are widely translated into foreign languages. Seferis dies in Athens in 1971.

1907: March 8   Birth of Constantine Karamanlis

Two-term Greek president and four-term prime minister Karamanlis is first elected to parliament in 1935. In 1952 he is appointed minister of public works in the government of Alexandros Papagos. When Papagos dies three years later, King Paul names Karamanlis to succeed him as prime minister. Karamanlis forms a new party, the National Radical Union, which wins 161 of 300 legislative seats in parliamentary elections in February, 1956. The Karamanlis government negotiates with Turkey to arrange the creation of an independent state on Cyprus in 1960. Karamanlis resigns his post in 1963 over a disagreement with King Paul and moves to Paris.

In 1974, after the fall of the military *junta* that has ruled Greece since 1967, Karamanlis once again becomes prime minister. He forms a civilian government and establishes the New Democracy Party, which wins a large majority in parliamentary elections. Under Karamanlis, the Greek government manages to avoid going to war with Turkey over their conflicting interests in Cyprus. Karamanlis also presides over Greece's entry into the European Economic Community. Karamanlis serves as president of Greece from 1980 to 1985 and from 1990 to 1995. He dies in Athens in 1998 at the age of 91.

1909   Coup topples Greek government

The Military League, a group of junior army officers, overthrows the government and seizes power. The following year, liberal leader Eleftherios Venizelos is appointed as prime minister.

1911: November 2   Birth of poet Odysseus Elytis

Nobel prize-winning poet Odysseus Elytis (1911–96) is born on Crete to a well-to-do family and studies law as a young man. He begins publishing poetry in the 1930s. These poems, first published in literary magazines, are collected in the 1940 volume *Prosanatolismoi* (Orientations). During World War II, Elytis joins the resistance and also writes a famous poem that becomes a rallying call for Greek patriots: *Heroic and Elegiac Song for the Lost Second Lieutenant of the Albanian Campaign.* Elytis publishes no poetry for fifteen years after the war. His next published work is the long poem *To Axion Esti* (Worthy It Is). Following the 1967 military coup, Elytis moves to Paris. He wins the 1979 Nobel Prize for Literature. Later volumes by Elytis include *The Sovereign Sun* (1967), *The Stepchildren* (1974), and *The Little Mariner* (1986). He dies in March 1996.

1912–13   Balkan Wars

Under nationalist leader Eleftherios Venizelos, and through its alliance with Serbia, Bulgaria, and Montenegro, Greece is able to win control of Crete, Epirus, Thrace, islands in the Aegean, and southern Macedonia, increasing its territory from 25,000 square miles (65,000 square kilometers) to nearly 42,000 square miles (110,000 square kilometers). The Ottoman Turks are permanently expelled from the Balkans (with the exception of a small territory in Thrace) in the course of this war.

1913   Constantine becomes king

When King George is assassinated, his son, Constantine (1868–1923), who has distinguished himself in military action in the Balkan Wars, becomes king of Greece.

The World Wars

1914–18   World War I

King Constantine, who is connected by blood to the German monarchy, supports neutrality although he favors Kaiser Wilhelm II of Germany, while Venizelos supports a policy of intervention on the side of the Entente. Finally, in 1915, Venizelos resigns and forms a pro-Allied government of his own in Thessaloniki. Two years later Constantine is forced to abdicate. Venizelos then returns to Athens and to the post of prime minister. At the conclusion of the war, Greece wins control of Thrace and the Smyrna region in Asia Minor. Nevertheless, the split between the two men divides the country into supporters and opponents of the monarchy and dominates Greek politics even after both leaders die.

1917   Constantine I abdicates

King Constantine I is forced to abdicate because of his pro-German position. He is replaced by his son, Alexander (1893–1920), who joins Venizelos in declaring war on the Central Powers.

1918–20   Greece wins territory in peace talks

Prime Minister Venizelos gains additional territory for Greece (including the Dodecanese and part of Anatolia) at the peace talks in Paris following World War I. However, Turkish nationalists under Mustafa Kemal begin a military campaign against the Greek forces. If Greece is to keep its gains it must be willing to defend them by force.

1919: February 5   Birth of Andreas Papandreou

Andreas Papandreou, prime minister of Greece and son of a prime minister, is educated in Athens and flees to the United States after being persecuted by the government for his leftist views. In the U.S. he completes a doctoral degree, becomes a citizen, serves in the navy, and accepts positions at several universities, eventually becoming the head of the economics department at the University of California at Berkeley. When his father, Georgios Papandreou, becomes prime minister of Greece in 1963, Andreas returns to his native land, winning a seat in parliament. Together with his father, Papandreou is imprisoned following the military coup of 1967. The elder Papandreou dies soon after being released, and the son goes into exile in Sweden and then Canada, coordinating resistance to the ruling junta from abroad.

When the military government falls in 1974, Papandreou returns to Greece and forms PASOK, a socialist political party that gains support over the decade. The party's decisive victory in the parliamentary election of 1981 makes Papandreou prime minister of the country. Once in office, he moderates the anti-U.S. platform of his campaign but implements liberal policies on religion and marriage laws, while expanding social welfare programs. PASOK wins another victory in 1985, and Papandreou serves a second term as prime minister. His party is swept from office at the end of the decade in the wake of both personal and financial scandals but regains power in 1993. The aging Papandreou once again assumes the prime ministership but failing health forces him to resign three years later, and he dies in June, 1996.

1920: October   King Alexander dies

King Alexander dies of a monkey bite at the age of twenty-seven. The death of the childless monarch once again raises the issue of Constantine.

1920: November   Constantine is recalled

Monarchist parties triumph in elections following the death of King Alexander, and the Greek people vote to restore Constantine to the throne. Venizelos goes into exile in Paris.

1921–22   Unsuccessful Turkish campaign

Warfare with Turkish nationalist forces continues under King Constantine with disastrous results, Greece loses its gains in Asia Minor and roughly 1.25 million Greeks in Asia Minor are exchanged for 400,000 Turks in Greece. The sudden influx of refugees burdens the Greek economy which is at the point of collapse.

The Greek defeat does have its positive aspects, however. The expulsion of the Greek population of Asia Minor lays to rest the Greek nationalist dream of restoring the Byzantine Empire. By eliminating the thorniest issue in Greek-Turkish relations, the population exchange sets the basis for improved Greek-Turkish ties in the 1930s.

1922   King Constantine abdicates

Following the disastrous war with Turkey, King Constantine II abdicates and is replaced by his eldest son, George II (1890–1947) (grandson of King George I).

1922: May 29   Birth of composer Iannis Xenakis

Xenakis (b. 1922) is born in Romania to Greek parents, and his family returns to Greece ten years later. He begins college-level studies in engineering, but they are interrupted by World War II, and Xenakis joins the anti-Fascist resistance. In 1945 he is severely wounded in the face, losing his sight in one eye. After the war, he settles in Paris and comes under the influence of contemporary composers including Milhaud, Honegger, and Messiaen. While pursuing a career as a composer, Xenakis retains his interest in engineering and architecture, working with famed architect Le Corbusier on the design of housing projects and other buildings. Xenakis's compositions also reflect the composer's interest in mathematics and structural complexity. Nevertheless, they retain a strong human and expressive element. His compositions include works for orchestra (including his first completed piece, *Metastatis* [1953]); chamber and instrumental works (such as the piano piece *Herma*; 1960–64); and choral works, including *Oresteia* (1956–66).

1923   King George II is forced to abdicate

A league of military officers forces the abdication of King George II and replaces him with a military regent. By the following year, Greece is declared a republic, and the monarchy is abolished.

1924: May 1   Republic is proclaimed

Following a plebiscite (a binding vote on an issue by the general population), a Greek republic is proclaimed. Eleftherios Venizelos returns from Paris to head the Liberal Party. However, political rivalries and economic problems foster continued political instability.

1925: July 29   Birth of composer Mikis Theodorakis

Theodorakis has little formal musical training and serves in the military until the age of nearly thirty. He then settles in Paris, where he studies at the Conservatoire and completes his first work, the ballet *Antigone,* which is performed at Covent Garden in London. In 1959 the composer returns to his native land, where he articulates an outspoken critique of both the musical establishment and also those of the other arts. Following the military coup of 1967 Theodorakis is imprisoned but released three years later as a result of international pressure on the governing junta. Like his early ballet *Antigone,* most of the composer's works are based on Greek historical or literary themes. His prolific output includes ballets, works for chorus and orchestra, film scores, and other works.

1928–32   Venizelos serves as prime minister

Venizelos's party wins an overwhelming victory in national elections, and the statesman serves another term as prime minister. However, his successes are clouded by the beginning of the Depression, and his government is ousted after four years.

1930   National Theater is founded

Greece's National Theater is established.

1934   Balkan Entente is created

Greece, Yugoslavia, Turkey, and Romania agree to a collective security agreement called the Balkan Entente. This alliance is aimed a halting Bulgarian attempts to regain territories lost following World War I.

1935   George II is reinstalled as king

Due to political turmoil and economic difficulties, the Greek monarchy is restored, and George II is recalled as king.

1936 Metaxas seizes power

With the support of King George II, Prime Minister Ioannis Metaxas suspends Greece's parliamentary government and begins ruling as a dictator, instituting a period of fascist-style rule.

1936: March 18 Death of Eleftherios Venizelos

Statesman and former prime minister Eleftherios Venizelos dies in Paris. (See 1864.)

1939–45 World War II

After first being invaded by Italy, Greece is occupied by the Germans for most of the war, which inflicts great economic hardship on the country.

1939: April Italy invades Albania

The Italians, under Benito Mussolini (1883–1945), invade Albania and then use it as a base from which to invade Greece.

1940: October Italy invades Greece

When Greece refuses to allow the building of Italian military bases on its soil, the Italians invade but in spite of their superior military power are driven back by the Greeks.

1941–44 German occupation

The Germans occupy Italy, and the Greeks organize a fierce and effective resistance movement to oppose them. King George II forms a government-in-exile in London. Within its own ranks, the Greek resistance is deeply divided between royalists, who support the king, and communists. By the time the Germans leave the country, this division has led to a civil war.

The Postwar Period

1944–49 Greek civil war

Royalists and communists battle each other in a long civil war that has begun during World War II. By providing weapons and financial aid, the United States plays a major role in the eventual royalist victory.

1945–51 U.S. helps rebuild the Greek economy

After the war, Greece is in dire economic straits, suffering from shortages of food, housing, and clothing, and destruction of its infrastructure. Over 700,000 Greeks have become refugees because of the war. The U.S. provides over $500 million in short-term relief aid and long-term development aid.

1947 Truman Doctrine

U.S. President Harry S Truman (1884–1972) pledges assistance to stop the spread of communism in Greece and Turkey. Known as the Truman Doctrine, this becomes part of the U.S. policy of the containment of communism.

1946–47 George II is restored

After World War II, King George II is restored to the throne following two periods of exile. However, he dies the following year.

1947–64 Reign of King Paul

Paul I (1901–64), the youngest son of King Constantine, succeeds to the throne following the death of his brother, King George II. Queen Fredericka, wife of King Paul I, sponsors organizations to provide food and health care assistance to Greek citizens after World War II.

1950s Greeks advocate *enosis* for Cyprus

The goal of *enosis* (union between Greece and the British colony of Cyprus, which has a predominantly Greek population) becomes a major political issue in Greece.

1952 New constitution is adopted

Following the end of Greece's lengthy civil war, the nation adopts a new constitution providing for election by a straight ballot system instead of proportional representation. With the combined effects of political reform and economic recovery, a period of political stability follows.

1952 Full female suffrage is achieved

Greek women win the right to vote on the same basis as men.

1952 Greece joins NATO

Greece becomes a member of the North Atlantic Treaty Organization (NATO), the military alliance between Western Europe and the United States formed to oppose the spread of communism in Europe.

1955–63 Karamanlis serves as prime minister

Constantine Karamanlis (1907–98), head of the anti-Communist Radical Union, serves as prime minister, presiding over a period of economic growth and political stability.

1957: October 26 Death of author Nikos Kazantzakis

Internationally acclaimed novelist Nikos Kazantzakis, author of *Zorba the Greek,* dies in Freiburg, Germany. (See 1885.)

1960 Cyprus achieves independence

Following an agreement between the Greek, Turkish, and Cypriot governments that balances political power between

The Cold War

The Cold War is the commonly used name for the prolonged rivalry between the United States and the Soviet Union that lasted from the end of World War II to the break-up of the U.S.S.R. in 1991. The Cold War also encompassed the predominantly democratic and capitalist nations of the West, which were allied with the U.S., and the Soviet-dominated nations of Eastern Europe, where Communist regimes were imposed by the U.S.S.R. in the late 1940s. Although the United States and the Soviet Union never went to war with each other, the Cold War resulted in conflicts elsewhere in the world, such as the Vietnam War, a prime example of the United States's "domino theory," which held that a Communist victory in one nation would result in the spread of Communism to its neighbors.

Following World War II, the political and military bond between the United States and its allies in Western Europe was cemented by the Marshall Plan, which helped the European countries recover from the war economically, and the 1949 creation of NATO (the North Atlantic Treaty Organization). Within the United States, an early consequence of the Cold War was McCarthyism, a political trend named for Senator Joseph McCarthy, who spearheaded the drive to expose and penalize Americans suspected of being Communist spies or sympathizers in the early 1950s, one of the most intense periods of the Cold War. McCarthyism, at its height between 1950 and 1954, ruined the careers and lives of many Americans who posed no actual threat of subversion to the country. Following the death of Soviet dictator Josef Stalin in 1953, the intensity of the Cold War abated for several years. However, the launch of the Soviet space satellite Sputnik in 1957 as well as Soviet efforts to extend its influence in the newly-independent (formerly colonial) states revived the rivalry by inaugurating a battle for supremacy in space exploration (the "space race") and the "Third World", respectively.

The single most serious political confrontation of the Cold War was the Cuban Missile Crisis of 1962, when the United States and the Soviet Union were brought to the brink of a nuclear confrontation over the stationing of medium-range nuclear missiles in Soviet-allied Cuba. Ultimately, the missiles were withdrawn, but both nations began a costly arms race that was particularly damaging economically for the Soviet Union.

In the early 1970s, under Soviet premier Leonid Brezhnev and U.S. president Richard Nixon, Cold War tensions began to ease, in a development known as "détente." During this decade the two nations signed the SALT (Strategic Arms Limitations Treaty) agreements limiting antiballistic and strategic missiles. However, outside the provisions of these agreements, both nations continued their massive arms build-up, and tensions increased in the 1980s during the presidency of staunch anti-Communist Ronald Reagan, who referred to the Soviet Union as the "evil empire."

The appointment of political reformer Mikhail Gorbachev as general secretary of the Communist Party in 1985 signaled the beginning of the end for the Cold War. The democratization and openness implemented by the Soviet government under Gorbachev culminated not only in the downfall of Communism but in the dissolution of the Soviet Union itself, which was preceded by the fall of Communism in the Eastern European countries of the former Soviet Bloc.

the Greek Cypriot majority and the Turkish minority, Cyprus becomes an independent nation. In addition, Britain, Greece, and Turkey become the island's guarantor powers with the right to intervene on the island in order to maintain stability. Independence satisfies neither Cypriot community, and in late 1963 the Turkish-Cypriots withdraw from the island's government and set up their own administration.

1961   Greece becomes an associate member of the EEC

Greece wins associate membership in the European Economic Community (EEC).

1963   Poet George Seferis wins the Nobel Prize

Poet, essayist, and diplomat George Seferis is awarded the Nobel Prize for Literature. (See 1900.)

1963: November   Center Union ousts Karamanlis government

The government of Constantine Karamanlis is defeated at the polls by Georgios Papandreou's Center Union party. The new liberal government enacts important social reforms, but Papandreou alienates many by appointing his son, Andreas—a socialist—as the top economic minister.

Territorial growth of
Greece, 1913–19.

1964–73 Reign of King Constantine

King Paul dies in 1964. He is succeeded by his son, Constantine II (b. 1940), who rules until 1973.

1965 Papandreou resigns

To quell rising unrest in the military, Prime Minister Papandreou asks King Constantine for permission to appoint himself as defense minister. When it is refused, he resigns, and a series of caretaker governments follows.

Military Rule

1967: April 21 Right-wing coup topples government

A coup by right-wing military officers overthrows Greece's civilian government. In December 1967 a counter-coup is attempted but it fails, and the royal family flees the country. Lieutenant General George Zoetakis is named as regent to take the place of the king, while the ringleader, George Papadapoulos (1919–99), becomes prime minister.

The Marshall Plan, an international plan to rebuild European nations after World War II, provides aid to Greece, assisted by Queen Fredericka. Here, two thousand loaves of bread, baked by a Greek army bakery from flour sent by the United States, are distributed to children in Athens. (EPD Photos/CSU Archives).

1968: September 29   Voters approve new constitution

Greek voters approve a new constitution put forward by the military junta, which has declared martial law. The constitution suppresses individual freedoms and gives the military broad political powers.

1970   U.S. aid is resumed

President Richard Nixon (1913–94) resumes U.S. aid to Greece, which has been suspended following the 1967 coup.

1971: September 20   Poet George Seferis dies

Nobel Prize-winning poet George Seferis dies in Athens. (See 1900.)

1973: May   Unsuccessful coup by the navy

The Greek navy, long a royalist stronghold, mounts an unsuccessful coup against the ruling junta aimed at returning Constantine to Greece.

1973: June   Greece is proclaimed a republic

In the wake of the failed royalist coup, a Greek republic is proclaimed, and the monarchy is abolished.

1973: July   Papadopoulos becomes president

George Papadopoulos is named president of the republic. He ends martial law and appoints a civilian cabinet. These are token concessions, however, since real power continues to rest with the military, particularly the chief of military security forces, Army Brigadier Dimitrios Ioannides.

1973: November 25   Papadopoulos is ousted by hardliners

Following protests by students and workers, right-wing elements of the military, led by Ioannides, who feel that Papadopoulos has been too liberal oust him in a coup, reimpose martial law, and postpone elections.

1974   Cyprus invaded, junta falls, Greece withdraws from NATO

On July 15, the military government in Athens mounts a coup against the Cypriot government and replaces it with a regime that favors *enosis*. Turkey invades Cyprus five days later. With the junta unable to mount military action against Turkey, it collapses and invites Karamanlis to return from exile to form a civilian government. When Turkey continues its invasion of Cyprus, Greece pulls out of the military wing of

NATO in protest but remains a member of the political branch of the organization.

1974   PASOK is founded

Following the downfall of the military junta, Andreas Papandreou returns from exile and forms the Panhellenic Socialist Movement (PASOK). Throughout the decade, the party gradually gains support in the national elections.

The Greek Republic

1974: July 23   Junta falls after Cyprus invasion

Charged with involvement in a coup by right-wing Greek army officers on Cyprus, Greece's military junta resigns, and its leaders are exiled. A civilian government is formed under former prime minister Constantine Karamanlis, and the 1952 constitution is restored. Constantine Tsatsos (1899–1987) becomes president of the new government.

1974: November 17   Parliamentary elections are held

The first free parliamentary elections since the 1967 coup are held. The party of Prime Minister Karamanlis, New Democracy (ND), wins a substantial majority of seats, and Karamanlis forms a new cabinet.

1974: December   Referendum abolishes the monarchy

A popular referendum abolishes the monarchy and formally deposes King Constantine, who has been in exile since 1967.

1975: June   New constitution is adopted

Greece adopts a new constitution abolishing the monarchy and creating a parliamentary republic. Constantine Tsatsos becomes the president of the new republic.

1976   National Museum building opens

The permanent home of the National Gallery opens on Vassileos Konstantinou Avenue in Athens.

1979   Elytis wins Nobel Prize for Literature

Poet Odysseus Elytis is awarded the Nobel Prize for Literature. (See 1911.)

1980   Karamanlis is elected president

Former prime minister Constantine Karamanlis is elected president.

1980   NATO participation resumes

Greece rejoins the military wing of the NATO.

1981   Language reform is mandated

The Greek government officially simplifies the system of accents used with the Greek alphabet.

1981: January 1   Greece becomes a full member of the EC

Greece's full membership in the European Community becomes effective.

1981: October   First socialist prime minister in office

After gradually building support over the previous decade, the Panhellenic Socialist Movement (PASOK) gains an overwhelming victory in parliament, winning 172 out of 300 seats. Andreas Papandreou, leader of the party and son of former prime minister Georgios Papandreou, becomes Greece's first socialist prime minister. Papandreou has campaigned on a strident anti-American campaign, pledging to pull Greece out of NATO and demand removal of U.S. military bases from Greece. Once in office, however, he moderates his stance but maintains his anti-American rhetoric.

1982   Civil marriages are legalized

The government passes legislation making civil marriages legal, ending the dowry system and allowing divorce by consent.

1982   Papandreou visits Cyprus

Andreas Papandreou visits Cyprus to demonstrate his support for its Greek population. He becomes the first Greek prime minister to visit the island.

1983   Government freezes wages

To combat the nation's inflation rate of twenty-two percent, the Greek government implements a wage freeze.

1983   Communists are granted amnesty

Under Papandreou, the government extends an amnesty to communists exiled since the end of the civil war in 1949.

1983: January 29   Women's rights law is passed

The Greek parliament passes a law intended to give women equal rights within the family, but it is expected to have only a limited effect on the traditional male dominance in the private sphere.

1985   PASOK wins a second term

The ruling PASOK party of Prime Minister Andreas Papandreou wins a majority in parliamentary elections and remains in power for four more years.

1985: March   Socialist is elected president

Christos Sartzetakis (b. 1929), the candidate of the leading socialist party, is elected president with the support of prime minister Papandreou.

1985–89   Troubles plague PASOK government

During Papandreou's second term as prime minister, Greece is beset by economic difficulties and political scandals that taint the PASOK party.

1987   Church property is nationalized

Greece's socialist government takes the controversial step of nationalizing the property of the Greek Orthodox Church. The church responds by excommunicating several government leaders.

1988: October   Financial scandal hits ruling party

Allegations of financial wrongdoing are leveled against PASOK. Three cabinet ministers resign or are dismissed. Greece's worst financial scandal, which also extends to the arms industry, ultimately leads to the ouster of the Papandreou government.

1989: June 18   Socialists are defeated in elections

ND and the Communist Party of Greece (KKE) defeat the PASOK government of Andreas Papandreou in parliamentary elections and form a coalition government. The decision of the center-right ND and the far left KKE—traditionally bitter enemies—to form a coalition indicates the deep hostility felt by these parties toward the center-left PASOK.

1990: April   ND leadership confirmed in new elections

The New Democratic Party wins another electoral victory, and its leader, Constantine Mitsotakis, becomes prime minister.

1992   Papandreou is acquitted on corruption charges

Former prime minister Andreas Papandreou is acquitted on charges stemming from the financial scandal that rocked his government in 1989.

1992: July   Greece approves Maastricht Treaty

Greece becomes the first nation to ratify the Maastricht Treaty laying the foundation of European economic and monetary union.

1993: October   PASOK triumphs in elections

The PASOK party wins a wide majority in national elections, and Andreas Papandreou becomes prime minister for a third time.

1995: March   Stephanopoulos is elected president

Parliament elects the socialist-backed candidate Costis Stephanopoulos (b. 1925) president. Although supported by PASOK, Stephanopoulos is a former member of ND who founded his own party, Democratic Renewal, in the mid-1980s and whose political philosophy is best characterized as center-right.

1996: January   Greek-Turkish war avoided

Greece and Turkey nearly go to war over conflicting claims on uninhabited islands in the Aegean Sea. The crisis is defused following United States mediation. The incident is only the latest in a series of disputes over air, sea, and mineral rights in the Aegean that stretch back to the early 1970s.

1996: June   Death of Papandreou

Three-term prime minister Andreas Papandreou dies six months after resigning from office.

1996: September 22   PASOK wins parliamentary elections

To confirm support for his government in the face of opposition to economic austerity measures, Prime Minister Costas Simitis calls new elections, in which his party, PASOK, wins a decisive victory.

1997: September 5   Greece is chosen to host the 2004 Olympics

The International Olympic Committee (IOC) announces that Athens has won its bid to host the 2004 Summer Olympics, edging out rival cities Rome, Buenos Aires, Stockholm, and Cape Town. The city plans to construct a new subway, airport, and main road before the games. Gianna Angelopoulos-Daskalaki, who heads the city's bid for the games, is also to chair the organizing committee for the 2004 Summer Olympics, becoming the first woman to head an Olympic organizing committee.

1998: April 23   Karamanlis dies

Former prime minister and president Constantine Karamanlis dies in Athens at the age of 91. (See 1907.)

1999   Popular opposition to NATO's campaign against Yugoslavia

The overwhelming majority of the Greek population opposes the NATO bombing campaign of Greece's traditional allies, the Serbs. Although the Greek government also opposes the bombing and refuses to use its own aircraft in the operation, it does permit its NATO allies use of Greek facilities for the operation. In addition, Greek troops participate as peacekeepers in the Serbian province of Kosovo once the air war is over.

1999: June 27   George Papadopoulos dies

Former junta leader George Papadopoulos dies of cancer at age eighty. Unrepentant and refusing to seek clemency till the end, he spends his last years in a guarded hospital room. His death is a low-key affair although about 1,500 junta supporters attend his funeral in Athens.

Bibliography

Clogg, Richard, ed. *Greece, 1981–89: The Populist Decade.* New York: St. Martin's Press, 1993.

Costas, Dimitris, ed. *The Greek-Turkish Conflict in the 1990s.* New York: St. Martin's Press, 1991.

Hadas, Moses. *History of Greek Literature.* New York: Columbia University Press, 1950.

Jouganatos, George A. *The Development of the Greek Economy, 1950–91.* Westport, Conn.: Greenwood Press, 1992.

Kourvetaris, George A., and Betty A. Dobratz. *A Profile of Modern Greece, in Search of Identity.*

Laisné, Claude. *Art of Ancient Greece: Sculpture, Painting, Architecture.* Paris: Terrail, 1995.

Lawrence, A.W. *Greek Architecture.* New Haven, Conn.: Yale University Press, 1996.

Legg, Kenneth R. *Modern Greece: A Civilization on the Periphery.* Boulder, Colo.: Westview Press, 1997.

Pettifer, James. *The Greeks: The Land and People Since the War.* New York: Viking, 1993.

Woodhouse, C. M. *Modern Greece: A Short History.* 5th ed. London: Faber & Faber, 1991.

Hungary

Introduction

Located at the intersection of Eastern and Western Europe and the Balkans, Hungary is both a modern industrial nation and home to a culture whose history spans centuries—the one thousandth anniversary of Hungary's conquest by the Magyars (ethnic Hungarians) was celebrated over a hundred years ago. Hungary's strategic location has brought long periods of foreign domination and territorial changes: the land that extended from the Adriatic Sea to the Black Sea in the Middle Ages lost three-fifths of its territory to Romania, Yugoslavia, and Czechoslovakia following World War I (1913–18). Whatever the difficulties of their history, though, the Hungarians have always retained a strong sense of national identity and a willingness to fight for their freedom, whether in 1848 against the Habsburg empire or in 1956 against the Soviet Union. In the 1990s Hungary has weathered the political and economic adjustments following the breakup of the Soviet Union. As elsewhere in Central and Eastern Europe, this period has brought old ethnic tensions to the forefront, but Hungary has found peaceful political ways of dealing with them, both internally and in relations with its neighbors.

A landlocked nation in Central Europe, Hungary occupies an area of 35,919 square miles (93,030 square kilometers) and consists of four major geographic regions: the Great Plain, which extends across more than half the country's area; the Little Plain to the northwest; Transdanubia, which consists of low hills and valleys and includes three of the country's largest lakes; and the northern highlands. According to the 1990 census, Hungary had an estimated population of ten million people, of whom roughly one-fifth lived in the capital city of Budapest.

History

The land that today is Hungary was first conquered by the Magyars (ethnic Hungarians) in A.D. 896, having previously been settled by Celtic tribes, Romans, and Germanic tribes. The Magyars established the Arpad dynasty, which lasted four hundred years. A milestone during this period was the conversion of the population to Christianity during the reign of King (later Saint) Stephen (r. 1001–38). After the reign of the last Arpad monarch, King Andrew III, Hungary fell under foreign rule, including that of the French Angevins, under whom the land flourished during the fourteenth century, becoming a major commercial and intellectual center. The first university was established at Pecs in 1367. Another golden era occurred in the latter half of the fifteenth century, during the reign of the enlightened monarch Matthias Corvinus, who combined foreign conquest with successful domestic administration and support of intellectual pursuits.

In the sixteenth century Hungary became the easternmost country to come under the influence of the Protestant Reformation: nearly the entire population converted to Protestantism, although the Protestant population decreased significantly with the counter-Reformation of the following century. Nevertheless, Hungary would continue to have the largest Protestant population of any Eastern European country. The sixteenth century also saw the conquest of Hungary by the Ottoman Turks, although their rule was soon challenged by Austria's Habsburg dynasty as well as the Hungarian nobility, and control of the region was divided among the three groups.

With the Austro-Turkish War at the end of the seventeenth century, the Habsburgs forced the Ottomans out of Hungary, and local resistance was overcome by 1713, although Hungary retained its own parliament (called the *Diet*). By the early decades of the nineteenth century, nationalism was flourishing in Hungary, with the more moderate activists advocating autonomy within the Austrian Empire, while the radicals called for complete independence. The political and intellectual ferment of this period climaxed in the revolution of 1848, when Lajos Kossuth (1802–94) and Ferenc Deak (1803–73) led their peers in declaring an independent Hungarian republic, which was crushed within months by the Austrians with the aid of Russian troops. By 1867, however, weakened by pressure from Prussia, Austria signed a compromise agreement establishing the Dual Monarchy, by which Austria and Hungary were to have separate and equal governments, and only selected government functions would be administered jointly.

In spite of the looser Dual Monarchy arrangement, Hungary was obliged to ally itself with Austria in World War I,

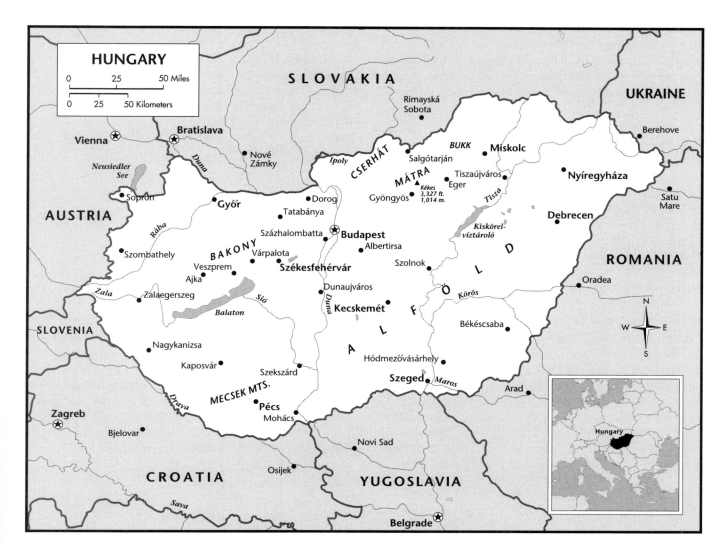

HUNGARY

0   25   50 Miles

0   25   50 Kilometers

whose outcome was disastrous for both countries. For Hungary, it resulted in the Treaty of Trianon, which divested the nation of three-fifths of its land and over half its population. By 1918 the Dual Monarchy was dissolved, and Hungary was proclaimed a republic. Within a few months, a communist government gained the ascendancy but was soon overthrown by an alliance of Hungary's neighbors who feared the spread of communism to their own countries. Between 1920 and 1944, Hungary's government was headed by Admiral Miklos Horthy (1868–1957), serving as regent. (Hungary had officially reverted to being a monarchy but had no king, its last one having abdicated.) In the 1930s, Hungary, beset by economic depression and hoping to regain its lost territories, established closer ties with Nazi Germany and was then obliged to enter World War II on the side of the Axis powers. After attempting to reestablish neutrality part way through the war, Hungary was invaded and occupied by Germany in 1944.

Liberated by Russia in the final months of the war, Hungary fell under the Soviet sphere of influence but still held democratic multiparty elections as mandated by the Yalta Agreement, and Hungary once again became a republic. However, Russian-backed communists gradually strengthened their position in the government until they assumed complete control by 1949, forming a single-party government modeled on that of the Soviet Union. In 1955 Hungary joined the Warsaw Pact, an alliance of Soviet satellite states formed by Russian premier Nikita Krushchev (1894–1971). However, popular discontent with communist rule and Soviet domination was growing and, encouraged by the success of a Polish attempt to win greater freedom from Soviet control, the Hungarian people staged an uprising during the last week of October, 1956. A democratic coalition government was formed under the leadership of former prime minister Imre Nagy (1896–1958), who vowed to implement moderate socialist economic policies within the framework of a multiparty government. Although initially appearing to accept the steps taken by the Hungarians, the Soviets sent in troop reinforcements and tanks and retook control of the country by force, installing Janos Kadar (1912–89) as prime minister and inflicting widespread retaliation on persons linked to the uprising. Nagy and his political associates were taken into custody and executed the following year, and around 200,000 Hungarians fled to other countries, robbing Hungary of many talented and well-educated professionals in a variety of fields.

Although serving only two terms as prime minister, Kadar remained in control of Hungary's government for over three decades as head of its Communist Party. During this time, the nation enjoyed one of the strongest economies in the Soviet bloc. In 1968 Kadar relaxed his government's adherence to the socialist economic model with a program called the "new economic mechanism" that introduced greater flexibility into the management of public enterprises, allowed for private ownership of small businesses, and led to increased trade with the West. Also in the 1960s, the Communist Party's harsh stance toward the Catholic church was relaxed. Relations between the two institutions improved following a 1964 accord, and by 1978 Hungary resumed diplomatic ties with the Vatican in Rome.

By the late 1970s, disruptions in energy supplies from the Middle East, economic problems suffered by Hungary's trade partners, and other factors were negatively affecting the nation's economy. In the 1980s, the Hungarian people became increasingly critical of Kadar's policies, and there was a growing trend toward a more democratic society. Parliamentary reforms were enacted in 1985. The following year there were mass student demonstrations, and for the first time the country's writers' union elected a totally non-communist leadership despite government pressure. In 1987 the longtime prime minister was replaced by a more reform-minded politician, Karoly Grosz, (b. 1930) and the following year Kadar retired from his post as party leader. Yet another important sign of changing times was the 1989 disinterment of the remains of Imre Nagy, the political leader who had been secretly executed and buried in an unmarked grave for his role in the 1956 uprising, and his re-burial with full public honors. On October 23 of the same year—the anniversary of the 1956 revolt—a new Hungarian Republic was proclaimed. Early in 1990 the nation held its first multiparty elections since the late 1940s, and a coalition government was formed.

As the decade progressed, support for the new government was eroded by the unwelcome economic effects of its privatization program—including high rates of unemployment and inflation—and political conflicts relating to tension among Hungary's varied ethnic groups. In 1993 the ruling government was forced to distance itself from its ultra-nationalist right wing, which then formed a new party. By 1994, the Hungarian electorate, dissatisfied with the government's economic liberalization program, elected a socialist government, which, ironically, speeded up privatization and reduced spending on social welfare programs. By 1997 the government's economic program, together with a credit arrangement by the International Monetary Fund and increased levels of foreign investment, had eased the country's economic situation.

In 1996 Hungary and Romania signed an agreement resolving a longtime border dispute and guaranteeing the rights of ethnic Hungarians living in Romania. For both countries, the pact was seen as a prelude to eventual membership in the European Union and NATO. In 1997 NATO announced that Hungary would be admitted to full membership in the alliance in 1999.

Timeline

Early History

1000 B.C.   Celtic tribes occupy Hungary

Present-day Hungary is inhabited by Celtic tribes, including the *Eravisci,* toward the end of the first millennium B.C.

9 B.C.   Romans conquer Hungary

During the reign of Augustus Caesar, the Romans occupy the western part of the Carpathian basin, which they incorporate into their provinces, Pannonia and Dacia.

4th century A.D.   Germanic tribes invade

Roman rule ends as Germanic tribes invade from the east, followed by the Huns from Asia.

406–453   Rule of Attila the Hun

Attila makes Hungary the center of his empire.

567–805   Rule by the Avars

The Avars, a Germanic tribe, rule Hungary for over two centuries.

The Magyars Conquer Hungary

896   The Magyars invade

The Magyars (Hungarians), a Finno-Ugric people whose culture and language have been influenced by the Turks, migrate from areas near the Ural mountains and conquer the region that will become Hungary. In Bulgaro-Turkic, their name is *onogur* (ten tribes), from which "Hungarian" is derived. They establish the Arpad dynasty.

955   Otto of Germany limits Magyar power

Otto the Great (912–73) of Germany, the Holy Roman Emperor, defeats the Magyars at Lechfeld near Augsburg and prevents further Magyar expansion into Eastern Europe.

1001–38 Reign of King Stephen

Stephen (977–1038), later St. Stephen, becomes Hungary's first king, instituting a centralized government. Although foreigners hold positions at court, the native Magyar language remains the official tongue. Under Stephen's reign the Hungarians are converted to Christianity, when he accepts a cross and crown from Pope Sylvester II.

1077–95 Reign of Ladislaus I

Under the reign of Ladislaus I Croatia is conquered.

1083 St. Stephen is canonized

King Stephen is canonized, becoming the patron saint of Hungary. August 20 is designated as his feast day.

1222 Golden Bull is signed

Nobles force King Andrew II to sign the Hungarian counterpart to England's Magna Carta, limiting the king's power, establishing a set of civil liberties, and laying the groundwork for the establishment of a parliament.

1241–42 Mongols invade

Mongol invasions, followed by famine, devastate Hungary, which loses half its population.

1301 Arpad dynasty ends

Hungary's first dynasty ends when King Andrew III dies. His death is followed by a period of civil war and foreign rule.

A Medieval Golden Age

14th century Period of economic and cultural achievement

Medieval Hungary, especially under Angevin rulers Charles Robert and Louis the Great, flourishes with the control of trade routes between the Baltic and Black seas and dominance in the European production of gold. Hungary becomes an international cultural and intellectual center, and crafts such as silversmithing flourish.

1308–42 Reign of Charles Robert

Charles Robert of Anjou establishes the Angevin dynasty in Hungary. Under Charles Robert, the monarchy becomes stronger and the power of the nobles shrinks.

1342–82 Reign of Louis I

Louis I, Hungary's second Angevin ruler, greatly enlarges Hungarian territory by gaining control of Dalmatia and Poland.

1367 First university is founded

Hungary's first university is established at Pecs.

15th century Ottomans stage series of attacks

Beginning in the mid-fifteenth century, the Ottoman Turks attack Hungary. At first they are driven back under the leadership of military hero Janos Hunyadi, but their attacks become more difficult to withstand as the kingdom is weakened by royal mismanagement, military decline, and a peasant revolt.

1432–72 Life of poet Janus Pannonius

The poems of Pannonius, Hungary's first notable author, are written in Latin. Hungarian as a written language does not become well established until the Reformation (sixteenth century).

1458–90 Reign of Matthias Corvinus

Matthias Corvinus, one of Hungary's most illustrious monarchs, successfully resists the Ottomans and conquers additional territory including Bohemia, Moravia, and Silesia. He further strengthens Hungary by undertaking administrative, judicial, and military reorganizations. An enlightened monarch, he founds universities and supports European humanistic culture. During his reign, many Italian artists and scholars come to the Hungarian court.

1472 First printing press operates

Hungary's first printing press is established by Andras Hess.

16th century Protestant Reformation influences Hungary

Hungary is the easternmost country affected by the Protestant Reformation. A large majority of Hungarians are converted to Protestantism, although many revert to Catholicism during the Counter-Reformation of the following century. However, Hungary retains the largest Protestant population of any Eastern European country.

1514 Peasant revolt

Thousands of Hungarian peasants, inspired by the leadership of Gyorgy Dozsa, revolt against their oppression by the Hungarian kings, overrunning feudal estates, killing members of the nobility, and burning their mansions. Royal troops suppress the rebellion, burning Dozsa to death and executing a total of around seventy thousand peasants in retaliation.

Ottoman Conquest

1526: August 29 Turks triumph at Battle of Mohacs

Hungary's rulers are decisively defeated by the Ottomans at the Battle of Mohacs. The Turks then march northward and capture the royal capital of Buda. Ongoing warfare over control of Hungary rages between the Hungarian nobles, the Ottoman Turks, and the Habsburgs of Austria. Ultimately, the Turks control the central part of the country, Austria controls the west, and the remaining region, consisting of Transylvania and the neighboring areas, becomes an autonomous principality ruled by Hungarian nobles.

1554: October 20   Birth of poet Balint Balassi

Hungary's greatest sixteenth century poet celebrates love, nature, and heroism in battle in lyric poems modeled after those of Ovid and influenced by the Western European Renaissance. Balassi's poems about war stem from his own experience as a soldier, and he dies (1594) after being wounded at the Battle of Esztergom.

1572–1608   Reign of Rudolph II

Rudolph II adopts a royal policy favorable to the Catholic Counter-Reformation, and the number of Protestants in Hungary declines from its high point in the fifteenth century, when it had reached 90 percent of the population.

1590   The Bible is translated into Hungarian

Gaspar Karoli produces the first complete Hungarian translation of the Bible.

1594: May 30   Death of poet Balint Balassi

Balint Balassi, the outstanding poet of his era, dies of wounds suffered at the Battle of Esztergom. (See 1554.)

1604–06   Bocskay leads revolt against Habsburgs

National hero Stephen Bocskay leads an uprising against the Habsburgs, winning political and religious concessions at the Peace of Vienna.

1635   Budapest University is founded

Budapest University is established.

1682–99   Habsburgs drive Turks out of Hungary

The defeat of the Ottomans in the Austro-Turkish War signals the ascendancy of the Austrians and the return of Catholic rule.

Habsburg Rule

1699   Peace of Karlowitz formalizes Austrian gains

At the Peace of Karlowitz ending the Austro-Turkish War, the Habsburg Empire formally annexes Hungarian lands that had been controlled by the Ottomans, as well as Croatia and Transylvania. However, most Hungarians do not accept Austrian domination and numerous uprisings are staged in the following years.

1703–11   Transylvanian prince leads revolt

Prince Ferenc Rakoczy II of Transylvania leads a major revolt against the Austrian Habsburgs, but it ultimately fails.

1713   Hungarians accept Habsburg rule

The Hungarian Diet (parliament) accepts the Pragmatic Sanction, effectively turning over rule of Hungary to the Habsburgs. However, the Diet, composed of two houses, called the Table of Magnates and the Table of Deputies, continues to function.

1740–48   War of the Austrian Succession

Hungary's Table of Magnates (one of two houses of parliament) supports Maria Theresa's claim to the Austrian throne, accepting her status as queen of Hungary.

1759: October 27   Birth of literary reformer Ferenc Kazinczy

Kazinczy is a scholar and author who works for the improvement of Hungarian literature by advocating reforms in both literary style and in the Hungarian language itself. Schooled extensively in foreign languages in his youth, he studies law and becomes a civil servant but continues to devote himself to literature, founding a literary review and becoming the center of a circle of writers. He corresponds with many European writers, translates plays into Hungarian, and introduces the sonnet form to Hungary, writing many sonnets of his own. Kazinczy introduces important language reforms—including reforms in grammar and spelling—that improve the quality of Hungarian literature. In addition to poetry and articles, Kazinczy's works include a collection of epigrams titled *Tovisek es viragok* (Thorns and Flowers; 1811). Kazinczy dies in 1831.

1782   Design school is founded

The *Institutum Geometricum* (later renamed the Technical University of Budapest) is established as a school for secular (as opposed to church) architecture and is the first institution in Europe to offer an engineering degree.

1801   National Szechenyi Library is founded

Hungary's largest library, containing about seven million volumes in the 1990s, is established.

1802   Hungary's first museum is established

The Hungarian National Museum is founded in Budapest to house the natural history, archaeological, and coin collections of Count Ferenc Szechenyi, which the count has donated to his country. Architect Mihaly Pollack designs a neoclassical building to house the collections.

1811: October 22   Birth of composer and pianist Franz Liszt

One of the great nineteenth-century composers and pianists, Franz Liszt (1811–86) is born in Hungary, where he lives until the age of ten, when his family moves to Vienna and then, two years later, to Paris. In his twenties, Liszt pursues a career as a pianist, settling in Geneva, Switzerland and touring Europe in 1839. Settling in Weimar in 1848, the musician ends his concert career and devotes himself to composing and conducting.

Liszt's compositions encompass many well-known orchestral works and piano pieces, including two piano concertos, the *Totentanz* for piano and orchestra, the *Faust Symphony,* the *Dante Symphony,* twelve symphonic poems for orchestra, and the very popular *Hungarian Rhapsodies* for piano. Liszt's late works are of a religious nature, including the *Legends* for piano and the *Christus* oratorio. Although many of Liszt's compositions are written in a popular, easily accessible Romantic style, his works also contain many harmonic and structural innovations that foreshadow the achievements of the great late-nineteenth-century and early-twentieth century French and Austrian composers. Liszt dies in 1886.

1815–48   Rise of Hungarian nationalism

Following the Napoleonic period, nationalism and reformism flourish in Hungary. Hungarian leadership is divided between radicals, such as Lajos (Louis) Kossuth, who advocate a complete break with Austria, and moderates, who want Hungary to remain a self-governing part of the Austrian empire.

1823: January 1   Birth of poet Sandor Petofi

Petofi, Hungary's national poet, is renowned both for the power of his poems and for his dedication to Hungarian national ideals and the revolution of 1848–49. Petofi's first published volume, entitled *Versek* (1844), establishes his reputation as a poet. He is a leading Hungarian literary and intellectual figure in the years preceding the revolution, when he serves as assistant editor of the periodical *Pesti Divatlap.* The poet enthusiastically espouses the principles of the French Revolution and attacks social injustice in Hungary.

Petofi's poem *Talpra magyar* (Rise, Hungarian) becomes the unofficial anthem of the 1948 revolution, in which he serves as an aide-de-camp of General Jozef Bem. The poet is captured by Russian forces at the Battle of Segesvar on July 31, 1849. Petofi is long thought to have died in battle, but new access to Soviet archives in the 1980s reveals that the poet is among a large group of Hungarian prisoners deported to Siberia, where he dies in 1856, reportedly of tuberculosis. His poetry is acclaimed for its expression of complex ideas in a passionate and easily accessible style. His epic poem *Janos vitez* (1845) remains one of his most popular works.

1823: January 21   Birth of poet and playwright Imre Madach

Madach is renowned as the author of the fifteen-act philosophical play *Az ember tragedidja* (The Tragedy of Man; 1861), which analyzes mankind's destiny through episodes in which the characters of Adam and Eve appear in different historical periods, constantly carrying out their conflict with Lucifer. Madach dies in 1864. Although written to be read rather than performed, *Az ember tragedidja* receives its stage premiere in 1883 at the Budapest National Theatre and remains a popular stage drama.

1825   Academy of sciences is founded

The Hungarian Academy of Sciences is established and plays an important role in integrating Western knowledge into the intellectual life of Hungary.

1831   Death of author and language reformer Ferenc Kazinczy

Leading literary figure Ferenc Kazinczy dies. (See 1759.)

1831: September 9   Birth of sculptor Miklos Izso

Hungary's most famous nineteenth-century sculptor, Izso takes part in the revolution of 1848 as a youth and is forced into hiding afterward to avoid retaliation. He later takes up stone carving and becomes a pupil of noted sculptor Istvan Ferenczy. Going abroad in 1857, he studies in Vienna and Munich before permanently returning to Hungary. Themes of Hungarian identity are central to Izso's sculpture, which includes such works as *Piping Shepherd* (1860), the *Dancing Peasant* series of terra-cotta and gypsum sculptures, and many busts of famous Hungarians in politics and the arts. Izso dies in Budapest in 1877.

1844: February 20   Birth of painter Mihaly Munkacsy

The most internationally acclaimed Hungarian painter, Mihaly Munkacsy (1844–1900) is born Mihaly Lieb, later taking as his professional name the region of his birth (Munkacs, now Mukachevo, Ukraine). Beginning in 1864, he studies abroad in Vienna, Munich, and Dusseldorf. His painting *Death Row,* which wins a gold medal at the Paris Salon, launches his career. Other paintings include *Lint Makers* (1872), *Woman Churning* and *Parting* (both 1872–73), *Pawn Shop* (1874), *Taking the Land* (1893), and *Parc Monceau at Night* (1895). Munkacsy moves to Paris in 1871. He dies near Bonn in 1900.

1848-49 Rebels declare independence from Austria

Hungarian nationalists led by Lajos Kossuth and Ferenc Deak revolt and issue a declaration of independence, forming their own parliamentary government. Austrian emperor Franz Joseph enlists Russian aid in ending the Hungarian rebellion and crushing the newly formed republic.

1848 March Laws grant Hungarian autonomy

Fearing a Hungarian revolt similar to those in Western Europe, Austria issues a directive granting Hungary virtual autonomy within the Habsburg empire.

1849: April Hungarian republic is proclaimed

Lajos Kossuth proclaims Hungary an independent republic.

1849: August 13 Hungarian resistance is defeated

After their defeat at the Battle of Temesvar, Hungarian nationalists surrender to Austrian and Russian troops. Thirteen Hungarian generals are hanged in retaliation, and rebel leader Lajos Kossuth goes into exile in Italy.

1856 Death of poet Sandor Petofi

Poet Sandor Petofi, captured at the Battle of Segesvar, dies after being transported to a Russian prison camp in Siberia. (See 1823.)

The Dual Monarchy

1867 Compromise agreement establishes a dual monarchy

Weakened by the Austro-Prussian War, Austria agrees to grant limited autonomy to Hungary, and the Austro-Hungarian empire, with a dual monarchy consisting of two separate and equal governments, is established. Each government has its own constitution and parliament, but foreign affairs and finances are administered jointly.

1868: June 18 Birth of Admiral Horthy

Miklos Nagybanya Horthy (1868–1957), the first head of state of modern Hungary, embarks on a naval career early in life, attending the naval academy at Fiume, and holding various commands before becoming the naval aide-de-camp of the Emperor Franz Josef between 1909 and 1914 and serving with distinction in World War I. In 1920 Horthy is chosen as regent (acting head of state in the absence of a monarch) of the postwar Hungarian government. In the 1930s Horthy takes a friendly stance toward Germany because of the Germans' promises to help Hungary regain the territory lost through the Treaty of Trianon (see 1920). He remains in office during World War II, which Hungary enters on the side of Germany, although it attempts to break away in 1943 and is occupied by the Germans. Horthy is imprisoned in Germany for the last year of the war after attempting to sign an armistice with Russia. He is freed by Allied forces at the conclusion of the war and spends his last years in Portugal, where he dies in 1957.

1870s Industrialization begins

Manufacturing becomes significant in Hungary. Peasants migrate to the cities to work in agriculture-related industries including flour processing, tanning, and brewing.

1873 Buda, Pest, and Obuda are united

The cities if Buda, Pest, and Obuda are merged into a major metropolitan area, which becomes a major commercial and artistic center.

1877: May 29 Death of sculptor Miklos Izso

Leading sculptor Miklos Izso dies in Budapest. (See 1831.)

1877: November 22 Birth of poet Endre Ady

An innovative and influential lyric poet, Endre Ady (1877–1919) works as a journalist in Hungary and Paris beginning in 1900. His first major success comes with the publication of the volume *Uj versek* (New Poems; 1906). Both the content and language of his poetry are revolutionary and controversial, and he inspires both acclaim and condemnation among his contemporaries, chiefly because he is also highly critical of his country, especially in his early poetry. Ady dies in his forties of alcoholism, after publishing ten volumes of poetry and many articles and short stories.

1878: January 12 Birth of author Ferenc Molnar

Ference Molnar (1878–1952), playwright, novelist, and short story writer, is born in Budapest and publishes his first short fiction at the age of nineteen. His play *Az ordog* (The Devil; 1907) establishes his reputation. A number of his plays are performed in other countries, including Austria, Germany, and the United States, and some are later made into films. Molnar is acclaimed for dealing with serious subjects using humor and irony. His short stories, particularly those collected in *Muzsika* (1908), have achieved great critical success. Molnar is also the author of novels including *A Pal utcai fiuk* (The Paul Street Boys; 1907). Molnar dies in New York City in 1952.

1881: March 25 Birth of composer Bela Bartok

Hungary's most renowned composer and an accomplished pianist, Bartok receives his early musical instruction from his mother and is performing his own compositions at the piano in public by the age of ten. After formal music studies at the Royal Hungarian Music Academy, he embarks on the joint pursuits of composition and collecting Central European folk music, which has a strong influence on his own work. He also joins the piano faculty at his alma mater and composes many works for piano, notably the *Mikrokosmos,* a series of rhythmically and harmonically innovative pieces that progresses from simple to increasingly advanced levels of study.

After the premiere of Bartok's opera *Bluebeard's Castle* (1918), the composer gains recognition throughout Europe and internationally, touring the United States and the Soviet Union. In 1940 he becomes a permanent resident of the United States, where he lives in New York and receives a grant from Columbia University to archive Yugoslav folk music. At the time of his death in 1945, he is completing his Third Piano Concerto. Bartok's best-known compositions include his three piano concertos, a violin concerto, six-string quartets, two sonata for violin and piano, the *Concerto for Orchestra,* the *Cantata profana* for chorus and orchestra, and a wide variety of piano works.

1882: December 16   Birth of composer Zoltan Kodaly

Hungary's most famous twentieth-century composer after his colleague Bela Bartok, Kodaly, like Bartok, works with Hungarian folk material, which the two composers jointly collect. Kodaly first achieves international recognition with his *Psalmus Hungaricus* (1923), a setting of the fourth Psalm commemorating the union of the cities of Buda and Pest fifty years earlier. In addition to his work as a composer, Kodaly is also a respected music critic and lecturer. His best-known composition is the opera *Hary Janos* (1926). Kodaly dies in 1967.

1884   National opera house is established

The State Opera House is founded. Opera becomes very popular throughout Hungary.

1886: July 31   Death of composer Franz Liszt

Internationally acclaimed Hungarian-born composer and pianist Franz Liszt dies. (See 1811.)

1896   Hungary celebrates its one thousandth anniversary

Hungarians commemorate the one thousandth anniversary of the Magyar conquest (see 896) in a giant celebration that includes fireworks and other festivities throughout the nation.

1896   Museum of Fine Arts is established

The Museum of Fine Arts is founded in Budapest to house Prince Miklos Esterhazy II's collection of Hungarian paintings, which the state has acquired through an act of parliament.

1896   Birth of political leader Imre Nagy

Born to a peasant family, Nagy becomes a communist when he is taken prisoner by the Russians in World War I and fights in the Bolshevik Revolution. He later returns to Hungary but by 1929 he is forced into political exile in Russia, where he remains until 1944. After the war, he is once again active in Hungarian politics and becomes prime minister in 1953 but is ousted by hard-line communists two years later. Nagy becomes the political head of the new government formed during the uprising of October, 1956, which outlines a liberal program combining private enterprise with a welfare state. The revolutionary government is crushed by Soviet troops in the beginning of November. Nagy and his colleagues are taken into custody by the Russians, who execute them two years later. In the late 1980s, a new liberal government has Nagy's remains exhumed and accords the slain leader the funeral of a national hero.

1900: May 1   Death of painter Mihaly Munkacsy

Hungary's most internationally renowned painter dies near Bonn, Germany. (See 1844.)

1901   Soccer association is formed

Hungary's first national soccer organization is formed.

1901   First film is made

The first Hungarian film, a newsreel, is produced in Budapest.

1902: October 12   First international soccer match is played

Hungarian soccer players participate in their first international match, against Austria.

1912: May 26   Birth of political leader Janos Kadar

Kadar, Hungary's longest-serving postwar communist leader, is born to a working-class family and works as a machinist. He joins the Communist Party in 1931 and is one of the organizers of the Hungarian resistance to the Nazis in World War II. After the war, he holds a succession of official posts, including minister of the interior and party secretary in Budapest. In the early 1950s he briefly falls out of favor, is imprisoned, and rehabilitated.

Although Kadar holds a position in the coalition government of Imre Nagy that comes to power during the uprising of October 1956, the Soviets sponsor him as the head of the government that is imposed on Hungary following the revolt. Kadar serves as prime minister from 1956 to 1958 and from 1961 to 1965 and as the head of Hungary's ruling party, renamed the Hungarian Socialist Workers' Party continuously until 1988. Under Kadar, the economic liberalization known as the "new economic mechanism" is introduced in

Territorial losses after World War I.
☐ Land lost

1968. Hungary prospers economically under Kadar's policies, and he retains a great degree of public support throughout the 1960s and 1970s. However, economic problems in the 1980s erode some of this support. Kadar resigns his post and retires in May, 1988. He dies the following year.

World War I and Its Aftermath

1914: June 28   Archduke Franz Ferdinand is assassinated

Franz Ferdinand, heir to the Austro-Hungarian throne, is assassinated in Sarajevo, an event that touches off World War I. Due to its political affiliation with Austria, Hungary enters the war as an ally of the Central Powers (Austria and Germany). In the course of the war, more than 380,000 Hungarian lose their lives in combat.

1916   Popular opposition to the war increases

Opponents of World War I break an internal truce and begin an active antiwar protest campaign.

1918: November   Communist party is formed

Returning Hungarian soldiers exposed to Marxism during World War I while in Russian prison camps establish the Communist Party of Hungary and work to spread communism among Hungary's workers and soldiers, many of whom are unemployed following the war.

1918: November 11   Dual monarchy is dissolved

Hungarian discontent aroused by World War I brings about the collapse of the dual monarchy. Its subjects in the various regions of Central Europe shift their allegiance to their own respective ethnic and nationality groups.

1918: November 16   Hungarian republic is proclaimed

A democratic republic is proclaimed under the leadership of Count Mihaly (Michael) Karolyi.

1919: January 27   Death of poet Endre Ady

Endre Ady, Hungary's first great modern poet, dies in Budapest. (See 1877.)

1919: February   Communist leaders are arrested

The government attempts to deal with the communist threat by arresting the party's leaders. Leaders of the communists and social democrats decide to merge the two parties, giving the communists a strong political base.

1919: March 21   Communist government is formed

Instead of aid from the West, war-torn Hungary receives an ultimatum demanding that it cede part of its territory, and Karolyi resigns his post in protest. Hungary looks to the new Soviet regime in Russia for help. The communists gain power locally under Bela Kun, who becomes the head of a Republic of Councils that rules Hungary for 133 days, instituting socialist policies.

1919   Kun government overthrown by foreign troops

Fearing the spread of communism to their own countries, Czechoslovakia, Serbia, and Romania, with help from France, send troops to overthrow Kun's government. Romanian troops occupy and loot Budapest.

1920–44   Horthy regime rules Hungary

After the downfall of the communist Kun government, Hungary comes under the control of a counterrevolutionary regime headed by naval commander Admiral Miklos Horthy. Because its last monarch, Charles II, has renounced power but not abdicated, Hungary is still nominally a monarchy, and Horthy becomes an elected regent acting as head of state, a position he holds until the final year of World War II.

1920–21   "White Terror"

Attacks known as the "White Terror" are carried out against supporters of the ousted communist government, the earlier democratic government, and Hungary's Jewish population.

1920   National Assembly is elected

Exercising universal suffrage and secret ballot rights, Hungarians elect a National Assembly. The majority of seats is won by the conservative Smallholders' Party, and the Christian National Party comes in second.

1920: June 4   Treaty of Trianon

The Treaty of Trianon, imposed on Hungary by the victorious Western powers, takes away over two-thirds of its territory, which is divided among Romania, Czechoslovakia, and Yugoslavia. The dismemberment of the country leaves its economy in shambles and large portions of its population living under foreign domination. Hungary later allies itself with Nazi Germany in hopes of regaining its lost territory.

1922–32   Bethlen in office as premier

Under the stewardship of Count Istvan Bethlen, who serves as premier for a decade, Hungary's two leading political parties are merged to form the United Party. Limited land reforms are carried out, rural education is improved, and the secret ballot is reinstituted in rural areas. Under a program of economic recovery, industry is expanded and inflation is halted. Hungary joins the League of Nations and later forms an alliance with Italy.

1926   First professional soccer games are played

Professionalism is introduced to Hungarian soccer.

1932   Gombos comes to power

The economic instability that accompanies the global Depression of the 1930s brings right-wing military leader Gyula Gombos to power as premier. Gombos supports fascism and Nazism and establishes closer ties with Germany.

World War II

1939–45   World War II

Hungary sides with Germany and the Axis powers in World War II and temporarily regains the territory lost to Romania, Czechoslovakia, and Yugoslavia. However, when Hungary later attempts to assume a neutral stance, it is invaded and occupied by the Germans, who, in turn, are driven out by Russia.

1940: November   Hungary signs Tripartite Pact

Hungary signs the Tripartite Pact committing it to support the Axis powers in World War II. In spite of a previous pact between Hungary and Yugoslavia, Germany crosses Hungarian territory to invade Yugoslavia, an act that Hungarian leader Pal Teleki protests by committing suicide.

1941   Hungary declares war on Russia

Claiming that Russian planes have bombed Kosice, Prime Minister Laslo Bardossy commits Hungary to joining the German offensive against Russia by declaring war on the USSR.

1944: March   Germany occupies Hungary

Germany invades and occupies Hungary. Political parties are disbanded, and the pro-German government begins deporting Hungarian Jews to concentration camps.

1944: October 15   Hungary seeks separate truce with Russia

Realizing that Germany cannot win the war, Admiral Horthy sends a diplomatic mission to Moscow to negotiate a separate peace, but the Germans force him to withdraw his offer and abdicate the regency.

1944–45   Hungary is liberated by Russia

Soviet troops battle the Germans. Besieging Budapest, they liberate Hungary.

1945: March   Land reform begins

Hungary's provisional postwar government launches an extensive land reform program under the supervision of Imre Nagy, then serving as minister of agriculture.

1945: November 4   Free elections are held

As mandated by the Yalta Agreement, secret-ballot elections are held, and a democratic government is elected. However the Russian military occupation forces the government to allot key political positions to communists in spite of their poor electoral showing.

1946: February   New republic is proclaimed

Hungary is once again declared a republic. Zoltan Tildy becomes president, Ferenc Nagy becomes prime minister, and a new constitution is adopted.

Communist Rule

1947–49   Communists gain control

The communists gradually strengthen their position in the coalition government by arresting opposition leaders. Religious leaders are also persecuted, notably Cardinal Mindszenty (see 1948).

1947: February   Paris Peace Conference reimposes Trianon borders

After the war, Hungary is forced to relinquish the territories it recaptured while allied with the Nazis and revert to the borders mandated by the Treaty of Trianon (see 1920). In addition, reparation payments to Yugoslavia, Czechoslovakia, and the Soviet Union are imposed.

1948   Cardinal Mindszenty is arrested and jailed

Cardinal Mindszenty, Roman Catholic primate of Hungary, is arrested and convicted in a communist show trial and imprisoned for eight years.

1949–53   Rakosi in office as prime minister

Matyas Rakosi, communist Hungary's first prime minister, institutes harsh, Stalinist-style rule, with the government taking control of nearly the entire economy and terrorizing suspected political opponents in a series of purges.

1949: May 15   Communist government is elected

After the Social Democrats have been forced to merge with the Communist Party, a single-ticket election is held, and a one-party state based on the Soviet model is formed.

1949: August 20   New constitution is adopted

Hungary's communist government adopts a new constitution, modeled on that of the Soviet Union, that establishes the Hungarian People's Republic. Industries are nationalized, agriculture is collectivized, and the Soviet Union becomes Hungary's major trade partner.

1952: April 1   Death of author Ferenc Molnar

Ferenc Molnar, known for his plays and short stories, dies in New York. (See 1878.)

1953   Nagy replaces Rakosi as prime minister

After the death of Russian leader Josef Stalin (1879–1953), the more moderate Imre Nagy replaces Matyas Rakosi as prime minister. His policies restore many civil liberties and improve the economy, but after two years he is ousted and Rakosi regains his post.

1955: May 14   Hungary joins the Warsaw Pact

Hungary becomes a member of the Warsaw Pact, an alliance of Soviet satellites formed by Soviet premier Nikita Krushchev to serve as a counterpart to the Western nations' North Atlantic Treaty Alliance (NATO).

The Uprising of 1956

1956: October 19–21   Protesters demand troop withdrawal

Encouraged by the success of Polish dissidents in extracting compromises from the Soviets, Hungarian intellectuals step up their demands for the withdrawal of Soviet troops from within Hungary's borders.

To produce paprika, an important agricultural export, red peppers are hung to dry outdoors. Workers then grade the peppers before sending them to be ground into the powdery spice. (EPD Photos/ CSU Archives)

1956: October 23   Police fire at students demonstrators in Budapest

A large student demonstration, at first prohibited by the government but then sanctioned, takes place in Budapest. Demonstrators demand the appointment of the moderate Imre Nagy as prime minister to replace the Erno Gero, the Soviet-picked official. They also issue a sixteen-point manifesto which they demand be broadcast on the radio. Police fire into the crowd, touching off a nationwide uprising. Revolutionary councils and committees are organized throughout the country, and the Hungarian army joins the uprising, but Soviet tanks remain stationed in Budapest.

1956: October 30   Rebels oust communist government

The nationwide uprising succeeds, and Hungary's communist government is ousted. Former moderate prime minister Imre Nagy forms a coalition government, promising free multi-party elections. His government outlines a political and economic program combining private ownership with a socialist welfare state. He withdraws Hungary from the Warsaw Pact, declares political neutrality, and vows to negotiate for the withdrawal of Soviet troops from Hungary. The Soviet Union appears ready to accept the Nagy government and begins to withdraw its troops.

1956: November 1–14   Soviets crush revolt and retake Hungary

The Soviet withdrawal stalls. Its soldiers regroup, and reinforcements enter the country from Russia, Romania, and Czechoslovakia. Soviet troops surround Budapest, tanks roll into the capital, and the city is bombarded by heavy artillery. Prime Minister Nagy broadcasts an appeal for help to the nations of the West, but world attention is concentrated on the Suez crisis in Egypt. French and British troops have just landed in Egypt, and the United States is occupied with working for a cease-fire in the region.

After three days and nights of heavy fighting, the rebels are vanquished, and the Soviets retake control of the country. Between 50,000 and 80,000 Soviet troops occupy the country, and the Soviets install Janos Kadar as prime minister. Many Hungarians go into exile as retaliation begins. Some 200,000 refugees flee to temporary camps in Austria in the weeks following the uprising, then move on to other destinations. Although an amnesty has been negotiated for the leaders of the revolt, it is violated and they are transported to Soviet headquarters. Some die in captivity; others, including Imre Nagy, are executed two years later.

1956: November 22   Leaders of uprising are interned by Soviets

Imre Nagy and other leaders of the uprising are taken into custody by Russia after promises of safety from Kadar.

1957   Hungarian National Gallery is formed

The Hungarian National Gallery is created from the Hungarian painting collection of the Museum of Fine Arts.

1957: February 9   Death of Admiral Horthy

Admiral Miklos Horthy, Hungary's head of state from 1920–44, dies in Portugal. (See 1868.)

1958: June   Nagy is executed

Imre Nagy, prime minister during the 1956 revolution, is secretly executed in Budapest together with several political associates. They are all buried in unmarked graves. (See 1896.)

The Kadar Era

1964   Compromise is reached with the church

Hungary's communist government and its Catholic church reach a compromise agreement to improve their relations.

1965   Kallai becomes prime minister

Gyula Kallai replaces Janos Kadar as prime minister, but Kadar continues to head the ruling Hungarian Socialist Workers Party.

1967: March 6   Death of composer Zoltan Kodaly

Composer Zoltan Kodaly dies in Budapest. (See 1882.)

1968   New economic program is introduced

The "new economic mechanism" (NEM) is instituted to reform Hungarian economic policy by increasing production and trade with the West and introducing greater flexibility and local control of economic policy. Public enterprises are given a greater voice in their own management, and many small, privately-owned businesses are launched with government approval.

1968: August 20–28   Hungarians help squash Czech uprising

Hungarian troops take part in the Soviet invasion of Czechoslovakia in response to democratic reforms in that country.

1972   Social welfare measures included in the constitution

State support for the sick, elderly, and disabled is written into Hungary's constitution.

1978   U.S. and Hungary settle World War II claims

War debt claims between Hungary and the United States are settled, and the U.S. returns royal treasures including St. Stephen's crown to Hungary.

1978   Diplomatic ties with the Vatican are restored

Church-state relations, disrupted since the beginning of communist rule, are strengthened with the resumption of diplomatic relations between the Hungarian government and the Vatican.

1979   Government imposes economic austerity program

Hungary's government imposes an austerity program to deal with mounting economic problems resulting from the rise in energy costs, the abandonment of the New Economic Mechanism due to political pressure, and recessions suffered by Hungary's trade partners. The country's economic woes include rising inflation and a growing foreign debt and trade deficit.

1980   Foundation organized to aid the poor

To help the nearly one-third of Hungarians estimated to be living in poverty, an unofficial foundation, SzETA, is formed.

1981   Hungarians win five-day work week

Fearing a duplication of the labor unrest rocking Poland, Hungary's government issues laws limiting the work week to five days. However, workers are required to maintain the same output within the shorter time frame.

1981   Hungarian film wins Academy Award

The film *Mephisto* by Istvan Szabo wins an Academy Award in the United States as best foreign film of the year.

1982   Hungary is admitted to international financial community

Membership in the International Monetary Fund and the World Bank enables Hungary to secure loans desperately needed to finance its foreign debt.

1984   Major economic reform enacted

The government sets up a system of elected councils to manage state-owned assets and oversee government enterprises.

1985   First parliamentary elections held under new system

As mandated by new democratic reforms, at least two candidates must be nominated for each parliamentary seat. However, the nomination of opposition candidates is blocked by the ruling politicians.

1986   Police break up student protests

Anti-government demonstrations by students on Hungary's national day are broken up by police.

1986   Hungarian soccer team in World Cup finals

Hungary's soccer team participates in the World Cup finals in Mexico but is eliminated in the preliminary rounds, which include a 6-0 loss to the Soviet Union.

1986   Writers' union elects non-communist leadership

In spite of government pressure, the Writers' Union Congress elects a completely non-communist leadership, and dissenting writers stage a walk-out.

1987: June   Karoly Grosz is appointed prime minister

Long-serving prime minister Lazar is replaced by Karoly Grosz, who is more open to reforms.

1988   Kadar is forced out

Janos Kadar, de facto head of Hungary's communist government since 1956, is forced out of office, and Prime Minister Karoly Grosz becomes the head of the Socialist Workers' Party.

1988   Budapest hosts figure skating championship

The World Figure Skating Championships are held in Budapest.

After Kadar: Political and Economic Reform

1989: May   Reformers end party control over appointments

Reformers within the ruling Socialist Workers' Party vote to end the *nomenklatura* system giving the party control over all government appointments, an important step in ending single-party rule of Hungary.

1989: June   Nagy's remains are reinterred

The remains of former prime minister Imre Nagy, executed for his role in the 1956 uprising, are buried with public honors.

1989: July 6   Death of Janos Kadar

Janos Kadar, head of Hungary's ruling party from 1956 to 1988, dies in Budapest. (See 1912.)

1989: October 23   New republic replaces communist regime

On the anniversary of the 1956 uprising, President Matyas Szuros proclaims the Hungarian Republic, ending communist domination of Hungary.

1989: November   Communist system is dismantled

The Communist Party is stripped of its militia and its workers' units in factories and required to account for its assets.

1990: March–April   Free multiparty elections are held

Hungary holds it first multiparty elections since the postwar communist takeover. A coalition government is formed by the Democratic Forum, Christian Democrats, and Smallholders' Party. Jozsef Antall becomes prime minister, and Arpad Goncz is president. Their government pursues a policy of privatization, but is criticized for the slow pace of economic transition. Antall also voices strong nationalistic sentiments, claiming to be the leader of the ethnic Hungarians living in Slovakian and Romanian borders as well as those within Hungary's borders.

1990 Hungary wins full membership in Council of Europe

Hungary becomes a full member of the Council of Europe.

1991 Monetary system is reformed

Reform of Hungary's monetary system gives the central bank a greater degree of independence from government control.

1993: January Coalition expels right wing

The governing coalition, led by the Democratic Forum, expels its right-wing members for their divisive ultra-nationalism, which inflames hostility toward groups including Jews and gypsies. Populist leader (and former president) Istvan Csurka forms his own political party, the Hungarian Justice and Life Party.

1994: May Democratic Forum suffers heavy electoral losses

The majority Democratic Forum Party loses nearly one-third of its seats in parliamentary elections that give the Hungarian Socialist Party 54 percent, calling into question the success of the government's program of privatization and economic liberalization. However, the new Socialist government continues many of the same economic policies to cope with Hungary's huge foreign debt.

1994: July Government takes greater control of the media

Socialist prime minister Gyula Horn arouses concern over preservation of the former coalition government's democratic reforms when his appointed heads of the state-run radio and television networks fire or suspend conservative journalists.

1995: June 16 World Bank approves loans

The World Bank approves loans to Hungary of up to $1.3 billion for government and economic reforms, education, environmental programs, and infrastructure improvements.

1995: July World War II restitution agreement signed by Jewish groups

The Hungarian government and Jewish groups sign an agreement providing compensation to Hungary's Jewish community for property seized during World War II. Under the agreement, a special fund is to be set up for this purpose, and restitution operations are to be administered by two committees.

1996: September 16 Border agreement reached with Romania

As a necessary step toward eventual European Union and NATO membership, Hungary and Romania settle a long-standing border dispute with an agreement that guarantees the rights of 1.6 million ethnic Hungarians living in Romania. A large number of Hungarians have resided within Romanian borders since the Treaty of Trianon (see 1920) ceded Transylvania to Romania.

1997: July NATO announces membership for Hungary

The North Atlantic Treaty Organization (NATO) announces that it will admit Hungary to full membership in 1999, along with Poland and the Czech Republic.

Bibliography

Bartlett, David L. *The Political Economy of Dual Transformations: Market Reform and Democratization in Hungary.* Ann Arbor: University of Michigan Press, 1997.

Berend, Ivan, and Gyorgy Ranki. *The Hungarian Economic Reforms 1953–1988.* New York: Cambridge University Press, 1990.

Burant, Stephen R., ed. *Hungary: A Country Study.* 2nd ed. Washington, D.C.: Government Printing Office, 1990.

Corrin, Chris. *Magyar Women: Hungarian Women's Lives, 1960s–90s.* New York: St. Martin's, 1994.

Gati, Charles. *Hungary and the Soviet Bloc.* Durham, N.C.: Duke University Press, 1990.

Hoensch, Jorg K. *A History of Modern Hungary, 1867–1994.* 2nd ed. New York: Longman, 1996.

Kun, Joseph C. *Hungarian Foreign Policy: The Experience of a New Democracy.* Westport, Conn.: Praeger, 1993.

Litvan, Gyorgy, ed. *The Hungarian Revolution of 1956: Reform, Revolt, and Repression, 1953–56.* Trans. Janos M. Bak and Lyman H. Legters. New York: Longman, 1996.

Sugar, Peter F., ed. *A History of Hungary.* Bloomington: Indiana University Press, 1990.

Szekely, Istvan P. and David M.G. Newberry, eds. *Hungary: An Economy in Transition.* Cambridge: Cambridge University Press, 1993.

Iceland

Introduction

Iceland, the "land of fire and ice" known primarily for its dramatic geographic contrasts, is also a country with a long and colorful history. An Icelander was the first European to land on the North American continent. Iceland established Europe's first parliamentary republic and created its most enduring body of medieval literature. In 1980 Iceland became the first country in the world to elect a female head of state. Today the standard of living and literacy rate enjoyed by the inhabitants of this sparsely populated nation are among the world's highest.

Lava fields, geysers, hot springs, and active volcanoes attest to the volcanic origins of Iceland, Europe's westernmost country and its second-largest island. As its name suggests, however, Iceland also has thousands of square miles that are covered by ice fields and glaciers. Located in the North Atlantic Ocean, it has an area of 39,769 square miles (103,000 square kilometers). The island has good natural harbors on its northern, eastern, and western coasts, which are indented by *fjords* (narrow sea inlets bordered by steep cliffs), bays, and inlets, but not on its sandy south and southeast coastlines. Iceland's many rivers are not navigable, but waterfalls and strong currents make them an important source of hydroelectric power. As of the late 1990s, Iceland is home to an estimated 280,000 people, of whom over half live in the capital city of Reykjavík on the southwest coast.

History

Iceland was settled in the late ninth and early tenth centuries by Norwegians leaving their homeland for a variety of reasons, including dislike of the power wielded by King Harald Fairhair. Many of these first settlers arrived after a sojourn in Scotland, Ireland, or the nearby islands, giving their descendants mixed Norwegian-Celtic ancestry since they brought with them Celtic wives or slaves. In 930 the *Althing,* a national assembly with both legislative and judicial functions, was established. It met for two weeks every summer in Thingvellir, establishing a legal code that was passed down through oral tradition. Around A.D. 1000, in the time when Leif Ericsson (the son of Eric the Red, who had discovered Greenland) sailed to North America, Christianity was introduced to Iceland by missionaries and adopted by the island's leaders. In addition to its religious impact, the establishment of the new religion played an important role in the rise of literacy in Iceland.

In the thirteenth century, a few families became very powerful and vied for control of the island, inaugurating a period of civil war that weakened Iceland, which came under Norwegian rule by 1262. Its native chieftains were replaced by Norwegian officials and the Althing, although permitted to continue in existence, was reduced in power. While this period witnessed a political decline in Iceland, culturally it was one of distinction, for the famous prose narratives called *sagas* were written at this time. Known for their realism, tragic dignity, and historic sweep, the greatest were the family sagas, fictionalized histories of Iceland's great tenth-century families. These prose epics have been recognized as great literature throughout the world and are still widely read and studied today. They also had an important influence on subsequent Icelandic literature.

In 1380, after Iceland had been ruled by Norway for over a century, Norway itself came under the domination of Denmark, and Iceland was obliged to join in the union of these two countries and Sweden. Iceland was to remain linked with Denmark politically for over five centuries, long after Denmark's ties with Sweden and Norway had been severed. Denmark imposed strict controls over Iceland's religious life in the fifteenth century and over trade in the seventeenth, when Iceland was forced to conduct all trade through a Danish company that took a large share of the profits for both imports and exports.

Iceland's worst natural disaster—and the worst volcanic eruption in recorded history—took place in 1783, when Laki, a fifteen-mile-long row of volcanoes, broke open and spewed twenty cubic miles of lava over the countryside and villages, causing the deaths of as many as one-third of Iceland's people. In addition to those whose villages were swallowed up in hot lava, others died from the smoke and fumes and from starvation, as much of the nation's fish and livestock was also killed in the disaster.

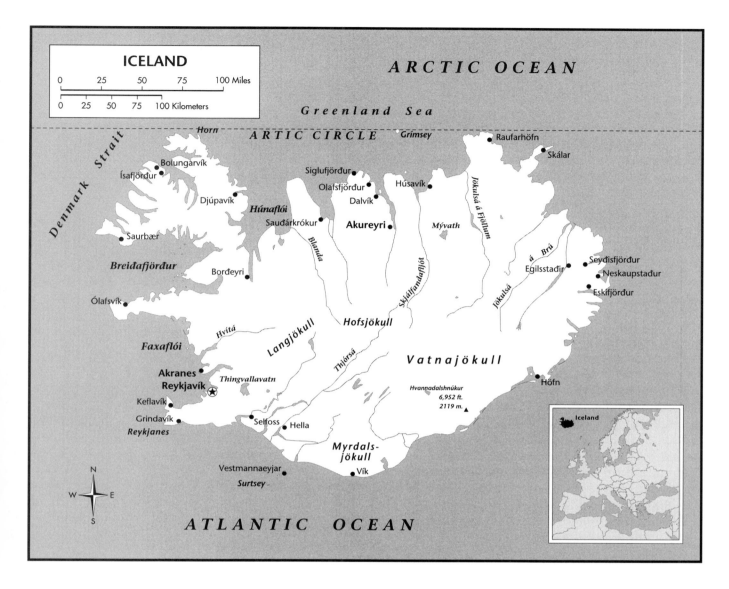

Like other countries throughout Europe, Iceland underwent a nationalist revival in the nineteenth century, given added impetus when Denmark finally abolished its national deliberative body, the Althing, which had been founded eight centuries earlier. Through the efforts of nationalists, most notably scholar and statesman Jón Sigurðsson (1811–79), the Althing was reestablished by 1843. The Danish ended their trade monopoly in 1854, and twenty years later a new constitution gave Iceland limited self-rule.

In 1918 Iceland signed a treaty with Denmark, making it an independent state that retained a connection with Denmark's king, who exercised control over the foreign affairs of both nations. Iceland was given the option of ending this connection after twenty-five years.

Like the rest of Scandinavia, Iceland was neutral in World War I. In World War II, however, the nation was forced into a growing involvement in international affairs. In the spring of 1940 British troops landed there following the Ger-

man invasion of Denmark and Norway. The British wanted to prevent Germans from occupying Iceland, whose location gave it strategic importance in the North Atlantic. When the United States entered the war the following year, Americans took over from the British, freeing English troops for deployment elsewhere. American troops remained in Iceland until 1947, and an agreement signed in 1951 established a continuing U.S. military presence in the country.

In 1943 Iceland's citizens voted to end the country's political union with Denmark and to become independent in both foreign and internal affairs. The Republic of Iceland was established on June 17, 1944 and soon demonstrated its intention of remaining an active participant in international affairs by joining the United Nations in 1946 and NATO in 1949. The major source of international tension for Iceland in the postwar years was the demarcation of its offshore fishing rights, which caused disagreements with Great Britain in the 1950s and 1970s.

In 1980 Vigdís Finnbogadóttir (b. 1930) was elected president of Iceland, becoming the world's first elected female head of state. She remained an immensely popular public figure, serving four four-year terms before retiring in 1996. In 1981 a constitutional amendment turned the bicameral Althing, the Icelandic parliament, into a unicameral (single-chamber) body. Four years later, Iceland's government declared the country a nuclear-free zone. Although Iceland had joined the European Free Trade Area (EFTA) when it was formed in 1970, in the 1990s it remained the only Western European nation to remain outside the European Union.

Timeline

c. 870–930 A.D.   Norwegians settle Iceland

Norwegians begin arriving in Iceland, seeking to escape the power of King Harald Fairhair, driven out by personal quarrels or leaving mountainous Norway to seek new farmland. Some come directly to Iceland, but many first migrate to the Celtic lands of Scotland, Ireland, England, the Hebrides, and the Orkney, Shetland, and Faroe Islands. Thus future Icelanders will have both Norwegian and Celtic blood, for Celts arrive as both wives and slaves with the Norwegians when they make the journey to Iceland. Fishing is the main means of subsistence for the settlers, but sheep farming soon grows in importance. By the end of this period, most of Iceland has been settled and divided into large landholdings.

A.D. 874   First recorded arrival of a Norwegian settler

Ingólfur Arnarson settles in Reykjavík, becoming the first recorded Norwegian to make his home on the island.

930   Legislative assembly is formed

A national Icelandic legislative assembly, the *Althing,* is established, and a uniform code of laws is promulgated. (Local leaders continue to wield power in their own districts.) The Althing meets for an annual two-week session at midsummer and serves both legislative and judicial functions. In the oral culture of the period, the laws are preserved through the assembly's chief officer, the "lawspeaker," who is responsible for reciting the entire legal code aloud in the course of his three-year term. Iceland is unique in having a national governing body but no king.

986   Greenland is settled

Eric the Red, exiled from Iceland for committing murder, lands on and settles Greenland.

1000   Leif Ericsson sails to North America

Leif Ericsson, the son of Eric the Red, discovers the North American continent, which he names Vinland for the grapes he finds growing there. Ericsson does not establish any settlements.

1000   Christianity is introduced

The first missionaries introduce Christianity, and it is adopted by Iceland's leaders in spite of opposition by the priests of the established pagan religion. A compromise is reached, and the priests build and run the churches. These local leaders, rather than the Vatican in Rome, exercise most of the religious power in Iceland. The establishment of the church is especially significant because of its role in the rise of literacy. The Althing opens special schools to train ministers and other church workers.

13th century   The Icelandic sagas are written

The sagas, a series of epic prose narratives, are Iceland's most acclaimed artistic achievement. Contrary to former theories that viewed them as recorded versions of earlier oral narratives, it is now thought that they are the work of individual authors. However, they still draw on earlier oral traditions (*saga* is related to the Icelandic word for *said*) and provide a detailed account of life in the period in which they are set (mostly between A.D. 930 and 1030, also called the "saga age"). There are several major types of saga: the kings' sagas, about the early leaders of Norway, the Icelanders' former homeland; the legendary sagas, which are concerned with Germanic and Scandinavian myths and legends; and the family, or Icelanders' sagas, historical and psychological descriptions of the human conflicts within families during the saga age.

The most famous ones, the family sagas, represent the highest level of artistic achievement. The stories they tell generally include tragic events involving rivalry, betrayal, and revenge. They far surpass the level of realism found in any other medieval literature and are renowned for the complex level of characterization achieved through the description of the characters' actions alone rather than the detailed descriptions of the characters' inner lives so often found in modern fiction. The most famous of all the sagas is *Njals Saga,* in which two virtuous heroes, Njall and Gunnar, both meet tragic fates because of implacable hatred caused by old injuries and feuds.

13th century   The Sturlunga Age

A few powerful families vie for control of Iceland in a period of civil war known as the Sturlunga Age after its best-known family, the Sturlusons, whose leaders include Snorri, Sighvatur, and Porthur.

13th century   The *Eddas* are compiled

Two great literary compilations are completed in thirteenth-century Iceland: the Prose *Edda* and the Poetic *Edda*. The Prose Edda, written by political leader and scholar Snorri Sturluson, probably in the first quarter of the century, is a guide to Icelandic poetry that summarizes and illustrates its poetic meters and retells some of the central Nordic myths that form its subject matter. The Poetic Edda, written later in the century, is an anthology of poems based on myths and heroic legends, including the cycle of tales on which the Germanic epic the *Nibelunglied* (Song of the Nibelungs) is based.

1241   Snorri Sturluson breaks with Norway

Snorri Sturluson, a major saga writer and historian as well as one of Iceland's leaders in the Sturlunga period, breaks off an alliance with King Haakon IV of Norway, and Haakon has him murdered by a rival Icelandic leader.

1262   Rule by Norway's king

Weakened by civil war, Iceland submits to rule by Norway, and the Icelanders swear allegiance to the Norwegian monarch, Haakon IV. The loss of independence begins a long period of decline for Iceland, which is soon subject to a trade monopoly by Norway. Although Iceland retains its governing assembly, the Althing, its power is gradually reduced.

1271   New law reduces Iceland's autonomy

Under a new legal system, Iceland's governing chieftains, or *gothar,* are replaced by officials appointed by the Norwegian court.

Danish Rule

1380   Union with Denmark

Together with Norway, Iceland comes under Danish rule when King Haakon VI dies and his widow, Margrethe of Denmark, becomes regent. The arrangement is consolidated by the Kalmar Union two decades later, uniting Denmark, Sweden, and Norway (and Iceland) under the Danish crown.

1402   The Black Death strikes Iceland

The Bubonic Plague becomes one of several forces that decimate Iceland in the fifteenth century, including volcanic eruptions and other natural disasters.

1484   Birth of Catholic martyr Jón Arason

Jón Arason, one of Iceland's national heroes, is martyred to the political and religious tyranny of Danish rule. In 1522 he becomes the bishop of Holar, Iceland's northern diocese. When Denmark's King Christian III begins his attempts to impose Lutheranism on Iceland, both Arason and his fellow bishop, Ogmundr, resist. Ogmundr is deported, but Arason continues his resistance. He is finally captured by the king's agents, and he and two of his sons are beheaded in 1550. In addition to being a religious leader, Arason is also the author of distinguished religious and satirical verse and introduces the first printing press to Iceland.

16th century   Lutheranism is imposed on Iceland

The Protestant Reformation reaches Iceland as Norway forces the Icelanders to accept Lutheranism.

1550   Last Catholic bishop is killed

As part of the campaign to convert Iceland to Lutheranism, its last Catholic bishop, Jón Arason, is beheaded (see 1484).

1589   First hymn book is printed

In connection with the Reformation, the first hymn book in Icelandic is published.

1602–1787   Denmark imposes trade monopoly

Denmark gives sole trading rights for Iceland to a Danish company, which charges the Icelanders high prices for imported goods while buying their own commodities cheaply. The monopoly is not completely ended until the mid-nineteenth century.

1614   Birth of poet Hallgrimur Pétursson

Iceland's greatest religious poet, Hallgrimur Pétursson, spends part of his youth in Denmark, where he receives a humanistic education. He later returns to Iceland, working first as a laborer and fisherman but serving as a parson between 1651 and 1669, when he contracts leprosy. The poems in which he expresses his suffering are published in *Passiusâlmar* (The Passion Hymns of Iceland), which is ranked with the world's greatest devotional poetry. The volume wins immediate popularity and continues to be widely read through the twentieth century, undergoing its sixty-fourth printing in 1957. Pétursson dies in 1674.

1662   Danish claim hereditary monarchy in Iceland

After proclaiming himself absolute monarch over Denmark two years earlier, Frederick III claims the same rights over Iceland, giving Iceland, like Denmark, an inherited king rather than one elected by parliament.

1674: October 27   Death of poet Hallgrimur Pétursson

Hallgrimur Pétursson, author of the greatest Icelandic book of religious poetry, dies. (See 1614.)

1728   Classic texts are lost to fire

Valuable printed poetry and saga manuscripts are lost in a great fire in Copenhagen after being removed to Denmark by scholar Arni Magnusson, who nevertheless manages to save most of those written on vellum. Upon Magnusson's death in 1830 the manuscripts become the property of the University of Copenhagen. More than two hundred years later, they are returned to Iceland. (See 1997.)

1783: January–June   Laki erupts

Laki, a fifteen-mile-long row of volcanic craters, has a series of eruptions over a six-month period in the worst volcanic activity in recorded world history. The lava covers 220 square miles (570 square kilometers); as many as 20,000 people— one-third of Iceland's population—are killed. The climax comes in June when Skaptar Jokul erupts, releasing a massive lava flow down the entire length of its open fissure. The lava, in two streams about forty miles long and one hundred feet deep, consumes entire villages as it sweeps to the sea. Covering an estimated twenty cubic miles, it is the largest recorded lava flow in history. Even in those villages not actually destroyed, inhabitants die from toxic fumes and ash or from starvation, as the disaster blights the landscape, destroys the nation's fish supply, and kills over two-thirds of its livestock over the next two years. Altogether, over nine thousand villages are destroyed.

1800   The Althing is abolished

The Danish king abolishes Iceland's national assembly, the Althing, which has been in existence since the Middle Ages.

Nationalist Revival

1811: June 17   Birth of nationalist and scholar Jón Sigurðsson

In contrast to Iceland's early Norse heroes, Jón Sigurðsson, the national hero of modern Iceland, is a scholar and statesman who spends a lifetime editing medieval sagas and other texts. Educated at the University of Copenhagen, he works for a foundation set up to preserve Icelandic culture. However, he is also an active advocate for Icelandic nationalism and takes part in the diplomatic campaigns that win the restoration of the Althing in 1843, the end of the Danish trade monopoly in 1854, and a revised constitution in 1874. Sigurðsson dies in 1879.

1833   Birth of artist Sigurthur Guthmundsson

Guthmundsson becomes the first native artist to established a career within Iceland. After studying art in Denmark at the Kongelige Danske Kunstakademi, he returns to Iceland, where he paints portraits and altarpieces and works as a theatre designer. Guthmundsson is active in Iceland's nineteenth-century nationalistic movement and works for the establishment of Iceland's National Museum (see 1863). He dies in 1874.

1843   Althing is reestablished

As a result of efforts by Jón Sigurðsson and other nationalists, a modern version of Iceland's assembly, the Althing, is established.

1847   Seminary is founded

Existing religious schools are consolidated into a theological seminary called the Prestaskoli Islands.

1853: October 3   Birth of poet Stephan Stephansson

Stephansson, who lives most of his life in Canada, produces some of the greatest Icelandic poetry ever written. The poet emigrates to the United States when he is twenty years old and, after living in Wisconsin and North Dakota, settles in Alberta, Canada, by the time he is thirty-six. He makes a living as a farmer and raises a large family, doing most of his writing at night. (For this reason, his most famous collection of poetry is called "Sleepless Nights.") Stephansson's poetry is written for Icelander readers, although he spends most of his life in North America. Stephansson dies in 1927.

1854   Trade monopoly ends

The 250-year-old Danish trade monopoly over Iceland is completely abolished.

1863   National museum is founded

The National Museum of Iceland is established.

1874   New constitution

A new constitution giving Iceland limited autonomy is adopted.

1874   National anthem is composed

Composer and pianist Sveinbjorn Sveinbjornsson (1847–1927) composes Iceland's national anthem.

1874: May 11   Birth of sculptor Einar Jónsson

Iceland's most famous sculptor, Jónsson studies in Copenhagen, where he is based until the early 1920s, making trips to Rome, Berlin, London, and the United States. He then returns to Iceland to stay and founds a museum, which he designs and the government finances. It opens in Reykjavík in 1923. Jónsson's many public sculptures include *Outlaws* (1898–1901), *Wave of Waves* (1894–1905), and *Ingólfur Arnarson* (1907). At his death in 1954, Jónsson leaves all his works to the Icelandic nation.

1879: December 7   Death of nationalist statesman Jón Sigurðsson

Jón Sigurðsson, Iceland's foremost nationalist advocate, dies in Copenhagen. (See 1811.)

1885: October 15   Birth of painter Johannes Kjarval

Kjarval, one of Iceland's leading twentieth-century painters, is educated in Reykjavík and Copenhagen, lives for a time in Italy, and also studies briefly in England, where he is influenced by the paintings of J. M. W. Turner (1775–1851). After returning to Iceland, he begins publishing an arts journal and becomes known for his pen-and-ink portraits of ordinary Icelanders, notably farmers. A trip to France in 1928 gets Kjarval interested in the modern French styles of painting, including Cubism and abstraction. Afterwards, he concentrates on landscape paintings in a distinct personal style, including a series of paintings depicting the historic Icelandic venue of Thingvellir, seat of the medieval Althing. After 1940 the scope of Kjarval's landscapes expands to encompass many parts of Iceland. Well-known paintings by Kjarval include *Moss and Lava* (1939), a painting of a lava field; the landscape painting *Mosfellsheithi* (1948); and the symbolic painting *Fantasy* (1940). Kjarval dies in Reykjavík in 1972.

1889: May 18   Birth of novelist Gunnar Gunnarsson

Gunnarsson, one of the foremost twentieth-century Icelandic writers, is the author of novels and short stories set in Iceland but written in Danish to reach a wider audience. He moves to Denmark as a young man and publishes his first best-seller at the age of twenty-three: the first volume of his novel *Af Borgslægtens Historie* (*The Borg Family Papers*; 1912). Gunnarsson remains in Denmark until 1939, when he returns to Iceland. In addition to writing forty novels and numerous short stories, Gunnarsson also authors articles and translations. He dies in 1975.

1898: October 10   Birth of author Guthmundur Hagalín

Hagalin is a well-known novelist, short story writer, and essayist whose works present a vivid picture of twentieth-century life in Iceland. As a young man, Hagalin travels and works as a journalist and librarian. He is known for his direct, economic style and strong characterization. Hagalin's best-known works include *Kristrun i Hamravik* (1933), *Sturla I Vogum* (1938), and *Mothir Island* (Mother Iceland; 1945). Hagalin also publishes several autobiographical volumes. He dies in 1985.

1902: April 23   Birth of author Halldór Laxness

Laxness, Iceland's greatest twentieth-century writer, is born Halldór Kiljan GuthdJónsson. In the 1920s he becomes a socialist, which is reflected in the emphasis on the rural and working-class poor in his novels over the following two decades. In 1955 he is awarded the Nobel Prize for Literature. Two of his most famous early novels are *Salka Valka* (1931–32), about a young woman living in a small fishing town, and *Sjalfstætt folk* (Independent People; 1945) about the struggles of poor farmers. Other works by Laxness include the historical trilogy *Islandsklukkan* (Iceland's Bell; 1943–46), set in the eighteenth century; *Gerpla* (1952), which satirizes mankind's use of violence by parodying the Icelandic sagas; *Brekkukotsannall* (The Fish Can Sing; 1957); and *Kristnihald undir Jokli* (Christianity at the Glacier; 1968). Laxness also publishes two volumes of memoirs, as well as poetry, plays, short stories, literary criticism, and translations. He dies in 1998.

1903   Autonomy is expanded

Iceland wins nearly total self-rule under the Danish crown.

1906: April   Birth of painter Thorvaldur Skulason

Skulason, one of Iceland's foremost modern painters, studies in Iceland and Denmark, where his teachers include former pupils of Henri Matisse (1869–1954). In 1931 he moves to Paris, where he exhibits paintings with a group of other Scandinavian artists known as the Colorists (*Koloristerne*). During the course of his career, Skulason's interest shifts from landscapes to paintings depicting Iceland's people, both urban dwellers and fishermen. He later embraces abstract art. In the late 1930s the Nazi military advance forces Skulason to leave Paris, and he returns to Iceland. After the war, Skulason becomes the focal point for a group of young artists called the Septembrists. Their annual exhibitions lay the foundation for modernist painting in Iceland. Skulason's work is also exhibited in many foreign countries. Skulason dies in 1984.

1911: June 17   University of Iceland is founded

Iceland's schools of theology, medicine, and law are consolidated to form the University of Iceland, which is officially established on the centenary of the birth of Jón Sigurðsson, Iceland's foremost national hero. Classes are held in the parliamentary buildings until 1940.

Independence

1914–18   World War I

Since Denmark declares its neutrality, Iceland remains neutral in World War I.

1918   Independence is won

Under the Icelandic-Danish Treaty of Union, Iceland becomes an independent state with its own flag but retains a connection with the Danish king, who is responsible for its foreign affairs. However, Iceland has the option of terminating this relationship in twenty-five years.

1927   First ceramic studio is established

Iceland's first ceramics workshop is set up in Reykjavík by sculptor and painter Guthmundur Einarsson (1895–1963). Much of his work features Viking themes.

1927: August 10   Death of poet Stephan Stephansson

Stephan Stephansson, one of the greatest Icelandic poets, dies in his longtime home of Alberta, Canada. (See 1853.)

1930   Reykjavík Musical Society is founded

As part of the cultural activities surrounding the commemoration of the Althing's one thousandth anniversary, the Reykjavík Musical Society is founded to promote classical music in Iceland. The society establishes a music school (the Reykjavík Conservatory of Music) and a symphony orchestra.

1930   Broadcasting service is established

The Icelandic State Broadcasting Service is founded. Its first music director is composer Pall Isolfsson.

1930   Birth of Vigdís Finnbogadóttir

Finnbogadóttir, Iceland's first female president and the world's first elected female head of state, is born in Reykjavík. She studies French literature and drama at the Sorbonne and the University of Grenoble and teaches at the high school and university levels when she returns to Iceland. Divorced in 1963, Finnbogadóttir adopts a baby girl in 1972, becoming one of Iceland's first single parents to adopt a child. From 1972 to 1980, Finnbogadóttir directs the Reykjavík City Theatre. Entering politics for the first time in 1980, she runs for and is elected to the non-partisan office of president, narrowly defeating three male candidates. Becoming extremely popular with the Icelandic people, she is reelected three times before retiring from the office in 1996.

1940: May 10   British troops land in Iceland

British troops enter Iceland to counter the German occupation of Denmark and Norway. Its strategic North Atlantic location makes it valuable as an air base for refueling and shipping. Iceland, which has declared neutrality, protests, but the British promise to respect Iceland's sovereignty, and an understanding is reached.

1941: July 7   U.S. troops arrive in Iceland

United States troops replace British forces in carrying out the Allied patrol of the North Atlantic, and the U.S. signs an agreement similar to that signed by the British (see 1940). The U.S. presence brings some benefits, including construction of new facilities and the provision of goods and services to the 60,000 U.S. troops.

1943: December   Iceland becomes fully independent

The 1918 Icelandic-Danish Treaty of Union linking Iceland to the Danish crown in matters of foreign affairs expires.

1944: May   Icelanders vote for independence

Although relations between Denmark and Iceland are good, more than 97 percent of Icelanders vote against renewing the Icelandic-Danish Treaty of Union, opting to form a republic of their own instead of remaining linked with the Danish crown.

The Republic of Iceland

1944: June 17   Icelandic republic is founded

The Republic of Iceland is established at Thingvellir. Sveinn Bjornsson becomes its first president.

1946: November 9   Iceland joins the United Nations

Iceland becomes a member of the United Nations.

1947   U.S. troops withdraw from Iceland

The U.S., whose request for a ninety-nine-year lease on three air bases has been denied, withdraws its remaining troops but wins permission to use the Keflavik air base.

1949   Iceland joins NATO

Iceland becomes a founding member of the North Atlantic Treaty Organization (NATO).

1950   National Theatre opens

Iceland's National Theatre opens in Reykjavík.

1951   Defense pact is signed with the U.S.

Iceland and the United States sign a defense agreement providing for U.S. troops to be stationed in Iceland.

1952   Iceland proclaims exclusive fishing zone

Iceland claims exclusive fishing rights in waters up to four miles offshore, expanding the existing limit by an additional mile. The action draws protests from Great Britain, which boycotts imported Icelandic fish.

1954: October 18   Death of sculptor Einar Jónsson

Einar Jónsson, Iceland's foremost sculptor, dies in Reykjavík. (See 1874.)

1955: October   Special Kjarval exhibit draws large crowds

A retrospective exhibit of paintings by Johannes Kjarval on his seventieth birthday draws 25,000 people—more than one-eighth of Iceland's population. (See 1885.)

The fishing grounds around Iceland yield a large catch. Here fish are unloaded at Westmanna Island, Iceland's busiest cod fishing center in 1940. (EPD Photos/CSU Archives)

1956: March   Government coalition collapses over NATO dispute

Conflicts over Iceland's membership in NATO break up the ruling coalition government of Progressive and Independents, and a new coalition of the Progressive, Labor, and People's Union-Socialist parties is formed following elections later in the year.

1958   Fishing zone is extended

Iceland unilaterally extends its offshore fishing zone to twelve miles. Like the previous expansion to four miles (see 1952), the action draws protests from Great Britain, which also fishes in these waters.

1970s   "Cod wars" with U.K. over fishing rights

Iceland continues to unilaterally expand the territory included in its exclusive offshore fishing zone, drawing protest and threatening action from Britain. By 1975 it has expanded its exclusive fishing rights to 200 miles offshore, up from 2 miles before 1952. Both countries send out gunboats to patrol the waters and take other threatening actions, including cutting the nets of some fishing trawlers, but there are no casualties.

Britain ultimately agrees to the 200-mile fishing limit with quotas for both countries. The United Nations Draft Convention on the Law of the Sea accepts Iceland's stand on its fishing rights, and it becomes the model for other countries as well.

1970   Iceland joins EFTA

Iceland joins the European Free Trade Area.

1972   Fischer-Spassky match focuses world attention on Reykjavík

The world championship chess match between U.S. Bobby Fischer and reigning champion Boris Spassky of the USSR is held in Reykjavík in the summer of 1972, attracting world-wide attention. Chess is a favorite activity in Iceland, which boasted six grand masters in the mid-1990s, a disproportionate number for a country with as small a population as Iceland's.

1972: April 13   Death of artist Johannes Kjarval

Kjarval, one of Iceland's foremost modern painters, dies in Reykjavík. (See 1885.)

1973   Volcanic eruption on Heimaey

The entire population of the island of Heimaey has to be evacuated when a volcano erupts without warning. The eruption continues for five months, producing a 500-foot- (150-meter-) wall of lava that threatens to bury the island and block its harbor.

1973: February   Tariff agreement with the EEC

Iceland ratifies a tariff agreement with the European Community.

1975   Iceland's women stage a one-day strike

Women in Iceland strike for one day to mark the beginning of a period that the United Nations has proclaimed Women's Decade. They strike again when it ends ten years later.

1975: November 21   Death of author Gunnar Gunnarsson

Author Gunnar Gunnarsson, best-selling author of novels and short stories exploring Icelandic life and history but written in Danish, dies in Reykjavík. (See 1889.)

1980   First woman is elected president

Vigdís Finnbogadóttir (see 1930) becomes Iceland's first female president and the first directly elected female head of state in the world. She serves four four-year terms before retiring from office in 1996.

1981   The Althing becomes a unicameral body

A constitutional amendment transforms the Althing, Iceland's legislative assembly, from a double- to a sixty-three-member single-chamber body.

1984: August 30   Artist Thorvaldur Skulason dies

Painter Thorvaldur Skulason dies in Reykjavík. (See 1906.)

1985   Iceland becomes a nuclear-free zone

Iceland's government declares the country a nuclear-free zone.

1985: February 26   Death of author Guthmundur Hagalin

Novelist and short story writer Guthmundur Hagalin dies. (See 1898.)

1986   Iceland hosts U.S.-Soviet summit meeting

Located between the United States and the Soviet Union, Iceland becomes the site of a nuclear disarmament summit meeting between U.S. president Ronald Reagan (b.1911) and Soviet leader Mikhail Gorbachev (b.1931), which is held in Reyjavík.

1987   Limited number of whales to be taken

In spite of a ban by the International Whaling Commission, Iceland announces it will capture 100 whales per year for scientific purposes.

1991   Iceland wins bridge championship

The Icelandic team wins the world bridge championship at a match held in Japan and broadcast live throughout Iceland.

1991: April   Center-right coalition is formed

General elections result in the formation of a center-right coalition of the Independence Party and the Social Democratic People's Party, led by Prime Minister David Oddsson.

1991: August 26   Iceland recognizes former Soviet republics

Iceland is the first country to grant formal recognition to the former Soviet republics of Latvia, Lithuania, and Estonia, which are the first to break with the USSR.

1994   U.S. military presence is renegotiated

Iceland negotiates a reduction in the number of U.S. troops stationed in the country.

1994   Fishing zone controversy is renewed

The controversy over Iceland's offshore exclusive fishing zone recurs, this time with Norway, and Norwegians cut the fishing nets of an Icelandic boat.

1995: April 8   Independence Party leads in general elections

The Independence Party is the victor in general elections, winning 25 seats compared to 15 for its nearest rival, the Progressive Party.

1996: June 27   Iceland legalizes gay marriage

Iceland becomes the fourth Scandinavian country to legalize marriage between homosexuals, who are allowed to marry in a civil ceremony and to share custody of any children they already have. However, they are not allowed to have a church ceremony, adopt additional children, or produce children through artificial insemination.

1996: June 29   Grimsson is elected president

Left-wing leader Olafur Ragnar Grimsson is elected to replace outgoing president Vigdís Finnbogadóttir. Grimsson receives 41 percent of the vote compared to his nearest rival, who wins only 29 percent. The highly popular Finnbogadóttir, who had been in office since 1980, was the world's first woman elected head of state.

1997: June   Last saga manuscripts are returned

Denmark returns the last of the saga manuscripts in its possession. Written versions of the medieval sagas, the greatest achievement in Icelandic literature, are taken to Denmark during the seventeenth and eighteenth centuries and stored there. Denmark starts returning them in 1971.

1998: February 8   Death of author Halldór Laxness

Nobel Prize-winner novelist Halldór Laxness, Iceland's greatest twentieth-century writer, dies is Reykjavík. (See 1902.)

1998: October   Hollywood returns celebrity whale to Iceland

Keiko, the orca whale who appears in the 1993 movie *Free Willy,* is brought to specially constructed facilities in Klettsvik Bay, Heimaey, for rehabilitation before a planned release into the wild. A special foundation has rescued the orca, taken from his native Iceland nineteen years earlier at the age of two, from unhealthy conditions in New Mexico and begun the rehabilitation effort in the United States, airlifting him to Oregon. In spite of the millions of dollars spent on this effort, it remains unclear whether the whale will be able to survive in the wild and be accepted by others of his species.

Bibliography

Auden, Wystan Hugh, and Louis MacNeice. *Letters from Iceland.* London: Faber & Faber, 1937.

Byock, Jesse L. *Medieval Iceland: Society, Sagas, and Power.* Berkeley: University of California Press, 1988.

Durrenberger, E. Paul. *The Dynamics of Medieval Iceland: Political Economy and Literature.* Iowa City: University of Iowa Press, 1992.

Hastrup, Kirsten. *Culture and History in Medieval Iceland.* London: Oxford University Press, 1990.

Jochens, Jenny. *Women in Old Norse Society.* Ithaca: Cornell University Press, 1995.

Jónes, Gwyn. *The Norse Atlantic Saga: Being the Norse Voyages of Discovery and Settlement to Iceland, Greenland, America.* London: Oxford University Press, 1986.

Lacy, Terry G. *Ring of Seasons: Iceland: Its Culture and History.* Ann Arbor: University of Michigan Press, 1998.

Roberts. David. *Iceland.* New York: H.N. Abrams, 1990.

Scherman, Katherine. *Daughter of Fire: A Portrait of Iceland.* Boston: Little, Brown, 1976.

Weingand, Darlene E. *Connections--Literacy and Cultural Heritage: Lessons from Iceland.* Metuchen, N.J.: Scarecrow Press, 1992.

Ireland

Introduction

Ireland is an island in the eastern part of the North Atlantic and covers an area of 32,595 square miles (84,421 square kilometers), with 27,136 square miles (70,282 square kilometers) belonging to the Irish Republic (Ireland) and the remaining land in Northern Ireland, part of the United Kingdom. The total area of the Irish Republic is slightly larger than the state of West Virginia. Ireland is bounded on the north by the North Channel, which separates it from Scotland; on the northeast by Northern Ireland; and on the south and southeast by the Irish Sea and St. George's Channel, which separate it from England and Wales. The entire western coast is bordered by the Atlantic Ocean.

Only ten percent of the land is arable (suitable for farming), while seventy-seven percent is meadow and pasture, eleven percent is used for rough grazing, and two percent is inland water. The center of the country is primarily a plains area, and there are mountain ranges in the east (Wicklow Mountains) and the southwest (Macgillycuddy's Reeks). The coastline is 1,969 miles (3,169 kilometers) long, and is heavily indented along the southern and western coasts. The most important of Ireland's many rivers is the Shannon, whose source is in the mountains along the Ulster border and which drains the central plains into the Atlantic Ocean.

Ireland, known for its lush greenery, has a very equable (steady) climate: it is rarely very hot or very cold. The average temperature in January is a mild 4°C (39° F), and the average in July is 15°C (59°F). This small range in temperature is due to the prevailing winds which come from the Atlantic; the ocean is warmer in winter and cooler in summer than the continental land masses.

Ireland has a population of 3.57 million people, and its largest ethnic group is Celtic, with an English minority. Throughout its history, however, Ireland has been inhabited by Celts, Norsemen, French Normans, and English, and the intermingling between these groups has left no pure ethnic divisions. There is no established (state) church, but the religions practiced are Roman Catholicism (95 percent), Church of Ireland (Protestantism) 4 percent, all other Protestants 1 percent, and Judaism 0.07 percent. The official languages spoken are English and Irish (Gaelic). There is a literacy rate of ninety percent, and children must attend school for at least nine years. About two-fifths of the people live in rural areas, and about one-third of Ireland's entire population lives within the greater metropolitan area of Dublin.

Until the 1950s, agriculture made the greatest contribution to Ireland's GNP (Gross National Product). After that time, liberal trade policies and industrialization stimulated economic expansion. Ireland's economy has been historically slower in its development than those of other Western European countries, and the worldwide recession in the early 1980s set economic progress back considerably. In 1986, unemployment was high at 18 percent, and as of 1993 the rate had fallen very slightly to 16.5 percent. Most industry is concentrated in the capital city of Dublin and the city of Cork. The most important products of manufacturing are food, metal and engineering goods, chemicals and chemical products, beverages and tobacco, nonmetallic materials, and paper and printing. The greatest recent industrial growth has been in high-technology industries such as electronics and pharmaceuticals. Tourism is also an important industry; a total of 3,535,000 tourists entered Ireland in 1991.

History

Ireland has a rich cultural history, but its past is also marked by years of strife and warfare. There has been strain not only between the Irish people and the British who controlled their country for hundreds of years, but also internal conflicts between Irish Catholics and the Protestants.

Early (pre-Christian) Irish history is recorded mainly through legend, though there is archaeological evidence of habitation of the island during the Stone and Bronze Ages. The roots of modern Ireland begin, however, with the arrival of the Celts from continental Europe in the fourth century B.C. These tall, red-haired people brought about the Iron Age in Ireland, settling down to form a Gaelic civilization and dividing the island into five kingdoms. After the arrival of St. Patrick in A.D. 432, Christianized Ireland became a center of Latin and Gaelic learning.

Near the end of the eighth century, a peaceful Ireland was invaded by the seafaring Norsemen, or Vikings, from Scandi-

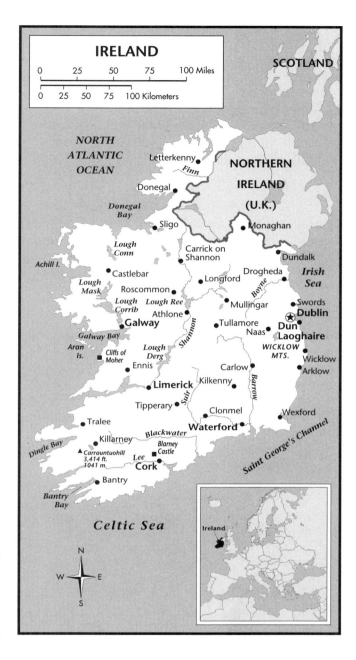

IRELAND

0 25 50 75 100 Miles

0 25 50 75 100 Kilometers

SCOTLAND

NORTH
ATLANTIC
OCEAN

Letterkenny

Finn

NORTHERN
IRELAND
(U.K.)

Donegal

*Donegal
Bay*

Sligo

Monaghan

*Lough
Conn*

Carrick on
Shannon

Dundalk

Achill I.

Castlebar

*Lough
Mask*

Roscommon

Longford

Drogheda

*Irish
Sea*

Boyne

*Lough
Corrib* *Lough Ree*

Athlone

Mullingar

Swords

Dublin

Shannon

Galway

Tullamore

Naas

**Dun
Laoghaire**

Galway Bay

*Aran
Is.*

Cliffs of
Moher

*Lough
Derg*

*WICKLOW
MTS.*

Wicklow

Ennis

Carlow

Arklow

Barrow

Limerick

Kilkenny

Suir

Tipperary

Clonmel

Wexford

Tralee

Waterford

Saint George's Channel

Killarney

Blackwater

Dingle Bay

▲ Carrauntuohill
3,414 ft.
1041 m.

Lee

Blarney
Castle

Cork

Bantry

*Bantry
Bay*

Celtic Sea

Ireland

N
W E
S

navia. The Vikings destroyed monasteries and wreaked havoc on the land but also began to integrate themselves into Irish culture through intermarriage. They established many prosperous coastal settlements which grew into Ireland's chief cities. Viking power was broken at the Battle of Clontarf in 1014, but only 150 years later the Anglo-Norman people from England began to invade the "emerald isle."

The Anglo-Norman invaders soon gained control of the entire country, and, like the Vikings before them, many became integrated into the Irish culture. The invasion, however, created a political attachment to the English Crown, which led to over 800 years of strife, as successive English monarchs sought to subdue both the Gaels and the Norman-

Irish. Mary I (r. 1553–58) was the first to confiscate Irish land and give it to English colonists, a practice which continued throughout the Elizabethan era, the Commonwealth period, and the reign of the English king William III (1650–1702). Poor treatment of the Irish reached a climax in the eighteenth century with the Penal Laws, which deprived Catholics and Dissenters (a majority of the population) of all legal rights.

Like the Americans who staged their revolution against the British in 1776, the Irish people resented the domination of the English Crown and their lack of self-rule. They demanded the formation of an independent Irish parliament in 1783, but it was abolished less than twenty years later by the 1800 Act of Union, which gave Ireland direct representation in Westminster, the British parliament. Meanwhile, Catholic oppression continued, and a movement for Catholic emancipation began, led by Daniel O'Connell (1775–1847). Emancipation was finally achieved in 1829, but the great potato famine of the 1840s highlighted the tragic condition of the Irish peasants and demonstrated the need for land reform.

The desire for complete Irish independence continued to grow, and the nineteenth century saw a series of uprisings and movements advocating home rule and independence. These movements culminated in the Easter Uprising of 1916, and in 1921 the Anglo-Irish Treaty was signed, establishing an Irish Free State with dominion (self-governing) status in the British Commonwealth. There was, however, violent opposition to dominion status, as well as to the separation of Protestant-dominated Northern Ireland from the primarily Catholic Free State, which had occurred in 1920. Civil war erupted, lasting close to a year. The Free State was officially established and a new constitution adopted in 1922, but the desire for reunification with the north remained strong, shown at its extreme by the terrorist acts of the Irish Republican Army (IRA). The IRA lost popularity after Éamon de Valera (1882–1975), a disillusioned supporter, took over the government in 1932, yet continued to use force, especially from the 1960s onward in the civil violence in Northern Ireland.

For many years, the Irish government continued to favor union with Northern Ireland, but only by peaceful means. In 1998, however, Ireland's prime minister signed an historic peace agreement that provided for continued British rule in the north but with a greater degree of autonomy.

Timeline

Middle Stone Age   First settlers

Ireland is the only country in Europe to have had no Old Stone Age. The nomadic hunting peoples living in what is now France and Spain cross over into England in search of game but never cross into Ireland, which remains uninhabited. The first settlers—primarily a group of hunters—arrive in Ireland during the Middle Stone Age. Their history is very shadowy, though it is known that they hunted, fished, domesticated animals such as oxen, goats, and sheep, and made rough pottery.

2200 B.C.–c. 600 B.C.   Irish Bronze Age

More settlers arrive in Ireland. These various clans are skilled in metalworking. The term Bronze Age is used to refer to the period when pure copper is used to make weapons and household objects until metalworkers learn how to alloy copper with other metals to create the tougher, more useful bronze. Among the artifacts left by these peoples are many stone circles, used for burial and ceremony, as well as Ireland's first swords.

4th century B.C.   Arrival of the Celts

The Celts—a tall, red-haired people from Gaul (a division of the Roman Empire in western Europe) or Galicia—arrive in Ireland. The Celts conquer the large Pict ("painted people") clan in the north and the Érainn tribe in the south, absorbing many of their traditions to establish a Gaelic civilization. The arrival of the Celts marks the beginning of the Iron Age. There is no written history of the Iron Age up to the fifth century A.D. because the Celts practice what is known as the oral tradition: they commit things to memory and pass them on orally. There is much fighting between the different Celtic clans.

A.D. 3rd century   Five Gaelic kingdoms established

By the third century A.D., five permanent Gaelic kingdoms exist: Ulster, Connacht, Leinster, Meath (North Leinster), and Munster. A high king, whose title is little more than honorary, presides over the five kingdoms at Tara in the northeast. The ruins of the king's castle at the Hill of Tara can be visited today.

A.D. 432   Arrival of St. Patrick

St. Patrick (c. 385–461) arrives on the island, bringing Christianity to the Gaelic people. Christian Ireland quickly becomes a center of Latin and Gaelic learning. Irish monasteries are home to not only the religious but also to intellectuals of the day, and Irish missionaries are sent to many parts of Europe.

c. A.D. 500   St. Brigid founds the first convent in Ireland

St. Brigid (453–523), or Bridgit, one of Ireland's patron saints, is alleged to have founded the first convent at Kildare. Though she is believed to have lived from 453 to 523, there is debate regarding whether she was actually a real person or merely a Christianization of the Celtic moon goddess Bridgit. Kildare is the former site of the shrine to the goddess Bridgit who was feasted with fire on February 1. St. Brigid's feast day is also February 1, and the nuns at the convent are said to tend a sacred fire (which no man is allowed to approach) for

many generations. St. Brigid and the goddess Bridgit correspond in many other ways. Regardless of the historical existence of a woman named Brigid who founds the convent at Kildare, the cult of St. Brigid is likely a continuation of the ancient worship of Bridgit.

Late 8th century   Viking invasions

Vikings from Scandinavia (northern Europe) begin to invade Ireland, destroying monasteries and pillaging the land. The Vikings, however, also intermarry with the Irish, adopting many of their customs, and establish many coastal settlements from which have grown Ireland's major cities. The Viking and Irish ways of life are not extremely different, except for the fact that the Irish are Christian, while the Vikings still worship pagan gods and goddesses.

1014   Battle of Clontarf

Viking power is finally broken at the Battle of Clontarf. The battle is fought between Brian Bóruma (d. 1014), king of Munster and one of the most successful early leaders in Ireland, and an alliance of Leinster and Viking forces. The success of Munster is considered a victory of Irish independence: though Brian dies in battle, his kingdom defeats the Vikings and the Leinstermen. Epic storytellers spread and glamorize the story of the battle. Before his death, Brian has succeeded in uniting all of Ireland, but in the following years this unity falls apart, with each of the provincial kings ruling their respective kingdoms.

c. 1160   Anglo-Norman invasions begin

The Anglo-Norman people from England begin to invade Ireland, eventually gaining control of the whole country. The Normans (also called the Northmen, or Norsemen), originate in Scandinavia and settle what is now called Normandy in France. Though they lose all connection with their Scandinavian homeland, they still fervently lust for adventure and expansion. First taking control of England, they look to Ireland for further expansion. Some of these invaders intermarry and adopt the Irish language, customs, and traditions. The invasions also result in a political attachment to the English Crown, which leads to almost 800 years of strife for the Irish people as successive English monarchs seek to subdue the various peoples of the island. Confiscation of lands and large plantations of English colonists begin under the rule of Mary I (Mary Tudor) (1516–58, r. 1553–58) and continues under Elizabeth I (1533–1603, r. 1558–1603), Oliver Cromwell (1599–1658, Lord protector of England 1653–58), and William III (1650–1702, r. 1689–1702).

1216   First government act of discrimination

Though the Normans do not invade Ireland with any thought of racial discrimination, it gradually comes about. The Norman (English) and Gaelic (Irish) cultures remain separate and distinct, and the Normans, as most conquerors do, feel their culture to be superior. In 1216, King John (1167–1216) grants special privileges within the Roman Catholic Church to Englishmen only.

1324 Woman accused of witchcraft

The wealthiest woman in Kilkenny, Alice Kyteler, is accused of witchcraft, most likely because of a property dispute. She resists authorities and finally flees to England, though her maid Petronella de Meath is burned at the stake.

1348 Black Death ravages Europe

The deadly disease known as the Black Death kills about forty percent of the people living in the densely packed towns and villages. However, most Irish live in the country and lead a wandering life, and thus the death toll is much lower than in other, more urban areas of Europe.

15th century Ireland ruled by deputies

Ireland is ruled by a series of deputies who are closely tied to the British Crown. In spite of this attachment to Britain, the Irish parliament declares in 1460 that it is entirely separate from the British parliament and that Ireland is its own, independent nation.

1534 Protestantism comes to Ireland

King Henry VIII of England (1491–1547) breaks away from the Roman Catholic Church when he is refused a divorce from his wife, Catherine of Aragon. Catherine (1485–1536) is unable to give birth to a male heir to the British throne, and Henry wants to wed his young mistress Anne Boleyn (1504–36) in order to produce such an heir. Thus, a divorce is necessary, and in order to win a divorce, Henry has to separate from the Catholic Church and form a new, Protestant church (the Church of England) with himself as its head. The new Protestantism soon spreads into different parts of Henry's realm, including Ireland.

1537 Church of Ireland founded

The Church of Ireland is the largest Protestant church, and from 1537 to 1870 it is the established state church, governed by the monarch. From 1870, it is an independent self-governing church.

1553–58 Mary I confiscates Irish land

Mary I (1516–58, r. 1553–58)—also known as "Bloody Mary"—begins persecuting the Irish by confiscating their lands and giving the property to English colonists. Many other English rulers will use the same technique against the Irish.

1558–1603 Reign of Elizabeth I of England

As the daughter of King Henry VIII (1491–1547) and the disgraced and beheaded Anne Boleyn (1504–36), Elizabeth I (1533–1603), the half-sister of Mary I (see 1553–58), is regarded all over Europe as an illegitimate ruler. Because her position in politics is so weak, Elizabeth views a disobedient Ireland as a menace to her reign. Hence, she makes Ireland an object of conquest and strengthens England's hold on the island by continuing "Bloody Mary's" practice of land confiscation and emigration of many English colonists. Yet the extent of the brutality that Elizabeth exerts in suppressing three Irish rebellions during her reign gives testament to her program of conquest.

1591 Founding of Trinity College and Trinity College Library

Trinity College Library in Dublin is the oldest and largest library in Ireland, with over three million volumes. Its many treasures include the Book of Kells (eighth century) and Book of Durrow (seventh century), two of the most beautiful illuminated manuscripts from the pre-Viking period.

1667 Birth of satirist Jonathan Swift

Born in Dublin, Jonathan Swift—a poet, satirist, and clergyman—is educated in Ireland, obtaining his degree from Trinity College only by 'special grace' in 1685. Thanks to family connections, Swift is able to become secretary to the renowned English diplomat Sir William Temple (1628–99). He writes his first two satires in 1704: *The Battle of the Books* and the powerful *A Tale of a Tub*, which attacks religious corruption. Active in politics, Swift is involved in a strenuous campaign for Irish liberties in the 1720s, for which he writes *Drapier's Letters*, a work concerned with the restrictions on Irish trade and the miserable state of Ireland's poor. Swift's most famous literary accomplishment, *Gulliver's Travels*, is published in 1726. *Gulliver's Travels* is a keen satire on the hypocrisy of the courts, political parties, and statesman, and it is the only work for which he receives any payment. Swift dies in 1745.

1685–88 Reign of James II

King James II (1633–1701) of England, who converts to Catholicism in 1669, spends his years as king trying to revive the Roman Catholic religion in the British Isles. James concentrates especially on Ireland, attempting to make the island a Catholic stronghold. From February 1687, the army, judiciary, and civil administration of Ireland are made overwhelmingly Catholic. A Scot himself, James is sympathetic only to the Protestant Scots who practice their own form of religion, which will become known as Presbyterianism. James, however, is overthrown in the revolution of 1688 when

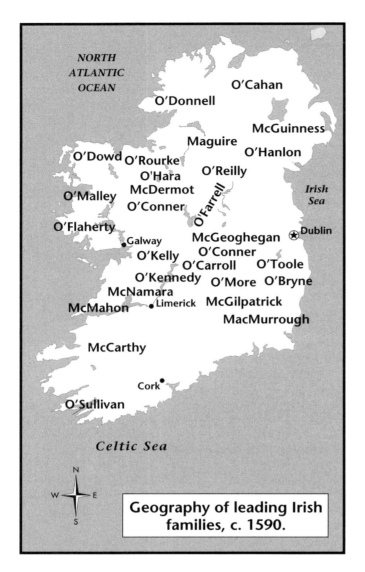

Geography of leading Irish families, c. 1590.

William III (1650–1702) of the Netherlands, or William of Orange, invades England. William III attacks England because he fears that James has been turning it into a "satellite" of France, his sworn enemy. Reluctant to offend the Catholics, William initially blocks the so-called "Penal Laws" (see 18th century), but eventually gives in to Protestant pressures and allows the laws to come into effect.

1685   Birth of philosopher George Berkeley

Both a philosopher and clergyman, Berkeley writes his most important works in his early years as a student and tutor at Trinity College, Dublin. In *Essay towards a New Theory of Vision* (1709), *A Treatise concerning the Principles of Human Knowledge* (1710), and *Three Dialogues between Hylas and Philonous* (1713), Berkeley develops the philosophy that "to be is to be perceived" (in other words, everything in the material world is an idea that only exists when a mind perceives

it.) Berkeley attempts to found a college in Bermuda, an endeavor which does not succeed. He is named Bishop of Cloyne and dies in England in 1752.

1688   "Glorious Revolution" in England

William III of the Netherlands invades England, taking over the throne from James II (see 1685–88). It is so-called due to its milestone achievement in the gradual process by which Parliament was to gain practical power over the monarch, and for the obvious importance even in the Glorious Revolution's other name, the "Bloodless Revolution."

18th century   Penal Laws put into effect

The persecution of the Irish reaches its peak in the eighteenth century with the Penal Laws, which deprive Catholics and Dissenters (those Protestants who refuse to accept the restored Episcopalian Church of Ireland)—who form a majority of the population—of all legal rights. The harsh laws are enacted partly as a Protestant reaction against the reign of James II (see 1685–88).

c. 1718   First female pirate (born in Ireland) active in the Caribbean

Anne Bonney (fl. 1720) as well as Englishwoman Mary Read (1690–1720) are the first female pirates on record. Bonney is born in Ireland and becomes the mistress of the pirate Captain "Calico Jack" Rackham. The two women work together as pirates and are tried and convicted as criminals in Jamaica in 1720. Both are pardoned, however, when they (falsely) plead that they are pregnant.

1731   Founding of Dublin Society

Inspired by the pleas of Jonathan Swift (see 1667) to improve the well-being of the Irish poor, many Protestants (who, as a result of the so-called "Protestant Ascendancy" movement, have reached the top of society by use of such measures as the Penal Laws) begin to take their high society responsibilities more seriously. Wealthy Protestant clergy found the Dublin Society, whose goal is to improve agriculture, manufacturing, and the fishing industry. Many landlords are inspired by this cause, and the lives of many countryfolk are improved. In 1733 special schools are established for the children of the poor. Though they have a Protestant bias (and therefore do not become popular), the schools are yet another sign of good intent.

1739   Death of soldier Kit Cavanagh

Soldier Kit Cavanagh (1667–1739)—also known as "Mother Ross" and "Christian Davies"—is buried with military honors. She fights with distinction in the British army for ten years before her sex is discovered, after which time she is only permitted to be a cook.

1751 Birth of dramatist Richard Brinsley Sheridan

Richard Brinsley Sheridan is best known for three comedies, all "comedies of manners": *The Rivals* (1775, very well received in the Covent Garden Theater), *The School for Scandal* (1777), and *The Critic* (1779). In 1776 Sheridan buys half the title to the Drury Lane Theater, London, which—after years of mismanagement and a fire in 1809—eventually brings about his financial downfall. Sheridan serves as a member of parliament and in various government roles in the late 1700s until 1812. He dies in poverty in 1816 but is given a lavish funeral at Westminster Abbey.

1775 Birth of Daniel O'Connell, Irish political leader

Known as "the Liberator," Daniel O'Connell is born near Cahirciveen, County Kerry. In 1798 O'Connell forms a very successful law practice. He becomes the leader in a movement advocating for the rights of Catholics, and in 1823 forms the Catholic Association, which successfully fights elections against the landlords. In 1828 O'Connell is elected as a member of Parliament, but as a Catholic he is prohibited from taking his seat. By the force of public opinion, the Catholic Emancipation Bill is passed in 1829, and he is reelected in 1830.

O'Connell continues to advocate for Irish and Catholic rights, leading agitation for abolition of tithes (money paid to the church), formation of an established church in Ireland, and opposition of poor laws (see 1838). In 1841 he becomes Lord Mayor of Dublin, where he revives an earlier movement to repeal the union between Ireland and Britain (he and his supporters are known as "Repealers"). O'Connell's "Repealer" agitation gains prominence, but the up-and-coming Young Ireland Party begins to become impatient with O'Connell's leadership. In 1847—after his power is broken by dissension, stress from the potato famine (see late 1840s), and ill health—O'Connell leaves Ireland and dies in Genoa on his way to Rome.

1782–1800 "Grattan's Parliament"

During this period—known as "Grattan's Parliament" after the Irish patriot leader in parliament, Henry Grattan (1746–1820)—Ireland enjoys an unprecedented prosperity. This is the longest time of stability (with no war or rebellion) since the time of the Viking invasions. As a result of this stability, the population rises from less than two million in 1700 to four million and continues to rise. The Irish economy booms. Trade restrictions are lifted, so that Catholic businessmen are able to invest in their home country instead of sending their money to France and Spain. In 1706 total Irish exports are worth £50,000. By 1796 they are worth over £5 million. Despite the prosperity of the times, Grattan's parliament is not truly representative of the Irish people: it is comprised of a handful of rich Protestants surrounded by disenfranchised (unable to vote) Catholics and Presbyterians.

1782 Birth of musician John Field

A child prodigy, John Field (1782–1837) is apprenticed to the Italian composer Clementi (1752–1832), with whom he travels in Europe. Field's nineteen *Nocturnes* and other keyboard music influence the future compositions of Polish composer Frederick Chopin (1810–49).

1783 Establishment of independent Irish parliament

By the end of the eighteenth century, many English colonists regard themselves as Irish and, like the English colonists in America, resent the domination of the British Crown and their own lack of self-rule. They force the establishment of an independent Irish parliament, which is abolished only seventeen years later by the Act of Union (see 1800).

1785 Beginnings of religious violence

Many Presbyterian farmers emigrate to America in the eighteenth century, leaving much land in Ulster (in northwestern Ireland) available for settlement. Catholics flood in and purchase the land by outbidding the Presbyterians in rent. Angry Presbyterians begin forming secret societies such as the "Peep o'Day Boys," who come to Catholic homesteads at dawn (the peep o'day) and terrorize the families living there. The Catholics react by forming similar organizations, one known appropriately as the "Defenders." In Ulster violence of this type has yet to cease.

1785 Founding of Royal Irish Academy

The Royal Irish Academy promotes study in science and the humanities and advocates for Ireland's membership in international scientific unions.

1792 Birth of Orientalist Edward Hincks

Hincks (1792–1866) is often credited with having discovered the true method of deciphering Egyptian hieroglyphics and is also said to have discovered the Sumerian language.

1793 Author defends education for women

British author Maria Edgeworth (1767–1849) advocates for female education in *Letters to Literary Ladies*. In 1798 she coauthors an adaptation of Rousseau's theories, *Practical Education*. Her 1800 novel *Castle Rackrent* condemns absentee landlords who keep the Irish peasants in poverty.

1799 Struggle for Catholic emancipation begins

Although the worst of the Penal Laws (see 18th century) are gone, Catholics continue to be excluded from many aspects of society when they are forced to take the Oath of Supremacy, swearing their loyalty to the monarch. The Oath not only keeps Catholics out of parliament, but prohibits talented middle- and upper-class Catholics from obtaining certain employments. For instance, no matter how good a lawyer, a

Catholic may never become judge; and no Catholic can obtain a scholarship or teach at Dublin University.

In 1799, British Prime Minister William Pitt (1759–1806) promises the Catholics that full emancipation will follow the Union between England and Ireland (see 1800), but when Pitt presses the issue at home, King George III (1738–1820) absolutely refuses to grant the emancipation. Pitt resigns in frustration, and Catholic freedom is postponed indefinitely.

1800   Act of Union

The Act of Union abolishes the independent Irish parliament, giving Ireland direct representation in the English parliament at Westminster.

1800   Height of the "cottage industries"

Only about one in ten Irish live in a town in 1800. Other Irish are primarily farmers, living in one-room cottages. Although agriculture is the main occupation of these rural people, many supplement their income by participation in "cottage industry," or small-scale manufacturing in the home. For example, the production of linen is often completed in the home, with all family members—even children—participating: the flax is rotted in water, then dried, then separated into strands and woven into linen cloth.

1808   Birth of composer Michael Balfe

Born in Dublin, Balfe (1808–70) begins to write music at age seven and makes his debut as a violinist at age nine. In 1823 he goes to London and studies music in Italy under the Italian composer Gioacchino Rossini (1792–1868). Balfe becomes inspired to sing opera, at which he succeeds. After returning to England, he becomes conductor of the London Italian Opera (1846). His most famous work is the opera *The Bohemian Girl*, which he composes in 1843.

1817–19   Fever in Dublin

Over 40,000 people are treated in Dublin hospitals for a fever which sweeps through the city's slums. Though the 200,000-person, bustling city has many prosperous and elegant areas, it is also home to great poverty and shocking slums, which by the end of the century are considered among the worst in Europe. The poor areas have tightly packed tenement houses, with no water or drains and little fresh air. Often, several families may occupy a single, one-room tenement. It is from crowded areas such as this that epidemics like the fever are spread.

1820   Beginning of industrialization in the North

Belfast begins to support the prosperous linen and shipbuilding businesses. The industrialization causes the city to expand quickly, and the population rises from 70,000 in 1840 to 350,000 in 1900. The overcrowding leads to tensions, especially between the Catholics and Protestants.

1822   Female boxing match

Irishwoman Martha Flaharty and Englishwoman Peg Carye compete in a boxing match. After downing half a pint of gin, Flaharty wins.

1822   Birth of Frances Power Cobbe

Frances Power Cobbe (1822–1905), (though technically British, she is born in Dublin) a reformer and philanthropist, spends her life advocating for the end of vivisection (operating on live animals for medical purposes), and for education for the working classes.

1823   Catholic Association founded

Daniel O'Connell (see 1775) founds the Catholic Association, established to campaign for Catholic emancipation. By making the membership fee affordable for a wide range of people, the association soon becomes a mass-based political movement, the first of its kind seen in Ireland.

1824   Founding of *Morning Register*

Michael Staunton founds the *Morning Register*, the first Irish newspaper to have reporters. The growth of papers such as this marks the emergence of a better and freer press. The "new" newspapers help to spread word of movements such as Daniel O'Connell's Catholic Association (see 1823).

1829   Catholic emancipation

Primarily through the efforts of Daniel O'Connell (see 1775), the Catholic Emancipation Bill is passed. The Emancipation Bill eliminates the portions in the Oath of Supremacy (see 1799) that are offensive to Catholics and replaces them with a simple oath of allegiance. O'Connell—the first to organize a peaceful political movement in Ireland—becomes a hero to democrats in England, America, and Europe. Unfortunately, the lower classes do not benefit from emancipation. The government counteracts emancipation by passing a bill which raises the voting fee from forty shillings to £10. The poorer classes, therefore, lose the right to vote because they cannot pay the higher fee and continue to pay high rents and tithes.

1830   Agitation for tithe reform begins

Most Irish farmers are opposed to paying the tithe, or tax, that supports their local Church of Ireland (Protestant) clergymen (see 1537). Farmers feel these taxes are unfair because most of them are either Catholic or Presbyterian. In 1838 an act reduces the tithe by twenty-five percent.

1831 Commissioners of National Education founded

After years of argument between Catholics and Protestants, the Commissioners of National Education are founded to ensure that Irish education is nondenominational. The commission demands that children of all ages be taught together in all subjects except religion. The schools are called the National Schools. However, many all-Catholic and all-Protestant schools refuse to join the National system, so a majority of Catholic and Protestant students continue to be taught in separate schools.

1838 Poor Law enacted

In response to growing worries over the number of poor in Irish cities, the Poor Law is enacted, partly as a response to the plight of the needy and partly from a fear that the unemployed Irish would flood into England and lower wages there. The law establishes workhouses to support the local poor. Whenever a person needs help, he goes to the workhouse, where conditions are uncomfortable so that only the most needy will go in. Families are separated (men, women, girls, and boys all live in different parts of the workhouse), and all persons are expected to earn their keep.

Early 1840s Beginning of the repeal movement

Daniel O'Connell (see 1775) begins a movement to repeal the union with Britain (see 1800) and plans to do so by using the same methods by which he won Catholic emancipation. He forms a Repeal Association and gradually gains support in the following years. In 1842 Thomas Davis (1814–45), Charles Duffy (1816–1903), and John Blake Dillon (1814–66)—soon members of a Repealer group which comes to be called Young Ireland—join the movement and found the weekly newspaper *The Nation*. The paper seeks to educate the people about the cause of Irish independence and to instill in them a knowledge of, and pride in, Ireland. The paper becomes tremendously popular. The repeal movement begins to win wide support beginning in 1843. An example of the movement's success is apparent in a meeting held at Tara on August 15, 1843: approximately 250,000 people attend to hear O'Connell and other Repealers speak.

1841 O'Connell serves as mayor of Dublin

Daniel O'Connell (see 1775) is named mayor of Dublin, the first Catholic to hold this post since 1688.

1841 First official census

The first reliable census of Ireland is completed in 1841, showing a population of 8,175,000. This population is almost twice of that reported in the 1970 census. The majority are farmers and have quite a low standard of living. They survive on potatoes, since an acre of land will grow enough potatoes to feed a family of five for most of the year. Farmers pay their workers by giving them a plot of land on which to grow potatoes. About three million of the population are directly dependent upon the potato.

1845 Robert Peel's reforms

Robert Peel (1788–1850), prime minister of England from 1841–46, begins to make social reforms in Ireland in an attempt to win support away from Daniel O'Connell's movement. Peel has O'Connell and others arrested in 1844 for a "conspiracy to overthrow the government." O'Connell is released from prison later that year and enacts such reforms as establishing new universities (Queen's Colleges) and completing a study on landlord-tenant relations.

1845–49 Irish Potato Famine

Introduced to Ireland by English navigator Sir Walter Raleigh (1552–1618) in 1610, the potato quickly becomes one of Ireland's most important crops, developing into the main source of nourishment for much of the Irish population. Potatoes cannot be kept fresh for very long, and even with the modern methods of the nineteenth century (when potatoes are stored in cool pits dug into the ground), they cannot last through the summer months. The danger of hunger in the summer months increases with every year as the growing population presses harder on the available land. Between 1831 and 1842, there are six seasons of poor potato crops, due principally to a blight, and in the 1840s millions die from starvation or emigrate for lack of potatoes. Meanwhile, wealthy landlords continue to export other crops to England for profit instead of keeping the food to stop the spreading starvation among peasants. The tragic famine emphasizes the great need for land reform and the impoverished condition of the Irish peasant, in one of the worst natural disasters in history.

1846 Repeal Association splits

O'Connell—whose tactics have become more conservative since his arrest—and the more radical Young Irelanders disagree on many issues regarding how the repeal movement should be handled. The Repeal Association (divided into pro-O'Connell and Young Ireland factions) splits just before the worst crisis in Irish history, the potato famine.

1848 Rebellion

Daniel O'Connell dies in 1847, and though his son John takes over leadership of the repeal movement, it soon loses strength due to the stress of the famine and the loss of the movement's founder. Meanwhile, however, the Young Irelanders have founded their own organization, the Irish Confederation. The Confederation agitates not only for repeal but for ownership of land, and many are soon convinced that the only way to reform is through rebellion. A successful rebellion in France in early 1848 gives the Confederation hope for their own movement's success, and they quickly begin to build up support among the people.

The rebellion begins in County Tipperary but is short-lived, as the government has many spies within the Irish Confederation and knows exactly what is being planned. A brief clash with police shows how weak the rebels are, and the followers disperse as the leaders are arrested.

Though the rebellion fails, the Young Irelanders leave a valuable legacy to future Irish Nationalists: their ideals are at the heart of the nationalist movement, and young rebels learn that disaster can result from lack of careful planning. Young nationalists are intent on having more success in the future.

1850s   Growth of the railways

In 1845 there are fewer than one hundred miles of railway; by 1865, there are over two thousand miles. The railroads encourage travel and tourism, as people may now travel at a steady pace of twenty miles an hour, as compared to the previous top speed (by stage coach or canal boat) of ten miles per hour. Towns on the railway prosper, while others decline, and railways also harm local industries by making imported goods cheaper. Cheap and quick transport, however, means that farmers get better prices for their produce.

1854   Birth of writer Oscar Wilde

A prolific playwright, novelist, essayist, and poet, Oscar Wilde is born in Dublin to Lady Jane Francesca Wilde (1826–96), a poet who writes under the pen name of 'Speranza,' and Sir William Wilde (1815–76), an occultist (believer in the mystical arts of alchemy, astrology, magic, etc.). Wilde attends Trinity College, Dublin, and Magdalen College, Oxford, where he gains a reputation for being outspoken and unconventional. Wilde becomes an accomplished classicist (a specialist in the Greek and Latin languages), and wins the Newdigate prize in 1878 for his poem "Ravenna." In 1881 his first volume of poetry—Patience—is published, and the next year Wilde leaves for a lecture tour in the United States. (Not known for his modesty, it is said that Wilde replied, 'Only my genius,' when asked by the U.S. customs agent if he had anything to declare.)

Married in 1884, Wilde has two sons, for whom he writes children's fairy tales. In 1890 Wilde publishes the novel The Picture of Dorian Gray, supposedly modeled on the poet John Gray, his presumed lover. After 1890 Wilde builds his reputation as a dramatist by publishing five plays, including his masterpiece The Importance of Being Earnest (1895). By this time, Wilde's homosexuality is commonly known Wilde files a lawsuit against the marquess of Queensberry (the father of his intimate companion) for libel concerning the marquess' flagrant accusations of homosexual offenses against Wilde. But the marquess files charges of his own under the Criminal Law Amendment concerning Wilde's homosexuality. Losing the case, Wilde is jailed for his homosexuality. Though he is released in 1897 and travels to France under an alias, he has become ill while in prison and dies in 1900.

1856   Birth of dramatist George Bernard Shaw

Born in Dublin, George Bernard Shaw inherits a love of music from his mother, which will have a profound impact on his work. Unhappy with both school and office work, Shaw leaves Ireland for good in 1871 to follow his mother and actress sister to London. The family's early years in London are full of poverty and struggle, and all of the five novels Shaw writes between 1879–83 are rejected by the more reputable publishers. Interested in socialism after reading the works of German theorist Karl Marx (1818–83), Shaw forms a 'kindly dislike' for capitalist society, which forms the backbone of all his work.

Shaw first makes his name between 1888–98 as a music and drama critic and marries the Irish heiress Charlotte Payne-Townshend in 1898. Shaw then turns to play writing, and some of his best-known early works include Man and Superman (1902), Getting Married (1908), and Misalliance (1910). Just before World War I (1914–18), Shaw writes some of his most charming plays: Androcles and the Lion (1912) and Pygmalion (1913), adapted in 1956 as the highly successful musical play My Fair Lady, filmed in 1964. Three of his greatest dramas are written after the war: Heartbreak House (1919), Back to Methuselah (1921), and Saint Joan (1923), which demonstrates most clearly Shaw's religious nature, his creative characterization, and his flair for drama.

Shaw is awarded the Nobel Prize for Literature in 1925 but donates the prize money to found the Anglo-Swedish Literary Foundation. Towards the end of his life, Shaw's plays—not as widely performed on stage—become more experimental and unique. He dies in 1950.

1858: St. Patrick's Day   Founding of the Irish Republican Brotherhood

James Stephens (1824–1901) and John O'Mahoney (1816–77)—both recently returned from exile after the 1848 rebellion—found the Irish Republican Brotherhood (or IRB), a group also known as the Fenians after the legendary heroes of ancient Ireland. The Fenians pledge to establish an Irish republic by force, and members of the group take a secret oath.

Two important groups are opposed to the Fenians: the old followers of O'Connell and the Catholic clergymen. Thus, it is difficult to spread the new movement, though by 1865 Stephens estimates there are some 80,000 Fenians in Britain and Ireland, as well as an American faction formed just after the American Civil War.

1865–67   The decline of Fenianism

James Stephens (see 1858: St. Patrick's Day) promises that 1865 will be the year of a revolt to gain independence from Britain and then (after no order to revolt comes) rashly promises that 1866 will be the year. Stephens, however, continues to hesitate, while Fenians continue to be arrested. Stephens's reputation therefore suffers irreparable damage, and the movement begins to fall apart. Finally, a coup overthrows Stephens, naming Colonel Thomas Kelly as their leader. Kelly attempts to mount rebellions in February and March 1867, both of which meet with failure. Hundreds are arrested in the March rebellion, and amnesty groups are formed to free the imprisoned Fenians. It is from these amnesty groups that the foundations for the next independence movement—the Home Rule Party—springs.

Fenianism never dies out completely, with many of those freed from prison going to America and starting movements there, in the 'greater Ireland beyond the seas.' It is the first movement that looks for support in American Irish.

1865   Birth of poet William Butler Yeats

William Butler Yeats, the son of artist John Butler Yeats (1839–1922), is born in the suburbs of Dublin. Yeats forms a life-long bond with the Irish land, culture, and history at a very young age; as a child, he spends much of his time in his mother's native County Sligo, the wild and naturally beautiful coastal area in the north.

Yeats first has his verse published while in college, where he forms a fascination with mysticism and the occult. He also pursues his interest in Irish mythology, which inspires much of his poetry. Yeats moves to London in 1887 and becomes friends with other well-known writers, such as George Bernard Shaw (see 1856) and Oscar Wilde (see 1854).

Homesick for Ireland, Yeats returns there in 1891, where he falls in love with the Irish nationalist Maud Gonne MacBride (1865–1953), who ultimately refuses to marry him. Inspired by Maud Gonne, Yeats writes *The Countess Kathleen* (1892). His best-known drama, *The Land of Heart's Desire* (1894), is the story of a young woman spirited away by a fairy child. Yeats gains his greatest fame as a poet with *Poems* (1895). In 1904 he founds the Abbey Theatre with Lady Gregory (see 1904). He marries Georgie Hyde-Lees in 1917, when he is fifty-two and she fifteen. Yeats is active in politics, becoming a member of the Irish senate in 1922 and receives the Nobel Prize for Literature in 1923. Yeats is a prolific writer as well as a public figure throughout his lifetime; he dies in 1939.

1868   Birth of astronomer Annie Russell Maunder

Annie Russell Maunder (1868–1948) becomes employed at the Greenwich Royal Observatory, where she specializes in sunspots and the relationship between earth's climate and changes in the sun.

1869: July 26   Irish Church Act

In spite of the anti-tithe reforms of the 1830s (see 1830), the Protestant Church of Ireland continues to have unfair powers: it is the official church of Ireland, though its members make up only one-eighth of the population. The Irish Church Act—part of a program to bring peace to Ireland—ends state involvement with the church, making it an independent organization.

1870   Gladstone's First Land Act

In 1870, 3 million of the 5,700,000 Irish are farmers, and unjust land laws are among the greatest grievances of the Irish people. Until 1870 these farmers get no relief from the government because most members of parliament are landowners and do not want to see their own interests ruined. After the Fenian uprisings, however, fears are aroused about Irish unrest, and Prime Minister William Gladstone (1809–98)—whose mission is to pacify Ireland—is able to pass his first Land Act. Though the act falls short of many expectations, it successfully provides compensation to tenant farmers who are unjustly evicted from their land.

1870   Home Government Association founded

Irishman Isaac Butt (1813–79) founds the Home Government Association, which strives to give Ireland her own parliament to look after internal matters (such as roads and schools), while allowing the British government to remain in control of international affairs. Though this is much less extreme than the repeal movement, Butt feels his Home Rule is the most that the English government would be willing to grant Ireland.

1870   Birth of artist Jack B. (John Butler) Yeats

The son of artist John Butler Yeats (1839–1922) and brother of William Butler Yeats (see 1865), Jack Yeats is best known for his comic strips and impressionist paintings. Educated in County Sligo, his first drawing is published at the age of eighteen. In 1894 he creates the first cartoon strip version of Sherlock Holmes, *Chubblock Holmes*. Yeats also writes and illustrates many children's books and draws comic strips until 1918, after which time he concentrates on painting and writing plays. He dies in 1957.

1873   The Home Rule Party

Isaac Butt (see 1870) turns his association into a political party (see 1865). The Home Rule Party is not very successful at first, as Butt is too gentle to truly fight for his ideal. However, some members of parliament who are Home Rulers (such as Charles Parnell, 1846–91) call attention to the movement by filibustering, or making long, dull speeches and debate in parliament in order to delay bills. One time, a Home Rule debate lasted twenty-six hours.

1879   Poverty and fears of famine

Immediately after the potato famine, farmers become more prosperous because the prices of products such as meat and butter rise rapidly. Consequently, landlords begin to demand higher rents. When prices fall again in 1876 and there are poor crops in the harvests of 1877–79, many farmers cannot pay their higher rents and are evicted. In the west, a new famine seems possible. The situation becomes critical in 1879.

1879 Founding of Land League

Former Fenian Michael Davitt (1846–1905) organizes protests demanding that landlords reduce their rents in the west, where famine is threatening. Davitt asks Charles Parnell (see 1873) to speak at a demonstration, where Parnell realizes that most Irishmen are more concerned with their land than with government matters, including Home Rule. Parnell sees, though, that if he can help these men gain control of their farms, they would support Home Rule. Thus, the Land League—which advocates for land rights—is founded, with Parnell as its president.

1881 Gladstone's second Land Act

Alarmed by the growth of the Land League (see 1879), Gladstone passes his second Land Act. The Act establishes special land courts that decide what is fair rent—and once the rent is settled by the courts, the tenant who pays it cannot be evicted. These courts reduce rents by an average of twenty percent.

1882–91 Parnell and the fate of the Home Rule Party

Between the years of 1882–85, Parnell (see 1873, 1879, and 1882: May 2) concentrates on reorganizing the Home Rulers, seeking a strong united party. He institutes reforms, and after the 1885 election, there are eighty-five determined Home Rule followers in parliament.

In 1886, a Home Rule bill—supported by Gladstone (see 1870 and 1881)—is introduced into parliament. The bill, however, fails because of a significant opposition, including the Protestant minority. They fear Home Rule because they feel that, in an independent Ireland, they would be persecuted by the Catholics. The party continues to gain strength, however, until a disaster in 1890.

Since 1880, Parnell has been in love with and the companion of Katherine O'Shea (1845–1921), the wife of former Home Ruler Captain William O'Shea (1840–1905). In 1890 Captain O'Shea exposes his wife's adultery and starts a tremendous scandal that sharply divides the Home Rule Party. Parnell is shamed and spends the next year attempting to gain support, but to no avail. An exhausted Parnell dies in 1891 at the age of forty-six and prospects of Home Rule die with him.

1882 Birth of Éamon de Valera, Irish statesman

Éamon de Valera is born in Brooklyn, N.Y., but is raised in County Limerick, Ireland, by a laborer uncle. Initially a mathematics teacher, he rises in the Irish Volunteers, leading men in the Easter Rebellion of 1916 ([see 1916:April 24] after which he is sentenced to death, saved only by intervention of the U.S. consul). He becomes affiliated with the Sinn Féin party (the political branch of the Irish Republican Army) and is elected to parliament in 1917. De Valera is opposed to the Anglo-Irish Treaty of 1921 and in 1926 forms the Irish Republican party, Fianna Fáil. In 1932 de Valera is appointed prime minister (a position he holds intermittently until 1959), until becoming president of Ireland for two consecutive terms, 1959 to 1973. De Valera dies in 1975.

1882: February 2 Birth of author James Joyce

Born in Dublin, James Joyce is an avid reader and completes his formal education at University College, Dublin. He corresponds with the well-known Norwegian dramatist Henrik Ibsen (1828–1906) and is influenced by such writers as the Italian poet Dante Alighieri (1265–1321) and the Irish poet William Butler Yeats (see 1865). Unhappy living in what he perceives to be a bigoted Roman Catholic society, Joyce leaves Ireland in the early 1900s with Nora Barnacle, who becomes his lifetime companion. The couple lives in Trieste, Italy; Zurich, Switzerland; and Paris, France, but Joyce continues to set his works in Ireland, especially Dublin. In 1998 two of his best-known works—*A Portrait of the Artist as a Young Man* (1914) and *Ulysses* (1922)—are named, respectively, the number three and number one works of fiction in the twentieth century in a survey by an independent board selected by Random House's Modern Library publishing arm. *Ulysses* ignites a flurry of controversy when it is first published in 1922 (and is banned from the United States on grounds of obscenity), but over the coming decades it becomes the focus of extensive literary criticism and analysis. Though many of his works are difficult to interpret, Joyce's readers enjoy his word play, sense of humor, and insight into the human condition. Joyce dies in 1941.

1882: May 2 The Kilmainham Treaty

The government suspects that Parnell is attempting to obstruct the second Land Act and jails him at Kilmainham prison. The Land League (see 1879), now without its leader, is virtually powerless, and there are widespread disputes about the new act. The new courts are packed with tenants seeking fair rents, but many who are dissatisfied or ignored resort to violence: in the six months Parnell is in jail, there are fourteen murders. Finally, the government realizes it cannot control the tenant crisis without Parnell's help and frees him under the so-called Kilmainham Treaty. The treaty releases Land League leaders, while in return Parnell is to work towards peace in the countryside.

1882: May 6 Phoenix Park murders

Two government officials (Lord Fredrick Cavendish, British secretary for Ireland, and Thomas Henry Burke, his undersecretary) are assassinated in Phoenix Park, Dublin, by a secret society called the "Invincibles." Britain and Ireland are shocked by the violence. The killings allow Parnell to break away from the Fenian-controlled Land League and form a new organization, the Irish National League, in 1882. Parnell is firmly in control of the new league, whose primary goal is to win Home Rule.

1883 Birth of actress Sara Allgood

Sara Allgood performs in Lady Gregory's (1852–1932) *Spreading the News* on the opening night of the Abbey Theater (see 1904). After many successes on stage in Great Britain and Australia, Allgood travels to the United States, becoming a U.S. citizen in 1945. Though she appears in over thirty American films—such as *Jane Eyre* (1943), *The Lodger* (1944), and *Between Two Worlds* (1944)—she is not given a high salary for her work and dies penniless in 1950.

1884 Birth of tenor John McCormack

After studying in Milan, Italy, McCormack makes his London debut in 1905, where he sings for the Covent Garden opera. He takes American citizenship in 1919 and begins to sing popular sentimental songs. McCormack dies in 1945.

1887 Birth of artist Sir William Orpen

Best known for his portraits, Orpen also depicts daily life in Ireland. Orpen does many sketches and paintings at the front during World War I (1914–18) and is present at the 1918 Paris Peace Conference as the official painter (the paintings from this conference may be found at the Imperial War Museum). Orpen dies in 1931.

1888 Birth of actor Barry Fitzgerald

Born William Joseph Shields, Barry Fitzgerald (1888–1961) begins his acting career on stage in Ireland before traveling to the United States to perform both on-screen and off. Some of his most successful films include Eugene O'Neill's (1888–1953) *The Long Voyage Home* (1940), Agatha Christie's (1890–1976) *And Then There Were None* (1945), as well as the highly successful *Going My Way* (1944). After acting in over forty films, Fitzgerald returns to Ireland in 1959, where he spends the last few years of his life.

1889 John Bury publishes history of the Romans

John Bagnell Bury (1861–1927) is only twenty-eight when he publishes his monumental scholarly work, *History of the Later Roman Empire*. A professor of modern history and Greek, Bury is best known for his histories of Greece and Rome.

1893 Founding of the Gaelic League

Douglas Hyde (see 1938) and Eoin MacNeill (see 1913) found the Gaelic League to try to revive Gaelic as the spoken language of the Irish people. Another object of the League is to encourage the writing of plays, poems, and novels in mod-

ern Irish. The two men hope that a love for the Irish language will bridge the gap between the different political affiliations.

1896 Birth of poet Austin Clarke

The early works of Austin Clarke (1896–1974), most notably *The Vengeance of Fionn* (1917), are heavily influenced by the poetry of Yeats (see 1865) and by Irish mythology and legend. His later poetry, however, becomes satirical and critical of many Irish attitudes. Clarke's first novel, *The Bright Temptation* (1932), is banned from Ireland until 1954.

1897 Birth of author Liam O'Flaherty

Liam O'Flaherty's (1897–1984) best-known works include *The Informer* (1926, a very popular novel which wins the James Tait Black Prize), *Spring Sowing* (1926), *The Assassin* (1928), *The Puritan* (1932), *Famine* (1937), and *Land* (1946). O'Flaherty fights for the British Army during World War I (1914–18) and then travels in North America and Latin America. He returns to Ireland in 1921 to fight on the Republican side in the Irish Civil War (see 1921), and then moves to London to write.

1900 Home Rule Party reunited

John Redmond (1856–1918) reunites the Home Rule (Liberal) Party, which continues to advocate for a completely autonomous Ireland. Opposition to the party includes the Unionists (who are tightly linked to the British Conservative Party). The Unionists are mostly Protestant and fear Home Rule because they feel that in an all-Irish parliament, the Catholic majority could deprive them of religious freedom.

1900 Daughters of Ireland founded

Actress and revolutionary Maud Gonne (1866–1953) founds the radical women's group Inghinidhe Na Eireann (the Daughters of Ireland) in Dublin. Devoted to the Irish Republican cause, Gonne supports the Easter Rebellion of 1916 (her husband John MacBride is executed for participating in the uprising) and is a relief worker during the troubled years following the rebellion. She is also a friend of the poet W. B. Yeats (see 1865).

1900 Birth of writer Seán O'Faoláin

Seán O'Faoláin (b. 1900) begins his literary career writing and translating the Gaelic language (as in 1938's *The Silver Branch*). His first novel, *A Nest of Simple Folk* (1933), is quite successful, while many of his later works are biographies, such as *Daniel O'Connell* (1938) and *De Valera* (1939).

1903 State forestry program begins

In an effort to restore Ireland's wooded areas, a state forestry program is inaugurated. Once a well-forested island, Ireland loses much of its timber in the seventeenth and eighteenth centuries, when absentee landlords make no attempt to refor-est stripped land. The spread of agriculture (with large plots of land needed for grazing and crops) has also contributed to deforestation.

1903   Birth of writer Frank O'Connor

A member of the Irish Republican Army as a teenager (1921–22), Frank O'Connor (pen name for Michael O'Donovan) fights against Great Britain and is imprisoned. After being released from prison, O'Connor becomes a librarian and writer, best known for his collections of short stories such as *Guests of the Nation* (1931), *Bones of Contention* (1936), *Crab Apple Jelly* (1944), and *Travellers' Samples* (1956). Yeats (see 1865) proclaims that O'Connor is "doing for Ireland what Chekhov (1860–1904, a highly influential Russian writer) did for Russia." He dies in 1966.

1904   Founding of national theater

The Abbey Theatre, Ireland's national theater, is founded by the poet William Butler Yeats (1865–1939; see 1865) and the playwright Lady Augusta Gregory (1852–1932). Many people consider the founding of the theater to be the beginning of Ireland's "cultural renaissance."

1905   Founding of Sinn Féin political party

Arthur Griffith (1872–1922), who in 1922 becomes the first president of the Irish Free State, founds a new political party. Sinn Féin (pronounced SHIN-FANE, which means 'We Ourselves') will come to be affiliated with the radical Irish Republican Army (IRA).

1906   Liberal party wins the general election

The Liberals, who support Home Rule, win in a general election by a large majority. They begin introducing social reforms, such as an old age pension. The Unionists (Conservatives) become alarmed by the growing influence of the Home Rulers.

1906   Birth of writer Samuel Beckett

Both an author and a playwright, Beckett is born in Dublin in 1906. From 1932 Beckett lives mostly in France, and becomes, for a time, secretary to James Joyce (see 1882: February 2). Beckett and Joyce share a fascination for language and for the pointlessness of life that humans strive to make purposeful. Beckett's early works are published in English, but his later works, namely the plays *En attendant Godot* (Waiting for Godot) and *Fin de partie* (End Game), first appear in French. *Waiting for Godot* is Beckett's most widely read and performed work, and it best exemplifies the Beckettian view of life: the human predicament and the loss of our hopes, philosophies, and endeavors. In 1969 Beckett is awarded the Nobel Prize for literature. He writes very infrequently towards the end of his life and dies in 1989.

1908   Marie Joseph Butler founds schools and colleges

Marie Joseph Butler (1860–1940), a nun, founds the Marymount schools and colleges in America and Europe. She opens her first school in 1908 in Tarrytown, New York. A year before becoming an American citizen in 1927, Butler is elected Mother General of the Congregation of the Sacred Heart of Mary, the first superior to head an American Catholic congregation whose motherhouse (headquarters of a religious organization) is in the Old World.

c. 1910   Writer and adventurer Beatrice Grimshaw travels in Africa

Beatrice Grimshaw (1871–1953), a travel writer who explores the South Pacific beginning in 1906, is the first woman to go up the Sepik and Fly rivers in Papua New Guinea. Grimshaw writes more than forty books about her adventures in the islands of Southeast Asia and Papua New Guinea.

1911   Parliament Act clears way for Home Rule

In the Parliament Act of 1911, the power of the British House of Lords is diminished. Typically, the more liberal House of Commons would pass a bill, only to have it stopped by the conservative Lords. The Parliament Act, however, states that the House of Lords may now merely delay bills by two years, instead of stopping them completely.

In 1912 a Home Rule Bill is proposed. Though rejected by the House of Lords, it is almost guaranteed to become law in 1914, after the two-year delay period.

1911   Birth of writer Flann O'Brien

Flann O'Brien (1911–66, pen name of Brian O'Nolan) writes his finest works while a college student at Blackrock College and University College, Dublin, completing an eccentric but fascinating novel, *At Swim-Two-Birds*, in 1935. O'Brien, though, does not publish the work until four years later, after his father dies, and he must supplement his income. The support of English writer Graham Greene (1904–91) does much to help *At Swim-Two-Birds* get published. O'Brien writes a column for the *Irish Times* for twenty years, and many of his works (such as 1940's *The Third Policeman*) are published after his death.

1912   School dismisses woman for suffrage activities

Hannah Sheehy-Skeffington (1877–1946), an Irish feminist and patriot, is the first woman to be dismissed from the Rathmines School of Commerce in Dublin for her militant suffrage activities. A founding member of the Irish Association of Women Graduates in 1901, Sheehy-Skeffington is also the cofounder of the Irish Women's Franchise League (1908) with Constance Markiewicz (see 1918).

1912: September 28   Signing of the Ulster Covenant

The conservative, anti-Home Rule Unionists are strongest in Ulster (northern Ireland), where they include farmers, laborers, and factory workers, and make up about half of the population. These Unionists fear Home Rule not only as a threat to their religious freedom, but also because they have prospered under the Union with Great Britain: the two great industries of Ulster—linen and shipbuilding—require the ability to sell these products freely to English markets. Thus, preparations are made to set up a provisional government in Ulster in order to take over should Home Rule become law. People are urged to sign the Ulster Covenant, which is a solemn promise of each individual to fight Home Rule. Over 400,000 people sign the Covenant.

1913   Founding of the "Irish Volunteers"

Eoin MacNeill (1867–1945), one of the founders of the Gaelic League (see 1893), urges the Irish Home Rulers to form a volunteer force to make sure the British government keeps its promise to grant Home Rule. MacNeill is encouraged to found the Irish Volunteers, and many who join the volunteer force are members of a brotherhood (the Irish Republican Brotherhood, or IRB) who hope to use the volunteer movement to organize a rebellion. Irish government officials become alarmed as the movement grows in numbers, and they take over the Volunteers in 1914.

1913: January   Ulster Volunteer Force formed

The Ulster Volunteer Force (UVF) is formed, recruiting over 100,000 members to defend Ulster should Home Rule become oppressive. The UVF sets up a defense fund and collects £1,000,000 to buy German armaments. Though illegal, the UVF has many supporters, and there is little the government can do to limit its power.

1914   World War I begins

Many members of both the volunteer forces of the Unionists (Ulster Volunteers) and the Home Rulers (Irish Volunteers) fight alongside the British in the First World War (1914–18). Both groups, however, hope that their service will give them preferential treatment after the war. Some Irish Volunteer leaders, however, believe that it is best to have a rebellion while the war is on and while Britain is at its weakest. By 1916, around 16,000 Volunteers support this cause. Members of the Irish Republican Brotherhood (IRB) begin to plan a revolution and decide to launch their rebellion on Easter Sunday, 1916.

1916: April 24   Easter Uprising

The IRB feels the greatest difficulty in beginning a successful rebellion is to make sure the commander of the Volunteers, Eoin MacNeill (1867–1945), gives them an order for a rising. MacNeill is opposed to the risky rebellion and would only give the Volunteers an order to rise if the government tries to disarm them. An attempt by the IRB to deceive MacNeill fails, and the remaining supporters try to muster as many men as possible and make an attempt at rebellion on Monday morning, a bank holiday. On Monday morning, the Volunteers march through Dublin and take over the General Post Office (GPO)—which becomes the designated rebel headquarters—as well as other public buildings. Only about 1,600 men turn out in the city, and by Wednesday they are outnumbered twenty to one by the British government forces. The city center of Dublin is shelled, and the GPO destroyed. On Saturday, April 29, the Volunteers surrender. Five hundred and fifty people have died and over two thousand are wounded in the rebellion.

1918   First woman elected minister of Parliament

Irish political activist Countess Constance Georgine Gore-Booth Markiewicz (1868–1927) is the first woman to be elected a minister of Parliament in the United Kingdom. Markiewicz had been serving a life term in prison for her contributions to 1916's Easter Rebellion (see 1916: April 24), and was sentenced to death before being pardoned in a general amnesty in 1917. She is elected to represent Sinn Féin in Parliament, but Markiewicz refuses to serve as a Nationalist protest. She is active in politics throughout her life. In 1909 she founds the radical youth group Na Finanna in Dublin. In 1919 she goes on to become the first female minister for labor in the illegal Irish parliament, Dáil Éireann.

1919   Declaration of Irish republic

Irish members of Parliament proclaim the existence of an independent Irish republic. They establish an illegal Irish parliament, known as Dáil Éireann, with Éamon de Valera (see 1882) as its head. These members of parliament plan on taking over the running of the country, leaving British institutions with less and less influence. Local governments accept the new parliament, and courts are set up. However, the ministers and members of parliament must frequently hide to avoid police raids.

1920s   Peig Sayers records Irish folksongs

Peig Sayers (1873–1958), an Irish storyteller and singer who lives on Great Basket Island in Ireland, is the first person to record over four hundred Irish stories and folksongs. Her songs and stories are collected by the Irish Folksong Commission in the 1920s. Sayers's two autobiographies (dictated in Gaelic in 1936 and 1970) vividly depict a harsh way of life and are now considered classics.

1920s   Consequences of war

The 1920s is a difficult time for the new Irish nation. The great social change of the era—marked by such shocking acts as dancing the new American Charleston and the publication of James Joyce's (see 1882: February 2) novel *Ulysses*—makes many Irish fearful that foreign influence might destroy Irish cultural pride, which holds the nation together in the early, uncertain years. Lawmakers are left to fine-tune the new constitution and laws and to try to revive the economy damaged by partition and war. The new state is left almost entirely without industry, as the now-separate northern counties of Ulster has been the center for such activity (see 1920: December 23).

1920: Easter   War of Independence begins

The Irish Volunteers—now referred to as the Irish Republican Army, or IRA—begin to attack British policemen and raid their barracks. The British, unable to defend all the barracks, withdraw into the towns and leave large areas to the control of the Volunteers. As a symbol of victory, Volunteers around the country burn empty barracks at Easter 1920. The British government responds by banning Dáil Éireann, Sinn Féin, and other nationalist organizations, and also imposes curfews.

1920   "Black and Tans" fight the Irish Republican Army

The government begins recruiting additional police forces in England. Because there are not enough uniforms for the new recruits at first, they wear a mixture of army khaki and the green-black police uniform, and are nicknamed the "Black and Tans." Given full power to search for suspects, the newcomers are ruthless and much less disciplined than the police. In response, the IRA stages ambushes; because they do not wear uniforms, members of the IRA can quickly attack and then disappear into a crowd. The Tans respond to this with terror, burning homes and shooting rebels.

1920: November 21   Bloody Sunday

In autumn 1920, the British begin to rebuild their network of spies, which has been destroyed by rebel Michael Collins (1890–1922). The British hire experienced spy agents, bringing them into Dublin to appear as ordinary Irish citizens. Collins, however, discovers the plan, and on the morning of November 21 groups of Volunteers force their way into the homes of the spies, killing eleven and wounding four. The British believe that the men guilty of the killings are among the crowd at a Gaelic football (soccer) match, and hundreds of Tans surround the game's spectators, killing twelve and injuring sixty. The day soon becomes known as Bloody Sunday.

1920: December 23   Division of Northern and Southern Ireland

As an effort for peace, the Government of Ireland Act creates the province of Northern Ireland, made up of six counties in Ulster. Northern Ireland remains part of the United Kingdom after the remainder of Ireland is made the Irish Free State (see 1921). Roughly two-thirds of the population of Northern Ireland are Protestant and Unionist. Many refugees begin flowing into the Free State to avoid the conflict-ridden North, where the militant IRA is very active.

1920   Irish sculptor invited to Russia

Clare Frewen Sheridan (1885–1970) is the first Irish sculptor to be invited to work in Russia. She is asked to visit Moscow by the Soviet Trade Commission in 1920 and during her stay makes busts of Zinoviev, Kamenev, Lenin and Trotsky.

1921   Signing of Anglo-Irish Treaty and civil war

The Anglo-Irish Treaty establishes the Irish Free State with dominion (self-governing capacity) status in the British Commonwealth. Northern Ireland, however, remains under the control of the British. Michael Collins (see 1920: November 21)—an Irish politician and Sinn Féin party leader—is largely responsible for the negotiation of the treaty. Violent opposition to dominion status and to the separate government in Northern Ireland leads to a civil war lasting almost a year (see 1922–23).

1922   Irish Free State and new constitution

The Free State is officially proclaimed, and a new constitution is adopted in 1922, but sentiment in favor of a reunified Irish republic remains strong. Extreme advocates for an Irish republic include members of the IRA (Irish Republican Army, formerly the Irish Volunteers), a terrorist organization. Arthur Griffith (1872–1922) is the first president of the Free State. The policies of Griffith and Michael Collins (first minister of finance and commander-in-chief of the army) become the basis for a new, moderate political party, Fine Gael. Fine Gael is the principal opposition party to Fianna Fáil, founded by Éamon de Valera (see 1882 and 1926).

1922–23   Civil war

The early months of 1922 are marked by confusion. The withdrawing British begin to hand over their barracks, some to pro-Treaty Irishmen, others to anti-Treaty leaders (see Anglo-Irish Treaty, 1921). Tensions become full-blown when some of the more extreme anti-Treaty forces seize the Four Courts and other public places in Dublin. The fragile peace between the two sides erupts into civil war.

In August 1922, pro-Treaty leaders Arthur Griffith and Michael Collins become some of the earliest victims of the civil war, Griffith perishing from overwork and Collins dying in an ambush. Yet the anti-Treaty forces are greatly outnumbered, and in May 1923—after months of raids, ambushes, and bloodshed—the anti-Treaty leaders (including Éamon de Valera, see 1882) decide that continued resistance is impossible. They order their men to surrender their weapons, and the war is over. Hundreds are dead, thousands in prison, and millions of pounds of damage has been done. Yet the worst con-

Rival leaders of the factions involved in civil war (1922–23) meet at a party given by the Archbishop of Dublin. From left, W.T. Cosgrave, former president of the Irish Free State; Rev. Dr. McQuaid, President of Blackrock College; Rev. Dr. Byrne, Archbishop of Dublin; and President Éamon de Valera. (EPD Photos/CSU Archives)

sequence of the war is the deep-seated bitterness among the people, which haunts Ireland for generations.

1923 William Butler Yeats wins Nobel Prize

The Nobel Prize for Literature is awarded to poet William Butler Yeats (see 1865). Yeats is the first Irish author to receive this honor.

1925 Death of pioneer Nelly Cashman

Nelly Cashman is the first white woman to settle in British Columbia, where she runs a series of boardinghouses, hotels, restaurants, and stores. Prospecting for gold in Alaska, she founds the Midnight Sun Mining Company, the first Catholic Church in Tombstone, the Miner's Hospital Association, and the Irish National League in Tombstone.

1926 Eamon de Valera breaks away from Sinn Féin

After the civil war, Éamon de Valera and the Sinn Féin party (affiliated with the IRA) continue to refuse to enter the Free State government and the parliament (Dáil) considering themselves the rightful leaders of Ireland. Arguments about the border make de Valera realize that the people represented by Sinn Féin need to have a voice in the Dáil. When Sinn Féin refuses to join the Dáil, de Valera breaks away and founds the new Fianna Fáil party, which promptly joins parliament.

1930s Irish Republican Army loses popularity

Though the IRA is powerful at first, it loses much of its support after Éamon de Valera (see 1882), a disillusioned supporter of the organization (see 1926), becomes prime minister in the Irish Free State government in 1932.

1931 Ninette de Valois founds Royal Ballet

Ninette de Valois (b. 1898) is the founder and first director of the Royal Ballet, which she continues to head until her retirement in 1963. A dancer of great energy and charm, de Valois also founds the National School of Ballet in Turkey (1947) and is the first woman to receive the Erasmus Prize Foundation Award (1974).

1931–32 First woman president of the Irish Trades Union Congress

Louie Bennett (1870–1956) is the first woman to become president of the Irish Trades Union Congress. Bennett works throughout her life for women's rights and suffrage.

1936 Birth of athlete Thelma Hopkins

Athlete Thelma Hopkins—skilled at high jumping, hockey, squash, long jumping, and the pentathlon—is born.

1937 New constitution

A new constitution replaces the governor-general with a president, who is elected by popular vote every seven years. The name of the nation is officially changed from the Irish Free State to Ireland (Éire in Irish). Among other new provisions, the constitution establishes a legislative branch of government. The Oireachtas (national parliament, pronounced "ir-ROCK-tas") sits in Dublin and is comprised of the president and two houses, the Dáil Éireann (House of Representatives, pronounced "DAW-il") and Seanad Éireann (Senate, pronounced "SHAN-ad"). Suffrage (the right to vote) is universal at age 18.

1938 Douglas Hyde becomes first president of the republic of Ireland

Active in politics, Douglas Hyde (1860–1949) becomes the first president of Ireland. Hyde, however, is also known for his contributions to the arts. Leaving his first career as a professor in England, Hyde returns to Ireland, leading the movement to revive the Irish language and literature (he uses an Irish pen-name, An Craoibhin). He is the founder and first president of the Gaelic League (1893–1915 [see 1893]) and helps to found the Abbey Theater (see 1904). Hyde writes a play, Casadh an tSúgán (The Twisting of the Rope; 1901), the first to be performed professionally in the Irish language.

1939–45 World War II

Ironically, the coming of the Second World War helps to heal wounds left by the Irish civil war. All parties within the Dáil agree to remain neutral, a rare occurrence of unity and agreement within the parliament. Though most are sympathetic to the Allies and dislike the German Nazi regime, they are still distrustful of Britain. Neutrality is also a way of proving the finality of Irish independence.

As the German threat increases, the Irish prepare themselves for attack, holding secret talks with Britain while remaining neutral. The American ambassador puts much pressure on Ireland to join the Allies, but to no avail, though the Irish quietly and secretly help the Allied cause (for example, by imprisoning crashed German airmen while returning British to Northern Ireland). Ireland is able to remain neutral because the British never feel it necessary to overtake the island, deciding it is better to have a friendly neighbor with food, labor, and army volunteers, rather than throwing that away with an invasion. The chief problem during the war is the shortages, especially of food.

1939 Birth of musician James Galway

Born in Belfast, flutist James Galway's unique musical style and talent attract large audiences around the world.

1939 Birth of poet Seamus Heaney

Heaney publishes his first collection of poetry—Eleven Poems—in 1965. Though he is born a Catholic in Northern Ireland, the violence there disturbs him greatly, and he moves to the republic of Ireland in 1975. Many of Heaney's works are inspired by the lush, beautiful, and sometimes menacing Irish countryside of his boyhood.

1942 Central Bank of Ireland established

The Central Bank is the monetary authority in Ireland. In 1971 its powers are increased by the Central Bank Act. In addition to giving the bank the ability to license and supervise other banks, the Central Bank Act makes the bank's status and powers comparable to those of central banks in other developed nations.

1945 First female appointed to Royal Society

Irish scientist Dame Kathleen Yardley Lonsdale (1903–71), who specializes in physical chemistry and the study of crystals, becomes the first female Fellow of the Royal Society in London, England. The youngest of ten children and daughter of an Irish postmaster, she is the first scientist to use Fourier analysis to study molecular structures. She is honored for research that leads to the development of x-ray crystallography.

1946 Institute for Industrial Research and Standards established

Along with the Agricultural Institute (founded 1958), the Institute for Industrial Research and Standards is the major organization doing scientific research in Ireland.

1948 Ireland leaves the Commonwealth

Ireland votes itself out of the British Commonwealth of Nations (whose other members include former British colonies and protectorates).

1949: April 18 Ireland declares itself a republic

1950s Agriculture no longer largest source of income

Until the 1950s, Ireland's economy is predominantly agricultural, with agriculture contributing the most to the Gross National Product (GNP). After the 1950s however, liberal trade policies, combined with industrialization, stimulate economic expansion. In 1958 agriculture accounts for 21 percent of the GNP, industry 23.5 percent, and other sectors 55.5 percent. By 1993 agriculture accounts for only 10.5 percent of the total GNP, while industry contributes 44.5 percent and other activities 45 percent.

1951 Ernest T. S. Walton wins Nobel Prize

Nuclear physicist Ernest T. S. Walton (b. 1903) wins the Nobel Prize for physics, for his work on transmutation of atomic nuclei.

1955: December 14 Ireland is admitted to the United Nations

The Republic of Ireland becomes a member of the United Nations (UN).

1958 First Program for Economic Expansion is launched

This program helps to develop the industrial sector of the economy. Part of this program has the Industrial Development Authority (IDA) administer incentives (such as low taxes, inexpensive land prices) to attract foreign investment. Other government agencies provide consulting services for research and development, marketing, exporting, and management.

1960s Civil rights movement

By the 1960s, it is apparent that many Catholics are willing to abandon their hopes for a united Ireland if they can instead have equality and an end to discrimination in Northern Ireland. The British army is periodically called in to restore order.

Early 1960s Mining industry grows

Ireland becomes a significant source of base metals thanks to an increase in mining exploration. (See late 1970s.)

1969 Playwright Samuel Beckett wins the Nobel Prize

Playwright and novelist Samuel Beckett becomes the second Irishman to be awarded the Nobel Prize for literature. (See 1906.)

1970 Women in Belfast protest

Women hold a protest march in Belfast, Northern Ireland, deliberately breaking the British curfew to protest for civil rights.

1973 Ireland becomes a member of the European Community

The goal of the European Community (EC) is to improve economic and political integration among Western European countries. Membership brings clear benefits to Ireland: farming is improved through the Common Agricultural Policy and industry is helped by the diverse markets. Membership in the EC also reduces Ireland's dependence on the British economy.

1974 Seán MacBride wins Nobel Peace Prize

Seán MacBride (1904–88), cofounder of Amnesty International, is the only person to win both the Nobel (1974) and Lenin (1977) Peace Prizes.

1974: April 1 Social insurance program implemented

All wage and salary earners between the ages of sixteen and sixty-eight become covered by retirement and old age pensions, widows' pensions, maternity benefits, and a death grant. The program is administered by the Department of Social Welfare.

1975 Equal-pay legislation enacted

Though "equal pay for equal work" legislation is enacted at the end of 1975, there is still a great disparity between men's and women's wages in the mid-1980s.

1976 Northern Ireland Peace Movement founded

Irish housewife Betty Williams (b. 1943) founds the Northern Ireland Peace Movement in Belfast. Williams begins her grass-roots organization in response to the violent deaths of three children in her neighborhood. Joined by Mairead Corrigan (b. 1944), Williams remains a leader of the organization until 1980. In 1976 she and Corrigan are awarded the Nobel Peace Prize, the first Irish women to win the award.

1976 Irish government gains "emergency powers"

From the late 1960s, the Irish government has been attempting to reduce the power of the "provisional wing" of the IRA, which has been using Ireland as a base for attacks in the north (Northern Ireland). In 1976 the government assumes emergency powers in order to deal better with IRA activities. Terrorist acts, however, continue.

1976 First female artist subject of retrospective exhibition

Eileen Gray (1879–1976), a professional architect and designer, is the first Irish female to have a retrospective exhibit dedicated to her at the Victoria and Albert Museum in London. Gray, a designer from the 1920s, is known for her modern style and use of innovative materials: steel, glass, and plastics.

Late 1970s Largest zinc-lead field in Europe begins production

In the late 1970s, the largest zinc-lead field in Europe—near Navan, County Meath—begins production. In 1991 the mine produces 187, 500 tons of zinc concentrates and 35,000 tons of lead concentrates (1992 improvements increase production to approximately 200,000 tons of zinc and 38,000 tons of lead). The country's zinc output is expected to double in the late 1990s, as two new mines are developed.

1977 First woman serves as parliamentary secretary

Marie Geoghegan Quinn (b. 1951), a member of the Fianna Fáil Party and former schoolteacher from Carna in County Galway, is the first woman to serve as a parliamentary secretary. She is appointed secretary to the minister of industry and commerce.

1979 Ireland joins the European Monetary System

Ending the 150-year-old tie to the British pound, Ireland joins the European Monetary System. This event culminates the growing independence of the Irish economy which had begun with its membership in the European Community (see 1973).

1979: August 12 Assassination of Earl Mountbatten

Earl Louis Mountbatten (1900–79), a distinguished British naval commander and statesman, as well as a member of British royalty, is murdered by an IRA bomb while sailing near his holiday home in County Sligo, Ireland.

Early 1980s Economic recession

Ireland's economy falls into recession. By 1986 unemployment soars to eighteen percent (compared to ten percent in 1981)—and over thirty percent for recent graduates of high school or college. The result of the recession is an increasing amount of emigration (leaving a country to live in another), including illegal residency in the United States and other countries. One estimate projects that emigration will reach 250,000 people—or seven percent of the population—by 1992.

Early 1980s Birth control becomes available to married couples

Because of the strong Roman Catholic influence in Ireland, the sale of contraceptives is entirely prohibited. However, in the early 1980s, birth control is available by prescription to married couples.

1983: April Antiabortion amendment passes the Irish parliament

Though approximately five thousand Irish women per year obtain abortions by traveling to England, an antiabortion amendment guaranteeing "the equal right to life of the unborn" passes the Irish parliament. In September the amendment also passes a national referendum.

1985 Birth control available without prescription

In 1985 contraceptives can be obtained without a doctor's prescription. The minimum age for marriage is also raised, from fourteen to eighteen for girls and sixteen to eighteen for boys. The fertility rate for 1985–90 goes down 20.6 percent from the previous five-year period.

1985: November Treaty between Ireland and the UK

Ireland and the United Kingdom ratify a treaty which enables Ireland to play a role in various aspects of Northern Ireland's affairs. The goal of the treaty is to promote peace between the two nations.

1986 Sinn Féin becomes a political party

Sinn Féin (We Ourselves), the political arm of the IRA, has been boycotting the Dáil (House of Representatives) for sixty-five years. However, in 1986 Sinn Féin ends the boycott and registers as a political party.

1986 Referendum for divorce fails

The Roman Catholic Church has a significant impact on the government's social legislation—both divorce and abortion are illegal in Ireland. A 1986 referendum, intended to provide for legalization of divorce under certain circumstances, is defeated by a large margin in parliament.

1990 Second Commission on the Status of Women established

The Second Commission on the Status of Women is established to encourage the participation of women in all aspects of Irish society. The commission works especially closely with issues such as child care, education, employment opportunities, and legal rights.

1990: November First woman elected president

Mary Robinson (b. 1944), an international lawyer, activist, and Catholic, becomes the first woman to hold the office of president. Robinson's election is especially notable because it comes during a time in Irish history when abortion and women's rights are very controversial issues. Robinson had promoted legislation that enables women to serve on juries and eighteen-year-olds to vote. While a member of the legislature in 1974, she shocks many Irish by advocating for the legal sale of contraceptives.

1992: November   Abortion referendum approved / More women in parliament

A referendum granting free access to abortion information and the right to obtain abortions outside the country is approved by a 3–1 margin.

The number of women in the 166-member Dáil (parliament) goes from thirteen to twenty members.

1993: January   Coalition formed to lower unemployment

The Fianna Fáil and Labor Party join to form a coalition to address the unemployment rate, which is among the highest in the European Community (EC). The coalition develops a five-year plan.

1993: June   President Robinson visits Northern Ireland

President Mary Robinson (see 1990: November) pays a private visit to Northern Ireland as part of a series of ongoing discussions with the British government.

1995   Sinn Féin Party opens office in Washington, D.C.

The first person to work for Sinn Féin in its newly opened office is a woman. Militant Sinn Féin leader Gerry Adams declares that while the British have 600 employees working at their American embassy, Sinn Féin has "one Irish woman. I think maybe the British are at a disadvantage."

1995   Poet Seamus Heaney wins Nobel Prize

Seamus Heaney (see 1939) is awarded the Nobel Prize for Literature, becoming the third Irish writer to win this honor. Previous recipients include William Butler Yeats (see 1865) and Samuel Beckett (see 1906).

1995: November   The Irish vote to legalize divorce

Overcoming opposition by the Catholic Church, the Irish electorate votes to end the country's constitutional ban on divorce by a narrow margin of 0.5 percent (9,000 votes) in a popular referendum.

1996: June 26   Prominent investigative reporter is murdered

Investigative reporter Veronica Guerin, who has won acclaim for her exposés of organized crime in Dublin, is shot and killed in her car by two men on a motorcycle. Guerin, thirty-six, a target of previous attacks, was a reporter for the *Sunday Independent* newspaper.

1997: June 26   Ahearn is chosen as prime minister

Bertie Ahearn, head of the Fianna Fáil party, is elected to the post of prime minister by the Irish parliament (the Dáil), becoming the youngest person in the history of the Irish republic to hold the office.

1998: April 10   Historic Northern Ireland peace agreement is reached

Political leaders meeting in Belfast announce a landmark agreement designed to bring peace to the troubled province of Northern Ireland. The province will remain under British rule but gain expanded autonomy, with many local matters decided by a newly mandated 108-member assembly. The peace talks at which the pact is announced are attended by Irish prime minister Bertie Ahearn, British prime minister Tony Blair, and leaders of both Northern Ireland's Unionist and Sinn Féin factions.

Bibliography

Breen, Richard. *Understanding Contemporary Ireland: State, Class, and Development in the Republic of Ireland.* New York: St. Martin's Press, 1990.

Daly, Mary E. *Industrial Development and Irish National Identity, 1922–1939.* Syracuse, N.Y.: Syracuse University Press, 1992.

Foster, R. F., ed. *The Oxford Illustrated History of Ireland.* New York: Oxford University Press, 1989.

Fulton, John. *The Tragedy of Belief: Division, Politics, and Religion in Ireland.* New York: Oxford University Press, 1991.

Hachey, Thomas E. *The Irish Experience: A Concise History.* Armonk, N.Y.: M. E. Sharpe, 1996.

Harkness, D. W. *Ireland in the Twentieth Century: Divided Island.* Hampshire, Eng.: Macmillan Press, 1996.

Inglis, Tom. *Monopoly: The Catholic Church in Modern Irish Society.* New York: St. Martin's Press, 1987.

Johnson, Paul. *Ireland, Land of Troubles—A History from the Twelfth Century to the Present Day.* London: Eyre Methuen, 1980.

Lee, Joseph. *Ireland, 1912–1985: Politics and Society.* New York: Cambridge University Press, 1989.

MacDonagh, Oliver, et al. *Irish Culture and Nationalism, 1750–1950.* New York: St. Martin's Press, 1984.

Sawyer, Roger. *"We Are But Women": Women in Ireland's History.* New York: Routledge, 1993.

Shannon, Michael Owen. *Northern Ireland.* Santa Barbara, Cal.: Clio Press, 1991.

Italy

Introduction

The Italian peninsula is simultaneously home to one of the world's oldest cultures and one of Europe's youngest countries. Italy's cultural heritage dates back two thousand years to the days of the Roman Empire, whose classical scholarship and arts—together with those of ancient Greece—enjoyed a rebirth during the Italian Renaissance of the fourteenth to sixteenth centuries, when Italy once again led Europe in art and learning. Due to the independence of its city-states and to centuries of foreign domination, however, Italy did not become a unified kingdom until 1861. In the postwar era, political unity has remained elusive in Italy, which has had over fifty elected governments since World War II. Nevertheless, the country has made a strong postwar economic recovery and is a leading industrial nation with one of the strongest economies in Western Europe. As home to the Pope and Vatican City, Italy is also the worldwide center of the Catholic church, which has played a major role in its affairs since about A.D. 300.

Most of Italy is located in the boot-shaped peninsula extending southeast into the Mediterranean Sea. The northern portion of the country borders France, Switzerland, Austria, and Slovenia. Roughly four-fifths of Italy's terrain consists of mountains, most belonging either to the Appenines that run from north to south, splitting the land in two lengthwise, or to the northern Alps, which include Europe's highest point, called Monte Blanco in Italian, or (in French) Mont Blanc. Italy also includes the Mediterranean islands of Sicily and Sardinia, as well as other, smaller islands, and two independent enclaves—Vatican City and San Marino—are found within the country's borders. Italy has an estimated population of roughly 57.5 million people; the capital city of Rome is home to over 2.5 million.

Even before the rise of Rome, the Italian peninsula was home to an advanced civilization, that of the Etruscans, who dominated the area between the ninth and sixth centuries B.C., leaving behind a wide range of skillfully crafted artifacts including terra-cotta vases, metalwork, and statues. During this period, according to legend, the city of Rome was founded—by the brothers Romulus and Remus. In the last

five centuries before the birth of Christ, Rome grew in importance, forming a republic and instituting a legal code that was to exert a major influence on Western civilization. Rome gradually expanded its control of the Italian peninsula and the Mediterranean region as a whole, defeating its major rival, Carthage, in the three Punic Wars. After a period of dictatorship under Julius Caesar (49–44 B.C.), the Roman Empire was inaugurated, bringing with it a two-hundred-year period of peace and stability known as the *Pax Romana.* Eventually the Roman Empire expanded to include most of Europe and reached to the Middle East and North Africa as well.

During the period of the Roman Empire, the new religion known as Christianity gained many followers but was harshly suppressed for over two centuries. By the fourth century A.D., however, it had found a champion in the person of Emperor Constantine (c. 274–337). By the end of the century it was the official religion of the Roman Empire, beginning the long and historically significant relationship between Rome and the Church, which would come to revolve around the pope and the—at one time considerable—land controlled by the papacy.

As the influence of Christianity grew, the Roman Empire declined. From the fourth century, it was split into two parts, with Italy claiming only the western half, and the eastern part centered in the Turkish city known as Constantinople, (previously Byzantium, now Istanbul). The same period saw the beginning of Germanic invasions from the east and north, inaugurating many centuries of foreign domination and internal fragmentation. The Goths and other early Germanic invaders were followed by the Lombards (eighth century), Saracens (ninth century), more Germans (tenth century), and the French Normans, who seized control of the south in the eleventh century.

As a power struggle began between the popes and the Holy Roman Emperors (a series of German monarchs at this time), a new political and commercial force sprang up around the tenth century: independent city-states, including Florence, Venice, and Milan, administered by guilds of artisans and other professionals. The growing prosperity and sophistication of these cities laid the foundation for the Renaissance, the great Italian cultural rebirth of the fourteenth to sixteenth centuries. Centered on the effort to revive the classical learning

of ancient Greece and Rome, the Renaissance resulted in the unprecedented artistic and intellectual flowering that gave the world the artistic masterpieces of Michelangelo (1475–1564), Raphael (1483–1520), and Leonardo de Vinci (1452–1519), the architecture of Filippo Brunelleschi (1377–1446), and the literary humanism of Petrarch (1304–74), Ficino (1433–99), and other authors.

By the end of Renaissance, however, colonization in the New World had brought new power to Spain, and Italy became the prize in a struggle between the Spanish and the French for dominance over Europe. At the end of the fifteenth century, France invaded Spain, and warfare between the two nations continued for decades, during which Rome itself underwent a fierce attack by the French. A 1559 treaty granted Spain territory in both the north and the south, and the Spanish remained in control of much of Italy until its territories went to Austria following the War of the Spanish Succession (1701–14). The 1748 treaty ending the War of the

Austrian Succession divided control of Italy between the Austrian Habsburgs and the Bourbons of Spain, an arrangement that continued virtually unchanged for over a century, except for the Napoleonic era at the beginning of the nineteenth century.

Although the Austrian monarchy regained control of Italy following the downfall of Napoleon (1769–1821), the period of French rule, during which republics had been formed in Milan, Genoa, and other cities, helped inspire a new spirit of revolution in Italy. The return of Austrian rule in 1815 spelled the beginning of the *Risorgimento,* a period of intense nationalism that culminated in the independence and unification of Italy. The new spirit of Italian nationalism was evident in many aspects of Italian life during this period, from the poetry of Ugo Fascolo to the operas of Giuseppe Verdi (1813–1901) and Gioacchino Rossini (1792–1868). The three leading political and military leaders of the Risorgimento were Giuseppe Mazzini (1805–72), who founded the "Young Italy" patriotic organization; Count Camillo de Cavour (1810–61), an aristocrat who mediated between the nationalists and King Victor Emmanuel II (1820–78); and the great general Giuseppe Garibaldi (1807–82), who led the Red Shirts to victory over Austria after spending twenty years exiled from Italy for revolutionary activities. The Kingdom of Italy, headed by King Victor Emmanuel II, was formally proclaimed on March 17, 1861.

Emerging from a long history of political fragmentation and foreign domination, the new nation struggled economically, burdened by an ever-growing national debt. The situation in the rural south was especially dire, and conditions became worse than ever in the 1880s, when farm prices declined throughout Europe. Poverty and civil unrest drove between five and six million Italians to emigrate in the decades preceding World War I. They also led to the rise of socialism in Italy.

Italy initially adopted a neutral stance in World War I but then joined the Allies, lured by the promise of winning back Austrian territories that it regarded as legitimately its own. The long and arduous war cost the country over half a million lives, and at the conclusion Italy did not receive all of the lands it had been promised. Reeling from the economic and psychological effects of the war, in 1922 the country came under control of Fascist leader Benito Mussolini (1883–1945). Over the succeeding years, Mussolini transformed Italy into a dictatorship, silencing all opposition and ruling by decree. In 1936, after annexing Ethiopia, he joined with Nazi Germany in forming the Rome-Berlin Axis and entered World War II on the side of Germany. The war did not go well for Italy, and when the Allies invaded in 1943, the Italians deposed Mussolini and joined the Allied side for the remainder of the war.

In 1946 Italians voted to end the monarchy, and Italy became a republic. However, its political life remained tumultuous in the postwar era, with prime ministers and governments changing frequently. One of the major constants was the longtime domination by the Christian Democrat party. The Italian economy, shattered by two world wars and a depression, revived dramatically after World War II and was one of Europe's strongest by the 1960s. By the 1970s, however, rising oil prices signaled the end of Italy's economic boom. The nation's stagnating economy spurred popular discontent, and violence by both right- and left-wing extremist groups proliferated, climaxing in the kidnapping and murder of former prime minister Aldo Moro by the Red Brigades in 1978.

In the 1980s the Italian government began a crackdown on organized crime that resulted in over one thousand suspects being brought to trial by the middle of the decade. The early 1990s saw an extensive campaign to root out the endemic corruption that had become an entrenched feature of Italian politics. The *mani pulite* (clean hands) campaign implicated many of the nation's top political leaders (including two former prime ministers) and businessmen in bribery scandals and, even more dramatically, spelled the end of political dominance by the Christian Democrats. In 1994 the Alliance for Freedom, a conservative coalition, won the first Italian elections held under electoral reforms enacted the previous year. The new government was headed by businessman Silvio Berlusconi. However, Berlusconi himself was soon brought down on corruption charges stemming from the business practices of his powerful Fininvest conglomerate, and Minister of Trade Lamberto Dini formed Italy's fifty-fourth postwar government.

Two years later, the country swung from the right to the left, with victory by the Olive Tree coalition, which formed Italy's first leftist government since the war.

Timeline

c. 1200 B.C.   Etruscans arrive in Italy

The Etruscans, the founders of Italy's first major civilization, arrive in northern Italy (present-day Tuscany). They establish walled city-states in the region that comes to be called Etruria, which eventually expands from the western plains to include the Po Valley and the Tyrrhenian coast.

9th–6th centuries B.C.   Etruria dominates central Italy

Etruscan civilization reaches its peak, as both commerce and art flourish. The Etruscans mine copper, tin, iron, bronze, and other metals. They create sophisticated gold jewelry and other artifacts, including terra-cotta vases, paintings, and statues. Their goods are traded widely traded throughout the Mediterranean region and Europe. The individual Etruscan city-states join together to form a federation.

8th century B.C.   Greeks settle in southern Italy

Greeks arrive in and colonize southern Italy, founding cities including Naples, Reggio, and Cuma. The administrative center of the region, called *Magna Graecia* (Greater Greece), is Sicily. Through trade and other types of contact, Greek art, religion, and other aspects of Greek culture are transmitted to the neighboring peoples.

753 B.C.   Founding of Rome

According to legend, the city of Rome is founded on April 21 of this year by the brothers Romulus and Remus.

6th–3rd centuries B.C.   Decline of the Etruscans

Etruscan civilization gradually declines, weakened by warfare with Celts in the north, Greeks in the south, and the tribes of Rome.

The Rise of Rome

509–10 B. C.   The Roman republic is formed

The city of Rome overthrows its Etruscan rulers and establishes a republic governed by a Senate. Its influence gradually extends throughout the Mediterranean over the next 500 years as it gains control, first of the Italian peninsula then of the entire region, defeating the rival city of Carthage in the Punic Wars. A rich cultural heritage is also established with the formation of a legal code making all citizens equal and the creation of great literary works by the poet Virgil (see 29 B.C.), the satirist Juvenal (c. 55–c.140), historians Livy (59 B.C.–A.D.17) and Tacitus (c. 55–120), and others.

272–250 B.C.   Italy is united under the Roman republic

All of the Italian peninsula comes under Roman rule.

264–146 B.C.   The Punic Wars

Rome is victorious in the three Punic Wars with the North African city of Carthage. In the course of these wars, the Romans win control of Sicily, Sardinia, Corsica, Greece, and Egypt, emerging as the dominant power in the Mediterranean region.

264–241 B.C.   First Punic War

Rome defeats Carthage in the First Punic War, triggered by Carthage's invasion of Sicily. Victory in the long conflict costs Rome about one-fifth of its population.

218–201 B.C.   Second Punic War

The Carthaginian general Hannibal (247–182 B.C.) crosses the Alps with 60,000 troops and attacks Rome, beginning the Second Punic War. The Carthaginians are ultimately defeated by the Roman general Scipio Africanus (237–183B.C.) and driven out of the Iberian Peninsula (present-day Spain and Portugal).

151–149 B.C.   Third Punic War

Following a Carthaginian revolt against Roman rule, Carthage is sacked and destroyed by Rome.

134–71 B.C.   Slave uprisings defeated

Recurrent slave revolts are contained under the leadership of the Roman generals Crassus (c.115–153B.C.) and Pompey (106–48 B.C.), who together with Julius Caesar (100–44 B.C.) form the *triumvirate* (three-member governing committee) to govern Rome.

50 B.C.   Caesar conquers Gaul

Julius Caesar conquers Gaul (present-day France).

49 B.C.   Caesar takes control of Rome

Following the death of Crassus, Caesar defeats Pompey, the remaining member of the triumverate, and takes control of Rome as a dictator.

44 B.C.: March 15   Assassination of Caesar

Julius Caesar is assassinated in the Roman Senate on the Ides of March.

29–19 B.C.   Composition of the *Aeneid*

The great Roman poet Virgil (70–19B.C.) writes the twelve-volume epic poem, the *Aeneid,* celebrating the exploits of the legendary hero Aeneas that ultimately led to the founding of Rome. The *Aeneid* is influenced by both the *Iliad* and *Odyssey* of the Greek poet Homer, as well as the local legends of Rome.

Empire and Conquest

27 B.C.–3rd century A.D.   Pax Romana

For two hundred years, the Roman empire—the greatest in European history—enjoys peace and stability, dominating commerce in the Mediterranean region and throughout Europe. A single currency is established, and great public works are erected in Rome, including roads, temples, aqueducts, the Colosseum, and the Circus Maximus.

27 B.C.   Augustus Caesar becomes emperor

Octavian (63B.C.–14A.D.), the nephew of Julius Caesar, becomes Rome's first emperor, ruling over an empire that includes most of Europe, part of North Africa, and much of the Middle East.

Hannibal and his troops, carried by elephants, march from Spain into Italy. (EPD Photos/CSU Archives)

Through his administrative skill Augustus strengthens Rome politically and economically and initiates the Pax Romana (see entry above).

3rd century A.D.   Rome threatened internally and externally

The stability of Rome is threatened internally by corrupt and inept emperors and rivalry between generals and externally by Asian and Germanic invasions in the east and north.

312–337   Reign of Constantine

Constantine (c. 274–337), the last of Rome's great emperors, moves the empire's capital from Rome to the ancient city of Byzantium, newly rebuilt and renamed Constantinople.

313 A.D.   The Edict of Milan

Following two centuries of Roman repression of Christianity and persecution of Christians, Constantine legalizes the new religion by the Edict of Milan and becomes a Christian himself.

391   Rome adopts Christianity

Christianity becomes the official religion of the Roman empire.

395   Division of Rome becomes permanent

Rome is permanently divided administratively into Eastern and Western empires with their respective capitals in Constantinople and Rome.

476   Germanic conqueror is proclaimed king of Rome

The last Roman emperor, Romulus Augustus (b. 461), is overthrown by Odeacer (434?–493), a leader of the Germanic Goths, who becomes the ruler of Rome, an event generally considered to mark the end of the Western Roman Empire. (The Eastern, or Byzantine, empire survives in Constantinople—located in present-day Turkey—until its destruction by the Ottomans one thousand years later.)

553   Reign of Justinian

The emperor Justinian (r. 527–565) reconquers Italy from the Goths, expelling them from the region.

568   The Lombards arrive in Italy

A Germanic tribe called the Lombards migrates to the Po River valley. With their capital at Pavia, they establish dominance in north and central Italy, creating their own kingdom.

590–604   Gregory strengthens the papacy

Pope Gregory (c.540–604) brings the papacy a new level of power and prestige, gaining power over religious officials in the Eastern empire and spreading Christianity abroad.

756   The Papal States are established

The Frankish king Pepin the Short (714?–768) defeats the Franks and forces them to give the pope control of territory in central Italy between Rome and Ravenna, called the Donation of Pepin. This land becomes an autonomous principality known as the Papal States.

800   Charlemagne is crowned Holy Roman Emperor

After reconquering the papal states from the Lombards, the Frankish king Charlemagne (742–814) is crowned Holy Roman Emperor by Pope Leo III (c. 750–816).

827   Saracens invade Sicily

The Saracens, Muslims from the Middle East and North Africa, conquer Sicily, ruling it for two centuries as part of their empire. Saracen power in Italy weakens and ultimately helps destroy the empire Charlemagne has bequeathed to his successors.

Rise of the Northern City-states

10th–14th centuries   Growth of medieval city-states

In the absence of a strong central government as Italy's emperors and popes vie for power, economically independent city-states grow and flourish in northern Italy. They include Venice, Florence, Genoa, Pisa, Siena, and Milan. Administered by craft, merchant, and other professional guilds, they also maintain political independence and eventually become great cultural centers and the birthplace of the Italian Renaissance. (See 13th century.)

951   Otto the Great invades Italy

The German ruler Otto the Great (912–73) invades Italy, inaugurating a period of rule by German monarchs.

962   Otto is crowned Holy Roman Emperor

To win German protection for the Papal States, Pope John XII (938?–964) crowns Otto the Great as Holy Roman Emperor, making him head of both church and state in Italy and, theoretically, ruler of the Christian countries of Europe.

1035–92   Normans conquer southern Italy

The French Normans invade and conquer southern Italy and the island of Sicily. They drive out the Saracens and the Lombards, and Pope Nicholas II (c. 980–1061) gives the Norman leader Robert Guiscard (c. 1015–85) title to the conquered lands. From this point, the region is administered separately from the city-states of the north.

1122   Concordat of Worms

A historic agreement governing the appointment of church leaders resolves a long-standing power struggle between Italy's popes and its German rulers.

1169   Eruption of Mount Aetna

The long-dormant volcano Mount Aetna erupts, killing about 15,000 people.

1183   Treaty of Constance

Through the Treaty of Constance, the Holy Roman Emperor is forced to recognize the independent city-states of the Lombard League.

1194–1250   Reign of Emperor Frederick II

Through a royal marriage, Holy Emperor Frederick II (1194–1250) gains sovereignty over Sicily and the kingdom of Naples. His long reign is marked by conflict with the papacy, which excommunicates him twice. This conflict leads to the formation of two competing political factions, the Guelphs (supporters of the pope) and the Ghibellines (supporters of the Holy Roman Emperors). Hostility between these two groups dominates Italian politics in the thirteenth and fourteenth centuries, creating extensive disruption in northern and central Italy, and allowing the local rulers of city-states to increase their own power. It also leads to the removal of the papacy to Avignon (see 1309).

1198–1216   Reign of Pope Innocent III

Pope Innocent III (1161–1216) greatly increases the power of the papacy, asserting his power over Holy Roman Emperor Otto IV (1174?–1218) and other European monarchs.

c. 1200   Normans lose southern Italy

The Normans are driven out of southern Italy by the rulers of the Holy Roman Empire.

1209   Franciscan Order is established

St. Francis of Assisi (1182–1226) founds the Franciscan Order, an institute devoted to living a life of poverty and being of service to the poor and sick.

1224   University of Naples is founded

Holy Roman Emperor Frederick II (r. 1194–1250) establishes the University of Naples.

1233 Inquisition is inaugurated

Pope Gregory IX (1147?–1241) launches the Office of the Inquisition in Italy.

1265 Birth of Dante Alighieri

Italy's greatest poet and one of the giants of Western literature, Dante Alighieri (1265–1321) is born and raised in Florence. He writes poetry from an early age and comes under the influence of the well-known Italian poet Guido Cavalcanti as a young man. In 1293 Dante publishes his first book of verse, *La Vita Nuova* (The New Life), which uses a prose narrative to link together poems written over a ten-year period. After 1295 Dante becomes active in Florentine politics, which results in his being permanently exiled from the city of his birth in 1302, when a rival political faction gains power. During the following period of nearly twenty years, the poet lives in various cities, ultimately settling in Ravenna, where he dies in 1321. During his exile, he writes *Il Convivio* (The Banquet; 1307), treatises on government and language, and his three-volume masterpiece, *La Divina Commedia* (The Divine Comedy), which describes an imaginary journey to the three realms of the afterlife: Hell, Purgatory, and Paradise (see 1308).

1279 State recognizes papal sovereignty

Habsburg king Rudolf I (1218–91) recognizes the sovereignty of the pope over the Papal States in central Italy and also over southern Italy.

The Renaissance

c. 14th–16th centuries   The Italian Renaissance

Fueled by the wealth of Italy's city-states and the heritage of its Roman past and facilitated by their political independence, the Italian Renaissance is a flowering of the arts, humanities, and sciences in a deliberate attempt to revive the classical culture of Greece and Rome while rejecting many of the beliefs and customs of the Middle Ages. The new emphasis on the ability of human beings to learn about, enjoy, and improve things in the world—rather than focusing mostly on God and the hereafter—is called humanism.

Hallmarks of the Renaissance are its emphasis on learning—particularly study of classical Greek and Roman texts; an enthusiasm for scientific investigation, personified by Galileo Galilei (see 1564) and Leonardo da Vinci (see 1452); and a new secular spirit in the art of such painters as Botticelli, (1444?–1510) Titian (1477–1576), Raphael (see 1483), and Michelangelo (see 1475).

14th–15th centuries   Rise of wealthy ruling families

Many of the achievements of the Renaissance are funded by the wealth of a handful of wealthy families who gain power in the cities of northern Italy. They include the Medici in Florence, the Sforza in Milan, the Este in Ferrara, and the Montefeltro in Urbino. The Medici, patrons of art and education, subsidize universities and churches and establish Europe's first public library. The Sforza are the patrons of Leonardo da Vinci (see 1452).

1304 Birth of Petrarch

Petrarch (Francesco Petrarca), the great Renaissance poet and humanist, is born in Arezzo, Tuscany. As a scholar, he is acclaimed for linking the ideals of Christianity to the culture of classical Greek and Rome and considered the founder of humanism. As a poet, he is renowned for his vernacular poems (*Rime*), especially his mastery of the fourteen-line form that is named for him—the Petrarchan sonnet, which has a major influence on European poetry. (In England it forms the basis for the Elizabethan sonnet.) Petrarch dies in Padua in 1374.

1308–21 Dante writes *The Divine Comedy*

Poet Dante Alighieri (see 1265) writes *La Divina Commedia* (The Divine Comedy), one of the masterpieces of Western Literature. Its three volumes (*Inferno, Purgatorio,* and *Paradiso*) tell the story of an imaginary journey the poet takes through Hell, Purgatory, and Paradise. In the first two realms, his guide is the Roman poet Virgil; in Paradise he is guided by the great love of his life, Beatrice Portinari (who has died in 1290). In addition to the beauty of its poetry and its vividly imaginative depiction of the afterlife, the work is a compendium of religious and philosophical ideas, history, and politics. It is also noteworthy for being written in Italian, and thus raising the vernacular to new heights at a time when literary works were commonly written in Latin.

1309–77 "Babylonian captivity"

Pope Clement V (1264–1314), a French prelate whose appointment as pope was secured through the influence of King Philip of France (1268–1314), moves the papacy to Avignon, in France. It is dominated by French kings for most of the century, a period referred to as the "Babylonian captivity."

1321 Death of Dante Alighieri

Poet Dante Alighieri dies in Ravenna shortly after completing his masterpiece, the epic *La Divina Commedia* (The Divine Comedy). (See 1365; 1308.)

1347 The Black Death reaches Italy

The Black Death, a plague starting in Asia and eventually affecting all of Europe, is spread to Italy by Genoese sailors returning from the East. It is estimated that a third of Europe's population dies of the Bubonic Plague between 1347 and 1351. Boccaccio's masterpiece, the *Decameron* (see 1351), is a collection of stories supposedly told by a group of aristocrats who flee to the countryside to escape the epidemic.

1349–53   Boccaccio writes the *Decameron*

Giovanni Boccaccio (1313–75) writes the *Decameron,* a collection of one hundred tales gathered in a framing narrative about a group of young aristocrats who flee from Florence to escape the plague. Taking refuge in the countryside, they gather every day for ten days and each person tells a story. Every day one of the tellers chooses a theme for all the stories to be told that day. Although many of the stories in the *Decameron* are well-known tales from folklore or mythology, Boccaccio's work is acclaimed for the sophistication of its structure and its use of language, which influences many other writers of the Renaissance.

1374   Death of Petrarch

The poet and scholar Petrarch (Francesco Petrarca) dies in Padua. (See 1304.)

1434   Medici gain control of Florence

Control of Florence by the Medici family begins, ushering in a golden era for the city in the late fifteenth century as the Medici become distinguished patrons of the arts.

1453   Fall of Constantinople

Constantinople, initially formed as the eastern capital of the Roman empire, falls to the Ottoman Turks.

1452: April 15   Birth of artist and scientist Leonardo da Vinci

Probably the single most representative figure of his age, Leonardo da Vinci (1452–1519) has been called the original "Renaissance man" in honor of the breadth of his talents and interests. Creator of some of the world's most famous paintings, he is also a skilled engineer and leaves behind notes documenting advanced scientific speculation in fields ranging from botany to aerodynamics. Much of Leonardo's adult life is spent in Milan working for Duke Lodovico "the Moor" Sforza (1451–1508) as a court artist, a period during which he paints *The Last Supper* (1495–98). Between 1500 and 1506, Leonardo resides in Florence, where he paints the *Mona Lisa.* He spends his final years in France in the employment of King Francis I (1494–1547). He dies in May, 1519, at the age of 67.

1454   Treaty of Lodi

Warring Italian city-states agree to the Treaty of Lodi granting the Sforza family the succession to the duchy of Milan and ending conflicts in the region for over forty years.

1469–92   Lorenzo Medici rules Florence

Medici rule of Florence reaches its high point under Lorenzo Medici, known as Lorenzo the Magnificent (1449–92). A statesman as well as a poet and a patron of the arts, Lorenzo is instrumental in keeping peace among the city-states as well as keeping foreign powers at bay.

1475: March 6   Birth of Michelangelo

Michelangelo Buonarotti (1475–1564), one of the giants of Western art, is born near Arezzo. The foundations of his training are laid during his adolescence when he becomes a protégé of Lorenzo Medici, the great Florentine statesman and patron of the arts. The early part of Michelangelo's career is spent in Florence, where he sculpts the *Pietà* and *David,* two of his greatest works, while still in his twenties. In 1505 he is summoned to the Vatican to fulfill commissions for Pope Julius II (1443–1513). The greatest of these is the fresco on the ceiling of the Sistine Chapel, a cycle depicting the creation of the world as narrated in the Bible. Other famous works by Michelangelo include the panel painting *The Holy Family* (1503), the sculpture of Moses (c. 1515), and the fresco *The Last Judgment* (1536–41). In addition to his achievements as a sculptor and painter, Michelangelo also works as an architect, designing various buildings in Rome and also completing the plans for St. Peter's Basilica, begun by Donato Bramante (1444–1514). Michelangelo dies in Rome in 1564.

1483: April   Birth of Raphael

Raphael (1483–1520), one of the greatest painters of the Italian Renaissance, is born Raffaello Sanzio, in Urbino. He first learns his craft in the studio of his father, who is also a painter. In 1504, the artist moves to Florence, where he comes under the influence of his contemporaries, Michelangelo and Leonardo da Vinci.

From 1509, Raphael lives in Rome, where Pope Julius II commissions him to paint frescoes in the Vatican, at the same time that Michelangelo is working on the ceiling of the Sistine Chapel. In addition to working for the pope, Raphael also receives commissions from prominent patrons in Rome, becoming known for his portraiture. Well-known portraits include those of author Baldassare Castiglione (1478–1529), Angelo Doni and his wife, and Pope Julius (1443–1513).

In addition to his paintings, Raphael is also known for his work as an architect, which includes churches and other buildings. Leaving behind a prodigious body of work, he dies in 1520 at the age of only 37.

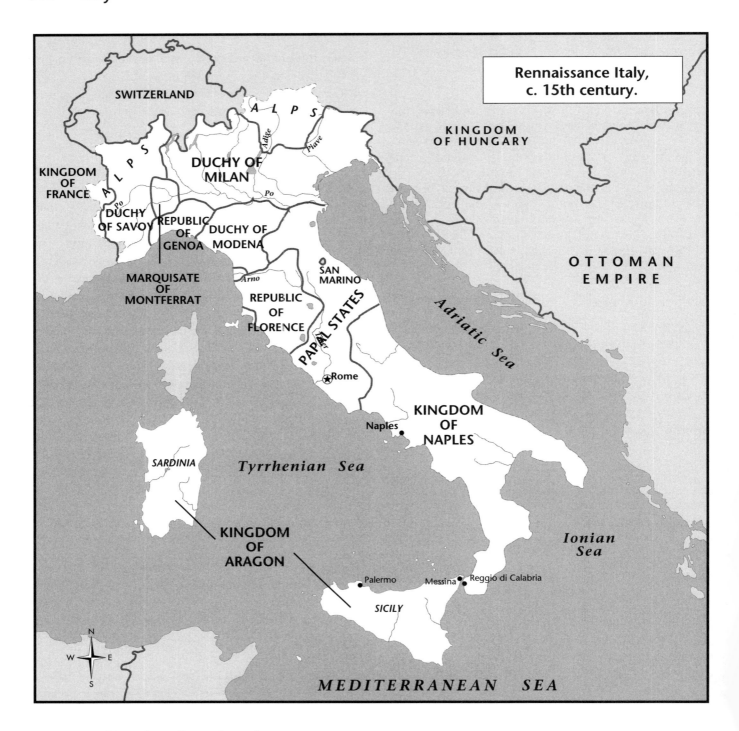

Rennaissance Italy, c. 15th century.

SWITZERLAND

A L P S

KINGDOM OF HUNGARY

KINGDOM OF FRANCE

A L P S

DUCHY OF MILAN

Adige

Piave

Po

DUCHY OF SAVOY

Po

REPUBLIC OF GENOA

DUCHY OF MODENA

OTTOMAN EMPIRE

MARQUISATE OF MONTFERRAT

Arno

SAN MARINO

REPUBLIC OF FLORENCE

PAPAL STATES

Adriatic Sea

⭑Rome

KINGDOM OF NAPLES

Naples

SARDINIA

Tyrrhenian Sea

Ionian Sea

KINGDOM OF ARAGON

Palermo

Messina Reggio di Calabria

SICILY

N W E S

MEDITERRANEAN SEA

Foreign Domination

1494   French invasion launches struggle for control of Italy

As discovery and colonization of the New World begins, Italy's former dominance over European commerce ends, and Spain's power increases. The Italian peninsula, divided into numerous political entities ruled by foreign powers, becomes a pawn in the power struggle between France and Spain. Over half a century of intermittent warfare begins when King Charles VIII (1470–98) of France invades Italy.

1503–13   Reign of Pope Julius II

The reign of Pope Julius II (1443–1513) is one of the high points of the Renaissance. A political as well as religious leader, Julius forms the Holy League, an alliance with Venice and Spain, to expel the French forces of King Louis XII

(1462–1515) from Italy. He also initiates ambitious projects for the restoration of the city of Rome, including the design and construction of St. Peter's basilica and the painting of the Sistine Chapel ceiling by Michelangelo.

1503   Da Vinci paints the *Mona Lisa*

Leonardo Da Vinci (see 1452: April 15) creates his most famous painting, the *Mona Lisa*.

1508–12   Michelangelo paints the Sistine Chapel

Creating one of the triumphs of Western art, Michelangelo paints the ceiling of the Sistine Chapel, covering it with frescoes representing scenes from the Old Testament, primarily Genesis. Altogether, the work contains nearly 350 larger-than-life figures.

1513   Machiavelli writes *The Prince*

Niccolo Machiavelli (1469–1527) publishes the classic political work, *Il Principe* (The Prince), a treatise on obtaining, exercising, and retaining political power. Based on his experience as a diplomat and dedicated to the former Florentine leader Lorenzo Medici, the treatise is born of the belief that only a strong leader can free Italy from foreign domination. Machiavelli's delineation of politics as an art divorced from morality and religion has made the term "Machiavellian" synonymous with the ruthless, cunning, and unprincipled quest for political power.

1519: May 2   Death of Leonardo da Vinci

Artist and scientist Leonardo da Vinci dies in France. (See 1452: April 15.)

1520: April 6   Death of painter Raphael

Raphael, one of the leading Italian painters of the Renaissance, dies in Rome. (See 1483.)

1521   Martin Luther is excommunicated

Pope Leo X (1475–1521) excommunicates radical German theologian Martin Luther (1483–1546) for allegedly heretical views, helping to trigger the Protestant Reformation. Italy becomes the center of the Counter-Reformation.

1527   Charles V invades Rome

Provoked by an alliance between Pope Clement VII (1478–1534) and the king of France, King Charles V (1500–56) of Spain launches an all-out attack on Rome, wreaking fierce destruction on the city and its inhabitants. The pope flees the Vatican.

1529   Treaties of Cambrai and Barcelona

Through the Treaty of Cambrai, France cedes all claims on Italian land to Spain. The same year, the pope and King Charles V settle their differences by the Treaty of Barcelona.

1555–59   Reign of Pope Paul IV

Pope Paul IV (1476–1559) dedicates himself to eradicating the Protestant heresy, expanding the activities of the Inquisition in Rome.

1559   Treaty of Le Cateau-Cambrésis

Over half a century of foreign struggles over Italy end as Spain is granted control of both Milan and the Kingdom of the Two Sicilies in the south.

1564: February 15   Birth of scientist Galileo Galilei

Galileo Galilei (1564–1642), famed for his discoveries in physics and astronomy, is born near Pisa and educated there. During his career he holds teaching posts at Pisa and Padua, later winning an appointment as court mathematician in Florence. Galileo is famous for his astronomical discoveries, using a telescope he constructs himself, and for his embrace of the heliocentric (sun-centered) theory of the Polish astronomer Nicolaus Copernicus (1473–1543), the first to claim that the earth revolves around the sun. Because of his beliefs, Galileo is brought before the Inquisition in 1633 and sentenced to life imprisonment. However, it is served in the form of house arrest, which allows him to pursue further some of his scholarly studies. In addition to his work in astronomy, Galileo researches a variety of other topics, including the principles of mechanics and motion and the laws governing the tides. The scientist dies in 1642.

1564: February 18   Death of Michelangelo

Sculptor, painter, and architect Michelangelo Buonarotti dies in Rome. (See 1475.)

c. 1567   Birth of Claudio Monteverdi

Acclaimed as the father of Italian opera, Claudio Monteverdi (1567?–1643) is also the major composer of the transition between Renaissance and Baroque music in Italy. In the course of his professional career, he holds positions at the courts of the Duke of Mantua and the cathedral of St. Mark's in Venice. The three types of compositions for which he is best known are madrigals, church music, and operas, the latter including *Arianna, L'Orfeo,* and *L'Incoronatione di Poppea.* Monteverdi dies in Venice in 1643.

1600   Giordano Bruno burned as a heretic

Dominican priest and freethinker Giordano Bruno (1548–1600) is burned at the stake as a heretic.

1626: July 30   Major earthquake levels Naples

The city of Naples is struck by a major earthquake that kills 70,000 people.

1633   Trial of Galileo

The scientist Galileo Galilei (see 1564) is tried by the Catholic Inquisition after publishing his book, *Dialogue Concerning the Two Chief World Systems,* which supports the Copernican theory that the earth revolves around the sun. Found guilty on "suspicion of heresy," he is sentenced to life in prison, but the sentence is commuted to house arrest.

1643: November 29   Death of composer Claudio Monteverdi

Claudio Monteverdi, known as the "father of Italian opera," dies in Venice. (See 1567.)

1647: July   Naples revolts against Spain

An unsuccessful revolt against Spanish rule is led by Masaniello (Tommaso Aniello, 1623–47) in Naples. Masaniello is a fisherman and patriot who leads a revolt by the people in Naples who are outraged by new taxes. The uprising results in the Spanish viceroy being driven out of the city, but a number of people are massacred in the process. Masaniello is murdered later the same year. Some accounts attribute his death to Spanish agents; other accounts report that he was murdered by his own followers who had lost confidence in his leadership.

1669: March 11   Mount Aetna erupts

At least 20,000 people are killed in the worst ever eruption of Mount Aetna (some estimates of the casualties range from 60,000 to as high as 100,000). The eruption follows three days of massive earthquakes that tear a twelve-mile-long fissure in the side of the mountain. The thousands of deaths are caused by thick rivers of molten lava, fire and ash, toxic vapor, and showers of boulders, some flung miles away. Some fifty towns are buried in floods of lava. The town of Catania, which has previously been a victim of Mt. Aetna'a eruptions, is buried again and once again is rebuilt.

1678: March 4   Birth of composer Antonio Vivaldi

Antonio Vivaldi (1678–1741), a leading Italian composer of the Baroque era, is renowned for his prolific instrumental works, especially concertos, of which the most famous is *The Four Seasons* (op.8). He takes holy orders in 1703 and serves for over thirty years as music director of the Ospedale della Pieta, a girls' orphanage in Venice. Altogether, Vivaldi writes over four hundred concertos for many combinations of instruments, as well as trio sonatas and other works. The composer dies in 1741.

c. 1700   Cristofori develops the first pianos

Bartolomeo Cristofori (1655–1731), curator of instruments at the Medici court in Florence, builds the first pianos, whose sophisticated actions (internal mechanisms) anticipate nineteenth-century instruments.

1701–14   War of the Spanish Succession

In the War of the Spanish Succession, touched off by the conflict over the succession to the Spanish throne, England and Austria effectively counter French military power. Most of Spain's possessions in Italy go to Austria's Habsburg rulers under the Treaty of Utrecht that ends the conflict. The duke of Savoy wins control of Sicily.

1707: February 25   Birth of dramatist Carlo Goldoni

Carlo Goldoni (1707–93), Italy's greatest comic playwright, is known for transforming Italian comic drama from the *commedia dell'arte* tradition to a more realistic and spontaneous form. His plays include a stage adaptation of *Pamela* (1740) the novel of English author Samuel Richardson; *Il Vero Amico* (1750–51); and *I rusteghi* (The Tyrants; 1760). In 1762 Goldoni is appointed director of the Comédie-Italienne in Paris, where he writes French that he later translates into Italian. Goldoni dies in Paris in 1793.

1726   Publication of *Scienza nuova*

Philosopher and historian Giambattista Vico (1668–1744) publishes the first edition of *Scienza nuova* (New Science), an intensive study of historical patterns that argues for a cyclical theory of history that progresses in stages from barbarism to civilization. Vico's work influences subsequent historians, political philosophers, and other thinkers.

1734   Spain controls Naples and Sicily

Spain seizes control of Naples and Sicily.

1741   Death of Antonio Vivaldi

Composer and violinist Antonio Vivaldi, known for his instrumental works, dies. (See 1678.)

1748   Treaty of Aix-la-Chapelle

The Treaty of Aix-la-Chapelle, which ends the War of the Austrian Succession, brings stability to Italy, establishing a political framework that will survive almost unchanged for over a century until the unification of the Italian states. The Spanish Bourbons win control of Sicily, Naples, Parma, and Piacenza; the Habsburgs control Tuscany, Milan, Mantua, and Modena.

1764   Famous treatise opposes capital punishment

The essay *Crimes and Punishments* by Cesare Conesana, the Marchesa di Beccaria, marks a milestone in the evolution of modern criminology and judicial procedure.

1778: February 6   Birth of nationalist poet Ugo Foscolo

Ugo Foscolo (1778–1827), an early literary figure associated with the Italian *risorgimento,* or resurgence of nationalism, is one of the country's most renowned poets and novelists. As a young man, his nationalism causes him to support Napoleon, but he is later disillusioned by the French leader, whose actions he criticizes in some of his works, including *Ultime lettere di Jacopo Ortis* (The Last Leters of Jacopo Ortis; 1802), hailed by some as the first modern Italian novel. Nevertheless, he takes the French side when Italy is invaded by Austria and Russia, serving in the French military. Following the fall of Napoleon and the return of Italy to Austrian rule, Foscolo leaves the country, ultimately settling in England. His patriotic works, which help inspire the movement toward Italian independence and unification, include the poem in blank verse *Dei sepolcri* (Of the Sepulchres); the tragedy *Aiace,* and the unfinished poem, *Le Grazie* (The Graces). Foscolo dies in London in 1827.

1792–1802   French revolutionary wars

Warfare erupts between the French revolutionary government and the rulers of Austria and Prussia. Many battles take place in Italy, which is eventually conquered by Napoleon.

1793: February 6   Death of dramatist Carlo Goldoni

Italy's greatest comic playwright dies in Paris. (See 1707.)

1797   Treaty of Campo Formio

Under the terms of this treaty, the French revolutionary army, which has invaded northern Italy under the leadership of Napoleon Bonaparte, creates republican governments in Milan (the Cisalpine Republic) and Genoa (the Ligurian Republic). Republics are later formed in Rome and Naples as well.

1799: April   French are expelled

Joint Austrian and Russian forces drive the French from northern Italy, spelling the end of the young republics and reprisals against supporters of France.

1805   Napoleon becomes king of Italy

After military victories over France's enemies, Napoleon retakes Italy and begins a political reorganization, appointing his brother, Joseph, king of Naples and annexing the Papal States.

1807: July 4   Birth of national hero Giuseppe Garibaldi

Giuseppe Garibaldi (1807–82), a central figure in the unification of Italy, is born in Nice, France, which becomes part of Italy's Piedmont kingdom a few years later. After a youthful career as a sailor on cargo ships, Garibaldi joins the "Young Italy" patriotic organization founded by Giuseppe Mazzini (1805–72) and attempts to promote revolution while a member of the navy but is exiled from the country in 1834. He takes refuge in South America, where he participates in revolutionary activities in Brazil and Uruguay, returning to Italy in 1848. Further insurgent activities result in a second exile, this time spent in the United States.

Having spent the better part of twenty years in exile, Garibaldi returns to Italy in 1854. Allying himself with Count Camillo di Cavour (1810–61), he organizes the volunteer Red Shirt forces and takes part in Piedmont's 1859 revolt against Austria, winning control of Lombardy. He later gains control of Sicily and Naples, which are added to the territory already under the control of King Victor Emmanuel, to form a new Italian kingdom. Garibaldi then organizes an aborted invasion of the Austrian-held region of Trentino, takes part in the campaign to win Venetia, and fights on the French side in the Franco-Prussian War. Before retiring from public life, Garibaldo serves in the Italian Parliament. The greatest military leader of Italy's battle for independence and unification, Garibaldo becomes a national hero. He dies in 1882.

c. 1808   The Carbonari society is founded

The best known of a number of secret societies, the Carbonari is founded as an anti-French group during the Napoleonic era and later becomes a focal point for liberal nationalists.

1813: October 10   Birth of composer Giuseppe Verdi

Giuseppe Verdi (1813–1901), generally considered Italy's greatest operatic composer, is born to a family of modest means in Parma. Demonstrating early talent, he receives his musical education through the aid of a wealthy benefactor, completing his training in Milan.

His first opera is produced at La Scala theater in Milan in 1839. After an unsuccessful second effort, the composer's reputation is established with the premiere of *Nabucco* in 1842. He goes on to write such classics as *Macbeth* (1847), *Rigoletto* (1851), *Il Trovatore* (1853), *La Traviata* (1853), and *Aïda* (1871). His reputation eventually reaches far beyond Italy, and he has works commissioned by the Paris Opera and St. Petersburg's Imperial Opera in Russia. In spite of his success on the international scene, Verdi remains strongly identified with his native land and the nineteenth-century rebirth (*risorgimento*) of nationalism that led to its unification. After a long and successful career, the composer dies of a stroke in 1901.

1814–15   Congress of Vienna restores former rulers

The Congress of Vienna that follows the downfall of Napoleon (1769–1821) restores the Spanish, Austrian, and papal rule that existed before the Napoleonic conquest of Italy. Austria once again controls most of northern Italy, Naples is controlled by Spain, the Papal States of central Italy are returned to the pope, and the duke of Savoy is given sovereignty over the Piedmont region.

The Risorgimento

1815–61   Risorgimento

With the restoration of Italy's former royal rulers, nationalist and pro-unification sentiment grows among the Italians, in a movement known as the *risorgimento*. In addition to its political focus, the spirit of the risorgimento appears in many areas of Italian life, including the poetry of authors such as poet Ugo Foscolo (1778–1827) and the music of Italy's great operatic composers Gioacchino Rossini, Gaetano Donizetti (1797–1848), Vincenzo Bellini (1801–35), and Giuseppe Verdi. Opposition to the restored rulers also has a substantial economic basis due to their refusal to share power with Italian landowners.

1827   Death of poet Ugo Foscolo

Nationalistic poet Ugo Foscolo, whose works are a source of inspiration for the risorgimento, dies in London. (See 1778.)

1831   Mazzini forms Young Italy movement

Nationalist leader Giuseppe Mazzini (1805–72), in exile in France, forms the Young Italy association, with the goal of unifying Italy and freeing it from foreign domination. Within two years, the association, which admits no one over the age of forty, has 60,000 members.

1832   Carbonari rebellion in the Papal States suppressed

A revolt by the Carbonari in the Papal States is crushed by Pope Gregory XVI (1765–1846).

1847   Benso di Cavour founds *Il Risorgimento*

Count Camillo Benso di Cavour (1810–61), an aristocrat who espouses the nationalist cause, founds the newspaper *Il Risorgimento*.

1848–49   Wave of unsuccessful revolts

Influenced by the tide of revolution elsewhere in Europe, Italians in Florence, Rome, Naples, and elsewhere stage a wave of revolts, but none is successful.

1859–60   Austrians defeated in Lombardy

Under the leadership of Count Cavour and King Victor Emmanuel II of Piedmont, and with the backing of Louis Napoleon, emperor of France, the Austrians are provoked into war, defeated, and driven out of Lombardy. However, the terms of the peace settlement with Austrian emperor Franz Joseph I (1830–1916) still leave Austria and the pope with control over most of Italy, as well as ceding Savoy and Nice to France.

1860: May–Sept   Garibaldi wins Sicily and Naples from Spain

A force of one thousand known as the Red Shirts, led by Giuseppe Garibaldi (see 1807), conquers the Kingdom of the Two Sicilies (Sicily and Naples), driving the Spanish out of Italy.

The Kingdom of Italy

1861: March 17   Kingdom of Italy is proclaimed

A unified kingdom of Italy is proclaimed under the leadership of Victor Emmanuel II (1820–78), who reigns until 1878.

1863: March 12   Birth of author Gabriele D'Annunzio

Gabriele D'Annunzio (1863–1938) is the most popular author and one of the most colorful personalities in Italy in the late nineteenth and early twentieth centuries. His works are known for their emphasis on the senses and for the author's fascination with the ideal of the Nietzschean "superman," a powerful figure who transcends traditional morality. D'Annunzio's greatest literary success is achieved between the 1890s and World War I. His most famous works include the novel *Il trionfo della morte* (The Triumph of Death, 1894); the poetry collection *Laudi del cielo del mare della terra e degli eroj* (In Praise of Sea, Sky, Earth, and Heroes, 1899); and the play *La figlia de Iorio* (The Daughter of Jorio, 1904).

In addition to his literary career, D'Annunzio plays a prominent, if controversial, role in public affairs. He speaks out in favor of Italy's entrance into World War I and joins the fighting himself once war is declared. After the war, he leads several hundred supporters in occupying the port of Fiume. He later becomes a zealous supporter of Mussolini's Fascist movement, but without exercising any real political influence. D'Annunzio dies in 1938.

1866   Northern territories are reclaimed from Austria

At the close of the Austro-Prussian War, the Kingdom of Italy, having allied itself with the victorious Prussians, reclaims Venetia from Austria.

1870 Rome is annexed by King Victor Emmanuel

King Victor Emmanuel II annexes Rome (which has remained under the control of the pope), virtually completing the unification of Italy. The following year, Rome becomes the capital of the Italian kingdom.

1878 Umberto I becomes king

Upon the death of his father, Victor Emmanuel, Umberto I (1884–1900) succeeds to the throne of Italy.

1880s Economy buffeted by falling farm prices

The Italian economy, largely dependent on farming, suffers from a general European decline in agricultural prices. Poverty in the agricultural south, as well as the accompanying riots and other civil disturbances, drive many Italians to emigrate and also fuel the rise of socialism. By the end of the decade, high tariffs are imposed on both agricultural and industrial products to protect the nation's economy.

1882: May 20 Treaty of the Triple Alliance is signed

Italy joins the Triple Alliance with the Central Powers (Germany and Austria) for mutual protection against possible French aggression. The treaty is renewed five years later.

1882: June 2 Death of Garibaldi

Military leader and national hero Giuseppe Garibaldi dies on Caprera (see 1807.)

1883: July 29 Birth of Benito Mussolini

Benito Mussolini (1883–1945), the Fascist dictator who leads the country from 1922 to 1943, is born in Dovia. He begins his political career as a socialist, becoming the editor of the party's newspaper, *Avanti*, in 1912. His emerging philosophy includes opposition to the church and the monarchy and advocacy of violence as a means of accomplishing political change. After serving in World War I, Mussolini organizes the first Fascist group, the *Fascio di Combattimento* (Union of Combatants), in 1919. It soon becomes a powerful political party, and by 1922 Mussolini is the prime minister and de facto ruler of Italy. Over the course of the decade, he transforms Italy into a totalitarian dictatorship, known as the Corporate (or Corporatist) State. In 1929 Mussolini wins the allegiance of the Catholic Church through the Lateran Treaty settling the unresolved question of the church's relationship with the Italian government.

Under Mussolini's leadership, Italy allies itself with Nazi Germany both before and during World War II. By the middle of the war, however, an armistice is signed with the Allies, and Mussolini is removed from power and arrested. Rescued by the Germans occupying Italy, the former dictator survives in hiding until close to the end of the war. On April 28, 1945, he and his mistress, Clara Petacci, are arrested and executed.

1887 Crispi becomes prime minister

Francesco Crispi (1819–1901), the dominant political figure of the latter part of the century, is appointed prime minister and presides over liberal social reforms at home and attempts at foreign colonization abroad.

1890s African colonization

Italy acquires African territories including Italian Somaliland and Eritrea.

1898 National soccer association is formed

Italy's first national soccer organization is established.

1900 Victor Emmanuel succeeds to the throne

Following the assassination of his father, King Umberto I, by an anarchist, thirty-year-old Victor Emmanuel III (1869–1947) inaugurates a nearly half-century-long reign as king of Italy.

1901 Marconi transmits first transatlantic radio signals

Italian physicist and inventor Guglielmo Marconi (1874–1937) receives transmission of the first radio-wave signal, sent from Cornwall, England to St. John's, Newfoundland.

1901: January 27 Death of composer Giuseppe Verdi

Famed opera composer Giuseppe Verdi dies in Milan. (See 1813.)

1903: November Giovanni Giolitti becomes prime minister

Giolitti, who serves as prime minister for most of the next eleven years, presides over significant economic improvements and social reforms. Industrial growth is encouraged by direct government backing and infrastructure improvements. Social reform measures include labor legislation and the passage of universal male suffrage.

1906 Simplon Tunnel is completed

Construction is completed on a tunnel through the Simplon Pass of the Alps measuring more than twelve miles (nineteen kilometers).

1907: November 28 Birth of author Alberto Moravia

Italy's best-known twentieth-century literary figure is celebrated for the critiques of modern middle-class life found in his novels and short stories. Born in Rome, he begins his career as a journalist and foreign correspondent. With the publication of his first novel, *Gli indifferenti* (Time of Indifference; 1929), he becomes an overnight literary success, although his works are later censored by both Mussolini and the Catholic Church. Moravia's major works include the novels *Agostino* (Two Adolescents; 1944), *Il conformista* (The Conformist; 1951), and *La ciociara* (Two Women; 1957), as well as the essays collected in *L'uomo come fine* (Man as an End; 1963). Moravia dies in Rome in 1990.

1908   Strikes suppressed by government

The government of prime minister Giovanni Giolitti (1842–1928) quells strikes by farm workers in Parma and Ferrara.

1908: December 28   Earthquake destroys Messina

The city of Messina, in Sicily, is destroyed by a massive earthquake that begins under the Strait of Messina and is accompanied by fifty-foot-high *tsunamis* (giant tidal waves). Besides Messina, the quake also levels towns and cities within 120 miles of its epicenter. Some statistics record the number of deaths at 160,000, but according to other accounts, it may be as high as 250,000.

1911: October 11   Italy invades Libya

After the failure of peaceful attempts to occupy Libya, Italian forces invade and annex the territory.

1912   Universal male suffrage is enacted

The Italian government, under liberal prime minister Giovanni Giolitti, enacts universal male suffrage.

The World Wars and Fascism

1914–18   World War I

In spite of its previous alliance with the Central Powers and an initial stance of neutrality, Italy enters World War I on the Allied side. In the secret Treaty of London, the Allies have promised if they are victorious to return Trentino, Trieste, and certain other regions currently under Austrian domination to Italy. Italy is also promised some of Germany's overseas colonies. Over half a million Italian lives are lost in the course of the country's long and difficult engagement in the war.

1919: June 2   Treaty of St. Germain

The Allies back out on part of their prewar agreement to award Austrian lands to Italy, although the Italians do reclaim Trieste, Istria, and Trentino.

1920: January 20   Birth of film director Federico Fellini

Federico Fellini (1920–94), Italy's most acclaimed film director, is born in Rimini, to a middle-class family. Exhibiting an early talent for drawing, he works as a comic-strip artist and writer before he enters the film industry as a writer of screenplays and assistant director. Early in his career, he works with well-known director Roberto Rossellini (1906–77). Fellini directs his first solo film in 1951. With his second film, *I*

Vitelloni (1953), he establishes an international reputation, which grows continually over the succeeding decades. Four of his movies win Academy Awards as best foreign film: *La Strada* (1954), *The Nights of Cabiria* (1957), the autobiographical *8½*, starring Marcello Mastroianni (1963), and *Amarcord* (1973). *La Dolce Vita* (1960) wins the grand prize at the Cannes Film Festival. In 1992 Fellini is honored with a lifetime achievement award by the U.S. Academy of Motion Pictures. The filmmaker dies in 1994.

1920: May 13   First international soccer match is played

Italy plays its first international soccer match, against France.

1921: November   Fascist party is formed

The Fascists, organized and led by Benito Mussolini, become an official political party.

1922: October   Mussolini takes control of the government

After Mussolini and his followers organize a march on Rome, the government of King Victor Emmanuel III resigns and turns power over to the Fascist leader, who becomes prime minister and is granted dictatorial powers for one year.

1924   Fascist election victory

The Fascists win a wide majority in national parliamentary elections, in a campaign marked by violence, and the Socialists charge them with electoral abuses.

1925   Formation of a totalitarian state begins

Mussolini begins turning Italy into a fascist dictatorship, called the Corporative State. His accountability to parliament is ended, and he claims the right to rule by decree. Political opposition is silenced, press censorship is instituted, and elective offices are abolished on the local level and replaced by appointed ones.

1929: February 11   Lateran Treaty

The Lateran Treaty, signed by Mussolini and the Roman Catholic Church, resolves the long-standing tension between the Italian government and the papacy by recognizing Vatican City as a sovereign enclave under the control of the pope and making Catholicism the official state religion of Italy. The pope also receives a financial settlement and declares support for the Fascist regime.

1934   Playwright Pirandello wins Nobel Prize

The Nobel Prize for Literature is awarded to playwright and fiction writer Luigi Pirandello (1867–1936), the leading voice in the Italian "theater of the grotesque," known for its ironic treatment of personal dramas and social themes. Pirandello's best-known plays are *Sei personaggi in cerca d'autore* (Six

Characters in Search of an Author; 1921) and *Enrico IV* (Henry IV; 1922).

1934 Italy hosts and wins World Cup soccer games

Italy, hosting its first World Cup soccer tournament, emerges as the winner, triumphing over runner-up Czechoslovakia.

1935: October 3 Italy attacks Ethiopia

Following a clash between border forces the previous year, Italian troops invade Ethiopia.

1936–39 Italy supports Franco in Spain

Italian troops are sent to Spain to aid the dictator Francisco Franco (1892–1975) in the Spanish Civil War.

1936: May Italy annexes Ethiopia

Defying censure by the League of Nations, Italian forces take the Ethiopian capital, and Italy officially annexes the country, uniting it with its other African possessions, Eritrea and Italian Somaliland, to form Italian East Africa.

1936: October Rome-Berlin Axis is formed

Italy allies itself with Germany to form the Rome-Berlin Axis.

1938 Italians score second World Cup victory

Italy earns its second consecutive World Cup victory at games held in France.

1938: March 1 Death of author Gabriele D'Annunzio

Leading turn-of-the-century literary figure Gabriele D'Annunzio dies. (See 1863.)

1939 Italy occupies Albania

After signing the Pact of Steel reaffirming its alliance with Germany, Italy invades and occupies Albania.

1939–45 World War II

Following an early stance of neutrality in World War II, Mussolini is persuaded by German leader Adolf Hitler (1889–1945) to enter the war on the side of Germany and the Axis powers. Italy does not fare well in the war, attacking Greece unsuccessfully from Albania, losing Libya to the Allies, and also losing its North African territories of Italian Somaliland, Eritrea, and Ethiopia.

1943: July Allied invasion of Italy

Allied forces invade Italy, joined by Italian partisans (resistance fighters).

1943: October Italy joins the Allies

After removing Mussolini from power and signing an armistice with the Allies, the Italian government renounces its alliance with Germany. Allied forces battle German troops continuing to occupy parts of the country.

1945: April Germans surrender in Italy

German troops occupying Italy surrender to the Allies.

1945: April 27–28 Mussolini is captured and executed

Mussolini is captured in his hiding place and executed. His remains and those of his mistress, executed along with him, are hung in a public square in Milan.

The Italian Republic

1946: June Italy becomes a republic

By a narrow margin, Italians vote to end the monarchy and form a republic. The king abdicates and leaves the country.

1947: February 10 Italy signs peace treaty with the Allies

Under the terms of the formal peace treaty signed with the Allies, who are still occupying the country, Italy agrees to pay $360 million in war reparations to Albania, the Soviet Union, and other countries and to cede all its overseas colonies and possessions. Trieste is placed under United Nations supervision but is later divided between Italy and Yugoslavia.

1948: January New constitution takes effect

The Italian republic adopts its first constitution.

1948 Christian Democrats win election

The Christian Democrats win control of the government.

1949 Italy joins NATO

Joining the West in the Cold War power alignment, Italy becomes a charter member of the North Atlantic Treaty Alliance (NATO).

1955 Italy joins the UN

Italy becomes a member of the United Nations (UN).

1955 *La Strada* wins Best Foreign Film Oscar

The movie *La Strada,* directed by Federico Fellini (see 1920: January 29), wins the Academy Award as Best Foreign Film. (It is the first of four Fellini films to win this award over the next two decades.)

The Cold War

The Cold War is the commonly used name for the prolonged rivalry between the United States and the Soviet Union that lasted from the end of World War II to the break-up of the U.S.S.R. in 1991. The Cold War also encompassed the predominantly democratic and capitalist nations of the West, which were allied with the U.S., and the Soviet-dominated nations of Eastern Europe, where Communist regimes were imposed by the U.S.S.R. in the late 1940s. Although the United States and the Soviet Union never went to war with each other, the Cold War resulted in conflicts elsewhere in the world, such as the Vietnam War, a prime example of the United States's "domino theory," which held that a Communist victory in one nation would result in the spread of Communism to its neighbors.

Following World War II, the political and military bond between the United States and its allies in Western Europe was cemented by the Marshall Plan, which helped the European countries recover from the war economically, and the 1949 creation of NATO (the North Atlantic Treaty Organization). Within the United States, an early consequence of the Cold War was McCarthyism, a political trend named for Senator Joseph McCarthy, who spearheaded the drive to expose and penalize Americans suspected of being Communist spies or sympathizers in the early 1950s, one of the most intense periods of the Cold War. McCarthyism, at its height between 1950 and 1954, ruined the careers and lives of many Americans who posed no actual threat of subversion to the country. Following the death of Soviet dictator Josef Stalin in 1953, the intensity of the Cold War abated for several years. However, the launch of the Soviet space satellite Sputnik in 1957 as well as Soviet efforts to extend its influence in the newly-independent (formerly colonial) states revived the rivalry by inaugurating a battle for supremacy in space exploration (the "space race") and the "Third World", respectively.

The single most serious political confrontation of the Cold War was the Cuban Missile Crisis of 1962, when the United States and the Soviet Union were brought to the brink of a nuclear confrontation over the stationing of medium-range nuclear missiles in Soviet-allied Cuba. Ultimately, the missiles were withdrawn, but both nations began a costly arms race that was particularly damaging economically for the Soviet Union.

In the early 1970s, under Soviet premier Leonid Brezhnev and U.S. president Richard Nixon, Cold War tensions began to ease, in a development known as "détente." During this decade the two nations signed the SALT (Strategic Arms Limitations Treaty) agreements limiting antiballistic and strategic missiles. However, outside the provisions of these agreements, both nations continued their massive arms build-up, and tensions increased in the 1980s during the presidency of staunch anti-Communist Ronald Reagan, who referred to the Soviet Union as the "evil empire."

The appointment of political reformer Mikhail Gorbachev as general secretary of the Communist Party in 1985 signaled the beginning of the end for the Cold War. The democratization and openness implemented by the Soviet government under Gorbachev culminated not only in the downfall of Communism but in the dissolution of the Soviet Union itself, which was preceded by the fall of Communism in the Eastern European countries of the former Soviet Bloc.

1957: March 25   EEC treaty is signed in Rome

Italy plays a leading role in drawing up the agreement creating the European Economic Community, or Common Market, which is signed in Rome.

1957   Poet Quasimodo wins Nobel Prize

Poet, literary critic, and translator Salvatore Quasimodo (1901–68) is awarded the Nobel Prize for Literature. Before World War II, he is associated with the Hermetic school of poetry that includes Giuseppe Ungaretti (1888–1970) and Eugenio Montale (1896–1981; see 1975), who are his major influences. Hermetic poetry is known for its idiosyncratic structure and symbolism and its violation of ordinary logic. After the war, Quasimodo's work becomes more concerned with social and political themes, including the legacy of fascism and the hardships of the war. Quasimodo's works include *Acque e terre* (Water and Land; 1930), *Ed è subito sera* (Suddenly It's Evening; 1942), *Giorno dopo giorno* (Day After Day; 1947), *La terra impareggiabile* (The Incom-

Construction of a subway begins to ease traffic congestion caused by the booming economy in Milan. A three-story subway station on the square in front of the city's famous Gothis cathedral, the Duomo, will feature a shopping center and restaurant. (EPD Phoitos/CSU Archives)

parable Earth, 1958), and *Dare e avere* (To Give and To Have; 1966).

Early 1960s   Economic boom

Italy enjoys a major economic boom, as it makes the transition to a fully-fledged industrial economy, with most manufacturing concentrated near the northern cities of Milan, Turin, and Genoa.

1968: Fall   Strikes and riots shake Italy

Widespread strikes and rioting give this period the label "The Hot Autumn." In response, the government makes major concessions to placate public sector employees and trade unions.

1970s   Italy faces economic crisis and terrorism

Italy's economy, which has been exceptionally strong in the early to mid-1960s, stagnates, mostly as a result of rising oil prices. Unemployment and inflation rise rapidly, and support for the ruling Christian Democrats erodes. Extremist right-

and left-wing terrorist groups emerge and begin a wave of political violence that gives the 1970s the nickname "the decade of the bullet" (*gli anni di piombo*), climaxing in the kidnapping and murder of Christian Democrat leader Aldo Moro (see 1978) by the Red Brigades (see 1970).

1970   The Red Brigades is formed

Italy's most notorious left-wing terrorist group is formed in Milan. The group commits robberies and kidnappings to fund its illegal activities, which include bombings and assassinations.

1974   Italians reaffirm divorce law

Italians vote against legalizing divorce.

1976   Eugenio Montale wins Nobel Prize

Poet Eugenio Montale (1896–1981) wins the Nobel Prize. His works include *Ossi di Seppia* (1925), *Le occasioni* (1939), *La bufera* (1956), *Satura* (1962), and *Xenia* (1966).

Published collections of his works in English include *The Second Life of Art* (1982), *Otherwise: Last and First Poems of Eugenio Montale* (1984), and *Collected Poems 1920–54* (1998). (See also 1957.)

1978   Red Brigades kidnap former prime minister

Former prime minister Aldo Moro is kidnapped by the Red Brigades, who execute him when their demands for prisoner releases are not met by the government. Following the incident, the Italian government institutes a serious crackdown on terrorism. However, political violence continues.

1978   Italy elects its first socialist president

Alessandro Pertini becomes the first socialist to be elected president of Italy.

1980: August 2   Bologna train station bombing

In the most serious terrorist incident attributed to right-wing extremists, a bomb is planted in the Bologna train station, killing eighty-four people. The perpetrators are arrested and convicted.

1980: November 23   Earthquake causes massive damage in the south

Southern Italy is rocked by its worst earthquake in sixty-five years. The quake, measuring 6.8 on the Richter scale, is responsible for some 3,000 deaths, and 200,000 people lose their homes. Rescue efforts are hampered by heavy fog in the Naples area and a lack of earth-moving equipment in the affected area, which is one of Europe's poorest regions.

1981   Restoration of the Sistine Chapel begins

The restoration of the ceiling of the Sistine Chapel, painted between 1508 and 1512 by Michelangelo, begins, with the goal of restoring the colors of the painting to their original brilliance.

1981: May   Government connections to secret lodge are revealed

The public learns about the existence of the Masonic lodge P2 ("Propaganda 2"), a secret right-wing organization aimed at undermining parliamentary democracy in Italy. Its membership is found to include hundreds of government officials—including three cabinet ministers—as well as military and religious leaders. The discovery causes a scandal that brings down the government of Christian Democrat Arnaldo Forlani.

1981: June 28   Republican Party leader becomes prime minister

Giovanni Spadolini, leader of the Republican Party, becomes Italy's first prime minister since 1945 who is not a Christian Democrat.

1982   Italy wins its third World Cup victory

For the third time, an Italian soccer team wins the World Cup championship tournament, defeating runner-up West Germany.

1983–87   Craxi serves as prime minister

Bettino Craxi (b. 1934) is Italy's long-serving prime minister since the end of World War II, as well as the first socialist to hold the office. Craxi's coalition government, which has strong Christian Democrat representation, introduces effective economic policies that cut inflation and reduce the nation's budget deficit while maintaining economic growth.

Bettino Craxi is born in Milan in 1934. He becomes actively involved in the Socialist Youth Movement, joining the Central Committee of the Italian Socialist Party in 1957.

1983: January   Moro kidnappers sentenced

Members of the Red Brigades are sentenced in connection with the 1978 kidnapping and murder of former prime minister Aldo Moro and other crimes. Twenty-three receive life sentences, and 36 others receive lesser sentences.

1985: February   Organized crime trials begin

Trials of over 1,000 suspects begin in the government's crackdown on organized crime.

1985: October   Hijacking of the *Achille Lauro*

An Italian cruise ship, the *Achille Lauro,* is hijacked by Palestinian terrorists, who kill Leon Klinghoffer, an American Jew, and throw his body overboard. The terrorists are freed by Egypt and put on a plane, but U.S. fighter pilots force it to land in Sicily. The Italian government refuses a U.S. request for extradition of the terrorists, who are tried, convicted, and imprisoned.

1987–88   Economy produces political instability

Inability to resolve the nation's chronic economic crisis leads to the rapid formation and dissolution of succeeding governments.

Early 1990s   Corruption scandals plague government

With the inauguration of the *mani pulite* (clean hands) campaign to uncover official corruption, bribery scandals plague Italy's government and its top business leaders, ending the political dominance the Christian Democrats have enjoyed since World War II. Over 1,000 political and business leaders, including former prime ministers Bettino Craxi and Giulio Andreotti, dozens of legislators, and top industrialists, are implicated in scandals involving kickbacks from a wide range of businesses and industries. The endemic bribery of political officials uncovered by the investigations is dubbed *Tangentopoli* (Bribe City). Further anti-corruption activities include investigation of government links to the Mafia and prosecution of top Mafia leaders.

1990 Sistine Chapel restoration is completed

The task of restoring Michelangelo's sixteenth-century Sistine Chapel ceiling is completed after nine years of work. (See 1981.)

1990 Italy hosts World Cup games

For the second time, Italy is the venue for the World Cup soccer championship, in which West Germany triumphs over runner-up Argentina.

1990: September 26 Death of author Alberto Moravia

Acclaimed novelist Alberto Moravia dies in Rome. (See 1907.)

1992: March Mafia crime wave begins

The execution of Christian Democrat leader Salvatore Lima touches off a wave of Mafia violence in Sicily.

1993: August Electoral reform law is passed

Legislation provides for a new electoral system combining simple majority voting with proportional representation for parties winning over 4 percent of the vote.

1994: March 27–28 Conservatives win national election

Following the Christian Democrats' extensive implication in corruption scandals, a center-right coalition, the Alliance for Freedom, wins the first national vote held under Italy's new electoral system. Media mogul Silvio Berlusconi, founder of the Forza Italia party, becomes the new prime minister.

1994: December 22 Berlusconi resigns following corruption charges

Weakened politically by internal disputes within the ruling conservative coalition and a corruption probe of his Fininvest media conglomerate, Prime Minister Silvio Berlusconi resigns after less than a year in office due to corruption charges and is replaced by Minister of Trade Lamberto Dini, who forms Italy's fifty-fourth postwar government.

1996: April 21 Communist-led coalition wins elections

With the victory of the center-left Olive Tree coalition, Italy's first leftist government since World War II is voted into office. Former economics professor Romano Prodi becomes the new prime minister.

1997: October 9 Fo wins Nobel Prize for Literature

Playwright and political satirist Dario Fo, age 71, wins the Nobel Prize for Literature. A long-time leftist, Fo is known for his plays satirizing capitalism and the Catholic Church. His best-known work is the 1970 play *Accidental Death of an Anarchist,* based on the actual arrest and imprisonment of Italian anarchist Giuseppe Pinelli.

1998: May 24 Italian film honored at Cannes

The Grand Prix at the Cannes Film Festival goes to *Life is Beautiful,* directed by Italian comic Roberto Benigni, who also stars in the film about Nazi persecution of the Jews in World War II.

1998: July Berlusconi convicted on corruption charges

In two separate trials, former prime minister Silvio Berlusconi is convicted of making illegal political contributions to the Socialist Party and of bribing government officials to obtain a favorable tax audit for his Fininvest conglomerate. A two-year-plus prison term handed down in the tax case is considered likely to be suspended.

Bibliography

Baranski, Zygmunt G., and Robert Lumley, ed. *Culture and Conflict in Postwar Italy: Essays on Mass and Popular Culture.* New York: St. Martin Press, 1990.

Burckhardt, Jacob. *The Civilization of the Renaissance in Italy.* 3rd ed. Oxford, England: Phaidon, 1995.

Duggan, Christopher. *A Concise History of Italy.* New York: Cambridge University Press, 1994.

Furlong, Paul. *Modern Italy: Representation and Reform.* New York: Routledge, 1994.

Ginsberg, Paul. *A History of Contemporary Italy: Society and Politics, 1943–1988.* London: Penguin, 1990.

Hearder, Harry. *Italy: A Short History.* New York: Cambridge University Press, 1990.

Moss, David. *The Politics of Left-wing Violence in Italy, 1969–85.* New York: St. Martin's Press, 1989.

Sassoon, Donald. *Contemporary Italy: Politics, Economy, and Society Since 1945.* New York: Longman, 1986.

Spotts, Frederic, and Theodor Weiser. *Italy: A Difficult Democracy.* New York: Cambridge University Press, 1985.

Trevelyan, George Macaulay. *Garibaldi and the Making of Italy.* New York: Longmans, Green, 1948.

Latvia

Introduction

The Republic of Latvia (Latvijas Republika) is the middle of three countries known as the Baltic states that border the Baltic Sea in the northeastern region of Europe, the other two being Estonia and Lithuania. With a total land area of 24,749 square miles (64,100 square kilometers), Latvia is just a little larger in size than the state of West Virginia. It is bordered by Estonia to the north, Lithuania to the south and south-west, Russia to the east, Belarus to the southeast, and the Baltic Sea to the west. Latvia consists of low-lying, undulating land that is part of the East European Plain. The largest of its many rivers is the Daugava River (known as the Dvina in Russian).

As of the late 1990s, Latvia has a total estimated population of nearly 2.5 million. Riga, its capital city, is located in the southeast corner of the Gulf of Riga, a deep inlet of the Baltic Sea. Founded in 1201, Riga is the largest city in the Baltic states and has a population of more than 830,000.

Latvian history has been strongly influenced by German, Russian Imperialist, and Soviet rule. As early as the twelfth century A.D., the Germans brought Christianity and feudalism to Latvia and spearheaded the development of trade and industry. By the early thirteenth century, much of the area today known as Latvia, which was previously fragmented politically, had been conquered by orders of crusading German knights. It was eventually organized into the state of Livonia, which also included portions of present-day Estonia. By the sixteenth century, control over the region was contested by Denmark, Sweden, Russia, and the political federation of Lithuania and Poland. Most of Latvia was divided between Sweden and Lithuania-Poland. After a period of rule by Sweden in the seventeenth century, Latvia was annexed by Russian czar Peter the Great (1682–1725) in 1710 and remained under Russian Imperialist rule until the last czar, Nicholas II, was overthrown in 1917.

During the nineteenth century, Russian political and social oppression sparked a national and cultural awakening unparalleled in Latvian history. The first Latvian-language newspaper was founded, and Riga University was established. Poet Krisjanis Barons (1823–1923) and others collected large numbers of Latvian folksongs, some of which were later incorporated into works by Latvian composers. Latvian artists, such as Janis Rozentals (1866–1916), brought new dignity to scenes from everyday life in their homeland and glorified its scenery with landscape paintings in a variety of styles. The Riga Latvia Association, founded in 1868, provided the impetus for numerous cultural activities, including the massive song and dance festivals that continue to this day. In response to this outpouring of nationalistic activity on the part of Latvians, their czarist rulers implemented a policy of Russification starting in the mid-1880s in an attempt to establish the Russian language and culture in the region, but the effort failed badly and was abandoned by the end of the century.

Latvia first established itself as a parliamentary democracy on November 18, 1918, although it was not really able to enjoy independence for two more years as rival Russian political factions were at war in the region, and Russia made an outright attempt to annex it. By 1920, however, Russia recognized Latvian sovereignty, and Latvia remained independent until the eve of World War II. It enjoyed democratic parliamentary government in the 1920s. In the following decade, a moderate dictatorship was imposed by its acting government in an attempt to head off victory at the polls by a pro-fascist group.

At the beginning of World War II, Latvia was annexed by the Soviet Union under the terms of the Molotov-Ribbentrop Pact signed with Nazi Germany in 1939. Soviet troops invaded the country in 1940. The Latvian Soviet Socialist Republic was created, and the mass deportation of Latvians to Siberia under brutal conditions was begun as a way to rid the country of any persons who might possibly pose a threat to the new rulers, especially well-educated and upper-class individuals. By 1941, however, Latvia had fallen to the Germans, who occupied it until 1945, forcing many Latvians into military service or labor camps, and spurring the formation of active resistance movements among both nationalist and communists.

At the end of World War II, Latvia reverted to control by the Soviet Union, which launched yet another mass deportation of Latvians in 1949, exiling about 40,000 people from

their homeland. In the following years, the U.S.S.R. adopted policies designed to integrate Latvia and her Baltic neighbors as fully as possible, politically, economically, and culturally. They changed the ethnic makeup of these countries by deporting many native residents while sponsoring large-scale immigration of ethnic Russians into the area, often with jobs in specific industrial enterprises. By 1953 one-third of Latvia's labor force consisted to ethnic Russians. Altogether, approximately 400,000 Russians emigrated to Latvia between the 1940s and the 1990s. Cultural integration was carried out largely by privileging Russian over Latvian as the language of government and education. In addition, religious organizations were closely scrutinized by state authorities, and religious believers were harassed and persecuted. Major churches in larger cities in Latvia were turned into museums or concert halls.

Nevertheless, Soviet rule did bring economic progress, based on the rapid growth of industry and the collectivization of agriculture. Latvia eventually came to have one of the highest standards of living among the Soviet republics. But

Latvians still seized the opportunity to proclaim their national identity and their sovereignty as a people when the leadership of Mikhail Gorbachev caused an easing of Soviet repression in the 1980s. The Soviet Union's collapse in August, 1991, gave the Baltic states and other Eastern and Central European nations the opportunity to become completely independent. Latvia reclaimed the independence it had known during the interwar years (1918–39) and officially became the Republic of Latvia.

Since then it has taken measures to restore the status of the Latvian language, as well as controversial measures stripping many ethnic Russians of their Latvian citizenship.

In 1993 Latvia held its first free parliamentary elections since the 1930s for seats in its newly restored one-hundred-member parliament, called the Saeima. Latvia was admitted to the Council of Europe in 1995 and the World Trade Organization in 1998. As the 1990s drew to a close, the nation looked forward to eventual membership in the European Union.

Timeline

9000 B.C.–3000 B.C.   First signs of humans in prehistoric era

The area currently known as Latvia first shows signs of human habitation around 9,000 B.C., after the Ice Age. Ancestors of the Baltic Finn peoples occupy the area beginning around 3,000 B.C. Belief in natural deities emerges as the oldest religion.

c. 2500–1500 B.C.   Tribal groups migrate to the coast

The Baltic coastal region attracts many different tribal groups migrating to the east. The proto-Baltic tribes, drawn by the maritime characteristics of the Baltic area, begin to occupy the region. The Baltic tribes (Selonians, Semgallians, Couronians, and Latgallians) and Finno-Ugric tribes (Estonians and Livonians) start to become culturally distinct.

The first peoples known to settle the area of Latvia are the Livonians. Ethnically, their closest relatives are the Old Prussian and Lithuanian tribes. These tribes remain independent of each other, and their ancient dialects form the foundation of the Baltic language group, of which only Latvian and Lithuanian survive.

c. A.D. 100   Romans record existence of Latvia

The inhabitants of Latvia are mentioned by the Roman historian Tacitus (c. A.D.56–c. A.D.120), but he offers no specific details about them.

8th–12th centuries   Independent Latvian kingdoms are formed

Living freely on their land, the Latvians develop several independent and culturally distinct kingdoms.

9th–10th centuries   Vikings raid Latvia

Vikings from Scandinavia launch raids down the Baltic coast, founding settlements along some of the region's rivers. They are interested only in opening a trade route from the Baltic Sea, and their control of the local inhabitants is limited to the imposition of taxes.

Conquest by Germany

13th–16th centuries   Germans control Latvia

Germans, traveling east to seek new trade routes, spread Christianity, and expand feudalism, are able to seize control of Latvia. They convert the populace to Christianity and impose a system of serfdom on them. The region that corresponds to present-day Latvia is divided into Courland in the south, Livonia in the north, and includes the southern part of present-day Estonia. The ethnic Latvians are treated as sub-

The State of Livonia, c. 1270.

jects and forced to pay taxes and provide free labor to their German rulers. The Latvians are able, however, to maintain their ethnic character through their oral folk tradition.

1201   City of Riga founded

The city of Riga is founded as an independent entity (the free city of Riga) and becomes one of the principal powers in the region, the other two being the Catholic church and the orders of crusading German knights.

1225–27   Teutonic Knights invade southern Estonia, Latvia, and Lithuania

The Latvians are conquered by the Knights of the Teutonic Order (a Germanic religious and military organization that emerges from the Crusades). The disunity prevalent among Latvian tribes and communities makes conquest by Western European Crusaders easier.

1237   Formation of the Livonian Order

The Livonian Order, an organization of Crusaders, is formed from the group formerly known as the Knights of the Sword.

1270s   State of Livonia is founded

The Crusaders establish the state of Livonia, a political union of territories belonging to the Livonian Order and the Catholic church and including the free city of Riga. The political and economic unity imposed by the Livonian Order leads to the unification of the Finno-Ugric and Indo-European peoples in the region as one Latvian linguistic community.

1282   Riga joins the Hanseatic League and becomes a trading center

The city of Riga is admitted to the Hanseatic League of northern Germany and grows to become one of the great trading cities of the Middle Ages. (The Hanseatic League includes a group of cities and towns in northern Germany allied to control long-distance trade in the Baltic and North Seas and thereby assume a central mediating role in east-west trade.) Trade and property rights of non-Germans living in Riga are severely restricted, and the Latvian people are increasingly enslaved to the land. Feudal estates develop, and German land barons establish themselves as the dominant influence in the Baltic territories. Over the next two centuries, an active musical life develops in Riga, with performing groups and musicians' guilds.

Rival Powers Vie to Control Latvia

1558–83   The Livonian War Period

The Germans are not alone in desiring to conquer Latvia and her Baltic neighbors. Due to the area's strategic location as a trading center, the German nobility faces increasing threats to their control from rivals to the north and east. Denmark, Prussia, Sweden, and Russia take turns dominating the state of Livonia, which includes southern Estonia and Latvia. The Livonian Wars begin when Ivan IV (the Terrible), czar of Russia from 1533–84, invades the region, seeking access to the Baltic Sea for trade expansion. When the wars end, Livonia is divided between Sweden and the Commonwealth of Poland-Lithuania, which takes the southern part of Livonia, including Latvia.

1561   Sweden seizes northern Livonia

After the Knights of the Livonian Order disband, Sweden seizes control of northern Livonia, which includes part of Estonia. The southern portion is claimed by the Commonwealth of Poland-Lithuania.

1587   Publication of the first religious music with Latvian texts

In Riga, Liturgical songbooks with Latvian texts are published for the first time.

1622   Sweden controls Livonia, including Riga

Sweden has by this time assumed control of all of Livonia, including Riga. The German nobility is able to retain its landed privileges and manors and rules over an enserfed Latvian peasantry in exchange for their support of Swedish rule.

1689   First Latvian-language Bible is printed

With support from the Swedish crown, Reverend Ernest Glueck publishes the first Bible printed in the Latvian language. The earliest examples of Latvian writing begin to emerge in the seventeenth century. The main religion in Latvia is Lutheranism

Czarist Rule

1710   Peter the Great begins his conquest of Latvia

The inclusion of Latvian territories in the Russian Empire begins with the new wave of Russian expansion in 1700. Peter I (the Great), czar of Russia from 1682–1725, begins the conquest of Latvia in the Great Northern War (1700–21), initiating the conquest of Livonia (the name given to southern Estonia and northern Latvia) from Sweden. The German nobility, which represents a small percentage of the population, thrives under Russian rule as they had when Sweden controlled the area.

1721   The Treaty of Nystad ends the Great Northern War

At the close of the Great Northern War, the Treaty of Nystad transfers control of Estonia and Livonia to Russia, and Russia thus becomes part of the community of European nations.

Late 18th century   Russian imperial rule

At the close of the eighteenth century all of Latvia comes under Russian Imperial control. A period of gradual change from feudalism to capitalist industrialism also begins. With the abolition of serfdom, industry begins to develop rapidly, and the population grows. Russian Orthodoxy replaces Lutheranism as the dominant religion. The era of Imperialist Russian rule spreads over two centuries and ends with the last czar, Nicholas II.

The Rise of Nationalism

19th–20th centuries   Latvian folk songs compiled

The compilation of Latvian folksongs is begun by writer/journalist Krisjanis Barons (1823–1923), known as the father of *dainas* or folk songs. He identifies more than 1.4 million dainas, four-line couplets reflecting the traditional ethics, morals, and life-styles of the Latvians. Folk music has a strong influence on Latvia's nineteenth-century composers, including Andrejs Jurgans, Jazeps Vitols, and Emilis Melngailis, who use these tunes in their compositions.

1816–19   Serfdom is formally abolished in the Baltic states

During this period peasants win the rights to their property and to farm their land without overlordship. Latvians are freed from serfdom in the countryside and, if they choose, can move to the city. Many move to Riga to find work in trades, business, and civil service and create intellectual circles.

1822   First Latvian-language newspaper is founded

German Latvians found the first newspaper to be published in Latvian, *Latviesu Avizes*.

1860s–70s   First Latvian novel, *Mernieku Laiki*, is published

The brothers Rainis and Matiss Kaudzitis write their masterpiece *Mernieku Laiki* (The Time of the Land Surveyors). The first Latvian novel depicts country life in Latvia the late nineteenth century.

1861   Riga University is founded

Riga University, Latvia's major institution of higher education, is established.

1865: September 11   Birth of author Janis Rainis

Janis Rainis (1865–1929), Latvia's most influential and prolific writer and a major advocate for Latvian independence, is born. As editor of the newspaper *Dienas Lapas,* he writes critically of Russian social and political oppression, which results in his being exiled. His leading works, all written in the early twentieth century, include the plays *Uguns un nakts* (Fire and Night) and *Jazeps un vina brali* (Joseph and His Brothers). His poetry volumes include *Gals un sakums* (The End and the Beginning) and *Piecas Dagdas skicu burtnicas* (Dagda's Five Notebooks). His wife, Aspazija, is a leading poet, playwright, and social critic in her own right, who is seen as an early advocate in the struggle for women's rights.

1866: March 18   Birth of artist Janis Rozentals

Rozentals, considered by many to be the founder of modern Latvian art, works as a painter in Riga in his youth and enters the St. Petersburg Academy of Arts in 1888. After two sojourns in Western Europe, he returns to Latvia permanently a decade later. Rozentals's art is closely tied to his native land, often featuring Latvian landscapes or scenes from everyday life, such as *Leaving Church: After the Service* (1894), *On the Porch,* and *The Market* (both 1896). Latvian mythology also serves as subject matter for his art. Rozentals becomes an acclaimed portrait painter and passes on his craft to a new generation of Latvian artists as an instructor at the Riga City Art School, where he directs the portrait-painting program from 1906–13. He also contributes artwork and essays to major journals. Rozentals dies in Helsinki in 1916.

1868   The Riga Latvian Association is formed

An increasing spirit of Latvian nationalism, or pride in one's people and culture and the desire for independence, grows and expresses itself in many ways. The Riga Latvian Association becomes the cornerstone of the Latvian national awakening and inspires the creation of a Latvian literature, an encyclopedia, a song festival, a national theater, and opera.

Latvians begin to consider themselves a viable separate nation. A group of the Latvian intelligentsia, referring to itself as the Young Latvians, proves instrumental in developing the Latvian literary style and Latvian culture.

1872: March   Birth of artist Vilhelms Purvitis

Purvitis is an influential painter, educator, and museum administrator who is at the center of artistic life in Latvia during the early part of the twentieth century. Born to a peasant family, he is educated at the St. Petersburg Academy of Arts. His paintings are closely identified with his homeland and include many landscapes. Many are patriotic in tone and pay tribute to the peasants' simple rural lives. Stylistically, Purvitis has been associated with French Impressionism because of the short brushstrokes that characterize many of his paintings. Purvitis serves as director of the Riga City Art School and founds the Proletarian Art Studios, also in Riga. He dies in 1945.

1873   Periodic mass song and dance festivals begin

Song and dance play an important role in distinguishing Latvian culture from Russian or German culture. Periodic mass song and dance festivals begin in 1873. (The tradition continues into the 1990s; a festival held in 1993 attracts 30,000 participants.) These festivals showcase Latvians in traditional folk costumes. Women wear long, colorful skirts, embroidered blouses, jackets or shawls, and elaborate headgear. Men wear less colorful clothing symbolic of their peasant past.

1880s   Russification

The new spirit of nationalism spreading throughout the Baltic states starts a trend of individualism, consciousness of self, and political activism that Imperial Russia finds intolerable. Beginning in the 1880s the Russian government institutes a program of deliberate Russification in the Baltics designed to undermine, by force if necessary, both the autonomy of the Baltic states and any growing nationalistic movements. Russification includes suppression of the Latvian language and coerced adherence to Russian cultural traditions.

1888   Epic poem, *Lacplesis*, is published

Poet Andrejs Pumpurs publishes the epic poem, *Lacplesis* (The Bear Slayer), ushering in the beginning of modern Latvian literature. The bear has traditionally symbolized Russia, and the poem is written to highlight rising Latvian patriotic sentiments. As Latvians continue to establish their literary independence, they face increased pressure from Russia to silence the national awakening.

1905   Russian Revolution spurs demands by Latvians

Czarist Russian authorities respond with violence and repression to Baltic demands for radical political change during the Russian Revolution of 1905. Latvians seek social and political liberation against Baltic German landowners and the Russian policy of assimilation (Russification).

1914–18   World War I

The First World War cripples Latvia economically as the German army occupies the western half of the region. One-fifth of Latvia's 2.5 million inhabitants become refugees, while most of Latvia's industry is moved to the Russian interior.

1916: December 26   Artist Janis Rozentals dies

Janis Rozentals, widely regarded as the founder of modern Latvian art, dies in Helsinki. (See 1866.)

1917   Czar Nicholas II abdicates; czarist regime collapses

As the Bolsheviks take control of Russia they make significant political gains in the Baltic region. The Russian Revolution together with Germany's defeat in the First World War pave the way for Latvian independence.

First Period of Independence

1918: November 18   Latvia declares independence

Latvia declares its independence and becomes the Republic of Latvia. Various Latvian political parties band together to form the Latvian People's Council which proclaims national independence and the creation of the Republic of Latvia, a parliamentary democracy. However, the achievement of true independence comes only after two years of warfare between competing Russian revolutionary forces in the region.

1919: September 28   Latvian State University opens

The new Latvian State University is officially opened by the Minister of Education, K. Kasparson.

1918–20   Baltic states wage war to defend independence

Soviet Russia attempts to permanently annex Latvia and the remaining Baltic states under the communist system.

1920–40   Riga Artists' Group is active

An association of modernist painters and sculptors called the Riga Artists' Group actively promotes *avant-garde* (experimental and modern) art in Latvia between the world wars.

Included among its members are Jekabs Kazaks, Romas Suta, and Uga Skulme. During its existence, the group sponsors thirteen exhibition of modern artworks.

1920: August 11   Peace treaty with Soviet Russia

A peace treaty is signed with Russia, which is forced to recognize the independence and sovereignty of Latvia and renounce all territorial claims in it.

1920   First Latvian opera is staged

Banuta, by J. A. Kalnins, becomes the first Latvian opera ever produced.

1920–22   Land reform

Land reform is carried out in the Baltic states; democratic constitutions are introduced.

1921   Opera is based on famous Latvian tragedy

Composer Janis Medins writes an opera based on Janis Rainis's classic work, *Fire and Night*.

1921   Latvia is admitted to the League of Nations

Latvia joins the League of Nations, an organization of states established at the close of World War I and dedicated to the preservation of peace.

1922: February 15   Constitution is adopted

The *Satversme* (Latvian Constitution) is adopted to provide basic rights and freedoms.

According to the terms of the new constitution suffrage is universal for Latvian citizens eighteen years or older. The government is divided into three branches: the Executive; the Legislative, which includes the *Saeima* or Parliament; and the Judicial branch. The legal system is based on civil law.

1929   Death of author Janis Rainis

Influential poet, playwright, and journalist Janis Rainis dies. (See 1865.)

1930s   Standard of living among highest in Europe

Latvia prospers and attains one of the highest standards of living in Europe. Many Latvians who emigrate to the West during Russian Imperial rule return. Under Russian rule the Latvian economy is concentrated in its coastal ports for trade purposes. As an independent nation, Latvia shifts its attention from the East as it seeks new markets in Western Europe.

1934: May   State of emergency declared

Karlis Ulmanis, one of the founders of independent Latvia, who has served as prime minister several times, declares Latvia to be in a state of emergency. He introduces an author-

League of Nations

Formed in the wake of World War I, the League of Nations—the forerunner of the United Nations—was the world's first international organization in which the nations of the world came together to maintain world peace. Headquartered in Geneva, Switzerland, the League was officially inaugurated on January 10, 1920. During the life span of the organization, over sixty nations became members, including all the major powers except the United States. Like the United Nations, the League of Nations pursued social and humanitarian as well as diplomatic activities. Unlike the United Nations, the League was primarily oriented toward the industrialized countries of the West, while many of the regions today referred to as the Third World were still the colonial possessions of those countries.

The League's structure and operations, which were established in an official document called the *League of Nations Covenant,* resembled those adopted later for the United Nations. There was an Assembly composed of representatives of all member nations, which met annually and in special sessions; a smaller Council with both permanent and nonpermanent members; and a Secretariat that carried out administrative functions. The League's procedures for preventing warfare included arms reduction and limitation agreements; arbitration of disputes; nonaggression pledges; and the application of economic and military sanctions.

Although the League of Nations solved a number of minor disputes between nations during the period of its existence, it lost credibility as an effective peacekeeper during the 1930s, when it failed to respond to Japan's takeover of Manchuria and Italy's occupation of Ethiopia. Ultimately, the organization failed in its most important goal: the prevention of another world war. However, it did make contributions in the areas of world health, international law, finance, communication, and humanitarian activity. It also aided the efforts of other international organizations.

By 1940 the League of Nations had ceased to perform any political functions. It was formally disbanded on April 18, 1946, by which time the United Nations had already been established to replace it.

itarian regime, resulting in the suspension of parliament, the Constitution, and the banning of all political parties in an effort to fend off attempts by the Soviet Union and Nazi Germany to seize control of Latvia.

World War II

1939: August 23   Nazi-Soviet nonaggression pact signed

The political challenges from Soviet Russia and Nazi Germany cannot be countered by Latvia's dictatorship. Under the terms of the Nazi-Soviet Nonaggression Pact (Molotov-Ribbentrop Treaty), Latvia is assigned to the Soviet sphere of influence along with Estonia, while Germany takes control of Lithuania. By October Lithuania is added to the Soviet sphere.

1939: October 5   Latvia signs the Pact of Defense and Mutual Assistance

Latvia and the other Baltic states are forced into signing treaties allowing Moscow to station troops on their soil. Latvia must accept occupation by 30,000 Soviet troops.

1940   First full-length film is produced

The first feature-length Latvian film, *Zvejnieka dels* (The Fisherman's Son), appears. Although the origins of Latvian cinematography can be traced to the 1920s, it is not until the Soviet years that Latvian film achieves success. The state-owned Riga Film studio produces several films yearly. Documentaries are especially popular. The most famous producers of documentaries are Juris Podnieks for *The Soviets,* Ivars Seleckis for *Crossroads,* and Herc Franks for *The Supreme Court.*

1940: June 17   Soviet troops invade Latvia

Latvia's de facto sovereignty and independence officially end when it is occupied by the Red Army. A pro-Soviet government is formed based on sham elections. A totalitarian reign of fear descends on each of the Baltic states during the first year of Soviet occupation.

1940: July 21   Soviet republic is established

The Latvian Soviet Socialist Republic is formed.

1941: June 13–14   Soviets deport Latvians to Siberia

Soviet authorities arrest and deport tens of thousands of Latvians to Siberia.

Many of those deported belong to the well-educated Latvian upper class and are considered a threat to communism. Women and children are herded into cattle cars riding to Siberia. Most of the deportees die. In all, Latvia loses over

35,000 people to deportations and executions during this short period.

1941: July–45   Nazi occupation

In July the invasion of Russia by Adolph Hitler's army places the Baltic states under total German occupation. The Nazi regime institutes compulsory draft of all Baltic peoples into labor or military service. Jews and Gypsies are annihilated. By the end of World War II Latvia has lost one-third of its population. Nationalist and communist resistance movements remain active.

1944–45   Soviet forces reoccupy Latvia

Hundreds of thousands of refugees flee to the West by way of the Baltic Sea or Gulf of Riga as Latvians face mass deportation to Siberia. As the Latvian population is systematically depleted, thousands of ethnic Russians are moved into Latvia to assume positions of power in government, business, and industry. Religious organizations are closely scrutinized by state authorities, and religious believers are harassed and persecuted. The clergy is persecuted, and church properties are confiscated. Major churches in important cities in Latvia are turned into museums or concert halls. Over time ethnic Latvians find themselves in the minority, particularly in cities like Riga. An estimated 400,000 Russians migrate to Latvia over the next forty years.

Latvia under Soviet Control

1945–52 Anti-Soviet guerrilla war continues in the Baltic states

For several years following the end of World War II, Latvian rebels wage an unsuccessful battle to oust its Soviet rulers. The Communist Party of Latvia (CPL) holds all political power, and the repression implemented by Soviet leader Josef Stalin (1879–1953) continues.

1945: January 14   Death of artist Vilhelms Purvitis

Influential painter and educator Vilhelms Purvitis dies in Germany. (See 1872.)

1947–51 Collectivization of agriculture in the Baltic states begins

Forced collectivization of agriculture is the benchmark of Soviet control. By 1949 Latvian farms are *collectivized* (removed from private ownership and control). Industry is also forcibly regulated by the Soviet government, and Latvia grows to be one of the most prosperous areas in the Soviet sphere of influence. Machine building and metal fabricating became two of the leading manufacturing industries.

1949: March 25   Mass deportations are carried out

To break the Latvians' resistance to Soviet control and agricultural collectivization, over 40,000 Latvians are sent to Siberia in the largest mass deportation carried out in Latvia.

1953   Repression eases after Stalin's death and economy grows

Ethnic Russians account for one-third of Latvia's population and provide a skilled labor force. Latvia continues to move ahead economically and to prosper. Manufacturing plants are built in smaller cities. Heavy industry builds ships, streetcars, power generators, diesel motors, and agricultural tools. Light industry concentrates on textiles, shoes, and clothing. Consumer goods such as refrigerators and washers increase, as do services.

Transportation services improve and expand throughout Latvia as extensive railroads and highways are constructed. Riga expands as a seaport and develops air links with Moscow and other Soviet cities. In all, Soviet occupation leads to Latvia's development as a fully industrialized nation.

1957–59   Local communist leaders launch Latvianization

With the era of Stalinist repression ended, a group of Latvian national communists, led by government minister Eduards Berklavs, attempts to counter the Russification of Latvia by imposing new measures that support Latvia's ethnic identity, including immigration restrictions and requirements that the Latvian language be used by government officials.

1959: July   Moscow cracks down on Latvian reformists

The Soviet government in Moscow disapproves of Latvia's new nationalist policies and institutes a purge of some 2,000 Latvian communists.

1982: September   NSRD artists' group is formed

Latvian artists, designers, and architects come together in the Workshop for the Restoration of Unfelt Sensations (Latvian acronym NSRD), an avant-garde group dedicated to introducing new forms of artistic expression, in visual, performance, and multidisciplinary art, to the Latvian public. Founded by Juris Boiko (b. 1954) and Hardijs Ledins (b. 1955), NSRD is associated with the political drive toward independence that gathers strength during the 1980s. In the seven years of its existence, the group produces three exhibitions; musical performances, recordings, and videos; and other projects.

1985   Soviet leader Mikhail S. Gorbachev introduces liberal policies

Beginning in the mid-1980s the Soviet Union enters a period of unprecedented economic and political crisis which threatens the very foundation of communist rule. *Glasnost,* a Russian term for public discussion of issues, and *perestroika,* which means restructuring, form part of Gorbachev's cam-

During Soviet occupation in the 1950s, production of consumer products increases. Washing machines await shipment at the Riga Electrical Engineering Works. (EPD Photos/CSU Archives)

paign in the Soviet Union to revitalize the economy, the Communist Party, and society by adjusting political, social, and economic mechanisms. The ramifications of these policies lead to the demise of the U.S.S.R. and freedom for the Baltic states.

1988: October   Popular front holds its first congress

The Popular Front of Latvia (*Latvijas Tautas Fronte,* or LTF), numbering more than 100,000 members throughout the republic, holds its first congress.

1989: May 5   Latvian Language Law is passed

Due to Soviet occupation and an intense policy of Russification, fewer than one-fourth of non-Latvians living in Latvia speak Latvian, and the Latvian language is in danger of becoming extinct. During the national reawakening of the 1980s a petition drive collects 354,000 signatures from Latvians and non-Latvians demanding that Latvian be named the official state language.

1989: July   Latvian Supreme Soviet adopts declaration of sovereignty

Under pressure by the Popular Front of Latvia, the Latvian Supreme Soviet decides to end the monopoly on political power held by the Communist Party in Latvia, clearing the way for the emergence of independent political parties and for Latvia's first free parliamentary elections since 1940.

1990: May 4   Transition to independence is approved

The Latvian Supreme Council votes for a transition to independence with the Declaration of the Renewal of the Independence of the Republic of Latvia.

Independence is Restored

1991: August 21   Latvian Supreme Council votes for full independence

Following the failed coup against the Gorbachev government in Moscow, the Baltic states restore diplomatic relations with many foreign countries. Latvia reestablishes de facto independence, and the authority of the constitution is recognized. Latvia moves to reestablish diplomatic relations with the West.

1991: September 6   Latvian independence is recognized

The U.S.S.R.'s central government in Moscow recognizes the independence of Latvia, along with that of the other two Baltic republics.

1991: October   Latvia is admitted to the UN

Latvia becomes a member of the United Nations.

1992: May 5   Language reforms take effect

The Language Center of the Cabinet of Ministers of Latvia is founded, and the Language Law goes into effect.

1993: June 5–6   Parliamentary elections are held

The first truly democratic elections since the 1930s are held for seats in the single-chamber Saeima, or parliament. It is composed of one hundred members who are elected in direct, proportional elections by citizens eighteen years or older. Political parties must receive at least five percent of the national vote to win seats in the Saeima.

1993: July 6   The Saeima (Parliament) restores 1922 constitution

Latvia's newly elected parliament restores the constitution that is in effect during the period of Latvian independence between the two world wars.

1993: July 7   Ulmanis is elected president

H. E. Guntis Ulmanis, of the Latvian Farmers Union, is elected president of Latvia. The president, elected by the Saeima for a three-year period, is Latvia's head of state.

1994: May   Nationalist party wins elections

The Latvian National Independence Movement finishes first in Latvia's first post-Soviet local elections.

1995: February   Latvia is admitted to the Council of Europe

After abandoning restrictive quotas on naturalization, Latvia is granted admission to the Council of Europe, which is founded in 1949 to oversee intergovernmental cooperation in areas such as environmental planning, finance, sports, crime, migration, and legal matters.

1995: Sept–October   Democratic party leads in parliamentary elections

The Democratic Party, *Saimnieks,* finishes first in Latvian parliamentary elections, followed closely by the far-right party, For Latvia.

1996: June 18   Ulmanis is elected to second term as president

President H .E. Guntis Ulmanis is reelected to a second three-year term as president by Latvia's parliament.

1998   Latvia becomes the first Baltic state to join the WTO

Latvia, which pursues a liberal foreign trade policy, joins the World Trade Organization. Plus it is an active member of the Council of Baltic Sea States (CBSS), working towards regional co-operation between the Baltic states. Latvia also promotes inter-Baltic co-operation and a free-market econ-

omy in a move to create a Baltic common market similar to the EU.

1998: October 3   People's Party leads in parliamentary elections

Over seventy percent of Latvian voters turn out for elections for the Saeima. With twenty-one percent of the vote, the People's Party wins twenty-four seats, followed by Latvia's Way with twenty-one. Andris Skele, the former prime minister who has resigned in 1997, regains his office. Altogether, six parties win enough votes to hold seats in the Saeima. Results for the remaining parties include LNNK (Fatherland and Freedom Party), seventeen seats; Latvian Social Democratic Alliance, fourteen; National Harmony Party, sixteen; and the New Party, eight seats.

1998: October 3   Citizenship laws are amended

In a national referendum held at the same time as parliamentary elections, Latvian citizenship laws enacted following independence in 1991 are modified in response to pressure by Latvia's ethnic Russians and the international community. Under the amended law, quotas on annual naturalizations are lifted, citizenship tests are simplified for persons over the age of sixty-five, and all persons born in Latvia after 1991 are automatically granted citizenship status.

1999: February 25   Joint military college opening is announced

Estonia, Latvia, and Lithuania officially announce the opening of a joint military college in Tartu, Estonia, in the hope that it will increase their chances of being accepted into NATO. Staffed by NATO officers, classes at the new Baltic defense college, informally called BALTDEFCOL, are slated to begin in August, 1999.

Bibliography

Gerner, Kristian, and Stefan Hedlund. *The Baltic States and the End of the Soviet Empire.* London and New York: Routledge, 1993.

Hiden, John and Patrick Salmon. *The Baltic Nations and Europe.* London and New York: Longman, 1994.

Iwaskiw, Walter R. *Estonia, Latvia, and Lithuania: Country Studies.* Washington, D.C.: Federal Research Division, Library of Congress, 1996.

Karklins, Rasma. *Ethnopolitics and Transition to Democracy: The Collapse of the USSR and Latvia.* Baltimore: Johns Hopkins University Press, 1994.

Lieven, Anatol. *The Baltic Revolution.* New Haven and London: Yale University Press, 1993.

Neimanis, George J. *The Collapse of the Soviet Empire : A View from Riga.* Conn.: Praeger, 1997.

Plakans, Andrejs. *The Latvians: A Short History.* Stanford, Calif: Hoover Institution Press, Stanford University, 1995.

Von Rauch, George. *The Baltic States: The Years of Independence.* Berkeley and Los Angeles: University of California Press, 1970.

Liechtenstein

Introduction

Liechtenstein is a small, landlocked principality lying along the Rhine River and located between Switzerland and Austria. In terms of area, at 160 square kilometers, it is the fourth-smallest country in Europe. It has a population of 31,000, of which over one-third is foreign-born. The inhabitants of Liechtenstein, called Liechtensteiners, are descendants of the Germanic Aleman invaders of the fifth century A.D. They speak a German dialect and are predominantly Roman Catholic. The government is a constitutional monarchy in which the prince wields considerable executive authority. The capital city of Vaduz has a population of under 5,000. The most populous city is Schaan with slightly over 5,000 inhabitants.

Although it occupies a very small area with little agricultural or natural resource wealth, Liechtenshtein has a prosperous economy and is one of the wealthiest states in Europe. The strengths of the principality's economy lie in the industrial and commercial sectors. Liechtenstein is a magnet for foreign investment due to a minimum of restrictions (and many tax breaks). Like neighboring Switzerland, Liechtenstein also offers secret bank accounts.

As a result of its small size, Liechtenstein has had to rely upon its neighbors for military protection and diplomatic representation. Its long association with the Holy Roman Empire caused Liechtenstein to contribute forces to the empire, and as a consequence, it often suffered from foreign invasion. Although the principality disbanded its military forces in 1868, its association with stronger neighbors has continued to play a leading role in its historical and economic development. Its association with losing Austria-Hungary during World War I nearly collapsed Liechtenstein's economy. In order to prevent such a calamity in the future, the principality turned to perpetual neutral Switzerland for an economic and diplomatic relationship. Liechtenstein avoided starvation during the Second World War and thrived within the booming postwar Swiss market. However, the onset of European integration in the 1990s has forced Liechtensteiners to question their continued association with a Switzerland that remains outside the European Union.

The continued existence of an independent Liechtenstein is an anomaly that is the result of peculiar historical circumstances. The region of modern Liechtenstein was unorganized until the Roman conquest in 14 B.C. Roman rule lasted for over four centuries until the Germanic Aleman tribes invaded the area in the fifth century A.D. These Germanic peoples were the direct ancestors of today's Liechtensteiners and have had the greatest cultural and historical impact upon the region. Frankish invaders took over the region in the sixth century. In the ninth century the territory of modern-day Liechtenstein became a part of the subcounty of Lower Rhaetia. The region around the present capital of Vaduz was not even established until the fourteenth century. Indeed, until the eighteenth century, a host of families ruled the area of modern Liechtenstein within the Holy Roman Empire. When war broke out, armies crossed the tiny area at will. Not until 1719 would the Principality of Liechtenstein be formed, having taken its name from Prince Adam of Liechtenstein who acquired the Schellenberg and Vaduz regions from the von Hohenem family in 1699 and 1712, respectively.

The Napoleonic Wars saw Liechtenstein invaded for the final time. Ironically, the campaigns of Napoleon did not serve as a catalyst for the development of nationalism among the principality's population. Unlike other German states which had become part of the Napoleonic-era Confederation of the Rhine and the post-Napoleonic Germanic Confederation, no strong German national organization emerged in the Principality of Leichtenstein. Perhaps this was due to the tiny state's geographic location. Situated between strongly multi-ethnic Switzerland and multi-national Austria, the Liechtensteiners were not in close contact with pan-Germanists from the north. Indeed, the Revolutions of 1848 passed nearly unnoticed.

During the second half of the nineteenth century, Liechtenstein embarked upon major internal improvements and modernization under the benevolent leadership of Prince Johannes the Good. A constitution was promulgated, the army abolished, and the arts patronized. In addition, peasants received financial relief in the abolition of the feudal system (which persisted in Liechtenstein despite the abolition of serfdom earlier in the century). It was also a period of pride for artistically minded Liechtensteiners. This period was the

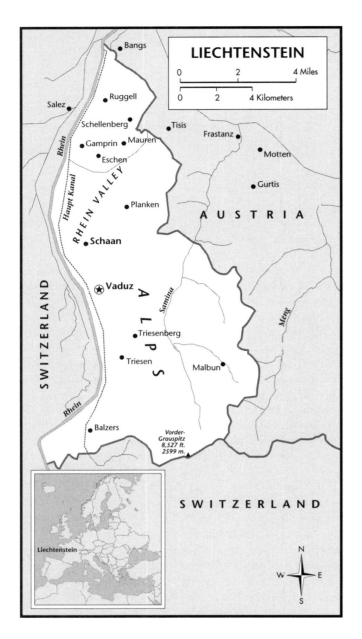

height of renown for Liechtensteiner composer and musician Josef Gabriel Rheinberger.

The continued economic relationship with Austria proved detrimental during the First World War. However, with the aid of a 75 million Swiss Francs gift from Prince Johannes and a shift in economic ties to Switzerland Liechtenstein set upon the path of recovery. Following the end of World War II, the principality's economy has grown rapidly.

Today, the industrial and tourist sectors form the largest segments of the principality's economy. Low tax rates—company taxes are no higher than 15 percent—entice many business to set up shop in the micro-state. Banking is another major business concern. Tourists also flock to Liechtenstein—partly, out of the novelty of visiting such a small coun-try, partly out of the state's attractions. Liechtenstein's stamps are world-renowned, and the principality's rulers have amassed a major art collection. Agriculture, once a mainstay of the economy, today accounts for less than five percent of all employment.

Timeline

c. 3000 B.C.   First settlement

Farmers and cattle raisers settle in the region of modern-day Liecthenstein. They become the first permanent inhabitants of the region.

14 B.C.   Roman conquest

Rome conquers the area of modern Liechtenstein. This region is incorporated into the Roman province of Rhaetia. During their five centuries of rule, the Romans introduce the grape-vine, better construction methods, and, in the later centuries of their rule, Christianity.

c. A.D. 450   Aleman invasions

Groups of Germanic Aleman invaders sweep through the region. The Alemani are the ancestors of today's Liechten-steiners. However, they transplant Roman influence slowly. Indeed, for centuries Romans and Alemanis coexist in the region, and Rhaetoroman remains the language of the area for hundreds of years. Not until the twelfth century does German become the language of the region.

600s   Frankish rule

The Franks, another Germanic people, enter the region and establish control over the area. Under their great leader of the eighth century, Charlemagne, the Frankish kingdom extends from modern-day France into central Europe and becomes known as the Holy Roman Empire.

814   Creation of subcounty of Lower Rhaetia

The area of modern-day Liechtenstein is incorporated into the newly-established subcounty of Lower Rhaetia. This region falls under the control of a Frankish count.

1342   County of Vaduz established

The county of Vaduz is established. It will become the most important city in Leichtenstein.

1434   Unification of Vaduz and Schellenberg regions

The Vaduz and Schellenberg regions are unified under the rule of the Holy Roman Emperor.

1510 Sale of Vaduz and Schellenberg

The counties of Vaduz and Schellenberg are sold to Rudolf von Sulz. Rudolf proves to be an able ruler during a period of widespread conflict. His leadership halts the spread of Protestantism in the area and prevents the spread of the German peasant wars into the region.

1613 Counties sold to Count Caspar von Hohenems

Count Carl Ludwig sells the counties of Vaduz and Schellenberg to Count Caspar von Hohenems, his son-in-law. Count Caspar comes from a military family, and his rule during a time of upheaval turns the counties of Vaduz and Schellenberg into a region of conflict.

1618–48 Thirty Years' War

Even tiny Liechtenstein does not escape the effects of the Thirty Years' War, a conflict that lays waste to most of Germany. Both Habsburg and Swedish forces march through Liechtenstein. Following the Swedish incursion, Liechtenstein pays a tribute to the Swedes.

The Thirty Years' War and its immediate aftermath are disastrous for Liechtenstein. War brings disease and an upsurge in religious fanaticism. In a three-decade-long witch-hunt, 300 Liechtensteiner men and women are killed. Liechtenstein's population at the time is roughly 3000.

1699 Purchase of Schellenberg

Prince Johann Adam Andreas (1657–1712) of Liechtenstein buys the region of Schellenberg from the von Hohenem family. Prince Johann Adam's reign marks the beginning of the Liechtenstein dynasty.

1712 Vaduz acquired, succession of Josef Wenzel

Prince Johann Adam buys Vaduz. The acquisition of Schellenberg and Vaduz establishes Leichtenstein's modern borders. At this time the Liechtenstein family also owns territory in Bohemia, a region under the Habsburg Empire.

Upon the death of Prince Johann Adam, Crown Prince Josef Wenzel (1692–1772) succeeds to the throne. Josef Wenzel has an illustrious career as both diplomat and soldier for the Habsburgs. In the former capacity, he serves with distinction in Berlin and Paris, while in his military capacity he serves against the Turks, Spanish, and French. Ultimately, he makes the rank of field-marshal.

1718 Prince Anton Florian receives Liechtenstein

In an exchange with his nephew Prince Josef Wenzel, Habsburg diplomat Anton Florian (1656–1721) receives Liechtenstein and becomes prince.

1719: January 23 Principality of Liechtenstein created

An act of Holy Roman Emperor Charles VI establishes the Principality of Liechtenstein. The ruler of the principality receives the title Prince von und zu Liechtenstein.

1721 Death of Prince Anton Florian,

Prince Anton Florian dies and is succeeded by Josef Johann Adam (1690–1732).

1732 Prince Josef Johann Adam dies, Johann Nepomuk Karl succeeds

Upon the death of Josef Johann Adam, Johann Nepomuk Karl (1724–48) succeeds to the throne.

1748 Prince Johann Nepomuk Karl dies, return of Josef Wenzel

Prince Johann Nepomuk Karl dies. Since he leaves no heir, the throne reverts to Josef Wenzel who had exchanged the principality with his uncle thirty-six years before.

1772 Prince Josef Wenzel dies, Franz Josef I succeeds to throne

Prince Josef Wenzel dies and Crown Prince Franz Josef I (1726–81) succeeds him.

1781 Accession of Alois I

Upon the death of Prince Franz Josef I, Crown Prince Alois (1781–1805) accedes to the Liechtensteiner throne as Prince Alois I.

1799 Last foreign occupations

French and Russian troops occupy Liechtenstein during the Wars of the Coalition. This is the last time foreign forces occupy the principality. These wars pit the conservative dynasties of Europe against revolutionary France.

Early 1800s Napoleonic Wars

At different points in the wars, both France and Russia invade the principality but do not stay.

1805 Treaty of Pressburg, Johannes I accedes to the throne

The Treaty of Pressburg assigns Liechtenstein to the Confederation of the Rhine, a collection of German states created by Napoleon for ease of French domination.

Upon the death of Prince Alois I, Crown Prince Johannes I (1760–1836) accedes to the principality's throne. Johannes I has an illustrious career as an officer in the Austrian army. He fights countless battles against the forces of Napoleon, and as recognition for his service, he is subsequently made a field-marshal.

1806   Confederation of the Rhine

France becomes a part of the Confederation of the Rhine. This is a grouping of the western Germanic states formed by Napoleon Bonaparte after his conquest of the region. The confederation results in a radical restructuring of German states that reduces their number from over 300 to 39. The reduction in the number of states also has the effect of promoting German nationalism. Liechtenstein, however, maintains its integrity and is not absorbed into a larger German state.

1808   Serfdom abolished

The principality abolishes serfdom. Following this reform, all peasants are free as they are no longer tied to landed estates.

1809   Land register created

The land register is created. The creation of this institution enhances the principality's credit system.

1815   Germanic Confederation

Napoleon's defeat leads to the dissolution of the Confederation of the Rhine. In its place, the great powers create the Germanic Confederation, a larger grouping of German states that includes parts of Prussia and Austria, as well as the lesser German states. Liechtenstein becomes a part of the Germanic Confederation and, once again, maintains its territorial integrity.

1818   Constitution adopted

Liechtenstein adopts its first constitution.

1836   Prince Johannes I dies, Alois II succeeds

Upon the death of Prince Johannes I, Crown Prince Alois II (1796–1858) accedes to the throne.

1839: March 17   Josef Gabriel Rheinberger born

Joseph Gabriel Rheinberger (1839–1901) is born in Vaduz. In the course of the nineteenth century, Rheinberger, a composer and musician, becomes Liechtenstein's most famous citizen. After completing early studies in music under Philipp Schmutzer in Feldkirch, he travels to Munich in 1851 to study theory under Julius Josef Mayer, the organ under Georg Herzog, and the piano under Emil Leonhard.

Rheinberger is a prodigy who, at the age of thirteen, becomes a deputy organist at the Court Church of St. Michael and at fifteen becomes *repetiteur* with the chorus of the Munich Oratorio Society. In 1859, he is named a teacher at his old school, and five years later becomes the Director of the Oratorio Society and is the soloists' coach at the court theater. International recognition follows along with a host of honors. In 1877 Rheinberger is made Chapel master to the Hight Court. Subsequently, he becomes a Knight of the papal order of St. Gregory, a Knight of the Bavarian crown, and a *Geheimrat* (Privy Councillor). In 1867 Rheinberger marries Fanny von Hoffnaas, a poet, and dies in 1901.

1842   Visit of Alois II

Prince Alois II becomes the first Liechtensteiner prince to visit the principality.

1848   Liechtenstein in the revolutionary year

Liechtenstein is represented in the abortive all-German parliament in Frankfurt. The principality's troops, however, are not used in a fashion sympathetic to the revolutionaries. At the behest of the Habsburgs, they are used to preserve order in Baden.

1852   Economic unity with Austria

Liechtenstein joins in a customs union with Austria. This relationship lasts until the Habsburg defeat in World War I.

1853   Prince Franz born

A son is born to Prince Alois and his wife. Prince Franz succeeds his brother as prince in 1929 and reigns for nine years.

1858   Prince Alois I dies, succeeded by Johannes II

Prince Johannes II (1840–1929) ascends to the throne. Known as Johann the Good, Father of the People, Helper of the Poor, Friend of Peace, Protector of the Arts, he rules for 71 years. Johannes proves to be an activist monarch. He plays a leading role in the constitutional development of Liechtenstein and helps promulgate two constitutions.

However, he is best remembered for his role as a benefactor. As prince, he promotes the arts and education. He also promotes the social well-being of the principality and donates 75 million Swiss francs to the country after World War I.

1862   New constitution

Liechtenstein adopts a second constitution. This document leads to the creation of a constitutional government in which the *diet* (assembly) gains the right to vote on the budget and initiate legislation.

1866   Dissolution of the Germanic Confederation, major newspaper founded

Following Prussia's triumph over Austria in the Austro-Prussian War, Prussia disbands the Germanic Confederation.

The principality's first major newspaper, the *Liechtensteiner Volksblatt* (Liechtenstein People's Paper), is founded in Vaduz. A conservative paper in the twentieth century, it becomes the newspaper of the Progressive Citizens' Party.

1868   Abolition of Liechtensteiner army

Liechtenstein permanently disbands its army.

1901: November 25   Josef Gabriel Rheinberger dies

Josef Gabriel Rheinberger, Liechtenstein's musical virtuoso, dies at the age of sixty-two.

1912: February 1   First stamps

The principality issues its first stamps. Eventually, Liechtensteiner stamps become one of the country's chief exports and sources of foreign currency. Philatelists the world over prize the unique stamps.

1914–18   World War I

Liechtenstein stays neutral during the World War I although it remains tied to Austria, a member of the defeated Central Powers, until the end of the conflict. The economic relationship with Austria during the war results in the economic collapse of the principality. As a result of this experience, Liechtenstein reconsiders its relationship with Austria after the war and embarks upon a closer relationship with Switzerland.

1918   Political parties formed; Bohemian lands of Liechtenstein house siezed

Wilhelm Beck, a lawyer, forms the first Liechtensteiner political party, the *Volkspartei* (People's Party), which opposes the government. The government's supporters respond in kind with the formation of the *Burgerpartie* (Citizens' Party).

The newly formed Czechoslovak state siezes the Liechtenstein holdings in Bohemia (620 square miles). It is a major loss for the family, particularly since these lands are of greater area than the Principality of Liechtenstein. The issue remains unresolved in the late 1990s.

1921: October 5   New constitution

Prince Johannes II promulgates a new constitution for Liechtenstein. Under this document, power is based both in the hands of the hereditary monarch and the people. The prince serves as head of state. In this capacity, he appoints state officials, sanctions laws, assembles and dissolves the Diet, and can pardon criminals. In addition, the prince has the right to issue decrees under emergency circumstances. The Diet ratifies treaties, nominates state officials, and enacts legislation. The members of the Diet are elected by universal male suffrage of Liechtensteiners (women do not get the vote until later).

In addition to the powers delegated to the people through their representatives in the Diet, the constitution allows for the right of initiative and referendum. The signatures of 600 voting-eligible citizens or resolutions by three communal assemblies are required for initiative, while 900 voting-eligible citizen signatures or resolutions by four communal assemblies is required for amending the constitution. A referendum is necessary for ratification of certain revenue bills as well as initiatives rejected by the Diet.

Vaduz is the site of a new coalition government that seeks to preserve the independence of the tiny principality ruled by the eighty-five-year-old Francis I. (EPD Photos/CSU Archives)

1923: March 29   Customs Treaty signed with Switzerland

Liechtenstein enters into a customs union with Switzerland. It takes effect on January 1, 1924. The Swiss franc becomes the currency of the principality. The Swiss also take over Liechtenstein's defense and diplomatic representation. These ties continue into the 1990s.

1929: February 11   Prince Johannes II dies, School system reorganized

Prince Johannes II dies after a seventy-one-year reign, by far the longest in Liechtenstein's history. He is succeeded by his seventy-nine-year-old brother, Crown Prince Franz I (1853–1938). Franz I, a former Austrian ambassador to tsarist Russia, is the first prince who resides in Liechtenstein for a considerable time.

The country's school system is reorganized. An eight-year period of compulsory education is established. Control of the schools remains under Catholic guidelines.

1938   Accession of Franz Josef II

Franz Josef II (1906–89) becomes prince. He is extremely popular and becomes Europe's longest reigning monarch.

1938: March   Liechtenstein threatened

Nazi Germany, under Adolf Hitler, annexes neighboring Austria. This action puts an expansionist Germany on Liechtenstein's doorstep. Fearing Nazi encroachment, a coalition government is formed in Vaduz. This new alignment seeks to preserve the principality's independence. The German annexation of Austria also forces newly crowned Prince Franz Josef II to set up permanent residence in the principality. The prince's move makes him the principality's first ruler to reside within its borders.

1939–45   World War II

Liechtenstein proclaims its neutrality during the Second World War. Unlike its experience during World War I, Liechtenstein avoids economic hardship during the second conflict. As a result of its ties to Switzerland, the principality is not subject to Allied blockade, and its population remains well-fed.

1945: February 14   Prince Hans Adam born

A son is born to Prince Franz Josef II and his wife, Princess Georgine Wilczek.

1961: October 5   National Library

A National Library is founded in Vaduz. The library serves as a repository of all books published in Liechtenstein as well as a center for technical and educational materials for the principality's population.

1968   Prince Alois born

A son is born to Crown Prince Hans Adam and his wife, Prince Alois.

1984   Crown Prince Hans Adam assumes executive powers

Crown Prince Hans Adam (b.1945) assumes the executive functions from his father, Prince Franz Josef II. A businessman by day, the crown prince seeks to attract ever more foreign capital to the principality.

1989: November 13   Death of Franz Josef II, accession of Hans Adam II

Prince Franz Josef II dies at age eighty-three after a reign of fifty-one years. Crown Prince Hans Adam succeeds as Prince Hans Adam II. The first decade of his reign features renewed Liechtensteiner assertiveness as he tries to guide his country along the path of European Unity. This policy alters Liechtenstein's relations with Switzerland.

1986   New penal code

Liechthenstein enacts a new penal code. Among the provisions is the aboltion of the death penalty.

1991   UN, EFTA membership

Liechtenstein forms EFTA (European Free Trade Area), a free trade grouping of non-European Community nations.

In September, Liechtenstein becomes a member of the United Nations (UN), an organizaiton to which Switzerland does not belong.

1992   Women's equality

A women's equality amendment is added to the constitution.

1995   EEA membership, transfer of diplomatic affairs

Liechtenstein joins the European Economic Area (EEA), another organization of which Switzerland is not a party.

Liechtenstein transfers the oversight of its diplomatic affairs from Bern to Brussells.

1995: May 24   Prince born

A son, named Prince Joseph Wenzel Maximilian Maria, is born to the Hereditary Prince Alois and his wife, Hereditary Princess Sophie of Liechtenstein, nee Duchess of Bavaria.

1995: June 6   Stamps issued to honor stamp designer and letter writing

Three commemorative stamps are issued to honor the 100th anniversary of the birth of Canon Anton Frommelt (1895–1975), stamp artist. The stamps depict Liechtenstein scenery and are part of a series of stamps entitled "Painters from Liechtenstein." Frommelt becomes a priest of Triesen in 1922. Six years later, he is elected to the Diet (Landtag), the Liechtenstein parliament. He serves as president of the Diet from 1928–45. When he retires from politics in 1945, Frommelt designs several stamp series and organizes stamp exhibitions.

Another series of stamps is entitled "Letters as Joyous Messages" to celebrate the tradition of letter writing. The present stamp issue is entitled "Fun With Letters," with individuals designs titled "Write Again Soon," "Bosom Friends," "Hurray, I'm Here," and "All the Best."

1996   Liechtenstein joins the EEA

The citizens of Liechtenstein vote to join the European Economic Area (EEA). Prince Hans-Adam II leads the movement to join the EEA. Bankers in Liechtenstein oppose joining the EEU, fearing the loss of traditional secrecy laws. EEA follow certain guidelines about sharing information which could discourage certain depositors who wish to keep financial assets secret from doing business in the country.

1996: October 17   Princess born

The Princess Marie Caroline is born to Prince Alois and Princess Sophie.

1997 Dispute pits prince against parliament

A rare constitutional debate rages over the rights to judicial appointments. The prince and the parliament both claim the right to appoint judges. This conflict is one of the few instances of serious constitutional debate in the principality. However, it exemplifies the controversial reign of Hans Adam II.

Bibliography

Background Notes: Liechtenstein. Washington, D.C.: U.S. Department of State, Bureau of Public Affairs, Office of Public Communication, Editorial Division, USGPO, 1989.

Bunting, James. *Switzerland, including Liechtenstein.* London: B.T. Batsford, 1973.

Carrick, Noel. *Let's Visit Liechtenstein.* London: Burke, 1985.

Duursma, Jorri C. *Fragmentation and the International Relations of Micro-States: Self-determination and Statehood.* Cambridge: Cambridge University Press, 1996.

Kohn, Walter S. G. *Governments and Politics of the German-speaking Countries.* Chicago: Nelson-Hall, 1980.

Moore, Russell Franklin. *Principality of Liechtenstein: A Brief History.* With an introduction by Baron Edward von Falz-Fein. New York: Simmons-Boardman Publishing Corporation, 1960.

The Principality of Liechtenstein: A Documentary Handbook. Vaduz: Press and Information Office of the Government of the Principality of Liechtenstein, 1967.

Raton, Pierre. *Liechtenstein: History and Institutions of the Principality.* Vaduz: Liechtenstein-Verlag AG, 1970.

Seger, Otto. *A Survey of Liechtenstein History.* Vaduz, n.d.

Who's Who in Switzerland, Including the Principality of Liechtenstein. Zurich: Central European Times Publishing Co., n.d.

Lithuania

Introduction

The Republic of Lithuania (Lietuvos Respublika) is one of three countries in northeastern Europe bordering the Baltic Sea that are known as the Baltic states, the other two being Estonia and Latvia. Formerly a part of the Soviet Union, Lithuania has been an independent republic since 1991 and has maintained strong ties both to Russia and to the West.

The largest of the Baltic states, Lithuania has a total land area of 25,174 square miles (65,200 square kilometers). It is bordered by Latvia to the north, Poland and Belarus to the south, Russia to the east, and Kaliningrad (part of the Russian Federation) and the Baltic Sea to the west. Lithuania's terrain alternates between highlands and lowlands with numerous rivers and lakes in the eastern regions. Over one-third of the land is arable, or farmable, and close to one-third is covered by forests and woodlands.

During the 1990s the Lithuanian population declined because of low natural growth rates and migration. In 1998 Lithuania's population totaled 3,600,158. Vilnius, the capital city, is located in the southeastern part of the country. It is Lithuania's largest city, with an estimated population of over 630,000. Only four other cities have populations exceeding 100,000: Kaunas, in central Lithuania, 424,000; Klaipeda, in western Lithuania, 205,000; Siauliai, 84 miles (140 kilometers) north of Kaunas, 147,000; and Panevezys, 140 kilometers north of Vilnius, with over 132,000.

Lithuania is a parliamentary democracy with universal suffrage at age eighteen. Adopted on October 25, 1992 the Lithuanian Constitution divided the government into three branches: executive, legislative, and judicial, with power delegated to the will of the people.

The executive branch consists of the president, elected by popular vote to a five-year term of office as head of state; the prime minister as the head of government who is appointed by the president and subject to parliamentary approval; and the council of ministers who comprise the fourteen-member cabinet appointed by the president on the nomination of the prime minister. The legislative branch consists of the *Seimas*, a one-chamber parliament with 141 members, who serve four-year terms.

The judicial branch includes the Constitutional Court, the Supreme Court, the Court of Appeals, circuit courts and district courts. Lithuania is divided into forty-four regions/administrative units and eleven municipalities for purposes of local self-government.

Lithuanian is the official language, but Polish and Russian are also spoken. There are some forty-seven religious denominations in Lithuania. Roman Catholicism is practiced by more than eighty percent of Lithuanians. Education in Lithuania is compulsory from age six to sixteen and is free at all levels. Lithuania offers a national social security system that provides social insurance and social benefits to all citizens from birth to death. Free medical care is provided through state owned and operated facilities. Private health care became legal and available in the late 1980s. Although the doctor-to-patient ratio is above average, medical equipment, supplies, and drugs are in short supply.

Environmental issues, including air and water pollution, are major concerns due to the heavy concentration of organic waste found in the Baltic Sea and the contamination of soil and groundwater with petroleum products and chemicals at former Soviet military bases. Lithuania is party to numerous international agreements designed to tackle such problems.

Lithuanian history has been influenced by Poland, Russia, and Germany. For nearly four hundred years, Lithuanian history was closely tied to that of Poland. In medieval times Lithuania was a powerful state known as the Grand Duchy of Lithuania. It first united with Poland in the fourteenth century, and the union was completed in the sixteeth century, when the Lithuanian-Polish Commonwealth was born. (The Klaipeda region of Lithuania, known as Lithuania minor, was dominated for seven hundred years by Germans who came to Lithuania via the Crusades.) During the eighteenth century Lithuania was assimilated by Imperial Russia, which sought to eliminate all traces of Polish influence and impose a Russian political framework and social structure. The Lithuanians staged major revolts in 1830 and 1863, but Russia maintained its grip on their country.

In 1918 Lithuania took advantage of Imperial Russia's collapse to seek and establish its independence, which lasted only for the duration of the interwar period. For twenty years, until the start of World War II (1939–45), Lithuania reclaimed its identity as a separate state. This period of Lithuanian independence is often called the first golden age of Lithuanian culture. The spirit of Lithuanian nationalism was mirrored in a three-volume novel of the 1930s titled *Altoriu Sesely* (In the Altars' Shadow) by Vincas Mykolaitis-Putinas (1893–1967), a former priest, regarded as a leading literary figure of the early twentieth century. Lithuanian art, language, literature, and traditions have survived because ethnic Lithuanians retained their cohesion through the centuries, and thus their culture was not diluted, as happened in neighboring Estonia and Latvia.

At the outset of World War II, however, Lithuania and the other Baltic states were annexed by the Soviet Union and later occupied by Nazi Germany. Following the war, they reverted to Soviet control, and Lithuania remained a Soviet republic until it regained its independence in 1991 as the collapse of the Soviet Union was imminent. The privatization of small and moderate sized businesses and a majority of large businesses since independence have improved Lithuania's economy. Among its major foreign policy goals are full membership in the North Atlantic Treaty Organization (NATO) and in the European Union (EU).

Timeline

9000–3000 B.C.   Prehistoric times

The area presently known as Lithuania first shows signs of human habitation around 9,000 B.C., after the Ice Age. Ancestors of the Baltic Finn peoples move westward from eastern and central Russia and begin to occupy the coastal area around 3,000 B.C. Belief in natural deities, or paganism, emerge as the oldest form of worship.

c. 2500–1500 B.C.   Tribal groups migrate east

The Baltic coastal region attracts many different tribal groups migrating east. The proto-Baltic tribes, lured by the maritime characteristics of the Baltic area, begin to occupy the region, and the culture of the Baltic tribes takes form. They begin to trade with the Roman Empire. These tribes (Selonians, Semi-gallians, Catgallians, Couronians, Livonians, Lithuanians, Latvians, and Prussians) become independent of one other, and their ancient dialects form the foundation of the Baltic language group of which only Latvian and Lithuanian survive.

7th–12th centuries   Early Middle Ages

From the seventh to the twelfth centuries, the Baltic peoples live freely as peasants and form many tribes and communities along the Baltic coastline.

The territory of Lithuania first appears after the seventh century A.D. from the merger of several separate tribes.

1009   First recorded mention of Lithuania

The first recorded mention of Lithuania occurs in the *Annals of Quedlinburg*. It is another two centuries before Lithuania can unite the scattered tribes in order to withstand conquest by Germanic tribes.

1236–63   The Mindaugas era

Grand Duke Mindaugas (r. 1253–63) is recognized as the individual responsible for bringing organization and unity to the scattered ethnic tribes occupying the territory known as Lithuania. He succeeds in banding the people together in order to create the state of Lithuania and, in so doing, to fight the eastward expansion of the Teutonic Order of Knights, a Germanic religious and military organization which grows out of the Crusades. The goal of the Teutonic Knights is to Christianize all of Europe. By this time the Teutonic Knights are successful in conquering neighboring Latvian and Prussian tribes.

1251   Grand Duke Mindaugas adopts Christianity

Grand Duke Minaugas is converted to Christianity and become a Roman Catholic.

1253: July 6   Mindaugas is crowned King of Lithuania by Pope Gregory IX

Pope Gregory IX crowns Grand Duke Mindaugas king of Lithuania. He is the first and only king in Lithuanian history.

1263   Mindaugas is assassinated

King Mindaugas is murdered, presumably by his nobles. Their motivation is twofold: they do not agree with his decision to co-exist with (rather than annihilate) the Teutonic Knights; and they do not share his vision of opening a trade route to western Europe. The nobility prefer to expand eastward. The question of whether Lithuania should develop western as opposed to eastern political, social, economic and cultural standards is one that follows all subsequent rulers. Of the original Baltic tribes only the Lithuanians succeed in creating an independent state, the Grand Duchy of Lithuania.

1316–41   The Gediminas era

When King Mindaugas is killed the Lithuanian monarchy ends. So, too, does the influence of Christianity. Paganism now re-emerges. Grand Duke Gediminas and his descendants are able to expand the Lithuanian territories south and east toward the Black Sea. Gediminas founds the city of Vilnius and starts the Gediminas dynasty. Members of this dynasty unite in marriage with many European monarchies. Lithuania becomes an empire and extends its boundaries from the Baltic Sea to the Black Sea.

1386   Jogaila is crowned king

Jogaila, the Grand Duke of Lithuania of the Gediminas dynasty, is crowned King of Poland.

During the reign of King Jogaila (r. 1377–1392) histories of Poland and Lithuania begin to intertwine, a connection which lasts for four hundred years. The Kingdom of Poland is a Roman Catholic country, and, during Jogaila's reign, Catholicism returns to Lithuania. (See 1387.) King Jogaila decides to expand Lithuania's interests westward which results in conflict with the Teutonic Knights who also covet western trade routes.

1387   Catholicism returns

As Lithuania becomes more closely aligned with Poland under the rule of Jogaila, Roman Catholicism becomes the dominant religion in Lithuania. It remains a primarily Catholic nation until the nineteenth century. Jogaila is tolerant of other religious groups and opens Lithuania's doors to Orthodox Christians, Muslims, and Jews.

1392–1430   The Vytautas era

During the reign of King Jogaila, Vytautas (r. 1392–1430), his cousin, becomes the Grand Duke of Lithuania. Together Jogaila and Vytautas develop Lithuania into one of the largest

The Polish-Lithuanian State, 15th century.

European states by assimilating significant Eastern Slav lands, including Belorussian, Russian, and Ukrainian territories, which fortify their expansion to the Black Sea.

1410   The Battle of Tannenberg

In this battle Grand Duke Vytautas succeeds in ending several centuries of expansion eastward by defeating the Teutonic Knights. Vytautas also deals a blow to any further attempts at westward expansion by the Mongol Tartar hordes. Vytautas is the first to attempt to sever Lithuanian ties to the Kingdom of Poland and create a separate crown and dynasty. But the Polish nobility prove too powerful and keep Lithuania tied to Poland.

1539   First higher education institution opens

The first Lithuanian school of higher learning is founded by Abraomas Kulvietis, a Protestant reformer. Tension between Roman Catholics and Protestants force the school's closing three years later.

1547   The earliest Lithuanian book is printed

The first book in the Lithuanian language, a Roman Catholic catechism, is printed.

The Lithuanian-Polish Commonwealth Era (1569–1795)

1569   Lithuania signs the Union of Lublin

Lithuania and Poland sign the Union of Lublin and formally unite to form the Polish-Lithuanian Commonwealth.

The Grand Duchy of Lithuania realizes it cannot successfully meet the growing pressure of eastern influences without help from the Kingdom of Poland. By signing the Union of Lublin, Lithuania loses its power, political independence and, in time, its social and cultural identity to Poland.

1579   The University of Vilnius (Alma Mater Vilnensis) founded

A former Jesuit college, established in 1570, becomes the University of Vilnius. When the University of Vilnius opens the Grand Duchy of Lithuania has a center of learning to rival those found in Moscow and Tartu. The University of Vilnius offers studies in science, the only university in northeastern Europe to do so.

1654–67   Imperial Russia engages Lithuania in war

During the second half of the seventeenth century a new threat emerges to challenge the Lithuanian-Polish Commonwealth: the Russian Empire.

1655   Russians capture Vilnius

Imperial Russia succeeds in capturing Vilnius. Although the occupation is short-lived, it marks the first time an invading army is able to penetrate so deeply into Lithuanian territory.

Late 17th century   Publication of first Lithuanian grammar

The first Lithuanian grammar book and dictionary, called *Lietuviu Kalbos Zodynas* (Dictionary of the Lithuanian Language), is published by Konstantinas Sirvydas. The *Biblija* (Bible) is translated into Lithuanian.

18th century   Decline of the Grand Duchy

Political decline and cultural fragmentation continue as the Grand Duchy of Lithuania declines.

1706   *Aesop's Fables* translated into Lithuanian

The eighteenth century ushers in a new era of secular Lithuanian literature. *Ezopo Pasakecios* (Aesop's Fables) is translated into Lithuanian.

1714–80   Kristijonas Donelaitis launches Lithuanian literature

Kristijonas Donelaitis, a narrative poet, is credited with establishing nonsectarian literature in Lithuania. Donelaitis's masterpiece titled *Metai* (The Seasons) is the first major work of fiction in Lithuanian classical literature. In it, Donelaitis uses poetry to depict serfs and village life. An important work, *Metai* has been translated into many languages.

1772–95   Lithuania is partitioned

Lithuania is partitioned three times (1772, 1793, and 1795) before being incorporated into Russia.

The Era of Imperial Russia

1795   Lithuania ceases to exist as a nation

The Grand Duchy of Lithuania is absorbed by Imperial Russia, and Poland is partitioned. The name of Lithuania does not appear again on the European political map until 1918.

At the close of the eighteenth century nearly all of Lithuania comes under Imperial Russian control when Poland is partitioned by Russia. Prussia and Austria, Germanic states allied to Russia, share in the partitioning of territories. Prussia gains control of a smaller, western part of Lithuania bordering the Baltic Sea which becomes known as Lithuania Minor. Vilnius becomes the third largest city in the Russian Empire, behind Moscow and St. Petersburg, and thus prospers from the annexation.

1803   The University of Vilnius becomes Imperial University

Vilnius University becomes one of the finest institutes of higher learning in Europe and is renamed Imperial University.

1816–19   Serfdom is abolished

Under Imperial Russia the area of Lithuania evolves from feudalism to capitalist industrialism. Serfdom is abandoned, cities grow, and the population increases. The era of Imperialist Russian rule continues over the next two centuries and ends with the last Russian tsar, Nicholas II.

1830   Lithuanian nobles revolt

Lithuanian nobles stage their first revolt against Imperial Russian rule. The Russian Empire responds quickly and continues to remove all traces of Polish influences on Lithuanian culture in favor of all that is Russian. Roman Catholicism faces increased opposition through the expansion of the Russian Orthodox Church and the Russian Old Believers Church.

1835–1916   Vincas Svirskis, Lithuanian folk artist

An interesting folk-art tradition involves carving large wooden crosses or suns, figures of saints, and/or weathercocks on tall poles (like totem poles) and placing them at crossroads, in cemeteries or village squares, or at the sites of extraordinary events. Vincas Svirskis masters this art form. Many of his works are held by the State Museum in Vilnius.

1860–61   Antanas Baranauskas creates his epic

Antanas Baranauskas (1835–1902) pens his epic poem, *Anyksciu silelis* (Anyksciai Pine Forest), representing a high point in nineteenth-century Lithuanian literature. The poem uses the forest as a symbol of Lithuania and bemoans its being cut down by foreign landlords (German and Russian).

1863   Nobles revolt

Lithuanian nobles stage a second revolt against Imperial Russian rule. The Russian Empire responds to this second attempt at revolution by restricting the use of the Lithuanian language. Publications written in the Lithuanian language and the Latin alphabet it uses are banned.

1864   Publishers forced to use Russian alphabet

Lithuanian literature is forced by the Russian Imperial government to use only the Russian Cyrillic alphabet. In addition to the publication restrictions, educational institutions like Imperial (Vilnius) University are closed.

1880s   Russification and National Awakening

A new spirit of nationalism begins to spread throughout Lithuania. It is a trend toward individualism, consciousness of self, and political activism which Imperial Russia finds increasingly intolerable. Beginning in the 1880s Russia introduces a policy of deliberate Russification in order to undermine, by force if necessary, the increasing autonomy of the Baltic states and any growing nationalistic movements. Lithuania becomes the first Russian territory to demand independence.

1883   First Lithuanian periodical, *Ausra*

Dr. Jonas Basanavicius publishes the first Lithuanian periodical, called *Ausra* (The Dawn). This publication is considered illegal because it is not approved by the Imperial government. It is part of a national movement of social and cultural rebirth or "awakening" taking place in Lithuania in the 1880s.

Lithuanians seeks to reclaim their language and culture through a movement called the Book-bearers which stresses self-education and home-schooling as a means to cultural preservation.

1895 Maciulis leads national revival

Poet Jonas Maciulis, known as Maironis, spearheads the Lithuanian national revival. His nationalist, romantic *Pavasario balsai* (Voices of Spring), published in 1895, is the start of modern Lithuanian poetry.

1896 The Social Democratic Party is founded

Lithuania's first political party, The Social Democratic Party, is founded. Under the totalitarian rule of the Soviet Union, the Social Democratic Party disappears and re-emerges a century later (see 1989: August 12) when the Soviet Union nears collapse.

Twentieth Century: Revolution, Occupation, and Independence

1905 Revolution in the Baltics

Russian Imperial authorities respond with violence and repression to the Baltic states' demands for radical political change. This revolution is a vain struggle for social and political freedom against foreign landowners (many of whom are German) and the Russian policy of national oppression and Russification.

1914–18 World War I

The First World War cripples Lithuania and her Baltic neighbors economically as the German army takes advantage of the political turmoil in Russia. In 1915 Germany succeeds in taking control of the Russian province of Lithuania and remains there until the end of the war.

1917 The Russian Empire collapses

The last tsar of Russia, Nicholas II (1868–1918), is forced to abdicate. As the Bolsheviks seize power in Russia they make significant political gains in the Baltics. The demands of the Russian revolution combined with Germany's defeat in WWI open the door for Lithuania, Estonia, and Latvia to declare independence.

Independence: 1918–39

1918: February 16 Statehood Day

Lithuania formally proclaims independence from Russia. On this date the Lithuanian Council issues a *de jure* declaration of independence, meaning it does not immediately take effect since much of the country is still occupied by the German army. Real or de facto independence comes only after several years of fighting against hostile powers, including Poland and Russia.

1918–20 Lithuania wages war to defend independence

The new Soviet Russia attempts to permanently annex the Baltic states under the communist system. Newly independent Poland seizes the eastern part of Lithuania, including the capital Vilnius. Lithuania does not regain the Polish-occupied territories until after World War II.

During the inter-war period the city of Kaunas replaces Vilnius as the center of higher learning when a university and several institutes are opened there. Kaunas becomes the capital city of Lithuania until Vilnius is re-claimed from Polish control.

1920: July 9 Peace settlement

Vladimir I. Lenin (1870–1924), leader of Soviet Russia, signs a peace treaty with Lithuania, as well as with neighboring Estonia and Latvia. The peace treaties force Moscow to recognize their independence and renounce all territorial claims.

1921 Lithuania is admitted to the League of Nations

Independent Lithuania becomes a member of the League of Nations.

1920–22 Land reform in Baltic states

Lithuania begins the gradual process of economic and social recovery. Agricultural reforms include a major redistribution of land from large estates to small and medium-sized private farms. The privatization of agriculture is the beginning of economic development in areas such as food processing, light industry, transportation, and foreign trade. These economic reforms continue until the Soviet occupation of 1940.

1923 Lithuania annexes Klaipeda region

Lithuania annexes the Klaipeda region and gains access to the Baltic Sea through Klaipeda harbor. When Lithuania takes control of the city of Klaipeda, over seven hundred years of German rule end. Klaipeda is an important seaport city on the shores of the Baltic Sea giving Lithuania direct access to the West.

1926–29 Military coup

The idea of a Republic and a democratic form of government are new concepts to the people of Lithuania. Antanas Smetona of the Nationalist Party becomes the acting president of the parliament. He censors the press and bans all political parties.

Independent Lithuania still flourishes in many areas. Primary and secondary schools and universities once closed under Imperial Russia are re-opened. Writers and artists find a new inspiration in independence. Land reforms and a stronger economy, including the introduction of a national currency, the *litas,* bring more stability to Lithuania.

1930s   Putinas publishes *Altoriu Sesely*

Vincas Mykolaitis-Putinas (1893–1967), a former priest, is regarded as a leading literary figure of the early twentieth century. His most noted work, penned after the 1930s, is a three-volume novel titled *Altoriu Sesely* (In the Altars' Shadow), which mirrors the spirit of nationalism evolving during the period between World War I and World War II.

1934   Baltic states sign 1934 Baltic Treaty on Unity and Cooperation

Faced with the dual threat of Soviet Russia and Nazi Germany, Lithuania and neighboring Estonia and Latvia determine that their individual national interests are best served and preserved by presenting a unified front.

World War II: 1939–45

1939: August 23   Nazi-Soviet Nonaggression Treaty is signed

Nazi-Soviet Nonaggression Treaty (Molotov-Ribbentrop Pact) is signed.

The political challenges from Soviet Russia and Nazi-Germany overpower the Lithuanian government. While Estonia and Latvia are assigned to the Soviet sphere of influence, Germany seizes control of Lithuania.

1939: October 10   Lithuania is assigned to the Soviet sphere of influence and Vilnius is given back to Lithuania

The Baltic states are forced into signing treaties allowing Moscow to station troops on their soil.

Soviet Occupation (1940–41)

1940: June 15   The Red Army of the Soviet Union occupies Lithuania

Pro-Soviet governments are elected using sham elections, and the Baltic states are formally annexed to the Soviet Union.

1940: August 6   Lithuania officially becomes the Lithuanian Soviet Socialist Republic

The Soviet government makes Lithuania a Soviet Socialist Republic.

1941: June 14–15   Mass deportation of Lithuanians to Siberia begins

A totalitarian reign of fear emerges under Joseph V. Stalin (1879–1953) as tens of thousands of men, women and children are forced from their homes to be sent by railcar to Siberia, never to return.

1941: June 22   Lithuania and neighboring Baltic states fall under Nazi-German occupation.

The invasion by Adolph Hitler's (1889–1945) army places the Baltic states under total Nazi-German occupation. The Nazi regime institutes compulsory draft of all Baltic peoples into labor or military service. Jews and Gypsies are annihilated. By the end of World War II Lithuania's population has suffered heavy losses. By 1953 estimates calculate Lithuania's population loss at nearly thirty percent. Nationalist and communist resistance movements remain active underground.

1944–45   Soviet forces re-occupy Lithuania and the Baltic states; reign of terror begins again

Thousands of Lithuanians flee to the West to escape Soviet repression. Many Western nations refuse to recognize the annexation of the Baltic states into the Soviet Union. The Soviet Union begins to eliminate the Polish influence on Lithuania's history through a system of cultural and social "cleansing" called *sovietization*. Religious repression begins, and Roman Catholicism in particular is targeted and replaced by the totalitarian policy of forced atheism. Vilnius Cathedral is closed, and the Archbishop of Vilnius banished.

1947–52   Agricultural collectivization is completed in Lithuania and her Baltic neighbors

Lithuania changes from an agricultural based economy to an industrialized based one during Soviet rule. Under the Communist policy of *collectivization* all private property is nationalized, and Lithuanians are forced to abandon their single family dwellings and farms, for multi-storied urban complexes. Forced agricultural collectivization is the benchmark of Soviet control. Farms are removed from private ownership, and industry is regulated by the Soviet government. Industrialization is forced, yet the Lithuanian economy begins to improve.

1950s–late 1980s   Lithuania under Communist rule

From the postwar years to the Gorbachev era (see 1985), Lithuanians continue to live under repressive Communist rule. Political opposition and personal freedoms are restricted, and religious activity is suppressed. Lithuania's needs are subordinated to the overall economy of the U.S.S.R. Growing numbers of Lithuanians join the Communist Party, as party membership is the ticket to well-paid jobs in government, business, and other fields.

Lithuanian troops march into Vilnius, restored in the 1950s to Lithuania by the Soviets. (EPD Photos/CSU Archives)

1953   The death of Joseph V. Stalin eases Soviet repression

Beginning in the 1950s the ranks of the Lithuanian Supreme Soviet gradually fill with an educated class of people (economists, writers, teachers, scientists, etc.) who have decided to use the system to help preserve ethnic Lithuanian language, culture, and traditions. Others join as political opportunists, realizing that feigned allegiance to the Communist Party attracts more favors than outright rebellion.

Late 1950s–60s   Lithuanians gain more local power

Ethnic Lithuanians gain increased power over some aspects of their own lives, including the pace of industrialization, as they make their way into the upper ranks of the local Communist hierarchy.

1964–82   Brezhnev advances Russianization policies

Under Soviet leader Leonid Brezhnev (1906–82), Lithuania, like the rest of the Soviet Union, comes under increased pressure to adopt the Russian language and culture. However, Lithuania, unlike its Baltic neighbors, has not had the ethnic character of its population diffused by massive emigration of ethnic Russians following World War II.

1968   Invasion of Czechslovakia provokes dissident activity

The Soviet Union's invasion of Czechoslovakia to halt the liberalization taking place under Alexander Dubcek (1921–92) is followed by even greater repression in the U.S.S.R. itself, leading to an increase in dissident activity. Underground publications multiply.

1970s–80s   Church-supported protest thrives

Dissident activity spreads, centered around the Roman Catholic Church. More dissident literature is published in Lithuania than in any other Soviet republic.

1972   Lithuanian student immolates himself as protest

Lithuanian student Romas Kalanta immolates himself (sets himself on fire) in protest against Soviet rule. The underground resistance movement in Lithuania which seeks independence and an end to Soviet repression continues, often with dramatic results.

1973 The Chronicle of the Catholic Church of Lithuania, an underground publication, begins

The persecution and destruction of all things relating to Roman Catholicism by the Soviet Union forces adherents underground as they seek to preserve the social and cultural identity of Lithuania. Through this publication (which continues to 1988) Lithuanians at home and abroad are encouraged to hold fast to their religious beliefs and their cultural heritage.

1985 Soviet leader Mikhail S. Gorbachev (b.1931) introduces policies of glasnost and perestroika

Beginning in the mid-1980s the Soviet Union enters a period of unprecedented economic and political crisis which threatens the very foundation of communist rule. *Glasnost*, a Russian term for public discussion of issues, and *perestroika*, which means restructuring, form part of Gorbachev's campaign in the Soviet Union to revitalize the economy, the Communist Party, and society as a whole by adjusting political, social, and economic mechanisms. The ramifications of such policies lead to the demise of the U.S.S.R. in favor of Old Russia and freedom for the Baltic states.

1987–88 Demonstrations and rebirth of nationalism

Baltic dissidents hold public demonstrations in Tallinn, Riga, and Vilnius (the Baltic capitals). A re-birth of nationalism begins in the Baltics begins.

Gorbachev's twin policies of *glasnost* and *perestroika* ignite the Lithuanian spirit of nationalism and the move for independence. Lithuanian is declared the official language as it has been once before, during the inter-war period of independence.

1988: June 3 Sajudis, the democratic Lithuanian Reform Movement, is founded in Vilnius

This organization includes Communist Party members and non-party leaders interested in capitalizing on Gorbachev's tenuous hold on Soviet Russia to reform the economic and political structure of Lithuania and to renew ethnic Lithuanian traditions and culture beginning with the restoration of Lithuanian as the official language.

1988: October Congress holds elections

The Sajudis congress elects as chairman Vytautas Landsbergis, a former music professor. Algirdas Brazaukas becomes Lithuanian communist leader.

1989: May Proclamation of sovereignty

The Lithuanian Supreme Soviet proclaims Lithuania's sovereignty and declares illegal its annexation by the Soviet Union.

1989: August 12 The Social Democratic Party is restored

Lithuania's first political party, The Social Democratic Party, re-emerges as the Soviet Union nears collapse.

1989: August 23 Human chain commemorates World War II

A human chain from Tallinn to Vilnius is a protest on the fiftieth anniversary of Nazi-Soviet Nonaggression Treaty. This human chain, extending some 600 kilometers in length, links nearly 2 million Lithuanians, Latvians, and Estonians hand-in-hand on the Vilnius-Tallinn road. This protest is called The Baltic Way.

1989: December The Communist Party of Lithuania splits from the Communist Party of the Soviet Union

The Communist Party of Lithuania evolves into a democratic, pro-independence group to become known as the Lithuanian Democratic Labor Party.

1990: March 11 The newly-elected Lithuanian Supreme Soviet (parliament) proclaims independence and elects Vytautas Landsbergis chairman

Beginning in the 1990s the new government starts a reform policy called *privatization* under which it seeks to make amends by restoring the rights of previous property owners or offering monetary compensation.

1990: April Moscow imposes economic blockade on Lithuania; Baltic agreement on economic cooperation signed by Estonia, Latvia, and Lithuania

In moving towards independence Lithuania realizes its economy is closely tied to that of Russia, its main import and export trading partner. The collapse of the Soviet Union means economic chaos for Lithuania also as it depends heavily upon Russia as a supplier of basic raw materials and energy.

1990: May Baltic states renew 1934 Baltic Treaty on Unity and Cooperation

In a move destined to antagonize Moscow's authority, the Baltic states renew the 1934 Baltic Treaty on Unity and Cooperation.

1990: June   Moratorium declared on independence

The Lithuanian Supreme Council agrees to a six-month moratorium on independence declaration; Moscow lifts economic blockade.

Independence Restored, 1991

1991: January 13   Lithuania's prime minister Kazimiera Prunskiene resigns after dispute with Vytautas Landsbergis

The Soviet Union resorts to force in order to oust the Lithuanian government from Vilnius in favor of Communist rule. The Soviet use of military intervention in Vilnius and Riga (capital of Latvia) results in the massacre of many civilians.

1991: February–March   Referenda in Estonia, Latvia, and Lithuania show overwhelming support for independence

This overwhelming show of strength in support of independence is bolstered by the decision of the United States and Europe to formally recognize the independent status of Lithuania.

1991: September 6   The Soviet Union recognizes the re-establishment of Lithuanian independence

In an about-face, the Soviet government recognizes the independence of Lithuania.

1991: September 17   Lithuania is admitted to the United Nations

Lithuania becomes a member of the United Nations.

1992: October 25   The Lithuanian Democratic Labor Party wins absolute majority of seats in Seimas; the Constitution of the new Republic of Lithuania is approved

For the first time since independence is restored elections are held for the 141 member state legislature or Seimas. Algirdas Mykolas Brazauskas of the Lithuanian Democratic Labor Party is elected chairman. A new Lithuanian constitution is approved by referendum and adopted by the Seimas. Lithuania is formally established as an independent democratic state.

1993: February 14   Algirdas Mykolas Brazauskas elected president

Algirdas Mykolas Brazauskas, chairman of Seimas, becomes the first democratically elected president of Lithuania

1993: May 14   Lithuania is admitted to the Council of Europe

The Council of Europe is organized in 1949 to provide intergovernmental cooperation in such areas as environmental planning, finance, crime, migration, law and sports.

1993: June 25   The litas is re-introduced as the official currency

The litas is re-introduces as the official currency of Lithuania. It replaces the Russian ruble.

1994: January 27   Lithuania joins the NATO Partnership For Peace program

Full NATO membership is a primary goal of Lithuania. The Baltic states create a joint peace-keeping force, a joint mine sweeping unit, and are working on creating a common air defense system. Russia is adamantly opposed to the admission of any Baltic state in NATO because it believes it would directly threaten its borders and security.

1997: December 21   Valdas Adamkus elected president

Valdas Adamkus is elected president of Lithuania; he is sworn in on February 25, 1998

1998: February 1   Lithuania becomes an associate member of the European Union (EU)

Lithuania pursues full membership in the European Union in order to acquire greater economic and political stability.

1999: February 25   Estonia, Latvia, and Lithuania announce the opening of joint military college

The Baltic states officially open a joint-military college in Tartu, Estonia, in order to improve the quality of their military and, hence, increase their chances of NATO membership. Staffed by NATO officers, classes at the new Baltic defense college, called BALTDEFCOL, are scheduled to begin in August 1999.

Bibliography

Gerner, Kristian and Stefan Hedlund. *The Baltic States and the End of the Soviet Empire.* London and New York: Routledge, 1993.

Gerutis, Albertas. *Lithuania: 700 Years.* New York: Maryland Books, 1969.

Hiden, John and Patrick Salmon. *The Baltic Nations and Europe.* London and New York: Longman, 1994.

Hiden, John. *The Baltic States and Weimar Ostpolitik.* Cambridge: Cambridge University Press, 1987.

———. *The Baltic States in Peace and War: 1917–45,* edited by V. Stanley Vardys and Romuald J. Misiunas. University Park and London: The Pennsylvania State University Press, 1978.

————. *The Baltic States: Years of Dependence, 1940–80,* edited by Romuald J. Misiunas and Rein Taagepera. Berkeley and Los Angeles: University of California Press, 1983.

Iwaskiw, Walter R., ed. *Estonia, Latvia, and Lithuania: Country Studies.* Washington, D.C.: Federal Research Division, Library of Congress, l996.

Jurgela, Constantine R. *Lithuania: The Outpost of Freedom.* The National Guard of Lithuania in Exile, Inc. and Valkyrie Press, Inc., 1976.

Kirby, David. *The Baltic World, 1772–1993.* London and New York: Longman, 1995.

Lieven, Anatol. *The Baltic Revolution.* New Haven and London: Yale University Press, 1993.

Lithuania Today. Press and Information Department, Ministry of Foreign Affairs of the Republic of Lithuania, 1998/1999.

Noble, John, Nicola Williams, and Robin Gauldie. *Estonia, Latvia, and Lithuania.* Australia: Lonely Planet Publications, 1997.

Page, Stanley W. *The Formation of the Baltic States.* New York: Howard Fertig, Inc., 1970.

Statistical Yearbook. Statistical Office of Lithuania, 1999.

Von Rauch, George. *The Baltic States: The Years of Independence.* Berkeley and Los Angeles: University of California Press, 1970.

Luxembourg

Introduction

Luxembourg is a landlocked nation in Western Europe, with an area of 998 square miles (2,586 square kilometers)—a size slightly smaller than the state of Rhode Island. Its eastern boundary with Germany is formed by the Our, Sûre, and Moselle Rivers. Luxembourg is bordered on the south by France and on the west and north by Belgium. Its capital city, Luxembourg, is located in the southern center of the country.

Luxembourg has two distinct geographical regions: the rugged, wooded uplands (*Oesling*) of the Ardennes mountains in the north, and the fertile southern lowlands, known as *Bon Pays* (Good Land). The average elevation in the north is 1,476 feet (450 meters) with the highest point at 1,834 feet (559 meters), while the average elevation in the south is 820 feet (250 meters). All of Luxembourg is criss-crossed by deep valleys with rivers, most draining into the Sûre in the east, which then flows into the Moselle.

Luxembourg's climate is mild and temperate: it is rarely very hot or very cold. Summers are generally cool, with an average temperature of 63°F (17°C), while winters are rarely severe, their average temperatures being 32°F (0°C). The peaks of the Ardennes shelter the lowlands from strong north winds, and the prevailing northwesterly winds have a cooling effect. Precipitation averages about 30 inches (75 centimeters) annually, with the most rainfall in the extreme southwest.

The most common trees in Luxembourg are the pine, chestnut, spruce, oak, linden, elm and beech, and fruit trees. There are also many shrubs, such as the blueberry and genista, as well as ferns and many varieties of wild flowers. There are many vineyards, found especially along the Moselle River in the east. Only a very few species of wild animals—deer, roe deer, and wild boar—may be found in Luxembourg, but there are many species of birds and also fish such as perch, carp, bream, trout, pike, and eel. Extinct animals are the wolf and European otter.

A 1991 census put the population at 384,062, with forty-seven percent living in cities. The capital city of Luxembourg has a population of 75,377, and the principal industrial city of Esch-sur-Alzette has a population of 24,012. As of 1992, approximately 114,700 residents of Luxembourg (about one-third of the population) were foreigners, mainly workers from Portugal, Italy, France, Belgium, Germany and other European nations. The indigenous Luxembourgers consider themselves to be a distinct nationality; despite a history of foreign occupation, the people have kept their individuality as a nation. Luxembourgers are trilingual, speaking not only their traditional dialect *Letzeburgesch* (Luxembourgish), but also French and German. Luxembourgish is Germanic, deriving from the Moselle Frankish tongue once spoken in western Germany. It has only recently gained more prominence as a written language. French is usually used for administrative purposes, while German is often used in other areas, such as religion. French is the most common language of instruction in secondary schools. Luxembourg has complete religious freedom, with ninety-five percent of the population being Roman Catholic and the remaining five percent being Protestant or other religions.

Luxembourg is one of the most highly industrialized countries in the world, with an extremely high standard of living. Steelmaking was once the nation's principal industry but has been in decline since 1974. It once contributed twenty-one percent to the nation's GDP (1974), whereas today the industry only contributes about 5.5 percent. Instead, plastics, rubber, and chemicals have been successfully developed, as well as service industries such as banking. Approximately 15.8 percent of the population is employed in industry, with 17.5 percent employed in trade, restaurants and hotels; 31.5 percent in services and public administration; 8.4 percent in construction; 9.2 percent in credit and insurance; 6.3 percent in transportation and communications; 3.2 percent in agriculture and forestry; and 8.1 percent in other sectors. Unemployment averages at less than 2 percent of the population, and there is virtually no adult illiteracy.

Natural resources include slate and nonmetallic minerals such as dolomite, gypsum, limestone, sand, and gravel. The traditional, historically-important natural resource has been iron ore, found in the southwest. However, mine depletion has resulted in a decline in production, from 2,079,000 tons in 1976 to 429,000 tons in 1981, after which time the mines closed their operations. Moreover, the country's lack of industrial fuels makes it entirely dependent upon imports of coke for steel production. About twenty-one percent of the land is forested, and there is a small commercial wood trade. In 1990, fifty percent of Luxembourg's land was devoted to agriculture and grazing, with the majority of that land used as meadows and pastures for livestock. There is a strong dairy exportation business, as well as other important exports such as wine, clover seeds, fruits (most notably apples, plums, and cherries), and specialty rosebushes (millions are exported annually).

Luxembourg has only been a completely independent, self-governing nation since 1867, though the country has existed for over 1,000 years, since A.D. 963. In its earliest history, Luxembourg was dominated by the Celts, then the Romans, and finally the Riparian Franks before it was proclaimed the County of Luxembourg in 963 by Siegfroid, Count of the Ardennes. The territory then tripled in size during the reign of Countess Ermenside (1196–1247). Count John (r. 1309–46)—known as "John the Blind" —became the national hero, for despite his blindness he laid the foundations of a strong, powerful dynasty. John's son Charles (1316–78), the second of four Luxembourgish princes to be named Holy Roman Emperor, made Luxembourg a duchy. Unfortunately, however, his successors ruined the country financially, and it came under the control of other European nations for the next 400 years.

Luxembourg fell under the control of the Burgundians in 1443 and did not regain its self-rule until the nineteenth century. Successively it passed to Spain (from 1506–1714, aside from 1684–97, when France briefly took power), Austria (1714–95), and France (1795–1815). In 1815, the Congress of Vienna—a group of all the most powerful leaders in Europe who met to redraw political boundaries and decide the fate of their nations after the fall of Napoleon Bonaparte (1769–1821)—made Luxembourg a Grand Duchy and part of the Kingdom of the Netherlands, after having given its territory east of the Our, Sûre, and Moselle Rivers to Prussia. Luxembourg lost territory once more, in 1839 to Belgium, but gained more independence in the bargain (the Dutch kings, however, continued to rule as Grand Dukes). Luxembourg was finally declared a completely independent and neutral state by the Treaty of London in 1867. The house of Nassau-Weilbourg became the ruling house of Luxembourg in 1890 when Adolphe (r. 1890–1905) became Grand Duke.

The country was occupied twice more in the twentieth century. During World War I (1914–18), German troops assumed control of the tiny nation. After the German loss, however, Grand Duchess Charlotte succeeded to the throne (r. 1919–64), with the overwhelming support of the Luxembourgish people. Luxembourgers resumed "business as usual"—forming an economic union with Belgium in 1921—until they were once again invaded by the Germans in May, 1940. The grand ducal family and other members of government were able to escape to England and form a government-in-exile in London, but the people suffered greatly under Nazi

rule. The country was liberated by Allied forces in September, 1944.

Luxembourg's 1000th anniversary as an independent state was celebrated in 1963. On November 12, 1964, Grand Duchess Charlotte abdicated her throne in favor of her son Jean, who remains Grand Duke as of 1999. His reign has been marked by continuing prosperity, as Luxembourg's economy has shifted from dependence on steel to a focus on services, most importantly, international banking.

Luxembourg has four main political parties: the Christian Social Party (CSP), Workers Socialist Party (WSP), the Democratic Party (DP), and the Green Alternative Party (GAP). The Roman Catholic oriented CSP enjoys the most popularity and is strongly in favor of NATO (the North Atlantic Treaty Organization) and its policies. The DP, also pro-NATO, is the oldest and smallest party, and draws its support from professions and the urban middle class. Industrial workers tend to support the moderately pro-NATO WSP, while the strongly anti-NATO GAP recently won ten percent of the vote in European elections.

For much of the twentieth century, Luxembourg has been working towards economic and political union with other European nations. The Benelux Customs Union (an economic alliance with Belgium and the Netherlands) was effected in 1948. In 1951, six nations formed the European Coal and Steel Community (ECSC). 1958 marked the establishment of The European Economic Community (EEC), the free trade organization whose membership had increased to fifteen by 1999. A 1992 treaty led to the creation of the European Union (now called the European Community). Alliance among the European states is very important to the small, trade-dependent nation of Luxembourg. Both the government and the people know that they must strive towards a united Europe in order to remain prosperous and productive for years to come.

Timeline

c. 5th century B.C.   Domination of the Celts

Some of the earliest known inhabitants of the land which is now Luxembourg are the Celts, an Iron Age people who are able to make more sophisticated and powerful weapons than their predecessors, who had used only stone and bone for tools. The Celts are strong warriors who drive chariots and engage in one-on-one combat. The Celts, however, are no match for the invading forces of ancient Rome (see 58 B.C.)

58 B.C.   Roman invasion led by Julius Caesar

Julius Caesar (102–44 B.C.)—Roman general and statesman—conducts military campaigns which extend Roman power in the west. Luxembourg is part of the area which falls under Roman control in 58 B.C.

58 B.C.–450 A.D.   Roman domination

For a period of approximately 500 years, the Romans control all of western Europe including Gaul (what is now France) and the land that is now Luxembourg. Luxembourg, however, is never home to any important Roman settlements, though many inhabit the land along the Moselle River and grow grapes (c. 300). Later, as farm buildings begin to be constructed from brick, tile, and stone rather than wood, the Romans use stone from the quarries of southern Luxembourg for construction. Yet the region in and around Luxembourg is used more as routes for going from one important settlement to another than as an actual center of Roman civilization.

Some historians believe it is Luxembourg's isolation beginning in the Roman period which has allowed it to remain separate, distinct, and independent despite its small size. It is quite unusual for an area as small as Luxembourg to maintain its autonomy; why was Luxembourg never absorbed into Belgium, Germany, or France? Some scholars believe that the location of the present-day Grand Duchy has much to do with its independence: the great posts of Rome in the general area of Luxembourg are Trier, Aachen, Arlon, and Virton, among others. Luxembourg lies within a rough circle formed by these important Roman settlements. Perhaps the inhabitants within this circle felt a sense of community which kept them settled there, and perhaps Luxembourg's "unimportance" allowed her to remain discreetly independent through centuries of tumultuous European politics.

c. A.D. 250   The Franks begin a series of invasions

The third to fifth centuries mark the domination of the *(Riparian)* Franks, a Germanic people who occupy most of the land which is now Luxembourg. By 459, the Franks control Luxembourg, aided in their invasion of the region by the well-built Roman roads. By the ninth century, the Frankish empire extends over what is now France, Germany, and Italy (see 800).

c. 300   First Christians arrive in region

Luxembourg today is a very religious nation, with over ninety-five percent of its population being Roman Catholic. The roots of Christianity can be found around the year 300, when the first Christians arrive in that region of Europe. The religion spreads quickly, as Christians enjoy great freedom from persecution. The Luxembourg area is almost completely Christian by the end of the sixth century.

A.D. 800   Charlemagne (Charles the Great) declared emperor

The pope names Charlemagne (747–814)—a Frankish king and great conqueror who expands the Frankish empire considerably—the Emperor of the Romans. Charlemagne's reign marks a period of great stability and cultural advancement. However, the empire does not last long beyond his death, for Charlemagne's sons lack both his authority and his vision.

8th–10th centuries   Unrest and instability

For approximately 200 years, the former imperial province of Rome (including what is now Luxembourg) is the source of continuous fighting between competing counts, lords, and dukes. Feudalism becomes the predominant way of life during this time, the height of Luxembourg's so-called Dark Ages.

855–963 Seven kings claim Luxembourg

Over a period of approximately 100 years, seven different rulers claim Luxembourg as their own, and for a time the area is divided in two. The provincial people's ties to their land become stronger because of the distant leadership, and the people need strong local leadership. No central power exercises effective control over the area.

963 Founding of County of Luxembourg

The Ardenne family is one of several rich and powerful landowning lineages which rises significantly above the local aristocrats. Siegfroid (r. 963–998), Count of the Ardennes, gains significant power not only for his family's wealth but also for his ties to the church. He founds the County of Luxembourg in 963, laying claim to six domains. Siegfroid rebuilds a small ruined fortress called *Lucilinburhuc* (Little burg) on the site of the present-day capital.

Luxembourg is able to remain independent until the fourteenth century, thanks primarily to a string of strong leaders.

11th century Manufacture of religious manuscripts

One of the most important monasteries in the Middle Ages is found at Echternach, located on the eastern border of present-day Luxembourg. The monastery is most famous for the Echternach Gospels, its lavishly decorated eleventh century manuscripts that are one of the best examples of Ottoman manuscript illumination (illustration).

1196–1247 Reign of Countess Ermesinde

Countess Ermenside (r. 1226–46) reigns for over twenty years, until her son comes of age. She extends the frontiers of Luxembourg not by war but by marriage alliances to other royal houses. Marriages are often more powerful tools than wars, as they create more allegiance and loyalty. Ermenside also marries to ensure her inheritance; she is widowed in 1214 at the age of 27 with four daughters and no male heir and, therefore, weds a German prince to whom she bears a son.

During the reign of Countess Ermesinde, the territory of Luxembourg triples in size. Moreover, Luxembourg's citizens gain much personal freedom, such as the right to sell possessions, organize themselves, and create institutions. Ermenside is often called the Founding Lady of Luxembourg.

c. 1200 Luxembourg isolated from its neighbors

A publication by the Ministry of Arts and Sciences proclaims that Luxembourg participates very little in the general prosperity that marks this time, known as the Gothic period. This is due to Luxembourg's location away from the great commercial routes.

1308 Henry VII is named Holy Roman emperor

Count of Luxembourg Henry VII (1274–1313) is named Holy Roman emperor in 1308. Henry—also king of Germany—helps his family gain even greater power by marrying his son John (see 1309–46) to Elizabeth, the heiress of Bohemia, in 1311.

1309–46 Reign of Count John (John the Blind)

John (1296–1346), the Count of Luxembourg (r. 1309–46) and king of Bohemia (r. 1310–46) becomes Luxembourg's national hero. Although he is blind for many years, John lays the foundations for a powerful ruling dynasty. He is killed in the Battle of Crécy, Northern France, during the Hundred Years' War (1338–1453).

John exemplifies all the characteristics of a chivalrous medieval knight, including honor, loyalty, and courage. As a ruler, he is also able to build Luxembourg's holdings, stimulate its economy, form a new defense system, and build new fortifications. It is for these reasons that John the Blind remains a national hero.

1347–78 Reign of Charles of Luxembourg (Charles IV)

Charles (1316–78), son of Count John, becomes king of Germany and Bohemia (now part of the Czech Republic) and Holy Roman emperor. He is the second of four Luxembourg princes to become leader of the Holy Roman Empire. Charles helps found the University of Prague (1348) and issues the so-called *Golden Bull* (1356), which lays down the constitutional framework of the Empire. The *Bull* contains procedure for the election of the monarch, decreases the influence of the Pope, and defines the rights of seven electors. Avoiding involvement in the conflicts in Italy, John instead builds up his empire around his hereditary domains—Bohemia and Moravia—with his capital at Prague. He makes Luxembourg a duchy, but under his successors the country is ruined financially.

1378–1400 Reign of Wenceslas IV

The son of Charles, Wenceslas (1361–1419) is crowned king of Bohemia and Holy Roman emperor in 1378. He is a poor ruler and allows Germany to collapse into anarchy. Wenceslas is deposed as emperor in 1400.

1443–1867 Luxembourg under foreign rule

Luxembourg and the other Low Countries (the Netherlands and Belgium) come to be controlled by the Burgundian (French) dynasty. It remains in foreign hands for the next 400 years. Successively, it is ruled by Spain (1506–1714, excepting 1684–97, when it is ruled by France), Austria (1714–95), and France under Napoleon (1795–1815).

By 1506, the Netherlands become disillusioned with the Spanish, Catholic rule, and the nation rebels in 1566, declaring itself an independent, Protestant state. Luxembourg and

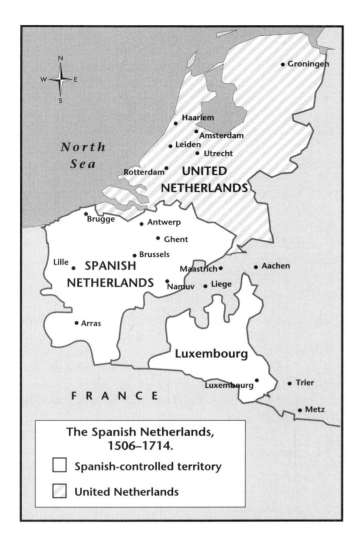

The Spanish Netherlands, 1506–1714.

- [ ] Spanish-controlled territory
- [ ] United Netherlands

Belgium, however, remain Catholic and under Spanish control.

1588–1631   Witchcraft trials

Between 1588–1631, as many as 100 women are put on trial for witchcraft annually. The accused are almost always executed for their so-called crimes.

1613   Cathedral Notre Dame constructed

The main center of worship for Luxembourg's Catholics is built in Luxembourg City. Named the Cathedral of Our Lady, or the Cathedral Notre Dame, the church is built in a Gothic manner with a Renaissance-styled door. Twin spires are added to the church in 1635.

1759   Birth of artist Pierre Redoute

Born in the town of St. Hubert in the Ardennes, Pierre Redoute (1759–1840) begins his artistic career at the age of thirteen. He becomes known as a highly influential botanical artist and works for the Emperor Napoleon's (1769–1821)

wife Josephine (1763–1814) in Paris, creating over 600 drawings that are still studied today.

1815   Luxembourg made a Grand Duchy

The Congress of Vienna (1815) raises Luxembourg from a Duchy to the rank of a Grand Duchy, giving it as an independent state to the king of the Netherlands. This arrangement is reached after Luxembourg cedes to Prussia all of its territory east of the Moselle, Sûre, and Our rivers. The new Kingdom of the Netherlands includes the land that is now Belgium, the Netherlands, and Luxembourg.

1839   Luxembourg loses territory

Luxembourg loses more than half its territory when Belgium declares itself an independent kingdom but gains more power and autonomy (ability to self-rule). The Dutch kings, however, continue to rule Luxembourg as grand dukes.

1845   National Museum of Historical Art founded

Luxembourg founds the National Museum of Historical Art.

1845   Birth of scientist Gabriel Lippmann

A professor of mathematical and physical sciences at the Sorbonne (1886), Gabriel Lippmann (1845–1921) invents the capillary electrometer and produces the first colored photograph of the spectrum. He is awarded the Nobel Prize in physics in 1908 for his pioneering work in color photography.

1850s   Luxembourg's iron industry booms

The quick development of the iron industry brings about Luxembourg's industrial revolution, brings wealth to the country, and changes villages into towns. Iron (and later steel) is the most important manufactured material in Luxembourg for more than a century, until 1975 when the world experiences a crisis in the iron and steel industries. (See 1975.)

1850–1900   Luxembourger emigration

During the second half of the nineteenth century, one out of every five Luxembourgers emigrates to the United States. The majority to Luxembourg immigrants now live in Illinois and Iowa, though the first groups to leave their home nation go to Wisconsin and Minnesota.

1867   Luxembourg declared an independent state

Under the Treaty of London, Luxembourg—though greatly reduced in its size (see 1839)—is declared an independent and permanently neutral state under the protection of the Great Powers (England, France, and Spain). Luxembourg, however, is required to dismantle its mighty fortress. Over a period of 400 years, the fortress of Luxembourg has been besieged, devastated and rebuilt more than twenty times. It takes sixteen years to complete the dismantling of the fort, a symbol of devastation and war.

Today, many tourists visit the remains of the fort, one of Europe's most powerful. Some of the ancient tunnels beneath the remains—once used to give troops shelter—have also been used in modern times, during the World Wars as bomb shelters and the Cold War as nuclear shelters.

1870s   Wave of Italian immigration

Many Italians come to Luxembourg in search of jobs in the steel industry. A similar wave of Italian workers arrives in the 1920s. For a discussion of immigration in Luxembourg, see 1970s.

1870   Iron deposits discovered

Rich deposits of iron ore are discovered in the south. A period of prosperity and industrialization follows, and many persons from neighboring countries migrate to Luxembourg.

1886   Birth of French statesman Robert Schuman

Though he serves as prime minister (1947–48) and foreign minister (1948–53) for France, Robert Schuman (1886–1963) is born in Luxembourg. Schuman serves on the French resistance during the Second World War and in 1950 initiates the Schuman Plan for the pooling of Western European coal and steel resources. Greatly esteemed by Luxembourgers, he is viewed as the architect of European unification.

1887   Birth of statesman and diplomat Joseph Bech

Joseph Bech (1887–1975) serves as prime minister from 1926 to 1937 and from 1953 to 1958. He also serves as foreign minister for thirty-three years, contributing to the effort to form the Benelux and European Common Market economic agreements following World War II. A modest and gentle personality, Bech does much to maintain Luxembourgers' morale during the war years and tumultuous following years.

1890   House of Nassau-Weilbourg comes to power

The union with the Netherlands ends when King William III (1870–90, r. 1849–90) dies, and Queen Wilhelmina (1880–1962, r. 1890–1948) assumes power. Luxembourg does not accept Wilhelmina as monarch because there is a male heir, Duke Adolphe, in a branch of the House of Luxembourg.

Once duke of Nassau (1839–66), Grand Duke Adolphe (1817–1905, r. 1890–1905) brings the house of Nassau-Weilbourg to power in Luxembourg. Adolphe's family line dates back to 1059.

1894   Birth of artist Joseph Kutter

Joseph Kutter (1894–1941) is a painter whose expressionist style is characterized by bright color. *Expressionism* seeks to give symbolic expression to inner thoughts and emotions.

1894   Birth of artist Nico Klopp

The son of a wine-grower along the Moselle River, Nico Klopp studies at the Royal Prussian Academy of Art in Dusseldorf, Germany, where he specializes in painting and engraving. His first paintings are marked by a tragic romanticism, while his later works are harsher with a powerful sense of light. Much of Nico's work meets with severe criticism at home, for the new expressionist style of Nico, Joseph Kutter (see 1894) and others is very controversial. Nico Klopp is one of the driving forces behind the first secessionist (expressionist) exhibition in Luxembourg in 1927, but his art is commercially unsuccessful, and he is forced to work as a local tax collector. He dies in 1930 at the age of thirty-six from meningitis, and his work only achieves a breakthrough after World War II (1939–45).

1900   Michael Theato wins Olympic gold

At the Paris Olympic Games, Michael Theato becomes the first Luxembourger to receive a gold medal when he wins the marathon with a time of two hours, fifty-nine minutes and forty-five seconds. However, no Olympic committee exists in the Grand Duchy at this time, so Theato has to register for the games with the French team and for many years is considered to be French.

1914   Luxembourg occupied by Germany

A neutral, pacifist nation, Luxembourg is occupied by German troops starting in the early stages of World War I (1914–18).

1918: November 11   World War I ends

Fighting ceases on the Western Front on November 11, 1918. After the war's end, all of Luxembourg's economic ties to Germany are severed, and the nation enters into an economic union with Belgium (see 1921).

1919: September 28   Grand Duchess Charlotte takes throne

In a referendum held to decide the country's future, a plurality of Luxembourgers support the succession to the throne of the Grand Duchess Charlotte (1896–1985, r. 1919–64).

1919: February 6   Constitution revised

The constitution is revised to include universal suffrage (women can vote) and proportional representation. The country is divided into four electoral districts, and sixty people are elected to the Chamber of Deputies (the legislative body) every five years. Legislation is introduced to the Chamber by the executive, and when a bill is passed it is sent to the Grand Duke for his signature, making the bill law.

1920 Joseph Alzin wins Olympic silver

Joseph Alzin becomes the second-ever Luxembourger to win an Olympic medal when he achieves a silver for weightlifting in Antwerp, Belgium.

1921 Economic union (BLEU) formed with Belgium

1928 Birth of statesman Gaston Thorn

Gaston Thorn is prime minister from 1974–79 and becomes president of the Commission of the European Community in 1981.

1933 Founding of Radio Luxembourg

Launched in 1933, the station Radio Luxembourg builds a huge transmitter to reach British audiences. This move is in response to the British domination of the radio, giving the listening public very little variety in radio shows. Radio Luxembourg responds to this by being the first station to give people what they want rather than what it is thought they should have. A major dispute breaks out between Radio Luxembourg and the British Broadcasting Company (BBC) over this move, but the station proves to be a huge success and broadcasts for over sixty years, closing in 1991. The station is truly international, with programs in German, French, and English. Radio Luxembourg is the first station ever to play a Beatles record over the air—1962's *Love Me Do*.

1940: May 10 Germans invade Luxembourg

Luxembourg is again occupied by the German army during World War II (1939–45). The grand ducal family (ruling Duke and Duchess) and most members of government escape to safety, establishing a government-in-exile in England. The Luxembourg people, however, suffer greatly under Nazi rule.

1942: August 30 Luxembourger revolt savagely repressed

Germany officially annexes Luxembourg, making it part of the German Reich. Many Luxembourgers revolt, engaging in a general strike to protest compulsory (required) service in the German army. The revolt is harshly repressed, with some Luxembourgers shot and others deported.

1944: September Luxembourg liberated by Allies

Allied forces march into Luxembourg and liberate it from the Nazis.

1944: December 16 The Battle of the Bulge

The Battle of the Bulge, one of the most decisive and bloody battles of the Second World War, is fought in northern Luxembourg. In early December, the Allied forces realize that Germany is launching a major offensive through the northern part of Luxembourg. U.S. General George Patton (1885–1945) anticipates a German surprise attack and has a contin-gency plan prepared: he orders the army to make a ninety-degree turn northwards, pushing through Luxembourg and attacking the advancing Germans at their flanks on the southern end of the Bulge (a term describing the geography of the land). After fierce combat in harsh winter conditions, Patton's army defeats the Germans at the end of January. The Battle of the Bulge depletes Germany's reserves and weakens them considerably; they surrender less than four months later (see 1945: May 8).

General Patton remains a national hero to Luxembourgers. After he dies from injuries in an auto accident, he is buried there, in the Hamm U.S. military cemetery among the men from the Third Army whom he commanded.

1944 Economic union with Belgium and the Netherlands

Though still in exile, the Luxembourg government agrees to form an economic union with the Netherlands and Belgium. The first phase of the union—the *Benelux* (Belgium-Netherlands-Luxembourg) Customs Union—is put into effect in 1948.

1945: May 8 V-E Day

World War II ends on V-E (Victory in Europe) Day, after the German surrender on May 7.

1945: October 24 Luxembourg joins the United Nations (UN)

Luxembourg is one of the founding members of the UN.

1950 Official Luxembourgish dictionary introduced

After several failed attempts to standardize the spelling of written Luxembourgish (*Letzeburgesch),* the government introduces an official dictionary of the language. This is a significant accomplishment, as it is rare for governmental legislation to succeed in imposing regularity on a formerly chaotic linguistic pattern. Luxembourgish is especially difficult to standardize because there are many dialects spoken all around the nation, and dialects are characterized by absent rules and norms. Luxembourgers take great pride in their language, however; after the introduction of the dictionary the language begins to become more widely used in its written form.

1951 Formation of European Coal and Steel Community

The European Coal and Steel Community (ECSC) is formed by Luxembourg, the other Benelux nations, Germany, France, and Italy. The six-member Community is founded with the understanding that if the individual nations do not control their industries for armaments, then they are no longer able to declare war. The ECSC's formation is the first step towards European integration. (See 1958, 1992.)

Grand Dutchess Charlotte shakes hands with Elenor Roosevelt at the White House during a two-day visit to Washington.
(EPD Photos/CSU Archives)

1952   Olympic Gold awarded to Josy Barthel

During the Summer Games in Helsinki, Finland, Josy Barthel beats the favorite from the United States against all odds to win the 1,500 meter race. His victory and personality win the hearts of the Luxembourger people, and Barthel goes on to head the Luxembourg Athletics Federation and then the Olympic and Sports Committee of Luxembourg. The main sports stadium in Luxembourg is named after him.

1958: February   Treaty of economic union signed

Effective in 1960, a treaty of economic union is signed by representatives of the three Benelux countries. The treaty establishes the European Economic Community (EEC), an institution whose role is to facilitate free trade of goods and services among the six member nations. By 1999, EEC membership increases to fifteen nations. The signing of the treaty and formation of the EEC is the second stage of European integration. (See 1951, 1992.)

1963: April   Luxembourg celebrates 1,000th anniversary

Luxembourg has been an independent state for 1,000 years.

1964: November 12   Grand Duchess Charlotte abdicates

Grand Duchess Charlotte abdicates, giving ducal powers to her son Jean. Jean's reign is marked by continued economic prosperity, as Luxembourg's economy shifts from dependence on steel to an emphasis on services, most notably international banking.

1967   Creation of volunteer military force

Conscription (the draft) is abolished, and newly-formed volunteer armed forces are made part of NATO (North Atlantic Treaty Organization) forces.

1969   Central University of Luxembourg founded

The Central University of Luxembourg is founded and has an average student population of 500. Most students, however, go to Belgium or France for their secondary education.

1970s   Wave of Portuguese immigration begins

Beginning in the 1970s, many Portuguese come to Luxembourg in search of jobs in the building and construction industries. Luxembourg is exceptional in that it has little intolerance or racial unrest. There is a positive attitude towards immigrants and a willingness to allow them to integrate and form culturally similar backgrounds; thus, Luxembourg is a melting pot, and its diversity is not perceived as a threat to national identity. An important factor contributing to this tolerance is the nation's low unemployment rate: social conflict is less likely to occur where jobs are plentiful. Moreover, the Luxembourger experience during World War II has taught the country about the dangers of discrimination; the country suffered greatly during that time, with most families losing members to German concentration camps, on the Russian front, or in the underground Resistance movement.

In 1993, immigrants comprise thirty-two percent of Luxembourg's resident population.

1970   First woman becomes mayor of Luxembourg City

Colette Flesch (b. 1937), educated both in France and the United States, becomes the first woman to achieve mayorship of Luxembourg City. She remains in the post of mayor for ten years, from 1970–80. In the 1980s, Flesch serves as Minister of Foreign Affairs, Minister of the Economy, and Minister of Justice.

1975   World iron and steel crisis

Iron and steel are the dominant force in Luxembourg's economy from the mid-nineteeenth century until 1975, when the world experiences a steel industry crisis. After that time, mining slows down and effectively ends in the mid-1980s, and many of the mining galleries are turned into museums. Currently, the entire steel industry has been restructured, with significant cutbacks in jobs. In the 1970s, ARBED–Luxembourg's steel company and Europe's third largest steelmaker–employs more than 27,000 people. At the end of the twentieth century, however, only about 6,000 are employed by ARBED.

1979   Death penalty abolished

1983   Formation of the Green Alternative Party

The Green Alternative Party (GAP) is formed in response to growing concerns about the environment. The party also opposes nuclear weapons and nuclear power, as well as Luxembourg's military policies.

The environment is a significant concern in Luxembourg. An especial concern is air quality: thick motor vehicle traffic and a concentration of industry has led to high levels of air pollutants. The nation now has almost forty sites to monitor the level of nitrogen dioxide in the air, double the number of such sites in Great Britain.

The GAP has been influential in Luxembourg, as a majority of Luxembourgers are willing to give up higher standards of living in order to keep their small nation clean. Luxembourg is one of the first European nations to split garbage into organic, plastics, and general, and also to make recycling common. Over one-fifth of Luxembourg's total land area is protected, in comparison to less than nine percent in the United States.

Late 1980s   High rankings in social insurance

During the late 1980s, Luxembourg is ranked second in the world after Sweden in the percentage of its national budget spent on social security and housing. The extensive system of social insurance covers practically all employees and their families.

1991   Archaeological site uncovered

Very few archaeological sites have been discovered in Luxembourg. In 1991, however, Bastendorf, a Celtic place of worship in the Ardennes, is uncovered. Excavations reveal that the religious shrine was used from the first century B.C. until the second half of the third century A.D., when it was probably abandoned in a rebellion.

1992   Entertainment data reported

According to a 1992 study by UNESCO (United Nations Educational, Scientific and Cultural Organization), Luxembourg has nearly a quarter of a million radio receivers and over 100,000 television sets.

1992   Skier Marc Giradelli wins Olympic medals

Though born in Austria, downhill skier Marc Giradelli is a citizen of Luxembourg when he wins two Olympic silver medals in Albertville, France, for the giant slalom and super-G. Giradelli also wins the World Cup skiing championship five times.

1992: July Maastricht Treaty approved

The signing of the Maastricht Treaty leads to the creation of the European Union (EU) in 1993, as well as setting goals for more ambitious political and monetary union in the future.

1993 Cape Verdeans receive aid from Luxembourg

In 1993, the largest portion of Luxembourg's international aid goes to the Cape Verde islands, financing such projects as the renovation of hospitals, provision of water to small villages, and building of schools. 11.6 million of the nearly 45 million francs of Luxembourg's direct food aid goes to the islands as well.

The Cape Verde islands and Luxembourg have close economic and humanitarian links. Several thousand Cape Verdean immigrants live in Luxembourg in the 1990s.

1995 New prime minister appointed

Jean-Claude Juncker becomes prime minister at the young age of forty. A leader in the Christian Socialist Party, Juncker is praised for his intellect and for the energy he brings to his job.

1995 Sixth Games of the Small European States

Under the patronage of Grand Duke Jean, the Olympic and Sports Committee of Luxembourg (C.O.S.L.) organizes the sixth Games of the Small European States. The games bring together the best sportsmen and women from European countries with a population of less than one million people—Andorra, Cyprus, Iceland, Liechtenstein, Luxembourg, Malta, Monaco, and San Marino. More than 1,000 athletes compete in sports including bicycling, basketball, judo, volleyball, tennis and swimming.

1995 Archaeological discovery

A farmer from the town of Vichten in the north discovers a seventy-two square yard (60 square meters) Roman mosaic on his land. The floor mosaic represents the nine muses, or mountain women, of Greek mythology. The mosaic is one of the largest and best-preserved to have been found north of the Alps. Its good condition is likely due to a landslide which covered the land with a layer of clay.

1995 Steps towards European monetary unity

The most difficult area within the European Union's negotiations towards European integration is the goal of monetary union; in other words, a single European currency and independent, autonomous central bank. Luxembourg is the only nation able to meet the initial 1995 criteria for a single currency.

There is great support in Luxembourg for European unity. Both the government and the people believe that there is no future outside of Union for a small nation like Luxembourg.

1997 Luxembourg is worst per capita carbon dioxide emitter

Luxembourg has the world's highest per capital level of carbon dioxide emissions, at thirty tons per person (versus twenty for the U.S.).

1998: May 3 Luxembourg to join European Monetary Union

European Union representatives meeting in Belgium announce plans to form an economic and monetary union (EMU) consisting of eleven European nations including Luxembourg. The centerpiece of the agreement is the adoption of a unified currency (the euro) in 1999.

Bibliography

Clark, Peter. *Luxembourg.* New York: Routledge, 1994.

Dolibois, John. *Pattern of Circles: An Ambassador's Story.* Kent, Oh.: Kent State University Press, 1989.

Eyck, F. Gunther. *The Benelux Countries: An Historical Survey.* Princeton, N.J.:Van Nostrand, 1959.

Gade, John Alleyne. *Luxembourg in the Middle Ages.* Leiden: Brill, 1951.

Hury, Carlo. *Luxembourg.* Oxford, England: Clio Press, 1981.

Newcomer, James. *The Grand Duchy of Luxembourg: The Evolution of Nationhood.* Luxembourg: Editions Emile Borschette, 1995.

Macedonia

Introduction

The Republic of Macedonia (recognized internationally as the Former Yugoslavia of Macedonia, or FYROM), which has been an independent republic since 1991, was one of the six republics that made up the former Yugoslavia following World War II (1939–45). The history of the country and region is complicated due to the multi-ethnic makeup of the region. Indeed, the country's name, Macedonia, has the same root as the French word, *macedoine*, which refers to a fruit salad. Throughout history, the geographic region of Macedonia has been the center of disputes between rival states. The very identity of the people has also been questioned. Although the inhabitants of the Republic of Macedonia refer to themselves as Macedonians, at various times Bulgarians, Greeks, and Serbs have challenged this assertion. Bulgarians claim that Slav-Macedonians (hereafter used to refer in a general sense to Slavic inhabitants of the geographic region of Macedonia) are merely Bulgarians who speak the western dialect of that language. In fact, well into the twentieth century, large numbers of—many observers would say, most—Slav-Macedonians considered themselves ethnic Bulgarians. Greeks link the term Macedonia to the ancient Macedonia of Alexander the Great and argue that one cannot be a Macedonian unless one is Greek. Slav-Macedonians who consider themselves Macedonian are referred to by Greeks as *Skopjians* (after the capital city of the Republic of Macedonia). On their part, Serbs traditionally referred to Slav-Macedonians as South Serbs.

The Republic of Macedonia accounts for the greater part of the land historically occupied by the Macedonian people (Vardar Macedonia); the remainder of geographical Macedonia comprises part of Bulgaria (Pirin Macedonia) and a province of Greece (Aegean Macedonia, which is the largest part of geographical Macedonia, the southern part of which has been populated by Greeks since antiquity). The relationship between the Greek and Yugoslavian regions has been a sensitive one diplomatically since the Republic of Macedonia declared independence, raising the specter of a unification of the two areas and prompting a trade embargo against the new nation by the Greek government. By 1995, however, many of

Greece's fears had been allayed through a compromise agreement, and the embargo had been lifted. However, Greece supported Macedonia's territorial integrity (although it still objects to the name) and became the largest source of foreign investment in the country.

Bulgaria, likewise had reservations about an independent Macedonia. Although Bulgaria was the first country to grant recognition to the Republic of Macedonia, it did not recognize the existence of a Macedonian nationality or Macedonian language.

Macedonia is located in the southwestern part of the Balkan Peninsula. Much of the republic of Macedonia is covered with mountains or high plains. The western part of the country is dominated by the Plakenska mountain chain; other mountain ranges are found along the Serbian and Bulgarian borders. The valley of the Vardar River, known as the Vardar Plain, dominates the center of the country. The capital, Skopje, is located in the northwest. As of 1999, the Macedonian republic has an estimated population of over two million people, about two-thirds of whom are ethnic Macedonians. Albanians comprise between twenty-five and thirty percent of the population, while Bulgarians, Greeks, Gypsies, Serbs, and Turks round out the remainder.

History

The Illyrians, claimed as the forerunners of today's Albanians, inhabited the geographic region of Macedonia as early as the second millenium B.C. The region of Macedonia, however, does not appear in history until the first millenium B.C. with the rise of a Macedonian kingdom. This kingdom reached its height under Alexander III (the Great, r. 336–323) who turned his state into an empire that stretched from Greece to Persia. Although historians continue to debate the origins of the ancient Macedonians, it is generally assumed that by the time of Alexander they had been Hellenized (Greekified) culturally, if they were not actually Greek in origin. Macedonia continued as an independent state for nearly two centuries after Alexander's death until the conquest of the Romans in 146 B.C. whereupon it became a province of the Roman Republic (later Empire).

MACEDONIA

0   20   40   60 Miles

0   2C   40   60 Kilometers

YUGOSLAVIA

BULGARIA

Luke

Rujen
7,388 ft.
2252 m.

Kumanovu

⊛ Skopje

Tetovo

Kočani

Dobrino

Veles

Bregalnica

Blatec

Korab
9,068 ft.
2764 m.

JAKUPICA

OGRAŽDEN

Debar

Treska

Kozjak
5,728 ft.
1746 m.

Strumica

Vardar

BELASICA

Sopotnica

Prilep

Struga

Gevgelija

Límni
Doïránis

Crno

NIDŽE

Idhoméni

Prespansko
Jezero

Bitola

Ohridsko
Jezero

Flórina

ALBANIA

GREECE

Macedonia

CRNA GORA

Today's Macedonians, however, are not descended from the ancient Macedonians. Rather, they are the descendents of Slavs who migrated to Macedonia in the sixth and seventh centuries A.D. as part of the great wave of migration that brought the South Slavic people to the Balkans. For much of the medieval period, the Byzantine Empire, Serbia, and Bulgaria competed for control of the region. During the ninth century, a period when Macedonia was ruled by Bulgaria, the Bulgarian ruler converted to Christianity and monks were invited into the country to Christianize the population, a process that was largely completed at this time. The religious frescoes painted in Macedonian churches in the Middle Ages, particularly the St. Sofia Church at Ohrid, remain one of the region's great artistic treasures.

Following the bitter defeat of the Serbian forces at the Battle of Kosovo Polje in 1389, the Ottoman conquest of Macedonia began; it was completed in about half a century, and the region came under Turkish rule for over three hundred years. Major rebellions occurred in the sixteenth and seventeenth centuries, and many Slavs fled northward to the Serb-Hungarian border region to escape Turkish retaliation at

the end of the seventeenth century. In the first half of the nineteenth century, a new wave of nationalism swept through the region inspired by the successful Serbian and Greek struggles for independence. The Slav-Macedonians, however, were among the last of the Balkan nationalities to undergo a national awakening. Nevertheless, by the middle of the century, a growing national consciousness was evident among the Macedonian Slavs, who had begun the publication of works in their own dialect. For all of the nineteenth and much of the twentieth centuries, however, most Slav-Macedonians who had developed a national consciousness identified themselves as Bulgarians. This fact was largely due to the creation of the Bulgarian Exarchate, an *autocephalus* (self-governing) Bulgarian Orthodox Church, in 1870 which set up Bulgarian schools throughout geographic Macedonia.

By the late nineteenth century, Turkish power in the region had declined to the point that the Ottomans lost control of Macedonia, which was briefly placed under Bulgarian rule as a result of the aborted Treaty of San Stefano in 1878. It was returned later that year by the Congress of Berlin, but over the next thirty years Ottoman control was contested by Bulgaria, Greece, and Serbia. By 1893 a Macedonian nationalist political group was formed, the Internal Macedonian Revolutionary Organization, or IMRO, which quickly gained support and created a considerable underground network that remained active until World War II. By 1895, IMRO had been divided into two factions, one favoring complete autonomy, the other, the Supreme Committee (or Supremists) seeking union with Bulgaria. Socialism had also come to Macedonia by the turn of the twentieth century.

Competing claims to Macedonia led to the first Balkan War in 1912, when the Balkan League, consisting of Serbia, Greece, Montenegro, and Bulgaria demanded that Turkey honor previous treaty provisions guaranteeing Macedonian autonomy. When the Turks refused, they were routed by the Balkan forces in a surprisingly rapid defeat, and the Ottomans ceded nearly all their European territory. However, the victors in the Balkan War were unable to agree on the division of Macedonia. Bulgaria, in particular, resented the allotment of only a fraction of the region and the failure to receive the Aegean port of Thessaloniki (which was captured by Greek forces a few hours ahead of the Bulgarians). A second war followed in June, 1913, initiated by a Bulgarian attack on Greece and Serbia. The second Balkan war ended in a month with the essential division of Macedonia that prevailed, with modification, for the rest of the twentieth century: the northern portion came under Serbian (later Yugoslav) control, while a smaller area to the east remained Bulgarian, and the southern portion became part of Greece.

Smarting over its failure to gain the majority of Macedonia, Bulgaria joined the Central Powers in 1915 and temporarily regained control of Macedonia. The Bulgarian defeat in 1918, however, forced it to withdraw. In December, 1918, the Serbian part of Macedonia became part of the newly formed

Kingdom of Serbs, Croats, and Slovenes (renamed Yugoslavia in 1929). In addition to political control, the Serbs also imposed their language and culture on the Macedonians. IMRO remained active between the World Wars, although now it was largely a terrorist organization which, among other acts, recruited the assassin who killed the Yugoslavian king, Alexander I, in 1934. During World War II, Macedonia was occupied by Bulgaria and Italian-occupied Albania, both of which were allied with Nazi Germany. In the midst of the war, Macedonia's Communists pledged their support for Partisan leader Josip Broz Tito (1892–1980), who promised to give Macedonia the status of a republic after the war. By 1944 the Macedonian Communists had proclaimed a republic, and in 1945 it became one of the six republics in the Yugoslavian federation formed by Tito.

From 1945 to 1991 Macedonia remained a Yugoslav republic and reaped some of the benefits of industrialization and material progress the country enjoyed under Tito, as well as the containment of the traditional rivalries among the region's ethnic and religious groups. Following Tito's death in 1980, however, these rivalries intensified until they brought Yugoslavia to the point of collapse within a decade. Like its fellow republics Slovenia and Croatia, Macedonia withdrew from Yugoslavia in 1991. Although Macedonia was not immediately confronted with a military crisis such as those that occurred in Croatia or Bosnia, its independence brought with it a unique problem: the hostile reaction of the Greek government, which feared future Macedonian claims against Greek Macedonia. Greece created diplomatic problems for the new republic and imposed a crippling trade embargo. By 1995, however, it had lifted the embargo in exchange for concessions by Macedonia, which included the removal of a Greek symbol from its national flag. The situation was explosive enough that the Macedonian concessions prompted an assassination attempt against the President Kiro Gligorov, who was seriously injured but survived and was eventually able to resume his duties. In 1999, Macedonia was caught up in regional politics as thousands of Albanian refugees from Kosovo fled the turmoil in that province. Many observers feared that the large influx of Albanian refugees into Macedonia which already had a large Albanian minority threatened the continued existence of an independent Macedonia.

Timeline

2nd millennium B.C.   Illyrians occupy Macedonia

The part of ancient Macedonia that today makes up the Republic of Macedonia is inhabited by Illyrians, who also occupy other parts of the Balkans, including present-day Serbia and Croatia.

4th century B.C.   Macedonia is unified by Philip and Alexander

After centuries of warfare between rival kings, Philip II (r. 359–336) of the Argeadae dynasty gains control of Macedonia. His son, Alexander (r. 336–323), extends Philip's kingdom into a vast empire.

146 B.C.   Roman occupation

The Romans annex Macedonia, turning it into a Roman province.

4th century A.D.   Macedonia comes under Byzantine control

When the Roman empire is divided into eastern and western halves, Macedonia becomes part of the eastern, or Byzantine, empire. By this time most of the population has converted to Christianity.

Slavic Migration and Conquest

6th–7th centuries   Slavic migration

Slavs migrate to Macedonia. Byzantine influence in the region is temporarily weakened, but it is restored during the reign of Heraclitus (610–641).

9th century   Bulgarian rule

The kingdom established by the Bulgars gains control of most of Macedonia. For the rest of the medieval period, control of Macedonia is contested by the Byzantine Empire, Serbia, and Bulgaria.

9th century   First church is built

Macedonia's oldest church, St. Panteleimon, is built. Saint Kliment is said to have preached the gospel in Slavic here for the first time, and he is buried near the church.

9th century   Cyril and Methodius are dispatched to Moravia

Two Greek monks from Salonika (Thessaloniki), Macedonia (in the twentieth-century Greek province), are sent to Moravia to teach the Holy Scriptures. They develop a Slav alphabet based on the Greek alphabet. Eventually known as Cyrillic, this alphabet is used in Russia and other parts of Eastern Europe. The two monks are later canonized.

865   Bulgarian ruler converts to Christianity

During the period when Macedonia is ruled by Bulgaria, the Bulgarian ruler, Khan Boris, converts to Christianity and invites two disciples of Cyril and Methodius, named Naum and Kliment (who is later canonized), to the region to convert the population. They translate the Scriptures into the local

Slavic dialect using a modified version of the alphabet devised by their teachers. (Scholars now believe that the original language was what is now called *Glagolitic,* while Cyrillic was actually the streamlined version devised by Naum and Kliment.) The Macedonian town of Ohrid becomes the religious center from which the Eastern Orthodox faith is spread throughout Bulgaria, Serbia, and Kievan Rus (present-day Ukraine).

976–1018   Macedonia at the center of a new empire

The former Bulgarian kingdom declines and a ruler named Samuilo comes to power. He builds a great but short-lived Slavic empire with Macedonia at its center. Its capital city is Ohrid, and the Macedonian church gets its own *patriarchate* (ecclesiastical jurisdiction).

1018   Byzantine rule

After Samuilo's empire falls to the Byzantine emperor Basil II, Macedonia reverts to Byzantine rule.

Early 11th century   St. Sofia Church is built

Macedonia's most treasured church, the St. Sofia at Ohrid, is built on ruins of a fifth-century basilica. Its historic frescoes will later be plastered over by the Ottoman Turks, who turn the building into a mosque, constructing a *minaret* (a tower from which the Muslim prayer call is made) at the site. (It, in turn, is torn down when Macedonia is liberated from the Ottomans in the twentieth century.)

Late 13th century   Serbia's Nemanja dynasty conquers Macedonia

The Nemanja princes of Serbia (then called Rascia) gain control of most of Macedonia during the reign of Stefan Dusan (r. 1331–55), the last great ruler of the dynasty. After Dusan's death, ten different Serbian lords claim portions of Macedonia.

14th century   Macedonia becomes a cultural center

Between the eras of Byzantine art and the Italian Renaissance, a unique artistic style flourishes in Macedonia, spreading from Salonika northward to Serbia. With support from Serbian royalty, a monastery in Hilander becomes a center for Slavic scholarship.

Ottoman Rule

1389   Turkish conquest begins

The Battle of Kosovo Polje marks the beginning of Turkish conquest of Macedonia, which continues for roughly 400 years.

15th–20th centuries   Ottoman rule

The Ottoman Turks rule Macedonia from the fifteenth century until the Balkan War of 1912. Much of the Christian nobility flees to the west and north. The Muslim population grows with the migration of Yuruk Turks to the region from Anatolia and is further augmented by a wave of conversions by Christians in the late seventeenth century. Additional Muslim emigration occurs when other areas of the Balkans, including Serbia and Bosnia, succeed in expelling the Ottomans, and their Muslim populations flee to Macedonia.

The Ottoman Period

15th century   Immigration by Sephardic Jews

Many Sephardic Jews expelled from Spain in the era of the Inquisition emigrate to Macedonia.

1560s   Rebellions against the Ottomans

Macedonians stage their first significant rebellions against the Turks during the reign of Ottoman Sultan Suleiman the Magnificent. Another major revolt takes place at the turn of the century.

1689   Habsburgs occupy Skopje

Austro-Turkish wars spur new Macedonian revolts. A major rebellion is led by a miner named Karpos as Austrian armies move toward the Macedonian city of Skopje. However, the Austrians retreat, the rebellion is put down, and its leader is executed. Large numbers of Macedonians are sold into slavery by the Turks. Many flee to the Hungarian border area of Vojvodina, decreasing the Slavic population of the region, which has already been diluted by Albanian immigration in the course of the century.

18th century   Growth of a Macedonian merchant class

Greek and Slavic Macedonians play a significant role in the development of a merchant class to trade the goods of landlords in the Balkans to Western Europe as the urban economy of their own region declines. These well-to-do Macedonians will help support the competing cultural revivals in the nineteenth century.

1792   First Macedonian primer is published

The first primer in the Slavic Macedonian language is published by Marko Teodorovic in Vienna.

1807–08   Serb revolt sparks Macedonian uprisings

The Serbian revolt against the Ottomans led by George Petrovic (Karadjordje) (1762–1817) leads to a wave of uprisings in Macedonia and helps further a Macedonian national consciousness.

The Ottoman Period

The Ottoman Empire began in western Turkey (Anatolia) in the thirteenth century under the leadership of Osman (also spelled Othman) I. Born in Bithynia in 1259, Osman began the conquest of neighboring countries and began one of the most politically influential reigns that lasted into the twentieth century. The rise of the Ottoman Empire was directly connected to the rise of Islam. Many of the battles fought were for religious reasons as well as for territory.

The Ottoman Empire, soon after its inception, became a great threat to the crumbling Byzantine Empire. Constantinople, the jewel in the Byzantine crown, resisted conquest many times. Finally, under the leadership of Sultan Mehmed (1451–81), Constantinople fell and became the capital of the Ottoman Empire.

Religious and political life under the Ottoman were one and the same. The Sultan was the supreme ruler. He was also the head of Islam. The crown passed from father to son. However, the firstborn son was not automatically entitled to be the next leader. With the death of the Sultan, and often before, there was wholesale bloodshed to eliminate all rivals, including brothers and nephews. This ensured that there would be no attempts at a coup d'état. This system was revised at times to the simple imprisonment of rivals.

The main military units of the Ottomans were the Janissaries. These fighting men were taken from their families as young children. Often these were Christian children who were now educated in the ways of Islam. At times the Janissaries became too powerful and had to be put down by the ruling Sultan.

The greatest ruler of the Ottoman Empire was Suleyman the Magnificent who ruled from 1520–66. Under his reign the Ottoman Empire extended into the present day Balkan countries and as far north as Vienna, Austria. The Europeans were horrified by this expansion and declared war on the Ottomans and defeated them in the naval battle of Lepanto in 1571. Finally the Austrian Habsburg rulers were able to contain the Ottomans and expand their empire into the Balkans.

At the end of the nineteenth century, the Greeks and the Serbs had obtained virtual independence from the Ottomans, and the end of a once great empire was in sight. While Europe had undergone the Renaissance, the Enlightenment, and the Industrial Revolution, the Ottoman Empire rejected these influences as being too radical for their people. They restricted the flow of information and chose to maintain strict religious and governmental control. The end of Ottoman control in Europe came in the First Balkan War (1912–13) in which Greece, Montenegro, Serbia, and Bulgaria joined forces to defeat the Ottomans.

During the First World War, (1914–18), the Ottoman Empire allied with the Central Powers and suffered a humiliating defeat. In the Treaty of Sévres, the Ottoman Empire lost all of its territory in the Middle East, and much of its territory in Asia Minor. The disaster of the First World War signaled the end of the Ottoman Empire and the beginning of modern-day Turkey. Although the Ottoman sultan remained in Constantinople (renamed in 1930, "Istanbul"), Turkish nationalists under the leadership of Mustafa Kemal (Ataturk) (1881–1938) challenged his authority and, in 1922, repulsed Greek forces that had occupied parts of Asia Minor under the Treaty of Sévres, overthrew the sultan, declared the Ottoman Empire dissolved, and proclaimed a new Republic of Turkey. That following year, Kemal succeeded in overturning the 1919 peace settlement with the signing of a new treaty in Lausanne, Switzerland. Under the terms of this new treaty, Turkey reacquired much of the territory—particularly, in Asia Minor—that it had lost at Sévres. Kemal is better known by his adopted name "Ataturk", which means "father of the Turks". Ataturk, who ruled from 1923 to 1938, outlawed the existence of a religious state and brought Turkey a more western type of government. He changed from the Arabic alphabet to Roman letters and established new civil and penal codes.

1821–29 Greek war of independence

Like the Serbian uprisings earlier in the century, the Greek struggle for independence increases the desire of all Balkan Christians to freedom from Ottoman rule.

Mid-19th century Growth of Macedonian nationalism

A national consciousness grows among the Macedonian Slavs, who have begun publishing works in their own central Macedonian dialect and agitating for its introduction into the schools. Not until much later do most Slavs in Macedonia adopt a Macedonian national identity. Well into the twentieth century, the majority of Slav-Macedonians identify themselves as Bulgarians.

1861 Macedonian folk poetry collection is published

An anthology of over 600 Macedonian folk poems collected by educator Dimitar Miladinov (1810–62) and his brother Konstantin is published in Zagreb under the title *Bulgarian Folk Poems.* In the same year, Dimitar Miladinov is jailed by the Turks for his nationalist activities and charged with treason. Attempting to aid his brother, Konstantin is also arrested, and both brothers die in prison in 1862.

1870 Bulgarian Orthodox Church is established

In an effort to undermine the growing Greek nationalist effort in Macedonia, the Ottoman government grants the Bulgarians their own branch of the Eastern Orthodox Church, called the Bulgarian *Exarchate,* which attempts to spread Bulgarian culture and religion. Bulgarian nationalists, in turn, use the Exarchate as a vehicle to unite the region with Bulgaria. Bulgarian bishops and schools replace those of the Greek patriarchate in Macedonia.

1878 Turkey loses and regains Macedonia

Macedonia changes hands twice in a year. Most of the region is briefly turned over to a newly created Greater Bulgaria in the Treaty of San Stefano (1878), but this decision is reversed by the Treaty of Berlin in the same year because it violates provisions of a previous treaty between Russia and Austria. Macedonia returns to Turkish control, but Greece, Bulgaria, and Serbia continue to claim rights to the region for the next thirty years.

1893 Nationalist group is formed

The Internal Macedonian Revolutionary Organization (IMRO) is formed by Slav-Macedonians seeking a popular mass uprising to free the region from Turkish rule. Many of its original members are craftsmen or students. It spreads throughout Macedonia, eventually forming an underground network divided into revolutionary regions and subdivided into districts and communes. It maintains an armed force of several thousand fighters active both in urban and rural areas.

IMRO also assumes some functions of a government, maintaining its own courts and revolutionary tribunals and operating an underground press, which publishes several books. Within a few years, the group divides into competing factions, one favoring incorporation into Bulgaria and the other supporting total autonomy. IMRO remains active as an underground group until World War II, and a political party still carries its name in the 1990s.

1894 Socialist group is established

Macedonia's first socialist movement takes root in Veles, founded by a carpenter named Vasil Glavinov.

1903: August 2 Macedonian uprising is crushed by the Turks

A revolt is organized by IMRO in the Monastir district, and the revolutionaries proclaim a socialist republic in the town of Krusevo. The uprising lasts two months before being crushed by 300,000 Turkish troops, who carry out massacres and torch villages.

1908 The Young Turks revolt in Macedonia

The Young Turks, a group of reformist Turkish officers, take control of Macedonia and attempt to modernize and develop the region. These officers succeed in overthrowing Sultan Abdul Hamid II and promote a new constitution for the Ottoman Empire. The Young Turks support a policy of Ottomanism which promises autonomy for the empire's subject peoples, but at the price of maintaining the territorial integrity of the Ottoman Empire.

1912: October The first Balkan war

The newly formed Balkan League, consisting of Greece, Serbia, Montenegro, and Bulgaria, demands that Turkey honor the provisions of the Treaty of Berlin mandating the creation of separate autonomous regions in Macedonia. When Turkey refuses, the League declares war and drives the Turks out of the Balkans.

1913: May 30 Turkey cedes European possessions

Following its defeat at the hands of the Balkan League, Turkey gives up nearly all its European possessions.

1913 Second Balkan war

Bulgaria attacks Serbia and Greece in a dispute over control of Macedonia and is defeated within a month by its former Balkan allies, with Romania and Turkey joining in as well. Macedonia is then divided between Greece, Serbia, and Bulgaria. These countries impose their own language and culture on their Macedonian populations.

As a result of these wars, population shifts begin in the region that continue intermittently for the next three decades.

Large numbers of Slav-Macedonians from Greece and Serbia with a Bulgarian national consciousness flee to Bulgaria where they form an important pressure group that advocates Bulgarian conquest of all of Macedonia.

1914–18   World War I

Induced by the prospect of territorial gain in Macedonia, Bulgaria joins the Central Powers (Austria-Hungary, Germany, and the Ottoman Empire) in 1915. During the course of the war, Bulgaria fights against Greece and Serbia, occupying parts of Macedonia but must cede these and other territories at the end of the war when the Central Powers are defeated.

1915   Yugoslav committee-in-exile is formed

Serb, Croat, and Slovene political leaders in exile form a Yugoslav committee to work toward the creation of an independent state for the South Slavs. The Yugoslav committee does not recognize the existence of a Macedonian nationality.

1917   Serbian government supports Yugoslav committee

The Serbian government-in-exile agrees to support the creation of a Yugoslav state in the form of a constitutional monarchy ruled by the Karadjordjevic dynasty.

1918: November 1918   End of World War I

By the time World War I ends in victory for the Triple Entente, the South Slavic lands have been liberated from Austro-Hungarian control. The new Kingdom of Serbs, Croats, and Slovenes retains Serbian Macedonia.

The Kingdom of Serbs, Croats, and Slovenes

1918: December 1   South Slavs win independence

Representatives of the South Slavic peoples meet with Regent Alexander Karadjordjevic, prince of Serbia, and proclaim an independent Kingdom of Serbs, Croats, and Slovenes. Alexander becomes the country's ruler, ascending the throne as Alexander I. (Ten years later he renames the country Yugoslavia.) As part of Serbia, the northern, Slavic portion of Macedonia becomes part of the new kingdom, which inherits a legacy of ethnic rivalry between Serbs, Croats, and Bosnian Muslims. Economically, the kingdom is faced with wartime damage, heavy debt, labor shortages, and a pressing need for land reform created by centuries of feudalism. Politically it is dominated by Serbia.

1920s   Macedonian refugees settle in Bulgaria

Large numbers of Slav-Macedonian refugees who identify themselves as Bulgarians settle in Bulgaria during the 1920s. Many leave following a population exchange between Greece

and Bulgaria. Greek Macedonia becomes thoroughly Greek in ethnic composition as a result of the population exchange with Bulgaria as well as a population exchange with Turkey. The latter exchange results in approximately 400,000 Turks in Greek Macedonia leaving for Turkey, while over 1, 250,000 Greeks from Turkey settle in Greece, mostly in Macedonia on land vacated by Turks or Slav-Macedonians.

1920s–30s   IMRO goes underground as a terrorist group

The Macedonian liberation group (see 1893) remains active between the world wars as a terrorist organization. The large refugee population forms the basis for a reinvigorated IMRO which comes under the control of Bulgarian leader Ivan Mihailov, who uses its fighters to assassinate his political opponents. Indeed, IMRO becomes the most powerful political force within Bulgaria for over a decade. Political rivals are gunned down in the streets of Bulgaria's capital, Sofia. In 1923 IMRO leads the bloody and successful coup against the agrarian prime minister of Bulgaria, Alexander Stambuliskii. An IMRO member, working for the Croatian Ustashe, also serves as the assassin of King Alexander I of Yugoslavia in 1934. Later that same year, however, a military coup in Bulgaria suppresses the organization whereby Mihailov is forced to reorganize.

1926   Natural history museum is established

The Natural History of Museum of Macedonia is founded in Skopje.

1929: January 6   Royal dictatorship is established

Using the assassination of Croat leader Stjepan Radic as justification for tighter government control, King Alexander imposes a royal dictatorship, overruling the constitution, abolishing the country's parliamentary government, and outlawing political parties. He also renames the country Yugoslavia.

1931   Constitution is adopted

The autocratic rule of King Alexander is modified by adoption of a constitution, and political parties are legalized. However, a number of other freedoms are still restricted.

1931   Global economic slump hits Yugoslavia

Yugoslavia feel the effects of the worldwide economic depression, which leads to bankruptcies and unemployment. The crisis is worsened by weather conditions that produce famine in rural areas.

1934   Balkan Entente is formed

Yugoslavia joins Greece, Turkey, and Romania in a defensive alliance known as the Balkan Entente. This treaty is aimed at stifling Bulgarian attempts at revising the World War I peace settlement.

1934: October   King Alexander is assassinated

King Alexander is murdered by an assassin working for the Ustashe, the Croatian pro-fascist liberation organization, and recruited through the Macedonian liberation group IMRO. Alexander's death brings on fears that Yugoslavia will collapse. Alexander's son, Peter II, becomes the country's regent. Three officials are appointed to rule for him while he is still a minor.

World War II

1941: March 25   Pact is signed with Germany

Under military pressure and surrounded by pro-Nazi countries, Yugoslavia's government, under Alexander's cousin, Prince Paul, agrees to join the Tripartite Pact with the Nazi allies in return for German guarantees of nonaggression.

1941: March 27   Yugoslavian government is overthrown

Because of his cooperation with Nazi Germany, the government of Prince Paul is overthrown, and sixteen-year-old Peter II, the regent and son of slain king Alexander, is declared king. In spontaneous demonstrations, the populace expresses its hostility toward the Nazis and their allies.

1941: April 6   Germany bombs and invades Yugoslavia

The Yugoslavian capital of Belgrade is bombed by the German *Luftwaffe* (air force), and ground forces invade the country. The government goes into exile, and the military surrenders unconditionally. Yugoslavia is divided among the Axis powers, except for Croatia, which is placed under the rule of the Independent State of Croatia (NDH), a nominally independent state controlled by the Nazis and led by Ante Pavelic, head of the Croatian Ustashe. The middle and eastern parts of Macedonia are occupied by Bulgaria, while Albanian takes the west.

Partisans (anti-Nazi freedom fighters) organized by long-time communist leader Josip Broz (1892–1980), popularly known as Tito, carry out rebellions against Croatia's Nazi-controlled puppet government run by the Ustashe. The Ustashe also faces armed opposition by Serbian Chetnik forces, led by General Draza Mihajlovic.

Bulgaria joins the German invasion and occupies most of Serbian and Greek Macedonia. The Bulgarian occupation is extremely harsh and antagonizes many Slav-Macedonians who still identify themselves as Bulgarians.

1943   Macedonian communists join Tito

Macedonian communists ally themselves with anti-Nazi Partisan leader Josip Broz Tito in return for his pledge to sponsor a Macedonian republic after the war.

1944   Communist republic is proclaimed in Macedonia

Macedonia communists proclaim the Macedonian People's Republic, with Macedonian as its official language.

1944   National library is founded

The Kliment Ohridski National and University Library is established in Skopje. By the 1990s its holdings include 2.16 million volumes.

1944   Orchestra is founded in Skopje

Following the city's liberation from the Nazis, Skopje's first philharmonic orchestra is established. After the war, it will go on to perform throughout Eastern and Western Europe and in other parts of the world, including Mexico.

1944: October 20   Red army liberates Belgrade

The Soviet army, aided by Partisan forces, marches into Belgrade and liberates Yugoslavia. Some 1.7 million Yugoslavians have died since the war began—more than half at the hands of other Yugoslavs. Cities are left in ruins and the countryside is also devastated.

1945   Art school is established

The Art School is established in Skopje through an effort spearheaded by artists Nikola Martinoski and Lazar Licenoski. (It is later renamed the School of Applied Arts.)

1945   Modern Macedonian alphabet is adopted

Macedonian scholars adopt a modern alphabet and grammar to turn the central Macedonian dialect into a formal written language. This is part of a greater effort to solidify the existence of a distinct Macedonian national identity that is not subject to foreign—especially, Bulgarian—influence.

Yugoslavia under Tito

1945: November 29   People's Republic of Yugoslavia is proclaimed

The Yugoslavian monarchy is officially dissolved, and the country is renamed the Federal People's Republic of Yugoslavia (later renamed the Socialist Federal Republic of Yugoslavia). Macedonia becomes one of six republics in a Soviet-style federation with a strong centralized government. Macedonian becomes one of the country's official languages and remains the official language of the Macedonian republic, used in government, schools, and the church.

For a time, Tito and Bulgarian Communist leader Georgi Dimitrov (1882–1949) discuss uniting Yugoslavia and Bulgaria to form a true Yugoslavia (land of the South Slavs). Under this scheme, the Republic of Macedonia would be enlarged to include Pirin Macedonia (the region of geographical Macedonia within Bulgaria). In the event of a Greek Communist Party (KKE) victory in the Greek Civil War (1946–49), Aegean Macedonia (the Greek portion of geographic Macedonia) would also join this enlarged Republic of Macedonia. Although the ouster of Yugoslavia from the Communist Information Bureau (*Cominform*) in 1948 and the defeat of the KKE in 1949 prevent the creation of an enlarged Macedonian republic, it remains the goal of Macedonian nationalists and causes problems in Yugoslavia's relations with Bulgaria and Greece.

1947–51 Five-year economic plan is adopted

A five-year plan modeled on those of the Soviet Union is inaugurated, emphasizing industrial development.

1948 Yugoslav-Soviet split

Following World War II, Yugoslavia allies itself closely with the Soviet Union. By 1948, however, growing tensions, begun with disagreement over the management of joint enterprise, create a rift between the two countries. Unlike the heads of other Eastern European satellites of the U.S.S.R., Tito does not owe his influence or position to the Soviets and will not allow them to dictate his policies. He becomes the only Eastern European communist leader to break with Stalin and remain in power. The break between the two countries is solidified when the Cominform denounces the Yugoslav Communist Party. Trade with the Soviet Union and other Communist bloc nations is reduced, and Tito is forced to turn to the West for new trade partners.

1948 Graphic arts program is inaugurated

The initiation of a graphic arts department at the School of Applied Arts in Skopje furthers the development of wood engravings, lithographs, and other forms of graphic art by Macedonian artists including Mira Spirovska (b. 1939) and Spase Kunovski.

1949 Greek Communists lose Greek Civil War

The KKE loses the Greek Civil War. This important defeat for Macedonian nationalists dashes hopes of incorporating Greek Macedonia into Yugoslavia. At the time of the KKE's defeat, ethnic Macedonians comprise about forty percent of the Democratic Army (the Communist forces). Following their loss, those Macedonians who had sided with the KKE flee north, and many of them settle in the Republic of Macedonia.

The defeat of the communists in the Greek Civil War helps complete the almost total Hellenization (Greekification) of Greek Macedonia that had begun in the aftermath of the Balkan Wars. While estimates vary, most independent observers place the number of ethnic Macedonians in Greece at under 100,000 in the aftermath of the Greek Civil War.

1949 Art gallery is established

The Art Gallery is founded in Skopje with a collection including artworks produced from the Middle Ages to the present by artists from the Yugoslav republics.

1949 Museums of archaeology and ethnography are founded

The Archaeological Museum of Macedonia and the Ethnographical Museum are established in Skopje.

1949 University of Skopje is founded

The University of Skopje is established. It offers programs in agriculture, forestry, veterinary medicine, engineering, geology, and other fields.

1950s Self-management economic system develops

Yugoslavia develops its communist economy through a system in which enterprises are locally managed by workers' councils, and economic goals are first formulated at the local level and coordinated centrally. Agricultural collectivization, one of the hallmarks of Soviet-style communism, is slowed and then abandoned altogether.

1950 Restoration of historic frescoes begins

Restoration work on medieval church frescoes, including many in Macedonia, begins. The most illustrious frescoes are those created between the eleventh and fourteenth centuries in St. Sofia Church at Ohrid.

1951 Yugoslavia forms defense pact with U.S.

Yugoslavia and the United States conclude a defense agreement.

1953 Constitution incorporates workers' councils

The constitution is modified to authorize the creation of workers' councils to run state-owned enterprises.

1960 Macedonian arts and sciences association is formed

The Association of Sciences and Arts is founded in Bitola.

1961: September Conference of nonaligned nations

A conference of nonaligned nations held in Belgrade confirms Yugoslavia's international leadership position in the group of nations that have proclaimed their neutrality in the Cold War, including India and Egypt. Their mutual goals include disarmament, peaceful coexistence among nations, and the elimination of rival political blocs.

1962   Struga poetry evenings are inaugurated

Struga Evenings of Poetry, Macedonia's annual international poetry festival, is launched in the city of Struga. Around two hundred poets and literary critics from Yugoslavia and other nations around the world gather to read and discuss poetry.

1963   Political prisoners are released

As part of a wider liberalization trend, the Yugoslavian government releases 2,500 political prisoners.

1965   Market socialism is introduced

Yugoslavian industry grows rapidly in the 1950s, but imports still exceed exports by a wide margin, and some industrial inefficiency remains. In response to an economic crisis of the early 1960s, the government modifies its economic policy, eliminating price controls and export subsidies while cutting import duties. The move toward decentralized economic self-management is stepped up as well.

1966   Tito purges top Serbian leader

Because the Serbs have once again risen to the position of political dominance they enjoyed following World War I, Tito dismisses Serbian leader Aleksandar Rankovic and disbands Rankovic's powerful secret police. During the period that follows, the balance of power among Yugoslavia's ethnic groups becomes more equal, and relations between the groups improve.

1966: September 10   Death of St. Kliment is commemorated

The one thousandth anniversary of the death of St. Kliment is commemorated at Ohrid by the Macedonian archbishop, representatives of other religions, and the prime minister of Macedonia.

1967–68   Yugoslavian republics gain more power

Constitutional reforms expand the power of the individual Yugoslavian republics relative to the central government.

1967   Academy of arts and sciences is formed

The Macedonian Academy of Arts and Sciences is established in Skopje. It has departments of medical and biological science, as well as the technical sciences and mathematics.

1967   Macedonian Orthodox Church is founded

Macedonians break away from the Serbian Orthodox Church to form their own church, although neither the Serbian nor any other Orthodox church recognizes it.

1968   Tito condemns Soviet invasion of Czechoslovakia

Yugoslav leader Tito condemns the U.S.S.R.'s forcible halt to Czech liberalization under Alexander Dubcek, straining Yugoslav relations with the Soviets. Tito prepares the Yugoslavian military to resist a possible Soviet invasion.

1970s   Experimental art thrives

The Macedonian art scene sees the growth of experimental art that combines video, film, and photography to reinterpret familiar objects and produce innovations in the use of materials and color. Two practitioners of the new art are Simon Semov and Nicola Fidanovski.

1971   Tito sets up collective presidency

Tito creates the framework for a collective presidency, intended to ease tensions when he retires. Intended to provide comparable representation to Serbs, Croats, and Muslims at the top level of government, it is to have two members from each of the country's three main ethnic constituencies and one additional member. Under his retirement, however, Tito remains president for life.

1971: December   Tito imposes crackdown on liberals

Tito purges Croatia's liberal Communist leadership. Thousands of officials are arrested and either jailed or expelled from the party. Included among them is future Croatian president Franjo Tudjman (b. 1922). The purges continue for over a year and spread to the other Yugoslavian republics also. They expand into a wider crackdown that includes press censorship, dissident arrests, and pressure on university professors to confirm with the party line.

1974   New constitution is adopted

A new constitution gives the individual republics and autonomous provinces a greater role in governing themselves. The role of the federal government is limited to foreign policy, defense matters, and certain economic powers. The constitution also formalizes and simplifies the structure of the collective presidency that Tito has created to assure the continuity of the Yugoslav federation after his death.

1979   University is founded at Bitola

Bitola University becomes Macedonia's second university.

Tito's Yugoslavia Collapses

1980s   Politics and the economy pull Yugoslavia apart

Following the death of Marshal Tito, the stability achieved over decades is threatened by age-old ethnic rivalries and affected by widespread economic problems in Russia and Eastern Europe, as well as Yugoslavia's inability to service its excessive foreign debt. The country is beset by inflation, labor strikes, food shortages, and financial scandals, as tensions grow between the individual republics. Serbian nationalism, which had been kept in check during the Tito years, gains strength in spite of government crackdowns.

1980: May 4   Death of Marshal Tito

Josip Broz Tito, who had greater successful in unifying the South Slavs than any other leader, dies. He is widely mourned at home, and forty-nine foreign dignitaries attend his funeral.

1987   Proposed constitutional changes would limit republics' power

Proposed changes to the 1974 constitution anger many in the republics, particularly Slovenia and Croatia, by increasing the power of the central government. Among the changes are a unified legal system throughout the country, one unified market for the country, central control of communications and transportation, and increased Serb control over the autonomous provinces of Vojvodina and Kosovo.

1990: November–December   Multiparty elections in Macedonia

With Slovenia and Croatia having paved the way, Macedonia holds its first free elections. More than twenty political parties field candidates.

1991: January   Macedonia claims sovereignty

The Macedonian Assembly declares sovereignty.

1991: June 3   Last proposal to prevent Yugoslav collapse

Macedonia joins with Bosnia and Herzegovina in a proposal to keep Yugoslavia together by forming a community of Yugoslav republics, but it is rejected by Serbia.

Independence

1991: November 20   Macedonia declares independence

Declining to remain in the "rump Yugoslavia" consisting solely of Serbia and Montenegro, Macedonia withdraws from the country and declares independence. Kiro Gligorov (b. 1917) is named president.

1993: April   Macedonia is admitted to the UN

The Macedonian republic is granted admission to the United Nations under the name Former Yugoslav Republic of Macedonia (FYROM) to distinguish it from the part of Macedonia that belongs to Greece.

1994: February   Greece imposes trade embargo in Macedonia

Fearing that Macedonia harbors expansionist designs against it, Greece imposes a trade blockade on Macedonia. The embargo is devastating to the economy of the young country, whose main trade partner, Serbia, is also the object of a trade embargo. The European Union (EU) initiates action against Greece for imposing the blockade.

1994: February 8   Macedonia is recognized by the U.S.

The United States grants diplomatic recognition to Macedonia under the compromise name, FYROM.

1995: October   Greece lifts embargo in Macedonia

Greece agrees to lift its trade embargo against Macedonia. In return Macedonia agrees to change its national flag, which contains a symbol that Greeks associate with their own history and culture. It also agrees to other concessions and pledges to make no territorial claims on land outside its borders. The lifting of the embargo proves advantageous to both countries. Indeed, by the end of the decade, Greece is Macedonia's largest foreign investor.

1995: October 3   President survives assassination attempt

Macedonian President Kiro Gligorov survives a car bomb attack that kills his driver and seriously injures him as his armored Mercedes drives past a parked car containing forty-five pounds (20 kilograms) of explosives. Gligorov, who is 78, suffers a skill fracture and loses an eye. Parliamentary speaker Stojan Andov is named acting president until Gligorov recovers. The political opposition condemns the attack, but extreme nationalists are known to be dissatisfied with the terms of the agreement under which Greece has suspended its trade embargo on Macedonia.

1996: April 8   Macedonia and Yugoslavia establish diplomatic ties

Macedonia establishes diplomatic relations with Yugoslavia, which renounces claims on any Macedonian territory.

1997: July   Ethnic tensions erupt in Gostivar

Tensions between Macedonians and Albanians erupt into violence in the town of Gostivar, where the flags of various nationalities are flown illegally outside the town hall. When police remove the flags, there is a standoff with thousands of protesters in which three ethnic Albanians are killed and several policemen are wounded.

1998: October–November   Ruling party ousted

Macedonia's ruling social-Democratic Union party is ousted in parliamentary elections by a coalition composed of two nationalist parties, the Democratic Alternative Party and Macedonia's traditional nationalist party, the Internal Macedonian Revolutionary Organization (IMRO), named after the famed nationalist group. IMRO's leader, Ljupco Georgievski, becomes the nation's new prime minister at thirty-two years of age.

1999   Kosovo refugees enter Macedonia

Large numbers of Albanian refugees flee from the crisis in Kosovo across the border to Macedonia, threatening to destabilize the already sensitive ethnic balance in the region. Ethnic Albanians already comprise between twenty-five and thirty percent of Macedonia's population, and the Macedonian government fears that the influx of Albanian refugees from Kosovo will result in further clashes between the two nationalities and jeopardize the continued independence of Macedonia.

1999: March–June   NATO bombing of Yugoslavia, influx of refugees

After efforts to reach an agreement fail between the warring Kosovo Liberation Army and the Yugoslav government, the North Atlantic Treaty Organization (NATO) begins a bombing campaign against Yugoslavia. Yugoslav forces in Kosovo, in turn, begin a mass campaign of "ethnic cleansing" against the Kosovar Albanians, many of whom flee into Macedonia. The international community, including relief agencies, send aid to Macedonia to deal with the crisis, but the Macedonian government is once again worried about its continued stability. Ultimately, the Macedonian government survives this crisis as Yugoslavia sues for peace and NATO ground forces enter Kosovo as peacekeepers. Albanian refugees soon follow and Macedonia is slowly relieved of its burden.

Bibliography

Banac, Ivo. *The Nationality Question in Yugoslavia.* Ithaca, NY: Cornell University Press, 1984.

Billows, Richard A. *Kings and Colonists: Aspects of Macedonian Imperialism.* New York: E.J. Brill, 1995.

Danforth, Loring M. *The Macedonian Conflict: Ethnic Nationalism in a Transnational World.* Princeton, NJ: Princeton University Press, 1995.

Jelavich, Barbara. *History of the Balkans.* 2 vols. Cambridge: Cambridge University Press, 1983.

Kofos, Evangelos. *Nationalism and Communism in Macedonia: Civil Conflict, Politics of Mutation, National Identity.* New Rochelle, NY: A.D. Caratzas, 1993.

Lange-Akhund, Nadine. *The Macedonian Question, 1893–1908.* Trans. Gabriel Topor. Boulder, Col.: East European Monographs; Distributed by Columbia University Press, 1998.

Perry, Duncan M. *The Politics of Terror: The Macedonian Liberation Movements, 1893–1903.* Durham: Duke University Press, 1988.

Poulton, Hugh. *Who Are the Macedonians?* Bloomington, Ind.: Indiana University Press, 1995.

Shea, John. *Macedonia and Greece: The Struggle to Define a New Balkan Nation.* Jefferson, NC: McFarland, 1997.

Vukmanovic-Tempo, Svetozar. *Struggle for the Balkans.* Trans. Charles Bartlett. London: Merlin Press, 1990.

Malta

Introduction

Malta is one of the smallest European nations. It consists of three main islands (Malta, Camino, Gozo and two uninhabited islands (Cominotto and Filfia). Malta's rich history and cultural contribution to the world is mainly attributed to its geographic location in the Mediterranean. Malta is located fifty-eight miles south of the island of Sicily and 179 miles north of Tunisia (Africa). The islands consist of a rocky, limestone base of deep harbors and rocky coves ideal for shipping and military strategy. The majority of the Maltese people are of Mediterranean, Italian and ancient Carthoginian stock. According to the *Columbia Gazetteer of the World '98,* there are nearly four hundred thousand people living on the islands. There are two official languages, English and Maltese, of which the latter is of Semitic origin. Roman Catholicism is the official religion of the country, but all faiths have the freedom to worship. Education is compulsory until the age of sixteen. Citizens are permitted to vote at the age of eighteen, and a majority of the citizenry exercise their right to vote. Women are gradually gaining more rights as citizens although once a woman is married she ceases to have any legal authority over her property, including her children.

Tourism accounts for one-fourth of the gross national product (GNP) of Malta. There are no natural resources except limestone. Only small quantities of potatoes, cauliflower, grapes, wheat barley, tomatoes, citrus fruit, cut flowers, hogs, poultry and eggs are produced due to the rocky soil conditions. Malta supplies twenty percent of its own food needs while fuel and raw materials for many of the items manufactured must be imported. In fact, desalination plants provide half of Malta's drinking water.

The other major industry besides tourism on Malta is ship repair. There are also light manufacturing enterprises in the clothing, textile, building, and food processing sectors but more than half of them are foreign owned.

There are two major political parties on Malta, and they permeate almost all aspects of Maltese life. The Labor Party represents the workers and the strong labor unions. It advocates, at times, protectionist policies and has had strong clashes with the Catholic church, especially when it comes to the issues of the separation of church and state. The Nationalist Party, on the other hand, professes a liberalized economy, away from governmental control and usually takes the side of the Vatican in issues regarding the government and church. Between 1971–84, The Labor Party had the majority in the government, but it has shifted now towards the Nationalist. The annual economic growth in the past seven years has been at five percent with a four percent unemployment rate, making Malta's economy favorable for future expansion.

The Nationalist Party, under Prime Minister Fenech-Adami, has been encouraging Malta to develop as a financial service area of the world. Malta favors membership in the European Union, but there is resistance because of its position of neutrality in foreign affairs and its size, which some feel would have a disproportional veto vote compared to larger countries in the European Union.

The history of Malta is reflected in festivals throughout the year. There is the Festival of St. Paul, which commemorates St. Paul converting Malta to Christianity. The traditional Christian holidays of Ash Wednesday, Easter and Christmas are also celebrated. Freedom Day celebrates the departure of the British troops in 1979 and St. Peter and St. Paul Day, also known as Imnarja, is a harvest festival coinciding closely with St. John Day, celebrating the departure of the Knights in the eighteenth century. Our Lady of Victories, which marks the defeat of the Turks by the Knights coincides with the celebration of the Blitz in 1942, and the Feast of the Virgin Mary. Independence Day and Republic Day festivals are also held. In addition to all these, there are local feasts in towns sometimes producing friction among the neighbors.

Timeline

5000–4000 B.C.   Discovered artifacts and stone monuments establish prehistoric man on Malta

Excavations in 1960s reveal that towns and temples are built during the Neolithic period. Much of the work discovered leads to evidence of early fertility cult worshipping.

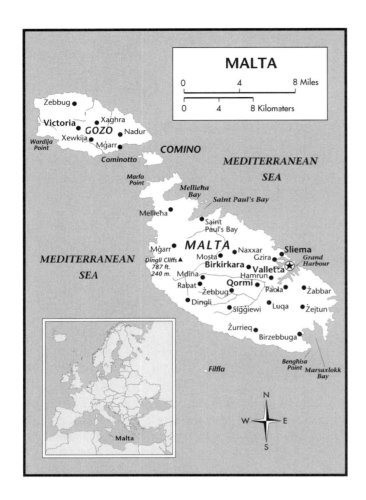

770–500 B.C.   Phoenicians colonize the islands

The Phoenicians, who like the Greeks never unite their cities into a single country, sail and trade and establish themselves on the islands of Malta.

550–218 B.C.   Period of the Carthaginian rule

From the city of Carthage, located three miles from the site of modern Tunis, the Carthaginians trade and rule the Mediterranean using Malta as a stepping stone.

218 B.C.   The Romans annex Malta in their strategy to win the Second Punic War

There are a series of three Punic Wars between Rome and Carthage. By this time Malta has a strategic significance.

60 A.D.   St. Paul is shipwrecked on Malta.

St. Paul is shipwrecked on Malta. Within two years, Malta is converted to Christianity.

533   The Byzantine Empire gains control.

Malta is part of the Byzantine Empire.

870   The Arabs conquer Malta

The Byzantine Empire consists of various sizes and strengths of territories around the Black and Mediterranean Seas.

300–600 A.D.   Byzantine Empire flourishes

The Byzantime Empire flourishes internally because of its ancient cultural and Christian practices. Eventually, the invasion of the barbarians—the Lombards from Germany, the Slavs, the Avars, and the Persians—erodes and weakens the Empire's borders. Byzantium divides around this time in the way it practices Christianity, the Roman as opposed to the Orthodox. Muslim Arabs take advantage of the circumstances and conquer Syria, Palestine, and Egypt. Malta becomes useful as a stepping stone for invasions, consequently putting it under Arab domination for 220 years.

1090   Malta is part of Sicily

Roger, the Norman, ruler of Sicily and parts of southern Italy, declares Malta as an appendage of Sicily. For the next 440 years, Malta is sold to various feudal lords and barons. It comes under the rulers of Swabia, Aquitaine, Aragon, Castile and Spain.

1429   Malta is sacked by the Muslims

Early 6th century   Malta barely exists

The islands are deprived economically and culturally and population dwindles to 20,000.

1523   Malta is given to the Knights of St. John of Jerusalem

Charles V inherits Malta as part of the Holy Roman Empire when he becomes king of Spain. He, in turn, gives Malta to the Knights of St. John because of their valor against the Turks during the Crusades. For the next 275 years, the Knights dominate the arts, culture and military architecture of Malta.

1526–86   Gerolamo Cassar, architect

Under the Knight's request, Cassar designs St. John's Cathedral, the Auberges, fortifications, and several churches in Valletta, promoting the baroque style.

1565   The Knights defeat Suleiman the Magnificant

The Knights hold out against the Turks (Siege of Malta) and defeat Suleiman. The town of Valletta, the capital city today, is named after the Grand Master Knight who leads the fight against the Turks.

The Feast of St. Paul is one of Malta's most important religious holidays. In the twentieth century, a ten-foot statue of St. Paul is carried through the streets to commemorate his shipwreck on Malta. (EPD Photos/CSU Archives)

1630–1710   Lorenzo Gafa, architect

Inspired by the Knights' valor, Gafa designs unique churches with huge domes and grandiose interiors with intricately ornate altars and walls covered with paintings and frescos.

1798   Napoleon invades Malta

Napoleon wants Malta as a base to invade Egypt. By this time, the Knights are weak because of their opulence and numbers and offer little if any resistance to Napoleon.

1800   Malta rebels

The people of Malta rebel against French rule and with the help of the British evict the French.

1813   Malta comes under British rule

Sir Thomas Maitland is appointed as the first British Governor of Malta. A constitution is written, and administrative reforms are enacted.

1814   The Treaty of Paris makes Malta part of the British Empire

According to the terms of the Treaty of Paris, Malta becomes a military and naval fortress and headquarters for the British Mediterranean fleet.

1853–56   Malta is a strategic supply station for British forces

During the Crimean War, England, Turkey, France, and Sardinia resist Russia's ambition to drive an outlet through the Mediterranean.

1869   The opening of the Suez Canal

The opening of the Suez Canal makes Malta an important port on the British route to India.

1914–18   World War I

Malta is a hospital for 20,000 troops. Immediately after the war, Malta is given internal autonomy by England

1933   Malta reverts to British Colony status

Malta is unsuccessful at existing as an autonomous nation, and reverts back to a British Colony.

1939–1945   World War II

In 1940, Italy enters the war, and Malta suffers its first air raid. In 1942, over 6,000 tons of bombs are dropped on Malta in April alone. This prompts King George of England to bestow the George Cross to "the island fortress of Malta—its people and defenders".

1947   Malta again is granted self government

The islands are granted self government, but there is a British appointed Governor who maintains control over foreign affairs, defense and currency.

1955   Proposal to integrate Malta into UK

In parliament, a radical proposal is put forward for full integration of Malta into the United Kingdom.

1956   Referendum on integration into UK

A referendum passes by three-fourths majority, but the integration idea disappears, and there is a new advocacy for independence from Britain.

1964: September 21   Malta becomes an independent

Malta becomes an independent parliamentary monarchy, within the Commonwealth of Nations. The English monarchy is represented by a governor-general. Malta signs a mutual defense treaty with Great Britain that allows British troops to remain on Malta. Dr. Borg Oliver is elected prime minister.

1969 The Labor Party is elected

The Labor Party, influenced by powerful labor unions, passes the Employment of Disabled Persons Act. This Act allows disabled persons to work for government agencies.

1970 The Nationalist Party returns

An Association Agreement is signed with the European Union. It allows Malta to export to these countries industrial products customs free. (Eighty percent of Malta's trade is with EU countries.)

1971 The Labor Party returns to power

The Labor Party returns to power, with Dominic Mintoff (b. 1916) as prime minister. The Labor Party puts the Association Agreement into cold storage because it wants to protect its unions. It also abrogates the Mutual Defense Treaty with Great Britain. From 1961–71, the Labor Party is aggressively nationalizing private institutions, i.e., banks, airlines, broadcast stations, oil and gas, and is in deep conflict with the Catholic church when it tries to nationalize the hospitals and schools. It reaches the point that the Church declares that any support or vote for the Labor Party is grounds for excommunication. Violence erupts. The Church lifts the ban. Ultimately this is settled by a Vatican-government agreement in 1985.

1972 Malta pressures Britain for more rent

The rental fee charges are tripled.

1974 Malta becomes a republic within the British Commonwealth

Malta becomes a republic within the British Commonwealth with executive authority invested in the president of Malta. Malta changes from an independent parliamentary monarchy to a republican parliamentary democracy, and the English Crown is no longer represented formally. Elections are every five years based on proportional representation. The president is head of the state, but the executive power is held by Parliament headed by the prime minister. The judiciary is independent.

1976 Industrial Relations Act is passed

A new industrial relations act is passed. Labor disputes may now be referred to an Industrial Tribunal, a government appointed body consisting of the government, and employer and employee groups or to binding arbitration.

1979 British forces leave

British forces leave after 179 years on Malta. This leaves an economic gap for the Country.

1979 Prime Minister Mintoff turns to Libya for help

After having failed to get a loan from France, Italy, or Algeria, Prime Minister Dom Mintoff of the Labor Party turns to Libya for help. Libyan leader Colonel Moamar Al Qadafi delivers oil and gas almost without charge and enters into a defense pact with Malta providing helicopters and coastal patrol boats. (One year later, however, a dispute over oil rights between the two states brings them to the International Court of Justice which it finally resolves in 1985.)

1980 Malta agress to remain neutral

Malta signs a military pact with Italy to remain neutral. In return, Italy promises Malta technical assistance and financial help.

1981 Malta agrees to store Soviet oil

Malta signs an agreement with the Soviet Union to store 300,000 tons of oil.

1984 Prime Minister Mintoff replaced by Karmenn Bonnici

Prime Minister Mintoff resigns and is replaced by Karmenn Bonnici, also from the Labor Party. The Labor Party states that it favors Malta's independence and neutrality. The party is anti-colonial in foreign policy and seeks closer ties with Third World nations and seeks foreign investment and economic help to industrialize and to boost their weak agriculture and harbor repair industry sectors. The Labor Party also believes in redistribution of property and income to create a more egalitarian society.

In Parliament, Prime Minister Bonnici announces that a defense aid agreement with Italy will lapse in favor of a new alignment with Libya.

1986 Former prime minister involved with Qadafi

Former Prime Minister Mintoff admits to having tipped off Libyan leader Moamar Qadafi minutes before the U.S. bombing raid on Libya.

1987 Nationalist Party returns to power

Prime Minister Fenech-Adami and his Nationalist Party return to power after sixteen years of Labor Rule. Prime Minister Adami enacts a series of laws liberalizing the economy. Government holdings in banks, insurance companies, and utilities are sold. Government work force is reduced and protectionist trade practices eliminated. A stock market and an autonomous money market are founded. The constitution is changed to entrench both non-alignment and neutrality and forbid foreign military bases on Malta. Fenech-Adami also declares that the military clause of the 1984 agreement with Libya will not be renewed.

1989 US-Soviet talks held off the coast of Malta

U.S. president George Bush and Soviet president Mikael Gorbachev meet on a ship moored off Malta. The purpose of this meeting comes about because of the events unfolding in Europe and elsewhere. The last of the Russian troops are leaving Afghanistan. The Communist Party is rewriting its manifesto, and the Soviet leadership is being seriously challenged within. President Bush is concerned about arms control, chemical and nuclear production, U.S. troop reduction in Europe, and the necessity that Russia liberalize its trade policies with the West. It is also the time of the Eastern European breakup, East Germany in particular. There is accusation at this time that the Soviets are supporting arms shipments to Central America, specifically to El Salvador. Although nothing of major significance is signed at this conference, it allows for dialogue and future cooperation between the two leaders. Malta is chosen as a sight for the conference because it has a pleasant Mediterrean climate and is known for its diplomatic traditions, imbued in protocol.

1990 Malta joins the UN and applies to the EU

Malta becomes a member of the United Nations (UN) and applies for membership in the European Union.

1991 A constitutional amendment is signed into law

The constitution obligates the government to promote equal rights for all persons regardless of sex. Redress in the courts for sexual discrimination is available.

1992 A new airport is built

A new airport in Malta opens, featuring one of the longest runways in the world. The airport handles 2.5 million passengers a year, and offers air service to all the major European cities.

1992 Telmalta is completed

Telmalta, one of the most advanced telecommunication systems in the world, begins operations.

1992 The Maltese Stock Exchange is established

1993 Child Protection Law is passed

Malta passes a law whereby children under sixteen may not be legally employed. During the summer service however, enforcement of the law is somewhat lax in domestic, restaurant kitchen help, and vendoring.

1994 Malta as a money service center of the world

National banks are privatized, and an amendment is passed to bring regulations and guidelines of the economy into line with the European Union practices. Joint ventures are initiated with Germany in insurance and investment banking.

1995 Malta joins NATO's Partnership for Peace Program

Malta joins Partnership for Peace, a forerunner for membership into NATO. It is introduced by NATO at the Brussels Summit in January 1994. The program focuses in particular on practical, defense-related and military cooperation activities.

1997: January 30   The Human Rights Report released

The Human Rights Report for Malta, 1966, is released by the Bureau of Democracy, Human Rights and Labor. The reports states that Malta is in compliance with civil liberties and freedom to address individual and institutional grievances. The report notes, however, that women's rights need attention because of the widespread pattern of family violence against women. This problem is addressed by a special police unit and several volunteer organizations around the country.

1997 Inglot receives international recognition

Reverend Peter Serracino Inglot, a Maltese, former head of the University of Malta, is recognized internationally for his outstanding achievements. Inglott has written two librettos for opera. He is also responsible for formulating a proposal that asserts the seas are common human property and should be administered by the United Nations for the benefit of all. He is recognized for his leadership of the 400-year-old University of Malta.

1998 Membership in EU renewed

Prime Minister Eddie Adami, of the Nationalist Party, renews membership into the EU.

Bibliography

Berg, Warren G., *Historical Dictionary of Malta*. Lanham, Jd.: Scarecrow, 1995.

Blouet, Brain, *The Story of Malta*. Rev. ed. London: Faber and Faber, 1972.

Caruana, Carmen M., *Education's Role in the Socioeconomic Development of Malta*. Westport, Conn.: Praeger, 1992.

Countries of the World and their Leaders Yearbook, 2 Vol, Detroit, Mich.: Gale Research Co., 1980.

Dobie, Edith, *Malta's Road to Independence*. Oklahoma: University of Oklahoma Press, 1967.

The Europa World Yearbook, Vol. 2, 39th Edition, 1998.

Elliot, Peter R., *The Cross and the Ensign: A Naval History of Malta 1798–1878*. Annapolis, Md.: Naval War College Press, 1980.

Evans, J.D., *The Prehistoric Antiquities of the Maltese Islands*. New York: Oxford University Press, 1971.

Jellison, Charles A., *Besieged: The World War II Ordeal of Malta 1940–42*. Hanover, N.H.: Published for the University of New Hampshire by University Press of New England, 1985.

Gregory, Desmond, *Malta, Britain, and the European Powers 1793–1815*. Madison: Fairleigh Dickinson University Press, 1996.

Muller, Thomas C., *Political Handbook of the World*. Edited by Arthur S. Banks. Binghampton, N.Y.: CSA Publications, 1997.

Spooner, Tony, *Supreme Gallantry: Malta's Role in the Allied Victory, 1939–45*. London: J. Murray, 1996.

Thompson, Wayne C., *Western Europe,* 15th Edition. Harpers Ferry, W.Va.:Stryker-Post Publications, 1996.

The World Factbook. Central Intelligence Agency, 1995.

Moldova

Introduction

Moldova is a small landlocked country of 13,012 square miles (33,700 square kilometers) located between Romania and Ukraine. Moldova is bordered by Ukraine to the north and east and Romania south of the Prut River. Prior to 1990, the country (then a part of the Union of Soviet Socialist Republics—U.S.S.R.) was known as Bessarabia or Moldavia. This has led to much confusion since the name Moldavia has also referred to the region of Romania running north-south along the eastern edge of the Carpathian mountains which sometimes included Bessarabia within its borders. In 1995, estimates placed Moldova's population at 4,489,657. Of these, nearly sixty-five percent were ethnic Romanians. Russians and Ukrainians comprised about thirteen percent each, Gaugauz (Turkic Christians) were 3.5 percent, while Bulgarians, Jews, and others were fewer than two percent.

An independent Moldova is a recent concept that is a direct result of the region's geographic position next to the former Soviet Union. The majority of Moldova's Romanian population, prior to 1991, would have chosen to be a part of Romania had they been given a free choice by Soviet authorities. Indeed, all of Moldova south of the Dniester river had been a part of Romania between the two World Wars.

History

In ancient times, parts of modern-day Moldova were part of the ancient kingdom of Dacia. During the years 105–271, much of Dacia (although none of modern-Moldova) came under Roman rule. Even the lands north of the Prut River came under Roman influence, and contacts continued after the Roman withdrawal from Dacia. The most significant Roman contributions were in the cultural realm; the Romans introduced the Latin language, which forms the basis of modern Romanian, as well as the Christian religion. Following the collapse of Rome, the region of modern Moldova was subjected to a series of barbarian invasions, including those of Slavs, Magyars (Hungarians), and Tatars (Mongols). Nevertheless, the proto-Romanians managed to keep a distinct culture by either taking to the hills or moving south of the Danube River.

By the fourteenth century, the invasions began to subside, and the Romanians began to reemerge. In 1349, Prince Bogdan established an independent principality in Bessarabia which he named Bogdania. Stephen the Great's Kingdom of Moldavia (which included Bessarabia) became a regional power and vanquished the Tatars from the region. However, Moldavia proved no match for the Ottoman Empire and became a tributary state in 1512 as did its sister Romanian state to the west, Wallachia.

Beginning in the eighteenth century, Moldavia became a battleground between the Russians and Ottoman Turks. In a series of wars between their two empires, Russian troops entered Ottoman territory through Bessarabia, and ownership of the strategic region changed hands several times during the course of the next couple of centuries. Russia gained all of the Transnistrian lands in 1792, and in 1812, the Ottomans ceded all of Bessarabia to Russia. The Russian defeat in the Crimean War in 1856 reversed the Russian annexation of 1812 and allowed the Romanians of Bessarabia to participate in the Romanian national movement based in the principalities, which led them to unite with an autonomous Romanian state in 1861. However, as a price of Romanian independence in 1878, Romania once again ceded southern Bessarabia to Russia.

Bessarabia remained a part of the Russian Empire until 1918 when, following the Russian Revolution, the Romanians of Bessarabia voted to unite with Romania. While the peace settlement recognized the Romanian annexation of Bessarabia, the Bolshevik regime in Moscow did not. In the early 1920s, as the basis for challenging Romanian rule over Bessarabia, the Soviet Union established the Moldavian Autonomous Soviet Socialist Republic (A.S.S.R.) in the region north of the Dniester River extending into the Ukrainian Soviet Socialist Republic. The Molotov-Ribbentrop Pact of 1939, which divided Eastern Europe into Soviet and German spheres of influence, sealed Bessarabia's fate; in June, 1940, Romania was forced to cede northern Bukovina and all of Bessarabia to the U.S.S.R. In turn, the U.S.S.R. redrew internal borders, abolished the Moldavian A.S.S.R., and created a new, landlocked Moldavian S.S.R. Although Romania tempo-

MOLDOVA

0   30   60 Miles

0   30   60 Kilometers

Mohyliv Podol's'kyy
Tul'chyn
Briceni
UKRAINE
Soroki
Răşcani
Floreşhti
Rybniţa
Bălţi
Falesti
Codri
Orhei
Mt. Balanesti
1,407 ft.
429 m.
Ungheni
Dubăsari
Chişinău
ROMANIA
Hills
Tighina
Tiraspol
Boţna
Cogalnic
Căuşenii
Leova
Ialpug
Basarabeasca
Bacău
Comrat
Step Pe
Bugeac
Prut
Cahull
Bilhorod
Dnistrovs'kyy
Galaţi
Moldova
Mouths
of the
Danube
Black
Sea

rarily recovered all of Bessarabia when it joined in the German invasion of the Soviet Union on June 22, 1941, the subsequent Romanian defeat reaffirmed Soviet control of the region.

As the Soviet Union began to disintegrate, national consciousness among the inhabitants of the Moldavian S.S.R. returned to the public sphere. The Romanian majority succeeded in changing the name of the republic from Moldavia to the Romanian Moldova in 1990. The name change, along with an effort to change the official language from Moldavian (a Soviet designation designed to quell Romanian *irredent-*

ism, the movement to unite all Romanians within one state) to Romanian symbolized the rise in national consciousness among the Romanians of the republic. This, in turn, antagonized the republic's ethnic Russian, Ukrainian, and Gagauz minorities who feared they would become second-class citizens in either an independent Moldova or a Greater Romania (most analysts believed that Moldova would eventually unite with Romania). As a result the Russian and Ukrainian minorities in Transnistria seceded and established their own republic, backed by Russian troops. The Gagauz also seceded. These actions resulted in bloody clashes between the secessionist minorities and Moldovans.

Following the unsuccessful coup against Soviet leader Mikhail Gorbachev in August 1991 Moldova declared its independence from the U.S.S.R.. The following year, a ceasefire ended the fighting in the troubled regions, and the Moldovan government took advantage of the return of stability to embark upon a program of economic reform. Largely as a result of the Transnistria and Gaugaz revolts, popular opinion swung away sharply from the desire for union with Romania. In 1994, Moldova and Russia reached an agreement for the withdrawal of Russian forces from Transnistria, while the Moldovan government promised these areas wide-ranging autonomy. As of the late 1990s a final solution had yet to be formulated. Nevertheless, Moldova continued with its economic reforms and is a stable republic.

Timeline

A.D. 105–271   Roman occupation of Dacia

Roman legions under the command of Emperor Trajan conquer Dacia; however, none of present-day Moldova comes under Roman rule. Nevertheless, the Roman conquest has a profound impact upon the Dacians. The Romans introduce their culture, including the Latin language, to the region. This contact with Rome forms the basis of the present claim by Romanians that they are a Latin people.

271   Romans abandon Dacia

For reasons of security, Roman garrisons leave Dacia and retreat to the southern banks of the Danube. Nevertheless, contact between the Dacians and the Roman Empire continue, thus furthering the process of Latinization in the former Roman province. These continuing contacts result in the Christianization of the Dacian region. In subsequent centuries, the region is overrun by countless invaders, including Slavs, Magyars (Hungarians), and Tatars. Although these groups leave their marks, the Roman legacy remains, particularly in the form of the Romanian language, a tongue directly descended from Latin.

Moldavia and Wallachia, c. 18th century.

1349   Independent Bessarabian principality established

Prince Bogdan establishes a state in the region of present-day Moldova which becomes known as Bogdania. His state stretches from the Carpathian Mountains to the Dniester River.

1400–32   Tatars pushed beyond Dniester River

Prince Alexander the Good leads the Tatars out of Bessarabia. The Tatars (also known as Tartars or Mongols), a group of people from what is now eastern Mongolia, first come to Europe in the thirteenth century.

1457–1504   Era of Stephen the Great

Stephen the Great rules Moldavia for over forty-six years and establishes his country as a leading regional power. In countless battles he vanquishes a host of Hungarian, Polish, and Turkish invaders. During his reign, Moldavia extends from the Milcovu to the Dniester and from the Carpathians to the Black Sea.

1512   Bessarabia becomes Ottoman tributary state

Bessarabia becomes part of the Principality of Moldavia. Along with the Principality of Wallachia, Moldavia holds a special status in the Ottoman Empire as a tributary state. In exchange for their annual tribute paid to the Ottoman Sultan in Constantinople and their pledge of neutrality, the two Romanian principalities retain their own princely governments and avoid an Ottoman military occupation.

1711   Peter the Great attacks Ottoman forces

In alliance with Prince Dimitrie Cantemir, Russian Tsar Peter the Great sends his army through Bessarabia on its way to attack the Ottoman Empire. Unfortunately for Tsar Peter and Prince Cantemir, the Sultan's forces vanquish the invaders. As a result, the Ottomans replace the Romanian princes with pro-Ottoman Phanariote Greeks who are called *hospodars*. These Greeks hail from Constantinople's Phanar (Lighthouse district) and serve in a variety of positions in the Ottoman administration.

1774   Treaty of Kutchuk Kainarji returns the Danubian Principalities to the Ottoman Empire

The Treaty of Kutchuk Kainarji ends the Russo-Turkish War. Under the terms of this treaty, Russia returns the Danubian Principalities of Moldavia and Wallachia to the Ottoman Empire. However, the inhabitants of the principalities are to retain a wide-ranging autonomy, including a reduction in the tribute owed the Ottoman sultan (emperor) and the election by the *boyars* (local nobility) of their own *voivod* (prince). In practice, however, the Ottomans violate these provisions regularly in the Principalities, particularly in the selection of *hospodars*.

1792   Treaty of Iasi signed

The Treaty of Iasi recognizes the Dniester River as the Russian frontier and reaffirms the privileges bestowed upon the Principalities.

1806–12   Russo-Turkish War

Yet another conflict breaks out between Russia and the Ottoman Empire. Russian forces seize the principalities.

1812   Russia annexes Bessarabia under the Treaty of Bucharest

The Treaty of Bucharest recognizes Russian annexation of Bessarabia but restores the Wallachia and the rest of Moldavia to the sultan under the provisions of the Treaties of Kutchuk Kainarji and Iasi.

1856   Treaty of Paris cedes southern Bessarabia to Moldavia

The Treaty of Paris ends the Crimean War. According to the Treaty, Russia cedes southern Bessarabia to the Principality of Moldavia. The treaty also places the Principalities under Great Power protection although they remain under Ottoman *suzerainty* (overlordship). However, the Ottoman promise to remain aloof from the principalities' internal affairs, provides an opening for Romanian nationalists.

1858   Alexander Cuza elected prince of both Moldavia and Wallachia

Assemblies from Moldavia and Wallachia elect Alexander Cuza as prince. This is the first step in the formation of a Romanian nation-state. The principalities are formally united in 1861, and Romania becomes independent in 1878. Following independence, Romania, like its Balkan neighbors, attempts to expand its borders to include all lands with Romanian speakers, such as Bessarabia. Diplomatic considerations (frequent Romanian alliances with Russia), however, prevent Romania from pursuing its aim.

1878: July 13   Congress of Berlin cedes southern Bessarabia to Russia; Romania declared independent

The Congress of Berlin restores southern Bessarabia to Russia. Romanians are outraged at this apparent Russian treachery. This Russian action happens in spite of a Russian guarantee the year before to respect Romanian territorial integrity in return for allowing Russian troops to pass through the country. Another provision of the Congress is the formal independence of Romania. As a result of the Russian annexation (as well as dynastic ties), Romania starts to align with Germany and Austria-Hungary (despite the presence of a large Romanian community in Austro-Hungarian-controlled Transylvania).

1914   Outbreak of World War I

On July 28, 1914, World War I breaks out when Austria-Hungary invades Serbia. By the following week the war spans the entire continent and pits the Entente and its allies (Great Britain, France, Russia, and Serbia) against the Triple Alliance (Austria-Hungary and Germany—alliance member Italy chooses neutrality for the time being).

Romania, although secretly a party to the Triple Alliance, also opts out of hostilities. Since Romania has territorial ambitions against members of both alliances (Transylvania from Austria-Hungary and Bessarabia from Russia), Romanian officials spend the next two years entertaining offers from both sides.

1916: August   Romania joins the Entente powers

After months of negotiation and having secured Entente support for the self-determination of Transylvania, Romania declares war on the Central Powers. The Romanians in Bessarabia are to remain under Russian rule.

1916: September–December   Central Powers defeat Romania

The Entente effort to smash through the Carpathian mountains into Austria-Hungary fails when Romanian forces are threatened along their southern frontier with Bulgaria. The Central Powers mount a counter-offensive that culminates in the capture of Bucharest on December 6. The line does not stabilize until January 10, 1917 to the north of the Romanian capital. The Romanian king and government flee to Iasi.

1917: April–May   National Council formed by Bessarabian politicians

In the wake of the February Revolution in which the Russian tsar, Nicholas II, abdicates, Romanians in Bessarabia begin forming their own political and cultural organizations. Meetings of Bessarabian intellectuals in April and May result in the formation of the *Partidul National Moldovenesc* (The Moldavian National Party) which agitates for autonomy.

1917   Romanian army regroups with Russian forces in Bessarabia

Despite the chaos in the aftermath of the February Revolution, the Romanian government continues to cooperate with the Provisional Government under Alexander Kerensky. One of its chief aims is the stabilization of the Eastern Front. As part of this effort, the remaining Romanian forces regroup with Russian units in the area of Moldavia (the Romanian province) and Bessarabia. The effort proves successful as Romanian and Russian forces repel successive German offensives in August and September.

1917: November   Bolshevik Revolution in Russia

The Bolsheviks, under Vladimir Lenin, seize control of the Russian capital of Petrograd and install themselves at the head of the Russian government. Their takeover and attempt to install a communist government results in massive unrest and chaos which soon lead to civil war that engulfs the country until 1920.

1917: December 15   Declaration of Moldavian Democratic Federated Republic

The *Sfat al Tarii* (the general assembly of Bessarabia) in Chisinau declares the establishment of the Moldavian Democratic Federated Republic. Bolshevik forces in the area move in to bring down the regime. The Romanian government responds by sending in forces of its own and evicting the Bolsheviks from Chisinau on January 26.

1918: February 6   Bessarabia declares independence from Russia

The Moldavian council declares the Moldavian Republic an independent state. Although this act is seen as the first step toward unification with Romania, the ongoing peace talks between the Bolshevik Russians and Austria-Hungary place Romania in a precarious situation. With Romanian armies defeated and most of Romania overrun, a peace settlement between Russia and the Central Powers would isolate Romania and leave it without allies.

1918: March 3   Treaty of Brest-Litovsk signed

Russia and the Central Powers sign a peace treaty at Brest-Litovsk. Under the terms of this treaty, Russia leaves the war and allows German and Austro-Hungarian armies to occupy large swaths of land in Eastern Europe that were part of tsarist Russian Empire. This loss of Russian territory allows the Central Powers to move into territory north of Romania and surround the remaining Romanian forces which have continued to resist.

1918: March 27   Sfat votes to unite with Romania

After consultation with Romanian officials, the Sfat votes to unite Bessarabia with Romania. Bessarabian officials insist upon and receive a large degree of autonomy.

1918: April 3   Constantin Stere elected president of Sfat

The Sfat elects Constantin Stere president and embarks upon a program of economic and social reform in Bessarabia.

1918: May 7   Treaty of Bucharest signed, recognition of Bessarabian union with Romania

Romania and the Central Powers sign the Treaty of Bucharest. According the provisions of this treaty, Romania loses territory in the Carpathian mountains to Austria-Hungary and control of its economy to Germany. In addition, the Central Powers occupy large portions of Romania.

As a consolation, the Treaty of Bucharest recognizes the union of Bessarabia with Romania.

1918: November 10   Romania reenters the war

With the Central Powers facing imminent collapse, Romania reenters the war on the side of the Entente. The following day, Germany, the last holdout among the Central Powers, sues for peace, and an armistice is declared. The defeat of the Central Powers combined with the collapse of Russia paves the way for the creation of a Greater Romania with the additions of Bessarabia (from Russia), the Banat, Bukovina, and Transylvania (from Austria-Hungary), and Southern Dobrudja (from Bulgaria).

1920: October 28   Romanian sovereignty over Bessarabia recognized

The victorious Allies finally recognize Romanian sovereignty over Bessarabia and agree to the Dniester River as Romania's eastern boundary. However, the Allies also stipulate that a final settlement must be negotiated between Romania and a Russian government with allied recognition (at the time, none of the Allies recognizes the Bolshevik regime). The Bolshevik refusal to recognize Romanian sovereignty over Bessarabia in the interwar period remains the cause of diplomatic unease between the two countries and contributes to the Soviet annexation of the region in 1940.

1924   U.S.S.R. forms Moldavian Autonomous Soviet Socialist Republic

The Soviet Union (as Russia is known after 1922), still smarting from its loss of Bessarabia in 1918, establishes the Moldavian Autonomous Soviet Socialist Republic (A.S.S.R.) opposite from Romania. This autonomous region has no historical basis, and aside from Transnistria (the area to the north of the Dniester River), few Romanians inhabit the region. The Soviets, however, do not recognize the Romanian presence and refer to Romanian speakers and their language as Moldavian. In addition, the U.S.S.R. claims that the Moldavian A.S.S.R. represents all Moldavians (i.e., those Romanians living across the border in Bessarabia). This attempt to create a distinct Moldavian region in the U.S.S.R. with its own Moldavian nationality provides the Soviet Union with a basis for reviving territorial claims against Bessarabia.

1929   Capital of Moldavian A.S.S.R. moved from Balta to Tiraspol

Soviet authorities move the capital of the Moldavian A.S.S.R. from Balta, where hardly any Romanian lives, to Tiraspol, a city in Transnistria, a region of mixed Ukrainian and Romanian settlement.

1939: August   Molotov-Ribbentrop pact signed

In an effort to secure his country from an eastern enemy just prior to his invasion of Poland, German leader Adolf Hitler sends his foreign minister, Joachim von Ribbentrop, to Moscow to sign a non-aggression pact with the Soviet Union. As part of the agreement, Ribbentrop and his Soviet counterpart, Vyacheslav Molotov, divide eastern Europe into spheres of influence. The U.S.S.R. receives the Baltic republics, eastern Poland, Bessarabia, and Northern Bukovina as part of its sphere of influence. Within a year, all of these regions are annexed to the Soviet Union.

1940: January 27   Petru Lucinschi born

Future Moldovan president Petru Lucinschi is born in Radulenii Vachi village in Bessarabia. In the 1960s he studies history at the Moldova State University and earns a Ph.D. Beginning in 1971, he holds a number of positions in the Communist Party of the Soviet Union (CPSU), including the following: Deputy Head of the Propaganda Department of the CPSU Central Committee, Secretary of the Central Committee of the Tajikistan Communist Party, Head of the Moldavian Komsomol (Young Communist League), and First Secretary of the Moldovan Communist Party (1991).

He continues his service after Moldovan independence and serves as Moldovan Ambassador to the Russian Federation (1992), and Chairman of the Moldovan Supreme Soviet of People's Deputies (1993). In February, 1994, he is elected to parliament as a member of the Agrarian Democratic Party and the following month is elected speaker of parliament. Finally, in 1996 Lucinschi, running as an independent, defeats incumbent President Mircea Snegur to become president of Moldova.

1940: June   The U.S.S.R. annexes Bessarabia and Northern Bukovina

Taking advantage of the sphere of influence accorded it under the terms of the Molotov-Ribbentrop pact, the Soviet Union forces Romania's King Carol to cede Bessarabia and North-

ern Bukovina to the U.S.S.R.. This action proves extremely unpopular among Romanians of all political persuasions except for the communists.

1940: August 2   Moldavian A.S.S.R. abolished, renamed Moldavian S.S.R.

In the aftermath of the annexation of Bessarabia, the Soviet Union abolishes the Moldavian A.S.S.R. and reorganizes the region. Bessaravia and Transnistria become the Moldavian Soviet Socialist Republic (S.S.R.) while the rest of the Moldavian A.S.S.R. reverts to the Ukrainian S.S.R..

1941: June 22   Romania joins the German invasion of the U.S.S.R.

The same day the German army violates the Molotov-Ribbentrop Pact and invades the Soviet Union, Romanian Prime Minister Ion Antonescu, without prior consultation with German authorities, sends Romanian troops across the border into Bessarabia and Northern Bukovina. Soon Romanian troops reconquer the lands lost to the Soviet Union. Although the reoccupation of Bessarabia and Northern Bukovina is popular with the Romanian public, Antonescu's decision to send Romanian troops beyond his country's 1940 borders is not. In three years as German allies, Romanian troops fight in many crucial engagements on the Eastern Front, including Odessa, Crimea, Stalingrad, and Kuban.

1941–44   The Holocaust in Bessarabia

In contrast to the vast majority of Jews in the rest of Romania, those in Bessarabia and Transnistria suffer bitterly. Summary executions break out, and Bessarabia serves as a dumping ground for Jews from the rest of Europe. In the rest of Romania, however, discriminatory measures against Jews are halted due to the intervention of King Michael, National Peasant Party leader Iuliu Maniu, and the Romanian Orthodox Church. Indeed, after 1942, Bucharest serves as a transit center for Jews emigrating to Palestine.

1944: August   Romania withdraws from German alliance and joins the Allies

In the latter stages of their great summer offensive, the Red Army overruns Romanian forces and the Romanian government to sue for peace. On August 23, King Michael ousts the pro-German Marshal Antonescu and sanctions a new government under the leadership of General Constantin Sanatescu. The Germans respond to the Romanian action by bombing Bucharest. Following the German action, King Michael declares war on the Axis. However, the Soviet Union shows no sign of relinquishing its control of Bessarabia and Northern Bukovina.

1947: February 10 Peace treaty signed between Romania and the U.S.S.R.

The Romanian delegation led by Gheorguiu Gheorghiu-Dej signs a peace treaty with the U.S.S.R. after six months of negotiations. According to the terms of this treaty, Bessarabia and Northern Bukovina revert to the Soviet Union (although Hungary returned the portions of Transylvania it had occupied in 1940). Although the treaty formally recognizes the Romanian loss of these regions, in practice, the Soviet Union has administered them as its own ever since it reoccupied them following the German and Romanian retreats in the summer of 1944.

The U.S.S.R. returns to its pre-war policies of supporting a Moldavian nationality with its own literary language in place of Romanian. As a result, Soviet authorities restore the Cyrillic alphabet and punish Moldavians who do not adopt the linguistic reforms in their speech and writings.

1945–49 Drought and famine plague Moldavian S.S.R.

A severe drought hits Moldavian agriculture in the first years of Soviet rule. In addition, communist agricultural policies prove disastrous for the region. Forced collectivization disrupts traditional agricultural patterns. Finally, the combined effects of the drought and collectivization result in a famine. Ironically, the Moldavian S.S.R. is one of the most fertile areas of the U.S.S.R. despite its being the smallest republic in area.

1949 Mihai Brunea born

Free-lance artist Mihai Brunea is born in Cucurazeni, Moldavian S.S.R.. Although fascinated with art as a youth, he gives up art to pursue other studies. In 1974, he graduates from the State University of Chisinau's Department of Physics and pursues a career as a physics teacher. However, his interest in art rekindles, and he soon embarks upon a career as an illustrator of books and creator of free-lance works. He eventually becomes vice-president of the Artists Union of the Republic of Moldova.

1950–52 Leonid Brezhnev serves as First Secretary of Moldavian Communist Party

Leonid Brezhnev, a Ukrainian, becomes the first secretary of the Moldavian Communist Party. Brezhnev's leadership lowers a severe repression on the republic. Mass arrests of dissidents and suppression of their organizations ensue. Brezhnev ultimately becomes the leader of the entire Soviet Union following his ouster of Premier Nikita Khrushchev in 1964. Brezhnev follows similar tactics on the entire country during his rule. His eighteen years in power (1964–82) are known collectively in communist parlance as "the years of stagnation".

1989 Moldovan Popular Front formed

In the wake of Soviet leader Mikhail Gorbachev's sweeping *glasnost* (openness) and *perestroika* (restructuring) attempts at reform, a collection of Moldavian cultural and political groups form the Moldovan Popular Front. This organization constitutes the major ethnic Romanian political reform movement in the Moldavian S.S.R..

1990: February 25 First democratic elections in Moldavian S.S.R.

The Moldavian S.S.R. holds its first democratic elections. Run-off elections are held the following month. The result is a sweeping victory for the Popular Front. Mircea Snegur is elected Chairman of the Supreme Soviet of Moldavia. In September, he becomes president.

1990: June Moldavia becomes Moldova

The official name of the republic is changed from Moldavian S.S.R. to Moldovan S.S.R.. The republic also claims that it is sovereign.

1990: August Gagauz announce secession from Moldavia

The Gagauz, Turkic Christians living in Moldova, secede from the republic. The Moldavian parliament refuses to recognize this action.

1990: September Slavs in Dniester region announce secession from Moldavia

The majority Slavic (Russian and Ukrainian) inhabitants of the Dniester region (Transnistria) declare their independence from the rest of Moldova. The insurgents take up arms and soon engage Moldovan Romanians in bloody clashes.

1991: May Moldovan S.S.R. becomes Republic of Moldova

The Moldovan parliament changes the name of the republic from Moldovan S.S.R. to Republic of Moldova. This move is yet another step toward independence from the Soviet Union. It is also taken to enhance the Moldovan government's position *vis-à-vis* the Gagauz and Transnistria separatists.

1991: August Attempted coup against Soviet leader Mikhail Gorbachev

Opponents of Mikhail Gorbachev, backed by segments of the military and security apparatus, mount a coup against the Soviet leader who is vacationing in the Crimea. After a few days, however, the coup fails, and Gorbachev returns to power, his authority irreparably damaged. The Moldovan government remains steadfast in its support for Gorbachev throughout the period of uncertainty.

1991: August 27   Moldova declares independence

Subsequent to the coup's failure, the Republic of Moldova declares its independence from the Soviet Union.

1991: October   Moldova organizes armed forces

Moldova begins to organize its own armed forces. Until this time, Moldova is completely reliant upon the Soviet military, particularly the Fourteenth Army, for its defense.

1992: March 2   Moldova joins the United Nations

Moldova becomes a member of the United Nations.

1992   Cease-fire in Transnistria

President Snegur and Russian president Boris Yeltsin negotiate a cease-fire agreement for Transnistria. According to the provisions of this agreement, a trilateral peace-keeping force of Moldovan, Russian, and Transnistrian forces is implemented. Moldova also agrees to give Transnistria a special status as well as to accord it the right of secession in the event Moldova reunifies with Romania.

1993: March   Privatization program begins

The Moldovan government embarks upon a program of *privatization* (the transfer of government-owned properties into private hands). As of 1998, nearly eighty percent of all housing and nearly 2,000 enterprises have been privatized.

1993: November   Moldovan *leu* replaces Russian ruble

As part of the country's economic reforms, the government introduces a new currency, the *leu*, to replace the Russian *ruble*.

1994: February 27   Agrarian Democratic Party of Moldova wins elections

In parliamentary elections, the Agrarian Democratic party captures a plurality with fifty-six of the 104 seats. The Socialist-Edistro Bloc finishes second with twenty-eight seats, while two pro-Romanian parties, the Peasants and Intellectuals Bloc and the Popular Front, finish with eleven and nine seats, respectively. Petru Licinschi is elected speaker of parliament while Andrei Sangheli becomes prime minister.

1994: March 16   Moldova sings Partnership for Peace treaty

Moldova signs on to the North Atlantic Treaty Organization's (NATO) Partnership for Peace. This program seeks to promote continental security and cooperation between the NATO members and those states of the former Warsaw Pact (the Soviet-led alliance system that serves as NATO's rival during the Cold War).

1994: April   Moldova joins CIS

Moldova joins the Commonwealth of Independent States (CIS), a grouping of former Soviet republics.

1994: July 28   New constitution ratified

The Moldovan parliament approves a new constitution which goes into effect on August 27, 1994 (the third anniversary of Moldovan independence). One of the chief factors of this document is the provision for autonomy for Transnistria and Gagauzia.

1994: October   Russian troop withdrawal agreement signed

Moldova and Russia sign an agreement for the withdrawal of Russian troops from the troubled Transnistria and Gagauzia regions.

1995: April 16   Democratic-Agrarian Party wins local elections

In local elections the Democratic-Agrarian Party captures a strong majority of 69.59 percent (643 mandates). The Communist Party wins 21.54 percent (199 mandates), while the Unity Movement picks up 8.87 percent (eighty-two mandates).

1995: April 27   President Snegur proposes language change amendment

President Snegur asks the Moldovan parliament to amend the constitution so that the official language of the country will be Romanian rather than Moldovan. Snegur, like many Moldovans, view the reference to a Moldovan language as a vestige of Soviet rule.

1995: August   Party of Renewal and Conciliation of Moldova formed

President Snegur forms the Party of Renewal and Conciliation of Moldova. This new party attracts ten parliamentary deputies of the Democratic-Agrarian Party.

1996: November 17   First round of presidential elections is indecisive

The first round of balloting in Moldova's presidential elections is inconclusive. Incumbent Snegur (Party of Rebirth and Conciliation of Moldova) wins a plurality with 38.75 percent. Petru Licinschi (Independent) finishes second with 27.66 percent, while Vladimir Veronin (Communist Party) garners 10.23 percent. Six candidates take the remaining 23.37 percent.

1996: December 1   Petru Licinshchi wins presidential run-off election

Parliament speaker and presidential challenger Licinschi defeats incumbent Snegur in the second round of presidential balloting. Licinschi receives 54.02 percent of the vote while Snegur garners 45.98 percent.

1998: March 22   Parliamentary elections

Elections for the Moldovan parliament give a plurality to the Communist Party of Moldova. Its 30.1 percent of the vote results in forty seats. Support for other parties is as follows: Democratic Convention of Moldova (CDM), formerly the Christian Democrats, 19.2 percent, twenty-six seats; Bloc for a Democratic and Prosperous Moldova (PMDP), 18.2 percent, twenty-four seats; Party of Democratic Forces (PFD), formerly the Congress of Intellectuals and Agrarians, 8.8 percent, eleven seats. The biggest losers in this election are the Agrarian Democratic Party (3.7 percent, zero seats) and the Socialist Unity Party (1.8 percent, zero seats). Despite their gains, the communists are unable to form a governing coalition. Instead, the CDM, PMDP, and PFD form a government with Ion Ciubic as prime minister.

1999   Sturza becomes Prime Minister

Ion Sturza replaces Ion Ciubic as prime minister of Moldova.

Bibliography

Belarus and Moldova: Country Studies. Washington, D.C.: Department of the Army, 1996.

Bruchis, Michael. *The Republic of Moldavia: From the Collapse of the Soviet Empire to the Restoration of the Russian Empire.* Trans. by Laura Treptow. Boulder, Col.: East European Monographs, 1996.

Georgescu, Vlad. *The Romanians: A History.* Transl. by Alexandra Bley-Vroman. Columbus: Ohio University Press, 1991.

Hitchins, Keith. *Rumania, 1866–1947.* Oxford: Clarendon Press, 1994.

Jelavich, Barbara. *History of the Balkans.* vol. 1, "Eighteenth and Nineteenth Centuries." Cambridge: Cambridge University Press, 1983.

———. *History of the Balkans.* vol. 2, "Twentieth Century." Cambridge: Cambridge University Press, 1983.

Lungu, Dov B. *Romania and the Great Powers, 1933–40.* Durham, N.C.: Duke University Press, 1989.

Papacostea, Serban. *Stephen the Great: Prince of Moldavia, 1457–1504.* Trans. by Seriu Celac. Bucharest: Editura Enciclopedica, 1996.

Rothschild, Joseph. *East Central Europe between the Two World Wars.* Seattle: University of Washington Press, 1974.

Treptow, Kurt W. *Historical Dictionary of Romania.* Lanham, Md.: Scarecrow, 1996.

Monaco

Introduction

Located on France's Mediterranean coast, Monaco is the world's smallest country (the only sovereign entity that is smaller is the Vatican). For over a century, it has been a famous gambling venue and tourist mecca and a magnet for wealth, glamour, and sophistication. Monaco is also home to the world's longest-ruling dynasty, the Grimaldis, who in 1997 celebrated their 700th anniversary as sovereigns of the tiny principality.

Like the neighboring resort areas along the French Riviera, Monaco is famous for the natural beauty of its beaches and coastline and its pleasant, sunny climate. With an area of less than one square mile, it is located entirely within the French department of Alpes-Maritimes. The principality is divided into three main areas. Monaco-Ville, located on the promontory that once gave Monaco the nickname "the Rock," is the oldest part of the principality and home to the ornate presidential palace. Monte Carlo is the location of its famed gambling casino, and La Condamine, the most modern section, is the business district and includes a modern industrial area (Fontvieille) built in land reclaimed from the sea. Together, Monaco's small size and its population of about 1,500 make it the world's most densely populated country.

Monaco's early history, like that of France itself, is one of repeated invasion and conquest—by the Roman Empire, by Germanic tribes, and, later, by Islamic invaders from Arabia (the *Saracens*). By the late twelfth century, Monaco had come under the control of the Italian city-state of Genoa, which built the fortress that was to be the principality's most prominent feature for many years. Control by the Grimaldis began one hundred years later with a colorful exploit that has passed into legend. François Grimaldi seized control of the fortress from the family's Genoese rivals by sneaking in disguised as a monk and then attacking its armed guards with a sword. (To this day, the Grimaldi coat of arms features two monks with drawn swords.)

The Grimaldis controlled Monaco intermittently until taking final possession of it in 1419. However, control of their tiny domain was still subject to the vicissitudes of European history over the following centuries. Monaco was dominated by the Spanish in the sixteenth century and then came under

French protection in the seventeenth. French revolutionary forces annexed it at the end of the eighteenth century, but Grimaldi rule was restored at the Congress of Vienna in 1815. In 1848 the towns of Menton and Roquebrune, which had formed part of the principality, were ceded to France, and Monaco suddenly found itself minus over eighty percent of its territory.

To guarantee the economic survival of the principality, Monaco's rulers decided to enter the lucrative casino and tourism business and built a casino in 1856. The foundations of Monaco's tourism business were laid in the 1860s with the establishment of the Société des Bains de Mer (SBM); the inauguration of a railway providing service between Monaco and Nice; and the construction of the elegant Hôtel de Paris. Also significant was an 1869 law abolishing direct taxation, which established Monaco's reputation as a tax haven. With the influx of wealthy and cultured visitors—and also through the efforts of Prince Albert I—came the growth of Monaco as a center for the arts, with its own world-famous opera house and ballet company. Prince Albert supports cultural and scientific achievements and also presided over the adoption of Monaco's first constitution, which established an elected national council to share legislative duties with the prince.

Although it was formally neutral in both world wars, Monaco could not avoid being touched by the conflicts. The nation was forced to renegotiate the terms of its sovereignty and its territorial rights at the Versailles conference following World War I, and its famed Hôtel de Paris was occupied by Gestapo troops at one point during the Second World War. In 1949 Rainier III, Monaco's first native-born (*Monégasque*) ruler since the eighteenth century, succeeded to the throne upon the death of his uncle, Louis II. Known to the world at large primarily through his 1956 marriage to American film star Grace Kelly, Rainier has been a shrewd and effective ruler. By promoting the establishment of light industry in Monaco's Condamine district, Rainier oversaw the creation of its Fontvieille industrial zone and reduced the nation's dependence on gambling revenues. His business acumen led to the birth of the phrase "Grimaldi Inc." as a nickname for Monaco.

Monaco was plunged into mourning in 1982 with the untimely death of Princess Grace in a car accident at the age

MONACO

0   .5   1 Miles

0   .5   1 Kilometers

Cinéma d'Été

Beausoleïl

Anglican Church

FRANCE

Monte-Carlo

Moneghetti Stadium

Casino of Monte Carlo Opera House

Port of Monaco

La Condamine

Exotic Garden

Monaco-Ville ✪

MEDITERRANEAN SEA

Palace of Monaco

Fontvieille

Fontvieille

Port of Fontvieille

Capd'ail

Monaco

of fifty-three. The family suffered yet another tragic loss when Stefano Casiraghi, the second husband of Princess Caroline, died in 1990. At the time the legal status of their three children was in question because Caroline's previous marriage to Philippe Junot had not formally been annulled by the Church. (This issue was of special concern because neither of Caroline's siblings had yet had children, and eventual succession to the throne hung in the balance.) In 1993 the Church annulled the marriage, legitimizing Caroline's marriage to Casiraghi. This action affirmed the eligibility of their heirs to succeed to the throne of Monaco, insuring continuity of the Grimaldi dynasty in future generations.

Timeline

221 B.C.　Hannibal docks at Monaco

During the Punic Wars between Carthage and Rome, the fleets of the Carthaginian general Hannibal use Monaco as a port.

58–51 B.C.　Roman invasion

The Romans, led by Julius Caesar, invade Gaul, which includes present-day France, including the Mediterranean coastal region where Monaco is located.

5th–6th centuries A.D.　Germanic tribes invade

Following the downfall of Rome, Monaco and the surrounding regions are invaded by Germanic tribes, including the Visigoths, Vandals, Franks, and Lombards.

9th century　Saracen invasions

The Muslim *Saracen* raiders plunder and dominate Mediterranean coastal areas, including Monaco.

1174: August 16　Genoa claims right to build fortifications

The Genoese draft a charter claiming the right to build a fortified castle in Monaco, popularly referred to as "the Rock."

1190　Genoa wins control of Monaco

Genoa, which has claimed Monaco for much of the medieval period, is formally awarded the title to it by Holy Roman Emperor Henry VI in exchange for exclusive use of any fortifications built there.

1215　Fortifications are built

The Genoese build a fortress on the site of what is today Monaco.

1247　Construction of a church is authorized

The Pope grants Monaco the right to build its own church. The Eglise St. Nicolas is completed in the latter half of the century.

Late 13th century　Grimaldis win control

The Grimaldis, a family of Genoese aristocrats (and ancestors of the current rulers), obtain control of Monaco. They belong to the Guelph political faction, which supports the papacy in its political rivalry with the Holy Roman Emperors.

1297: January 8　Grimaldis wrest fortress from political foes

François, a nephew of sea trader Rainier Grimaldi, infiltrates and seizes Monaco's fortress disguised as a monk. To commemorate this exploit, the Grimaldis later adopt a coat of arms showing two monks with drawn swords. After the fortress is taken, the family allies itself with France and makes war on Genoa but still loses control of Monaco four years later.

1338 Grimaldis formally gain title to Monaco

Charles, the son of Rainier Grimaldi, purchases the title to Monaco from Genoa and founds the House of Grimaldi, the royal line of Monaco.

1355–57 The Genoese retake Monaco

Fearing an alliance between the Grimaldis and France, Genoa attacks Monaco with four thousand troops, overpowering the forces of the Grimaldis.

1419 Grimaldis purchase Monaco again

Ambroise, Antoine, and Jean, the sons of Rainier II, purchase Monaco from Queen Yolande of Aragon and take turns ruling it.

1454 Grimaldi line of succession is established

Jean Grimaldi's will establishes the lines of succession for the Grimaldi dynasty. The eldest son will inherit the throne. It there is no male heir, the title passes to the eldest daughter on condition that her husband adopt the Grimaldi name and coat of arms.

1480s Monaco is recognized by France

King Charles VII of France and the Duke of Savoy formally recognize the sovereignty of Monaco.

16th century Monaco is under control of Spain

For most of the sixteenth century, Monaco is controlled by Spain, whose exploration and colonization of the New World make it the most powerful country in Europe.

1525 Treaty of Burgos

The Treaty of Burgos, signed by Augustin Grimaldi, brings Monaco under Spanish influence for over a century.

1635 Honoré II wins French protection against Spain

Prince Honoré II and King Louis XIII of France sign a treaty guaranteeing French protection and aid in ousting the Spanish from Monaco. However, France soon becomes embroiled in the Thirty Years' War and does not have the resources to actively aid Monaco against Spain.

1641: November 17 Honoré storms Spanish garrison

Honoré II mounts an attack on Monaco's Spanish garrison, defeating the troops stationed there.

1642: April Honoré recognized by France

King Louis XIII formally recognizes the sovereignty of Honoré II as Prince of Monaco.

1713 Treaty of Utrecht affirms Monaco's sovereignty

The Treaty of Utrecht concluding the War of the Spanish Succession reaffirms the sovereignty of the principality of Monaco.

1793: February Monaco is annexed by France

France's revolutionary government deposes the Grimaldis and annexes Monaco.

Prince Honoré III is thrown into prison, his daughter-in-law is guillotined, and the royal family's property is confiscated. Monaco is renamed Fort d'Hercule.

1815 Congress of Vienna

The Grimaldi principality is restored at the Congress of Vienna following the downfall of Napoleon Bonaparte, but it is placed under the protection of Sardinia.

1848 Menton and Roquebrune secede from Monaco

The towns of Menton and Roquebrune declare independence from Monaco but are immediately occupied by Sardinian troops.

1848: November 13 Birth of Prince Albert I

Albert, one of Monaco's most distinguished princes, succeeds his father, Charles III, in 1889. Carrying on the tradition of his seafaring ancestors, he is known for his love of and contributions to oceanography. He develops new oceanographic equipment and techniques and founds an oceanographic museum in Monaco (see 1910) and an oceanographic institute in Paris. He also serves as a patron of other sciences, including anthropology, botany, and paleontology. During Albert's reign, Monaco becomes a constitutional monarchy. Its first constitution, adopted in 1911, creates an elected national council. After a short first marriage that ends in divorce, Albert marries the widowed, American-born Duchess de Richelieu (Monaco's first American princess), who shares many of his intellectual and cultural interests. Together, they play an important role in making Monaco a cosmopolitan European cultural capital. After roughly a decade, however, this marriage ends unhappily as well. Albert dies in 1922 and is succeeded by his son from his first marriage, Louis II.

1856 Gambling concession is chartered

In its first attempt to attract tourists, Monaco allows two French entrepreneurs to build a gambling casino and tourist facilities and establish steamboat service between Monaco and Nice. The Société des Bains de Monaco ends in failure.

By the 1930s, gambling brings people from all over the world to Monaco. They play games of chance like roulette, shown here. (EPD Photos/ CSU Archives)

1860: July 17   Sardinian troops withdraw from Monaco

With the withdrawal of Sardinian troops from Monaco, the Sardinian protectorate ends, and Monaco returns to the protection of France.

1861: February 2   Menton and Roquebrune become part of France

Monaco and France sign a treaty providing for the reversion of Menton and Roquebrune to French control, decreasing Monaco's territory by over eighty percent. In return, France agrees to pay over four million francs to Monaco's prince, Charles III, and also to build a road for carriage travel from Nice to Monaco, as well as including Monaco in the route for the projected Nice–Genoa railway.

1863   Société des Bains de Mer is formed

Monaco's failed tourist concession, renamed and placed under new management, prospers and launches Monaco's first tourism boom. Streets, squares, hotels, a park, and other public facilities are built.

1864: January   Hôtel de Paris opens

The Hôtel de Paris, modeled on the famous Grand Hôtel in Paris, is inaugurated for well-to-do vacationers. Its lavish decor makes it one of the world's leading deluxe lodgings.

1866   Monte Carlo is named

The concentration of gambling and tourism facilities built by Société des Bains de Mer is named Monte Carlo, after Monaco's current prince, Charles.

1868: October   Railroad is completed

Monaco's first railroad is launched, hastening the economic development of the principality. Train service to Nice takes fifteen minutes.

1869   Taxes are abolished

Monaco's Prince Charles eliminates direct taxation, making Monaco even more attractive to potential wealthy residents.

1879: January   Opera house is opened

The first performance—a play starring the great Sarah Bernhardt—is held in Monaco's opera house, the Theatre de

Monte Carlo, designed by the prominent architect Charles Garnier. Works by Massenet, Fauré, and Ravel later receive their world premieres at the theater.

1900–14 Monaco becomes a cultural center

With the aid of its growing financial resources and the efforts of Prince Albert I, the principality becomes a European mecca for the arts. The Monte Carlo opera features performances by great singers including Enrico Caruso, Feodor Chaliapine, and Nelly Melba, and famous operas receive their premiers in the city. In addition, the opera house becomes the home of the famed Ballet Russe de Monte Carlo founded by Russian dancer and choreographer, Serge Diaghilev.

1910 Oceanographic Museum is completed

The Oceanographic Museum is built by Prince Albert I, who also establishes an oceanography institute in Paris. Noted oceanographer Jacques Cousteau later serves as director of the museum, which contains an aquarium, a display of rare ocean specimens, and a 50,000-volume library.

1911: January 7 Monaco adopts its first constitution

Previously under absolute rule by its princes, Monaco adopts a constitution that provides for a national council elected by universal male suffrage.

1918 Treaty of Versailles

Monaco's is obliged to renegotiate its sovereignty and territorial rights following World War I.

1922 Death of Prince Albert

Prince Albert I dies in Paris at the age of seventy-four and is succeeded by his son, Louis II (1870–1949).

1923: May 31 Birth of Prince Rainier

Prince Rainier III is born to Charlotte, daughter of Prince Louis II and Count Pierre de Polignac. Rainier is educated in France, Switzerland, and England and serves with the French in World War II in 1944. Rainier succeeds to the throne just before the death of Prince Louis II in 1949 and works to reduce economic dependence on the gambling industry. In 1956 he focuses world attention on his tiny principality when he marries American film star Grace Kelly, who becomes Princess Grace de Monaco. The couple have three children, Caroline, Albert, and Stéphanie. In 1982 Princess Grace is killed in a car crash and deeply mourned by her husband and the nation.

1929 Auto race makes it debut

The first Monaco Grand Prix is held with sixteen participants going up to eighty kilometers per hour.

1930: December National Council is dissolved

Prince Louis dissolves the National Council and rules by decree.

1939–45 World War II

Monaco formally remains neutral in World War II, but Prince Louis II is sympathetic to the Vichy government.

1943 Germans occupy the Hôtel de Paris

Monaco's premier hotel is occupied by Gestapo forces.

1945 Women can vote in municipal elections

Monaco's women win the right to vote in local elections.

1949: May 5 Prince Rainier ascends the throne

Prince Rainier III, grandson of Louis II, succeeds to the throne of Monaco, becoming the first native-born Monégasque to rule the principality since Honoré IV in the eighteenth century.

1951 Tourism academy is founded

Prince Rainier establishes the International Academy of tourism, which publishes a dictionary of tourism and a quarterly publication, *Revue Technique du Tourisme*.

1956 Prince Rainier weds Grace Kelly

Prince Rainier III marries American film star Grace Kelly, who brings a new level of international glamour, as well as an element of Americanization, to Monaco. The births of their three children, Caroline, Albert, and Stephanie, guarantee the continuation of the Grimaldi line.

1962 Tax agreement reached with France

After tensions between Monaco and France over taxes and customs duties lead French president Charles de Gaulle to seal off Monaco's borders with France, Monaco and France sign a new tax agreement that restricts the ability of French citizens to use the principality as a tax haven.

1962: December 17 New constitution is adopted

Monaco adopts a new constitution providing for a single-chamber National Council whose eighteen members are elected to five-year terms by direct popular vote; a minister of state with French citizenship to assist the prince with executive duties; and a three-member Council of Government to assist the minister of state.

1963: February   Women vote in National Council elections

Women's suffrage is expanded to include voting for the first time in elections for the National Council.

1970s   Industrial zone is established

To diversify Monaco's economic base, Prince Rainier establishes the Port de Fontvieille, a zone for light industry, including electronics and perfumes.

1972   National Museum is founded

Monaco's National Museum is established.

1982: September 14   Princess Grace dies in car crash

Princess Grace dies when she suffers a stroke while driving and her car plunges over a cliff. Her daughter, Stéphanie, who is in the car with her mother, survives the accident with no permanent injuries. The popular princess is deeply mourned by her adopted country.

1992: January   Citizenship laws for women are modified

Laws governing Monégasque citizenship for women are amended to end the practice of automatic citizenship for women marrying Monégasques, who must now wait five years to become citizens. The government also rules that women can transfer Monégasque citizenship to their children.

1993   State-owned company is investigated

The Société des Bains de Mer, the company that runs Monaco's gambling and tourism facilities, is investigated at the orders of Prince Rainier after charges of strong-arm tactics to recover debts incurred by gamblers at the casinos.

1993: February 23   Pope declares Caroline's children legitimate heirs

Pope John Paul II issues an edict making Princess Caroline's three children by her late husband, Stefano Casiraghi, legiti-mate heirs to the throne of Monaco. The status of the children has long been in limbo because Caroline's marriage to her first husband, Philippe Junot, has never been formally annulled, making her 1983 civil marriage to Casiraghi invalid in the eyes of the church (and thus making her children illegitimate and disqualifing them from royal succession).

1993: May 28   Monaco joins the UN

Monaco becomes a member of the United Nations.

1997: January   Grimaldis celebrate 700 years of rule in Monaco

The Grimaldi family—the world's oldest ruling dynasty—inaugurates festivities marking the 700th anniversary of the original Grimaldi takeover of Monaco in the Middle Ages (see 1297). A ten-month-long $270-million celebration is planned.

Bibliography

Bernardy, Françoise de. *Princes of Monaco: The Remarkable History of the Grimaldi Family.* London: Barker, 1961.

Duursma, Jorri. *Self-determination, Statehood, and International Relations of Micro-states: The Cases of Liechtenstein, San Marino, Monaco, Andorra, and the Vatican City.* New York: Cambridge University Press, 1996.

Edwards, Anne. *The Grimaldis of Monaco.* New York: William Morrow, 1992.

Jackson, Stanley. *Inside Monte Carlo.* Briarcliff Manor, N.Y.: Stein & Day, 1975.

Sakol, Jeannie and Caroline Latham. *About Grace: An Intimate Notebook.* Chicago: Contemporary Books, 1993.

Sherman, Charles L. *Five Little Countries of Europe: Luxembourg, Monaco, Andorra, San Marino, Liechtenstein.* Garden City, N.J.: Doubleday, 1969.

The Netherlands

Introduction

The Netherlands—often called Holland, which is only one region of the country—is a nation in northwestern Europe, bordered by Germany on the east, Belgium on the south, and the North Sea to the west and north. The name *Netherlands* (literally, "low lands") refers to the fact that much of the country is built on low-lying land reclaimed from the sea. The hard work and tenacity that went into reclaiming this land through an extensive system of dikes and dams gave rise to the popular saying, "God made the world, but the Dutch made Holland."

Not long after winning their own independence from Spain, the Dutch built the greatest maritime empire of the seventeenth century, an achievement paralleled by the great flowering of artistic talent and scientific discovery that made the era a cultural as well as a commercial Golden Age. Although the Dutch colonial empire now belongs to the past, the Netherlands continues to prosper as a highly industrialized country and to play an active role in European economic and political affairs.

With a total area of 14,413 square miles (37,330 square kilometers), the Netherlands consists of four major geographical areas. To the extreme west lies a narrow band of dunes, created over thousands of years by wind and water. Apart from dunes, the western half of the country consists of lowlands, also called polders (the name for artificially reclaimed land areas surrounded by dikes or walls). Both the country's richest agricultural land and its largest cities—including Amsterdam and The Hague—are found in this lowland region. The land to the east, called the sand plains region, is drier and hillier. In contrast to the polders, which are characterized by an excess of water, the sand plains region actually needs more water, and many farms in this region rely on irrigation systems. The remaining region of the Netherlands, a small upland section to the southeast, is the highest in the country.

For its small size, the Netherlands is one of the most densely populated countries in Western Europe, with a population estimated at about 15.5 million people in 1999. The capital city of Amsterdam is home to over a million people.

With its small size and strategic seacoast location, the Netherlands was overrun and occupied by a variety of foreign invaders, including the Romans, Franks, Saxons, and Vikings, in the first centuries of its history. The Romans introduced the construction of dikes to keep the sea at bay and drain the region's marshy coastal lands. Under the Frankish king Charlemagne (742–814), the populace was converted to Christianity. During the twelfth and thirteenth centuries, the Netherlands—now a group of semi-autonomous kingdoms—grew as more land was reclaimed from the sea. Agriculture flourished, and cities were built.

History

In the fourteenth century, the Netherlands came under the control of the Dukes of Burgundy, who united the scattered kingdoms into seventeen provinces covering the territory that today comprises the three Benelux countries (Belgium, the Netherlands, and Luxembourg). When Mary of Burgundy married Maximilian I of Austria in 1477, the Habsburg dynasty attained sovereignty over these provinces. Another royal marriage uniting two dynasties brought the region under the rule of King Charles I of Spain and his son Philip II. Philip's harsh rule and persecution of Dutch Calvinists gave rise to a rebellion, led by William of Orange, that turned into the Eighty Years' War (1568–1648). Early in this period, there was a decisive break between the northern and southern provinces that at that time made up the Netherlands. The seven largely Protestant northern provinces, which had essentially driven out the Spanish by the early 1570s, declared themselves the United Provinces in 1579 and asserted their religious freedom and their political freedom from Spain. The mostly Catholic provinces of the south (later to become Belgium and Luxembourg) remained loyal to Spain and became known as the Spanish Netherlands (after 1714, the Austrian Netherlands). Spain did not recognize the independence of the United Provinces until 1648 and made intermittent attempts to win back the region until that point.

At the end of the sixteenth century, the Netherlands claimed the island of Java, in present-day Indonesia, and its merchants entered the lucrative spice trade, forming a number of trading companies that were consolidated into the Dutch

NETHERLANDS

| 0 | 25 | 50 Miles |
| 0 | 25 | 50 Kilometers |

East India Company by 1602. This enterprise—one of the first joint-stock companies ever formed—laid the foundation for the Netherlands to become the world's foremost trading nation in the seventeenth century, with possessions extending from Asia and Africa to the New World. It was in the seventeenth century that the Dutch gained control of present-day Indonesia and bought Manhattan Island, in North America, from the local Indians for $24 (later ceding it to the British). The 1600s are now considered the Golden Age of the Netherlands, both in terms of commerce and of the cultural and scientific advances that accompanied the prosperity generated by the Dutch overseas possessions. The century that produced the paintings of Rembrandt and Vermeer also encompassed the later career of composer and organist Jan Sweelinck, the philosophical works of Baruch Spinoza, and the invention of the microscope by Anton van Leeuwenhoek.

However, the seventeenth century was also a period of virtually continuous warfare for the Dutch, who fought three wars with the British and two against France. By the eighteenth century the cost of these wars (and the long War of the Spanish Succession from 1701 to 1714), together with declining profits due to increased trade competition from France and Britain, brought about the decline of the Netherlands as a colonial power. In 1795 French Revolutionary forces invaded the Netherlands, beginning a twenty-year period of French occupation, first under the newly formed Batavian Republic, then under the reign of Napoleon Bonaparte's brother, Louis. At the Congress of Vienna following the downfall of Napoleon, the European powers reunited the northern provinces of the Netherlands with the southern, or so-called Austrian, Netherlands for the first time since the late sixteenth century, hoping that a larger and more powerful Dutch nation would serve as a buffer to prevent any possible expansionism by France in the future. However, the religious and cultural differences between north and south proved too strong, and by 1830 the southern provinces had revolted and formed the Kingdom of Belgium. (By 1890, Luxembourg would also become independent of the Netherlands.)

In 1848 the Netherlands became a constitutional monarchy with increased powers accorded to its legislative body. Toward the close of the nineteenth century, industrialization expanded rapidly, transforming the Netherlands' economy from a primarily agricultural one to that of a modern industrial nation, and also leading to the growth of trade unionism and to social reform in a variety of areas. The 1860 novel *Max Havelaar* by Multatuli (the pseudonym for Eduard Douwes Dekker) led to the abolition of the colonial system of forced labor on Java. In 1870 the Netherlands became one of the first nations to outlaw capital punishment. At the turn of the century, education was made compulsory, and a new, more enlightened policy was adopted toward the nation's overseas colonies.

The Netherlands remained neutral in World War I, but even without active involvement, the nation still suffered as supply routes were blocked and normal trade was disrupted. The global depression of the 1930s brought new hardships, and unemployment was particularly severe. In the Second World War, the Dutch were unable to remain neutral. On May 10, 1940 the country was invaded by Nazi Germany. With its queen and her government in exile, the Netherlands was occupied by the Germans until 1945. During this period, numerous Dutch resistance fighters were killed, the port city of Rotterdam was virtually destroyed, and other cities, such as The Hague, suffered major damage. In the last year of the war, an artificial, Nazi-imposed famine starved about 10,000

people in the western part of the country and came close to causing an even more massive tragedy.

Immediately following the war, the Netherlands was confronted with the demise of its colonial empire when Indonesia, its last major colonial possession, declared independence. Four years of alternating negotiations and armed conflict followed before Indonesian independence was recognized internationally through a negotiated agreement. (The small Dutch colony of Suriname in South America became independent in 1975.)

On the night of January 31, 1953, the reputedly impregnable dikes of the Netherlands gave way to the North Sea for the first time in hundreds of years following days of heavy rain and wind. Over one hundred towns and villages were submerged within minutes. Over 1,800 people were killed, and more than 72,000 were evacuated from their homes. In the wake of the disaster, the Dutch government laid out plans for a massive new flood control project, the Delta Project, which dammed the arms of the delta facing the North Sea. The project, which was begun in 1958 and completed in 1986, reduced the total length of the coastline by several hundred miles.

In the new postcolonial world in which the Netherlands has found itself since World War II, it has cooperated closely with its European neighbors. In 1948, it formed a customs union—the Benelux Union—with Belgium and Luxembourg, and it was a founding member of the European Economic Community ten years later. In 1991 the Netherlands hosted the conference that drew up the Maastricht Treaty providing for the creation of the European Union and the eventual integration of the economies of the member nations, including the adoption of a common currency.

Timeline

55 B.C.   Julius Caesar conquers the Lowlands

Roman emperor Julius Caesar conquers the region that makes up the present-day Netherlands, as well as Belgium and Luxembourg, a region collectively known as the Lowlands (or Low Countries). Most of the native Celtic and Germanic peoples are subdued. The Romans build dikes and an extensive system of roads.

4th–8th centuries A.D.   Germanic tribes overrun the Lowlands

The Germanic Franks and Saxons conquer the Lowlands following the waning of Roman power in the region. Most of the conquerors claim and farm stretches of land, and German becomes the common language of the area.

768–814   Reign of Charlemagne

The Frankish king Charlemagne rules over the Lowlands as part of the Holy Roman Empire. During his reign, the population is converted to Christianity.

9th–11th centuries   Vikings raid the Low Countries

The Vikings, seamen from Scandinavia, make repeated raids on the Low Countries, plundering settlements and capturing or killing their inhabitants.

12th–13th centuries   Land is reclaimed from the sea

The Netherlands grows as increasing areas of land are reclaimed from the sea to be used for agriculture and settlement.

12th century   First courtly epic is written

Henric van Veldeke (c. 1140–c. 1210) authors the *Eneide,* a courtly epic in the French style, later translated into German. Born in Limburg, van Veldeke is the first poet of the Low Countries whose name is known to present-day scholars. Van Veldeke is also the first known author of troubadour lyrics in the Netherlands.

c. 1200–40   Life of poet Hadewijch

One of the greatest writers of secular courtly lyrics, this female poet left 45 poems that are acclaimed for their lyricism and range of emotion. Her prose works, including letters and accounts of mystical visions, are also highly regarded.

Burgundian and Habsburg Rule

14th–16th centuries   Burgundian rule

The French dukes of Burgundy reintegrate and control most of the semi-autonomous states that have formed in the Low Countries following the decline of the Frankish empire.

1384   Philip becomes first Burgundian ruler

Philip the Bold of Burgundy (1342–1404), the brother of the French king, becomes the ruler of Flanders and Artois, beginning the Burgundian rule of the Low Countries, which is later expanded to include Holland, Zeeland, Hainaut, Namur, Limburg, and Luxembourg.

1421   Flood disaster kills 100,000

After days of heavy rains and high winds, the dikes surrounding the city of Dort give way and the city is destroyed in the Netherlands' second worst flood. Seventy-two nearby villages are also obliterated. Altogether, approximately 100,000 lives are lost. The flood detaches the city of Dordrecht from the mainland, permanently surrounding it with water.

c. 1450   Birth of artist Hieronymus Bosch

Bosch, a Flemish painter born in the province of Holland, is known for his terrifying visions of human folly, temptation, and sin and for the unique semi-human images (often called devils) with which he peopled these visions. He is active as a painter for about forty years and dies in 1516. Among his most famous works are *Carrying of the Cross, The Hay Wain, The Temptation of Saint Anthony,* and *The Garden of Earthly Delights.*

1477   Habsburgs gain control of the Low Countries

Mary of Burgundy marries Holy Roman Emperor Maximilian I, bringing the Low Countries, now consisting of seventeen provinces (the area that currently comprises the Benelux countries), under the control of the Habsburg dynasty.

1482   Maximilian becomes regent

Mary of Burgundy dies, and her widower, Maximilian, of Austria's Habsburg line, rules as regent for their son, Philip.

Spanish Rule

1516   Low Countries fall under Spanish rule

Charles (1500–58), the grandson of Mary of Burgundy and son of Juana of Castile and Aragon, inherits the Spanish throne, becoming King Charles I of Spain and Holy Roman Emperor Charles V, and the Low Countries come under Spanish rule.

1516: August 9   Death of Hieronymus Bosch

Artist Hieronymous Bosch dies. (See 1450.)

1530: November 1   Worst flood in Dutch history kills thousands

The North Sea, agitated by a gale, bursts the system of protective dikes reclaiming some forty percent of Dutch land from the sea, causing the worst flooding in the history of the Netherlands. Entire towns are wiped out, and an estimated 400,000 people are killed.

1533: April 24   Birth of William of Orange

William of Orange, also known as William the Silent, becomes Prince of Orange at the age of eleven, inheriting large portions of land in the Netherlands. Emperor Charles V directs his education, and the young prince shows a talent for politics. Although a Catholic himself, he disapproves of the persecution of Dutch Calvinists by Philip II when Charles makes him ruler of the Netherlands in 1556. In response to repressive measures instituted by the Duke of Alva, whom Philip appoints as governor of the Netherlands, William leads a Dutch revolt against Spain. However, the initial attempt fails, and William organizes additional forces to try again. Under his leadership, the northern provinces break free from Spanish control but not those in the south.

In 1580 King Philip places William under a death sentence (called a ban). The same year William publishes a historic document, the *Apology,* defending Dutch freedom. After surviving one assassination attempt in 1583, William succumbs in 1584 to a second assassin, who shoots him in his house at Delft. His sons, Maurice and Frederick Henry, complete the battle for independence begun by William, whose services to his country also include founding the first Dutch university (the University of Leiden) and instituting the use of the Dutch language in official communications. William is revered to this day as the father of his country.

1539–40   Rebellion in Ghent is put down

A revolt against Spanish rule takes place in Ghent and is crushed by King Charles.

1549   Union of Spain and the Netherlands

The Netherlands are formally joined to Spain by Charles I.

1556   Philip becomes king of the Netherlands

After the abdication of his father, King Charles, Philip II of Spain (1527–98) becomes king of the Netherlands. He attempts to impose Catholicism on the population, harshly persecuting those who have embraced Calvinism and arousing much hostility among the populace. In addition to religious persecution, the Dutch also resent Philip's use of the resources of their homeland to finance Spain's wars with France.

1562: May   Birth of organist and composer Jan Sweelinck

Sweelinck plays a leading role in the development of keyboard music in the Baroque era. He serves as the organist of the Oude Kirk (Old Church) in Amsterdam for most of his life, taking over the post from his father. He is known primarily for his keyboard compositions, which include fantasias, chorale variations, and toccatas. However, he is also the composer of a number of choral works—including settings of all 150 psalms—that are the culmination of the great Renaissance tradition of Dutch and Flemish choral music. Through both his compositions and his teaching he has a major influence on Dutch and German organ playing and keyboard composition for generations to come. His teaching is so influential that he is known to many of his contemporaries as "the organist maker." Sweelinck dies in 1621.

1567   Duke of Alva sent as governor of the Netherlands

After Protestant crowds attack churches and demolish their decorations as idolatrous, Philip II sends Fernando Alvarez de Toledo, Duke of Alva (1507–82) to govern the unruly province and keep rebellion there in check. Alva further incenses the population by trying and executing members of the Dutch nobility. However, William of Orange escapes and begins organizing a revolt while still in exile in Germany.

The United Provinces

1568   William of Orange leads revolt against Spain

William, Prince of Orange, leads the Low Countries in rebelling against political, religious, and economic domination by Spain. They are eventually aided by the British. Although the Spanish are effectively driven from the seven northern provinces within the first years of the war, Spain continues its efforts to retake them. Full Spanish recognition of Dutch independence does not come for eighty years. Thus the rebellion is known as the Eighty Years' War.

1570   Flooding kills 50,000

Exactly forty years after the worst flood disaster in the history of the Netherlands (see 1530), North Sea breaches its newly rebuilt dikes, with a resultant loss of an estimated 50,000 lives. The city of Friesland in the north is completely destroyed.

1574: October 1–2   Spanish are driven out of Leiden

In a major victory for the Dutch, the Spanish are driven back from the city of Leiden, which they have besieged for months, when dikes miles away are broken and the Spanish encampment is flooded, killing some 20,000 of its troops. Some accounts attribute the flooding to a storm; others claim that the dikes are broken deliberately at the order of William of Orange to drown the Spanish and allow the allies of the Dutch to sail to their rescue over the floodwaters.

1575   University of Leiden is founded

To reward the people of Leiden for their courage during the Spanish siege of the city, William of Orange establishes the Netherlands' first university there.

1576   Pacification of Ghent

All seventeen provinces of the Low Countries, united under the leadership of William of Orange, petition the Spanish government for greater religious tolerance and more liberal government.

1579: January 23   Union of Utrecht

Although the northern and southern provinces of the Netherlands have been allies in fighting the Spanish, the Protestant north and the Catholic south remain divided by religion. Through the union of Utrecht, the seven northern provinces (Holland, Zeeland, Utrecht, Gelderland, Groningen,

Friesland, and Overijsel) affirm their break with Spain and their right to religious freedom by forming the United Provinces of the Netherlands. The southern provinces, known as the Spanish Netherlands, remain under Spanish control.

1580   Philip II bans William

King Philip II of Spain places William of Orange under a ban, inviting his assassination and calling him a "plague upon the human race."

1581: March 16   Birth of poet Pieter Hooft

Pieter Corneliszoon Hooft, a great poet and dramatist of the Dutch Golden Age, is also a prose stylist whose influence extends to the nineteenth century. Hooft is strongly influenced by the styles of French and Italian Renaissance authors after spending time in their countries. His early works include love lyrics and a pastoral play (*Grandida*; 1605). Hooft spends nineteen years writing a monumental history of the Netherlands in the period between 1555 and 1584, which celebrates the greatness of William of Orange, who is treated as an epic hero. Hooft dies in 1647.

1581: July 26   Dutch independence is declared

In retaliation for the ban placed on William of Orange by Philip II, the northern provinces of the Netherlands break with Spain completely, declaring their independence in the Act of Abjuration. This also represents a break with the southern provinces, which remain loyal to Spain.

1584: July 10   William is assassinated

William of Orange is assassinated. He becomes a Dutch national hero for his role in winning independence for the Netherlands.

1585   The British ally themselves with the Dutch

Seeking foreign protection against Spain, the United Provinces offer Queen Elizabeth sovereignty over the region. She refuses it but sends her adviser, the Earl of Leicester, to the Netherlands with a small army.

1587: November 17   Birth of poet Joost van den Vondel

Vondel is the most acclaimed poet of the Dutch Golden Age and regarded by many as the single greatest poet in the history of the Netherlands. His dramatic tragedies are his most famous works. Vondel's writing is strongly grounded in the Greek and Latin classics, some of which he translates into Dutch, including Sophocles' *Elektra*. Many of his works have Biblical themes, including his first major play, *Het Pascha* (*The Passover*; 1612), a dramatization of the Jews' exodus from Egypt in which the Egyptians' subjugation of the Jews is meant to parallel the Spanish persecution of Dutch Calvinists. Vondel's greatest work is a trilogy of religious plays

written between 1659 and 1667: *Lucifer, Adam in ballingschap* (Adam in Exile), and *Noah*. Vondel dies in 1679.

1595   Spanish troops withdraw from the north

Following the defeat of its Armada by England in 1588 and other reverses, a weakened Spain withdraws its forces from the northern provinces of the Netherlands.

1596   Java is claimed by the Dutch

Dutch explorer Cornelius de Houtman lands on the island of Java with several vessels and claims it for the Netherlands. Only 89 of the original 258 members survive the two-and-a-half-year expedition, but it returns to the Netherlands laden with valuable spices, encouraging further exploration of the region. A number of merchants form trading companies, which are soon consolidated in the Dutch East India Company (see 1602). The Indonesian archipelago becomes known as the Dutch East Indies.

The Golden Age

17th century   Dutch Golden Age

In spite of almost continual warfare with the French and English, the United Provinces becomes Europe's leading trading nation and builds a profitable overseas empire that spans Asia, Africa, the Caribbean, and the Americas. Together with commercial success come cultural achievements in fields including painting, crafts, architecture, literature, philosophy, and science. The great painters of the era include Rembrandt (see 1606), Jan Vermeer (see 1632), Jacob van Ruisdael, and Frans Hals. Philosophers Baruch Spinoza (see 1632), Hugo Grotius, and the Frenchman Rene Descartes (now living in the Netherlands) are active during this period. Great scientists of the period include Jan Swammerdam and Anton von Leeuwenhoek.

1602   Dutch East India Company is founded

The government charters the Dutch East India Company to reduce costs and risk by consolidating the trading operations of Dutch merchants, thus laying the framework for a monopoly on the world spice trade and the formation of an overseas empire. The firm is one of the first joint-stock companies ever formed.

1606   Birth of Rembrandt

Rembrandt Harmenszoon von Rijn, one of the greatest painters in Western art, is born in Leiden, the son of a miller. After a brief period as a student at the University of Leiden, he apprentices himself to a local artist. Around 1623 he leaves Leiden to study with a prominent painter in Amsterdam. Returning to his birthplace, he opens his own studio there in 1625, soon winning professional success and accepting pupils. In 1631 the painter returns to Amsterdam, becoming one of the country's most renowned artists and the capital's most sought-after portrait painter. His financial security increases further when he marries the daughter of a wealthy public official.

In 1642 the painter's fortunes begin to undergo a reversal when his wife dies, leaving him alone with a year-old son. Later in the decade, Hendrickje Stoffels becomes his common-law wife, and they have a daughter. However, the painter, whose work has become less popular with the Dutch public, experiences financial difficulties and is forced to declare bankruptcy in 1656. Before his death in 1669, Rembrandt also undergoes the loss of his second wife and his son by his first marriage.

Rembrandt creates about 600 paintings and almost 300 etchings. Although his paintings encompass different genres, including landscapes and biblical scenes, Rembrandt is best known for the psychological depth of his portraits, including a remarkable series of self-portraits done at different periods of his life. He is also renowned for the quality of his drawings, of which some 1,400 survive.

1609–21   Twelve years' truce

A truce with Spain assures the independence of the United Provinces of the Netherlands for twelve years, after which Spanish attacks resume.

1609   Henry Hudson explores the North American coast

The Atlantic seaboard of North America is explored by Henry Hudson, an English seaman hired by the Dutch to find a passage to the Far East by sailing across the Atlantic Ocean. Hudson sails down the waterways in present-day New York State and Canada that are later named for him.

1611   Stock exchange is established

The Amsterdam Stock Exchange—the world's oldest—is founded.

1619–67   Dutch gain control of the Malay Archipelago

The Netherlands gains control of Java, Sumatra, the Cape of Good Hope, and other parts of the Malay Archipelago (present-day Indonesia).

1621   Dutch West India Company is formed

The Dutch West India Company is established to carry out commerce and colonization in the New World and Africa, including participation in the slave trade. It founds settlements in the Caribbean and the colony of New Netherland in North America, later ceded to the English.

1621: October 16   Death of Jan Sweelinck

Organist and composer Jan Sweelinck dies in Amsterdam. (See 1562.)

1626   Manhattan Island is purchased

The Dutch purchase Manhattan Island from native American leaders for $24, renaming it New Amsterdam. It is later ceded to the British, who name it New York, in an exchange that includes Dutch control of present-day Suriname.

1630s   Tulip demand fuels wild speculation

Rapidly rising demand for tulips, newly imported from Turkey, makes prices skyrocket, fueling a wave of frenzied speculation that makes and breaks fortunes before subsiding.

1632   Birth of painter Jan Vermeer

One of the greatest artists of the Netherlands' Golden Age, Jan Vermeer is the son of an innkeeper and art dealer. Producing a relatively small number of paintings, even for an artist who dies young (age 43), Vermeer is a meticulous and painstaking worker, living in modest quarters with his wife, who bears eleven children. To bring in additional income, he tries dealing in art but without great success. He dies in 1675, having lived to see only three of his children reach adulthood, and is buried at Delft. His paintings—of which there are only about 40 in existence—are renowned the world over for their subtle use of light and perspective. Through their concentration on everyday domestic activities, they also provide a detailed record of Dutch daily life during the prosperous seventeenth century.

1632: October 24   Birth of Anton van Leeuwenhoek

Leeuwenhoek, famed as the inventor of the microscope, is born in Delft. Receiving no formal scientific training, he works as a civil servant while developing and producing some 250 microscopes during his lifetime and making major biological discoveries about humans and other life forms. Through his use of the microscope, he is the first person to view bacteria and other micro-organisms. Leeuwenhoek makes important contributions to the study of blood circulation in humans and insect development. Scientifically, his discoveries are significant for helping to refute the popularly accepted theory of spontaneous generation. Leeuwenhoek becomes a Fellow of the Royal Society in 1680 and becomes a member of the French Academy of Sciences in 1697. He dies in 1723.

1632: November 24   Birth of philosopher Baruch Spinoza

Baruch Spinoza (also called Benedict de Spinoza) is born in Amsterdam to Jewish emigrants from Spain. He is brought up as an Orthodox Jew but excommunicated by the religious authorities for his views, which are deemed heretical. He pursues his philosophical writing and correspondence with leading intellectual figures while working as a lens grinder and polisher for telescopes and microscopes. He is offered a professorship at the University of Heidelberg in Germany but declines to avoid conflicts with clerical opponents of his views. His most famous works are the *Tractatus Theologico-Politicus* (1670) and the *Ethics* (completed in 1675 and published posthumously in 1677). Spinoza is known for his study of human nature in the framework of the Copernican Revolution. He applies scientific principles to arrive at a philosophy about the nature of God and the place of human beings in the cosmos. Spinoza dies in 1677.

1641   The Dutch capture Malacca

The Dutch capture Malacca, ending Portuguese domination of the straits between Malaya and Sumatra and gaining control of important shipping lanes.

1642   Tasmania and New Zealand are discovered

Dutch explorer Abel Tasman (1603–59) sails to present-day New Zealand and Tasmania.

1647   Death of Pieter Hooft

Renowned poet, dramatist, and historian Pieter Hooft dies in The Hague. (See 1581.)

1648   Treaty of Westphalia

In a peace agreement signed at Münster, Spain formally recognizes the Netherlands as an independent nation as part of the Treaty of Westphalia.

1652–54   First Anglo-Dutch war

Competing commercial interests lead to war between the Netherlands and the Commonwealth of England, led by Oliver Cromwell (1599–1658). In addition to the trade rivalry between the two nations, the English fear Dutch support for the restoration of the Stuart line in England. With the advantage of a newly reorganized navy, the English soundly defeat the Dutch.

1667   Second Anglo-Dutch War

The Netherlands and England go to war a second time following the restoration of Charles II. Dutch admiral Michiel de Ruyter carries out a daring raid that brings his forces nearly all the way to London. At the conclusion of the war, the Netherlands gains control of Dutch Guiana (present-day Suriname) on the northern coast of South America, in exchange for lands including New Amsterdam (New York).

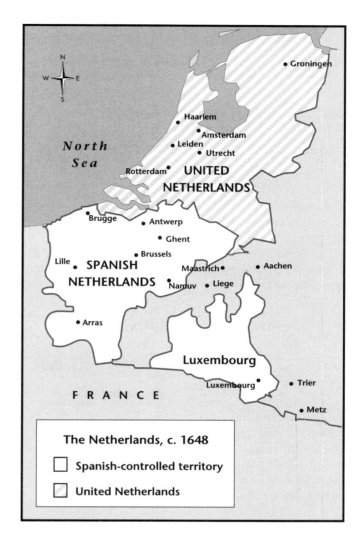

The Netherlands, c. 1648

☐ Spanish-controlled territory

▨ United Netherlands

1668   War with France

The Netherlands joins with England and Sweden in forming the Triple Alliance to drive back a French invasion of the Spanish Netherlands to the south, in the War of Devolution.

1669   Death of Rembrandt

Renowned painter Rembrandt van Rijn dies in Amsterdam. (See 1606.)

1672–78   Third Anglo-Dutch War

After forming a secret alliance, Charles II of England and Louis XIV of France launch a new war against the Dutch. They invade the United Provinces, but their forces are driven back when dikes are opened, causing flooding.

1677   Death of Spinoza

Philosopher Baruch Spinoza dies. (See 1632.)

1679: February 5   Death of author Joost van den Vondel

The great dramatist and poet Joost van den Vondel dies in Amsterdam. (See 1587.)

1688-97   War of the League of Augsburg

The Dutch join England in the first of two wars against France. (See 1701.)

1688   Dutch monarch installed in England

At the request of the British parliament, William III, grandson of William of Orange, and his wife, Mary II, replace the Catholic monarch James II on the British throne, in what comes to be known as the Glorious Revolution. Together with a new dynasty, the House of Orange, constitutional monarchy is introduced to England.

Decline of Colonial Supremacy

18th century   The Netherlands declines as a colonial power

Faced with competition for trade and maritime supremacy by the French and British, the Netherlands becomes engaged in costly wars and is not able to maintain its colonial and economic supremacy. With increased competition, profits from the spice trade begin to fall, and revenues from the Dutch colonies decline. The country becomes increasingly dependent on its alliance with England.

1701–14   War of the Spanish Succession

A long war with France drains Dutch resources and marks the end of the seventeenth-century Golden Age of commerce and culture.

1714   Spanish Netherlands are ceded to Austria

The southern provinces of the Netherlands, which have remained linked to Spain, are ceded to Austria as part of the Peace of Utrecht.

1723–47   Rule of the regents

Four of the eight northern provinces that now comprise the independent Netherlands reject William IV, heir of the House of Orange, as governor (*stadholder*) and choose to remain under local rule, in what comes to be called the "rule of the regents."

1723: August 26   Death of Anton van Leeuwenhoek

Anton van Leeuwenhoek, inventor of the microscope, dies at Delft. (See 1632.)

1747   All provinces accept William IV

Fearing an invasion by the French, the Netherlands unites under William IV of the House of Orange, who is previously accepted as governor of only half the provinces.

1756: September 7   Birth of poet Willem Bilderdijk

Bilderdijk is a central figure in the introduction of Romanticism into Dutch literature, through both his work and his theories of poetry. He studies law in Leiden and is a practicing attorney in The Hague until 1795, when he is forced into exile for refusing to swear allegiance to the French government that has taken over the country. In his poem *De kunst der poezij* (The Art of Poetry), Bilderdijk introduced the Romantic emphasis on feeling and emotion to Dutch poetry. His best-known single work is the unfinished Biblical epic *De Ondergang der eerste wareld* (The Destruction of the First World), which dramatizes the fate of Cain's children. Bilderdijk dies in 1831.

1795   France invades the Netherlands

The Dutch kingdom collapses following conquest by French revolutionary forces, who found the Batavian Republic.

1800   Rijksmuseum is founded

The Rijksmuseum, the Netherlands' outstanding museum and one of the world's finest, is founded in Amsterdam.

1806   Napoleonic kingdom is founded

The French emperor Napoleon I establishes the Kingdom of Holland. His brother, Louis, becomes its king.

1810   Dutch territories are annexed by France

After Napoleon's brother Louis is overthrown, all Dutch territories are annexed by France.

Reunion and Division

1815   Reunited kingdom is formed

French occupation of the Netherlands and its territories is ended at the Congress of Vienna, which reunites the United Provinces in the north with the Austrian Netherlands in the south, as well as the Grand Duchy of Luxembourg, to strengthen the Low Countries against possible future attacks by France. A new monarchy ruled by King William I (1772–1843), a descendant of William of Orange, is installed to rule the reunified and expanded Kingdom of the Netherlands.

1820: March 2   Birth of author Multatuli

Multatuli (the pen name of Eduard Douwes Dekker), author of the famous novel *Max Havelaar,* is a great nineteenth-century Dutch prose writer and campaigner for social justice. As a young man working for the Dutch government in the East Indies, he becomes aware of injustices suffered by the Javanese, resigning his post in 1856 because of the government's refusal to support his recommendations. After adopting the pseudonym *Multatuli* (Latin for "I have suffered greatly"), he publishes *Max Havelaar,* his most famous work (see 1860). Other works include *Minnebrieven* (1861)*,* which portrays a fictional correspondence between two lovers and his masterwork, the seven-volume *Ideen* (Ideas; 1862–77), consisting of essays on a variety of subjects, that includes an autobiographical novel, *Woutertje Pieterse.* Dekker dies in 1887.

1824 Treaty of London

Colonial territories in present-day Malaysia are divided between the Netherlands and Britain. The British claim to newly founded Singapore is accepted by the Dutch, and the Netherlands cedes Malacca in exchange for British-controlled ports in Sumatra.

1825–30 Revolt in the East Indies

A revolt breaks out in the East Indies, whose control has been transferred from the Dutch East India Company to direct rule by King William I. An estimated 200,000 Javanese and 15,000 Dutch are killed. On the brink of bankruptcy from the joint effects of the Javanese revolt and the secession of Belgium (See 1830), the Netherlands launches a system of forced labor (*cultuurstelsel*) in Java. Javanese peasants are required to labor over two months of every year growing export crops for the colonial government. Eventually, the export income produced through this system accounts for one-third of the total Dutch budget.

1830 Southern provinces revolt

Northern rule creates deeply held grievances in the southern provinces. King William tries to impose the Dutch language on the French-speaking Walloons, and the Catholics of the south resent being governed by Protestants. Spurred by revolutionary activities in France, the southern provinces revolt and secede from the Netherlands. The kingdom of Belgium is formed, and Luxembourg is divided between Belgium and the Netherlands.

1831: December 18 Death of poet Willem Bilderdijk

The single most important figure in Dutch Romanticism dies in Haarlem. (See 1756.)

1839 The Netherlands recognizes Belgium

King William of the Netherlands formally recognizes Belgian independence and neutrality at the London Conference.

1848 New constitution is adopted

Under King William II (1792–1849), a new liberal constitution is promulgated, creating a constitutional monarchy. The king remains head of state, but the powers of the States-General (parliament) are expanded. In addition, the king loses his exclusive control over the East Indies, which increasingly becomes the province of the legislature.

1853: March 30 Birth of painter Vincent Van Gogh

Although he sells only one painting in his lifetime, Van Gogh is considered one of the world's great artists. He is born in Groot-Zundert to the family of a clergyman. In 1869 he goes to work for a firm of art dealers, working at branches in The Hague, London, and Paris. However, due to Van Gogh's lifelong eccentricity, he is dismissed by the company in 1876, after which he studies theology and goes to work as a missionary among Belgian coal miners. During this period, he produces charcoal drawings of miners and begins considering art as a career. He studies anatomy and perspective in Brussels in 1880-81. After spending two years in The Hague, the artist joins his parents at Nuenen, where he paints his first major work, *The Potato Eaters* (1885).

Van Gogh next lives in Paris for two years, with the help of his younger brother, Theo. There he meets such artists as Henri Toulouse-Lautrec, Paul Gauguin, and Georges Seurat and is influenced by impressionistic painting and Japanese art. From Paris, Van Gogh leaves for the more peaceful rural environment of Arles, where he produces some of his greatest paintings, including *The Postman Roulin* and *House at Arles.* With his mental and physical health deteriorating, the artist voluntarily enters an asylum at Saint-Rémy, where he continues painting when he is well enough. The artist shoots himself on July 27, 1890 and dies shortly afterward. Van Gogh's vivid, emotionally intense paintings, with their bold brush strokes, are generally considered the beginning of the modern Expressionist movement in Western art.

1860 Publication of *Max Havelaar*

This landmark Dutch novel by the author Multatuli (see 1820) is an indictment of Dutch colonial exploitation of the Javanese in Indonesia through the forced labor system (*cultuurstelsel*) in place since 1830. *Max Havelaar* tells the story of a well-meaning liberal official whose enlightened efforts to introduce a more humane administration in Java fail. The novel's effect on the Dutch public is similar to that of a contemporary American work, *Uncle Tom's Cabin,* on U.S. readers in bringing the evils of slavery to the forefront of national consciousness in the United States. In the case of *Max Havelaar,* the exploitation it describes draws attention not just in the Netherlands but even in England and Germany. The forced labor system in Java is discontinued for all products except sugar and coffee soon after the book's publication.

1863: June 10   Birth of novelist Louis Couperus

A distinguished novelist associated with the Dutch literary revival of the 1880s, Couperus is born in the Dutch East Indies and travels widely, publishing newspaper essays about his experiences. His first and best-known novel is *Eline Vere* (1889), a detailed account of life in The Hague that is influenced by French realism. Other works include *Extase* (Ecstasy; 1892) and *De berg van licht* (The Mountain of Light; 1906). Some of Couperus's works reflect the author's interests in Oriental philosophy and the occult. Couperus dies in 1923.

1864: December 3   Birth of playwright Herman Heijermans

Heijermans is a Dutch playwright and novelist whose work engages a variety of contemporary social issues. His novel *Kamertjeszonde* (Petty Sin; 1898) criticizes the bourgeois sexual repressiveness and hypocrisy of his society. *Op hoop van zegen* (The Good Hope; 1901) takes up the social problems of fishermen, and *Glück auf* (Good Luck; 1911) explores the lot of miners. Heijermans also authors satirical sketches under the pen name of Samuel Falkland. The author dies in 1924.

Late 19th century   Growth of industrialization

The Netherlands is transformed from an agricultural to an industrial economy as the textile, chemical, and other industries grow. Industrialization is accompanied by the growth of trade unionism.

1870   Capital punishment is banned

The Netherlands becomes one of the first countries to abolish the death penalty.

1872   Birth of artist Piet Mondrian

Mondrian is born in Amersfoort. After becoming interested in the works of Pablo Picasso and Georges Braque, he moves to Paris in 1912 and begins painting in the Cubist style. Gradually, representative elements, as well as curved lines, disappear from his work, which becomes increasingly abstract and geometric. Mondrian becomes the leader of an artistic movement known as *de Stihl* (the Style), which abandons realism and embraces the most basic aspects of painting—primary colors, straight lines, and right angles. Mondrian continues his artistic development in New York City, where he moves in 1940. He dies in 1944.

1880s–1890s   Cultural revival takes place

Dutch commercial and industrial growth is accompanied by an outburst of cultural and intellectual activity, whose most famous single figure is painter Vincent Van Gogh (see 1853.)

1882   Amsterdam Concertgebouw is built

A state-of-the-art concert hall, the Amsterdam Concertgebouw, is constructed.

1888   Concertgebouw Orchestra is founded

The Amsterdam Concertgebouw Orchestra marks the beginning of a modern orchestral tradition in the Netherlands. The orchestra soon acquires an international reputation, turning Amsterdam into one of Europe's major musical centers.

1890   Luxembourg gains its independence

Because of an old law by which the duchy cannot recognize a female monarch, Luxembourg's union with the Netherlands is ended when Queen Wilhelmina (1880–1962) becomes the Dutch queen.

1899   International court is set up at The Hague

The International Court of Arbitration (also known as The Hague Tribunal) is established at The Hague. The same year, The Hague hosts the first international peace conference.

Early 20th century   "Ethical policy" adopted toward colonies

A new government under Abraham Kuyper inaugurates an enlightened policy (called the "ethical policy") of improving social and economic conditions in the Dutch colonies. Railways are built in Java and Madura, educational facilities are improved, and banking services are expanded.

1900   Education becomes compulsory

The first legislation is passed making education compulsory in the Netherlands.

1902   First game of korfball is played

The Dutch game of korfball is developed by Amsterdam schoolteacher Nico Broeukheuysen, who takes part in the first game with his students. The game is known for its educational advantages, including the fact that it is coeducational and involves no opportunities for contact or collision that can injure players.

Combining elements of basketball and handball, it is played by two teams of twelve players. The object, as in basketball, is to get a round ball into the opposing team's basket while attempting to prevent the opponents from scoring. The playing field is 44 yards (40 meters) by 99 yards (90 meters), and the players of each team are stationed in three zones, within which they must play until they rotate to another zone. The ball can only be moved with one's hands. Games are 90 minutes long.

Today there are some 600 korfball clubs in the Netherlands, with a total of more than 100,000 players. Korfball is also played in other countries, including the Czech Republic, Hungary, Suriname, India, and Taiwan.

1907 Royal Dutch Shell is formed

The Royal Dutch Shell oil company, which will become one of the world's leading corporate giants, is formed by a merger of the Royal Dutch Petroleum Company and Shell Transport and Trading.

The World Wars

1914–1918 World War I

The Netherlands remains neutral in World War I. However, it is adversely affected by the resulting decline in international trade, which forces the country to rely almost entirely on its own resources to feed its population.

1917 Pacification agreement gives state aid to religious schools

A long struggle over state funding for religious schools culminates in the government's decision to fund both public and religious schools. The normally liberal Catholics join forces with Protestants in opposing the secular liberals' opposition to state support for church-operated school.

1919 Female suffrage is enacted

Dutch women win the right to vote.

1919 Major labor legislation is passed

The work week is limited to 45 hours, and old age pensions are introduced.

1923: July 16 Death of novelist Louis Couperus

Louis Couperus, one of the leading Dutch novelists of the late nineteenth and early twentieth centuries, dies in De Steeg. (See 1863.)

1924: November 22 Death of playwright Herman Heijermans

Playwright and social critic Herman Heijermans dies in Zandvoort. (See 1864.)

1928 High Court of Labor is established

The High Court of Labor is set up to arbitrate labor disputes.

1929 Depression brings economic decline

The Netherlands suffers economic hardships with the onset of global depression. Unemployment is particularly severe. Public discontent leads to some support for a pro-Nazi political party.

1930 Research organization is founded

The Netherlands Organization for Applied Scientific Research is established.

1938 Korfball receives royal charter

Queen Wilhelmina grants a royal charter to the Netherlands' official korfball association. (See 1902.)

1939–45 World War II

The Netherlands is invaded by Germany, and its government, headed by Queen Wilhelmina, flees to England, where it is reestablished in exile for the duration of the war. Underground resistance is punished by mass executions, and the country is ravaged by the Nazis. Rotterdam, The Hague, and other cities suffer major damage. Some 240,000 Dutch lives are lost during the war. About 400,000 Dutch are sent abroad to provide labor in Germany. In addition, the country's infrastructure is heavily damaged, and it is looted for livestock and equipment. An artificially imposed famine kills thousands in the western part of the country and comes close to causing deaths on an even larger scale.

1940: May 10 Germany invades the Netherlands

The Netherlands is invaded by Nazi Germany in a surprise attack, and after five days the Germans occupy the country. Air raids destroy 35,000 buildings in the city of Rotterdam, virtually leveling the country's main seaport.

1940: May 13 Queen Wilhelmina goes into exile

The Dutch queen flees the country with her retinue, establishing a government in exile in England.

1944–45 Nazis create famine in the western part of the country

In the final year of World War II, the Nazi occupiers of the Netherlands deliberately create a famine that kills as many as 10,000 people in the western part of the country. Having confiscated stockpiled food in 1940 and commandeered over half of all Dutch produce, the Germans have already created a scarcity of food, especially in the industrialized western part of the country. In the fall of 1944, the Germans cut off food shipments to the west to counter a rail strike imposed by the Dutch government in exile and to retaliate for the Allied air attack on Arnheim. In addition, food is taken from warehouses and moored cargo ships and sent out of the country.

Within a short period of time, famine conditions set in. Thousands are threatened with starvation, and about 10,000 die. Eventually, the commanding officer of the Nazi occupation allows medical relief teams from the West to enter to country just prior to liberation, preventing hundreds of thousands of additional deaths.

1944   Death of painter Piet Mondrian

Piet Mondrian, one of the foremost twentieth-century Dutch painters, dies in New York City. (See 1872.)

1945: May   The Netherlands is liberated

Allied troops liberate the Netherlands.

The Postcolonial Era

1945: August 17   Indonesia declares independence

Newly freed from wartime occupation by Japan, Indonesia unilaterally declares independence. Postwar Allied occupation of Indonesia is assigned to the British, who are not interested in helping the Dutch maintain political control of their colony.

1945: December 10   The Netherlands joins the UN

The Netherlands becomes a founding member of the United Nations.

1946: November   The Dutch meet with Indonesian nationalists

A meeting between the Dutch and representatives of independent Indonesia at Linggadjati produce *de facto* Dutch recognition of jurisdiction over Java, Madura, and Sumatra by the Indonesian republic. The Dutch accept the formation of an Indonesian federation affiliated with the Netherlands. Nevertheless, the Dutch send troops to the region and skirmishes with republican forces take place.

1947: July   First Dutch "police action" against Indonesia

With 150,000 troops in the region, the Dutch launch a police action against republican troops in Java and Sumatra, capturing territory in Java.

1948   Queen Wilhelmina abdicates

Queen Wilhelmina abdicates the throne in favor of her daughter, Juliana (b. 1909).

1948: January 1   Benelux union is formed

The Netherlands forms a joint customs union with Belgium and Luxembourg.

1948: December 19   Dutch launch second police action

Dutch forces launch a second offensive against republican troops in Indonesia but are halted by UN calls for a cease-fire and the pressure of world opinion.

Queen Wilhemina, who is abdicating the throne in favor of her daughter Juliana, waves to a crowd of citizens from a balcony. (EPD Photos/CSU Archives)

1949   The Netherlands helps found NATO

The Dutch are founding members of the North Atlantic Treaty Organization (NATO).

1949: November   Dutch agree on Indonesian independence

At the Hague Round Table Conference, the Dutch agree to the formation of a federal United States of Indonesia that will be part of an Indonesian-Dutch union politically. The Netherlands receives a guarantee of safety for Dutch citizens and investments in Indonesia and retains control of Western New Guinea (West Irian).

1949: December   Indonesia is granted independence

Following Japanese occupation of the East Indies during World War II and four years of intermittent warfare, the Netherlands recognizes Indonesia as an independent country. Indonesian independence represents the loss of the Netherlands' main legacy from its Golden Age of maritime supremacy in the seventeenth century, as well as the elimination of what had been a major economic base up to the eve of World War II.

1951: October   Television broadcasts begin

The first television broadcasts take place.

1952   Dutch join coal and steel group

The Netherlands joins the European Coal and Steel Community (ECSC), an economic union among the coal and steel producing states.

1953: February 1   Flooding kills 1,800

A heavy North Sea storm overwhelms the Netherlands' fifty primary dikes, which have been impregnable under the worst storm conditions for almost 400 years. After days of heavy rain and winds up to 100 miles (161 kilometers) per hour, they all suddenly give way at once, catching the terrified population unprepared and causing catastrophic flooding—the third worst in the nation's history (see 1421 and 1530). Over 100 towns and villages are inundated within minutes. More than 72,000 people are evacuated from their homes, and over 1,800 are killed.

When the flooding is over, more than half a million acres of land (and almost 50,000 homes) are under water, and over 50,000 head of livestock have died. The water breaks through the country's dikes in sixty-seven places, inflicting major damage on farm land and industrial facilities. Emergency rescue and food relief arrives from twenty-five foreign countries. The Dutch government subsequently plans a major flood control project, called the Delta Project, to protect its citizens from similar future disasters by damming off the country's delta arms that lead to the sea, except for the channel used for shipping. Concrete caissons are to be sunk and boulders dumped into the gaps between islands, reducing the total length of the coastline several hundred miles and turning former sea inlets into freshwater lakes.

1954: December   Two territories are granted self-rule

Through the Charter for the Kingdom, Dutch Guiana (present-day Suriname) and the Netherlands Antilles (Aruba, Bonaire, Curacao, Saba, St. Eustatius, and St. Maarten) become equal in status with the Netherlands as part of a single kingdom and are granted full autonomy over their domestic affairs.

1956: April   Indonesia ends union with the Netherlands

Indonesia nullifies the agreement linking it politically with the Netherlands, defaulting on a debt of over four million guilders and eventually expelling Dutch nationals.

1957   Indonesia nationalizes Dutch assets

The government of Indonesia begins to nationalize Dutch properties, which represent a total investment of over $1 billion.

1957   Civil rights gains for Dutch women

Dutch women gain the right to open their own bank accounts, and female government employees are no longer dismissed from their jobs when they marry. (However, as late as 1969, fifty-five municipalities still retain the right to fire married women from government jobs.)

1958   European Economic Community is formed

The Netherlands is one of six founding members of the European Economic Community (EEC), newly created to allow free trade among member nations.

1958   International organ competition is launched

The St. Bavo Cathedral holds its first annual International Organ Competition.

1958   Work begins on the Delta Project

The Delta Project, a series of dikes and dams designed to protect the southwestern part of the country from North Sea flooding, is begun.

1959   Natural gas is discovered

Natural gas is discovered in the Slochteren Field, located in the province of Groningen. With known reserves of 58,245,000 million cubic feet (1,650,000 million cubic meters), it is one of the world's largest gas fields.

1960   Benelux Economic Union is launched

The Benelux Economic Union, intended to lead toward economic integration of the Netherlands, Belgium, and Luxembourg, is inaugurated.

1960   Indonesia ends diplomatic ties with the Netherlands

Indonesia severs diplomatic relations with the Netherlands.

1963: May   Talks held on future of West Irian

Amid Indonesian threats to take West Irian (the remaining Dutch possession in Indonesia) by force, the Dutch and the Indonesians agree to let the area's residents decide their own fate in open elections.

1966   Democracy 66 party is formed

The progressive political party, Democracy 66 (D66), is established.

1969   West Irian votes to join Indonesia

In UN-supervised elections, residents of West Irian vote to become part of Indonesia.

1971   Environmental agency is established

The Dutch government establishes the Ministry of Health and Environment.

1975   Suriname wins independence

The territory known as Dutch Guiana, on the northern coast of South America, becomes the independent nation of Suriname.

1976   Prince is involved in Lockheed scandal

Prince Bernhard, the husband of Queen Juliana, is obliged to give up his official posts when he is linked to a scandal involving illegal payments by U.S. aircraft maker Lockheed to obtain Dutch military contracts. The popularity of the Dutch monarchy enables it to successfully survive the crisis.

1979   Van Gogh museum is founded

The Rijksmuseum Van Gogh is established in Amsterdam. It houses 700 paintings and drawings by the Dutch artist.

1980: April 30   Queen Beatrix is crowned

Queen Beatrix ascends the throne following the abdication of her mother, Queen Juliana.

1981   Abortion is legalized

A law legalizing abortion in the Netherlands is passed.

1983   Strike by public employees

Government employees go on strike to protest wage cuts.

1984   Proposed NATO missiles spur protests

The proposed deployment of NATO cruise missiles spurs large demonstrations, delaying final government action on the plan.

1985   Pope visits the Netherlands

Pope John Paul II visits the Netherlands. Protesters demonstrate against the pope's power over the Dutch Catholic Church.

1986   Final barrier of Delta Project is completed

The final part of the construction project to protect the country from North Sea flooding is completed.

1986   Immigrants vote in elections

For the first time, recent immigrants are allowed to vote.

1986: February   Cruise missile deployment is approved

The proposed deployment of NATO cruise missiles in the Netherlands wins parliamentary approval.

1987: August   Rotterdam port operations are disrupted by labor disputes

Plans to cut the work force at the Rotterdam port lead to labor unrest, which is resolved through government intervention.

1988   Equal pay for women mandated by law

Women workers are guaranteed equal pay for equal work and protected from being fired because of pregnancy.

1989   Major environmental program is unveiled

The government announces an ambitious environmental program designed to reduce air, water, and soil pollution in the densely populated Netherlands by seventy percent within twenty years. It calls for reductions in energy consumption and waste and increased use of public transportation.

1991   Maastricht Treaty

The Netherlands hosts the conference that draws up the Maastricht Treaty creating the European Union with eventual adoption of a common currency and elimination of all trade barriers.

1991   Economic austerity program draws protests

A program to slash the budget deficit spurs mass walkouts by private and public sector employees. However, the government adheres to the program.

1992: February   Maastricht Treaty is approved

The Netherlands approves the Maastricht Treaty creating the European Union.

1993: November 1   European Union is inaugurated

The European Union (EU) is launched. A more comprehensive version of the EEC, it is intended to insure European cooperation on both security and economic matters and in other areas as well.

1994: May   Ruling party ousted in historic general election

The ruling Christian Democratic Appeal (CDA) party is ousted from power, along with its leader, Prime Minister Ruud Lubbers, by a coalition composed of the Labor, Liberal, and Democracy '66 parties. Lubbers has headed the government for twelve years, as head of both center-right and center-left coalitions. After losing twenty seats in the lower house of parliament, the CDA becomes the opposition party for the first time in recent history. Labor Party leader Wim Kok becomes the new prime minister.

1995: January 1–February 30   Floods force evacuation of 250,000

Flooding caused by melting snow in the Alps and heavy rain forces the evacuation of 250,000 people in the Netherlands—the most extensive peacetime evacuation in the nation's history. In spite of the massive Delta Project, completed in 1986, to shore up the nation's coastal barrier, some dikes begin to crumble. Most evacuees return to their homes by February 5. Estimates place flood damage and the cost of evacuation at about two billion guilders ($1.18 billion). Some criticize the Dutch government for concentrating too heavily on coastline protection and ignoring inland river dikes. The government promised to upgrade these dikes by the year 2000.

1996: October 24   Insider trading scandal rocks stock exchange

Allegations of insider trading and fraud at Dutch investment companies lead to the arrest of Han Vermeulen, the director of the brokerage company Leemhuis & Van Loon. At the same time, government officials raid the firm's offices and the Amsterdam Stock Exchange. The probe, which also includes British firms, involves alleged trading violations dating back as far as 1985.

1997: June 10   Dutch students excel at math in international survey

The results of an international survey of achievement in math and science at the grade school level rank primary students in the Netherlands fifth in the world in math, behind students from four Asian nations (Singapore, South Korea, Japan, and Hong Kong).

1997: July 30   Dutch scientists discover distant galaxy

Dutch and U.S. astronomers announce the discovery of a galaxy thirteen billion light years away—the most distant sighting ever made from Earth.

1998: May 6   Ruling coalition retains majority in general elections

The ruling government coalition retains its parliamentary majority in general elections. Labor, the largest party, increases its representation in the 150-seat lower house from 37 to 45 seats. The coalition's second-largest party, the Liberals, see their total rise from 31 to 38. However, the D66 party results are disappointing—down from 24 to 14. The opposition Christian Democratic Appeal (CDA), ousted in 1994 after decades in power, suffers further losses in the current election, with its total seats down from 34 to 29.

1998: July   Dutch play in the World Cup semifinals

The Netherlands participates in the World Cup Tournament elimination rounds, defeating Yugoslavia 2–1 in the second round of competition and Argentina 2–1 in the quarterfinals. However the Dutch lose to Brazil in the semifinals, in a game that is tied at 1–1, with a final score of 4–2 decided on penalty kicks. (Brazil loses 3–0 to host team France in the final round.)

Bibliography

Andeweg, R.B. *Dutch Government and Politics.* New York: St. Martin's, 1993.

Fuykschot, Cornelia. *Hunger in Holland: Life During the Nazi Occupation.* Amherst, N.Y.: Prometheus Books, 1995.

Hoogenhuyze, Bert van *The Dutch and the Sea.* Weesp: Maritiem de Boer, 1984.

Israel, Jonathan Irvine. *Dutch Primacy in World Trade, 1585–1740.* New York: Oxford University Press, 1990.

Schama, Simon. *The Embarrassment of Riches: An Interpretation of Dutch Culture in the Golden Age.* New York: Knopf, 1987.

Schilling, Heinz. *Religion, Political Culture, and the Emergence of Early Modern Society: Essays in German and Dutch History.* New York: E.J. Brill, 1992.

Slive, Seymour. *Dutch Painting 1600–1800.* New Haven, Conn.: Yale University Press, 1995.

Wee, Herman van der. *The Low Countries in Early Modern Times.* Brookfield, Vt.: Variorum, 1993.

Wolters, Menno and Peter Coffey, ed. *The Netherlands and EC Membership Evaluated.* New York: St. Martin's, 1990.

Zanden, J.L. van. *The Rise and Decline of Holland's Economy: Merchant Capitalism and the Labour Market.* New York: St. Martin's, 1993.

Norway

Introduction

Norway (derived from *Nordweg*, or "Northern way") is aptly named: half the country lies above the Arctic Circle, and Norway is the location of the world's northernmost town (Hammerfest) as well as the northernmost point in Europe. Famed for the beauty of its mountains and fjords, Norway is also known for the courage and independence its people have displayed throughout their history, from the days of the Vikings to the World War II era, when a country that had planned on remaining neutral mounted one of the most remarkable resistance efforts of the war. In recent years, the issue of Norwegian independence has come to the fore in the national controversy over joining the European Union. Norway is one of the few Western European countries remaining outside the organization.

Norway occupies the western part of the Scandinavian peninsula, bordering Sweden, Finland, and Russia. To the north and west lie the Arctic Ocean and the North Sea. Most of its 125,182 square miles (324,220 square kilometers) are mountainous, with mountain ranges extending for almost the entire length of the country. Norway has traditionally been divided into four main regions: Vestlandet and Ostlandet in the wider southern part of the country, and Trondelag and Nord Norge to the north. In recent years, Norway's southernmost area, Sorlandet, has been recognized as a separate region as well. As of 1999, Norway has an estimated population of approximately 4.5 million people.

Historical

Historical accounts of Norway begin with the first Viking conquests in the late eighth century A.D. During the same period, Harald Fairhair (d. c. 940) became the first king to unite a substantial portion of the country, which had previously been divided into many small kingdoms. In the eleventh century, two kings—Olaf I (c. 964–1000) and Olaf II (C. 995–1030) —attempted to introduce Christianity into Norway, and Olaf II succeeded. He was canonized after his death and later became Norway's patron saint. From the mid-elev-

enth to the mid-thirteenth centuries, rival dynasties vied for control of the kingdom, unrest which worsened when the Black Death spread to Norway from England in 1349, killing at least half the population, destroying much of the region's wealth, and weakening its nobility. By the late fourteenth century, Denmark had taken de facto control of Norway, and in 1397 the formation of the Kalmar Union formally united Norway, Sweden, and Denmark.

Norway was to remain under Danish domination for over four hundred years. This domination became even more complete when Sweden withdrew from the Kalmar Union in 1523; thirteen years later Norway was declared a province of Denmark. Together with the political control, the Danes also imposed the Lutheran religion on Norway, replacing Norway's Catholic bishops with Lutheran clergy and requiring Norwegians to use the Lutheran Bible. However, Norway did enjoy some economic progress during the long period of Danish control. Silver, copper, and iron mining were introduced in the seventeenth century, and the fish and timber industries thrived in the eighteenth. By the end of the eighteenth century, a revival of Norwegian nationalism was under way.

The Norwegians were freed from Danish rule by the Napoleonic wars, when a British blockade weakened Denmark's hold on Norway, which was allowed its own provisional government. After the downfall of Napoleon Bonaparte (1769–1821), however, the European powers put Norway under the control of another foreign country, Sweden. In the meantime, though, the Norwegians had declared independence and adopted their own constitution. Although unable to retain the full independence they sought in the face of the superior might of Sweden, the Norwegians nevertheless won an important compromise: they were able to keep their constitution and remain largely autonomous domestically, while allied diplomatically with Sweden under a single monarch (the Swedish king).

For the rest of the nineteenth century, nationalism played an important part in Norwegian life. It was evident in the art and activities of such cultural figures as poet Henrik Wergeland (1808–45), violinist and composer Ole Bull (1810–80), composer Edvard Grieg (1843–1907), and Bjørnstjerne Bjørnson (1832–1910), poet, playwright, and creator of Nor-

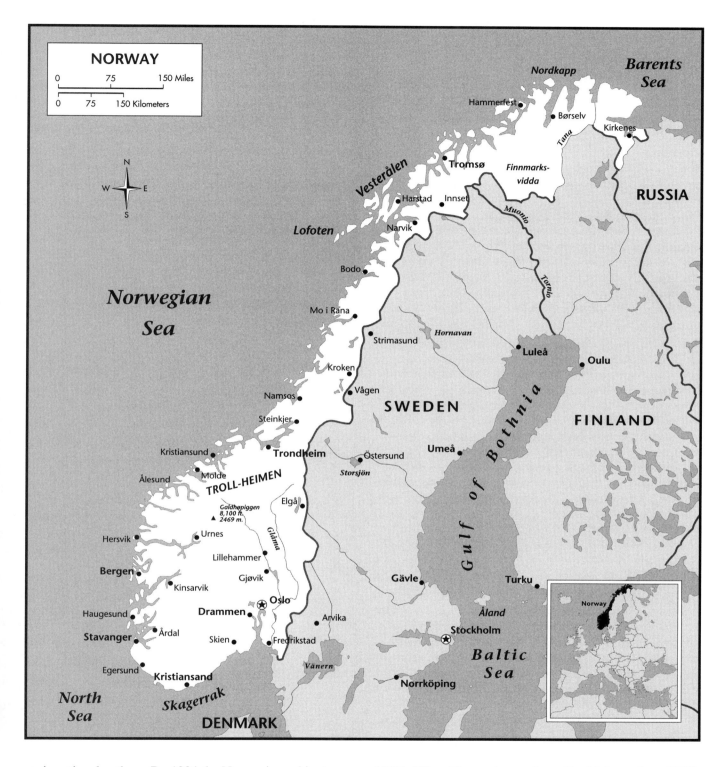

way's national anthem. By 1884 the Norwegian cabinet was made responsible to Norway's own parliament, the Storting, rather than to the Swedish king. In 1905 the union between the two countries was dissolved, and Norway was formally declared an independent constitutional monarchy with a newly appointed king, the Danish-born Haakon VII. (1872–

1957). When Norwegian explorer Roald Amundsen (1872–1928) became the first person to reach the South Pole six years later, it was the Norwegian flag that he planted there, naming the surrounding plateau for his king.

Norway remained neutral in World War I but still lost many merchant ships—and 2,000 sailors—to German submarines. Between the wars the most significant development was the rise to power of the Labor Party, which controlled the government by 1935 and was to do so for nearly thirty years more, implementing a broad network of government-funded health, housing, and social welfare programs. As Europe prepared for possible warfare in the late 1930s, Norway planned on remaining neutral once again, but this option was eliminated by a surprise German invasion on April 9, 1940 that resulted in a five-year occupation of the country. Aided by French and British troops, the Norwegian military mounted a vigorous two-month resistance effort but was ultimately overpowered. King Haakon VII refused to collaborate with the Nazis and, together with his cabinet ministers, fled to England, where he set up a government-in-exile. Norway's remaining military personnel also left the country to fight with Allied forces abroad. The Norwegians at home created an extensive network of clandestine activities to thwart the collaborationist government of Norwegian Nazi leader Vidkun Quisling(1887–1945). On June 7, 1945 King Haakon VII returned to Norway in triumph. Quisling and twenty-four other Nazi sympathizers were tried for treason and shot.

Warfare and occupation had severely depleted the country's resources, leveling towns, destroying communications networks, and producing a severe housing shortage. The first postwar priorities were reconstruction and economic recovery, and Norwegians enjoyed strong economic growth in the 1940s and 1950s under Labor Party leadership. Norway became a founding member of the United Nations in 1945, and the country's foreign minister, Trygve Lie (1896–1968), became the first UN secretary general. Disillusioned following the Communist takeover of Czechoslovakia in 1948, Norway abandoned its former policy of neutrality and joined the North Atlantic Treaty Organization (NATO) the following year. In 1957 King Haakon VII died. His son succeeded him as King Olaf V (1903–91) and, like his father, became a highly popular monarch.

An already prosperous economy was boosted further with the discovery of North Sea oil off the coast of Norway in the 1960s and the development of the oil and petrochemical industries in the 1970s. However, the growing importance of oil also had a down side, making the Norwegian economy sensitive to fluctuations in world oil prices. When prices fell in 1986, Norway was strongly affected. Inflation rose, the government's extensive network of social welfare programs was threatened, and the nation was crippled by mass strikes.

The Labor Party's long continuous domination of Norwegian politics ended in the early 1960s, and a nonsocialist coalition government was formed in 1965. In the following decades, the country alternated between periods of Labor and Conservative rule. Beginning in the 1980s, the dominant figure in the Labor Party was Gro Harlem Brundtland (b. 1939), who served briefly as prime minister in 1981, becoming the first woman in Norway's history to hold that post. She returned to it in 1986 after a five-year period of Conservative rule.

A central political and economic issue of the postwar period has been the controversy over membership in the European Union (EU), formerly the European Economic Community (EEC). After centuries of foreign domination, many Norwegians are wary of any arrangement that seems to threaten their nation's prerogative to exercise sovereignty over its own affairs. In addition, there is the fear that EU membership may threaten Norway's generous farm subsidies and the special restricted zones enjoyed by its fishing industry. Proponents of EU membership fear the effects of any policy that results in economically isolating Norway in the world's current global economy and point out the advantages of the lower tariffs enjoyed by EU members. Referendum votes have defeated proposals for EU membership in 1972 and 1994.

Timeline

c. 9000 B.C.   First humans inhabit Norway

Human habitation of present-day Norway begins after the ice-age glaciers have receded.

c. 3000 B.C.   Agriculture is introduced

Agriculture is introduced in the eastern regions, where barley is grown and cows and sheep are raised.

A.D. 793–1050   Viking conquests

Beginning with the first recorded Viking attack (in Scotland), the Norse Vikings inaugurate two centuries of exploration and plunder, invading Ireland, Scotland, parts of England and France, the Faero Islands, Iceland, and Greenland.

9th century A.D.   Norway is divided into districts

Norway is divided into districts called *fylker,* with their own assemblies and laws.

A.D. 870   Settlement of Iceland

Norwegians begin to settle Iceland. The 529-mile (850-kilometer) journey can take as long as three weeks. Between 15,000 and 20,000 settlers arrive within a sixty-year period. Most are farmers looking for new land.

Early Kingdoms

A.D. 872–930   Reign of Harald Fairhair

The son of a minor king, Harald defeats neighboring rulers to form a more extensive kingdom, uniting most of western Norway.

c. 981   Erik the Red lands on Greenland

After being exiled from both Norway and Iceland, Erik the Red (950?–1004?) becomes the first to land on and explore Greenland. Settlement begins shortly afterward.

995–1000   Reign of Olaf I

Olaf I, a descendant of Harold Fairhair raised in England, reigns over Norway and attempts to forcibly introduce Christianity. His enemies, allied with Sweden and Denmark, defeat him at the Battle of Svold and divide the country among themselves.

1016–30   Olaf II reigns

After driving the Swedes and Danes from Norway, Olaf II reigns. Like Olaf I, he tries to Christianize Norway, still meeting with heavy resistance. Nevertheless, he succeeds in establishing Christianity in Norway. Olaf is canonized after his death and later becomes the patron saint of Norway.

1030   Danes gain control of Norway

The enemies of King Olaf II conspire with the Danes and defeat Olaf at the Battle of Stiklestad. The Danish Canute II becomes Norway's ruler.

1035   Magnus I succeeds to the throne

Magnus I (1024–47), the son of Olaf II, becomes king upon the death of Canute. He unites Norway and Denmark and inaugurates three centuries of Norwegian rule by native kings.

Mid-11th to mid-13th centuries   Civil wars by feuding nobility

For much of these two centuries, Norway falls prey to feuds by rival dynasties and civil conflict.

1184–1202   Reign of Sverre

Sverre (c. 1149–1202) reestablishes a strong monarchy supported by the peasants. He asserts control over the church, resulting in the Crosier War (see 1196.)

1196–c. 1202   Crosier War

The attempt by Sverre to assert the power of the monarchy over the church leads to warfare against Norway's bishops.

1217–63   Reign of Haakon IV

Haakon IV (1204–63) reunites Norway, organizing and strengthening the central government. A royal council is established, and the king appoints judicial officers and local governors. Iceland and Greenland recognize the kingdom of Norway for the first time, and trade expands.

1250–1300   Hanseatic merchants take over trade

Norway's economy suffers when the Hanseatic League of German merchants establishes a presence in Bergen. By the end of the century they dominate grain imports into the country.

1274–76   Legal code is established

A national legal code is adopted, as well as rules governing succession to the throne.

1319–43   Reign of Magnus VII

Norway comes under foreign domination when the throne passes to the Swedish king Magnus VII. (1319–74)

1349   The Black Death

The Plague strikes Norway, killing up to half the population and crippling the nation's dairy farming business.

Danish Rule

1380–1814   Denmark rules Norway

Denmark rules Norway for over four hundred years, referred to by Norwegian playwright Henrik Ibsen as "four hundred years' night." The political independence and cultural identity of the Norwegian people are suppressed, and political power is held by Danish and German nobles. Norway loses its overseas territories, Denmark's Lutheran religion is forced on the Norwegians, and Danish is declared the official language of Norway. In addition, Norwegian men are required to serve in the Danish navy.

1380   Danish monarch gains control of Norway

When King Haakon VI dies with no direct heirs, a joint Danish-Norwegian throne is created with the infant prince, Olaf Haakonsson, named as king. As regent, his mother, Queen Margrethe of Denmark (1353–1412) holds the real power in Norway, beginning more than four hundred years of Danish rule. She remains in power when the young King Olaf dies at the age of seventeen, eventually conquering Sweden as well.

1397   Kalmar Union

Through the Kalmar Union, Sweden, Denmark, and Norway are united politically under the Danish crown.

1450   Danish-Norwegian treaty is signed

Denmark signs a treaty making Norway an equal political partner, but the Norwegians still do not achieve real equality.

1469   Norway loses the Orkney and Shetland islands

Norway's territories, the Orkney and Shetland islands, are lost when King Christian I mortgages them to Scotland.

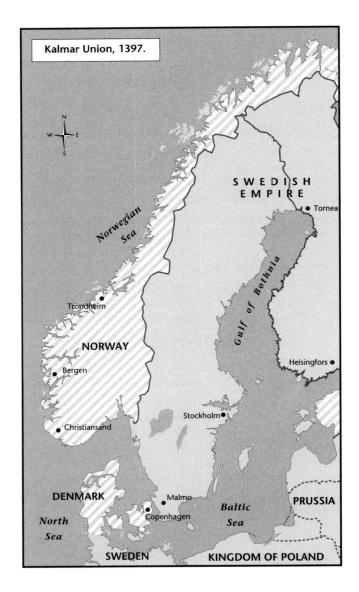

Kalmar Union, 1397.

1523   Sweden withdraws from the union

Sweden withdraws from the Kalmar Union, leaving only Denmark and Norway.

1536   Norway becomes a Danish province

Norway is declared a province of Denmark. Religious domination goes hand-in-hand with political control, as Lutheranism spreads northward and a revolt by the Catholic archbishop of Trondheim is defeated. Norway's Catholic bishops are replaced by Lutheran bishops from Denmark, and the Catholic Church's property is confiscated. Norwegians are required to use Denmark's Lutheran Bible and liturgy in religious services.

1550   Sawmill is introduced

With the introduction of the sawmill, Norway's timber industry develops. Timber producers send logs downriver to the coast to be sawn in the mill and exported to the Netherlands and other destinations.

1588–1648   Reign of Christian IV

Danish king Christian IV (1577–1648) helps the Norwegian economy develop by launching silver, copper, and iron mining. Under Christian's rule, a separate Norwegian army is established.

1643   First printing press

Norway's first printing press is put into operation.

1661   Norway is designated as an equal kingdom

A new constitution makes Denmark and Norway "twin kingdoms," giving them official legal equality.

18th century   Norway prospers economically under Danish rule

Although still under Danish rule, Norway maintains a sound economy with exports of fish and timber. For the most part, political stability and internal peace reign.

1700–20   Great Northern War

Denmark and Sweden fight the third in a series of wars over which country will become the major power in Scandinavia. In the course of this war, Sweden invades Norway but meets with strong resistance at the border fortress of Frederiksten.

1718   Siege of Frederiksten

The Norwegians resist a Swedish attack on this border fortress, and the Swedes retreat after their king, Charles XII, is fatally wounded.

1743   Artists' guild is formed

An artists' guild is established in Bergen, which at this time is the largest city in Norway.

1760   Royal Scientific Society is founded

The establishment of the Royal Scientific Society is part of the beginning of Norway's nationalist revival.

1796   Lay preacher prompts national reawakening

Lay preacher Hans N. Hauge (1771–1824) undertakes a campaign to bring spiritual renewal to the Norwegian peasantry, sparking the beginnings of a nationalist revival that the Danish cannot stop even though they imprison Hauge.

Napoleonic Wars

1807–14   Britain blockades Norway during war with Napoleon

During the Napoleonic Wars, the British blockade Norway, cutting off all shipments of food to the country, creating hardship that reaches the starvation stage during the two years when harvests are bad. However, the British inadvertently grant the Norwegians greater freedom because the blockade makes it impossible for Danish government administrators to reach the country.

1807   Norway gets a provisional government

Under military pressure from the British during the Napoleonic wars, Denmark grants Norway its own provisional government administration headed by Christian Frederick, the Danish prince.

1808   Birth of author Henrik Wergeland

Wergeland is Norway's first major modern literary figure and is widely considered its greatest lyric poet. He has been compared to the contemporary English poet Percy Bysshe Shelley for his lyric poetry, commitment to social causes, and brief life (he dies at the age of thirty-seven). Deeply involved in the causes of Norwegian nationalism and social justice, Wergeland is celebrated as Norway's national poet. In addition to his poetry and essays, his support of humanitarian causes includes giving lectures, distributing printed materials, establishing lending libraries, and teaching classes in Norwegian in his own home. He also founds and edits two journals and lobbies strenuously to abolish the constitutional provision banning Jews from Norway. His best-known poetry includes *Poems, First Cycle* (1829) and the narrative poems *Jan van Huysums blomsterstykke* (Jan van Huysum's Flowerpiece; 1840) and *Den engelske lods* (The English Pilot; 1844). Wergeland dies in Oslo in 1845.

1810: February 5   Birth of violinist and composer Ole Bull

Ole Bull, a great violinist and Norway's outstanding nineteenth-century cultural icon, is a child prodigy who plays in a string quartet by the age of eight and debuts as a soloist a year later. Initially destined for a career in the church, he is appointed to an orchestral conducting position by the age of twenty and also embarks on a career as a composer. He also begins touring abroad in 1829 , becoming a cultural ambassador for Norway. Bull performs with some of the greatest nineteenth-century musicians, including Robert Schumann (1810–56), Felix Mendelssohn (1809–47), and Franz Liszt (1811–86). In addition to his other accomplishments, he modifies his own violin, bringing it closer to a Norwegian folk instrument called a Hardanger fiddle and making possible certain virtuosic feats not matched by any other violinist. Bull

plays for enthusiastic crowds in Paris and London and tours Germany, Austria, Russia, and the United States.

Ole Bull is a prominent promoter of Norwegian independence. His best-known composition is *Den 10. December.* Written to commemorate a patriotic speech given in that year, it evokes the Norwegian mountains in the summertime. Bull also promotes Norwegian folk music and promotes the creation of a national theater (the Norwegian Theatre in Bergen), which brings him into close contact with the famed dramatists Henrik Ibsen (see 1828) and Bjørnstjerne Bjørnson (see 1832). In addition, he serves as a mentor of the composer Edvard Grieg (see 1843) when Grieg is a young student. In the latter part of his life, Bull divides his time between Norway and the United States, where the daughter of a Wisconsin senator becomes his second wife. He continues concertizing until his death in 1880.

1811   University of Oslo is founded

Norwegians become increasingly dissatisfied with their lack of a university, which forces those who want to receive an advanced education to study at the University of Copenhagen. The demand for a Norwegian university becomes part of the growing feeling of national identity in Norway. Under pressure from the Napoleonic wars, Denmark grants Norway permission to establish its own university, and the University of Oslo is founded with money previously collected for this purpose.

1814   National library is founded

The Universitetsbiblioteki Oslo, Norway's most extensive library collection of art materials, is both the library of the University of Oslo and the country's national library.

1814: January   Sweden annexes Norway

After the defeat of Napoleon Bonaparte, Denmark, which supported him, is forced to cede Norway to Sweden by the Treaty of Kiel. It is thought that Norway will serve as a buffer for Sweden's western border and compensate Sweden for having lost Finland to Russia in 1809.

1814: February   The Norwegians declare independence

The Norwegians try to resist annexation by Sweden, declaring their independence and electing the Danish prince Christian Frederik as their regent at a meeting of Norwegian leaders. Norwegians nationwide then elect representatives to a national assembly.

1814: April 11   Delegates meet at Eidsvoll

Representatives elected by the Norwegian people meet at Eidsvoll to debate the future of their country. The majority vote for independence from both Denmark and Sweden

1814: May 17   A constitution is adopted

The delegates assembled at Eidsvoll adopt a democratic constitution modeled on those of the United States and revolutionary France. It provides for a government with separate executive, judicial, and legislative branches, whose legislative body will be called the Storting, a name derived from Old Norse.

1814: July–August   Sweden attacks Norway

Refusing to accept Norway's declaration of independence, Sweden sends in its troops and attacks Norway.

Union with Sweden

1814: November   Norway accepts Swedish rule

In a compromise agreed on following the Swedish military attack, Norway agrees to Swedish rule and, in return, Sweden agrees to accept the constitution adopted at Eidsvoll (see 1814:May). The Storting, Norway's elected assembly, accepts the Swedish king as its head of state. Although united by a single king and pledged as military allies, Norway and Sweden are considered two separate countries in charge of their own internal affairs. Nevertheless, a nearly century-long political power struggle ensues between the two countries as Norwegian nationalism grows.

1818   Art institute opens in Oslo

Norway's first formal institution for training artists, the Kongelige Tegne- og Kunstskole, is founded in Oslo, fostering the creation of a professional artists' class. Most artists go on to study abroad for a period after their graduation.

1821   Norway abolishes its nobility

The Storting votes to abolish Norway's nobility to prevent the Swedish king from gaining more support within the country.

1828: March 20   Birth of dramatist Henrik Ibsen

Norway's most famous literary figure, Henrik Ibsen revolutionizes the European theater of his day by introducing social and psychological realism into his plays. The realistic contemporary drama he pioneers portrays recognizable middle-class characters whose stable lives unravel when they confront a crisis set in motion by threatening forces underlying their superficially comfortable existence.

Ibsen begins working in the theater as a young man, appointed at the age of twenty-three by Ole Bull (see 1810) to run the national theater at Bergen and write a play for it every year as part of the job. From 1857–1862 he occupies a similar position with the Norwegian Theater in Oslo, struggling to create plays within the conventions of the day. In 1864 Ibsen leaves Norway and lives abroad for over twenty-five years, mostly in Italy and Germany. With the creation of *Pillars of Society* in 1877 and *A Doll's House* in 1879, he pioneers a new kind of play that shows realistic people struggling with controversial contemporary problems. His new plays create a sensation with theater-goers and critics. So do the ones that follow: *Ghosts* (1881), *An Enemy of the People* (1882), *The Wild Duck* (1884), and *Rosmersholm* (1886). Psychological insight is combined with symbolism in the plays of the 1890s, including *Hedda Gabler* (1890), *The Master Builder* (1892), and *When We Dead Awaken* (1899), which is Ibsen's last play. Henrik Ibsen is crippled by strokes in 1900 and 1901 and dies in 1906.

1830s   Farmers begin serving in the Storting

Norway's government becomes more democratic when farmers start electing other farmers to protect their own interests, instead of electing legislators solely from the professional classes.

1832: December 8   Birth of author Bjørnstjerne Bjørnson

Bjørnstjerne Bjørnson, who will become the third person ever to win the Nobel Prize for Literature, is a prominent writer of fiction, plays, and poetry, as well as a journalist, editor, theater director, and outspoken participant in the public debates of his day, in which he is renowned as a defender of the rights of minorities. He is also the author of the Norwegian national anthem. His earliest published works are rural narratives, including *Arne* (1858) and *En glad gut* (The Happy Boy; 1860), and saga plays based on medieval Norse history. During this period, Bjørnson becomes the artistic director of the National Theater at Bergen, succeeding Henrik Ibsen in that post.

Bjørnson's *Poems and Songs* volume is published in 1870. During the 1860s, Bjørnson's involvement in political controversies and literary feuds is so intense that he spends part of the following decade in Rome in order to concentrate exclusively on his writing. His two plays from this period establish his reputation as a dramatist: *En fallit* (The Bankrupt; 1875) and *Redaktoren* (The Editor; 1875). Bjørnson resumes his active role in Liberal Party politics in the 1880s while continuing to write both novels and plays. In 1903 he receives the Nobel Prize for Literature. Bjørnson dies in Paris in 1910.

1837   Local governments are created

The Storting (the Norwegian parliament) sets up a system of local self-government, getting more Norwegians interested in participating in the political process.

1842   First art museum is founded

Norway's first art museum, the Nasjonalgalleri, opens in Oslo. Its holdings include a collection of paintings donated by Norwegian painter J.C. Dahl.

1843: June 15   Birth of composer Edvard Grieg

Grieg, Norway's most famous composer, receives early musical training at home and then attends the Leipzig Conservatory in Germany. The young composer Rikard Nordraak (1842–66) has a decisive influence on Grieg's work, leading to his interest in integrating Norwegian folk elements into his compositions and develop a distinctively Norwegian composing style. Grieg's most famous piece, his Piano Concerto in A minor (1868) is praised by the famous pianist and composer Franz Liszt. Starting in 1873, Grieg has enough financial security to devote himself nearly full-time to composing. He is awarded honorary doctorates in music from both of England's most prestigious universities, Cambridge and Oxford, and conducts concerts throughout Europe. The entire nation mourns Grieg's death in 1907. In addition to the Piano Concerto, other famous pieces by Grieg include the *Peer Gynt* suite, written as incidental music for a play by Norwegian dramatist Henrik Ibsen (see 1828) and the *Lyric Pieces* for piano, short compositions that are still widely played by pianists and piano students in countries throughout the world.

1845: July 12   Death of poet and patriot Henrik Wergeland

Wergeland, Norway's first great modern poet and a lifelong Norwegian patriot, dies in Oslo. (See 1808.)

1849: January   National theater is founded

Musician and nationalist Ole Bull presides over the establishment of the Norwegian National Theatre in Bergen, intended to encourage the composition and performance of dramas in the Norwegian language. The playwright Henrik Ibsen is later recruited by Bull to take charge of the theater.

1850–1920   Large-scale immigration to America

To escape rural poverty, some 800,000 Norwegians immigrate to the United States, where most settle in the Midwest. Only Ireland surpasses Norway in immigration to the U.S. during this period.

1855   Publication of *The Governor's Daughter*

The novel, *The Governor's Daughter*, an indictment of a social system that forces women into unhappy marriages, is published by Camilla Collett (1813–95), the sister of famed poet Henrik Wergeland. It is later considered Norway's first feminist novel.

1859: August 4   Birth of author Knut Hamsun

Nobel Prize-winning novelist, poet, and playwright Knut Hamsun is born into a farming family in northern Norway. As a young man, he holds many different jobs, both in Norway and in the United States, where he spends two periods of time in the 1880s. His first successful published work is the novel *Hunger* (1890). Other novels of the 1890s include *Mysteries* (1892) and *Victoria* (1898). After 1900, Hamsun's novels expand their scope beyond the problems of the individual to address broader social issues. Hamsun is critical of modern society, and his works are among the first to express the anxiety and alienation of modern existence. In 1909 he and his second wife, Marie, attempt to return to a simpler existence by moving to a farm near Hamsun's boyhood home, but the experiment only lasts a short time.

Hamsun's 1917 novel *The Growth of the Soil* helps win him the Nobel Prize for Literature in 1920, and the author's renown continues to increase over the next two decades. An impressive array of major authors, including Thomas Mann, André Gide, and H. G. Wells, contribute essays to a volume celebrating Hamsun's seventieth birthday in 1929. However, the author's reputation in Norway is severely damaged by his support for the Nazis during their occupation of the country during World War II. Following the war, he is forced to undergo mental examinations and required to pay a large fine for his behavior during the war. Hamsun's last book, *On Overgrown Paths,* is published in 1949 and he dies three years later.

1860s   Elementary schools are established

A nationwide system of elementary schools replaces education by traveling tutors.

1861: October 10   Birth of explorer Fridtjof Nansen

Explorer and humanitarian Fridtjof Nansen is born to a distinguished family of leaders and explorers. In 1888, at the age of 26, he organizes a six-man expedition and becomes the first person to cross the Greenland ice cap. Already a national hero, he plans to test a theory about Arctic currents by crossing the Arctic Ocean in a ship (the *Fram*) specially designed to withstand ice. Setting out in 1893, he drifts with the ship as planned for three years, but then becomes impatient and sets out for the North Pole on a sledge with one companion. Although forced to turn back eventually, he reaches a point 86°14'N—closer to the Pole than anyone has come before. His ship and the rest of his crew also return safely along a course that validates Nansen's original theory, and Nansen spends years summarizing the scientific results of his journey.

In the twentieth century, Nansen pursues his interest in oceanography but also turns his attention to political and humanitarian matters. He helps broker the agreement that leads to Norway's independence from Sweden in 1905, serving as envoy to Great Britain for the new government. He organizes relief and repatriation efforts following World War I and is appointed the League of Nations high commissioner for refugees. In 1922 Nansen is awarded the Nobel Peace Prize. Fridtjof Nansen dies in 1930. The following year, the Nansen International Office for Refugees is founded in Geneva, Switzerland, to carry on his work, and this organization wins its own Nobel Peace Prize in 1938.

1863: December 12   Birth of artist Edvard Munch

Painter and printmaker Edvard Munch, Norway's most renowned artist, is exposed to the harsh side of life in his youth, when his mother and sister die of tuberculosis. While an art student, he becomes active in intellectual circles in his native city of Christiania (later Oslo). In Paris, Munch works with lithographs and woodcuts and is exposed to postimpressionist painting, which influence him to abandon realistic representation in favor of emotional expression. He also studies etching in Germany. Munch's paintings are known for their portrayals of emotional trauma through the distortion of familiar forms and colors. Munch's most famous work is the painting *The Scream,* which has become such a familiar image throughout the world that it is even copied in parodies and other forms. Other well-known works include *Ashes, Death in the Room, Summer Night,* and *Jealousy.* After suffering an emotional breakdown in 1908, Munch returns to work in Oslo and continues to support himself with his painting. He dies in 1944.

1866   First recorded ski competitions are held

Norwegians hold their first recorded competitions in which skiers compete for prizes.

1869: April 11   Birth of sculptor Gustav Vigeland

Norway's most famous twentieth-century sculptor—and the first Norwegian sculptor to win international acclaim—is born to a furniture maker and displays an early talent for wood-carving. He is apprenticed to a wood-carver and sculptor in Oslo (then called Christiania) and takes evening classes as well. In 1889 Vigeland's first sculptures are shown in the National Art Exhibition in Oslo. In 1893 Vigeland visits Paris, where the work of the great French sculptor Auguste Rodin (1840–1917) has a major influence on him.

In the early twentieth century, Vigeland produces busts of prominent Norwegians, including playwright Henrik Ibsen (see 1828). In addition to more traditional sculptures, he also creates some unconventional works, including a nude statue of the composer Ludwig van Beethoven (1770–1827). In 1915 Vigeland begins producing woodcuts. In 1921 the authorities of Oslo offer to provide him with a regular salary and a studio if he leaves all his works to the city, and the artist agrees. In addition he creates a sculpture park in Oslo, producing the sculptures over a period of time. A highlight of the park is a series of figures called *Cycle of Life* that includes figures representing all ages of mankind. The single most dramatic piece in the park is *The Monolith,* which portrays 121 figures climbing upward toward its peak, which is 55 feet (17 meters) high. Gustav Vigeland dies in 1943.

1870s–1880s   Parliament divides over Norway's future

Two major factions divide the Storting. The conservatives—mostly landowners, merchants, and politicians—support the existing government structure. The liberals, who are mostly farmers and radical intellectuals, favor reducing Sweden's influence by taking legislative privileges away from ministers of the Swedish king.

1872: July 16   Birth of explorer Roald Amundsen

Amundsen, famous as the first man to reach the South Pole, is born in southeast Norway. Abandoning early plans for a career in medicine, he gains sea-going experience working on a merchant ship in the Arctic. Amundsen's first great achievement is crossing the Northwest Passage between Europe and Asia, which he accomplishes between 1903 and 1906. In the process he gathers a wealth of scientific data as well as information about the native Inuit population, some of whose survival techniques he adopts.

Amundsen's second great journey is the voyage to the South Pole, undertaken in a last-minute decision made when he is beaten to the North Pole by American Robert Peary in 1909. He decides to reroute his original expedition and race the Englishman Robert Falcon Scott to the South Pole. On December 14, 1911 Amundsen becomes the first to reach his goal, beating Scott by five weeks. He plants the Norwegian flag and names the plateau leading to the Pole King Haakon VII's Plateau. In the 1920s Amundsen explores the Arctic Ocean by seaplane. He dies in 1928 trying to rescue a fellow explorer, Umberto Nobile. Nobile is found alive by a second search party, but Amundsen's group never returns.

1872: August 3   Birth of Haakon VII

The future king of Norway is born Prince Carl of Denmark and becomes a naval officer. He marries Princess Maud, the daughter of Edward VII, Prince of Wales (the future King Edward VII of England). When Norway wins independence from its union with Sweden in 1905, Prince Carl is offered the crown of the new kingdom, an offer supported by a popular referendum in Norway. On June 22, 1906 he becomes the first ruler of an independent Norwegian kingdom in more than five hundred years, taking the historic Norwegian name Haakon. Under Haakon VII, Norway becomes a constitutional monarchy and a democratic society.

The Danish-born king is extremely popular with his subjects, and he grows even further in stature with his resistance to the Nazis in World War II. Following the German invasion of Norway in 1940, Haakon refuses to become part of a puppet government and leads an armed resistance in the north of the country. When this effort fails after two months, Haakon and his cabinet ministers flee to England, where they form a government-in-exile whose authority is recognized by the Norwegian people for the duration of the war instead of the collaborationist administration headed at home by Vidkun Quisling. Haakon VII returns to Norway in triumph in 1945. He dies in 1957 and is succeeded by his son, Olaf V.

1874–77   Sweden's king rejects parliamentarianism for Norway

Successive bills proposing parliamentarianism—a system linking the country's executive and legislative branches—are passed by the Storting but rejected by Sweden because they would make Norway's cabinet answer to the Storting rather than to the Swedish king.

1880: August 17   Death of violinist and patriot Ole Bull

Violinist and composer Ole Bull, one of Norway's most famous and most beloved nineteenth-century cultural figures, dies near Bergen. (See 1810.)

1882   First woman is admitted to the University of Oslo

Ida Cecille Krogh is the first woman to gain admission to the University of Oslo.

1882: May 20   Birth of author Sigrid Undset

Nobel Prize-winning author Sigrid Undset, best known for her novel *Kristin Lavransdatter,* grows up in a family impoverished by her father's death at a young age. She is not able to attend a university, instead taking a secretarial job at the age of sixteen to support herself and contribute to the support of her mother and sisters. Writing in the evenings and weekends, she launches her literary career with the publication of her first novel at the age of twenty-five. Until 1919 Undset writes realistic novels exploring the lives of ordinary people in contemporary Norway. During this time, she leaves her secretarial job and meets her husband, the Norwegian painter Anders Castus Svarstad, while staying in Rome. She and Svarstad have three children and also make a home for the three children from his previous marriage.

By 1920 Undset is divorced and living with her children in a house she has had built in Lillehammer. At this point she returns to her first interest—the historical novel—and writes two major works set in medieval Norway: the three-volume *Kristin Lavransdatter* and the four-volume *Olaf Audunssonn,* between 1920 and 1927. Both win critical acclaim and become bestsellers, and Undset receives the Nobel Prize for Literature in 1928. In the 1930s Undset returns to the contemporary novel, also writing historical works and literary essays and doing translations. In 1940 she is forced to flee the country after the Nazi invasion, in which her eldest son is killed serving in the Norwegian army. She spends the duration of the war in the United States, returning to Norway in 1945 and dying four years later.

1883   Constitutional crisis leads to impeachment of cabinet

In the midst of a crisis over the relationship between the executive and legislative branches, Norway's high court impeaches all of the nation's cabinet ministers for refusing to accept legislation making them responsible to the Storting.

Although the Swedish king has refused to sign the bill, Liberals argue that it is still valid because it has been passed and rejected three separate times.

1884   Norwegians win parliamentary standoff

Following an intense political showdown, the Norwegians win the right to a government that is responsible to its own parliament (the Storting) rather than to the Swedish monarch. Sweden's king appoints a cabinet led by Liberal leader Johan Sverdrup.

1884   Women's rights association is founded

The Norwegian Association for the Rights of Women is established and works to improve the education of women and the rights of married women.

1885   First speed-skating contest

A fjord near the Akershus fortress in Oslo is the site of Norway's first speed-skating competition.

1888   First women's ski races are held

The first ski races for women are held in Lillehammer.

1895: July 12   Birth of soprano Kirsten Flagstad

Internationally renowned opera star Kirsten Flagstad is born into a musical family and begins singing professionally by the age of eighteen. For nearly twenty years she pursues a successful career in Scandinavia. In 1933 she is invited to sing at the Bayreuth Festival in Germany. Her 1935 Metropolitan Opera debut as Wagner's Sieglinde wins her international recognition, and she goes on to perform major roles at Covent Garden in England.

During World War II Flagstad's husband becomes a supporter of the Nazi collaborationist Quisling government, compromising his wife's political reputation, and after the war her popular support in the United States cools. However, English audiences are still welcoming, and she continues to perform major operatic roles at Covent Garden and other venues to great acclaim until her retirement from operatic performance in 1953. Afterward, she serves as director of the Norwegian State Opera for several years. Kirsten Flagstad dies in 1962.

1898   Universal male suffrage is granted

All Norwegian males are granted voting rights.

1899   Oslo's National Theatre opens

Norway's main theater, the National Theatre in Oslo, opens. Playwright Henrik Ibsen (see 1828) and his wife are seated in the front row for the first performance.

1901 First Nobel Prizes are awarded

The Nobel Institute in Oslo announces the first Nobel Prize awards, inaugurated through the will of Alfred Nobel, the Swedish inventor of dynamite.

1903 Bjørnson receives the Nobel Prize

Playwright, novelist, and poet Bjørnstjerne Bjørnson becomes the third person to receive the Nobel Prize for Literature.

Independence

1905: June 7 Norway dissolves its union with Sweden

Following a strategy devised by Prime Minister Christian Michelson (1857–1925), Norway's entire cabinet resigns following a legislative veto by Sweden's king. Based on the principle that this leaves the Swedish king no agency through which to rule Norway, the Storting declares the union between the two countries dissolved.

1905: August Referendum is held on independence

Sweden agrees to a referendum on independence for Norway. The Norwegians choose independence almost unanimously, and Sweden recognizes Norway as a sovereign nation.

1905: November Norway votes to remain a monarchy

In a referendum, Norwegians overwhelmingly favor a constitutional monarchy over a republic, and Denmark's Prince Carl is enthroned as King Haakon VII, a name that links the new sovereign to the medieval Norwegian kings.

1906: May 23 Death of playwright Henrik Ibsen

Pioneering dramatist Henrik Ibsen dies. (See 1828.)

1907: September 4 Death of composer Edvard Grieg

The composer Edvard Grieg, who evolved a distinctly Norwegian musical style, is mourned by the nation. (See 1843.)

1910: April 26 Death of author Bjørnstjerne Bjørnson

Bjørnstjerne Bjørnson, one of Norway's leading nineteenth-century literary figures, dies in Paris. (See 1832.)

1911 First woman is elected to parliament

Anna Rogstadt becomes the first woman elected to Norway's parliament (the Storting).

1911: December 14 Amundsen reaches the South Pole

Explorer Roald Amundsen (see 1872) leads the first expedition to the South Pole.

1913 Female suffrage is granted in national elections

Norway becomes the second European nation (after Iceland) to enact female suffrage when it gives women over the age of 25 the vote.

1914–18 World War I

Norway remains neutral in World War I. However, hundreds of its merchant ships are sunk by German submarines, killing over 2,000 sailors. Its trade is also disrupted by Allied blockades and embargoes.

1914 Birth of explorer and anthropologist Thor Heyerdahl

Thor Heyerdahl, unlike previous Norwegian explorers, takes Polynesia rather than the polar regions as his domain and is also unique in his focus on anthropology. On Heyerdahl's first visit to Polynesia in 1937 he notices connections between Polynesian culture and that of South America. However, the prevailing scientific view is that it would have been impossible for primitive people to make the journey between the two regions on the craft available in earlier ages and that the Polynesians are of exclusively southeast Asian ancestry. Ten years later, Heyerdahl challenges that notion with the now-famous voyage of the *Kon Tiki,* a balsa wood raft designed to simulate an Incan craft. In three months he and a six-man crew sail the 5,000 miles (8,045 kilometers) from Peru to Polynesia, proving that the connection he theorizes is, if not certain, at least possible. Heyerdahl publishes a book presenting his theory and supporting it with the evidence of his own journey.

Heyerdahl makes other voyages to support theories about connections between early civilizations, traveling to the Galapagos Islands in 1953 and Easter Island in 1955–56, and sailing from North Africa to the Caribbean in 1969 and 1970 in the *Ra* and *Ra II,* papyrus reed boats.

1917 Foreign investment is regulated

The Storting passes legislation regulating foreign investment in Norway, reflecting national concern that nearly one-third of the shares in the nation's businesses are held by foreign interests. Restrictions on foreign investors include the payment of production fees to the Norwegian government.

1920s Labor gains power in Norwegian politics

Norway's Labor party, which begins among fishermen and tenant farmers and later gathers industrial support, becomes a powerful force in national politics. The party undergoes various alignments, including a break with the Communists, after which it emerges as a social reform party.

1920   Knut Hamsun wins the Nobel Prize

Author Knut Hamsun is awarded the Nobel Prize for Literature.

1922   Nansen wins Nobel Peace Prize

Explorer and humanitarian Fridtjof Nansen is awarded the Nobel Peace Prize for his relief work in Russia.

1928   Sigrid Undset receives the Nobel Prize

The Nobel Prize for Literature is awarded to Sigrid Undset, author of *Kristin Lavransdatter,* a popular multivolume novel set in medieval Norway.

1935   Labor government voted into office

A moderate Labor government is elected and implements social programs advancing housing, public health, social welfare, and cultural activities. Labor will dominate the Norwegian government until 1963.

1939: April 20   Birth of Gro Harlem Brundtland

Gro Harlem Brundtland, Norway's first woman prime minister, studies medicine in Oslo and at Harvard University and works in public health administration after receiving her degree. She enters politics in 1969, joining the Labor Party. She rises through the ranks in the 1970s, serving in the government as environment minister from 1974 to 1979. By 1981 she heads the party and briefly becomes prime minister in 1981 when the incumbent Labor prime minister, Odvar Nordli, resigns for health reasons. However, she is forced to cede the position when the Conservative Party comes to power in elections later that year.

Brundtland becomes prime minister once again in 1986 with a minority Labor cabinet and, except for a one-year period in 1989, continues to occupy that position until 1996, when she retires from government service amid speculation that she may be appointed secretary general of the United Nations. In 1998 Brundtland is named Director General of the World Health Organization. Brundtland's other accomplishments include chairing an international commission on development and the environment that produces the report *Our Common Future*. She is also the recipient of a Third World Foundation prize for leadership on environmental issues.

World War II

1940: April 9   Germany invades Norway

Norway is prepared to remain neutral in any coming conflict, but Germany launches a surprise invasion of both Denmark and Norway. Denmark capitulates within the day, but the Norwegians sink the German ship *Blücher* and launch a two-month-long resistance effort. The king and other members of the government manage to leave Oslo before it is overrun by the Germans and flee northward, where members of the Storting, meeting in emergency session, pass legislation recognizing the legitimacy of a government-in-exile in case they have to flee the country.

1940: April–June   Norway fights back

The Germans demand that King Haakon VII surrender to them and appoint Vidkun Quisling, head of the Norwegian Nazi Party, as prime minister. The king refuses, and the Nazis bomb the area where he and the other government officials are hiding. However, they escape and remain on the move, eventually setting up a provisional capital in the northern city of Tromsø. With help from French and British troops, Norwegians continue fighting in the northern part of the country for two months, but they are ultimately unable to keep Germany from occupying all of Norway. Eventually, there are some 400,000 German troops stationed in the country.

1940: June 7   Norway's king and cabinet escape to England

King Haakon VII and Norway's cabinet ministers escape to England, where they declare a government-in-exile, which is endorsed by the Storting and by the Norwegian people, once they feel confident that England will not be invaded as well. When the government leaders escape, they take with them all of the nation's gold reserves, strengthening their position in London and making them able to finance the activities of the Norwegian Armed forces units that flee with them.

In Norway, a pro-Nazi government is formed, headed by former defense minister Vidkun Quisling (whose name later becomes synonymous with "collaborator"). Resistance activities continue within Norway and include clandestine military training, sabotage, and anti-German propaganda. By 1943 there are sixty underground newspapers in Norway. Norwegians also undermine the German occupation by resigning from their posts in government, religious, and other institutions. When the Germans attempt to organize Norwegian sports, members of all sports organizations resign in a national sports strike. Schools are temporarily closed when more than a thousand teachers are arrested for refusing to sign a loyalty oath. Escapes are organized for persons who are about to be arrested or who want to join the Norwegians fighting with Allied forces overseas.

1943: February 27   Norwegians sabotage German plant

In the most important Norwegian resistance action of the war, nine saboteurs break into a German "heavy water" plant at Rjukan designed for the development of nuclear weapons. Eluding German guards, they plant explosives inside the plant, destroying it, and flee to safety, evading a 3,000-man search effort mounted by the Germans. Five of the men ski to safety in Sweden—a journey of 250 miles (400 kilometers).

Crowds gather to celebrate Norway's Day of Independence. (EPD Photos/CSU Archives)

1943: March 12   Death of sculptor Gustav Vigeland

Famed sculptor Gustav Vigeland dies in Oslo. (See 1869.)

1944: January 23   Death of painter Edvard Munch

Edvard Munch, Norway's most famous painter and the creator of the painting *The Scream,* dies near Oslo. (See 1863.)

The Postwar Period

1945–61   Labor Party holds electoral majority

The Labor Party wins an electoral majority in the first elections following World War II and retains it until the early 1960s. Labor directs Norway's postwar economic recovery by instituting a planned economy in which the government has a central role, distributing subsidies in many areas, enacting permanent wage and price controls, and establishing state-owned enterprises.

1945   Norway joins the United Nations

Norway, which is active in the League of Nations following World War I, becomes a founding member of the United Nations (UN). The Norwegian foreign minister, Trygve Lie, becomes the first UN secretary general.

1945: June 7   King Haakon returns to Norway

Following the end of World War II, King Haakon VII, in exile in England, returns to Norway amid great public celebration. Vidkun Quisling, head of the wartime Nazi puppet government, is convicted of treason and shot along with twenty-four other Nazi collaborators. Thousands more receive prison sentences.

1949: April   Norway joins NATO

Disillusioned about the benefits of neutrality, which prove unreliable in World War II, and distrustful of the Soviet Union, Norway becomes a founding member of the North Atlantic Treaty Organization (NATO).

1949: June 10   Death of Sigrid Undset

Nobel Prize-winning author Sigrid Undset dies. (See 1882.)

1952: February 19   Death of author Knut Hamsun

Novelist, poet, and playwright Knut Hamsun, winner of the Nobel Prize for Literature, dies near Grimstad. (See 1859.)

1957   King Haakon dies

King Haakon VII dies and is succeeded by his son, Olaf V.

1959   Norwegian Opera is founded

The Norwegian Opera is founded. Its first director is soprano Kirsten Flagstad. She is followed by Odd Gruner-Hegge and Lars Runsten.

1962: December 7   Death of soprano Kirsten Flagstad

Kirsten Flagstad, one of the leading opera singers of her time, dies in Oslo.

1963   Munch museum opens

The Munch Museum, containing an art collection donated posthumously by painter Edvard Munch, opens in Oslo.

1964   First Ski for Light event is held

Erling Stordahl organizes the first Ridderrennet (Ski for Light) ski race for the blind, which uses beeping sounds along the ski track. Eventually, Norway's Sports Organization for the Disabled encompasses seventeen different sports.

1965   Nonsocialist coalition takes office

Norway's Labor government is voted out of office for the first time since 1935, and a coalition of nonsocialist parties forms a new government.

1967   Government-financed concert foundation is created

The Rikskonsertene (Norwegian State Foundation for the Nationwide Promotion of Music), supported by the government ministry of education, is founded to promote concerts.

1969   Oil is struck in the North Sea off Norway

Phillips Petroleum oil venture finds petroleum in a part of the North Sea now known as the Ekofisk oil fields.

1970s   Statoil is created

Statoil, a state-owned oil company, is formed following the discovery of oil in the North Sea. The company is involved in the exploration, processing, and distribution of petroleum and natural gas.

1972   EEC entry is rejected by voters

Norwegians defeat a proposal to join the European Economic Community (EEC) in a popular referendum despite support for the measure by the ruling Labor government. Opposition to membership stems from fears that it would harm Norway's major industries, including shipbuilding and fishing, by making them less competitive.

1981: February   First woman prime minister is chosen

Gro Harlem Brundtland of the Labor Party is elected Norway's first woman prime minister, replacing Odvar Nordli, who resigns for health reasons.

1981: September   Conservative government is elected

Following the decline of support for the Labor Party through most of the 1970s, Norway elects its first Conservative government since 1928, headed by Conservative Party leader Kare Willoch. The Conservatives attempt to reduce the government's role in regulating the economy.

1984   Statoil is reorganized

Statoil, Norway's state-owned oil company, is reorganized to increase government control over it.

1986   Reindeer harmed by Chernobyl disaster

Norwegian reindeer herds are affected by radioactive contamination from the Chernobyl nuclear accident in Russia.

1986   Falling oil prices drive inflation

Falling world prices for oil hurt Norway's economy, driving up inflation and threatening the country's extensive social welfare programs.

1986: April   Nation is crippled by mass strikes

Economic problems brought about by falling oil prices lead to Norway's worst strikes in decades, as about 100,000 workers in various industries strike for a week.

1987   Weapons export to U.S.S.R. spurs controversy

Public discovery of weapons export to the Soviet Union by Konigsberg Vapenfabrikk, Norway's state-owned weapons manufacturer, leads to controversy and scandal.

1988   Committees are required to appoint women

Norway's equal rights legislation is amended to include provisions requiring that women constitute at least 40 percent of all publicly appointed boards and committees.

1990   Rules of succession are changed

The rules governing succession to the throne of Norway are changed to permit women to inherit the crown. Under the new system, the first-born child succeeds to the throne regardless of gender.

1991: January   King Olaf dies

Olaf V dies and is succeeded by his son, Harold V. Harold's wife, Sonja Haraldsen, born a commoner, becomes Norway's queen. She also becomes Norway's first Norwegian-born queen—the wives of the two preceding monarchs had been born in, respectively, England and Sweden.

1991: October 17   Bank collapse is averted

Following the collapse of Norway's second-largest bank, the government funds a costly bailout to save the nation's banking system from collapse.

1994: February 12–28   Norway hosts Winter Olympics

The seventeenth annual Winter Olympics are held in Lillehammer, a town of 23,000 located north of Oslo. A total of 1,920 athletes, representing sixty-six nations, compete in the games.

1994: February 12   *The Scream* is stolen

Edvard Munch's world-famous painting *The Scream* is stolen from Oslo's National Art Museum, where it is on display as part of a cultural festival linked with the Winter Olympics in Lillehammer. The entire burglary, which is completed by two men in 50 seconds, is filmed by the museum's surveillance cameras, which show the men breaking into the gallery through a window. A reward of 200,000 kroner ($26,700) is offered by the Norwegian government for the safe return of the painting.

1994: May 7   *The Scream* is found

Edvard Munch's painting *The Scream,* stolen during the Winter Olympics, is recovered in a hotel south of Oslo, and three men are arrested. The suspects have been tracked down by police through their attempts to sell the painting. Prominent figures in Norway's antiabortion movement have attempted to link the theft to an antiabortion film called *The Silent Scream,* but police have ignored their claims.

1994: November 28–29   New referendum on EU fails

Norwegian voters narrowly defeat a second proposal to join the European Union (EU) (formerly the EEC), this time by a vote of fifty-two percent to forty-two percent. Although the vote is nonbinding, the Storting is expected to respect the results and vote against EU membership. Opponents of EU membership, many of whom live in rural areas, fear that it will threaten Norway's generous farm subsidies and that the country will be forced to open its fishing zones to other countries. These voters also associate EU membership with foreign domination, a sensitive subject among Norwegians after their years of domination by Denmark and Sweden and the long German occupation during World War II.

Those in favor of joining the EU fear that economic isolation will hurt Norway's economy and argue that the lower tariffs accompanying EU membership would increase business investment. They also point out that Norway already observes most trade laws of the EU through its membership in the European Economic Area.

1996: August 29   Arctic crash is Norway's worst air disaster

A Russian airliner crashes into a mountain on Spitsbergen Island, a Norwegian territory in the Arctic Circle. All 141 passengers and crew aboard the plane are killed in the crash, which occurs six miles from the nearest airport. The cause of the air disaster—the worst in Norway's history—remains undetermined.

1996: October 23   Brundtland steps down as prime minister

Gro Harlem Brundtland announces her decision to retire as prime minister. She is replaced by Labor Party leader Thorbjoern Jagland, who is sworn in two days later. Brundtland first serves as prime minister in 1981—becoming the youngest person and the first woman elected to the post—but her tenure ends within a year when a Conservative government was elected. When Labor returns to power in 1986, she regained her former post, serving as prime minister almost continuously for ten years and becoming the dominant figure in Norwegian politics during that period. Brundtland says she intends to consider serving in the Storting and seek reelection in 1997. However, she is rumored to be under consideration for a top post at the United Nations.

1997: October 13   Jagland steps down as prime minister

Thorbjoern Jagland resigns as Norway's prime minister following the September parliamentary elections, in which the Labor Party wins roughly thirty-five percent of the vote, down slightly from its results in the previous election four year earlier, but still twice as high as the total for any other party. Jagland had vowed to resign unless the party's support was at least equal to the 36.9 percent garnered in the 1993 election.

1997: October 17   New government is formed

Norway's new prime minister, Christian Democrat leader Kjell Magne Bondevik, forms a new center-right coalition that includes the Christian Democrats, the Center Party, and the Liberal Party. These parties hold a total of only forty-two of the 165 seats in Norway's parliament.

Bibliography

Baden-Powell, Dorothy. *Pimpernel Gold : How Norway Foiled the Nazis.* New York: St. Martin's Press, 1978.

Berdal, Mats R. *The United States, Norway and the Cold War 1954–60.* New York: St. Martin's, 1997.

Cole, Wayne S. *Norway and the United States, 1905–55 : Two Democracies in Peace and War.* Ames: Iowa State University Press, 1989.

Galenson, Walter. *A Welfare State Strikes Oil : The Norwegian Experience.* Lanham, Md.: University Press of America, 1986.

Greve, Tim. *Haakon VII of Norway: The Man and the Monarch.* New York: Hippocrene, 1983.

Heide, Sigrid. *In the Hands of My Enemy: A Woman's Personal Story of World War II.* Trans. Norma Johansen. Middletown, Conn.: Southfarm Press, 1996.

Jochens, Jenny. *Women in Old Norse Society.* Ithaca, N.Y.: Cornell University Press, 1995.

Jonassen, Christen Tonnes. *Value Systems and Personality in a Western Civilization: Norwegians in Europe and America.* Columbus: Ohio State University Press, 1983.

Kersaudy, Francois. *Norway 1940.* New York: St. Martin's Press, 1991.

Petrow, Richard. *The Bitter Years: The Invasion and Occupation of Denmark and Norway, April 1940–May 1945.* New York: Morrow, 1974.

Shaffer, William R. *Politics, Parties, and Parliaments: Political Change in Norway.* Columbus: Ohio State University Press, 1998.

Vanberg, B. *Of Norwegian Ways.* New York: Harper & Row, 1984.

Poland

Introduction

Poland's history as an independent, sovereign nation dates back a thousand years, but its strategically central location and level terrain have led to long periods of invasion and domination by its powerful neighbors, especially Russia to the east. Through these periods, however, the Poles have retained their cultural identity, united by their common language and history and their strong allegiance to the Roman Catholic Church. In the late 1980s, Poland was in the forefront of national renewal and political reform that led to the end of communism in Eastern Europe. Like its neighbors, it has grappled with the problems inherent in the transition to a democratic society and a free-market economy and, together with Czechoslovakia and Hungary, ended the decade on a historic note with formal admission in 1999 to the North Atlantic Treaty Organization (NATO).

With an area of 120,726 square miles (312,680 square kilometers), Poland is divided into three principal geographic regions running east to west in roughly parallel bands: a low-lying plain in the north, the hilly central lowland of central Poland and upper Silesia, and the Sudeten and Carpathian mountains to the south. Poland's longest river is the Vistula, and about a quarter of its area is covered by forest land. As of the late 1990s, Poland has an estimated population of close to 39 million people. Warsaw, the capital and largest city, has an estimated metropolitan population of over 2.2 million.

History

The Slavs living in present-day Poland were first united politically in the tenth century, and their sovereignty was recognized in AD 999 by Holy Roman Emperor Otto III. The division of the kingdom of Poland upon the death of Boleslaw III in 1138 ushered in two centuries that were dominated by political fragmentation and included Mongol invasions in the thirteenth century. However, by the mid-fourteenth century, Poland was once again on the ascendant under one of its greatest monarchs, Casimir III, and in 1386 a long and illustrious period in Polish history was launched with the founding of the Jagiellonian dynasty and the reign of its first king, Ladislaw II. Military victories over the knights of the Teutonic Order in the fifteenth century led to territorial expansion, and the historic legislative assembly known as the Sejm was convened for the first time in 1467. Poland also made a great contribution to learning in the fifteenth century when the theories of astronomer Nicolaus Copernicus challenged the entrenched belief that the sun and other heavenly bodies revolved around the earth.

The sixteenth century, particularly the reign of Sigismund II (1548–72), brought a golden age in scholarship and the arts, influenced by the humanism and architecture of the Italian Renaissance. Krakow became an intellectual and cultural center, and Jan Kochanowski became the first major poet to write in Polish, making an important contribution in developing literary forms and conventions where none existed before. He also wrote the first Polish-language drama.

Poland's fortunes began to decline in the seventeenth century. Much of the power formerly held by the monarch had been ceded to the nobles, and the government engaged in a series of mostly unsuccessful wars against Sweden and Russia. In addition, much of Poland was ravaged by the rebellion of Cossacks in Ukraine. By 1667, Poland had ceded the eastern portion of Ukraine to Russia, which, together with Austria and Prussia, posed a growing threat to Polish sovereignty throughout the eighteenth century. Stanislaw II, the last king of Poland, relied heavily on the support of Russian empress Catherine II and was dominated by Russian interests. By the end of the century, a series of agreements had partitioned the country among its three adversaries: central and eastern Poland, including Ukraine, were taken over by Russia; the northwestern region was claimed by Prussia; and Austria-Hungary gained control of Galicia. By 1795 Poland as a sovereign nation had ceased to exist.

Throughout the nineteenth century, the Poles sought to regain their independence. As part of the Napoleonic conquest of Europe, a Polish state, the Grand Duchy of Warsaw, was created, but the 1815 Congress of Vienna that followed the downfall of Napoleon Bonaparte (1769–1821) largely returned Poland to its partitioned borders. The Poles subsequently waged unsuccessful rebellions in 1830 and 1863–64. Nationalism came to the forefront of Polish culture in the literary works of authors, including Adam Mickiewicz (1798–1855) and Henryk Sienkiewicz , the music of Frédéric Chopin (1810–49), and the paintings of Jan Matejko (1838–93).

The first real breakthrough for the Poles came with World War I, when the powers occupying Poland went to war with each other. In the midst of the war, while the Central Powers (Germany and Austria) had the upper hand, they authorized the creation of a Polish state on territory they had won from Russia. After their defeat, the Polish lands they had occupied were ceded to the newly formed independent Polish republic that was proclaimed on November 11, 1918. Nationalist leader Josef Pilsudski was proclaimed its first head of state, and its boundaries were established officially at the Versailles conference in 1919. In 1926, when Pilsudski's presidential term was to end, he engineered a *coup d'état* (armed overthrow of the government) and took over the government, ruling through a hand-picked successor. Pilsudski's autocratic government remained in power until World War II.

On September 1, 1939, Germany invaded Poland, launching World War II. German soldiers were to remain in the country until the end of the war, carrying out a repressive occupation that permeated all aspects of Polish life. In addition to forming a government-in-exile abroad, the Poles resisted the German occupation by forming the largest underground resistance movement in Europe. When Germany invaded the Soviet Union in 1941, violating the non-aggression pact between the two countries, Poland became a battleground between the two countries. The last Germans were expelled from the country in March, 1945 by Soviet troops.

In 1944 a Soviet-backed Committee of National Liberation was established in the city of Lublin. After the war, the Allied powers agreed to recognize this political organization as the government of Poland, providing that it expanded to include non-communists. However, by 1947 the Polish communists had formed a Soviet-style people's republic and begun a program of repression against non-communists. The Soviet Union retained strong control over Poland into the 1950s, and, in accord with the communist position against religion, the government waged a campaign against the Catholic Church, but the Polish people still remained devoutly Catholic even though they couldn't express their religious convictions as openly as before.

Government repression lessened after mass demonstrations and rioting in 1956, when Wladislaw Gomulka (1905–82) came to power, and strict government adherence to communist policies was relaxed. In the 1960s, major economic reforms were launched, but repression of individual freedom continued. Widespread student unrest at the end of the decade led to the resignation of Gomulka as head of the government. However, popular discontent with the government continued, highlighted by mass strikes and demonstrations in 1976 in response to cuts in food subsidies, and in 1978 opposition to communism in Poland received a strong boost with the appointment of the first Polish pope, Pope John Paul II.

In the 1980s the Solidarity labor union federation, led by Gdansk shipyard worker Lech Walesa (b. 1943), led the fight for democratic freedom in Poland. By 1988, the nation's respect for Walesa, who won the Nobel Peace Prize in 1983, had reached a point where the government was compelled to call on him for help in quelling new outbreaks of civil unrest. In return, the labor leader demanded, and received, major political concessions including the legalization of Solidarity and a promise of free elections. On June 4, 1989, Poland held its first free elections since the beginning of the communist era, and political candidates allied with Solidarity won nearly every parliamentary seat they contested. A new government was formed, and the People's Republic of Poland became the Republic of Poland. The Solidarity government launched major economic reforms to institute a free-market economy. Government enterprises were privatized, government subsidies reduced, and price controls abolished.

Like their Eastern European neighbors, Poles found that the transition to a Western-style economy and a democratic society posed serious challenges and that they disagreed about how best to meet them. Due to policy differences between Lech Walesa and Prime Minister Tadeusz Mazowiecki (whom Walesa had backed), Walesa ran for president in 1990 and won a five-year term. By 1993, however, the Polish parliament was controlled by the a leftist party, and in 1995 Walesa was narrowly defeated for reelection by voters critical of the sometimes painful effects of government economic reforms and the growing political power of the Catholic Church. In 1997 the North Atlantic Treaty Alliance (NATO) announced that it would invite Poland to become a member in two years, and official membership documents were signed in March, 1999.

Timeline

c. 1st century A.D.   Slavic tribes begin settling in Poland

The first Slavic settlers are thought to arrive in present-day Poland.

c. A.D. 960   Slavs united by the Piast monarchs

The Piast dynasty establishes control over the Slavic tribes living between the Vistula and Oder rivers, most notably the tribe known as the Polanie, from which modern Poland gets its name.

A.D. 966   The Poles embrace Christianity

The Piast ruler Mieszko I accepts the Christian religion, linking Poland with the lands to the west rather than to the Slavs of Russia and other eastern regions, who follow the Eastern Orthodox religion and use the Cyrillic alphabet. The Poles adopt the Latin alphabet used in the West, but with additional letters for Slavic sounds that can't be represented by the western alphabet.

999   Polish sovereignty is recognized

Holy Roman Emperor Otto III recognizes Polish sovereignty during the reign of Boleslaw I, son of Mieszko I. Boleslaw is the first Piast ruler to officially be crowned and recognized abroad as king of Poland. He presides over the expansion of Polish territory as far as the Dnieper River and the Carpathian Mountains.

1138   Poland is divided

Upon the death of Boleslaw III, Polish territory is divided among his five sons, beginning a long period of fragmentation and decline. The power that had formerly belonged to the unified kingdom is taken over by feudal lords and the Church.

1177–94   Reign of Casimir II

Casimir II restores the unity of much of Poland. He establishes a central government assembly and supports peasants' rights.

1241   Mongol invasion

Invasion by Mongols and Tatars from the east destroys the Polish unity that was restored under Casimir II. At the same time, Lithuanians and Prussians invade from the north.

1320–33   Ladislaw I rules

Much of Poland's former territory is reunified under Ladislaw I, who is crowned king of Poland.

1333–70   Reign of Casimir III

Poland's monarchy reaches a high point under the rule of Casimir III, also known as Casimir the Great. He wins new territories in the east, including Galicia, Volhynia, and Podolia, introduces administrative reforms, including a code of laws (see 1347), and founds Poland's first university (see 1364). During Casimir's reign, living conditions of the peasantry are improved, and Jews fleeing persecution in Western

The Polish-Lithuanian State, 15th century.

Europe are allowed to settle in Poland. Casimir has no sons, and the throne goes to Louis I of Hungary upon his death.

1347   Statutes of Wislica

A legal code known as the Statutes of Wislica is adopted by Casimir III.

1364   University is founded

One of Europe's oldest universities (today known as Jagiellonian University) is established at Krakow.

The Jagiello Dynasty

1386–1572   Rule by the Jagiello dynasty

Founded by Ladislaw Jagiello, the Jagiellonian dynasty rules Poland, presiding over a period of political expansion and cultural advancement.

1386   Ladislaw II becomes king

Upon marrying into Poland's ruling family, the grand duke of Lithuania becomes king and rules as Ladislaw II, inaugurating the long-lived Jagiellonian dynasty.

1410   Battle of Tannenberg

The combined forces of Poland and Lithuania defend the growing territory of the Jagiellonian dynasty by defeating the knights of the Teutonic Order in battle at Tannenberg (also known as Grunwald).

1454–66   War with the Teutonic Knights

Poland once again goes to war with the Teutonic Knights (and wins) during the reign of Casimir IV.

1454   Casimir IV grants charter to nobles

Casimir IV, unable to curb the power of the Church or the nobles, grants the nobles a charter at Nieszawa, an action comparable to the signing of the Magna Carta in England by King John.

1466: October 19   Treaty of Torun

With the Treaty of Torun, the long war between Poland and the Teutonic Knights ends in victory for Poland, which takes control of Pomerania and Danzig (Gdansk).

1467   First parliament is convened

The first Polish parliament (the Sejm) is convened.

1473: February 19   Birth of astronomer Nicolaus Copernicus

Copernicus, celebrated as the father of modern astronomy, pursues studies at the University of Krakow and in Bologna and Padua, Italy, receiving degrees in both religion and medicine. He returns to Poland by 1506, first working as a physician and then in an appointed post within the Catholic church. A multi-talented individual, Copernicus also serves as a diplomat as well as an artist, linguist, and mathematician. However, it is for his contribution to astronomy that he is known today.

Copernicus works steadily throughout his life on a treatise challenging the prevailing *geocentric* view of the universe inherited from the second-century astronomer Ptolemy of Alexandria. According to this view, the earth is at the center of the cosmos, and the other heavenly bodies revolve around it. In contrast, Copernicus pursues a *heliocentric* (sun-centered) view of the universe that is gradually accepted by scientific authorities over the next century and a half. First, however, it is condemned by church authorities, including the Roman Inquisition, which sentences the Italian scientist Galileo Galilei (1564–1642) to prison for endorsing it. Copernicus is not affected by the controversies surrounding his theory because he dies soon after it is published in 1543, in the volume *De libris revolutionum narratio prima*.

Poland's Golden Age

1505   Nihil Novi statute expands parliament's powers

Under pressure by the Polish nobles, King Alexander signs a statute giving parliament an equal voice in government, including a veto in certain matters.

1526   Battle of Mohacs

Poland is defeated by the Ottoman Turks at the Battle of Mohacs, leading the Jagiellonian monarchy to lose control over Bohemia and Hungary.

1530   Birth of poet Jan Kochanowski

Kochanowski is the leading literary figure of Poland's cultural golden age during the Renaissance. He is exposed to the humanist culture of Renaissance Italy while studying in Padua and is active in spreading its ideas upon his return to Poland, where he serves as a royal secretary in Krakow. Although he starts out by writing poetry in Latin, the common literary language of his time, Kochanowski soon switches to Polish. In the absence of any established literary version of the language, he is obliged to create his own poetic forms and linguistic conventions. His greatest work is *Treny* (Laments; 1580), a sequence of poems written in response to the death of his baby daughter. Kochanowski also achieves renown as the author of the first drama written in Polish, the tragedy *The Dismissal of the Greek Envoys* (1578). Kochanowski dies in Lublin in 1584.

1543: May 24   Death of Nicolaus Copernicus

Nicolaus Copernicus dies in Frauenberg. He is the first modern astronomer to challenge the geocentric theory.

1548–72   Reign of Sigismund II

The reign of Sigismund II, the last Jagiellonian monarch, is a golden age for scholarship and the arts. Krakow becomes a center of humanist learning, Polish literature, and Renais-

sance architecture, as well as a prominent site of Protestant religious reform.

1569: July 1   Lublin agreement strengthens Polish-Lithuanian union

Poland and Lithuania are united under a single monarch, as well as a single *Diet* (parliament), a single set of laws, and a common currency.

1572   Pacta Conventa weakens the monarchy

Through the *Pacta Conventa* (also called the Henrician Articles), the Polish nobility gain sweeping governmental powers, in effect turning Poland into a parliamentary republic, in which the king is reduced to a weak chief executive. Legislators gain the right to choose each new king in a special election, the right to veto legislation. They also have the final say in the imposition of taxes and the waging of war. The king is also obliged to retain a council of legislators as his advisors, and the parliament even reserves the right to choose the king's wife.

1584: August 22   Death of poet Jan Kochanowski

Jan Kochanowski, the leading poet of Poland's golden age, dies in Lublin. (See 1530.)

1587–1668   Rule by the Vasa dynasty

The reign of Sigismund III begins the tenure of the Vasa dynasty as rulers of Poland.

1587–1632   Reign of Sigismund III

The half-Swedish Sigismund III tries to regain the Swedish crown but holds it only briefly and begins a series of wars against Sweden.

1610   Poland invades Russia

During a period of Russian turmoil, Polish troops invade Russia and occupy Moscow.

1648–68   Reign of John II Casimir

Poland's power is progressively weakened during the reign of John II Casimir as it faces rebellion and invasion.

1648–49   Cossack rebellion

Cossacks in the Ukraine stage a brutal rebellion against the Polish nobility, causing devastation over a large area of the country.

1667   Treaty of Andrusov

The Treaty of Andrusov ends hostilities with Russia, to which Poland cedes the eastern portion of Ukraine.

1682–99   Poland joins Austria against the Ottomans

Poland joins forces with Austria in fighting the Ottoman Turks.

1683   Poles defeat Turks at Vienna

Led by their king, John III Sobieski, the Poles win a famous victory over the Ottoman Turks at Vienna. However, Sobieski is forced to grant territory to Russia in exchange for protection against other foreign enemies.

1697–1763   Reigns of Augustus II and Augustus III

German kings rule Poland for most of the eighteenth century when Augustus II of Saxony assumes the throne and is succeeded by his son. During this period, a major threat emerges in the person of Frederick I, who unites the Prussian state, which will be one of the three powers that carries out the coming division of Poland. Together with Prussia, Russia and Austria exercise considerable power over Poland during these decades.

c. 1700–25   Dominance by Russia

During the first part of the eighteenth century, Russia's tzar, Peter the Great, virtually controls the Polish monarchy.

1746: February 12   Birth of national hero Tadeusz Kosciuszko

Tadeusz Kosciuszko, a patriot and freedom fighter, attends military schools in Warsaw and France, acquiring a specialty in military engineering. Drawn by the struggle for freedom in the American colonies, he travels to North America in 1776 and aids the colonists with fortifications and other strategic projects. He is honored by the U.S. Congress in 1783. Returning to Poland, Kosciuszko is awarded the rank of major general. In 1794 he leads Polish forces against Russia after the second partition of the country, but the campaign in unsuccessful, and Kosciuszko is captured and held prisoner in Russia until 1796. He later returns to the United States, where he forms a friendship with Thomas Jefferson, whom he appoints executor of his will (in which he specifies that his American property should be used to buy freedom for slaves and provide them an education). Settling in France in 1798, Kosciuszko continues to work for the liberation of his homeland until his death in 1817.

1764–95   Reign of Stanislaw II

With the backing of Empress Catherine II of Russia and the rulers of Prussia, Stanislaw August Poniatowski becomes the last king of Poland. His reign is dominated by foreign influence.

1768–72   Confederation of Bar

Polish Catholics form the Confederation of Bar to oppose Russian power, but the confederation is crushed within four years.

Partition and Occupation

1772–95   Poland is partitioned

The partition of Poland in stages by neighboring Russia, Prussia, and Austria spells the end of the country as an independent political entity. Austria-Hungary gains control of Galicia, Prussia wins northwestern Poland, and central and eastern Poland, including Ukraine, go to Russia.

1772   First partition of Poland

Poland loses one-third of its territory to Russia, Prussia, and Austria in the first of three partitions. Ceded lands include Galicia, part of Krakow, Podolia, eastern Belarus, and other lands east of the Dnieper River. The victors create a new constitution for the conquered lands.

1788–92   The Great Sejm

A wave of Polish patriotism triggered by the First Partition (see 1772) leads to the convening of the Great Sejm, a special assembly led by nationalists Stanislaw Malachowski, Ignacy Potocki, and Hugo Kollontaj.

1791: May 3   New constitution is adopted

While Russia is distracted by foreign wars, the Great Sejm promulgates a new constitution giving Poland a hereditary monarchy rather than one elected by parliament. In sweeping nationalist reforms, Polish towns obtain greater autonomy and political representation, peasants' rights are expanded, and the unpopular parliamentary veto is abolished.

1792   Confederation of Targowica is formed

In opposition against the new liberal Polish constitution, conservatives form the Confederation of Targowica, which is joined by the king and supported militarily by Russia.

1793: January 23   Second partition of Poland

In violation of a prior military alliance, Prussia joins with Russia in attacking Poland. Polish resistance is overcome, and more of the country is annexed by both powers.

1794   Kosciuszko leads uprising

The Poles, led by national hero Tadeusz Kosciuszko (see 1746), rise up against the occupying foreign powers but are ultimately unsuccessful.

1795: October 24 Third partition ends Polish sovereignty

Poland as a united and independent political entity is abolished with the third partition, in which Austria joins Russia and Prussia.

1798: December 24 Birth of poet Adam Mickiewicz

Mickiewicz is one of Poland's greatest poets. As a student, his ardent nationalism leads him to join a secret society *Filareci* dedicated to achieving Polish independence. His first volume of poetry, *Poezja I* (1822), is an attempt to introduce Western European poetic forms into Polish literature. His next collection, *Poezja II* (1823), includes experiments in dramatic form that involve the use of Polish folklore. In 1823 Mickiewicz is arrested for his activities in the secret society and sent to Russia, where he is detained for four years, during which time he meets some of the major Russian literary figures. After his release from Russia, Mickiewicz settles in Paris, where he publishes the epic poem *Pan Tadeusz* (see 1834). He also teaches literature at the Collège de France and the University of Lausanne in Switzerland. Mickiewicz dies in 1855.

1807 Grand Duchy of Warsaw is created

As part of his conquest of Europe, Napoleon Bonaparte (1769–1821) reconstitutes a greatly reduced Polish state, called the Grand Duchy of Warsaw, from territories annexed by Prussia during the partition of Poland. The duchy, which is expanded two years later, greatly encourages Polish hopes of a return to national sovereignty.

1810: February 22 Birth of composer Frédéric Chopin

Chopin, one of the greatest composers of music for the piano, is born near Warsaw to a French father and a Polish mother. Demonstrating his musical gifts at an early age, he has his first composition published at the age of seven. He receives his formal musical education at the Warsaw Conservatory, graduating at the age of nineteen. He then begins giving concerts throughout Europe, establishing a base in Paris, where he becomes a popular piano teacher to the rich and fashionable. In 1836 he begins an eight-year affair with the eccentric Baroness Dudevant, who becomes famous as a writer under the pseudonym George Sand. During this period, the composer's already fragile health deteriorates rapidly, and he dies of tuberculosis in 1849.

Chopin is unusual among the great Western composers in having written almost exclusively for the piano. However, his works for this instrument span a wide range of expression, encompassing waltzes, preludes, etudes, nocturnes, scherzos, ballades, rondos, sonatas, and concertos. Chopin is identified with the beginning of the Romantic Movement in the arts, and his piano compositions are noted for both lyricism and virtuosity. Chopin was also ahead of his time in introducing nationalistic elements into his music. His mazurkas and polonaises in particular are associated with the music of his homeland.

1815 Congress of Vienna

The Congress of Vienna returns Poland in large part to its partitioned borders. Krakow is designated a free city-republic but is still a protectorate of the three foreign powers that control Poland. The Polish lands under Prussian control become the Grand Duchy of Poznan, and a part of Poland that is under Russian rule is designated as Congress Poland.

1830: November November Revolution

The Poles stage an unsuccessful rebellion against Russia. Czar Nicholas I retaliates by making Russian rule even more restrictive. Poland is reduced to the status of a Russian province.

1834 Publication of *Pan Tadeusz*

The great Polish poet Adam Mickiewicz publishes his epic masterpiece, *Pan Tadeusz*. In telling the story of the rivalry between two noble families, it provides a detailed account of nineteenth-century life among the Polish gentry.

1838: June 24 Birth of painter Jan Matejko

Matejko, one of Poland's greatest painters, is known for devoting his art to the development of nationalism among the Polish people. After studies in Vienna and Munich, he returns to Krakow, where he settles permanently. Matejko is known for his interest in Polish history and contemporary events, both of which provide subjects for his paintings. The most famous include *Stanczyk* (1862), in which he gives the court jester of King Sigismund I his own features; *Battle of Grunwald*, which portrays one of the great military victories in Polish history; *Year 1863—Polonia*, which depicts an unsuccessful Polish uprising against Russia in which he lost his own brother; and *Union of Lublin*, which celebrates the political union between Poland and Lithuania. Aside from the popularity of his paintings with the general public, Matejko also wields a strong influence on successive generations of artists as director of Krakow's School of Fine Arts in the last twenty years of his life. By the end of his life, the artist is a nationally revered figure whose house is turned into a museum three years after his death in 1893.

1846: May 5 Birth of author Henryk Sienkiewicz

Nobel Prize-winning author Henryk Sienkiewicz is a journalist and critic as well as the writer of acclaimed and widely popular short stories and novels. In the 1870s, he begins publishing fiction and spends two years in the United States as a special correspondent for a Polish newspaper. In the 1880s he is coeditor of the daily paper *Slowo*, in which some of his novels are serialized. During this decade he writes a renowned trilogy of historical novels set in late-seventeenth-century Poland: *With Fire and Sword* (1884), *The Deluge* (1886), and *Pan Michael* (1887–88). Sienkiewicz establishes his international reputation with the 1896 publication of the novel *Quo Vadis?*, set not in Poland but in Rome during the days of the Roman Empire. In 1905 Sienkiewicz is awarded the Nobel Prize for Literature, becoming the first Pole to receive a Nobel Prize. During World War I he works for Polish independence and assists with relief efforts for Polish war victims. He dies in Switzerland in 1916.

1849: October 17   Death of Frédéric Chopin

Composer and pianist Frédéric Chopin dies in Paris. (See 1810.)

1855: November 26   Death of poet Adam Mickiewicz

Nationalist poet and dramatist Adam Mickiewicz dies in Constantinople. (See 1798.)

1860: November 18   Birth of musician and statesman Jan Paderewski

Ignace Jan Paderewski is a world-famous musician and a respected Polish patriot. Beginning in his late twenties, his virtuosic piano performances win him a level of international fame enjoyed only by fellow performer/composer Franz Liszt during this period. His magnetic personality at the keyboard draws great crowds in London, Paris, Vienna, New York, and elsewhere. Paderewski is also the composer of the opera *Manru* (1901), the *Symphony in B Minor* (1909), and the *Minuet in G,* and produces a respected edition of the works of another famous Polish-born composer, Frederick Chopin.

Paderewski is also known as a lifelong Polish patriot, and the nonpartisan respect commanded by the musician leads to his being named to prominent government posts at key moments in twentieth-century Polish history. His first period of political service is during World War I, when he serves on the Polish National Committee and acts as its representative to the United States. After the war, he briefly becomes prime minister of the newly independent nation. Paderewski is once again called on to serve his country during World War II, when he is active in the Polish government-in-exile in Paris until the Nazi invasion, when he flees to the United States, where he dies in 1941.

1862   Fine arts museum is opened

The Museum of Fine Arts (later the National Museum) is established in Warsaw. In addition to its other holdings, it develops a collection of Polish art dating from the twelfth century.

1863–64   January Insurrection

The Poles rebel against Russia and are defeated again after two years of guerrilla warfare. Russia moves to intensify its presence and institutions in Poland through *Russification* (a policy of forcibly promoting Russian national identity).

1866   Galicia wins home rule

Galicia, which belongs to the area ruled by Austria, obtains administrative autonomy, with the right to run its own courts, schools, and government offices.

1869: January 15   Birth of artist and author Stanislaw Wyspianski

Wyspianski is both a major artist and one of the foremost figures in the history of Polish drama. During the early part of his short career, he concentrates mostly on his art, producing stained-glass windows and other decorations and turning from oil paintings to pastels when he develops an allergy to oil paint. Returning to Poland, he makes his home in Krakow. Among his most famous artworks are the stained-glass windows for the city's Franciscan church. He also designs windows for Wawel Cathedral that feature figures from Polish history. Wyspianski produces portraits, domestic scenes, landscapes, self-portraits, and book illustrations and serves as artistic director for a weekly magazine.

Most of Wyspianski's plays, for which he designs the sets and costumes, are written in the last decade of his life. Combining Greek mythology and Polish history, they place him in the forefront of early-twentieth-century European theater and modern Polish drama. His most famous work is *Wesele* (The Wedding), which premieres in 1901 and encompasses themes from Polish history and politics. First staged in Krakow, it becomes popular throughout the country. Wyspianski dies in Krakow in 1907.

1879   National museum is founded in Krakow

Krakow's National Museum is founded and becomes home to the country's most important collection of Polish artifacts.

1893: November 1   Death of painter Jan Matejko

Acclaimed painter and patriot Jan Matejko dies in Krakow. (See 1838.)

1905   Nobel Prize awarded to author Henryk Sienkiewicz

Novelist Henryk Sienkiewicz becomes the first Pole to win a Nobel Prize when he receives the award for literature. (See 1846.)

1905: February 5   Birth of political leader Wladyslaw Gomulka

Wladyslaw Gomulka, who plays a prominent role in Polish politics during the communist era, first becomes active in the Communist Party while working as a mechanic. He soon devotes himself full-time to political activism and is arrested numerous times before World War II, during which he is a

Nazi resistance fighter. He is part of the wartime communist government set up in Lublin by the Soviet Union and becomes one of its three most powerful leaders after the war. By 1948, however, Gomulka is in trouble with the Soviets for promoting Poland's freedom to choose its own style of socialism. He is removed from his government post and, in 1951, sent to prison.

Gomulka's fortunes improve following the death of Soviet leader Josef Stalin (1879–1953). By the end of 1956 he holds the highest political office in the land. Despite his own earlier persecution, Gomulka's policies become increasingly repressive and pro-Soviet, sparking outbreaks of civil unrest in 1968 and leading to his ouster in 1970 following mass rioting by workers. Gomulka dies in Warsaw in 1982.

1907: November 28   Death of artist and playwright Stanislaw Wyspianski

Stanislaw Wyspianski, an artist and playwright considered the founder of modern Polish theater, dies in Krakow. (See 1869.)

1911: June 30   Birth of poet Czeslaw Milosz

Nobel Prize-winning poet Czeslaw Milosz is born in Lithuania, which belongs to Poland at the time. He is educated in the Lithuanian capital of Wilno and publishes his first book of poetry at the age of twenty-one. He is a socialist in his youth and participates actively in the Polish resistance to the Nazis during World War II. He serves for a time as a diplomat in the postwar communist regime but later defects to France, ultimately settling in the United States, where he teaches at the University of California at Berkeley and becomes a naturalized U.S. citizen. In 1980 he is awarded the Nobel Prize for Literature. In addition to numerous volumes of poetry, Milosz also publishes an autobiography (*Native Realm*; 1959); novels, including *The Seizure of Power* (1955); literary criticism; and *The Captive Mind* 1953), a widely read volume of essays on the situation of intellectuals in communist regimes. Among Milosz's later works are a volume of poetry entitled *Bells in Winter* (1978) and *The Witness of Poetry* (1983), a series of lectures.

1916: November 9   Wartime kingdom is established in Poland

With a large part of Poland captured by Austria and Germany, the Central Powers approve the formation of an independent Polish kingdom in the region formerly controlled by Russia.

1916: November 15   Author Henryk Sienkiewicz dies

Nobel Prize-winning novelist Henryk Sienkiewicz dies in Switzerland. (See 1846.)

1918: January 8   Wilson voices support for Poland

U.S. president Woodrow Wilson (1856–1924) includes Polish independence in his historic Fourteen Points.

1918: November 11   The Republic of Poland is proclaimed

With the defeat of the Central Powers in World War I, an independent Polish republic is proclaimed, with nationalist leader Joseph Pilsudski as its first head of state.

1919   Polish borders are decreed at Versailles

Boundaries of the new Polish state are decided at the Versailles conference in Paris following World War I. Poland is given access to the sea and special rights in Gdansk (Danzig), which is declared a free city.

1919: October 9   Poland occupies Vilnius

Polish troops occupy the disputed Lithuanian city of Vilnius (Wilno), whose population is made up of a majority of ethnic Poles although it is historically under Lithuanian rule.

The occupation turns into a de facto annexation that wins international recognition in the 1923 Treaty of Riga.

1920: April–October   Poles attempt to retake Ukraine

Polish leader Pilsudski and his Ukrainian counterpart launch an invasion to retake Ukraine from Russia. It fails, and Poland itself is nearly invaded by the Russians, who are turned back at Warsaw.

1920: May 18   Birth of Pope John II

Pope John Paul II (b. 1920) is born Karol Wojtyla in Wadowice, Poland. He begins studying for the priesthood in Krakow during World War II and is ordained in 1946. He continues his studies in Rome, receiving a doctorate in 1948. During the 1950s and 1960s, he rises through the ranks of the priesthood, becoming archbishop of Krakow in 1964 and being named a cardinal in 1967. Wojtyla is elected pope in October 1978 and enthroned the same month as John Paul II.

The papacy of John Paul II is distinguished by the high energy level and personal magnetism of the pope, who thrives on personal contact with the faithful and makes numerous public appearances throughout the world and at the Vatican. The bullet of a Turkish would-be assassin in 1981 does not curb his normal activities, which are resumed within a year. Coupled with his outgoing manner is a conservative philosophy on many issues, including such controversies as abortion and the ordination of women. He issues many encyclicals, as well as a new Code of Canon Law (1983) and a new Catechism (1992). John Paul is a strong supporter of the Solidarity movement from its inception in the 1980s because of its opposition to communism. In the 1990s he becomes increasingly frail but continues to travel and maintain as active a schedule as possible.

1920: September 12   Birth of author Stanislaw Lem

Internationally recognized science fiction author Stanislaw Lem studies medicine in order to pursue a career as a biolo-

Thousands protest in Warsaw, and Joseph Pilsuski takes control of the government. (EPD Photos/CSU Archives)

gist but turns to writing, publishing essays, poetry, and fiction. In the 1950s he turns to science fiction with the novel *Astronauts* and after that devotes his career to short stories and novels in this genre. Among the best known are the collection *The Cyberiad* (1965); the series of stories involving a time traveler named Ijon Tichy (some of which appear in English as *The Star Diaries*); *Solaris* (1961); and *Memoirs Found in a Bathub* (1961), about the battle between a man and a computer. Lem has also written television dramas, literary criticism, and serious scientific essays.

1926: March 6   Birth of Andrzej Wajda

Poland's best-known filmmaker is strongly affected by the political events of his youth. The rise of fascism in Europe looms over his childhood, and his father is killed in World War II, in which he participates as a resistance fighter while still a teenager. After studying painting, Wajda turns to film and begins his directorial career in 1952 with *A Generation,* which instantly establishes his reputation. This movie and his next two, *Kanal* (1957) and *Ashes and Diamonds* (1958), form a trilogy about World War II.

Over the course of his career, Wajda directs a wide range of films. In the West, he is perhaps best known for two movies dealing with communist repression and the struggle against it: *Man of Marble* (1977) and *Man of Iron* (1981). The latter film, made during the period when mass labor strikes threatened communist rule and the Solidarity labor federation was born, expands international awareness of political events in Poland. Wajda is persecuted by the Polish

government after the film's release and goes into exile in France but is able to return to Poland by 1989, the year that communism's grip on Poland is broken. While continuing to pursue his art, he is also named head of the leading theater in Warsaw and serves as a member of Poland's newly created Senate.

1926: May 12   Pilsudski engineers a coup

Polish leader Joseph Pilsudski launches a successful *coup d'état* and takes over the government, ruling through Ignacy Moscicki, who is elected president with his backing. Pilsudski remains the real power behind the new and increasingly repressive government.

1928   National Library is established

Poland's National Library is founded in Warsaw. By the end of the century, it holds close to five million volumes. In addition to books and periodicals, its holdings include manuscripts, maps, and music.

1933: November 23   Birth of composer Krzysztof Penderecki

One of the world's leading *avant-garde* (experimental) composers, Penderecki studies music in Krakow and holds teaching posts in Poland, Germany, and the U.S., where he is on the faculty of Yale University. Penderecki is known in particular for expanding the repertoire of sounds available to composers by using unusual instrumental effects and unorthodox vocal sounds. His first major success was with the 1960 *Threnody in Memory of the Victims of Hiroshima* for stringed instruments. Other works include *Polymorphia* (1961); settings of religious texts, including the *St. Luke Passion* (1966) and the *Magnificat* (1974); and an opera based on the epic poem *Paradise Lost* by English poet John Milton (1608–74).

1935: April 22   Constitution is adopted

The Polish government adopts a constitution that increases the power of the central government, limits the functions of parliament, and places restrictions on political parties.

1938: October   Poland occupies Teschen

Following the Munich Agreement that paves the way for the German break-up of Czechoslovakia, Poland occupies Teschen, a disputed territory previously claimed by both Poland and Czechoslovakia.

1939: March   Germany demands Polish territory

Nazi Germany demands that Poland cede the city of Gdansk and makes other demands for territorial concessions that lead to a diplomatic stand-off between the two countries.

1939: August   Germany and U.S.S.R. agree to partition Poland

As part of the German-Soviet non-aggression pact, the two countries agree to divide Poland between them.

World War II

1939: September 1   Germany invades Poland

Germany launches a surprise land and air attack on Poland, beginning World War II. Some of Poland is incorporated directly into Adolf Hitler's (1889–1945) Third Reich, while the rest becomes a German protectorate under a separate administration. Poland's legitimate leaders flee, forming a government-in-exile. Polish resistance forces organized abroad fight with the Allies. At home, Poles organize the largest anti-Nazi resistance movement in Europe.

Poland is subjected to a harsh German occupation, in which all aspects of daily life are controlled by the Nazi occupiers, and large numbers of people are sent to labor camps and concentration camps. Polish Jews are persecuted and confined to ghettos. Millions of Poles, including three million Jews, are killed in attacks and in concentration camps.

1939: September 17   Soviets attack Poland

Following the German invasion, the Soviet Union invades Poland from the east to claim its share of the country.

1940   Katyn Massacre

Soviet troops secretly murder about 15,000 Polish soldiers and bury them in a mass grave. When the bodies are discovered three years later, Poland breaks off diplomatic relations with the U.S.S.R.

1941: June   Germany invades Russia

Germany invades the Soviet Union, breaking the nonaggression pact between the two countries, and the Germans seize the part of Poland previously occupied by the Soviets.

1941: June 29   Death of musician and patriot Jan Paderewski

Pianist, composer, and statesman Jan Paderewski dies in New York City. (See 1860.)

1941: July 30   Poland signs pact with Soviets

After Germany invades the Soviet Union, Poland and the U.S.S.R. sign an agreement restoring previously disrupted diplomatic relations, reestablishing the border between the two countries, and allying the two countries in the effort against Nazi Germany. The Soviet Union subsequently backs out on its promise to return all Polish prisoners and recognize prewar boundaries between the two countries.

1942   Deportation of Jews begins

The Nazis begin implementation of the *Final Solution* (elimination of the Jews) by deporting Poland's Jews from ghettos to concentration camps.

1943: April   Warsaw ghetto uprising

The 350,000 Jews of the Warsaw ghetto stage a one-month uprising against their Nazi captors rather than be deported. The camp is destroyed and its remaining inhabitants deported.

1943: September 29   Birth of Lech Walesa

Lech Walesa, the labor leader who heads the opposition to communist rule and becomes Poland's first post-Communist president, begins his career as an electrician at the Gdansk shipyard. He loses his job as a result of his union activities during the strikes of 1976 but continues his activism, leading a strike at the shipyard in 1980. He becomes a founder and leader of the Solidarity labor federation that is formed as an alternative to the existing government-controlled unions. When the government cracks down on the organization, Walesa is imprisoned for a year and Solidarity is outlawed. In 1983 Walesa is awarded the Nobel Peace Prize, to the consternation of Poland's communist leaders, who henceforth are limited in the degree of pressure they can apply against him due to the international attention this would attract.

Walesa's moral authority within the country is such that he is called in by the government to negotiate an end to strikes in 1988. In return, the government agrees to legalize Solidarity and give it a voice in the government. Solidarity candidates sweep the free elections held in June 1989, and Walesa is elected to a five-year presidential term in popular elections held in 1990. Once in office, he arouses opposition by his attempts to strengthen executive power, the rapid pace of transition to a free-market economy, and his support for measures demanded by the Roman Catholic Church. In 1995, Walesa is defeated by a narrow margin in his bid for reelection, with victory going to leftist leader Aleksander Kwasniewski.

1943: November–December Allies agree on postwar Polish border

At a meeting in Teheran, Iran, the heads of state of the Allied governments agree to draw Poland's eastern border according to the Curzon Line, which favors the Soviet Union.

1944: January Soviets rout Germans across the Polish border

Soviet troops cross the Polish border in pursuit of fleeing German troops.

1944: July 22 Soviet-backed political body is formed in Lublin

The Soviet-backed Polish Committee of National Liberation is set up in the city of Lublin.

1944: August–October Resistance fighters besiege Warsaw

Polish resistance fighters led by Tadeusz Komorowski lay siege on Warsaw but are finally defeated by the occupying Germans, who nearly level the city.

1945: February Yalta Conference

The U.S. and Great Britain approve the Soviet annexation of eastern Poland and agree to recognize the existing authority in Lublin as the basis for Poland's postwar government. However, it is supposed to expand and admit non-communist members.

Communist Rule

1945: July 5 Lublin government is recognized abroad

The United States formally recognizes the communist-controlled Lublin government as the provisional government of Poland and withdraw diplomatic recognition from the government-in-exile formed during the war.

1945 Poland joins the UN

Poland becomes a member of the United Nations.

1947: January 19 First postwar elections are held

Poland holds its first elections since before the war. Communists (the Polish Workers' Party) win 382 out of 444 seats in the parliament.

1947: February 5 Bierut is elected president

Communist leader Boleslaw Bierut is elected president by the members of the Sejm. A Soviet-style people's republic is instituted and a wave of Stalinization—repression like that under Josef Stalin (1879–53) in the Soviet Union—begins against non-communists, including members of Poland's wartime government resistance movement.

1948: September Gomulka is ousted

Communist Party general secretary Wladislaw Gomulka (1905–1982) is "purged" from his top party and government posts and later arrested.

1948: December Communist and Socialist parties merge

The Polish Socialist Party and the Polish Workers' Party merge to form the Polish United Workers' Party (PZPR) with Boleslaw Bierut as its general secretary.

1949 Poland joins COMECON economic group

Poland becomes a member of the Council for Mutual Economic Assistance (COMECON), the economic alliance of communist countries.

1949: November Russian assumes top defense post

Soviet field marshal Konstantin Rokossovsky is named Poland's defense minister and commander-in-chief.

1952: July 22 New constitution is adopted

Poland adopts a new constitution providing for a government modeled on that of the Soviet Union. The office of prime minister replaces that of president.

1953: September Top church leader is arrested

Poland's primate (highest-ranking religious leader), Cardinal Wyszynski, is arrested in the most dramatic development in the communist campaign against the strong influence of the Catholic Church.

1955 Poland joins the Warsaw Pact

Poland becomes a founding member of the Warsaw Pact alliance of nations belonging to the communist bloc.

1956: June   Students and workers protest at Poznan

Some 50,000 students and workers stage mass demonstrations and riots at Poznan protesting food shortages, government repression, and control of their government by the Soviet Union.

1956: October   Gomulka becomes party leader

Government leadership changes in response to civil unrest the previous spring, as well as liberalization in the Soviet Union following the 1953 death of Josef Stalin and the "de-Stalinization" policies of Stalin's successor, Nikita Krushchev (1894–1971). Wladislaw Gomulka becomes general secretary of the Communist Party and institutes a series of government reforms that reduce the influence of the Soviet Union and adherence to communist economic doctrine. Most collective farms are dissolved, relations with the Catholic Church are improved, and freedom of expression is expanded.

1964   Economic reform program

The Gomulka government launches an economic reform program but becomes more repressive of individual freedoms.

1968: March   Students demand freedom of speech

Widespread student demonstrations are held to call for more relaxation of government controls on free speech. The result is a crackdown that leads many intellectuals and dissidents to leave the country.

1968: August   Poland joins in Soviet invasion of Czechoslovakia

Poland, whose government has opposed the liberalization taking place in neighboring Czechoslovakia during the "Prague spring" earlier in the year, aids in the Soviet invasion of Czechoslovakia aimed at removing the liberal Dubcek regime.

1970: December 7   Treaty is signed with West Germany

Poland and West Germany sign a non-aggression pact mutually recognizing the line of the Oder and Neisse Rivers as the border between the two nations. The treaty is expected to pave the way for West German aid for Poland's beleaguered economy.

1970: December 15   Food price increases spur riots

Steep government price increases on food, fuel, and other basic commodities lead to rioting in Gdansk, Gdynia, and Sczcecin. Armed military intervention results in 70 deaths and 1,000 injured, and Party Secretary Gomulka is forced from office.

1970: December 20   Gierek replaces Gomulka

Edward Gierek replaces Wladislaw Gomulka as head of Poland's Communist Party. Gierek retains this post for ten years.

1976: June 25   Workers strike over food prices

Polish workers stage strikes and demonstration to protest government cuts in food subsidies.

1979: June   Pope visits Poland

Polish-born Pope John Paul II visits Poland, becoming the first pope to visit a Communist country.

1980   Poet Czeslaw Milosz receives Nobel Prize

The Nobel Prize for Literature is awarded to poet Czeslaw Milosz. (See 1911.)

1980: July–August   Gdansk strikes spread widely

In response to another round of food price increases, strikes sweep the northern part of the country, beginning in the shipyards of Gdansk and spreading to other Baltic cities, Silesia, and elsewhere. A workers' committee headed by organizer Lech Walesa (b.1943) and two colleagues submits a list of 22 demands to the government. Following negotiations, the government meets most of the demands, including the right to form independent trade unions and the right to strike.

1980: September   Solidarity is formed

Having just won the right to form independent unions, Polish workers create a nationwide union federation, headed by Lech Walesa, called Solidarity. It soon has ten million members.

1981: January   Solidarity strikes for five-day work week

The government agrees to Solidarity's demands for a five-day work week following a strike.

1981: April 10   Strikes are temporarily banned

The government bans all strikes for two months.

1981: April 27   Nation narrowly avoids bankruptcy

With agriculture and industry stagnating and food shortages becoming critical, the Polish government narrowly escapes declaring bankruptcy as Western nations agree to refinance its debt.

1981: October 18   Jaruzelski is named to top post

General Wojciech Jaruzelski, the commander-in-chief of Poland's armed forces, is appointed to head the Communist Party. Jaruzelski takes a harsher stance toward labor unions than his predecessors.

1981: December 13   Martial law is imposed

In response to new demands by Solidarity, the Polish government declares martial law and arrests thousands of union leaders, including Walesa.

1982: September 1   Death of communist leader Wladislaw Gomulka

Longtime communist leader Wladislaw Gomulka dies in Warsaw. (See 1905.)

1982: October 1982   Solidarity is outlawed

Following the refusal of Solidarity leaders to compromise with the government position, Solidarity is banned and replaced by government-controlled unions.

1983   Walesa wins the Nobel Peace Prize

Lech Walesa, leader of the banned Solidarity movement, is awarded the Nobel Peace Prize over the protests of the Polish government.

1984: July   Government declares a general amnesty

The fortieth anniversary of the Polish republic is marked by a general government amnesty for political prisoners.

1986   Government moves to reconcile with Solidarity

A new Consultative Council created by the government includes former members of the banned Solidarity trade union federation.

1988: April–May   New wave of strikes breaks out

Government price hikes spur a new and crippling wave of strikes, and the government turns to Walesa for help in containing them. In return, Walesa demands political reforms and the legalization of Solidarity.

1988: May   Workers take over the Lenin shipyard

Thousands of Polish workers occupy the Lenin shipyard calling for reinstatement of Solidarity.

1989   Government agrees to major political reforms

The government makes major concessions, including the legalization of Solidarity and authorization of free elections to a newly created upper house of parliament. In addition, opposition candidates will be allowed to contest more than one-third of the seats in the lower house, and the presidency will no longer be restricted to members of the Communist Party.

1989: May   Catholic Church is recognized

In a major move away from the communist position, the Polish government gives its official recognition to the Roman Catholic Church.

Post-Communist Poland

1989: June 4   Solidarity triumphs in general elections

In the first elections since World War II that provide for the selection of non-communist candidates, Solidarity scores a major triumph, winning practically all competitions it is allowed to contest, including all but one seat in the upper house of parliament.

1989: June   Gorbachev declares support for reforms

Soviet president Mikhail Gorbachev (b. 1931) visits Poland, proclaims support for the government reforms, and promises that the Soviet Union will not intervene.

1989: June   Jaruzelski chosen to remain president

By a narrow margin, Woyciech Jaruzelski, Poland's last leader under communism, is elected as head of state (president) by the nation's parliament.

1989: August 24   Mazowiecki becomes prime minister

With the support of Lech Walesa, Solidarity leader Tadeusz Mazowiecki becomes head of the new government.

1989: December   Poland's name is changed

Poland's name is officially changed from the People's Republic of Poland to the Republic of Poland. In addition, all references to the country's Communist Party are removed from the constitution.

1990s   Catholic Church gains political power

With the overthrow of communism in Poland, the country's traditionally strong allegiance to Catholicism becomes a prominent factor in political affairs. The church uses its growing influence to win passage of legislation asserting its authority over education, the media, and women's rights.

1990: January   Economic reform plan begins

The new government launches a program to rescue the nation's faltering economy through privatization, abolition of price controls and reductions in government subsidies to industry, and changes in fiscal policy. Although unemployment figures rise, many Poles open small businesses in the new atmosphere of entrepreneurship. In addition, inflation is reduced and the supply of consumer goods grows.

1990: May Local elections are held

Solidarity makes further gains in free elections held locally throughout the country.

1990: September Jaruzelski resigns

Elected by parliament in 1989 to a six-year term, General Wojciech Jaruzelski resigns as president of Poland, setting off a contest for that position between two top Solidarity leaders: Party Chairman Lech Walesa and Prime Minister Tadeusz Mazowiecki, whose differing positions on government policy indicate a growing split within Solidarity itself.

1990: November 25 Walesa is top vote-getter in preliminary round

Lech Walesa garners 40 percent of the votes in the preliminary round of presidential voting, less than many had expected but still more than any other candidate. Because no candidate wins a clear majority, a second round of voting is required.

1990: December 9 Walesa is elected in run-off election

In a run-off election, Solidarity leader Lech Walesa is chosen to serve as president of Poland, making the head of state and head of government (i.e. prime minister) both members of Solidarity.

1991: October Political divisions surface in national elections

With their recent liberation from single-party rule, Poles field candidates from dozens of parties, beginning the formation of a series of coalition governments.

1992: October 17 "Little Constitution" is approved

Parliament adopts a shortened constitution that guarantees democratic freedoms and institutions.

1993 Walesa signs new abortion legislation

Reflecting the strong influence of the Catholic Church in post-communist Poland, President Lech Walesa approves an abortion law that is one of the most restrictive in Europe. It allows abortions only if the mother's health is seriously threatened, if she has been raped or subjected to incest, or if serious fetal defects can be determined.

1993: September Leftist party leads in elections

Alliance of the Democratic Left, created in 1989 by former Communist Aleksander Kwasniewski, leads the voting in national elections, winning 173 parliament seats. The Peasants' Party comes in second with 128 seats, and the two parties form a ruling coalition.

1995: November 19 Walesa is unseated as president

Aleksander Kwasniewski, leader of the formerly Communist Alliance of the Democratic Left, is elected president, defeating incumbent candidate Lech Walesa by a voting margin of 3 percent in runoff elections. A record 68 percent of eligible voters go to the polls. Walesa's loss of popular support is attributed to a lack of consensus-building ability. The forty-one-year-old Kwasniewski has strong support among younger voters, while Walesa's support is stronger among older ones. Like Walesa, Kwasniewski is committed to economic privatization, democratic reforms, and Poland's admission into the North Atlantic Treaty Alliance (NATO).

1996: February 7 New prime minister is sworn in

Wlodzimierz Cimoszewicz is sworn in as prime minister, replacing Jozef Oleksy, who resigns after being charged with supplying state secrets to the former Soviet Union.

1996: October 3 Nobel Prize is awarded to Polish poet

Poet Wislawa Szymborska receives the Nobel Prize for Literature. She becomes the fifth Polish author to be given the award, and the ninth woman. Although her work is not well known outside Poland, within the country it is widely read.

1997 Prospective NATO membership announced

The North Atlantic Treaty Alliance (NATO) announces that Poland, along with the Czech Republic and Hungary, will formally be invited to join the organization in 1999. This will be the first expansion of NATO since Spain joined in 1982, as well as the first admission of nations formerly belonging to the Warsaw Pact, the alliance of Soviet Bloc nations that was NATO's adversary during the Cold War.

1997: September 21 Solidarity coalition wins victory at the polls

A coalition of groups affiliated with Solidarity wins over 33 percent of the vote in parliamentary elections, compared with 27 percent for the ruling Alliance of the Democratic Left.

1997: October 20 New coalition government is formed

Solidarity Electoral Action (AWS) and Freedom Union, the parties with the strongest showing in September's election, agree to form a right-of-center coalition government. Jerzy Buzek serves as the new prime minister.

1999: March 12 Poland is formally admitted to NATO

Together with Hungary and Czechoslovakia, Poland is formally admitted to NATO ten years after the collapse of communism in Eastern Europe. Foreign ministers from all three nations sign documents of accession in a ceremony at the Harry S Truman Library in Independence, Missouri. The addition of three new members expands total NATO membership from sixteen nations to nineteen.

Bibliography

Bernhard, Michael H. *The Origins of Democratization in Poland: Workers, Intellectuals, and Oppositional Politics, 1976–80*. New York: Columbia University Press, 1993.

Blazyca, George and Ryszard Rapacki, eds. *Poland into the 1990s: Economy and Society in Transition*. New York: St. Martin's, 1991.

Bromke, Adam. *The Meaning and Uses of Polish History*. New York: Cambridge University Press, 1986.

Engel, David. *Facing a Holocaust: The Polish Government-in-Exile and the Jews, 1943–45*. Chapel Hill: University of North Carolina Press, 1993.

Goodwyn, Lawrence. *Breaking the Barrier: The Rise of Solidarity in Poland*. New York: Oxford University Press, 1991.

Staar, Richard F., ed. *Transition to Democracy in Poland*. New York: St. Martin's, 1993.

Steinlauf, Michael. *Bondage to the Dead: Poland and the Memory of the Holocaust*. Syracuse, N.Y.: Syracuse University Press, 1997.

Tworzecki, Hubert. *Parties and Politics in Post-1989 Poland*. Boulder, Colo.: Westview, 1996.

Walesa, Lech. *The Struggle and the Triumph: An Autobiography*. New York: Arcade, 1992.

Wedel, Janine R., ed. *The Unplanned Society: Poland During and After Communism*. New York: Columbia University Press, 1992.

Portugal

Introduction

By the end of the twentieth century, Portugal held on to small sliver of land in southeast China, the last remain of what was once a mighty empire that stretched across three continents. Macao, ruled by the Portuguese for more than four hundred years, was scheduled to return to Chinese control in December, 1999.

The return of Macao to the Chinese and Portugal's negotiations with Indonesia to determine the future of the former Portuguese colony of East Timor underscored the dramatic political and social changes that took place in the second half of the twentieth century in this small European nation.

Portugal is the oldest nation in Europe—with its borders nearly unchanged since the middle of the thirteenth century. During the Age of Discovery, Portugal became one of the wealthiest nations in the world. Yet, by the 1960s, the country languished in mediocrity, led by a repressive dictatorship that stubbornly held on to its colonial past. It would take a nearly bloodless revolution, led by a group of young army officers, in 1974 to dramatically alter the course of the nation.

Along the western edge of the Iberian Peninsula, the Portuguese have historically turned their backs to the rest of Europe. Theirs is a tiny nation, with about 10 million people living in an area slightly smaller than the state of Indiana. Portugal shares its northern and eastern borders with Spain, once considered its most dangerous foe. The Atlantic Ocean laps against its southern and western shores.

Portugal's interior is mountainous, while plains lie along the Atlantic. The country has four major rivers, the Tagus, Douro, Minho, and Guadiana. More than thirty percent of the country is covered with forests.

Nearly twenty percent of the population live in Lisbon, the capital city. The *Azores* Archipelago, a group of nine islands about eight hundred miles west of Portugal, and the islands of *Madeira* and *Santo Porto,* more than six hundred miles off the coast of Morocco, belong to Portugal.

The Portuguese population is one of the most homogeneous in Europe, with no distinct national minority. Its people are an amalgamation of many ethnic groups, including Celtic, Arab, and Berber. Through the centuries, Lusitanians, Phoe-nicians, Carthaginians, Romans, Visigoths, and Jews left their mark on the Portuguese people. Today, African immigrants from the former colonies continue to add to Portugal's ethnic mix. About 100,000 gypsies live in the Algarve region.

It is, by all measures, a devout Roman Catholic nation, where superstition and Catholicism are profoundly intertwined. The church and the state were formally separated only in 1976, yet Catholicism has shaped the nation's mores, laws, schools, and many other aspects of society. While only one-third of the population attends church regularly, nearly all Portuguese are baptized and married within the church and receive last rites before they die. Many priests are venerated as heroes, and several of the nation's major holidays are religious.

Generally, the Portuguese are formal and conservative, more so in the northern part of the country. They often greet each other with kisses to both cheeks. By the 1990s, Portugal remained a patriarchal society, with women expected to stay home and care for children. Despite the staunch conservatism, the position of women improved dramatically following the ouster of the military dictatorship in 1974. The 1976 constitution gave women equal rights, and by the 1990s, more than half of those enrolled in higher education were women.

Economic development slowed down after the government nationalized major sectors of the national economy in 1975. The government targeted for takeover banks, insurance companies, petroleum refineries, steel industry, public transportation, and heavy industries. Many of the country's largest privately owned lands were expropriated. Many peasants illegally seized other lands. The government shifted gears dramatically in 1986, when Portugal joined the European Community. State intervention was reduced, and the country received many loans to modernize its infrastructure. By 1989, Portugal began gradual privatization of public enterprises. Portugal, which traditionally relied on an export-import economy, moved to develop an overseas business policy focused on investment.

Portugal is the world's leading producer of cork, with wine stoppers accounting for more than fifty percent of cork exports. But demand for cork has declined in recent years, and the Portuguese have began planting more Eucalyptus trees for the pulp industry. The planting of Eucalyptus trees,

however, has become a major controversy. Opponents claim the trees damage the soil and water table, and displace traditional farming

The early history of Portugal is profoundly connected to the rest of the Iberian Peninsula, now home to the nations of Spain and Portugal. The early peoples of Portugal were known as Celtiberians and Lusitanians. Over time, they withstood invasions by Celts, Romans, Visigoths, and Moors. The Romans left an indelible mark in early Portugal. And so did the Moors of North Africa, who invaded the peninsula starting in 711. The Moors brought with them Islam, a competing religion with Christianity. Islam tolerated other religions, and many Christians prospered during Moorish occupation. Many others converted. But ultimately, the Moors were considered interlopers. The small Christian kingdoms in the north soon launched the *reconquest* of the peninsula. Nearly 800 years went by before the Moors were pushed back across the Mediterranean.

Portugal derived its name from ancient *Portucale*, now the northern city of Oporto. It was here, at the mouth of the Douro River, where the Portuguese monarchy rose to power during the Christian Reconquest. *Portucale* was a small province of Castile and León (in present-day Spain). It declared its independence in 1139, after King Afonso Henriques defeated the Moors at *Ourique*. Henriques consolidated his power over Portugal and continued to push the Moors south. The capture of Faro in 1249 marked the end of Moorish occupation in Portugal.

With the Moors out of the way, the Portuguese turned their attention to their powerful neighbors, the Spanish kingdoms that fought to control the rest of the peninsula. The neighboring Castilians had claims to the Portuguese throne and barely tolerated Portugal's sovereignty claims. Marriages and family ties further complicated the relationship between the monarchies. Portugal expanded its navy, strengthened its ties to Britain, and built castles to defend itself from Castilian threats. The dispute was settled with a major battle at Aljubarrota, when Portuguese and British forces routed the Castilians in 1385. For the next 200 years, Spanish kingdoms did not pose a threat.

Starting in the fifteenth century, in what came to be known as the Age of Discovery, the small Portuguese kingdom began an astonishing period of conquest and exploration. In 1415, Portugal captured Ceuta in North Africa. By 1500, Portuguese sailors had explored the West Coast of Africa, rounded the Cape of Good Hope, and reached India and Brazil. By 1513, they had trading posts in Macao and Canton. In the sixteenth century when Portugal reached Brazil and was profiting from its monopoly on the spice trade from Asia, only Spain rivaled Portugal's mighty empire.

Overconfident, the Portuguese attempted to invade Morocco in 1578 and were soundly defeated. The Spaniards claimed the Portuguese throne two years later and ruled for the next sixty years. Spain recognized Portugal's independence in 1668, after decades of fighting.

Portugal's Age of Discovery was tainted by shameful acts. The Portuguese kingdom played an important role in the enslavement of Africans. The first modern slave market in Europe was held in Portugal. When the native peoples of Brazil refused to work for the Portuguese, the empire brought more than 4 million slaves to the South American colony before slavery was abolished in 1888. Portugal extracted the wealth of its colonies giving little in return.

Throughout its history, Portugal had most of its wealth accumulated in the hands of the monarchy, the military aristocracy, nobles, and the powerful Catholic Church. With its riches, Portugal built exquisite palaces and cathedrals, yet did little to improve the national economy or create industries.

The nation remained backward and isolated, and the bulk of its population struggled to survive.

By the end of the eighteenth century, the American and French Revolutions inspired Europeans to challenge the absolute power of the monarchies. These liberal ideas of representative government, civil liberties, and equality before the law eventually reached Portugal, which resisted change. Portugal finally abolished its monarchy in 1910 and replaced it with a republic. But deep political differences and instability led to a military takeover in 1926. The mid-twentieth century was marked by the rule of António de Oliveira Salazar, an economics professor who became the country's prime minister but ruled as virtual dictator until 1968.

While other European nations gave up their colonies after World War II (1939–45), Portugal under Salazar stubbornly held on to its possessions. In the 1960s, Portugal's repressive military dictatorship killed thousands to halt the independence movement in its profitable African colonies. While the rest of Europe prospered, Portugal barely got by, in part thanks to wealth generated by its African colonies. Under Salazar, emigration reached a zenith, with 100,000 people leaving the country each year in the 1960s. By the end of the twentieth century, more than four million Portuguese, nearly thirty percent of the population, lived outside the country.

By 1974, young military officers became tired of the African wars and overthrew the dictatorship. All the colonies but Macao were given independence, and the young officers embarked the nation on painful and chaotic democratic reforms that threatened a return to a repressive regime.

Yet by the 1990s, democracy had taken root, the country was a committed member of the European Community, the economy was healthier, and Portuguese migration had slowed.

Timeline

7000 B.C.–1500 B.C.   Hunting and fishing tribes populate Portugal

What is now known as the nation of Portugal is inhabited by hunting groups who move along its river valleys. By 7000 B.C., hunters and fishermen are living in the valley of the Tagus River (*Rio Tejo*). Twentieth century anthropologists find many artifacts dating to this era, including kitchen middens that contain the remains of shellfish and the bones of animals.

By 3000 B.C., Neolithic peoples are building primitive dwellings and practicing agriculture. They use stone tools, make ceramics, and erect large funerary monuments known as *dolmens,* many of which survive into the twentieth century. More than 130 of the *pedras talhas*, hewn stones, near Evora, date to 4000 and 2000 B.C. The tall megaliths suggest fertility rites while *cromlechs*, carved monoliths spaced in groups, appear to have religious significance.

1500 B.C. More people colonize the Iberian Peninsula

Many groups of people move into the Iberian Peninsula (what is now Portugal and Spain) between 1500 and 1000 B.C. Among them are the *Lígures*, who remain a mystery, and the Iberos, who appear to arrive from North Africa. The *Iberos* practice agriculture and use tools, including a primitive plow and wheeled carts. They have writing and make offerings to the dead as part of their social customs.

1200 B.C. Phoenicians expand empire into Iberian Peninsula

The Phoenicians establish trading posts at Cádiz, Málaga, and Seville. People from the interior of the peninsula supply them with silver, copper, and tin. By 800 B.C., large numbers of Celtic people from central Europe begin to move to the western part of the Iberian Peninsula and begin farming. In time, they blend with the Iberos, forming a new group known as the *Celtiberians*.

700 B.C.–500 B.C. Invaders fight for control of Iberian Peninsula

By the seventh century B.C, Greeks have settled several colonies on the Iberian Peninsula. They are followed by the Phoenicians and later the Carthaginians. The Carthaginians close the Straits of Gibraltar to the Greeks and attempt to conquer the peninsula. They only manage to control the southern coast.

218 B.C. Romans invade Iberian Peninsula

Roman troops march into the Iberian Peninsula in 218 B.C. and fight their arch-enemies, the Carthaginians, for supremacy of the western end of the Mediterranean.

For the Romans, the Carthaginians are only part of the problem. After conquering all of Carthage's territories in present-day eastern Spain, Romans sweep west across the peninsula and attempt to subdue the Celtiberian tribes living along the Atlantic.

139 B.C. Lusitanian resistance ends with death of chieftain

The tribes put up fierce resistance to the Roman invasion but eventually fall one by one to the greater force of the invaders. The Lusitani, in present-day Portugal, hold the Romans at bay for more than two decades. They give up only after their leader, the feared Viriato, is murdered by three of his own people who are bribed by the Romans. His death in 139 B.C. marks the end of major resistance in the peninsula. Viriato later becomes a symbol of Portuguese sovereignty against foreign invaders, especially against Spain.

The Romans consolidate their power and rule the local people by establishing several administrative centers throughout the Iberian Peninsula. By A.D.300, Emperor Diocletian reorganizes the entire Roman Empire and divides the Iberian Peninsula into five provinces. The province of Lusitania occupies an area roughly equal to present-day Portugal. At first it is ruled from Olissipo, present-day Lisbon, and later from Emerita (present-day Merida in Spain). Julius Caesar briefly rules from Lisbon in 60 B.C.

19 B.C. Romans pacify Portugal

By 19 B.C., nearly two hundred years after they arrive in the Iberian Peninsula, the Romans pacify Lusitania. Their presence profoundly changes the way local people live. To control the local population, the Romans force the Lusitanians to give up their hilltop fortifications and settle them in Roman towns (*citânias*). In these towns, the Lusitanians are romanized. They learn Latin, the basis of modern Portuguese, and adopt many Roman social customs, including religion.

The Romans open Greek and Latin schools and develop brick, tile, and iron smelting industries; they build roads and bridges to connect the administrative centers. Roman landlords plant olive-groves and vineyards and begin to export their products to Italy.

They also institute a highly stratified social order, with wealthy Romans at the top of the power structure. Known as *senatores* (senators), they monopolize civic and religious posts and own the great states. At the bottom are indentured Iberian laborers and slaves.

A.D. 200 Christianity becomes firmly established

Christianity takes hold in the Iberian Peninsula.

A.D. 409 Germanic tribes invade Portugal

As the Roman Empire collapses and loses its control of its provinces, Germanic tribes begin to invade the Iberian Peninsula starting around A.D. 406. The Vandals, the Alani, and the Suevi peoples reach Portugal around A.D. 409, settling in different parts of the region.

The Suevi settle between the Minho and Douro rivers and establish a kingdom.

415 Visigoths invade Peninsula

The Romans customarily use troops from one part of the empire to control the inhabitants in another. They commission the Visigoths, a romanized German tribe, to restore the lost provinces. The Visigoths sweep through the peninsula and force the Vandals, 80,000 men, women and children, to sail to North Africa in 429.

In 457, the Visigoths kill Rechiarus, the Suevi king, and consolidate their power over the peninsula. At first, the Visigoths rule in the name of Rome, but they later establish their own empire, with an elected kingdom that rules from Toledo (in present-day Spain).

Moorish Domination of the Iberian Peninsula

A.D.711   Muslim armies invade Iberian Peninsula

The emergence of Islam in present-day Saudi Arabia under the Prophet Mohammed (570–632) has a profound impact on the Iberian Peninsula. Islam, like Christianity and Judaism, is a monotheistic religion that acknowledges the absolute sovereignty of God.

Conquering Arabs quickly spread in all directions, taking Islam with them. After sweeping through North Africa and converting the local Berber tribes, the Arabs and Berbers, collectively known as Moors, cross the Mediterranean and invade the Iberian Peninsula starting in 711.

The name Moor derives from Mauri, a North African Berber people. Europeans use the term to refer to anyone from North Africa, Berber or Arab.

The Moors crush the Visigoths. Within five years, they control all but a tiny region in the northern part of the peninsula. Asturias, in present-day Spain, remains under Christian control. The Iberian Peninsula becomes a province of the *Caliphate* (Islamic state) of Damascus.

A.D.756   Peninsula becomes independent Muslim kingdom

In 756, at the battle of Al Musara, Abd al Rahman defeats the provincial governor and founds the independent kingdom of Al Andalus, with its capital city at Córdoba (present-day Spain).

For the next 250 years, Al Andalus flourishes into one of the wealthiest and most powerful kingdoms in Europe. The caliphs encourage cultural growth, open schools and libraries, and support the study of the sciences, with an emphasis in mathematics. On the western edge of the peninsula, the Muslims add more than 600 Arabic words to the Portuguese language.

Muslims introduce architecture, develop industries, and import irrigation systems that transform many parts of the arid south into orchards. In present-day Portugal, Muslims divide the land and begin to grow crops. Yet Muslims who migrate to northern Portugal do not stay there long. Unaccustomed to the cooler and wetter climate, they return to the more familiar, drier south, where they leave a long lasting legacy. Many Christians find refuge in the north, where they are protected from Moorish excursions by the mountainous terrain.

The golden age of Al Andalus comes to an end in the eleventh century, when local nobles begin to carve the caliphate into independent city-states. As the caliphate breaks down, Christians go on the offensive, beginning the reconquest of the Iberian Peninsula.

Christian Reconquest of Iberian Peninsula and the Formation of Portugal

1000–1139   Christian Reconquest of Portugal

The Christian struggle to recover the Iberian Peninsula from the Moors dates to 722, when a Visigothic nobleman, Pelayo, defeats a small Muslim contingent at Covadonga in the northwestern edge of the Peninsula.

Pelayo emerges as King of Astúrias (later León in present-day Spain) following the minor skirmish. The push to dislodge the Muslims from Portugal begins in earnest in the eleventh century, when the kingdom of Astúrias-León becomes a major player in Portuguese developments.

From this northwestern corner of the Peninsula, Visigothic kings who rule after Pelayo's death retake several towns in present-day northern Portugal, including Braga and Oporto. This area remains a buffer zone between the Christian and Moors for more than two hundred years. The territory becomes known as *Portucale,* or Portugal in the vernacular. The name develops from two Roman settlements, Portus and Cale, across from each other on the Douro River. In time, Portucale becomes a small province of the kingdom of León. Rugged mountains stand between the kingdom and the province. Its isolation forces the king to place the region under the control of his counts.

1109   King's death leads to war and puts Portugal on road to independence

In 1096, the crusader-knight Henry is given a hereditary title to Portucale and Coimbra as a dowry for marrying Teresa, the illegitimate daughter of Alfonso VI, king of León. In his subordinate position, Henry remains loyal to the king until his death in 1109.

Civil war follows the death of the king, as Aragonese, Galician, and Castilian (territories in present-day Spain) barons fight for the crown. Henry remains neutral, and so does his wife after his death in 1112.

1127–28   Queen Teresa loses power struggle

Alfonso VII emerges as the winner in the civil war and attempts to bring Portucale back into his realm. But Teresa, now accustomed to her independence, refuses to become a vassal of the king. She is forced into submission after a six-week war in 1127.

Her son, Afonso Henriques (1105–85), supported by barons and other nobles, continues the fight. He rebels against his mother, defeats her allies, and exiles her to Galicia in 1128.

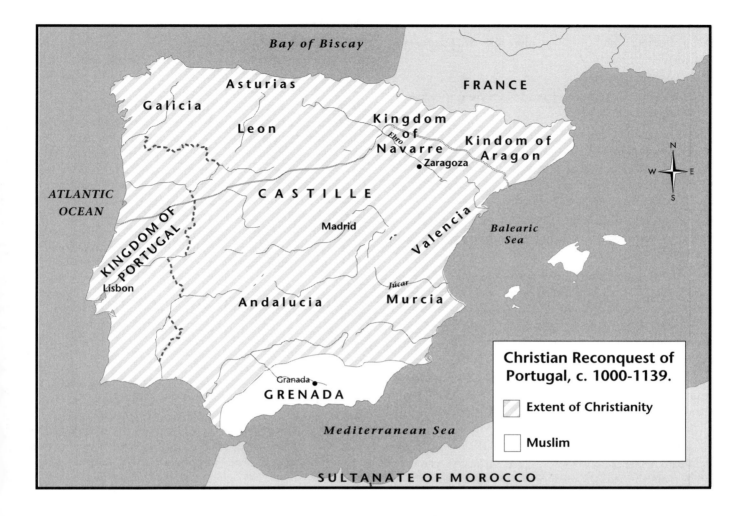

Christian Reconquest of Portugal, c. 1000-1139.

Extent of Christianity

Muslim

1139–43   Afonso Henriques declares himself king of Portugal

Afonso Henriques, the so-called founder of what is now Portugal, resists pressures to rejoin Alfonso VII's kingdom and goes to battle to secure independence. In 1139, he begins to go by the title of king. In 1143, he signs a peace treaty with Alfonso VII, who recognizes him as king of Portugal (and first monarch of the House of Burgundy). Later, the pope, as reward for his successes against the Moors, officially declares Afonso Henriques king of Portugal.

Henriques pushes his troops south, conquering more Muslim territory than any other Christian king in the Iberian Peninsula. With the help of English, Flemish, and German Crusaders on their way to Palestine, he takes Lisbon in 1147. In his later years, he tightens his grip on power and consolidates his kingdom.

His heirs continue to fight the Moors. A victory against a small Muslim contingent in 1249 seals Portugal's southern border at the Atlantic.

1279–1325   Portuguese becomes national language under King Dinis

With the consolidation of Portugal and the defeat of the Moors completed by earlier kings, King Dinis turns his attention to the economy and culture. His kingdom develops policies to boost agriculture and the economy, encourages maritime trade, and negotiates a trade agreement with the English monarchy in 1303.

Dinis (r. 1279–1325) founds one of the world's oldest and most prestigious universities at Coimbra in 1290 and decides that all official documents are to be written in the Portuguese language, instead of Latin.

Spoken Latin, brought by the Romans to the Iberian Peninsula, slowly evolves into Portuguese, Castilian, French and Galician. Later, invading Germanic tribes and the Arabic language of the conquering Moors influence and modify Portuguese.

By the middle of the sixteenth century, when the kingdom establishes an overseas empire in Asia, Africa, and the Americas, Portuguese grammar is well established.

By the end of the twentieth century, Portuguese is the official language of seven nations—Portugal, Brazil, Angola,

Cape Verde, Guinea-Bissau, Mozambique, São Tomé and Príncipe islands—and spoken by more than two hundred million people.

1325–1411 Portugal struggles for sovereignty against Spanish kings

Starting with the reign of Afonso IV (r. 1325–57), wars of succession in neighboring Castile threaten Portugal's sovereignty. The difficult relationship is complicated by the many marriages between members of the House of Burgundy and members of various Spanish kingdoms.

1355 A victim of the succession wars becomes literary figure

One of the most tragic tales that comes out of the succession wars belongs to Inês de Castro (?–1355), a Galician lady of the court who secretly starts a relationship with the Infante Pedro, son of King Afonso IV. Reasons of state force Pedro to marry Constanza of Castile. After her death, de Castro begins to live with him in 1345.

A group of noblemen in Afonso's court fear de Castro's family will meddle in succession affairs if she is allowed to maintain her relationship with Pedro. They convince Afonso to order her murder.

Afonso dies in 1357, two years after de Castro's murder, and Pedro becomes king. He takes revenge on de Castro's two killers by having their hearts torn out and declares that he has been married to de Castro. Pedro has her body exhumed and crowned. He forces his court to kneel in front of her and kiss her decomposed hand before laying her to rest.

Their story of romance and death is retold frequently in the literature of many nations.

1385: August 14 The Battle of Aljubarrota

Pedro rules for ten years. At his death, his son Fernando I takes the throne. Succession rivalries reach a climax in 1383, when Portugal's direct line of descent ends with Fernando's death. Pedro's illegitimate son, João, claims the throne. Juan of Castile opposes the claim and goes to war.

On August 14, on a small plateau near Aljubarrota in central Portugal, 7,000 Portuguese, with the aid of 800 English longbowmen, gather under the command of Nuno Álvares Pereira to defend Dom João's claim to the throne. In a quick battle, they defeat 30,000 Castilian troops, ensuring the independence of Portugal for the next 200 years.

The battle brings an end to the House of Burgundy. João I (r. 1385–1433) becomes the first king in the House of Avis. Hostilities continue between Castile and the House of Avis until a peace treaty is reached in 1411.

1386: May 9 Portugal signs treaty of perpetual alliance with England

During the fourteenth century, the Portuguese foster strong relations with England. King Dinis signs a maritime agreement with the British in 1303. Later, after England helps Portugal in its war against Castile, King João I and Richard II of England sign the treaty of Windsor. It remains in effect into the twentieth century, making it the oldest alliance between two nations.

A year later, João I further strengths relations with England by marrying Philippa of Lancaster, the daughter of John of Gaunt (who at one time has claims to the throne of Castile). Philippa introduces English customs to the Portuguese court and fosters English commercial interests in her adopted country. João I and Philippa help revive the fortunes of Portugal under the House of Avis.

Voyages of Discovery: Portugal's Golden Age

1415 Portuguese troops strike in Muslim territory

With its independence assured, Portugal begins to expand its territory beyond the Iberian Peninsula. In 1415, the Portuguese capture Ceuta in Morocco; in 1419, they take the archipelago of Madeira; by 1427, they control the archipelago of the Azores; and in 1434, they round Cape Bojador and explore the mouth of the Gambia River (present-day Senegal).

1418 Keeper of royal archives chronicles royal family's fortunes

In 1418, Fernão Lopes (b. 1380) becomes the official keeper of the Royal archives. Lopes chronicles the history of Portugal with a direct and vigorous style. The royals are treated reverentially, but Lopes also spends considerable space chronicling the lives of ordinary people. His accounts of the years 1383–85 are considered among the masterpieces of medieval literature.

1437 Prince Henry the Navigator allegedly opens navigation school

Early historians give Prince Henry (1394–1460), the third son of João I and Philippa, much of the credit for Portugal's voyages of discovery. But some contemporary historians have re-evaluated his contributions.

Prince Henry is said to open a maritime school in 1437 at Sagres, where he trains Portuguese captains and plans voyages until his death. But the existence of such a school remains questionable.

While Henry crosses the Straits of Gibraltar in 1415 to fight the Moors, he never sails. He leaves that to his subordinates, financing expeditions along the African coast and monopolizing all trade south of Cape Bojador (Western Sahara). Henry, Verlinden argues, is a businessman who grows rich with his monopolies. Henry does not discover the Atlantic islands of Madeiras and the Azores, but he brings them into the Portuguese fold and fosters early colonization. One of Henry's explorers, Nuno Tristão, brings with him the first slaves from Africa in 1441. A plaque in the town of Lagos marks the site of the first modern slave market in Europe.

1481–95   João II's reign and voyages of exploration

With an eye on profits, João II begins to fund maritime voyages during his reign, laying the foundations of the Portuguese Empire. The maritime expeditions are driven by many factors, including commercial gain and traditional Christian hostility against Muslims.

Portugal expands its maritime activities in the Atlantic when pirates and Turks threaten commercial shipping in the Mediterranean. During this time, valuable spices from Asia make their way to the Levant over several overland routes. The cargoes are loaded aboard Genoese and Venetian ships and taken to Europe. The Turks close the overland routes and heavily tax the spices. The Europeans begin to look for other routes to Asia in order to bypass the troubled areas.

In time, the Portuguese believe it is possible to reach Asia by traveling around Africa, then unknown territory to the Europeans. With royal patronage, Portuguese sailors explore the mouth of the Congo (now Zaire) River and the coast of Angola. By 1487, Bartolomeu Dias rounds the Cape of Good Hope, and ten years later, Vasco da Gama sails up the east coast of Africa to Mombasa. From there, Arab pilots guide him to India.

In 1500, a flotilla of thirteen *caravels* (small ships) under the command of Pedro Álvares Cabral is blown off course and reaches the coast of Brazil. By 1503, da Gama is building several coastal fortresses between Mozambique and India. By 1512, the Portuguese reach the Moluccas (Spice Islands), and a year later, they open trading posts at Macao and Canton in China. By 1543, they reach Japan.

1495–1521   Reign of Manuel I, The Fortunate

Pepper, cloves, nutmeg, and cinnamon are considered exotic and expensive spices. They turn Portugal into a mercantile superpower, rivaled only by neighboring Spain.

Dubbed "The Fortunate" king, Manuel I (1469–1521) heads the richest monarchy in Europe thanks to the monopoly on spices and luxury goods from Asia. Yet, most of the profits end up in the hands of the monarchy, the nobility, and the church. Lavish cathedrals and palaces are built during this era. Lisbon, the port of departure for the voyages, disproportionately receives more at the expense of the rest of the country. The wealth does not transform Portugal's basic social structure and does little to create a middle class. The nation remains tightly controlled by the military aristocracy, the crown, and the church.

1515–21   Torre de Belém, a jewel of Manueline architecture

With so much wealth generated by trade, Portugal embarks on lavish and extremely elaborate construction projects, giving rise to a style known as *Manueline* architecture.

Named after Manuel I, Manueline architecture is a Portuguese variant of Late Gothic and its inspiration comes from Portugal's Golden Age of Discovery. It is typified by elaborate decoration on all surfaces of a building, including windows and doors. Its most striking characteristic is the use of maritime detail, including twisted ropes and braided *colonnettes*, swaths of seaweed, coral-encrusted masts and anchor chains. Important details include the *armillary* (arm bracelet) spheres, a navigational device, and the Cross of The Order of Christ, an emblem of a military order that finances some of the early voyages.

There are many fine examples of Manueline architecture throughout Portugal. One of them is the *Torre de Belém*, commissioned by Manuel I and built between 1515 and 1521. The tower is on the Tagus, the starting point for the voyages of discovery. The Portal of the church *Conceição Velha* in Lisbon, also commissioned by Manuel, is decorated with flowers, angels, beasts, *armillary* spheres, and a figure of Manuel.

1540   Portuguese Inquisition claims its first victim

To counter the Protestant Reformation in the first decades of the sixteenth century, the Roman Catholic Church establishes the Inquisition, a formal tribunal in charge of suppressing heresy. Portugal establishes its first Court of Inquisition in 1536.

The Portuguese Inquisition focuses its attention on the New Christians, Jews who convert to Christianity after Manuel I orders all Jews, including those who are forced out of Spain in 1492 and seek refuge in Portugal, to accept Christianity or leave the kingdom in 1497. The Portuguese believe the Jews have only converted in name and continue to practice Judaism in their homes. The first public burning of a heretic takes place in Lisbon in 1540. In the next 150 years, more than 1,400 people are executed.

1572   Epic poem a treasure of Portuguese language

The publication of Luis de Camões' *Os Lusíadas* (The lusiads) in 1572 marks the coming of age for the Portuguese language and the country's literary tradition.

The epic recounts Vasco Da Gama's voyage to India and events and legends from Portuguese history. The reckless and passionate Camões (1524–80) has been described as a soldier and adventurer, a widely read humanist scholar and a well-traveled explorer. Banished from court for his behavior, he enlists in 1547 and sets sail for North Africa, where he loses an eye. He is imprisoned after another fight and agrees to serve Portugal in India.

His experience—in a fleet of ships, his vessel is the only one that survives the stormy seas and reaches India—ignites his poetry. Yet he dies in obscurity after his epic is published. Later, Portugal enshrines his memory (National Day or Camões day is celebrated each year on June 10) and recognizes him as its greatest writer.

1578 The Battle of the Three Kings

Events in Morocco lead to the fall of the House of Avis. Abdel Malik deposes his nephew, Sa'adi king Mohammed Al-Mutawwakil, because he contravenes the family rule that the eldest male in the family should succeed, not the eldest son. Al-Mutawwakil escapes and asks King Philip II of Spain to assist him regain power. He declines, but Portuguese King Sebastião, seeing an opportunity to reestablish rule in Morocco, agrees to help.

Sebastião assumes the throne at age fourteen. He has been described as sickly, mentally unstable and poorly educated, and obsessed with conquering the Muslims of North Africa. He lands in Morocco with an army of 20,000 men, where Abdel Malik waits with 50,000 troops. At day's end, the Portuguese troops are crushed, Sebastião and Al-Mutawwakil have perished in battle, and Malik dies shortly after. A young Sa'adian prince, Ahmad al-Mansur, emerges as the sole ruler of Morocco.

The crushing defeat brings Portugal's monarchy to its knees and drains its coffers. Spain, which has been waiting for 200 years to reassert its authority over Portugal, prepares for an invasion.

1580–1640 Spanish occupation

A large Spanish army invades Portugal and marches on to Lisbon in 1580. Spain annexes Portugal, and Philip II is declared Philip I of Portugal (under the new dynasty of the House of Austria). Philip decides to give Portugal some degree of autonomy. He names Portuguese to a six-member council to rule Portugal and keeps many of its institutions untouched.

The Portuguese royal court has been using the Spanish language and etiquette since the fifteenth century, and the nobility readily accepts Spanish rule. The same is not true outside the privileged circles.

The seemingly harmonious relationship soon deteriorates, as Portugal's autonomy begins to erode under Spanish occupation. By the 1630s, rebellions break out in response to higher taxes and orders to recruit Portuguese to aid Spain in its battle against France. By 1637, the nobility and the lower classes rally behind João, duke of Bragança, and proclaim him king.

1640: December 1   Restoration of Portuguese rule

João's supporters enter the royal palace and arrest Portugal's Spanish governor, the duchess of Mantua, a cousin of the king of Spain. Five days later, João IV (r. 1641–56) becomes the first king in the new ruling dynasty of the House of Bragança.

João presides over a much smaller empire. During the Spanish occupation, Portugal is dragged into Spain's wars with England and Holland, losing its trade monopoly in the Far East and India. The shrinking empire is sustained for the next 180 years by Brazil's wealth (See Brazil section for history of Portuguese colonialism).

1703 Treaty with English boosts Portugal's wine industry

In 1703, Portugal and England strengthen their commercial relations by signing the Methuen Treaty, giving Portugal preferential duty on wine export to England. In return, Portugal removes restrictions on the importation of English-made goods. In time, the treaty proves more advantageous to England, which dominates the Portuguese economy.

The English develop a fondness for Port, a sweet wine that dates from the seventeenth century. Early British merchants "discover" port when they begin to treat wine from the Douro region with brandy to keep it from turning sour. They soon find that a stronger and sweeter wine makes for better flavor. Over the years, producers refine the classic after-dinner drink, and Portugal becomes known around the world for its port. Officially port only comes from a small region in the upper Douro valley, east of Oporto. Throughout its history, English and Scottish firms have controlled trade. For Portugal, port is important to its economy and culture.

One of the celebrated figures of port is Joseph James Forrester (1809–61), the Port Baron. In 1831, the Englishman travels to Oporto to join his uncle's wine company and ends up reforming the industry. He battles bad port with his 1844 treatise, *A Word or Two on Port,* which criticizes shippers who adulterate the wine. Pedro V bestows on him the title of baron in 1855 for his contributions. In 1862, he drowns when his boat capsizes in the Douro, and he is dragged to the bottom by a heavy money belt.

1732 Master tile maker dies

António de Oliveira Bernardes (1660–1732) is considered one of the greatest masters of painted ceramic tiles, or *azulejos,* as they are known in Portugal. Oliveira produces the bulk of his work during the zenith of tile making in Portugal. By the eighteenth century, Portugal is a leading producer of *azulejos* in Europe.

The Moors introduce ceramic tiles to the Iberian Peninsula. The decorative tiles are used to cover walls, floors and ceilings. By the sixteenth century, Portugal begins to produce its own style. In time, *azulejos* become important decoration for the interior and exterior architecture of Portuguese buildings.

Oliveira's best-known tiles are at the Misericórdia Church in Évora. Here, he uses blue and white story-telling *azulejos* that are remarkable for their detail. Story-telling tiles begin to appear around 1690. Before their introduction, most tiles are geometrical in design, a Moorish legacy. Carpet tiles, so called because they imitate the patterns of Moorish rugs, are popular during the seventeenth century. So are tiles that incorporate Hindu motifs. The inspiration for the design comes from printed calicoes and chintzes imported from India. Spain also influences tile design in Portugal. Most twentieth century Portuguese artists have worked with tiles, which remain an important part of the country's artistic heritage. A convent built in 1509 in Lisbon now houses the National Tile Museum.

1726: September   The *Santa Rosa* sinks off the coast of Brazil

The *Santa Rosa,* the mightiest ship in Portugal's fleet, sinks, killing 700 men, women and children. Only seven survive.

The *Santa Rosa* sets sail for Portugal from the Brazilian port of Salvador in late August. On September 6, as the ship passes Recife, the gunpowder in the hold blows up. In the 1990s, a group of corporate treasure hunters attempts to find the *Santa Rosa,* which is believed to have carried 5 tons of gold, valued at more than $500 million, in its hold.

1755: November 1   Earthquake shatters Lisbon

Shortly after 9:30 a.m. on All Saints' Day, with crowded churches to celebrate the holiday, an earthquake reduces over half of Lisbon to rubble. More than twenty churches collapse, crushing the devotees who have gathered for early mass. Fires follow a third shock. Huge waves roll in from the Tagus and flood the lower part of the city an hour later. In Lisbon alone, more than 15,000 people lose their lives.

Philosophically, the earthquake is devastating as well, and has a profound effect on European thought. People wonder if the earthquake is a natural phenomenon or divine wrath. Pre-earthquake Lisbon is a marvelous city, decadently wealthy, with a well-earned reputation for its part in the horrible Inquisition executions. Preachers interpret the earthquake as warning of more punishment to come. Even French author Voltaire, who writes a small poem about the disaster, argues that evil is real, and man is powerless, doomed to an unhappy life.

In Lisbon, recovery is more important than philosophizing. Reconstruction falls on Sebastião José de Carvalho e Melo, later Mârques de Pombal (1699–1782). As prime minister to the indolent José I, Pombal restores order and prepares a plan to rebuild Lisbon. Within a month, with the aid of architect Eugénio dos Santos and the engineer Manuel da Maia, Pombal devises a modern grid of parallel streets and reconstruction gets under way.

José I doesn't like to bother with matters of the state and allows Pombal a free hand in running his government. Pombal gains enormous respect for rebuilding Lisbon and sets out to Portugal. Virtually operating as a dictator, he attempts to reform the economy and modernize the country. He arrests the nobles who oppose him, tortures and executes many of them, and he exiles the Jesuits. While some changes are seen as positive, his despotism is finally curbed when José I dies. His queen, Maria I (r. 1777–1816) banishes Pombal to his estate and dismantles his government.

1807   Portuguese royalty seeks refuge in Brazil

During the Napoleonic wars, Portugal remains faithful to England despite French pressures to end the relationship. Portugal becomes a base of operations for the British in the Iberian Peninsula. In 1807, Napoleon sends General Junot (1771–1813) and his troops to invade Portugal. He only does so, Napoleon claims, to liberate the Portuguese from British economic domination. Maria I and her son, the future João VI, sail to Brazil, and establish a government in exile.

Through 1810, French forces invade Portugal three times, and each time they are defeated by British or British-Portuguese forces. They are finally thrown out of the country in 1813.

1809   Huge walls are built to save Lisbon

To save Lisbon from a Napoleonic invasion, Arthur Wellesley (1769–1852) orders the construction of defensive lines to protect the city. Two walls, known as the Lines of Torres Vedras, are built, with more than six hundred guns and 152 masonry fortresses stretching twenty-nine miles from the sea to the Tagus River. A second wall runs behind the first for twenty-four miles. Within a year, workers move tons of dirt, dam many rivers, and demolish homes and farms.

On October 14, 1810, the defensive walls surprise General Massena (1758–1817) and his 65,000 troops. They soon realize they are impregnable. By November, they fall back and, in 1811, after suffering heavy losses, retreat to Spain.

Wellesley, born in Dublin, later becomes the Duke of Wellington. He serves England in India before being sent to Portugal with an expeditionary force.

1816   Maria I dies in exile

With the death of Maria I, who is insane for the last twenty-four years of her life, her son João VI (r. 1816–26) becomes king. João is popular with Brazilians and does not immediately return to Portugal, where liberals, inspired by the French and U.S. Revolutions, demand an end to the monarchy.

1820–22   Revolt leads to constitutional monarchy

With an absent king, regiments in Oporto revolt in 1820, and the military establishes a provisional government until a new constitution is written.

The constitution of 1822 establishes a hereditary monarchy with limited powers. It creates a strong central government with executive, legislative, and judicial branches. Members of the unicameral Chamber of Deputies are directly elected by literate males. The king has no representation in the chamber and no power to dissolve it. João accepts his greatly diminished powers and returns to Portugal in 1822.

His son, Pedro IV, stays in Brazil. With British aid, he declares Brazil's independence and becomes emperor in 1822.

1822–34   Constitutional crisis leads to civil war

The new constitution is challenged immediately by the nobility and the clergy. They believe their king has been humiliated and demand a return to absolutism, which vests all power in the monarch and his advisers. João's young son, Miguel, refuses to accept the constitution and joins rebels calling for a return to monarchy in 1824. His father exiles him to France after the movement fails.

The death of João in 1826 complicates royal succession. Pedro IV, now emperor of Brazil, is the rightful heir, but Brazilians and Portuguese have no interest in a unified monarchy. Pedro abdicates in favor of his daughter, Maria da Gloria, who is seven, with the condition that she will marry his brother Miguel.

Pedro gives Portugal a new constitution (known as the Constitutional Charter) and returns to Brazil, leaving the throne to Maria, later Maria II (1834–54), and his brother Miguel as regent. The new constitution creates four branches of government, giving both liberals and absolutists a role. But Miguel, who returns from exile in 1828, does not follow his brother's wishes. Backed by supporters of absolutism, Miguel claims the throne and nullifies the charter.

1831   Deposed emperor returns to fight his brother

Revolts against Miguel break out throughout Portugal. In the next six years, thousands of liberals are killed, imprisoned or forced to leave the country. In the meantime, Pedro IV, who is deposed in Brazil in 1831, returns to Portugal to lead the liberals against his brother.

It is a complex struggle with many fronts and many factions. Some Portuguese want a complete return to the monarchies of the past; others want a king with limited power; and yet a third group wants wide democratic reforms with an elected parliament.

By July, 1833, Pedro IV controls Oporto, and his troops march toward Lisbon. Miguel's troops are finally defeated in May, 1834. He is exiled, and Pedro restores the constitutional charter.

Portugal's constitutional crisis does not end with Miguel's exile. For the next several decades, Portuguese continue to fight over how best to rule the nation. Gradually, minor democratic reforms are approved, and more people are allowed to participate in government. Yet by the end of the 1800s, only one percent of the population is eligible to vote.

1836   One of Portugal's greatest musicians leaves long-lasting legacy

One of the greatest singers of *fado*, Maria Severa (1810–36), dies after a short and prodigious career, but her legacy lives on.

Fado, like the blues, expresses longing and sorrow. It has its roots in indigenous music, with Moorish and African influences. The music is said to express the unique Portuguese concept known as *saudade*, which expresses melancholy and nostalgia for what has been lost and for what has never been attained.

The *fado* (which roughly translates to fate) is an expression of Portuguese society, but it is most widely practiced in Lisbon. Severa is considered the first great *fadista*. She dies young, after what is considered at the time a scandalous life. Her life inspires poems, novels, plays, and *fado*, as well as Portugal's first sound movie in 1931. All female *fadistas* today, including the renowned Argentina Santos and Amália Rodrigues, pay tribute to Severa by wearing a black shawl in her memory during a performance.

Fado is sung by women and men, often dressed entirely in black, and accompanied by two guitarists. One of the instruments is the viola, and the other is a twelve-string *guitarra*, an instrument peculiar to Portugal. The *guitarra* evolves from a simple nineteenth century design and later becomes a finely decorated instrument, heavily inlaid with mother-of-pearl. Its sound is an essential ingredient of a good *fado*.

The music, in its authentic form, is only played at *fado* houses, where listeners can get drinks and a meal. Most modern *fado* houses are owned by *fadistas*.

1884–85   European conference slices up African territory

In the second half of the nineteenth century, Portugal turns its attention to Africa, especially to the Congo region. Portugal has long been established on the shorelines of Guinea, Angola, and Mozambique but does not colonize the interior, today the countries of Zambia, Malawi, Zimbabwe, and Zaire.

Britain, France, and Germany, three countries with immense colonial ambitions, challenge Portugal's historical claims to vast areas of Africa. Portugal proposes a conference to settle the claims and meets with the other European nations in Berlin in 1884–85. The conference awards the Congo to the king of Belgium and declares that a claim to African territory is only valid when a nation can demonstrate "effective occupation," not simply historical claims. The Berlin Conference awards Mozambique, Angola, and Guinea to Portugal.

In 1886, Portugal signs two treaties marking the boundaries between Portuguese territories and those of France and Germany. Those two nations recognize Portuguese sovereignty in the interior territories between Mozambique and Angola.

1890: January 11 England threatens war over disputed African territory

Shortly after its border agreements with France and Germany, Portugal publishes a so-called pink map claiming the lands between Mozambique and Angola.

The map ignites a crisis with Britain, which claims the same area for itself. The British want to build a railroad line through central Africa to connect Cape Town (in South Africa) and Cairo. The British government demands the immediate withdrawal of Portuguese forces from the disputed area. Portugal, unprepared to wage war against a more powerful nation, gives in to British demands.

The Portuguese feel humiliated by the foreign policy defeat and focus their anger on the monarchy. Sentiments grow against the monarchy and nobility, institutions that are now seen as a danger to the country. The Portuguese begin to call for a republic to replace the monarchy. Republican sentiments grow with calls for democratic reforms, universal suffrage, the separation of church and state, and the abolition of the monarchy and nobility.

1907: May–1910 Monarchy attempts to hold on

By May, 1907, the political situation in Portugal reaches a crisis. King Carlos (1889–1908) dissolves Parliament and gives Prime Minister João Franco the power to govern by decree.

The military, supported by the Republicans, attempts a coup in January, 1908. In February, two Republicans assassinate the king and his youngest son. Manuel II, who is only 18, becomes the new king.

Prime Minister Franco steps down as a last effort to save the monarchy and calls for new elections. But attempts to form a stable government fail six times. On October 3, 1910, the army refuses to step in when the crews of two Portuguese ships mutiny. One of the ships begins to shell the royal palace, forcing Manuel II and the royal family to seek refuge in Britain. On October 5, a provisional republican government takes office.

1911: August 21 New constitution abolishes monarchy

The new constitution abolishes the monarchy and inaugurates Portugal's first republican government. The document calls for the separation of church and state, prohibits religious instruction in the public schools, and prohibits the military from taking part in religious observances. Workers are allowed to strike. The early years of the new Republic are marked by political and economic crisis.

Forty-five different governments rule in the first 15 years of the republic. During that time, eight presidential and seven congressional elections are held. Petty arguments and the clash of personalities sometimes spark a government crisis. A greater danger to the young republic, however, comes from those who want to bring back the monarchy.

1916 Portugal enters World War I (1914–18)

In 1916, in the midst of World War I, the Portuguese government seizes seventy German ships anchored in Portuguese harbors. At the time, the Portuguese fear that a German victory would lead to the loss of Portugal's colonies of Mozambique and Angola. A force of 40,000 men is sent to Flanders (Belgium) to fight alongside the Allies. But the poorly trained and equipped Portuguese suffer devastating casualties.

1917: May 13 Timely apparition

Three peasant children, ten-year-old Lucia Santos and her young cousins, Francisco and Jacinta Marta, claim to see the Virgin Mary in a holm oak tree in Fátima, central Portugal. The apparition energizes Christian faith and the Portuguese Catholic Church, which is increasingly under attack by the republican government.

According to the children, the virgin orders them to return to the tree on the same day for the next six months. By October, more than 70,000 pilgrims accompany the children. The Virgin, the child claims, speaks of peace—as World War I rages on—and Russia, although there's no explanation on that issue. A prophesy made by the Virgin is never divulged. By the end of the twentieth century, the Sanctuary of Fátima is a massive devotional shrine and one of the most important pilgrimage destinations in Europe.

1926 Military assumes reigns of power

The political crisis spins the nation into serious conflict. Many Portuguese lose faith in republicanism. The frequent strikes and changes of government and the political assassinations, including the murder of President Sidónio Pais in 1918, have Portuguese clamoring for a return to the monarchy or a more authoritarian government. In 1926, a coup puts the military in charge.

1933 The New State

The year 1933 marks the beginning of a new period for Portugal. The *Estado Novo* (New State) will reshape Portuguese society, newly appointed prime minister António de Oliveira Salazar says. Backed by the army and police, Salazar rules dictatorially until 1968.

Salazar (1889–1970) comes from a poor and very religious family. He is a good student and considers the priesthood before deciding to attend the University of Coimbra, where he studies law and economics.

Salazar is mildly in favor of a monarchy and stands out for his opposition to the republicans who spouse anticlerical views. He is briefly suspended from teaching at the university, where he is investigated for propagating monarchist ideas.

Thousands of pilgrims gather each year at the shrine commemorating the apparition of the Virgin Mary believed to have come to three shepherd children in Fátima. (EPD Photos/CSU Archives)

His teachings on political economy at Coimbra and attempts at political office under the banner of the Catholic Church attract notice. In 1928, military leaders offer him the post of minister of finance. Salazar quickly brings the national budget under control and manages a small surplus.

By 1932, Salazar wins many important people to his side with his anti-Communist and anti-liberal views. The church considers him an ally; and monarchists believe he will in time return the nation to monarchical rule. Others are impressed by his authoritarian and hierarchical views. On July 5, 1932, he is appointed prime minister.

1933   Salazar shapes a new constitution

Nationalism and religion are two of the main features in Salazar's life. He is heavily influenced by the writings of Saint Thomas Aquinas (1225–74) and contemporary church leaders. Aquinas sees the world organized hierarchically according to natural law and God's just ordering of the universe. According to natural law, the social order is fixed and immutable, and inequalities among men are simply part of the

natural order. A fisherman should be happy being a fisherman and not aspire to more, for example.

The 1933 constitution is founded on some of these ideas. The constitution proclaims that Portugal is a corporate and unitary republic. In theory, a corporate state represents interest groups rather than individuals. In essence, an authoritarian and hierarchical state is above the social groups, which are above the individual.

With the constitution in place, Salazar begins to consolidate his power. He censors the press, changes the school curriculum to reflect his government's philosophy, strengthens relations with the church, orders the installation of crucifixes in the classrooms, and establishes the post of teacher of moral education.

1934   St. Antony becomes patron saint of Portugal

Pope Pius XI declares Santo Antonio (St. Antony, 1195–1231) the patron saint of Portugal. St. Antonio joins the Franciscan Order in 1220 and soon becomes known for his devotion to the poor and his deft touch for converting heretics. While he lives most of his life in Portugal, he dies in Padua, Italy, where he is canonized a year after his death. To many, he is known as St. Antony of Padua, the finder of lost property.

1939–45   Portugal's role in World War II

Portugal officially remains neutral during World War II but is forced to sell minerals to Germany when it threatens the country's shipping. In 1942, Salazar meets Spanish dictator Francisco Franco to sign a pact of nonaggression. By 1943, Portugal allows the British and Americans to establish military bases in the Azores.

1949   Portuguese wins Nobel Prize

Ántonio Caetano Egas Moniz (1874–1955) receives the Nobel Prize in physiology for his pioneering work in neurosurgery. He trains at Coimbra University, and in 1911, he holds the chair of Neurology at Lisbon.

1955   Multimillionaire leaves wealthy trust to Portugal

Calouste Sarkis Gulbenkian (1869–1955) leaves a $275 million trust to support cultural activities in Portugal. Born into a wealthy Armenian family in Turkey, he becomes a British citizen in 1902. He acquires his fortune by organizing companies for the exploitation of oil in the Middle East and earning five-percent royalties on all dealings. A resident of Paris, he moves to Lisbon when the Germans invade France during World War II. His trust funds numerous cultural activities. It has its own ballet company, orchestra, libraries, concert halls, and museums.

During his lifetime, Gulbenkian acquires a large and valuable collection of artifacts spanning more than 4,000 years of history, from ancient Egypt to modern art. The Gulbenkian Collection in a small museum in Lisbon is considered one of the finest in the country.

1961 Portugal loses its Indian colonies, fights to keep others

By 1961, Portugal, unlike some of its European allies, attempts to hold on to its colonies against mounting world pressure to give them up. In 1961, India forcibly takes back Goa, Daman, and Diu.

That same year, the Union of Angolan Peoples begins guerrilla warfare against Portugal. In 1963, revolts break out in Guinea, and a year later, guerrilla warfare erupts in Mozambique. With warfare on three fronts, Portugal mobilizes forces on a large scale.

While some high-ranking members of the armed forces believe the country is in no position to fight and even less win a colonial war, Salazar is convinced Portugal must remain in Africa to defend Western and Christian civilization.

Portugal believes its colonial holdings are simply extensions of the nation overseas. In Africa, they claim, they are only carrying a non-racist, Christian mission to civilize the peoples of Mozambique, Guinea, and Angola. Portugal claims the Africans living in the colonies have the same rights as Portuguese citizens. But in practice, an African must abandon traditional ways of living and adopt the Portuguese language, culture, and religion to become a Portuguese citizen.

Even then, Portugal makes it nearly impossible for Africans to become Portuguese citizens, or *assimilados* (assimilated into Portuguese culture). Most Africans remain classified as *indigenas* (natives). Under Portugal's racist policies, the majority of the indigenas are poor, and many of them are subjected to forced labor. By 1974, Portugal commits 140,000 troops to the African colonies.

1968 Salazar's reign ends

In 1968, Salazar is incapacitated by cerebral hemorrhage and is unable to rule. Marcello Caetano (b. 1906), a close associate of Salazar, becomes the new prime minister. Caetano announces he will seek "evolution within continuity," a clear signal that he will not deviate too far from Salazar's policies.

Caetano improves living conditions for the working poor and attempts to restructure the economy. Supporters of the New State countered many of his proposed social changes, launching a new political crisis.

1974: April 25 The Carnation Revolution ends dictatorship

Ultimately, the colonial wars bring an end to the New State. By 1973, young army officers openly begin to challenge Caetano's policies. On February 22, 1974, Gen. Antonio Sebastião Ribeiro de Spínola publishes a book, *Portugal e o Futuro* (Portugal and the Future). The book advocates a political rather than a military solution to the African wars.

By 1974, Portugal finds itself increasingly isolated, as the United Nations continues to support independence for the African colonies, and some of the world powers voice political support for the guerrilla movements. The war demoralizes Portugal's armed forces, which face an increasingly powerful guerrilla movement.

Young army officers, collectively known as the Armed Forces Movement, stage a coup on April 25, 1974, to stop the bloodshed in the colonies. The nearly bloodless coup comes to be known as the Carnation Revolution because citizens reward the rebellious soldiers with red carnations.

Spínola emerges as the new leader of a provisional government. He announces the abolition of the secret police and the remaining institutions created by Salazar's regime. He releases all political prisoners, abolishes censorship, and lifts restrictions against political parties and labor unions. Yet the euphoria is short-lived. Soon after the coup, the nation breaks into near anarchy when Spínola is forced to resign. Left-and right-wing groups fight for control of the nation.

1975 Crisis reaches boiling point

In March 1975, Spínola and many officers flee the country after a right-wing coup attempt. The attempted coup leads to a leftist counter movement. Banks, businesses, and farms are nationalized. But attempts to turn Portugal into a socialist state are met with resistance in the more conservative north.

Several provisional governments fail to bring stability. Adding to the crisis is the return of 800,000 Portuguese settlers from the former colonies.

The military announces a new constitution in April. The document defines Portugal as a republic "engaged in the formation of a classless society." It creates the Military Council of the Revolution, which is supposed to protect democracy. The council, along with the president, have the power to veto any legislation that threatens revolutionary actions.

The document provides for a strong, popularly elected president with the power to appoint a prime minister and cabinet. The constitution is substantially revised in 1982 and in 1989, giving more power to the president and eliminating the Military Council of the Revolution. Portugal is ruled by a parliamentary system.

1984: August 5 Portuguese runners earn Olympic medals

Portuguese runner Rosa Mota joins fifty competitors from twenty-eight nations in the inaugural women's marathon in Olympic competition in Los Angeles. While Mota wins the marathon in the European Championships in 1982, she is not considered a favorite to win a gold medal. Yet Mota proves her ability by taking the bronze medal, the first Olympic medal for a Portuguese woman, in a personal best of two hours, 26 minutes and 57 seconds. American Joan Benoit wins the gold, and Norwegian Grete Waitz finishes in second place. The marathon is a breakthrough for women's competition. Until the 1980s, there is no long distance races for women in the Olympics.

Four years later, Mota stands on the starting line of the Seoul Olympics focused on winning the race. She almost does not compete. While considered popular among her marathoning peers for her graciousness and good-natured manners, she is not well liked by Portuguese running officials. She is too independent and defiant, and they consider leaving her out of the Olympics. Mota goes on to win a gold medal in a time of 2:25:40.

In the 1984 men's Olympic Marathon, Portugal's Carlos Lopez is not considered a favorite despite an splendid running resume that includes a silver medal in the 10,000 meters at the Montreal Olympics in 1976 and two world cross-country titles. At 37, he is considered too old as well. But Lopez overcomes his age, heat and smog of Los Angeles to win the gold medal in a new Olympic record of 2:09:21. Later, he sets a new a world record of 2:07:12 in the marathon before he retires.

1986   Portugal's first elected president in sixty years

In the 1986 elections, four candidates run for president, but none is able to win a majority. In the runoff election, former prime minister Mário Soares (b. 1926) becomes the first civilian president in sixty year. Soares, a socialist, is elected to a second five-year term in 1991. Portugal also joins the European (Economic) Community in 1986.

The socialists, with the election of President Jorge Sampaio in 1996, continue to dominate national politics. Sampaio announces his administration is ready to meet the challenges of a global economy and will begin to privatize public companies and invest abroad.

1997   Despite democratic changes, police brutality continues in Portugal

The death of five people in the custody of police—with one suspected robber shot in the head and then decapitated—shakes the nation and touches off a national debate on the shortcomings of Portugal's police and court system, Tony Smith writes for the Associated Press (January 3, 1997). Amnesty International, a human-rights group based in London, criticizes Portugal in a 1997 report and asks authorities to ensure full investigations into allegations of police brutality.

1997   Portuguese Americans hold on to heritage with mock bullfights

Portuguese Americans in the San Joaquin Valley in California remember their heritage with bloodless bullfights, reports the Associated Press (August 10, 1997).

Bullfighters use a spear with velcro on its tip and attach it to a patch on the bull's back. Many bullfights are held in California each year, in conjunction with religious festivals. As many as eight bullrings are in the San Joaquin Valley, where many Portuguese dairy farmers emigrated generations ago.

1998: February–June   Portuguese struggle with abortion issues

Portugal's parliament approves a landmark measure to give women greater access to abortion. Portugal has one of the toughest anti-abortion laws in Europe, and more than 16,000 illegal abortions are performed here each year.

The measure is highly unpopular in Portugal, a Roman Catholic nation. Parliament in 1984 allows women with serious health problems to abort and until the sixteenth week if the fetus is not viable. The law also allows women who have been raped to abort until the twelfth week.

On February 4, parliament agrees to legalize abortion through the twelfth week of pregnancy in all cases and through the twenty-second week of "serious illness or deformation." But the law is immediately challenged, and voters are asked to decide the abortion issue through referendum, which asks whether women should be allowed to terminate a pregnancy in the first ten weeks.

But only thirty-two percent of the electorate turn out to vote on June 28—too low to validate the referendum—with 50.9 percent of the votes cast against abortion. The abortion issue is far from over, and the country's politicians continue to debate the issue and the importance of the referendum.

The abortion issue underscores women's rights issues. Until 1969, women are not allowed to obtain a passport without the permission of their husbands or male relatives. Spousal abuse in the home and sexual harassment at work are widespread problems despite laws to protect women.

1998: October   Portuguese writer wins Nobel Prize

Jose Saramago (b. 1922) wins the 1998 Nobel Literature Prize for his many novels and an award of $978,000.

With the publication of *Baltasar and Blimunda* in 1982, Saramago becomes one of the most popular contemporary writers in Portugal, and his novels are published in at least twenty different languages. He is best known for *The Stone Raft,* a novel in which the Iberian Peninsula breaks off from Europe and drifts into the Atlantic.

While often magical, his novels are deeply political and critical of modern society, earning him praise and criticism for his views. The Vatican denounces his prize as a political ploy, showing the church's displeasure for Saramago's Communist ties and his book, *The Gospel According to Jesus Christ.*

Saramago is described as an outspoken nonconformist, a writer with a soft spot for the common man. His frequent commentaries on radio and in newspapers reflect his concern for society. That preoccupation is echoed in his latest novel, *Blindness.* Saramago writes about a blindness epidemic that sweeps through an unnamed city. The first victims are confined in a vacant mental hospital. Civil society breaks down, and anarchy rules, even among the blind.

1999   In death, Pessoa comes alive

Virtually unknown outside his country for many decades, Fernando Pessoa (1888–1935) comes to be known in the 1990s as one of the greatest poets of the twentieth century, more than sixty years after his death.

Pessoa is born in Lisbon and spends most of his adult life in that city. Fluent in English and French, he earns a modest living as a commercial writer and translator. He publishes very little of his work during his lifetime and dies in obscurity. Many of his poems are published in Portugal in the 1940s, and he is quickly recognized as one of the country's most important literary figures.

He is best known for writing under many fictitious names: *heteronyms* or literary alter egos with unique identities and writing styles. Three of his best known alter egos are Alberto Caeiro, Álvaro de Campos, and Ricardo Reis, each with his own poetic voice, ideology and biography.

1999: March   Highway of death

With twenty-nine traffic fatalities per 100,000 residents, Portugal's roads are the most dangerous in Europe. A campaign develops to improve road safety after the well-publicized death of a five-year-old girl in a traffic accident. The country's main highway and link to the rest of Europe is known as the death highway, with more than 300 fatalities in a decade.

Portugal's socialist government adopts a road safety campaign and announces plans to rebuild some of the country's most dangerous roads.

1999: April   Portuguese radio station to broadcast from U.S.

The University of Massachusetts-Darmouth announces it will become the North American headquarters for a major Portuguese public broadcasting network. In an agreement with the university, *Radiolevisao Portuguesa*, the equivalent of the Public Broadcast System, commits to set up a twenty-four-hour international television station that will broadcast programs on current events, music, sports and culture.

1999: April 24   Portugal mediates in East Timor peace agreement

The foreign ministers of Indonesia and Portugal agree to hold a vote in East Timor that could determine whether the tiny province is integrated into Indonesia or granted independence. Two days earlier, after twenty-four years of conflict, East Timor's two warring factions sign a peace agreement. But international observers remain pessimistic that East Timor can reach peace any time soon.

1999: April 25   Discovery in Portugal leads to new human evolution theories

Paleontologists who discover the 24,500-year-old skeleton of a young boy in a shallow graveyard in Portugal, about ninety miles north of Lisbon, believe his remains show that Neanderthals and modern humans not only coexisted for thousands of years but also cohabitated.

Neanderthals live in Europe and western Asia from 300,000 years ago until 28,000 years ago, when the last of them vanish from the Iberian Peninsula. Modern humans, according to prevailing theory, appear in Africa less than 200,000 years ago and appear in Europe about 40,000 years ago. While the theory of cohabitation is not widely accepted by scholars, the discovery of the bones is considered an important addition to continued research on human evolution.

1999: April 25   Country reflects on anniversary of the Carnation Revolution

Most Portuguese hold the Carnation Revolution in high regard. Yet on its twenty-fifth anniversary, many of them reflect on the consequences of the revolt that toppled a forty-year dictatorship.

Portugal is troubled by the legacy of civil wars and deep poverty in its former colonies, even though the country appears close to breaking all of its ties to its colonial past. Portugal is scheduled to return Macao to China in December, 1999, and has been negotiating with Indonesia over its seizure of East Timor.

1999: May   Macao gets its first Chinese leader

After 442 years of Portuguese rule, a forty-four-year-old Chinese banker, Edmund Ho, is named to administer the colony after it reverts to Chinese rule in December, 1999.

Macao, on the southeast coast of China, has been a Chinese territory under Portuguese administration since 1975. The Portuguese settle in Macao in 1557 to open trade with China. The Portuguese declare that Macao is separate from China in 1849, and Beijing recognizes the annexation in 1887. In 1974, Portugal's new government attempts to return the territory to the Chinese, but Beijing declines because Macao is a major source of foreign exchange. In 1987, Portugal and China sign an agreement that declares Macao a Special Administrative Region of China by 1999. Under the arrangement, Macao is supposed to function as a semi-autonomous region, with its own legal system.

As Portugal prepares to leave, Macao has entered a deep recession. The economy is aggravated by violent gang war for control of Macao's lucrative gambling industry.

Bibliography

Birmingham, David, *A Concise History of Portugal*, Cambridge: Cambridge University Press, 1993.

De Figueiredo, Antonio, *Portugal: Fifty Years of Dictatorship*, New York: Holmes & Meier Publishers, 1976.

Gallagher, Tom, *Portugal: A Twentieth-century Interpretation*, Manchester: Manchester University Press, 1983.

Harvey, Robert, *Portugal: Birth of a Democracy*, New York: St. Martin's Press, 1978.

Hermano Saraiva, Jóse, *Portugal: A Companion History.* Manchester: Carcanet Press, 1995.

Herr, Richard ed. *The New Portugal: Democracy and Europe.* Berkeley: University of California at Berkeley, 1992.

Machado, Diamantino P. *The Structure of Portuguese Society: The Failure of Fascism.* New York: Praeger, 1991.

Opello, Walter C., *Portugal: From Monarchy to Pluralist Democracy.* Boulder, Col.: Westview Press, 1991.

Russell-Wood, A.J.R. *A World on the Move: The Portuguese in Africa, Asia, and America, 1415–1808,* New York: St. Martin's Press, 1993.

Winius, George D. ed., *Portugal, The Pathfinder: Journeys from the Medieval toward the Modern World, 1300–ca.1600,* Madison: The Hispanic Seminary of Medieval Studies, 1995.

Romania

Introduction

The eastern European country of Romania is located in the north of the Balkan Peninsula. Bound on the North and Northeast by Ukraine and Moldova, on the east by the Black Sea, on the south by Bulgaria, on the southwest by Serbia, and on the west by Hungary, Romania occupies a total area of 91,699 square miles (237,500 square kilometers). The Carpathian Mountains, which include the Transylvanian Alps, arch across the southeastern and western portions of the land. Mt. Moldoveanu, rising to 8,343 feet (2,543 meters), is the highest peak. The wide plains of Moldavia and Wallachia, two of the three historic Romanian homelands, stretch across the eastern and southern fringes of the Carpathians to the Prut River in the North and the Danube River in the South. The eroded plateau of the Dobrudja in southeastern Romania has elevations of 1,310–1,970 feet (400 to 600 meters).

The population is about 21.7 million, a little over half of which live in urban areas, especially the capital city of Bucharest (population 2.1 million). Romanians make up the largest percentage of the population. Hungarians and Germans are the most important minority groups, but there are significant numbers of Gypsies and Greeks living in the country. The Romanian Orthodox Church, one of the autocephalous Eastern Orthodox churches, is the dominant religion. Other denominations include the Greek Catholic or Uniate Church, the Reformed (Calvinist) Church of Romania, the Evangelical Church of the Augsburg Confession, and the Unitarian Church. There are also small numbers of Baptists, Seventh-Day Adventists, Muslims, and Jews.

History

The area now occupied by the country of Romania has been continuously inhabited for thousands of years. Primitive cultivation of wheat and rye began as early as the sixth millennium B.C.. Three thousand years later, the Cucetini people were producing fine ceramics decorated with a distinctive spiral motif. Greek literary and historical works of the sixth and fifth centuries B.C. placed the Getae, a northern branch of the Thracian people, in the region. By the first century A.D. the Dacians, as they were now called, had united into a powerful kingdom. The Roman Emperor Trajan conquered Dacia in A.D. 106 and instituted a program of Romanization in the territory. Repeated attacks by nomadic tribes led to the withdrawal of the Romans in 271 A.D. By that time, the Daco-Romans were partially Christianized and spoke Latin.

The Daco-Romans vanished from the historical record after the fourth century A.D. Three hundred years later, the term *Vlach* was used to denote the Roman population of the region south of the Danube River. In the late thirteenth and fourteenth centuries, the principalities of Wallachia and Moldavia emerged as independent states but quickly fell under the suzerainty of the Ottoman Empire. Transylvania, formerly under Hungarian control, came under Turkish influence in 1526. The Siege of Vienna in 1683 marked a turning point in the fortunes of the Ottoman Empire. By the terms of the Treaty of Karlowitz in 1699, Austria (later Austria-Hungary) annexed Transylvania. The 1812 Treaty of Bucharest awarded Bessarabia, part of Moldavia, to Russia. In 1856, the Congress of Paris secured the autonomy of Moldavia and Wallachia and returned a portion of Bessarabia to Moldavia. The two principalities of Moldavia and Wallachia were unified in 1859 under the rule of Prince Alexandru Ioan Cuza. Seven years later, Cuza was forced to abdicate in favor of Carol I (1830–1914) of the house of Hohenzollern-Sigmaringen. The new king ruled as a constitutional monarch.

Romania gained full independence from the Ottoman Empire in 1878 at the Congress of Berlin but was forced to return southern Bessarabia to Russia.

Romania participated in the Second Balkan War in 1912–13, taking Dobrudja from Bulgaria. In World War I (1913–18), the country joined the Allied Powers (the major allied powers were Russia, France, England, and later the United States). Territorial gains included Bessarabia, Bukovina, and Transylvania. The Treaties of St. Germain, Neuilly, and Trianon guaranteed the newly expanded state. In 1927 Corneliu Zelea Codreanu organized the Legion of the Archangel Michael, an extreme-nationalist party also known as the Legionary Movement. The return of Carol II (1893–1953) in 1930 after abdicating five years earlier in favor of his son Michael (b. 1921) led to a governmental crisis. The former king deposed his son and set up a royal dictatorship.

ROMANIA

0   50   100 Miles

0   50   100 Kilometers

The country joined the Axis Powers in 1940 (together with Germany, Italy and Japan) after ceding Bessarabia and northern Bukovina to the Soviet Union and northern Transylvania to Hungary. The resignation of Carol II the same year brought Michael back to the throne. In 1941, Romania joined the German invasion of the Soviet Union. The wartime regime of General Ion Antonescu was overthrown after Soviet troops entered the country in August. In 1942 a Communist government under Petru Groza (1884–1958) was established. The Romanian People's Republic was declared in 1947 following the abdication of King Michael. The Paris Peace Treaty restored northern Transylvania to the Romanian state.

During the late 1940s and 1950s, Romania followed a pro-Soviet line, nationalizing the economy and collectivizing

agriculture. Nicolae Ceaucescu (1918–89) emerged in 1965 as the leader of the nation and the Communist Party. He steered Romania on a more independent course by denouncing Soviet intervention in Czechoslovakia and Afghanistan and signing economic treaties with the United States.

Internally, Ceaucescu used the Securitate (his own, widely-feared police force) to quell political opposition. The worst strike of the Communist period occurred in 1977 when miners in the Jiu Valley, west of Bucharest took to the streets to protest Romania's ailing economy. Ceaucescu emerged from the incident more powerful than ever.

In the early 1980s, the Romanian economy was in a state of collapse. Ceaucescu announced in 1987 that he would not follow the reform plans of Mikhail Gorbachev in the Soviet Union. In March of the following year, the regime attempted to forcibly urbanize the peasantry by instituting the "systemization plan." Romania's Hungarian minority saw the plan as an attack on their cultural autonomy. Ethnic tensions continued to mount throughout 1989 and in mid-December spontaneous demonstrations broke out in the city of Timisoara. The Securitate attempted to end the protests by arresting Laszlo Toekes, a popular clergyman and spokesman for local Hungarians. Thousands were killed in the two days of rioting that followed.

The riots in Timisoara sparked further demonstrations throughout Romania. After returning from a three-day trip to Iran, President Ceaucescu organized a mass rally intended to boost the government's position. The rally turned into a riot directed against his regime. Ceaucescu and his wife attempted to flee the country but were quickly arrested and tried for crimes against Romania. On December 25, 1989, they were executed. The Council of National Salvation took power under Ion Iliescu, a former Communist, and a government dominated by ex-Communists.

In the elections of May 1990, the National Salvation Front, the political wing of the Council of National Salvation won the majority of the vote and Ion Iliescu (b. 1930) was returned to the presidency. Popular discontent continued into the next year. In June 1992, miners went on strike in Bucharest to protest government economic policies. Bowing to public pressure, Iliescu dismissed Prime Minister Peter Roman, the leading supporter of radical economic reform. The new prime minister, Theodor Stolojan, delayed a move on the question of economic reform until after the general elections of September 1992. Iliescu was again returned to the presidency. The National Salvation Front was the largest party with 28 percent of the vote. The strong showing of the Democratic Convention, an anti-Communist opposition coalition, further demonstrated the continuing hostility of the populace to Iliescu and other ex-Communists still in power.

In November 1992, Iliescu dismissed Prime Minister Stolojan and replaced him with Nicolae Vacaroiu, a pro-reform official who had no links to the Ceaucescu government. The move pleased the international financial community, but Romania's fitful economy could not keep up with the stringent reforms demanded by the West. Foreign investment remained low.

Since the dissolution of the Warsaw Pact in 1991, Romania has joined several Western organizations. In 1993, Romania signed an association agreement with the European Union. Romania was the first Eastern European country to join North Atlantic Treaty Organization's Partnership for Peace program (NATO). In 1995, Romania became an associate member of the European Union.

In November 1996, Emil Constantinescu became Romania's first post-Communist leader after winning 54 percent of the vote in second-round elections. Romania's ailing economy was the most important issue facing the new government.

Timeline

c. 6000 B.C.   Agriculture introduced

Primitive agricultural techniques are used to cultivate wheat and rye.

c. 4000 B.C.   Penentration of Neolithic tribes

Neolithic peoples dominate the Carpathian-Balkan region until the second millennium.

c. 3000 B.C.   Emergence of Cucetini civilization

The Cucetini dominate the Danubian area. Distinguished by highly decorated ceramics, Cucetini culture is also noted for its copper and gold weapons.

c. 2200–1800 B.C.   Penetration of Indo-Europeans

Nomadic Indo-European tribes penetrate the territory north of the Danube River. These tribes mix with the native population to form a distinct ethnic group called the Thracians.

c. 600s B.C.–500s B.C.   Founding of Greek colonies

Greek settlers found colonies in southern Moldavia on the shores of the Black Sea. The Greeks call the Thracians with whom they come into contact Getae.

c. 500s–400s B.C.   Greek writings mention Getae

The Getae are discussed in the works of Greek authors Hecateus, Sophocles, and Thucydides. The *Histories of Herodotus* mentions the Getae warding off an expeditionary force sent by the Persian King Darius I (521–486 B.C.).

480 B.C.   Kingdom of Odris Thracians founded

Under rule of Teres and his successors, the kingdom of the Odris Thracians becomes an important regional power in the Carpathian-Balkan area.

341 B.C.   Invasion of the Macedonians

King Philip II of Macedon (382–336 B.C.) conquers and annexes the Thracian kingdom of Odris.

340 B.C.   Alliance between the Getae and the Macedonians

The Getae and the Macedonians forge an alliance to drive the Scythians, a nomadic Iranian people, from Dobrudja. The marriage of Meda, daughter of the Getic king Colthas, and Phillip II of Macedon cements the new friendship.

335 B.C.   Campaign of Alexander the Great

Alexander the Great (356–323 B.C.), son of Philip II, leads an army against Thracians living near the Danube River.

82 B.C. Unification of Geto-Dacian tribes

Burebista (82 B.C.–44 B.C.), unites the Geto-Dacian tribes into a powerful kingdom. Assisted by his high priest and advisor, Decaeneus, Burebista extends his rule over the region comprising modern-day Romania. Julius Caesar (c.100–44 B.C.) plans an invasion of Dacia but his assassination in 44 B.C. prevents him from carrying it out. The death of Burebista the same year results in the disintegration of the kingdom.

A.D. 1st century Latin sources mention Dacians

Roman writings note the Getae for the first time, calling them Dacians. The name Dacia as a territorial designation for the Danubian region appears for the first time in the writings of Pliny the Elder (A.D. 23–79).

A.D. 46 Annexation of Dobrudja

Dobrudja, located in southeastern Romania between the Danube River and the Black Sea, becomes part of the Roman province of Moesia.

A.D. 87–106 King Decebal

Decebal (87–106) reunites the former kingdom of Burebista after defeating Roman expeditions led in 86–89 by the Emperor Domitian (Titus Flavius Domitianus, A.D. 51–96). By the terms of the peace treaty signed between them, Roman engineers build several fortresses for the Dacians. The Roman Emperor Trajan (Marcus Ulpius Trajanus, c. 53–117 A.D.) leads two campaigns against the Dacians, the second of which results in the destruction of Dacia and the death of Decebal.

A.D. 100s–800s Development of Romania

The intermingling of the Daco-Romans and various nomadic tribes leads to the emergence of the modern Romanian people as a distinct ethnic group. The Romanian language develops as a Romance language, the sole heir of Latin in Eastern Europe.

A.D. 101–02 First Roman-Dacian War

The first of the two wars leading to the conquest of Dacian breaks out. Despite important victories, the Roman Emperor Trajan fails to take the kingdom. A peace treaty ends the conflict.

A.D. 105–06 Second Roman-Dacian War

The Roman Emperor Trajan defeats the Dacian armed forces and annexes Dacia. King Decebal commits suicide to avoid capture. A massive Romanization effort is undertaken. Roads and towns are built on Roman models, and Roman citizens are encouraged to colonize the area. Free Dacians living to the north and east of the Carpathians are also gradually Romanized over time.

A.D. 117–1242 Nomadic invasions

For over one thousand years, Daco-Romans face waves of invasion by various Germanic, Slavic, and Turkic tribes. The invading peoples are either absorbed or repulsed by the local population and do not significantly change Daco-Roman culture.

A.D. 271 Withdrawal of the Romans

The Romans partially withdraw from Dacia after repeated attacks by migrating tribes. Although troops and administrators from the province south of the Danube River, the Roman Empire retains several fortified cities on the left bank of the Danube and the region remains in the orbit of the Eastern Roman Empire for several centuries.

A.D. 395 Annexation of Dobrudja

The Byzantine Empire annexes Dobrudja.

c. A.D. 400s Adoption of Christianity

The Daco-Romans adopt Christianity, becoming one of the first Eastern European people to profess that faith.

A.D. 600s People of the Danube become known as Vlachs

The term Vlach appears for the first time in historical sources as a designation for the Romanized people of the Danubian (Danube River) area.

A.D. 800s Gesta Hungarorum

An anonymous Hungarian chronicle, the *Gesta Hungarorum,* notes the resistance of three Transylvanian principalities to Hungarian expeditionary forces.

1054: July 16 The Great Schism

Political and theological differences split eastern and western Christendom leading to the eventual development of Orthodox Christianity and Roman Catholicism. The Romanians fall under the authority of the patriarch of Constantinople and thus adhere to Orthodox Christianity.

1185–86 Asan Dynasty

The Vlach brothers, Asan and Petru, rebel against the Byzantine Empire and found the Asan dynasty, also known as the Vlacho-Bulgarian Empire.

1241–42   Nomadic invasions

Great invasions of Petchnegs, Cumans, and Mongols devastate the Danubian region. The Crimean Khanate threatens Romania until the end of the eighteenth century.

1264 –65   Conquest of Transylvania

King Bela of Hungary conquers Transylvania after Voivode (Prince) Stefan refuses to pay tribute.

1330: November 9–10   Battle of Pasada

Basarab (1310–52), Voivode (ruler) of Wallachia, defeats the army of Charles Robert of Anjou (1288–1342), king of Hungary, at the Battle of Pasada, ending Hungarian suzerainty. His victory marks the emergence of the principality of Wallachia as an independent state. The Basarab Dynasty produces such notable Wallachian rulers as Mircea the Old (see 1391) and Vlad Tepes, the historical Dracula (see 1448).

1348–1386   Emergence of Dobrudja

Under the rule of Dobrotici, medieval Dobrudja emerges as a powerful Danubian state, capable of interfering in Byzantine and Genoese politics and fighting off expeditionary forces sent by the Ottomans.

1359   Emergence of Moldavia

Prince Bogdan (r. 1359–65) rejects Hungarian suzerainty and proclaims himself sole ruler of Moldavia. This act marks the emergence of Moldavia as an independent state.

1369   Turkish defeat

Prince Vladislav I Vlaicu of Wallachia defeats the first Ottoman attempt to invade the area.

1391   Ottoman tribute

Mircea the Old (1386–1418) becomes the first Wallachian ruler to agree to pay tribute to the Ottoman Empire.

1395: May 17   Battle of Rovine

Mircea the Old defeats the forces of the Ottoman Turks at the Battle of Rovine. His victory fails, however, to prevent further attacks. A coalition of French, Burgundian, German, and English knights led by the king of Hungary, Sigismund of Luxembourg, join forces with the Romanians under Mircea to repulse the Turks. In 1396, Sultan Bayezid I defeats the coalition forces at Nicopolis.

1400–32   Reign of Alexander the Good

Alexander the Good (1400–32) reigns in Moldavia. An able ruler, Alexander promotes ties of friendship and commerce with neighboring states and helps Poland repulse attacks by

Teutonic knights. In 1420, Alexander defeats an Ottoman expeditionary force.

1417   Tribute

After a temporary lull, Mircea the Old is again forced to pay tribute to the Ottoman sultan.

1419   Conquest of Dobrudja

The Ottomans conquer Dobrudja and retain possession of it until the Romanian War for Independence in the late nineteenth century.

1432–56   Rule of John Hunyadi

Janos (John) Corvinus Hunyadi (c.1387–1456), voivode of Transylvania (governor of Hungary from 1444) repulses a Turkish advance. Hunyadi conceives of a coalition of knights, mercenaries, and peasants as an anti-Ottoman defense. A victory against the Turks in Belgrade in 1456 is cut short by his death after the battle from the plague.

1437–38   Bobalna Revolt

Romanian and Hungarian peasants in Transylvania revolt against the oppressive measures of the Catholic Bishop Gheorghe Lepes. The first great peasant rebellion in Romanian history, the rebels defeat an early effort by noble armies to put down the revolt but are crushed by a unified force of Hungarians, Szecklers, and Saxons. In the years following the revolt, Romanian serfdom becomes more deeply entrenched than ever before.

1448   Reign of Vlad Tepes

Vlad Tepes (the Impaler, r. 1448, 1456–62, 1476) seizes power in Wallachia. Unable to hold the throne, Vlad flees and spends several years wandering in Moldavia and Wallachia. (See 1456 and 1476.)

1456   Vlad Tepes retakes throne

Vlad retakes the Wallachian throne, assisted by Hungarian warrior Janos (John) Corvinus Hunyadi (c.1387–1456; see 1432–56).

1462   Vlad stops paying Ottomans

Strong relations with Hungary convince Vlad to stop paying tribute to the Ottomans. In response, Sultan Mohammed II (Mehmet, the Conqueror 1432–81) drives Vlad from the throne and replaces him with his brother Radu the Handsome. Vlad languishes in Hungarian prisons (see 1476).

1457–1504   Rule of Stefan the Great

Stephen the Great comes to power in Moldavia through an alliance with Vlad Tepes of Wallachia. The Ottoman Turks remain throughout his reign his greatest foe. A great builder of monasteries and churches, Stephen is an able ruler who bases most of his laws on Wallachian customs.

1459 Refusal of Tribute

Vlad Tepes of Wallachia refuses to pay tribute to the Turks.

1459: September 20 Bucharest

Bucharest is recorded for the first time in writing as the residence of Vlad Tepes.

1462 Flight of Vlad Tepes

Sultan Mohammed II campaigns against Wallachia. Vlad appeals to Matthias Corvinus, the king of Hungary but is later imprisoned at Vishegrad by him. The sultan does not alter Wallachia's status as an autonomous province within the Ottoman Empire. Vlad's brother Radu the Handsome succeeds him.

1467: June Revolt in Transylvania

A revolt breaks out in Transylvania against the rule of Matthias Corvinus, son of John Hunyadi and king of Hungary.

1467: December 15 Battle of Baia

Stephen the Great defeats an attempt by Matthias Corvinus, king of Hungary, to extend his rule in Moldavia.

1471 Invasion of Wallachia

Stephen the Great invades Wallachia in an attempt to stop the passage of Turkish armies through the area. He replaces Radu the Handsome with Basarab-Laiota.

1473 Alliance

An alliance between Stephen the Great of Moldavia and Basarab-Laiota of Wallachia results in the total repulsion of the Ottomans from the Danubian region.

1474 Escape of Vlad Tepes

Vlad Tepes escapes from imprisonment in Hungary. Two years later, he temporarily re-occupies the throne of Wallachia. He is killed by Turkish soldiers one month later.

1476 Withdrawal of the Ottomans

Stephen the Great faces defeat by the armies of Mohammed and flees to northern Moldavia. A cholera outbreak leaves the Ottoman army in a weakened state. Stephen is able to push them out of the area with the help of forces from northern Transylvania, forcing the sultan to withdraw.

1476 Vlad seizes throne

Vlad Tepes again seizes the throne of Wallachia. He is killed one month later. The name "Dracula," which Vlad affixes to several documents, means "son of Dracul" (the Dragon) after his father, a member of the Order of the Dragon. Bram Stoker's novel turns the Wallachian ruler into a vampire.

1486 Withdrawal of the Ottomans

Sultan's forces withdraw from Moldavia after Stephen defeats them in battle. Two fortress cities on the Danube remain under Turkish occupation. During the last years of his reign, Stephen faces a new threat from Poland. He advises his son to submit to the Ottomans in preference to the Hungarians and the Poles if the terms are light enough.

1512–21 Rule of Neagoe Basarab

Neagoe wins the throne of Wallachia in 1512 and takes the dynastic name of Basarab to legitimize his action. His books of statecraft are among the first original works in Romanian literature. During his reign Wallachia becomes an important center of Orthodox Christianity.

1526: August 29 Annexation of Transylvania

The Ottomans defeat the Hungarians at the Battle of Mohacs. Hungary comes under the domination of the Ottoman Empire.

1538 Moldavia under Ottoman domination

Moldavia becomes a permanent tributary state of the Ottoman Empire and like Wallachia holds the status of a self-governing province under Ottoman overlordship. The status of Moldavia remains unchanged until Phanariot rule in the early eighteenth century.

1541 Ottoman suzerainty in Transylvania

The Diet at Debrecen in Transylvania recognizes Ottoman suzerainty over the Principality of Transylvania, ending over five hundred years of Hungarian rule.

1552 Conquest of Banat

The Ottomans conquer and annex the Banat and part of Crisana.

1572–74 Reign of John the Brave

Prince John the Brave buys the throne of Moldavia from the Ottoman sultan. He employs measures designed to enhance the powers of the prince at the expense of the nobility and tries to assert Moldavian independence. The last year of his reign is dominated by war with the Ottoman Empire after the prince refuses to continue paying tribute to the sultan.

1574 Death of John the Brave

John the Brave, Prince of Moldavia, defeats the Turks at Jilistea. Attacks on Turkish forces at Nistru and Burgeau fail to produce victories and ultimately lead to his capture and execution by decapitation.

1593–1601   Reign of Michael the Brave

Michael the Brave takes the throne of Wallachia with the full support of the Ottomans. He later embarks on an anti-Ottoman campaign, ordering the massacre of Turkish forces in Wallachia and attacking Ottoman positions along the Danube. Despite a peace treaty of 1598, Michael continues his anti-Ottoman policy, eventually conquering and uniting the three Romanian principalities of Moldavia, Wallachia, and Transylvania. Michael is assassinated at Campia Turzii in 1601 on the orders of the Habsburg General George Basta.

1599: November 1   Conquest of Transylvania

Michael the Brave conquers Transylvania and is crowned prince of that region.

1600   Unification of Romania

Michael the Brave unifies the three Romanian lands of Wallachia, Moldavia, and Transylvania.

1601: August 19   Assassination of Michael the Brave

Michael the Brave is assassinated on the orders of the Habsburg General George Basta.

1613–29   Reign of Gabriel Bethlen

Gabriel Bethlen overthrows Prince Gabriel Bathory (1608–13) and takes the throne of Transylvania. He steers an independent course between Habsburg and Ottoman dominance.

1659   Bucharest

Bucharest emerges as the capital of Wallachia.

1683   Siege of Vienna

The Siege of Vienna ends in withdrawal of Turks from Central Europe.

1686: June 26–July 6   Treaty of Vienna

The Treaty of Vienna concludes peace between the Habsburg and Ottoman Empires. As part of the terms of the treaty, Transylvania accepts the protection of the Habsburg Empire.

1693, 1710–11   Reign of Dimitrie Cantemir

Dimitrie Cantemir reigns in Moldavia. A scholar, historian, philosopher, and writer, Cantemir spends his later life in exile in Russia.

1697: March 27   Creation of Uniate Church

A portion of the Romanian Orthodox Churches in Transylvania unites with the Roman Catholic Church, resulting in the creation of the Greek-Catholic or Uniate Church.

1699: January 26–February 6   Treaty of Karlowitz

By the terms of the Treaty of Karlowitz, the Ottoman Empire recognizes the Austrian annexation of Transylvania.

1703–11   Rebellion in Transylvania

The nobles of Transylvania, led by Francis II Rackoczi (1676–1735), rebel against Habsburg occupation.

1711   Exile of Cantemir

Dimitrie Cantemir, ruler of Moldavia from 1710 to 1711, joins the anti-Ottoman campaign of Peter the Great. After the battle of Stanilesti, he lives out the rest of his years in exile in Russia writing historical works. In 1714 he is elected as a member of the Brandenburg Scientific Academy (Berlin Academy) and later writes *The History of the Ottoman Empire* (1743). After Cantemir's rebellion, the Turks decide to disband the armed forces of Wallachia and Moldavia and to take away regional autonomy by having the two principalities governed by Greek officials from the Phanar or Lighthouse District of Constantinople. The first Phanariot, as these officials come to be known, is Nicolae Mavrocordat, a descendant of a noble Greek family on his father's side and a noble Moldavian family on his mother's side. He rules in 1711, between 1715 and 1716, and between 1719 and 1730.

1746: August 5–16   Abolition of serfdom

The Phanariot Prince Constantin Mavrocordat abolishes serfdom in Wallachia and Moldavia.

1775: May 7–18   Annexation of Bukovina

The Habsburg Empire annexes Bukovina, formerly a part of Moldavia.

1784   Revolt against Hungarians

Romanian serfs join poor Hungarians in Transylvania in a massive peasant uprising against Magyar rule. The Habsburg Emperor Joseph II suppresses the revolt.

1812: May 16–28   Treaty of Bucharest

The Treaty of Bucharest awards Russia the Ottoman regions of Bessarabia and northern Moldavia.

1821   Revolt of Tudor Vladimerescu

Tudor Vladimerescu, commander of a unit of panduri (soldiers serving in return for tax exemptions) leads a rebellion against Ottoman rule and asks for support from Moldavia and

Wallachia. Vladimerescu's Proclamation of Pades expresses for the first time in a public document the idea of the national unity of all Romanians living in the two principalities.

1822   Greek War of Independence

Phanariot rule in Wallachia and Moldavia is abolished after the outbreak of the Greek War of Independence.

1829: September 2–14   Treaty of Adrianople

The Treaty of Adrianople concludes the Russo-Turkish War of 1828-29. By the terms of the document, the Ottomans agree to end their commercial monopoly in Wallachia and Moldavia.

1830   Organic Statutes

The *Reglement Organique* (Organic Statutes) is promulgated in Wallachia and Moldavia as part of the terms of the Treaty of Adrianople. The document is intended to modernize the administrations of the two principalities with the eventual goal of incorporating both regions into the Russian Empire. Except for a brief period during the 1848 Revolutions, the Organic Statutes remain in effect until 1858.

1848   Revolution in Moldavia

Revolution breaks out in Moldavia. Nicolae Balcescu (1819–52) and Mihail Kogalniceanu (1817–91) lead the revolution in Moldavia by submitting with other young men a petition for reform to Prince Mihai Sturzaa. The revolutionaries of 1848 are inspired by the example of the French Revolution.

1848: June   Revolution in Wallachia

Led by Nicolae Balcescu, a participant in the Moldavian Revolution, the revolution of 1848 breaks out in Wallachia. The Proclamation of Islaz (Oltenia) provides for a provisional government. Prince Gheorghe Bibescu abdicates and flees the principality. Among the revolutionary decrees adopted are the abolition of aristocratic titles, corporal punishment, and Gypsy slavery. The Organic Statutes are burnt in the streets. The revolution is crushed by a joint Turko-Russian effort.

1848: May   Revolution in Transylvania

At a rally in Blaj, Transylvania, 40,000 Romanians declare their independence, demanding equality with other nationalities in the region and the union of the two principalities of Moldavia and Wallachia. The Russian and Habsburg Empires join forces to put down the revolt.

1849: April 19–May 1   Convention of Balta Liman

The Russian and Ottoman Empires agree through the Convention of Balta Liman to suppress the Revolutions of 1848 in Moldavia and Wallachia.

1856   Congress of Paris

The Congress of Paris guarantees the autonomy of Wallachia and Moldavia but does not grant the two principalities complete independence from Ottoman rule. Part of the territory apportioned to the Russian Empire in 1812 is reincorporated into Moldavia. Romanian Assemblies called Ad Hoc Divans are organized to revise the Organic Statutes.

1858   United Principalities

The Ad Hoc Divans set up by the Congress of Paris continually call for the union of Moldavia and Wallachia. Great Britain, France, and Russia (Great Powers) reject the notion of a unified Romanian state because of the effect it might have on the balance of power in the region. Instead, the Great Powers decide that the two principalities should have common legislation and be titled the United Principalities of Moldavia and Wallachia. The Convention of 1858 replaces the Organic Statutes until 1864 as the law of the principalities and provides that each country should have its own government and legislative assembly.

1859: January   Union of the two principalities

The two principalities of Wallachia and Moldavia are unified. Colonel Alexandru Ioan Cuza (1859–66) is elected prince. A new constitution proclaims Romania a limited monarchy. The modern Romanian state dates from the unification of the Ottoman principalities of Moldavia and Wallachia.

1862: January 24–February 5   International recognition

Opening the first Parliament of Romania in Bucharest, Prince Alexandru Cuza declares the Union of the Two Principalities. The new state immediately receives international recognition.

1864: June 16–28   The Developing Stature of the Convention of 1858

Prince Alexandru Ioan Cuza submits to the people for ratification by plebiscite The *Developing Stature of the Convention of 1858*, a new constitution accepted later by the Great Britain, Russia, and France. The document greatly increases the powers of the executive at the expense of the legislature and increases the number of electors.

1865: November 24–December 6   Hungary incorporates Transylvania

The Diet of Cluj votes to merge Transylvania with Hungary.

1866: February 11   Abdication of Cuza

Prince Alexandru Ioan Cuza is forced to abdicate after his reforms arouse the hostility of both conservatives and radicals in Romania.

Territorial growth of Romania, 1856–1920.

U . S . S . R

BUKOVINA (1919)

Budapest

HUNGARY

Somes

Iaşi

BESSARABIA

MOLDAVIA (1856)

Chisinau

Szeged

TRANSYLVANIA (1920)

Mureş

Siret

Timişoara

Resiţa

WALLACHIA (1856)

Siret

Galati

Belgrade

Jiu

Olt

Arges

Ploieşti

Ialomiţa

Bucharest

Dunărea (Danube)

YUGOSLAVIA

Dunărea

Craiova

Dunărea (Danube)

Constanţa

SOUTHERN DOBRUDJA (1913)

Black Sea

(Danube)

BULGARIA

1866: February 25–March 9   Literary debut of Mihai Eminescu

Poet, writer, and journalist Mihai Eminescu (1850–89) makes his literary debut with the publication of a poem in the journal *Familia*. He is later deemed Romania's greatest poet.

1866: May 10–22   Rule of Hohenzollern-Sigmaringens

An interim governing body sets up after the abdication of Prince Alexandru Ioan Cuza offers the throne to Prince Carol (1839–1914) of Hohenzollern-Sigmaringen. The Hohenzollern-Sigmaringens are related to the major ruling families of Europe. The move is supported by the Ad Hoc Divans, which argue that a prince of foreign blood would prevent dissension and provide dynastic links among Romania, Prussia, and France.

1866: July 1–13   New constitution adopted

A new constitution is adopted and remains in effect until 1923. The document reaffirms the name of county and the rights of the prince but does not declare full independence from the Ottoman Empire.

1867: February 5–17   Austro-Hungarian Empire formed

The Austro-Hungarian Empire (Habsburg Empire) is formed. Hungary annexes Transylvania.

1867: May 25–June 8   Francis Joseph I

The Austrian Emperor Francis Joseph I receives the title of King of Hungary and upholds the annexation of Transylvania.

1877: April 4–16   Russian Romanian Convention

By the terms of the Russian Romanian Convention, Russian troops are allowed to cross Romanian territory in time of war.

1877: April 12–24   Russia declares war

Russia declares war on the Ottoman Empire.

1877: April 29–May 11   Romanian Declaration of Independence

The Assembly of Deputies in Bucharest declares independence from the Ottoman Empire. The Chamber of Deputies seconds the motion a few weeks later.

1877: May 9–21   War against the Ottomans

Romania allies with Russia against the Ottoman Empire.

1877–78   Romanian War for Independence

Romanian armies battle Ottoman forces for independence.

1878: February 19–March 3   Treaty of San Stefano

The Treaty of San Stefano concludes the Russo-Turkish War. Romania is awarded Dobrudja and the Danube Delta, provisions that will become official in October. Russia is given southern Bessarabia.

1880: February 8–20   Recognition of Independence

Germany, Great Britain, and France officially recognize the independence of Romania from the Ottoman Empire

1881: March 14–26   The kingdom of Romania

Romania is declared a kingdom. Since many Romanians still live outside the new state's borders in Russian Bessarabia and Bulgaria, the government's main goal is *Romania Mare* or the union of all Romanian-inhabited lands. This goal is achieved for a short time after World War I when the Allies sanction the Romanian acquisition of Transylvania, Bessarabia, northern Bukovina, and southern Dobrudja. Romania will cede most of these lands after World War I.

1883: October 18–30   Treaty of Vienna

By the secret Treaty of Vienna, Romania allies with the Habsburg and German Empires (Central Powers).

1885: May 1–13   Approval of Tomos

King Carol I recognizes the autocephalous nature of the Romanian Orthodox Church.

1891–94   Memorandist Movement

The founding of the Romanian National Party in Transylvania leads to a movement to bring the grievances of Romanian nationals living in the region to the attention of the Austro-Hungarian Emperor. *The Memorandum*, a document specifying grievances, is in 1893 submitted to Vienna. The ensuing trial of its signers results in a temporary ban on the party and the condemnation of the accused. All of the signers are later acquitted.

1913   Second Balkan War

Romania campaigns against Bulgaria and annexes southern Dobrudja.

1914: July 15–28   World War I

World War I begins when Austria-Hungary declares war on Serbia.

1914: July 21–August 3   Declaration of neutrality

Romania declares neutrality.

1916: August   Romania enters World War I

Romania joins the Entente Powers of Great Britain, France, Italy, and Russia, and enters World War I after declaring war on Austria-Hungary.

1917: Summer   Battle of Maraseti

The Romanian army attempts to remove the country from World War I by securing Moldavia.

1918: February 20–March 5   Preliminary peace treaty

A preliminary peace treaty between the Entente and the Central Powers is reached at Buftea.

1918: April 24–May 7   Peace of Bucharest

The Peace of Bucharest is signed and imposes heavy territorial and economic sanctions on Romania.

1918: November 15–28   National Congress meets

In Bukovina, the National Congress of Romania meets and decides to unify with Romania.

1918: November 18–December 1   Unity proclaimed

At an assembly in Alba Iulia, Romanians proclaim the union of Transylvania and Banat with Romania.

1918: October–November   Ultimatum issued

An ultimatum issued by the Romanian government instructs the occupation forces of the Central Powers to leave the country within twenty-four hours. The following day Romania re-enters the war when King Ferdinand I declares war on the Central Powers a second time. From October 29 to November 11, hostilities end after an armistice between Germany and the Entente Powers is signed.

1918: April 24–May 7   Treaty of Bucharest

Germany declares the Treaty of Bucharest null and void and withdraws its forces from Romania.

The opening session of the Romanian parliament after the death of King Ferdinand I. (EPD Photos/CSU Archives)

1918: November 15–28   Union with Romania

The Federal Congress of Bukovina declares the union of Bukovina with Romania.

1918: December 11–24   Union with Romania

King Ferdinand I declares the union of Transylvania with Romania.

1919: September 10   Treaty of St. Germain-en-Laye

The Treaty of St. Germain-en-Laye concludes hostilities between the Entente Powers and Austria, ending World War I. By the terms of this treaty, Austria recognizes the union of Bukovina with Romania. Romania signed the document on December 10, 1919.

1919: March 5–18   Introduction of the Gregorian Calendar

The Gregorian Calendar is introduced by statute in Romania to begin April 1, which will become April 14.

1919: April 15–May 1   Hungarians attack

An alliance of Romanian and Czechs prevents the Hungarian Red Army from joining Soviets attacking from the east.

1919: June 28   League of Nations

Romania enters the League of Nations as a founding member.

1919: July 20–August 4   Hungarians attack

Hungarian troops attack Romanian troops on the east bank of the Tisa. A Romanian counterattack leads to the occupation of Budapest. The Romanians withdraw in March of the following year.

1919: December 10   Peace treaties signed

Romania signs the treaties of Saint Germain-en-Laye and Neuilly-sur-Seine.

1919: December 29   Union approved

The Romanian parliament approves the union of Transylvania, Bukovina, and Bessarabia with Romania.

1920: June 4   Treaty of Trianon

The Treaty of Trianon recognizes the frontier between Romania and Hungary.

1922: October 15   Coronation of Ferdinand and Marie

King Ferdinand and Queen Marie are crowned rulers of Greater Romania (Romania Mare).

1923: March 23   New constitution

A new governing document declares Romania a unitary and indivisible state.

1924: April 5   Banning of Communist Party

The Romanian Communist Party is banned.

1925: December   Abdication of Prince Carol II

Prince Carol II (1893–1953) abdicates in favor of his son Michael (b. 1921) and agrees not to return to Romania for ten years.

1927: July 20   Death of Ferdinand I

Ferdinand I dies and is succeeded by his six-year-old grandson Michael I. Iuliu Maniu of the Peasant Party acts as regent.

1927: June 24   Legion of the Archangel Michael

Corneliu Zelea Codreanu founds the Legion of the Archangel Michael, also known as the Legionary Movement, an extreme-nationalist party.

1930: June 6   Carol returns

Carol of Hohenzollern-Sigmaringen clandestinely returns to Romania to reclaim his throne, breaking the promise he made at the time of his abdication (see 1925: December) not to return for ten years.

1930: June 7   Resignation of the Government

The government and the Regency Council resign.

1930: June 8   Carol II proclaimed king

Carol II is proclaimed king of Romania. Michael becomes a hereditary prince.

1933: December 29   Ministers assassinated

Members of the Legionary Movement assassinate Prime Minister I.G. Duca at the train station in Sinaia.

1937: November   Non-Agression Pact

A non-aggression pact designed to guarantee free elections is signed by leaders of the National Peasant Party, the Legionary Movement, and the dissident branch of the National Liberal Party.

1937: December 20   Elections held

Elections are held in Romania. Despite the fact that they received less than ten percent of the vote, Carol II orders Octavian Goga and the anti-semitic National Christian Party to form the new government.

1938   Royal dictatorship established

Carol II dissolves all political parties and sets up a single party of his own, the National Revival Front. He restricts the powers of parliament, establishes a personal dictatorship and promulgates a new constitution.

1938: May   Show trial of Codreanu

Corneliu Zelea Codreanu is put on trial and found guilty by way of fabricated evidence. He is condemned to ten years of hard labor.

1938: November   Assassination of Codreanu

Carol II and Interior Minister Armand Calinescu order the assassinations of Codreanu and thirteen other members of the Legionary Movement.

1939: September 1   World War II begins

Germany invades Poland, beginning World War II.

1939: September 21   Calinescu assassinated

Interior Minister Armand Calinescu is assassinated by a group of Legionaries in revenge for the death of Codreanu. In response, King Carol II orders the deaths of 252 leading members of the Legionary Movement, all of whom are executed without charge or trial in a single night.

1940: June   Soviet ultimatum

The Soviet Union demands the immediate cession of Bessarabia and northern Bukovina. Romania complies by ceding Bessarabia and northern Bukovina to the Soviet Union, northern Transylvania to Hungary, and southern Dobrudja to Bulgaria.

1940: July 4   Alliance with Axis powers

Romania concludes an alliance with the Axis powers of Germany, Italy, and Japan.

1940: August 30   Diktat of Vienna

The Diktat of Vienna, issued by the Axis powers, awards northwestern Transylvania to Hungary.

1940: September 4   Antonescu in power

After the resignation of the government headed by Ion Gigurtu, King Carol II orders General Ion Antonescu to form a new government. Antonescu sets up a fascist dictatorship by bringing the Legionaries into the government.

1940: September 6   Carol II abdicates

King Carol II abdicates in favor of his son, Michael. German troops enter the country.

1940: September 7   Treaty of Craiova

The Treaty of Craiova restores southern Dobrudja to Bulgaria.

1940: September 14   National Legionary State

Romania is proclaimed a National Legionary State. King Michael I appoints Ion Antonescu head of state.

1940: November 28   Tri-Partite Pact

Romania becomes a member of the Tri-Partite Pact.

1941: January 21–23   End of National Legionary State

Members of the Legionary Movement fail to topple the wartime regime of General Ion Antonescu, who takes complete power and declares an end to the National Legionary State.

1941: June   Invasion of Soviet Union

Romania participates in the German invasion of the Soviet Union.

1943: August 1   Bombing of Ploesiti

American and British air forces bomb the city of Ploesiti in an attempt to damage German oil supplies.

1944: March 17   Secret negotiations

Prince Barbu Stirbei, leader of the political opposition in Romania, begins secret negotiations in Cairo with Allied diplomats on the question of armistice.

1944: April 4   Bombing of Bucharest

American and British air forces bomb Bucharest.

1944: August 20   Soviets enter Romania

Soviet troops enter the country.

1944: August 23   Antonescu overthrown

King Michael engineers a wartime coup that results in the overthrow of the Antonescu regime. General Constantin Sanatescu sets up a new government. Romania joins the Allied forces against Germany.

1944: August 24–28   Bombing of Bucharest

Hitler bombs Bucharest. The Germans are ultimately pushed out of the city.

1944: September 12   Armistice signed

Romania signs an armistice convention with the government of the United Nations, confirming its withdrawal from alliance with the Axis Powers. The armistice nullifies the Vienna Diktat and northern Transylvania is returned.

1945: March 6   New government formed

A Communist-led coalition under Petru Groza (1884–1958) forms a new government.

1945: August 21   Royal strike

King Michael I asks for the resignation of the Groza government and when refused, initiates a royal strike.

1945: October 16–22   Gheorghiu-Dej elected

The National Congress of the Romanian Communist Party elects Gheorghe Gheorghiu-Dej (1901–65) secretary-general of the Central Committee. Gheorghiu-Dej later holds the office of premier and serves as president of the State Council.

1946: February 5   Romania recognized

Great Britain and the United States recognize Romania and reestablish diplomatic relations.

1946: May 17   Bloc of Democratic Parties

The Bloc of Democratic Parties, a Communist front organization, is created.

1946: November 19   Elections held

The Bloc of Democratic Parties receives 79.86 percent of the vote and 376 of 414 seats in Parliament in an election marred by intimidation and fraud.

1947: December 30   Abdication of King Michael

King Michael is forced to abdicate. The Romanian People's Republic is declared.

1947: February 10   Paris Peace Treaty

The Paris Peace Treaty sets Romania's borders to those of January 1, 1941 and restores northern Transylvania. The treaty returns Bessarabia and northern Bukovina to the Soviet Union and nullifies the Vienna Diktat.

1948: June   Nationalization of key industries

The government nationalizes major industrial, mining, banking, insurance industries.

1948: August   Securitate established

The General Direction of Popular Security or the Securitate is established to eliminate opposition.

1949   Council for Mutual Economic Assistance

Romania becomes a member of the Council for Mutual Economic Assistance.

1951–1975   Five-Year Plans

Five-year Plans after the Soviet model are introduced in 1951 to speed the work of industrialization and collectivization. The first lasts from 1951 to 1956, the second from 1961 to 1965, the third from 1966 to 1970, the fourth from 1971 to 1975.

1952   New constitution adopted

A new constitution, more directly patterned after that of the Soviet Union, replaces the Communist Constitution of 1948.

1952–55   Rule of Gheorghiu-Dej

Gheorghe Gheorghiu-Dej serves as prime minister, basing his government on the Soviet model. His rule is marked by harsh political oppression.

1955 Romania joins organizations

Romania becomes a member of the United Nations and a founding member of the Warsaw Pact.

1956 UNESCO membership

Romania becomes a member of UNESCO.

1961–65 State Council presidency for Gheorghiu-Dej

Gheorghe Gheorghiu-Dej serves as president of the State Council and holds the position until his death in 1965. Towards the end of his life, he attempts to move away from the Soviet orbit and lessen political repression.

1965: March Death of Gheorghe Gheorghiu-Dej

Gheorghe Gheorghiu-Dej and is succeeded by Nicolae Ceaucescu. The new leader's reliance on the Securitate leads to the arrest and persecution of tens of thousands.

1965: July Ceaucescu elected General Secretary

The Ninth Congress of the Romanian Communist Party elects Nicolae Ceaucescu (1918–89) General Secretary of the Central Committee of the Romanian Communist Party.

1965: August Socialist Republic proclaimed

A new constitution proclaims Socialist Republic of Romania.

1967: June 11 Six-Day War

Romania is the only Communist country to keep diplomatic relations with Israel during the Six-Day War.

1968: August 21 Soviet intervention denounced

Romania denounces Soviet intervention in Czechoslovakia, the only Warsaw Pact country to do so.

1969: August 2–3 Visit of Richard Nixon

Richard Nixon becomes the first U.S. president to visit Romania.

1970: January 31 Treaty ratified

Romania ratifies the Treaty for the Non-Proliferation of Nuclear Arms.

1971: November 14 GATT signed

Romania becomes a member of the General Agreement on Tariffs and Trade or GATT.

1972: December 9 Accords signed

Romania adheres to the accords of the International Monetary Fund and the World Bank for Reconstruction and Development.

1970s Attempts to secure aid

Ceaucescu attempts to secure foreign aid and fails. He adopts a policy of forced repayment of external debt that imposes severe privation on the country.

1973: December 4–7 Ceaucescu visits the United States

On a visit to the United States, Nicolae Ceaucescu signs an agreement on economic, industrial, and technical cooperation.

1974: March 28 Election of Ceaucescu

Nicolae Ceaucescu is elected the first president of the Socialist Republic of Romania.

1975: April 2 Agreement signed

Romania becomes the first communist country to receive most favored nation status by the United States.

1976: July–August Olympic gold

At the Olympic Games in Montreal, fourteen-year-old Nadia Comaneci receives seven perfect scores of ten and wins three gold medals, including individual all-around champion.

1977: August 1–3 Miners strike

Miners in the Jiu Valley, west of Bucharest, strike to protest the ailing Romanian economy. Two of the strike leaders are assassinated. It is the largest strike of the communist era.

1980s Ceaucescu dominates Communist Party

Ceaucescu and his family dominate the Romanian Communist Party. A cult of personality springs up around the president.

1982 Calls for withdrawal

Ceaucescu calls on the Soviet Union to withdraw from Afghanistan.

1984: July 28–August 12 Los Angeles Olympic Games

Romania is the only Soviet-Bloc country to participate in the Summer Olympic Games in Los Angeles.

1987 Romania's hard-line policy, election riots

Ceaucescu announces that Romania will not follow Mikhail Gorbachev's reform program. Parliamentary elections are marred by a popular revolt in Brasov, later squashed by the Securitate and the army.

1988: March Systemization plan adopted

Ceaucescu initiates his "systemization plan," an attempt to forcibly urbanize about half of the country's peasantry.

1989: November   Defection of Comaneci

Olympic gold-medallist Nadia Comaneci defects to the West.

1989: December 16–22   Tensions build

Ethnic tensions follow the introduction of the systemization plan, especially among Romania's 2.5 million Hungarians. These tensions, coupled with many economic grievances, lead to a spontaneous demonstration in the western city of Timisoara. The Securitate, Romania's secret police, attempts to deport Laszlo Toekes, a popular clergyman and leading spokesman for local Hungarians. Thousands take to the street in protest. Troops are summoned to quell two days of rioting. News of the riot leads to new demonstrations across the country. Ceaucescu leaves for a three-day tour of Iran. Upon his return, he organizes a rally that degenerates into an anti-government demonstration. On December 22, Ceaucescu and his wife are arrested and tried after a failed attempt to flee the country. The Council of National Salvation takes power under president Ion Iliescu (b. 1930) and Prime Minister Peter Roman.

1989: December 25   Execution of Ceaucescu

Nicolae Ceaucescu and his wife Elena are executed for crimes against Romania.

1990: May   Elections held

Ion Iliescu is elected president of Romania.

1991: July 1   Warsaw Pact dissolved

The Warsaw Pact is dissolved.

1991: September 24–28   Tensions mount

Miners from the Valea Jiului occupy several buildings in the capital city of Bucharest to protest deteriorating economic conditions. Prime Minister Peter Roman, an advocate of radical reform, resigns. Iliescu appoints Theodor Stolojan as Prime Minister.

1991: December   New constitution adopted

A new constitution based on Western European models is adopted.

1992: September 27–October 11   Elections held

Ion Iliescu is re-elected president of Romania.

1992: November   New constitution adopted

A new constitution is adopted by parliament. On December 8, the document is ratified by referendum. The new constitution affirms a multiparty system, a free-market economy, and respect for human rights. Legislative power is vested in a bicameral parliament composed of a 341-seat Chamber of Deputies and a 143-seat Senate. Members are elected by universal suffrage, using proportional representation, for four-year terms. The president is elected to a four-year term with a two-term limit.

1993: February 1   Association agreement signed

Romania signs an Association agreement with the European Union.

1993: March 24   Council for National Minorities established

The government establishes the Council for National Minorities, a governmental division under the secretary-general, composed of representatives from public institutions and recognized ethnic groups living in Romania.

1993: September 28   Council of Europe

Romania becomes a member of the Council of Europe.

1993: October   MFN status for Romania

The United States reinstates Most Favored Nation (MFN) status for Romania.

1994: January 26   Partnership for Peace

Romania becomes the first Eastern European Country to join the Partnership for Peace program of the North Atlantic Treaty Organization (NATO).

1994: September 2   Status renewed

The United States renews Romania's MFN status.

1994: November   Comaneci's visit

Five years after her defection, Olympic gold-medallist Nadia Comaneci returns to visit Romania.

1995: February 1   European Union

Romania becomes an associate member of the European Union (EU).

1995: March   Aviation disaster

Sixty people are killed when an Airbus A-310 (TAROM flight from Bucharest to Brussels) explodes over Balotesti, making it the worst aviation disaster in Romanian history.

1996: September   Basic treaty signed

Romania and Hungary sign a basic treaty upholding existing borders and the rights of ethnic minorities in both countries.

1996: November   Elections held

Voters elect Emil Constantinescu president of Romania. The new government advocates full membership for the country in the EU and NATO.

1997: January   Economic reforms

The government launches a fast-track economic reform program that wins the support of the International Monetary Fund (IMF) and the World Bank.

1997: July   Exclusion of Romania

Romania is excluded from first-round talks for NATO and EU enlargement.

1998: February   IMF dissatisfaction with economic reform

The IMF refuses to disburse the third installment of a $410 million standby loan, citing dissatisfaction with Romania's stalled reform program.

1998: October–November   Crisis in Bucharest

Public services in the city of Bucharest are near a state of collapse as sewers fail, garbage goes uncollected, and packs of stray dogs roam the streets.

1998: November   Abuses in Romania

A European Union report concludes that Romania has failed to root out government corruption and protect the individual liberties of its citizens.

1999   Economic crisis

The slow pace of reform leads to the increasing destabilization of economy. Rising support for extremist parties sparks ethnic tensions. The recognition of Hungarian and German as official languages remains an important issue.

Bibliography

Bachman, Ronald D., ed. *Romania: A Country Study.* 2d ed. Washington, D.C.: Library of Congress, 1991.

Castellan, Georges. *A History of the Romanians.* New York: Columbia University Press, 1989.

Commission on Security and Cooperation in Europe. *Human Rights and Democratization in Romania.* Washington, D.C.: Commission on Security and Cooperation in Europe, 1994.

Economist Intelligence Unit. *Country Report: Romania.* London: The Economist Intelligence Unit, 1999.

———. *Romania: Country Profile.* London: The Economist Intelligence Unit, 1998

Fischer, Mary Ellen. *Nicolae Ceaucescu: A Study in Political Leadership.* Boulder, Colo.: L. Rienner Publishers, 1989.

Fischer-Galati, Stephen A. *Twentieth Century Romania.* 2d ed. New York: Columbia University Press, 1991.

Otetea, Andrei and Andrew MacKenzie, eds. *A Concise History of Romania.* New York: St. Martins, 1985.

Shafir, Michael. *Romania: Politics, Economics, Society.* Boulder, Colo.: L. Rienner, 1985.

Treptow, Kurt W. and Marcel Popa, eds. *Historical Dictionary of Romania.* Lanham, Md.: Scarecrow Press, 1996.

Russia

Introduction

Russia is the largest country in terms of area in the world. At 6,592,771 square miles (17,075,200 square kilometers), Russia spans two continents, Europe and Asia, and one-eighth of the total land area of the earth. Although Russia suffers from severe economic problems, much of the country is rich in natural resources, including iron ore, copper, bauxite, nickel, and zinc. Much of the country, particularly the tundra above the Arctic Circle, is barren and uninhabitable.

Estimates in 1998 placed Russia's population at 146,861,022 of which 81.5 percent were ethnic Russians. Other groups included 3.8 percent Tatars, 3 percent Ukrainians, and 1.2 percent Chuvash. In recent years the country's population has been declining, and the trend is expected to continue into the next century. Seventy-three percent of the population is urban.

The modern Russian state has its origins with the settlement of the Slavs in the European plain north of the Black Sea beginning in the eighth century A.D. The vast flat land served as an easy migration route, and the Slavs were only the latest in a long line of peoples who used this route to travel between Europe and Asia. The earliest evidence of human habitation in the area extends back about 10,000 years. By c. 1000 B.C. the Cimmerians, an iron-working people related to the Thracians, settled just north of the Black Sea. The Cimmerians were supplanted by the Scythians (a group whose origin is disputed), who were followed by the Greeks. The Greeks established colonies in the Crimea in the seventh century B.C. linking the region with the rest of the Greek world. Over the course of the next millenium, Iranian cattle-breeding Sarmatians, Germanic Goths, and Central Asian Huns and Avars all swept through the region. The arrival of the Turkic Khazars in the seventh century brought a measure of stability to the region and established trading contacts with the Arab and Byzantine empires. As Slavs began settling in the region in the eighth century, Varangians (Normans) from Scandinavia explored, traded, and settled along much of the Eurasian plain. Out of the interaction of these two peoples originated Rus, the first Russian state, in the ninth century.

At its height, Rus stretched north to south from the Baltic to the Black Seas and east to west from the Volga to the Nieman Rivers. After consolidating its power in the European plain, Rus attempted to expand southward and mounted two unsuccessful attacks against the Byzantine capital of Constantinople. After Rus's second unsuccessful attack in 944, it negotiated a trade agreement with Byzantium thereby opening the way for Byzantine cultural penetration of Russia. The most significant impact of Russian contact with Byzantium was the conversion of Rus to Orthodox Christianity in the tenth century. After the fall of Byzantium, Russia used its Byzantine cultural heritage to claim for itself the title of Third Rome (the Byzantine Empire originated as the eastern half of the Roman Empire).

By the eleventh century, Kievan Rus (centered on the city of Kiev) began its decline as centralized authority evaporated and power devolved toward regional princes. This loss of a strong central authority coincided with the rise of the Mongols, a group of nomadic warriors from Central Asia. Beginning in the eleventh century, under the leadership of Genghis Khan, the Mongols began their westward expansion and dominated Rus. Not for over a century did a new Russian state emerge under the leadership of Prince Dimitri Ivanovich "Donskoy" of Moscow.

Ivan III (the Great) succeeded in consolidating power centrally in the fifteenth century. In addition, Ivan III solidified the image of Moscow as the Third Rome through his marriage to Sophie Paleologue, niece of the last Byzantine emperor. The strength of the royal family got a major boost under Ivan IV (the Terrible) in the sixteenth century. Ivan took the title *tsar* (ceasar) and embarked upon a policy of eastward expansion that would ultimately give Russia control of Siberia. Ivan's reign is also significant for his destruction of the power of the *boyars* (landholding nobles), which led to the establishment of serfdom, an institution that lasted until 1861, as well as the tradition of *autocracy* (unlimited rule).

Until the late seventeenth century, the major influences on Russia's development were Orthodox Christianity, the Mongol Yoke (as the period of Mongol rule was called), and Russia's expansion to the east. As a result, Russia was isolated from Western Europe and fell further behind in development. The reign of Peter I (the Great, 1689–1725) attempted

RUSSIA

to open up Russia to the West. Peter, an admirer of Western Europe, traveled extensively, studied in the West, and sought to apply the Western model of development to Russia. Thus, he stressed Western education, manners, dress, and governmental organization. To symbolize this new Western outlook, Peter built a new capital on the newly-acquired Baltic coast which he named after himself, St. Petersburg.

The foundations laid by Peter permitted a further expansion of Russian power. Throughout the eighteenth century, Russia continued its expansion and gained significant territories to the west and in the south along the Black Sea coast. Russian military power had grown so much that by 1814 Russia was the greatest land power on the European continent. Yet the growth in Russian power came at a price. Ninety percent of the Russian population remained peasants (mostly serfs) tied to the soil. In an age of increasing industrialization and urbanization, Russia remained overwhelmingly agrarian and rural. Moreover, the autocrat refused to grant liberalizing measures as had been done in much of Western Europe. The battle between autocracy and reformers would dominate the last century of Russian history.

Russia's weaknesses became apparent following its humiliating defeat in the Crimean War (1853–56). Russia's loss spurred Tsar Alexander II to action. He pursued a modernization program (including the abolition of serfdom in 1861). Yet he did not go far enough in alleviating peasant poverty, nor did he show interest in granting a constitution. By the 1870s opposition to autocracy became widespread among Russia's tiny but growing *intelligentsia* (a group of well-educated individuals sharing the same lifestyle and cultural tastes). Students played a major role in radical terrorist groups such as *Narodnaya Volya* (People's Will). Indeed, this group assassinated Alexander II in 1881.

When Alexander III ascended to the throne after the assassination, Russia entered a period of harsh political repression coupled with a renewed drive toward industrialization. The *okhrana* (tsarist secret police) conducted investigations and raids against revolutionary groups. The government also pursued a program of Russification of its significant non-Russian population. Jews were particularly singled out for *pogroms* (organized persecutions). The industrial drive brought major advances to Russian development. During the

1890s, the Trans-Siberian Railroad connected Vladivostok to European Russia. Peasants began migrating to the cities. By 1900 Russia was the fifth leading industrial power in the world.

The politically and socially tumultuous nineteenth century also coincided with the greatest period of artistic and scientific flourishing in Russian history. Writers such as Alexander Pushkin, Fyodor Dostoyevsky, Leo Tolstoy, and Ivan Turgenev received worldwide acclaim. Some writers, notably Alexander Herzen and Nikolai Chernyshevsky, attached themselves to revolutionary movements. In music, Pyotr Tchaikovsky, Modest Musorgsky, and Mikhail Glinka composed masterpieces. Dimitri Mendeleyev's invention of the periodic table of the elements, and Ivan Pavlov's work with conditioned reflexes in dogs brought worldwide recognition.

The Twentieth Century

At the onset of the twentieth century, therefore, Russia's condition represented a paradox. On the one hand, Russia was a modernizing great power. On the other hand, Russia remained an unreformed autocracy on the verge of revolution. The defeat in the Russo-Japanese War (1904–05) brought revolution to the empire that resulted in the promulgation of an imperial constitution by Tsar Nicholas II. Yet the 1905 "October Manifesto" which created a *Duma* (parliament) was a token gesture. The prime minister lacked power which was still retained by the autocrat. Moreover, nothing was done to alleviate the continuing social inequalities among the empire's subjects.

The ordeal of World War I (1914–18) brought the end of tsarism and left the empire in tatters. A series of military reverses in the field and misery at home resulted in Nicholas's abdication in February 1917 (he and his family were subsequently executed by the Bolsheviks in July 1918). When the interim Provisional Government failed to consolidate its control domestically and proved ineffective militarily, it too, faced revolt. In October 1917, the Bolsheviks (a faction of the Russian Socialist Democratic Labor Party), under Vladimir Lenin, seized control of Petrograd (as St. Petersburg had been known since 1914). They quickly established an authoritarian regime based on mass repression by the party-controlled secret police. Although in control of the capital, the Bolsheviks faced opposition throughout most of the empire even after concluding a peace with the invading Germans. It took two years of civil war before the Communists (as the Bolsheviks were known after 1918) regained the vast majority of the Russian Empire. The civil war cost Russia millions of lives (including some of its most productive citizens, who emigrated) and devastated its industry. The Communists then faced the task of rebuilding.

Seeing themselves as the vanguard of a world socialist revolution, the Communists originally expected a quick trans-

formation to a society run by the working class. Instead, they faced almost universal hostility externally and chaos internally. Thus, the Communists spent most of the 1920s consolidating their regime. Opposition parties were banned as were factions within the ruling Communist party. In 1922, the country was reorganized and named the Union of Soviet Socialist Republics (U.S.S.R.) and consisted of theoretically autonomous Soviet Socialist Republics (S.S.R.'s) from among the country's major nationalities. Within the ruling Communist Party, a power struggle ensued after Lenin's death in 1924 which concluded with the ascendancy of Josef Stalin to the leadership. Under Stalin, the U.S.S.R. began a policy of "Socialism in One Country" with a program of crash industrialization as its centerpiece.

The Stalin era (c. 1928–53) was the most brutal period in the country's history. The policy of collectivization (the forced replacement of private holdings by government-run collective farms) brought famine and disorganization. It took generations for Soviet agriculture to reach its pre-1914 levels of production. In the 1930s, Stalin's own paranoid personality led him to purge (eliminate) the political and military leadership of the country. Paradoxically, however, the country continued to function, and the crash industrialization did increase the country's productive capacity.

In the wake of the turmoil of the 1930s followed the Second World War in which an estimated twenty-seven million Soviet citizens died. Capitalizing on the destruction of the Soviet officer corps, the German invasion of 1941 threatened to take Moscow (the capital, once again, since 1918) but improved Soviet resistance as well as the Russian winter prevented this disaster. By 1944, the U.S.S.R. had driven the invader from its territory, and, in 1945, the Red Army captured the German capital, Berlin.

The Soviet victory in World War II left it second only to the United States in terms of military power. The U.S.S.R. also extended its control into its Eastern European neighbors where it established friendly Communist regimes which were also hostile to the United States. The postwar fallout in relations with its wartime American allies plunged the Soviet Union into a Cold War which it could ill afford (see sidebar).

Stalin's successor, Nikita Khrushchev, did much to ease the excesses of the Stalin years. Khrushchev denounced Stalin's purges and attempted to pursue peaceful coexistence with the United States. The Khrushchev years also featured Soviet advances in the area of space exploration as the U.S.S.R. beat the United States with the first satellite in 1957 and the first man in orbit in 1961. The reforms under Khrushchev remained limited, however, The Cold War almost became hot in 1962 when Khrushchev placed missiles in Cuba, ninety miles south of the Florida coast. At home, the Soviet Union continued to be a police state.

A return to a semi-Stalinist repression occurred under Leonid Brezhnev's rule (1964–82). Brezhnev reversed some of Khrushchev's reforms and cracked down harder on dissent.

Abroad, the U.S.S.R. took active interest in exporting Communist revolution to newly-independent countries. Within the Soviet Union, however, the economy began to stagnate due to corruption, the inefficiencies of the Communist system, and the inability to keep pace with the United States in an escalating arms race.

By the early 1980s, the U.S.S.R. faced a perilous economic situation. Living standards dropped. An unpopular war in Afghanistan fueled the fires of a quiet opposition. The rise of Mikhail Gorbachev to the leadership of the Soviet Union in 1985 brought these problems to the fore through his introduction of *glasnost* (openness) and *perestroika* (restructuring). He introduced these policies in an effort to reform the Soviet system. While his leadership proved instrumental in easing and ending Cold War tensions, his efforts to reform Communism failed. *Glasnost*, which included free elections, opened the door to widespread public opposition to his policies; *perestroika* could not find a suitable substitute for the inefficient Communist system of centralized planning, and many rising nationalist tensions threatened the unity of the country.

In 1991, prior to the signing of a new union treaty, hardline Communists mounted a coup against Gorbachev. When the coup failed a couple of days later, the Soviet Union began to dissolve as republics began to secede. By year's end, the U.S.S.R. had ceased to exist. An independent Russia, composed of the former Russian Soviet Federative Socialist Republic (R.S.F.S.R.) emerged as the successor to the former Soviet Union, under the leadership of President Boris Yeltsin. Yeltsin moved quickly to establish major economic and political reforms. Nevertheless, he too, faced serious opposition from nationalists and Communists who viewed him as pro-Western.

The Yeltsin years have been marked by continued turmoil. He resorted to force against his parliament in 1993 and against ethnic Chechen separatists in 1994. Moreover, the Russian economy remains in crisis and organized crime continues to rise. Warm relations with the West remain tenuous as evidenced by the strong Russian opposition to the NATO bombing of Serbia and Montenegro. Nevertheless, Russia is nearing the end of one decade of democratic rule. Yeltsin won reelection over nationalist and communist rivals in 1996. The 2000 presidential elections (in which Yeltsin is barred from running) will be a major crossroads for Russia's future.

Timeline

c. 8000 B.C.   Paleolithic (Old Stone Age) habitation

Humans inhabit the Eurasian plain. Hunter-gatherers, they use primitive tools and weapons of crude chipped stone.

c. 4000 B.C.   Neolithic (New Stone Age) habitation

Eurasian peoples develop more sophisticated polished stone tools and weapons and begin domesticating animals and cultivating the land.

c. 1st millenium B.C.   Invaders strike through the Eurasian plain

The Eurasian plain is the site of western invasion routes by peoples originating in Asia and the Middle East.

c. 1000 B.C.   Cimmerians inhabit region north of Black Sea

The Cimmerians, an iron-working people related to the Thracians, settle in the plains to the north of the Black Sea.

7th century B.C.   Scythians supplant Cimmerians

The Scythians, a group whose origin is disputed, replaces the Cimmerians as the leading group in the Black Sea coastal region.

7th century B.C.   Greeks in the Crimea

The Greeks establish colonies in the Crimea and found the city of Olbia at the mouth of the Bug River in 644. The establishment of a Greek city links the Eurasian plain to the rest of the Greek world via trade.

3rd century B.C.   Sarmatians settle in the Eurasian plain

The Sarmatians, Iranian cattle-breeding nomads, arrive in the Eurasian plain from Central Asia. The Sarmatians preserve the old trade routes in the Eastern Mediterranean world.

1st century A.D.   Goths move south into Eurasian plain

The Goths, Germanic people divided into several tribes, begin moving south into the Eurasian plain.

4th century   Hun invasions

The Huns, a fearsome nomadic tribe from central Asia, sweep westward and displace those who stand in their way, including the Goths who flee west and help bring about the decline and collapse of the Western Roman Empire. At the same time, another group, the Slavs, begin their migration from central and eastern Europe. Meanwhile, the Huns, under their greatest leader, Attila (406–53), rampage through Europe, including the Roman Empire. Following his death, Hun power declines amid a series of wars between his successors.

6th century   Avars displace Huns

The Avars displace the Huns. The Avars are yet another barbarian group to emerge from central Asia and are a mixture of Chinese, Mongolian, and Turkish. By the late sixth century, the Avars control a large eastern European state between the Volga and Elbe Rivers. In the early seventh century, the Avars suffer a catastrophic defeat at the hands of the Byzantine (Eastern Roman) Empire, and their state quickly falls apart.

7th century   Rise of the Khazars

The Khazars, a Turkic group from the area between the Black and Caspian Seas, emerge as a leading military power that comes into commercial contact with the Arab and Byzantine Empires. Many Slavs inhabit the region under Khazar control.

One of the important characteristics of Khazar rule is the cosmopolitan nature of the region. Christianity, Islam, and Judaism seek converts in the area—indeed, the Khazar dynasty eventually converts to Judaism.

8th century   Slavs in the Dnieper Basin

The eastern branch of Slavs settle in the area of the Dnieper River. This branch of Slavs are the ancestors of the modern-day Great Russians, Belorussians (White Russians), and Ukrainians (Little Russians). The other branches of Slavs are the South Slavs (modern-day Bulgars, Croats, Macedonians, Slovaks, and Wends) and the West Slavs (modern-day Czechs, Poles, and Slovaks).

From Rurik to the Romanovs, Ninth Century–1613

Ninth century   Rurik and the emergence of the first Russian state

The first Russian state, Rus, emerges in the ninth century under the leadership of Rurik, a Varangian (Norman) from present-day Sweden. Rus is centered in the area of Novgorod. Rurik's retainers extend the state's boundaries southward to include the city of Kiev, situated on the Dnieper River.

860   Attack on Constantinople

A Russian attack on Constantinople, the capital of the Byzantine Empire, fails.

882   Union of Novgorod and Kiev

Oleg the Wise (r. 879–912) unites the Novgorod and Kiev regions. The expanded state—as its name, Kievan Rus, indicates—is centered politically and culturally on the city of Kiev.

944   Second attack on Constantinople

A second attempt to take Constantinople fails. Following their defeat, the Rus sign a trade agreement with the Byzantine Empire. This treaty allows for Byzantine penetration into Kievan Rus ultimately paving the way for conversion to Orthodox Christianity.

988   Conversion to Christianity

Prince Vladimir (r. 978–1015) seeks to convert his subjects to a religion and is deeply impressed by the form of Christianity practiced in the Byzantine Empire (known as Orthodox Christianity following the split in the Christian Church in 1054). At the same time, the Byzantine Empire is in need of allies and welcomes the conversion of Vladimir. Thus, the Byzantines offer Vladimir the emperor's sister, Anna, in marriage in return for military support. Once the threat to Byzantium passes, however, the Byzantines renege on their offer and Vladimir attacks Byzantine territory in the Crimea. When the Byzantines lose, they agree to let Vladimir marry Anna in return for Vladimir's conversion. Following his baptism, Vladimir orders the rest of his population baptized.

10th–11th centuries   Rise of Kievan Rus

Kievan Rus becomes a major power. Under the leadership of Sviatoslav in the tenth century, the Rus destroy the Khazar state and the Danubian Bulgarian kingdom (whose inhabitants are Turkic and bear no relation to present-day Bulgarians). By the eleventh century, the Kievan Rus dominates the Eurasian plain north of the Black Sea, and the Slavs supplant the Varangians in the upper class of the state. A hereditary prince heads the state, but his power is shared with *boyars* (landowning nobility) and urban assemblies.

11th–13th centuries   Decline of Kievan Rus

Beginning in the mid-eleventh century, centralized authority dissolves as power increasingly moves toward regional princes.

1223   Battle of Kalka River

The Mongols, a nomadic people from central Asia, win a major victory against Kievan forces on the Kalka River, just north of the Sea of Azov. Under the leadership of Genghis Khan (1162–1227), the Mongols create an empire that stretches from the Sea of Japan to eastern Europe. Even after Genghis Khan's death, his empire continues to expand westward. Mongol rule over Russia lasts over two centuries and the period of Mongol rule is commonly referred to as the Mongol Yoke.

1240   Prince Alexander Nevsky defeats the Swedes on the Neva

Prince of Novgorod, Vladimir Alexander Nevsky defeats Swedish forces on the Neva River. Two years later, his forces drive back an invading force of Teutonic Knights. The Orthodox Church subsequently canonizes Nevsky.

1251   Batu establishes himself at Sarai

Batu (d. 1255), a grandson of Genghis Khan, establishes the Khanate of the Golden Horde at Sarai along the banks of the

Volga River. Under Batu the Mongols continue moving westward into Europe and reach as far as the Adriatic Sea.

1325–41 Rule of Ivan I Kalita ("Money Bags")

Ivan I, known as Kolita, or "Money Bags," due to his large tributary payments to the Mongols, receives the title Prince of Vladimir. Ivan's close relationship with the Mongols increases the significance of Moscow in comparison with other Russian cities.

1359–89 Rule of Dimitri

Prince Dimitri Ivanovich (later known as "Donskoy") extends Moscow's power through his victories over the rival states of Tver and Ryazan. He subsequently defeats the Empire of the Golden Horde (Mongols).

1380 Battle of Kulikovo

Under Dimitri, Muscovy (emanating from Moscow) forces defeat those from the Empire of the Golden Horde. This is the first Russian victory over the Mongols and establishes Muscovy as the center of Russian civilization.

1439 Russian Orthodox Church rejects the Union of Florence

Surrounded by the Ottoman Turks and desperate for foreign assistance from Western Europe, the Byzantine Empire agrees to place its church under the jurisdiction of Rome, the center of Roman Catholicism. The Russian Orthodox Church views this act, known as the Union of Florence, with disdain and breaks with Constantinople. Subsequently, many in the Russian Orthodox Church argue that the fall of Byzantium to the Ottomans in 1453 is punishment for the union.

1462–1505 Rule of Ivan III (the Great)

Ivan III succeeds to the throne and consolidates the Russian possessions by securing the loyalty of all princes to the grand prince of Moscow. In addition, Ivan begins cultivating the image of Russia as the Third Rome. He marries Sophia Paleologue, niece of the last Byzantine Emperor, Constantine XI (1404–53) and adopts Byzantine symbols such as the double-headed eagle and the title *autocrat* (independent ruler, although it later comes to mean ruler with unlimited power).

1478 Novgorod conquered

Ivan III conquers Novgorod. He later breaks the city's continued resistance through deportations and massacres.

1480 Moscow freed from the Golden Horde

Moscow officially becomes independent of the Empire of the Golden Horde. The Mongol Empire, in turn, disintegrates.

1547 Ivan IV Grozny (the Terrible) is crowned tsar

Ivan IV, grand prince of Moscow at the age of three in 1533, is crowned *tsar* (Caesar). His rule brings mixed results. On the one hand, he carries out reforms of the administration, the law, and the army, and encourages Siberian exploration and expansion. On the other hand, the later years of his rule feature an arbitrary rule that alienates the boyars (landholding nobility) and brings the country to the brink of ruin in a series of wars, particularly his unsuccessful twenty-five year war in Livonia.

1552 Expansion into Kazan and Astrakhan begins

Ivan IV conquers Kazan and later moves against Astrakhan. These new territories extend Russian rule to the Volga River and the Caspian Sea.

1558 Stroganov family gets rights to Siberia

Ivan IV grants the rights to Siberia to the Stroganov family of merchants contingent upon the colonization of the area. This sets the stage for the greatest territorial expansion and exploration by a single state in history.

1565–72 Partition of imperial territories between tsar and the public

Ivan IV institutes a policy known as the *oprichina* which is a division of the imperial lands between the tsar and the public. The policy effectively breaks the power of the boyars, many of whom own territory claimed by Ivan IV. In addition to breaking the power of the boyars, Ivan's policy also ties the peasants to the land, thus setting the foundation for serfdom, which lasts until 1861.

1571 Crimean Tatars attack Moscow and burn its suburbs

The Crimean Tatars attack Moscow, burn the city, and take 100,000 Russians prisoner.

1581–84 Hetman Yernak crosses Western Siberia

The Cossack Hetman Yernak (d. 1584) crosses into Western Siberia in the service of the Stroganov family. He conquers the Khanate of Sibir, and the territory is annexed to Russia.

1584 Port of Archangel is founded

Russia founds the port city of Archangel on the White Sea in order to facilitate trade.

1588 Boyar Boris Godunov becomes regent

Ivan IV dies in 1584 and is succeeded by his son Fedor, who is mentally unfit. As a result, Boris Godunov becomes regent (one who governs when the sovereign is a minor).

1589   Moscow becomes an independent patriarchate

The Russian Orthodox Church becomes formally independent of Constantinople with the proclamation of the Patriarchate of Moscow.

1591   Tsar's son is murdered

Dimitri, a son of Ivan IV is murdered under mysterious circumstances.

1598   Boris Godunov is elected tsar

Tsar Fedor dies without any heirs and the *semski sobor* (a special assembly of boyars, clergy, gentry, and merchants) elect the regent Godunov, tsar.

1601–03   Impostor Dimitri

Disaster strikes Russia as crops fail and famine follows causing widespread discontent. A pretender to the throne claiming to be Dimitri (d. 1606) emerges from Poland and tries to take advantage of the chaos. Dimitri marches into Moscow and, after Godunov's death in 1605 and the murder of Godunov's son, is crowned emperor.

1605–13   The Time of Troubles

In the wake of the False Dimitri's coronation comes eight years of chaos that includes civil war with intervention by Poland and Sweden. The period ends with the election of Mikhail Romanov as tsar in 1613.

1606–07   Second Impostor

Polish and Swedish forces help overthrow the first False Dimitri who is replaced by the boyar Vasili Shuyski in 1606. Shuyski allies with the Swedes, but a second False Dimitri (d. 1610), allied with Poland, appears. The second impostor claims the throne in 1610 when Polish forces capture Moscow. Soon thereafter, however, Russians rebel against the Poles and drive the second False Dimitri from the throne.

Romanov Period, 1613–1917

1613   Michael Romanov becomes tsar

The Time of Troubles comes to an end when the new *zemski sobor* elects Mikhail Romanov, a boyar, tsar. This event marks the beginning of the three hundred year Romanov dynasty.

1682–1725   Reign of Peter the Great

Peter I (1672–1725) becomes co-tsar at age ten with his half brother while his half-sister Sofia runs the government. In 1689 Peter succeeds in ousting his sister and, upon the death of Ivan in 1696, becomes sole tsar. Peter, one of the most significant tsars, tries to westernize Russia. During his reign he

wages a successful war against Sweden and in the early eighteenth century founds a new capital city, St. Petersburg (later Petrograd, Leningrad), on the Baltic Sea. In addition, he imports Western European scientific and technological experts and urges Russian nobles to adopt Western dress, mannerisms, and taste.

1700–21   Great Northern War

Peter I fights a long and successful war against King Charles XII of Sweden. Known as the Great Northern War, the Russian victory results in significant Russian expansion in the Baltic, including Livonia and Finland. Russia becomes the leading Baltic power.

1709   Battle of Poltava

In the most significant battle of the Great Northern War, Russian forces inflict a decisive defeat on the Swedes at Poltava in Ukraine. The Russians destroy the Swedish army, and Charles escapes to the Ottoman Empire.

1711   Mikhail V. Lomonosov is born

Future academic Mikhail V. Lomonosov (1711–65) is born. After receiving his education at the Slavo-Greco-Latin Academy in Moscow he studies in Germany. Upon his return to Russia he is a professor at the Academy of Science in St. Petersburg where he studies matter, writes a history of Russia, a grammar of Russian, and founds the country's first chemical laboratory.

1725   Academic University opened in St. Petersburg

The Academic University (currently known as St. Petersburg University) opens in St. Petersburg. It is the first university in Russia.

1735–39   War against the Ottoman Empire

War against the Ottoman Empire brings a Russian victory that results in the capture of a small part of the Crimea and the port of Azov on the Black Sea.

1749   Alexander Radischev is born

Liberal author Alexander Radischev (1749–1802) is born. His critical account of conditions under serfdom, *A Journey from St. Petersburg to Moscow* (1789), earns him the wrath of Catherine the Great who sentences him to death. The sentence is later changed to Siberian exile. Radischev commits suicide in 1802.

1755   Moscow University is established

Moscow University is established. Although at first few students attend and instruction is in Latin, the university soon becomes a center of intellectual activity. In the nineteenth and twentieth centuries it becomes the premier institution of higher learning in Russia.

1762–96   Reign of Catherine the Great

Following the murder of her weak husband, Peter III, the German-born Catherine becomes tsaritsa. She expands Russian territory and strengthens autocratic power, particularly during her suppression of the Pugachev peasant rebellion in 1773.

1766   Nikolai Karamzin is born

Romantic author Nikolai Karamzin (1766–1826) is born. He is best known for his *Letters of a Russian Traveler* (1791), and *Poor Liza* (1792).

1768–74   Russo-Turkish War

Russia fights another victorious war against the Ottoman Empire and makes further gains in the Crimea. The Treaty of Kutchuk Kainarji (1774) ends the conflict and also recognizes Russia as the "protector of the Balkan Christians" thus, paving the way for future Russian intervention in the Balkans on behalf of its fellow Orthodox.

1768   Ivan A. Krylov is born

The critic, playwright and teller of fables Ivan A. Krylov (1768–1844) is born. He is best known for his moral fables, the tone of which antagonizes many in St. Petersburg.

1772   Partition of Poland

Austria is disturbed by the Russian victories against the Turks and fears that an extension of Russian power will destroy the delicate balance of power among the leading European states. In an effort to prevent Austrian action against Russia, King Frederick II (the Great) of Prussia (1712–86) proposes a partition of roughly twenty-five percent of Poland among Austria, Prussia, and Russia. The three states agree and under this plan, Russia gains territory in Belorussia which is populated primarily by Belorussians rather than Poles.

1773–74   Pugachev Rebellion

The Cossack (a group of frontier settlers given extensive privileges by the tsar in exchange for providing troops for the imperial army) Emilian I. Pugachev (1742–75) leads a peasant revolt in the region between the Volga River and Ural Mountains. Pugachev has widespread support that reflects popular discontent among serfs with the government of Catherine the Great.

1783   Annexation of the Crimea

Russia annexes the remainder of the Crimean peninsula.

1787–92   Russo-Turkish War

The Ottoman Empire accuses Russia of violating the Treaty of Kutchuk Kainarji and attacks Russia. Despite simultaneously fighting a war against Sweden, Russian forces win a convincing victory, and Russia gains more territory along the Black Sea including the port of Odessa. Russia's borders now extend to the Dniester River.

1792   Nikolai Ivanovich Lobachevsky is born

Nikolai Lobachevsky (1793–1856) is born in Nizhni-Novgorod. He invents non-Euclidian geometry and serves as a professor and rector at the University of Kazan.

1793, 1795   Second and Third Partitions of Poland

In two further partitions, Austria, Prussia, and Russia eliminate Poland from the map of Europe. As a result of these annexations, Russia gains the rest of Belorussia as well as territory that is ethnically Polish.

1799   Alexander Pushkin is born

Poet Alexander Pushkin (1799–1837) is born. During his short life, he becomes the greatest Russian poet of all time. His works reflect French Enlightenment influence; his criticism of serfdom and the authorities in works such as *Ode to Liberty*, *The Village*, and *Hurrah! He's Back in Russia Again* land him in exile in southern Russia for a time. Following his return in 1824, he writes a series of acclaimed works including: *The Gypsies*, *Boris Godunov*, and *Eugene Onegin*. He dies in a duel in 1837.

1801   Tsar Alexander I ascends to the throne

Following the murder of his father, Paul I (1754–1801), Alexander I (1775–1825) becomes tsar. His reign coincides with the peak of Russian power. The years of Alexander's rule are divided into roughly two periods: the era of the liberal period corresponding with the Napoleonic Wars, and a conservative era (c. 1814–25). In the early years of his reign, Alexander considers continuing the modernizing reforms pursued by his father. As his reign continues, however, he grows more conservative and generally opposes revolutions. For instance, he backs a return of Spanish rule in Latin America.

1804   Mikhail Glinka is born

Composer Mikhail Glinka (1804–57) is born. As a youth, he studies piano under John Field and opera in Italy. His best known works include *Ivan Susanin* (a.k.a. *A Life for the Tsar*) and *Ruslan and Liudmilla*.

1812   Alexander Herzen is born

The radical Alexander Herzen (1812–70) is born. The son of a nobleman, Herzen becomes the leading Russian proponent of socialism in the mid-nineteenth century. Using the fortune he inherits, he moves to London where he edits the anti-tsarist newspaper *Kolokol.*

1812 Napoleon invades Russia

As a result of the Russian refusal to comply with the Continental System (a European-wide blockade of Britain imposed by French emperor Napoleon Bonaparte 1769–1821), the Franco-Russian alliance disintegrates, and French forces invade Russia in 1812. At first the French advance is swift and victorious. In response, the Russian forces adopt a "scorched earth" policy which lays waste to evacuated territories. In September, French forces seize Moscow only to find it destroyed. In October, after Alexander refuses to negotiate peace terms with Napoleon, the French leader orders his troops out of Moscow. The retreat is disastrous for Napoleon as the cold Russian winter and attacks by Russian forces batter the French forces. Only 30,000 of Napoleon's nearly 600,000 troops survive the retreat.

1814: March 30 Alexander I rides into Paris

After a series of French defeats, Alexander I enters Paris at the head of the anti-Napoleonic coalition. This moment best represents the Russian's position as the strongest land power in Europe.

1818 Ivan Turgenev is born

The novelist Ivan Turgenev (1818–83) is born. The son of a noble, he is arrested in 1852 following publication of his obituary of Gogol which the state deems unsatisfactory. After his pardon he settles in St. Petersburg and later moves to Paris where he writes most of his novels. His works include: *Rudin* (1856); *Nest of Gentlefolk* (1859); *On the Eve* (1860); *Fathers and Sons* (1862), his most well-known work; and *Smoke* (1867).

1821 Fyodor Dostoyevsky is born

The novelist Fyodor Dostoyevsky (1821–81) is born to a wealthy family. Unlike many of his contemporary writers, Dostoyevsky is not a radical, nor is he a conservative (although he briefly joins a radical discussion group which causes the tsarist police to arrest and exile him). His works focus on his strong religiosity. His best known novels include *Poor Folk* (1845), *The Double* (1846), *Notes from the House of the Dead* (1861), *Crime and Punishment* (1866), *The Idiot* (1868–69), and *The Brothers Karamazov* (1879–80).

1825 Nicholas I becomes tsar amidst an uprising (Decembrist Revolt)

Upon the death of his older brother, Alexander I, Nicholas I (1796–1855) becomes tsar in December 1825. His succession is marred by a revolt among young officers and intellectuals who favor the promulgation of a constitution. The rebels, known as Decembrists, number only about 3,000 and are

poorly organized. The revolt collapses almost as soon as it begins, but it acquires legendary status among future Russian revolutionaries who view the Decembrists as martyrs for the cause of reform.

Following the suppression of the revolt, Nicholas begins a thirty-year reign characterized by unswerving conservatism and opposition to the most rudimentary reforms as symbolized by his motto, "Autocracy, Orthodoxy, and Nationalism." During Nicholas's reign, Russian military power drops markedly, and his refusal to reform leaves Russia even farther behind its European rivals economically, socially, and politically. Nicholas dies in 1855, toward the end of the disastrous Crimean War.

1828 Alexander Mikhailovich Butlerov is born

The Scientist Alexander Mikhailovich Butlerov (1826–86) is born. As a professor at Kazan University and the University of St. Petersburg, Butlerov discovers formaldehyde and devises a theory of the structure of organic matter.

1828 Nikolai Chernyshevsky is born

Radical writer Nikolai Chernyshevsky (1828–89) is born in Saratov. He is best known as the author of *What Is to Be Done?*, a political novel he writes while in prison in 1863. The novel serves as an inspiration to future Russian revolutionaries, including Vladimir Lenin. In addition to novels, Chernyshevsky also writes for the periodical, *Sovremennik.*

1828 Count Leo (Lev) Tolstoy is born

Acclaimed novelist Count Leo Tolstoy (1828–1910) is born. Tolstoy is best known for his 1869 novel *War and Peace.* A historical novel set during Napoleon's invasion of Russia, the work is extremely complex and views the period from the perspective of different groups and social classes. Other well-known works include *Anna Karenina, A Confession* (1882), and *The Kreutzer Sonata* (1889).

1833 Treaty of Unkiar Iskelessi

In a rare foreign policy victory for Nicholas I, Russia signs the Treaty of Unkiar Iskelessi with the Ottoman Empire. According to the terms of this treaty, Russia promises to support the maintenance of the Ottoman Empire while the Porte (a term used to refer to the Ottoman government) agrees to close the Straits to foreign warships. This treaty marks the high point of Russian influence with the Porte.

1834 Dimitri I. Mendeleyev is born

Scientist and industrialist Dimitri I. Mendeleyev (1834–1907) is born. After completing his education, Mendeleyev embarks upon a multi-faceted career as a scientist, professor, and industrialist. He is best known for discovering the periodic table of the elements. In addition, he conducts private and public scientific research and publishes his findings.

1839 Modest Musorgsky is born

Composer Modest Musorgsky (1839–81) is born. He is primarily renowned for his operas of which the best known is *Boris Godunov*.

1840 Pyotr Ilyich Tchaikovsky is born

Composer Pyotr Ilyich Tchaikovsky (1840–93) is born. Although trained as a lawyer, he attends the St. Petersburg Conservatory and subsequently becomes a professor at the Moscow Conservatory. Tchaikovsky's well-acclaimed works include *The Sleeping Beauty* (1889), *The Nutcracker* (1892). In 1893 he composes the Sixth Symphony (*Pathétique*), widely considered his best work.

1846 Karl Fabergé is born

Karl Fabergé (1846–1920) scion (offspring) of one of St. Petersburg's leading jewelers is born. Eventually, he becomes jeweler to the imperial family and is best-remembered for the jeweled eggs he designs for them.

1848–49 Russia helps suppress revolution

Revolutions sweep through Europe (but not Russia) and threaten several conservative regimes. In Austria, the revolutions force the resignation of Prince Clemens von Metternich (1773–1859), the very symbol of conservatism while the empire's Hungarians revolt. In response to a request from the Austrian emperor, Franz Josef (1830–1916) Nicholas I sends Russian troops in May 1849, and they crush the revolt.

1849 Ivan Petrovich Pavlov is born

Renowned scientist Ivan Petrovich Pavlov (1849–1936) is born. Following his studies at the University of St. Petersburg, Breslau, and Leipzig, Pavlov becomes a professor of physiology at the Russian Academy of Sciences. While there, he conducts conditioning experiments in dogs which win him international acclaim, including a Nobel Prize in 1904.

1852 Vera Figner is born

Revolutionary Vera Figner (1852–1942) is born to a noble family. While a student in Zurich, Switzerland she becomes a revolutionary and joins the revolutionary movement, *Narodnaia Volia* or "People's Will" (see 1870s), the group which assassinates Tsar Alexander II in 1881 (see 1881). Figner is eventually captured and spends twenty years in prison. Upon her release, she leaves Russia but returns in 1917.

1853–56 Crimean War

A trivial religious dispute in the Ottoman-controlled Holy Land between Orthodox and Catholic monks causes Russia (acting in its role of "protector" of the Orthodox in the Ottoman Empire) to pressure the Porte for greater influence. However, Britain and France fear an extension of Russian influence into the eastern Mediterranean and offer backing to the Porte to resist the Russian demands. The diplomatic situation deteriorates and war breaks out. Sardinia also joins the anti-Russian coalition. In 1855, Austria declares that it, too, will enter if Russia refuses to surrender. The Austrian decision to oppose Russia is viewed by the latter as a betrayal in the wake of Nicholas's aid in crushing the Hungarian revolt.

The outcome of the war is a humiliating defeat for Russia. The British and French send forces to the Crimea in 1854 and, following the Austrian threat to intervene in 1855, Russia agrees to cede the principalities of Moldavia and Wallachia (both part of present-day Romania) to the Porte and to dismantle its Black Sea fleet and its bases. The most important outcome of the defeat is the realization that Russia is militarily, economically, and technologically inferior to its European rivals and needs to embark upon serious reform to catch up.

1855–81 Reign of Alexander II (the "Tsar Liberator")

Near the end of the Crimean War, Nicholas I dies and is succeeded by his son, Alexander II. Realizing that Russia needs major modernizing reforms if it is to continue as a major European power, Alexander concludes the Crimean War and begins a policy of modernization, the Great Reforms. His most ambitious reform is the abolition of serfdom in 1861. Other reforms include changes in the judiciary, military, educational system, and local government. As his reign continues, however, the *intelligentsia* (educated Russians of different social backgrounds with a shared political and cultural outlook) oppose him, and he is assassinated by terrorists in 1881.

1856 Georgi Plekhanov is born

Georgi Plekhanov (1856–1918), the "Father of Russian Marxism," is born in St. Petersburg. He becomes a revolutionary in the 1870s and leaves Russia for Western Europe in 1880 where he adopts his Marxist ideology. Although he returns to Russia in 1917, he opposes his former revolutionary pupil, Lenin.

1857 Konstantin Eduardovich Tsiolkovsky is born

Space flight theoretician Konstantin Eduardovich Tsiolkovsky (1857–1935) is born. When flight and rocketry are in their early stages, Tsiolkovsky devises theories on manned space flight and space colonies.

1859 Alexander Stepanovich Popov is born

Inventor Alexander Stepanovich Popov (1859–1905) is born. Popov is a pioneer in radio who develops the first receiver in 1895 and the first transmitter the following year.

1859–95   Expansion into the Caucasus and Central Asia

Russia expands its frontiers southward into the Caucasus and Central Asia. Russians subdue the Caucasus by the early 1860s. By contrast, the expansion process in Central Asia is a gradual one based on the continued acquisition of desert oases. By the mid-1880s, Russian control of Bukhara, the Kazakh Steppe, Transcaspia, and Turkestan stirs a renewed rivalry with Britain, which fears the loss of its supply line to India. Russia's frontiers in Central Asia are settled in 1895 although its rivalry with Britain over spheres of influence continues until 1907.

1860   Anton Chekhov is born

Playwright Anton Chekhov (1860–1904) is born in Taganrog. Although trained in medicine, Chekhov turns his attention to writing (which, through freelance work pays for his education) and soon emerges as one of Russia's foremost writers. Among his most influential plays are: *The Sea Gull* (1895), *Uncle Vania*, *The Three Sisters*, and *The Cherry Orchard* (all written between 1899 and 1903). In poor health from tuberculosis, he dies in 1904.

1861   Serfdom abolished

Alexander II institutes his most significant reform, the abolition of serfdom. Prior to this reform, the majority of Russia's peasants were serfs, peasants tied to the land they worked. The abolition of serfdom sets them free. Yet their emancipation leaves them in serious difficulty because the land they gain is limited, they owe dues for several years after emancipation, and they are tied to the *mir* (peasant commune).

1863   Konstantin Stanislavsky (Alekseyev) is born

Artistic leader Konstantin Stanislavsky (Alekseyev) (1863–1938) is born. He founds the Society of Art and Literature and, in 1897, forms the Moscow Art Theater.

1866   Vasily Kandinski is born

Modernist artist Vasily Kandinski (1866–1944) is born in Moscow. Following a period of working abroad, he returns to Russia to found the Museum of Painting in St. Petersburg. After the Bolshevik Revolution he leaves Russia for good and lives in Germany, France, and the United States.

1868   Maxim Gorky is born

Writer Maxim Gorky (1868–1936) is born. Some of his works include *Makar Chudra* (1892), *Mother* (1907), and *Among Strangers* (1915). Gorky's works reflect a broad humanitarianism that reflects his own experiences as an impoverished youth. His own poverty makes him a convert to Marxism, and he is an early supporter of Lenin. Exiled during the late tsarist period, he returns in 1913 and works for the Bolsheviks. He criticizes some of Lenin's actions after the revolution and leaves Russia only to return in 1931.

1870s   Rise of radical movements

Increasing discontent among the intelligentsia leads to a rise in radical opposition movements, including those that espouse violence. The most famous of these terrorist groups is *Narodnaia Volia* (People's Will), a socialist group that assassinates Tsar Alexander II in 1881.

1870: April 9   Vladimir Ilyich Ulyanov (Lenin) is born

Revolutionary and Soviet leader Vladimir Ilyich Ulyanov (Lenin) is born in Simbirsk. Ulyanov becomes a revolutionary after his brother is arrested and executed by the tsarist government for attempting to assassinate Alexander III. Ulyanov becomes a Marxist whereupon he joins the Russian Socialist Democratic Labor Party and adopts the name Lenin. His revolutionary activity earns him three years of exile in Siberia.

Upon his release, Lenin leaves Russia and in 1903 he becomes the leader of the *Bolshevik* (majority) faction of the party (see 1903). Although he returns to Russia during 1905, he is not a leader in the revolution (see 1905) and leaves the country once again in 1907. He spends a decade in Western Europe and eventually settles in Switzerland. While in exile, he begins to form the core of a tightly-knit revolutionary party. After the February Revolution (see 1917), he returns to Russia and begins organizing his followers. In October 1917, he mounts a successful coup against the Provisional Government and becomes the new Russian leader.

He inherits a country deeply divided and on the brink of civil war (which breaks out in 1918). Although his plans for world revolution and a quick economic transformation for his own country fail, he does manage to exert Communist control on nearly all of the former Russian Empire (known after 1922 as the Union of Soviet Socialist Republics) by 1920–21 (see various entries for the period (1917–24). Lenin suffers a stroke in 1922 and dies in 1924. Following his death (and against his wishes), Lenin's body is embalmed and displayed in a specially-constructed mausoleum in Moscow. Statues of him appear throughout the Soviet Union and his name graces many prizes and honors throughout the Soviet era.

1870   Ivan Bunin is born

Poet Ivan Bunin (1870–1953) is born. Like many Russian intellectuals, Bunin travels and works abroad where he receives most of his acclaim. An opponent of the Bolsheviks, he leaves Russia for good after the revolution and continues to work abroad. In 1933, his work garners him a Nobel Prize for literature.

1872 Sergei Diaghilev is born

Artist and musician Sergei Diaghilev (1872–1929) is born. After a stint as an editor of an art journal and art exhibitor, he begins working as a Russian concert organizer in Paris in 1905. From 1911 to 1929 he is the director of the *Ballets Russes* (Fr., Russian Ballet).

1872 Alexandra M. Kollontai is born

Revolutionary, feminist, and diplomat Alexandra M. Kollontai (1872–1952) is born. An early convert to the revolutionary cause, she later joins the Bolsheviks and becomes Lenin's collaborator and lover while he is in exile in Switzerland. Although she opposes Lenin's decision to sign a peace with Germany (see 1918), she remains with the Bolsheviks and as Soviet Ambassador to Norway, Mexico, and Sweden serves as the first woman ambassador in the world.

1873 Fyodor Chaliapin is born

The great opera basso Fyodor Chaliapin (1873–1938) is born. He begins his career in 1896 and is famous for his performances as Boris Godunov and Ivan the Terrible. He leaves the Soviet Union in 1921.

1873 Vsevolod Y. Meyerhold is born

Actor and theater director Vsevolod Y. Meyerhold (1873–1942) is born. An actor at the Moscow Art Theater, he joins the Communists in 1918 and is later made director of Moscow theaters. He subsequently falls out of favor with the Soviet government which closes his own theater in 1938 and exiles him to a labor camp.

1873 Sergei Rakhmaninov is born

Composer Sergei Rakhmaninov (1873–1943) is born. After the Bolshevik revolution he leaves Russia for Sweden and eventually settles in the United States. *Rhapsody on a Theme of Paganini* is among his major works.

1874 Serge Koussevitzky is born

Famed double-bass player and orchestra conductor Serge Koussevitzky (1874–1951) is born. He founds the Moscow Symphony Orchestra, but leaves Russia during the civil war. While in the United States he becomes the conductor of the Boston Symphony Orchestra (1924–49).

1877–78 Russo-Turkish War

In response to an uprising in the Ottoman territory of Bosnia, Russia attacks the Ottoman Empire and fights a short, victorious war. However, the Treaty of San Stefano, which ends the fighting, alarms the other European powers. These states fear that the creation of a large Bulgarian state allied to Russia will give the Russian navy access to the Mediterranean Sea and therefore upset the balance of power. As a result, the powers convene the Congress of Berlin which eliminates the "Great Bulgaria." The results of this conference demonstrate the limits of Russian power in Europe.

1878 Kasimir Malevich is born

Abstract painter Kasimir Malevich (1878–1935) is born. Malevich devises his own style, Suprematism.

1879: November 8 Lev Davidovich Bronstein (Leon Trotsky) is born

Lev Bronstein (Leon Trotsky) (1879–1940) is born. Originally a Menshevik (see 1917), he switches his support to the Bolsheviks in 1917 and is one of Lenin's key allies in the revolution. As Commissar of Foreign Affairs, he concludes the Treaty of Brest-Litovsk with Germany in 1918. During the civil war, he serves as Commissar of War and organizes the victorious Red Army. Trotsky loses the factional power struggle after Lenin's death and is exiled. From exile he opposes Stalin's theory of "socialism in one country" and founds the Fourth International to oppose the Comintern (see 1919). On Stalin's orders, a NKVD agent assassinates Trotsky in Mexico in 1940.

1879: December 21 Josef V. Djugashvili (Stalin) is born

Josef V. Djugashvili (Stalin) (1879–1953) is born in Georgia. While studying to be a priest, Djugashvili becomes a radical who joins the Russian Socialist Democratic Labor Party. Following the factional split in 1903, he joins the Bolshevik faction. He later adopts the pseudonym, Koba and changes his last name to *Stalin* (man of steel). His role in the party's central committee gets him exiled but he is released following the February Revolution. After the Bolshevik Revolution, he serves as Commissar of Nationalities. Lenin subsequently promotes him to secretary-general of the Communist Party, a position that he uses to enhance his own power base. In the wake of Lenin's death, Stalin defeats his opponents and establishes a one-man dictatorship.

Stalin's regime is one of the most brutal in all of world history. He maintains control over the country through the often arbitrary use of terror by the security services. Periodic purges (removals) of party officials and ordinary citizens, alike, prevent opponents from threatening his power. The worst purge, the "Great Terror" of the late 1930s claims the lives of millions. He also establishes a cult of personality around himself while minimizing the roles of others in the functioning of government or even basic day-to-day life. Indeed, his leadership often contributes to extreme inefficiency due to the bloated security apparatus and fear among officials and the public. The collectivization of agriculture best typifies Stalinist ineptness.

At the same time, his rule—in spite of its excesses—succeeds in industrializing the U.S.S.R. and making it the second strongest military power in the world. As a result, subsequent

Soviet governments limit their repudiation of his methods because to do so completely would be to discredit the whole Soviet system. A thorough discrediting of Stalin occurs only on the eve of the U.S.S.R.'s collapse.

1881 Alexander F. Kerensky is born

Future prime minister Alexander Kerensky (1881–1970) is born. After earning a law degree, he enters politics and joins the Duma (see 1905) as a right-wing Socialist Revolutionary (S.R.). Following the tsar's abdication, he becomes minister of justice and then war minister of the Provisional Government. In July 1917, he becomes prime minister. His refusal to negotiate an end to the war and his failure to convene a constituent assembly cost him his government support. The Bolsheviks overthrow his government in a bloodless coup in October 1917 and Kerensky flees Russia. He dies in the United States.

1881–84 Reign of Alexander III

Alexander III (1845–94) succeeds his father following the latter's assassination at the hands of terrorists. Unlike his father, Alexander III is a most conservative—some would say, reactionary—autocrat. He concentrates his efforts on restoring order through the suppression of liberal and revolutionary movements, Russifying the non-Russian population of the Empire (particularly Jews), and rapid industrialization. Under the direction of Count Sergei Y. Witte (1849–1915) Russia begins construction of major industries and railroads (including the Trans-Siberian, that links the Russian Far East to European Russia). Industrial progress is rapid; by the turn of the century, Russia is the fifth largest industrial power in the world. Unfortunately, for Russia, however, all of its leading rivals are more industrialized.

1881 Anna Pavlova is born

Famed ballerina Anna P. Pavlova (1881–1931) is born. She studies in St. Petersburg, but begins performing abroad in 1909. She creates her own ballet company in London in 1914.

1889 Vaclav F. Nijinski is born

Famous ballet dancer Vaclav F. Nijinsky (1889–1950) is born. He performs at the Marynski Theater and then joins the Ballets Russes. Mental illness later strikes ending Nijinski's career.

1890 Boris L. Pasternak is born

Poet Boris L. Pasternak (1890–1960) is born. Pasternak studies abroad but remains in Russia after the revolution where he becomes one of the most prolific Russian poets of the century. He also translates the writings of Goethe and Shakespeare into Russian. Pasternak's *Dr. Zhivago* (published abroad in 1958), earns him the Nobel Prize for literature.

Under pressure from the Soviet government, Pasternak refuses the award.

1891 Sergei S. Prokofiev is born

Concert pianist, conductor, and composer Sergei S. Prokofiev (1891–1953) is born. Although he emigrates after 1917, he returns to the Soviet Union in 1933 where he continues to compose. In 1948, however, his work is criticized by the authorities. As a result, his subsequent work suffers from the pressure to conform to party guidelines.

1894–1917 Reign of Nicholas II

Nicholas II (1868–1918) succeeds his father, Alexander III. Nicholas II is best known for being the last tsar of Russia. Temperamentally, he is ill-suited to rule a country facing perpetual revolutionary agitation; a warm, generous man on a personal level, he is indecisive, and a staunch conservative who opposes further reform. Although he preserves his throne during the Revolution of 1905, the disastrous effects of World War I force him to abdicate in 1917. He and his family are held captive in Russia following the Bolshevik takeover and are murdered by the Bolsheviks in 1918.

1894: August 17 Nikita S. Khrushchev is born

Soviet leader Nikita S. Khrushchev (1894–1971) is born in Ukraine to Russian parents. He joins the Communists in 1918 and rises through the ranks. By 1939 he becomes a member of the Politburo. After Josef Stalin's death, Khrushchev wins a power struggle for the country's leadership.

1897 Marc Chagall is born

Painter Marc Chagall (1887–1985) is born in Vitebsk. He studies in Paris, but returns to Russia in 1914. Although the Bolsheviks make him commissar of fine arts in Vitebsk and allow him to open an art academy, he leaves Russia in 1922 and settles in France where his work remains popular.

1898 Sergei M. Eisenstein is born

Film director Sergei M. Eisenstein (1898–1948) is born. During the early 1920s, he becomes a legendary filmmaker and achieves international recognition with *Battleship Potempkin*. Subsequent films include *Alexander Nevsky* and *Ivan the Terrible*.

1903 Russian Socialist Democratic Labor Party splits

At its second party congress in London, the Russian Socialist Democratic Labor Party splits into two factions: Bolshevik (majority), and Menshevik (minority). The Bolsheviks, under Lenin, represent the radical wing of the party while the Mensheviks are moderates. Ultimately, the Bolsheviks lead the revolution that sweeps them to power in 1917. Following the

The portrait of Tsar Nicholas II, ruler of Russia from 1894to 1917, hangs in the Russian Duma. (EPD Photos/CSU Archives)

revolution they are known as Communists. The Mensheviks stay out of government until Lenin outlaws them in 1922.

1904–05   Russo-Japanese War

War breaks out between Russia and Japan, the result of growing rivalry between the two countries over influence in China. Japan mounts a sneak attack against the Russian naval base at Port Arthur (China) and follows with an invasion of Russian-occupied Manchuria. The Russian fleet takes heavy losses while the Russian army loses several battles. In a desperate move, Nicholas II orders the Baltic Fleet to aid besieged Port Arthur. In May 1905, the Baltic Fleet meets with disaster in the Straits of Tsushima between Korea and Japan. In the meantime, Port Arthur falls before the fleet even arrives. A peace is soon signed in which Russia cedes Port Arthur, Dairen, Manchuria, and the southern half of Sakhalin Island to Japan. It is the first time in the modern era that a European

power is defeated by a non-European power. The defeat is a major humiliation for Russia and results in major domestic upheaval.

1905   Revolution

Following the Russian defeat, a wave of strikes hits the empire. In response, Nicholas II issues a constitution (see 1905: October).

1905: October   "October Manifesto" gives Russia its first constitution

Faced with widespread calls for reform, on October 17, 1905, Nicholas II reluctantly issues a manifesto that guarantees fundamental civil liberties, and creates a *Duma* (parliament). Yet it does not create a constitutional monarchy. The first Duma meets in 1906 and has a liberal majority. The conservative tsar dissolves the body and calls for new elections in 1907. When the second Duma proves more radical than the first, Nicholas begins ignoring the body rendering it ineffective except as a platform for debate.

1906   Dimitri D. Shostakovich is born

Composer Dimitri D. Shostakovich (1906–75) is born in St. Petersburg. He enjoys a mixed relationship with the Soviet authorities, particularly during the Stalinist period. Later his works were often criticized but he enjoys a strong international reputation.

1906: December 19   Leonid I. Brezhnev is born

Soviet leader Leonid I. Brezhnev (1906–82) is born. After earning a degree from the metallurgical institute in 1935, Brezhnev begins his rise in the party hierarchy as a Red Army political officer. He later serves as a party chief in Moldavia and in Kazakhstan. His success in these posts earns him a position in the Politburo. In 1964, he uses his power in this body to oust Nikita Khrushchev from leadership.

Brezhnev rules the Soviet Union for eighteen years until his death in 1982, the second-longest rule of any Soviet leader. This period is one in which the U.S.S.R. continues supporting Communist expansion into the Third World even while facing economic stagnation at home.

1907   Triple Entente

After settling disputes with Britain over spheres of influence in Persia, Russia concludes an understanding with the former. This cements an informal alliance, known as the Triple Entente, that also includes France. This alliance plays a key role in World War I (see 1914–18).

1907–11   Petr Stolypin attempts reforms

Prime Minister Petr Stolypin attempts the most significant agricultural reforms since the abolition of serfdom. Stolypin allows peasants to leave the *mir* to settle elsewhere and the right to acquire more land.

1914–18   World War I

Following the assassination of the heir to the Austro-Hungarian throne by a Serb nationalist, Austria-Hungary declares war on Serbia. Russia, a Serbian ally, mobilizes its army to aid the Serbs. This sets off a chain reaction in Europe that leads to a general European war that extends to Europe's colonies, World War I (1914–18). The war pits the Allies, Britain, France, Serbia, and Russia (later joined by Italy, the United States, and other minor countries) against the Central Powers (Austria-Hungary, Germany, the Ottoman Empire, and, after 1915, Bulgaria). The war proves disastrous for Russia, resulting in foreign occupation, revolution, and civil war.

1914   Battle of Tannenberg

The Russians begin the war by invading the German province of East Prussia but are thrown back at Tannenberg. This defeat begins a general Russian retreat against the German armies that results in the loss of Poland in 1915 and parts of Belorussia and Ukraine by the following year.

1916   Brusilov Offensive

Russian forces under the command of General Aleksei A. Brusilov (1853–1926) on the Austro-Hungarian front along the Carpathian mountains launch a major offensive that succeeds in driving back enemy forces. Yet its long-term effects are limited. Russian forces are low on supplies and morale among troops remains low.

1917: February   Revolution brings abdication of Nicholas II

Protests in Petrograd (renamed in 1914 because St. Petersburg sounded too German) beginning on International Women's Day result in a full-fledged uprising that causes the collapse of tsarism. Tsar Nicholas II abdicates in favor of his younger brother, Grand Duke Mikhail Romanov who declines the throne, thus bringing the imperial period to a close. Governmental authority is divided between the Petrograd Soviet (council) of soldiers and workers which is Menshevik and Socialist Revolutionary in orientation and the rival Provisional Government backed primarily by the liberal Kadet (Constitutional Democrat) and Octobrist parties under the prime ministership of Prince Lvov (1861–1925).

1917: July   Kerensky government

Lvov's government is replaced by a new coalition headed by the S.R. (Socialist Revolutionary) Alexander F. Kerensky (1881–1970). Unrest plagues the new government almost from the start. In September, an attempt by right-wing general Lavr Kornilov (1870–1918) to overthrow Kerensky fails, but the prime minister relies upon Bolsheviks to help him sup-press the uprising. This Bolshevik aid only increases their power (and that of the Petrograd Soviet where the Bolsheviks are gaining strength) while further undermining that of Kerensky.

Soviet Period, 1917–91

1917: October   Bolsheviks take power in Petrograd

The Bolsheviks mount a successful coup in Petrograd. They face opposition throughout much of the country, and it takes years, and a civil war, to subdue the opposition sufficiently so that they can become masters of the entire country.

1918: January   Constituent Assembly meets and is dissolved

The Constituent Assembly, elected only three weeks after the Bolshevik takeover, meets in Petrograd but is dispersed after only one day. The body has a non-Bolshevik majority, and Lenin fears that it can be used as a weapon against the revolution. Many of its leaders are subsequently arrested.

1918   Capital moved back to Moscow

Fearing an imminent German takeover of Petrograd, the Bolsheviks move the capital of Russia back to Moscow.

1918: March   Treaty of Brest-Litovsk pulls Russia out of World War I

With the country weary after nearly four years of war and the new Bolshevik regime facing a precarious future, Lenin signs a peace treaty with the Central Powers at Brest-Litovsk. Under the terms of this agreement, Russia loses the Baltic region, Finland, Poland, and Ukraine. In addition, the Russian government owes reparations to the Central Powers. Although the terms of this agreement are harsh, they provide the Bolshevik government with an opportunity to consolidate its power. In addition, many Bolsheviks believe that a world Communist revolution is imminent and have no qualms about surrendering territory for a short while. For their part, the Germans largely ignore the provisions of the treaty and continue their advance into Russia.

1918–20   Civil war

Civil war begins in the winter of 1918 and lasts for over two years. The Bolsheviks (Reds) are opposed by a broad spectrum of groups (referred to as Whites) that includes monarchists, liberals, and moderate socialists. In addition, Russia's former Entente allies, including Britain, France, Japan, and the United States also intervene in the conflict on behalf of the Whites and in order to preserve supplies they had sent to the old government.

Under the leadership of Trotsky, the Red Army triumphs over its divided and poorly organized White opponents. Many who fight on the side of the Whites flee into exile or are captured and killed by the Reds. The civil war is disastrous for the country. Most industry is destroyed, and communications and trade routes are disrupted. Agriculture is also largely devastated. However, the unity of the state is preserved since the Reds manage to recapture most of the territories of tsarist Russia.

1918   Alexander I. Solzhenitsyn is born

Dissident writer Alexander I. Solzhenitsyn is born. He is best known for his novels about life in forced labor camps where he, himself, is a prisoner from 1945 to 1963. His works include, *One Day in the Life of Ivan Denisovich*, and *The Gulag Archipelago* (see 1971, and 1974). His subsequent writings display a strong Russian nationalism and faith in Russian Orthodoxy.

1918: July   Royal family executed

The Bolsheviks execute Nicholas and his family who are held prisoner in the Ural town of Ekaterinberg. The Bolsheviks order the execution as White forces approach the town. Lenin fears that the White capture of the royal family would serve as a rallying point for anti-Bolshevik forces throughout Russia.

1919: March   Comintern formed

The leaders of the Bolshevik (now called Communist) regime announce the creation of the Communist (or Third) International (Comintern), an international organization of communist parties. This grouping is designated the successor to the Second International, a collection of socialist parties that divides during the First World War when most of its leaders prove to be patriots and support the war efforts of their respective countries. Although the Comintern is ostensibly an international organization independent of any specific Communist party, the Comintern is headquartered in Moscow and is a foreign policy instrument of the Russian Communists. The real purpose of its creation is to provide support for the embattled Russian Communist regime in the early years of the revolution.

1919–20   Russo-Polish War

An independent Poland reemerges at the end of World War I. In an effort to extend Polish borders eastward, the Poles attack Red forces in Ukraine. Although Red Army troops repulse them and drive toward Warsaw, the Poles counterattack and eventually force a harsh peace on the Reds by taking the western Ukraine. This Russian loss remains a point of contention between the two governments into the 1940s.

1921: March 7–18   Kronstadt revolt suppressed

An uprising by sailors of the naval garrison on the island of Kronstadt outside of Petrograd in the Gulf of Finland is crushed by the Red Army. For opponents of the Communists, this revolt symbolizes the ruthlessness of Lenin's regime. In the early stages of the revolution, the Kronstadt Sailor's *Soviet* (council) had proven the most loyal of Bolshevik allies. When they demand more freedom, the party turns against them.

1921   New Economic Policy (NEP) introduced

The civil war leaves the Russian economy in tatters. In an effort to jump-start the economy, Lenin moves away from some of the Communist measures of expropriation and state ownership and allows a return of limited capitalism. Small-scale enterprises reopen and the economy improves. For Lenin, this is a temporary step that will ease the way for full implementation of Communism.

1921   Andrei D. Sakharov is born

Andrei D. Sakharov (1921–89) is born. After earning a doctorate in physical science, he begins work on the Soviet hydrogen bomb in the early 1950s. Beginning in the 1960s, his opposition to the Soviet regime lands him in trouble with the Communist authorities and makes him a worldwide symbol in the struggle for human rights in the U.S.S.R. After he wins the Nobel Peace Prize in 1975, the Soviet authorities send him to internal exile in Gorky (Nizhny Novgorod) where he remains for eleven years.

1922   Treaty of Rapallo signed

Germany and Russia reach an agreement at Rapallo. According to the provisions of this treaty, both countries agree to restore diplomatic relations and renounce war reparations. In addition, they reach a secret understanding on military cooperation. The final clause is particularly important since both countries are international outcasts in the 1920s. The secret military cooperation clause allows Germany to train in Russia while the Russians benefit from German military training. The experience gained in this training goes a long way toward helping Germany re-arm quickly under Adolf Hitler (1889–1945) in the 1930s.

1922: December   Formation of the Union of Soviet Socialist Republics

Communist Russia is reorganized into the Union of Soviet Socialist Republics (U.S.S.R.). The state consists of the following republics: Russian Soviet Federative Socialist Republic (R.S.F.S.R.), Ukrainian S.S.R., Belorussian S.S.R., and the Transcaucasian S.F.S.R. The T.S.F.S.R. is later divided into several republics, while the Turkmenian, Tajikistan, and Uzbek S.S.R.s are created and join years later.

1923   New constitution drafted

A new constitution is adopted for the U.S.S.R. The constitution distinguishes between republic and federal powers and makes no mention of the Communist Party. In reality, however, power rests in the party through the Council of People's Commissars. In addition, all republic governments are subservient to Moscow, where all regional appointments originate.

1924: January 21   Lenin dies

Lenin, the leader of the revolution and the U.S.S.R. dies. In poor health as a result of a major stroke in 1922, his sudden death results in a power struggle over succession. Lenin, himself, is turned into a cult figure by his successors and his body is preserved—against his wishes—and put on public display in a special mausoleum erected for him.

1924: February   Great Britain recognizes the U.S.S.R.

Great Britain grants diplomatic recognition to the Soviet Union. This is a major victory for the U.S.S.R. which is still viewed as a pariah state lacking international legitimacy.

1925   Trotsky removed as Commissar of War

Stalin, with help from his allies Lev Kamenev (1881–1936) and Grigory Zinoviev (1882–1936), ousts Trotsky from his position as People's Commissar of War and thus removes the key obstacle to Stalin's authority. Joseph Stalin, People's Commissar of Nationalities, and Secretary-General of the Central Committee of the Russian Communist party had begun maneuvering his way to the top of the leadership of the Soviet Union in the early twenties. Using his position in the central committee, Stalin began appointing his allies to key positions and playing off different factions within the party against each other. By the late 1920s he controls all of the key positions within the party structure.

1928   First Five-Year Plan introduced

Stalin announces the first Five-Year Plan for the industrial development of the U.S.S.R. The plan is part of Stalin's efforts to achieve "socialism in one country" first rather than world revolution. Thus, the emphasis is placed on economic self-sufficiency through the exploitation of resources. Stalin's methods are brutal and take a toll on the standard of living (see 1929–1930).

1929–30   Collectivization of agriculture

As part of Stalin's effort to industrialize the country, he forcibly eliminates private agriculture. Stalin believes that the abolition of private farms through the creation of great state-run collective farms will result in greater efficiency. Instead, chaos results. Farmers refuse to join collectives except at gunpoint, and many slaughter their livestock rather than give them to the state. Particularly hard hit are the *kulaks*, "well-to-do" farmers who are targeted by state authorities for elimination as a "class." In many cases, crops go unharvested and famine strikes. Of the crops that are harvested many are earmarked for export in order to provide foreign exchange from which Stalin can subsidize his drive towards industrialization. Agricultural production levels, already reduced by World War I and the civil war, drop even lower. It is decades before Soviet agriculture reaches pre-war levels.

1930s   Socialist Realism

Socialist Realism becomes the art form officially sanctioned by the Communist Party. Socialist Realism is best described as an artistic style that depicts a socialist utopia in a present-day setting. Artists not conforming to Socialist Realism are frequently denied government sponsorship and often face repression.

1931   Mikhail S. Gorbachev is born

The last Soviet leader, Mikhail S. Gorbachev is born in Stavropol. While studying at the University of Moscow he joins the Communist Party in 1952. As he rises through the ranks, he serves in a variety of posts including Stavropol party leader, a member of the Central Committee, and a member of the Politburo.

Gorbachev becomes leader of the Soviet Union in 1985 amid a serious economic and infrastructure crisis. A believer in socialism, Gorbachev nevertheless begins a program of reform intended to restructure the Soviet system in order to make it more efficient. His policies of liberalization lead to chaos, however, as the socialist system cannot be saved. The parallel rise of nationalism among the U.S.S.R.'s republics leads to the final collapse of the Soviet Union. Gorbachev moves into retirement after the dissolution of the U.S.S.R., but remains a prominent international figure.

1931   Boris N. Yeltsin is born

First Russian President Boris N. Yeltsin is born near the Urals. Originally, an ally of Mikhail Gorbachev, Yeltsin breaks with the Soviet leader in the late 1980s over the pace of economic and political reforms. Eventually, Yeltsin becomes president of the R.S.F.S.R. and challenges the Soviet leader for authority by demanding a greater role for the republics. Once the Soviet Union dissolves in the 1990s, Yeltsin becomes president of an independent Russia (see multiple entries for the 1990s).

1933   The United States recognizes the Soviet Union

The United States opens diplomatic relations with the U.S.S.R.

1934   The U.S.S.R. joins the League of Nations

In what is seen as another step aimed at ending its diplomatic isolation, the Soviet Union joins the League of Nations.

1934   Yuri Gagarin is born

Yuri Gagarin (1934–68) is born in the Smolensk region. Upon completing his studies at the Polytechnic, he joins the Red Air Force and becomes a fighter pilot. When the Soviet Union begins its space program, Gagarin makes history as the first man in space aboard Vostok I in 1961 (see 1961). He is later killed in a crash during a training flight in 1968.

1936   "Stalin Constitution" replaces 1923 document

Stalin adopts a new constitution for the U.S.S.R. which replaces that promulgated (put in force) in 1923. Known as the "Stalin Constitution," the new document provides for the creation of a state of eleven soviet socialist republics (with the right of secession), united at the federal level. Separation of powers among the branches of government is another key feature. Other clauses, however, underscore the central role played by the Communist Party in government. In practice, the document proves to be a sham since Stalin's control (along with that of his henchmen) remains key to any exercise of power.

1935–38   Great Purge

The December 1934 murder of Leningrad party chief Sergei Kirov (1886–1934), possibly at the instigation of Stalin, ushers in a massive purge of party members, military and security commanders, and general citizenry of the Soviet Union. Millions are repressed and either executed or sent to work in the *GULag* (Chief Administration of Camps, forced labor camps that housed both criminals and political prisoners). "Old Bolsheviks," i.e., prominent party members from the revolution, are arrested and executed on trumped up charges such as "wrecking" and conspiring with foreign powers to betray the revolution. In addition to "Old Bolsheviks," the purges also eliminate most Red Army generals as well as leading members of the NKVD (People's Commissariat of Internal Affairs, the secret service), the same organization that performs the arrests. According to later accounts, the NKVD places arrest quotas on city populations, which in some cases are exceeded.

The purges are particularly devastating in their effect on the Soviet Armed forces. The purges plunge captains and majors into divisional commands—commands they are not trained to handle. Valuable military experience is also wasted. The disruptive effects of the purges are largely responsible for the heavy losses suffered by the Red Army against Finland in 1939–40 and the Werhmacht (German Armed forces) in 1941–42.

1937   Valentina Tereshkova is born

Future cosmonaut Valentina Vladimirovna Tereshkova is born. She later becomes an amateur parachutist before becoming the first woman in space aboard Vostok VI in 1963

League of Nations

Formed in the wake of World War I, the League of Nations—the forerunner of the United Nations—was the world's first international organization in which the nations of the world came together to maintain world peace. Headquartered in Geneva, Switzerland, the League was officially inaugurated on January 10, 1920. During the life span of the organization, over sixty nations became members, including all the major powers except the United States. Like the United Nations, the League of Nations pursued social and humanitarian as well as diplomatic activities. Unlike the United Nations, the League was primarily oriented toward the industrialized countries of the West, while many of the regions today referred to as the Third World were still the colonial possessions of those countries.

The League's structure and operations, which were established in an official document called the *League of Nations Covenant*, resembled those adopted later for the United Nations. There was an Assembly composed of representatives of all member nations, which met annually and in special sessions; a smaller Council with both permanent and nonpermanent members; and a Secretariat that carried out administrative functions. The League's procedures for preventing warfare included arms reduction and limitation agreements; arbitration of disputes; nonaggression pledges; and the application of economic and military sanctions.

Although the League of Nations solved a number of minor disputes between nations during the period of its existence, it lost credibility as an effective peacekeeper during the 1930s, when it failed to respond to Japan's takeover of Manchuria and Italy's occupation of Ethiopia. Ultimately, the organization failed in its most important goal: the prevention of another world war. However, it did make contributions in the areas of world health, international law, finance, communication, and humanitarian activity. It also aided the efforts of other international organizations.

By 1940 the League of Nations had ceased to perform any political functions. It was formally disbanded on April 18, 1946, by which time the United Nations had already been established to replace it.

(see 1963). Following her mission she becomes a Soviet goodwill ambassador.

1939: August   Molotov-Ribbentrop Pact

Germany's foreign minister, Joachim von Ribbentrop (1893–1946) and his Soviet counterpart, Vyacheslav Molotov (1890–1986) conclude a non-aggression treaty. This treaty between two sworn enemies stuns the world. In a secret clause, the two countries agree to partition Poland and divide Eastern Europe into German and Soviet spheres of influence. This provision is the basis for the subsequent Soviet annexation of parts of Finland and Romania, as well as the total absorption of Estonia, Latvia, and Lithuania. The Germans agree to the terms since it spares them having to fight the Red Army when the Wehrmacht invades Poland on September 1, 1939 and begins World War II (1939–45). The U.S.S.R. invades three weeks later and annexes eastern Poland to the Soviet Union.

1939: November 30   Soviet Union invades Finland

After the Finnish government rejects Stalin's demands for territory, the Red Army invades Finland. The vastly outnumbered Finns put up a strong resistance, and the Soviets are halted for months before finally breaking through in the winter of 1940. The Winter War, as the conflict is known, demonstrates the purge-induced ineffectiveness of the Red Army to the world. Following the Soviet victory, Stalin orders a thorough reorganization of his military to make up for any inefficiencies.

1939: December   U.S.S.R. expelled from League of Nations

As a result of its aggression against Finland, the Soviet Union is expelled from the League of Nations.

1940: March 12   Treaty of Moscow ends Winter War

The Winter War comes to an end with a Soviet victory. Finland is forced to cede large parts of land on the Karelian peninsula, including an outlet on the Arctic Ocean as well as land north of Leningrad, including the city of Vipuri (Vyborg).

1940   Baltic Republics, Bessarabia, and Northern Bukovina annexed by the Soviet Union

As allowed under the secret provisions of the Molotov-Ribbentrop Pact, Stalin takes the opportunity to forcibly annex the Baltic republics of Estonia, Latvia, and Lithuania. In addition, the U.S.S.R. also annexes the Romanian regions of Bessarabia (present-day Moldova) and Northern Bukovina.

1941: April 13   Non-aggression treaty signed with Japan

The Soviet Union and Japan conclude a non-aggression treaty that proves mutually beneficial. The Japanese fear a renewal of tension with the U.S.S.R. (in 1939, the Russians had inflicted a major defeat on Japanese forces) while the Soviets wish to avoid a conflict in the Far East with war still raging in Europe.

1941: June 22   Germany invades the U.S.S.R.

In violation of their non-aggression pact, the Wehrmacht invades the Soviet Union. According to Nazi ideology, the invasion of Russia is necessary in order to acquire *lebensraum* (living space) for the German people. The Slavs of Russia are viewed as racial inferiors who are to be reduced to slavery for their German overlords.

The Germans are subsequently joined by Finland, Hungary, Italy, Romania, and a Spanish volunteer division. The first weeks of the invasion are a total disaster for Soviet forces. Rather than falling back and regrouping, Red Army divisions are encircled and decimated. The Soviet Air Force, the largest air force in the world, loses half its planes in the first week and soon ceases to be an effective fighting force. By early December the advance units of the Wehrmacht are within sight of the Kremlin in Moscow and have besieged Leningrad. Nevertheless, the Russian population unites against the invader. Indeed, the Germans squander an opportunity to turn mass discontent against the Soviet regime, particularly in Ukraine, to their advantage. Instead of seeking the support of the local population, German troops antagonize them through the application of Nazi ideology. Ultimately, the German forces are halted before Moscow and thrown back in a counteroffensive that lasts into early winter 1942.

1942–43   Battle of Stalingrad

Following their reversals in early 1942, the Wehrmacht regroups and begins a drive in the south toward the Volga River that summer with their key objective the Caucasus oil fields. A major city that needs to be captured along the way is Stalingrad (formerly Tsarytsin, present-day Volgograd). Although German troops take ninety percent of the city by October, Soviet resistance is stubborn. Moreover, the Red Army is being reinforced from factories still operating in the uncaptured part of the city as well as from behind the lines while German forces are at the end of a precarious supply line. In early 1943, Soviet forces counterattack, cut off the German Sixth Army in the city and force the Germans to surrender in early February. Out of over 250,000 men of the Sixth Army, only 90,000 remain. It is a major victory for the Red Army and a decisive defeat for the Wehrmacht, which, in addition to its material losses, loses the Sixth Army's commander, Field Marshal Friederich von Paulus, to captivity.

1943 Comintern dissolved

In an effort to improve relations with his non-Communist allies, Stalin dissolves the Comintern.

1943: July Battle of Kursk

The Germans mount their last major offensive on the Eastern Front when they strike against the industrial city of Kursk over 200 miles southwest of Moscow. In the largest tank battle in history, the Wehrmacht is repulsed, and the Red Army begins its own counteroffensive that by year's end liberates large sections of eastern Ukraine.

1943: November Tehran Conference

Stalin meets with his wartime allies, Prime Minister Winston S. Churchill (1874–1965) of Great Britain, and President Franklin D. Roosevelt (1882–1945) of the United States in Tehran, Iran to discuss wartime strategy. Known collectively as the "Big Three," the Western leaders agree to mount an invasion of France the following year in order to open up a western front against the Germans and, thus, relieve the Red Army of some of the burden it faces.

1944: January Siege of Leningrad lifted

The Red Army continues its advance all along the front and finally lifts the German siege of Leningrad after over two years.

1944: July–August Destruction of Army Group Center

In the largest Soviet offensive of the entire war, the Red Army destroys the Wehrmacht's Army Group Center, the chief German army group on the eastern front. The Soviet summer offensive leaves Red Army troops at the gates of Warsaw and brings about the surrender of Romania and Bulgaria. Once finished, however, the U.S.S.R. does not mount another major offensive until January 1945. As a result, an underground Polish uprising against German forces in Warsaw is crushed by the Germans. The Soviets contend that their supply lines are outstretched, while Western observers subsequently note that Stalin is content to allow the Polish forces to be destroyed since they are non-Communist and are allied to the very same Polish government that the Soviet Union helped destroy in 1939.

1945: February Yalta Conference

The Big Three meet in Yalta in the Crimea to discuss the pending end of the war and postwar reconstruction. The Soviet Union agrees to free elections for Poland and other Eastern European states and to enter the war against Japan within three months of the end of the European War.

1945: May 8 Germany surrenders

The Red Army captures Berlin and the German high command surrenders more than a week after Hitler's suicide. This event brings the European war to an end. It is a costly victory for the U.S.S.R. which has lost over twenty-seven million lives in the conflict, most of them civilian. In addition, vast tracts of land in the western Soviet Union are devastated. Nevertheless, the U.S.S.R. emerges from the war as the second most powerful military in the world.

1945: June 26 U.S.S.R. is a founding member of the United Nations

The Soviet Union becomes one of the founding members of the United Nations.

1945: July Potsdam Conference

The Big Three meet for a final time outside of Berlin, in Potsdam. By now, however, the United States is represented by President Harry Truman (1884–1972) who succeeds Roosevelt upon the latter's death, and Britain's place is occupied by Prime Minister Clement R. Atlee (1883–1967) who defeats Churchill in elections midway through the conference. The Allies continue to pledge cooperation, but there is increasing tension between Britain and the United States on the one hand and the Soviet Union on the other. Eventually, these disputes result in mutually suspicious and hostile relations.

1945: August Soviet forces attack Japan

In accordance with its pledge at Yalta, the Soviet Union renounces its 1941 non-aggression treaty and attacks Japanese forces in Manchuria. The Japanese are quickly routed. The U.S.S.R. attacks occur between the U.S. atomic bombings of the Japanese cities of Hiroshima and Nagasaki. Japan quickly surrenders, and the Red Army gains control of large parts of Manchuria. In addition, the Red Army annexes southern Sakhalin Island and the Kurile Islands. The present-day occupation of these is a major source of tension between Russia and Japan.

1946 Government reorganization, new round of Stalinist repression begins

With the war over, Stalin begins a major reorganization of the Soviet government and resumes the repression of the late 1930s. Agencies are regrouped, and government controls tightened. In addition, the Soviet government launches a campaign against "cosmopolitanism," (i.e., anything of Western or capitalist influence) which results in the arrest and/or execution of leading artists, writers, party officials, and ordinary citizens.

1946 Churchill gives Iron Curtain speech, Cold War emerges

Speaking to an audience in Fulton, Missouri, Churchill proclaims that "an Iron Curtain" divides the European continent between the Western, free countries and the Eastern, Communist satellites of the U.S.S.R. The speech symbolizes the

The United Nations

The United Nations (UN) is the international organization created by the major Allied powers at the end of World War II to maintain peace and encourage cooperation among the world's sovereign nations. The UN officially came into being on October 24, 1945, when its charter was ratified by a majority of the fifty nations whose representatives had drafted the document at a conference in San Francisco the previous April. The organization replaced the older League of Nations that had been in existence between the world wars. Since its inception, the United Nations has been headquartered in the UN Building in New York City. In 1998 it had 185 members.

As specified in its charter, the UN consists of six main bodies, of which the most prominent are the General Assembly and the Security Council. All member nations send representatives to the General Assembly, which holds regular sessions annually, as well as special sessions. Each nation may send up to five delegates but still has only one vote. The General Assembly passes resolutions on issues of concern to the international community, amends the UN charter, elects judges to the International Court of Justice (described below), and handles budgetary and other matters. The Security Council is, as its name suggests, concerned with international security. It has five permanent members—the United States, Britain, Russia, China, and France—and ten nonpermanent members elected to two-year terms by the General Assembly. The nonpermanent memberships are distributed so that all the different regions of the world are represented on the Council. The Security Council can recommend ways of resolving international disputes, impose sanctions, and pass binding resolutions. It has deployed peacekeep-ing forces to regions around the world, including Bosnia, Haiti, Northern Ireland, and Somalia. The Security Council also nominates the UN Secretary General.

The remaining four UN bodies are the Secretariat, the UN's administrative arm; the International Court of Justice (or World Court), headquartered in the Hague, which settles disputes; the Economic and Social Council, which oversees economic, humanitarian, and cultural programs; and the Trusteeship Council. The Secretariat is headed by the UN Secretary-General. As the single most prominent UN official, the Secretary-General often acts as a UN spokesman and as a conciliator in international disputes in addition to his duties as an administrator.

The United Nations also sponsors a number of additional agencies and programs, including the United Nations Children's Fund (UNICEF); the United Nations Educational, Scientific and Cultural Organization (UNESCO); the World Health Organization (WHO); the International Monetary Fund (IMF); and the International Bank for Reconstruction and Development (the World Bank).

Twice in its history, major political developments around the world have led to "growth spurts" in UN membership. In 1960 sixteen new African nations were admitted to the world body, reflecting the end of Europe's colonial empires. In the 1990s, the break-up of the Soviet Union into separate republics and developments in other parts of Eastern Europe brought nearly two dozen new member nations into the UN, bringing the total number of members to 185.

---

quick breakdown of wartime relations with the Soviet Union and the beginning of the Cold War.

1947   Cominform created

The Soviet Union returns to open support for international Communism with the creation of the Communist Information Bureau (Cominform) which replaces the defunct Comintern.

1948–49   Berlin blockade

In an effort to sabotage American, British, and French control of their portions of occupied Berlin, the Soviet Union blockades the city from all ground sources of supply. In response, the Western allies resupply their part of the city by air, and the Soviet blockade fails.

1949: January   Anniversary of Lenin's death

Russian leaders gather to commemorate the twenty-fifth anniversary of Lenin's death.

1949: August   First Soviet atomic bomb

The U.S.S.R. explodes its first atomic bomb over Siberia. In addition to their own scientists, the Soviets are aided by Western spies including the American Communists Julius (1918–53) and Ethel (1915–53) Rosenberg who provide them with

The Cold War

The Cold War is the commonly used name for the prolonged rivalry between the United States and the Soviet Union that lasted from the end of World War II to the break-up of the U.S.S.R. in 1991. The Cold War also encompassed the predominantly democratic and capitalist nations of the West, which were allied with the U.S., and the Soviet-dominated nations of Eastern Europe, where Communist regimes were imposed by the U.S.S.R. in the late 1940s. Although the United States and the Soviet Union never went to war with each other, the Cold War resulted in conflicts elsewhere in the world, such as the Vietnam War, a prime example of the United States's "domino theory," which held that a Communist victory in one nation would result in the spread of Communism to its neighbors.

Following World War II, the political and military bond between the United States and its allies in Western Europe was cemented by the Marshall Plan, which helped the European countries recover from the war economically, and the 1949 creation of NATO (the North Atlantic Treaty Organization). Within the United States, an early consequence of the Cold War was McCarthyism, a political trend named for Senator Joseph McCarthy, who spearheaded the drive to expose and penalize Americans suspected of being Communist spies or sympathizers in the early 1950s, one of the most intense periods of the Cold War. McCarthyism, at its height between 1950 and 1954, ruined the careers and lives of many Americans who posed no actual threat of subversion to the country. Following the death of Soviet dictator Josef Stalin in 1953, the intensity of the Cold War abated for several years. However, the launch of the Soviet space satellite Sputnik in 1957 as well as Soviet efforts to extend its influence in the newly-independent (formerly colonial) states revived the rivalry by inaugurating a battle for supremacy in space exploration (the "space race") and the "Third World", respectively.

The single most serious political confrontation of the Cold War was the Cuban Missile Crisis of 1962, when the United States and the Soviet Union were brought to the brink of a nuclear confrontation over the stationing of medium-range nuclear missiles in Soviet-allied Cuba. Ultimately, the missiles were withdrawn, but both nations began a costly arms race that was particularly damaging economically for the Soviet Union.

In the early 1970s, under Soviet premier Leonid Brezhnev and U.S. president Richard Nixon, Cold War tensions began to ease, in a development known as "détente." During this decade the two nations signed the SALT (Strategic Arms Limitations Treaty) agreements limiting antiballistic and strategic missiles. However, outside the provisions of these agreements, both nations continued their massive arms build-up, and tensions increased in the 1980s during the presidency of staunch anti-Communist Ronald Reagan, who referred to the Soviet Union as the "evil empire."

The appointment of political reformer Mikhail Gorbachev as general secretary of the Communist Party in 1985 signaled the beginning of the end for the Cold War. The democratization and openness implemented by the Soviet government under Gorbachev culminated not only in the downfall of Communism but in the dissolution of the Soviet Union itself, which was preceded by the fall of Communism in the Eastern European countries of the former Soviet Bloc.

crucial information. The Soviet explosion of an atomic bomb ends the American monopoly on these weapons and intensifies the Cold War.

1950: June   Korean War breaks out

Cold War tensions between the United States and the Soviet Union intensify when Communist North Korea invades South Korea. After World War II, the territory once belonging to Japan has been divided along the thirty-eighth parallel. The North, under Kim Il Sung, adopts a Communist government and keeps close ties with the U.S.S.R., while the South, under Syngman Rhee, allies with the United States. Although it is unclear whether Stalin orders the invasion, he certainly supports it. Soviet support for North Korea is part of general backing for Communist movements opposed to the United States and its allies throughout Asia. The United States quickly mobilizes support from the United Nations (UN) for military intervention against North Korea. The UN intervention goes forward due to the absence of the Soviet Union

During a January, 1949, commemoration of the twenty-fifth anniversary of Lenin's death, Joseph Stalin, in the center of the group, is seated beneath a towering likeness of Lenin. (EPD Photos/CSU Archives)

from Security Council sessions. The U.S.S.R., which can block any UN military action through its veto power, is boycotting meetings of the UN Security Council due to that body's refusal to seat the Communist People's Republic of China.

The UN forces succeed in repulsing the North's invasion, but Soviet-backed Chinese intervention on behalf of the North halts the UN advance and raises the possibility of further escalation. Ultimately, the opposing sides sign a ceasefire in 1953. The Soviet failure to prevent the UN action is a costly mistake for its foreign policy. By contrast, Communist China gains prestige among more radical Communists throughout the world through its defense of North Korea.

1953: January   "Doctor's Plot" signals the start of a new purge

The secret police announce the existence of a conspiracy among Soviet doctors, many of whom are Jewish, to murder leading party officials. Most observers view this as the start of a new purge similar to that of the 1935–38 period. Only Stalin's death in March prevents it. Shortly thereafter, the new leadership announces that the "Doctor's Plot" is a hoax.

1953: March 5   Stalin dies

Josef Stalin, dictator of the Soviet Union for over a quarter century dies under circumstances that remain questionable. His death unleashes a new power struggle.

1953: July   Khrushchev, Malenkov, and Molotov emerge as new leaders

A new *troika* (trio) of leaders emerges consisting of Nikita Khrushchev, Grigory M. Malenkov (1902–88), and Vyacheslav Molotov. The *troika* are united in their effort to rid themselves of KGB (as the secret police are now named) chief Lavrenti Beria (1899–1953) who is arrested and executed. By 1957 they have a falling out, and Khrushchev emerges as the sole leader. Unlike the custom in the Stalin era when rivals are eliminated, Khrushchev gives his opponents minor posts. Malenkov becomes the manager of a hydroelectric plant in Kazakhstan while Molotov becomes ambassador to Mongolia.

Khrushchev's rule (1953–64) is characterized by a relaxation of some of the repressive conditions prevalent under his predecessor, which Khrushchev denounces in 1956. In foreign affairs, Khrushchev pursues a policy of "peaceful coexistence" with the West. Cold War tensions remain, however, as indicated by the invasion of Hungary, the erection of the Berlin Wall, and the Cuban Missile Crisis (see 1956, 1961, 1962). Domestically, through his "Virgin Lands" project, the Soviet leader concentrates his energies on making the Soviet Union the leading grain producer in the world. Yet the grandiose scheme fails due to poor planning and inflated expectations. Indeed, the U.S.S.R. begins importing grain. After a succession of bad harvests, and criticism of his foreign policy, the Presidium (as the Politburo is called under Khrushchev) removes him from office in October 1964. Khrushchev lives out the rest of his years in retirement and writes his memoirs.

1953: August   Soviet Union tests a hydrogen bomb

The U.S.S.R. conducts its first test of a hydrogen bomb.

1955: May   Warsaw Pact created

In response to the decision of the North Atlantic Treaty Organization (NATO, an alliance of Western European nations, including the United States and Canada) to admit the Federal Republic of Germany (West Germany), the Soviet Union creates the Warsaw Treaty Organization (commonly known as the Warsaw Pact). The Warsaw Pact is an alliance of the Soviet Union with the following satellites: Albania (stops participation in 1962 and withdraws in 1968), Bulgaria, Czechoslovakia, German Democrat Republic (East Germany), Hungary, Romania, and Poland.

1956: February   Khrushchev delivers "secret speech"

At the twentieth party congress of the Communist Party of the Soviet Union, Khrushchev delivers a "secret speech" in which he exposes the brutal repressions of the Stalin era. This speech signals the start of a policy of "de-Stalinization" (the removal of the cult of personality surrounding the former dictator) and the "rehabilitation" (restoration of respect) of many of Stalin's victims.

1956: November   Hungarian revolution crushed

In Hungary, a popular revolt leads the government of Imre Nagy to announce his country's withdrawal from the Warsaw Pact and multiparty elections. In response, the new First Secretary of the Hungarian Communist Party appeals to the Soviet Union to crush the rebels. Khrushchev orders an invasion of Hungary which crushes the uprising and installs a Communist government loyal to the Soviet Union.

1957: August   First ICBM launched

The U.S.S.R. launches its first Intercontinental Ballistic Missile (ICBM). This allows the Soviet Union to launch nuclear weapons against the United States and its allies without the risk of sacrificing bombers on long missions.

1957: October   Soviets launch first satellite

Taking advantage of their new missile technology, the U.S.S.R. launches Sputnik I, the world's first satellite, into earth orbit. The success of this mission takes the world by surprise and results in the "space race" between the Soviet Union and the United States.

1959: October   Pasternak wins Nobel Prize for literature

Boris Pasternak wins the Nobel Prize for literature for his *Dr. Zhivago*, a novel that focuses on rebirth throughout a man's life. The Communist party, however, which refuses to allow publication of the novel in the U.S.S.R., forces Pasternak to reject the prize. This incident, together with the crackdown in Hungary, indicates that there are limits to Khrushchev's "thaw."

1960: May   U-2 incident

An American U-2 spy plane is shot down over the Soviet Union and the pilot, Francis Gary Powers (1929–77) is taken prisoner. Soviet relations with the United States take a downward turn.

1961: April 12   Gagarin becomes first man in space

The Soviet Union stuns the world again when it sends cosmonaut Yuri Gagarin on a one-orbit flight aboard Vostok I.

1961: August   Berlin Wall erected

In an effort to stop the flow of refugees fleeing Communist East Germany through West Berlin, Khrushchev permits the East German government to build a wall around the city. The Berlin Wall effectively halts the flow of East Germans to the West although some continue to try to escape (and many are killed in the process). The Wall comes to symbolize the Iron Curtain until its removal in 1989–90.

1962: October   Cuban Missile Crisis nearly results in war against the United States

In need of a major foreign policy victory, Khrushchev deploys medium-range nuclear missiles fitted with nuclear warheads to the Caribbean island of Cuba. Cuba, a recent addition to the Communist community, fearful of an American invasion, welcomes the Soviets. When he learns of the deployment, U.S. President John F. Kennedy (1917–63) decides to "quarantine" (a diplomatic way of saying "blockade") Cuba with American warships and threatens an invasion. Ultimately, Khrushchev agrees to withdraw the missiles in return for an American pledge not to invade Cuba and a secret agreement to remove United States missiles based in Turkey. For Khrushchev, the missile crisis is a major foreign policy defeat that contributes to his ouster two years later.

1963: June 16   Tereshkova becomes first woman in space

The Soviet space program scores another first when cosmonaut Valentina Tereshkova becomes the first woman in space when she flies in Vostok VI.

1963: August   Limited Test Ban Treaty signed

In a major breakthrough to halt the spread of nuclear weapons, Great Britain, the Soviet Union, and the United States sign and ratify a treaty that bans atmospheric testing of nuclear weapons. It is the first agreement signed between the Superpowers on nuclear weapons.

1964: October   Khrushchev ousted

Citing their leader for incompetence and rash, adventurist policies, the Presidium (the main governing body of the Communist Party) removes Khrushchev from office. He is replaced by Leonid Brezhnev and Alexei N. Kosygin (1904–80). By the late 1960s, Brezhnev establishes himself as the dominant leader of the regime while Kosygin is relegated to the background. Under Brezhnev government controls tighten even more, and there is a partial restoration of Stalinism.

1968: September   Soviet forces invade Czechoslovakia, "Brezhnev Doctrine" announced

Soviet forces invade Czechoslovakia and put an end to the liberalization introduced by Czechoslovak Communist leader Alexander Dubcek (1921–92). Dubcek is ousted and replaced by pro-Soviet Communist leaders. Following the invasion, Brezhnev announces that the sovereignties of the U.S.S.R.'s Eastern European allies is restricted; the Soviet Union will not tolerate changes in their political and economic system.

1969: March   Soviet and Chinese forces engage in border clashes

Soviet and Chinese border forces engage in serious clashes along their frontier. Relations between the two countries, once closed, become bitter by the mid-late 1960s as Chinese Communist Party chairman Mao criticizes the Soviet leadership for timidity and accuses it of abandoning socialist goals.

1970: December   Solzhenitsyn awarded Nobel Prize for Literature

Soviet dissident writer Alexander I. Solzhenitsyn (b. 1918) wins the Nobel Prize for literature (see 1918).

1971: May–June   Salyut I becomes first space station

The Soviet Union achieves another space first when it launches Salyut I, the first space station. However, the first mission to the station ends in tragedy when the three-man crew perishes as a result of an oxygen leak during reentry.

1972: May   U.S.-Soviet summit in Moscow, arms treaties signed

United States President Richard M. Nixon (1913–94) visits Moscow for a summit meeting with Brezhnev. The two leaders sign the SALT I (Strategic Arms Limitations Treaty), the first treaty limiting nuclear weapons as well as the ABM (Anti-Ballistic Missile) treaty, which limits the two countries to the construction and deployment of one anti-ballistic missile defense system, each. The conclusion of these treaties signals the start of *détente* (reduction of tension) between the two Superpowers.

1974   Solzhenitsyn exiled

Hounded by the authorities for his dissident writings, the KGB arrests Solzhenitsyn and deports him.

1975: July   Apollo-Soyuz space flight

In a sign of the thaw in United States-Soviet relations, the two countries conduct a joint space flight during which an American Apollo spacecraft docks with a Soviet Soyuz in earth orbit.

1975: August   Helsinki Accords signed

The Soviet Union signs the Helsinki Accords, a series of documents that pledge European governments to support human rights and respect international frontiers. Although the U.S.S.R. signs the treaties, it continues to violate its citizens' civil liberties, and it still interferes in the internal affairs of other states.

1975: December   Sakharov wins Nobel Peace Prize

Andrei Sakharov (1921–89), the father of the Soviet hydrogen bomb program, and later a dissident, wins the Nobel Peace Prize for his opposition to the arms race. He is subsequently exiled to the city of Gorky.

1977: October   New constitution replaces the Stalin Constitution of 1936

The Communist Party promulgates a new constitution to replace that of 1936. Like its predecessor, the new document reaffirms the central position of the Communist Party in the government. It also includes provisions for the protection of fundamental rights and grants the republics the right of secession. In practice, however, these rights continue to be suppressed.

1979: June   SALT II Treaty signed

Brezhnev and American president Jimmy Carter (b. 1924) sign the SALT II treaty. Like its predecessor, this treaty commits the superpowers to limitations on nuclear arms. However, the United States Senate refuses to ratify the treaty following the Soviet Union's invasion of Afghanistan in December 1979.

1979: December 25–26   Soviet forces invade Afghanistan

Soviet forces enter Afghanistan at the "invitation" of one of the factions of the Afghan Communist government. By the spring of 1980 over 85,000 Soviet troops are in the country and face heavy opposition from an Afghan Muslim resistance group, the Mujahadin. The war continues for nearly a decade as Mujahadin forces receive aid from Islamic countries and the United States. The high casualties suffered by Soviet forces results in popular opposition to the war in the U.S.S.R.

1980: August   Moscow hosts summer Olympics

Moscow hosts the summer Olympic games. The Olympiad is blighted, however, by the American-sponsored boycott of the games in protest of the Soviet invasion of Afghanistan.

1982: November   Brezhnev dies

Leonid Brezhnev, leader of the U.S.S.R. for eighteen years, dies of natural causes. His death results in quiet maneuvering for succession. Within days, former KGB head Yuri Andropov (1914–84) emerges as the new Soviet leader. Andropov wishes to institute major reforms in the Soviet system, but his poor health hampers his efforts.

1983: September   Soviets shoot down Korean Airlines jet

The Soviet Air Force shoots down a Korean Airlines Boeing 747 when it strays over Sakhalin Island. The world condemns the Soviet action. The U.S.S.R. insists that the plane did not heed warnings and was on a spy mission. All passengers and crew perish.

1984: February   Andropov dies

After less than two years in power, Yuri Andropov dies of kidney failure. He is succeeded by Constantin Chenenko (1911–85), a hard-liner who opposes Andropov's reforms.

1985: March   Chernenko dies

After only eleven months at the helm, Chernenko dies. He is replaced by Mikhail S. Gorbachev (b. 1931), a protégé of Andropov who ultimately unleashes a series of reforms that culminate in the collapse of the Soviet Union and the end of the Cold War.

1985: November   Gorbachev-Reagan summit in Geneva

Gorbachev and U.S. President Ronald Reagan (b. 1911) hold a summit meeting in Geneva, Switzerland. It is the first of several summits the two leaders will hold and lays the foundation for a close personal relationship that eventually leads to the end of the Cold War.

1986: April   Nuclear accident at Chernobyl

An explosion at a nuclear plant at Chernobyl in Ukraine spreads radiation throughout much of Eastern Europe. The accident reveals the poor safety at Soviet nuclear facilities. The plant is closed, and much of the area surrounding the plant is evacuated.

1986–87   Gorbachev launches *glasnost* and *perestroika*

In an effort to reinvigorate the Soviet system, Gorbachev announces the policies of *glasnost* (openness) and *perestroika* (restructuring). Now, Soviet citizens can openly criticize their government and debate its policies. *Glasnost* and *perestroika* reveal to the world the gross inefficiencies of the Soviet system. Although Gorbachev dismantles the centrally planned and administered Soviet system, he erects no new structure in its place. The result is chaos and the ultimate collapse of the Soviet Union.

1987: December   INF Treaty signed

The United States and Soviet Union sign the Intermediate Nuclear Forces (INF) treaty that eliminates intermediate range ballistic missiles in Europe.

1988   Ethnic tensions begin rising

Tensions between the various ethnic groups living in the Soviet Union begin rising. Resentment toward Russian domination fuels their anger. As more openness is tolerated in Soviet society, nationalists begin raising their demands, ultimately calling for independence. The decision of the U.S.S.R.'s constituent republics to secede from the Soviet Union, brings about its collapse.

1988: December   Supreme Soviet dissolved

The Supreme Soviet, the chief Communist legislative body, authorizes the establishment of a 2,250-member Congress of People's Deputies to be chosen by free election. At the same time a smaller-member Supreme Soviet is to serve as a second house of parliament. Once these measures are enacted the old Supreme Soviet dissolves itself.

1989: March–April   Free elections for Congress of People's Deputies

In free elections for Congress of People's Deputies, about 300 reformers win seats while the Communists retain about eighty-seven percent of the seats. Nevertheless, this is seen as a big victory for reformers.

1989: Spring–Fall   Collapse of Communism in Eastern Europe

Following the lead of the Soviet Union, the other Communist satellites of Eastern Europe adopt liberalizing reforms. In May, Poland holds free elections in which the Communists are defeated. In the fall, Communist regimes collapse in Bulgaria, Czechoslovakia, East Germany, Hungary, and (after much bloodshed) in Romania.

1989: December   Gorbachev and Bush declare Cold War over

At a summit meeting between Gorbachev and the new U.S. president George Bush (b. 1924) the two leaders announce that the Cold War is over.

1990   Continued ethnic tension in the U.S.S.R. threatens unity, Soviet republics begin to secede

By 1990, ethnic tensions in the U.S.S.R. threaten its very existence. In March, Lithuania declares independence.

Throughout the year ethnic riots occur in Azerbaijan, Tajikistan, Kyrgyzstan, and Moldavia, while the pro-independence parties win a majority in the Georgian parliament.

1990   Yeltsin emerges as a rival to Gorbachev

Within the R.S.F.S.R., Boris Yeltsin (b. 1931) emerges as the leading opponent to Gorbachev's policies. Once an ally of Gorbachev, Yeltsin breaks with him over the pace of reform, which Yeltsin argues is too slow. In addition, Yeltsin capitalizes on growing Russian discontent with the Soviet Union. Ethnic Russian opposition to the U.S.S.R. grows as many Russians resent having to pay a large share of the union's expenses while seeing their culture placed on an equal level with those of the other republics. In 1990, Russia follows the lead of the Baltic republics and declares its own sovereignty, therefore indicating that it considers Russian law supreme to Soviet law.

In spite of his support for Russian sovereignty, Yeltsin does not want to break up the Soviet Union. Rather, he favors reorganizing it with Russia at its head. Events over the course of 1991 prevent the creation of a new union.

1991: January   Soviet troops crack down on Baltic independence movements

Gorbachev sends in the Red Army against independence agitators in the Baltic republics. This has little practical effect, however, since Yeltsin signs agreements on behalf of Russia that recognize the sovereignty of the Baltics.

1991: March   Warsaw Pact dissolved

The Warsaw Pact is formally dissolved.

1991: March   Union referendum

Gorbachev holds a union referendum. All of the republics except the Baltics, Armenia, Georgia, and Moldavia participate. The voters from the participating republics approve the plan to preserve the union.

1991: June   New union treaty drafted, to be signed by seven republics

In an effort to save the Soviet Union, Gorbachev drafts a new union treaty which is to be signed by seven republic leaders in August.

1991: July   START I Treaty signed

The United States and the Soviet Union sign the START I (Strategic Arms Reductions Treaty), the first treaty which reduces, rather than limits, nuclear arms between the two countries.

1991: August   Communist hard-liners mount an unsuccessful coup

One day before the scheduled signing of the new union treaty, a group of communist hard-liners mount a coup against Gorbachev, while the Soviet president is on vacation in the Crimea. After several days of stand-offs between protestors, led by Boris Yeltsin, and soldiers, the coup collapses. In the aftermath of the coup, Gorbachev resigns from the Communist Party and tries to remove hard-liners from their posts, and by November he bans the party. His efforts do not save his own position, however. Institutions and organizations throughout the U.S.S.R. collapse as republic after republic declares independence. By year's end the Soviet Union collapses.

1991: December   CIS formed, Soviet Union dissolved

After a last-ditch effort to sign a union treaty fails in November, the Russian Belarus (formerly Belorussia) and Ukraine sign a treaty forming the Commonwealth of Independent States (CIS). Later that month, Armenia, Azerbaijan, Kazakstan (formerly spelled Kazakhstan), Kyrgyzstan, Tajikistan, Turkmenistan, and Uzbekistan join. In later years, Georgia and Moldova (formerly Moldavia) enter into the CIS. The creation of the CIS dooms the U.S.S.R. which ceases to exist on December 25, 1991. The CIS does not take its place but serves as a cooperative unit instead and efforts to cement closer ties are strongly resisted by the member states.

Post-Soviet Russia, 1991–present

1991: December 25   Russia becomes independent

With the dissolution of the Soviet Union, Russia becomes an independent state.

1992: March   Russian Federation union treaty signed

Yeltsin signs a new treaty that reinforces the unity of Russia. Russia, itself, is a federation consisting of many autonomous regions that consist of non-Russian ethnic groups, many of whom wish to secede from Russia.

1992: April   Russia becomes Russian Federation

In accordance with the terms of the Russian Federation Treaty the R.S.F.S.R. officially becomes the Russian Federation.

1992: December   Yeltsin and parliament debate economic reforms

In the first year after independence, Yeltsin pursues a program of radical privatization and the lifting of nearly all economic controls. This course of action causes widespread opposition among Communists and others who want the reforms halted or reversed. At the same time, Yeltsin and parliament clash over the extent of presidential powers which are

Break-up of the Soviet Union

The Soviet Union, formed following the Russian Revolution of 1917, was one of the two great superpowers of the post-World War II era. Its founders established a system of Communist government, based on the writings of Karl Marx and Friedrich Engels, in theory designed to eradicate extremes of social inequality by eliminating private property and giving economic control to the government. In practice, however, the Soviet government resorted to authoritarian repression throughout its existence in order to maintain control over an inefficient system and a discontented citizenry.

In the 1980s, many Soviet citizens were dissatisfied with living conditions in the U.S.S.R., which was in the midst of an economic crisis, and with the further curtailment of civil liberties that accompanied a government crackdown on dissent. Mikhail Gorbachev, who became general secretary of the Communist Party in March 1985, responded to the widespread discontent in Soviet society by instituting a series of reforms that allowed for greater political and economic freedom, eliminating censorship and instituting free-market economic measures.

In spite of popular support for his policies, Gorbachev was criticized both by conservatives for weakening the Communist Party and by other reformist politicians, such as Russian president Boris Yeltsin, for not going far enough. On August 19, 1991, conservative elements in the government, army, and KGB (secret police) attempted a coup to take over the government, placing Gorbachev under house arrest. The failure of the coup within two days demonstrated the lack of popular—and even military—support for Communist hard-liners and set the stage for the ascendancy of Yeltsin as the country's most powerful government leader. Yeltsin gained worldwide attention when he climbed atop a tank in front of Russia's parliament building and addressed a crowd assembled there, calling for mass resistance to the coup.

Only days after the failed coup, Gorbachev disbanded the central committee of the Communist Party. In the fall, he issued decrees outlawing the party and confiscating its property. Russia and the remaining Soviet republics unilaterally declared themselves independent of the Soviet Union. (The Baltic republics of Latvia, Lithuania, and Estonia had already declared their independence the previous year.) On December 8, 1991, Russia, Ukraine, and Belarus laid the groundwork for the Commonwealth of Independent States (CIS), effectively dissolving the Soviet Union. Formal dissolution of the U.S.S.R. came on December 25; on the same date Mikhail Gorbachev resigned as head of the government.

largely undefined. The two sides reach a compromise; Yeltsin appoints Viktor Chernomyrdin prime minister and the issue of presidential powers are to be put forward in a referendum.

1993: July   Clashes between Yeltsin and parliament continue

Throughout 1993 Yeltsin and parliament renew their clashes. The referendum on presidential powers is canceled and parliament proceeds to limit Yeltsin's authority. The subject of presidential authority is also a key issue in the draft constitution adopted by the constitutional assembly. Although the draft constitution limits some of the president's powers, parliament rejects the document.

1993: August–October   Yeltsin calls for new elections

Yeltsin again calls for a new referendum and also dissolves parliament, which refuses to comply. Instead, parliament begins impeachment proceedings against the president, and its members refuse to leave their building. In response, Yeltsin sends in military units to surround the building and, after a two-week standoff, dispatches soldiers to occupy the parliament.

1994: June   Russia joins NATO's Partnership for Peace

Russia joins NATO's Partnership for Peace, a program between NATO and non-members (most of which are either former Warsaw Pact satellites or former Soviet republics) which provides for military cooperation in preparation for regional security.

1994: September   Chechnya war begins

Separatists in the Russian autonomous province of Chechnya begin a war of independence. Yeltsin quickly dispatches Russian forces to suppress the revolt, but demoralized Russian troops are unable to quell the rebellion. The war drags on intermittently for nearly three years and is further indication of the sad state of Russia's armed forces.

1996: June–July   Yeltsin wins presidential elections

After two rounds of voting—the Russian constitution demands a runoff between the two leading candidates in the event no one receives a majority—Boris Yeltsin wins.

1997: March   Treaty with Belarus

Russia signs a treaty with Belarus that stresses the continued integration of the two republics. The close relations between the two countries is part of the Russian effort to re-exert its influence in the "near abroad," former Soviet republics.

1997: May   Crimean and Black Sea Fleet disputes with Ukraine resolved

The Crimea is populated mostly by Russians but is transferred to Ukraine by Khrushchev. When the Soviet Union dissolves, ethnic Russians in the peninsula favor a return to Russia, but Ukraine insists on retaining the region. In addition, the breakup of the Soviet Union leaves the distribution of the Black Sea Fleet and the use of Ukrainian ports in dispute. Under this 1997 treaty, Russia accepts Ukrainian sovereignty over the Crimea while Ukraine agrees that Russia keep eighty percent of the Black Sea Fleet. In addition, Russia receives a twenty-year lease on the port of Sevastopol.

1997: August   Ruble reform

In the wake of a major currency crisis, Yeltsin announces plans to reform the ruble the following January. He revalues the currency by dropping three zeros from all denominations.

1999: March   Russia suspends all cooperation with NATO

As an expression of its opposition to NATO air strikes against its traditional ally, Yugoslavia (Serbia-Montenegro), Yeltsin suspends all of its cooperation with NATO. Russia opposes NATO's decision to intervene militarily in order to restore autonomy to the Albanian-majority Serbian province of Kosovo. Russian opposition is based on the fear that foreign intervention in Serbia will set the precedent for foreign interference in Russia should any more of its autonomous regions decide to secede.

1999: June   Russian peacekeepers sent to Kosovo

Following a cease-fire agreement between NATO and Yugoslavia, Russian peacekeepers rush to Kosovo where they arrive at the airport of the Kosovar capital, Pristina, ahead of NATO forces. The action is seen as a major public relations victory for Yeltsin who is able to get NATO to agree to Russian participation in the Kosovo peacekeeping force.

Bibliography

Barner-Barry, Carol, and Cynthia A. Hody. *The Politics of Change: The Transformation of the Former Soviet Union.* New York: St. Martin's Press, 1995.

Channon, John, and Robert Hudson. *The Penguin Historical Atlas of Russia.* London: Viking, 1995.

Clements, Barbara Evans. *Bolshevik Feminist: The Life of Aleksandra Kollontai.* Bloomington: Indiana University Press, 1978.

Conquest, Robert. *The Great Terror: A Reassessment.* New York: Oxford University Press, 1990.

———. *Harvest of Sorrow: Soviet Collectivization and the Terror-Famine.* New York: Oxford University Press, 1986.

Curtis, Glenn E., ed. *Russia: A Country Study.* Washington: Library of Congress, 1998.

Daniels, Robert V., ed. *The Stalin Revolution: Foundations of the Totalitarian Era.* 4th ed. Boston: Houghton Mifflin, 1997.

Dmytryshyn, Basil. *A History of Russia.* Englewood Cliffs, NJ: Prentice-Hall, 1977.

Fitzpatrick, Sheila. *The Russian Revolution.* New York: Oxford University Press, 1994.

———. *Stalin's Peasants: Resistance and Survival in the Russian Village after Collectivization.* New York: Oxford University Press, 1994.

———, Alexander Rabinowitch, and Richard Stites, eds. *Russia in the Era of NEP: Explorations in Soviet Society and Culture.* Bloomington: Indiana University Press, 1991.

Getty, J. Arch, and Roberta Manning, eds. *Stalinist Terror: New Perspectives.* Cambridge: Cambridge University Press, 1993.

Jelavich, Barbara. *A Century of Russian Foreign Policy, 1814–1914.* Philadelphia: Lippincott, 1964.

Lincoln, W. Bruce. *Red Victory: A History of the Russian Civil War.* New York: Simon and Schuster, 1989.

MacKenzie, David, and Michael W. Curran. *A History of Russia, the Soviet Union, and Beyond.* 5th ed. Belmont, CA: West/Wadsworth, 1999.

Rabinowitch, Alexander. *The Bolsheviks Come to Power: The Revolution of 1917 in Petrograd.* New York: W.W. Norton, 1976.

———. *Prelude to Revolution: The Petrograd Bolsheviks and the July 1917 Uprising.* Bloomington: Indiana University Press, 1968.

Raymond, Boris, and Paul Duffy. *Historical Dictionary of Russia.* Lanham, MD: Scarecrow, 1998.

Robinson, Geroid Tanquary. *Rural Russia under the Old Regime.* Berkeley: University of California Press, 1959.

San Marino

Introduction

San Marino is the oldest Republic in the world. It is located on a three-tiered mountain on the northern part of the Italian penninsula. Surrounded by Italy, San Marino's situation is comparable to that of the Vatican, which is also an independent state surrounded by Italy. San Marino consists of an area of twenty-four square miles stretching in length from between five and five and one half miles across on all sides. The *World Almanac* places its 1998 population at 24, 714 inhabitants.

The citizens of San Marino are referred to as Sanmarinese. They include a mixture of Mediterranean, Alpine, Adriatic, and Nordic origins. They resent being called Italian, but they speak Italian with a local dialect called Romagnolo. San Marino is 95 percent Roman Catholic. There is no military conscription, but the citizens are subject for call to defend the country. Children are required to attend school up to the age of fourteen, similar to that of Italy. The literacy rate is 97 percent. There are some technical schools in San Marino, but university study is pursued in Italy or elsewhere.

The geography of San Marino has a strong influence on the life style of the Sanmarinese. Mount Titano virtually rises out of the ground as a giant seam of white limestone. It has three peaks that have a commanding view of the Adriatic coastline. On each of the mountain peaks is an ancient castle. The mountain has many valleys and woods with rugged terrain. It is virtually a safe haven from modern, as well as ancient, invading armies. The Maraus and the Ausa rivers run through the mountain and has supplied, until very recently, the Sanmarinese with their fresh water.

History

According to legend, the Republic was founded by Father Marino, a Christian stonecutter who established a monastery on the mountain in 301 A.D. San Marino's republican form of government dates back to 951 with the *Diploma di Berengario*, a document that provided for the establishment of governmental institutions responsible to the local population. The first governing body enshrined by the *Diploma* was the *Arengo*, a group of the Sanmarinese family heads of households.

San Marino has maintained its continued independence despite turbulent events in subsequent centuries. Throughout the Middle Ages, the tiny republic served as a battleground. Beginning in the late thirteenth century, the greatest threat to Sanmarinese freedom came from the Papacy which demanded that the tiny republic pay taxes to the Holy See. San Marino refused and submitted its case to a court in Rimini which found in the republic's favor. Nonetheless, San Marino was not spared from the interminable warfare of Renaissance Italy. In 1503, San Marino even fell (albeit briefly) to foreign occupation under Cesare Borgia. Papal claims to the republic continued into the seventeenth and eighteenth centuries in the face of Sanmarinese resistance. In the late 1730s, Cardinal Giulio Alberoni, Papal Legate for the Province of Romanga, ordered troops to occupy San Marino. Once more, the Sanmarinese appealed—this time to the pope, himself—for assistance. The pope's decision to side with the republic's citizens and withdraw his forces in early 1740 marked the end of the papal threat to San Marino's independence.

The Napoleonic Wars and the unification of Italy did not threaten Sanmarinese freedom. When Napoleon Bonaparte's forces marched into Italy, they bypassed the republic, and Bonaparte even offered to increase its territory. Fearing future resentment from their larger neighbors, the citizens of San Marino refused. In the wake of the failed revolutions of 1848, Sanmarinese gave refuge to Italian nationalist Giusseppe Garibaldi as he fled from Austrian and Papal troops (although Garibaldi fled when Austria threatened to invade). In 1862, Italy and San Marino signed a friendship and customs union treaty. Although amended in 1939 and 1971, this agreement has served as the basis of continued Sanmarinese independence.

San Marino remained neutral during both world wars. Despite this neutrality, however, San Marino did not escape entirely unscathed the turmoil of those years. Sanmarinese volunteered to fight with the Italian army during World War I; a fascist government ruled the republic during Benito Mussolini's dictatorship in Italy (1922–1943); and Allied planes mistakenly bombed the republic on June 26, 1944. In spite of with Italy, San Marino began issuing its own currency once more, while in 1984 the tiny republic instituted its first income tax. Educational opportunites increased as well. The government founded the University of Degli Studi, a technical institute with departments of science and technology. San Marino also began joining international organizations. By the early 1990s, the republic belonged to the Council of Europe and the United Nations.

Timeline

500–400 B.C.   Pieces of flint on Mt. Titano

The pieces of flint uncovered over two millenia later are from the Etruscan period. The Etruscans rule for a period of time the northern part of Italy. They are a highly complex civilization who first establish the city as we know it today. The language from this period has not as yet been translated.

430–354 B.C.   Roman coins left behind with the imprint of Augustus I

Roman coins are left between two towers on one of the peaks on Mount Titano. Although the Romans rule during this period, their empire is on its decline when San Marino is founded in 301.

10–9 B.C.   A bronze razor is in use

This artifact uncovered nearly two millenia later is used during the Villanovian period. Its civilization centers around the modern day city of Bologna. It appears that from other artifacts later discovered (no weapons or parts of fortresses) that this period is relatively peaceful.

A.D. 301   The founding of San Marino

According to legend, a stonecutter born on the island of Arbe, off the Dalmatian coast in the Adriatic Sea moves to this area to avoid persecution or simply to find a better way of life. This Christian stonecutter sails westward to the city of Rimini on the Italian coast. Finding the Roman Emperor Diacletian persecuting Christians there, Father Marino leaves for the nearby woods and Mount Titano, for the solace and protection in which to practice his Christian beliefs. Others soon follow, and eventually Father Marino establishes a monastery on top of one of the peaks. As legend has it, there is a wealthy woman who owns the mountain and has a son, Verissimo. One day Verissimo draws his arrow and points it at Father Marino and the boy becomes paralyzed. When Verissimo's mother learns of her son's plight, she begs Father Marino to cure him. He does and for her gratitude, she gives Father Marino ownership of the mountain.

There exists in the archives a letter by a monk named Eugippis who writes to a cleric at the time that on Mount Titano there is "a community of monks, shepherds, peasants,

its own hardships, San Marino still offered refuge to over 100,000 people who fled the Second World War.

Since the end of World War II, Sanmarinese politics has swung back and forth from left to right. From 1945 to 1957, communists and left-wing socialists formed a governing coalition. When an Italian-supported bloodless revolution ousted this government in 1957, Christian Democrat-Social Democrat coalition began which lasted until 1973. Christian Democrats then aligned with the socialists. Following the May 1978 elections, however, the communists returned to power in coalition with the Socialist Unity Party. Although this coalition lasted until 1986, the communists continued in power as they formed an alliance with the center-right Christian Democrats.

Regardless of political change, a constant theme in the republic's postwar history has been social, political, economic, and institutional reform. In the 1950s, San Marino initiated a social development program which included disability and old-age pensions. In 1975, a state hospital opened to deal with minor medical problems. Women's suffrage became law in 1960. Following the collapse of communist governments throughout Eastern Europe, the Sanmarinese Communist Party renamed itself the Demcratic Progress Party. Under the 1971 revision of the 1862 treaty artisans, hunters and woodsmen." According to some historians, this is sufficient evidence that San Marino exists.

885 The *Placito Feretrano* Document

A torn document in the archives in 1749 at San Marino relates an argument between Stefano, the Abbot of the Monastery at San Marino, and Deltone, Bishop of Rimini about the ownership of the monastery and the surrounding territory. A cleric rules that Bishop Rimini has never owned the monastery and that "the land owed no loyalty, religion or civil to anyone".

951 *Diploma di Berengario*

This document explains the outline for a government for San Marino. Dated September 26, 951, the document states that a considerable population exists on San Marino and that it has constructed its first fortification of towers and walls. Clearly a government (civil authority) is in place on San Marino which has democratic principles and in which citizens meet and discuss how best to regulate their affairs. Although it has undergone many changes, the government remains the same as it is at the beginning.

Today, there are five basic institutions by which San Marino is governed: the *Arengo*, The Grand and General Council, the Council of Twelve, the Congress of State, and the Captains Regent. The *Arengo*, by the beginning of the tenth century, consists of all heads of families in San Marino who meet regularly to discuss matters of importance. The outcome of these discussions determines final decisions. The Arengo elects a Rector to chair the meetings and together with the Captain of Defense, who is also elected by the Arengo, conducts the affairs of the state. However, as time passes this group grows; consequently, the arrangement becomes impractical. In 1906, the Arengo decides to delegate its powers to a body which becomes known as the Grand and General Council.

The Council of Twelve protects citizens from injustices that a government agency may have taken against them, including the power to annul government actions. The Council of Twelve is also the guardian of Sanmarinese property. If a foreigner wishes to build or purchase property in San Marino, this body must approve it.

Because San Marino is small, it does not have local municipal governments in the sense that other local communities have. Its communities may be likened to townships. San Marino is divided into nine regions called Castles: Acquaviva, Borgo-Maggiore, Chiesanuova, Domagnano, Faetano, Fiorentino, Montegiardino, and San Marino and Serravalle. Each castle has an auxiliary council elected to a four-year term headed by an official, called the Captain of the Castle, who is elected every two years.

1243 Arengo elects two consuls

In 1243, the Arengo elects the first two Captains Regent (originally called Consuls). To avoid abuse of power the

Nine Castles (Regions) of San Marino

Acquaviva was known in the Middle Ages. Its name comes from the water it draws from its rocks. The Monte Cerreto Natural Park, a beautiful green park and recreation facilities, is located here.

Borgo-Maggiore has had an ancient market since 1244. It is an historical center with museums, churches, monuments, and typical court roads.

Chiesanuova includes a peak with the medieval castle from the Busignano's court. From this beautiful country one can have a spectacular view of the San Marino Capital and Mount Titano.

Domagnano is a small village in 1300, but after the 1463 victorious battle against the Malatesta Court of Rimini, the Montelupo Fortress is expanded.

Faetano is the ancient Malatestian Territory, the last San Marino territorial expansion in 1463. There is a church here built in the eighteenth century.

Fiorentino is an ancient Malatestian fortress conquered by San Marino in 1463. Today, a public service building and a football field for summer evening sports are on this location.

Montegiardino is an ancient village, dating back even further than the Longobards period. In 1463, it joins San Marino. Church and castle architecture, along with other art, is of interest here.

San Marin Citta is the site where Father Marino organized the small community on Mount Titano in 301 AD. Millions of tourists visit every year. Stores line the way for souvenir shoppers. Important annual events are held here: the Captains, Regency Ceremony (every six months) and the Republic Foundation's Day crossbow contest (on September 3).

Serravalle was first mentionned in a manuscript written in 962. It became part of San Marino in 1463, and is a modern urban center.

Arengo devises two heads of state, with equal power and a brief term in office. Every six months, on April 1 and October 1, the Grand and General Council elects two new Captains

Regent. These individuals serve as joint Heads of State; they are the commanders of the army and presidents of the Grand and General Council, the Council of Twelve and the Congress of State. No matter can be discussed without the approval of both Captains Regent. They can take different sides on government decisions, and each has the right to veto the other, but their input is absolutely essential. Over time and because of the short terms, all classes of citizens have participated as government heads. Once elected, the Captains Regent enact traditional San Marino ceremonies beginning at the Valloni Palace. Afterwards, the Captains Regent lead a procession to the cathedral for mass, even if the elected party is from the Communist Party. Following mass, the procession returns to the palace where the two Captains Regent are sworn in by the Secretary of State.

1247 The Pope excommunicates all Sanmarinese citizens

In a contest between two families, the Ghibellino and the Guelph, who seek the throne of Germany, the Sanmarinese side with the Ghibelline family against the Pope who aligns with the Guelph. The pope excommunicates the Sanmarinese. Pope Innocent IV revokes this two years later.

1291 The Sanmarinese resist Papacy efforts to dominate them

In 1291, a priest named Theodorico is sent to San Marino to persuade the inhabitants to declare themselves subjects of the pope and pay their taxes. At this time, the pope is more than a spiritual head and closer to a monarch. The dispute is submitted to a lawyer from Rimini for a decision. The Judge, Palamede, rules that the Sanmarinese are "free and independent." As a consequence to this ruling, intermittent fighting occurs from 1304 until 1463, but the decision clearly defines San Marino as a separate and independent state.

1463 San Marino defeats Sigismondo Malatesta and gains territory

Sammarine forces, with help from the Duke of Urbino, defeat Sigismondo Pandolfo Malatesta, Lord of Rimini, and gain territory for their little republic. Pope Pius II awards San Marino the towns of Fiorentino, Montegiardino, and Serravalle. That same year, the town of Faetano decides the join the Republic. With these additions, the present-day borders of San Marino are fixed.

1503 Cesare Borgia conquers San Marino

An Italian family and its army under the leadership of Cesare Borgia conquer San Marino for a brief period. Borgia occupies the region around San Marino but leaves after six months. He finds the Sanmarinese uncooperative and leaves voluntarily.

16th Century Giambattista Belluzzi, a famous military engineer, is born in San Marino

Giambattista Belluzzi, a famous military engineer in the service of Florence, is born in San Marino.

17th century Neighboring principalities (states) plot to take possession of San Marino

On June 4, as legend goes, an army from Rimini sets out in two directions to invade San Marino. A dense fog descends upon one of the columns and they become lost. Eventually, the fog lifts and the Sanmarinese see them, prepare against them, and prevent a disastrous invasion. San Marino is saved. Shortly afterwards, an attempt by another neighbor to invade the Republic meets the same fate.

The Sanmarinese realize that in order to preserve their independence they need allies. They sign a defense treaty with a nobleman, Guidubaldo, Duke of Urbino. In 1631, San Marino is surrounded by Papal States and the duke dies. The pope absorbs the duke's principalities and declares San Marino is under his domain. San Marino refuses and continues to govern itself.

1739 Cardinal Giulio Alberoni subjects San Marino to Papal authority

Cardinal Alberoni (1664–1752) lives in San Marino and is the Papal Legate for the Province of Romagna. The Sanmarinese bicker among themselves in regard to coming under Papal authority. Without getting permission from the pope and believing that the Samarinese are willing, Alberoni orders 500 soldiers into San Marino to take over and subject the citizens to papal authority. There is heavy resistance, and the Sanmarinese appeal directly to the pope. The pope sends Monsignor Enrico Enriquez, Governor of Perugia, to investigate. It is decided that the Cardinal Alberoni "exceeded his authority" and that "the Pope had no intention to change the Rebulic's system of government". San Marino is set free. The last soldier leaves on the Feast of Agatha, February 5, 1740, and St. Agatha becomes a joint patron saint of the Republic alongside Saint Marino.

1743 Giuseppe Balsamo is born

Giuseppe Balsamo (1743–95) is born in San Marino. Also known as Count Alessandro Cagliostra, he is considered to be an adventurer, imposter and alchemist.

1781 Bartolommeo Borghesi is born

Bartolommeo Borghesi (1781–1860) resides in San Marino from 1821-60. He is known as an antiquarian, epigrapher, and numismatist.

1797 **Napoleon marches by San Marino crushing all Italian states which oppose him**

When Napoleon arrives to San Marino, he by-passes the republic. He sends an emissary to tell the Sanmarinese that he will not invade because "It is our duty to preserve San Marino as an example of freedom".

As the population of the republic expands down the mountain to the valley, Napoleon offers the Sanmarinese more land. The Sanmarinese refuse Napoleon's offer and respond by explaining that "San Marino is small in size and poverty protects it from its greedier neighbors". Expansion, they believe, is an invitation for war.

1815 **Congress of Vienna**

Following the defeat of Napoleon at the battle of Waterloo, the Congress of Vienna is called to reshape Europe. It draws up a list of European nations and includes San Marino among them.

1849 **Giuseppe Garibaldi finds refuge in San Marino**

Garibaldi (1807–82), a leader and founder of modern-day Italy, is pursued by the Austrian and Papal troops. He takes refuge in San Marino and later repays the citizens of the republic by recognizing them in the Treaty of 1862.

1861 **U.S. President Abraham Lincoln praises San Marino**

In a letter to one of the Captains Regents, U.S. President Abraham Lincoln writes: "Although your dominion is small, nevertheless your State is one of the most honoured throughout history."

1862 **Treaty of Friendship with Italy**

Italy becomes a new nation. It signs with San Marino a Treaty of Friendship that recognizes the independence, liberty, and sovereignty of this small nation-state.

1864 **Money minted**

San Marino begins minting its own money. This practice continues until 1938 and is then resumed in 1971.

1879 **Telegraph service opens with Italy**

A telegraph communication service opens linking San Marino with Italy.

1904 **Telephone service with Italy**

San Marino establishes telephone connection with Italy.

Italian soldier Giuseppe Garibaldi (1807–82) finds refuge in San Marino. (EPD Photos/CSU Archives)

1906 **The "peaceful revolution"**

Citizens are face economic and financial hardships. A Grand and General Council forms which serves as an elected parliament. It has sixty members who are elected for a five-year term. Elections are conducted according to a proportional representation. Any party that gains more than two percent of all votes cast is guaranteed a seat in the Council which brings many political parties together over the years. It is common for the Council to include the Christian Democratic Party, various socialist parties, the Communist Party, and a Marxist-Lenninist Party, along with a local party, such as the Defense of the Republic Party. The function of the parliament is to pass laws and regulations. In addition, the parliament also has the right to grant mercy or amnesty to criminals and appoint government officials and magistrates. The Sanmarinese celebrate March 25 as a public holiday.

1914–18 **Sanmarinese citizens volunteer to fight on the side of Italy**

Although San Marino declares itself neutral, volunteers enroll in the Italian army.

1922–43 Benito Mussolini rules Italy

Benito Mussolini (1883–1945) adopts the title *il Duce* (leader) and rules Italy as a fascist state. Under Mussolini, Italy becomes a one-party state, and society is divided into a number of "corporations" which come under the ultimate control of the Fascist Grand Council. In foreign affairs, Mussolini adopts an expansionist policy, seeking to carve out Mediterranean and African empires and an alliance with Nazi Germany. Although il Duce seeks Italian expansion, he does not move against San Marino, which maintains its independence and neutrality in foreign affairs.

1938 San Marino adopts the Italian *lire*

San Marino abandons its own currency and adopts the Italian *lire*. It keeps the *lire* as its official currency until 1971.

1939: March The Treaty of 1862 is amended

The Treaty of Friendship between Italy and San Marino is amended to read as "a protective friendship". This treaty reflects the fear among Sanmarinese of losing their independence to Mussolini's Italy.

1939–45 World War II

During the Second World War, San Marino adopts a fascist government but declares itself neutral. On June 26, 1944, San Marino is bombed by British planes. The bombing kills many people and destroys a railway line. It also hits the Valloni Palace where many pieces of art, including Stozzi's 'St. John the Baptist", are damaged or destroyed. San Marino is caught between two worlds. It declares itself to be a free and neutral country but does not refuse asylum or help to those persecuted by misfortune or tyranny. During the Second World War, San Marino takes in more that 100,000 refugees.

1945 Congress of State created

The Congress of State is the newest institution. Its members are called Representatives, with the exception of those who head the Department of Affairs, the Department of the Interior, and the Treasury who are called Secretaries of State. They are appointed for five years and can be reappointed. The ten members of this Congress are presided over by the two most powerful persons in San Marino, the Captains Regent.

1945 A coalition of communists and left wing socialists gains control of the government

From 1945 to1957, a communist and socialist-led coalition rules the Republic of San Marino. In 1956, Hungary revolts against the Soviet Union, and its repercussions are felt within the ranks of the left wing parties. There is dissension within the communist and socialist parties. The Socialist and Christian Democratic Parties quickly form a coalition and, recognized by Italy, take over the reigns of the government. The government, however, remains unstable for a number of years. In 1978, the communists, along with the Socialists and a Unity Party, regain control of the government but by only one seat. San Marino remains unstable until 1986, when the communists align themselves with the Christian Democrats to form yet another—but more stable—coalition.

1956 Social insurance program created

The government incorporates a comprehensive social insurance program that includes disability, family supplement payments, and old age pensions. The creation of a social insurance program is part of a long-term social development plan initiated by the Sanmarinese government during the postwar era. In subsequent years, San Marino develops medical and educational facilities, including a state hospital and the Institute of Cybernetics and the Degli Studi, both of which specialize in science and technology.

1960 Women gain the right to vote

In 1960, women in San Marino gain the right to vote and hold office. Since then, they have been elected to the Grand General Council. The equal rights of citizens, however, remain a contested issue. Only in 1982 were women who marry foreign citizens given the right to keep their citizenship. They still may not transmit their citizenship to a husband or children—a right accorded to men.

1971: September The Treaty of 1939 between Italy and San Marino is amended

The 1939 treaty is amended to read, "The Republic of San Marino confirms its neutrality and is certain that the warm friendship and deep cooperation of the Italian Republic. for the preservation of San Marino's ancient liberty and independence, will never fail. For this purpose the governments of Italy and San Marino will consult regularly on problems of common interest." This treaty establishes a regular consulting relationship between both countries. In 1987, it allows for San Marino to establish its own television station with an Italian loan and reinstates the San Marino right to operate a casino. It also enables San Marino banks to deal directly with foreign banks instead of going through Italian financial operations.

1971 San Marino returns to circulating its own money

In 1971, the government of San Marino begins circulating its own currency once again but has a major problem with the process. Tourists, which number over three million per year, use the money for souvenirs and take the money home with them. This forces the government to constantly replenish to compensate for the supply of money taken out of circulation by tourists. Although the Italian *lire* remains legal tender in San Marino, Sanmaranese coins are not accepted outside of the tiny republic.

1975   A state hospital is opened in San Marino

This hospital is more a public health institution. Serious matters, such as surgery, are still directed to Italy. This institution is primarily a dispensary for the poor. However, surgery excepted, it provides free and comprehensive medical care for all citizens.

1984: October   The first income tax

The first income tax passed by the Grand and General Council is progressive for its citizens and a flat tax for the corporate institutions. Other levies are also initiated at this time, for instance, stamp duties, registration, and mortgage taxes.

1988   San Marino joins the Council of Europe

In 1988, San Marino joins the Council of Europe. Founded in 1949, the Council of Europe is an international organization whose aim is to protect human rights and pluralist democracies. It has a Parliamentary Assembly whose members are elected by the national parliaments of the respective member states. Except for defense, its agenda covers human rights, mass media, legal cooperation, social and economic issues, health education, culture, heritage, sports, youth, local and regional self-government and the natural environment.

1992   San Marino joins the United Nations

Up to this time, San Marino participates with only "Observer Status". In 1972, it sends a delegation to the United Nations Conference on Human Environment, but it cannot vote or speak unless invited.

The United Nations is an organization of sovereign nations. Founded in 1948, its central purpose is to preserve world peace. It does not legislate but rather provides the machinery for solving problems that face people around the world. The United Nations has six main parts: The General Assembly, the Security Council, the Economic and Social Council, the International Court of Justice and the Secretariat. In addition, there are fourteen specialized agencies in health, finance, agriculture, civil aviation and telecommunications.

Bibliography

Bent, James Theodore. *A Freak of Freedom; or, The Republic of San Marino.* Port Washington, N.Y.: Kennikat Press, 1970.

Carrick, Noel. *San Marino.* New York: Chelsea House Publishers, 1988.

Duursma, Jorri C. *Fragmentation and the International Relations of Micro-States.* Cambridge: Cambridge University Press, 1996.

Johnson, Virginia Wales. *Two Quaint Republics: Adorra and San Marino.* Boston: Estes, 1913.

Packett, Neville. *Guide to the Republic of San Marino.* Bradford, England: Lloyd's Bank Chambers, 1964.

United States Department of State. Bureau of Public Affairs; Background Notes: *San Marino.*

World Reference Atlas: San Marino. London: Darling Kindersley Publishing, Inc., 1998.

Slovak Republic

Introduction

With the creation of the Slovak Republic following the break-up of Czechoslovakia in 1993, Slovakia became an independent political entity for the first time. The Slovaks had been ruled by Hungary through most of their history, winning independence in 1918 only as part of a political union with the Czechs—a union in which they were the subordinate member, both in its early years and during the Communist regime that followed World War II (1939–45). Once the authoritarian rule of Communism was removed in 1989, separatist sentiments came to the fore, and the Slovaks demanded control over their own destiny. As a sovereign nation, they have faced the challenges of balancing government control with respect for human rights and settling on the best pace for the introduction of free-market economic reforms.

Bordered by the Carpathian Mountains to the north and northeast, Slovakia is a landlocked country occupying 18,923 square miles (49,011 square kilometers)—roughly twice the area of the state of New Hampshire. To the south are hills and river valleys, and in the southwest is the Hungarian Plain, where Bratislava, Slovakia's capital and largest city, is located. The Slovak Republic has an estimated population of 5.4 million people.

Hungarian and Habsburg Rule

The early history of Slovakia was similar to that of the Czech lands to its west. Slavic settlers arrived in the region in the fifth century A.D., settling in agricultural villages. The first unified state in the region was founded in 625, but it was short-lived. In the ninth century the Great Moravian Empire was founded, becoming the first political entity to include both Czechs and Slovaks. The decisive split between the Czech and Slovak peoples occurred with the Magyar (Hungarian) invasions of the tenth century, which began nine centuries of Hungarian rule for the Slovaks. The Czech territories, meanwhile, saw the emergence of the Kingdom of Bohemia, which fell within the Frankish sphere of influence and that of the Holy Roman Empire. Thus the two groups began a long period of separate cultural and political development, emerging as two distinct ethnic entities.

For a long time the Slovaks remained largely a rural peasant population ruled by Hungarian nobles. However, the Mongol invasions of the thirteenth century improved their situation by reducing the power of the Hungarians. Cities grew, and a Slovak merchant class emerged. Slovakia also benefited by the arrival of new immigrants from Germany and other areas. With the military victory of the Ottoman Turks over the Hungarians in the sixteenth century, Slovakia came under the control of the Habsburg Empire.

The first important development in the emergence of a political consciousness among the Slovaks was the period of enlightened Habsburg rule in the eighteenth century, which, ironically, led to waves of both Austrian and Hungarian nationalism that resulted in the repression of the Slovak language and culture. By the 1840s, the first Slovak-language periodical was being published. In spite of their growing sense of national identity, the Slovaks did not make substantial progress toward political autonomy until the idea of an independent political union with the Czechs took hold early in the twentieth century.

Czechoslovakia, 1918–91

With the dissolution of the Habsburg Empire in World War I (1914–18), this idea became a reality, and Czechoslovakian independence was declared on October 28, 1918. Czech leader Tomas Masaryk (1850–1937), became the first president of the new nation, which had a strong industrial base but struggled to create unity among its diverse ethnic population. The Slovaks, who had expected to have internal autonomy, found themselves under the direct rule of a strong central government, which declared Czech the only official national language. The new government did not act to remedy imbalances in education and economic development between the two regions, and many Slovaks felt like second-class citizens. By 1925 a Slovak separatist political party scored a strong showing in national elections, and separatist feeling remained strong up to the eve of World War II.

Between the disastrous Munich Pact of 1938 and the spring of 1939, Czechoslovakia was effectively dismembered

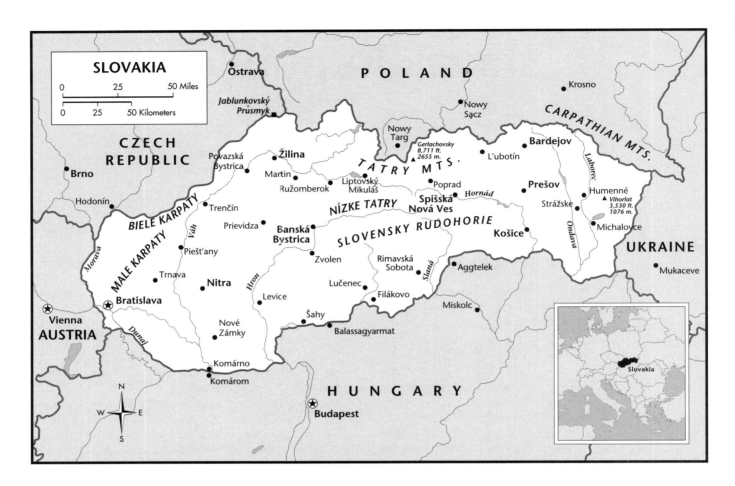

SLOVAKIA

0   25   50 Miles

0   25   50 Kilometers

by Nazi Germany, which sponsored the creation of a Slovak puppet state that complied with Nazi directives throughout the war (and whose leader was later executed for treason). Slovak leaders in exile in London, however, declared their opposition to the Nazi-controlled government, and resistance fighters at home participated in an unsuccessful revolt against it.

Czechoslovakia was reunited after the war, and Slovakia was once again promised autonomy, this time under a new government allied with the Soviet Union. But prospects of autonomy vanished with the imposition of exclusive Communist rule in 1948. Slovak leaders protesting government interference in their region were labeled "bourgeois nationalists" and persecuted in the political purges of the 1950s. In 1956 mass demonstrations in Slovakia, inspired by the rebellion in Hungary that year, had to be crushed by both Czechoslovakian and Soviet troops. By the 1960s, however, the worst of the repression was over, and in 1968 Alexander Dubcek (1921–1992), the head of the Slovak Communist Party, was promoted to first secretary of the party for the entire nation— the first time this post had been held by a Slovak. Dubcek instituted a series of reforms aimed at creating "Socialism with a human face."

The period that followed was characterized by unprecedented freedom of expression, as media censorship was lifted and plans were made for economic reforms also. However, the government leaders of the Soviet Union and Czechoslovakia's other Communist neighbors felt threatened by the growing freedom so near their borders and criticized the new Czech policies. On August 20, 1968, the brief period of freedom known as the "Prague Spring" ended when Soviet tanks rolled into Czechoslovakia. Dubcek and other top leaders were taken into Soviet custody and pressured to reverse their policies to conform with the wishes of the Soviets.

A new government was formed under another Slovak, Gustav Husak (1913–91), who, in spite of his imprisonment during the purges of the 1950s, had developed into a hard-line Communist. Husak was appointed party secretary in 1969 and also served as president for fourteen years, beginning in 1975. Husak resisted the Soviet liberalization of the 1980s that took place under Soviet leader Mikhail Gorbachev (b. 1931), and anti-government feeling grew in Czechoslovakia throughout the decade. In November 1989 the brutal government response to a large student demonstration in Wenceslas Square triggered the "Velvet Revolution," a ten-day period of protests, rallies, and strikes that led to the resignation of the country's Communist leaders in December and the parliament's election of dissident playwright Vaclav Havel (b. 1936) to the presidency.

The country was renamed the Czech and Slovak Federal Republic, and the first free elections since 1946 were held in June 1990. Tensions between the Czechs and Slovaks grew, however. Many Slovaks wanted more autonomy than that provided under the central government and disapproved of the rapid pace with which the country was moving toward a free-market economy. Efforts at reconciliation by President Havel were made more difficult by the fact that the new Slovak leader, Vladimir Meciar, was a strong separatist. By 1992 representatives of both sides agreed to separate, popularly known as the "Velvet Divorce." As of the beginning of 1993, Czechoslovakia ceased to exist and was replaced by separate Czech and Slovak republics.

Except for a brief period in 1994, Meciar remained in office until 1998. His government's treatment of the country's ethnic Hungarian minority aroused controversy at home and censure abroad and contributed to Slovakia's being passed over in the first round of Eastern European admissions to the European Union (EU, which included the Czech Republic). However, it was anticipated that the country would join the EU as part of the next round of admissions.

Timeline

c. A.D. 500   Arrival of the Slavs

Slavic tribes who are the ancestors of today's Slovak people migrate to present-day Slovakia from the east.

625–658   Samo unifies the Slavs

The first unified Slavic state in the region is founded by Samo, a member of a Germanic tribe called the Franks. Centered in Bohemia, it dissolves upon the death of its founder.

9th century   The Great Moravian Empire

A kingdom is established in Moravia by a Slavic leader named Mojmir. Later expanded to include parts of Bohemia, Slovakia, Poland, and Hungary, it becomes known as the Great Moravian Empire. It unites the Czechs and Slovaks politically for the first time.

850–870   Cyril and Methodius convert Czechs to Christianity

Prince Rostislav of the Moravian Empire asks the Byzantine Emperor to send holy men to teach his people about the Eastern Orthodox religion in their own language. The famous Orthodox monks Cyril (c. 827–69) and Methodius (c. 825–884) convert many Czechs to the eastern church, also introducing a new script (the Cyrillic alphabet) to render religious texts in Slavic. Ultimately, however, Roman Catholicism becomes the predominant religion in the region.

850   First Christian church is built

The first Christian church in Slovakia is built in the western part of the region, in present-day Nitra.

Hungarian Rule

10th century   Hungarian (Magyar) invasions

The Great Moravian Empire breaks up as the Hungarians (Magyars) invade from the east, subjugating the Slovaks, most of whom remain peasants (and eventually serfs) ruled by the Hungarian nobility. Meanwhile, the Kingdom of Bohemia, under the influence of the Franks and the Holy Roman Emperor, emerges in the Czech lands of Bohemia. The Czech and Slovak peoples are thus separated politically and culturally, encouraging their development as two distinct ethnic groups.

13th century   Economic progress

Mongol invasions, which weaken Hungary, lead to greater freedom and economic progress for the Slovaks, who also benefit from the arrival of immigrants from other lands, particularly Germany. A Slovak merchant class emerges, and cities begin to grow.

15th century   Hussite movement affects the Slovaks

Religious divisions created by the Hussite religious movement in Bohemia bring new immigrants to Slovakia and spread the religious doctrine of Czech church reformer and martyr Jan Hus (c. 1369–1415).

1526   Battle of Mohacs

Following the defeat of the Hungarians at the hands of the Ottoman Turks at the Battle of Mohacs, the Hungarian Empire is divided into three regions. Slovakia is included in the area ruled by the Habsburgs, and the city of Bratislava in western Slovakia becomes the Habsburg capital until the expulsion of the Turks from Hungarian lands at the end of the seventeenth century.

Nationalist Revival

Late 18th century   Enlightened rule brings nationalism

There are major changes in Habsburg rule in the eighteenth century thanks to the enlightened policies of Empress Maria Theresa (r. 1740–80) and Emperor Josef II (r. 1780–90). Together with needed social reforms and an increase in religious toleration, however, comes a loss of national identity, as the use of the German language is institutionalized throughout the Habsburg Empire. This action leads to a wave of Hungarian nationalism which, in turn, touches off a corresponding reaction on the part of the Slovaks, who react as negatively to having Hungarian forced on them as they do to Germanification. Slovak authors Jan Kollar (1793–1852), and Jozef Safarik (1795–1861), who write in classical Czech, play a prominent role in nationalist revivals among both the Czechs and Slovaks.

1793: July 29   Birth of nationalist poet Jan Kollar

Kollar is an important figure in the revival of Slavonic nationalism and cultural identity in the early nineteenth century. He is educated at the University of Jena, and in the course of his varied career, he serves as a pastor and a professor of Slavonic archaeology, as well as writing poetry that contributes to the spirit of Slavic cultural solidarity. Among his best-known works is the lyric-epic poem *Slavy dcera* (The Daughter of Slava; 1824).

1836   Hungarian is declared the official language of Slovakia

The Hungarians decree that their language must be the only official language of the Slovaks. In protest, some Slovak authors who had written in Czech begin to write using the Slovak dialect.

1840s   Slovak literary language is created

Author Ludovit Stur (1815–56) develops a Slovak literary language called *sturovcina*.

1845   First Slovak periodical begins publication

Ludovit Stur (see 1840s) begins to publish the first Slovak-language periodical, *Slovenskje Narodnje Novini.*

1846   National epic poem is published

Marina, the Slovak national epic, is published by Andrej Sladkovic.

1848   Slovaks draw up a list of demands

During the Hungarian Revolution, the Slovaks draw up the "Demands of the Slovak Nation," which include the formation of a Slovak assembly and the use of the Slovak language in schools, local government, and the courts. The Hungarians ignore the Slovak demands.

1849: February 2   Birth of poet Hviezdoslav

The great Slovak poet Hviezdoslav (pen name of Pal Orszagh) starts out by writing in Hungarian but switches to the Slovak language early in his career. His poetry, which is a landmark in the development of a Slovak poetic tradition, is notable for combining realism with Slavic folk material. Major works include "The Gamekeeper's Wife" (1886) and "Blood-Red Sonnets" (1919), a book of World War I poetry.

1850s–1860s   First plays are written in Slovak

The comedies of Jan Palarik become the first Slovak-language plays.

1852   Slovak grammar is published

Martin Hattala publishes a Slovak grammar based on the literary language developed by Ludovit Stur and his contemporaries (see 1840s). Hattala's book is accepted as the standard linguistic authority.

1861   Slovaks again petition for expanded rights

Slovak nationalists again demand autonomy from the Hungarians and are refused.

1863   Slovak National Museum is founded

The Slovak National Museum is established at Martin.

1867   Creation of the Dual Monarchy

With the creation of the Austro-Hungarian Dual Monarchy, Hungary intensifies its efforts at "Magyarization" (spreading the Hungarian language and culture) and its repression of Slovak nationalism. Slovak high schools, the Slovak National Museum, and the cultural center Matica Slovenska are forced to close. Politically, the Slovaks are second-class citizens, rarely holding high elective office.

1886   The Bratislava National Theatre opens

Construction is completed on the Bratislava National Theatre.

1913: January 10   Birth of Gustav Husak

Communist leader Gustav Husak is born near Bratislava, where he studies and practices law. In the 1930s he joins the Communist Party and becomes one of its leaders during World War II (1939–45), when he also is one of the leaders of an anti-Nazi Slovak uprising. After the war, he is active in both the national Czechoslovakian Communist Party and the regional Slovak party. He loses his party standing in the government purges of the early 1950s. Labeled a Slovak nationalist, he is imprisoned from 1954 to 1960. In the 1960s he once again rises through the ranks of the party and is head of the Slovak Communist Party by 1968, when his fellow Slovak, Alexander Dubcek (1921–92; see 1921: November 27) becomes head of the Czechoslovakian party. Following the Soviet invasion of the same year, Husak is named to the top Party position and presides over the dismantling of the liberal reforms of Dubcek's "Prague Spring." In addition to his party post, Husak becomes president in 1975 and holds both posts until the late 1980s, heading a regime characterized by harsh repression of political dissidents. Husak is forced out of office in 1989 by the "Velvet Revolution" and dies on November 18, 1991.

Territorial losses after World War I.

☐ Land lost

1914–18   World War I

During World War I, Czech and Slovak leaders lay the framework for a unified, independent state to be created when the war ends. Slovak nationalist leader Milan R. Stefanik (1880–1919) spends the war years in the West, working for Czechoslovak independence together with Czech leaders Tomas Masaryk (1850–1937) and Eduard Benes (1884–1948). Father Andrej Hlinka (1864–1938) and other Slovak patriots work for independence at home.

1916   Czechoslovak National Council is formed

Czech nationalist leader Tomas Masaryk, together with fellow Czech Eduard Benes and Slovak leader Milan Stefanik, founds the Czechoslovak National Council, headquartered in France (see 1914–18). The three leaders lobby for recognition by the Western powers while their compatriots at home form underground resistance groups.

1918   The Allies recognize the Czechoslovak council

The Allied powers recognize the Czechoslovak National Council and approve the formation of a Czechoslovak state.

1918: May 30   The Pittsburgh Declaration is signed

Slovak and Czech leaders in wartime exile in the United States sign an unofficial document agreeing to jointly form an independent state with autonomy for Slovakia, which is to have its own parliament and judiciary and enjoy full language rights.

The Czechoslovak Republic

1918: October 28   Masaryk declares Czech independence

A provisional Czech government is set up, with Tomas Masaryk named as president. Still in exile in the United States, Masaryk issues a declaration of Czechoslovak independence. With a population of over 13.5 million, the new nation will encompass a variety of ethnic groups, including Czechs, Slovaks, Moravians, Ruthenians, and Sudeten Germans. It will have a large part of the industrial base that formerly belonged to the Austro-Hungarian empire, including china, glass, automobile, and beer manufacturing. However, far more of the industrialization is in the Czech part of the country than in Slovakia.

1918: October 30   Slovaks declare independence

Two days after the Czechs declare Czechoslovak independence, Slovak leaders issue the Declaration of Turciansky Svaty Martin, announcing the secession of Slovakia from Hungary and officially approving the Czech-Slovak union.

1919: December 10   First Slovak opera is performed

Czech composer Bedrich Smetana's opera *The Bartered Bride* becomes the first operatic work to be performed in Slovak.

1920s and 1930s   Tensions between Czechs and Slovaks

Through the 1920s, Czechoslovakia maintains a stable central government with a vigorous party system, and its strong industrial base brings prosperity to the new nation. The conflicting interests and expectations of its diverse ethnic constituency, however, create serious problems for the country, especially between the central government and the Slovaks. Many Slovaks have expected full autonomy within the new nation, but the government retains strict central control of the country, and many Slovaks feel like second-class citizens. There are fewer than two hundred Slovaks among some eight thousand government employees. Even within Slovakia itself, the top civil service positions are largely held by Czechs. Slovak is not declared a national language, and the government does not attempt to redress the economic imbalance between the Czech and Slovak regions by supporting economic development in Slovakia.

1920–21   Formation of the Little Entente

Czechoslovakia, Yugoslavia, and Rumania form the Little Entente (understanding or agreement) for mutual protection against a possible restoration of the Habsburg monarchy.

1920 Constitution is adopted

Czechoslovakia adopts its first permanent constitution, which provides for a parliamentary democracy with a bicameral (having two legislative chambers) national assembly and a president elected to a seven-year term who shares executive power with an appointed cabinet. Slovaks are disappointed that the constitution contains no provisions for Slovak autonomy.

1920: March 1 Opera society sponsors first performance

A performance of *The Kiss* by Bedrich Smetana (see 1919: December 10) launches the Slovak National Opera Society.

1921: November 8 Death of poet Hviezdoslav

Major Slovak poet Hviezdoslav dies. (See 1849.)

1921: November 27 Birth of Alexander Dubcek

Dubcek is famed as the liberal Communist leader who inaugurated the "Prague Spring" of 1968 (see 1968) that leads to the Soviet invasion of Czechoslovakia the same year. Dubcek serves in the underground resistance to the Nazis during World War II (1939–45) and works his way up the hierarchy of the Communist Party in the postwar period. After his prominent opposition to longtime prime minister Antonin Novotny (1904–75), Dubcek is appointed first secretary of the Communist Party of Czechoslovakia in January 1968 and institutes wide-ranging liberal reforms, including greater freedom of expression and rehabilitation of political figures prosecuted during the purges of the early 1950s. Uneasy with the newfound political freedoms in one of its "satellite" countries, the Union of Soviet Socialist Republics (U.S.S.R.) sends troops into Czechoslovakia in August 1968. Dubcek and other top Czech leaders are taken to the Soviet Union, where they are coerced into abandoning much of their reform program.

Dubcek is allowed to return to Czechoslovakia but removed from his former position in April 1969 and expelled from the Communist Party in 1970. He remains in political obscurity until the "Velvet Revolution" of 1989 (see 1989: November 17), when the Communists lose their hold on Czechoslovakia. At this time, Dubcek is elected chairman of the national assembly. In 1992 he becomes the head of the Slovak Social Democratic party. Alexander Dubcek dies in Prague in 1992 as a result of injuries suffered during a car crash.

1923 Government bans demands for autonomy

A law prohibits protests against the "unitary structure" of the Czech government by those who favor a looser federalist government with greater autonomy for individual regions. Under this law, demands for Slovak autonomy are viewed as subversion.

1924 Defense pact is signed with France

Czechoslovakia and France sign a mutual defense treaty aimed at protection against possible German aggression in the future.

1924 National museum is founded

The Slovak National Museum is launched in Bratislava.

1925 Slovak separatist party scores large electoral showing

The People's Party, led by Slovak nationalists Andrej Hlinka and Monsignor Jozef Tiso, wins thirty-two percent of the Slovak vote in national elections.

1926 Scientific library is established

The State Scientific Library is founded in Banska Bystrica, eventually growing to house nearly two million volumes.

1927 Slovak provincial administration is formed

Slovakia gets its own provincial administration, but Slovak leaders continue to agitate for full autonomy.

1928 Slovak professor is jailed for advocating autonomy

Under a law that bans demands for regional autonomy (see 1923), Professor Vojtech Tuka is sentenced to fifteen years in jail for a newspaper article calling for a popular vote on Slovakia's role in the Czech-Slovak union.

1933 Slovak newspaper is shut down

The government shuts down the newspaper of the nationalist Slovak People's Party and arrests its editor following a mass demonstration calling for Slovak autonomy.

1935 Masaryk resigns

Tomas Masaryk, president of Czechoslovakia since 1918, resigns due to age and fragile health, and is succeeded by Eduard Benes.

1938: June Bill for Slovak autonomy is introduced

As Czechoslovakia nears a crisis over German demands for cession of the Sudetenland, Slovak demonstrations and demands for autonomy are renewed. The Slovak People's Party introduces a bill in the Czech parliament providing for Slovak autonomy, and about 120,000 Slovaks demonstrate in Bratislava.

World War II

1938: September 29–30   Munich Pact

In response to threats by Nazi leader Adolf Hitler (1889–1945), the major Western powers pressure Czechoslovakia to cede the Sudetenland to Germany. Representatives of Britain, France, Italy, and Germany meet in Munich to sign an agreement with Germany, with no representation by Czechoslovakia. Hoping to avert a war, they sign the Munich Pact, which deprives the Czechs of the Sudetenland and other territories. Czechoslovakia loses its most industrialized and prosperous region, as well as a large percentage of its population.

1938: October 15   Benes resigns and leaves the country

Czech president Eduard Benes (see 1914–1918, and 1935) resigns his post and goes into exile in London.

1938: November   New government is formed

A new government is formed under Emil Hacha and named the Federated Republic of Czechoslovakia (also known as the Second Republic). Hacha's right-wing government attempts to appease the Germans by cooperating with their demands, adopting a new constitution that weakens the nation yet further by granting substantial autonomy to Slovakia and Subcarpathian Ruthenia.

1939: March 15   German takeover creates Slovak puppet state

Bohemia and Moravia become German protectorates, effectively ending the existence of Czechoslovakia as a sovereign nation. Following previous negotiations between Nazi and Slovak leaders and a declaration of Slovakian independence, the Independent Slovakian Republic is formed with Monsignor Josef Tiso as president. The Slovakian government signs the Anti-Comintern Pact, and its policies are dictated by Nazi Germany. The Anti-Comintern Pact is an anti-Soviet alliance that takes its name from the Communist International, an ostensibly independent association of communist parties that is headquartered in Moscow. In reality it is controlled by the Soviet Union. As a result of Slovakia's adherence to this alliance, Slovakian forces are sent to take part in the German invasion of the Soviet Union.

1940–42   Slovak Jews are deported to Poland

Following German directions, the Slovak government deports Jews to Nazi-occupied Poland for resettlement. Slovak president Tiso stops the deportations when he learns that many of the deported Jews are being killed.

1943   New Czech-Soviet pact is signed

Feeling that they have been betrayed by the West and anticipating that it is the Soviets who will liberate Czech lands from their German occupiers, Czech leaders strengthen their ties to the Soviet Union throughout the war years. This includes the signing of a twenty-year mutual defense pact.

1943: December   Slovaks in exile form provisional government

Liberal Slovak leaders in London form the Slovak National Committee, an underground resistance group, and sign the Christmas Agreement declaring their intention to take control of Slovakia back from the Nazi-controlled Tiso government.

1944: August   Slovak Revolt

Partisans with Communist ties lead a rural revolt in eastern Slovakia against the Nazi-controlled government. The London-based Slovak National Committee declares support for the rebels, and President Tiso requests German military assistance. The revolt collapses, but German troops occupy the country until the end of the war.

1944: October 18   Russians enter Slovakia

The first Russian troops enter Slovakia to liberate the country from occupying German troops and are met with a Nazi counter-offensive.

1945: March   Benes sets up base in Slovakia

Czech leader Eduard Benes returns from exile in the United States to set up a provisional government from a base in Kosice, Slovakia. The Kosice Program outlines his plans for reunifying Czechoslovakia and punishing participants in the wartime Slovak government for collaborating with the Nazis. Most of these government officials flee westward but are extradited back to Czechoslovakia by the United States and made to stand trial for treason.

1945: April   Germans are expelled from Slovakia

Slovakia is completely freed of German troops by the Soviet army.

1945: April   New government is formed

A new Czech government is formed with Eduard Benes (see 1938: October 15) as president. The government is controlled by parties friendly to the Soviet Union and Communists hold important posts, but Czechoslovakia remains a democracy, and the Soviet Union pledges to respect its independence. The new Czech government promises autonomy to Slovakia, which is to have its own parliament and cabinet. A separate Slovak Communist Party is also organized.

1946   Communists do poorly in Slovak elections

Communist candidates have a poor electoral showing in heavily Catholic Slovakia.

1946–47   Wartime Slovak leaders tried for treason

Officials of the German-controlled wartime Slovak government are tried before a National Tribunal in Bratislava.

1947: April   Tribunal sentences Slovak leaders

After sitting for almost a year, the National Tribunal sentences Slovak leaders tried for collaborating with the Nazis during World War II (1939–45). Most are sentenced to serve prison terms. The leader of the wartime government, Monsignor Josef Tiso (see 1939: March 15), is condemned to death by hanging.

Communist Rule

1948   Slovak National Gallery is founded

The Slovak National Gallery opens in Bratislava, displaying Slovak art dating back to the twelfth century, as well as works by Dutch, Flemish, and Italian masters.

1948: February   Communists take control of government

Czechoslovakia's twelve non-Communist cabinet ministers resign to protest the actions of the ministry of the interior, which include alleged police protection of Communist terrorists. Instead of raising support for their position, the move consolidates the power of the Communists. The twelve ministers are replaced by Communists and Communist sympathizers, putting the nation's government entirely under Communist control.

1948: May   New Czech constitution restricts Slovak autonomy

Czechoslovakia adopts a new constitution setting up a people's republic based on the Soviet model. In order to maintain strict control of the country, the Communists set up a strongly centralized government, revoking the autonomy earlier promised to Slovakia. The Slovak Communist Party is merged into the Czechoslovak Communist Party, becoming a branch of the larger organization.

1949   Slovak Philharmonic orchestra is founded

The Slovak Philharmonic Society is established as a professional orchestra with its base in Bratislava. Composed of one hundred instrumentalists and an eighty-member chorus, it presents a regular concert series and also performs on tour throughout Czechoslovakia and abroad. Conductors include Vaclav Talich, Ludovit Rajter, and Ladislav Slovak.

1950: April   Slovak religious orders are dissolved

To carry out the policies of the central government, Czech police enter Slovakia to break up religious orders and expel their members from the country, an action that causes resentment within the Slovak community. When Slovak officials protest, they are charged with "bourgeois nationalism." (According to the Communist philosophy, nationalism violates the international solidarity of workers.)

1951   Slovak leaders purged

Slovak Communist leaders are purged from the Communist Party and punished. Foreign Minister Vladimir Clementis is executed in 1952, and others are sentenced to prison, including future Czech president Gustav Husak (1913–1991; see 1913: January 10), who receives a life sentence. As part of a general nationwide purge that continues through 1954, many other Slovaks are removed from government and Communist Party positions.

1954   First television station goes on the air

The first television station in Slovakia is launched in Bratislava.

1956   Hungarian revolt sparks Slovak protests

The unsuccessful Hungarian revolt against Soviet domination leads to demonstrations in Slovak cities. Troops are sent in by both the Czechoslovakian government and the Soviet Union to maintain control.

1959   Safarik University founded

The Pavel Josef Safarik University, named for a prominent nineteenth-century Slovak poet and nationalist, is established in Kosice.

1959   Bratislava Municipal Gallery is founded

A new municipal gallery in Slovakia's capital features Slovak, as well as Czech and other Central European art of the seventeenth through twentieth centuries.

1960s   Construction of nuclear power station

The Jaslovske Bohunice nuclear power plant is built, eventually supplying more than half of Slovakia's electricity.

1960   Husak is released from prison

Following an inquiry into the purges of the early 1950s, Slovak leader Gustav Husak is released from prison, where he is serving a life sentence.

1962   Party congress overturns convictions of Slovak leaders

As part of a larger program of de-Stalinization (repudiation of repressive measures inspired by the policies of Soviet leader Josef Stalin [1879–1953]), the Twelfth Congress of the Czechoslovak Communist Party vindicates hanged foreign minister (and Slovak) Vladimir Clementis (see 1951) and nullifies the convictions of other Slovak government leaders.

1962   Liberal Slovaks replace unpopular leaders

Relations between the Slovaks and the central government improve when the government replaces the unpopular leaders it had appointed to top positions in Slovakia with two who have popular support, including Alexander Dubcek (see 1921: November 27), who becomes the first secretary of the Slovak Communist Party.

1965: October   The Bratislava Festival is founded

The annual two-week Bratislava Festival makes its debut. It features orchestral and chamber music, solo recitals, operas, ballets, and a musicology convention. A series of concerts featuring young performers sponsored by the United Nations Education, Scientific, and Cultural Organization (UNESCO) later becomes part of the festival.

Prague Spring and Soviet Invasion

1968: January   Dubcek is named head of the Communist Party

Alexander Dubcek, first secretary of the Slovak Communist Party, replaces Antonin Novotny as leader of Czechoslovakia's Communist Party, becoming the first Slovak to hold that post (see 1921: November 27).

1968   "Prague Spring"

Under the leadership of Alexander Dubcek, wide-ranging liberal reforms are implemented with the goal of creating "Socialism with a human face." Media censorship is lifted, and freedom of expression is allowed to an unprecedented degree. Plans for economic liberalization are announced as well. In addition, Dubcek introduces a federalization plan that grants greater autonomy to Slovakia. Dubcek's reforms are strongly criticized by the Soviet Union and Czechoslovakia's Eastern European neighbors.

1968: August 20–21   Soviets invade Czechoslovakia

The Soviet Union, aided by troops from other Warsaw Pact nations, invades and occupies Czechoslovakia. Top government officials, including Alexander Dubcek, are seized and taken to the Soviet Union. However, there is strong opposition to the invasion among the Czech people, and the remaining Czech leaders refuse to cooperate by forming a puppet government whose policies will be dictated by the Soviets.

1968: October   Pact permits stationing of Soviet troops

Czechoslovakia is forced to sign an agreement permitting Soviet troops to be stationed in the country indefinitely. The Dubcek government is nominally allowed to remain in office but has to abandon its liberalization program. Nevertheless, much of the "Prague Spring" leadership is removed from office within a year by resignation, dismissal, or demotion.

1968: December 31   New federal republic increases Slovak autonomy

As part of its effort to weaken Czechoslovakia's central government following the "Prague Spring," the Soviet Union backs the transformation of the nation into a loosely structured federal state consisting of separate Czech and Slovak republics. The Slovaks thus gain a long-sought measure of autonomy by virtue of Soviet repression.

1969: April   Husak replaces Dubcek as party secretary

Alexander Dubcek resigns as secretary of Czechoslovakia's Communist party and is replaced by Gustav Husak, a fellow Slovak who was imprisoned in the 1950s for "bourgeois nationalism." Husak follows policies designed to maintain good relations with the Soviet Union and ends the political reforms initiated under Dubcek (see 1913: January 10).

1975: May   Husak becomes president

Husak assumes the office of president while remaining secretary of the Communist Party, holding both positions for the twelve remaining years of Communist domination. Increasingly conservative, he resists the spirit of reform that originates from the Soviet Union when Mikhail Gorbachev (b. 1931) becomes party secretary in the 1980 (see 1987).

1977: January   Dissidents publish manifesto

A group of over two hundred Czechoslovakian intellectuals calls for government compliance with the human rights provisions of the Helsinki Agreement in a manifesto (Charter 77), which is published in the West. It is not taken seriously as a threat by the Czech government. Czech police arrest many of those who signed the document. Prominent among the signers is future president Vaclav Havel (b. 1936; see 1989: December 28–29).

1980s   Widespread environmental damage

Government economic policies lead to excessive burning of lignite, which harms the environment through acid rain and serious deforestation. Documented adverse effects on the nation's health include shorter life expectancy and higher infant mortality.

1983: October   Deployment of Soviet missiles is announced

The deployment of Soviet tactical nuclear missiles in Czechoslovakia is announced and triggers antinuclear demonstrations in 1983–84.

1986: April   Fallout from Chernobyl affects Czechoslovakia

The Chernobyl nuclear accident occurs in the Soviet Union, causing radioactive nuclear fallout over many parts of Czechoslovakia.

1987   Husak belatedly endorses Soviet liberalization

After initial resistance, Gustav Husak reluctantly declares his support of Soviet leader Mikhail Gorbachev's liberal policies of *perestroika* (restructuring) and *glasnost* (openness). He resigns as head of the Communist Party but remains president of Czechoslovakia for two more years.

The "Velvet Revolution"

1989: January–October   Growing anti-government protests

Increasingly larger anti-government protests take place in Prague throughout the year. The government reacts by suppressing the demonstrations and arresting their leaders.

1989: November 17   Youths stage mass protest in Wenceslas Square

Some three thousand young people demonstrate in Wenceslas Square, calling for free elections. Violent government suppression of the protest sparks meetings, demonstrations, and a general strike—the "velvet revolution"—that lead to the resignation of the government of Prime Minister Milos Jakes and the retirement of Gustav Husak as president.

1989: November 20   The Civic Forum is formed

Led by Vaclav Havel, the Czech dissident groups join together to form the Civic Forum, and the Slovaks form the Public Against Violence, or VPN.

1989: December   Communist leaders resign

Czechoslovakia's Communist leaders apologize to the country and resign, and the constitution is amended to abolish the power of the Communist Party.

1989: December 28–29   Havel is named as president

Dissident playwright Vaclav Havel becomes president by acclamation (overwhelming approval but not an actual count). Alexander Dubcek, the Slovak who was the nation's leader during the "Prague Spring" of 1968, is named chairman of the national assembly.

1990: February   Soviets pledge to withdraw troops

The Soviet Union apologizes to the Czechs for the 1968 invasion and occupation of their country and promises to withdraw its troops from Czechoslovakia by July 1991.

1990: June 20   Free elections are held

Czechoslovakia (now renamed the Czech and Slovak Federal Republic) holds its first free elections since 1946, with a turnout of ninety-six percent of eligible voters. The Civic Forum and the Slovak VPN win a decisive victory, and the government begins a program of economic reform.

1991–92   Slovak separatist sentiment grows

Tension between Czechs and Slovaks becomes increasingly evident, and pressure for separation builds. The Czechs want a strong central government and a rapid shift to a free-market economy, while the Slovaks desire greater autonomy and a slower pace of reform. President Vaclav Havel makes extensive efforts to mediate between the two groups. The top Slovak official, Vladimir Meciar, backs the separatist cause.

1991: November 18   Death of Gustav Husak

Former Communist leader Gustav Husak dies in Bratislava (see 1913: January 10).

1992: July   Slovak constitution is adopted

The Slovak National Council adopts a constitution providing for the creation of the Slovak Republic, with a 150-seat national assembly that elects the president, who serves as head of state.

1992: August 26   Czechs and Slovaks agree to separation

Czech and Slovak leaders sign an agreement providing for separation into two republics by the beginning of 1993.

1992: November 7   Death of Alexander Dubcek

Political leader of Czechoslovakia during the "Prague Spring" liberal reforms of 1968, Alexander Dubcek dies in Prague (See 1921: November 27.).

The Slovak Republic

1993: January 1   Slovak Republic is declared

The political union of Czechs and Slovaks ends as the Czech and Slovak Federal Republic is dissolved and replaced by two separate republics. Vladimir Meciar (see 1991–92) and his Movement for a Democratic Slovakia (HZDS), in power before the split, continue to rule Slovakia. Meciar drastically slows the pace of privatization and moves to censure media criticism of his policies, losing both public support and the support of his own party, as a number of members resign.

1994: February   Ruling party splits

The ruling party of Prime Minister Vladimir Meciar (Movement for a Democratic Slovakia) splits with the withdrawal of party leaders opposed to Meciar's policies, including Foreign Minister Jozef Moravcik.

1994: March 11   Meciar ousted as prime minister

A no-confidence vote in parliament ousts Prime Minister Vladimir Meciar. He is replaced by former foreign minister Jozef Moravcik, who heads an interim government. Moravcik returns to the economic reforms abandoned by Meciar, effecting a dramatic improvement in economic growth within six months.

1994: September–October   Meciar rebounds in first national elections

The Slovak Republic holds its first national elections since the break-up of Czechoslovakia. The HZDS (Movement for a Democratic Slovakia) party of ousted Prime Minister Vladimir Meciar rebounds to score the highest percentage of votes (thirty-five percent). Meciar forms a coalition government and resumes the office of prime minister, which he had vacated six months earlier.

1995: August 31   President's son is kidnapped

The adult son of President Michal Kovac is kidnapped near Bratislava and transported to Austria, where he is taken into police custody for alleged illegal financial dealings. Kovac's supporters charge that the kidnapping is related to the ongoing feud between Kovac and Prime Minister Meciar. A government secret service agent admits to participation in the kidnapping, but Meciar denies all involvement.

1995: November 15   Controversial language law is passed

The Slovak parliament passes a new law making Slovak the country's only official language, meaning that other languages cannot be used in government communications. The law, which is passed by an overwhelming majority of legislators (108–17), has the greatest consequences for Slovakia's Hungarian minority, which accounts for roughly ten percent of the country's population. Hungarian prime minister Gyula Horn protests the adoption of the law, and Hungary recalls its ambassador to Slovakia.

1996: July 4–5   Hungarians advocate autonomy within Slovakia

An international summit of ethnic Hungarians held in Budapest issues a statement in favor of Hungarian autonomy within other countries in which they are a minority population. The communiqué strains newly mended relations between Hungary and Slovakia, which is home to approximately 600,000 ethnic Hungarians.

1997: May   Referendum on NATO membership is declared invalid

The Slovak government nullifies the results of a national referendum on joining NATO because an additional question about direct election of the Slovak president had been omitted from the vote. The action is part of a power struggle between Slovak prime minister Vladimir Meciar and President Michal Kovac.

1998: March   Preliminary talks on EU membership launched

Preliminary talks begin in Brussels on the eventual admission of the Slovak Republic to the European Union (EU), along with four other Eastern European countries. The EU has not included Slovakia among the first five Eastern and Central European nations to be admitted (a group that includes the Czech Republic) because of its objections to the suppression of democracy under Prime Minister Vladimir Meciar.

1998: September 26–27   Ruling party is defeated

The Movement for a Democratic Slovakia (HZDS) party of Slovakian prime minister Vladimir Meciar is defeated by opposition groups in national elections. The four leading opposition parties, which collectively win fifty-eight percent of the vote, announce they will form a coalition government.

1998: October 30   Dzurinda is sworn in as prime minister

Mikulas Dzurinda becomes Slovakia's second prime minister since the formation of the Slovak Republic in 1993, succeeding Vladimir Meciar. Dzurinda is selected by the four-party coalition that defeated Meciar's Movement for a Democratic Slovakia (HZDS) Party in the September elections. He is the head of the dominant party, the center-right Slovak Democratic Coalition (SDK). The other three parties are the Party of Civic Understanding (SOP), the Party of the Democratic Left (SDL), and the Hungarian Coalition Party (SMK).

Bibliography

Brock, Peter. *The Slovak National Awakening: An Essay in the Intellectual History of East Central Europe.* Toronto: University of Toronto Press, 1976.

Goldman, Minton F. *Slovakia Since Independence: A Struggle for Democracy.* Westport, Conn.: Praeger, 1999.

Jelinek, Yeshayahu A. *The Parish Republic: Hlinka's Slovak People's Party: 1939–1945.* New York: Columbia University Press, 1976.

Johnson, Owen V. *Slovakia, 1918–1938: Education and the Making of a Nation.* New York: Columbia University Press, 1985.

Kirshbaum, Stanislav J. *A History of Slovakia: The Struggle for Survival.* New York: St. Martin's Press, 1995.

Leff, Carol Skalnik. *The Czech and Slovak Republics: Nation Versus State.* Boulder, Colo.:Westview Press, 1997.

Oddo, Gilbert Lawrence. *Slovakia and its People.* New York: R. Speller, 1960.

Palickar, Stephen Joseph. *Slovakian Culture in the Light of History, Ancient, Medieval and Modern.* Cambridge, Mass.: Hampshire Press, 1954.

Steiner, Eugen. *The Slovak Dilemma.* Cambridge: Cambridge University Press, 1973.

Slovenia

Introduction

Slovenia is a small country, 7,836 square miles (20,296 square kilometers), located on the northern coast of the Adriatic Sea. It borders Italy to the west, Austria to the north, and Croatia to the east. The coastal areas are relatively flat, while the interior of the country is Alpine in character. As of 1991, Slovenia's inhabitants numbered 1,965,986. Of these, eighty-eight percent were Slovene, three percent were Croats, two percent were Serbs, and one percent were Muslim. Other minorities included Italians and Germans. Important cities include Ljubljana, the capital (with 278,000 people), Maribor, (population 105,000), and Koper, Slovenia's port.

Until 1991, Slovenia had never existed as an independent state. Stronger neighbors have ruled over Slovenia for most of its history. Only in 1918 did Slovenes become part of a country in which they were theoretically on an equal footing with other peoples.

Slavic Invasions and Germanic Influence

The Slovenes trace their history to the Slavic invasions of the sixth century A.D. Specifically, the Slovenes were part of the South Slav branch that settled throughout the Balkan Peninsula from the Adriatic coastline to Greece. The modern Slovenian language is further evidence of this legacy, a language of the South Slavic branch from the Slavic group of Indo-European languages.

A major feature of their history that separates the Slovenes from their South Slav neighbors is the heavy Germanic influence on the region. This began as early as 748, when the area fell to the Franks. By the tenth century, the territory of modern-day Slovenia had become an integral part of the German-dominated Holy Roman Empire. By the fifteenth century, the rule of the German Habsburg dynasty extended over nearly all of Slovenia and resulted in the establishment of a feudal system of control over the Slovenes. German overlords ruled their Slovenian peasants. Moreover, the dominance of the Habsburgs prevented the emergence of an independent Slovenian state during this period.

Slovene national consciousness emerged slowly over the course of centuries. Indeed, a distinctly Slovene nationalism did not appear until the 1800s. As with most instances of national awakening, the emergence of a Slovenian sense of identity paralleled a linguistic revival. The growth of printing in the wake of the Protestant Reformation (1517–c. 1600) set the stage for a rise in the number of Slovenian-language books. In addition to the printing of religious works such as the catechism and Holy Bible, sixteenth century publishers provided readers with grammar books as well. However, it was the educational reforms of Habsburg empress Maria Theresa (1717–80) and Emperor Joseph II (1741–90) in the eighteenth century that finally provided a solid foundation for a sizable literate Slovene population. The late eighteenth century saw the publication of the first Slovenian-language grammar, the first Slovenian-German-Latin dictionary, and the first Slovenian newspaper.

Slovenia Part of Illyrian Provinces

The Napoleonic Wars had a profound impact upon the development of Slovenian nationalism. For four years (1809–1813) Slovenian territories were part of the French-administered Illyrian Provinces, which included fellow South Slav Croats and Serbs. Some Slovene intellectuals were drawn into the Illyrian Movement, a program that advocated cultural unity of all South Slavs in the Habsburg Empire. Although these territories reverted to Habsburg control in 1813, the ideas of the Illyrian Movement maintained a following among many South Slav intellectuals.

Moreover, the liberal ideas of the French Revolution (1789–99) manifested themselves in the abortive United Slovenia manifesto issued during the Revolutions of 1848. This document, which called for the creation of a kingdom of Slovenia to include Carniola, Carinthia, Styria, and the Littoral (coastal area), represented the most radical of Slovene nationalist ideas. Most Slovene intellectuals did not entertain notions of an independent Slovenia. Believing their nation to be too small to survive on its own, they sought autonomy within Austria. Yet once revolution had exhausted itself-throughout the continent, the prospect of autonomy had died too. Not until after 1867 would Slovene nationalism resurface

SLOVENIA

as a major force—and then as part of a larger (though not always cooperative) South Slav movement.

Writers (est. 1872) also advocated increased rights for Slovenes.

Dual Monarchy

The Austrian defeat at the hands of Prussia in 1866 destroyed the German Confederation and threatened the very existence of Austria. The empire was reorganized as a Dual Monarchy with Hungarians placed on an equal footing with the Austro-Germans. The South Slavs emerged as the biggest losers in this scheme, as they were now subject to systematic discrimination by both the Austrian and Hungarian halves of the empire. In the face of renewed oppression, many Slovene nationalists returned to the policies of the United Slovenia manifesto. Others joined renewed calls for a South Slavic state within the Habsburg Empire. With education widespread, Slovenia had the base for the establishment of mass political parties. The creation of the Catholic People's Party (1892) and the National Progressive Party (1894) provided the constitutional means for airing Slovene grievances. Significantly, both sought a reorganization of the Dual Monarchy. Cultural organizations, such as the Society of Slovene

World War I

At the outset of the First World War (1914–18), most Slovenes still supported the Habsburg Empire. As the war dragged on, this feeling changed, and many Slovenes opposed fighting Serbs and Russians, both of whom were fellow Slavs. By 1918 many Slovene political leaders openly advocated secession from Austria-Hungary and union with an independent South Slav state. This aim became reality on October 29, 1918 when the National Council of all Slavs of the former Habsburg lands proclaimed separation from the Dual Monarchy. On December 1, these lands merged with the Kingdom of Serbia (and ultimately, the Kingdom of Montenegro) to form the Kingdom of Serbs, Croats, and Slovenes under the rule of King Alexander Karadjorjevic (1806–85) of Serbia.

Problems plagued this new South Slav state from the start. It had territorial disputes with every one of its neighbors (except Greece). Furthermore, rather than being a kingdom in which each of the nationalities enjoyed equal rights, the kingdom was, in reality, a "Greater Serbia." Political life became so acrimonious (bitter) that in 1929 King Alexander suspended the constitution and proclaimed martial law. That same year, he also changed the name of the state to Yugoslavia (Land of the South Slavs). These actions did little to stabilize the country. Indeed, they only contributed to further instability. Those opposed to the king's rule resorted to terrorist actions, and Alexander himself became a victim when he was assassinated in 1934 while on a state visit in Marseilles, France.

World War II and Emergence of Tito

The perpetual internal conflict within the country could not continue much longer without the kingdom's implosion (bursting inward). Therefore, the regent (Prince Peter, the heir to the throne, was a minor) who succeeded Alexander sought to strengthen the kingdom by belatedly offering autonomy to non-Serbs. By the time World War II broke out across the continent (1939–45), the government also sought security from external threats which culminated in the signing of the Tripartite Pact in March 1941, a treaty that aligned Yugoslavia with the Axis powers. This alliance proved only temporary when, on March 27, a popular coup led by air force officers ousted the prince regent and replaced him with Peter. The new government denounced the Tripartite Pact and signed an agreement with the Soviet Union instead. Infuriated at the dramatic turn of events in Belgrade, Adolf Hitler (1889–1945) hastily ordered an invasion of Yugoslavia.

"Operation Retribution" began on April 6, and lasted less than one month. By that time, Yugoslavia ceased to exist, and the country was partitioned. Slovenia disappeared entirely and was divided between Italy and the Greater German Reich. Opposition to the new regime began almost immediately. Many Slovenes flocked to resistance organizations. Others, however, served in collaborationist Home Guard units and fought the Communist-dominated Partisans.

The liberation of Slovenia in 1945 brought a radically changed political situation. The victorious Partisans under Josip Broz Tito (1892–1980), himself half-Slovene, abolished the monarchy and established a Communist state. In addition, the country was reorganized politically into six autonomous republics. For the Slovenes, this meant the creation of the Socialist Republic of Slovenia. For its role in the struggle on the side of the Allies, Yugoslavia received new territory. Slovenia added most of the Istrian Peninsula, including Rijeka (but excluding Trieste, which became a point of contention plaguing Yugoslav-Italian relations, and the Klagenfurt region of Austria).

Tito Breaks Away From U.S.S.R.

The start of the Cold War between the United States and the Soviet Union placed Yugoslavia in a unique situation. Tito broke with the Union of Soviet Socialist Republics (U.S.S.R.)

in 1948 and pursued a path of nonalignment. He became a major world figure as many new nations adopted non-aligned policies. Moreover, Tito's split with the Soviet Union also opened Yugoslavia to foreign trade and investment. Of all republics, Slovenia profited most from this arrangement. In fact, by the 1980s, the republic's economic success led many leading Slovenes to question their continued association with Yugoslavia.

Slovenian Independence

Upon Tito's death in 1980, many foreign observers questioned the continued survival of Yugoslavia. The lack of a unifying national figure exacerbated already strong ethnic divisions in the country. Economic grievances aggravated these historical splits. Many Slovenes felt that their continued economic association with the poor regions of Yugoslavia hindered their advancement. The collapse of Communism in Eastern Europe in the fall of 1989 encouraged those who sought either a reorganization of Yugoslavia or secession. On June 25, 1991, after repeated attempts at salvaging Yugoslav unity had failed, Slovenia became the first republic to secede. European mediation halted the war against the Yugoslav Federal Army (JNA) after only ten days and ensured for a JNA withdrawal by October. International recognition followed shortly thereafter. The country drafted a new constitution in December 1991.

Since independence, Slovenia has become a full-fledged member of the international community. Its economy remains strong and growing. It is also a potential candidate for future membership in the European Union (EU) and the North Atlantic Treaty Organization (NATO).

Timeline

c. 80,000-40,000 B.C.   Neanderthal settlement in Slovenia

Evidence unearthed beginning in A.D. 1980 places Neanderthal man in the region of modern-day Slovenia. The evidence discovered includes hollowed-out bones that may have been used as musical instruments.

300–200 B.C .   Roman conquest

Roman forces occupy the region of modern-day Slovenia and incorporate the area into the Roman Republic. Roman rule continues into the fourth century A.D.

c. A.D. 550   Early Slavic settlement

The first settlement of Slavs in the plains near the Adriatic Coast begins.

745 Frankish conquest

The region of modern Slovenia falls to the Franks. It ushers in a period of external domination that lasts until the twentieth century. Unlike their Croat neighbors to the south, the Slovenes never develop an independent kingdom.

863 Cyril and Methodius

The Greek missionary monks, Cyril and Methodius, begin translating the Scriptures into Slavic. To this end, they create a Slavic alphabet, known as Cyrillic. Although the Cyrillic script does not take hold in Slovenia, the missionary efforts of Cyril and Methodius are significant because they result in the large-scale conversion of Slavs to Christianity.

900s Holy Roman Empire

Slovenia is incorporated into the Holy Roman Empire. Along with German rule comes heavy Germanic influence which is felt into the 1990s.

1054 Catholic-Orthodox split

Centuries of increasing tension over political and theological disputes result in the break between the eastern and western branches of Christianity centered in Constantinople and Rome, respectively. With Slovenia under German control and away from Byzantine influence, the region remains loyal to the pope in Rome. From this point on, Roman Catholicism in Slovenia is seriously challenged only by the Protestant Reformation.

1400s Habsburg control begins

The Habsburg dynasty gains control over the Slovene lands which become part of the crown lands of Carinthia and Carniola. Germanic influence increases. Feudalism under the Habsburgs results in German overlordship of the native Slovene population.

From the Reformation to the End of World War I

The effects of the Protestant Reformation on the Slovene regions are primarily cultural rather than political. Like most of the Habsburg regions, Protestantism never gains a large following in Slovenia. However, the growth of printing in the Reformation's wake results in first steps toward the creation of a Slovenian literature four centuries later.

1551 Translation of catechism

Privoz Trubar translates the catechism into Slovenian.

1552: Grammar book

Trubar publishes *Abecedarium*, a basic grammar of Slovenian, though it was not published in the Slovene language.

1575 Publishing house is formed

The first Slovenian publishing house is established in Ljubljana.

1578 Bible translated

Jurij Dalmatin is the first to translate the Bible into Slovenian. He makes his work readable to any Slovene or Croat by using an inclusive vocabulary.

1595 College founded

As part of the Catholic Counter-Reformation, a Jesuit college is founded in Ljubljana. The Counter-Reformation is a vigilant Catholic attempt to fight the Protestant Reformation.

1673 Academy established

An academy of arts and sciences is established. This is the first large institution of non-religious learning in Slovenia.

1740–90 Reforms of Maria Theresa and Joseph II

Habsburg monarchs Maria Theresa (1717–80) and her son Joseph II (1741–90) embark upon a period of significant internal reform intended to strengthen the empire. Their reforms are liberal in nature and stress education (in German) and religious toleration. For the non-Germanic peoples of the empire, however, the reforms have a limited effect since they do not achieve an increase in political rights. Nonetheless, a flowering of Slovenian-language works occurs. By the late nineteenth century, the effects of the birth of this writing, together with the influence of the French Revolution, give rise to Slovenian nationalism.

1765 *Abecedika*

Marko Poblin publishes *Abecedika*, the first grammar of the Slovenian language published in Slovenian.

1781 Slovenian-German-Latin dictionary

Poblin publishes the first Slovenian-German-Latin dictionary.

1797 First Slovenian newspaper

Valentin Vodnik (1758–1819) publishes the first Slovenian newspaper.

1809 Napoleonic conquest

Slovenia is conquered by Napoleon Bonaparte (1769–1821) and administered as part of the Illyrian Provinces of his empire. Although Napoleon's control over the region lasts only four years, the period of French rule is significant. Napoleon's soldiers bring with them the nationalist ideas of the French Revolution (1789–99) which influence generations of European nationalists.

1813   Return of Habsburg authority

After Napoleon's forces are driven out of the Illyrian Provinces, the region reverts to Habsburg control. Nonetheless, the ideas of the French Revolution remain.

1848: May   The Revolution of 1848 in Slovenia

Slovene nationalists issue the United Slovenia manifesto. They demand a Kingdom of Slovenia extending over Carniola, Carinthia, Styria, and the Littoral (coastal area). The government is to be a parliamentary monarchy. However, the demands for nationalist programs are rejected.

1867   Creation of Austria-Hungary

With the proclamation of the *Ausgleich* (division of the empire into two equal halves), the Habsburg Empire becomes a Dual Monarchy known as Austria-Hungary. Under this arrangement, the Slovene lands fall into the Austrian half of the empire, while their fellow South Slavs, the Croats and Serbs, are under Hungarian rule.

Late 1800s   Realism

Slovene art and literature embraces the concept of Realism. While the artists are best known for the realistic style in their works, the Slovene literary Realists are best known for their politics. The literary movement splits into two mutually opposed factions. The Old Slovenes fight for Slovene rights within the Habsburg monarchy, while the Young Slovenes seek to apply the principles of nationalism within the context of the Illyrian Movement, a balkan peninsular movement that stresses self-rule of the peoples on it.

1872   Society of Slovene Writers formed

This group is the first organization of Slovenia's writers. It comes to play a leading role in Slovenian cultural development until the First World War (1914–18). In the latter twentieth century it serves as a forum for those opposed to communist rule.

1892   Creation of Catholic People's Party

The Catholic People's Party is created under the leadership of Janez Evangelist Krek (1865–1917). Krek, a Christian Socialist, advocates closer ties with Croats and Serbs within the Dual Monarchy. He favors a "Trialist" (the creation of a South Slav state in the Habsburg monarchy) solution to the Habsburg nationalities problem. After the outbreak of World War I, however, Krek advocates an independent South Slav state.

1892: May 25   Tito born

Josip Broz is born to Slovene and Croat parents. He serves in the Austro-Hungarian army in World War I, is taken prisoner by the Russians, and becomes a Communist. Better known by his adopted name Tito, he goes on to lead the Partisan resistance forces, who fight the Axis within Yugoslavia, during World War II (1939–45) and becomes the postwar leader of Yugoslavia until his death in 1980.

1894   Slovene liberal party founded

As a counter to the clerical Slovene People's Party, anticlerical middle-class elements rally around the program of the National Progressive Party. This party is a liberal formation characterized by a Pan-Slavic orientation. They wish to secure Slovene rights through the creation of a federal Austria, which would give more Slovene representation in the Dual Monarchy. Among their members are two mayors of Ljubljana, Ivan Hribar (r. 1896–1910) and Ivan Tavcar (1911–21).

1908   Annexation of Bosnia-Herzegovina

Austria-Hungary annexes Bosnia-Herzegovina, which it has occupied since 1878. In addition to causing a major diplomatic crisis in Europe, this annexation increases the numbers of South Slavs in the empire and brings the issue of the future of the South Slavs in the Dual Monarchy to the fore.

1908–14   Neo-Illyrism

The annexation of Bosnia-Herzegovina rejuvenates the Illyrian Movement. The new program (now referred to as Neo-Illyrism) calls for South Slav unity in culture and language, as well as politics. Thus, some Slovene intellectuals who support the movement argue that Slovenes should adopt the Serbo-Croatian language.

1912–13   Balkan Wars

Balkan states fight two wars over the delineation of Ottoman (Turkish) territory in the southern part of the peninsula. Serbia, which gains most of the territory it desires to the south, now casts its eyes north to the Austro-Hungarian Empire.

1914: June 28   Outbreak of World War I

The assassination of the heir to the Austro-Hungarian throne, Archduke Francis Ferdinand (1863–1914), and his wife, Sophie, while on a visit to Sarajevo, by a nationalist Serb, leads to war in a little over a month. The First World War strains the relationship of Slovenes toward the Dual Monarchy. While many Slovenes serve in the Austro-Hungarian army, many others, including intellectuals, oppose fighting their Serbian brethren.

1914–1918   World War I

Many Slovenes fight in the First World War, the vast majority in the Austro-Hungarian army, and battle in Serbia, Russia, and (after 1915) Italy. Their attitude toward fighting the Serbs is mixed. Many genuinely believe in maintaining the Habsburg Empire, while others show little desire to fight fellow

Slavs (the same is true of those on the Eastern Front). However, most Slovenes do view the Italians as a threat and, as a result, fight well on the Italian Front.

1918: August 12   National Council for Slovene Lands formed

In Ljubljana, Slovene nationalists form the National Council for Slovene Lands.

1918: October 12   National Council of all Slavs founded

In Zagreb, leaders of the South Slavs of the Habsburg Empire form the National Council of all Slavs. This organization is led by Monsignor Anto Koresec, head of the Slovene People's Party, and aims to provide united leadership for the South Slavs now that Austria-Hungary is disintegrating.

1918: October 29   Separation of South Slavs from Austria-Hungary

The National Council of all Slavs declare the separation of South Slav areas from Austria-Hungary and the creation of a Kingdom of Serbs, Croats, and Slovenes.

From the First Yugoslavia to Independence

The euphoria that surrounds the incorporation of the Slovene lands into the newly formed Kingdom of Serbs, Croats, and Slovenes quickly fade as the new state turns out to be a "Greater Serbia." The 1920s witnesses political repression that culminates in the establishment of a royal dictatorship by King Alexander (1888–1934). The 1930s feature continued turmoil which foretells the collapse of Yugoslavia during the German invasion of April 1941. The successful wartime resistance of the Communist Partisans under Tito forms the basis for the second Yugoslavia in 1945. However, Tito keeps the peace through suppression of all nationalities. After his death in 1980, nationalist tensions slowly resurface. The end of the Cold War provides the impetus for the final collapse of the second Yugoslavia, as wars break out in Slovenia, Croatia, and Bosnia and Herzegovina.

1918: December 1   Kingdom of Serbs, Croats, and Slovenes declared

The South Slav lands of the former Habsburg monarchy join with Serbia to form the Kingdom of Serbs, Croats, and Slovenes under King Alexander Karadjordjevic. Although all three nationalities ostensibly have equal rights, the kingdom is, from the start, essentially a "Greater Serbia" in which Serbs dominate all levels of administration. This leads to disillusionment and antagonism from Croats and Slovenes and eventually to the fall of the first Yugoslavia (state of the South

Slavs), in the aftermath of the Axis invasion of 1941 (see 1941: April 6).

1919–20   Border questions

The extent of Slovenian borders remains an open question as the Italian nationalist poet, Gabriele D'Annunzio (1863–1938) marches into Rijeka (Fiume) at the head of a motley collection of Italian nationalist thugs and claims the city for Italy. Further Slovene national aims are stifled when the Klagenfurt region of Austria votes to remain in Austria rather than join the Kingdom of Serbs, Croats, and Slovenes.

1920s   Modernism

Already the leading art form in the era before the First World War, Modernism becomes the chief form of art in interwar Slovenia and continues to play a role half a century later. Modernism includes artistic styles such as postimpressionism, expressionism, and cubism—styles made popular by Vincent Van Gogh, Paul Gauguin, and Pablo Picasso. Leading Slovene modernists include Anton Gojmir Kos, Stane Kregar, Marij Pregelj, Lojze Spacal, and Gabrijel Stupica.

1920   Treaty of Rapallo

The Treaty of Rapallo delineates Yugoslavia's borders with Italy. The treaty assigns Italy all of the Istrian peninsula, including Rijeka. It is a difficult loss for many Slovenes to accept since they constitute a majority in the region.

1929   Royal dictatorship

Following a decade's worth of tense political bickering, King Alexander declares the constitution suspended and invokes martial law. He also changes the name of the country to Yugoslavia (Land of the South Slavs).

1934   Alexander assassinated

King Alexander is assassinated in Paris along with French foreign minister Louis Barthou. The assassination is done by the *Ustacha* (a Croatian ultranationalist terrorist group that favors Croatian independence) together with the Internal Macedonian Revolutionary Organization (IMRO), a pro-Bulgarian terrorist group opposed to Serbian control of part of Macedonia. Since the heir to the throne is a minor, a regency takes over the monarch's duties.

1936   Academy founded

The Slovenian Academy of Arts and Sciences is founded in Ljubljana.

1941: March   Yugoslavia signs Tripartite Pact

Yugoslavia aligns with the Axis powers (Germany, Italy, and Japan). Prince Regent Paul (b. 1893) takes this step not only as a result of his own pro-German leanings. but also as a result of Yugoslavia's diplomatic isolation. This alliance, however, is short-lived.

1941: March 27   Coup ousts Prince Paul

In a bloodless coup led by Yugoslav Air Force officers, Prince Regent Paul is ousted and replaced by the minor, Peter (1923–70), who is declared king. The action is popular, particularly with the Serbs. Peter quickly abrogates (abolishes) the Tripartite treaty.

1941: April 6   Axis forces invade Yugoslavia

German, Italian, and Hungarian forces invade Yugoslavia in an operation code-named "Retribution." Yugoslav defenses collapse. Indeed, non-Serbs (especially Croats) surrender without a fight to the invader. Before the end of the month, the Yugoslav military surrenders to the Axis. The king and his government flee into exile, first to Crete, then to Cairo and London.

German forces occupy parts of the country until the end of the war (although they are driven out of most of Yugoslavia by late 1944). Yugoslavia ceases to exist. The Italians set up a puppet kingdom of Croatia (which also includes Bosnia-Herzegovina) with *Ustacha* support, while the Germans create a puppet Serbian protectorate. Kosovo and parts of Yugoslav Macedonia are ceded to Italian-controlled Albania, Bulgaria occupies most of Macedonia, Hungary takes the Vojvodina, while Slovenia is partitioned between the Greater German Reich and Mussolini's Italy.

The period of occupation is the setting for a Yugoslav civil war, partly to settle scores from the interwar period. Heavy fighting takes place between the forces of Josip Broz (Tito) (1892–1980) and Draza Mihailovic (d. 1946). Tito's multiethnic Communist Partisans battle Mihailovc's Serb-nationalist Chetniks, as well as German and Italian occupation forces. Ultimately, the forces of Tito (who is half-Croat, half-Slovene; see 1892: May 25) triumph and remake Yugoslavia.

1941: April 27   Slovene Anti-Imperialist Front established

Edvard Kardelj and Franc Leskosek form the Anti-Imperialist Front. This is the first anti-Axis organization formed in occupied and dismembered Slovenia. The group later changes its name to the Liberation Front (*Osvobodilna fronta*) (OF) and cooperates with Tito's forces.

1944   Tito victorious

By 1944 the Partisans (with Soviet help) have driven Axis forces out and begin to reorganize the country. In an effort to avoid the Serbian-dominated interwar state, Tito creates autonomous republics for each of the country's ethnic groups. Multiethnic Bosnia-Herzegovina also receives its own state.

1945: November 11   General election

The Communists win overwhelming support in the general election. Tito uses this mandate to proceed with the creation of a People's Republic of Yugoslavia consisting of six autonomous republics. Domestically, agricultural reform begins with the breakup of large land holdings. Private enterprise is also taken over by the state. The new regime imprisons political opponents and many disappear. Later revelations indicate that nearly ten thousand are executed.

1945: November 29   Monarchy abolished

Tito abolishes the monarchy and declares Yugoslavia to be a republic.

1945–54   Trieste dispute

Yugoslavia and Italy dispute control over the Istrian city of Trieste. Although the city's population is predominately Italian, Tito claims the city for Yugoslavia on the basis of the Slovenian majority in the hinterland (an inland region claimed by the state that controls the coast). The Western allies, led by the United States, disagree and force Yugoslav forces out of the city in 1945, only weeks after they liberate it. As a compromise, the powers create the Free Territory of Trieste. Italy and Yugoslavia divide control of the surrounding region.

1946   More schools founded

Both the Ljubljana Geological Institute and the Institute for Karst Research are founded this year.

1948   Cominform expulsion

Yugoslavia is expelled from the Communist Information Bureau (Cominform), the leading international body of Communist states and parties. This is a significant step, for now Yugoslavia is on its own in foreign affairs. As a result of its expulsion, Yugoslavia becomes the first Communist state to break with the Soviet Union.

1954   London Agreement

The London Agreement abolishes the Free Territory of Trieste. Under this arrangement, Trieste remains Italian, while Italy and Yugoslavia incorporate those areas under their control from the outlying regions. The Italian Parliament, however, fails to adopt the measure.

1961   Non-Aligned conference

Tito hosts the first summit of the Non-Aligned Movement in Belgrade. This meeting of the group of states not allied to either superpower (the United States and the USSR) reinforces Tito's position as a major world leader.

1965 Market Socialism

In an effort to increase production and living standards, Tito allows for some economic competition and private enterprise through his program of market socialism. Slovenia, in particular, benefits tremendously from this effort to create a small mixed economy.

1970s Postmodernism

Postmodernism starts to influence Slovene literature. Leading Slovene postmodernists inlcude Lojze Kovacic, Gregor Strnisa, Drago Jancar, Milan Jesih, and Ivo Svetina. Often described as a cynical intellectual movement, postmodernist works are characterized by a questioning of reality and efforts to probe behind the "dominant discourse"—that is, behind the commonly perceived way of looking at the world that has been imposed upon a society.

1971 Slovenian Spring

Both the Slovenian and Croatian Communist parties pursue liberalizing reforms. Press and speech restrictions are relaxed. This opening reveals pent up frustration on the part of many Yugoslavs. Slovene and Croat nationalism is expressed by many, inluding party leaders. In response, Tito mounts a massive party purge which ousts and jails liberal reformers.

1974 New constitution

Yugoslavia ratifies a new constitution. This is largely in response to the nationalist grievances aired three years previously (see 1971).

1975 Osimo Agreement

This agreement provides for a final settlement of the Trieste question based upon the London Agreement of 1954, which the Italian Parliament fails to ratify.

1980: May 4   Tito dies

Josip Broz Tito, Yugoslavia's only postwar leader, dies in Ljubljana at the age of eighty-seven. He is mourned as a great international leader, and his death poses great concern. He has no successor; the constitution of 1974 provides for a presidency that rotates among the six republics. Most importantly, observers wonder whether Yugoslavia can continue to exist as a functioning multinational state.

1987: February   The journal *Nova Revija* expresses Slovene nationalism

Nova Revija publishes a series of articles dealing with the many problems confronting Slovenia. Such topics include the current role of the Communist Party, the status of the Slovenian language, and most importantly, Slovenian independance. In actuality, the *Nova Revija* becomes a declaration for Slovenian secession from Yugoslavia and finds much acclaim among a growing number of Slovenes.

1988: March   The Yugoslav government reports secessionist activity

A secret report is submitted by the Yugoslav army's Military Council to the Federal Presidency alledging Slovenian secessionist planning. The army calls for abrupt measures against the liberal secessionists. However, quick intervention by Slovenian leaders succeeds in halting army action.

1990: January 23   Slovenian withdrawal from Yugoslav Communist meeting

Slovenia withdraws its delegation from the meeting after its calls for reform go unheeded. Soon afterward, the Slovenian Communist Party changes its name to the Party for Democratic Renewal. Other parties soon form.

1990: April 10   Free elections in Slovenia

Free elections in Slovenia result in a victory for a six-party coalition known as *Demos*. While the victors gain a total of fifty-five percent of the vote, the Party for Democratic Renewal garners only seventeen percent. The Socialist Party wins five percent, while the Liberal Democrats gain fifteen percent. Dr. Lojze Peterle is named prime minister, while Milan Kusan is elected president with fifty-four percent of the vote. He is a former head of the League of Communists of Slovenia.

1990: October   Proposal to save Yugoslavia

Croatia and Slovenia draft a proposal to preserve Yugoslavia as a federation, but it fails.

1990: December 23   Plebiscite for independence

A plebiscite (a direct vote by the people on an issue) calling for independence wins eighty-three percent approval from Slovenes. It calls for secession from Yugoslavia if there is no solution to the federal problem.

1990: December 26   Declaration of Sovereignty

As a preliminary step toward the possibility of independence, Slovenia adopts a Declaration of Sovereignty.

1991: February 20   Federal laws annulled

In a further step toward dissociation from Belgrade, Slovenia declares all federal laws null and void.

1991: June 25   Declaration of independence

Slovenia becomes the first Yugoslav republic to declare its independence from Belgrade. This sets the stage not only for armed opposition from the Federal Yugoslav Army (JNA), but also for increased demands for independence from neighboring Croatia.

1991: June 27   War in Slovenia

JNA forces try to seize Slovenia by force. Some fighting also occurs in cities and border crossings, but casualties are limited. Surprisingly, the Slovenes put up a strong resistance and over 3,200 JNA soldiers surrender. The war ends in ten days when the European Community (EC) brokers a cease-fire. According to the provisions of this arrangement, the JNA is to withdraw by October 1991. Thus, unlike Croatia and Bosnia, Slovenia avoids prolonged conflict on its territory. The most probable reason for this, however, is the relatively small number of Serbs in Slovenia.

1991: December 18   First recognition of Slovenian independence

Germany becomes the first country in the world to recognize the independence of Slovenia (the Germans also recognize Croatia). Austria soon follows.

1991: December 23   New constitution

Slovenia adopts a new constitution.

1992: January 15   EC recognition

Following the lead of Germany, the entire EC recognizes the independence of Slovenia and Croatia.

1992: April 7   U.S. recognition

After months of hesitation, the United States recognizes Slovenian and Croatian independence. Believing that its failure to give prompt recognition resulted in war between these two republics and JNA forces, Washington also extends diplomatic recognition to Bosnia and Herzegovina. Unfortunately, this action does not prevent the outbreak of armed conflict there.

1992: April 23   United Nations membership

Slovenia, Croatia, and Bosnia and Herzegovina, join the United Nations (UN).

1992: December 6   Elections

A coalition government sweeps to power in these first elections held under the new constitution. The winning grouping consists of the Liberal Democrats, the Christian Democrats, and the United List Group of Leftist Parties. Milan Kusan is again elected president, while Dr. Janez Drnovsek becomes prime minister.

1993   Council of Europe membership

Slovenia joins the Council of Europe.

1994   Association with the North Atlantic Treaty Organization (NATO)

Slovenia, along with most former Communist countries in Eastern Europe, joins NATO's Partnership for Peace Program.

1995: Spring   Major art show

A major Slovenian art show, "House in Time," takes place under the direction of Zdenka Badovinac. This presentation features the works of a variety of Slovenian artists, and its theme reflects the country's historical experience. *Art in America* describes the exhibition subject: the conditions under which one inhabits a site determine one's sense of belonging or displacement. Such issues are obviously of daily, often mortal consequence in the new nations formerly subsumed by Yugoslavia."

1997   Continued progress brings Slovenia closer to Western Europe

Slovenia continues its remarkable progress in trying to become a full-fledged member of Western Europe. NATO cites Slovenia, along with Romania, as a potential candidate for the next round of the alliance's expansion. In addition, Slovenia's name is proposed as a candidate for the next round of European Union expansion.

Privatization efforts, in which Slovenia had lagged behind other former communist states, pick up momentum. In Ljubljana, plans are drawn for several multiplex theaters to be located in shopping malls.

1998   High credit ratings

Slovenia continues to earn high marks for its post-Communist economic recovery. Leading investment agencies give the republic high marks on its borrowing risk.

Bibliography

Cohen, Lenard. *Broken Bonds: The Disintegration of Yugoslavia*. Boulder, Colo.: Westview Press, 1993.

Fine, John V. A., Jr. *The Early Medieval Balkans: A Critical Survey from the Sixth to the Late Twelfth Century*. Ann Arbor, Mich.: University of Michigan Press, 1983.

Fink-Hafner, Danica, and John R. Robbins, eds. *Making a New Nation: The Formation of Slovenia*. Brookfield, Vt. Aldershot: Dartmouth Publishing, 1997.

Glenny, Misha. *The Fall of Yugoslavia: The Third Balkan War*. New York: Penguin, 1993.

Jelavich, Barbara. *History of the Balkans (vol. 1). Eighteenth and Nineteenth Centuries*. Cambridge: Cambridge University Press, 1983.

_____, *History of the Balkans (vol. 2). Twentieth Century*. Cambridge: Cambridge University Press, 1983.

Magocsi, Paul Robert. *Historical Atlas of East Central Europe*. Seattle: University of Washington Press, 1993

O'Brien, John, "Ljubjana Goes International," *Art in America* 84 (June 1996) 6: 41-45.

Owen, David. *Balkan Odyssey*. New York: Harcourt Brace, 1995.

Rogel, Carole. *The Breakup of Yugoslavia and the War in Bosnia*. Westport, Conn.: Greenwood Press, 1998.

Rothschild, Joseph. *East Central Europe between the Two World Wars*. Seattle: University of Washington Press, 1974.

Silber, Laura, and Allan Little. *Yugoslavia: Death of a Nation*. New York: TV Books, 1996.

Zimmerman, Warren. *Origins of a Catastrophe: Yugoslavia and its destroyers—America's last ambassador tells what happened and why*. New York: Times Books, 1996.

Spain

Introduction

The region of southern Europe known as Iberia or the Iberian peninsula is comprised of Spain and Portugal. The southernmost end of the Iberian peninsula is only about ten miles (sixteen kilometers) away from the African coast. Linked to Europe by the Pyrenees mountains, Spain borders Portugal to the west, the Mediterranean Sea to the east, and the Atlantic Ocean to the west and north. This geographic location explains the Iberian Peninsula's function as a bridge between the east and the west for centuries.

Spain comprises approximately four-fifths of the Iberian peninsula, and covers an area of 194,880 square miles (504,750 square kilometers). The Mediterranean coastal area as well as the Tajo, Ebro, and Guadalquivir river valleys enjoy a warm climate and fertile lands. The country has many rivers. The largest—the Miño, the Douro, the Tajo, the Guadiana, and the Guadalquivir—flow into the Atlantic Ocean, while only the Ebro flows into the Mediterranean.

Ancient Spain

The earliest signs of human life on the Iberian peninsula are at least 28,000 years old. Archeologists believe Neanderthal people lived in the region during the Paleolithic period. After 10,000 B.C., hunters and gatherers left drawings in the eastern and southern parts of the peninsula. Celtic peoples migrated across the Pyrenees around 800 B.C., bringing with them knowledge of the use of metals.

The Tartesians were the first inhabitants of the Iberian peninsula with an advanced political organization and a rich and sophisticated economy, probably the result of their contact with the Phoenician and Greek colonizers. The Tartesians were absorbed by the Carthaginians (inhabitants of Carthage on the northern coast of Africa near modern-day Tunis) by the end of the sixth century B.C.

Iberian art reveals Greek and Eastern influences, especially in sculpture (Dama de Elche, Dama de Baza), pottery, and metal work (votive bronze offerings from Cerro de los Santos, Albacete). The Iberians occupied present-day Andalucia and the Mediterranean, although they extended as far as the Rhone. The Celtiberians were the ethnic result of alliances and intermarriage between the Iberians and the Celts. They settled in *Celtiberia,* a region in the northern area of the central plateau. They lived in urban settlements like Numancia (Soria) and mined iron from the region of the Moncayo mountain.

The development of sailing brought people—Phoenicians, Carthaginians, and Greeks—to Iberia. Carthage and Rome both became mighty military powers and soon confronted each other for control of the Mediterranean world. The Punic Wars between the two lasted more than a century (see 264–146 B.C.) and ended with the victory of Rome. The Iberian peninsula was then Romanized and became known as *Hispania.* People living there became part of a universal civilization and language and continued under Rome's cultural influence, even under the Visigoths, until the arrival of the Arabs.

The Middle Ages, 414–1492

In the sixth century the Visigoths established their capital in Toledo, and kept themselves apart from the Hispano-Romans and had different laws. At the first Council of Toledo in 589, the Visigoths became Roman Catholics and, around the same time, adopted Hispano-Roman language, culture, and laws. In 711 Muslim invaders landed in Spain and defeated the last king of the Visigoths, Rodrigo. By 718 they conquered the whole peninsula.

The Hispano-Muslims followed closely the cultural trends coming from Damascus and Baghdad and developed a flourishing culture that made Cordoba the most advanced city of Europe. Other cities—Seville, Zaragoza, Valencia, and Toledo—soon reached an advanced scientific and cultural level. The Hispano-Muslim economic prosperity stemmed from active trade both with the East and the West, the development of agriculture (including a sophisticated system of irrigation), and the development of mining techniques.

Mozárabes (Arabized) was the name given to the Christian minorities who lived under Muslim rule. They enjoyed religious freedom, kept their own customs and were governed by a Christian nobleman. *Mudéjares* were Muslims who incorporated Hispano-Muslim decorative elements into buildings. Their style flourished from the second half of the twelfth to the end of the fifteenth century.

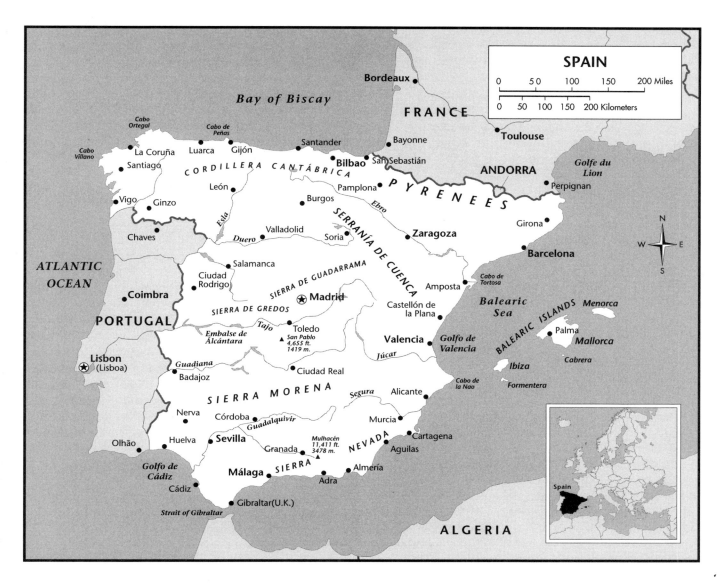

Jews had been present in the peninsula since Roman times. At the time of the Muslim invasion Jews were being persecuted and consequently saw a common cause with the invaders. Contrary to the Mozárabes, the Jews integrated quickly into Muslim culture and occupied positions of responsibility. It was a "golden era" for Sephardic culture (Sefarad was the Hebrew name for Spain), without equal in other times or nations. Rifts between the Christian and the Muslim communities and the Jews increased in the eleventh century, erupting into confrontations and riots.

At the beginning of the eleventh century, after the destruction caused by Almanzor, Christian rulers asked the French Cistercian monastic order of Cluny to help restore religious life. In the following century, the Cistercian order, also from France, provided similar assistance with sustaining religious life. In the thirteenth century, the Franciscan and the Dominican religious orders were founded. By the fifteenth century, the Catholic monarchs King Ferdinand and Queen Isabella enforced a policy of religious unity, expelling all non-Christians from the country.

Aragón and Navarra became independent from the Muslims at the end of the ninth and the beginning of the tenth centuries respectively. Catalonia evolved from the *Marca Hispánica,* a territory subject to the Franks from the beginning of the tenth century; it adopted the feudal model of government of the rest of Europe. Aragón was united with Catalonia in 1164. The Aragonese crown created a vast mercantile empire in the Mediterranean thanks to the help provided by the Catalonian bourgeoisie (middle class people) who had great social and economical influence in this era.

Al-Andalus (or Andalusia—the Arabic name for Muslim Spain) was originally a province of the Damascus caliphate until the arrival of the Omayad prince Abderraman I. He declared Al-Andalus an independent emirate. During the reign of Abderraman III (r. 912–61), Córdoba became the richest and most cultivated city in Europe. After the death of the military leader Almanzor, the caliphate broke into independent kingdoms called *taifas*. North African kingdoms provided military assistance to the taifas in their conflict with the Christians. The last African invaders were the *Benimerines,* who were finally defeated in 1340.

From the twelfth century on, the kings repaid the services rendered by noblemen and the military orders with land grants. Throughout the Middle Ages, diverse types of laws govern society: the Hispano-Romans use Roman Law until the promulgation of the Visigothic *Fuero Juzgo,* and the Muslim communities used Islamic Law. The *Fueros* or local laws were also important in some cases.

Literary developments in this era included *mester de juglaría,* poetic compositions of epic (heroic) character transmitted orally. *Mester de Clerecía,* on the other hand, were the work of skilled poets (most monks like Gonzalo de Berceo, the author of *Milagros de Nuestra Señora*—Our Lady's Miracles) who narrated the miracles of the Virgin Mary or the life of the saints. *Cantigas de Santa Maria* (Saint Mary's Songs), composed by the Castilian poet-king Alfonso X, *el Sabio,* (also known as the Wise, 1221–84; r. 1252–80) belong to the same genre.

The Renaissance, 1492–1598

The Catholic monarchs Ferdinand and Isabelle maintained a system of authoritarian and centralized government that protected the bourgeoisie and kept the nobility under control. They organized a modern state by taking power away from the Cortes, creating a professional army, reforming the administration, and recruiting civil servants from the educated bourgeoisie. Ferdinand and Isabelle wanted a unified Roman Catholic Spain. Members of the religious minorities—Jews and Muslims—either converted or left the country. Jews were frequently more highly educated than members of other groups and were therefore employed in political and administrative posts; they practiced skilled professions and trades and were active in commerce. Their exodus from Spain thus had a negative effect on the economy.

Once the Reconquest of Granada was completed in 1492, Spain was ready to turn its attention from the Mediterranean to the Atlantic. Missionary and commercial interests, pushed the Spaniards to discovery and conquest. In October 1492 Christopher Columbus arrived at Guanahani (present day Dominican Republic) in search of the Indies.

Francisco de Vitoria and Francisco de Suárez (1548–1617) created the basis of the *Tractatus de Legibus as Deo Legislatore* (the future international law), formulating for the first time in Europe the concepts of national sovereignty and of the illegality of war. They also formulated the principle that all citizens are equal. In the same spirit were the *Leyes de Indias* (Laws for the Indies) that prohibited the enslaving of the Indians and applied to them the same laws as to the rest of the Spaniards. For Fray Bartolomé de Las Casas (1474–1566), the staunchest defender of the Indians, only missionary work justified the presence of the Spaniards in the New World.

During the sixteenth century new institutions emerged: universities and hospitals, the *Casa de Contratación* in Seville (both a scientific institution and a court for trade) and the Madrid Academy of Mathematics. The *Consejo de Indias* (Council for the Indies), founded in 1524, sponsored geographical explorations; the study of astronomy, cartography and mathematics benefited nautical knowledge, and the *Arte de Navegar* (Art of Sailing) by Pedro de Medina became a standard textbook in Europe. Dr. Miguel Servetus (1511–53) discovered the blood circulatory system. Mining techniques as well as civilian and military engineering were notably advanced.

The Reformation, the protest movement against the leaders of the Roman Catholic Church headquartered at Rome, Italy, was started by German theologian Martin Luther (1483–1546). It split Europe between Catholicism and Protestantism. The Counter-Reformation was the Catholic reaction against the Protestant doctrines, and Spain was the staunchest defender of Roman Catholic orthodoxy.

Erasmus of Rotterdam (1466?–1536), the greatest of the Renaissance humanists, favored a religiosity based on the free examination and interpretation of the Holy Scriptures, which was contrary to prevailing orthodoxy. His Spanish followers were persecuted. Among them was the humanist Alfonso de Valdés, who criticized the relaxed customs of the Roman clergy in his book *Diálogo de Mercurio y Caron,* prohibited by the Inquisition.

Economically, the influx of New World gold provoked inflation, increased the price of land and thus the economic power of nobility while ruining the bourgeoisie. Products made in Spain were not competitive with the increasingly numerous imports and many entrepreneurs went out of business. The *hidalgos,* the lowest class of nobility whose status did not allow them to do manual work or engage in commerce, were impoverished. Meanwhile, the artisans and other members of the working class were looked down upon for doing work formerly done by Moriscos and Jews. American gold was mainly used to pay the Spanish armies that were fighting in America, against the Protestants in Northern Europe, and the Turks in the Mediterranean.

The sixteenth and seventeenth centuries were called the Golden Age or *Siglo de Oro* because of the extraordinary development of Spanish culture. The Gothic and Classical architectural styles appear, combined during the transition from the Middle Ages to the Renaissance. During the reign of Ferdinand (1452–1516) and Isabelle (1451–1504), flamboyant Gothic incorporated ornamental *Mudéjar* elements and a new architectural style, *Isabelino,* was born. (An example is the *Colegio de San Gregorio* in Valladolid.) Another style known as *Plateresco* was a combination of Mudéjar and Italian elements (as seen in the facade of the University of Salamanca). Later, Classicism brought a severe and functional use of decorative elements (examples include the Alcázar of Toledo).

The Gothic style lasted longer among the painters but by the end of the fifteenth century a Spanish school emerged that combined realism with idealism, and included many excellent artists. Spanish composers were paramount in polyphonic music.

Popular narrative poetry, *Romances*, were originally transmitted orally. At the beginning of the sixteenth century, Juan Boscán and Garcilaso de la Vega (1503–36) introduced the new Italian Renaissance poetic style. Ferdinand de Rojas's work, *La Celestina*, was the first novel written in Spanish. With dramatists Juan del Encina and Lope de Rueda, theater came of age.

The Baroque, 1598–1700

In his testament, Carlos II (r. 1665–1700) appointed Felipe d'Anjou, grandson of Louis XIV of France, as his heir to the crown of Spain. Disregarding the condition that Felipe could not reign over two countries, Louis XIV sustained his grandson's right to the French throne. The Austrian Habsburgs' hopes to place one of their members as king of Spain having failed, they allied themselves to England, who felt threatened by a powerful Bourbon block. Thus started the War of Succession, lasting from 1701–14 and affecting the whole of Europe.

The history of Spain under the Habsburg dynasty can be divided into two periods: one of military and political might under Carlos V and Felipe II lasting the main part of the sixteenth century, and another period of decadence under the remaining kings of the same dynasty lasting until the end of the seventeenth century.

Neither Felipe III (r. 1598–1621) nor his *privado* (his favorite), the Duke of Lerma, were able to rule the country, although both continued to pursue an imperialistic policy. Upon the king's death, his young son, Felipe IV, left power in the hands of the Count-Duke of Olivares, who was an able politician in favor of economic and social reforms that were opposed by the *Cortes* and the grandees.

The Baroque period saw two opposing literary currents. *Culteranismo* preferred poetry and used a metaphorical and obscure language. Its main representative was Luis de Góngora y Argote (1561–1637), author of *Soledades*. *Conceptismo* expressed itself mainly through a prose style rich in allegories and symbols, mythological allusions, and puns.

Representative of the Baroque vision of life was the picaresque novel. In autobiographical form, it depicted contemporary society seen through the eyes of a rogue who lives by his wits, an anti-hero who lacks moral, religious, and social values. The most notable novelists of this era are Mateo Alemán (1547–1614); Vicente Espinel (1550–1624); and Francisco de Quevedo (1580–1645). All reflect disenchantment, pessimism, and a lack of faith in a hypocritical and decadent society.

Baroque drama and comedy generally defended the status quo of a highly structured society. Tirso de Molina (1584?–1648?) wrote religious and historical plays as well as comedies of intrigue. Pedro Calderón de la Barca (1600–81) used language abounding in symbols and allegories. Calderón sees honor as not depending on social status but on the innate dignity of a person as a human being and as a Christian.

After the truce with the Dutch expired in 1621, Spain decided to declare war in order to protect its overseas empire from the constant attacks of the Dutch pirates. Allied to the Austrian Habsburgs, Spain embarked on the Thirty Years War that yielded catastrophic results. In spite of the initial victories of the Imperial armies, France, Sweden, and later England, sided with the Dutch and in 1648 Spain had to recognize the independence of Holland.

The reign of Carlos II (r. 1665–1700) a sickly, weak and childless king, was characterized by intrigue between the pro-Austrian and the pro-French parties. The decline of Spain continued during this period, attributable mainly to the decrease of the population due to emigration to the colonies, plagues, and war. The amount of the religious orders increased, as did the number of beggars.

Eighteenth Century

A united Spain was not achieved until the eighteenth century under the new Bourbon dynasty. The *Cortes* is a modern-day version of the *Curia Regia*. Both were representative bodies made up of members of the nobility, the church, and the people, and were advisory bodies to the king. The king needed authorization from this body to impose or raise taxes.

During the eighteenth century Europeans experienced the Enlightenment, substituting reason for tradition as the source of knowledge. The *Ilustrados* (the Enlightened) constituted a minority of non-conformist freethinkers. In Spain they endeavored to renew the country but had to face the opposition of traditionalists and conservatives. The Spanish *Ilustrados* combined new ideology with national spirit and were respectful of Catholic dogma while criticizing abuses of the Church.

The eighteenth century saw the birth of *Juntas de Comercio* (Merchants' Guilds) and *Sociedades Económicas de Amigos del País* (Societies for the Economical Improvement of the Country). Businessmen and merchants started to invest, and Spain's industry developed, notably with iron works in the Basque country and textile industries in Catalonia. The bourgeoisie (middle class), largely living in urban areas, acquired more economic power. The number of priests, monks, and nuns decreased from 220,000 to 170,000; members of the nobility decreased as well, from 700,000 to 400,000. At the same time, there was improvement in the social status of peasants and tradesmen.

The university system was greatly improved. Universities had been in the hands of the Jesuit Order, with the students mainly coming from the upper classes, but they were now open to the middle classes. In order to foster and regulate the progress of the arts and sciences, Royal Academies were created. Observatories, botanical gardens, libraries, and archives were instituted as well. The presses actively printed books dealing with a diversity of scientific and literary subjects.

Yet the results of the French Revolution slowed enthusiasm for reform and provoked a backlash. The clergy had held great power and had always considered the new ideas coming from France to be dangerous. Part of the nobility had embraced the ideas of the Enlightenment, but the majority remained faithful to the ideology of the old regime and affected an enthusiasm for bullfighting, popular music and dance, dress and customs of *lo castizo* (the lower classes) which they considered a patriotic and nationalistic statement. The end of the century witnessed the birth of "the two Spains," one liberal and progressive and another, its opposite, traditionalist and conservative.

With the new Bourbon dynasty came French arts and culture, protected by the court at the expense of Spanish tradition. The majority of Spaniards did not abandon their preference for their own comedy and drama, however. New authors like Ramón de la Cruz (1731–94) filled the playhouses with works exalting traditional values and satirizing French tastes and fashions. Other authors wrote in the Neoclassical vein, among them, Vicente García de la Huerta (1734–87), author of *Raquel,* a tragedy inspired by Spanish history. Later, Nicolás Fernández de Moratín (1737–80) attacked Spanish Golden Age theater in his *Desengaños al teatro español.* Nicolas's son, Leandro (1760–1828), criticized the ills of contemporary society and tried to educate the public with comedies.

In architecture, Spanish Baroque lasted through the eighteenth century, mostly in churches and religious buildings (*Obradoiro* facade in the Santiago de Compostela cathedral, by Pedro de Casas y Novoa; facade of the Hospicio de San Fernando, by Pedro de Rivera). The Churriguera family of architects gave its name to the *churrigueresco* style that enriched and exaggerated the decorative elements of Baroque. One of the Churrigueras, Alberto, built the Plaza Mayor of Salamanca. The Spanish Baroque was gradually superseded by the academic and functional French Baroque. Best examples of this style are the Royal Palaces of Madrid and Aranjuez.

Nineteenth Century

During the nineteenth century the *dos Españas* (two Spains—liberal and conservative) fought each other. As a consequence there were frequent *pronunciamientos* (military uprisings), revolutions, and changes of governments, A liberal minority seized power several times during the century and promulgated ten different constitutions, favoring a parliamentary government with a division of power. The conservatives, mainly landowners and members of the clergy, supported the absolute powers of the king.

Carlos IV and his wife Maria Luisa's favorite was Manuel Godoy, originally a Palace Guard, who dictated Spanish policy for several years, with disastrous results. After the death of the French king Louis XVI, Spain declared war on the French Republic. With the San Ildefonso Treaty (1796), Spain and France became allies, only to be defeated by the English in the naval battle of Cape Trafalgar, near Cádiz, in 1805. The considerable weakening of the Spanish navy affected afterwards its capacity to defend the American colonies.

In 1808, with the pretext of crossing Spain to invade Portugal, Napoleon's army occupied the main strategic points of Spain. The dissension between Carlos IV, his wife and Godoy, on one side, and prince Fernando (later Fernando VII), on the other, escalated and gave rise to the *Aranjuez mutiny* (1808) which brought a violent end to Godoy's power. Napoleon lured Carlos IV and his son Fernando to Bayonne, in France, forced both of them to abdicate, and replaced them with his own brother Joseph. On May 2, 1808, the people of Madrid revolted against the French in a bloody reprisal . The revolt spread to other parts of the country and the War of Independence began.

As a counterpart to the Madrid government of José I, imposed by Napoleon, the Spanish patriots established another government in Cádiz with the Cortes (Parliament) to rule in the name of the exiled Spanish king. They enacted the Constitution of 1812 reflecting the ideology of the French Enlightenment. The King shared his powers with the Cortes, freedom of the press was established, and the Inquisition and the privileges of the nobility were abolished.

The War of Independence lasted until 1814, ending with the victory of the Spanish army and guerrillas, aided by British troops. After the battle of Vitoria in June 1813, the French army retreated from Spain, and Fernando VII, the new king of Spain, was freed. Upon his return from France, Fernando abolished the Constitution of 1812 and reverted to absolute rule. Colonel Rafael del Riego, commander of one of the regiments scheduled to embark in Cádiz to fight in the American colonies, headed a successful military coup and reasserted the validity of the 1812 Constitution. After a brief three-year period, the monarchy was reinstated and a new period of repression began.

Fernando VII died on September 29, 1833, leaving his wife Maria Cristina as regent, and his young daughter, Isabel as queen. Fernando's brother, Prince Carlos, the late king's reactionary brother, started an unsuccessful seven-year war to claim the crown for himself.

The reign of Isabel II (also known as Isabella II, 1830–1904; r. 1843–68) was characterized by the struggle for power between liberal conservatives, *moderados,* and *progresistas* who had the backing of the masses.

By the middle of the nineteenth century, the Industrial Revolution and the concomitant social and economic reforms produced a rapid growth of the economy. The Catalan textile industry was prosperous, as were the Basque country iron works. Prosperity created an upper class of bankers, industrialists, and businessmen, as well as successful military officers. Isabel II was driven into exile as a result of a revolution headed by Serrano, Topete, and Prim in September of 1868.

A republic was proclaimed in the absence of government (1873–74). The short-lived regime was defended by a group of Republicans calling themselves "volunteers of liberty," and opposing military conscription and centralized rule. However, their weak decentralized structure led quickly to their demise in January 1874 with a military coup by General Manuel Pavia. Alfonso XII, the son of the deposed Queen Isabel II, was declared to be king. The period that followed, known as the *Restauración,* was marked by political stability. After defeat in the Spanish-American War in 1898, Spain lost control of Philippines and Cuba.

The loss of colonies in America and of the trade they provided in the first half of the century proved detrimental to Catalonian and Basque business which accused the central government of inefficiency. To increase their earnings, factory owners imposed long hours of work for little pay, thus alienating the workers, who had neither social security nor retirement pay. In time, they acquired considerable political power, organized themselves and joined socialist and anarchist trade unions.

Twentieth Century

By the dawn of the twentieth century, Spain still had not solved the persistent problems that afflicted it of old: uneven distribution of wealth, lack of a well-established middle class, illiteracy (half of the population could not read or write at the beginning of the century), meddling of the military in politics and of the Church in society; a mostly rural economy; weak industry, and growing power of trade unions. The army allied itself with the oligarchy in power (landowners and industrialists) and imposed dictatorial regimes between 1923–29 under General Primo de Rivera and from 1939 on under General Francisco Franco. The Church sided with them, weary of an intellectual leftist bourgeoisie and of the revolutionary working class. Regionalist movements, only complicated matters further.

Franco Era, 1939–1975

During the years that dictator Francisco Franco (1892–1975, dictator 1939–75) was in power, the Roman Catholic Church asserted excessive control over society. By the 1990s, the Church saw most of its former power and influence erode. Church attendance was down, many openly opposed the position of the Church in matters concerning divorce, abortion, and family planning, and few young people were willing to enter the priesthood. The Constitution stipulated the separation of church and state and recognized religious freedom (*libertad de conciencia*) of Spaniards. As a result, the Church evolved a new role in twentieth-century Spanish society .

Economic development was rapid after 1959. By the end of the twentieth century, Spain was among the ten most industrially developed countries in the world. Its main economic pillars are industry, agriculture, and tourism. Major ports include Barcelona and Cartagena on the Mediterranean Sea, and Cádiz, La Coruña, Santander, and Bilbao on the Atlantic Ocean. Spain has an excellent system of public transportation with forty-two airports, a modern network of roads, and a national railroad system (*Renfe*) with an increasing number of superspeed trains. Madrid, Valencia, and Barcelona have modern subway systems.

In November 1973, Admiral Luis Carrero Blanco, chosen by Franco to be his successor was murdered in Madrid by ETA. Francisco Franco died on November 20, 1975, and two days later, Prince Juan Carlos de Borbón, Alfonso XIII's grandson, was proclaimed king of Spain.

Modern Spain, 1975–2000

Franco's death marked the beginning of a new era. The *Transición,* the gradual transition from a dictatorship to a constitutional and parliamentary monarchy, was peaceful under the guidance of the *Unión de Centro Democrático* Party (*UCD*) with Adolfo Suárez in power from 1977 to 1982.

Since 1975, Spanish life has undergone a rapid evolution towards modernity, and the new society has European values and a commitment to democracy. For some time, the trend was to move from rural to urban areas (mainly to Madrid, Barcelona, Seville, Zaragoza, and Bilbao), and by the 1970s, over half of the population lived in cities with more than 50,000 inhabitants. By the late 1990s, the birthrate was the lowest in the history of the country. The population is getting older and is unevenly distributed: the periphery of the country and cities are more densely populated than the interior.

The Constitution of 1978 creates seventeen autonomous regions in Spain (see 1978). It states that the autonomous communities have equal rights but do not supersede "the indissoluble unity of the Spanish Nation, the country of all Spaniards." Four languages are spoken. Three—Castilian, Catalan, and Gallegan—derive from Latin and were developed in the Middle Ages. The fourth, Basque, is of unknown origin. Although Castilian is the official language of Spain, the other languages are also official in their own autonomies. Castilian, referred to as generic Spanish in the United States, is also spoken in eighteen American states. *Ladino* (Sephardic Spanish) is spoken by about 1.2 million people. (*Sefarad* is the Hebrew name given to the Iberian peninsula.)

In the 1980s, the government implemented an income tax system similar to that of the United States. In response, an *economía sumergida* (underground economy) developed that amounted to an estimated twenty-five percent of the gross national product.

Until well into the twentieth century, the position of women was relegated mostly to that of wife and mother. By the 1980s, the general liberalization experienced by the country radically changed women's rights and roles, both legally and socially. By the end of the twentieth century, women had the same job possibilities as in other European countries, and they frequently occupied key posts in politics, government, industry, and academia. Divorce and abortion were legalized.

The majority of Spaniards (an estimated eighty percent) report little interest in politics although they prefer democracy to any other system of government. Spain is no longer isolated from the world, and the modernization of information systems, mass media and infrastructure is a reality. The economy has undergone substantial changes and has shown a solvency and a capacity to grow that were unknown in the last two centuries, and that has allowed the country to integrate easily within the framework of the European Community.

The main problems of contemporary Spanish society are Basque terrorism, unemployment, drugs and the resulting insecurity and street crime, as well as illegal immigration from Africa, Latin America, and Eastern Europe. Yet, in spite of these problems, Spaniards are politically mature, confident of the future, and proud of their present democratic system.

Timeline

c. 15,000–10,000 B.C.   Cave paintings at Altamira and Castillo, Santander

Archeologists have found a variety of weapons, tools, and ornaments of the Neanderthal people who lived during the Paleolithic period (between 25,000 and 10,000 B.C.). Most important were their artistic achievements found in the interior of caves in Southern France and in Northern Spain where they left realistic polychrome paintings of animals (Altamira, Cantabria).

c. 10,000–5,000 B.C.   Mesolithic cultures

After 10,000 B.C., with warmer temperatures following the end of the last Ice Age, a group of hunters and gatherers leave drawings in the eastern and southern parts of the Iberian peninsula. These drawings are in the open, usually beneath rock overhangs (abrigos). The drawings are stylized and monochromatic, representing humans fighting, dancing, or hunting animals. They are found near Alpera, Albacete; Valltorta; and Castellón.

c. 1100 B.C.   Phoenician traders found Gádir (Cádiz)

The development of sailing brings people to the Iberian Peninsula. The Phoenicians come from Tir (in present-day Lebanon) and found Gádir (present-day Cádiz) on the Southern Atlantic coast, which soon becomes a prosperous center of trade. They are not only active merchants who deal in gold, silver, and copper, but are also manufacturers and sailors.

The Carthaginians originate from Carthage, a former Phoenician colony in North Africa (present-day Tunisia) that

becomes an independent military power. They are a commercial people who found Carthago Nova (present-day Cartagena), the center of their area of influence.

1000–800 B.C.   Celtic peoples enter the area

At the beginning of the first millennium B.C., Celtic peoples migrate across the Pyrenees mountains. Around 800 B.C., they settle in the northwest regions of the peninsula. They have a rudimentary knowledge of the use of metals.

c. 700–600 B.C.   Tartesians reside in the area

Archeological findings in the lower Guadalquivir valley confirm the existence of the Tartesians, an ancient people who are the first inhabitants of the Iberian Peninsula with an advanced political organization and a rich and sophisticated economy. This is probably due to their contact with the Phoenician and Greek colonizers. The Tartesians monopolize the tin trade until they are absorbed by the Carthaginians (inhabitants of Carthage on the northern coast of Africa near modern-day Tunis) at the end of the sixth century B.C.

The Greeks arrive in the seventh century B.C., also attracted by commercial gain, and founded several colonies on the Mediterranean coast, among them Emporion (Ampurias), where archeologists in modern times discover statues, pottery, coins, jewelry and other remains of their rich civilization.

6th century B.C.   Tartesians absorbed by Carthage

The Carthaginians, residents of Carthage, a city on the north coast of Africa, absorb the Tartesians into their society.

4th century B.C.   Iberian and Celtiberian people thrive

The Iberians occupy present-day Andalucia and the region near the Mediterranean Sea, although they extend as far north as the Rhone River. The people known as the Celtiberians are the ethnic result of alliances and intermarriage between the Iberians and the Celts; they settled in Celtiberia, a region in the northern area of the central plateau where they live in urban settlements like Numancia (Soria) and mine iron from the region of the Moncayo mountain.

The Iberians must resort to war in order to subsist. Among their values are military courage and fidelity to their leader. They believe in an after life and in deities like the God of War, and they worship a Goddess of Fertility, who reigns over life and death. The Iberians have a writing system, derived from Greco-Phoenician, that has not yet been deciphered. They are the first to cultivate orchards and use irrigation systems, later perfected by Romans and Muslims. They are organized in independent tribes united by means of regional alliances, and their settlements are surrounded by walls made of large stones (known as *ciclopean* walls, such as in the city of Tarragona).

Iberian art reveals Greek and Eastern influences, especially in sculpture (Dama de Elche, Dama de Baza), pottery, and metal work. (Examples include votive bronze offerings as seen in Cerro de los Santos , Albacete).

264–146 B.C.   Punic Wars between Rome and Cartage

Carthage and Rome become mighty military nations and confront each other for the control of the Mediterranean world. The Punic Wars last more than a century and end with the victory of Rome and the destruction of Carthage. As a result, the Peninsula is Romanized and becomes known as *Hispania*. Its inhabitants, now called *hispanos*, become part of a universal civilization and language. They continue under Rome's cultural influence, even under the Visigoths, until the arrival of the Arabs.

Under the Romans, *Hispania* becomes a linguistic, cultural, social, and political unit, with a solid municipal administration. There are more than five hundred cities thriving in Hispania during the era of Roman influence. In return, Hispania gives Rome four emperors: Trajan (see c. A.D. 53), Hadrian (see A.D. 76), Marcus Aurelius (A.D. 121–180), and Theodosius I, the Great (see c. A.D. 347).

197 B.C.   Hispania is divided into Citerior and Ulterior

Hispania is divided into *Citerior* and *Ulterior*. The emperor Augustus (63 B.C.–A.D. 14) divides *Ulterior* into *Bética* and *Lusitania*.

A.D. 69–70   Roman law introduced in Hispania

Vespasian (A.D. 9–79) gives Roman law to Hispania (A.D. 69–79). During this period, Hispania contributes writers and thinkers. Lucio Anneo Seneca (c. 4 B.C.–c. A.D. 65) is a distinguished author of tragic plays. The stoic philosopher, Lucan (Marcus Annaeus Lucanus, A.D. 39–65) is Seneca's nephew. In A.D. 60, he publishes the epic poem *Pharsalia* about the civil war between Pompey and Julius Caesar.

c. A.D. 53   Trajan is born

Trajan (c. A.D. 53–117), also known as Marcus Ulius Trajanus, is born near Seville. He becomes sole ruler of the Roman Empire in A.D. 98. During his reign, he increases the territory controlled by the Empire, adding Armenia and Mesopotamia. He builds canals, bridges, harbors, and roads, and enjoys great popularity. His ward and trainee is Hadrian (see A.D. 76) who becomes emperor upon Trajan's death.

76   Hadrian is born

Hadrian (A.D. 76–138) is born of Spanish parents in Rome and becomes the ward of Trajan. Upon Trajan's death in 117, Hadrian becomes Emperor of Rome. From around 120, he spends his time traveling around the empire, including visits to Germany, Britain (where he builds a wall that now is known by his name), Spain, Egypt, and Greece.

211   Hispanos become Roman citizens

Roman emperor Caracalla (A.D. 188–217) grants Roman citizenship to all inhabitants of the Empire. The Hispano-Roman society has a hierarchical organization: free men, who are either Roman citizens, Latins, or foreigners; people with a semi-free status; and slaves.

Late 3rd century   Tenant farmers replace slaves

When the price of slaves increases, free tenant farmers are substituted for slaves for farm labor. Tenant farmers' relationship to their masters develops over time in such a way that the farmers remain forever attached to the land by the *colonato*, a system that is half-way between slavery and feudalism.

313   Constantine proclaims Edict of Milan

Constantine (307-37) proclaims the Edict of Milan extending religious freedom to Christians.

347   Theodosius I, the Great, is born

Theodosius I, the Great (c. A.D. 347–95) is born in northwest Spain. He is named co-ruler of the Roman Empire with Gratian in 379. Theodosius is a devout Christian and makes Christianity the official religion of the empire. He affirms the Nicene Creed, drafted by Osio, an important document of the Christian Church, in 381, and torments pagans and heretics. He closes all non-Christian temples and bans all pagan rituals.

414–1492   Middle Ages

The Spanish Middle Ages (or Medieval period) is generally considered to begin in 414 with the arrival of the Visigoths and to end with the reconquest of Granada from the Moors in 1492 and the discovery of America in the same year.

Throughout the Middle Ages, the Spaniards are ruled by diverse types of laws: the Hispano-Romans continue to use Roman Law until the promulgation of the Visigothic *Fuero Juzgo* (see 653); the Muslim communities use Islamic Law; the Castilian-Leonese use the Levantine and Pyrenaic laws that appear in the eleventh and thirteenth centuries. Besides these, there are the *Fueros* or local laws that in time acquire great importance.

Medieval Christian society is divided into nobility (high and low nobility and high clergy); free men (clergymen and monks and the bourgeois class); and serfs (an intermediate class between free men and slaves who depend on the lord who owns the land they farm). While the nobility and the high clergy increase their privileges, the kings protect the bourgeoisie, formed by enterprising free men who live in the cities and form trade associations that first appear in Catalonia.

414   Arrival of the Visigoths

The Visigoths are a Germanic people who arrive in Hispania as allies of the imperial Roman army. They readily accept

Roman cities prosper, with public works such as roads, bridges, and aqueducts (pictured), many of which survive to modern times. (EPD Photos/CSU Archives)

Roman culture and institutions and rule the country for three centuries.

c. A.D. 417   Paulo Orosio publishes seven-volume history

Paulo (or Paulus) Orosio (or Orosius) publishes a seven-volume history of the world, *Historiarum adversus Paganos,* from its creation to A.D. 417. He is the first to consider Hispania a national entity.

500   Cities prosper

Romans build an extensive network of roads connecting the provinces of Spain with Rome for their military and commercial purposes. The *Via Augusta* enters through the Pyrenees, reaches Cartagena, and from there goes to Córdoba and Cádiz. By this time, more than 500 cities exist, among them many as prosperous as Barcelona, Valencia, Zaragoza, Seville, and Lisbon. In the late twentieth century, the ruins of Roman cities like Italica (near Seville) and Mérida provide evidence of theaters, baths, public markets, temples, and monuments, as well as public works like triumphal arches, bridges, and aqueducts.

576   Toledo proclaimed capital of Leovegildo's Visigothic kingdom

The Visigoths establish their capital in Toledo. A military minority, the Visigoths keep themselves apart from the Hispano-Romans and have different laws, which they maintain for over a decade (see A.D. 589). They subscribe to the *Arrian* sect of Christian beliefs (that is, they do not accept the dogma of the Christian Trinity and the divinity of Christ).

589   First Council of Toledo is held

The First Council of Toledo meets, proclaiming Roman Catholicism to be the official religion of the state after the conversion of King Recaredo. The Visigoths become Roman Catholics, renouncing the Arrian sect (see A.D. 576), also at this Council. They assume Roman culture and Latin as their language. Saint Isidore of Seville (570–636), the most illustrious scion of the new society, writes his *Etimologías* and *History of the Goths* in Latin.

653   Death of Chindaswinth

Chindaswinth, Visigoth king of Spain (642–53) dies. Chindaswinth rules jointly with his son Recceswinth from 649. Chindaswinth begins compiling the Visigoth *Fuero Juzgo* laws, integrating the diversity of the Hispano-Roman and Visigoth judicial systems. This system, issued by Recceswinth about 654, is known as the *Liber iudiciorum* (later as the *Liber* or *Forum iudicum*).

Visigothic architecture is mostly religious, combining the classical Greco-Roman and Byzantine traditions. It uses the horseshoe arch that later becomes one of the characteristic elements of Muslim art. The churches of San Juan de Baños in Palencia, Santa Combe de Bande in Orense, and San Pedro de la Nave in Zamora date from the seventh century.

711   The Muslims land in Spain

Muslim invaders land in Spain near Algeciras (on the southern coast of Spain) to take part in the civil wars of the Visigoths. They defeat the last king of the Visigoths, King Rodrigo, at the battle of Guadalete and soon conquer the Peninsula. Groups of Visigothic and Hispano-Roman survivors find shelter in the northern mountains of Asturias. There they organize resistance to the invaders.

718   Battle of Covadonga

According to legend, Pelayo, a Christian chieftain, defeats the Muslims in Covadonga, marking the beginning of the Reconquest. From then on, small groups of Christians battle the Muslims for more than seven centuries.

756   Abd-ar-Rahman I declares his Emirate

Abd-ar-Rahman I (731–88), the first of the Muslim Ummayyad dynasty rulers, declares an independent emirate in Córdoba.

756–1200 Muslim Spain thrives

Al-Andalus, the Arabic name for Muslim Spain, is originally a province of the Damascus caliphate. When the Ummayyad prince Abd-ar-Rahman I arrives, he declares Al-Andalus an independent emirate. This lasts from 756 to 912.

9th century Asturians establish separate kingdom

The Asturians establish a separate kingdom by the ninth century with their capital in Oviedo.

899 The apostle Santiago's remains discovered

The remains of the apostle Santiago are discovered in Galicia. By this time the cult of the Apostle Santiago is established. The apostle's tomb in the city of Santiago of Compostela becomes one of the most popular destinations of medieval pilgrimage, an important factor in establishing unity among Christians on the Iberian peninsula and linking them to Christians in Europe.

912–61 Reign of Abd-ar-Rahman III

During the reign of Abd-ar-Rahman III (r. 912–961), Córdoba became the richest and most cultivated city in Europe.

10th century Asturians move capital to León

At the beginning of the tenth century the Asturians moved their capital to León, resettling the area with small landowners who form rural municipalities and popular armies.

961 The County of Castile becomes independent

The County of Castile accepts independence from the Kingdom of León. It soon becomes the strongest power among the Christians.

985–88 Almanzor conquers cities

The military leader Almanzor conquers Barcelona in 985 and Santiago de Compostela in 988.

Late 900s–early 1000s Aragón and Navarra gain independence

Aragón and Navarra became independent from the Muslims.

1007 Birth of poet Ibn Zaydun

The poet Ibn Zaydun (1007–70) is born. His love poetry is considered among the finest in the Arab world. Characteristic of Hispano-Arabic poetry are the *jarchas,* lyric poems that incorporate local linguistic elements into the Arabic text.

1031 End of the caliphate

After Almanzor's death (see 985–88), the caliphate breaks apart into a series of independent kingdoms called *taifas.* At times the taifas numbered as many as twenty-three.

c. 1043 Birth of El Cid

Rodrigo Díaz de Vivar (c. 1043–99), known as El Cid, is born in Burgos. He is a national hero who is constantly involved in battles. He becomes known as El Cid (The Lord) and El Campeador (The Champion). He captures Valencia from the Moors in 1094 and becomes the ruler there. His exploits inspire many epics, ballads, and legends.

c. 1050 Christian rulers seek aid from France

After the destruction caused by Almanzor, the Christian rulers ask for help from the French monastic order at Cluny to reorganize religious life in their domains. They introduce French culture and a universalistic spirituality, the devotion to the Virgin Mary, the cult of Santiago, whose sepulcher became a center of European pilgrimage, and the idea that the Reconquest is a holy war against Islam. In the following century, the *Cistercian* Order also comes from France with a similar purpose.

1070 Construction of the cathedral at Santiago begins

The building of the Santiago de Compostela cathedral begins. (See 899.) It is the most important building of Spanish Romanesque architecture.

Romanesque art is fundamentally religious and appropriate to a rural feudal society. Brought to Spain in the eleventh century and found along the road to Santiago, its defining elements are barrel vaults, semicircular or Roman arches, decoration with geometric figures, and columns and facades decorated at times with human and animal figures, frequently with a symbolic meaning. The first Romanesque churches appear in Catalonia, and by the eleventh century they can be found all over Christian-held territory.

1085 Alfonso VI conquers Toledo

Alfonso VI, a Christian, conquers Toledo. After his conquest, the city becomes a center of learning where scholars of the three religions (Christian, Muslim, and Jewish) translate and comment on works written in Arabic and Hebrew.

1097 Alfonso VI of Castile gives land to his daughter

Portugal is the result of a donation made by King Alfonso VI of Castile to his daughter Theresa in 1097. Catalonia evolves from the *Marca Hispánica,* a territory subject to the Franks from the beginning of the tenth century; it adopts the feudal model of government of the rest of Europe.

12th century Kings make land grants

Beginning in the twelfth century, kings repay the services rendered by noblemen and honor members of military orders with land grants. Because of this, some institutions of medieval origin later become obstacles to economic development. These are known as entailed states and *manos muertas* (mortmain, or properties belonging to a judicial person, or a religious order, which cannot dispose of them).

12th century   Gothic style of architecture

Gothic style, the art of the bourgeoisie and of the cities, is brought to Spain by the Cistercian monks. Its architects increase the height of the buildings and idealizes its forms. Sculpture and painting abandon Romanesque symbolism and reproduce naturalistic and popular elements.

1126   Birth of scholar Ibn Rusd

Hispano-Arabic scholar Averroes (also known as Ibn Rusd; 1126–1198) is born in Cordoba. A leading Hispano-Arabic scholar, Averroes studies and comments on Aristotle; his writings and ideas about reason and faith revolutionized European scientific thought from the twelfth century on.

c. 1150   *Poema de Mio Cid* is written

The greatest epic poem in Castilian literature, *Poema de Mio Cid*, recalls the deeds of the Castilian knight Rodrigo Diaz de Vivar (c. 1043–99), called El Cid (see c. 1043).

c. 1150–1500   Mudéjar decorated style flourishes

Mudéjares are the Muslims who live in territories conquered by the Christians. They incorporate Hispano-Muslim decorative elements into buildings with a Christian-style architecture. The Mudéjar style flourishes from the second half of the twelfth to the end of the fifteenth century.

1164   Aragón is united with Catalonia

Aragon and Catalonia are united. The Aragonese crown creates a vast mercantile empire in the Mediterranean thanks to the help provided by the Catalonian bourgeoisie.

1195   The *Almohades* defeat the Castilians

The *Almohades* of North Africa provide military support to the Muslim taifas and defeat the Castilians at Alarcos (see also 1031, 1212, and 1340).

13th century   Military and religious orders founded

With the idea of fighting the Muslims and protecting the pilgrims, the Military Orders of Santiago, Calatrava, Alcántara, and Montesa are founded. Two religious orders—the Franciscans and the Dominicans—are also born. Both religious orders exert great influence on the social and cultural life of the country.

Mester de juglaria are poetic compositions of epic (heroic) character transmitted orally. *Mester de Clerecía*, on the other hand, are the work of skilled poets who narrate the miracles of the Virgin Mary or the life of the saints. Most authors are monks like Gonzalo de Berceo, the author of *Milagros de Nuestra Señora* (Our Lady's Miracles).

1212   Defeat of the Almohades by the Christians

Christian armies defeat the Almohades at the battle of Las Navas de Tolosa.

1221   Birth of Alfonso X

Alfonso X (1221–84), king of Leon and Castile, is born in Burgos. He ascends to the throne when his father, Fernando III (also known as Ferdinand III) dies (see 1252). He captures Cádiz and the Algarve from the Moors. He manages to repel an uprising by one son, Philip (see 1271), only to be defeated in another insurrection led by another son, Sancho (see 1282). Alfonso X is regarded as the father of Castilian literature, since he sponsors the first Castilian-language history of Spain and a translation of the Old Testament. His *Siete Partidas* (Code of Laws) are of major significance. He is known as "The Wise" because he adapts charts of the planets from those developed by Arabic astronomers. As a poet, he composes *Cantigas de Santa Maria* (Saint Mary's Songs).

1232   Inquisition launched by Gregory IX

Pope Gregory IX (1148–1242), pope from 1227, issues a bull sending inquisitors to Aragon, thus launching centuries of the Inquisition. The Inquisition is the name given to harsh religious court judgments levied against citizens who are believed to be heretics or guilty of crimes against the Roman Catholic Church. (See 14th century.)

1243   University of Salamanca founded

King Fernando III founds the University of Salamanca.

1246   Seville is conquered

King Fernando III conquers Seville.

1252   Reign of Alfonso X

Alfonso X (see 1221) ascends to the throne.

1265   Barcelona governed by Council of the One Hundred

Barcelona is governed by the *Consejo de Ciento* (Council of the One Hundred) composed of one hundred free men from all social classes. The *Consulat del Mar* (Sea Guild) is a professional association that has legislative power in mercantile matters; it mints money, has its own fleet, and maintains consulates abroad.

1271   Philip attempts to overthrow his father

King Alfonso X (see 1221) quells an uprising led by his son, Philip.

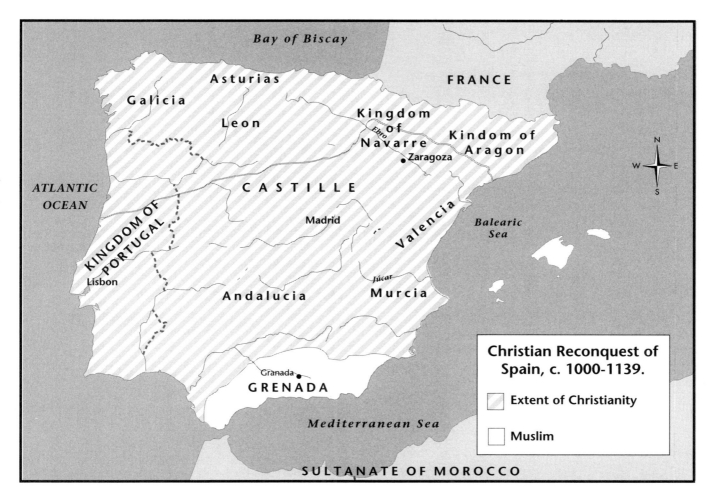

Christian Reconquest of Spain, c. 1000-1139.

Extent of Christianity

Muslim

1273   Feuding cattlemen unified by King Alfonso X

Castilian cattlemen always win favor from the rulers over common farmers because of the substantial taxes they pay for the trade and export of wool. Cattlemen belong to community associations. King Alfonso X (1221–84) unified these diverse associations under the name *Honrado Concejo de la Mesta* (Honorable Mesta Council). One of the privileges granted to members of this Council is the use of preferred rights of way (cañadas) for the cattle to move to winter pastures.

c. 1275   King Alfonso X writes poetry

King Alfonso X (1252–1280), the Castilian poet-king, composes the *Cantigas de Santa Maria* (Saint Mary's Songs).

1281   King Alfonso X defeated

King Alfonso X is defeated by his son, Sancho. Sancho leads a successful insurrection against his father. The coup is in response to Alfonso's unstable, weak and obstinate rule coupled with the constant squabbles with his nobles.

14th century   Inquisition leads to massacres

The Inquisition (see 1232) is a system of harsh religious court judgments against those violating the laws and teachings of the Roman Catholic Church. At Valencia, over 10,000 persons are forced to be baptized with the threat of death. (See 1480.)

1340   The defeat of the *Benimerines*

The North African *Benímerines* invade to assist the taifas in fighting the Christians. They are defeated (see 1031 and 1195).

1348   The Black Death invades Europe

The Black Death, a plague that sweeps across the continent during the fourteenth century, first enters Europe, probably at Genoa, Italy. (Although no one is certain, it seems likely that what is referred to as Black Death was bubonic plague.) In parts of Europe, historical accounts report deaths of two-thirds to three-fourths of the population. One historian, Hecker, writing in the 1800s in his work, *Epidemics of the*

Middle Ages estimates that one-fourth of the total population of Europe dies in the Black Death.

1350   Pedro of Castile ascends to the throne

Pedro I of Castile (1334–69; r. 1350–69) succeeds his father, Alfonso XI.

1397   Birth of Ausias March

Poet Ausias (Auzias) March (1397–1459) is born in Valencia. He leads the movement among poets away from the style of lyrical ballads to a more realistic approach. His main themes are drawn from the major experiences of life, such as love and death. He follows a trend begun by the earlier mystic poet, Raimundo Lulio (1235–1315).

1469   Marriage of Ferdinand and Isabella

The monarch, King Ferdinand "The Catholic" (1452–1516) and Isabella of Castile (1452–1504, also known as Isabella the Catholic) are married. During their reign they strengthen the monarchy, curb the power of the nobles, and organize a police force called the *santa hermandad* (holy brotherhood). They enforce a policy of religious and political unity, achieved by the expulsion of the non-Christians. (See 1480.)

1480   Ferdinand and Isabella and the Spanish Inquisition

The Inquisition (see 1232) has been passing religious judgments on people throughout Europe for over two centuries. Four popes—Sixtus IV (1414–84; pope from 1471–84), Innocent VIII (1432–92; pope from 1484–92), Alexander VI (1431–1503; pope from 1492–1503), and Julius II (1443–1513; pope from 1503–13)—support this effort. The popes believe Spain was lending powerful government support to preservation of Roman Catholic purity.

1485   Birth of Hernán Cortés

Conquistador and explorer Hernán (or Hernando) Cortés (1485–1547) is born. He enrolls in the University of Salamanca but leaves after two years. He sails for San Domingo in 1504 and in 1511 joins the expedition of Diego Velázquez; they are successful in conquering Cuba. In 1518, Velázquez appoints Cortés to head an expedition of over five hundred men to attempt to conquer Mexico. They are successful in a series of conquests; their final victory over the city of Tenochtitlán in 1521 completes their conquest of Mexico. Cortés destroys the city and builds Mexico City in its place. In 1522, he is appointed governor and captain-general of New Spain. In 1528, he returns to Spain and is created a marquis by Charles V. He dies in Spain, near Seville, in 1547.

The Renaissance

1492   Surrender of Granada

The Moors are defeated and surrender at Granada.

1492   Jews leave Spain

Ferdinand and Isabelle establish the Catholic religion as the state religion. The Jews leave Spain, but they are, in reality, being expelled since members of religious minorities are being forced to convert to Catholicism if they remain in Spain.

1492: October 12   Exploration of America

Ferdinand and Isabelle commission Christopher Columbus to find a route to the so-called "Indies" spices.

1511–13   Balboa explores the Pacific Ocean

Vasco Nuñez de Balboa (1475–1519) is believed to be the first European to see the Pacific Ocean. In 1511, he stows away on an expedition to Central America. He founds a colony there and explores the area. While exploring, he climbs a high peak and spots the Pacific Ocean. He claims it for Spain. He undertakes many expeditions for the governor, Pedrarias Dávila, and eventually marries his daughter. When Dávila and Balboa have a major disagreement in 1519, Balboa is beheaded.

1516   Reign of Carlos V

Carlos V is crowned king of Spain.

1524   Council for the Indies founded

The *Consejo de Indias* (Council for the Indies) is founded. The Council sponsors a plan for explorations in the distant lands of the Americas.

1534   Founding of the Jesuit order

The Jesuit religious order is founded in part to respond to the Reformation. Spain, the staunchest defender of Roman Catholic Church, contributes with theologians to the Council of Trent (1545–1563), called to fight the spread of the Lutheran Reformation.

1542   Leyes Nuevas enacted

The Leyes Nuevas (New Laws of the Indies), the new colonial body of laws for the protection of the Indians, are enacted. They are based mainly on the theories of the Dominican friar Bartolomé de las Casas (1474–1566), known as the Apostle of the Indians.

Bartolomé de las Casas is born in Seville and joins the third voyage of Christopher Columbus in 1498. In 1510, he becomes a Roman Catholic priest and the next year accompanies an explorer to Cuba. He travels as a missionary in Mexico, Nicaragua, Peru, and Guatemala, and is appointed Bishop of Chiapas (México) in 1544.

1547   Birth of writer Miguel de Cervantes

Miguel de Cervantes (1547–1616) is born near Madrid. He begins publishing his writings in his early twenties. He enlists as a soldier and serves in Tunis in northern Africa. When he attempts to return to Spain he is captured by Algerians and held in prison. Although he makes a number of attempts at escape, he is held for about five years in Algiers. Friends and sympathetic monks negotiate his ransom, and he is released in 1580. He returns to Spain, where he marries Catalina de Salazar y Palacios (1565–1626). He begins publishing, attempting to earn his living as a writer.

In 1594, Cervantes accepts a position as a tax collector for Granada. Three years later, when his collections fall short of the total the government is owed, Cervantes is imprisoned. The first part of his most famous work, *Don Quixote,* is published in 1605. He completes the second half in 1615. His other works include *Novelas Ejemplares* (Exemplary Novels; 1613); *Viage al Parnasa* (1614), a three-thousand line poem; and *Persiles y Sigismunda* (Travels of Persiles and Sigismunda; 1619).

1554   Publication of report on destruction of Indians

Dominican friar Bartolomé de las Casas (1474–1566) publishes a report exposing the abuse of the Indians by European colonists, especially Spanish explorers and conquerors.

c. 1562   Birth of dramatist Lope de Vega Carpio

Lope de Vega Carpio (1562–1635), considered the creator of Spanish national theater, is born. Immensely popular in his time, his plays mix tragic and comic elements and glorify honor, the monarchy, and religion. Lope deals with the theme of honor, previously considered inherent only to nobility, as also attributable to commoners.

1566   Reign of Felipe II

Felipe II is crowned king of Spain.

1571   Battle of Lepanto

Spanish, Venetian, and Papal forces achieve victory over the Turks in the Battle of Lepanto. This is significant because the Turks are considered the greatest naval power in the Mediterranean at the time.

c. 1575   El Greco arrives in Spain

Domenico Theotocopoulos (1541–1614), known as El Greco (The Greek), arrives in Spain. Born in Greece, he begins painting in the early 1570s. He settles near Toledo in 1577, where he receives a commission to decorate the church of Santo Domingo el Antiguo. Among the paintings El Greco creates for this church is *Assumption of the Virgin.*

1580   Felipe II inherits the Portuguese crown

Felipe II (also known as Philip II, 1527–1598) becomes king of Portugal.

1588   The Spanish Armada defeated

The English Navy defeats the Spanish Armada, signifying the decline of Spanish domination of the seas.

Baroque Era

1598   Felipe III (1598–1621) becomes king of Spain and Portugal

1600   Plague

More than 800,000 Spanish citizens die in a devastating plague.

1605   Miguel de Cervantes publishes *Don Quixote*

Miguel de Cervantes (1547–1616) publishes the first part of his most famous work, *Don Quixote*. It is widely believed that he writes part of the work while in prison in La Mancha. (See 1547.)

1609   Birth of Ferdinand, Prince of Spain

Ferdinand, Prince of Spain, Cardinal-Infante (1609–41) is born the fifth son of King Philip III and Margarita of Austria. He is designated a cardinal in the Roman Catholic Church by Pope Paul V in 1619 and designated lieutenant cardinal of Catalonia in 1632. During the Thirty Years' War, he joins forces with the army of Austria, sharing command with his cousin Ferdinand, King of Hungary and Bohemia (later Emperor Ferdinand III). They defeat the Swedish-Weimarian army of Marshal Gustav Horn and Bernhard of Saxe-Weimar at a battle in Nordlingen on September 6, 1634. In 1636, Ferdinand leads an invasion of France. He dies in November 1641.

1618   Painter Velázquez established studio

Court painter Diego de Silva y Velázquez (1599–1660) establishes his studio. Velázquez is a court painter with a realistic style and is one of the greatest European artists. Besides his most famous work, *Las Meninas* (Maids of Honor), he painted *The Surrender of Breda*, and *Las hilanderas* (The Spinweavers). He also paints portraits of people from all walks of life.

1621   Reign of Felipe IV

Felipe IV (1621–65) becomes king of Spain.

1626 Francisco de Quevedo writes about Spanish society

Spain's political challenges are analyzed by political thinker, Francisco de Quevedo, in his work, *Vida del Buscón.*

1639 Revolt in Catalonia

Felipe IV's policy of taxing all the territories of the Crown gives rise to secessionist movements in Andalucía, Naples, and Sicily. Catalonia revolts against the central government, switches allegiance to the French, and only in 1652 reverts to Spain.

1640–68 Independence of Portugal

Portugal becomes an independent state. A feeble monarchy together with the weakening caused by revolt and war allows Portugal to slip from Spanish control without even needing to strike a blow. In 1668, Spain officially recognizes Portugal's independence.

1640 Holland becomes independent

A truce with the Dutch expires in 1621, and Spain is at war with them in an effort to protect Spanish overseas territories from attacks by Dutch pirates. Spain and Austria ally to win early victories (see 1609). But France and Sweden, and later England, side with the Dutch, and in 1648 they win control. Spain is forced to recognize the independence of Holland.

1665 Reign of Carlos II

Carlos II (r. 1665–1700) is a sickly, weak, and childless king, easily manipulated by priests and favorites. His reign is marked by palace intrigues between the conflicting pro-Austrian and the pro-French factions. Spain's population decreases due to emigration to the colonies, plagues, and war.

Eighteenth Century

1700 War of Succession

The War of Succession is fought between two pretenders to the crown of Spain: king Felipe V, grandson of Luis XIV of France, and Archduke Carlos of Austria.

1712 Opening of the National Library

The National Library is officially opened in Madrid. It remains the only national library and steadily grows as a repository for scholarly materials, by 1999 holding three million volumes.

1713 Peace of Utrecht

Under the terms of the Peace of Utrecht, Felipe V (1700–46) becomes king of Spain.

1714 Creation of the Royal Spanish Academy of the Language

The *Academia de la Lengua* (Royal Spanish Academy of the Language) is the first of many academies which fosters development of arts and sciences.

1733 *Pacto de Familia*

An agreement, *Pacto de Familia,* is negotiated between the Bourbon kings of Spain and France to counter British expansionism.

1746 Fernando VI, king of Spain

Fernando VI's reign (r. 1746–59) is marked by a policy of neutrality in Europe and of internal prosperity.

1746 Birth of Goya

Francisco (José) de Goya y Luciente (1746–1828) is born near Saragossa. He works as a designer for the Royal Tapestry Works beginning in 1775. In the first years of his career, Goya paints scenes full of color and gaiety reflecting characters and contemporary society. From 1786 when he is appointed Court painter, he executes numerous portraits as well as the famous *Maja Desnuda* (The Naked Maja) and *Maja Vestida* (The Clothed Maja), painted during the years 1797–1800. About 1799, ill and deaf, Goya publishes his *Los Caprichos,* a collection of etchings in which he satirized modern institutions and society. In the same vein is another collection of etchings inspired by the War of Independence, *Los desastres de la guerra* (The Disasters of War). Goya was not only a revolutionary in the arts of his times, but he also influences later artistic movements like romanticism, impressionism and postimpressionism.

1759 Carlos III, king of Spain

King Carlos III (r. 1759–78) is a representative of enlightened despotism, surrounding himself with excellent ministers. He and his advisors, known as the *ilustrados,* are in favor of *Despotismo Ilustrado* (Enlightened Despotism) based on the principle of improving the economic and educational situation of the people although denying them any participation in government. This process is summed up in the formula *Todo para el pueblo pero sin el pueblo* (Everything for the people, but without the people). Carlos's reign is an innovative period of reforms aimed at the improvement of cultural and economic levels of the Spanish people. In the second half of the century, the economy and demographics are both improving: by the end of the century the population has increased from eight to twelve million.

1767 Jesuits expelled

Carlos III expels the Jesuits from his kingdom and opens the universities, formerly under Jesuit control, to the middle classes, thus providing the country with many enlightened *technocrats.*

1778   Carlos IV ascends to the throne

Carlos IV (r. 1778–1808) succeeds his father on the throne, but lacks both the talent and the inclination to lead. He marries the energetic Maria Luisa of Parma, and he soon leaves the management of affairs to his wife and to the favorite, Manuel Godoy.

1785   Prado opens

The Prado, Spain's most famous museum, is built in Madrid. Included in its collections are masterpieces by Spanish artists such as Velazquez, Goya, and El Greco.

Nineteenth Century

1805   Battle of Trafalgar

The Spanish navy allied to the French is defeated by the English in the battle of Cape Trafalgar, near Cádiz, Spain. As a result of this action, Spain loses most of its navy and, consequently, is unable to protect its colonies in America.

1808: May 2   Revolt against French invasion

The people of Madrid revolt against the French and the War of Independence begins.

1812: March 19   The Constitution is proclaimed

1810–24   Independence for colonies

The majority of the Spanish colonies in the Americas gain independence.

1814   Return of King Fernando VII

Fernando VII is released by Napolean when the War of Independence ends with Spanish victory. Fernando is acclaimed as *El Deseado* (The Desired One), and he authorizes a military coup, abolishes the Constitution of 1812, reverts to an absolute rule, reinstates the Inquisition, imprisons liberal members of the Cádiz Cortes, and suppresses several liberal military plots.

1820–23   The Liberal Three Years

Colonel Rafael del Riego, commander of a regiment scheduled to fight in the American colonies, heads a successful military coup and reasserts the validity of the 1812 Constitution. The resulting era is known as the Liberal Three Years.

1823   The *Santa Alianza* (The Holy Alliance) reinstates monarchy

The Holy Alliance, a conservative league formed by the European monarchies, sends a powerful army named Saint Louis's One Hundred Thousand Sons that reinstate the Spanish King to his full powers and launches "the ominous decade," a ten-year period of repression.

1828   Death of painter Francisco de Goya

1833   *Oda a la Patria* is published

Buenaventura Carlos Aribau (1798–1862) publishes *Oda a la Patria* (Ode to the Motherland), an influential early example of Catalonian poetry. Among poetic works it inspires are the two epics *La Atlántida* and *Lo Canigó* by Mosen Jacinto Verdaguer (1845–1902), the greatest Catalonian poet of the nineteenth century. Of special importance is the work of the philologist Pompeu Fabra, whose grammar and other works set the rules for modern Catalonian.

1833: September 29   Death of Fernando VII and beginning of the Carlist War

Fernando VII dies, leaving his wife Maria Cristina as regent, and his young daughter, Isabel as queen. Prince Carlos, the late king's reactionary brother, starts a war (Carlist War) to claim the crown for himself. His *Carlista* party is made up of clergy, rural landowners, and peasants. The Basque provinces of Navarre and Catalonia supported Don Carlos in exchange for the reinstatement of their suppressed *Fueros* (local fiscal and legal privileges). The army and the majority of the country support the young queen, Isabel. María Cristina, the queen regent, initiates numerous reforms, among them the granting of amnesty of the exiled liberals and the definitive abolishment of the Inquisition. The first Carlist War lasts seven years and ended with the defeat of the Carlist faction and a peace agreement.

1834   Angel Saavedra, Duke of Rivas , publishes *El moro expósito*

This narrative poem marks the beginning of Romanticism in Spain.

1843–68   Reign of Isabel II

The reign of Isabel II (also known as Isabella II, 1830–1904; r. 1843–68) is characterized by the struggle for power between liberal conservatives, *moderados,* and *progresistas,* who have the backing of the masses.

c. 1845–1900   Galician poetry movement

Nineteenth-century Galicia produces a group of distinguished poets, beginning with the publication of the poem *Alborada* by Nicomedes Pastor Díaz. Three other notable poets are Rosalía de Castro (1837–1885) and whose *Cantares gallegos* (1873), *Folhas novas* (1880) reveal the author's melancholy, her love of her native land, and her condemnation of poverty, emigration, and social injustice; Eduardo Pondal (1835–1917); and Manuel Curros Enriquez (1851–1906).

1848 The Barcelona-Mataró railway is built

The Barcelona-Mataró railway, the first to be built in Spain, is begun.

1849 First realist novel published

Cecilia Bohel de Faber (1796–1877), writing under the pen name Fernán Caballero, publishes *La gaviota*. Set in Andalusia, the work is considered the first Spanish realist novel.

1852 Writers Pardo Bazán and Leopoldo Alas are born

Feminist and naturalist writer Emilia Pardo Bazán (1852–1921) is born. A countess, she writes *Los pazos de Ulloa*, which describes the decadence of an aristocratic family living in a Galician village. In its sequel *La madre Naturaleza*, nature ultimately destroys the lives of the protagonists. Her writings sustain the idea that the individual is endowed by God with free will and thus is able to fight successfully against the influence exerted by genetics and by social environment.

Another naturalist writer, Leopoldo Alas ("Clarín", 1852–1901), is also born this year. His novel, *La regenta*, depicts the intrigues and miseries of daily life in Vetusta, a stagnant provincial capital.

1860 Birth of composer Isaac Albéniz

Isaac Albéniz (1860–1909), composer and pianist, is born. He is a child prodigy, studying with Franz Liszt (1811–86; see Hungary) and travelling extensively. He begins composing music, employing traditional rhythms and harmonies of Spain. His collected works for piano, known as *Iberia*, reflect the musical traditions of the various regions of Spain.

1868: September Monarchy exiled

Queen Isabel is exiled with her family to France. Although she is a popular queen, she is not able to control the rival factions in Spain, and she is driven out of power. General Juan Prim, the leader of the Progressive Party, heads a coalition that drives Isabel II into exile in a revolution called *la Gloriosa* (the Glorious one).

1869 Period of popularity of *zarzuela* begins

The *zarzuela,* a play usually in one or two acts accompanied by music, is created to provide entertainment for the monarchy and their guests. These short plays are extremely popular until around 1910. An essentially Spanish musical genre, zarzuela deals with problems of everyday life of the lower classes—love, money, and politics—seen from a comical and ironic perspective. Outstanding composers and librettists are Francisco Asenjo Barbieri (1823–94), author of *Pan y toros;* Tomás Bretón (1849–1920), author of *La verbena de la Paloma;* and Ruperto Chapí (1851–1919) author of *La revoltosa.*

1869 Constitution of 1869 grants universal suffrage

1869 Birth of writer Ramón María del Valle-Inclán

Ramón María del Valle-Inclán is born. Valle-Inclán settles in Madrid, where he writes novels, plays, and poetry. Among his works are *La guerra carlista* (The Carlist War; 1908), *Aguila de blasón* (The Eagle of Heraldry; 1907), and *Romance de Lobos* (published in English as *Wolves! Wolves!;* 1957). Valle-Inclán triggers a renewal of Spanish theater with *esperpento,* a new dramatic genre with high critical and satirical content and puppet-like characters.

1870–73 Amadeo I is chosen king of Spain

Italian Amadeo I is chosen to become king of Spain. He is opposed by both Conservative and Liberal parties, and abdicates after a brief reign. He returns to Italy.

1873 The First Spanish Republic

The first republic is proclaimed after Amadeo I abdicates. From its beginnings it is troubled by dissension among those favoring a federal republic, who revolt in Cartagena and elsewhere. A military coup led by General Manuel Pavía ends the republic after only two years.

1874 Alfonso XII becomes king

Brigadier Arsenio Martínez Campos declares Alfonso XII (1857–85), son of Isabel II, the new king. He is exiled with his parents in 1868, and is educated in Austria and England. Supported by Martinez Campos and Canovás del Castillo, he consolidates the monarchy, winning greater popularity for it than it has enjoyed under his mother or grandfather, Fernando VII. He dies in a cholera epidemic, and his widow, Maria Christina (1858–1929), is regent during the minority of his son, Alfonso XIII (see 1886).

1876 A new constitution

A new constitution is created, largely due to the influence of Antonio Canovás del Castillo (1828–97). He serves four terms as premier: 1875–81, 1884–85, 1890–92, and 1895–97). His leadership during his first term as premier brings an era of political stability. Canovas believes in *Doctrinarismo* (sovereignty shared jointly by the King and the Cortes) and he establishes a political system based upon the rotation of power between the Liberal and the Conservative parties (*turno pacífico de partidos*). This artificial balance is maintained by means of trumped-up elections (*pucherazo*) backed by the government itself, which corrupts and demoralizes the electoral system. Nevertheless, under his leadership, Spain enjoys a period of economic prosperity lasting until the agricultural crisis of the 1880s.

The Spanish government's lack of flexibility and its slowness to grant its colonies political and fiscal reforms results in a new flare up in Cuba, inspired this time by José Martí in 1895, and in the Philippines under Andrés Bonifacio, both backed by the United States. The sinking of the American battleship *Maine* gives rise to the armed intervention of the United States, resulting in the loss of the Spanish navy, The Philippines, and Cuba. With the Treaty of Paris (1898) Spain loses its last overseas colonies: Cuba and Puerto Rico in the Caribbean, and Guam and the Philippines in the Pacific.

As a result widespread discontentment spreads to different sectors, among them Catalonian and Basque industrialists, the working classes and the intellectuals later known as the Generation of 1898. The loss of the mainland colonies in America and of the trade they provide in the first half of the century has proven detrimental to Catalonian and Basque business whose owners accuse the central government of inefficiency. They drift towards Carlism, and later, towards nationalism. By the second half of the century the proletariat or urban working class achieves significant numbers. To increase their earnings, factory owners imposed long hours of work for little pay, thus alienating the workers, who have neither social security nor retirement pay. In time, they acquire considerable political power, organize themselves and join socialist and anarchist trade unions.

1876   The *Institución Libre de Enseñanza* is founded

The Institución Libre de Enseñanza, an alternative to the university, is founded by the Krausists, a group of philosophers who follows German thinker Karl Christian Friedrich Krause (1781–1832). For Krause, God is an essence that contains the whole universe, and man is an integral part of the perfection of universal life. The *Institución Libre de Enseñanza* is designed to promote cultural projects and was responsible for the intellectual training of many intellectuals in Spain.

1876   Birth of composer Manuel de Falla

Composer Manuel de Falla (1876–1946) is born in Cádiz. He plays piano as a child and wins both piano competitions and contests for original compositions. His most famous work is *El sombrero de tres picos* (Three-Cornered Hat; 1919). His works include the opera *El Retablo de Maese Pedro* (Master Peter's Puppet Show; 1922), *El amor brujo,* and *Noches en los jardines de España* (Nights in the Gardens of Spain, 1909–16). When the Spanish Civil War erupted (see 1936: July 18–1939: April 1), he flees Spain and settles in Argentina.

1876   Birth of cellist Pablo Casals

Cellist Pablo Casals (1876–1973) is born in Vendrell, a town in the Tarragona region. He studies at the Royal Conservatory in Madrid. He is a professor of cello in Barcelona and an orchestra cellist in Paris before beginning a career as a solo artist. (See also 1919.)

1882   Building of Gaudí's *Sagrada Familia* church begins

Construction of the huge *Sagrada Familia* church designed by Antonio Gaudí (1852–1920) begins in Barcelona. Gaudí mixes different styles in order to create his own style.

1884   Landmark novel, *La Regenta,* published

Leopoldo Alas, known as Clarín, publishes his novel *La Regenta.*

1886   Alfonso XIII becomes king

Although he is not born until after his father's death, the infant Alfonso XIII (1886–1941) becomes king. He rules Spain from 1886–1931, but his mother, Maria Christina (1858–1929) is regent on his behalf from 1886 until 1902 (when Alfonso XIII reaches age 16). Alfonso XIII soon has to face unrest as the people increasingly demand political and social rights.

1887   Benito Pérez Galdós publishes *Fortunata y Jacinta*

Novelist Benito Pérez Galdós (1843–1920) publishes his best-known work, *Fortunata y Jacinta* (Fortunata and Jacinta). It portrays Madrid society during the last third of the nineteenth century. In addition to novels, his writings encompass memoirs, plays, travel diaries, and criticism.

1888   Birth of the *Unión General de Trabajadores*

The *Unión General de Trabajadores* (UGT), the Socialist party's trade union, is formed.

1892   Birth of Francisco Franco, soldier and dictator

Francisco Franco (1892–1975), future leader of Spain, is born in Galicia. Educated at military academies in both Spain and France, becomes chief of staff of the army in Spain in 1935 and governor of the Canary Islands the following year. He becomes chief of the army and leads the insurgents in the Spanish Civil War (see 1936: July 18–1939: April 1). The Spanish Civil War ends when the insurgents capture Madrid, and Franco becomes dictator, ruling harshly until his death in 1975. During his rule, the country's economy shifts from agrarian to industrial, the standard of living increases, and for the first time in history there is a sizable middle class, which is hard-working, consumer-oriented, and politically moderate.

1893   Birth of artist Joan Miró

Joan Miró (1893–1983), painter, sculptor and ceramist, is born in Montroig. He studies art in Paris and Barcelona. He begins exhibiting with the Surrealists around 1925. In his work, he uses bright colors and curvy lines to suggest an ingenuous primitivism and dreamlike settings. His works include *Catalan Landscape* (1923–24) and *Maternity* (1924).

1894 Birth of Andrés Segovia

Guitarist Andés Segovia (1894–1987) is born in Linares. He develops a unique technique and earns an international reputation as a guitar virtuoso.

1895 Revolt in Cuba and the Philippines

The Spanish government's lack of flexibility and its slowness to grant its colonies political and fiscal reforms result in a flare-up in Cuba, inspired by Cuban writer and activist José Martí (1853–95). In the Philippines, the activists are led by Andrés Bonifacio. Both are backed by the United States.

1897 Angel Ganivet writes *Idearium español*

Angel Ganivet (1862–1898) publishes his book *Idearium español*. It is an attempt to define the character of the Spanish people.

1898 Spanish-American War

The sinking of the American battleship *Maine* near Cuba gives rise to the armed intervention of the United States in the uprisings in Cuba. As a result, the Spanish navy suffers defeat, and Spain loses control of the Philippines and Cuba.

1898 Treaty of Paris

With the Treaty of Paris, Spain loses its last overseas colonies. These include Cuba and Puerto Rico in the Caribbean, and Guam and the Philippines in the Pacific.

1898 Generation of 98 writers

The Generation of 98 is a group of young writers and thinkers who are young adults at the time of the Spanish-American War of 1898. While these writers react in different ways to the defeat, all of them share a concern for the decadence of Spain and search for the country's spiritual essence.

Twentieth Century

1902 Birth of composer Joaquín Rodrigo

Joaquín Rodrigo (1902–99) is born in a town near Valencia. At the age of three, he loses his eyesight. He learns to play the piano and studies in Paris before returning to Spain to live. His works include the *Concierto de Aranjuez* (Aranjuez Concerto; 1940), featuring the guitar, and *Cinco Piezas Infantiles* (Five Pieces from Childhood; 1925), for which he wins a Spanish competition. He becomes one of the best-known Spanish composers.

1906 Biologist Dr. Santiago Ramón y Cajal wins Nobel Prize

Biologist Dr. Santiago Ramón y Cajal (1852–1934) is awarded the Nobel Prize in medicine. He is born in Petilla de Aragón and is educated at Saragossa University. He earns his degree in 1873 and joins the Army Medical Service. He enters academic life as a professor of anatomy at Valencia (1883–86), and then as professor of histology, first at Barcelona (1886–92) and then at Madrid (1892–1922). His research focuses on the nervous system. He shares the Nobel Prize with Italian Camillo Golgi.

1906 Alfonso XIII marries

Alfonso XIII marries Princess Victoria Eugenie of Battenberg, granddaughter of Queen Victoria of Great Britain. An attempt is made to kill the couple on their wedding day, the first of several assassination attempts. Although Alfonso enjoys some personal popularity, the monarchy is threatened by social unrest in the newly industrialized areas, by Catalan agitation for autonomy, by dissatisfaction with the constant fighting in Morocco, and by the rise of socialism and anarchism.

1909 Francisco Ferrer Guardia is executed

The government was widely attacked for the execution of the radical publicist Francisco Ferrer Guardia following an uprising in Barcelona.

1909 Protests in Barcelona

The *Semana trágica* (Tragic Week) takes place in Barcelona. Begun with an uprising against the sending of troops to fight in a war in Morocco, it is followed by a general strike.

1913–18 Spain stays out of World War I

Alfonso XIII does not involve Spain in World War I.

1917 Poet Juan Ramón Jiménez publishes *Platero y yo*

Poet Juan Ramón Jiménez (1881–1958) publishes his poetic prose story of a poet and his donkey, *Platero y yo* (published in English as *Platero and I*). He later wins the Nobel Prize (see 1956).

Juan Ramón Jiménez, aiming to write *poesía pura* (a concise poetry), had a great influence on a group of poets known as the *Generacion del 27,* formed by Vicente Aleixandre (Nobel Prize for Literature, see 1977), Miguel Hernández, Rafael Alberti, Dámaso Alonso, Jorge Guillén, Luis Cernuda, Pedro Salinas and Federico García Lorca (1898–1936). Lorca is a neopopularist whose work turns to his Andalusian roots (*Romancero gitano*, 1928).

1919 Barcelona Orchestra is founded

Pablo Casals (see 1876) founds the Barcelona Orchestra.

1921 Founding of the Spanish Communist Party (PCE)

1921 Disaster of Annual in Morocco

Its colonial empire already severely weakened by the outcome of the Spanish-American War, the Spanish expeditionary force suffers a bloody defeat by Moroccan forces pushing for separation from Spain.

1921   Novecentistats focus on European trends

The *Novecentistas* are a group of prose writers who want to acquaint Spaniards with the scientific and cultural trends going on in Europe. Paramount among them is José Ortega y Gasset (1883–1955), a thinker of European repute. Like his predecessors of the Generation of 98 (see 1898), he writes extensively about the decadence of Spain.

1922   Playwright Jacinto Benavente is awarded the Nobel prize in literature

Jacinto Benavente (1866–1945) receives the Nobel Prize for literature. Benavente is born in 1855 in Madrid. His is a best-selling playwright for several decades with such works as *La malquerida* and *El nido ajeno* (The Other Nest; 1893). In addition to his masterpiece *Los intereses creados* (Human Concerns; 1907), he writes a number of plays for children.

1923–29   Primo de Rivera's dictatorship

In agreement with the king, General Miguel Primo de Rivera (1870–1930) acquires dictatorial powers. He restores law and order, and devises a plan of public works that ended unemployment.

Primo de Rivera serves in Cuba and the Philippines during the Spanish–American War (1898) and Morocco (1909–13). He holds the position of Military Governor of Cadíz (1915–19), Valencia (1919–22), and Barcelona (1922–23). He leads as a dictator until 1930, when he loses the support of the army and the king and resigns.

1923   Military government takes over

After keeping Spain out of World War I, Alfonso is dissatisfied with the functioning of parliamentary government. He decides to lend his support to General Miguel Primo de Rivera (1870–1930) in establishing a military dictatorship. Primo de Rivera is a veteran of the Spanish–American War (1898), and of the military action in Morocco (1909–13).

1924   La Cierva builds the first helicopter

Inspired by the plans of Leonardo da Vinci in the sixteenth century, Juan de la Cierva considers the helicopter as one method of flight. He works as an engineer on the project envisioned by French inventor Louis Breguet. La Cierva finishes building his helicopter at about the same time as Igor Sikorsky in the United States.

1925   Primo de Rivera lands at Morocco

General Primo de Rivera lands with an expeditionary force at Alhucemas, Morocco, leading to a durable peace in the Spanish Protectorate in Morocco.

1927   Moroccan War ends

General Miguel Primo de Rivera successfully ends the Moroccan War.

1928   Salvador Dalí joins Surrealists

Salvador Dalí (1904–89) joins the Surrealists, becoming one of its leading figures. He studies abnormal psychology and dream symbolism and paints seemingly incongruent objects into landscapes with detailed precision and realism. His 1931 painting, *The Persistence of Memory,* popularly known as *Limp Watches,* hangs in the Museum of Modern Art in New York. At the end of his career, he turns his attention to the depiction of religious subjects.

1928   Luis Buñuel finishes *Un chien andalou,* a surrealist film

Film director Luis Buñuel (1900–83) finishes the surrealist film, *Un chien andalou* (An Andulusian Dog), which he codirects with artist Salvador Dalí (1904–89). Buñuel is the most dynamic and innovative Spanish director, and he enjoys an international reputation for his body of work, which includes *Viridiana* (1960) and *Tristana* (1970).

1928   *Opus Dei* is founded

Monsignor José María Escrivá founds *Opus Dei,* a religious institute for Roman Catholic laymen and laywomen. *Opus Dei* members strive to resist the trends toward secularism in Spanish society and aim for Christian perfection.

1930   Primo de Rivera loses power

Primo de Rivera resigns as dictator after he loses the support of both his army and the ruling class, including Alfonso XIII. Discontent runs high in Spanish society.

1931: April   Municipal elections held

The municipal elections show an overwhelming lack of support for Alfonso and corresponding support for a republican majority. When it becomes obvious from the election results that the country is divided, Alfonso XII responds by "suspending the exercise of royal power" and leaves Spain for exile in Italy to avoid a civil war. A few weeks before his death in Rome he renounces his claim to the throne in favor of his third son, Don Juan de Borbon.

1931: April 14   Birth of the Second Republic

The Second Republic is established. The new Constitution acknowledges popular sovereignty, separation of powers, religious freedom, and the Spanish regions' right to autonomy. The triumph of the Second Republic means the triumph of the leftist bourgeoisie, a small class feeble enough to be attacked by an aggressive and violent working class, by the anarchist, socialist and communist parties, on one hand, and by the animosity of the reactionary alliance of the oligarchy, the Church and the Army, on the other. Lawlessness is rampant and there are strikes, terrorism and burning of churches.

1933: October   Founding of the *Falange* (Fascist Party)

José Antonio Primo de Rivera (1903–36), a son of General Primo de Rivera (see 1923–29), founds the *Falange* (Fascist) political party. He is executed in 1935 by the Republicans.

1933   Birth of soprano Montserrat Caballé

Soprano Montserrat Caballé (b. 1933) is born in Barcelona. She performs at all the major opera houses of the world during her long performing career and contributes to many opera recordings.

1934: October   Miners seize control of mines

Asturian miners, most of whom belongs to left-wing trade unions, take violent control of the mines (*Revolución de Octubre*) and the republic is obliged to put the revolution down forcibly.

1935   Birth of Teresa Berganza

Mezzo-soprano Teresa Berganza (b. 1935) is born in Madrid. She makes her singing debut in 1955 and earns her reputation in roles with major opera companies around the world. In 1994, she is elected to the Spanish Royal Academy of Arts, the first woman to be so honored.

1936   Playwright García Lorca assassinated

Playwright and poet Fernando García Lorca is assassinated in Granada by agents of the government. His writings are considered dangerously influential.

1936: July 13   Right-wing politician murdered

The right-wing politician José Calvo Sotelo is murdered, setting in motion the events that lead to the Spanish Civil War.

1936: July 18–1939: April 1   The Spanish Civil War

General Francisco Franco's uprising on July 18, the last one in a long series of military *pronunciamientos* (uprisings) starts the Spanish Civil War. It lasts until 1939. In September Franco is appointed head of the Spanish state and commander-in-chief of the army. Franco's Nationalist forces are helped by Germany and Italy, both opposed to a Communist victory in Spain. At the same time, Soviet Russia backs the Republic, while France and Britain and the United States remain neutral, although these and other countries send volunteer units to fight for one side or the other. As the war progresses, the Nationalists become more disciplined and united while the Republic suffers from the dissidence among the different political parties and from the bloody fight between communists and anarchists. The war ends with the defeat of the Republic on April 1, 1939.

Franco Era, 1939–75

1939–45   Spain maintains neutrality in World War II

Spain maintains a position of official neutrality during World War II, although Franco (see 1892) does respond to pressure by Germany's Adolph Hitler and provides troops to assist Nazi Germany in battles against the Russians.

1941   Birth of tenor Plácido Domingo

Plácido Domingo is born in Madrid. He becomes one of Spain's most celebrated operatic tenors. His family moves to México when Plácido is still a child; he studies at the National Conservatory of Music in Mexico City. His operatic debut is in 1959; his first major role as a tenor is Alfredo in *La Traviata* (in 1960). He performs at all the major opera houses of the world. In 1990, he sings with tenors Luciano Pavarotti and José Carreras (see 1946) in an internationally acclaimed concert in Rome, billed as the "Three Tenors." His autobiography, *My First Forty Years,* is published in 1983.

1941: May   Agreement between Franco and the Vatican

Under the terms of an agreement between Spanish dictator Francisco Franco (see 1892) and the Vatican, Franco designates individuals to become bishops in the Spanish Roman Catholic Church, and the Vatican has power to ratify these appointments.

1945   Soprano Victoria de los Angeles makes debut

Victoria de los Angeles (b. 1923) makes her opera debut in her home city of Barcelona. During her career, she performs at all the major opera houses of the world—Paris Opera; La Scala in Milan, Italy; Covent Garden in London, England; and at the Metropolitan Opera in New York. She retires in 1969.

1945   Enactment of the *Fuero de los Españoles*

The declaration of the *Fuero de los Españoles* officially pronounces the rights and duties pertaining to the Spanish citizen. This is enacted at about the same time as the *Ley de Referendum Nacional* (the National Referendum Law), which states the Spanish people's right to elect by vote their own representatives.

1946   José Carreras is born

Tenor José María Carreras is born in Barcelona. In 1970, he makes his singing debut, and goes on to appear at all the major opera houses. In 1990, he performs with Plácido Domingo (see 1941) and Luciano Pavarotti in the concert, "Three Tenors," in Rome. In 1992, Carreras performs at the summer Olympic Games hosted by the city of Barcelona.

Over 4,000 children gather to receive communion in front of the Holy Family shrine during the thirty-fifth Eucharistic Congress held in Barcelona in 1953. (EPD Photos/CSU Archives)

1950s Eroding of political and religious ideology

Along with the progressive growth in Spain's economy is a progressive eroding of political ideologies in the 1950s. The Roman Catholic Church also experiences a backlash. During the years that dictator Francisco Franco is in power, the Roman Catholic Church asserts excessive control over society.

1953 Eucharistic Congress in Barcelona

The thirty-fifth International Eucharistic Congress is held in Barcelona, with Roman Catholics from over thirty nations participating.

1953 Agreement with the United States and Concordat with the Holy See

During the Cold War between Russia and the Western powers, U.S. president Dwight Eisenhower signs a military pact (which still stands) between Spain and the United States. In

1959, he visits Spain, thus putting an end to international sanctions against Spain.

1956 Spain is admitted into the United Nations

1956 Juan Ramón Jiménez is awarded the Nobel Prize

The Andalusian poet Juan Ramón Jiménez (1881–1958), already universally known by his poetic work and his prose book *Platero y yo,* is awarded the Nobel Prize in Literature. He locates the story in the town of his birth, Moguer. He leaves Spain at the outset of the Civil War (see 1936: July 18– 1939: April 1), and settles in Florida in the United States.

1957 Birth of golfer Seve Ballesteros

Severiano Ballesteros (b. 1957), known as Seve, is born in Santander. He begins his association with golf as a caddy. He is the youngest person in the twentieth century to win the British Open golf championship, in 1979. (He wins it again in 1984 and 1988.) He is the youngest person—and the second European—ever to win the U.S. Masters tournament in 1980. He is captain of the Ryder Cup team when the tournament is held in Spain (see 1997).

1959 Period of economic development

Intense economic development is stimulated by the Economic Stabilization Plan. The plan liberalizes the economy, thus preparing for the integration of Spain into the international markets.

1963 *Planes de Desarrollo* (Economic Development Plans)

Planes de Desarrollo (Economic Development Plans) mark the beginning of a period of increased economic prosperity. The so-called *economic miracle* that follows implementation of these plans is mainly due to a combination of foreign investment, tourism, and the money sent home by Spaniards working in Western Europe.

1964 Birth of Spanish cyclist Miguel Indurain

Spanish cyclist Miguel Indurain is born in Villava. In the early 1990s, he becomes the fourth competitive cyclist to win five Tour de France tournaments (1991–95). He leads the team, Banesto, and becomes wealthy from cycle racing prize money and endorsements. He is a national hero in Spain, but he retires from cycle racing in 1997.

1968 ETA begins violent attacks

The terrorist organization Basque Homeland and Freedom (known as ETA) begins its campaign of violence. It carries out terrorist acts, usually resulting in deaths, to call attention to the cause for independence for the Basque region. Under the new system of regional autonomy, the four Basque provinces now enjoy a large measure of autonomy.

1973: December   Admiral Luis Carrero Blanco assassinated

Admiral Luis Carrero Blanco, designated to succeed Franco, is assassinated by ETA.

Modern Spain, 1975–2000

1975   Death of Franco; Juan Carlos de Bourbon becomes King of Spain

Franco's (see 1892) dies. Juan Carlos de Bourbon ascends to the throne.

1976   *Interview* begins publication

Interview, a weekly magazine, begins publication. At its peak it sells a million copies per week, offering a mixture of sensationalism, sex, and political scandals.

1976: May   *El País* begins publication

El País, a daily publication with a center-left political orientation, begins publication. It exerts considerable influence on Spanish culture.

1977   Nobel Prize awarded to poet Vicente Aleixandre

The Nobel prize is awarded to poet Vicente Aleixandre (1898–1984). Aleixandre is born in Seville, and contracts a form of tuberculosis in childhood. He began publishing in the late 1920s, with *Ambito* (Ambit). Other works include *La Destrucción o el amor* (Destruction or Love) and *Pasión de la Tierra* (Passion of the World), both published in 1935. In 1937, his *Mis Poemas Mejores,* a collection of poems, establishes his reputation. He also publishes *Presencias* (Presences; 1965) and *Antologia Total* (Complete Works; 1976).

1977: December 1   Censorship of movies lifted

A new law lifts censorship of films. This allows the movie industry to venture into the areas of political satire and social criticism. This is seen in such films as Mario Camus's *Los santos inocentes* (1984); and Carlos Saura's *Ay Carmela!* (1990). The best known Spanish cinema director is Pedro Almodóvar whose movies *(Mujeres al borde de un ataque de nervios* and *Atame)* are a mixture of parody, social satire, sex, violence and humor.

1978   The Constitution creates seventeen *Autonomías*

The Constitution of 1978 establishes *Autonomías* (autonomous regions) of Andalucía, Aragón, The Canary Islands, Cantabria, Castile-La Mancha, Castile-León, Catalonia, The Balearic Islands, the Valencian Community, Extremadura, Galicia, Rioja, the Autonomous Community of Madrid, the Foral Community of Navarre, the Basque Country, the Principality of Asturias, and the Region of Murcia. The *Autonomías*

are of different sizes, have different resources and number of inhabitants. Catalonia, Galicia, and the Basque country have two official languages—their own language and Castilian, the official language of greater Spain. In some of these autonomies there are pro-independence groups that have resorted to terrorism, notably *ETA* in the Basque country and *Terra Lliure* in Catalonia.

1980s   *La movida madrileña* develops

La movida madrileña (Madrid action) is a term used to describe the eccentricities and extravagances of groups of young adults in Madrid who combine fun, excitement, and artistic creativity with business-minded activities. Musical groups and decadent and bizarre fashions are part of the frenzy that overtakes life in Madrid. It reflects the hedonistic and narcissistic values of a society where money, success, sex, and consumerism have become every day values.

1980   Anti-terrorism law passed

The government passes the *Ley contra el terrorismo* (Antiterrorist Law) to quell the civil uprisings in Catalonia and the Basque region.

1981: February 23   Colonel Tejero stages failed coup

Coronel Tejero, representing several generals, stages a coup d'etat, entering Congress and holding congressmen at gunpoint while they are in session. The firm attitude of the King condemning the coup results in Tejero's giving up the coup attempt and saving the country from a new military dictatorship.

1981: September 10   Picasso's *Guernica* returned to Spain

The New York Museum of Modern Art returns to Spain the *Guernica* painting by Spanish painter Pablo Picasso (1881–1973). Picasso is a major figure in twentieth-century art. He spends much of his career painting in France. Picasso serves as director of the Prado in Madrid from 1936–39. *Guernica* (1937) expresses Picasso's shock at the bombing of the town of Guernica during the Spanish Civil War.

1982: October 25   PSOE election victory

The Spanish Socialist Workers Party (PSOE) wins the General Elections by a landslide and controls the government under the presidency of Felipe González from 1982 to 1996. González focuses on improving the economy. Most of his reforms are positive although not always well received by the trade unions, which stage frequent strikes. Among these reforms are the closing of any unprofitable state-sponsored factories and enterprises and the ending of *pluriempleo* (holding two or more jobs at the same time).

1982   International Fair of Contemporary Art is inaugurated

A yearly art fair, the International Fair of Contemporary Art (*Arco*), is established in Madrid.

1983: December 25   Death of surrealist painter Joan Miró

Joan Miró (1893–1983), painter, sculptor and ceramist, dies.

1986: January 1   Spain joins the European Economic Community

By joining the European Economic Community, known as the Common Market, Spain ensures positive economic development.

1986: March   Spain votes to stay in NATO

In a referendum, the country votes to remain a member of the North Atlantic Treaty Organization (NATO).

1989   Camilo José Cela wins Nobel Prize for literature

Novelist Camilo José Cela (b. 1916) wins the Nobel Prize for literature. He attended Madrid University. His first novel, published in 1942, is entitled *La familia de Pascual Duarte* (The Family of Pascual Duarte). It creates a new style of fiction, known as *tremendismo* (*tremendo* in the sense of terrible, harsh), includes vivid scenes of violence, and is banned in Spain. His best-known work, *La Colmena* (The Hive), is published in 1951. It depicts life in Madrid following the Spanish Civil War.

1992   Barcelona hosts the Olympic Games

The city of Barcelona hosts the summer Olympic Games.

1993   Emergency repairs made to the Prado

After centuries of operation, the Prado (see 1785), Spain's premier art museum, is in need of repair. The roof and walls are leaking, and priceless are treasures are threatened with water damage. Emergency repairs are made to avoid losses, but a permanent solution has not been developed yet.

1995: March 18   Princess Elena weds

Infanta (Princess) Elena María Isabel Dominica de Silos de Borbón y Grecia, thirty-one, daughter of King Juan Carlos and Queen Sofia, marries Jaime de Marichalar y Saenz de Tejada, a thirty-one-year-old banker. The couple takes the titles of Duke and Duchess of Lugo.

1996: March   *Partido Popular* defeats the Socialists

In the elections, *Partido Popular* the right-center party, defeats the Socialists. Their leader, José María Aznar, becomes head of the government. From 1990 on, unemployment, disenchantment with the economic situation and increasing cases of corruption scandals at all levels of government contributes to the decrease of credibility and popularity of the Socialist party, all of which lead to their defeat.

1996: June   ETA calls a halt to bombing

The terrorist organization, ETA, calls a one-week halt to the bombing attacks they have been carrying out since 1968. When bombing resumes, bombs are exploded in Granada, Jaén, and Pamplona.

1996: July 10   Bomb explodes at the Alhambra

The Alhambra, a Moorish palace in Granada that dates to the thirteenth century, is a popular tourist attraction. ETA explodes a bomb there, but damage is slight and there are no injuries.

1996: July 11   Bomb explodes at hotel

Basque separatists set off a bomb at a hotel in the Santa Catalina castle outside Jaén. There are no injuries.

1997   Ryder Cup golf tournament held in Spain

The Ryder Cup golf tournament is held in Spain, and Spanish champion golfer Seve Ballesteros (see 1957) is captain of the European team.

1997: October   Guggenheim Museum opens in Bilbao

The Guggenheim Museum, designed by Canadian-born architect Frank Gehry, opens in Bilbao. The $210-million museum is constructed of flowing metallic curves and is located on an abandoned industrial site.

1997: October   Princess Cristina weds

Princess Cristina, thirty-two-year-old daughter of King Juan Carlos and Queen Sofía, marries twenty-nine-year-old Olympic athlete Iñaki Urdangarín. Princess Cristina is the first member of the Spanish royal family to attend university.

1998: October 25   Regional elections held

Elections are held in the Basque region. More Basque citizens vote than have ever cast ballots in any previous election. Many separatists cast votes in support of ETA's political wing. It is renamed Euskal Herritarok (EH) for these elections. EH wins fourteen of the seventy-five seats in the regional parliament, up from eleven in 1994.

1999: July 7   Death of Joaquín Rodrigo

Composer Joaquín Rodrigo (1902–99) dies at the age of ninety-seven (see 1902).

Bibliography

Cantarino, Vicente. *Civilización y Cultura de España,* 3rd. ed., Englewood Cliffs, NJ: Prentice Hall, l995.

Carr, Raymond. *The Republic and the Civil War.*. London, Eng.: Macmillan, 1971.

———. *Modern Spain, 1875–1980.* New York: Oxford University Press, 1980.

———. *Spain 1808–1975.* London: Oxford University Press, 1982.

——— and Juan Pablo Fussi, *Spain: Dictatorship to Democracy.* London, Eng.: George Allen and Unwin, 1979.

Castro Lee, Cecilia, ed., "The Literature of Democratic Spain 1975–1992" *The Literary Review.* Volume 36, Number 3, Spring 1993.

Fishman, Robert, *Working-Class Organization and the Return to Democracy in Spain.* Ithaca, N.Y.: Cornell University Press, 1990.

Gillespie, Richard. *The Spanish Socialist Party. A History of Factionalism.* Oxford, Eng.: Clarendon Press, 1989.

Gilmour, David. *The Transformation of Spain..* London: Quartet Books, 1985.

Herr, Richard. *The Eighteenth Century Revolution in Spain.* Princeton, N.J.: Princeton University Press, 1958, reissued 1969.

———. *Spain.* Englewood Cliffs, NJ: Prentice Hall, 1971.

Hopper, John. *The New Spaniards.* Suffolk, Eng.: Penguin, l995.

Jordan, Barry. *Writings and Politics in the Franco's Spain.* London, Eng.: Routledge, 1990.

Leahy, Philippa. *Discovering Spain.* New York: Crestwood House, 1993.

Lieberman, Sima. *The Contemporary Spanish Economy: A Historical Perspective.* London: George Allen and Unwin, 1892.

Lynch, John. *Bourbon Spain, 1700–1808.* New York: Blackwell, 1989.

Maravall, José. *Dictatorship and Political Dissent: Workers and Students in Franco's Spain.* New York: St. Martin's Press, 1979.

———. *The Transition to Democracy in Spain.* New York: St. Martin's Press, 1982.

Orwell, George. *Homage to Catalonia.* New York: Harcourt, Brace and World, 1952.

Payne, Stanley G. *A History of Spain and Portugal.* 2 vols. Madison: University of Wisconsin Press, 1973.

———, ed., *Politics and Society in Twentieth-Century Spain.* New York: Viewpoints, 1976.

Pitt-Rivers, Julian. *The People of the Sierra.* Chicago: University of Chicago Press, 1963.

Preston, Paul. *The Coming of the Spanish Civil War.* London: Macmillan, 1978.

Thomas, Hugh. *The Spanish Civil War.* New York: Harper & Row, 1963.

Wernick, Robert. "For Whom the Bell Tolled. (the Spanish Civil War)." *Smithsonian.* April 1998, vol. 28, no. 1, pp. 110+.

Sweden

Introduction

The Scandinavian country of Sweden has been an independent kingdom since 1523, when the Swedes ended more than a century of Danish rule. In the following two hundred years they turned their country into one of the major European powers. Since the Napoleonic era, Sweden has been known internationally as a peaceful nation, remaining neutral through the two world wars of the twentieth century. Peace was accompanied by prosperity, as Sweden grew into a modern industrial state with one of the highest standards of living in Europe while managing to look after all its citizens through an enviable network of social welfare programs. As the end of the twentieth century approached, difficult economic times forced a reassessment of the nation's commitment to these programs and a vigorous national debate on how best to continue them while maintaining a sound economy.

With an area of 173,732 square miles (449,964 square kilometers), Sweden is the largest Scandinavian country and the fourth largest country in Europe. The northern and southern parts of the country make up two distinct natural regions, each characterized by a variety of topographical features. The north has areas where mountains predominate, a region of marshes and bogs, and an area of coastal plains. To the south are the central lowlands, the Småland highlands, and the plains of Skåne. Sweden's capital, Stockholm, is situated on the Baltic sea coast. In the late 1990s, Sweden had an estimated population of 9 million people.

History

The ancestors of today's Swedes first enter recorded history in the first century A.D. in the works of the Roman historian Tacitus (ca. A.D. 56–120), who refers to them as the *Svears*, or *Suiones*. Thought to have been a Germanic people, they established the first documented state in Sweden. Beyond this, little is known of the Swedes until the ninth and tenth centuries, during the Viking era. While the Danish and Norwegian Vikings were raiding Western Europe, their Swedish counterparts mostly headed eastward across Russia. A Swedish Viking named Rurik founded the city of Novgorod in Kievan Rus; other Vikings founded a dynasty in Kiev. By the ninth century, Christianity had been introduced to Sweden, although it took several centuries for the new religion to replace native pagan worship.

The twelfth and thirteenth centuries saw the growth of towns and new levels of material wealth as trade expanded and the merchant class became more powerful. In the middle of the twelfth century, Swedes began the conquest of Finland, part of a general expansion across the Baltic Sea. By the late fourteenth century, however, Sweden came under Danish rule when the Swedish nobles invited Denmark's regent, Margaret, to assume control of their country. The resulting Kalmar Union united Sweden, Norway, and Denmark politically. However, Danish rule of Sweden was challenged repeatedly throughout the fifteenth century. When Christian II of Denmark attempted to consolidate control of Sweden—attempts culminating in the infamous Stockholm Bloodbath of 1520—the Swedes rose up under Gustavus Vasa and drove the Danes out of their country for good. Gustavus was established as king of an independent Sweden and the founder of Sweden's royal dynasty.

In the seventeenth century, under the leadership of Gustavus Adolphus (Gustavus II), Sweden became a major military power, undertaking foreign conquests in the Baltic region and winning additional territory as a result of the Thirty Years War. Beginning with the Great Northern War (1700–21), however, Sweden's military supremacy declined, although Swedish culture flourished during the reign of Gustavus III (1771–92), when the Swedish Royal Academy of Music, the Royal Opera, and the Royal Dramatic Theatre were founded. In spite of Gustavus's support for the arts, his imposition of absolute monarchy on the Swedish nobles led to his assassination—which, appropriately enough, later formed the subject matter for an opera by Italian composer Giuseppe Verdi (*Un Ballo in Maschera*).

In 1805 Sweden became involved in the Napoleonic Wars. When they were over, the Swedish had lost Finland to Russia but gained control of Norway in a dynastic union that lasted until 1905. It also gained an heir to succeed the childless Charles XIII on the throne of Sweden in the person of Jean Baptiste Bernadotte, a French marshal. As Charles XIV, he reigned from 1818–44, inaugurating a period of peace and prosperity. Toward the end of the nineteenth century, Sweden's largely rural economy foundered when competition from North America and Russia drove grain prices down. As a result, there was a mass wave of emigration, much of it to the United States. By the end of the century, however, the Industrial Revolution was gaining strength and emigration slowed.

Sweden inaugurated the twentieth century with the awards in 1901 of the first Nobel Prizes. The fund for the prizes had been established by the estate of inventor Alfred Nobel (1833–96) to afford annual recognition to men and women who made a substantial contribution to the good of humanity. The prizes would remain a respected tribute to excellence throughout the century. In 1909 Swedish novelist Selma Lagerlöf was awarded the prize for literature, becoming both the first Swede and the first woman to win a Nobel Prize.

Although Sweden remained neutral in World War I, its citizens suffered food shortages and other privations in the final year of the war. The first postwar decade, the 1920s, brought a level of prosperity that made Sweden one of Europe's wealthier nations. However, the global Depression of the 1930s caused an economic downturn that brought a coalition of the socialist and agrarian parties to enact social welfare and farm support legislation. Sweden was once again neutral in the Second World War, although its citizens supported the Allied cause in a variety of ways, from harboring refugees to aiding Russian troops in Finland. Although Sweden refrained from joining any defense alliance after the war, it took an active role in non-defense-related international cooperation, joining the United Nations in 1946 and helping found the Nordic Council in 1953 and the European Free Trade Association in 1960.

Ecological issues came to the forefront as a political and social issue in the 1970s, most notably the issue of nuclear power, over which a vigorous and long-running debate was waged. In 1980 the Swedish electorate voted to maintain but restrict the growth of nuclear energy, keeping the country's maximum number of reactors at twelve and later reducing this number gradually. The first reductions, slated for completion by 2001, were agreed on in 1997.

By the 1970s controversy had arisen over the high level of taxation needed to maintain Sweden's welfare state, which had continued to grow in the postwar period. Reflecting voter discontent over this issue, the Center Party was voted into power in 1976, ending a long era of Social Democratic rule. The Social Democrats were returned to power in 1982 and then ousted again in the early 1990s with the onset of a recession that created Sweden's worst economic crisis since the 1930s and launched a thoroughgoing reassessment of its economic policies. An austerity plan was implemented, entailing cuts in government spending and employment, and taxes were reduced for most Swedes. By 1993 an economic recovery had begun.

During the same period, Sweden's foreign policy was also undergoing a reassessment following the demise of Communism in Russia and Eastern Europe. With the Cold War over, Sweden's government decided that its traditional policy of neutrality could be reinterpreted, and the country applied for membership in the European Union in 1991. Three years later, Swedish voters approved entry into the organization, and on January 1, 1995, Sweden officially became a member of the EU. However, it was not among the original group of nations who joined the monetary union and adopted the euro in 1999.

Timeline

A.D. 98   Early Swedes are documented by Tacitus

The Roman historian Tacitus (c. A.D. 56–120) mentions the *Suiones* (also called *Svears*), who come to prominence during the decline of the Roman Empire, establishing the first documented state in Sweden, with its capital near present-day Uppsala. Thought to be of Germanic descent, the Svears are known as warriors and sailors. They conquer the Götar to the south, establishing control over most of what today is southern and central Sweden.

9th–10th centuries   Viking era

Some Swedish Vikings raid parts of Western Europe with their Danish and Norwegian counterparts. Most, however, venture eastward through Russia in search of trade with the East, looting and pillaging as far as the Caspian seas. They found a dynasty in Kiev. The Russians call them the Varangians. Some Swedish warriors reach the imperial courts of Turkey and Arabia.

A.D. 829   First Christian missionary arrives in Sweden

The first Christian missionary to arrive in Sweden is Ansgar (Anschar). He attempts to introduce Christianity in the region, but it takes several centuries until the conversion of the Swiss people is complete.

A.D. 862   Novgorod is founded by Swedes

A Swedish Viking named Rurik founds the city of Novgorod in Kievan Rus.

993–1024   Reign of Olaf Skötkonung

Olaf Skötkonung, the last Svear king, is converted to Christianity, becoming Sweden's first Christian monarch.

12th–13th centuries   Growth of an urban merchant class

An expanding merchant class leads to the growth of towns. The greatest trade center is the city of Visby on the island of Gotland, which serves as a base for the Hanseatic League, an association of German merchants. The Swedish monarchy develops new administrative systems, Latin education is introduced, and material progress is accompanied by cultural advances.

1150–1300   Sweden expands to Finland and the Baltic

The Swedes begin their expansion into the lands across the Baltic Sea and occupy Finland.

1150–60   Reign of Eric IX

Eric Jeduarsson rules Sweden as Eric IX. He furthers the spread of Christianity in Sweden and leads Swedish troops into Finland, beginning the conquest of that region. Eric is killed in 1160 by Prince Magnus Henrikson of Denmark and later becomes a patron saint of Sweden.

1275–90   Reign of Magnus I

Magnus I deposes his brother, Waldemar, founder of the Folkung dynasty. He expands the king's control of the nobles but exempts the nobility from taxation in exchange for military service.

c. 1303–73   Life of St. Bridget

St. Bridget, later declared a patron saint of Sweden, supports the overthrow of Magnus II by the nobility. Her writings, collected in *Revelations,* are widely considered the greatest work of medieval literature produced in Sweden.

1319–65   Reign of Magnus II

Magnus II, already king of Norway, is elected to the Swedish throne at the age of three, uniting Sweden and Norway. His reign begins under a regency.

The Kalmar Union

1397   Kalmar Union

The Swedish nobles invite Margaret, regent to the king of Denmark and Norway and widow to the heir of the Swedish throne, to govern their country, uniting all three countries under one crown.

15th century   Rule by regents

Queen Margaret rules Sweden from 1397 to 1412. The real power behind the government moves to Denmark, and Sweden is ruled by a series of regents under Danish control in the fifteenth century.

1477   University of Uppsala is founded

Sweden's first university is founded at Uppsala.

1496: May 12   Birth of Gustavus I

Gustavus I, born Gustavus Vasa, is Sweden's national hero, revered for freeing the Swedes from Danish rule. Born in Lindholm, he is educated at the University of Uppsala. Soon

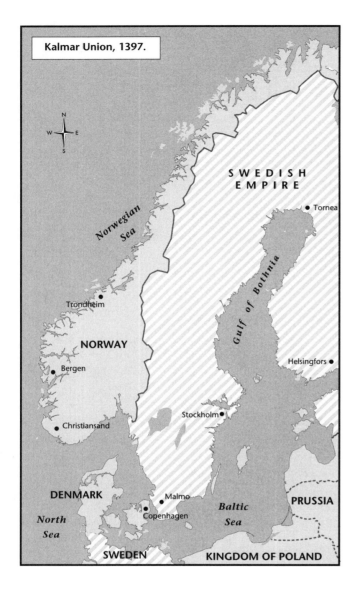

Kalmar Union, 1397.

1520: November Stockholm Bloodbath

Christian II has eighty-two nationalist Swedish nobles executed in a massacre that comes to be known as the Stockholm Bloodbath.

1521–23 Swedes rebel

With the support of the Swedish peasants, Gustavus Vasa organizes a rebellion, fought mainly by German mercenaries, against King Christian II. The Danes are expelled from the country, and the Kalmar Union is dissolved.

1523–60 Reign of Gustavus I

Rebel leader Gustavus Vasa is elected king and rules Sweden as King Gustavus I, establishing a new hereditary dynasty. Under Gustavus, Sweden begins the transition to a modern state. A national army and navy are established, and Lutheranism becomes the state religion. Sweden becomes a dominant force in Baltic trade as the power of the Hanseatic League declines.

1527 Church lands are seized

Sweden comes under the influence of the Protestant Reformation. Following a long dispute between Gustavus I and the Catholic Church over the appointment of bishops, the king urges confiscation of lands owned by the Church.

1544 Hereditary monarchy proclaimed

Gustavus I proclaims a hereditary monarchy, beginning his own dynasty.

1560–92 Reign of Eric XIV

Eric XIV, son of Gustavus I, rules Sweden. He seeks to expand Swedish power into the Baltics following the collapse of the Livonian state and initiates the conquest of Estonia.

1560: September 20 Death of Gustavus I

Gustavus I (Gustavus Vasa), the liberator of his country and founder of its royal dynasty, dies in Stockholm. (See 1496.)

1569–92 Reign of John III

John III marries a Polish princess, uniting the two countries politically in the following generation, when his son, Polish king Sigismund III, becomes the ruler of Sweden

1586 First music is printed

A hymn book contains the first music printed in Sweden.

1592–99 Sigismund reigns and is deposed

Polish king Sigismund III, whose father, John III married into the Polish royal family, succeeds to the throne of Sweden. A

afterwards, he spends time in Denmark as a hostage under the rule of Christian II but escapes. When his father is killed in the Stockholm Bloodbath (see 1520), Gustavus vows to take revenge on the Danes and organizes a rebellion in southern Sweden. The Danes are defeated, and Gustavus is crowned king of Sweden on June 6, 1523, beginning the dynasty that will rule Sweden through the twentieth century. As monarch, Gustavus unites and strengthens the Swedish state and supports the growth of commerce and industry. He dies in 1560, revered as his nation's liberator. His life becomes the subject of numerous works of art throughout Swedish history, including poems, operas, paintings, and a play by August Strindberg (see 1849).

1513–23 Reign of Christian II

Christian II attempts to strengthen Danish control of Sweden, invading Stockholm.

Catholic, he antagonizes the Swedish Protestants and is deposed.

1594: December 9   Birth of Gustavus Adolphus

Sweden's greatest military hero is born in Stockholm, the grandson of Gustavus I. He ascends the throne at the age of sixteen, when Sweden is at war with Denmark and Poland and entangled in an alliance with Russia. In addition, the aristocracy is challenging his authority at home. Gustavus, with the aid of his long-time chancellor, Axel Oxenstierna, manages to obtain the cooperation of the nobles and achieves a harmonious relationship with the *Riksdag* (National Assembly), effecting important administrative reforms. He also initiates significant reforms in secondary and university education and attracts foreign investment to Sweden's copper industry. Above all, Gustavus is known as a military genius, who introduces new linear tactics to Sweden's armed forces and achieves a string of military successes beginning in the 1620s. As a strong defender of Protestantism, Gustavus leads his country into the Thirty Years War in 1630 but is killed leading a cavalry charge at the Battle of Lützen two years later.

Sweden becomes a Military Power

1611–32   Reign of Gustavus Adolphus

Gustavus Adolphus, famed as a great military strategist and leader, ascends the Swedish throne as Gustavus II. (He is the grandson of Gustavus I.) Gustavus inaugurates an era of military conquest, adding Ingermanland, Karelia, and most of Livonia to Swedish territory. Under Gustavus, Sweden enters the Thirty Year War on the side of the Protestants. Gustavus reforms government administration at the national and local levels, creating more efficient bureaucratic structures. The constitution is amended to provide for a Riksdag. The nation prospers through the expansion of iron and copper exports. Gustavus Adolphus is killed at the Battle of Lützen in 1632.

1611–13   Kalmar War

Sweden and Denmark go to war, and Denmark wins. Denmark retains control of the strategic fortress of Alvsborg, the sole port on the west coast of Sweden and the future site of the city of Göteborg, pending the receipt of a large indemnity which is not paid until 1619.

1630   Sweden enters Thirty Years War

In response to pleas for help from German Protestants, Gustavus Adolphus leads Sweden into the Thirty Years War. He scores several victories but is killed two years later at the Battle of Lützen.

1632–54   Reign of Queen Christina

Sweden enjoys the position of dominant Baltic power during the reign of Queen Christina, during which the Thirty Years' War is concluded by the Peace of Westphalia. Christina's reign, which is inaugurated when she is six years old, begins under the regency of Axel Oxenstierna. At the age of eighteen, she takes over control of government herself. In 1654 Christina abdicates in favor of Charles of Zweibrüken, her cousin, who rules as Charles X.

1632: November 6   Death of Gustavus Adolphus

King Gustavus Adolphus, Sweden's greatest military hero, is killed at the Battle of Lützen. (See 1594.)

1648   Peace of Westphalia

The Peace of Westphalia ends the Thirty Years War. Sweden gains control of western Pomerania and wins land at the mouths of the Weser and Elbe rivers.

1655–60   First Northern War

Under Charles X Gustavus, Sweden wages war against Poland, which is forced to cede Livonia. The Swedes also win control of territory to the south previously ruled by Denmark.

1660–97   Reign of Charles XI

Charles XI establishes an absolute monarchy, overcoming the power of the nobles with the support of other segments of society. He confiscates the royal estates, which greatly increase government revenue. Charles also establishes a policy of neutrality that helps expand Swedish trade in the Baltic region at the end of the seventeenth century.

1666   University of Lund is founded

Sweden's second university is established at Lund.

1667–68   War of Devolution

After France attempts to seize control of the Spanish Netherlands, Sweden, England, and the Netherlands form the Triple Alliance against Louis XIV of France and drive out the French. The War of Devolution ends with the first Treaty of Aix-la-Chapelle.

1688–97   War of the League of Augsburg

Sweden joins the League of Augsburg, consisting of England, the United Provinces of the Netherlands (the northern, Protestant provinces not under Spanish rule), and the Habsburgs of Austria. The League opposes the expansionist goals of France's Louis XIV. A long, grueling struggle that stretches over nine years, the war ends in a stalemate. The same underlying conflict leads to the War of the Spanish Succession (see 1701).

1688: January 29   Birth of philosopher Emanuel Swedenborg

Scientist, philosopher, and theologian Emanuel Swedenborg is born in Stockholm and studies at Uppsala. He travels widely in his youth, pursuing interests in many areas of science and creating innovative inventions, including navigational equipment and machinery for transporting boats on land. He is recognized for his knowledge of minerology and appointed an assessor of Sweden's Royal Bureau of Mines. Three years later he is made a member of the House of Nobles. Swedenborg also publishes many scientific volumes and edits a periodical focusing on inventions and mechanical theories.

Beginning in 1743, Swedenborg's interests shift from science to religion and mysticism. On the basis of what he perceives as a divine revelation, he begins expounding a new doctrinal system in theological writings, of which the best known is *Heaven and Hell* (1758). After his death in 1772, proponents of these doctrines form the Church of the New Jerusalem in London, and Swedenborgian beliefs also takes root in two separate religious groups the United States.

Military Decline

1700–21   Great Northern War

When Charles XII ascends the throne at the age of fifteen, Sweden's rivals, Denmark, Russia, and Poland, attempt to take advantage of his youth and inexperience by launching an offensive against Sweden. Charles proves to be an exceptionally able leader even at a young age and defeats his adversaries in the early campaigns of the war. However, Swedish fortunes decline after the disastrous loss to Russia at Poltava (1709). By the end of the war, Sweden has lost almost all its overseas possessions, and its days as one of the great European powers are over.

1707: May 23   Birth of botanist Carolus Linnaeus

Carolus Linnaeus, the originator of the modern system of botanical classification, is born in Rashult. He studies at Lund and Uppsala, where he works with Olaf Celsius. As a young man, he is commissioned to survey the flora of Lapland and spends three months compiling a comprehensive survey of the region. Linnaeus obtains a medical degree in 1735 in Holland, where he begins publishing the works that lay out his system for classifying plants based on their reproductive structures. Going even further, his *systema naturae* undertakes the classification of all objects in nature—plants, animals, and minerals. These are arranged by class, order, genus, and species.

Linnaeus helps found the Royal Swedish Academy of Sciences and serves as its first president. He obtains a university appointment in 1741 and spends his mature years teaching and writing. He also becomes a popular public lecturer, thanks to his gifts as a speaker. In all, Linnaeus publishes over 180 books as well as numerous scientific papers. In addition to his system of classification, he is also famous for devising the binomial Latin names for plant and animal species that have been universally used since his lifetime. Linnaeus dies in Uppsala in 1778.

1709   Defeat at Poltava

Sweden is defeated by the Russians at Poltava in the Ukraine. The Swedish army surrenders and Charles XII flees to Turkey, where he seeks support from the sultan. Following this defeat, Prussia and Hanover join the forces allied against Sweden.

1718   Charles XII is killed in battle

Charles XII is killed in battle in Norway. His sister Ulrika Eleonora ascends the throne with her husband, Frederick of Hesse. However, they must grant concessions to the Swedish nobles.

1721–38   Arvid Horn has power as chancellor

Following the Great Northern War, the government is dominated by the nobles of the Riksdag, whose chancellor, Arvid Horn, leads the country in recovering from twenty years of warfare. He is forced out by a pro-French political faction called the Hats. (Their rivalry with the pro-Russian Caps dominates Swedish politics for much of the century.)

1740: February 4   Birth of poet Carl Michael Bellman

Carl Michael Bellman, a poet and musician, is the foremost Swedish literary figure of the eighteenth century. He is known for his humorous songs, which become popular throughout Scandinavia. They include drinking songs, biblical parodies, and songs celebrating the common man. His best known song collection is *Fredmans epistlar* (1790), whose early songs parody the Pauline Epistles. Other works by Bellman include *Fredmans sanger* (1791) and *Bacchi tempel* (1783), which contains songs, poetry, and engravings. Bellman dies in 1795.

1741–43   Russo-Swedish War

Under the pro-French political faction, the Hats, Sweden goes to war with Russia and suffers a disastrous defeat. The war ends with the Treaty of Abo, which cedes Sweden's Finnish territory to Russia.

1771–92   Reign of Gustavus III

Gustavus III, the son of Gustavus Adolphus, stages a coup against Sweden's constitutional government and reimposes absolute monarchy on Sweden. He effects major improvements in the nation's military, judiciary, civil service and stabilizes its currency. Gustavus, a gifted man who is a playwright and a patron of the arts, supports opera and drama and founds the Swedish Academy. However, his imposition of absolute rule angers the nobility, and he is assassinated in 1792 at a masked ball, an incident that provides the subject matter for the opera, *Un Ballo in Maschera,* by Giuseppe Verdi (1813–1901).

1771   Music academy is founded

The Swedish Royal Academy of Music is founded.

1772: March 29   Death of Emanuel Swedenborg

Scientist and theologian Emanuel Swedenborg dies in London. (See 1688.)

1773   Opera company is established

The Royal Opera, which becomes one of Europe's premier opera companies, is founded in Stockholm.

1778: January 10   Death of botanist Carolus Linnaeus

Linnaeus, founder of the modern system of naming and classifying plants, dies in Uppsala. (See 1707.)

1782: November 13   Birth of poet Esaias Tegnér

Esaias Tegnér is the most popular Swedish poet of the early nineteenth century. He is a professor of Greek at the University of Lund and later bishop of Växjö. Early in his career he is associated with the Romantic movement but later adopts more classical ideals. His best-known works include the poetic cycle *Frithiofs saga* (1825), based on one of the Icelandic sagas, and two works of narrative poetry: *Children of the Lord's Supper* (1820) and *Axel* (1822). Tegnér dies in 1846.

1787   Royal Dramatic Theatre is established

The Royal Dramatic Theatre, which becomes Sweden's leading drama company, is founded in Stockholm.

1795: February 11   Death of Carl Michael Bellman

Sweden's most popular poet dies in Stockholm. (See 1740.)

1796: July 23   Birth of composer Franz Berwald

Franz Berwald, Sweden's best known composer, receives little formal musical training. In his youth he studies violin with his father and gives public concerts. In 1812 he joins the court orchestra in Stockholm, in which he continues to play until 1828 except for two short interruptions. In 1829 he goes abroad and tries to establish himself as a composer in Berlin, without significant success. In 1835 he opens a successful medical institute for orthopedic problems using a popular treatment method. In the 1840s, Berwald lives in Vienna and Paris and travels widely. He also composes his most highly regarded works during this decade, most notably his four symphonies.

Berwald returns to Sweden in 1849 and continues composing music although he fails to win a professional appointment and finds work managing a glass factory. During the 1850s he also begins publishing essays on current social issues. In the final years of his life he is named a fellow of the Royal Academy of Music and a professor of composition. He dies of pneumonia in 1868. In addition to his symphonies, Berwald's works include operas, choral and vocal works, and chamber music. Although Berwald does not receive recognition as a composer until the twentieth century, his four symphonies come to be regarded as significant contributions to nineteenth-century orchestral literature.

The Napoleonic Era

1805   Sweden opposes France

Sweden joins Russia, Austria, and Great Britain in opposing France in the Napoleonic wars.

1808–09   Russo-Swedish conflict

After Russia turns against the coalition arrayed against France, Sweden and Russia go to war, and Russia seizes Finland.

1809   Gustavus IV is overthrown

King Gustavus IV (r. 1792–1809) is overthrown by a military coup. Gustavus's uncle, Charles, is named king on condition that he agree to adopt a more democratic constitution restoring the power of the Riksdag.

1810   Bernadotte comes to power

Since the reigning king, Charles XIII, has no children, the Swedish nobles invite Jean Baptiste Jules Bernadotte, a French marshal, to become the heir to the Swedish throne in hopes that France will help Sweden regain Finland from Russia. Taking the name Charles John, he rapidly becomes an influential force in French politics, although he does not ascend the throne until 1818.

c. 1810   Scandinavianist movement begins

Scandinavianism, a nationalist and cultural movement influenced by German Romanticism, becomes influential in Sweden, launching a renewed interest in Swedish culture. Among the figures associated with this movement are the poets Esaias Tegnér (1783–1846) and Erik Geijer (1783–1847).

1812–13   War against France

Charles John allies Sweden with Russia against France and Denmark to gain control of Norway.

1814 Union of Sweden and Norway

The Congress of Vienna unites Sweden and Norway in a dynastic union under the Treaty of Kiel. At first the Norwegians resist and proclaim their independence; they eventually agree to a union under terms that grant them a degree of autonomy.

Peace And Prosperity

1818–44 Reign of Charles John

Charles Jean (Jean Bernadotte) reigns over Sweden as Charles XIV during a period of peace, prosperity, and progress. Liberalism gains strength as the middle class becomes more influential.

1833: October 21 Birth of inventor Alfred Nobel

Alfred Nobel, the inventor of dynamite and founder of the Nobel Prize, is born in Stockholm to an arms developer and manufacturer. Nobel spends part of his youth in Russia, where his father works on submarine explosives for the Russian government. Upon the family's return to Sweden in 1859, the Nobels establish an explosives laboratory near Stockholm. In the course of experimentation on nitroglycerin, there are several explosions in the laboratory, including one on September 3, 1864, that kills one of Alfred Nobel's brothers.

In the course of searching for explosives that would be safer to work with and transport, Alfred Nobel invents the blasting cap in 1863 and dynamite in 1866. A talented entrepreneur as well as an inventor, Nobel establishes a business empire of dynamite plants and offices throughout the world. In addition to explosives, Nobel also experiments with a variety of materials, including synthetic rubber and silk. He stipulates that most of his estate will go into a fund for the Nobel Prizes. The fund is established after his death in 1896.

1844–59 Reign of Oscar I

Oscar I, the son of Charles XIV, reigns. Like his father, he is king of both Sweden and Norway. He is sympathetic to middle-class demands for liberal reforms and to the Scandinavianist movement seeking to unite the Scandinavian countries more closely culturally and politically.

1846: November 2 Death of poet Esaias Tegnér

Esaias Tegnér, the foremost Swedish poet of the early nineteenth century, dies. (See 1782.)

1849: January 22 Birth of playwright August Strindberg

August Strindberg (1849–1912), one of the greatest figures in the history of Swedish literature, is born in Stockholm and experiences a childhood marred by poverty and neglect. His first published novel, *Röda rummet* (The Red Room; 1879), paints a satirical portrait of contemporary Swedish society. It is a success and makes its author famous. However, it takes eighteen years until Strindberg's first play, *Master Olof* (1872), is produced, although later it is widely considered the first modern Swedish drama. In the 1880s Strindberg spends six years living in different parts of Europe, publishing his first volume of short stories (*Giftas*; 1884–85), which leads to his prosecution for blasphemy. Strindberg is acquitted, but his mental health suffers from the incident.

Strindberg continues writing plays, including *Fadren* (The Father; 1887) and *Fröken Julie* (Miss Julie; 1888), and novels (The People of Hemsö; 1887). His plays demonstrate the characteristics of naturalism in drama—everyday speech, spare sets, and the use of symbolic props. Strindberg's personal life deteriorates in the 1890s: he is divorced in 1891 and suffers from alcoholism. Two marriages later in life are short-lived. Well-known later plays by Strindberg include *Ett drömspel* (A Dream Play; 1902) and *Spöksonaten* (The Ghost Sonata; 1907). The dramatist dies in Stockholm in 1912.

1858: November 20 Birth of novelist Selma Lagerlöf

Nobel Prize-winning novelist Selma Lagerlöf authors works often based on Swedish folkore, including legends and sagas. Her first novel, *Gösta Berlings Saga* (1891), whose memorable title character is a fallen priest, is part of a wider revival of Romantic literature in Sweden. In 1894 she publishes the short-story collection *Osynliga länkar* (Invisible Links). With the publication of the two-volume novel *Jerusalem* (1901–02), after a winter spent in the Middle East, Lagerlöf becomes Sweden's leading novelist. She is awarded the Nobel Prize for Literature in 1909, the first Swede—and the first woman—to win this honor. Among her later works are memoirs, including *Ett barns memoarer* (Memories of My Childhood;1930), and the Värmland trilogy (1925–28). Lagerlöf dies in Marbacka in 1940.

1859–72 Reign of Charles XV

King Charles reigns over Sweden and Norway. He supports the creation of a bicameral legislature (see 1866).

1860: February 18 Birth of artist Anders Zorn

Zorn, one of Sweden's most prominent painters, is born in Mora, which—throughout his life-long travels and foreign sojourns—remains his permanent home in Sweden. He begins his career painting watercolors. From 1881 he primarily lives abroad, especially in London, and travels widely, although still summering in Sweden. In the late 1880s Zorn turns to oil painting and moves to Paris, where he becomes known for his portraits and nudes. His painting displays the influence of Impressionism, particularly the works Edouard Manet (1832–1883), and the Old Masters are also an important influence.

In 1893 Zorn travels to the United States to oversee the Swedish art exhibit in the Chicago World's Fair. He returns to the U.S. frequently and paints portraits of numerous prominent Americans, including U.S. president William Howard Taft (1857–1930). In 1913 Zorn builds a permanent residence in his hometown of Mora. The building, which houses his art and crafts collections, is later turned into a museum. At various points in his life, Zorn takes up etching and sculpture. In 1903 he produces a statue of Gustavus Vasa, the founder of Sweden's royal dynasty, in Mora. Zorn dies in 1920.

1866 Riksdag becomes a bicameral body

The Riksdag, the Swedish parliament, which previously consists of several estates representing different classes of society, is transformed into a modern two-chamber legislative body.

1875: June 23 Birth of sculptor Carl Milles

One of Sweden's foremost artists, Milles is born near Uppsala. His studies in Stockholm are followed by sojourns in Paris and Munich. Milles returns to Sweden in 1908. In the 1920s he serves as a professor of art at the Konsthogskola. In 1931 he accepts a teaching appointment in the United States at the Cranbrook Academy of Art in Bloomfield Hills, Michigan, where he remains until 1945, when he becomes a U.S. citizen. Milles's Swedish commissions include *Europa and the Bull* (1926) in Halmstad and *Orpheus* (1936) in Stockholm. His works in the U.S. include the Delaware Monument (1938) in Wilmington, Delaware and the Resurrection Fountain (1950–52) in Falls Church, Virginia. Milles dies near Stockholm in 1955.

1877 University of Stockholm is founded

The University of Stockholm is founded as a private university (it becomes a public institution in the twentieth century).

1880s Large wave of emigration

After grain imports from Russia and North America create hardship for farmers by driving prices down, many Swedes emigrate. However, emigration slows in the following decade as industry grows.

1889 Social Democratic Party is formed

With the growth of industry comes the introduction of socialism, and Sweden's Social Democratic Party is founded.

1891: May 23 Birth of author Pär Lagerkvist

Lagerkvist, a Nobel Prize winner, is Sweden's most prominent twentieth-century literary figure. A poet, playwright, and novelist, Lagerkvist is concerned with the loss of traditional religious and social values in the modern world. This theme is prominent in works including the story *Det eviga leendet*

(The Eternal Smile; 1920) and the autobiographical work *Gäst hos verkligheten* (Guest of Reality; 1925). Lagerkvist expresses increased faith in humanity in the prose monologue *Det besegrade* (The Triumph over Life; 1927). The 1936 play *Mannen utan sjal* (The Man Without a Soul) is a condemnation of fascism.

Widespread acclaim does not come to Lagerkvist until relatively late in his career, with the publication of his 1944 bestselling novel *Dvärgen* (The Dwarf). The novel *Barabbas* establishes his reputation internationally, and he is awarded the Nobel Prize for Literature the following year. Lagerkvist dies in Stockholm in 1974.

1896: December 10 Death of inventor Alfred Nobel

Inventor Alfred Nobel dies. His estate is used to set up the fund for the Nobel Prize. (See 1833.)

Early 20th century First orchestras are established

Sweden's first symphony orchestras are formed, including the Stockholms Konsertförening (1902) and the Göteborgs Orkesterförening (1905). Smaller groups are founded in Helsingborg and other cities.

1901: December 10 First Nobel Prizes are awarded

The first Nobel Prizes are awarded on the fifth anniversary of the death of Alfred Nobel, the inventor who founded the awards as recognition for those who "during the preceding year, shall have conferred the greatest benefit on mankind." Nobel also donates his estate to establish the fund from which the prize money is to come. According to Nobel's instructions, there are awards in five areas: physics, chemistry, physiology or medicine, literature, and the promotion of peace. (A sixth award—for economics—will be added by the Riksbank in 1969.) All awards except the Peace Prize are to be awarded in Stockholm, and each may be shared by up to three persons. Award winners receive a diploma and a medal in addition to a cash prize.

1904: May 6 Birth of author Harry Martinson

Harry Martinson is a poet, novelist, and essayist of working-class origins who becomes one of Sweden's foremost poetic voices of the early twentieth century and a Nobel Prize winner. His experiences as a merchant seaman are chronicled in his first book of poetry, *Spökskepp* (Ghost Ship; 1929) and in two autobiographical novels, *Nässlorna blomma* (Flowering Nettle; 1935) and *Vägen ut* (The Way Out; 1936). Other well-known works include the poetry collection *Passad* (Trade Wind; 1945); the 1948 novel *Vägen till Klockrike* (The Road; 1948); *Aniara* (1956), a long poem about space travel; and the poetry collection *Tuvor* (Tussocks; 1976). In 1974 Martinson is awarded the Nobel Prize for Literature jointly with Eyvind Johnson. Martinson dies in 1978.

A procession passes before the king of Sweden at the dedication of the Olympic stadium in 1912. (EPD Photos/CSU Archives)

1905   Union with Norway is dissolved

Following a period of strained relations between Norway and Sweden, Norway declares its independence, and the dynastic union between the two countries is terminated.

1907–50   Reign of King Gustavus V

King Gustavus V reigns for nearly the entire first half of the twentieth century, maintaining Swedish neutrality through two world wars.

1909   General strike

Labor conflicts arising from the ups and downs of the business cycle lead to a general strike.

1909   Novelist Selma Lagerlöf is awarded Nobel Prize

Novelist Selma Lagerlöf becomes the first Swede and the first woman to win the Nobel Prize for Literature.

1910   Liberal Party is formed

The Liberal Party is formed, as Sweden moves toward a true multiparty system of parliamentary government.

1911   Retirement pensions are introduced

Old-age pensions are introduced, marking the beginning of Sweden's extensive network of progressive social welfare programs.

1912   Sweden hosts the Olympic Games

The fifth modern Olympic Games are held in Stockholm, Sweden. The city constructs a stadium for the event, whose size and international representation brings a new level of prestige to the games. Some 2,500 athletes from twenty-eight countries—and every continent—compete in 102 events. For the first time, the competitions are regulated by officials from the international federations of the individual sports represented rather than by local officials from the host country. In addition to sports, the Stockholm Olympics include competitions in music, literature, painting, sculpture, and city planning.

1912: May 14   Death of dramatist August Strindberg

August Strindberg, the preeminent figure in modern Swedish literature, dies in Stockholm. (See 1849.)

Neutrality

1914–18   World War I

Sweden remains neutral in World War I. In the final year of the war, however, the Swedes suffer food and fuel shortages as a result of German submarine warfare and are forced to resort to rationing.

1918   Composers' society is formed

The *Förening Svenska Tonsättare* is established to support the work of Sweden's composers.

1918: July 14   Birth of film director Ingmar Bergman

Bergman, widely regarded as one of the great directors of the twentieth century, is born in Uppsala to a Lutheran clergyman and receives a strict upbringing. While attending the University of Stockholm he becomes involved in theater productions and afterwards works as theater director. He enters the Swedish film industry in the 1940s, directing his first feature film in 1945. However, Bergman's reputation as a film director is not established until the release of *Smiles of a Summer Night* (1955) and *The Seventh Seal* (1957). The latter film, in particular, brings Bergman international recognition, including multiple Academy Award nominations. Throughout the 1960s and into the 1970s, Bergman continues his exploration of the serious philosophical themes of religious faith and mortality, also exploring the female psyche in films such as *Persona* (1966).

In 1976 Bergman suffers a nervous breakdown following his arrest on tax-evasion charges, and he vows never again to work in Sweden. He establishes a base in Munich and begins releasing films in the United States. By 1978, however, he has returned to his Swedish island retreat of Faro and once again begins directing stage productions at the Royal Dramatic Theater in Stockholm. Bergman's international reputation is revitalized with the release in 1983 of *Fanny and Alexander,* which wins multiple international honors including an Academy Award in the U.S. for best foreign film. Bergman announces his retirement from filmmaking but goes on to produce *After the Rehearsal* the following year and to author the screenplay for *The Best Intentions* in 1992. He is also the author of two autobiographical works, *The Magic Lantern* (1988) and *My Life in Film* (1993).

1920s   Prosperity reigns

Although no political single party has enough clout to form a government with staying power, Sweden enjoys a period of prosperity that makes it one of the wealthier nations in Europe.

1920: August 22   Death of painter Anders Zorn

Zorn, an internationally recognized portrait painter and one of Sweden's greatest artists, dies in his hometown of Mora. (See 1860.)

1921   Universal suffrage is adopted

Sweden enacts universal male and female suffrage.

1930s   Depression strengthens the Socialists

When the global economic depression strikes Sweden, the Socialist and Agrarian parties join forces to enact social legislation and support agriculture.

1932   Social Democrats are voted into office

Serious unemployment caused by the Depression strengthens support for the Social Democrats, who form a government under Per Albin Hansson. Lacking a parliamentary majority, they form a coalition with the Agrarian Party the following year.

1939–45   World War II

Sweden remains officially neutral but aids the Danish and Norwegian resistance movements and harbors refugees from the Nazis. Raoul Wallenberg, a wealthy Swede working in Budapest, saves the lives of about 100,000 Hungarian Jews in the closing months of the war by distributing Swedish passports and through other strategems.

1940: March 16   Death of novelist Selma Lagerlöf

Nobel Prize-winning novelist Selma Lagerlöf dies in her hometown of Marbacka. (See 1858.)

The Postwar Era

1946–69   Tage Erlander serves as prime minister

Social Democrat Tage Erlander presides over the Swedish government during the postwar decades. His administration maintains Swedish neutrality throughout the Cold War and expands the social welfare program at home, increasing child welfare allowances, old age pensions, and rent subsidies.

1946   Sweden joins the UN

Sweden becomes a member of the United Nations. However, due to its policy of neutrality, it does not join the North Atlantic Treaty Alliance (NATO).

1950–73   Reign of Gustavus VI

During the reign of Gustavus VI, Sweden's social welfare programs expand.

1951   Author Pär Lagerkvist receives Nobel Prize

The Nobel Prize for Literature is awarded to Swedish poet, dramatist, and novelist Pär Lagerkvist. (See 1891.)

1951   Swedes are granted right to withdraw from state church

Swedes born into the state church are granted the right to withdraw from it later.

1953   Nordic Council is formed

Sweden, Denmark, Norway, and Iceland form the Nordic Council. They are later joined by Finland.

1955: September 19   Death of sculptor Carl Milles

Sculptor Carl Milles, one of Sweden's best-known artists, dies near Stockholm.

1958   Women win right to ordination in state church

Women are approved for ordination in Sweden's state church.

1960   Sweden helps form the EFTA

Sweden is a founding member of the European Free Trade Association (EFTA).

1968   Social Democrats score major electoral victory

For the first time since the eve of World War II, the Social Democrats win majorities in both houses of the Riksdag.

1968   Riksbank establishes a sixth Nobel Prize

Sweden's central bank, the Riksbank, establishes an economics prize, the Nobel Memorial Prize in Economic Science, to accompany the five existing Nobel Prizes awarded annually. The first economics prize is awarded the following year.

1969: October   Olof Palme becomes prime minister

Social Democrat Olof Palme, a young politician, becomes prime minister of Sweden, taking over from long-time leader Tage Erlander.

1970s   Ecological issues gain importance

Issues affecting the environment gain prominence in Swedish politics, notably nuclear power, which pits the centrists and communists against the liberals and moderates.

1971: January   Riksdag becomes a single-chamber body

The Riksdag, the Swedish parliament, is transformed from a bicameral body to a single-chamber legislature.

1972   Unions win right to name corporate boards

Legislation gives unions the prerogative of appointing boards of directors for corporations that employ more than one hundred persons.

1973: September 15   Reign of Carl XVI

Following the death of his father, Gustavus VI, Carl XVI becomes king of Sweden.

1974   Swedish authors share Nobel prize

Swedish authors Eyvind Johnson and Harry Martinson (see 1904) share the Nobel Prize for Literature.

1974: July 11   Death of author Pär Lagerkvist

Nobel Prize-winning author Pär Lagerkvist dies in Stockholm. (See 1891.)

1976   Centrists end long Social Democratic era

Reflecting voter discontent with the high taxes needed to maintain Sweden's welfare state, the Center Party ousts the Social Democrats from the preeminent position they have enjoyed in Swedish politics for over forty years. Thorbjörn Fälldin replaces Olof Palme as prime minister.

1978: February 11   Death of author Harry Martinson

Nobel Prize-winning poet, novelist, and essayist Harry Martinson dies in Stockholm. (See 1904.)

1980: March   Swedes vote to retain nuclear power

In a special referendum, the Swedish electorate votes to expand Sweden's nuclear power plants to no more than twelve by the middle of the decade but eventually to phase out nuclear power.

1982–86   Social Democrats return to power

With Sweden facing a slow-down in economic growth, budget deficits, inflation, and unemployment, the Social Democrats are returned to power with Olof Palme as prime minister. Palme attempts to retain Sweden's social welfare program and still curb public spending through a "Third Way" program that places tax revenues into "wage-earner funds."

1986–91   Carlsson faces continuing economic problems

The administration of Ingvar Carlsson, who succeeds Olof Palme as prime minister, faces an economy that continues to deteriorate, endangering the social welfare system and causing labor unrest. Environment issues also continue to be a major concern.

1986: February 28   Palme is assassinated

Prime Minister Olof Palme is shot and killed by a lone gunman while walking unescorted on a Stockholm street. The Swedish public is shocked by this violent event. The sole suspect, a mentally unbalanced drifter, is convicted of the crime and sentenced to life imprisonment in 1989 but released on appeal. Palme's assassin is never brought to justice, and no motive is established for the killing.

1987   Bofors scandal

A scandal erupts when officials of the Bofors arms manufacturer are charged with selling weapons to a nation at war and bribing foreign officials to gain arms contracts abroad.

1988: January   Sweden wins border dispute with U.S.S.R.

A dispute over territorial rights in the Baltic Sea is settled in Sweden's favor.

1990   Sweden faces recession

Sweden confronts its worst recession since the 1930s. Its normally low unemployment level soars, and many workers hold temporary jobs. Changing economic and political conditions force a reassessment of the country's traditional economic policies, and the government institutes an austerity plan, cutting government spending and employment and privatizing some state-owned enterprises.

Neutrality after the Cold War

1991   Sweden applies to join the EU

With the Cold War ending, the strict neutrality that has dictated Sweden's economic relations with its European neighbors becomes less relevant, and the nation applies for admission to the European Union, although the move still faces strong opposition at home.

1991: January   Income taxes are abolished for most Swedes

Income taxes for all but the wealthiest Swedes are eliminated, with the revenue made up for by sales tax.

1991: September   Coalition government elected

Swedes elect a four-member nonsocialist coalition government headed by moderate politician Carl Bildt as prime minister. Facing a serious economic crisis, the new government makes substantial cuts in public spending.

1993: May   Neutrality policy is modified

Sweden's traditional diplomatic policy of neutrality is modified by the Riksdag to allow for the possibility of joining defense alliances.

1994   Social Democrats return to power

National elections return a minority Social Democrat government to power with Carl Bildt reassuming the post of prime minister.

1994: November   Swedish votes approve EU membership

A popular referendum on membership in the European Union is approved by Swedish voters.

1995: January 1   Sweden joins the EU

Sweden officially becomes a member of the European Union.

1997: February 2   Agreement is reached on shutting down nuclear plants

Sweden's major political parties agree on a schedule for decommissioning two nuclear power plants by 2001, raising a variety of objections from economists, labor leaders, and environmentalists. Sweden currently relies on nuclear power for about half its electricity.

1997: August 8   Olympic stadium is bombed

Stockholm's Olympic stadium, the site of the 1912 Olympic Games, is bombed by an extremist group opposed to the city's bid to host the games in 2004. The bombing occurs one month before the International Olympic Committee is scheduled to choose from among four cities under consideration.

1998: September 20   Social Democrats win parliamentary elections

The ruling Social Democratic Party wins the largest percentage of the vote in nationwide elections. However, its thirty-six percent plurality is down significantly from the forty-five percent tally in the 1994 election, forcing it to form a coalition with two other leftist parties, the former Communist Party (renamed the Left Party) and the Green Party. The Social Democrats have promised to reverse the recent trend toward decreased public spending on Sweden's social welfare programs.

1998: October 30   Deadly discotheque fire kills sixty

A rapidly spreading fire in a second-floor makeshift discotheque kills sixty and injures some 180 others. Most of those killed are teenagers from immigrant families who have been attending a Halloween party when the fire breaks out and are unable to leave the building, where some four hundred people are jammed into a space with a legal capacity of one hundred fifty. Arson or a short circuit are suspected as possible causes of the blaze, which is Sweden's worst in decades.

Bibliography

Cole, Paul M. *Sweden Without the Bomb: The Conduct of a Nuclear-Capable Nation Without Nuclear Weapons.* Santa Monica, Calif.: Rand, 1994.

Elstob, Eric. *Sweden: A Popular and Cultural History.* Totowa, N.J.: Rowman & Littlefield, 1979.

Heclo, High, and Henrik Madsen. *Policy and Politics in Sweden: Principled Pragmatism.* Philadelphia: Temple University Press, 1987.

Milner, Henry. *Sweden: Social Democracy in Practice.* New York: Oxford University Press, 1989.

Moberg, Vilhelm. *A History of the Swedish People: From Renaissance to Revolution.* New York: Pantheon Books, 1973.

Palmer, Alan. *Bernadotte: Napoleon's Marshal, Sweden's King.* London: Murray, 1990.

Puffendorf, Samuel F. *The Compleat History of Sweden,* 2 vols. Folcroft, Penn.: Folcroft Library Editions, 1977.

Roberts, Michael. *From Oxenstierna to Charles XII: Four Studies.* New York: Cambridge University Press, 1991.

Rothstein, Bo. *The Social Democratic State: The Swedish Model and the Bureaucratic Problem of Social Reforms.* Pittsburgh: University of Pittsburgh, 1996.

Sundelius, Bengt, ed. *The Committed Neutral: Sweden's Foreign Policy. Boulder,* Col.: Westview Press, 1989.

Switzerland

Introduction

Located in west central Europe, Switzerland is a small nation with several distinctive characteristics, including the picturesque Alpine scenery that once drew foreign visitors to health spas and now attracts tourists to popular ski resorts. A landlocked country ringed by larger nations—three of which have at one time or another been the dominant powers of Europe—Switzerland has guarded the integrity of its borders by a policy of permanent neutrality since the sixteenth century. This stance has led to a unique role in world affairs. Switzerland has been the site of numerous international peace negotiations and a mobilizing point for various humanitarian efforts, of which the best known is the International Red Cross. In the twentieth century, the adoption of strict secrecy policies by Switzerland's banking industry have helped make the country an international financial center.

With an area of 15,942 square miles (41,290 square kilometers), Switzerland is about twice the size of the state of New Jersey. It is composed of three major geographic regions: the Jura Mountains along the Swiss-French border in the northwest; the fertile central plateau, or Mittelland; and, to the south, the snow-covered Alps, which account for roughly half of Switzerland's territory. Switzerland has an estimated population of 7.2 million people, of which over 120,000 live in the capital city of Bern.

History

The first known group to inhabit present-day Switzerland were called the Helvetii (or Helvetians), who settled there several hundred years before the Christian Era. It is from this Celtic group that the Swiss people's name for their own country (Helvetia) comes. The region came under the control of the Roman Empire in 58 B.C. and was overrun by Germanic tribes between the third and sixth centuries A.D. The Franks, who were the last of these groups, occupied the area until after the death of their greatest monarch, Charlemagne, in the ninth century, after which their empire broke up and various noble families vied for control of the territory. The most powerful were the Habsburgs of Austria, who controlled much of central Europe by the end of the thirteenth century.

In 1291, three of the Swiss cantons (provinces) laid the foundations for what would become the nation of Switzerland by forming an alliance for mutual protection against the Habsburgs. (The date of this pact—August 1—is still the Swiss national holiday.) This alliance grew into the Swiss confederation, which had eight member cantons by 1353 and adopted its name from the largest of the original three cantons (Schwyz). The confederation survived through a series of military victories against would-be conquerors in the late fourteenth and fifteenth centuries, and the Swiss army gained so formidable a reputation that its soldiers later hired themselves out as mercenaries to other countries for hundreds of years until the practice was outlawed in the nineteenth century. By the end of the fifteenth century, Austria had recognized the independence of the confederation, which kept growing and included thirteen cantons by 1513.

The Protestant Reformation played a prominent role in Swiss affairs in the sixteenth century, when religious reformer Ulrich Zwingli formed his own Protestant sect in northern Switzerland. After Zwingli's death in the Kappel Wars between Swiss Protestants and Catholics (1529–31), the French-born John Calvin, based in Geneva, emerged as the country's main Protestant leader, and a new spirit of religious toleration enabled the Swiss to avoid further bloodshed over religion. (Three centuries later, however, religious conflict returned when several Catholic cantons formed a breakaway alliance called the Sonderbund and a brief civil war ensued.) During the sixteenth century, Switzerland also adopted a policy of neutrality in international affairs, which enabled it to avoid taking sides in the next century's Thirty Years' War. The one subsequent period when the Swiss came under foreign domination was the period following the French Revolution, when forces led by Napoleon Bonaparte invaded and occupied the country, creating the Helvetic Republic. Following Napoleon's downfall and the 1815 Congress of Vienna, however, the Swiss Confederation was restored.

Switzerland, which had remained a loose confederation of cantons since the thirteenth century, formed its first centralized federal government under the constitution of 1848, which authorized the government to conduct foreign diplo-

SWITZERLAND

0   25   50 Miles

0   25   50 Kilometers

macy, issue currency, and take on a variety of other functions. However, the cantons still retained a large measure of autonomy over their internal affairs. Internationally, Switzerland served as the site of a series of international treaties (the Geneva Conventions, held between 1864 and 1949) that established a code of behavior for civilized nations in wartime, including guidelines covering the treatment of political prisoners. Another historic step taken by Switzerland during this period was the establishment of the International Red Cross in 1864.

As the nineteenth century progressed, Switzerland's industrial sector grew, and the country became known worldwide for the production and export of high-quality manufactured goods. Among other items, the Swiss became famous for watches and chocolate. Another field in which the Swiss

distinguished themselves was the emerging science of psychology. Switzerland was home to two famous figures in twentieth-century psychology, both of whom began their careers in the early part of the century: Carl Jung, who broke with Freud to develop his own theory, called Analytical Psychology, and Jean Piaget, who laid the foundations for the field of child psychology. In addition, the famous inkblot test was developed by Swiss psychiatrist Hermann Rorschach in the 1920s.

In the twentieth century, Switzerland remained neutral in both world wars, although its people were more strongly united in World War II, in their opposition to Nazi Germany than the time in the first world war when they experienced a degree of competing allegiances between their French- and German-speaking populations. In both wars, Switzerland participated in humanitarian activities and welcomed refugees fleeing persecution and conflict in their homelands. Following World War I, Geneva became the home of the League of Nations, and Switzerland maintained a special neutral status within the peacekeeping organization. In 1934 a new law mandated strict secrecy guidelines for Swiss bank accounts and prevented banks from releasing information about accounts without the explicit permission of depositors. The law gave a boost to Switzerland's financial sector as Swiss bank accounts gained new international popularity.

Switzerland has maintained its political neutrality in the postwar era. It has declined membership in the European Union, NATO, and the United Nations (participating in UN sanctions or possible military commitments would violate its neutrality). However, the Swiss have cooperated with other international organizations and agencies that do not pose the danger of forcing it to breach its neutrality policy. In 1991 Switzerland celebrated the seven hundredth anniversary of the "Perpetual Alliance" that united the first three Swiss cantons in 1291. As the twenty-first century approached, Switzerland faced the challenge of reconciling its traditional neutrality with the growing movement toward European economic integration.

Timeline

c. 500 B.C.   The Helvetii migrate to Switzerland

The Helvetii, a Celtic people, settle in present-day Switzerland (whose name, in Swiss—Helvetia—comes from this group). The Rhaeti, who originated in northern Italy, also populate the area during this period.

58 B.C.   Caesar conquers the Helvetii

Fleeing Germanic invaders, the Helvetii push westward toward Gaul (present-day France) but are repulsed at Geneva by the Romans, led by Julius Caesar, who want them to stay where they are so they can serve as a buffer between the Romans and the invading tribes. They are forced to return but provided with Roman military protection, and Helvetia becomes a semi-autonomous part of the Roman empire. The Romans introduce their legal code as well as improved agricultural methods, and the region prospers for several hundred years, until the Romans withdraw their forces to protect other parts of the empire.

3rd–5th centuries A.D.   Germanic tribes invade Switzerland

Present-day Switzerland is invaded and occupied by two Germanic tribes, the Alemanni and the Burgundians, whose differing dialects—which evolve into German and French— lay the foundation for Switzerland's linguistic diversity.

6th century   Frankish conquest

The Franks, another Germanic tribe, displace the Alemanni and Burgundians, controlling much of Switzerland as well as the eastern part of France. Under the Frankish Merovingian dynasty, Christianity is introduced to the region.

9th century   Reign of Charlemagne

As part of the Frankish empire, Switzerland is ruled by the great king Charlemagne (742–814), who is named Holy Roman Emperor. After Charlemagne's death, his vast kingdom is divided, and rival noble families battle for control of Swiss territory. The two most successful are the Habsburgs and the Savoy line.

1032   Switzerland again becomes part of the Holy Roman Empire

For the first time since the death of Charlemagne, Switzerland once more is united under a single ruler, the Holy Roman Emperor Conrad II.

Late 13th century   Habsburg rule is consolidated

The Austrian Habsburgs, who are now the predominant power in central Europe, strengthen their control over eastern Switzerland, appointing governors to rule over the region and establish a system of taxation.

The Swiss Confederation

1291: August 1   Perpetual alliance links Swiss cantons

The Swiss forest cantons (communities) of Schwyz, Uri, and Unterwalden form an alliance aimed at retaining their local independence and resisting the power of the Habsburgs. The alliance is the forerunner of the Swiss confederation (*Eidgenossenschaft*), and the date of its signing becomes the Swiss national holiday. William Tell, a native of Uri, becomes a folk hero for his legendary resistance to Habsburg rule.

1315: November 15   Habsburgs are defeated at Morgarten Pass

In one of the most important military victories in Swiss history, a peasant force of 1,400 from the allied cantons of Schwyz, Uri, and Unterwalden defeats 20,000 invading Habsburg troops led by Duke Leopold.

1332–53   Swiss confederation expands

The group of independent Swiss cantons expands to form an eight-member confederation with the addition of Luzern (1332), Zurich (1351), Glarus (1352), Zug (1352), and Bern (1353). The members are collectively known as Schwyzer (Swiss), based on the name of the largest of the original cantons (Schwyz).

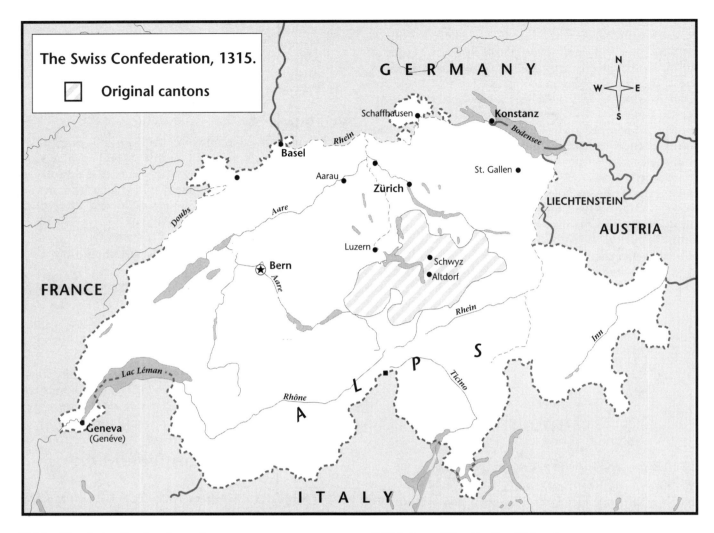

The Swiss Confederation, 1315.

☐ Original cantons

1339   The Swiss flag is adopted

Troops from the Swiss cantons fight under the national red-and-white banner for the first time.

14th–15th centuries   Swiss independence is assured by military victories

The confederation of Swiss cantons maintains its independence by a series of military victories against foreign powers, including the Habsburgs, the Burgundians, the dukes of Savoy and Milan, and King Francis I (1494–1547) of France. The Swiss military gains an international reputation for its fierce and skillful fighting.

1386: July 9   Battle of Sempach

The Swiss win a military victory against the Habsburgs. Swiss hero, Arnold von Winkelried, dies in battle.

1388: April 9   Battle of Nafels

The Swiss win a second victory against Austria at the Battle of Nafels in central Switzerland.

1476: June 22   Battle of Morat

Allied with France and Austria, the Swiss defeat the forces of the Burgundian duke Charles the Bold at the Battle of Morat. The Swiss army begins its long-lived tradition of hiring itself out to foreign governments as a mercenary force. Soldiers become Switzerland's chief export.

1484: January 1   Birth of religious reformer Ulrich Zwingli

Zwingli, Switzerland's most important Protestant leader aside from John Calvin, studies philosophy and theology in Bern and Vienna and is ordained as a Catholic priest in 1506. He serves as parish priest for several congregations before settling permanently at Gross-Munster. Gradually coming to accept the Holy Scripture as his primary source of religious authority rather than the holy fathers of the Church, Zwingli breaks with the Roman Catholics and founds his own religious sect in Zurich in 1523. He publishes his ideas in the book *On the True and False Religion* in 1525 and also translates the Bible into Swiss German. Zwingli serves as an army chaplain in the religious wars between Protestants and Catholics known as the Kappel Wars (1529–31) and dies on the battlefield at Kappel on September 11, 1531. After his death, the French-born reformer John Calvin becomes Switzerland's major Protestant leader.

1493: December 17   Birth of physician and alchemist Paracelsus

Paracelsus is the first physician to link chemistry and medicine, challenging medical principles accepted since the days of the ancient Greeks. He is born Theophrastus Bombastus von Hohenheim and later adopts the name "Paracelsus" to indicate that he has gone beyond ("para") the knowledge of the renowned Roman physician Celsus. As a young man, he spends years studying medicine at various universities in Europe, as well as alchemy in the Middle East and also serves as an army surgeon.

After 1524 Paracelsus's reputation as a physician and teacher spreads throughout Europe, and he becomes a lecturer at the University of Basel at the age of 33. However, he antagonizes the medical and civil authorities with his controversial beliefs and brash behavior, including publicly burning the books of the most revered Greek and Middle Eastern medical authorities. Facing threats from his enemies, he is forced to leave Switzerland and lives abroad for eight years, continuing his medical discoveries. He returns to Switzerland in 1538 but dies three years later at the age of 48. Among his many achievements are the identification of the cause of silicosis in miners; the development of theories that help lay the foundation for the school of medical treatment known as homeopathy; and his emphasis on the body's natural healing powers.

1499: September 22   Treaty of Basel

Following Switzerland's victory over the Austrian armies of Holy Roman Emperor Maximilian I (1459–1519) at Graubunden, Austria recognizes the independence of the Swiss cantons.

1505   The Swiss Guard is founded

Pope Julius II forms the Swiss Guard, a 110-member armed guard, recruited from Switzerland's Catholic cantons, that is stationed at Vatican City and serves as bodyguards for the pope.

1513   Three cantons are added

Basil, Appenzell, and Schaffhausen join the Swiss confederation, bringing its total number of cantons to thirteen. The Federal Union of the Thirteen Cantons remains a stable political entity until the end of the eighteenth century and the French Revolution.

1515: September 13–14   Military defeat at Marignano

As part of the conflict over the Po Valley in the north of Italy, Swiss forces are defeated by France, suffering some 8,000

losses, including losses incurred by Swiss mercenaries killing each other while fighting for opposing sides. Following this calamity, the Swiss adopt a policy of neutrality which they maintain through the twentieth century.

The Protestant Reformation

1519   The Reformation comes to Switzerland

Ulrich Zwingli (see 1484), a priest from Zurich, breaks with the Catholic Church and forms his own branch of Protestantism in northern Switzerland. Together with the French-born John Calvin (1509–64), he is one of Switzerland's two great religious reformers of the sixteenth century.

1523   Zurich bans Catholic services

With the help of the city's church council, the rites of the Catholic Church are banned from Zurich as part of Zwingli's program of religious and civic reform. This controversial move helps trigger religious warfare between Switzerland's Catholic and Protestant cantons.

1529–31   Reformation leads to war (Kappel Wars)

Armed conflict erupts between Swiss Protestants and Catholics. Swiss Reformation leader Ulrich Zwingli is killed in battle in 1531. The Swiss adopt a policy of religious toleration in the wake of this conflict.

1536   Geneva revolts against Savoy

Geneva, under the influence of Reformation leader John Calvin (1509–64), revolts against the Catholic Duke of Savoy. The city becomes a center of Calvinism, with the Protestant church exercising a strong influence over its government and laws. Under Calvin's leadership, a twelve-member tribunal is set up to rule on the behavior of local residents. In the severely puritanical spirit of the times, there are penalties for such infractions as wearing brightly colored clothing or iceskating.

1541: September 24   Death of Paracelsus

The pioneering physician Paracelsus dies in Austria at the age of forty-eight. (See 1493.)

1547   Publication of the *Dodecachordon*

Henricus Glareanus (1488–1563) publishes the *Dodecachordon,* one of the most authoritative musical treatises of the Renaissance.

1648   Treaty of Westphalia

Under the Treaty of Westphalia ending the Thirty Years' War (in which the Swiss have remained neutral), Switzerland is recognized as an independent, neutral state.

1741: February 7   Birth of artist Henry Fuseli

One of Switzerland's most original artists, Fuseli (born Johann Heinrich Füssli) is raised in Zurich but later lives in Germany, Italy, and England, where he settles permanently. His works, which have been associated with those of the English poet and engraver William Blake, are mostly based on literary works of Homer (c. 9th century B.C.), Dante (1265–1321), Shakespeare (1564–1616), and other classic authors. Fuseli's work is known for its bold lines and imaginative qualities. The artist dies in London in 1825.

1746: January 12   Birth of educator Pestalozzi

Johann Heinrich Pestalozzi, a prominent educational reformer, is born and educated in Zurich. His early political interest in helping the disadvantaged is later channeled into educational reform. He tests his theories and puts them into action in the schools he heads, most notably that at Yverdon, which he directs for twenty years beginning in 1805. His emphasis on careful observation and other principles have a major influence on educational reform in the nineteenth century. Pestalozzi dies in 1827.

1798   France invades Switzerland

Disregarding traditional respect for Swiss neutrality, French revolutionary forces led by Napoleon Bonaparte (1769–1821) invade Switzerland, replacing the loose federation of cantons with the newly created "Helvetic Republic." New democratic laws end the power of the ruling aristocracy and weaken control by the old feudal families.

1802   Act of Mediation

Napoleon confers a new constitution (the Act of Mediation) on Switzerland, which restores much of the local autonomy previously removed by the French and expands the number of Swiss cantons to nineteen. The central government is given power over foreign affairs and the military.

1806: September 2   Avalanche kills hundreds in the Swiss Alps

The top of Rossberg Peak suddenly crumbles, causing an avalanche as it carries an entire forest down the slopes with it, killing eight hundred people in four mountainside villages. The calamity culminates in a gigantic forest fire triggered by friction and steam geysers.

1815   Swiss Confederation is restored

Following the downfall of Napoleon, the Swiss Confederation is restored at the Congress of Vienna with a total of twenty-two cantons, and its permanent neutrality is recognized internationally.

1818: May 25   Birth of cultural historian Jakob Burckhardt

Burckhardt, a pioneering cultural scholar and critic, is born in Basel and studies history and theology in Bonn and Berlin. He teaches on the university and secondary school levels in Basel throughout his career, while authoring major works on cultural history. He is best known for his 1860 book *Die Kultur der Renaissance in Italien* (translated into English as *The Civilization of the Renaissance in Italy*), in which he argues that the Italian Renaissance was the driving force behind the birth of the modern way of regarding the world. His other major works are *The Age of Constantine the Great* (1853) and *Cicerone* (1855), an overview of Italian painting. Burckhardt's view of the Renaissance is central to the work of cultural historians for decades and remains influential in the twentieth century. Jakob Burckhardt dies in 1897.

1825: April 16   Death of artist Henry Fuseli

Swiss-born artist Henry Fuseli dies in London. (See 1741.)

1827: February 17   Death of educator Pestalozzi

Teacher and educational reformer Johann Pestalozzi dies. (See 1746.)

1845   Sonderbund is established

As elsewhere in Europe, liberal sentiment grows in Switzerland in the first half of the nineteenth century. In reaction to calls for secular education and taxes on church property by the more liberal (and mostly urban) cantons, seven Catholic cantons (Lucerne, Uri, Schwyz, Unterwalden, Zug, Fribourg, and Valais) form an alliance called the Sonderbund and secede from the Swiss confederation.

1847   Civil war

In a brief civil war, the confederation government forces the dissolution of the Sonderbund, a breakaway alliance of seven Catholic cantons.

Federal Government

1848   Constitution adopted

A new constitution replaces the confederation with a centralized federal government and balances liberal and conservative interests. Central government functions are expanded to include foreign diplomacy, customs regulation, and the issuing of currency. A bicameral legislature and a seven-member executive council are created as the organs of government. The Jesuits are banished from Switzerland, but religious tolerance is mandated in all other matters.

1857: November 17   Birth of linguist Ferdinand de Saussure

Ferdinand de Saussure, widely considered the founder of modern linguistics, is born in Geneva to a family of French refugees. He attends college in Leipzig, Germany, receiving a doctorate at the age of twenty-three, by which time he has already published an academic paper that has established his professional reputation. From 1881 to 1891 he teaches linguistics in Paris, also serving as secretary of the city's linguistic society. He later lectures at the University of Geneva, and dies in Geneva in 1913. Saussure revolutionizes the study of language by shifting its focus from historical changes over time (diachronic) to the underlying structure of languages as they exist in the present (synchronic). Saussure's theories become widely known after his death when two of his students publish their lecture notes in the form of a book called the *Course in General Linguistics* (1915).

1864–1949   The Geneva Conventions

Henri Dunant (1828–1910), a young Swiss who organized ad hoc relief services to aid thousands of unattended wounded men at the Battle of Solferino (1859) in Italy, plays a leading role in inaugurating the Geneva Conventions, a series of international treaties setting basic ground rules for humanitarian measures in wartime. The first convention (1864) provides for the neutrality of medical personnel and civilians who aid those wounded in warfare. The following conventions extend the provisions of the original agreement to cover naval warfare (1906); prisoners of war (1929); and civilians in time of war (1949).

1864: August   Red Cross is established

Measures laying the foundation for the establishment of the International Red Cross are agreed to at the meeting that produces the first Geneva Convention—the Geneva Convention for the Amelioration of the Condition of the Wounded and Sick of Armies in the Field (see 1864–1949). The agreement provides for humane treatment of the wounded in wartime by neutral medical personnel and civilians identified by the now-familiar Red Cross emblem of a red cross on a white background (a Swiss flag with its colors reversed), adopted in honor of Henri Dunant, the Swiss man who is a guiding force behind the Convention. After his experiences observing and aiding the wounded in battle, Dunant proposes the formation of voluntary societies by countries throughout the world to aid victims of war and other disasters.

1874   Constitution revised

The Swiss constitution is modified to expand the powers of the federal government.

1875: July 26   Birth of psychiatrist Carl Jung

Carl Jung, a famous psychiatrist and founder of the school of Analytical Psychology, is born to a family of theologians and clergymen and is raised and educated in Basel, where he studies medicine. In 1900 Jung begins working at the Burgholzli Psychiatric Clinic of the University of Zurich under the direction of Eugen Bleuler. In 1907 he meets Sigmund Freud (1856–1939), beginning a collaboration that lasts until 1913. The collaboration ends due to professional differences which cause Jung to sever other professional ties as well as; he develops his own system of psychoanalysis. The principles underlying this system are first detailed in the book *Symbols of Transformation* (1912).

Probably Jung's most famous formulation is the concept of the collective unconscious containing archetypes representing psychic patterns potentially present in all people. Also widely influential is Jung's theory of types, most notably the introverted and extroverted types. Jung's interest in the psychic elements common to all people leads him to investigate the myths and rituals of non-Western peoples, including the Pueblo Indians and Kenyan tribes, as well as the symbolism and teachings of the major Eastern religions. The C. J. Jung Institute in Zurich is founded in 1948 to perpetuate Jung's teachings and train analysts in his system. (Jungian analysis is also taught at other institutions in Europe and the United States.) Considered one of the major figures of twentieth-century psychology, Carl Jung dies in Zurich in 1961.

1876   Forestry law is passed

The Swiss parliament passes a forestry bill that becomes one of the world's first environmental laws.

1877: July 2   Birth of writer Hermann Hesse

German-born novelist Hermann Hesse moves to Switzerland during World War I. After the war he settles there permanently, becoming a Swiss citizen in 1924. He is known for novels that explore the conflict between individual identity and social expectations. His best-known works include *Siddhartha* (1922), which re-creates the early life of the Buddha; *Demian* (1919); *Steppenwolf* (1927); and *The Glass Bead Game* (1943). Hesse wins the Nobel Prize for Literature in 1946. He dies in 1962.

1879: December 18   Birth of artist Paul Klee

Paul Klee is one of the best-known, most innovative and influential twentieth-century artists. His paintings are known for their frequent use of abstract linear shapes, simple, primitive-looking figures, and, in his mature period, bright, vivid color. Klee is born near Bern and demonstrates early gifts for both the visual arts and music. He studies art briefly in Munich and Italy but soon rejects the conventionality of academia and develops an interest in the avant-garde. He returns to Switzerland for several years before moving to Munich in

1906. His first one-man shows are held in Switzerland in 1911–12. Klee serves in World War I as a noncombatant. In 1920 he is appointed to the staff of the Bauhaus, a famous school of design that promoted the production of art shaped by its eventual use. He develops his own educational method and becomes a highly successful teacher.

With the rise to power of the Nazis in 1933, Klee leaves Germany for Switzerland, where he settles in Bern. Retrospective exhibits of his work are shown in Bern and Basel in 1935. Suffering from the disease scleroderma in his last years, Klee continues working, often in brilliant colors. He dies of a heart attack in 1940.

1880   Children's story *Heidi* is published

One of the most famous Swiss books—*Heidi* by Johanna Spyri (1880)—is published.

1880: July 24   Birth of composer Ernest Bloch

Ernest Bloch, one of Switzerland's most distinguished twentieth-century composers, is born in Geneva, where he studies music and later teaches composition at the Geneva Conservatory. He settles in the United States after World War I, obtaining U.S. citizenship in 1924. Among Bloch's achievements is the integration of traditional Hebrew music into symphonic and choral compositions such as *Schelomo* (1915–16), a rhapsody for cello and orchestra, and "Avodath Hakodesh" (1934), a piece for choir and orchestra. Bloch also composes music in many different styles, and his works include *Macbeth* (1903), an early opera that was widely performed in Paris; the choral symphony *America* (1917); and *Helvetia*, a symphonic poem commemorating his birthplace (1929). After teaching in New York and Cleveland, and on the West Coast of the United States, Bloch returns to Switzerland from 1931 to 1941, devoting himself solely to composing, with financial support from a wealthy patron. After World War II, he returns to the United States and settles in Oregon, where he dies in 1959.

1887: October 6   Birth of architect Le Corbusier

Le Corbusier, born Charles Edouard Jeanneret, is one of the most famous architects of the twentieth century. His building designs, which are highly influential, include the Ronchamp Chapel in Vosges, France, a structure with no straight lines; the controversial seventeen-floor Radiant City apartment complex in Marseilles, France; and the Center for the Visual Arts at Harvard University, Le Corbusier's only building in the United States. The architect dies in France in 1965.

1890: September 15   Birth of composer Frank Martin

Martin, one of Switzerland's foremost twentieth-century composers, is born in Geneva and studies music in Zurich and Munich, Germany. In the 1920s he begins an association with the music educator and developer of Dalcroze Eurhythmics,

Emile Jacques-Dalcroze (1865–1950). Martin first studies and later teaches at the music institute founded by Jacques-Dalcroze. During World War II, Martin serves as president of the Swiss musicians' union. After the war, he moves to the Netherlands, and his works receive increasing recognition. Martin dies in Naarden (the Netherlands) in 1974.

Influences on Martin's works vary from Bach (1785–1850) and Chopin (1810–1849) to Arnold Schoenberg (1874–1951), whose twelve-tone system is adopted and modified by Martin in works written after 1930. In his lifetime, Martin composes dozens of works, including operas and choral, orchestral, and instrumental pieces. Individual works by Martin include the *Rhapsodie for Five Strings* (1935), the choral work *Le Vin Herbé* (1938–41), the oratorio *Golgotha* (1945–48), *Petite Symphonie Concertante* (1946), and *the Requiem* (1970–71).

1892: March 10   Birth of composer Arthur Honegger

Composer Arthur Honegger is born in France to Swiss parents and educated at the Zurich Conservatory. His teachers include Vincent d'Indy (1851–1931), and the Group of Six including Darius Milhaud (1892–1974) and Francis Poulenc (1899–1963) are early colleagues. Honegger, who spends his professional life in France, is a prolific and successful composer best-known works include the oratorio *King David* and the experimental orchestral piece *Pacific 231*, both written in 1923; the opera *Judith* (1926); and *Joan of Arc at the Stake* (1938). Honegger dies in Paris in 1955.

1896: August 9   Birth of child psychologist Jean Piaget

The work of psychologist Jean Piaget has been central to the study of child psychology in the twentieth century. Piaget is born in Neuchâtel and displays scientific precocity at an early age, publishing his first professional paper at the age of ten and more papers as a teenager. After receiving a Ph.D. in natural history at the age of twenty-two, Piaget works briefly with psychiatrist Carl Jung (see 1875) and psychological testing pioneer Theodore Simon. While administering psychological tests to children, Piaget becomes interested in their thought processes, and this interest determines his subsequent career path.

Noting that there are significant differences between the thought processes of adults and children, Piaget constructs a framework of cognitive development in children composed of four stages that move gradually toward abstract thought: sensorimotor, preoperational, concrete, and formal operations. Piaget serves as director of research at the Jean-Jacques Rousseau Institute in Geneva and as a faculty member of the University of Neuchatel, the Sorbonne, and the University of Geneva. He also directs the Center for Genetic Epistemology at the University of Geneva, which is founded in 1952. Piaget also authors a number of books, including *The Language and Thought of the Child* (1926), *The Origin of Intelligence in*

Children (1954), and *The Development of Thought* (1977). Jean Piaget dies in 1980.

1897: August 8   Death of cultural historian Jakob Burckhardt

Jakob Burckhardt, one of the most influential cultural historians of the nineteenth century, dies in Basel. (See 1818.)

1898   National museum opens

Switzerland's national museum, the Schweizerisches Landesmuseum, opens in Zurich.

1901   Dunant receives the Nobel Peace Prize

Henri Dunant, the leading force behind the first Geneva Convention and the founding of the International Red Cross, is awarded the first Nobel Peace Prize.

1901: October 10   Birth of sculptor Alberto Giacometti

The son of a successful Impressionist painter, sculptor and painter Alberto Giacometti becomes involved with art at an early age. His career includes sojourns in Paris and Rome, but his birthplace of Stampa, Switzerland remains his principal residence. He is known primarily for his bronze sculptures of slender, elongated human figures, sometimes portraying single individuals and sometimes groups. During his lifetime, retrospectives of his work are shown at the Museum of Modern Art in New York and at the Zurich Museum of Fine Arts. Giacometti dies on January 12, 1966.

1907   The Swiss National Bank is founded

The Swiss National Bank is established, becoming the nation's official bank of issue.

1909   Kocher wins Nobel Prize for Medicine

Physician and pathologist Emil Theodor Kocher (1841–1917) is awarded the Nobel Prize for Medicine.

1911: May 15   Birth of author Max Frisch

Playwright and novelist Max Rudolf Frisch is one of Switzerland's leading twentieth-century authors. As a young man he works as a journalist and architect but devotes himself full time to writing after 1955. His plays, which are known for their portrayal of skepticism about contemporary society, include *Santa Cruz* (1947); *Die Chinesische Mauer* (The Chinese Wall; 1947); *Als der Krieg zu Ende war (*When the War Was Over; 1949); and *Don Juan oder die Liebe zur Geometrie* (Don Juan, or The Love of Geometry; 1962). Among Frisch's novels are *Stiller* (1954); *Homo Faber* (1957); and *Der Mensch erscheint im Holozän* (Man in the Holocene; 1979). Frisch dies in Zurich in 1991.

1913: February 22   Death of linguist Ferdinand de Saussure

Influential linguistic pioneer Ferdinand de Saussure dies in Geneva. (See 1857.)

The World Wars

1914–18   World War I

Switzerland remains strictly neutral during World War I but still must arm to defend its neutrality. The nation suffers shortages of food and other raw materials and rising unemployment, as well as internal dissension between supporters of France and Germany.

1916   Publication of the *Course in General Linguistics*

The linguistic theories of Ferdinand de Saussure (see 1857) are published posthumously based on the lecture notes of his students.

1918: November   General strikes

Discontent stemming from the hardships imposed by war erupts in a series of general strikes, but order is restored.

1919   Geneva is chosen to house League of Nations

Geneva becomes the headquarters of the newly formed League of Nations, the first international organization dedicated to preserving world peace. Switzerland joins the League, having been granted special status due to its political neutrality. It is later declared exempt from the organizations trade sanctions against Germany.

1920s   Rorschach develops the inkblot test

Swiss psychiatrist Hermann Rorschach (1884–1922) develops a personality assessment based on the subject's reactions to a series of ten inkblot pictures, which form an essentially neutral stimulus for the projection of the patient's emotions. Subjects look at the cards one at a time and describe what each inkblot resembles. Rorschach subscribes to the method of analysis pioneered by another Swiss psychiatrist, Carl Jung (see 1875), and his experiments with inkblot tests were used to assess the Jungian categories of extraversion and intraversion.

1921: January 5   Birth of dramatist Friedrich Dürrenmatt

Friedrich Dürrenmatt is educated in Zurich and Bern. He writes novels and essays but is best known for the comic absurdity of his dramas, which play a leading role in the resurgence of German-language theater following World War II. Dürrenmatt's best-known plays include *Die Ehe des Herrn Mississippi* (The Marriage of Mr. Mississippi; 1952); *Der Besuch der alten Dame (*The Visit; 1956); *Die Physiker* (The

Swiss citizens are firmly behind their government's position of neutrality. Twenty thousand citizens gather to pledge neutrality at the start of World War II. (EPD Photos/CSU Archives)

Physicists; 1962); and *Porträt eines Planeten* (Portrait of a Planet; 1970). Dürrenmatt, who also writes literary criticism, detective novels, and radio plays, dies in 1990.

1932–34   Geneva Conference

The Geneva Conference fails in its goal of arms reduction when German leader Adolf Hitler (1889–1945) withdraws from the talks.

1934   Banking law protects depositors' privacy

The Swiss government implements a banking law imposing the strict secrecy guidelines that have since given Swiss bank accounts their international popularity among depositors seeking privacy in their financial dealings. Under this law, banks may not provide any information about accounts to third parties without authorization by the account holders.

1939–45   World War II

The Swiss, while united in their opposition to Nazi Germany, preserve their neutrality but arm to prevent attacks by aggressors. Nearly one in every four citizens is mobilized to protect the neutrality of Switzerland's borders. The Swiss also carry on humanitarian activities, notably medical assistance to war victims through the International Red Cross. As it has done in

World War I, Switzerland becomes a haven for political refugees. Surrounded by the Axis powers, Switzerland shares in the economic hardships of war and is forced to ration food and freeze prices.

1940: June 29   Death of artist Paul Klee

Painter and graphic artist Paul Klee dies. (See 1879.)

1946   Novelist Hermann Hesse receives the Nobel Prize

German-born novelist Hermann Hesse, who is a Swiss resident and citizen, is awarded the Nobel Prize for Literature. (See 1877.)

1947   Early patriot is canonized

The fifteenth-century national leader Nicholas of Flue is canonized and becomes the patron saint of Switzerland.

Postwar Neutrality

1950s   Switzerland leads postwar recovery effort

Having suffered relatively little wartime damage, Switzerland enters the postwar period with a sound economy and is able to help other European nations recover from the war with loans and other types of assistance.

1951   Female suffrage bill is defeated

Legislation to introduce female suffrage is defeated in the Swiss parliament.

1951: January 20   Worst avalanches in 35 years hit the Alps

The Swiss, Italian, and Austrian Alps suffer the worst avalanches since World War I, triggered by snow and rain conditions and hurricane-force winds. Over 45,000 people are trapped for weeks in towns and villages including the famous ski resorts of Davos and St. Moritz. Communications and rail lines are cut off, and one Swiss village, Vals, is entirely destroyed.

1955: November 27   Death of composer Arthur Honegger

Swiss-French composer Arthur Honegger dies in Paris. (See 1892.)

1959   Switzerland joins European Free Trade Association

While remaining aloof from the newly formed European Economic Community (EEC, later the European Union), Switzerland becomes a founding member of the European Free Trade Association.

1959: July 15   Death of composer Ernest Bloch

Highly regarded composer Ernest Bloch dies in Portland, Oregon. (See 1880.)

1960s   Rising immigration causes tension

With a booming economy and nearly full employment, Switzerland has hired large numbers of foreign workers from southern European countries—over one million between 1950 and 1970. The presence of these workers, who come to comprise some seventeen percent of the population, eventually causes social and political tensions and is linked to problems such as housing shortages. The Swiss government responds by restricting the number of foreign workers who may be employed in the country. Proposed constitutional amendments are introduced to effect further limits on foreigners in the work force.

1960   Women are allowed to vote in local elections

Women win the right to vote in municipal elections.

1961: June 6   Death of Carl Jung

Psychiatrist Carl Jung dies in Zurich. (See 1875.)

1962: August 9   Death of author Hermann Hesse

German-born Swiss novelist Hermann Hesse dies. (See 1877.)

1965: August 27   Death of architect Le Corbusier

Le Corbusier, one of the twentieth century's foremost architects, dies at Roquebrune Cap-Martin, France. (See 1887.)

1966: January 12   Death of sculptor Alberto Giacometti

Famed Swiss sculptor Alberto Giacometti dies in Chur, Switzerland. (See 1901.)

1970s   Swiss economy is slowed by oil crisis

The global increase in oil prices puts a damper on Switzerland's economy, leading to rising inflation, growing unemployment, and labor unrest.

1971: February 7   Female suffrage enacted

Legislation guaranteeing women the vote in most of Switzerland's cantons is passed.

1972   Free-trade pact with the EC

Swiss voters ratify a free-trade agreement aimed at eliminating tariffs on manufactured products traded with EC countries by 1977.

1973   Ban on Jesuits is repealed

The banishment of Jesuits from Switzerland by the constitution of 1848 is overturned by voters in a national referendum.

1974: November 21   Composer Frank Martin dies

Distinguished Swiss composer Frank Martin dies in the Netherlands. (See 1890.)

1979: January 1   New canton is formed

The predominantly French-speaking canton of Jura becomes Switzerland's twenty-third canton. Carved out of the surrounding German-speaking canton of Bern, it is Switzerland's first new canton in one hundred fifty years.

1980: September 16   Death of Jean Piaget

Jean Piaget, the single most influential figure in child psychology, dies. (See 1896.)

1980: September   St. Gotthard tunnel is completed

The St. Gotthard tunnel—at 10.6 miles (17 kilometers) the longest road tunnel ever constructed—opens in the Ticino.

1981 Constitution is amended to guarantee women's rights

A constitutional amendment is approved that outlaws discrimination against women in education and employment and requires all cantons and municipalities to allow women to vote.

1984 First woman is elected to the federal council

Elisabeth Kopp becomes the first woman elected to Switzerland's federal executive body, the seven-member Federal Council.

1985 Legislation mandates equal rights in marriage

Men and women gain equal rights in marriage under Swiss law, which has previously entitled husbands to make all family decisions, including whether wives can work outside the home.

1985 Road tax is aimed at reducing emissions

The Swiss government imposes a road tax as an environmental measure aimed at reduction of vehicle emissions.

1986: March Swiss voters reject UN membership

In a special referendum, Swiss voters reject a proposal for full membership in the United Nations after it has been approved by its parliament. (Switzerland has had special observer status in the organization since its formation and has participated in its humanitarian activities and is represented in its special agencies.)

1986: November 1 Toxic waste spills into the Rhine

A fire in a chemical warehouse sets off a major environmental accident, as some thirty tons of toxic waste spill into the Rhine River. It is estimated that about half a million fish and eels are poisoned.

1987: April Voters limit the extension of political asylum

Fearing negative economic consequences from large numbers of political refugees, Swiss voters approve a measure to limit the number of refugees who may receive political asylum in the country.

1990 Demonstrations protest secret files

Mass demonstrations are held to protest the existence of secret files kept on 200,000 persons by a federal prosecutor's office.

1990: December 14 Death of playwright Friedrich Dürrenmatt

Distinguished Swiss dramatist Friedrich Dürrenmatt dies in Neuchâtel, Switzerland. (See 1921).

1991 Women demonstrate to protest discrimination

Half a million Swiss women hold a mass demonstration to protest gender-based discrimination.

1991 Swiss mark 700th anniversary

Switzerland observes the 700th anniversary of the founding of the Swiss Confederation.

1991 Women gain the vote in final canton

Following a federal court ruling, the half-canton of Appenzell Inner-Rhoden becomes the last local region to fully enfranchise women, which includes allowing them to participate in the traditional annual assemblies called *Landsgemeinde*.

1991: April 4 Author Max Frisch dies

Playwright and novelist Max Frisch dies in Zurich. (See 1911.)

1992: May Membership in financial organizations wins approval

Swiss membership in the International Monetary Fund (IMF) and the World Bank is approved in a referendum.

1992: December Voters reject economic cooperation initiatives

Swiss voters defeat a government measure proposing membership in the European Union (EU) and participation in the European Economic Area (EEA) linking the European Free Trade Area with the EU in a free-trade zone.

1994: October 5 Cult members found dead

The bodies of forty-eight members of the Order of the Temple of the Sun are discovered in two Swiss villages (five cult members are also found dead in Canada). They have been shot as part of a mass suicide or murder and their residences then set on fire. Among the dead are the cult's founder and leader, Luc Jouret, a Belgian practitioner of alternative medicine. The group, founded in 1984, combined New Age philosophies with traditional Christian apocalyptic beliefs. It is thought that Switzerland attracted them because its banking laws allowed the group's financial practices to go unexamined.

1998: August 12 Swiss banks agree to settlement with Holocaust survivors

The two largest Swiss banks, Credit Suisse and UBS, agree to pay $1.25 billion to Holocaust victims to compensate for unclaimed assets placed in Swiss bank accounts during the World War II era. The banks had been under international pressure to provide reparations, especially from the U.S., where many state and local governments have threatened to boycott the banks involved. A portion of the funds will go directly to Holocaust survivors and their heirs who can document the existence of unclaimed deposits. The rest will be used to aid survivors around the world.

1998: September 2   Swissair jet crash kills 229

All 229 passengers and crew aboard a Swissair jetliner die as the plane plunges into the Atlantic Ocean off the Nova Scotia coast a little over an hour after takeoff from Kennedy International Airport in New York City. The plane had been bound for Geneva, Switzerland. Rescue efforts are carried out by the Royal Canadian Mounted Police and fishermen from the village of Peggy's Cove near the crash site. Authorities report no evidence of terrorist sabotage and blame the crash on mechanical malfunction.

Bibliography

Bacchetta, Philippe, and Walter Wasserfallen, ed. *Economic Policy in Switzerland.* New York: St. Martin's Press, 1997.

Diem, Aubrey. *Switzerland: Land, People, Economy.* Kitchener, Ont.: Media International, 194.

Eu-Wong, Shirley. *Culture Shock!: Switzerland.* Portland, Or.: Graphic Arts Center Pub. Co., 1996.

Hilowitz, Janet Eve, ed. *Switzerland in Perspective.* New York: Greenwood Press, 1990.

Linder, Wolf. *Swiss Democracy: Possible Solutions to Conflict in Multicultural Societies.* Houndsmills, U.K.: Macmillan, 1994.

Martin, William. *Switzerland from Roman Times to the Present.* New York: Praeger, 1971.

Meier, Heinz K. *Switzerland.* Santa Barbara, Calif.: Clio Press, 1990.

New, Mitya. *Switzerland Unwrapped: Exposing the Myths.* New York: I.B. Tauris, 1997.

Ross, John F. L. *Neutrality and International Sanctions: Sweden, Switzerland, and Collective Security.* New York: Praeger, 1989.

Schimel, Carol L. *Conflict and Consensus in Switzerland.* Berkeley: University of California Press, 1981.

Schwarz, Urs. *The Eye of the Hurricane: Switzerland in World War II.* Boulder, Colo.: Westview, 1980.

Steinberg, Jonathan. *Why Switzerland?* 2nd ed. Cambridge: Cambridge University Press, 1996.

Turkey

Introduction

Although modern-day Turkey was founded in 1923, its history dates to civilizations inhabiting the area from 10,000 B.C. Twentieth-century Turkey is located mostly in Asia Minor and partially in Europe and is geographically bordered by Greece, Bulgaria, Iran, Iraq, Armenia and Syria. Its main border, however, is the Mediterranean Sea. It is strategically located because it controls two straits that connect the Black Sea to the Mediterranean—the Bosporus and the Dardanelles. An interior lake, Lake Van, is a salt lake. The topography is varied with two main mountain chains, the Taurus and the Pontic, and many plateaus. The highest point is Mt. Ararat which is 16,853 feet (5140 meters) above sea level. According to the 1998 census, the capital city of Ankara had 2,846,000 people, while the most populous city, Istanbul, had close to 8,000,000 people. The total population was 64, 566,511.

Turkey has been populated by many ancient civilizations and was often referred to as Anatolia, from its Greek name *anatolé*, meaning sunrise. The term Asia Minor was coined in the fifth century A.D. to distinguish it from the larger adjacent region of Asia. Anatolia is also used to describe the region of Asia Minor and Thrace, a region of European Turkey. Turkey has also often been referred to as the bridge between Europe and Asia because many different peoples have crossed through this area and have lent their cultural influences to it. All of the many peoples who crossed the borders of Anatolia emerged changed by the experience. The proximity of Greece and Greek culture is also part of the history of ancient Turkey.

Turkey is so ancient that history and mythology have been interwoven for centuries. Archaeological finds have verified what was once thought to be legend. The ancient city of Troy, now rediscovered, as well as the reality of the Trojan War, mentioned in Greek poems thought to be legends, blend real history with literature. The land of the poet Homer and the *Iliad* is a part of history. Mount Ararat, from the Biblical story of Noah, is located in present-day Turkey. The history of Turkey can be divided into five main periods: the Anatolian period (Hittites, Phrygians and Lydians, Armenians and Kurds, Greeks); the Roman period; the Turks and the Seljuks;

the Ottoman Empire; and finally modern-day Turkey from 1923. Most of the secrets of ancient Anatolia have come to light only since the end of the Ottoman Empire and the establishment of modern-day Turkey.

The earliest written record of inhabitants in the area of ancient Anatolia dates from about 6500 B.C. when possibly the first human settlement was founded at Çatalhüyük. Handmade pottery and other artifacts depicting a matriarchal society with an Earth Mother as goddess have been unearthed. Artifacts dated earlier than those at Çatalhüyük have been found at Ashikli Höyük (8000 B.C.) and at Nevali Cori , but Çatalhüyük was the oldest agrarian community. Around 3000 B.C. bronze weapons and artifacts were introduced into society, and that produced a trade for tin with neighboring Assyria. Tin was necessary in the production of bronze. The Anatolians had copper but traded for woolen cloth and tin which the Assyrians supplied. While this trade existed, it contributed to the prosperity of Anatolia. The decline of Anatolia led to the rise of the Hittites who established themselves in the ancient city of Troy.

The Hittites crossed the Caucasus Mountains in the third millennium B.C. They spoke an Indo-European language and subjugated the indigenous peoples of Anatolia called the Hatti. They took their name from these peoples and adopted their religion. They had a cuneiform alphabet and are related to the Hittites in the Old Testament. Their form of government was a feudal system. Their capital was Hattusas. They were very successful until their defeat at the hands of the Phrygians in 1200 B.C. The first king of the Hittites was Hattushili I who expanded his kingdom into present-day Syria. His reign was punctuated by frequent attempts to depose him. He installed his nephew, Murshili, as successor, but Murshili was soon assassinated. The kingdom was in turmoil until 1465 B.C. when Telipinu took control and provided for an orderly succession through laws of heredity. Telipinu is recognized for imposing a system of fair and just laws which required restitution instead of punishment for wrongdoings. The Hittites had many wars with the neighboring Egyptians (the longest war, culminating at the Battle of Kadesh, came to a draw in 1275 B.C.) but finally came to a lengthy peace agreement established by the marriage of the Hittite king's daughter to Ramses II of Egypt in 1246 B.C. Many cuneiform

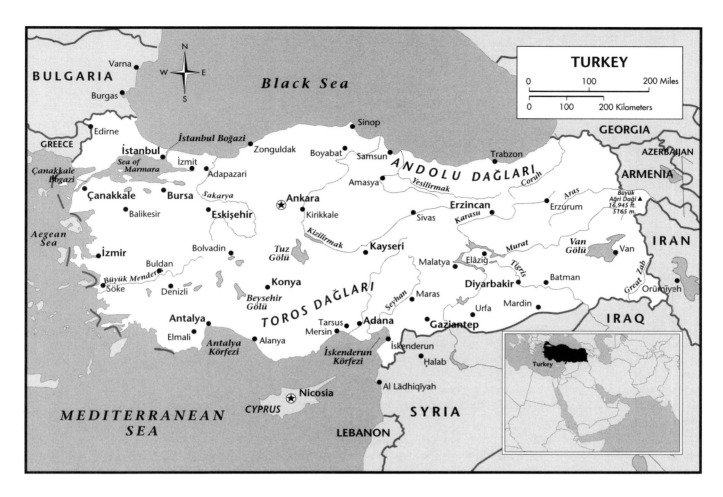

texts document the history of the Hittites. The Hittite civilization came to a sudden halt around 1200 BC. when the city of Hattusas was levelled by fire. The arsonists were thought to be the Phrygians, but there are many theories that natural catastrophes could have caused the demise of the Hittite empire.

The next people to inhabit the area were the Phrygians who are probably known more for what were previously thought to be legends. Their two most famous kings were Gordius (of the famed Gordian Knot) and Midas (known for his golden touch). The capital, Gordium, is where present-day Ankara is located.

The Phrygians ruled the area until about the sixth century B.C. They in turn gave way to invasions by the Cimmerian nomads from the Caucasus. Finally the Lydians, a warrior group from Thrace, became the dominant rulers from about 700 B.C. to 550 B.C. The most famous of these kings was Croesus, of Greek legend, who was a real person. The Greek legend described him as the richest man ever. The Lydians panned and mined gold and minted the world's first coins. The expression "richer than Croesus" has a basis in fact.

Paralleling these communities were neighboring Greek coastal settlements. Around 1000 B.C., a group of these peoples established a Panionian League that consisted of differ-

ent Greek settlers who populated the western Anatolian coast. The first of these groups to arrive were the Ionians, soon followed by the Aeolians and the Dorians. These Greeks were soon to spread their ideas of Western civilization through the arts, education, and politics. Most of these newcomers kept their own identities but got along peaceably with their neighbors. The result of their refusal to band together was their eventual conquest by foreign invaders. First the Lydians, controlled the area and ruled from their court city of Sardis. Along with their king Croesus, they ruled until their defeat at the hands of the Persians in 546 B.C. under Cyrus the Great. Most of the city states of Greece, except for Athens and Sparta, remained under Persian domination until Philip of Macedon united them in an effort to gain independence. His son, Alexander the Great, continued the fight and defeated the Persians at the Granicus River in 334 B.C. He cut the Gordian Knot and became the ruler of a kingdom which eventually stretched from the Nile River in Egypt to the Indus River in India. He ruled this area until his death in 323 B.C.

After the death of Alexander the Great, many of his successors fought for control of his vast kingdom. One of them, Seleucus Nicator, did manage to control parts of Anatolia, Thrace, and Syria, Mesopotamia, and Persia. Other groups, such as the Gauls (from present-day France) set up city states.

The Gauls established the kingdom of Galatia whose capital was at present-day Ankara. The spread of Greek culture was at its height, and Anatolia became the center of western civilization. The two great cities of Pergamum and Antioch were models of Hellenistic (Greek-style) culture. This period of Hellenistic rule was short-lived. In 133 B.C. the Romans began invading and taking over cities and towns in Anatolia. These became part of the vast Roman Empire that lasted until the end of *Pax Romana* in about 330 A.D.

The Roman Empire brought peace and great wealth to Anatolia. The Romans called this province Asia and established Ephesus as its capital. Historical greats such as Julius Caesar, Antony and Cleopatra, Hadrian, and Diocletian were the new inheritors of this culture and added their knowledge to this growing history. The Romans ruled until their decline in the fourth century A.D. Their empire had grown so large it was impossible to rule from Rome, so the empire was divided into the eastern and western sectors, each ruled by a separate emperor and an assistant. The eastern half of the empire, which had its capital at Byzantium (later Constantinople and still later Istanbul) on the Bosphorus, became known as the Byzantine Empire. The Roman emperor Constantine converted to Christianity and transferred his capital out of Rome to Byzantium. Later the city was renamed Constantinople in his honor. For the next 700 years, this would be the heart of the Byzantine Empire.

In the sixth century A.D. Turkish tribes began migrating out of central Asia and into Asia Minor, home of the land that would eventually bear their name. The Oguz Turks, who had converted to Islam in the tenth century, began migrating in many directions and were united under their leader and known as Seljuks. The Seljuk rule eliminated other Arabs and and made Sunni Islam the main religion, as opposed to Shia Islam. The Turks slowly edged the Romans out of the area previously carved by the Byzantine Empire and finally destroyed the Byzantine Army in 1071 at the Battle of Manzikert. After this battle the Seljuks had control of Anatolia. They called their new government the Sultanate of "Rum" which was Rome or the old Byzantine Empire. Sovereignty was decided by ruling families and tribal loyalties. The leadership was constantly challenged and depended on which tribe had the most relatives. The new Turkish government was basically the same as the old Roman government, but now the bureaucracy was headed by Muslim Turks instead of Christians.

The Sultanate of "Rum" soon gave way to what was to become the great Ottoman Empire (1300–1922). Founded by Osman I, an Anatolian ruler in the thirteenth century, the Ottoman Empire would soon embrace much of the present-day Balkan countries as well as present-day Turkey and last into the early twentieth century. The Ottoman Empire embraced many diverse ethnic peoples such as Persians (present-day Iranians), Turks, Mongols, and other Mediterranean societies. It was largely Muslim, and the Ottoman leaders forcibly converted their conquered peoples. However, Jews, Greeks, and some Christians were allowed some degree of autonomy in business because it was profitable for the empire. The Ottomans often took children of Christian families to be raised in Muslim homes and trained them to be Janissaries, or paid soldiers. The zenith of the Ottoman Empire came to an abrupt halt with the defeat of the Ottomans at the Battle of Lepanto in 1571. This stopped the expansion of the empire into Vienna and further into the Hapsburg Empire. Perhaps the greatest Ottoman ruler was Suleiman the Magnificent, (c. 1494–1566), who made Istanbul a beautiful capital by building mosques, bridges, and palaces. He promoted the arts and music but, in the end, he was wary of Western civilization because it was a threat to Islam. Thus, the Enlightenment period of science and technology that swept Europe in the eighteenth century bypassed the Ottoman Empire. Eventually, the tight hold of the successors of Suleiman fell to ill-prepared sultans. It was customary for all of the sultan's sons but the eldest to be killed when the son took the throne. While they were awaiting their fates, most of them led decadent lives that ill-prepared them to lead. Most of those who followed Suleiman were either incompetent or mentally deranged, and the Ottoman Empire was badly weakened before its eventual demise after World War I (1914–18). Before the actual dissolution of the Ottoman Empire, a committee known as the Young Turks deposed the sultan in 1909. They installed a leadership under Enver Pasa that allowed more liberal reforms such as the establishment of political parties, permission to discuss religious ideas, and the admittance of women to public life. However, during World War I, Pasa allied his country with the Russians and Colonel Mustafa Kemal led his troops into battle. The Turks were defeated and, after the War, Anatolia was divided into French, Italian, and Greek spheres of influence. Mustafa Kemal (1881–1938) organized a guerrilla movement which managed to clear all foreign troops from Turkish land. In 1923, the National Assembly declared Turkey a republic, and Mustafa Kemal was elected its first president. The grateful people gave him the nickname of *Atatürk,* which means father of the Turks. Without this remarkable leader, Turkey would not have entered the twentieth century as an independent country, carrying with it its heritage of over 10,000 years.

Mustafa Kemal (Atatürk) planned to design his country's political and intellectual future on that of western Europe. Atatürk first changed his country from an Islamic-based society to a secular culture. This did not come without resistance, but he persisted and closed religious schools, changed to the Gregorian calendar, and instituted a European legal code. He also encouraged people to adopt a more western look by outlawing the women's veil and the men's fez. Perhaps the most dramatic was the change from Arabic to Roman script for the Turkish language. This made the Turkish language accessible to other people who used the Roman alphabet.

After Atatürk's reign, Turkey began its troubled journey towards self-determination. The most pressing problem seemed to be whether Turkey could remain a secular country or would return to its Islamic roots. On several occasions it was necessary for government troops to intervene to repress the opposition. Also the small Kurdish minority began mounting its demands for independence from Turkey and staging guerrilla attacks. During the 1960s, contrary to Atatürk's decree that the military should stay out of politics, the armed forces staged three coups against the sitting government. The elected government was replaced by the Committee of National Unity which basically acted like a dictatorship. General Cemal Gürsel assumed the duties of president, prime minister, and defense minister. This set the stage for political unrest which continued into the 1970s. The main reasons for political dissent were whether to have a secular or Islamic state; whether to have authoritarian or democratic rule; and whether to have a capitalistic, western economy which embraced free trade. By the 1980s other western nations were beginning to doubt Turkey's resolve to enter the world of free trade and democracy. The government was constantly plagued by the opposition which wanted to return to an Islamic state. Another coup in 1980 was deemed necessary to control terrorism and establish security. In 1981 the military took the drastic measure of disallowing political parties and forbidding their leaders from participating in any political activity. The military junta confiscated all property previously owned by these political parties. Many western democracies were appalled at the draconian tactics used by the military. Finally the military agreed to elections but only under severe restrictions. It tried to ban as many political parties as possible to avoid having a coalition government. Several previously involved politicians were placed under house arrest for violating rules not to participate in politics. The military finally transferred power to an elected parliament in 1983. In the 1990s, an Islamic front formed a coalition government and had enough seats to elect a female prime minister of the True Path Party. However, the pro-Islamic government was forced to resign in 1997 under pressure from the military. Since that time the military has increased pressure to thwart the inroads of religion into government.

Culturally, Turkey is a treasure trove of riches. Art, music, and poetry dating back thousands of years have come to light as a result of the opening up of research during Atatürk's regime. Slowly the western world has come to discover the hidden treasures of Turkey's past. Before Atatürk's regime, most literature came from Islam. However, during the 1850s, Turkey began to experiment with literary forms from Western Europe, such as the novel. Some famous writers from that period were Halid Uakligil, Huseyin Gurpinar, and the poet Tevfik Tikret. Nationalistic fever also gripped Turkey. Writers from that period include Ziya Gokalp, Reat Guntekin, Kemal Tahir, and Orhan Kemal. Some famous present-day writers are Yaar Kemal, Mahmut Makal. And the woman writer Nehziye Meric. Turkish writing has combined a blend of themes from ancient and modern cultures into a type of literature called *Eyubbî*. Another socio-political type of literature called *Tahirî* reflects elements of anti-Kemalist sentiment, glorifying the ancient Ottoman past. Islam forbade the representation of the human form in art, but modern-day Turkey has many artists. Much of this art is folk and tribal art, Turkish carpets, vases, and other pottery forms. Most Turkish music reflects its Middle Eastern heritage. At the close of the twentieth century, the secular government supports the arts, and tourism has expanded the knowledge of Turkey's contributions to civilization. Economically, Turkey has tried to ally itself with the West, but Islamic factions are always at the forefront attempting to curtail Turkey's openness to other cultures.

At the close of the Ottoman Empire, the economy of Turkey was very weak. For the most part people engaged in subsistence agriculture. The new government did much to try to stimulate the economy by developing a manufacturing industry, mostly subsidized by the state. In the 1980s, the president Türgüt Özal succeeded in opening Turkish markets to other nations. Other nations were interested in Turkey's development of food processing, textiles, and the manufacture of automobiles. Turkey has applied for inclusion in the European Community, but, as of 1999, it has not proven itself economically stable enough to be admitted. Present-day Turkey's main exports include cotton, tobacco, sugar beets, fruits, and figs. It is vigorously promoting tourism which is of great interest because of all the classical ruins. Most industries are state controlled. Mining, small production of trucks and machine tools, and petroleum refineries complete this fledgling economy.

From its inception as a new nation in 1923, Turkey has endeavored to create a secular society in which all people's rights would be respected. This also involved changing the status of women. In the previously religious based society, women did not have many civil rights and were often relegated to roles as wives and mothers. Modern-day Turkey has eliminated conservative dress styles and has encouraged the education of all of its citizens. A recent prime minister was a woman, and many women have since entered politics. However, the religious political parties are still trying to force the government to return to its Islamic roots. This remains a constant source of friction at the close of the twentieth century.

Timeline

10,000 B.C.   Evidence of earliest inhabitants

There is evidence of early agrarian societies in what is now present-day Iraq and southern Turkey from about 10,000 to 8,000 B.C. This site at Nevali Cori is discovered in the 1980s where archaeologists find structures for grain storage and a room with a terrazzo floor which is thought to be a temple. Ashliki Höyük, an ancient city, reveals burial of the dead.

6500 B.C.   First urban settlement at Çatalhüyük

There is archaelogical evidence of a civilization which leaves remains of buildings and artifacts. Some are thought to be religious structures. The people raise livestock as well as plant crops. There is evidence of small manufacturing and trade. The population estimate is between 5,000 and 10,000 inhabitants.

2371–2316 B.C.   Reign of King Sargon of Akkad

King Sargon begins an Assyrian dynasty and rules from Mesopotamia to what is now present-day Turkey.

2200 B.C.   Hittites invade

The Hittites, an Indo-European people, cross the Caucasus mountains and invade Anatolia. They choose their name from the Hatti, indigenous people whom they conquer, and settle in the area. They are warriors from the Bronze Age who adopt the religious deities of the Hatti. They even write in the cuneiform alphabet. They impose an almost feudalistic societal system on the Hatti with peasants supporting the highly ranked warrior caste. Their capital city is Hattusas.

1275 B.C.   Hittites battle Ramses II of Egypt

The Hittites do battle with the pharaoh of Egypt, Ramses II (r. 1292–1225 B.C.). They fight his troops to a standstill at the Battle of Kadesh. This battle stops the advancement of Egypt into Mesopotamia. Hattushili III (of Anatolia) finally forms a peace agreement with Ramses by sending his daughter to wed him. This is verified by a carving on the temple at Abu Simbel in Egypt.

1184 B.C.   End of the Trojan War

Excavations by Heinrich Schliemann (1822–90) in 1870 uncover the ruins of the ancient city of Troy. The city is immortalized in the Greek poem, *The Iliad,* by the poet Homer who supposedly lived around 850 B.C. Scholars are divided as to whether Homer was an actual person or a conglomerate of poetic writers. The real war is about the control of trade routes. The poem by Homer recounts the kidnapping of Helen Queen of Sparta by a Trojan prince named Paris and the war that ensues from her capture. When the Trojans open the gates of the city, the Greek warriors emerge and conquer them. Until recently, this war was thought to be fictitious, but discoveries of archaeologists prove the existence of the ancient city.

1200 B.C.   Collapse of the Hittite Empire

The Phrygians, a competing population, conquer the Hittites and burn their capital city Hattusas. This comes on the heels of a long period of aggression by many different civilizations in Anatolia. The Phrygians basically adopt the religious beliefs of the Hittites and absorb their civilization. They are architects, builders, and iron workers. Two great and perhaps legendary kings of the Phrygians are King Gordius and King Midas. King Gordius is famous for the *Gordian Knot.* Supposedly King Gordius ties a knot so complicated that he declares whoever shall untie it shall be fit to rule the Phrygians. This is finally accomplished by Alexander the Great who does not untie the knot but cuts it with his sword. King Midas is famous for the *Midas Touch.* According to legend, everything the king touches turns to gold. The capital city of the Phyrigians is Gordium which is close to present-day Ankara.

1000 B.C.   Asia Minor (Anatolia) invaded

The west coast of Anatolia is systematically invaded by Aeolians, Ionians, and Dorians from Greece. They populate the Turkish mainland and fight with existing civilizations. From about 1200–900 B.C., Ionian Greek refugees flee to the west coast of Anatolia. They establish commerce with the Lydians and Phrygians and found the great city of Byzantium (657 B.C.) on the Bosporus strait.

699–546 B.C.   Lydians rule Anatolia

The Lydians, a warrior people from Thrace, take over the Phrygian society. They are goldsmiths, deriving their gold from the Hermus River. The Lydians are one of the first societies to use coinage to conduct trade. They make coins out of gold and silver. This gives importance and power to their king, Croesus, who is considered to be the richest man in the world. Croesus (d. 546 B.C.) rules from his court at Sardis, and the Lydians control most of western Anatolia until the invasion of the Persians (present-day Iranians).

546 B.C.   King Cyrus invades Lydian kingdom

King Cyrus of Persia, also known as Cyrus the Great (600–529 B.C.), invades the kingdom of Croesus and expands his empire. He conquers the Babylonians and frees the Jewish people from captivity. He controls the cities on the Anatolian coast, a move the Greeks find repressive. However, what the Greeks lose in name, they gain in civilization. By the time of the arrival of Alexander the Great, the expansion of Greek civilization is complete.

334–323 B.C.   Reign of Alexander the Great

Alexander III of Macedon (356–323 B.C.) succeeds his father, Philip II, to become one of the greatest leaders of the western world. He conquers all of Greece and extends his power to the Persian empire. He occupies northern Egypt and founds the city of Alexandria, home of the world's greatest library (lost to modern civilization). By liberating the population from Persian control, he spreads Greek language, literature, and philosophy more than ever before. His untimely death from a fever in 323 B.C. creates havoc in his empire, and the era after his death is marked by infighting among cities. Finally the Romans gain control of Alexander's empire.

200 B.C.   First references to the Turks

The first written references to Turks appears in Chinese writings naming the tribes *Hsiung-nu*, a form of the western word for Hun. These are believed to be the ancestors of modern-day Turks.

334–133 B.C.   The Hellenistic period

The Hellenistic period is noted for the spread of western thought and culture. The reign of Alexander the Great takes Greek culture throughout the region, including Anatolia. The cities of Ephesus, Pergamum, and Antioch become centers of culture and prosperity. With the death of Alexander comes a search for leadership. There is much quarrelling and infighting. However, the Hellenic way of life is securely established, and those who come after Alexander adopt its customs and ideas. The lack of leadership leads to the invasion of the Romans who next rule Anatolia but eventually succumb to its culture.

190 B.C.   Last of the Persian kings

One of Alexander's generals, Seleucus (358–280 B.C.), gives his name to a dynasty that lasts until the death of Antiochus III (242–187 B.C.) who, after his defeat by the Romans at the Battle of Magnesia near Ephesus (190 B.C.) cedes all his territory to Rome. The territory that is modern-day Turkey becomes known as the Roman province of Asia Minor. The city of Byzantium (later called Constantinople and now called Istanbul) eventually becomes the center of the Roman Empire and later the capital of the Eastern Roman or Byzantine Empire.

The Roman Period

133 B.C.   Rome consolidates its holdings in Asia Minor

The area previously known as Anatolia, the Lydian Empire, the Persian Empire, and the Seleucid Dynastic Empire is now the beginnings of the Roman Empire. All of Anatolia, except for Armenia, is incorporated into the Roman Empire by 43 A.D. The peace of the Roman Empire brings prosperity and security to the area. The inhabitants speak more Greek than Latin but are part of a Roman system of government with representative delegates sent to Rome. Roman citizenship is extended to all free men and women.

A.D. 285–305   Reign of Emperor Diocletian

Diocletian (245–313) decides to reorganize the Roman Empire. Because of its size and different peoples (Greek and Roman) the Empire is divided into Latin-speaking and Greek-speaking halves. He declares Nicomedia as an Eastern capital city.

306–337   Reign of Emperor Constantine

Constantine establishes his capital city at Byzantium and calls it the "New Rome." It is strategically located on the western side of the Bosporus in the middle of Europe and Asia Minor. This crossroads of civilization remains important throughout history.

330   Beginning of organized Christianity

Anatolia is not only the crossroads of civilization, but the birthplace of all of the world's greatest religions. Christianity is no exception. Even though Jesus was born in Bethlehem, his main ministry was preached through Paul of Tarsus. Paul preached in Ephesus and spreads the Gospel throughout the area. By 380, Christianity is the official religion of the Roman Empire. By the time of this official decree, the majority of the people in the area are already Christian; under the Edict of Milan (313), Christianity is granted "toleration."

330–1071   The Byzantine Empire

The Byzantine Empire is actually the Eastern Roman Empire. The Western Roman Empire collapses in about 500, but the Eastern Empire stays intact. It inherits all the trappings of the old Roman Empire, with a large bureaucracy and a centralized government. It establishes Orthodox Christianity as the official religion. During the reign of the emperor Justinian I (527–563), the Byzantine Empire reaches a cultural zenith.

600–1071   Constantinople head of huge empire

The city of Constantinople becomes the head of an empire that includes Greece, Anatolia, Syria, Egypt, Sicily, Italy, and the Balkans. The Byzantine Empire spends enormous amounts of money to defend these vast frontiers which are continuously under attack. The Goths, Slavs, Persians, and Arabs all seek to expand their empires at the expense of the Byzantines. By the eighth century, continuous raids seriously weaken the empire. Constantinople is beseiged and most trade is halted.

The Seljuk Turks

1071   Beginning of Seljuk rule

The Seljuks are basically responsible for the origins of modern Turkey. The Oguz, a nomadic Turkish people, are Islamic. Slowly they migrate out of Pakistan and into present-day Iran and Iraq. They travel into Anatolia, converting people to

Islam. In 1071 at the Battle of Manzikert, they capture the Byzantine emperor Diogenes (Romanus IV, d. 1071). He cedes part of Anatolia to the Seljuks, and they set up an empire they call the "Sultanate of Rum." *Rum* is a corruption of Rome or the Roman Empire.

1081   The Sultanate of Rum

The capital city of this new empire is at Konya (Iconium) in present-day Turkey in the central Anatolian plateau. The *gazis* or Turkish warriors carve out their territories and establish feudal states, the strongest of which is Rum. The Seljuks adopt the social and economic structure of the former Byzantine Empire, except now it is no longer Christian, but Muslim. They impose their language and religion on the inhabitants and intermarry with former subjects of the Byzantine Empire. They are famous for their architecture and stone bridges and also for their poet Celaleddin Rumi (C. 1207–73), the founder of the Whirling Dervishes, known for their wild religious dances.The Seljuks quarrel among themselves and eventually split into two groups—the Rum Seljuks and the Great Seljuks of Syria and Iraq.

1243   Fall of the Seljuk Empire at Rum

The Mongols defeat the Seljuks and their emperor Sultan Keyhüsrev II at the Battle of Kösedag.

The Ottoman Empire

1300   Rise of the Ottoman Empire in Turkey

The Ottoman Turks derive their name from their Seljuk leader Osman I (1259–1326). As Anatolia becomes disorganized, the Turkoman tribes move in and begin to organize themselves into small emirates. They lay claim to the remnants of the Sultanate of Rum.

1326   Beginning of Ottoman Expansion

Osman's son Orkhan (1279–1359) takes over and captures Bursa in 1326. He marries a Byzantine princess to increase his power. Next he adds Nicaea and Nicodemia (Ismit), crosses into Europe, and conquers Thrace. He organizes a corps of mercenary soldiers known as Janissaries. They are young boys taken from Christian families and trained in fighting and Islam. They live at Topkapi Palace and study mathematics, philosophy, and the arts. Orkhan's son Murad adds Bulgaria, Albania, and Serbia to his kingdom.

1453   Conquest of Constantiople

The jewel of the Seljuk empire, Constantinople, becomes the prize of the Ottomans. Mehmet the Conqueror (1432–81) builds a siege castle on the Bosporus (Rumeli Hisar). Soon after are the conquests of the Peloponnese, Greece, and Trebizond in Asia Minor. Mehmet wins the Battle of Otranto in

Italy but never incorporates Italy into the Ottoman Empire. The Ottomans are now an established force, and their main enemies are Poland, Hungary, Austria and Venice. Mehmet is a great ruler, intelligent, and well-read. The main goal of the Ottomans is to establish themselves in Europe (Vienna). This they fail to do after numerous attempts.

c. 1460–c. 1570   Effects of the Ottoman Empire on the development of Turkey

Turkey is now a melting pot of Greeks, Bulgarians, Serbs, Arabs and Jews expelled from Spain and Portugal, and Persians. Their language is enriched by Greek, Arabic, and French. They become a prosperous force. The Ottoman Empire has Turkish roots and is Islamic in faith. Ethnicity is determined by religious affiliation, not language or culture. Muslims are all considered one group. Other religions are grouped into *millets,* largely autonomous religious communities. Under their most famous leader, Süleiman the Magnificent, they increase learning and the arts and reach their height as an empire. However, the sultans who follow Süleiman never come close to his greatness. They are affected by the perverse system of killing all rivals to the throne. Those who survive are not always the best and the brightest. They are often given disparaging names such as Selim "the Grim," and Ibrahim "the Mad." The decline of the great Ottoman Empire begins with its defeat at the Battle of Lepanto in 1571.

1571   Battle of Lepanto

The Ottoman naval fleet is destroyed by Spanish and Venetian forces at the Battle of Lepanto. No longer are they considered to have the military might they formerly exerted on the Mediterranean area. The Ottomans disdain the Europeans and their culture and refuse to recognize that they are no longer dealing from a position of strength. Even in matters of trade, they become increasingly indebted to Europe. Their leaders are not capable, and the Ottoman Empire begins its slow decline which lasts until the twentieth century.

1699   Treaty of Carlowitz

The Treaty of Carlowitz ends the advancement of the Ottoman Empire into Europe. The Ottomans are forced to accept defeat.

18th century   Age of Enlightenment

Turkey, by virtue of its defeat by the Europeans, begins to import some ideas of the Enlightenment period. The printing press and French designs first appear in Turkey. The French are infatuated with Turkey and many things "oriental." The Turks also import tobacco from Europe. They send envoys to European capitals and receive ambassadors as well. The French are interested in stopping Russia's advance into the Ottoman Empire.

The Ottoman Period

The Ottoman Empire began in western Turkey (Anatolia) in the thirteenth century under the leadership of Osman (also spelled Othman) I. Born in Bithynia in 1259, Osman began the conquest of neighboring countries and began one of the most politically influential reigns that lasted into the twentieth century. The rise of the Ottoman Empire was directly connected to the rise of Islam. Many of the battles fought were for religious reasons as well as for territory.

The Ottoman Empire, soon after its inception, became a great threat to the crumbling Byzantine Empire. Constantinople, the jewel in the Byzantine crown, resisted conquest many times. Finally, under the leadership of Sultan Mehmed (1451–81), Constantinople fell and became the capital of the Ottoman Empire.

Religious and political life under the Ottoman were one and the same. The Sultan was the supreme ruler. He was also the head of Islam. The crown passed from father to son. However, the firstborn son was not automatically entitled to be the next leader. With the death of the Sultan, and often before, there was wholesale bloodshed to eliminate all rivals, including brothers and nephews. This ensured that there would be no attempts at a coup d'état. This system was revised at times to the simple imprisonment of rivals.

The main military units of the Ottomans were the Janissaries. These fighting men were taken from their families as young children. Often these were Christian children who were now educated in the ways of Islam. At times the Janissaries became too powerful and had to be put down by the ruling Sultan.

The greatest ruler of the Ottoman Empire was Suleyman the Magnificent who ruled from 1520–66. Under his reign the Ottoman Empire extended into the present day Balkan countries and as far north as Vienna, Austria. The Europeans were horrified by this expansion and declared war on the Ottomans and defeated them in the naval battle of Lepanto in 1571. Finally the Austrian Habsburg rulers were able to contain the Ottomans and expand their empire into the Balkans.

At the end of the nineteenth century, the Greeks and the Serbs had obtained virtual independence from the Ottomans, and the end of a once great empire was in sight. While Europe had undergone the Renaissance, the Enlightenment, and the Industrial Revolution, the Ottoman Empire rejected these influences as being too radical for their people. They restricted the flow of information and chose to maintain strict religious and governmental control. The end of Ottoman control in Europe came in the First Balkan War (1912–13) in which Greece, Montenegro, Serbia, and Bulgaria joined forces to defeat the Ottomans.

During the First World War, (1914–18), the Ottoman Empire allied with the Central Powers and suffered a humiliating defeat. In the Treaty of Sévres, the Ottoman Empire lost all of its territory in the Middle East, and much of its territory in Asia Minor. The disaster of the First World War signaled the end of the Ottoman Empire and the beginning of modern-day Turkey. Although the Ottoman sultan remained in Constantinople (renamed in 1930, "Istanbul"), Turkish nationalists under the leadership of Mustafa Kemal (Ataturk) (1881–1938) challenged his authority and, in 1922, repulsed Greek forces that had occupied parts of Asia Minor under the Treaty of Sévres, overthrew the sultan, declared the Ottoman Empire dissolved, and proclaimed a new Republic of Turkey. That following year, Kemal succeeded in overturning the 1919 peace settlement with the signing of a new treaty in Lausanne, Switzerland. Under the terms of this new treaty, Turkey reacquired much of the territory—particularly, in Asia Minor—that it had lost at Sévres. Kemal is better known by his adopted name "Ataturk", which means "father of the Turks". Ataturk, who ruled from 1923 to 1938, outlawed the existence of a religious state and brought Turkey a more western type of government. He changed from the Arabic alphabet to Roman letters and established new civil and penal codes.

The Ottoman Empire, c.1812

1734 First European military school opens in Constantinople

A school of military engineering opens in Üsküdar, a suburb of Constantinople. This is followed by other schools sponsored by the French to help the Ottomans resist the Russians. This is not entirely successful, and the Ottomans are forced into another treaty ceding control of the Black Sea.

1774 Treaty of Küçük Kaynarca

The Treaty of Küçük Kaynarca with Russia cedes control of the Black Sea to Russia. It also grants Russia the right to interfere on behalf of Balkan Christians in the Ottoman Empire.

1792 New order under Selim III

Selim III (1761–1808) comes to power and attempts to re-organize his army under European models. He dismisses the Janissaries and tries to modernize taxes and administration. The Janissaries revolt and overthrow the Sultan in 1807.

1806 Russia invades Turkey

Russia begins a series of military attempts to control the Black Sea area.

1808 Rule of Mahmut II

Mahmut II (1785–1839) comes to power and massacres the Janissaries in 1826. He forms a new army and tries to organize it in a Western European manner. Other European ideas, such as nationalism, an idea born of the French Revolution in 1789, also come into the Ottoman Empire, and factionalism begins to divide people who were once united. The ideas of the French Revolution are the beginning of the undoing of the old Ottoman order.

19th century Religious and ethnic conflict begins

Many groups within the Ottoman Empire begin breaking away along lines of religion and ethnicity. The Serbians (1804), Egyptians (1811), Greeks (1821), and Bulgarians (1826) all break free. In 1853 the Russians declare war on the Ottomans (the Crimean War) in an attempt to take over the Balkans.

1839 Tanzimat reforms

The Turkish reformers, Mehmed Emin Âlî Pasa (1815–71) and Mehmed Fuad Pasa (1815–69), want to introduce the ideas of the French Revolution to Ottoman society. They want to free the society from the old bureaucracy and implement the ideas of liberty and equality. This is considered to be the beginning of the reforms later adopted by Mustafa Kemal (Atatürk).

1854 Siege of Sevastapol

The Russians under Czar Alexander II (1818–81) attempt to annex the Balkan countries now under Ottoman control. The Russian Czar feels Christians are not being fairly treated in Jerusalem, now under Turkish rule. The Turks enlist the aid of Britain and France during the war. The siege of Sevastapol during the Crimean War lasts a year. During this time, English-born Florence Nightingale (1820–1910) heads a group of thirty-eight nurses and sets up a hospital at Scutari (1854) where her stringent sanitation rules reduce the death rate for cholera, typhus, and dysentery from 50 to 2 percent.

1856 Treaty of Paris

The Turks drive the Russians from the Balkans. The Treaty of Paris concludes the war, but the problem of the Balkans remains and will again surface during World War I (1914–1918). At the end of the Crimean War, the Ottomans grant concessions to the non-Muslims living in the Ottoman territories and seek more interaction with Europe.

20th century Emergence of Turkey as a modern republic

1876–1909 Reign of Abdül Hamid II

In the 1870s, Abdül Hamid tries to modernize the government, but it proves to be his undoing. In 1908, a group called the Young Turks insists on a return to the constitutional government first instated by Abdul Hamid and then reneged upon. The Young Turks force Abdül Hamid to renew the constitutional form of government.

1878 Russia again threatens Turkey

Russia again tries to invade a greatly weakened Ottoman Empire. It forces Turkey to sign the Treaty of San Stefano, which pushes the Russian border almost to Constantinople.

Later that year, European powers try to save the Ottomans, and the Congress of Berlin restores Turkey's old borders.

1908   Military coup

The Young Turks, promising a new democratic form of government, force Abdül Hamid II to abdicate. However, their good intentions are not well-received by the various factions in Turkey. Each ethnic group fears it will be sucked into a new form of Ottoman rule. Albanians, Slavs, and Greek and Armenian nationalists rise in protest. By 1912, all the Balkan possessions are lost to the Ottoman Empire. The Young Turks form a government under the leadership of Enver Pasa (1881–1922). The ideas of the Young Turks are those of the French Revolution, but they are elitist and do not appeal to the masses. They form the Committee of Union and Progress, later known as the Party of Union and Progress.

1914   Beginning of World War I

Enver Pasa decides to enter the war on the side of the Germans in order to defeat his old enemy, Russia. This proves to be disastrous. The Germans are defeated and the Ottoman Empire collapses. The Turks must accept the occupation of Constantinople.

1919–22   Greeks invade Turkey

Greece, which receives Ottoman territory in Thrace and Anatolia, invades Turkey in an effort to suppress the nationalist movement of Mustafa Kemal (1881–1938). In 1922, Kemal's forces defeat the Greeks and result in a revised peace treaty..

1919: May–1922: September   Mustafa Kemal leads war for independence

Mustafa Kemal, known as Atatürk, starts the Turkish war for independence. In April 1920, he is elected to the presidency of the Grand National Assembly. With the abolition of the Ottoman dynasty, an armistice is signed (see 1922) and Turkey becomes a new Turkish state. The republic is proclaimed (see 1923: October 23).

Mustafa Kemal is born in 1881 in Salonica, in present-day Greece, but then part of the Ottoman Empire. At this time, Turks do not have last names. Mustafa comes from a poor family, but his widowed mother sends him to a military high school where his teacher gives him a second name, Kemal, which means perfection. Mustafa Kemal begins his career by siding with military officials who topple the sultan's regime. He again is a hero during World War I, helping to liberate some provinces in eastern Turkey.

Mustafa Kemal is eventually given the surname of Atatürk (Father of the Turks) when he proclaims everyone must now have a last name. He begins major reforms in the new republic. The most important reform is to make the new country completely secular instead of religious. He abolishes religious schools and reforms education, giving women equal rights. He abolishes the old form of dress for women and the wearing of the fez (a red hat) for men. His new philosophy is known as Kemalism and later as Atatürkism. His new Civil Code is based on European models from Switzerland, Italy, and Germany. All people are equal under the law. People of all religions are free to practice their faiths without any persecution or special favors. He instates the use of the Western calendar instead of the Islamic one.

One of his most dramatic changes is to use the Latin alphabet for the Turkish language instead of the old Arabic script, following foreigners to pronounce and read the signs. Modern Turkish has twenty-nine letters (eight vowels and twenty-one consonants). He also introduces a new, more modern vocabulary and grammar meant to purify the new Turkish language from its old Arabic and Ottoman roots. This totally revolutionizes education and makes more people literate in a shorter period of time. Women benefit most from these reforms which enables them to achieve an equal education and opens the door to representative politics. Atatürk encourages the study of the ancient civilizations which gave rise to modern Turkey. He also supports the study of Western culture and the arts. He opens museums and showcases opera and classical music. To disseminate knowledge to all people, he creates "People's Houses," designed to provide access to new culture to all the people. The activities provided are: language and literature; fine arts; theater; sports; social assistance; public education; libraries and publications; rural development; and history and museums. There are also smaller branches of these houses called "People's Rooms." Even though they are for the general public, they are operated by party members only. Atatürk stabilizes life in Turkey by securing its borders and bringing peace to a region used to invasions and war. On the economic front, the overriding policy is *etatism* or the formation of private enterprise. This helps stimulate a fledgling economy.

1922   Mudania armistice

This grants total sovereignty of the Turks over Anatolia. General Mustafa Kemal emerges as the new leader of Turkey. He abolishes the sultanate and Mehmet VI goes into exile.

1923: July 24   Treaty of Lausanne

Turkey signs the Treaty of Lausanne which establishes peace with Great Britain, France, Italy, Japan, Greece, Romania, and the Serb-Croat-Slovenian governments. Great Britain is given permission to annex Cyprus and proclaim British nationality for its citizens. Turkish citizens may refuse such automatic citizenship.

The ruling family of Turkey includes, from left, Ismet Pash, Mrs. Mustafa Kemal, and Mustafa Kemal (known as Atatürk). (EPD Photos/ CSU Archives)

Modern Turkey

1923: October 29 Turkish Republic proclaimed

Mustafa Kemal is its first president. The official political party is the Republican People's Party (RPP). Ankara is chosen as the capital city because it is more protected than Constantinople.

1930 Istanbul officially replaces Constantinople

In keeping with the modernization of the country, Turkey changes the name of one of its oldest cities. Constantinople is a leftover from the old Byzantine Empire.

1932 Turkey joins League of Nations

Turkey beomes a member of the League of Nations.

1938 Atatürk dies

Just as the new way of life is getting started, the founder and hero of Turkey, Mustafa Kemal Atatürk, dies.

1936 Treaty of Montreux

This treaty gives Turkey the right to fortify the borders of the Dardanelles.

1939–45 World War II

Atatürk is succeeded by his prime minister Ismet Inönü (1884–1973). He manages to keep Turkey neutral during the war. This is important because the new country is now stable and free from outside conflicts for the first time in thousands of years. Inönü also wants to keep Turkey from being absorbed into the Soviet Empire. The Marshall Plan, developed by former U.S. President Harry S. Truman (1884–1972) and George C. Marshall (1880–1959), gives aid to Greece and Turkey (among other countries) to prevent them from being wooed by the Soviets. This economic aid gives stability to the area. Inönü also welcomes a real parliamentary government after Atatürk's solitary leadership.

1946 Inönü recognizes opposition political parties

In order to remain in good standing with the charter of the United Nations and adhering to individual principles of rights and privileges, Turkey allows the formation of multiparty

League of Nations

Formed in the wake of World War I, the League of Nations—the forerunner of the United Nations—was the world's first international organization in which the nations of the world came together to maintain world peace. Headquartered in Geneva, Switzerland, the League was officially inaugurated on January 10, 1920. During the life span of the organization, over sixty nations became members, including all the major powers except the United States. Like the United Nations, the League of Nations pursued social and humanitarian as well as diplomatic activities. Unlike the United Nations, the League was primarily oriented toward the industrialized countries of the West, while many of the regions today referred to as the Third World were still the colonial possessions of those countries.

The League's structure and operations, which were established in an official document called the *League of Nations Covenant*, resembled those adopted later for the United Nations. There was an Assembly composed of representatives of all member nations, which met annually and in special sessions; a smaller Council with both permanent and nonpermanent members; and a Secretariat that carried out administrative functions. The League's procedures for preventing warfare included arms reduction and limitation agreements; arbitration of disputes; nonaggression pledges; and the application of economic and military sanctions.

Although the League of Nations solved a number of minor disputes between nations during the period of its existence, it lost credibility as an effective peacekeeper during the 1930s, when it failed to respond to Japan's takeover of Manchuria and Italy's occupation of Ethiopia. Ultimately, the organization failed in its most important goal: the prevention of another world war. However, it did make contributions in the areas of world health, international law, finance, communication, and humanitarian activity. It also aided the efforts of other international organizations.

By 1940 the League of Nations had ceased to perform any political functions. It was formally disbanded on April 18, 1946, by which time the United Nations had already been established to replace it.

activity. The Republican People's Party is the main political party, but now the Democrat Party (DP) is formed mainly at first as the loyal opposition. It is headed by former Prime Minister Celâl Bayar (1883–1986) and Adnan Menderes (1899–1961). General elections in 1946 give the DP 61 seats out of 465 in Parliament. The DP is a middle-class party and wants to lift restrictions on the private economic sector. They basically oppose the state's control of business.

1950: May   New elections

The DP gains widespread popularity, and these elections give the DP 408 seats in Parliament versus sixty-nine to the RPP. Bayar is now president, and he names Menderes as his prime minister. The DP enacts laws prohibiting and dissolving all religious political parties.

1950s   Crisis in Cyprus

Ethnic Greeks from the island of Cyprus (then a British colony) demand union with Greece. The island's Turkish community opposes the Greek-Cypriots, demands a partition of the island between Greece and Turkey, and seeks help from Turkey to avoid becoming part of Greece.

1952: February 18   Turkey gains admission to NATO

Turkey solidifies its alliance with the West by gaining admittance to the North Atlantic Treaty Organization, a unit of military support.

1954: May   Elections give DP large gains

The DP continues to increase its parliamentary majority. The government is increasingly sensitive to the attacks by the RPP and limits freedom of the press and right to assembly.

1957: October   People disenchanted with DP

There is much political unrest due to the strict controls of the DP. In these elections opposition parties win many seats, thus diluting the DP's strength.

1960   Inönü encounters discontent

RPP leader Inönü takes a tour of central Anatolia and encounters outbreaks of resistance to the Bayar regime. Menderes retaliates by imposing martial law and limiting political activity. There are demonstrations, and in April, 1960, students in Istanbul are fired upon by the police and several are killed. The army is now at odds with its citizens.

1960: May 27   Military coup

The Turkish army under General Cemal Gürsel takes control of many government buildings and arrests President Bayar and Prime Minister Menderes. They accuse Bayar and Menderes of creating a despotic regime. There is no violence, and General Gürsel becomes the acting president, prime minister, and defense minister. The acting government is replaced by the CNU (Committee of National Unity) composed of the officers who are responsible for the coup. Gürsel announces that this is just an interim government, and he will hold elections soon. The main benefit of this coup is that Turkey becomes a country of various political parties.

1960: August Cyprus gains its independence

Cyprus receives its independence from Britain and becomes a member of the British Commonwealth. Under the agreements signed by Britain, Greece, and Turkey, the island is to remain indpendent: union with Greece or Turkey, or a partition of the island is prohibited. In addition, Greeks and Turks are to share power on the island. However, neither Greek- nor Turkish-Cypriots relinquish their respective goals of union with Greece, or partition between Greece and Turkey. As a result the Cypriot constitution breaks down in 1963 and plunges the island into civil war.

1960: October Trial of Bayar and Menderes

About 600 former government officials are tried, and most are found guilty of corrupting the government. Fifteen death sentences are ordered. Bayar is sentenced to life in prison but Menderes is hanged.

1961: January Second Turkish Republic

This new government submits constitutional reforms but remains faithful to Kemalism. The new constitution is approved by 60 percent of the voters.

1961: October New elections

In these elections, fourteen new political parties offer candidates for seats in Parliament. Only four parties gain seats. A coalition government is formed. The Justice Party is formed from the ruins of the old DP. New parties emerging are the New Turkey Party and the Nationalist Action Party. The new legislature elects General Gürsel as its president. He asks former President Inönü to form a coalition government. The goal of the new Justice Party (JP) is to get amnesty for former DP party members. After much dissension, Inönü grants amnesty to 283 of those convicted in the 1960 trials. Two years later, former President Bayar is released.

1963–64 Civil War in Cyprus

Following a breakdown of the island's constitution, civil war breaks out between Greek- and Turkish-Cypriots. The Turkish-Cypriots, supported by Turkey, gather in heavily defended enclaves throughout the island. A United Nations peacekeeping force is sent to the island in 1964, and a tense standoff ensues. Yet, with the exception of fighting in 1967, the island remains at peace until 1974.

1964 Inönü again elected

Inönü again is the leader of Turkey but resigns shortly thereafter in a vote of no confidence. Another interim government takes the helm.

1965: October Süleyman Demirel new leader

Süleyman (b. 1924), of the JP, forms a new government. He has enough votes to form a single party government. The JP is a continuation of the former DP, and Süleyman is a sympathizer of the now executed Menderes. His main support comes from rural farmers. The JP still formally supports secularism but shows more tolerance for open support of Islam. This also appeals mainly to traditional farmers. Under Süleyman the government develops a solid economic policy, expands education, and creates positions in the government to give more people jobs. Süleyman favors a free economic market which the more traditional wing of the JP opposes. His party keeps pressing him to install Islam as a part of government. This sets up the opposition party, the RPP, to begin the fight to maintain a secular government.

Late 1960s The rise of Bülent Ecevit

Ecevit has government experience from being part of Inönü's cabinet. He represents the RPP and has an authoritarian voice within the party. He is also an intellectual and a popular poet. He represents a return to Kemalism and directs his appeal to the younger voters. He is one of the first to recognize the power of television to promote party politics. He serves as prime minister several times in the 1970s and again in the late 1990s.

1969: October Election results

The Justice Party becomes the majority party in these elections. However, it is evident that both parties are pulling away from centrist politics. Emerging from this election is Alparslan Türkes, an organizer of the Republican Nation Party, later renamed the Natinal Action Party (NAP). This party is ultra authoritarian and nationalistic. The party line is to re-incorporate Islam into party politics. He organizes a youth movement which is known as the "gray wolves." These young people are basically gang thugs who are weak in ideology but strong in physical violence. They succeed in causing many demonstrations. This election sets the tone for the volatile nature of 1970s politics to come.

1970 Turkey enters period of political turmoil

Social changes contribute to the instability of the government. Many people are moving to urban areas. These are formerly rural people who have more fundamental Islamic leanings. Their arrival in the cities heralds the push for a return to a more religiously oriented form of government. New political parties emerge with religious agendas. They are often led by terrorist factions and there are many episodes of violence.

The new parties—the Turkish Communist Party, the National Salvation Party, and the Nationalist Action Party—engage in terrorist acts against the government and its own citizens. The weak coalition government cannot control these groups and Turkey slowly descends into a state of civil war. This lasts until the 1980s when again the military is forced to take control of the government.

1974: July–August   Turkey invades Cyprus

After a coup supported by the military government in Greece installs an anti-Turkish government on Cyprus, Turkish troops invade the island. The Turkish invasion nearly sparks a Greek-Turkish war. As a result of the fighting, Turkish-Cypriots move to the northern part of the island protected by Turkish troops. Greek-Cypriots congregate in the southern part of the island. Turkish-Cypriots occupy roughly thirty-seven percent of the island and demand that the republic's government become a federation. Peace talks with the Greek-Cypriots continue intermittently, but over twenty-five years later, the island remains divided.

1974–75   Emergence of the Sovietists

Many Leftist groups emerge in this period of political unrest. Some favor the Chinese and the philosophies of Mao Zedong, and some favor the Soviet regime. These groups are faced with opposition from the ultra nationalist groups who favor a return to Kemalism. The violent and radical National Movement Party (NMP) is led by Türkes and speaks of a return to a great Turkish state. It hides its racist views by proclaiming all people of Turkey true Turks. However the NMP is hostile to minorities such as the Kurds. The NMP also embraces a return to Sunni Islam. Its main ideology is violence.

1978: December 26   Ecevit proclaims martial law

Bülent Ecevit institutes martial law.

1980   Kurds organize for independence from Turkey

Abdullah Ocalan, head of the PKK (Kurdish Workers' Party) begins forming opposition troops in nearby Syria. These terrorist groups are trained to attack the Turkish government.

1980: September 12   Military rule proclaimed

All property of former political parties is confiscated.

1981: April   General Kenan Evren takes control

In order to control the outbreaks of violence, Chief of Staff General Kenan Evren outlaws all political activities. Previous political parties are banned, and former politicians are precluded from holding any political office for ten years. Those who are members of the Grand National Assembly (parliament) are banned for five years.

1982   New constitution grants repressive powers to military

With this constitution, the military continues to exert undue influence over politics. Most voices of the opposition are shut down. This is not in keeping with a true democracy. The government justifies this by saying that pro-Islamic and pro-Kurdish groups are a direct threat to Turkey. However, nonviolent journalists and writers are arrested and often jailed. Many publications are shut down. Human rights watchers are alarmed at prison conditions and abuses.

1983   Turkish zone in Cyprus declares independence

Turkish-Cypriots formally declare their independence and call their state the Turkish Republic of Northern Cyprus. However, the United Nations condemns the move and only Turkey grants the breakaway state formal diplomatic recognition.

1983: April   Government slowly re-instates political activity

In anticipation of new upcoming elections, the government permits some political activity. However, the government reserves the right to investigate all new political parties. Also, in order to prevent more chaos, the political parties must receive at least ten percent of the vote to have a representative in the Grand National Assembly.

1983: August   Fifteen political parties form

Concerned about the possibility of another coalition government, the government bans most of these parties. They ban the Grand Turkey Party which is really a disguise for the banned JP and has ties to Süleyman Demirel. They arrest Demirel. His supporters form the True Path Party (TPP) with lawyer Husamettin Cindoruk as its leader. Former members of the RPP establish the Social Democratic Party with Erdal Inönü as its leader. He is the son of former president Ismet Inönü. During this period the government regularly bans any party with the slightest connection to past politics. Three parties eventually stand for election: the National Democratic Party (NDP) whose leader is Turgut Sunalp; the Populist Party (PP) whose leader is Necdet Calp; and the Motherland Party (MP) whose leader is Turgut Özal. The NDP is rightwing; the PP is leftist; and the MP is center-right. Most of the people support the right and center right parties. Former deputy prime minister Özal becomes the prime minister and gets a vote of confidence.

1983–87   Turgut Özal wins confidence of people

Turgut Özal (1927–93) becomes prime minister and gains great popularity among the Turkish people. He is the founder of the Motherland Party.

1984   Ocalan attacks Turkish police

Ocalan steps up his terorist activities and attacks the Turkish police and other government representatives. This is to enforce his demands for Kurdish independence.

1985: November   Birth of the Social Democratic Populist Party

The PP and the Social Democratic Party join forces to form the SDPP.

1986: April   Government relaxes control

The government now permits Demirel and Ecevit and some other former leaders to re-enter the political arena. They are still not allowed to run for office, but they may express their views.

Late 1980s   Kurdish unrest

The Kurds are the largest ethnic minority in Turkey and have been occupying the territory loosely known as Anatolia since the fifth century B.C. They are mainly located in the southeastern provinces of Turkey. In the cities they often live in *gecekondu* (ghettos) where they foment hopes of separatism. They speak their own Kurdish language. In the late 1980s, this language is not taught in any Turkish schools, and there are no Kurdish newspapers at this date, although there have been publications in the past. The Kurds are not united on religious principles, some following the Sunni rites and others the Shafii rites. There are also members of a secret Yazidi sect who combine elements of paganism, Islam, Christianity, Judaism, and Zoroastrianism. The Kurds lean toward disruption of authority, and in the late 1980s they organize the Workers' Party of Kurdistan (the PKK), lead by Abdullah Ocalan. The PKK is a Maoist guerrilla organization which stages terrorist raids in order to force the Turkish government to the Kurds on independent homeland.

1989–1993   Turgut Özal becomes president

Former prime minister Turgut Özal, head of the Motherland Party, becomes president of Turkey. He creates an economic austerity program which is successful in reviving the inflated economy. Özal's government is the first civilian administration since the 1970s to rule without a coalition.

1990: August   Iraq invades Kuwait

Turkey offers military aid to the United States to stop Iraq's aggression in the Gulf War. Iraq is the source of pro-Kurdish rebels and supplies military aid and shelter to them.

1993   Suleyman Demirel becomes president

Demirel (b. 1924) takes the leadership of Turkey upon the death of Özal. A coalition government with the first woman prime minister, Tansu Çiller of the True Path Party, leads the country. Demirel is an engineer and has been involved with the government since 1965.

1995: March 20   Turkey battles the PKK

Turkey is engaged with the pro-Kurdish terrorist group, the PKK. Turkey invades Iraq in an attempt to quell support and destroy military installations making attacks from the Iraqi border. Turkey sends 35,000 troops into battle. The government of Turkey issues promises to the Kurds living in Turkey, granting them broadcast privileges on radio and television. It also promises to permit the Kurds to open their own schools and teach the Kurdish language.

1995: April 27   Turkey tries to halt Kurdish violence

Turkey vows to crush all Kurdish opposition to the government and secures the borders between Turkey and Iraq.

1995: June 1   Greece ratifies Law of Sea Treaty

Greece ratifies the Law of the Sea Treaty which further undermines Greek-Turkish relations. Under this treaty, Greece has the right to extend its territorial waters in the Aegean Sea from six to twelve nautical miles. Such a move would mean that any Turkish vessels sailing into the Mediterranean from Istanbul or Turkey's Aegean coast would have to sail through Greek waters—a move that Turkey vows to resist with force. Although Greece signs the treaty, it does not extend its territorial waters but reserves the right to do so in the future. In addition, continued tension over Cyprus mars Greek-Turkish relations.

1995: September 20   Çiller forced to resign

Tansu Çiller, the first woman prime minister, is forced to resign because she cannot win a vote of confidence. Finally she forms a coalition government. Her True Path Party merges with the MP leader Mesut Yilmaz, leaving out Necmettin Erbakan and the Welfare Party, an Islamic group.

1996: July   Çiller accused of corruption

Charges of corruption against Çiller force her resignation. Erbakan becomes the first constitutional Islamic party prime minister. He is anti-West and anti-secular. This causes Western economic investments in Turkey to decline sharply. Western democracies fear a return to a closed Islamic government.

1997: January   High court disbands Welfare Party

In an effort to rid the government of Islamic influences, the high court of Turkey bans the Welfare Party and Erbakan is forced to resign. He is banned from politics for five years.

1997: July 12   Yilmaz and Ecevit win control of parliament

Mesut Yilmaz (b. 1947) of the Motherland Party and Bülent Ecevit of the Republican Peoples' Party form a coalition government.

1998: April   Turkey captures Kurdish rebel

Turkey invades Iraq and captures Kurdish rebel leader, Semdin Sakik.

1998: November   Yilmaz government gets vote of no confidence

President Suleyman Demirel asks Bülent Ecevit to form a coalition government. Ecevit appeals to the secular government because he refuses to deal with the Islamic-leaning parties. He is a former poet and journalist.

1999: January 12   Ecevit's message to Parliament

Ecevit speaks to the Grand National Assembly (parliament) and tells them of his firm support for a secular government. He orders a crackdown on Islamic radicals and states that no anti-secular propaganda may run on television or radio. He also forbids the wearing of Islamic clothing (the chador for women). If Islamic supporters get many votes, they probably will not be allowed to get a seat in Parliament and their party will be banned.

1999: February 16   Capture of Abdullah Ocalan

Abdullah Ocalan is a known terrorist for the Kurdish separatists. His capture calms some of the tensions between Turkey and terrorist hideouts in Iraq and Syria. Ocalan has been seeking exile in other European countries, but he is too much a political liability for anyone to want to grant him shelter. Finally, Ocalan is forced out of asylum at the Greek embassy in Nairobi. He is seized and extradited to Turkey. At his trial which ends in June, he is convicted and sentenced to death.

1999: March 8   Virtue Party

The resurgence of an Islamic based party sends shock waves through the government. The Virtue Party is backed by Islamic groups in Moscow. European countries oppose Turkey's entry into the European Union because they fear radicals will take over the Turkish government.

1999: March 20   Pro-Islamics try to unseat Ecevit

Pro-Islamics from the Virtue Party want to force Ecevit to resign. Vural Savas, the chief prosecutor, is now seeking the death penalty for Erbakan's supporters for trying to install an Islamic state.

1999: March 29   Rebels plan election protest

Rebels plan to disrupt elections scheduled for April 18, 1999. They also press for the investigation of Yilmaz and Çiller.

They want to force parliament to remain in power so they can stage demonstrations.

1999: April 18   Elections

Elections give a sweeping victory to Bulent Ecevit's social democratic movement as well as right-wing nationalists. The two groups form a coalition government.

1999   Turkey faces the year 2000

As Turkey enters the next millennium, the country faces severe crises. Since the presidency of Atatürk, Turkey has vacillated on whether to accede to an Islamic government. The ruling parties have always outlawed a religious government, but the Islamic fundamentalists have managed to create unstable situations in which the military has had to take control of the government. This is not healthy for a newly-created democracy. Three times in the past forty years, the military has had to intervene. The rest of the elected governments have been mostly coalition governments which have collapsed on a regular basis. Political parties have been banned as well as political figures, only to resurface in another guise. Turkey continues to battle the Kurdish separatists. The upcoming trial of Abdullah Ocalan should prove whether Turkey can mollify the Kurds or whether it will try to suppress them with violence. The human rights watch will surely criticize the government if it disregards the rights of its Kurdish citizens. Also there remains the unsolved problem of Turks living in Cyprus and the control of Aegean waters. Although Turkey and Greece are at peace for the moment, threats of intrusion into each other's territory could signal military intervention. At the same time, new elections threaten to promote more violence. The current president, Demirel, and the current prime minister, Ecevit, have vowed to eliminate a return to a religious-based government. It is now a question of whether Atatürk's strong legacy will prevail.

1999: August 17   Earthquake

A devastating earthquake, one of the worst natural disasters of the twentieth century, strikes west of Istanbul. The death toll reaches about 13,000 by the end of August, with an estimated 60,000 building destroyed over an area that extends 150 miles. As bulldozers clear the rubble, observers predict that the precise number of deaths from the quake will never be tabulated.

Bibliography

Cole, Simon. *Coping With Turkey.* New York: Basil Blackwell Inc., 1989.

Fodor's Turkey. New York: Fodor's Travel Publications, 1999.

Macqueen, J.G. *The Hittites and Their Contemporaries in Asia Minor.* Boulder, Col.: Westview Press, 1975.

Mango, Andrew. *Discovering Turkey.* New York: Hastings House Publishers, 1971.

Mitchell, Stephen. *Anatolia: Land, Men, and Gods in Asia Minor.* New York: Oxford University Press, 1993.

Renda, Günsel and C. Max Kortepeter, ed. *The Transformation of Turkish Culture the Atatürk Legacy.* Princeton, N.J.: The Kingston Press, Inc., 1986.

Schick, Irvin Cemil and Ahmet Tonak Ertugrul. *Turkey in Transition.* New York: Oxford University Press, 1987.

Stoneman, Richard. *A Traveller's History of Turkey.* New York: Interlink Books, 1998.Vryonis, Speros, Jr. *The Turkish State and History Clio Meets the Grey Wolf.* New Rochelle, N.Y.: Aristide D. Caratzas, Publisher, 1991.

Time-Life Books. *Anatolia: Cauldron of Cultures.* Alexandria, Va.: Time-Life Books, 1995.

Ukraine

Introduction

Although it was an independent state by the ninth century A.D., Ukraine has been under the domination of foreign powers since about 1054. Known as Kievan Rus (a name associated with its capital city, Kiev) in its original days of independence, it has mostly been controlled by two powerful neighbors—Russia to the east and Poland to the west (in the nineteenth and early twentieth centuries, control of the western portion was transferred to Austria). For most of the twentieth century, Ukraine was a constituent republic of the Soviet Union. During this period, it was the site of the worst nuclear accident in history, which took place in 1986 at the Chernobyl nuclear power plant. With the demise of the Soviet Union in 1991, Ukraine became an independent republic.

Ukraine is known for its fertile farmland: approximately fifty-five percent of its 233,090 square miles (603,700 square kilometers) is under cultivation. The land mainly consists of steppes and plateaus, with higher elevations found in the Carpathian Mountains to the west and the Crimean Mountains to the south. The eastern and western regions are divided by the Dnieper River, one of Europe's longest rivers and an important source of hydroelectric power. The major city in eastern Ukraine is Kiev, the capital, which overlooks the Dnieper; the principal western city is Lviv (formerly Lvov). Ukraine has an estimated population of 50.4 million people.

Slavs first settled in present-day Ukraine in the sixth century A.D., and the city of Kiev was founded by the seventh century. By 882 it was the capital of Kievan Rus, the largest state in Europe. At the end of the tenth century, under Prince Vladimir (c. 956–1015), Orthodox Christianity became the official state religion. When Yaroslav the Wise (980–1054) died, Kievan Rus was divided and began a long period of decline. Mongol invasions in the thirteenth century decimated Kiev, and many Ukrainians moved westward to the area known as Galicia-Volhynia, maintaining their distinct ethnic identity. In the fourteenth century the region came under the control of Poland and Lithuania. When these two countries formed a political union two hundred years later, Ukraine came under the direct control of the Polish monarchy, which gave parcels of land to Polish nobles who imposed a system of serfdom on the peasantry and also tried to impose their Roman Catholic religion on the Orthodox Ukrainians.

The most significant resistance to the Poles came from frontiersmen called Cossacks who fled from Polish control and established their own strongholds in the southwest. Between 1648 and 1654 they staged a major uprising, led by their most famous leader (called a *hetman*), Bohdan Khmelnytsky (c. 1595–1657). In 1654 Khmelnytsky signed an agreement forming an alliance with Russia to protect his people from the Poles. Nevertheless, after thirteen years of warfare, Ukraine was still divided between the two powers, with Russia ruling the land east of the Dnieper and Poland ruling in the west. When Poland was partitioned at the end of the eighteenth century, most of Ukraine fell to Russia, although the westernmost part came under the control of the Austrian Habsburg Empire. Under Empress Catherine the Great (1729–96), most of Ukraine was so completely merged into Russia that the name *Ukraine* was no longer used to identify it.

Emergence of Ukraine Nationalism and Culture

In the nineteenth century, a vigorous nationalist movement emerged, concentrated mainly in the western region of Galicia, which was under Habsburg rule. An important touchstone for this movement was the secret Brotherhood of Saints Cyril and Methodius, even though it lasted only a year (1846–47) before being banned by Russia. Also associated with Ukrainian nationalism were prominent cultural figures such as the poet and painter Taras Shevchenko (1814–1861), who was born into serfdom and later freed, and the composer Mykola Lysenko (1842–1912), who made Kiev an important musical center. Other cultural activities during this period included the establishment of a Ukrainian-language journal and a scientific society. However, in 1876 Russian czar Alexander II (1818–81) cracked down on Ukrainian nationalism, including the use of the Ukrainian language, which led to even more resentment against Russian rule.

Soviet Control

World War I (1914–18), which brought down both of the powers controlling Ukraine—Czarist Russia and the Austro-Hungarian Empire—served as a catalyst for Ukrainian independence. Following the Russian Revolution in 1917, an independent Ukrainian republic was declared at Kiev. A year later, after the defeat of the Central Powers in World War I, a western republic was declared at Lvov, and the two republics were merged early in 1919. True Ukrainian independence, however, was thwarted once again, as the region was once again buffeted by warfare between opposing forces, including Poles and competing Russian political factions. By 1921 Ukraine was once again essentially divided between Russia and Poland. The Ukrainian Soviet Socialist Republic controlled the eastern portion of Ukraine, while the west was mostly controlled by Poland, with smaller areas annexed to Romania and the newly formed Czechoslovak state. In the 1920s Ukrainians in both the eastern and western areas enjoyed relative cultural freedom, but the 1930s brought

purges and repression in the east as part of a "Russification" program implemented by Soviet leader Josef Stalin (1879–1953). In addition, drought conditions produced a famine that the Soviet government tried to hide, resulting in the deaths of as many as seven million Ukrainians—many of which could have been prevented by international relief efforts.

Post-World War II and Independence

Invaded and occupied by both Germany and Russia, Ukraine suffered heavy casualties and extensive destruction in World War II (1939–45). Six million Ukrainians died, and thousands of villages were destroyed. Ukrainian resistance forces who had battled the Nazis during the war fought the Soviets for several more years until they were finally overpowered by 1953. In the 1940s and 1950s extensive reconstruction was undertaken, and Ukraine's natural resources—notably its rich farmland—as well as the rapid development of industry and mining resulted in a good economic recovery. By the late 1970s, however, the region's collectivized economy began a downturn that led to increasing dissatisfaction with the Soviet government. In 1986 Ukraine became the site of the worst nuclear accident in history when one of four reactors at the Chernobyl nuclear plant exploded, contaminating a wide region with nuclear radiation.

In the late 1980s the Soviet liberalization under Mikhail Gorbachev (b. 1931) helped spur a revival of Ukrainian nationalist activity. In 1990 Ukraine took important steps to assert its sovereignty, and the following year, after the breakup of the Union of Soviet Socialist Republics (USSR), Ukraine became an independent republic. The new state confronted important political and diplomatic questions in the 1990s, as well as the economic dislocation resulting from the collapse of Communism in the Soviet Union and Eastern Europe.

Timeline

4500–2000 B.C. Trypillian civilization

The earliest known civilization in Ukraine, identified in late-nineteenth-century archaeological digs, is built by hunter/gatherers who make artifacts out of smelted bronze.

1000 B.C.–A.D. 1 Many groups populate Ukraine

As a crossroads of the ancient world, Ukraine is populated by nomadic groups including Scythians, Sarmatians, Cimmerians, and Goths.

6th century A.D. First Slavs settle in Ukraine

Eastern Slavic tribes first settle in Ukraine.

Kievan Rus

7th century Kiev is founded

Kiev, the major city in Ukraine, is first settled and becomes an important trading center.

882 Oleg the Wise occupies Kiev

Viking prince Oleg the Wise (d. c. 912) comes to Kiev from Novgorod, and the city becomes the capital of Kievan Rus, the largest state in Europe.

988 Christianity is adopted

Orthodox Christianity is adopted during the reign of Prince Vladimir (r. 980–1015).

1054 Empire is divided on the death of Yaroslav

When Yaroslav the Wise (r. 1019–54) dies, he divides Kievan Rus, leaving portions of it to all of his sons, an act that begins a period of decline, civil unrest, and aggression by foreign powers, including Poland, Sweden, and the Teutonic Knights (a religious and military order of German knights). The region also becomes less central to trade because of new routes resulting from the First Crusade to the Holy Land to recover it from the Muslims. However, Kievan Rus survives as a political entity for two more centuries.

1223–40 Mongol invasion

Ukraine is invaded by Mongols under the leadership of Batu Khan (d. 1255). Kiev is destroyed, and Moscow becomes the new capital of a larger, unified Mongol state known as the Golden Horde. The Ukrainians of the former Kievan Rus form a new state in Galicia-Volhynia to the west. Removed to the furthest edges of their former empire, they call their new territory *Ukraine* ("borderlands").

Polish Rule

1340 Galicia-Volhynia is conquered by Poland and Lithuania

The Ukrainian region of Galicia-Volhynia comes under the control of Poland and Lithuania.

16th–17th centuries Emergence of the Cossacks

Ukrainian peasant frontiersmen called Cossacks become the major source of opposition to Polish rule, escaping from their landlords and setting up strongholds in the steppes of southwestern Ukraine.

1569 Union of Lublin

Poland and Lithuania form a political union, placing all Ukrainians under the unified rule of the Polish monarchy and attempting to control the southwestern areas inhabited by the Cossacks. In order to settle this area, they distribute the land to Polish noblemen, who turn the Ukrainians into serfs. The Poles also try to impose their Catholic religion on the Eastern Orthodox Ukrainians.

1648–54 Cossack uprising

The Cossacks stage a major uprising against Poland, led by their most famous *hetman* (leader), Bohdan Khmelnytsky (c. 1595–1657). They are victorious and form their own state, called the *hetmanate*. They also form an alliance with Russia. Eventually, however, the lands claimed by the Cossacks are divided between the two major powers in the region, Poland and Russia.

1654 Treaty of Pereyaslav

Ukrainian leader Bohdan Khmelnytsky (see 1648–54) offers to help bring his homeland under Russian control to prevent a Polish takeover. The Ukrainians fear that Poland may impose Roman Catholicism on the Eastern Orthodox Ukrainians and

The Polish-Lithuanian State, 15th century.

turn the peasants into serfs. The Ukrainian Cossacks sign a treaty with Russia, inaugurating thirteen years of warfare between Russia and Poland.

Russia and Poland Divide Ukraine

1667 Truce of Androsovo

Poland and Russia end their conflict over Ukraine by agreeing to Russian control of all Ukrainian lands east of the Dnieper River and Polish rule west of the river.

1677 Publication of music manual

Musical scholar Nikolay Diletsky (c. 1639–90) publishes a theory manual, *Mouskiyskaya hramatyka,* that lays out principles of polyphony (composing with multiple independent yet harmonious, melodic lines).

1709 Battle of Poltava

Cossack leader Ivan Mazepa (1687–1709) sides with Sweden against Poland and Russia in the Great Northern War (1700–21) and is defeated in battle by Russian czar Peter I (Peter the Great; 1672–1725).

Russian-Austrian Rule

1772–95 Partition of Poland

When Poland is partitioned in three stages, the western Ukrainian lands under its control are divided between the Habsburg Empire of Austria, and Russia, which now controls the great majority of Ukrainian lands. Russian empress Catherine the Great (1729–96) merges Ukraine directly into Russia, ending the governing organizations of the Cossacks and replacing them with Russian administrative units called *guberniyi.* After the third partition, the name *Ukraine* is no longer used for this region.

19th century Growth of Ukrainian nationalism

A nationalist movement emerges in Ukraine, beginning with the region's urban intellectuals and concentrated mainly in Galicia, the western region under Habsburg control, which is more liberal than the Russian rule of the areas to the east. The Habsburg monarchy encourages Ukrainian nationalism, which it sees as a way to counter the power of Russia and pan-Slavic nationalism to the east.

The Ukrainian nationalists emphasize the historic ties between Ukraine and the earlier state of Kievan Rus and its cultural separateness from the rest of Russia, advocating the creation of an independent state in the region. Russia reacts harshly to the nationalist movement, imposing language restrictions and carrying out armed attacks (pogroms) on Ukrainians. In addition to the efforts of prominent nationalist writers, including Taras Shevchenko (1814–61; see 1814: February 25) and Mykhailo Drahomanov (1841–95), the nationalist movement is promoted by a secret religious brotherhood (see 1846). By the end of the century, nationalist and socialist parties are formed.

1809 Ukrainian symphony is premiered in Odessa

Symphony no. 21 by Ukrainian composer Mykola Ovsianiko-Kulikovsky receives its premiere performance at the opening of the Odessa Theater in Russia. The symphony incorporates Ukrainian folk melodies into a classically composed work.

1814: February 25   Birth of poet and artist Taras Shevchenko

A poet, painter, and influential Ukrainian nationalist, Shevchenko is born a serf. When his owner moves to St. Petersburg in 1831, Shevchenko becomes acquainted with artists and writers who buy his freedom in 1838, when he enrolls at the Academy of Arts. Shevchenko's earliest poetry romanticizes the Ukrainian Cossack past. Although his approach to Ukrainian history soon changes, notably in the long poem *Haydamaky* (1841), both his poetry and paintings continue to express romanticism and nationalism. Sometimes they are directly linked, as in *Katerina* (1842), a painting of a serf girl betrayed by an officer, a figure who also serves as the subject for a poem from the same period.

In 1846 Shevchenko joins the secret nationalist Brotherhood of Saints Cyril and Methodius. He is arrested and exiled from Ukraine when the group is broken up by the Russian authorities the following year. He is forbidden to write or paint and forced to serve in the military. However, he continues to write in secret. Released in 1857, he returns to St. Petersburg, where he produces etchings and continues to write poetry until his death four years later. In addition to his influence on later artists and writers, Shevchenko is celebrated as a nationalist hero.

1842   Birth of composer Mykola Lysenko

Mykola Lysenko, Ukraine's foremost nineteenth-century composer, studies at the Leipzig Conservatory (in Germany) and also takes lessons with famed Russian composer Nikolay Rimsky-Korsakov (1844–1908) in St. Petersburg. A pianist, composer, and concert impresario, he is active as a performer and conductor. In 1876 he settles in Kiev, where he founds a popular choir that concertizes throughout Ukraine. In 1904 he founds the music institute that bears his name. Lysenko is instrumental in making Kiev an important musical center and in reviving interest in Ukrainian music. He dies in 1912.

1846   Secret nationalist brotherhood is formed

The nationalist Brotherhood of Saints Cyril and Methodius, whose thirty members include author Taras Shevchenko (see 1814: February 25), is founded in Kiev. After only one year, the group is disbanded by Russia, and its members are exiled from Ukraine.

1856: August 27   Birth of author Ivan Franko

Franko is born in a small town in Galicia, a portion of western Ukraine under Austrian rule at the time of his birth. (It is later renamed Ivana-Franka in his honor.) He studies at the university in Lvov (which is also named for him later) and in Vienna, becoming politically active as a socialist. Starting out

as a writer in the romantic tradition of the early and mid-nineteenth century, Franko turns to realism, publishing novels that provide a detailed portrait of contemporary Galicia. He also writes plays, poetry, articles, short stories, translations, and children's poems. Among Franko's many works are the long poem *Moysey* (Moses; 1905), the poetry collections *Miy izmarahd* (My Emerald; 1897) and *Iz dniv zhurby* (From the Days of Sorrow; 1900), and the novels *Boryslav smiyetska* (Boryslav Laughs; 1882) and *Osnovy suspilnosti* (Pillars of Society; 1895). Franko dies in 1916.

1861   Ukrainian journal is launched

A Ukrainian journal, *Osnova,* begins publication with permission of Russian czar Alexander II (1818–81).

1861: March 10   Death of Taras Shevchenko

Taras Shevchenko, nationalist poet and painter born into serfdom, dies in St. Petersburg. (See 1814: February 25).

1872   Ukrainian scientific society is founded in Lvov

The Shevchenko Scientific Society, named for the prominent Ukrainian nationalist author, Taras Shevchenko (see 1814: February 25), is founded in Lvov with the permission of its Habsburg rulers.

1876   Russia cracks down on Ukrainian cultural activities

Czar Alexander II bans the publication of all Ukrainian-language books and journals and closes Ukrainian-language schools in a crackdown on nationalism in the region. Many Ukrainian nationalists in Russian-controlled areas emigrate to Lvov in Austrian-ruled Galicia or to Switzerland.

1882: October 30   Birth of painter Mykhaylo Boychuk

A major influence on twentieth-century Ukrainian art, Boychuk studies art in Krakow, Poland, Munich, Germany and Vienna, Austria, and also spends time in Paris, France, and Italy. Returning to Ukraine, he restores medieval paintings in Lvov, and then settles in Kiev. Following World War I (1914–18), Boychuk is one of the nationalist artists who flourishes under the relatively tolerant Soviet rule of the 1920s. He is the leader of a group of painters who focus on "monumental painting" (frescoes on walls and other public spaces) as a way to connect art with the common people and express Ukrainian nationalistic aspirations. He and his colleagues jointly create frescoes in a number of public spaces, including an army barracks, a sanitarium, an agricultural school, and a theater. During the Stalinist repression of the 1930s, Boychuk's paintings fall out of favor. He is accused of "bourgeois nationalism," and his public works are destroyed. He dies in 1939.

1890 Regional studies university chair is established

Reflecting the growth of nationalism in Ukraine, a professorship in southeastern European history is established at the University of Lvov.

1912 Death of composer Mykola Lysenko

Composer and conductor Mykola Lysenko, the foremost figure in nineteenth-century classical music in Ukraine, dies. (See 1842.)

1916: May 16 Death of author Ivan Franko

Poet and novelist Ivan Franko dies in Lvov (See 1856: August 27).

World War I and Postwar Independence

During the First World War, Ukraine is overrun by the Central Powers (Germany and Austria-Hungary) in their advance through the Russian Empire. The First World War brings very limited damage to the Ukraine, in comparison to the horrors of destruction brought by the Second World War. For the Central Powers, large-scale destruction of Ukraine's infrastructure is unnecessary, as the Russian army is outmaneuvered and thoroughly decimated by conventional warfare alone. Near the end of the First World War, Ukraine originally declares itself a republic within the framework of a federated Russia. Ukraine proclaims complete independence in 1918, after the Bolshevik Revolution. Soviet troops are sent into the Ukraine, but the Central Powers, having acknowledged Ukrainian independence, then overrun the territory with their own soldiers as the Treaty of Brest-Litovsk (signed in March, 1918) forces the Red Army to withdraw. However, the World War I armistice is signed in November, 1918, and it forces the Central Powers to themselves withdraw from the Ukraine.

1917: April Ukrainian national assembly meets

A Ukrainian national congress meets in Kiev and elects a council, called the Rada.

1917: November 20 Independent republic is proclaimed

Taking advantage of the revolutionary upheaval in Russia following the overthrow of the Russian czar, the Ukrainian Rada (national council) declares independence at Kiev, proclaiming the Ukrainian People's Republic.

1917: December 26 Bolsheviks proclaim separate republic

A Ukrainian Soviet Republic is established by the Bolsheviks at Kharkov as a rival to the republic proclaimed the previous month in Kiev.

1918: January Ukraine declares its sovereignty

The Ukrainian People's Republic declares itself a sovereign nation with its own constitution, government, and currency.

1918: November 1 Western republic is proclaimed

Following the defeat of Germany and Austria in World War I (1914–18), a separate Ukrainian state, the West Ukrainian People's Republic, is proclaimed at Lvov, in the part of Ukraine formerly ruled by the Habsburg Empire.

1919: January 22 Ukrainian republics are merged

The eastern and western Ukrainian republics are merged to create a new state. It is recognized by more than forty nations.

1920–21 Ukraine falls to Poland and Russia

Poles, Bolsheviks, and Russian royalists attack Ukraine from all sides, and peasant anarchists compound the chaos. Even with the former imperial powers gone, Ukraine once again emerges under the control of Poland and Russia. The Ukrainian Soviet Socialist Republic emerges in the east, while Poland controls most of western Ukraine, including Galicia and Volhynia. Northern Bukovina comes under the control of Romania, and Subcarpathian Ruthenia is annexed by the newly formed nation of Czechoslovakia.

Soviet Republic

1920s Soviets follow liberal Ukrainization program

Under the leadership of Mykola Skrypnyk, the new Soviet republic of Ukraine is hospitable to the expression of Ukrainian culture and identity, leading to a national literary and artistic revival.

1920s Ukrainian nationalism flourishes in Polish-controlled areas

In the Polish-controlled parts of Ukraine, Ukrainian nationalism thrives in both legal forums and in underground political and military organizations, which attract militant student support. The Ukrainians enjoy freedom of religion and of the press and have an active business community.

1923: August 1 Language act favors Ukrainian

A language act in the Russian-controlled Ukraine gives the Ukrainian language priority over Russian as the first official language.

1928–39 Soviet Ukrainians suffer Stalinist reign of terror

Soviet leader Josef Stalin (1879–1953) adopts a policy of repressing all forms of nationalism, and it is harshly applied to Ukraine. Language rights are modified, artists and intellectuals are persecuted, and politicians known or suspected to harbor strong nationalist feelings are purged from the Communist Party.

1928   Status of Russian language is elevated

Russian is made the second official language of Ukraine. This underscores Stalin's attempt to Russianize Ukraine, even in language.

1930   Ukrainian intellectuals are arrested for nationalism

Forty-five prominent Ukrainian intellectuals are arrested by the state for "nationalist deviation" and tried for treason. Thirteen receive death sentences, and the rest are sent into exile. More purges are carried out in the next three years, and three top government officials or former officials are shot.

1932–33   Famine claims millions of lives

Agricultural collectivization is imposed on Ukraine by the Soviet government. Private farms are abolished, and farmers are required to give up their property and work on government-owned collective farms created by combining a number of smaller private plots. These changes cause massive disruption of the region's agricultural production. Farmers who resist are exiled or executed. When drought produces a famine in 1932–33, the government keeps it a secret to protect the reputation of its collectivization program. Famine-stricken areas are closed off, and relief cannot reach them. An estimated five to ten million Ukrainians die.

1934: January 9   Birth of political leader Leonid Kravchuk

Leonid Makarovych Kravchuk serves as the first president of the Ukraine after it breaks away from the Soviet Union. Kravchuk is born in Volhynia, a part of western Ukraine under Polish control during his youth. Earning a degree in political economy at Kiev State University in 1958, he joins the Communist Party in the same year and rises through its ranks over the next thirty years. For much of that time, he develops propaganda for the Communist government. By 1990 he is chairman of the Ukrainian ruling body (Supreme Soviet) under Soviet rule. In keeping with Mikhail Gorbachev's policy of *glasnost* (openness), Kravchuk is tolerant toward Ukrainian nationalism and sympathetic to the newly formed nationalist front, the Rukh.

When Ukraine becomes fully independent after the demise of the Soviet Union in late 1991, Kravchuk wins its first presidential election. In his official position, he takes part in the formation of the Commonwealth of Independent States (CIS) with other former Soviet republics. As president, he is a staunch defender of Ukrainian independence and an advocate of economic reform and nuclear disarmament. He also works to consolidate relationships with other Eastern European countries. Due to popular dissatisfaction with Ukraine's continuing post-independence economic crisis, Kravchuk is nar-

rowly defeated at the polls by Leonid Kuchma in his 1994 bid for reelection (see 1994: July 10).

1939   Death of painter Mykhaylo Boychuk

Nationalist painter Mykhaylo Boychuk dies (See 1882: October 30).

World War II

1939: September 1   Germany invades Poland

Germany launches a surprise land and air attack on Poland, beginning World War II (1939–45). Some parts of Poland are incorporated directly into Adolf Hitler's (1889–1945) Third Reich. Others become a German protectorate under a separate administration.

1939: September 17   Soviets attack Poland

Following the German invasion, the Soviet Union invades Poland from the east, annexing those parts of Ukraine (Galicia and Volhynia) formerly under Polish control.

1940: June   Soviets annex more territory in the western Ukraine

Romania cedes northern Bukovina and Bessarabia to the Soviet Union.

1941: June 22   Germany invades the Soviet Union

Germany overruns Ukraine in the early days of its invasion of the Soviet Union. At first many Ukrainians, happy to be freed from Soviet rule, welcome the Germans. However, when they learn that they are to be equally enslaved under the Nazis, they begin to organize resistance efforts. Many Ukrainians join with the Soviet Army and the Soviet Partisans to fight the Germans. Repeated invasion and occupation takes a heavy toll on Ukraine: in the course of the war, hundreds of cities and thousands of villages are destroyed, and six million Ukrainians are killed. Large-scale economic recovery and reconstruction efforts are undertaken in the 1940s and 1950s.

Post-World War II Soviet Rule

1945: June 29   Ukraine regains former Czechoslovak territory

The Soviet Union annexes Subcarpathian Ruthenia, which had been ceded to Czechoslovakia following World War I and fell under Soviet control during World War II.

1945–53 Ukrainian military resistance to the Soviets continues

Ukrainian nationalist troops active during World War II (1939–45) continue fighting against the Soviet Union following the war. They are not fully contained until 1953.

1950s Industrialization and mining boost economy

Ukraine becomes a vital part of the Soviet economy as its traditional agricultural economy is bolstered by the addition of industrialization (including steel and weapons production) and mining.

1953 Stalin's death heralds liberalization

From the end of World War II until the death of Soviet leader Josef Stalin (1879–1953), Ukrainian nationalism is ruthlessly suppressed. With the liberalization initiated by Stalin's successor, Nikita Khrushchev (1894–1971), Ukrainian intellectuals and artists gain more freedom of expression.

1954: February The Crimea is annexed to Ukraine

Soviet leader Nikita Khrushchev (see 1953) makes the Crimea (the south-western peninsula that extends into the Black Sea) part of the Ukraine in honor of the 300th anniversary of the 1654 alliance between the Ukrainian Cossacks and the Russian empire.

1964 Brezhnev cracks down on dissidents

Under Leonid Brezhnev (r. 1964–82), the Soviet government imposes strict controls on Ukrainian dissident writers, including Ivan Dziuba (b. 1931), Vyacheslav Chornovil (b. 1938), and Valentyn Moroz (b. 1936). Freedom of expression is generally curtailed.

1972 Hard-liner becomes Ukraine party chief

Hard-liner Volodymyr Macherbyhtsky replaces Petro Shelest as head of the Communist Party in Ukraine.

Late 1970s Economic decline begins

After years of economic strength, the Ukrainian economy declines as years of inefficient management result in shortages of food and consumer goods, and the Communist government's insistence on a collectivized economy it can control leads it to resist needed economic reforms.

Late 1980s Nationalism grows under *glasnost*

Mikhail Gorbachev's policy of greater openness (*glasnost*) helps foster the revival of Ukrainian nationalism. The Ukrainian language is once again taught in the schools, and a nationalist movement, the Rukh, is formed. The head of the Ukrainian Communist Party, Vladimir Shcherbitsky, dies and is replaced by Leonid Kravchuk, a more liberal politician who is more tolerant of nationalist aspirations (see 1934: January 9).

1986: April 26 Chernobyl nuclear accident

One of four reactors in the Chernobyl nuclear plant about seventy miles north of Kiev explodes after engineers illegally shut off all its emergency safety systems while conducting a test to prevent the shut-off reactor from starting back up. Operators ignore signs that pressure from residual energy is reaching dangerous levels, and the reactor blows up, sending flames 1,000 feet into the air and releasing approximately nine tons of radioactive matter that spreads contamination to the surrounding area, throughout Europe, and eventually, to a lesser degree, throughout the world.

The city of Pripyat, where Chernobyl is located, has to be evacuated, and neighboring countries take precautions to reduce exposure to radiation in food, water, and air. A report from the Lawrence Livermore Laboratory in the United States asserts that the Chernobyl accident released more long-term radiation into the air, water, and topsoil than all previous nuclear bombs and tests combined. By the early 1990s 150,000 people (including 60,000 children) who used Kiev's contaminated water system are suffering from thyroid disorders. At the end of the decade, the full extent of casualties from radiation has yet to be determined. Because of Ukraine's energy needs, the three remaining Chernobyl reactors are kept in operation throughout the 1990s in spite of protests by the other former Soviet republics.

1990 Freer elections are held

Semi-free legislative elections are held in Ukraine, producing a new assembly with nationalists accounting for one-quarter of its members.

1990: July 16 Ukraine asserts its sovereignty

Ukraine proclaims its sovereignty.

1990: November 21 Russia-Ukraine treaty is signed

Ukraine and Russia sign a treaty recognizing each other as sovereign states.

The Post-Soviet Era

1991: August 24 Ukraine declares independence

Following the abortive right-wing coup that puts into motion the demise of the Soviet Union, Ukraine declares full independence pending a popular referendum.

1991: December 1 Ukrainian voters support independence

Over ninety percent of Ukrainian voters support their government's declaration of independence from the Soviet Union. They also elect Leonid Kravchuk (see 1934: January 9) the first president of democratic Ukraine with sixty percent of the vote. Most countries worldwide recognize Ukrainian independence.

1991: December 8 Commonwealth of Independent States is created

Ukraine, Russia, and Belarus form the Commonwealth of Independent States (CIS) to help coordinate political, military, and economic policy. Eight other republics join later in the month. CIS unity, however, is impeded by conflicts between Ukraine and Russia in several areas, notably the control and phaseout of nuclear weapons.

1991: December 25 Gorbachev resigns

Mikhail Gorbachev, the last leader of the Soviet Union, resigns.

1992: May Nuclear weapons shipped to Russia

Shipment of tactical nuclear weapons to Russia for dismantling is completed after becoming stalled earlier in the year when Ukraine charged that Russia was failing to properly carry out the dismantling process.

1992: May 6 Crimea declares independence

The Crimea, ceded to Ukraine in 1954 by the Soviet Union, declares independence. Ukrainians only constitute about a quarter of the Crimea's population, of which another quarter are Tatars. The Russian government declares the 1954 annexation illegal and void. (see 1954: February.)

1992: October 1,000 percent inflation

Consumer inflation is recorded at over 1,000 percent. Ukraine shortly withdraws from the ruble block (those countries using the ruble as their official currency) and begins issuing its currency in coupons.

1994: February START I is ratified

Ukraine ratifies the Strategic Arms Treaty I (START I) nuclear disarmament agreement.

1994: March Independents and Communists lead in legislative elections

In response to the new government's failure to remedy the country's growing economic problems, a revamped Communist Party and other left-wing groups do well in parliamentary elections, although the highest number of votes is won by independent candidates, many of them prominent local leaders. Independents win 163 out of 338 contested parliamentary seats, and the Ukrainian Communist Party wins 86.

1994: July 10 Kuchma is elected president

President Leonid Kravchuk's bid for reelection is defeated, and Leonid Kuchma, leader of a centrist party, is elected in a second-round contest, winning fifty-two percent of the vote, compared with forty-five percent for Kravchuk. Most of Kuchma's support comes from the Russian-speaking eastern part of Ukraine.

1995: December 20 Ukraine agrees to close Chernobyl plant by 2000

Ukraine signs a pact with the Group of Seven (G-7) industrialized nations, promising to close the two remaining reactors in operation at the Chernobyl nuclear plant in northern Ukraine by 2000. The reactors have continued functioning despite the nuclear accident at the site in 1986 because Ukraine badly needs the energy they produce—five percent of all energy produced in the country. The G-7 nations pledge $2.3 billion in aid to pay the expenses of closing down the plant and to help Ukraine develop alternative energy sources.

1996: June New constitution is adopted

A new Ukrainian constitution provides for implementation of a five-year presidential term, the establishment of the Supreme Council, a unicameral (having one legislative chamber) parliament with members elected for four years.

1996: June 1 Last strategic warheads are sent to Russia

In keeping with an accord signed early in 1994, Ukraine completes the transfer of its nearly two thousand strategic warheads to Russia for demolition. In return, Ukraine will receive fuel rods for its nuclear power plants and financial compensation.

1997: March 29 Leftists lead in national elections

The Communist Party and other leftist groups maintain their dominance in Ukraine's parliament. The Communists win 123 of the assembly's 450 seats. Independent candidates remain an important force, winning nearly as many seats as the leftists—114. The nationalist Rukh Party wins 46 seats in an election marred by violence and government media manipulation.

1997: June New civil code is adopted

The Supreme Council, Ukraine's newly formed unicameral legislative body, adopts a new code of laws.

1997: July 9   Agreement reached with NATO

A charter of cooperation is signed by Ukrainian president Leonid Kuchma (see 1994: July 10) and the North Atlantic Treaty Organization (NATO) leaders.

Bibliography

Goncharenko, Alexander. *Ukrainian-Russian Relations: An Unequal Partnership.* London: Royal United Services Institute for Defense studies, 1995.

Gordon, Linda. *Cossack Rebellions: Social Turmoil in the Sixteenth-Century Ukraine.* Albany: State University of New York Press, 1983.

Hosking, Geoffrey A., ed. *Church, Nation and State in Russia and Ukraine.* New York: St. Martin's Press, 1991.

Kuzio, Taras. *Ukraine under Kuchma: Political Reform, Economic Transformation and Security Policy in Independent Ukraine.* New York: St. Martin's Press, 1997.

Marples, David R. *Ukraine under Perestroika: Ecology, Economics and the Workers' Revolt.* New York: St. Martin's Press, 1991.

Motyl, Alexander J. *Dilemmas of Independence: Ukraine after totalitarianism.* New York: Council on Foreign Relations Press, 1993.

Rudnytsky, Ivan L., ed. *Essays in Modern Ukrainian History.* Edmonton: Canadian Institute of Ukrainian Studies, 1987.

Shen, Raphael. *Ukraine's Economic Reform: Obstacles, Errors, Lessons.* Westport, Conn.: Praeger, 1996.

Subtelny, Orest. *Ukraine: A History.* Toronto: University of Toronto Press, 1994.

Wilson, Andrew. *Ukrainian Nationalism in the 1990s: A Minority Faith.* New York: Cambridge University Press, 1997.

United Kingdom

Introduction

The United Kingdom (also known as Great Britain) has played a role in world history that belies the small size of this island nation. Its maritime achievements and colonial expansion between the sixteenth and twentieth centuries produced an empire that, at its peak, covered a quarter of the globe. The Industrial Revolution of the eighteenth century made Great Britain the world's first (and for many years the foremost) great manufacturing nation. Britain also influenced the lives of people throughout the world through its cultural achievements and government institutions.

The twentieth century has seen the decline of the empire on which "the sun never set," as colonial independence movements gradually spelled the end of the U.K.'s imperial power. Two world wars and a number of economic crises have made heavy demands on the resilience of the British people. The postwar era has seen the erosion of the nation's traditional manufacturing base, resulting in poverty and unemployment, especially in the north, and immigration from its former colonies has produced social tensions with the growth of new ethnic minorities. Nevertheless, Britain retains a significant role on the world stage as a leading financial and industrial center and a respected and influential member of the international community.

The United Kingdom of Great Britain and Northern Ireland (commonly called Great Britain, or simply Britain) comprises the island of Great Britain (consisting of England, Scotland, and Wales) and the northern portion (known as Northern Ireland) of the neighboring island on which the Republic of Ireland is located. The combined total area of the country is 94,526 square miles (244,820 square kilometers). The terrain of England consists of highlands to the north, west, and southwest and fertile lowlands to the east and southeast. The Pennine mountain range, extending north to south roughly halfway across England divides the central portion into the East and West Midlands, a fertile area of hills and valleys. Scotland's major geographical regions are the northern Highlands, Central Lowlands, and Southern Uplands, and much of Wales's topography is accounted for by the Cambrian mountains. By contrast, Northern Ireland con-sists of low-lying plateaus. As of the late 1990s, the United Kingdom has an estimated population of 58.8 million people, including 49.1 million in England, 2.9 million in Wales, 5.1 million in Scotland, and 1.7 million in Northern Ireland.

History

The island of Britain was overrun by successive waves of invaders in the first millennium and a half of its history, most of them contributing to the culture of the future nation—the Celts (fifth century B.C.); Romans (55 B.C.); Germanic Angles, Saxons and Jutes (fifth century A.D.); Danes (eighth through eleventh centuries); and Normans (eleventh century). Christianity, introduced at the end of the sixth century, was a major influence on the culture and history of Britain. The Norman Conquest in 1066 established the first English dynasty and brought the language and culture of France to a dominant position in England.

The medieval period in England saw the emergence of three important areas of conflict that would recur over the coming centuries: religious controversy (the most famous resulting in the murder of Thomas à Becket in 1170); rivalry between England and France; and conflicts over how much power should be wielded by the king (resulting, during this period, in the signing of the Magna Carta in 1215). The four-teenth century was particularly tumultuous, with the onset of the Hundred Years' War, the Peasants' Rebellion, and the Black Death, which wiped out roughly one-third to one-half of England's population. However, it was also the period that saw the composition of Geoffrey Chaucer's masterpiece, *The Canterbury Tales,* which painted a vivid portrait of the entire spectrum of English society, from nobles and clerics to farmers and tradesmen.

By the end of the fifteenth century, the Wars of the Roses (1455–85) between the house of York and the reigning house of Lancaster had ended with the accession to the throne of Henry VII and the beginning of the Tudor period. The wool trade had grown into a staple of the English economy, and England's first books (including Thomas Malory's *Morte d'Arthur*) had been printed by William Caxton. Under Henry's rule, the balance of political power swung back to the king, who increased his power not only over the English

611

UNITED KINGDOM

0   50   100   150 Miles

0   50   100   150 Kilometers

nobles but also over Wales. The merchant marine was developed and England became a major center for trade. In addition, John Cabot's exploration of the North American coast opened an era of maritime exploration.

Religious conflict came to the fore under the reign of Henry VIII who in 1531 broke with the Catholic Church following the pope's refusal to annul Henry's marriage to Cathe-

rine of Aragon and declared himself head of the Church of England. Henry's daughter Elizabeth I, by his second wife, Anne Boleyn, ascended the throne in 1558, becoming one of England's strongest and most popular rulers. She presided over an era of flourishing commerce, naval power, and colonial expansion, accompanied by a cultural flowering that included the plays of William Shakespeare and Christopher

Marlowe, and the poetry of Philip Sidney and Edmund Spenser.

Conflicts between Parliament and the monarchy rose to new heights in the seventeenth century with the civil wars of the 1640s and the beheading of King Charles I in 1649, followed by the decade of the Commonwealth, headed by Oliver Cromwell, and the restoration of the monarchy in 1660. The poet John Milton, contemplating the downfall of the Puritan cause to which he had devoted so much effort, and seeking to "justify the ways of God to man," wrote his epic *Paradise Lost,* one of the towering achievements in English literature. The tumultuous period of the Restoration included a second major outbreak of the Black Death and the Great Fire of London, which leveled more than two-thirds of England's capital, by then the largest city in the world. The interplay between politics and religion once again became a determining factor in the course of British history with the removal of James II, a Catholic, in favor of the Protestant monarchs William III and Mary II in the "Glorious Revolution" of 1688.

The eighteenth century saw the official creation of Great Britain through the Act of Union (1707) uniting England and Scotland. A German dynasty, the House of Hanover, succeeded to the throne of England following the death of Queen Anne, in the person of King George I. Culturally, this was the great period of English wit and satire, whose foremost literary proponents were Alexander Pope and Jonathan Swift, and which also found expression in the highly popular engravings of William Hogarth. Political activity by the supporters of the Catholic "Pretender" (the son of deposed monarch James II) continued in the first half of the eighteenth century, and the Seven Years' War (1756–63) consolidated Britain's growing colonial supremacy globally. Perhaps the most important development of the eighteenth century, however, was the advent of the Industrial Revolution, in which technological advances, most notably the development of water and steam power and of the smelting process, led first to modernization of textile, iron, and steel production, and then to far-ranging changes in factory production of all types, and to the basic structure of the economy itself.

Revolution

Due to developments in the American colonies and France, political as well as industrial revolution was in the air in the late eighteenth century, leading to another period of war with France. The Democratic ideals of the French Revolution were also a rallying point for a new generation of poets—the Romantics—dedicated to throwing off established conventions in both literature and life. Conflict with France continued into the beginning of the nineteenth century with the Napoleonic Wars, which established British military supremacy abroad but led to neglect of pressing social and economic problems at home. The Industrial Revolution had spurred widespread migration from the country to increasingly overcrowded cities and created a new class of urban poor.

The year of Napoleon's defeat at Waterloo—1815—was also the year of the Corn Law, which raised the price of bread and intensified calls for reform, a major theme in nineteenth-century British politics and social theory. The need for reform was eloquently illustrated in the novels of Charles Dickens, with their portraits of urban poverty. Three major reform acts (1832, 1867, and 1884) vastly enlarged the size of the electorate and gave the new urban working and middle class a voice in the government. Other nineteenth-century laws effected reforms in working conditions, sanitary conditions, education, and treatment of prisoners and the indigent.

The great symbol of British unity, prosperity, and virtue for much of the nineteenth century was Queen Victoria, the longest-reigning monarch in the country's history (r. 1837–1901). Governing with the aid of her consort, Prince Albert, until his death in 1861 and then alone, she won the respect and admiration of the British public, which held two great celebrations of her reign, the Golden (1887) and Diamond (1897) jubilees. However, amidst the confidence engendered by prosperity and imperial dominance, some Victorians struggled to reconcile the Christian view of the earth and its history with the principles set forth in Charles Darwin's *On the Origin of Species* (1859) and scientific discoveries in other fields such as geology.

World Wars

World War I (1914–18), which exposed the British people to the horrors of modern mechanized warfare and decimated a generation of young men, is widely considered the end of a long era of confidence and prosperity and a watershed in British history. The war was followed by a period of economic depression, with high unemployment leading to civil unrest and general strikes. The great representative British literary work expressing the pervasive sense of fragmentation and dislocation in the postwar years is T.S. Eliot's 1922 poem, *The Waste Land.*

By the 1930s, Britain, like the rest of the world, was witnessing the rise of fascism in Europe, and responding with a policy of appeasement that culminated in the signing of the Munich Pact by Prime Minister Neville Chamberlain in 1938 in hopes that war could be avoided. By the following year, however, the nation entered World War II (1939–45), sending troops abroad even as its population coped with the German bombing "Blitz" aimed at its own cities. Britons rallied under the leadership of Winston Churchill, averting the dreaded German invasion, although heavy losses were still incurred both at home and on the battlefield.

The aftermath of war brought further hardships, including food shortages and rationing, and Britain's postwar Labour government laid the foundations for the nation's current social welfare system with the establishment of the

National Health Service and the nationalization of other institutions and services. The postwar period also saw the final dissolution of a formerly vast empire as Britain's remaining colonies in Asia and Africa won their independence. Elizabeth II succeeded to the throne in 1952. Elizabeth II, one of Britain's longest-reigning monarchs, was still in power as the twentieth century reached its final years, when questions were being raised about the role image of the monarchy in modern British society.

Beginning in the 1970s, the "troubles" in Northern Ireland drew increasing attention to that region's religious and cultural divisions, especially when the Irish Republican Army carried its campaign of terrorism to the British mainland, with numerous bomb explosions, including the bombing of the Parliament buildings. In the spring of 1998 the leaders of Great Britain and the Irish Republic and representatives of Northern Ireland's Protestants and Catholics met and drafted a historic agreement aimed at bringing peace to the region.

On a broader diplomatic stage, the 1990s saw a historic departure from the traditional insularity that has often characterized the island nation of Britain—Parliamentary ratification of the Maastricht Treaty providing for greater economic integration with the countries of the European Community in the proposed European Union, with a common currency (the Euro) to be adopted by the millennium. Controversy over the Euro was heated enough to threaten the standing of Conservative prime minister John Major, who ultimately retained leadership of his party, although that party was unseated two years later in a victory that brought the Labour Party, under the leadership of Tony Blair, back to power for the first time in eighteen years.

Timeline

5th century B.C.   Celts invade Britain

The Celts—speakers of the common language that is the basis for Breton, Irish, Welsh, and Gaelic—invade Britain. Organized into small tribes, in which all members claim a common ancestor, they live in settlements, growing crops and raising livestock. They practice an animistic religion, guided by spiritual leaders called Druids, who are not only priests but also teachers, judges, and healers. They produce an abstract late-Bronze Age style of artwork referred to as *La Tène*.

Roman Rule

55–54 B.C.   Julius Caesar invades Britain

Roman leader Julius Caesar invades the British Isles during the conquest of Gaul, but the Roman armies withdraw. It is Caesar who first uses the name Britannia.

A.D. 43–84   Romans conquer Britain

Under the emperor Claudius, the Romans conquer Britain, beginning with the eastern and southeastern lowlands and eventually occupying most of present-day Britain. Roman culture spreads, especially in the southeast. The network of roads built by the Romans helps unify the province.

78–85   Agricola expands the Romans' territory

Gnaeus Julius Agricola (40–93), who serves as Roman governor of Britain during this period, consolidates and nearly doubles the area controlled by Rome. Pushing northward, he conducts six campaigns in Scotland.

120–136   Construction of Hadrian's Wall

Roman Britain's most impressive monument is built to protect Roman settlements from attack from the north. Crossing the Tyne-Solway isthmus from Wallsend, Northumberland, to Bowness, Cumbria, it spans 80 miles (128 kilometers).

410   Last Roman troops withdraw from Britain

Pressured by barbarian invasions beginning in the late fourth century, the Romans are unable to maintain a presence in Britain and gradually withdraw their troops.

5th century   Invaders and natives fight for control of Britain

After the withdrawal of the Romans, native Celtic groups vie for dominance, as Picts and Scots attack from the north and the Germanic Angles, Saxons, and Jutes invade from the east. Eventually, the Celts are pushed westward to present-day Wales. Saxon and Anglian victories gain new territory that separates the Celts of Wales from those of Scotland to the north. The Germanic invaders retain their native culture and language, renaming the land England—home of the Angles.

The Anglo-Saxon and Danish Invasions

6th–8th centuries   Anglo-Saxon kingdoms are established

The invading Germanic tribes establish the following kingdoms, known as the Heptarchy: Northumbria, Mercia, East Anglia, Essex, Sussex, Wessex, and Kent. The dominant ones are Northumbria, Mercia, and Wessex. Wars between rival kingdoms are frequent.

597   Saint Augustine brings Christianity to England

Saint Augustine heads a group of monks sent to England by Pope Gregory I to reintroduce Christianity, which disappears when the Celts are driven out. Its influence is bolstered by the conversion of King Ethelbert of Kent.

664   Synod of Whitby

The Synod of Whitby convenes to choose between the two competing versions of Catholicism practiced in the British Isles: Celtic and Roman Christianity. The synod decides in favor of Roman Christianity, thus strengthening England's religious ties with the European continent.

787–878   Scandinavians raid England

Different Scandinavian groups—collectively referred to as Danes—initiate raids on England, first for plunder and later for conquest.

871–899   Reign of Alfred the Great

Alfred of Wessex (849–99) gains renown by being the first English monarch to resist Danish conquest. He is also famed for advancing education and law.

878   Alfred defeats the Danes at Edington

King Alfred's forces defeat the Danes at the Battle of Edington, preventing them from making a complete conquest of England. The victory leads to a division of England that creates a Danish area in the northeast called Danelaw, with Alfred retaining control in the south and west. Later, Alfred's successors drive the Danes out of Danelaw.

927–36   Athelstan becomes first king to unite England

Athelstan (d. 939), a grandson of King Alfred of Wessex, conquers Northumberland and Cornwall, bringing all of England under one government for the first time.

994–1012   New round of invasions by the Danes

The Danes invade England again and succeed in conquering it and driving out the reigning king. Sweyn, the leader of the invasion, becomes king and is succeeded by his son, Canute.

1042–66   Edward the Confessor restores Anglo-Saxon rule

Edward the Confessor (c. 1003–66) returns to England from Normandy and regains the throne for the house of Wessex. Many French-speaking Normans return with Edward to serve him, introducing the French language and Norman culture to the court.

The Norman Conquest

1066   The Norman Conquest

King Edward dies. He leaves no heir but names his brother-in-law, Harold (c. 1022–66), as his successor. William the Conqueror, duke of Normandy (c. 1028–87), makes a claim to the throne and invades England, in what becomes known as the Norman Conquest.

1066: October 14   Battle of Hastings

In the decisive battle of the Norman Conquest, King Harold II is killed, and William becomes king. He is crowned in Westminster Abbey. Anglo-Saxon nobles lose their titles, their lands, and in some cases, their lives. The land is distributed to barons of the Norman nobility.

1087   William dies

William the Conqueror dies. William's reign is followed by those of his two sons, William II Rufus (c. 1056–1100) and Henry (Henry I, 1068–1135).

1135–54   Civil war follows the death of Henry I

After Henry I dies, the rivalry over the succession to the throne leads to an anarchic and economically damaging period of civil war. Stephen (1097–1154), a grandson of William the Conqueror, occupies the throne but lacks the authority to govern effectively. His rule is challenged first by Matilda (1102–67), the daughter of Henry I, and later by her son, Henry Plantagenet (1133–89), heir to the house of Anjou. Stephen agrees that Henry will succeed him when he dies.

1154–89   Henry II rules England

Henry Plantagenet, heir to the house of Anjou and grandson of Henry I, becomes king of England following the death of King Stephen. With Henry's ascent to the throne, the Angevin dynasty replaces that of the Normans. Henry centralizes the power of the government in a number of ways and makes important reforms in judicial procedures.

c. 1160   Oxford University is founded

Oxford, England's oldest university, has its beginnings when English students are unable to study at the University of Paris due to a royal quarrel involving King Henry II. Scholars at Oxford decide to form a similar institution, and lectures are documented by 1186. The first colleges, including Balliol and Merton College, are founded in the thirteenth century.

1164   Constitutions of Clarendon limit the power of the church

Henry II challenges the power of the church by issuing the Constitutions of Clarendon, which restricts the power of the church in a number of ways, including making priests subject to state law and barring appeals to papal courts.

1170   Thomas à Becket is murdered

Following a long conflict between Henry II and Thomas à Becket (c. 1118–70), the archbishop of Canterbury appointed by Henry, Becket is murdered in his cathedral by allies of the king. Due to public outrage over this incident, Henry modi-

Jesus College at Oxford University. Oxford, England's oldest university, was established in the late twelfth century. (EPD Photos/CSU Archives)

fies some of the terms of the Constitutions of Clarendon (see 1164).

1189–99   Reign of Richard the Lionhearted

Richard (1157–99), son of Henry II, rules England for ten years while living abroad and visits the country only twice. However, order is maintained thanks to the elaborate legal system established by his father. Richard is known for his role in the failed Third Crusade to retake Jerusalem from the Muslims.

1199–1216   Reign of King John

King John (1167–1216), son of Henry II and brother of Richard, has a troubled reign, with challenges from the French, from Pope Innocent III, and from the English barons. The latter leads to the signing of the Magna Carta (see 1215).

1204   England goes to war with France

King John goes to war against Philip II of France. England loses, forfeiting Normandy and its other possessions on the European continent.

1209   King is excommunicated by the pope

King John is excommunicated from the church by Pope Innocent III as the result of a dispute. John is later forced to sign the English crown over to the authority of the pope.

1209   Beginning of Cambridge University

The foundation is laid for Cambridge University by clerics from Oxford who leave for Cambridge after a run-in with townspeople (the original quarrel from which the phrase *town and gown* is derived). The new institution, whose first college is Peterhouse, is largely modeled on that of Oxford.

1215: June 15   Signing of the Magna Carta

The English barons demand to meet King John at Runnymede, where he is pressured into signing the Magna Carta ("Great Charter"), a document with sixty-three provisions in their favor. The overall significance of the historic signing is to make the king answerable to the law. It is seen as paving the way for the establishment of the parliamentary system.

1258   Provisions of Oxford

King Henry III (1209–72) is forced by the barons to accept the Provisions of Oxford, a series of reforms that limit his power. The Provisions are later annulled by the Mise of Amiens.

1264–67   English barons cause a civil war

Civil war breaks out among the barons, and the king is taken prisoner.

1265   The first parliament meets

Simon de Montfort, earl of Leicester, the leader of the English barons, issues a writ summoning the first parliament, which includes representatives of towns and boroughs as well as members of the nobility.

1272–1307   Reign of Edward I

After the gains made by the barons earlier in the thirteenth century, Edward (1239–1307) gains back power for the monarchy. He also begins a concerted English effort to conquer Scotland and Wales.

1284: March 19   England annexes Wales

Following the death in battle of the last Welsh king, England annexes Wales, enacting the Statute of Rhuddlan. Welsh resistance continues into the sixteenth century.

1295   Model parliament meets

A second parliament convenes, modeled on that of 1265 but closer to the modern British parliament.

1311   Barons force limitations on the king

A group of twenty-one barons force King Edward II (1284–1327) to accept ordinances yielding most of his power to them. The ordinances are revoked eleven years later.

1314   England defeated by Scots at Bannockburn

English control of Scotland ends with the victory at the Battle of Bannockburn by Scottish forces under the command of Robert I.

c. 1329   John Wycliffe is born

Religious reformer John Wycliffe (c.1329–84) is born. After a career as a theologian at Oxford, he begins writing works critical of the papacy which articulate the principles that are later at the center of the Protestant Reformation. He claims that the Scriptures themselves, rather than the clergy, are the only true source of religious authority, and that Christians should have access to the Bible in their native language, and he rejects the doctrine of transubstantiation (the belief that the bread and wine in communion is changed into the body and blood of Christ). In England his ideas are adopted by a group called the Lollards; they are also carried forward by the Czech religious reformer Jan Hus.

1337–1453   Hundred Years' War

Begun during the reign of Edward III, the Hundred Years' War with France extends intermittently to the middle of the fifteenth century. Occasioned by Edward's claim to the throne of France, it is also fought over control of the extensive English possessions in France. Although England wins major victories at Sluy (1340), Crécy (1346), and Calais (1347), the war continues through the reigns of Edward III, Henry IV, and Henry V, and the English are eventually driven out of France.

c.1343   Birth of poet Geoffrey Chaucer

England's most famous medieval poet, Geoffrey Chaucer, is born to a middle-class London wine merchant and receives a good education. As a teenager he serves as a page to Lionel of Antwerp, a son of King Edward III, beginning a long and close association with the highest circles of England's nobility. He later serves as a personal assistant to King Edward himself and to John of Gaunt, another son of the king. Chaucer serves in the military and later carries out diplomatic and trade missions, including a trip to Italy. He is appointed Controller of Customs on wool for London (1374–86) and serves in Parliament in 1385–86. Chaucer dies in London in 1400.

Chaucer produces poetry throughout his varied career. His works include the *Book of the Duchess* (1369–70); *The Parlement of Foules* (c. 1380); and the long romance *Troilus and Criseide* (c. 1385). However, the work for which he is most famous is his masterpiece, *The Canterbury Tales*, probably written between 1386 and 1400 (Chaucer did not live to complete as many tales as he had planned). The work consists of twenty-two stories supposedly told by a group of pilgrims who find themselves journeying together to the shrine of the murdered archbishop Thomas à Becket (see 1170). Between the tales, a "frame narrative" describes the pilgrims themselves and depicts their behavior toward each other. Bringing together characters from all levels of society, *The Canterbury Tales* provides a vivid portrait of English society in the fourteenth century and also serves as a compendium (and in some cases a satire) of the literary genres current at the time.

1348–49   The plague strikes England

The Black Death (thought to be a form of bubonic plague) ravages England, together with the rest of Europe. Brought from Central Asia by sailors and Crusaders, it is transmitted to humans by the fleas on rats and causes an extremely painful death within three days of onset. Carried to England on a ship from Calais in the summer of 1348, it begins on the southern coast and spreads throughout the country within a year, killing an estimated one-third to one-half the population. There is a second attack of the plague in 1361, and intermittent further outbreaks into the seventeenth century.

1381   The Peasant Rebellion

Led by Wat Tyler, the Peasant Rebellion threatens London but is effectively suppressed, and it does not result in any long-term gains for the insurgents.

1384   Death of John Wycliffe

John Wycliffe, a theologian and critic of the established church, dies. (See 1330.)

c. 1386–1400   *The Canterbury Tales* is written

Geoffrey Chaucer writes *The Canterbury Tales*. (See 1343.)

1399–1413   Reign of Henry IV

After a successful rebellion against Richard II, Henry Bolingbroke, Duke of Hereford (1366–1413), ascends the throne of England as Henry IV, establishing the House of Lancaster.

1400   Death of poet Geoffrey Chaucer

Geoffrey Chaucer, author of *The Canterbury Tales,* dies in London. (See 1343.)

1407   Wool trading company is formed

The Merchants Adventurers trading company is established, as the wool trade becomes an important part of England's economy.

1413–22   Reign of Henry V

Henry V renews England's war against France, winning the Battle of Agincourt and other victories. Henry places himself in the line of succession to the French throne by becoming regent of France during the lifetime of Charles VI. He also marries Charles's daughter, Catherine of Valois. However, Henry dies two months before Charles, and England loses its chance to claim the French throne.

1455–85   Wars of the Roses

The Wars of the Roses are fought between the houses of Lancaster and York due to a complicated series of disputes. (The two groups have different colored roses as their symbols: white for the Yorks and red for the Lancasters.) They end with the death of Richard III in battle at the hand of Henry Tudor (1457–1509) and lead to the accession of Henry to the throne. Henry marries the daughter of the York king Edward IV, finally uniting the houses of York and Lancaster.

c. 1470   Malory completes *Morte d'Arthur*

Sir Thomas Malory completes the first version of the Arthurian legend to be written in Middle English prose. Malory's version of the adventures of King Arthur and the Knights of the Round Table differs from his French models in placing greater emphasis on the brotherhood among the knights than on courtly love.

1476   Caxton establishes the first printing press in England

William Caxton (c. 1420–92) is the first to successfully run a printing press in England. Caxton, who studies the printing craft abroad, has already published the first printed book in the English language—a translation of *Le Recueil des histoires de Troye*—while living in Belgium. His press at Westminster publishes almost 100 books, including works by Chaucer, Malory, and poet John Gower.

Tudor England

1485–1509   Reign of Henry VII

The Welshman Henry Tudor ascends the throne of England as Henry VII and inaugurates the Tudor period, customarily regarded as the beginning of the modern era in England. Henry restores the king's power over the barons and brings Wales under firm English control through legal reforms. The Court of Star Chamber is created to deal with civil unrest, corruption, and other forms of disorder that thrived during the chaotic years of war with France. The development of shipping leads to expanded trade, and commercial treaties are signed with Denmark, Venice, and Flanders. The condition of the royal treasury and of the country's economy as a whole is improved.

1497   John Cabot explores the coast off North America

Italian-born John Cabot is commissioned by Henry VII to explore the newly discovered lands across the Atlantic Ocean, and sails down the coastline of present-day Nova Scotia and Newfoundland, beginning the development of the British Empire.

1509–47   Reign of Henry VIII

Henry VIII (1491–1547) ascends to the throne of England, succeeding his father, Henry VII. To form an alliance with Spain, he marries Catherine of Aragon, the daughter of the Spanish king and queen, Ferdinand and Isabella. Catherine's failure to bear him a son and the church's refusal to annul the marriage, cause Henry to separate the Church of England (or Anglican Church) from the Church of Rome and to declare himself its head.

1513: September 9   Scots are defeated at Flodden Field

The Scots invade England while Henry VIII is fighting a war in Europe. The invasion is crushed at the Battle of Flodden, the last major border conflict between Scotland and England. The Scottish king James IV and many of his highest-ranking nobles are killed.

1529   Henry VIII seeks annulment of his marriage

King Henry VIII seeks to have his marriage to Catherine of Aragon, his wife of over twenty years, annulled because she has not borne him a male heir.

1531   Henry declares himself the head of the Anglican Church

Henry VIII breaks with the Catholic Church following the pope's refusal to annul his marriage to Catherine of Aragon.

1533: September 7   Birth of Elizabeth I

Elizabeth, one of England's greatest monarchs, is born to Henry VIII and Anne Boleyn, a former lady-in-waiting to Henry's first wife, Catherine of Aragon. (Henry marries Anne after divorcing Catherine in defiance of the Catholic Church. Anne, who produces no male heir, is, in turn, beheaded when Elizabeth is three years old.) At the age of twenty-one Elizabeth is imprisoned in the Tower by her half-sister, Mary Tudor (1516–58), a devout Catholic who has become Queen and fears the Protestant Elizabeth as a threat to her rule. For five years, Elizabeth is in danger, but she survives and succeeds to the throne on November 17, 1558 to great public acclaim.

Elizabeth's moderate, pragmatic policies give England a period of stability that also encompasses a cultural golden age boasting the works of such figures as William Shakespeare, Sir Philip Sidney, and Edmund Spenser. Elizabeth pursues a middle course with regard to both religion and foreign policy, making England secure but also attempting to avoid involving the nation in costly wars. Although using her hand in marriage as a form of political leverage for some time, she ultimately remains single. Using pomp and pageantry as a means of reinforcing her public image, she retains the allegiance and respect of the public throughout a forty-five-year reign. She dies on March 24, 1603.

1534   Parliament passes the First Act of Supremacy

The First Act of Supremacy, passed by Parliament, ends papal jurisdiction over England. The Church of England is created, and Henry VIII is named as its leader. The Act brings together a series of laws already enacted over several years that have effectively severed England from Rome, including a bar on paying dues to the Pope and making appeals to the papal court.

1536–37   Religious uprisings in northern England

Uprisings protesting King Henry's break with the Roman Catholic church take place in northern England are suppressed and lead to the imposition of martial law.

1536   Wales and England are formally united

Wales is officially integrated into England.

1547–53   Reign of Edward VI

Edward VI (1537–53), the ten-year-old only son of Henry VIII, succeeds to the throne upon his father's death. While Edward is a minor, actual power in the government is held first by his uncle, Edward Seymour, the Duke of Somerset (c. 1500–52 named "Protector of the kingdom") and later by John Dudley, the Earl of Warwick (1502–53). Struggles begin over who should succeed Edward, who is in frail health and dies in his teens. Warwick's plot to keep the throne from the rightful heir, Mary, the daughter of Henry VIII and Catherine of Aragon, by promoting the succession of Lady Jane Grey fails and the throne goes to Mary.

1549   The Book of Common Prayer is published

Thomas Cranmer, the first Archbishop of Canterbury within the reformed Church of England, compiles the Book of Common Prayer.

1553: July 10–19   Nine-day reign of Lady Jane Grey

Lady Jane Grey (1537–54), a great-granddaughter of Henry VII, succeeds to the throne following the death of Edward VI, with the support of the Earl of Warwick, who wants to keep Mary, the daughter of Henry VIII, from becoming queen. Jane Grey's reign lasts only nine days, after which she is held in the Tower of London and executed the following year.

1553–58   Reign of Mary I

Mary, daughter of Henry VIII and Catherine of Aragon, succeeds to the throne of England following the death of Edward VI and the overthrow of Lady Jane Grey, who ruled for only nine days. Mary, who opposes the break with the Catholic church, acts to reinstate Catholicism in England. Religious reforms that depart from the teachings and practices of the church are rescinded, papal authority over England is reestablished, and, once resistance is aroused, Protestants are persecuted. Nearly 300 persons are burned at the stake for heresy. This religious persecution, coupled with Mary's unpopular marriage to Philip II of Spain, arouses widespread resistance among the people, and she becomes known as "Bloody Mary." She is nearly overthrown by force, and her death on November 17, 1558 is greeted with relief and rejoicing.

1554   Mary weds Philip of Spain

Queen Mary marries Philip of Spain, who is her second cousin. The marriage is highly unpopular with the English people.

1558–1603   Reign of Elizabeth I

Elizabeth I, daughter of Henry VIII and Anne Boleyn, succeeds Mary as ruler of England. She is a strong ruler, successful in both domestic and foreign policy and shrewd in her choice of advisors. Under Elizabeth England becomes a major world power with a strong navy and a growing empire. Commerce expands and culture flourishes. Most of William Shakespeare's plays are written during Elizabeth's reign.

1559   Second Act of Supremacy is passed

The Second Act of Supremacy reestablishes the Anglican Church, following the reversion to Catholicism under Queen Mary. Like her father, Henry VIII, Elizabeth is named head of the Church of England.

1560   Treaty of Edinburgh

England and France agree to withdraw their troops from Scotland to mitigate hostilities between Catholics and Protestants.

1564: April   Birth of William Shakespeare

Commonly regarded as England's greatest author, William Shakespeare (1564–1616) is born in Stratford-on-Avon to a prominent local official. He is assumed to have attended the local grammar school, but does not attend either Oxford or Cambridge. In 1582 he marries Anne Hathaway and they have three children (including twins) between 1583 and 1585. By 1592 Shakespeare is active on the London stage and becomes an actor and the main playwright for the Chamberlain's Men, which is formed two years later. By 1597 Shakespeare is successful enough to purchase a stately house in Stratford.

The playwright's history plays, based on actual historical accounts from such sources as Raphael Holinshed's *Chronicles,* focus on struggles for political power. Most of these are written in the 1590s, including *Richard II, Richard III, Henry IV Parts I and II,* and *King John.* The comedies (*As You Like It, Twelfth Night, The Merchant of Venice, and Much Ado About Nothing*), written around the turn of the century, focus on the resolution of misunderstandings and other barriers to romantic love. Shakespeare writes his great tragedies (*Hamlet, Macbeth, King Lear,* and *Othello*) in the first decade of the seventeenth century. The quality of their poetry and their psychological insight into the human spirit place them among the masterpieces of Western literature. Shakespeare's final plays, written in 1610 and 1611, are called "romances" or "tragicomedies" and include *The Tempest, Cymbeline,* and *The Winter's Tale.* Shakespeare is also celebrated for his nondramatic poems, especially his sonnets.

In 1610 the playwright retires from London to his home in Stratford, where he dies in 1616. His contemporary, Ben Jonson, eulogies him using the famous phrase, "He was not of an age, but for all time!"

1568–87 Mary, Queen of Scots is imprisoned and executed

Mary (1542–87), the daughter of James V (1512–42) of Scotland and Mary of Guise, and a claimant to the English throne, is imprisoned in England for nearly twenty years by Queen Elizabeth before finally being executed.

1577–80 Sir Francis Drake sails around the world

Sir Francis Drake becomes the first English explorer to circumnavigate the globe.

1583 English lay claim to Newfoundland

Newfoundland becomes England's first overseas possession.

1585 First colonists travel to Virginia

England's first colonial settlers arrive in Roanoke, Virginia.

1588 Defeat of the Spanish Armada

Angered by the piracy of Sir Francis Drake, by England's break with the Catholic Church, and by its support for a revolt in the Netherlands, Spain sends an armada of 130 ships to invade England. The English navy, aided by inclement weather, halts the armada in a victory that is considered conclusive evidence of English naval superiority.

1590–1609 Publication of *The Faerie Queene*

Edmund Spenser's epic poem is one of the great artistic achievements of the Elizabethan period, bringing together Protestant religious allegory, British nationalism, and the literary tradition of the Italian romantic epic. Like epics of all kinds, *The Faerie Queene* is organized around an elaborate structure, with the quests of twelve different knights illustrating twelve major virtues. In one of the work's many layers of allegory, its heroine, Queen Gloriana, symbolizes Queen Elizabeth. The poem is written in the distinctive nine-line Spenserian stanza and utilizes a unique archaic vocabulary. Books I–III are published in 1590; books IV–VI in 1596; and the first complete edition in 1609.

1594 The Chamberlain's Men theatrical group is formed

The Chamberlain's Men, a theater troupe that includes William Shakespeare, is organized.

1599 The Globe Theatre is built

England's most famous theater is built as a home for its premier acting troupe, the Chamberlain's Men (later the King's Men), with which William Shakespeare is closely associated.

1599: April 25 Birth of Oliver Cromwell

Cromwell, Lord Protector of England during the Puritan revolution, is born in Huntingdon to members of the English gentry. As a young man, he pursues the life of a country gentleman and farmer, although he is active in civic affairs. His career in national politics begins with his election to Parliament in 1628, just before the eleven-year period when Charles I dismisses the legislative body. A fervent Puritan, Cromwell is elected to the turbulent parliaments of the 1640s and serves with distinction as a leader of the cavalry in the two civil wars. He also plays a leading role in organizing the tribunal that condemns Charles I to death in 1649. When a republic is declared, Cromwell becomes the chairman of its Council of State. When the Commonwealth is formed in 1653, he becomes its first and only Lord Protector. At his death in 1658 he names his son, Richard, as his successor, but by 1660 the monarchy is restored. Royalists later exhume and desecrate Cromwell's remains, which have been buried at Westminster Abbey.

1600 British East India Company is established

The British East India Company is formed and enters the spice trade, gradually expanding its interests in India.

1603: March 24   Death of Queen Elizabeth

Elizabeth I dies, ending a long and successful reign. (See 1533.)

1603–25   Reign of James I

King James VI of Scotland, the son of Mary, Queen of Scots, succeeds to the throne after the death of Queen Elizabeth, becoming James I and beginning the Stuart line. He attempts to keep both Puritans and Catholics happy, as well as accommodate Parliamentary demands with regard to taxation and foreign affairs. Under both James and his son Charles I (r.1625–49), Parliament becomes more powerful, and the middle class gains increasing influence.

1605: November 5   The Gunpowder Plot is crushed

A group of Catholics plots to overthrow the government by bombing Parliament during its opening session, thus also killing King James I, who will be in attendance for official ceremonies. They store thirty-six barrels of gunpowder in a cellar beneath the House of Lords. Word of the plot is widely spread throughout London's Catholic community, and one member leaks the news to warn a relative who is a member of Parliament. The police are notified, and the conspirators are arrested and hanged. Restrictions on England's Catholics become even worse after the failed plot, and Guy Fawkes, the Catholic convert assigned to set off the explosives, is hanged in effigy every year on November 5 from then on.

1607   Jamestown is founded

Jamestown, Virginia, becomes the first permanent English settlement in North America.

1608: June 30   Birth of poet John Milton

John Milton (1608–74), famed for his epic poem *Paradise Lost,* is born in London to a businessman and displays a precocious gift for learning, mastering Latin, Greek, and several other foreign languages even before beginning his studies at Cambridge University. After receiving two degrees from Cambridge, he studies on his own for six years and then travels in Europe for a year (in 1638). Early poetic works include *Comus,* a masque (a type of drama) and *Lycidas,* a famous poem eulogizing a classmate who dies in a boating accident. During the 1640s, Milton, an ardent supporter of the Parliamentarians and the Cromwell regime, publishes a number of tracts arguing his views on political and religious issues. With the downfall of Cromwell and the restoration of King Charles II, Milton, now blind, is arrested and fined and loses most of his property.

Retiring from public life, the poet writes his masterpiece *Paradise Lost* (published in 1667), intended to "justify the ways of God to men." In blank verse (seldom used before other than in dramas), the lengthy poem, in twelve books, narrates the fall of Adam and Eve and their expulsion from the garden of Eden. Its multitude of classical and religious allusions places it in the Renaissance tradition, but its ethic of personal responsibility and internal repentance and conversion also make it a work of the Reformation. Milton's other major works, written after *Paradise Lost,* include *Paradise Regained* (1671) and *Samson Agonistes* (1674). The poet dies in Buckinghamshire on November 8, 1674.

1611   The King James Bible is published

Commissioned by King James I, the best-known and most popular English translation of the Christian Bible is mostly based on the text of an earlier translation by William Tyndale. It is also known as the Authorized Version of the Bible.

1616: April 23   Death of William Shakespeare

The great poet and playwright William Shakespeare dies in Stratford-on-Avon. (See 1564.)

1620   Plymouth Colony is founded

The Pilgrims, seeking a haven from religious persecution, arrive in New England and establish a colony.

1625–49   Reign of Charles I

Charles I becomes increasingly dependent on Parliament for the funds to maintain his government, and Parliament becomes progressively stronger. He antagonizes Parliament by marrying a Roman Catholic and provokes a war with Scotland by trying to impose Anglican Church practices on the Scottish church.

1628: May   Petition of Right

The Parliament presents the Petition of Right, guaranteeing freedom from arbitrary arrest and imprisonment, to King Charles.

1629   King Charles dissolves Parliament

Charles I dissolves Parliament and rules on his own for eleven years.

1635   Post Office is founded

England's post office is established. It later becomes the world's first post office to use adhesive stamps.

1637   Charles provokes Scottish rebellion

King Charles attempts to force the Scots to use the Anglican liturgy, and Scottish Presbyterians rebel.

1642–48   English Civil War

King Charles's attempt to have members of Parliament arrested sparks a civil war between Parliamentary forces and royalists supporting the king. Wales supports the king, while

The English Civil War, c. 1645.

☐ Parliamentary forces

▨ Royalist support

most Scots side with Parliament. Charles's forces suffer a decisive defeat in 1645, and Charles takes refuge in Scotland. However, fighting resumes in the spring and summer of 1648.

1645   Royalists are defeated at Naseby

The royalists suffer a decisive defeat at Naseby in the summer of 1645.

1646: May   Charles takes refuge in Scotland

After the defeat of his troops at Naseby, Charles takes refuge in Scotland and negotiates with the Parliamentarians.

1647: February   Charles is turned over to the English

King Charles is turned over to his enemies in England.

1649: January 20   Charles's trial begins

Charles I is tried for treason and convicted.

1649: January 30   Charles is beheaded

Charles I is beheaded.

The Commonwealth

1649: February   England becomes a commonwealth

The kingship is abolished, and England becomes a commonwealth ruled by a forty-one-member council of state. Oliver Cromwell becomes the head of state and sends armies into Ireland and Scotland to subdue revolts over the next three years. Three-fourths of Irish land is confiscated to pay England's debts and reward its army.

1653: December 16   Cromwell becomes Lord Protector

England becomes a protectorate, according to guidelines spelled out in a written constitution (the only one in the country's history). As Lord Protector, Oliver Cromwell becomes head of the governments of England, Scotland, and Ireland.

1658   Cromwell dies

Lord Protector Oliver Cromwell dies (see 1599.) His death plunges England into political chaos for two years. His son Richard becomes Lord Protector but cannot govern effectively and resigns after less than one year.

1659   Birth of composer Henry Purcell

Purcell is associated with the Chapel Royal and Westminster Abbey in a variety of capacities, beginning as a choir boy and culminating in his appointment as chapel organist. He writes a large volume of music commissioned by the royal family on special occasions, including songs, anthems, and services. His compositions also include incidental music for dramas and operas. Purcell is best known for *Dido and Aeneas* (c.1689), generally regarded as the first true English opera. More generally, Purcell is known for integrating European musical styles with the English native tradition. Purcell dies in London in 1695.

1660–1703   Period covered by Samuel Pepys's *Diary*

The diary kept by naval official Samuel Pepys (1633–1703) is a classic of English letters. It provides a frank and detailed account of aristocratic life in England during the Restoration and describes the major events of the period such as the outbreak of the plague (see 1664) and the Great Fire of London (see 1666). It is first published in 1825.

The Restoration

1660–88 The Restoration

Based on the vote of a specially convened parliament, England once again becomes a monarchy, governed by a king and by parliamentary houses of Lords and Commons. The term "restoration" comes to refer not just to the political events of 1660 (the restored throne) but to all aspects of English history (social, economic, and cultural) over a period of nearly thirty years from the time when the new monarch, Charles II (1630–85), is crowned.

1660: May 8 Charles II is crowned

Charles II, the eldest son of Charles I, is crowned.

1664 Navigation Act is passed

The Navigation Act places strict controls on trade with the colonies. It arouses intense opposition among the colonists and eventually helps lead to the American Revolution.

1664–65 The Black Death recurs

There is a second major outbreak of the Black Death that sweeps through England in the fourteenth century (and has never been totally eradicated). Beginning in the winter of 1664–65, it reaches its height the following summer, with the worst outbreak in London, where thousands die.

1666: September 2–6 Great Fire of London

Dry conditions following a prolonged drought allow an ordinary fire to sweep rapidly through London, demolishing more than two-thirds of the world's largest city in five days. The fire begins in the chimney of a bakery, spreads to neighboring houses and other buildings, and on to the wharf of the Thames River, where it kindles combustible goods stored there. Firefighting techniques are so poor that the major "remedy" applied is to tear down houses in the path of the fire, and the blaze ends only when the wind direction changes. Some 13,000 homes, 87 churches, and dozens of public buildings are destroyed, including the Royal Exchange and Guild Hall. Amazingly, only eight deaths are recorded. Within a year, the world's first fire insurance company is founded. The fire is also thought to have eradicated the plague, whose return has haunted the English for decades, and which continues to spread on the Continent for over a century more.

1667 Paradise Lost is written

Poet John Milton (1608–1674), who has become blind, dictates the text of Paradise Lost. (See 1608.)

1674: November 8 Death of John Milton

John Milton, one of England's greatest poets and author of Paradise Lost, dies. (See 1608.)

1685–88 Reign of James II

James II (1633–1701), the younger brother of King Charles II, ascends the throne following Charles's death. A Catholic, he is an unpopular king. When an heir is born to him, fears are raised that he will be succeeded by yet another Catholic monarch, and James's opponents ask Mary and William of Orange, James's Protestant son-in-law, to take over the throne.

1685: February 23 Birth of composer George Frederick Handel

The greatest Baroque composer besides J.S. Bach, Handel is born Georg Friedrich Haendel, in Saxony, Germany. He begins his musical career in 1702 as a church organist in his native city of Halle. He soon settles in Hamburg, where he begins working as a performer and composer. By 1710 Handel is appointed *Hofkapellmeister* (chief music director) to George of Hanover, who becomes king of England upon the death of Queen Anne in 1714. Handel spends increasing amounts of time in England over the following years, and eventually settles there permanently, becoming a naturalized British citizen in 1726.

Handel first becomes successful as a composer of operas, including *Floridante, Ottone, Giulio Cesare,* and *Rodelinda.* After 1737 he turns his attention to oratorios, producing works such as *Saul, Israel in Egypt, Samson,* and *Judas Maccabaeus,* all written in English and based on religious texts. Handel's best-known oratorio is *Messiah,* famous for its "Hallelujah Chorus." Among the composer's best-known instrumental works are the *Water Music* and *Fireworks Music.* Handel goes blind by 1753 but continues to compose with the aid of a transcriber. He dies on April 14, 1759 and is buried in Westminster Abbey.

The Glorious Revolution

1688 The Glorious Revolution

The Catholic monarch James II is removed from the throne and replaced by Protestants William III (1650–1702) and Mary II (1662–94). England and Wales support the new monarchs, but in James's homeland of Scotland, resistance leads to the Battle of the Boyne (see 1690). Also known as the Bloodless Revolution, the Glorious Revolution introduces constitutional monarchy to England.

1689 Parliament enacts the Bill of Rights

England's Bill of Rights consolidates the balance of power favoring Parliament. Among its provisions are bans on royal suspension of laws, levying of taxes without parliamentary approval, a guarantee of the right to petition the crown, a provision for frequent sessions of parliament, and the requirement that parliamentary approval be obtained for forming a standing army in time of peace.

1690   James is defeated at the Battle of the Boyne

James II attempts to retake the throne with support from the French and Irish but is thwarted by King William, who proceeds to harshly subjugate the Irish.

1695: November 21   Death of Henry Purcell

Composer Henry Purcell dies in London. (See 1659.)

1697: November 10   Birth of William Hogarth

Hogarth, a painter and engraver, is a pioneer of English painting, notable not only for his own paintings but for his role in improving the status of artists in England. By publishing engravings of his own paintings, he leads the way for other artists to achieve financial success without relying on the patronage of the wealthy. The paintings for which he is best known, including *A Harlot's Progress* and *A Rake's Progress,* are famous for their realistic details and satirical view of contemporary society. Hogarth dies in 1764.

1701–14   War of the Spanish Succession

The War of the Spanish Succession is triggered when the Spanish throne is bequeathed to Philip of Anjou, grandson of Louis IV of France. Louis has recognized the claim to the throne of James, Stuart, the Old Pretender (the son of James II), and opposed English trade interests, so England allies itself with Holland and the other forces opposing Spain and its allies. Also known as Queen Anne's War, it is fought on the European continent and in the New World colonies.

1702–14   Reign of Queen Anne

The throne is occupied by Queen Anne (1665–1714), the last Stuart monarch. Although relying strongly on ministers because of her health problems, Anne is capable of governing effectively and tries to steer a middle course between the conflicting political interests of the newly powerful Whig and Tory parties.

1707: May 1   Act of Union unites England and Scotland

The Act of Union unites England and Scotland politically, forming Great Britain, governed by a single monarch and parliament.

1714   Alexander Pope writes *The Rape of the Lock*

Alexander Pope (1688–1744) writes the mock epic *The Rape of the Lock,* one of the defining literary works of Enlightenment neoclassicism. The wit and satire characteristic of much eighteenth-century English literature is showcased in this mock epic that places a personal incident revolving around the theft of a lock of hair into the framework of a classic epic, complete with battles, muses, invocations, guiding spirits, a descent to the underworld, and allusions to the heroes of famous Greek and Latin epics.

The House of Hanover

1714–27   Reign of King George I

Queen Anne dies without leaving any heirs to the throne, so George I (1660–1727), a great-grandson of James I, is brought from Hanover, Germany, to succeed to the throne, establishing the House of Hanover.

1715   Jacobite revolt

The supporters of the deposed King James II become known as Jacobites. After James's death in 1701, they shift their allegiance to his heirs, including his son, James, who is in exile in France and becomes known as the Old Pretender.

1719–20   Daniel Defoe writes *Robinson Crusoe*

The adventure narrative *Robinson Crusoe,* regarded by many as the beginning of the English novel tradition, is written.

1723   Birth of painter Joshua Reynolds

Widely regarded as England's greatest portrait painter, Reynolds is also known for his theories on the nature of art, especially as collected in his *Discourses.* In 1768 he is appointed the first director of the Royal Academy. Reynolds's paintings are extremely popular within his lifetime, and he has as many as 150 people a year sit for portraits. Knighted in 1769, Reynolds dies in 1792 and is buried in St. Paul's Cathedral, the first artist to be granted this honor since Van Dyck.

1726   Publication of *Gulliver's Travels*

Jonathan Swift's (1683–1760) great satirical work is published anonymously. Its four books describe the voyages of the shipwrecked surgeon and seaman Lemuel Gulliver to the lands of the diminutive Lilliputians, the giant Brobdingnagians, the flying island of Laputa and the continent of Lagado, and the land of the rational horses known as Houyhnhms (and the degraded humanoid Yahoos).

1727–60   Reign of George II

George II (1683–1760) succeeds to the throne, becoming the second monarch in the Hanover line.

1742   Handel composes *The Messiah*

German-born composer George Frederick Handel composes the oratorio, *The Messiah,* one of the masterpieces of Western choral literature.

1744   Cricket rules are adopted

The rules for cricket, the first modern ball game, are codified.

1745–46   Second Jacobite uprising

The Jacobites (supporters of the Catholic Stuart line) launch a new revolt, led by Prince Charles Edward (1720–88) (known as Bonnie Prince Charlie), the son of James, the Old Pretender, and grandson of King James II.

1755   Publication of Johnson's dictionary

Critic and essayist Samuel Johnson (1709–84) publishes his *Dictionary of the English Language* after working on it for eight years. It is the first English dictionary to illustrate word usage with historical quotations.

1756–63   The Seven Years' War

Fought against France, Austria, and Russia, the Seven Years' War establishes Britain's colonial supremacy in India and North America (where it is called the French and Indian War). The war is ended by the Treaty of Paris.

1759   Death of George Frederick Handel

Composer George Frederick Handel dies. (See 1685.)

1760–1820   Reign of George III

The controversial monarch whose actions provoke the American Revolution becomes insane late in life.

1764: October 25   Death of William Hogarth

Painter and engraver William Hogarth dies. (See 1697.)

Mid-18th century   The Industrial Revolution begins

Technological advances coupled with the growing demand for goods by an increasing population bring about the inauguration of modern industry, as small-scale, home-based production shifts to factory-based mass production. Progress in technology is led by advances in the textile, metals, and coal industries. Important inventions include James Hargreaves's spinning jenny, Richard Arkwright's water-powered frame for spinning cotton, and Abraham Darby's development of the coke smelting process.

1768   The Royal Academy is founded

The Royal Academy is founded to further the art of painting in Britain by offering artists training and a place to exhibit and sell their work. Since its founding, it has been housed in Somerset House, a National Gallery building in Trafalgar Square, and Burlington House.

1770: April 7   Birth of William Wordsworth

The major poet associated with the beginning of English Romantic poetry, Wordsworth is born in the Lake District in northern England and attends Cambridge. He makes two trips to France between 1790 and 1792, becoming an ardent supporter of the French Revolution and fathering a daughter, Catherine, with a Frenchwoman, Annette Vallon. After Wordsworth's second return to England, he experiences several years of emotional and financial turmoil before settling down with his sister, Dorothy, in Dorsetshire. He meets Samuel Taylor Coleridge (1772–1834), and the two poets begin the close professional association that results in the publication of *Lyrical Ballads* (1798), which opens with Coleridge's *Rime of the Ancient Mariner* and closes with Wordsworth's *Tintern Abbey.* This work is widely regarded as launching the Romantic period in English poetry. The direct, personal, lyrical quality of these poems, and their emphasis on the joys and solace of nature, contrast with the more elaborate, erudite neoclassical poetry associated with the Restoration period.

Wordsworth next begins writing his great autobiographical poem *The Prelude,* which he continues working on until his death in 1850. Most of Wordsworth's great poetry is written in the decade between 1798 and 1808. However, he enjoys professional recognition, public acclaim, and financial stability for the rest of his life and is appointed poet laureate of England in 1843. *The Prelude* is published after his death and is regarded as his masterpiece.

1775–83   The American Revolution

The growing divergence of British and colonial interests leads to increasing tensions between the mother country and its North American colonies. These are intensified by events including the passage of the Stamp Act and the "Intolerable Acts," new import duties, and the Boston Tea Party. Once hostilities are begun in Concord, Massachusetts, the British military strategies of driving a wedge between the colonies or conquering the South fail, and the war is essentially lost by 1781. It is formally ended by the Treaty of Versailles.

1775: April 23   Birth of Joseph Turner

Joseph M. W. Turner is one of England's premier landscape painters, known for both his oil and watercolor paintings. He is instrumental in gaining recognition not just for his own talents but for the validity of landscape painting as a serious art. After first establishing his reputation within the Royal Academy, he starts his own gallery in Harley Street in 1804, although he continues to participate in exhibitions at the Academy, where he is also appointed professor and begins to lecture in 1811. Turner dies in 1851, bequeathing many of his sketches and paintings to the British public.

1775: December 16   Birth of Jane Austen

Jane Austen, widely regarded as one of the England's greatest novelists, is born in Hampshire to a rector and schoolteacher, the seventh of eight children. She grows up in the world of the landed country gentry, a world she portrays in realistic detail in her novels. She receives five years of formal education and is then schooled at home by her father. By 1787 Austen is writing plays, poetry, and parodies of popular romance fiction. After Austen's father dies in 1805, Austen, her mother, and her sister live in a succession of temporary lodgings before settling in a cottage on her brother's estate at Chawton in 1809.

Austen begins revising drafts of novels she has written earlier. Six novels are published anonymously in rapid succession between 1811 and 1817 (the last two posthumously): *Sense and Sensibility* (1811), *Pride and Prejudice* (1813), *Mansfield Park* (1814), *Emma* (1815), and *Persuasion* and *Northanger Abbey* (both in 1817). After Austen's death on July 18, 1817 of a long illness thought to be Addison's disease, her brother reveals her identity to the public. The quality of Austen's prose, and her wit and insight into human nature give her novels a permanent place in the canon of English literature. The advent of feminist literary criticism over the past quarter century has drawn more scholarly attention than ever to Austen's work and provided additional perspectives for appreciation of it.

1776   *The Wealth of Nations* is written

Adam Smith (1723–90) writes *The Wealth of Nations,* which revolutionizes economic theory.

1783–1801   Pitt serves as prime minister

William Pitt (1759–1806) the younger expands the power of the prime minister and restores the strength of the Tories. He provides strong leadership during the difficult period of war with France.

1783   Treaty of Versailles

The Treaty of Versailles ends the American War for Independence, giving the thirteen American colonies the liberty to decide their own political future. The loss of its most valuable group of colonies is a major blow for Britain, and it redirects its energy toward the colonization of India, later settling Australia and New Zealand as well.

1787   Marylebone Cricket Club is founded

England's most prestigious cricket organization is established.

1789–94   Publication of Blake's *Songs of Innocence* and *Songs of Experience*

William Blake's great double set of lyrical poems is published with hand-colored illustrations by the author. *Songs of Innocence* appears by itself in 1789, and is followed by *Songs of Innocence and Experience: Shewing the Two Contrary States of the Human Soul* in 1794. The poems celebrate the innocence of nature and childhood and decry the corrupting influences of organized religion and modern industrialized society.

1792   *Vindication of the Rights of Woman* is published

Mary Wollstonecraft (1759–97) publishes one of the first famous feminist tracts, in which she challenges misogynist misconceptions about women and calls for improved female education and equal political rights.

1792   Death of painter Joshua Reynolds

Portraitist Joshua Reynolds dies. (See 1723.)

1793–1802   French Revolutionary Wars

Britain is forced to join the countries at war on the Continent by a declaration of war by the French revolutionary government.

1798   Publication of *Lyrical Ballads*

William Wordswoth and Samuel Taylor Coleridge jointly publish *Lyrical Ballads,* containing poems by both authors. It is a landmark in the development of English Romantic poetry. (See 1770.)

The United Kingdom

1801: January 1   Act of Union

Great Britain and Ireland unite, forming the United Kingdom. The Irish gain representation in Parliament but are still economically oppressed by absentee landlords. A pronounced split grows between the northern and southern Irish counties.

1803–15   Napoleonic Wars

Coming right on the heels of the French Revolutionary Wars, England's engagement in the struggle against the French leader Napoleon results in the country being at war for over a generation. England emerges as the leading European military power but faces economic and social problems at home.

1812–15   War of 1812

American forces attack sites in British-controlled Canada, partly for purposes of expansion and partly to retaliate for the British navy's impressment of Americans on the high seas and Britain's blockade of Europe in the Napoleonic wars. In response to the American victory at York (present-day Toronto), the British occupy Washington and burn down the White House. Although grievances of both sides are unresolved, peace is imposed by the Treaty of Ghent.

1812: February 7   Birth of novelist Charles Dickens

Generally considered the greatest Victorian novelist, Charles Dickens (1812–70) is born in Portsmouth to a government clerk. When he is twelve, his father is thrown into debtors' prison, and he is forced to leave school and go to work in a factory to help support his family. Although the family fortunes later revive, this experience leaves an indelible mark on the novelist and is reflected in the many unhappy childhoods and prison images found in his works. Dickens begins his professional career as a journalist and starts publishing essays and short fiction in 1833. He marries in 1836 and begins a family that will eventually include nine children.

Dickens achieves immediate success as a novelist with the publication of *The Pickwick Papers* in 1837. This novel is followed by *Oliver Twist* (1838), *Nicholas Nickleby* (1839), *The Old Curiosity Shop* (1841), and *Barnaby Rudge* (1841). Dickens then makes a highly successful tour of the United States. His classic *A Christmas Carol* is published in 1843, followed by *Dombey and Son* (1848) and the semiautobiographical *David Copperfield* (1850). Dickens's later novels, including *Bleak House* (1853), *Hard Times* (1854), and *Great Expectations* (1861), are darker in tone and more suffused with serious social criticism. Between 1850 and 1859 the novelist also publishes a popular weekly collection of essays, fiction, and poetry called *Household Words* and achieves a growing reputation for the quality of his public readings. Dickens dies in Kent on June 9, 1870. His novels are acclaimed for their vivid characters and dialogue, their illumination of social injustice, their detailed and colorful portrayals of London, and the emotional pathos of their plots.

1815   The Corn Law is passed

By restricting imports of foreign grain, the Corn Law raises bread prices, creating hardship and resentment among the working classes and instigating calls for reform.

1819–24   Byron writes *Don Juan*

The poet George Gordon, Lord Byron (1788–1824) writes his masterpiece, *Don Juan,* a satirical narrative in verse. He turns the historical figure of Don Juan into a picaresque hero, portraying his activities in love, politics, and war, and using the events of the story to satirize his own society. Although the poem has classical elements, such as the use of the *ottava rima* stanza, its celebration of the individual, "natural" man against the mores of society also locates the work within the Romantic tradition.

1819: May 24   Birth of Queen Victoria

The longest-reigning monarch in British history, Victoria is an effective and popular ruler, whose name has come to symbolize an entire era of British prosperity, expansion, and confidence in the destiny of its empire. The daughter and only child of Prince Edward, Duke of Kent, Victoria succeeds to the throne at the age of eighteen after two uncles, who are next in succession, die and leave no heirs. She is crowned at Westminster Abbey on June 28, 1838. She quickly gains approval and respect for her native intelligence and common sense. In 1840 she marries her first cousin, Prince Albert, on whose judgment she comes to rely heavily for both private and public matters. They have a close and happy marriage, producing nine children.

In 1861 Prince Albert dies of typhoid fever, and the queen goes into a state of profound mourning, from which she never fully emerges. There is some discontent with her absence from public performance of her duties, but by the 1870s her popularity is once again secure. There are great public celebrations of her golden (fiftieth) and diamond (sixtieth) jubilees in 1887 and 1897. The queen dies on January 22, 1901 and is buried near Windsor in the final resting place that also houses Prince Albert.

1819: August 16   Peterloo Massacre

British cavalry arrest demonstrators calling for parliamentary reform and repeal of the Corn Laws, killing several people and injuring hundreds.

1821   Publication of Shelley's elegy *Adonais*

Percy Bysshe Shelley (1792–1822) publishes the poem *Adonais* commemorating the death of fellow Romantic poet John Keats (1795–1821). Tragically, Shelley himself is killed in a boating accident the following year.

1825   First railroad locomotive is built

The first steam locomotive (Stephenson's Rocket) is built by George Stephenson, leading to extensive railroad development by the mid-nineteenth century.

1829   Catholic Emancipation Act

The Catholic Emancipation Act ends the official penalization of Catholics by the state. Catholics are allowed to serve in Parliament.

1829   Scotland Yard is established

London's first metropolitan police headquarters is set up.

1831   Faraday introduces principles of electromagnetism

Michael Faraday (1791–1867) demonstrates his theory of electromagnetic induction. Faraday, who delivers royal lectures before Prince Albert, is also a leader in the development of atomic particle theory.

1832 Historic reform bill enacted

The Reform Act of 1832 is the first major attempt to remedy the serious social ills caused by urbanization following the enclosure of grazing land and the growth of the factory system. The rights of the poor as well as the rising middle class are guaranteed by the reform acts of 1832, 1867, and 1884. Working class and middle class people win the vote, doubling the size of the voting public.

1832–33 Cholera outbreak strikes London

Contracted through passengers on a ship arriving from Hamburg, Germany, England's first cholera epidemic ravages London in two separate outbreaks, killing 6,800 people. Medical progress based on bacteriological discoveries is not yet advanced enough to aid in fighting the disease.

1833–50 Composition of the poem *In Memoriam*

Alfred, Lord Tennyson (1809–92) works on the poem *In Memoriam,* begun in response to the death at a young age of his closest friend, Arthur Hallam. Describing the successive stages of Tennyson's own grieving process, the poem dramatizes Victorian conflicts over scientific advances in fields such as anthropology and geology that call into question basic tenets of traditional Christian religious belief. The poem's great popular success is partly responsible for Tennyson's appointment as England's poet laureate in 1850.

1833: July 14   Keble sermon launches the Oxford Movement

The Oxford Movement, aimed at restoring High Church traditions within the Church of England, is launched with a sermon by clergyman John Keble. Prominent figures associated with the movement include Cardinal John Henry Newman, who, along with many of its other supporters, eventually joins the Roman Catholic Church.

1834 Slavery is abolished

Slavery is abolished throughout the British Empire, with major social and economic repercussions in many of Britain's possessions throughout the world.

The Victorian Era

1837–1901   Reign of Queen Victoria

Victoria's long reign marks the height of the British Empire as a world power, and a period of unprecedented commercial prosperity and technological advancement. With its superior merchant fleet, England gains a world market for its cotton and other manufactured goods, pouring the profits into investments both at home and in its expanding overseas empire. In addition to the steam-powered inventions that fuel industrial growth, the country's rapid technological progress during this period encompasses the introduction of the telegraph, photography, and anesthesia. Important social advances include universal compulsory education.

1837   Publication of *The Pickwick Papers*

Novelist Charles Dickens (see 1812) enjoys his first major literary success with the publication of *The Posthumous Papers of the Pickwick Club.*

1838–48   Chartism flourishes

Chartism is a popular working-class movement for democratic reforms. It is centered around the Six Points articulated in the "People's Charter" drawn up by cabinet-maker William Lovett in 1838: 1) universal (male) suffrage; 2) annual elections; 3) secret ballots; 4) fair electoral districting; 5) end to property requirements for members of Parliament; and 6) salaries for members of Parliament. Chartism eventually becomes diffused and is associated with a diverse group of working-class and radical movements.

1839–42   First Opium War

Britain wins important trade concessions from China as a result of the first Opium War.

Five ports are opened to British trade, the island of Hong Kong is ceded to the British, tariffs are limited, and Britain gains most-favored-nation trade status. British nationals are also granted extraterritoriality (exemption from Chinese laws).

1841   Conservative Party gains power

Under Prime Minister Robert Peel, the newly formed Conservative Party comes to power and is responsible for repealing the heavily disputed Corn Laws.

1846   Repeal of the Corn Laws

Long-standing import restriction laws protecting landowners are repealed in a major victory for England's urban middle-class constituency.

1848–53   Rise of the Pre-Raphaelite movement

The Pre-Raphaelite movement begins as a protest against the perceived lack of aesthetic honesty, directness, and spirituality in the style of painting encouraged by the Royal Academy. Its leaders include Dante Gabriel Rossetti, John Everett Millais, and William Holman Hunt, who look to pre-Renaissance Italian painting as their model and begin a new vogue for medievalism in both the visual arts and poetry. Besides Rossetti, other authors linked to the Pre-Raphaelites include William Morris, Algernon Charles Swinburne, and Christina Rossetti.

1851   The Great Exhibition

A landmark and symbol of Victorian grandeur, the mammoth Great Exhibition, held in Hyde Park, is housed in the specially constructed Crystal Palace, composed of prefabricated glass and metal sections and measuring 1,600 by 384 feet (488 by 117 meters).

1851: December 19   Death of painter Joseph Turner

Renowned landscape artist Joseph Turner dies. (See 1775.)

1854–56   Crimean War

Britain goes to war with Russia in the Black Sea region to protect its interests in Turkey and access to Asia. Conditions on the field are so bad that more soldiers die of disease than in fighting, in spite of advances in nursing by Florence Nightingale. The war is notable for the first use of rifles by the army and the use of steam power by the navy.

1859   *On the Origin of Species* is published

Charles Darwin (1809–82) publishes *On the Origin of Species.* His principles of natural selection and biological evolution gain universal acceptance and are widely influential in many fields.

1863   Football Association formed

Soccer's governing organization is formed to standardize the game, which is already being played according to a set of provisional rules by competing teams at the nation's top public [English private] schools, including Eton, Harrow, Rugby, and Winchester.

1867   Reform bill is passed

The Reform Act of 1867 reapportions seats in Parliament and doubles the size of the electorate by extending the franchise.

1868–69   Publication of *The Ring and the Book*

Poet Robert Browning (1812–89) publishes his greatest work, a series of twelve dramatic monologues based on a Roman murder trial that takes place in 1698. Each monologue is presented from the viewpoint of a different character. This work of over 12,000 lines is written in the blank verse characteristic of other Browning poems in the dramatic monologue form, such as *My Last Duchess* and *Fra Lippo Lippi.*

1870   Charles Dickens dies

Famed novelist Charles Dickens dies. (See 1812.)

1871   The Rugby Football Union is established

Dissenters who favor different rules from the ones adopted by the Football Association for the game of soccer form their own organization.

1874: November 30   Birth of Winston Churchill

Winston Churchill, one of twentieth-century Britain's most important leaders, is born in Oxfordshire to Lord Randolph Churchill and Jenny Jerome, an American. Throughout his life, Churchill pursues joint careers as a writer and politician. As a young man, he covers wars in Cuba, India, and South Africa as a foreign correspondent. He also participates in a military campaign in Egypt (1897) and later takes part in the trench warfare of World War I, when he fights in Flanders. Churchill first enters politics as a Conservative member of parliament in 1900 but soon defects to the Liberal party (later in life, he once again allies himself with the Conservatives).

Beginning in 1907, Churchill accepts government positions of increasing responsibility, including first lord of the admiralty, secretary of war, head of the Colonial Office, and, in 1924, Chancellor of the Exchequer (a post once held by his father). By now a Conservative again, Churchill is forced out of office by the Labour victory of 1929. With the onset of World War II Churchill becomes prime minister and rallies the English people for the war effort against Germany. After a period of Labour Party rule, Churchill and the Conservatives are voted back into office in 1951. He is knighted in 1953 and retires from politics two years later. Already the Nobel Prize-winning author (1953) of an extensive body of historical and biographical works, Churchill publishes his four-volume *History of the English-Speaking Peoples,* between 1956 and 1958. He dies on January 24, 1965 and is buried at his birthplace and ancestral home after a state funeral at St. Paul's Cathedral.

1875   Britain buys stock in the Suez Canal

Britain's purchase of an interest in the Suez Canal, negotiated by Prime Minister Benjamin Disraeli, begins the nation's involvement with Egypt.

1875   *Trial by Jury* is produced

Trial by Jury becomes the first of many successful operettas created by librettist William Gilbert (1836–1911) and composer Arthur Sullivan (1842–1900). Others include *HMS Pinafore* (1878), *The Pirates of Penzance* 1879), and *The Mikado* (1885).

1877   First Wimbledon matches are held

The first tennis championship matches are held at Wimbledon.

1880s   School soccer play is introduced

Soccer (association football) is added to the curriculum of England's elementary schools.

1885   Birth of novelist D. H. Lawrence

David Herbert Lawrence, one of the most important twentieth-century British novelists, is born in Nottinghamshire and attends University College in Nottingham. He publishes his first major novel, *Sons and Lovers,* in 1913, as well as a volume of poems, *Love Poems and Others.* During World War I, Lawrence is ostracized for his pacifism and for the German ancestry of his wife, Frieda. His novel *The Rainbow* (1915) is banned on grounds of obscenity. After the war, Lawrence travels widely, to destinations that include Italy, Germany, Austria, Ceylon, Australia, New Mexico, and Mexico. His other major novels include *Aaron's Rod* (1922), *Women in Love* (1922), and *Lady Chatterley's Lover* (1928), which is legally banned in both England and the U.S. until the 1960s. Lawrence suffers from tuberculosis in the final years of his life and dies in 1930 in France. His novels are known for their passion, sensuality, and frank treatment of eroticism, which is ahead of its time.

1886   Irish Home Rule bill fails

Reforms in Ireland fail to stem the growing nationalist sentiment there. Parliament defeats a bill to grant the Irish Home Rule, and liberal prime minister William Gladstone is forced out of office for his support of the bill.

1888   Early soccer championship is awarded

A league of twelve soccer clubs from the north of England and the Midlands plays a season of scheduled games and awards a championship to the winning team.

1888: September 26   Birth of T.S. Eliot

Acclaimed modernist poet and Nobel prize winner Thomas Stearns Eliot is born in the United States, in St. Louis, Missouri, to a cultured family with strong roots in New England. After graduating from Harvard, Eliot spends a year in France and then returns for further study. After meeting the poet Ezra Pound, Eliot moves to England in 1914. Three years later he publishes *Prufrock and Other Observations,* which marks the start of a campaign by Eliot and Pound to free English poetry from Victorian conventions. After briefly working as a teacher and as a bank clerk, Eliot is able to devote himself solely to literary pursuits, publishing literary criticism, poetry, and plays, editing the journal *The Criterion* (from 1922 to 1939), and working as an editor at the publishing house *Faber & Faber.* The publication of *The Waste Land* in 1922 wins heightened acclaim for the poet, as does his later work *Four Quartets* (1943). His plays include *Sweeney Agonistes* (1932), *The Cocktail Party* (1950), and *Murder in the Cathedral* (1935). Eliot wins the Nobel Prize for Literature in 1948. He dies on January 4, 1965 in London.

1899–1902   Boer War

British imperial expansion in Africa leads to war with settlers of Dutch descent in South Africa. England is the victor, winning the capitulation of the two Boer provinces of Transvaal and the Orange Free State, but the British pay a high price in casualties: 22,000 men die in the conflict.

1900   Labour Party is founded

A coalition of socialists and trade unionists led by a Scottish miner, James Keir Hardie, forms a new political party, originally known as the Labour Representation Committee and renamed the Labour Party in 1903. From the 1920s on, it becomes one of Britain's two dominant parties, the other being the Conservative Party.

1901: January 22   Death of Queen Victoria

Queen Victoria, England's longest-reigning monarch, dies at the age of 81. (See 1819.)

1901–10   Reign of King Edward

Upon the death of Queen Victoria, Edward VII (1841–1910), her eldest son, ascends the throne, ruling until his death nine years later.

1906   Thomson wins Nobel Prize for Physics

Sir Joseph John Thomson (1856–1940) is awarded the Nobel Prize for Physics for discoveries involving the electron.

1907   Kipling receives the Nobel Prize

Rudyard Kipling (1865–1936), author of poetry, short stories, and children's books, is awarded the Nobel Prize for Literature. His work is known for its patriotic (and, according to some later views, racist) glorification of the British empire.

1912: April 14   Sinking of the *Titanic*

The British luxury ocean liner *Titanic* strikes an iceberg and sinks on her maiden voyage. With its watertight compartments, the *Titanic* is touted as unsinkable. However, its owners in fact know that it will sink if more than five compartments are ruptured, but expect that this will never happen. The disaster is caused largely by a failure to heed numerous iceberg warnings from other ships, partly because passengers are flooding the ship's radio room with personal messages to the U.S., and partly due to other mishaps.

Compounding the tragedy is the common practice of providing lifeboats only for the minimum rather than the maximum possible number of passengers. The passengers and crew total 2,206, but the 20 lifeboats can accommodate only 1,178. Even these lifeboats are not filled to capacity as they leave the ship, and 1,507 passengers remain aboard as the *Titanic* sinks. Of these, some dive into the freezing water of the North Atlantic, while others go down with the ship, which sinks in a final dive beginning at 2:20 a.m. when the boilers explode. The *Carpathia,* arriving on the scene in response to distress signals, rescues 705 survivors aboard the lifeboats between 4:45 and 8:30 a.m. Of those left in the water, none survives.

1913: November 22   Birth of Benjamin Britten

Benjamin Britten is one of England's foremost twentieth-century composers, noted especially for his vocal music, including the cantatas *Hymn to St. Cecilia* (1942) and *A Ceremony of Carols* (1942); the operas *Peter Grimes* (1945) and *Billy Budd* (1951); the chamber operas *Albert Herring* (1947) and *The Turn of the Screw* (1954), and the *War Requiem* (1962), whose text integrates the World War I poetry of Wilfred Owen with Latin religious texts. Well-known instrumental works by Britten include the *Simple Symphony* (1934) and the *Young Person's Guide to the Orchestra* (1946). Britten is made a peer in 1976, which is also the year of his death.

Two World Wars

1914–18   World War I

Britain plays a major role in World War I, which marks the beginning of modern mechanized battle. Heavy losses are suffered in trench warfare, especially at the Battle of the Somme in France (1916) and at Passchendale in Belgium (1917). Over a million British lives are lost, and a generation of young men is decimated.

1917   Balfour Declaration

The Balfour Declaration pledges British support for the creation of a Jewish state in Palestine.

1918   Limited female suffrage granted

Women over thirty win the right to vote. Unlimited adult female suffrage is granted ten years later (see 1928).

1920   Britain helps found the League of Nations

Britain is a founding member of the League of Nations.

1920   Irish Home Rule is granted

Britain grants home rule to Ireland in the Government of Ireland Act, which also separates the six northern counties, creating Northern Ireland. However, civil unrest in Ireland continues, and the Anglo-Irish Treaty is signed the following year providing for Ireland's twenty-six southern counties to separate from the U.K. and form the Irish Free State.

1920   Premiere of *The Planets*

The orchestral suite *The Planets* by Gustav Holst (1874–1934), one of the most popular twentieth-century British symphonic compositions, is first performed. Each of its seven movements portrays a planet (only Earth and Pluto—not yet discovered—are missing). Its modernist elements place it ahead of its time, compared with contemporary British compositions.

1922   T.S. Eliot writes *The Waste Land*

Poet T.S. Eliot writes the modernist masterpiece, *The Waste Land,* one of the most influential English literary works of the twentieth century. It is acclaimed for its powerful articulation of postwar disillusionment and dislocation. The five-part, 433-line poem, in free verse, is loosely organized around the theme of the search for the Holy Grail and includes numerous literary allusions and other scholarly references, many of which are explicated in footnotes by the author.

1922   Publication of *Ulysses*

Irish author James Joyce (1882–1941) publishes his masterpiece, one of the great works of English literature, and one of the greatest books of the twentieth century. Using a multitude of styles, Joyce narrates a single day in the life of Leopold Bloom, an ordinary middle-class Dublin resident. *Ulysses* is distinguished by its use of stream-of-consciousness technique (showing what the characters are thinking), its imaginative structure, the poetic quality of its language, and its vivid, humorous, and complex portrayal of the human condition.

1925   Birth of Margaret Thatcher

Britain's first woman prime minister (born Margaret Roberts) is a native of Lincolnshire. She attends Oxford University, where she joins the Conservative Party. After earning a degree in chemistry, she obtains a research job. She marries Denis Thatcher, a businessman, and later earns a law degree and increases her involvement in politics. She is elected to Parliament in 1959 and gradually rises in the Conservative Party hierarchy. She becomes the head of the party in 1975, when Labour is in power and Conservatives are the opposition party.

With the Conservative victory in 1979, Thatcher becomes prime minister, the first woman in British history to hold this position. Thatcher presides over a Conservative revolution that reverses the Labour policies of the previous government. She is reelected twice, serving the longest tenure in office of any British prime minister in the twentieth century. In 1990 she resigns as a result of an internal power struggle within the Conservative party, and John Major becomes prime minister. Two years later she is awarded a peerage, becoming Baroness Thatcher of Kesteven.

1926: May 4   General Strike is called

Years of high postwar unemployment and related civil unrest reach a peak when the Trade Union Congress calls a general strike of workers in all industries in support of demands by mine workers. The strike ends after nine days under government pressure, although the miners remain on strike for the rest of the year. The following year Parliament passes the Trades Disputes Act making further general strikes illegal.

1928   Women gain full suffrage rights

Women win the right to vote on the same basis as men, at the age of twenty-one.

1929   The Great Depression

The British economy, already suffering from depression and high unemployment in the 1920s, plunges yet further with the onset of the worldwide economic depression of the 1930s. Million of workers are jobless.

1930: March 2   Death of novelist D. H. Lawrence

D. H. Lawrence dies in France of tuberculosis. (See 1885.)

1931: December 11   The British Commonwealth is created

The Statute of Westminster forms the British Commonwealth, composed of the United Kingdom and self-governing dominions of equal status.

1936   Abdication of Edward VIII

Edward VIII (1894–1972) abdicates the throne to marry American divorcée Wallis Simpson. Edward's brother ascends the throne as George VI (1895–1952), husband to Elizabeth (today known as the Queen Mother) and father of the future Queen Elizabeth II.

1938   Britain signs the Munich pact

As part of a policy of appeasement toward Nazi Germany, Prime Minister Neville Chamberlain approves a pact that leads to the dismantling of Czechoslovakia.

1939–45   World War II

Declaring war on Germany following the German invasion of Poland, Britain suffers heavy losses, both from the German air attacks on civilians and from military campaigns in Europe, Africa, and the Middle East. Over 900,000 lives are lost, and massive property damage is inflicted by German bombing, known as the "Blitz." Many residents of London sleep in the subway stations to escape the bombing, which flattens much of the city. Under the inspirational leadership of Winston Churchill, the British people mobilize fully for the war effort, both on the battlefield and at home, through defense measures and increased defense production.

1940   The Battle of Britain

British fighter planes engage German aircraft over the English Channel and succeed in warding off a German invasion of England.

The Postwar Period

1945   Labour government is elected

A Labour government is elected, and Clement Atlee becomes the prime minister. Industries are nationalized, and other Socialist reforms are enacted, including socialized medicine.

1945   Britain joins the United Nations

The U.K. is a founding member of the UN.

1947   Electricity and transportation are nationalized

Electricity production is placed under the direction of the Electricity Authority. A newly authorized Transportation Company takes charge of rehabilitating the nation's railroads and reorganizing its trucking industry.

1947: August 15   India becomes independent

The separate, independent Hindu and Muslim states of India and Pakistan are established in the former British India.

1948   Britain hosts the Olympic games

The Olympic games are held in London. Britain wins gold medals in rowing and yachting events.

1948   National Health Service is inaugurated

Following the passage by Parliament of the National Health Service law two years earlier, Britain's National Health Insurance program goes into effect, making free medical care available to all Britons. Among the major effects of nationalized health care are dramatic declines in infant mortality and maternal death in childbirth.

1951–55   Churchill's second term as prime minister

Winston Churchill, the U.K.'s wartime leader, is elected prime minister again. Churchill, a Conservative, reverses the policies of the previous Labour government, privatizing nationalized industries and the Bank of England.

1952   Ascension of Queen Elizabeth II

Elizabeth (b. 1926) succeeds her father, George VI, as England's monarch upon his death.

1956   The Royal Ballet is chartered

The former Sadler's Wells Ballet receives a royal charter and is renamed the Royal Ballet.

The coronation of Queen Elizabeth II in London. (EPD Photos/CSU Archives)

1956: November   The Suez Crisis

British and French troops invade Egypt and take over the Suez Canal after it is nationalized by Egyptian leader Gamal Abdel Nasser. The attack draws international condemnation from other countries.

1957–63   Macmillan becomes prime minister

Prime Minister Anthony Eden is forced to resign because of the Suez Canal crisis of the previous year. However, the Conservatives stay in power under the leadership of Harold Macmillan.

1959   *Australopithecus* is major anthropological discovery

British anthropologists Louis and Mary Leakey discover the skull of *Australopithecus*, a species of early man, in northern Tanzania. The skull is believed to be 1.75 million years old.

1962   Rudolf Nureyev begins dancing with the Royal Ballet

Acclaimed Russian dancer Rudolf Nureyev, who defects from the Soviet Union the previous year, begins his association with the Royal Ballet, which becomes his base. Its star dancer, Margot Fonteyn, becomes his dance partner.

1963: April   Beatles' first No. 1 hit

The Beatles, the most successful British pop music group of all time, have one of their songs (*From Me to You*) reach the top of the pop charts for the first time. It is followed by three more top hits over the coming year (*She Loves You, I Want to Hold Your Hand,* and *Can't Buy Me Love*), and Beatlemania is launched, both at home and internationally. The group remains together for the remainder of the decade, producing two hit movies, *A Hard Day's Night* and *Help!,* and a series of best-selling albums and singles.

1963: August 5   Britain signs nuclear test ban treaty

The U.K. signs the nuclear test ban treaty with other major world powers.

1964–70   Labour regains power

A new Labour government is elected, led by Prime Minister Harold Wilson. Wilson launches a campaign for British membership in the Common Market (EEC), but it is thwarted by opposition from French president Charles de Gaulle (1890–1970).

1965: January 4   Death of poet T.S. Eliot

T.S. Eliot, Nobel Prize winning poet and one of the great figures of literary modernism, dies at the age of 76. (See 1888.)

1966   England hosts and wins the World Cup games

The World Cup soccer championship is held in England, which wins the title, with West Germany coming in second.

1968   Abortion is legalized

An 1861 law making abortion a crime under all circumstances is overturned by a new law legalizing abortions when the pregnancy threatens the life or health of the mother. Critics of the law predict that abortion rates will rise to 400,000 a year, but actual rates the first year the law is in effect are closer to 30,000.

1968: October 5–6   Era of violence begins in Northern Ireland

Roman Catholic protesters marching in Londonderry clash with police, setting off a new era of violence in Northern Ireland. Catholics hold demonstrations protesting religious discrimination and inadequate representation in government. The Irish Republican Army (IRA) launches a campaign of terrorism in Northern Ireland and England to further its demands that Northern Ireland be integrated into the Irish Republic. British troops are sent to Belfast and Londonderry.

1970–74   Conservative government is elected

The Conservatives regain power, and Edward Heath becomes prime minister.

1970s   England is plagued by strikes

The British economy is hobbled by major strikes by dockworkers, postal workers, coal miners, and other groups.

1970   Britain wins membership in the Common Market

Britain is admitted to the European Common Market.

1972: January 20   Bloody Sunday in Northern Ireland

Fourteen civilian demonstrators are killed by British troops in Northern Ireland, in an occurrence that becomes known as Bloody Sunday.

1974: June 17   IRA bombs Parliament

The Houses of Parliament are hit by an IRA bomb, and eleven are injured. In the coming decade, the terrorist campaign between the Irish Republican Army and Protestant paramilitaries is repeatedly carried to the British mainland.

1975   North Sea oil development begins

The development of England's North Sea oil reserves begins.

1976: December 4   Death of composer Benjamin Britten

Britten, one of the foremost figures in twentieth-century British music, dies after receiving a life peerage earlier in the year. (See 1913.)

1979–90   The Thatcher era

Conservative Margaret Thatcher (see 1925) becomes Britain's first woman prime minister, serving the longest term of any single prime minister in the twentieth century. She reverses the nationalizations of the preceding Labour government, privatizing nearly half of Britain's state-owned enterprises by the end of the decade. Her government also implements cuts in social spending and increases in defense allocations. U.S. president Ronald Reagan (b. 1911) becomes a close ally.

1980: October   IRA prisoners go on hunger strike

Irish Republican Army members held prisoner in Ulster begin a series of hunger strikes that continue for a year, by which time ten men have died.

1981: July 20   Marriage of Prince Charles and Diana Spencer

The Prince of Wales marries Lady Diana Spencer in a televised ceremony seen by millions of people around the world via communication satellites.

1982: April 2   Argentina invades the Falkland Islands

Argentina invades the sparsely inhabited Falkland Islands (the Malvinas), which the British have held since 1833. The British send troops to the South Atlantic to reclaim the islands.

1982: June 14   Falklands recaptured

After more than two months of air, sea, and land battles, the British recapture the Falklands. The seventy-two-day war costs roughly 2,000 British and Argentine lives. Britain continues to maintain 2,500 troops on the islands. Britain and Argentina do not resume full economic and diplomatic ties for another seven years.

1983   Author William Golding wins the Nobel Prize

The Nobel Prize for literature is awarded to British author William Golding (1911–93), best known as the author of *Lord of the Flies* (1954). His other novels include *The Inheritors* (1955), *Pincher Martin* (1956), *The Spire* (1964), and *Rites of Passage* (1980).

1984: September 26   Britain and China agree on return of Hong Kong

Britain agrees to return its crown colony of Hong Kong to Chinese rule by 1997, and China agrees to retain the current free enterprise system for at least fifty years.

1985: March   Coal miners end strike

British coal miners end their year-long strike over mine closings.

1985: November   Anglo-Irish accord on Northern Ireland negotiations

An agreement signed by British Prime Minister Margaret Thatcher and Ireland's leader, Barrett FitzGerald, gives the Irish Republic a role in deciding the future of Northern Ireland.

1987   Thatcher wins second reelection

Prime Minister Margaret Thatcher is reelected for the second time since 1979, heading a nationwide wave of Conservative victories. She pledges to continue her program of government spending cuts and privatization.

1988: February   *The Satanic Verses* author condemned by Iranian leader

Indian-born British author Salman Rushdie publishes the satiric work *The Satanic Verses*. The book is condemned by Iran's religious and political leader Ayatollah Khomeini, who places Rushdie under a *fatwa* (death warrant). The author goes into hiding.

1988: December 21   Pan Am jet downed over Scotland by terrorist bomb

Pan Am passenger flight 103 from London to New York is downed over Lockerbie, Scotland, after an explosion by a terrorist bomb. All 256 persons aboard the plane are killed, as well as eleven on the ground, in the worst air disaster in British history. The bomb, contained in a tape recording device in the baggage compartment, sends the flaming plane plummeting earthward, where it crashes near a gas station, setting fire to the station and neighboring houses. After initially suspecting involvement by Palestinian or Iranian terrorists, the United States demands the extradition of two Libyan terrorists, but Libya's leader Moammar Qaddafi refuses, drawing censure from the international community.

1990   Diplomatic ties with Argentina are resumed

Britain renews diplomatic relations with Argentina, severed since the Falklands War (see 1982).

1990: November   Margaret Thatcher resigns

Margaret Thatcher ends her unprecedented eleven-year tenure as prime minister by resigning in the face of growing unpopularity. She is succeeded by fellow Conservative John Major, who confronts the nation's growing unemployment and inflation problems and ends the unpopular poll tax system instituted by the Thatcher government.

1991   Britain sends troops to the Persian Gulf

Censuring Iraq's invasion of Kuwait, Britain joins the U.S.-led multinational coalition, sending 42,000 troops to the region. There are twenty-six British casualties in the war.

1991: May   Queen Elizabeth addresses U.S. Congress

Queen Elizabeth, on an official visit to the United States, becomes the first British monarch to address the Congress.

1991: July   BCCI offices are closed

The British offices of the Bank of Commerce and Credit International (BCCI) are shut down on grounds of illegal activities by order of the governor of the Bank of England.

1992   First woman is named speaker of the House of Commons

In a bipartisan vote, Labour Party member Betty Boothroyd is named speaker of the House of Commons, the first woman to hold the post in the institution's 700-year history.

1993: July 23   Parliament ratifies Maastricht Treaty

A 339–299 vote in the House of Commons provides final British ratification of the Maastricht Treaty providing for a proposed European Union with closer political, defense, and economic ties between the nations of the European Community and the eventual creation of a single European currency.

1994: February 16   Penguin withdraws book based on Stephen Spender's life

Viking Penguin announces it will withdraw from publication American author David Leavitt's novel *While England Sleeps,* based on the life of British poet Sir Stephen Spender (b. 1909) in response to a copyright infringement suit by Spender, who has objected to Leavitt's use of material from his own autobiography, as well as to the sexually explicit nature of some passages in the book. A new edition of the book omitting the offending passages is to be issued.

1994: May 5   Tories suffer major losses in local elections

The Conservative Party chalks up its worst electoral results in local elections in the twentieth century—about 27 percent nationwide. The poor showing is yet another setback for Prime Minister John Major, already beleaguered by controversy over the proposed European Union and government scandals.

1994: May 6   Channel Tunnel is officially inaugurated

An inauguration ceremony is held for the newly completed Channel Tunnel between England and France, with Britain's Queen Elizabeth and French President Francois Mitterand presiding. The tunnel—already dubbed the "Chunnel"—will carry passengers between Folkestone, England and Calais, France, making it possible to travel between London and Paris in three hours via railway. The thirty-one-mile (fifty-kilometer) tunnel will also provide a car ferry service. The tunnel, which still must undergo safety tests, is not expected to be operational until October. Final construction costs for what is called one of the major achievements in modern engineering are reportedly $15–16 billion.

1995: June 22   Major resigns as leader of Conservative Party

In response to divisions within the Conservative Party over the adoption of a unified currency by members of the European Union, Prime Minister John Major resigns as head of the party, opening the way for other Conservatives to challenge his leadership.

1995: July   Conservative Party reaffirms support for Major

The Conservative Party votes to retain John Major as its leader, rejecting a challenge by Welsh Secretary John Redwood.

1995: December 13   Hundreds riot in Brixton over alleged police brutality

Hundreds of working-class youths riot in the Brixton area of greater London, protesting the death of a black man the previous week while in police custody. The disturbances cause about $1.5 million in property damage. Brixton has previously been the scene of race riots in 1981 and 1985.

1996: March 13   Gunman murders 17 in Scottish kindergarten

In one of the most horrific mass murders in modern British history, a gunman enters a kindergarten class in Dunblane, Scotland, and opens fire on its occupants, killing 16 children and their teacher. Dozens of others are wounded either in the classroom or outside on the school grounds. No motive is determined for the action by Thomas Hamilton, known to have a history of mental instability.

1996: June 15   IRA bomb injures over 200 in Manchester

A homemade bomb explodes in a crowded shopping district in Manchester, injuring 206 people. Four days later the Irish Republican Army (IRA) claims responsibility for the blast, which is seen as an attempt to disrupt the current peace process in Northern Ireland. The bomb is set off in a van parked outside the shopping center, which is unusually crowded because of a soccer tournament. Containing between 2,000 and 3,000 pounds (1,900–1360 kilograms) of explosives, it is the largest bomb the IRA has ever exploded in England.

1996: August 28   Charles and Diana are divorced

A high court decree finalizes the divorce of the Prince and Princess of Wales. Diana, said to have received a lump sum payment of over $22 million, is stripped of the title "Her Royal Highness." Charles claims he has no plans to remarry, but is said to be continuing his ongoing relationship with Camilla Parker-Bowles.

1996: November 3   British Telecom and MCI announce they will merge

In one of the largest mergers in the telecommunications industry, British Telephone and the U.S. firm MCI Communications announce they will merge to form the first transatlantic telephone company, to be known as Concert Global Communications PLC. The firm will serve about 43 million customers, with annual sales projected at $42 billion.

1997: May 1   Labour Party ends 18-year period of Conservative rule

The Labour Party returns to power after eighteen years of Conservative rule, winning the national elections with a 179-seat majority in the House of Commons. After campaigning from a centrist position, Tony Blair becomes Britain's new prime minister, pledging to rein in government spending and refrain from tax hikes for five years.

1997: July 1   Hong Kong reverts to Chinese rule

After 156 years as a British territory, Hong Kong reverts to Chinese rule and is renamed the Hong King Special Adminis-

trative Region. Under the "one country, two systems" plan drawn up by Chinese leader Deng Xiaoping, Hong Kong will become part of the People's Republic politically while maintaining its current capitalist economy and its status as a leading global financial center.

1997: August 31   Diana killed in Paris car crash

Diana, Princess of Wales, and her companion Emad Mohamed (Dodi) Fayed are killed when Fayed's car veers out of control in an underpass beneath a bridge over the Seine river in Paris. Fayed's driver, Henri Paul, who is also killed, is said to have been traveling at speeds of up to 100 miles (160 kilometers) per hour to elude pursuing *papparazzi* (tabloid photographers). Paul is also reported to have high levels of alcohol and prescription medications in his blood. The only survivor of the crash is Trevor Rees-Jones, a Fayed family security guard, who is seriously injured.

The death of Diana, age thirty-six, draws expressions of sympathy and mourning from throughout the world and unprecedented demonstrations of grief in England, where thousands of bouquets pile up outside Kensington Palace in London and other royal residences and many wait in line for hours to sign condolence books at St. James's Palace. Diana, divorced from Prince Charles a year earlier (see 1996), is survived by her two sons, Princes William and Harry. In addition to widespread sympathy for her difficult marriage to Prince Charles, Diana has won public admiration and respect for her involvement in charitable causes, including efforts to combat AIDS and the campaign to end the use of land mines. There is widespread speculation that her brief but intense romantic involvement with Fayed, an Egyptian-born film producer and scion of a wealthy family, was about to culminate in an engagement.

1997: September 6   Diana's funeral is seen by millions around the world

The funeral of Diana, Princess of Wales, killed in a car accident (see 1997: August 31), is attended by 2,000 invited guests from around the world, including prominent political figures, such as U.S. First Lady Hillary Clinton, and celebrities from the worlds of entertainment and fashion. Crowds of over a million gather outdoors to hear the memorial service on loudspeakers, while billions worldwide watch the ceremony on television. Before the funeral, Diana's coffin is transported from Kensington Palace to Westminster Abbey in a procession that includes her sons, Princes William and Harry.

Diana is eulogized by, among others, Prime Minister Tony Blair and her brother, Charles, Earl Spencer, who blasts the tabloid press for its relentless pursuit of his sister during her lifetime and its widely perceived role in the accident that caused her death. At the conclusion of the service, a minute of silence is observed throughout the nation. Afterwards, the coffin is taken to the Spencer family home, Althorp House, in Northamptonshire for burial and a private ceremony.

1997: September 15   Sinn Fein takes part in Northern Ireland peace talks

For the first time, Sinn Fein, the political wing of the IRA, takes part in peace talks to decide the future of Northern Ireland.

1998: April 10   Tentative settlement is reached on Northern Ireland

A historic agreement, involving extensive home rule by Northern Ireland, is reached by political leaders at peace talks aimed at ending the violence that has plagued the region for thirty years and also spread to mainland Britain. Among those attending the talks are British Prime Minister Tony Blair, Irish leader Bertie Ahearn, David Trimble of Northern Ireland's Protestant Ulster Unionist Party, and Gerry Adams of Sinn Fein. The agreement is subject to approval by the parliaments of Britain and Ireland and to referendum votes in Ireland and Northern Ireland.

Bibliography

Abrams, M. H., ed. *The Norton Anthology of English Literature.* 2 vols. 6th ed. New York : Norton, 1993.

Brittain, Vera. *Testament of Youth: An Autobiographical Study of the Years 1900-1925.* New York: Penguin Books, 1989.

Cannon, John and Ralph Griffiths. *The Oxford Illustrated History of the British Monarchy.* New York: Oxford University Press, 1988.

Cook, Chris. *The Longman Handbook of Modern British History, 1714–1995.* 3rd ed. New York: Longman, 1996.

Delderfield, Eric F. *Kings & Queens of England & Great Britain.* New York: Facts on File, 1990.

Figes, Kate. *Because of Her Sex: The Myth of Equality for Women in Britain.* London: Macmillan, 1994.

Foster, R.F. (ed.) *The Oxford History of Ireland.* New York: Oxford University Press, 1992.

Fussell, Paul. *The Great War and Modern Memory.* New York:Oxford University Press, 1975.

Glynn, Sean. *Modern Britain: An Economic and Social History.* New York: Routledge, 1996.

Havighurst, Alfred E. *Britain in Transition: The Twentieth Century.* 4th ed. Chicago: University of Chicago Press, 1985.

Hobsbawm, E. J. *The Age of Empire, 1875–1914.* 1st American ed. New York : Pantheon Books, 1987.

Jenkins, Philip. *A History of Modern Wales, 1536–1990.* New York: Longman, 1992.

Judd, Denis. *Empire: The British Imperial Experience from 1765 to the Present.* London : HarperCollins, 1996

Kavanagh, Dennis. *British Politics: Continuities and Change.* 2d ed. New York: Oxford University Press, 1990.

Kearney, Hugh F. *The British Isles: A History of Four Nations.* New York: Cambridge University Press, 1989.

Kenner, Hugh. *The Pound Era.* Berkeley: University of California Press, 1971.

The Oxford History of Britain. New York: Oxford University Press, 1992.

Powell, David. *British Politics and the Labour Question, 1868–1990.* New York: St. Martin's, 1992.

Stone, Lawrence. *The Family, Sex and Marriage in England, 1500–1800.* London: Weidenfeld & Nicolson, 1977.

Strachey, Lytton. *Eminent Victorians: The Illustrated Edition.* New York: Weidenfeld & Nicholson, 1988.

Vatican

Introduction

Vatican City, which is also called simply the Vatican, is situated in Rome, Italy, near the right bank of the Tiber River. It is the smallest state in the world with a fixed area of 109 acres (0.5 square kilometers). Often referred to as the State of the Vatican City, it is the official residence of the pope, who is the worldwide sovereign head of the Roman Catholic Church. The pope, who exercises supreme legislative, executive, and judicial power over the Holy See and the State of the Vatican City, is an absolute monarch elected to a life term by the College of Cardinals.

Vatican City is trapezoidal in shape, entirely urban, and has no colonies or territories. Its boundaries are the Colonnade of Saint Peter's Square, Via di Porta Angelica, Piazza del Risorgimento, Via dei Bastioni di Michelangelo, Viale Vaticano, and Via della Stazione Vaticana–architectural masterpieces built primarily during the Medieval and Renaissance periods. Temperatures in Vatican City are pleasant. Summer days are warm and evenings cold. The average temperature in July is 75 degrees Fahrenheit (24 degrees Celsius). Winter temperatures average 45 degrees Fahrenheit (7 degrees Celsius). Rainfall is light from May to September. The wettest months are October and November. The environment of Vatican City resembles Rome. Due to its urban character Vatican City compensates for the lack of natural flora and fauna with its famed gardens which are filled with orchids and exotic flowers.

The population of Vatican City is approximately 750, and includes Italian and Swiss nationals, as well as cardinals, bishops, and other clergymen from all over the world who live inside the walls. An estimated 400 persons are citizens. The official language of Vatican City is Italian, while the official language of the Holy See is Latin, which is also used for papal encyclicals and other official declarations. French is also spoken in Vatican City. The work force is comprised of 3,000 lay workers who reside outside the Vatican.

The Vatican is a major training center for clergy and operates about sixty-five papal educational institutions for higher Roman Catholic learning throughout Rome. Among the more familiar insitutes are the Gregorian University, the Biblical Institute, the Institute of Oriental Studies, the Institute of Christian Archaeology, and the Institute of Sacred Music.

Land-use allocation in the Vatican is strictly urban. Residents of Vatican City do not pay taxes. The Vatican instead derives income from travel and tourism by charging fees to visit the many art galleries, museums, the gardens, and the Vatican itself. The Vatican accepts income in the form of voluntary charitable contributions and is paid interest on its extensive valuable real estate, industrial, and artistic holdings. The Vatican is home to a few income-producing industries, including a studio manufacturing mosaic works, a sewing business which makes uniforms, and its publishing house—Vatican Polyglot Press.

Vatican City is recognized as an independent state under international law and may enter into international agreements. The state was established to maintain the spiritual mission of the Church and to administer all real estate properties belonging to the Holy See in Rome, including the surrounding territories. The Vatican has the authority to issue its own passports, stamps, and coinage.

The Vatican legal system is based on ecclesiastical or canon law. When canon law cannot be applied, the laws governing the city of Rome apply. For example, criminal cases are handled in the Italian courts. The pope delegates the day-to-day internal administration of Vatican City to the Pontifical Commission for the State of the Vatican City. This commission, comprised of seven cardinals, one special lay delegate, and a board of twenty-one lay advisers is responsible for the following: a central council to oversee various administrative departments; the museum(s) directors, technical services, economic services such as the post office and telegraph systems; medical services; the Swiss Guard; Vatican radio and television systems; the Vatican Observatory; the villa at Castel Gandolfo director; and other papal residences.

Vatican City is not to be confused with the Holy See, as they are separate and distinct entities united in the person of the pope. Vatican City is under the jurisdiction of the Papacy, an ecclesiastical government, and the administrative capital of the Roman Catholic Church. The *Holy See*, which is also called the See of St. Peter or the Apostolic See is the seat of authority, jurisdiction, and sovereignty vested in the pope

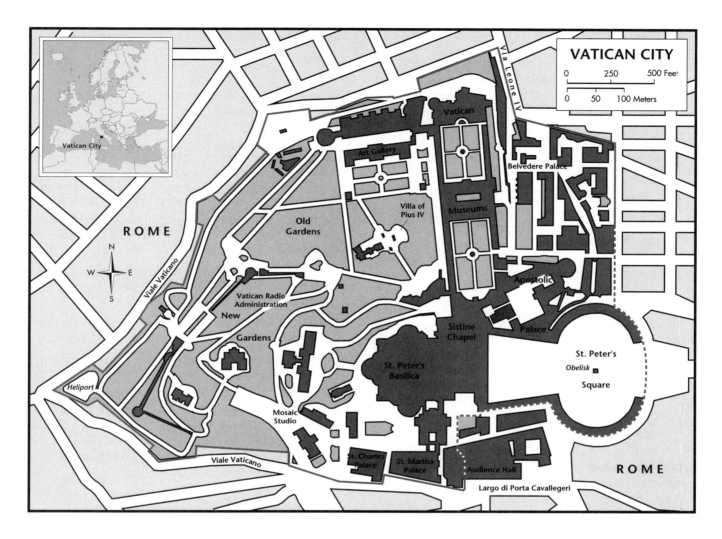

(and his advisers) as the spiritual head of the Church worldwide. Its independence and sovereignty are provided for by the Lateran Treaty, which the Vatican signed with the Republic of Italy in 1929. Political parties and forms of local government have no standing in the Vatican.

The boundaries of the temporal or secular power of the papacy have shifted through the centuries; today's Vatican is the sole remaining part of what used to be a much larger area called the Papal States, formed in the eighth century AD from territories donated to the reigning pope by the French king. At various times, they occupied large parts of Italy and even southern France. The most dramatic upheaval in the history of the papal territories was the Avignon Papacy (1309–77), when Pope Clement V moved the papacy to Avignon, France, for some seventy years, temporarily giving the French, rather than the Italians, control of the Church.

In the middle of the nineteenth century, the Papal States accounted for a large area extending across central Italy. However, most of it came under the control of King Victor Immanuel II during the unification of Italy in the latter half of the century. By 1860, only the city of Rome and surrounding coastal regions were under papal control. When French troops who were helping defend the city departed in 1870 to fight in the Franco-Prussian War, Rome itself fell to the kingdom of Italy and became its capital. The relationship between the Italian government and the papacy remained unresolved until the Lateran Treaty of 1929, under which the State of the Vatican City was created. A new concordat replaced the Lateran Treaty in 1984.

The United States maintained consular relations with the Papal States from 1797–1870 and diplomatic relations with the Pope, as head of the Papal States, from 1848–68. These relations ceased in 1870 when all papal territories were annexed to the Kingdom of Italy. From 1870–1984 the United States did not maintain formal diplomatic relations with the Holy See. Several presidents, including Richard M. Nixon, Gerald Ford, Jimmy Carter, and Ronald Reagan, assigned personal envoys to visit the Holy See periodically to discuss issues of international humanitarian and political appeal. In 1984 the United States established formal diplomatic relations with the Holy See.

The Holy See maintains an especially active role in various international organizations. It has been granted permanent observer status (as opposed to member status) at the United Nations in New York; the UN Office in Geneva; the Educational, Scientific, and Cultural Organizations of the UN in Paris, and the UN Food and Agriculture Organization in Rome. It also has a member delegate at the International Atomic Energy Agency and at the UN Industrial Development Organization in Vienna and maintains diplomatic relations with the European Union in Brussels .

Timeline

440–461 Reign of Leo I, the Great

Leo I, the Great, is the first of only two popes (the other being Gregory I) to be called "the Great". Leo I earns the title because he succeeds in persuading Asian conqueror Attila the Hun not to invade Rome. This accomplishment is significant for the papacy. The Roman Empire is in decline and can not raise a defense against the Hun. It is the Church who has to defend the central territories of the Italian peninsula using papal forces. Leo's victory strengthens the papacy's role as the major influence in western European affairs.

590–604 Reign of St. Gregory I, the Great

Gregory is influential in matters of doctrine, liturgy, and missionary work and is regarded as a truly great leader of the Church. He is the first pope recognized as the leader in western Europe of all spiritual and religious matters. Like Leo, Gregory is a fierce military leader who defends central Italy from northern invaders.

754–55 Emergence of the States of the Church or Papal States

The Papal States refers to that part of the Italy ruled directly by the pope as a temporal kingdom. These states are created through the Donation of Pepin, the French king who gives the territories as a gift to Pope Stephen III (reign 752–757).

By giving the papacy control of the states, backed by French troops, Pepin hopes to protect central Italy from future attacks.

The Middle Ages

1073–85 Reign of St. Gregory VII

Gregory (b. Hildebrand, c. 1015) debates with Holy Roman Emperor Henry IV (king of Germany) over whether the pope or the emperor has the right to nominate bishops. Gregory believes in the absolute power of the pope over all lay or secular authority, including the emperor. Henry and his German bishops depose Gregory and installed an antipope. Gregory excommunicate Henry and forced him to do public penance at the village of Canossa. But Henry feigns penitence and marches on Rome, driving Gregory into exile. Such opposition between religious and secular rule plagues papal successors through the Middle Ages.

1179 College of Cardinals as electors of pope initiated

The election of the Pope by the College of Cardinals, who serve as the electors and the principal advisers of the Pope, is formally instituted. The College of Cardinals grows from the ancient tradition of selecting Roman priests as papal advisers. To elect a pope the cardinals meet in a conclave, an assembly behind locked doors, in the Sistine Chapel. The reigning pope, Alexander III, declares that a new pontiff must be elected by two-thirds plus one cardinal votes.

1309–77 Pope Clement V moves the papacy to Avignon, France

Pope Clement V (r. 1305–14) relocates the papal headquarters to Avignon, France. This Avignon papacy lasts for over seventy years and through six successor popes. Clement is invited back to his native France by the French king when the politics of Rome threaten to dismantle the papacy. While in Avignon Clement fills the cardinal ranks with loyal French cardinals. When the papacy finally returns to Rome its power and influence is French rather than Italian.

1370–78 Reign of Pope Gregory

During his reign, Pope Gregory XI returns the papacy to its henceforth permanent location in Rome. When Gregory returns the papacy to Rome he finds that seventy years in Avignon has done little to end the political battles and intrigue that threaten Rome's continuance as the center of western Christianity. Gregory is significant in that he is the first pope to establish residence in the Vatican palace, which has remained the home of the pope and the papal court.

The Renaissance

1450 Apostolic Library of the Vatican conceived

The Apostolic Library of the Vatican is first conceived by Pope Nicholas V (r. 1447–55). The Renaissance, a period of great artistic achievement, introduces opulent construction and decoration in the Vatican. Nicholas V is credited with rebuilding the Vatican and making it the center of all papal administrative activity. Leon Battista Alberti and Bernardo Rossellino are the Vatican Library architects; Fra Angelico (Giovanni Di Fiesola) is commissioned to paint the frescoes.

1475   Apostolic Library of the Vatican is established

The Apostolic Library, established after the death of Pope Nicholas V, is one of the most famous worldwide with a collection including over 1.1 million books, 72,000 manuscripts, 8,300 *incunabula* (works of art), 80,000 archival files, and 100,000 engravings.

1492–1503   Reign of Alexander VI (Rodrigo Lanzol y Borgia)

Born in Spain, Pope Alexander is noted for dividing colonial territories in the New World between Spain and Portugal. He is also credited with the revitalization of the University of Rome.

1503–13   Reign of Pope Julius II

During the reign of Pope Julius II (r. 1503–13), the Vatican Museums are established. The reconstruction of St. Peter's Basilica in Rome is also carried out.

Julius II concerns himself more with art and warfare than with spiritual matters. His concept for a museum in the Vatican begins with the collecting of notable monuments of classical culture in the courtyard of the Palazzetto del Belvedere. Julius then commissions some of the finest artistic talent of the day to realize his dreams for Rome. Michelangelo paints the ceilings of the Sistine and Pauline Chapels. Raphael is commissioned to paint frescoes in the Stanze.

As a result the Vatican has an impressive twelve museums, housing some of the world's finest masterpieces. Some of these museums are the Pius Clementine, the Chiaramonti, the Gregorian Etruscan, the Gregorian Egyptian, and the Collection of Modern Religious Art.

1506   The Swiss Guard is established as the papal patrol force

The guard are recruited from several Roman Catholic cantons of Switzerland. They carry ceremonial weapons and wear colorful striped uniforms supposedly designed by Michelangelo.

Reform and Reformation

1542   The Congregation of the Holy Office is created by Pope Paul III

Pope Paul III (r. 1534–49), often referred to as the founder of the Catholic Reform Movement, establishes a sacred congregation designed to align the cardinals and bishops of various departments with papal objectives for a more divine ministry.

1572–85   Pope Gregory XIII creates the Vatican Observatory

During his reign, Pope Gregory III establishes the Vatican Observatory—one of the most sophisticated in the world. By the twentieth century, it includes an astrophysics laboratory and a 33,000-volume library.

1586   Limit set on College of Cardinals membership

Pope Sixtus V sets the maximum number of members of the College of Cardinals at seventy. Among his other accomplishments is the reorganization of the *Roman Curia* (administrative structure) and the requirement that all bishops make a pilgrimage to Rome every five years.

1603   The Pontifical Academy of Sciences is created

Under Pope Clement VIII (r. 1592–1605), the Pontifical Academy of Sciences is established. The purpose of the Academy is to promote the study of science and mathematics.

1611–14   The Vatican Archives are established

During the reign of Pope Paul V, the archives are founded with the purpose of preserving papal titles to land, along with other rights and privileges. Also during Paul's reign, the facade to St. Peter's Basilica is completed.

1655–67   Reign of Pope Alexander VII

Alexander is noted for having St. Peter's colonnade built.

The Age of Enlightenment and the Napoleonic Era

18th century   Enlightenment spurs anti-religious sentiment

The Age of Enlightenment ushers in a new era of anti-Christian and anti-religious feeling, partially in reaction to the Church's past abuse of power. The rulers of Europe no longer regarded the papacy as a major European power.

1798   The Papal States are replaced by a new Roman republic

During the reign of Pope Pius VI (r. 1775–99), Napoleon Bonaparte of France begins his conquest of Europe. He believes that the papacy has become obsolete and, when Napoleon's forces invade Italy and sack Rome, Pius is taken prisoner and sent to France, where he dies. His successor, Pope Pius VII (r. 1800–23) is also taken prisoner by Napoleon and exiled to France for four years.

1815   The Papal States are restored

Following the downfall of Napoleon, the Papal States are restored, and Pope Pius VII finishes his reign. However, the damage to the papacy inflicted by Napoleon is irreversible. Never again does the papacy enjoy the standing it once had in Europe. Its military strength is gone and the general attitude in Europe toward the papacy is negative, despite attempts to involve laymen in the administration of papal affairs

The 19th Century

1846–78   Italy wins control of the Papal States

During the reign of Pius IX, the papacy loses the Papal States to the kingdom of Italy. Pius IX has the longest reign of any Roman pontiff in history; it includes the convening of the First Vatican Council (see 1869). However, Pius's attempts to liberalize and reform the papacy fail.

1848   Pius IX is driven into exile

Pope Pius IX is driven from Rome by Italian revolutionaries.

1849: February 9   Roman republic is proclaimed

With Pius in exile, a new government is declared. The secular power of the pope is abolished, and Church property is confiscated. But by July Pius is able to return to Rome, aided by French troops.

1860   Victor Emmanuel unifies Italy and assumes control of the Papal States

Italy is unified under King Victor Emmanuel II. The spirit of nationalism that spreads throughout Italy leads to the annexation of the Papal States by the newly formed kingdom of Italy. Pope Pius IX is to rule only the Vatican and the area surrounding it, as his papal successors have ever since.

1861   L'Osservatore Romano, a daily newspaper, is founded

A semi-official newspaper, *L'Osservatore Romano*, is founded. *L'Osservatore Romano* is published daily in Italian by Roman Catholic laymen and carries official church information. It also appears weekly in English, Spanish, Portuguese, German, and French. (Following the election of Pope John Paul II, a native of Poland, in 1978, a monthly edition is printed in Polish.) As of 1996 it has a circulation of 70,000.

1869–70   The First Vatican Council is convened

Vatican I, convened by Pope Pius IX, is the first ecumenical council held in St. Peter's Basilica. Significantly, it establishes the doctrine of papal infallibility, meaning that on all matters of faith and morals the pope is never wrong.

1870: September 20   Rome falls to the Italians

The besieged city of Rome falls to the Italians when Napoleon III withdraws the French troops supporting Pius in order to participate in the Franco-Prussian War.

1870: October 2   Rome officially becomes the capital of Italy

The city of Rome is formally annexed to the kingdom of Italy and made the national capital. The Vatican rejects Italian legal and financial guarantees, and its pontiffs begin calling themselves "prisoners of the Vatican." The legal and political relationship between Italy and the Vatican becomes a long-standing unresolved issue known as the "Roman Question".

1871: May   Law of Guarantees is adopted

The new Italian government establishes the Law of Guarantees to define relations between the Italian kingdom and the papacy. Under the Law of Guarantees, (1) the pope is to be held inviolate; (2) full liberty in religious functions and in the conduct of diplomatic relations is guaranteed the pope; (3) an annual indemnity is given the pope to compensate for income lost through the annexation of the Papal States; and (4) the pope wins the right of extraterritoriality over the Vatican and the papal palaces. Extraterritoriality applies to numerous churches and palaces in Rome. Outside Rome it applies to the papal villa and its environs at Castel Gandolfo and to Santa Maria di Galeria, where the Vatican's radio broadcast facilities will later be located.

Pius refuses to accept the law or the money allowance and, instead, he and his successors choose to become "prisoners of the Vatican." Until 1919, Roman Catholics are prohibited by the papacy from participating in the Italian government.

1880–81   Secret Archives are opened to scholars

The Secret Archives of the Vatican, containing the strictly private records of Vatican affairs, is opened to scholars. Pope Leo XIII (r. 1878–1903) decrees that the Archives should be opened to all serious students, making it a world center for historical research. The materials found in the Archives, however, date no further back than 1200. A School for Archivists is also part of the Vatican Archives.

The 20th Century

1922–39   Reign of Pope Pius XI

Pope Pius XI (Achille Damiano Ratti, 1857–1939) is the reigning pope when the Vatican City State is created (see 1929). Pius renovates the Vatican Library, the papal palaces, and the papal summer palace at Castel Gondolfo, where the papal Observatory is relocated. He also builds a post office, a railroad station, a studio for mosaic construction, and the buildings housing the tribunals and other internal services.

1927   Religious press service is established

The international Agenzia Internazionale Fides (Religious Press Service–AIF) is founded to distribute news of missionary activity and publish Information (multilingual weekly), Documentation (irregular), and Photographic Service (weekly).

1929: February 11   Lateran Treaty is signed

The "Roman Question" is finally resolved with the signing of the Lateran Treaty between the Vatican and Italy. The State of the Vatican City is established, settling the relationship between the Vatican and the kingdom of Italy. Under the Lateran Treaty (actually three treaties in one), the Vatican recognizes the kingdom of Italy and, in turn, the Vatican is recognized as an independent neutral state. The first treaty recognizes the full sovereignty of the Vatican and establishes its territorial boundaries. The second treaty is a concordat establishing the Roman Catholic Church as the state church of Italy. The third treaty awards the Vatican 750 million "old lire" in cash and 1 billion "old lire" in interest-bearing state bonds in lieu of all financial claims against Italy for annexing the Papal States.

1930   The Vatican starts minting coins again

Following a lapse of sixty years, the Vatican resumes issuance of its own coinage, the lira (L). The Vatican agrees to issue no more than 300 million lire in any year. There are coins worth 10, 20, 50, 100, and 500 lire. Italian notes are also available. The currencies of Italy and the Vatican are mutually convertible.

1931   Radio Vatican is founded

Radio Vatican has two facilities, one in Vatican City and the other outside Rome at Santa Maria di Galeria. Pius XI builds the radio station in the Vatican Gardens, assisted by Guglielmo Marconi (1874–1937), the father of the wireless. Radio Vatican, the official radio station, remains one of the most influential in Europe.

1934   Weekly newspaper is launched

The Vatican publishes a weekly paper called *L'Osservatore della Domenica* (The Sunday Observer).

1936   The Pontifical Academy of Sciences is founded

Pope Pius XI founds this institute, recognized worldwide for experimental sciences and mathematics. The pope chooses from among the most prestigious scientists of all nations to promote the fields of research developed by the academy.

1942   The Vatican Bank is founded

Officially known as the *Instituto per le Opere di Religione* (Institute for Religious Works—IOR), Vatican Bank carries out fiscal operations and invests and transfers the funds of the Vatican and the Roman Catholic religious communities worldwide. The Administration of the Patrimony of the Holy See manages the Vatican's capital assets.

1947   Lateran Treaty terms are incorporated in Italian constitution

The Italian Republic adopts a constitution that includes the terms of the Lateran Treaty of 1929.

1957   Vatican TV station is established

Pope Pius XII (r. 1939–58) founds a television station. He also builds a more powerful radio station for the Vatican at Santa Maria di Galeria.

1958–63   Reign of Pope John XXIII

Pope John XXVIII (Angelo Giuseppe Roncalli, 1881–1963) wants to modernize the Church in the wake of an increasingly uncertain atomic age fueled by the Cold War. In an effort to re-connect the Church with those who have left and with non-Catholics, Pope John convenes the Second Vatican Council, also known as Vatican II (see 1962).

1962–65   The Second Vatican Council

The Second Vatican Council, the first worldwide ecumenical council since 1870, is begun under Pope John. Pope John XXIII closes the first session of Vatican II on December 8, 1962. He dies before Vatican II reconvenes. Nine months later, Pope Paul VI, Pope John's successor, carries on his work.

Vatican II leads to alterations of the Church's leadership in spiritual and social matters. Pope John makes history by altering the text of the canon of the Mass for the first time since the seventh century and by strongly defining the position of the Church on problems of labor and social progress in his encyclical (a papal document addressed to the bishops) *Mater et Magistra* of June, 1961.

Pope Paul VI creates a new form of collaboration between the pope and bishops, called the Synod of Bishops, which is presided over by the pope himself.

1963–78   Reign of Pope Paul VI

Pope Paul VI (Giovanni Battista Montini, 1897–1978) continues Pope John's effort to bring unity to the Christian world. Paul is the first pope to travel extensively worldwide.

1963: April 10   Pope John pleads for world peace

In his eighth encyclical, *Pacem in Terris,* Pope John XXIII makes a profound plea for peace and hails the United Nations as a defender of human rights.

1964   Paul VI meets Eastern Orthodox patriarch

Pope Paul VI meets in Jerusalem with Athenagoras, the ecumenical patriarch of the Eastern Orthodox Church. This meeting illustrates the desire of the Roman Catholic and the Orthodox churches to reconcile their differences.

Pope Pius XII, who introduces television and radio to the Vatican, holds an audience from a balcony. (EPD Photos)

1965   Pope denies Jewish guilt for death of Jesus

Pope Paul VI attempts to establish better relations between Roman Catholics and Jews by officially declaring the Jews innocent of Jesus's death. This helps to stifle religious animosity between Catholics and Jews.

1965: October 4   Paul VI addresses UN General Assembly

Pope Paul VI addresses the United Nations General Assembly (the first pope ever to do so). He appeals for world peace and international cooperation.

1965: December 7   Congregation of the Holy Office is renamed

The Congregation of the Holy Office, created by Pope Paul III in 1542, is renamed the Sacred Congregation for the Doctrine of the Faith.

1965: December 8   Vatican II concludes

Pope Paul VI presides over the concluding sessions of the second Vatican Council. He then travels to the Holy Land (another papal first).

1967: August 15   Paul VI establishes the Curia in its modern form

In his apostolic constitution *Regimini Ecclesiae Universae,* Paul VI reorganizes the Roman Curia (the collective administrative structure of congregations, tribunals, and offices through which the Pope runs the Church). Department heads, members (cardinals and bishops), secretaries and consultors are to be named by the Pope for a five-year period.

1970s   Decommercialization Policy enacted by Pope Paul VI

In an effort to defuse criticism of the Holy See as being too secular-oriented, Paul takes steps to emphasize the religious nature of the Vatican. One of those steps involves reducing the size of the papal armed forces. Also, a bakery, butcher shop, and fabric shop are closed. Income is instead derived from the sale of religious literature, stamps, and souvenirs and from admission to the many Vatican museums.

1970: September 14   Pope Paul VI disbands the papal patrol force

The Swiss Guard remains the only papal patrol force in existence, numbering approximately 100 members. The Vatican also employs a civilian security force to maintain order and protect Vatican personnel and property and the invaluable art treasures belonging to the Church. This security force reports directly to the Central Office of Security. The Vatican has its own jail.

1971: January 1   Age limits placed on the activities of cardinals

Pope Paul VI decrees that cardinals will cease to be members of departments of the Curia upon reaching the age of eighty and lose the right to meet in conclave and participate in papal elections.

1971   The Vatican backs nuclear treaty

The Holy See announces its decision to adhere to the Nuclear Non-Proliferation Treaty in order to give its moral support to the principles it is based on.

1972: September   Paul VI upholds traditional doctrines

Pope Paul VI, conservative on doctrinal questions, reaffirms the Catholic institution of the all-male, celibate priesthood as well as the discipline of dissident priests. He also opposes artificial methods of contraception and reaffirms papal infallibility.

1973   Paul VI meets with other religious leaders

Pope Paul VI confers with the Coptic Orthodox patriarch of Alexandria and, later in the year, meets with the exiled Dalai Lama in the first-ever meeting between a pope and a Buddhist leader.

1974: May   Italy legalizes divorce

In a special referendum, some sixty percent of the Italian people vote to retain legislation permitting divorce, in spite of strong opposition from the Church.

1978: August 26   John Paul I becomes pope

Pope John Paul I succeeds Paul VI, who dies at Castel Gandolfo on August 6.

1978: September 28   John Paul I dies suddenly

Pope John Paul I (Albino Luciani, 1912–78) dies suddenly after a thirty-four-day reign. John Paul is the first pope to assume a double name, and the first who refuses to be crowned. He vows to uphold the decrees set forth in Vatican II.

1978: October 16   John Paul II becomes pope

Polish Cardinal Karol Wojtyla (b. 1920) is elected pope. As John Paul II, he is the first non-Italian to serve as pope in 456 years and the most widely traveled pontiff since Paul VI.

1981: May 13   John Paul survives assassination attempt

Pope John Paul II survives an assassination attempt in Vatican Square by a Turkish gunman, who is sentenced to life in prison. The pope later visits the gunman and offers him forgiveness. Six accomplices in the shooting are held, but released for lack of evidence.

1982   *Opus Dei* attains the status of personal prelature

Opus Dei, an international organization of over 72,000 laity and priests known for its doctrinal fidelity, is elevated to the status of personal prelature by Pope John Paul II.

1983   Vatican Television Center is established

The Vatican Television Center is founded to produce and distribute religious programming.

1984   New concordat supersedes the Lateran Treaty of 1929

A new agreement between the Vatican and the Italian government replaces the Lateran Treaty. Under this concordat the pope retains temporal authority over Vatican city: he is the absolute sovereign and undisputed head of the Church worldwide. He is represented by the cardinal secretary of state in civil governance of Vatican City. The concordat also ends the status of Roman Catholicism as the official state religion of Italy.

1984: January 10   Diplomatic ties established with the U.S.

The United States and the Holy See announce the establishment of diplomatic relations.

1984: March 7   Wilson confirmed as first U.S. ambassador to the Holy See

The U.S. Senate confirms William A. Wilson, President Ronald Reagan's personal envoy to the pope since 1981, as ambassador to the Holy See. The Holy See names Archbishop Pio Laghi as *Apostolic Nuncio* (equivalent to ambassador) to the United States.

1986: March 29   Assassination accomplices are acquitted

Three Bulgarians and three Turks, the alleged accomplices in the attempted assassination of John Paul II, are acquitted of conspiracy charges due to lack of evidence.

1987: June   John Paul II grants controversial audience to Kurt Waldheim

Pope John Paul II invites criticism and controversy when he grants Austrian President Kurt Waldheim an audience amid accusations that Waldheim participated in war crimes as a German officer during World War II.

1993   Papal encyclical urges adherence to moral doctrine

John Paul II urges Catholic bishops to uphold traditional moral doctrine (celibacy, anti-abortion) in a papal encyclical.

1993: December   Diplomatic ties with Israel and the PLO

The Vatican and the Israeli government formally establish diplomaitc relations. And for diplomatic balance, a year after signing the mutual recognition agreement with Israel, the Vatican establishes diplomatic relations with the Palestinian Liberation Organization (PLO).

1995: October   Vatican employees are required to sign statement

New lay employees of the Vatican must sign a statement binding them to observe the moral doctrines of the Roman Catholic Church.

1997   Diplomatic ties are set with African nations

The Holy See establishes diplomatic relations with Libya, Guyana, and Angola.

1997   The Holy See joins the WTO

The Holy See becomes a member of the World Trade Organization.

Bibliography

Accattoli, Luigi. *Life in the Vatican with John Paul II*. Trans. Marguerite Shore. New York : Universe, 1998.

Chadwick, Owen. *Britain and the Vatican during the Second World War*. New York: Cambridge University Press, 1986.

Duffy, Eamon. *Saints and Sinners:A History of the Popes*. New Have, Conn.:Yale University Press, 1997.

Hebblethwaite, Peter. *In the Vatican*. Bethesda, Md.: Adler & Adler, 1986.

Hofmann, Paul. *O Vatican! A Slightly Wicked View of the Holy See*. New York: Congdon & Weed. Distributed by St. Martin's Press, 1983.

Holmes, Derek. *The Papacy in the Modern World 1914–18*. New York: Crossroad, 1981.

Packard, Jerrold M. *Peter's Kingdom: Inside the Papal City*. New York: Scribner, 1985.

Reese, Thomas J. *Inside the Vatican: The Politics and Organization of the Catholic Church*. Cambridge, Mass.: Harvard University Press, 1996.

Rosa, Peter de. *Vicars of Christ: The Dark Side of the Papacy*. New York: Crown Publishers, 1988.

Volpini, Valerio. *Vatican City: Art, Architecture, and History*. New York, Portland House: Distributed by Crown Publishers, 1986.

Yugoslavia

Introduction

The modern Federal Republic of Yugoslavia is the third political entity to bear the name Yugoslavia. The first was a kingdom created in the aftermath of World War I, the second a communist-ruled federation of six republics. Four of these republics withdrew from the federation in 1991 and 1992, leaving only Serbia and Montenegro, which currently make up the country. In spite of official borders, however, the fates of all the Balkan republics remain closely intertwined by a long history of foreign domination, contrasting cultures, and ethnic and religious divisions. The violent upheavals in the region since 1992—in Croatia, Bosnia, and Kosovo—have underscored the importance of this history and the persistence of the emotional scars that are its legacy.

Located on the Balkan peninsula in southeastern Europe, the Federal Republic of Yugoslavia has an area of 39,479 square miles (102,250 square kilometers). Serbia, by far the larger of its two constituent republics, has an area of 34,136 square miles (88, 412 square kilometers), compared with Montenegro's 5,299 square miles (13,724 square kilometers). Included within Serbia's current borders are two regions that were autonomous provinces from World War II until the 1990s: Vojvodina to the north and Kosovo to the south. Vojvodina's terrain consists of low-lying plains that are among the most fertile in Europe. The rest of Serbia is mostly mountainous, as is that of Montenegro, through which both the Dinaric and Albanian Alps extend (and whose name means black mountain). Yugoslavia's capital is the Serbian city of Belgrade, located in the northern portion of the country. Yugoslavia has an estimated population of over 10.5 million people, of whom roughly ninety-five percent live in Serbia.

The Serbs were one of the South Slavic peoples who emigrated to the Balkans in the seventh century A.D., settling in present-day Bosnia and Herzegovina as well as Serbia, Kosovo, and Montenegro. Organized into tribes, they lived a settled existence as farmers and shepherds. Led by a Slavic leader named Vlastimir, the Serbs founded their first principality, although it was under the control of the Byzantine Empire, one of the major powers in the region during this period. The Serbs converted to the Eastern Orthodox branch of Christianity, which became an important component of their national identity. Along with their conversion to Orthodoxy, the Serbs adopted the Greek-based Cyrillic alphabet. Their fellow South Slavs to the West, the Croats, adhered to Roman Catholicism and adopted the Latin alphabet. Serbia attained its greatest glory as an independent kingdom during the reign of the Nemanja dynasty founded by Stefan Nemanja in the 1160s. Expanding the domain of Rascia (as Serbia was then called), Stefan also broke free of Byzantine rule. It was during Stefan's reign that Kosovo was incorporated into Serbia, eventually becoming a major political and religious center. In 1196 Stefan turned over the throne of Serbia to his two sons, and the kingdom thrived for another century and a half, with a growing population, expanded trade, and cultural achievement. Under its most illustrious prince, Stefan Dusan (r.1331–55), Serbia reached the proportions of an empire, stretching from the Aegean Sea to the Danube, and along the Adriatic coast. During this period, it also received its first legal code.

By the final years of the Nemanja dynasty, the threat of Ottoman invasion was also looming over the horizon, and it became a reality by the end of the fourteenth century. In 1389 Serb forces suffered a bitter defeat at the hand of the Ottoman Turks in the Battle of Kosovo Polje, a hard-fought struggle in which the leaders of both forces were killed. The battle—and Kosovo itself—came to symbolize Serbian heroism and honor and remained a powerful symbol in Serbian consciousness long after the ethnic makeup of the region had shifted and it was largely populated by Albanians. By 1459 the Ottomans had completed their conquest of Serbia, which they then occupied for more than three centuries. Ottoman repression of Serbian Christians led to numerous unsuccessful rebellions. Many Serbs emigrated to other lands, while others took to the forests, living as outlaws.

After forming part of the medieval Serb kingdom, Montenegro came under Italian rule by the late fourteenth century. The fierce fighters of this small but mountainous region managed to stave off Ottoman conquest altogether despite being surrounded on all sides by Turkish-occupied land, although battles against the Turks raged intermittently for centuries. From the sixteenth to the mid-nineteenth centuries, Montenegro became a *theocracy*—all of its rulers were clerics. Being

YUGOSLAVIA

celibate, they had no natural heirs, so the succession was determined by vote or went to a nephew of the retiring ruler.

Meanwhile, large-scale migrations in Serbia in the seventeenth and eighteenth centuries strongly affected the futures of two border regions. In 1690 the Serb patriarch (religious leader) led thousands of Serbian families northward toward Hungary to escape Turkish reprisals for Serb uprisings. They settled in a military frontier region later named Krajina, which was to retain a mix of Serbs and ethnic Croats. During this migration and another one in 1736–39, many Serbs left Kosovo, and the Turks encouraged Albanians—the vast majority of whom, like themselves, were Muslims—to settle there. By the late twentieth century, only 10 percent of Kosovars were Serbs.

In the early nineteenth century, the Serbs succeeded in putting an end to Turkish rule of their homeland. The most prominent figure in the struggle for independence was George Petrovic (1762–1817), known as Karadjordje, or Black George. In 1811 he was granted hereditary royal rights, inaugurating the Karadjordjevic dynasty, which was to rule Serbia intermittently until the monarchy was abolished in 1945. Alternating with the Karadjordjevices was a second dynasty begun by Karadjordje's rival (and the force behind his assassination in 1817), Milos Obrenovic. Although nominally still under Turkish control, Serbia was recognized by the Turks as an autonomous principality and allowed to establish independent relations with other governments. In Montenegro, Prince Danilo II (r. 1851–60), who wished to marry, had ended the three-hundred-year tradition of clerical rule and instituted a hereditary monarchy. However, Nicholas, who succeeded Danilo, was to be the last Montenegrin monarch.

Serb nationalism combined with the Romantic movement to produce a literary and cultural revival in Serbia in the nineteenth century. The philologist Vuk Karadzic (1787–1864) made important linguistic innovations and scholarly contributions to the development of Serbian as a literary language. Literary periodicals were published in Novi Sad, and Serbian poets such as Laza Kostic (1841–1910) and Djura Jaksic (1832–78) were active. Serbs also became involved in the pan-Slavic Illyrianist movement founded by Croat author Ludevit Gaj (1809–72) in the 1830s, whose cornerstone was the advocacy of a joint Serbo-Croatian language.

The early twentieth century saw the growth of both pan-Slavism and radical nationalism among the Serbs, who wanted to end Austro-Hungarian influence in the Balkans. In 1914 a radical Serb student, Gavrilo Princip, touched off World War I by assassinating the Austrian Archduke Franz Ferdinand. A South Slavic council was active during the war years, and at the end of the war an independent Kingdom of Serb, Croats, and Slovenes was declared with King Alexander Karadjordjevic (1888–1934) of the Serbian Karadjordjevic dynasty as its first monarch. Tensions between Serbs and Croats provided a pretext for Alexander I to replace the country's parliamentary government with a royal dictatorship in 1929, at which time he renamed the nation Yugoslavia. Alexander was assassinated in 1934, and his son, Peter II (1923–1970), was named regent. Peter's government was brought down in 1941 after pledging cooperation with Nazi Germany, and Germany invaded Yugoslavia in April of that year.

During World War II, Yugoslavia was divided among the Axis powers. Croatia was supposedly an independent state, but it was governed by a puppet Croat fascist party, the *Ustashe,* which persecuted Serbs and other ethnic minorities. Two Slavic forces were formed to resist the Nazis: the Serbian *Chetniks* and the Partisans, headed by communist leader Josip Broz Tito (1892–1980). By the end of the war, the Partisans were recognized as the leading resistance group, and,

following liberation by the Red Army, Tito formed a new Yugoslavian government—a Soviet-style federation of six republics and two autonomous provinces, Vojvodina and Kosovo. In spite of his early alliance with the Soviet Union, Tito split with the U.S.S.R. by 1948 over his refusal to be controlled by Soviet leader Josef Stalin (1879–1953). In the postwar years, Yugoslavia pursued its own style of communism, with greater government decentralization than the countries of the Soviet bloc and less repression of personal freedoms. Nevertheless, Tito maintained a tight grip on power to keep the country's traditional ethnic rivalries in check.

After Tito's death in 1980, these rivalries, together with an economic crisis, brought about the collapse of the Yugoslav federation in little more than a decade. In the late 1980s, Slobodan Milosevic (b.1941) rose to power in Serbia and spearheaded a resurgence of nationalism in the republic, which in 1990 claimed direct control of Kosovo and Vojvodina by abolishing the autonomy of these regions. In the same year, the Communist parties of Slovenia and Croatia suffered defeats in free elections in those republics and both withdrew from Yugoslavia in June, 1991, followed by Macedonia in September. Bosnia and Herzegovina withdrew the following year, and the remaining republics of Serbia and Montenegro proclaimed a new Yugoslav federation.

With the break-up of the old Yugoslavia, the Balkans erupted into violence. In 1991, the Serbs of Croatia's Krajina region were backed up by Serb-dominated Yugoslav federal forces in their attempt to form a separate state. A three-year civil war began in Bosnia the following year. In 1998 the Yugoslavian government, led by Milosevic, began a campaign against ethnic Albanian separatists in Kosovo that by mid-1999 had decimated dozens of Kosovar villages, created nearly a million refugees, and resulted in a protracted campaign of NATO air strikes.

Timeline

7th century A.D.   Serbs migrate to the Balkans

The Serbs, a South Slavic people, migrate to the river valleys of present-day Bosnia, Kosovo, Montenegro, and southern Serbia, where they settle and live as farmers and shepherds, loosely organized in tribes ruled by leaders called *zupans.* They drive out the existing romanized populations, and rival leaders struggle for control over the following centuries.

850 A.D.   First Serb state is formed

A Slavic leader named Vlastimir founds the first Serb principality, under the control of Byzantium. The Serbs convert to Eastern Orthodoxy religion. Between the ninth and twelfth centuries, more Serb principalities are formed.

1077 Zeta becomes a kingdom

The principality of Zeta, which later forms part of Serbia, is raised to the status of a kingdom when its ruler, Mihajlo, is anointed king by Pope Gregory VII.

The Medieval Serbian Kingdom

1160s Nemanja dynasty is inaugurated

Stefan Nemanja becomes the grand zupan (head ruler) of Rascia under Byzantine rule and expands his territory through political alliances. He eventually grows powerful enough to declare independence from Byzantium. He allies himself with the Holy Roman Emperor and eventually wins back the support of the Eastern Orthodox Church by suppressing heresy in his lands.

1180–90 Kosovo incorporated into medieval Serbia

Serbian ruler Stefan Nemanja incorporates Kosovo into his kingdom, and it becomes a major political and religious center.

1196–1355 Serbia thrives under the Nemanja dynasty

Stefan Nemanja turns over the throne of Serbia to his two sons, Stefan and Vukan, and Serbia remains a powerful state under the succeeding Nemanja rulers for over one hundred fifty years. Its population grows, trade and literacy expand, and the arts flourish, supported by wealthy patrons.

1331–55 Reign of Stefan Dusan

By the reign of Stefan Dusan (also known as Stefan the Mighty), Serbia has expanded into an empire stretching from the Aegean Sea to the Danube River and along the Adriatic coast. The territorial expansion, while seemingly a sign of strength, actually dilutes the power of the central government at a time when the threat of Ottoman invasion is arising in the east. Following the death of Stefan Dusan, the Nemanja empire declines due to internal power struggles and growing pressure by the Ottomans.

1349 Legal code is devised

Serbia receives its first legal code during the reign of Stefan Dusan. It introduces an early version of constitutional monarchy by making the emperor subject to the rule of law. The code also includes provisions affecting commerce and serfdom.

The Nemanja Dynasty, 1196–1355.

1360–1421 Montenegro ruled by Italian dynasty

After forming part of Serbia for most of the Nemanja dynasty, Montenegro comes under the rule of the Balsa family.

1371 Last Nemanja leader dies

Stefan Uros, the last ruler of the Nemanja dynasty, dies.

Turkish Conquest

1389 Battle of Kosovo Polje

Serb forces are defeated by the Ottomans at Kosovo Polje, in a fierce battle in which the leaders of both sides are killed. The battle becomes a rallying point for Serb nationalism for centuries and is commemorated in epic poems and songs, as Kosovo becomes a symbol of Serb nationalism and solidarity. It continues to serve this function even after most ethnic Serbs have been driven from the region during the Ottoman occupation. Militarily the Battle of Kosovo Polje is significant in opening the way for the Ottoman conquest of the region.

The Ottoman Period

15th–19th centuries   Montenegro resists Turkish rule

Although completely surrounded by Turkish-occupied territory, Montenegro's mountainous terrain helps the region's tenacious fighters resist Turkish conquest.

1459   Ottomans complete conquest of Serbia

The Ottoman Turks conquer Smederjevo, the last Serb stronghold, inaugurating a three-and-a-half-century occupation and setting off a Serb diaspora, as inhabitants of the region flee to other lands. Although the Serbs that remain under Ottoman rule are granted internal autonomy, they still suffer mistreatment and exploitation. Most of the Christian Serbs evolve into a rural peasant class, while the upper rungs of society are occupied by Muslims. Many Serbian Christian youths are taken from their families and drafted into government service, where they are converted to Islam. Church lands are appropriated, and the Greek patriarchate is placed in control of the Serbian Christians until the middle of the sixteenth century. Serb peasants stage rebellions in response to Ottoman persecution, and many take to the woods and become outlaws known as *hajduci*.

A number of Serbs flee to Hungary, where they are granted an autonomous military frontier region to serve as a buffer against a Turkish invasion. The region is eventually named Vojvodina. Others later settle in a similar region in Croatia, which becomes known as the Krajina, forming a self-contained ethnic enclave in that province.

1493   Print shop set up in Montenegro

Montenegro's leader, Ivan the Black, launches a printing shop in the Montenegrin capital of Cetinje.

1516–1851   Montenegro is ruled by clerics

The Bishop of Cetinje takes over control of Montenegro, which becomes a theocracy for the next three centuries, with rulers either elected from the ranks of Cetinje's clerics or designated by their predecessors.

1557   Autonomy is restored to Serbian church

The Serbian church, which has been overseen by the Greek patriarchate (a man heading an organization which is ruled exclusively by men) since the beginning of Ottoman rule, has its own *patriarchate* restored through the efforts of the Serbian-born Grand Vizier Mehmed Pasha Sokolovic, who is turned over to the Muslims as a boy and rises through their ranks. Under the Ottoman *devshirme* system, Christian youths are often taken by the government as a tax and reared as Muslims in preparation for military or administrative service to the state. Grand Viziers are the sultan's leading administrators.

1680s   Serb rebellions against Ottoman rule

Following the Siege of Vienna, Christian armies led by Polish monarch Jan Sobieski inflict a series of military defeats on the Turks and come close to driving them out of Europe. The Serbs seize this opportunity to rebel against their Ottoman rulers, but the Ottomans ultimately prevail.

1690   Serb patriarch emigrates to Hungary

Fearing Turkish reprisals following unsuccessful Serb uprisings, Serb patriarch Arsenije III Carnojevic leads some 36,000 Serb families over the border into the Austrian Habsburg Empire. They settle in the military frontier region later known as Vojna Krajina. In Kosovo, the areas vacated by Serbs are populated by Muslim Albanians, who gradually come to comprise the province's dominant ethnic and religious group.

18th–19th centuries   Serbian cultural renaissance takes place

Serbians' pride in their national culture undergoes a rebirth through the efforts of scholars and literary figures, such as philologist and folklorist Vuk Karadzic (1787–1864), who compiles folksongs and poetry and modifies the Serbian alphabet. Dositej Obradovic also introduces language reforms and translates the New Testament into Serbian.

1702: December 24   Massacre of Muslims in Montenegro

To end suspected collaboration with the Ottoman Turks, Montenegro's ruler, Danilo Petrovic Njegos, orders Muslim men of all ethnicities murdered on Christmas Eve.

1736–39   Ottoman victories spur new wave of emigration

Turkish victories over Austria spur a new round of emigration, in the course of which many Serbs emigrate to northern Serbia from Kosovo, where the Turks encourage Muslim Albanians to settle. Albanians eventually account for some ninety percent of the province's population, but the Serbs maintain their emotional attachment to the region.

1762: November 3   Birth of national hero Karadjordje

George Petrovic, better known as Karadjordje, is born to a peasant family. He is nicknamed Karadjordje (black George) because of his dark coloring. In 1787 he joins the Austrian army to fight the Ottoman Turks and distinguishes himself as a military leader. After a period as a civilian, Petrovic heads a successful Serb rebellion against the elite Turkish Janissary

The Ottoman Period

The Ottoman Empire began in western Turkey (Anatolia) in the thirteenth century under the leadership of Osman (also spelled Othman) I. Born in Bithynia in 1259, Osman began the conquest of neighboring countries and began one of the most politically influential reigns that lasted into the twentieth century. The rise of the Ottoman Empire was directly connected to the rise of Islam. Many of the battles fought were for religious reasons as well as for territory.

The Ottoman Empire, soon after its inception, became a great threat to the crumbling Byzantine Empire. Constantinople, the jewel in the Byzantine crown, resisted conquest many times. Finally, under the leadership of Sultan Mehmed (1451–81), Constantinople fell and became the capital of the Ottoman Empire.

Religious and political life under the Ottoman were one and the same. The Sultan was the supreme ruler. He was also the head of Islam. The crown passed from father to son. However, the firstborn son was not automatically entitled to be the next leader. With the death of the Sultan, and often before, there was wholesale bloodshed to eliminate all rivals, including brothers and nephews. This ensured that there would be no attempts at a coup d'état. This system was revised at times to the simple imprisonment of rivals.

The main military units of the Ottomans were the Janissaries. These fighting men were taken from their families as young children. Often these were Christian children who were now educated in the ways of Islam. At times the Janissaries became too powerful and had to be put down by the ruling Sultan.

The greatest ruler of the Ottoman Empire was Suleyman the Magnificent who ruled from 1520–66. Under his reign the Ottoman Empire extended into the present day Balkan countries and as far north as Vienna, Austria. The Europeans were horrified by this expansion and declared war on the Ottomans and defeated them in the naval battle of Lepanto in 1571. Finally the Austrian Habsburg rulers were able to contain the Ottomans and expand their empire into the Balkans.

At the end of the nineteenth century, the Greeks and the Serbs had obtained virtual independence from the Ottomans, and the end of a once great empire was in sight. While Europe had undergone the Renaissance, the Enlightenment, and the Industrial Revolution, the Ottoman Empire rejected these influences as being too radical for their people. They restricted the flow of information and chose to maintain strict religious and governmental control. The end of Ottoman control in Europe came in the First Balkan War (1912–13) in which Greece, Montenegro, Serbia, and Bulgaria joined forces to defeat the Ottomans.

During the First World War, (1914–18), the Ottoman Empire allied with the Central Powers and suffered a humiliating defeat. In the Treaty of Sévres, the Ottoman Empire lost all of its territory in the Middle East, and much of its territory in Asia Minor. The disaster of the First World War signaled the end of the Ottoman Empire and the beginning of modern-day Turkey. Although the Ottoman sultan remained in Constantinople (renamed in 1930, "Istanbul"), Turkish nationalists under the leadership of Mustafa Kemal (Ataturk) (1881–1938) challenged his authority and, in 1922, repulsed Greek forces that had occupied parts of Asia Minor under the Treaty of Sévres, overthrew the sultan, declared the Ottoman Empire dissolved, and proclaimed a new Republic of Turkey. That following year, Kemal succeeded in overturning the 1919 peace settlement with the signing of a new treaty in Lausanne, Switzerland. Under the terms of this new treaty, Turkey reacquired much of the territory—particularly, in Asia Minor—that it had lost at Sévres. Kemal is better known by his adopted name "Ataturk", which means "father of the Turks". Ataturk, who ruled from 1923 to 1938, outlawed the existence of a religious state and brought Turkey a more western type of government. He changed from the Arabic alphabet to Roman letters and established new civil and penal codes.

corps (see sidebar). In 1805 he leads a rebellion against the Ottoman rulers and liberates Serbia from Turkish control. In 1808 he is recognized as hereditary ruler of Serbia, beginning a dynasty that will last—but rule only intermittently—until 1945.

Political machinations between Serbia, Austria, Russia, and Turkey result in the temporary loss of Serb autonomy in 1813 and the exile of Petrovic to Hungary. Autonomy is regained two years later by a rival Serb leader, Milos Obrenovic, who has Petrovic murdered when he returns to the country in 1817 and sends his head to Constantinople to win favor with the Turkish sultan. The dynasties begun by Petrovic and Obrenovic begin a fierce rivalry for the political leadership of Serbia that lasts until the end of World War I in 1918. However, Petrovic, the first to liberate his people from Turkish rule, is revered permanently as Serbia's national hero.

1787: November 6   Birth of linguist Vuk Karadzic

Vuk Stefanovic Karadzic, folklorist and language reformer, works as a schoolteacher and scribe early in the nineteenth century. He later studies Slavonic languages in Vienna and devises a simpler way for Serbian to be written in the Cyrillic script, eliminating eighteen letters for which there are no sounds in Serbian and adding six new letters. At first, his innovation draws heavy opposition. The first book allowed to be published in his reformed alphabet appears five years before his death, which occurs in 1864. Karadzic is also renowned as a collector of Serbian folk literature and for his 1818 work *Srpski rjecnik* (Serbian Lexicon), a mammoth Serbian-Latin-German dictionary with scholarly observations on Serb folklore. Karadzic also publishes a collection of stories and translate the New Testament into Serbian.

1788–92   Period of Austrian rule

Austria takes over control of Serbia for a brief period and rallies Serb forces to fight the Turks. Serbian soldier Djordje Petrovic, known as Karadjordje, or black George, eventual founder of a Serb royal dynasty, first rises to prominence during this period.

1790   Assembly of Temesvar

Serb settlers in Hungarian border region of Vojvodina demand autonomy from the Habsburg government.

1791   First Serbian-language newspaper is published

The first Serbian-language newspaper begins publication in Vienna.

1798   First Montenegrin code of laws is formulated

Under the rule of Peter I (r. 1782–1830), Montenegro receives its first legal code. During this period, the Ottomans formally recognize Montenegrin independence.

19th century   National literary revival

Under the dual influences of Romanticism and Serb nationalism, the region's literature undergoes a revival. The city of Novi Sad, in Vojvodina, becomes the primary literary center and home to the literary journals *Danica* (Morning Star), *Matica,* and *Javor.* In addition to philologist Vuk Karadzic (see 1787), leading literary figures include poets Laza Kostic (1841–1910) and Djura Jaksic (1832–78).

Serbian Autonomy is Regained

1804   Serbian uprising against the army

Djordje Petrovic, known as Karadjordje, leads the first large-scale Serb revolt against the Ottoman Turks, who rise up against the elite Turkish Janissaries, incited by the murder of Serbian leaders. They then demand that the sultan grant them autonomy.

1805   Serbs revolt against the sultan

When the sultan refuses to grant the Serbs autonomy, Karadjordje leads them in a war against the Ottomans, and the Serbs prevail.

1806: September   Serb autonomy enacted

The Ottoman sultan officially grants the Serbs political autonomy.

1811   Karadjordje is confirmed as Serbian monarch

A Serbian assembly makes Karadjordje a prince with hereditary rights, inaugurating the Karadjordjevic dynasty, which will rule Serbia intermittently until 1945.

1813–51   Life of author and Prince Peter II

Montenegro's Prince Peter II, author of epic verse celebrating the region's struggles against Turkish oppression, is widely regarded as the founder of the Serb and Montenegrin national literatures. Among his best-known works is the poem *Mountain Wreath* (1847).

1813   Turks reoccupy Serbia

After the European powers fail to support Serb independence, Serbia reverts to Ottoman rule, and the Turks exact a brutal revenge for prior Serb victories, plundering Serbian villages and going on killing sprees. Karadjordje flees across the border to Hungary.

1815   New Serbian uprising

A second major Serb uprising against the Turks is led by Milos Obrenovic, who wins the support of Russia and inaugurates his own dynasty, rivaling that of exiled leader Karadjordje.

1817 Karadjordje is assassinated

To solidify his hold on Serb leadership Obrenovic has Karadjordje assassinated. To gain favor with the Ottoman sultan, he has Karadjordje decapitated and sends his head to Constantinople. The rivalry between the Karadjordjevic and Obrenovic dynasties lasts into the twentieth century.

1824 *Matica Srpska* cultural society is founded

A literary and cultural society, *Matica Srpska,* is founded and plays an important role in the growth of Serb nationalism.

1826 Portrait collection spurs artistic renewal

A book containing portraits of well-known Serbs is published by Josif Milovuk and spearheads a resurgence of artistic activity in Serbia.

1830s Language-based Slavic nationalist movement gains influence

Croatians, led by nationalist writer Ludevit Gaj (1809–72), develop their own pan-Slavic nationalist movement, *Illyrianism,* to counteract the efforts of ethnic Hungarian administrators in the Krajina and Vojvodina at imposing Hungarian culture *(Magyarization)* on the South Slavs within the Habsburg Empire. To give their movement greater credibility, they attempt to involve other South Slavs, especially Serbs and Slovenes, and they promote the adoption of the Zagreb dialect, used by both Serbs and Croats, as a national language.

1830 Serbia regains autonomy

Aided by the Greek war for independence, Serbia wins autonomy for the second time. It is recognized by the Turks as a principality, with revolutionary leader Milos Obrenovic as its prince. Although nominally under Ottoman rule, Serbia is generally treated as an independent state, and the Turks allow foreign governments to establish relations with the Serbs. Although he is an autocratic ruler, Obrenovic brings progress to Serbia, improving education, pursuing land reform, and expanding trade.

1838 Constitutional government

Turkey and Russia impose a constitutional government on the Serbian principality, with a legislative council, ministers, and plans for the formation of an assembly.

1839 Obrenovic abdicates

Serbia's ruler, Milos Obrenovic, abdicates the throne when the Turks impose constitutional government. He is succeeded by his son, Mihajlo.

1843–58 Karadjordjevices regain power

The Karadjordjevices overthrow the Obrenovic dynasty and seize control of the government. Alexander, son of Karadjordje, becomes the new king of Serbia, ruling until 1858.

1843 National Museum is founded

The National Museum is established in Belgrade. By the twentieth century, it has a collection ranging from prehistory to contemporary times and including art by Serbians and artists throughout Europe, including French masters Auguste Renoir (1841–1919) and Edgar Degas (1834–1917).

1844 Greater Serbia concept is introduced

Prime Minister Ilija Garasanin introduces the concept of a Greater Serbia incorporating all Serbs within Serbia. Regions eyed by Serbian nationalists include the Krajina and Vojvodina, the military frontier regions on Serbia's border with Austria. He takes as his model the historical reconstruction of Stefan Dusan's medieval Serbia. Greater Serbia becomes an influential twentieth-century political goal of the Serbs, from World War I through the end of the century.

1847 Serbian translation of the New Testament appears

Linguist Vuk Karadzic (see 1787) translates the New Testament into Serbian.

1848 Revolution threatens autonomy of Vojvodina

During the revolutionary activities that sweep Europe in 1848, Hungary revolts against Austrian control and threatens to impose Hungarian culture on the Serbs in the Vojvodina region (a policy known as *Magyarization*). Fearing that they will be subject to Hungarian rule, Serbs ally themselves with Austria, hoping that the South Slavic provinces will later by united under Austrian rule. However, conditions only become more repressive under Austria, which imposes absolutist rule and silences all dissent.

1852 Hereditary succession proclaimed in Montenegro

Breaking a three-hundred-year tradition of clerical rule, Danilo II (1851–60) institutes a hereditary succession to the throne.

1860–1918 Reign of the last Montenegrin monarch

Nicholas, nephew of Danilo II, reigns as the last independent Montenegrin ruler. In a series of wars with Turkey, including the Balkan wars of 1912–13, the extent of Montenegro's territory more than doubles. Improvements are made in education, finance, and the infrastructure, and educated Montenegrins are drawn to the cause of Serbian nationalism.

1860–68   Reign of Mihajlo Obrenovic

The Serb assembly restores the Obrenovic dynasty, and Mihajlo Obrenovic ascends the throne following the death of his father, Milos. Although his personal life scandalizes the country and leads to his assassination eight years later, he is a progressive ruler, strengthening parliamentary monarchy, beginning industrial development, and bolstering the Serb military in preparation for a war of liberation.

1864: February 6   Death of scholar Vuk Karadzic

Linguistic innovator, scholar, and folklorist Vuk Karadzic dies in Vienna. (See 1787.)

1867   Formation of Austria-Hungary

After decades of demands by the monarchy's ethnic Hungarians for equal status with the ruling Austro-Germans, the Habsburgs restructure their empire into a dual monarchy, Austria-Hungary. With the formation of this state, Vojvodina reverts to Hungarian jurisdiction.

1873: October 11   Birth of painter Nadezda Petrovic

Petrovic, a painter, art critic, and promoter, is born in Cacak to a drawing teacher who provides her earliest instruction. She later studies in Munich before returning to Serbia. She consciously integrates the techniques of European art—both contemporary and traditional—into a uniquely Serbian style, characterized by vibrant colors and bold brushstrokes. Petrovic dies in 1915.

1876: July   Serbia declares war on the Ottoman Empire

In a show of support for Serbs in the neighboring Ottoman province of Bosnia and Herzegovina, Prince Milan Obrenovic declares war against the Ottoman Empire. His army is unprepared, and by October, Ottoman forces are threatening Belgrade. Russian mediation results in an armistice in November which restores the status quo.

1878   Serbia receives independence

Following the Ottoman defeat in the Russo-Turkish War (1877–78), the peace Congress of Berlin awards the Serbs with independence and some additional territory from the Ottoman Empire. However, Serbian gains come nowhere near the desire for a Greater Serbia. As part of the settlement, Austria-Hungary receives the right to occupy Bosnia and Herzegovina (although it remains nominally a part of the Ottoman Empire), a region with a substantial Serbian population. Serbia's weak military position precludes any stand against the Habsburgs.

1880s   Serbia comes under the Austrian sphere of influence

Having won independence from the Ottomans, Serbia becomes dependent on Austria as a result of military aid and a commercial treaty.

1884   Choral society is founded

The Academic Obilic Choral Society, the most important of numerous such groups in Serbia, is established.

1885   Serbia defeated by Bulgaria

In an attempt at territorial expansion, King Milan declares war against Bulgaria which is wracked by unrest. Unfortunately for Milan, his forces suffer a quick and humiliating defeat. Austro-Hungarian mediation on behalf of Serbia prevents the loss of Serbian territory.

1886   Arts and science academy is founded

The Serbian Academy of Science and Art is founded in Belgrade.

1895   First art school is founded

Serbia's first academy of drawing and painting is established by Kirilo Kutlik (1869–1900). By training students who later study abroad in such cultural centers as Paris and Munich, the school makes a major contribution toward the progress of modern artistic techniques in Serbia.

1897: August 6   Birth of painter Milo Milunovic

Milunovic, born in Montenegro, studies painting in Florence with Alberto Giacometti (1901–66) and is exposed to the works of Paul Cézanne (1839–1906) and the Cubist painters in France following World War I. Among his best-known works—most painted in the late 1920s and early 1930s—is *Still-life with Violin* (1930). Milunovic's work is known for its geometric composition and anti-illusionistic devices. In the later part of his career, he focuses on seascapes in his native Montenegro. He dies in 1967.

1899   Music school is founded

The Mokranjac State Music School is established.

1900–1920   Belgrade becomes artistic center

The arts center of Serbia shifts from Vojvodina to Belgrade, where artists including Kosta Milicevic (1877–1920) and Milan Milovanovic (1876–1946) introduce Impressionism to the region.

1903: June   Bloody coup ousts Obrenovic, radical change in Serbian foreign policy

In a bloody coup led by army officers, the Karadjordjevices regain the throne as Peter I becomes king of Serbia, ending the last period of rule by the Obrenovic dynasty. The plotters kill King Alexander Obrenovic, Queen Draga Masina, the war minister, and the queen's two brothers.

The return of the Karadjorjevic dynasty brings a change in Serbian foreign policy. Unlike the Obrenovic dynasty, Peter Karadjorjevic is anti-Habsburg and pro-French and pro-Russian. In addition, the conclusion of economic agreements with Bulgaria antagonizes the Austro-Hungarians who view the shift to new political and economic partners with alarm and retaliate by imposing a ban on the transport of Serbian livestock. This dispute, known as the Pig War, proves disastrous for Austria-Hungary since it forces the Serbs closer to their French and Russian allies (both of which are enemies of the Dual Monarchy).

1907: July 27   Birth of painter Peter Lubarda

Lubarda, a Montenegrin, studies painting in Belgrade and Paris. He is known particularly for his portrayal of the people and landscape of his native land.

1908   Austrian annexation of Bosnia and Herzegovina

Austria-Hungary annexes Bosnia and Herzegovina, which it has administered since 1878, making its de facto rule official. The annexation frustrates Serbian ambitions in the region.

1910   Early film is produced

The movie *Karadjordje* describes the life of Serbia's national hero. (See 1762.)

1912–13   First Balkan War

Serbia, Montenegro, Greece, and Bulgaria form the Balkan League and join together to expel the Ottoman Turks from the Balkans. Serbia gains control of the southern Adriatic coastline but is forced to cede it when Austria and Italy back the creation of an independent Albanian state. The Serbs also liberate Kosovo from Turkish rule, although by now the area is inhabited primarily by Albanians.

1913   Second Balkan War

After Bulgaria attacks its former South Slavic allies, Serbia and Greece, to win control of Macedonia, they respond and the Bulgarians are defeated. Serbia retains control of northern and central Macedonia.

World War I

1914: June 28   Serbian assassinates Austrian archduke, triggering World War

Austro-Hungarian Archduke Franz Ferdinand and his wife, the archduchess, are assassinated by a Serbian student, Gavrilo Princip, during a state visit to Sarajevo to view mili-

tary maneuvers. Princip is associated with the nationalist Serb secret society, the Black Hand. Secret societies formed by Serbian radicals have already made several assassination attempts on Austrian officials.

1914: July 23   Ultimatum by Austria-Hungary

Austria-Hungary threatens to attack Serbia unless it can participate in bringing the archduke's killer to justice, and Serbia agrees to ban secret societies.

1914: July 28   War is declared

In spite of Serbian attempts to meet the conditions laid out by the Austrians, Austria-Hungary declares war on Serbia and begins bombing Belgrade. Together, Austria and its allies—Germany, the Ottoman Empire, and after, 1915, Bulgaria—are known as the Central Powers. Serbia allies itself with France, Britain, and Russia, collectively known as the *Triple Entente*. The war pits ethnic Serbs in territories ruled by Austria-Hungary—including Bosnia—against Serbs in Serbia and Montenegro. In spite of ethnic and religious division, the idea of an independent state uniting the South Slavs gains force during this period.

1914: August 11   Austria-Hungary invades Serbia

Austria-Hungary launches a ground attack on Serbia, which refuses to surrender despite great losses and a devastating typhus epidemic. Following the Bulgarian invasion in 1915, the Serbs later retreat across the mountains of Albania and Montenegro to the Adriatic Sea and the Greek island of Corfu.

1915: April 3   Death of painter Nadezda Petrovic

Painter, exhibitor, and art critic Nadezda Petrovic, a pioneer of modern Serbian art, dies in Valjevo. (See 1873.)

1916   Serbs join the Allies at Salonika

Following the dangerous trek across Albania and Macedonia, Serbs join the Allied forces at the Salonika front in Greek Macedonia.

1917: July   Corfu Declaration

Serbian Prime Minister Nikola Pasic and the Croat leader of the Yugoslav Committee (a group dedicated to the formation of a greater South Slav state), Ante Trumbic, reach an agreement for the postwar creation of a Kingdom of Serbs, Croats, and Slovenes united under the Karadjorjevic dynasty.

1918: November   End of World War I

By the time World War I ends in victory for the Triple Entente, the South Slavic lands have been liberated from Austro-Hungarian control.

The Kingdom of Serbs, Croats, and Slovenes

1918: December 1   South Slavs win independence

Representatives of the South Slavic peoples meet with Regent Alexander Karadjordjevic, prince of Serbia, and proclaim an independent Kingdom of Serbs, Croats and Slovenes. Alexander becomes the country's ruler, ascending the throne as Alexander I. (Eleven years later he renames the country Yugoslavia.)

The new nation is troubled politically and economically from the outset. It inherits a legacy of ethnic rivalry, and the balance of power within the new government favors the Serbs. The capital, Belgrade, is located in Serbia, and Croatia is under-represented in the new country's balance of power. In Kosovo, the Serbs actively pursue the assimilation of the Muslim Albanian majority, abolishing Albanian schools, and turning Albanian land over to Serbs. Currency reforms favor the Serbs at the expense of the Croats. Serbs and Croats also clash over the issue of government centralization versus federalism. As the largest group in the country, the Serbs favor central administration from Belgrade while the Croats favor decentralization. Economically, the kingdom is faced with wartime damage, heavy debt, labor shortages, and a pressing need for land reform created by centuries of feudalism.

1920   Opera company is founded

An opera company is established at the National Theater through the efforts of composer S. Binicki.

1921–22   Little Entente is formed

At the instigation of France, the South Slavic kingdom forms a military alliance with Romania, Czechoslovakia, and France to prevent attempts at treaty revision by the defeated Austrians, Bulgarians, Germans, and Hungarians. The alliance is known as the Little Entente.

1921: June 28   Constitution is adopted

A constituent assembly approves a constitution for the country. The Serb position prevails, and a strong centralized government is created over the objections of the Croatian Peasant Party, which withdraws from the assembly.

1923   Belgrade Philharmonic Society is founded

The Belgrade Philharmonic Society is established.

1925   Croat Peasant Party is outlawed

The government outlaws the Croat Peasant Party, which opposes provisions of the constitution enacted in 1921. The party's leader, Stjepan Radic, is imprisoned and forced to agree to a coalition with the ruling Serbian party. However, he later resumes his dissident stance.

1927   Birth of animator Dusan Vukotic

One of the major contributors to the internationally recognized Zagreb Film School of animators, Montenegrin-born Vukotic wins numerous awards, including an Academy Award for his cartoon *Ersatz* in 1961. His other work includes *The Disobedient Robot* (1956), *Concerto for Sub-Machine Gun* (1959), *A Stain on the Conscience* (1969), and the full-length animated feature film *The Seventh Continent* (1966).

1928   Opposition leader Radic is assassinated

Croat Peasant Party founder and leader, Stjepan Radic, and several other Croats are assassinated by a Serb representative at a session of the Yugoslavian parliament.

1929: January 6   Royal dictatorship is established

Using the assassination of Croat leader Stjepan Radic as justification for tighter government control, King Alexander imposes a royal dictatorship, abrogating the constitution, abolishing the country's parliamentary government, and outlawing political parties. He also renames the country Yugoslavia.

Ante Pavelic forms the Croatian Liberation Organization, or *Ustashe,* to oppose the dictatorship of King Alexander. He later flees to Italy, where his organization becomes allied with fascist leader Benito Mussolini (1883–1945).

1930s   Composers' circle is formed

A group of Serbian composers who studied in Prague is active in Belgrade and adds sophistication to Serbian musical life. The composers include Milan Ristic, Stanojlo Rajicic, and Ljubica Maric.

1931   Constitution is adopted

The autocratic rule of King Alexander is modified by adoption of a constitution, and political parties are legalized. However, a number of other freedoms are still restricted.

1931   Global economic slump hits Yugoslavia

Yugoslavia feels the effects of the worldwide economic depression, which leads to bankruptcies and unemployment. The crisis is worsened by weather conditions that produce famine in rural areas.

1932: October 13   Birth of film director Dusan Makavejev

Makaveyev, a leading figure of Yugoslavian cinema in the 1960s, is born in Belgrade and attends Belgrade University and the Belgrade Academy of Theater, Radio, Film and Television. His first film, *Man Is Not a Bird* (1966), features the

Over five hundred thousand mourners line the streets as the funeral cortege of King Alexander passes. Kneeling peasants wear arm bands to signify their grief. (EPD Photos/CSU Archives)

humor and cinematic experimentation that are to characterize his films during this period, including *Love Affair: Or The Case of the Missing Telephone Operator* (1967). Makaveyev incurs the displeasure of the communist authorities with his 1971 film *WR: Mysteries of the Organism* about the pioneering psychoanalyst and advocate of sexual freedom, Wilhelm Reich. The film is banned and Makaveyev is unable to find work, so he directs films abroad until the late 1980s, when he returns to Yugoslavia. His more recent films include *Manifesto* (1988) and *Hole in the Soul* (1994).

1934   First graphic arts exhibition

The first graphic arts exhibition in Serbia features the work of sixteen artists. The Serbs are pioneers in the development of photography since the development of the daguerreotype in the nineteenth century. Kodak cameras are issued to soldiers in the Balkan Wars (1912–13) to document events of the period.

1934   Balkan Entente is formed

Yugoslavia joins Turkey, Greece, and Romania in a defense alliance known as the Balkan Entente. This treaty is aimed at stifling Bulgarian attempts at revising the World War I peace settlement.

1934: October   King Alexander is assassinated

King Alexander is murdered in Marseilles, France, by a Bulgarian assassin working for the *Ustashe,* the Croatian pro-fascist liberation organization. His death brings on fears that Yugoslavia will collapse. Alexander's son, Peter II, becomes the country's regent. Three officials are appointed to rule for him while he is still a minor.

1939   Autonomous region is ceded to the Croats

In an agreement (Sporazum) negotiated between Serb and Croat leaders to improve relations between the Yugoslav government and the Croats, an autonomous Croatian territory is formed.

World War II

1941: March 25   Pact is signed with Germany

Under military pressure and surrounded by pro-Nazi countries, Yugoslavia's government agrees to join the Tripartite Pact (the Axis alliance concluded among Germany, Italy, and Japan) in return for German guarantees of nonaggression.

1941: March 27   Yugoslavian government is overthrown

Because of its cooperation with Nazi Germany, the Yugoslavian government is overthrown in a coup led by Yugoslav Air

Force officers and sixteen-year-old Peter II, the regent and son of the slain king Alexander, is declared king. In spontaneous demonstrations, the populace expresses its hostility toward the Nazis and their allies.

1941: April 6   Germany bombs and invades Yugoslavia

The Yugoslavian capital of Belgrade is bombed by the German *Luftwaffe* (air force), and ground forces invade the country. The government goes into exile, and the military surrenders unconditionally. Yugoslavia is divided among the Axis powers, except for Croatia, which is placed under the rule of the Independent State of Croatia (NDH), a nominally independent state controlled by the Nazis and led by Ante Pavelic, head of a Croatian fascist group, the Ustashe. All of Serbia except for Vojvodina (occupied by German ally Hungary) is placed under German control and administered by General Milan Nedic.

Serbs in Ustashe-controlled Croatia are singled out for elimination through deportation or extermination, together with Jews and gypsies. The atrocities committed by the Ustashe during the war add yet more fuel to the hostility between Serbs and Croats and will be repeatedly invoked by the Serbs during the Serb-Croat war and the Bosnian civil war of the 1990s.

1941   Partisans and Chetnik forces organized to oppose the Ustashe

Partisans (anti-Nazi freedom fighters) organized by long-time communist leader Josip Broz (1892–1980), popularly known as Tito, carry out rebellions in the countryside and remain active throughout the war in spite of German retaliation. Although most Partisans are Serbs, all ethnic groups are represented (Tito himself is Croatian-born). The Ustashe also faces armed opposition by Serbian Chetnik forces, led by General Draza Mihajlovic. Unlike the pro-Communist Partisans, the Chetniks recognize the authority of the prewar Serbian government-in-exile. Although Mihajlovic and Tito meet early in the war in an attempt to forge cooperation between their respective forces, they disagree over Tito's Communist orientation, and a three-way struggle eventually develops between the Ustashe, the Partisans, and the Chetniks. After recognizing the Chetniks as the main resistance army early in the war, the Allies in 1943 throw their support behind the Partisans, which have become the largest and most active group.

1941: August 29   Birth of Slobodan Milosevic

Future Serbian president Slobodan Milosevic is born in Pozarevac. Although a member of the Communist Party throughout his adult life, he first enters politics in his forties. Prior to that, he heads a state-owned gas company and serves as bank president. In 1984, he is named head of Belgrade's Communist Party and soon rises to the top party position in Serbia. He develops a populist style meant to sway the masses while bypassing the party hierarchy and plays on nationalist sentiments by demanding the return of the autonomous regions of Kosovo and Vojvodina to Serb control. By the late 1980s, Kosovo's population is approximately ninety percent Albanian, while the Vojvodina's population is a mixture of Hungarians, Serbs, and Slovaks.

In 1989, Serbia's national assembly elects him president of that republic. When Croatia, Slovenia, Macedonia, and Bosnia secede from Yugoslavia in 1991–92, Molosevic becomes the president of a reduced nation consisting of Serbia and Montenegro. Throughout the decade, Milosevic's tenure as president is marked by ethnic division and political repression. In the late 1990s, Milosevic directs a concerted effort to return formerly autonomous province of Kosovo to Serbian rule, instituting a brutal policy of "ethnic cleansing" by evacuating and terrorizing ethnic Albanians, who form the ethnic majority in Kosovo. In early 1999, international censure of his actions leads to NATO bombings of Serbia.

1943: September   Italy surrenders

The surrender of Italy bolsters the Partisan effort by providing access to the coast as well as a supply of arms and a supply route. The Partisans win control of much of the country.

1944: October 20   Red army liberates Belgrade

The Soviet army, aided by Partisan forces, marches into Belgrade and liberates Yugoslavia. Some 1.7 million Yugoslavians have died since the war began—more than half at the hands of other Yugoslavs. Cities are left in ruins, and the countryside is also devastated.

Yugoslavia under Tito

1945: March 7   Provisional government is installed

A communist-dominated provisional government, led by Tito, takes office.

1945: November 29   Yugoslavia's monarchy is dissolved

The new Yugoslav parliament meets for the first time. The monarchy is officially dissolved, and the country is named the Federal People's Republic of Yugoslavia (later renamed the Socialist Federal Republic of Yugoslavia). Serbia becomes one of six republics in a Soviet-style federation with a strong centralized government. In an effort to dilute Serbia's power within the federation, Macedonia and Montenegro are recognized as republics, Kosovo is established as an autonomous Albanian province, and Vojvodina becomes an ethnically mixed autonomous province. Tito also keeps ethnic rivalries in check by setting up a centralized, single-party Communist government and suppressing political opposition and organized religion.

Josip Broz Tito and his wife. (EPD Photos/CSU Archives)

other Eastern European satellites of the U.S.S.R., Tito does not owe his influence or position to the Soviets and does not allow them to dictate his policies. He becomes the only Eastern European communist leader to break with Stalin and remain in power. The break between the two countries is solidified when the *Cominform* (Communist Information Bureau) denounces the Yugoslav Communist Party. Trade with the Soviet Union and other Communist bloc nations is reduced, and Tito is forced to turn to the West for new trade partners.

1950s Self-management economic system develops

Yugoslavia develops its communist economy through a system in which enterprises are locally managed by workers' councils, and economic goals are first formulated at the local level and coordinated centrally. Agricultural collectivization, one of the hallmarks of Soviet-style communism, is slowed and then abandoned altogether.

1950s Postwar literature divided between realists and modernists

Two major trends distinguish postwar Yugoslavian literature. The social realists, including Montenegrin Radovan Zogovic and Dobrica Cosic, focus on social issues, while modernists such as Radomir Konstantinovic pioneer new literary techniques. Works from this period include the novels *Breakthrough* (1952) by Branko Copic and *Kaleja Mountain* (1957) by Mihajlo Lalic.

1951 Yugoslavia forms defense pact with U.S.

Yugoslavia and the United States conclude a defense agreement.

1953 Constitution incorporates workers' councils

The constitution is modified to authorize the creation of workers' councils to run state-owned enterprises.

1955 *Rapprochement* with the Soviets under Khrushchev

Following the death of Josef Stalin, relations between Yugoslavia and the U.S.S.R. under Nikita Khrushchev thaw. The two leaders sign an agreement pledging mutual respect and cooperation.

1955 Tito visits India and Egypt

Tito travels to India and Egypt to extend cooperation among the nonaligned nations.

1958 Contemporary art museum opens

The Museum of Contemporary Art is established to exhibit the work of Balkan artists in all visual media, including painting, sculpture, and graphic arts.

1945–66 Serbs dominate Yugoslavia

In spite of the measures taken by Tito to neutralize the power of Yugoslavia's ethnic constituencies, Serbs come to dominate Yugoslavian politics in the postwar years.

1946: March Wartime Serb leader is captured

Serbian Chetnik leader General Draza Mihajlovic is captured in Bosnia and tried for treason and collaboration.

1946: July 17 Mihajlovic is executed

General Draza Mihajlovic is executed.

1947–51 Five-year economic plan is adopted

A five-year plan modeled on those of the Soviet Union is inaugurated, emphasizing industrial development.

1948 Yugoslav-Soviet split

Following World War II, Yugoslavia allies itself closely with the Soviet Union. By 1948, however, growing tensions, begun with disagreement over the management of joint enterprise, create a rift between the two countries. Unlike the heads of

1961   Yugoslavian animated film wins Academy Award

Ersatz, an animated film directed by Dusan Vukotic (see 1927), wins an Academy Award.

1961: September   Conference of nonaligned nations

A conference of nonaligned nations held in Belgrade confirms Yugoslavia's international leadership position in the group of nations that have proclaimed their neutrality (the Non-Aligned Movement) in the Cold War, including India and Egypt. Their mutual goals include disarmament, peaceful coexistence among nations, and the elimination of rival political blocs.

1963   Political prisoners are released

As part of a wider liberalization trend, the Yugoslavian government releases 2,500 political prisoners.

1965   Market socialism is introduced

Yugoslavian industry grows rapidly in the 1950s, but imports still exceed exports by a wide margin, and some industrial inefficiency remains. In response to an economic crisis of the early 1960s, the government modifies its economic policy, eliminating price controls and export subsidies while cutting import duties. The move toward decentralized economic self-management is stepped up as well.

1966   Tito purges top Serbian leader

Because the Serbs have once again risen to the position of political dominance they enjoy following World War I, Tito dismisses Serbian leader Aleksandar Rankovic and disbands Rankovic's powerful secret police. During the period that follows, the balance of power among Yugoslavia's ethnic groups improve as do relations between the groups.

1967: February 11   Death of painter Milo Milunovic

Milunovic, a leading twentieth-century Yugoslavian painter, dies in Belgrade. (See 1897.)

1967–68   Yugoslavian republics gain more power

Constitutional reforms expand the power of the individual Yugoslavian republics relative to the central government.

1968   Tito condemns Soviet invasion of Czechoslovakia

Yugoslav leader Tito condemns the U.S.S.R.'s forcible halt to Czech liberalization under Alexander Dubcek, straining Yugoslav relations with the Soviets. Tito prepares the Yugoslavian military to resist a possible Soviet invasion.

1971   Tito sets up collective presidency

Tito creates the framework for a collective presidency, intended to ease tensions when he retires. Intended to provide comparable representation to Serbs, Croats, and Muslims at the top level of government, it is to have two members from each of the country's three main ethnic constituencies and one additional member. Under this arrangement, however, Tito will remain president for life.

1971   Croatian nationalism antagonizes Serbia

Those in the vanguard of the Croatian leadership begin demanding even greater economic and political autonomy, including the establishment of a Croatian army, the restoration of the traditional national assembly (the Sabor), and even complete independence. The Serbs begin urging Tito to restrain the growing nationalist fervor in Croatia.

1971: November   University students strike

The Croatian government does nothing to restrain mass student demonstrations.

1971: December   Tito imposes crackdown on liberal Croats

Tito purges Croatia's liberal Communist leadership. Thousands of officials are arrested and either jailed or expelled from the party. Included among them is future Croatian president Franjo Tudjman (b.1922). The purges continue for over a year and spread to the other Yugoslavian republics also. They expand into a wider crackdown that includes press censorship, dissident arrests, and pressure on university professors to conform with the party line.

1972–73   Liberal Serbian leaders are purged

On the heels of his purge of liberal Croatian leaders, Tito carries out a similar campaign in Serbia.

1974   New constitution is adopted

A new constitution gives the individual republics and autonomous provinces a greater role in governing themselves. The role of the federal government is limited to foreign policy, defense matters, and certain economic powers. The constitution also formalizes and simplifies the structure of the collective presidency that Tito has created to assure the continuity of the Yugoslav federation after his death.

Tito's Yugoslavia Collapses

1980s   Politics and the economy pull Yugoslavia apart

Following the death of Marshal Tito, the stability achieved over decades is threatened by age-old ethnic rivalries and affected by widespread economic problems in Russia and Eastern Europe, as well as Yugoslavia's inability to service its excessive foreign debt. The country is beset by inflation, labor strikes, food shortages, and financial scandals, as tensions grow between the individual republics. Serbian nationalism, which is kept in check during the Tito years, gains strength in spite of government crackdowns. Another source of tension is the economic disparity between the standard of living in the wealthier republics of Croatia and Slovenia and that of the republics to the south and east.

1980: May 4   Death of Marshal Tito

Josip Broz Tito, who has greater success in unifying the South Slavs than any other leader, dies. He is widely mourned at home, and forty-nine foreign dignitaries attend his funeral.

1984   Winter Olympics are held in Sarajevo

The Winter Olympic Games are held in Sarajevo, the capital of Bosnia and Herzegovina.

1987   Proposed constitutional changes would limit republics' power

Proposed changes to the 1974 constitution anger many in the republics, particularly Slovenia and Croatia, by increasing the power of the central government. Among the changes are a unified legal system throughout the country, one unified market for the country, central control of communications and transportation, and increased Serb control over the autonomous provinces of Vojvodina and Kosovo.

1987   Slobodan Milosevic becomes top Serb Communist leader

Slobodan Milosevic is appointed head of the Serbian League of Communists. In his quest for more power, he becomes a strong advocate of Serb nationalism and a strong central government. While retaining the governing structure of the Communist Party, he institutes a nationalist ideology.

1989   Milosevic seizes control of Kosovo and Vojvodina

Slobodan Milosevic institutes Serbian control over the autonomous provinces of Kosovo and Vojvodina. In Kosovo, political purges allow Milosevic to place his own followers in key government posts, and persecution of ethnic Albanians intensifies. In response, a nonviolent Albanian separatist movement headed by Ibrahim Rugova sets up separate Albanian institutions outside those run by the Serb government.

1990: January   Communist congress rejects liberal reforms

Liberal reforms proposed by Slovenia are rejected at a congress of Yugoslavia's League of Communists. Following this rebuff, the Slovenians distance themselves increasingly from Yugoslavia, preparing for secession.

1990: March–April   Slovenia and Croatia hold free elections

The nationalist Croat Democratic Union (HDZ) edges out the Communist Party in free elections, forty percent to thirty percent, and its leader, Franjo Tudjman, becomes president of Croatia. Slovenia holds its first free elections since World War II and elects a non-Communist government headed by Dr. Lojze Peterle as prime minister.

1990: September   New Serbian constitution limits Kosovo's autonomy

Under the provisions of a newly adopted constitution, the administration of Slobodan Milosevic ends the autonomous status of Kosovo and Vojvodina, abolishing their provincial governments and limiting the rights of ethnic minorities in both regions. Milosevic cracks down on Kosovo's Albanian majority, which accounts for some ninety percent of the province's population. The Milosevic government removes thousands of Albanians from Kosovo's government, firing them from administrative jobs and shutting down the Kosovar assembly. Serbia takes over Kosovo's security forces and educational system, and Albanian activists are persecuted and imprisoned.

1991–95   Secession and warfare

After the failure of last-ditch efforts to restructure the federal government, four of the six Yugoslavian republics—Croatia, Slovenia, Macedonia, and Bosnia and Herzegovina—declare independence and secede from the country. Once Milosevic and Serbian nationalists resign themselves to the breakup of Yugoslavia, they set out to redraw borders by incorporating all land inhabited by Serbs in a Greater Serbia. In response, Milosevic attacks Slovenia, Croatia, and Bosnia, costing thousands of lives and widespread destruction of property.

1991: June 25   Slovenia and Croatia declare independence

Slovenia and Croatia declare independence and secede from Yugoslavia. Slovenia is invaded by Yugoslav forces two days later. However, the Slovenians hold their ground, and their strong resistance results in the surrender of over 3,200 Yugoslav troops. In a memorable move, the Slovenian government calls for the parents of the captured soldiers to come and take them home. In Croatia, the Serb-populated Krajina region declares its independence from the rest of Croatia and hostilities begin, with the Serb militia supported by the Serb-dominated Yugoslav federal army of 180,000.

1991: July–December   Heavy fighting in Croatia

Fighting increases in Croatia. The cities of Vukovar and Dubrovnik are heavily damaged, and altogether an estimated 10,000 persons lose their lives. The Serbs of the Krajina gain control of about one-third of Croatia's territory.

1991: July 5   Cease-fire in Slovenia

Hostilities in Slovenia end after ten days with a cease-fire negotiated by representatives of the European Community.

1992: January   Yugoslav forces withdraw from Croatia

Under the terms of a UN-sponsored cease-fire, Yugoslavia withdraws its federal forces from Croatia, and the cease-fire mostly holds. By this time, however, the city of Vukovar has been almost totally destroyed. The UN peace plan calls for local Serb militias to disarm, refugees to be repatriated, and the return of the Krajina to Croatia.

1992: April   Hostilities break out in Bosnia

Following the declaration of independence by Bosnia and Herzegovina, Bosnian Serbs declare a separate state. With military equipment provided by the Yugoslavian army (Yugoslavia at this point consists of Serbia and Montenegro), Bosnian Serb forces start a civil war, taking control of seventy percent of Bosnia's territory. Over the next three years, the country is shattered by the conflict, as cities are bombed and cease-fires repeatedly fail. Serb atrocities against Bosnian Muslims and Croats (so-called ethnic cleansing) are reported, including rape and mass execution. Serbian forces lay siege to Sarajevo for over a year and a half, until they are turned back by the threat of NATO air strikes.

The Federal Republic of Yugoslavia

1992: April 27   New Yugoslavian republic is formed

Serbia and Montenegro form the Federal Republic of Yugoslavia.

1992: May   UN imposes sanctions

The United Nations imposes strict sanctions on Yugoslavia for its actions in Bosnia and Herzegovina.

1992: September   Yugoslavia is expelled from the UN

As world opinion condemns Yugoslavian atrocities in Bosnia, the country is expelled from the United Nations.

1992: December 22   Milosevic is elected president of Serbia

Slobodan Milosevic is elected president of Serbia, and his Socialist Party also wins a majority in parliamentary elections.

1993   Moderate Serb leaders are ousted

As Serb moderates protest Milosevic's policies, their leaders, including Milan Panic and Dobrica Cosic, are ousted, and opposition leader Vuk Draskovic is beaten.

1994   War crimes tribunal is established

The United Nations sets up a tribunal in the Hague for war crimes in the former Yugoslavia.

1994: February   NATO ultimatum

Following a heavy shelling of Sarajevo that kills sixty-six people, NATO issues an ultimatum to the Serbs, and they agree to the presence of UN peacekeeping forces, and the pressure on Sarajevo subsides. However, the Serbs now focus their offensive on the cities of Tusla and Gorazde, continuing to attack in spite of UN protests.

1994: July   Milosevic endorses new UN peace plan

With serious economic problems demanding attention in Serbia itself, Milosevic tries to get the Bosnian Serbs to end hostilities there. He endorses a new UN peace plan that proposes to divide the country and give the Serbs forty-nine percent. Milosevic pledges to end assistance to the Bosnian Serbs until they endorse the new peace plan, and he closes the border between Serbia and Bosnia. The Bosnian Serbs begin attacking UN peacekeepers.

1994: September   UN sanctions are eased

In exchange for Milosevic's withdrawal of support for the Bosnian Serbs, the United Nations reduces the scope of its sanctions against Yugoslavia.

1995: February 19   Milosevic rejects contact group peace plan

Following negotiations at Wright-Patterson Air Force Base in Dayton, Ohio, a peace plan for Bosnia by the five-nation contact group is rejected by Serb leader Milosevic.

1995: November 21   Dayton peace accord is drawn

A peace agreement is framed by the presidents of Bosnia, Serbia, and Croatia. It maintains Bosnia as a single nation but divided into two parts: one governed by the Bosnian Serbs (Republica Srpska) and another by Bosnians and Croats (the Federation of Bosnia and Herzegovina).

1995: December   Dayton accords are signed

The Dayton peace accords ending the Bosnian civil war are signed in Paris. Estimates of the number of deaths in the three-and-a-half-year war range from 25,000 to 250,000, and the number of refugees is estimated at three million.

1995: December   UN lifts trade sanctions

The United Nations lifts its trade embargo against Yugoslavia.

1996: March   Anti-Milosevic demonstrators protest in Belgrade

Some 20,000 protesters demonstrate against the policies of the Milosevic government, which has brought devastation to the Serb economy with years of warfare.

1996: August 7   Tudjman and Milosevic meet in Athens

Croatian president Franjo Tudjman and Serbian president Slobodan Milosevic meet in Athens, Greece and agree to establish diplomatic relations for the first time since the beginning of hostilities between their countries in the spring of 1991.

1996: September 9   Croatia and Serbia renew diplomatic ties

Croatia renews diplomatic ties with Serbia following an agreement between the presidents of the two countries.

1996: November 3   Milosevic-led coalition wins parliamentary elections

The leftist coalition led by Slobodan Milosevic and his wife, Mirjana Markovic, wins a majority in parliamentary elections.

1996: December   Independent radio station is closed

The government draws protests by shutting down Belgrade's independent radio station.

1997: February   Milosevic reverses decision on local elections

Reversing his previous action, Milosevic recognizes the results of local elections that are won by opposition party candidates.

1997: July   Milosevic elected president of Yugoslavia

With his term as president of Serbia nearing an end, Milosevic is elected to the presidency of Yugoslavia by the federal assembly in an eighty-eight–ten vote, assuring him of continued political control. He resigns the presidency of Serbia on the day he is sworn in, as 1,000 demonstrators gather across the street from parliament.

The Kosovo Crisis

1998: February–March   Violence in Kosovo escalates

The province of Kosovo sees its worst violence since the Serb elimination of its autonomous status nine years earlier. Fifty people are killed in one week. Unrest is fueled by the newly formed Kosovo Liberation Army (KLA), which has taken over towns, attacked police stations, and assassinated a Serb official. The KLA's guerrilla forces are not allied with the

separatist movement of Ibrahim Rugova, but he does not denounce their activities.

1998: May 15   Milosevic meets with Rugova

Yugoslavian president Slobodan Milosevic and Kosovar Albanian leader Ibrahim Rugova meet for the first time, but no significant agreement is reached between the two.

1998: May 22   Serbia steps up Kosovo operations

Serbia expands operations against Albanian separatists, in the worst fighting in the Balkans since the end of the Bosnian war in 1995. Dozens of Albanians are killed, and between five and ten thousand flee to Albania in the first week of June alone.

1998: June 15   NATO exercise conducted near Yugoslav border

In a show of force designed to push Yugoslav leader Slobodan Milosevic into halting attacks on Albanian separatist in Kosovo, NATO forces hold airborne military exercises over Albania and Macedonia. Milosevic agrees to some concessions.

1998: September 23   UN Security Council calls for cease-fire in Kosovo

The United Nations Security Council passes a resolution calling for a cease-fire in Kosovo.

1998: October   Cease-fire follows NATO approval of air strikes against Serbia

U.S. envoy Richard Holbrooke and Yugoslavian president Slobodan Milosevic agree on the terms of a cease-fire for Kosovo after NATO approves possible air strikes against Serbia. The NATO vote reflects reaction to reports of massacres in several Kosovar towns in September that raise charges that the Serbs are attacking not only separatists but the Albanian population as a whole. Nearly one-fifth of Kosovo's population is reported to have been driven from their homes, creating a growing refugee crisis, as Serb forces bomb, loot, and set fire to scores of Albanian villages.

1999: January 29–30   Contact group orders peace talks

The six-nation contact group attempting to push Yugoslavia into ending its attacks on Kosovo threatens military intervention unless representatives of Yugoslavia and the Kosovo Liberation Army hold peace talks. NATO votes to authorize air strikes unless a resolution is reached.

1999: February 6–23   Rambouillet peace talks

Peace talks on the Kosovo crisis are held in Rambouillet, France, and an agreement in principle between the two sides is reached.

1999: March 15–19 Albanians sign peace agreement; Yugoslavia refuses

The peace agreement drafted at Rambouillet is signed by ethnic Albanians. Under the agreement, an international peace-keeping force will stay in Kosovo for a three year period, after which a referendum will determine the final status of Kosovo. Objecting to the use of foreign peacekeepers and fearing that this means the loss of Kosovo to Serbia, Milosevic rejects the agreement.

1999: March 24 NATO launches air strikes against Yugoslavia

NATO forces strike Yugoslavia in air attacks intended to disrupt Yugoslav operations in Kosovo.

1999: April Air strikes heighten refugee crisis

According to the UN High Commissioner for Refugees, the Serbian attacks in the wake of NATO air strikes have displaced over 400,000 ethnic Albanians, some of whom have fled Kosovo while others remain in the province. Many villages have been torched by the Yugoslavians after being emptied of inhabitants.

1999: May International implications widen in Kosovo crisis

The diplomatic implications of the NATO bombing of Yugoslavia grow as U.S. NATO forces mistakenly bomb the Chinese embassy in Belgrade, killing three persons and injuring twenty others. The bombing also causes a rift between the United States and Russia, which opposes the military action by NATO.

1999: May 27 Milosevic is indicted for war crimes

The international tribunal in the Hague indicts Yugoslavian President Slobodan Milosevic for war crimes in Kosovo, claiming he is responsible for the death and destruction wrought by Serbian troops even if he is not directly involved. The indictment—the first time a sitting head of state has been indicted for war crimes—could pose problems for Western nations attempting to negotiate an end to the Kosovo crisis. However, it is deemed unlikely that Milosevic will actually be arrested. Two top Bosnian Serb leaders indicted for similar crimes in 1995—Radovan Karadzic and Ratko Mladic—are still at large even though their whereabouts are widely known to their own government and to Westerners.

Bibliography

Bennett, Christopher. *Yugoslavia's Bloody Collapse: Causes, Course and Consequences.* London: Hurst & Company, 1995.

Cohen, Lenard J. *Broken Bonds: Yugoslavia's Disintegration and Balkan Politics in Transition.* Boulder, Colo.: Westview, 1995.

Denitch, Bogdan Denis. *Ethnic Nationalism: The Tragic Death of Yugoslavia.* Minneapolis: University of Minnesota Press, 1996.

Dyker, David A., and Ivan Vejvoda, ed. *Yugoslavia and After: A Study in Fragmentation, Despair and Rebirth.* New York: Longman, 1996.

Jelavich, Barbara. *History of the Balkans.* 2 vols. Cambridge: Cambridge University Press, 1983.

Judah, Tim. *The Serbs: History, Myth, and the Destruction of Yugoslavia.* New Haven, Conn.: Yale University Press, 1997.

Lampe, John R. *Yugoslavia as History: Twice There Was a Country.* New York: Cambridge University Press, 1996.

Pavkovic, Aleksandr. *The Fragmentation of Yugoslavia: Nationalism in a Multinational State.* New York: St. Martin's Press, 1997.

Ramet, Sbrina P. *Balkan Babel: The Disintegration of Yugoslavia from the Death of Tito to Ethnic War.* Boulder, Colo.: Westview, 1996.

West, Richard. *Tito and the Rise and Fall of Yugoslavia.* New York: Carroll & Graf, 1995.

Woodward, Susan L. *Balkan Tragedy: Chaos and Dissolution After the Cold War.* Washington, D.C.: Brookings Institution, 1995.

Glossary

abdicate: To formally give up a claim to a throne; to give up the right to be king or queen.

aboriginal: The first known inhabitants of a country. A species of animals or plants which originated within a given area.

allies: Groups or persons who are united in a common purpose. Typically used to describe nations that have joined together to fight a common enemy in war.

In World War I, the term Allies described the nations that fought against Germany and its allies. In World War II, Allies described the United Kingdom, United States, the USSR and their allies, who fought against the Axis Powers of Germany, Italy, and Japan.

Altaic language family: A family of languages spoken in portions of northern and eastern Europe, and nearly the whole of northern and central Asia, together with some other regions. The family is divided into five branches: the Ugrian or Finno-Hungarian, Smoyed, Turkish, Mongolian, and Tunguse.

amendment: A change or addition to a document.

Amerindian: A contraction of the two words, American Indian. It describes native peoples of North, South, or Central America.

amnesty: An act of forgiveness or pardon, usually taken by a government toward persons for crimes they may have committed.

animal husbandry: The branch of agriculture that involves raising animals.

Anglican: Pertaining to or connected with the Church of England.

animism: The belief that natural objects and phenomena have souls or innate spiritual powers.

annex: To incorporate land from one country into another country.

anti-Semitism: Agitation, persecution, or discrimination (physical, emotional, economic, political, or otherwise) directed against the Jews.

apartheid: The past governmental policy in the Republic of South Africa of separating the races in society.

appeasement: To bring to a state of peace.

arable land: Land that can be cultivated by plowing and used for growing crops.

archipelago: Any body of water abounding with islands, or the islands themselves collectively.

archives: A place where records or a collection of important documents are kept.

arctic climate: Cold, frigid weather similar to that experienced at or near the north pole.

aristocracy: A small minority that controls the government of a nation, typically on the basis of inherited wealth.

armistice: An agreement or truce which ends military conflict in anticipation of a peace treaty.

ASEAN *see* Association of Southeast Asian Nations

Association of Southeast Asian Nations: ASEAN was established in 1967 to promote political, economic, and social cooperation among its six member countries: Indonesia, Malaysia, the Philippines, Singapore, Thailand, and Brunei. ASEAN headquarters are in Jakarta, Indonesia. In January 1992, ASEAN agreed to create the ASEAN Free Trade Area (AFTA).

asylum: To give protection, security, or shelter to someone who is threatened by political or religious persecution.

atoll: A coral island, consisting of a strip or ring of coral surrounding a central lagoon.

atomic weapons: Weapons whose extremely violent explosive power comes from the splitting of the nuclei of atoms (usually uranium or plutonium) by neutrons in a rapid chain reaction. These weapons may be referred to as atom bombs, hydrogen bombs, or H-bombs.

austerity measures: Steps taken by a government to conserve money or resources during an economically difficult time, such as cutting back on federally funded programs.

Australoid: Pertains to the type of aborigines, or earliest inhabitants, of Australia.

Austronesian language: A family of languages which includes practically all the languages of the Pacific Islands—Indonesian, Melanesian, Polynesian, and Micronesian sub-families. Does not include Australian or Papuan languages.

authoritarianism: A form of government in which a person or group attempts to rule with absolute authority without the representation of the citizens.

autonomous state: A country which is completely self-governing, as opposed to being a dependency or part of another country.

autonomy: The state of existing as a self-governing entity. For instance, when a country gains its independence from another country, it gains autonomy.

Axis Powers: The countries aligned against the Allied Nations in World War II, originally applied to Nazi Germany and Fascist Italy (Rome-Berlin Axis), and later extended to include Japan.

Baha'i: The follower of a religious sect founded by Mirza Husayn Ali in Iran in 1863.

Baltic states: The three formerly communist countries of Estonia, Latvia, and Lithuania that border on the Baltic Sea.

Bantu language group: A name applied to the languages spoken in central and south Africa.

Baptist: A member of a Protestant denomination that practices adult baptism by complete immersion in water.

barren land: Unproductive land, partly or entirely treeless.

barter: Trade practice where merchandise is exchanged directly for other merchandise or services without use of money.

bicameral legislature: A legislative body consisting of two chambers, such as the U.S. House of Representatives and the U.S. Senate.

bill of rights: A written statement containing the list of privileges and powers to be granted to a body of people, usually introduced when a government or other organization is forming.

black market: A system of trade where goods are sold illegally, often for excessively inflated prices. This type of trade usually develops to avoid paying taxes or tariffs levied by the government, or to get around import or export restrictions on products.

bloodless coup: The sudden takeover of a country's government by hostile means but without killing anyone in the process.

boat people: Used to describe individuals (refugees) who attempt to flee their country by boat.

Bolshevik Revolution: A revolution in 1917 in Russia when a wing of the Russian Social Democratic party seized power. The Bolsheviks advocated the violent overthrow of capitalism.

bonded labor: Workers bound to service without pay; slaves.

border dispute: A disagreement between two countries as to the exact location or length of the dividing line between them.

Brahman: A member (by heredity) of the highest caste among the Hindus, usually assigned to the priesthood.

Buddhism: A religious system common in India and eastern Asia. Founded by and based upon the teachings of Siddhartha Gautama, Buddhism asserts that suffering is an inescapable part of life. Deliverance can only be achieved through the practice of charity, temperance, justice, honesty, and truth.

buffer state: A small country that lies between two larger, possibly hostile countries, considered to be a neutralizing force between them.

bureaucracy: A system of government that is characterized by division into bureaus of administration with their own divisional heads. Also refers to the inflexible procedures of such a system that often result in delay.

Byzantine Empire: An empire centered in the city of Constantinople, now Istanbul in present-day Turkey.

CACM *see* Central American Common Market.

canton: A territory or small division or state within a country.

capital punishment: The ultimate act of punishment for a crime, the death penalty.

capitalism: An economic system in which goods and services and the means to produce and sell them are privately owned, and prices and wages are determined by market forces.

Caribbean Community and Common Market (CARICOM): Founded in 1973 and with its headquarters in Georgetown, Guyana, CARICOM seeks the establishment of a common trade policy and increased cooperation in the Caribbean region. Includes 13 English-speaking Caribbean nations: Antigua and Barbuda, the Bahamas, Barbados, Belize, Dominica, Grenada, Guyana, Jamaica, Montserrat, Saint Kitts-Nevis, Saint Lucia, St. Vincent/Grenadines, and Trinidad and Tobago.

CARICOM *see* Caribbean Community and Common Market.

cartel: An organization of independent producers formed to regulate the production, pricing, or marketing practices of its members in order to limit competition and maximize their market power.

cash crop: A crop that is grown to be sold rather than kept for private use.

caste system: One of the artificial divisions or social classes into which the Hindus are rigidly separated according to the religious law of Brahmanism. Membership in a caste is hereditary, and the privileges and disabilities of each caste are transmitted by inheritance.

Caucasian or Caucasoid: The white race of human beings, as determined by genealogy and physical features.

ceasefire: An official declaration of the end to the use of military force or active hostilities, even if only temporary.

censorship: The practice of withholding certain items of news that may cast a country in an unfavorable light or give away secrets to the enemy.

census: An official counting of the inhabitants of a state or country with details of sex and age, family, occupation, possessions, etc.

Central American Common Market (CACM): Established in 1962, a trade alliance of five Central American nations. Participating are Costa Rica, El Salvador, Guatemala, Honduras, and Nicaragua.

Central Powers: In World War I, Germany and Austria-Hungary, and their allies, Turkey and Bulgaria.

centrist position: Refers to opinions held by members of a moderate political group; that is, views that are somewhere in the middle of popular thought between conservative and liberal.

cession: Withdrawal from or yielding to physical force.

chancellor: A high-ranking government official. In some countries it is the prime minister.

Christianity: The religion founded by Jesus Christ, based on the Bible as holy scripture.

Church of England: The national and established church in England. The Church of England claims continuity with the branch of the Catholic Church that existed in England before the Reformation. Under Henry VIII, the spiritual supremacy and jurisdiction of the Pope were abolished, and the sovereign (king or queen) was declared head of the church.

circuit court: A court that convenes in two or more locations within its appointed district.

CIS *see* Commonwealth of Independent States

city-state: An independent state consisting of a city and its surrounding territory.

civil court: A court whose proceedings include determinations of rights of individual citizens, in contrast to criminal proceedings regarding individuals or the public.

civil jurisdiction: The authority to enforce the laws in civil matters brought before the court.

civil law: The law developed by a nation or state for the conduct of daily life of its own people.

civil rights: The privileges of all individuals to be treated as equals under the laws of their country; specifically, the rights given by certain amendments to the U.S. Constitution.

civil unrest: The feeling of uneasiness due to an unstable political climate, or actions taken as a result of it.

civil war: A war between groups of citizens of the same country who have different opinions or agendas. The Civil War of the United States was the conflict between the states of the North and South from 1861 to 1865.

Club du Sahel: The Club du Sahel is an informal coalition which seeks to reverse the effects of drought and the desertification in the eight Sahelian zone countries: Burkina Faso, Chad, Gambia, Mali, Mauritania, Niger, Senegal, and the Cape Verde Islands. Headquarters are in Ouagadougou, Burkina Faso.

CMEA *see* Council for Mutual Economic Assistance.

coalition government: A government combining differing factions within a country, usually temporary.

Cold War: Refers to conflict over ideological differences that is carried on by words and diplomatic actions, not by military action. The term is usually used to refer to the tension that existed between the United States and the USSR from the 1950s until the breakup of the USSR in 1991.

collective bargaining: The negotiations between workers who are members of a union and their employer for the purpose of deciding work rules and policies regarding wages, hours, etc.

collective farm: A large farm formed from many small farms and supervised by the government; usually found in communist countries.

collective farming: The system of farming on a collective where all workers share in the income of the farm.

colonial period: The period of time when a country forms colonies in and extends control over a foreign area.

colonist: Any member of a colony or one who helps settle a new colony.

colony: A group of people who settle in a new area far from their original country, but still under the jurisdiction of that country. Also refers to the newly settled area itself.

COMECON *see* Council for Mutual Economic Assistance.

commerce: The trading of goods (buying and selling), especially on a large scale, between cities, states, and countries.

commission: A group of people designated to collectively do a job, including a government agency with certain law-making powers. Also, the power given to an individual or group to perform certain duties.

common law: A legal system based on custom and decisions and opinions of the law courts. The basic system of law of England and the United States.

common market: An economic union among countries that is formed to remove trade barriers (tariffs) among those countries, increasing economic cooperation. The European Community is a notable example of a common market.

commonwealth: A commonwealth is a free association of sovereign independent states that has no charter, treaty, or constitution. The association promotes cooperation, consultation, and mutual assistance among members.

Commonwealth of Independent States: The CIS was established in December 1991 as an association of 11 republics of the former Soviet Union. The members include: Russia, Ukraine, Belarus (formerly Byelorussia), Moldova (formerly Moldavia), Armenia, Azerbaijan, Uzbekistan, Turkmenistan, Tajikistan, Kazakhstan, and Kyrgyzstan (formerly Kirghiziya). The Baltic states—Estonia, Latvia, and Lithuania—did not join. Georgia maintained observer status before joining the CIS in November 1993.

Commonwealth of Nations: Voluntary association of the United Kingdom and its present dependencies and associated states, as well as certain former dependencies and their dependent territories. The term was first used officially in 1926 and is embodied in the Statute of Westminster (1931). Within the Commonwealth, whose secretariat (established in 1965) is located in London, England, are numerous subgroups devoted to economic and technical cooperation.

commune: An organization of people living together in a community who share the ownership and use of property. Also refers to a small governmental district of a country, especially in Europe.

communism: A form of government whose system requires common ownership of property for the use of all citizens. All profits are to be equally distributed and prices on goods and services are usually set by the state. Also, communism refers directly to the official doctrine of the former U.S.S.R.

compulsory: Required by law or other regulation.

compulsory education: The mandatory requirement for children to attend school until they have reached a certain age or grade level.

conciliation: A process of bringing together opposing sides of a disagreement for the purpose of compromise. Or, a way of settling an international dispute in which the disagreement is submitted to an independent committee that will examine the facts and advise the participants of a possible solution.

concordat: An agreement, compact, or convention, especially between church and state.

confederation: An alliance or league formed for the purpose of promoting the common interests of its members.

Confucianism: The system of ethics and politics taught by the Chinese philosopher Confucius.

conscription: To be required to join the military by law. Also known as the draft. Service personnel who join the military because of the legal requirement are called conscripts or draftees.

conservative party: A political group whose philosophy tends to be based on established traditions and not supportive of rapid change.

constituency: The registered voters in a governmental district, or a group of people that supports a position or a candidate.

constituent assembly: A group of people that has the power to determine the election of a political representative or create a constitution.

constitution: The written laws and basic rights of citizens of a country or members of an organized group.

constitutional monarchy: A system of government in which the hereditary sovereign (king or queen, usually) rules according to a written constitution.

constitutional republic: A system of government with an elected chief of state and elected representation, with a written constitution containing its governing principles. The United States is a constitutional republic.

Coptic Christians: Members of the Coptic Church of Egypt, formerly of Ethiopia.

Council for Mutual Economic Assistance (CMEA): Also known as Comecon, the alliance of socialist economies was established on 25 January 1949 and abolished 1 January 1991. It included Afghanistan*, Albania, Angola*, Bulgaria, Cuba, Czechoslovakia, Ethiopia*, East Germany, Hungary, Laos*, Mongolia, Mozambique*, Nicaragua*, Poland, Romania, USSR, Vietnam, Yemen*, and Yugoslavia. (Nations marked with an asterisk were observers only.)

counterinsurgency operations: Organized military activity designed to stop rebellion against an established government.

county: A territorial division or administrative unit within a state or country.

coup d'ètat or coup: A sudden, violent overthrow of a government or its leader.

criminal law: The branch of law that deals primarily with crimes and their punishments.

crown colony: A colony established by a commonwealth over which the monarch has some control, as in colonies established by the United Kingdom's Commonwealth of Nations.

Crusades: Military expeditions by European Christian armies in the eleventh, twelfth, and thirteenth centuries to win land controlled by the Muslims in the middle east.

cultivable land: Land that can be prepared for the production of crops.

Cultural Revolution: An extreme reform movement in China from 1966 to 1976; its goal was to combat liberalization by restoring the ideas of Mao Zedong.

customs union: An agreement between two or more countries to remove trade barriers with each other and to establish common tariff and nontariff policies with respect to imports from countries outside of the agreement.

cyclone: Any atmospheric movement, general or local, in which the wind blows spirally around and in towards a center. In the northern hemisphere, the cyclonic movement is usually counter-clockwise, and in the southern hemisphere, it is clockwise.

Cyrillic alphabet: An alphabet adopted by the Slavic people and invented by Cyril and Methodius in the ninth century as an alphabet that was easier for the copyist to write. The Russian alphabet is a slight modification of it.

decentralization: The redistribution of power in a government from one large central authority to a wider range of smaller local authorities.

declaration of independence: A formal written document stating the intent of a group of persons to become fully self-governing.

deficit: The amount of money that is in excess between spending and income.

deficit spending: The process in which a government spends money on goods and services in excess of its income.

deforestation: The removal or clearing of a forest.

deity: A being with the attributes, nature, and essence of a god; a divinity.

delta: Triangular-shaped deposits of soil formed at the mouths of large rivers.

demarcate: To mark off from adjoining land or territory; set the limits or boundaries of.

demilitarized zone (DMZ): An area surrounded by a combat zone that has had military troops and weapons removed.

demobilize: To disband or discharge military troops.

democracy: A form of government in which the power lies in the hands of the people, who can govern directly, or can be governed indirectly by representatives elected by its citizens.

denationalize: To remove from government ownership or control.

deportation: To carry away or remove from one country to another, or to a distant place.

depression: A hollow; a surface that has sunken or fallen in.

deregulation: The act of reversing controls and restrictions on prices of goods, bank interest, and the like.

desalinization plant: A facility that produces freshwater by removing the salt from saltwater.

desegregation: The act of removing restrictions on people of a particular race that keep them socially, economically, and, sometimes, physically, separate from other groups.

desertification: The process of becoming a desert as a result of climatic changes, land mismanagement, or both.

détente: The official lessening of tension between countries in conflict.

devaluation: The official lowering of the value of a country's currency in relation to the value of gold or the currencies of other countries.

developed countries: Countries which have a high standard of living and a well-developed industrial base.

dialect: One of a number of regional or related modes of speech regarded as descending from a common origin.

dictatorship: A form of government in which all the power is retained by an absolute leader or tyrant. There are no rights granted to the people to elect their own representatives.

dike: An artificial riverbank built up to control the flow of water.

diplomatic relations: The relationship between countries as conducted by representatives of each government.

direct election: The process of selecting a representative to the government by balloting of the voting public, in contrast to selection by an elected representative of the people.

disarmament: The reduction or depletion of the number of weapons or the size of armed forces.

dissident: A person whose political opinions differ from the majority to the point of rejection.

dogma: A principle, maxim, or tenet held as being firmly established.

dominion: A self-governing nation that recognizes the British monarch as chief of state.

dowry: The sum of the property or money that a bride brings to her groom at their marriage.

draft constitution: The preliminary written plans for the new constitution of a country forming a new government.

Druze: A member of a Muslim sect based in Syria, living chiefly in the mountain regions of Lebanon.

dual nationality: The status of an individual who can claim citizenship in two or more countries.

duchy: Any territory under the rule of a duke or duchess.

due process: In law, the application of the legal process to which every citizen has a right, which cannot be denied.

dynasty: A family line of sovereigns who rule in succession, and the time during which they reign.

Eastern Orthodox: The outgrowth of the original Eastern Church of the Eastern Roman Empire, consisting of eastern Europe, western Asia, and Egypt.

EC *see* European Community

ecclesiastical: Pertaining or relating to the church.

ecology: The branch of science that studies organisms in relationship to other organisms and to their environment.

economic depression: A prolonged period in which there is high unemployment, low production, falling prices, and general business failure.

elected assembly: The persons that comprise a legislative body of a government who received their positions by direct election.

electoral system: A system of choosing government officials by votes cast by qualified citizens.

electoral vote: The votes of the members of the electoral college.

electorate: The people who are qualified to vote in an election.

emancipation: The freeing of persons from any kind of bondage or slavery.

embargo: A legal restriction on commercial ships to enter a country's ports, or any legal restriction of trade.

emigration: Moving from one country or region to another for the purpose of residence.

empire: A group of territories ruled by one sovereign or supreme ruler. Also, the period of time under that rule.

enclave: A territory belonging to one nation that is surrounded by that of another nation.

encroachment: The act of intruding, trespassing, or entering on the rights or possessions of another.

endemic: Anything that is peculiar to and characteristic of a locality or region.

Enlightenment: An intellectual movement of the late seventeenth and eighteenth centuries in which scientific thinking gained a strong foothold and old beliefs were challenged. The idea of absolute monarchy was questioned and people were gradually given more individual rights.

epidemic: As applied to disease, any disease that is temporarily prevalent among people in one place at the same time.

Episcopal: Belonging to or vested in bishops or prelates; characteristic of or pertaining to a bishop or bishops.

ethnolinguistic group: A classification of related languages based on common ethnic origin.

EU *see* European Union

European Community: A regional organization created in 1958. Its purpose is to eliminate customs duties and other trade barriers in Europe. It promotes a common external tariff against other countries, a Common Agricultural Policy (CAP), and guarantees of free movement of labor and capital. The original six members were Belgium, France, West Germany, Italy, Luxembourg, and the Netherlands. Denmark, Ireland, and the United Kingdom became members in 1973; Greece joined in 1981; Spain and Portugal in 1986. Other nations continue to join.

European Union: The EU is an umbrella reference to the European Community (EC) and to two European integration efforts introduced by the Maastricht Treaty: Common Foreign and Security Policy (including defense) and Justice and Home Affairs (principally cooperation between police and other authorities on crime, terrorism, and immigration issues).

exports: Goods sold to foreign buyers.

external migration: The movement of people from their native country to another country, as opposed to internal migration, which is the movement of people from one area of a country to another in the same country.

faction: People with a specific set of interests or goals who form a subgroup within a larger organization.

Fascism: A political philosophy that holds the good of the nation as more important than the needs of the individual. Fascism also stands for a dictatorial leader and strong oppression of opposition or dissent.

federal: Pertaining to a union of states whose governments are subordinate to a central government.

federation: A union of states or other groups under the authority of a central government.

fetishism: The practice of worshipping a material object that is believed to have mysterious powers residing in it, or is the representation of a deity to which worship may be paid and from which supernatural aid is expected.

feudal society: In medieval times, an economic and social structure in which persons could hold land given to them by a lord (nobleman) in return for service to that lord.

final jurisdiction: The final authority in the decision of a legal matter. In the United States, the Supreme Court would have final jurisdiction.

Finno-Ugric language group: A subfamily of languages spoken in northeastern Europe, including Finnish, Hungarian, Estonian, and Lapp.

fiscal year: The twelve months between the settling of financial accounts, not necessarily corresponding to a calendar year beginning on January 1.

fjord: A deep indentation of the land forming a comparatively narrow arm of the sea with more or less steep slopes or cliffs on each side.

folk religion: A religion with origins and traditions among the common people of a nation or region that is relevant to their particular life-style.

foreign exchange: Foreign currency that allows foreign countries to conduct financial transactions or settle debts with one another.

foreign policy: The course of action that one government chooses to adopt in relation to a foreign country.

Former Soviet Union: The FSU is a collective reference to republics comprising the former Soviet Union. The term, which has been used as both including and excluding the Baltic republics (Estonia, Latvia, and Lithuania), includes the other 12 republics: Russia, Ukraine, Belarus, Moldova, Armenia, Azerbaijan, Uzbekistan, Turkmenistan, Tajikistan, Kazakhstan, Kyrgizstan, and Georgia.

free enterprise: The system of economics in which private business may be conducted with minimum interference by the government.

fundamentalist: A person who holds religious beliefs based on the complete acceptance of the words of the Bible or other holy scripture as the truth. For instance, a fundamentalist would believe the story of creation exactly as it is told in the Bible and would reject the idea of evolution.

GDP *see* gross domestic product

genocide: Planned and systematic killing of members of a particular ethnic, religious, or cultural group.

Germanic language group: A large branch of the Indo-European family of languages including German itself, the Scandinavian languages, Dutch, Yiddish, Modern English, Modern Scottish, Afrikaans, and others. The group also includes extinct languages such as Gothic, Old High German, Old Saxon, Old English, Middle English, and the like.

glasnost: President Mikhail Gorbachev's frank revelations in the 1980s about the state of the economy and politics in the Soviet Union; his policy of openness.

global warming: Also called the greenhouse effect. The theorized gradual warming of the earth's climate as a result of the burning of fossil fuels, the use of man-made chemicals, deforestation, etc.

GMT *see* Greenwich Mean Time

GNP *see* gross national product

grand duchy: A territory ruled by a nobleman, called a grand duke, who ranks just below a king.

Greek Catholic: A person who is a member of an Orthodox Eastern Church.

Greek Orthodox: The official church of Greece, a self-governing branch of the Orthodox Eastern Church.

Greenwich (Mean) Time: Mean solar time of the meridian at Greenwich, England, used as the basis for standard time throughout most of the world. The world is divided into 24 time zones, and all are related to the prime, or Greenwich mean, zone.

gross domestic product: A measure of the market value of all goods and services produced within the boundaries of a nation, regardless of asset ownership. Unlike gross national product, GDP excludes receipts from that nation's business operations in foreign countries.

gross national product: A measure of the market value of goods and services produced by the labor and property of a nation. Includes receipts from that nation's business operation in foreign countries

guerrilla: A member of a small radical military organization that uses unconventional tactics to take their enemies by surprise.

gymnasium: A secondary school, primarily in Europe, that prepares students for university.

harem: In a Muslim household, refers to the women (wives, concubines, and servants in ancient times) who live there and also to the area of the home they live in.

harmattan: An intensely dry, dusty wind felt along the coast of Africa between Cape Verde and Cape Lopez. It prevails at intervals during the months of December, January, and February.

heavy industry: Industries that use heavy or large machinery to produce goods, such as automobile manufacturing.

Holocaust: The mass slaughter of European civilians, the vast majority Jews, by the Nazis during World War II.

Holy Roman Empire: A kingdom consisting of a loose union of German and Italian territories that existed from around the ninth century until 1806.

home rule: The governing of a territory by the citizens who inhabit it.

homeland: A region or area set aside to be a state for a people of a particular national, cultural, or racial origin.

homogeneous: Of the same kind or nature, often used in reference to a whole.

Horn of Africa: The Horn of Africa comprises Djibouti, Eritrea, Ethiopia, Somalia, and Sudan.

human rights issues: Any matters involving people's basic rights which are in question or thought to be abused.

humanist: A person who centers on human needs and values, and stresses dignity of the individual.

humanitarian aid: Money or supplies given to a persecuted group or people of a country at war, or those devastated by a natural disaster, to provide for basic human needs.

hydroelectric power plant: A factory that produces electrical power through the application of waterpower.

IBRD *see* World Bank

immigration: The act or process of passing or entering into another country for the purpose of permanent residence.

imports: Goods purchased from foreign suppliers.

indigenous: Born or originating in a particular place or country; native to a particular region or area.

Indo-Aryan language group: The group that includes the languages of India; also called Indo-European language group.

Indo-European language family: The group that includes the languages of India and much of Europe and southwestern Asia.

infanticide: The act of murdering a baby.

infidel: One who is without faith or belief; particularly, one who rejects the distinctive doctrines of a particular religion.

inflation: The general rise of prices, as measured by a consumer price index. Results in a fall in value of currency.

insurgency: The state or condition in which one rises against lawful authority or established government; rebellion.

insurrectionist: One who participates in an unorganized revolt against an authority.

interim government: A temporary or provisional government.

interim president: One who is appointed to perform temporarily the duties of president during a transitional period in a government.

International Date Line: An arbitrary line at about the 180th meridian that designates where one day begins and another ends.

Islam: The religious system of Mohammed, practiced by Moslems and based on a belief in Allah as the supreme being and Mohammed as his prophet. The spelling variations, Muslim and Muhammad, are also used, primarily by Islamic people. Islam also refers to those nations in which it is the primary religion.

isthmus: A narrow strip of land bordered by water and connecting two larger bodies of land, such as two continents, a continent and a peninsula, or two parts of an island.

Judaism: The religious system of the Jews, based on the Old Testament as revealed to Moses and characterized by a belief in one God and adherence to the laws of scripture and rabbinic traditions.

Judeo-Christian: The dominant traditional religious makeup of the United States and other countries based on the worship of the Old and New Testaments of the Bible.

junta: A small military group in power in a country, especially after a coup.

khan: A sovereign, or ruler, in central Asia.

khanate: A kingdom ruled by a khan, or man of rank.

labor movement: A movement in the early to mid-1800s to organize workers in groups according to profession to give them certain rights as a group, including bargaining power for better wages, working conditions, and benefits.

land reforms: Steps taken to create a fair distribution of farmland, especially by governmental action.

landlocked country: A country that does not have direct access to the sea; it is completely surrounded by other countries.

least developed countries: A subgroup of the United Nations designation of "less developed countries;" these countries generally have no significant economic growth, low literacy rates, and per person gross national product of less than $500. Also known as undeveloped countries.

leftist: A person with a liberal or radical political affiliation.

legislative branch: The branch of government which makes or enacts the laws.

less developed countries (LDC): Designated by the United Nations to include countries with low levels of output, living standards, and per person gross national product generally below $5,000.

literacy: The ability to read and write.

Maastricht Treaty: The Maastricht Treaty (named for the Dutch town in which the treaty was signed) is also known as the Treaty of European Union. The treaty creates a European Union by: (a) committing the member states of the European Economic Community to both European Monetary Union (EMU) and political union; (b) introducing a single currency (European Currency Unit, ECU); (c) establishing a European System of Central Banks (ESCB); (d) creating a European Central Bank (ECB); and (e) broadening EC integration by including both a common foreign and security policy (CFSP) and cooperation in justice and home affairs (CJHA). The treaty entered into force on November 1, 1993.

Maghreb states: The Maghreb states include the three nations of Algeria, Morocco, and Tunisia; sometimes includes Libya and Mauritania.

majority party: The party with the largest number of votes and the controlling political party in a government.

Marshall Plan: Formally known as the European Recovery Program, a joint project between the United States and most Western European nations under which $12.5 billion in U.S. loans and grants was expended to aid European recovery after World War II.

Marxism *see* Marxist-Leninist principles

Marxist-Leninist principles: The doctrines of Karl Marx, built upon by Nikolai Lenin, on which communism was founded. They predicted the fall of capitalism, due to its own internal faults and the resulting oppression of workers.

Marxist: A follower of Karl Marx, a German socialist and revolutionary leader of the late 1800s, who contributed to Marxist-Leninist principles.

Mayan language family: The languages of the Central American Indians, further divided into two subgroups: the Maya and the Huastek.

Mecca (Mekkah): A city in Saudi Arabia; a destination of pilgrims in the Islamic world.

Mediterranean climate: A wet-winter, dry-summer climate with a moderate annual temperature range.

mestizo: The offspring of a person of mixed blood; especially, a person of mixed Spanish and American Indian parentage.

migratory workers: Usually agricultural workers who move from place to place for employment depending on the growing and harvesting seasons of various crops.

military coup: A sudden, violent overthrow of a government by military forces.

military junta: The small military group in power in a country, especially after a coup.

military regime: Government conducted by a military force.

militia: The group of citizens of a country who are either serving in the reserve military forces or are eligible to be called up in time of emergency.

minority party: The political group that comprises the smaller part of the large overall group it belongs to; the party that is not in control.

missionary: A person sent by authority of a church or religious organization to spread his religious faith in a community where his church has no self-supporting organization.

monarchy: Government by a sovereign, such as a king or queen.

Mongol: One of an Asiatic race chiefly resident in Mongolia, a region north of China proper and south of Siberia.

Mongoloid: Having physical characteristics like those of the typical Mongols (Chinese, Japanese, Turks, Eskimos, etc.).

Moors: One of the Arab tribes that conquered Spain in the eighth century.

mosque: An Islamic place of worship and the organization with which it is connected.

Muhammad (or Muhammed or Mahomet): An Arabian prophet, known as the "Prophet of Allah" who founded the religion of Islam in 622, and wrote *The Koran,* the scripture of Islam. Also commonly spelled Mohammed.

mujahideen (mujahedin or mujahedeen): Rebel fighters in Islamic countries, especially those supporting the cause of Islam.

mulatto: One who is the offspring of parents one of whom is white and the other is black.

municipality: A district such as a city or town having its own incorporated government.

Muslim: A follower of the prophet Muhammad, the founder of the religion of Islam.

Muslim New Year: A Muslim holiday. Although in some countries 1 Muharram, which is the first month of the Islamic year, is observed as a holiday, in other places the new year is observed on Sha'ban, the eighth month of the year. This practice apparently stems from pagan Arab times. Shab-i-Bharat, a national holiday in Bangladesh on this day, is held by many to be the occasion when God ordains all actions in the coming year.

NAFTA (North American Free Trade Agreement): NAFTA, which entered into force in January 1994, is a free trade agreement between Canada, the United States, and Mexico. The agreement progressively eliminates almost all U.S.-Mexico tariffs over a 10–15 year period.

nationalism: National spirit or aspirations; desire for national unity, independence, or prosperity.

nationalization: To transfer the control or ownership of land or industries to the nation from private owners.

NATO *see* North Atlantic Treaty Organization

naturalize: To confer the rights and privileges of a native-born subject or citizen upon someone who lives in the country by choice.

neutrality: The policy of not taking sides with any countries during a war or dispute among them.

Newly Independent States: The NIS is a collective reference to 12 republics of the former Soviet Union: Russia, Ukraine, Belarus (formerly Byelorussia), Moldova (formerly Moldavia), Armenia, Azerbaijan, Uzbekistan, Turkmenistan, Tajikistan, Kazakhstan, and Kirgizstan (formerly Kirghiziya), and Georgia. Following dissolution of the Soviet Union, the distinction between the NIS and the Commonwealth of Independent States

(CIS) was that Georgia was not a member of the CIS. That distinction dissolved when Georgia joined the CIS in November 1993.

Nonaligned Movement: The NAM is an alliance of third world states that aims to promote the political and economic interests of developing countries. NAM interests have included ending colonialism/neo-colonialism, supporting the integrity of independent countries, and seeking a new international economic order.

Nordic Council: The Nordic Council, established in 1952, is directed toward supporting cooperation among Nordic countries. Members include Denmark, Finland, Iceland, Norway, and Sweden. Headquarters are in Stockholm, Sweden.

North Atlantic Treaty Organization (NATO): A mutual defense organization. Members include Belgium, Canada, Denmark, France (which has only partial membership), Greece, Iceland, Italy, Luxembourg, Netherlands, Norway, Portugal, Spain, Turkey, United Kingdom, United States, and Germany.

nuclear power plant: A factory that produces electrical power through the application of the nuclear reaction known as nuclear fission.

OAPEC (Organization of Arab Petroleum Exporting countries): OAPEC was created in 1968; members include: Algeria, Bahrain, Egypt, Iraq, Kuwait, Libya, Qatar, Saudi Arabia, Syria, and the United Arab Emirates. Headquarters are in Cairo, Egypt.

OAS (Organization of American States): The OAS (Spanish: Organizaciûn de los Estados Americanos, OEA), or the Pan American Union, is a regional organization which promotes Latin American economic and social development. Members include the United States, Mexico, and most Central American, South American, and Caribbean nations.

OAS *see* Organization of American States

oasis: Originally, a fertile spot in the Libyan desert where there is a natural spring or well and vegetation; now refers to any fertile tract in the midst of a wasteland.

occupied territory: A territory that has an enemy's military forces present.

official language: The language in which the business of a country and its government is conducted.

oligarchy: A form of government in which a few people possess the power to rule as opposed to a monarchy which is ruled by one.

OPEC *see* OAPEC

open market: Open market operations are the actions of the central bank to influence or control the money supply by buying or selling government bonds.

opposition party: A minority political party that is opposed to the party in power.

Organization of Arab Petroleum Exporting Countries *see* OAPEC

organized labor: The body of workers who belong to labor unions.

Ottoman Empire: A Turkish empire founded by Osman I in the thirteenth century, that variously controlled large areas of land around the Mediterranean, Black, and Caspian Seas until it was dissolved in 1923.

overseas dependencies: A distant and physically separate territory that belongs to another country and is subject to its laws and government.

Pacific Rim: The Pacific Rim, referring to countries and economies bordering the Pacific Ocean.

pact: An international agreement.

panhandle: A long narrow strip of land projecting like the handle of a frying pan.

papyrus: The paper-reed or -rush which grows on marshy river banks in the southeastern area of the Mediterranean, but more notably in the Nile valley.

paramilitary group: A supplementary organization to the military.

parliamentary republic: A system of government in which a president and prime minister, plus other ministers of departments, constitute the executive branch of the government and the parliament constitutes the legislative branch.

parliamentary rule: Government by a legislative body similar to that of Great Britain, which is composed of two houses—one elected and one hereditary.

partisan politics: Rigid, unquestioning following of a specific party's or leader's goals.

patriarchal system: A social system in which the head of the family or tribe is the father or oldest male. Kinship is determined and traced through the male members of the tribe.

per capita: Literally, per person; for each person counted.

perestroika: The reorganization of the political and economic structures of the Soviet Union by president Mikhail Gorbachev.

periodical: A publication whose issues appear at regular intervals, such as weekly, monthly, or yearly.

political climate: The prevailing political attitude of a particular time or place.

political refugee: A person forced to flee his or her native country for political reasons.

potable water: Water that is safe for drinking.

pound sterling: The monetary unit of Great Britain, otherwise known as the pound.

prime meridian: Zero degrees in longitude that runs through Greenwich, England, site of the Royal Observatory. All other longitudes are measured from this point.

prime minister: The premier or chief administrative official in certain countries.

privatization: To change from public to private control or ownership.

protectorate: A state or territory controlled by a stronger state, or the relationship of the stronger country toward the lesser one it protects.

Protestant: A member or an adherent of one of those Christian bodies which descended from the Reformation of the sixteenth century. Originally applied to those who opposed or protested the Roman Catholic Church.

Protestant Reformation: In 1529, a Christian religious movement begun in Germany to deny the universal authority of the Pope, and to establish the Bible as the only source of truth. (*Also see* Protestant)

province: An administrative territory of a country.

provisional government: A temporary government set up during a time of unrest or transition in a country.

purge: The act of ridding a society of "undesirable" or unloyal persons by banishment or murder.

Rastafarian: A member of a Jamaican cult begun in 1930 as a semi-religious, semi-political movement.

referendum: The practice of submitting legislation directly to the people for a popular vote.

Reformation *see* Protestant Reformation

refugee: One who flees to a refuge or shelter or place of safety. One who in times of persecution or political commotion flees to a foreign country for safety.

revolution: A complete change in a government or society, such as in an overthrow of the government by the people.

right-wing party: The more conservative political party.

Roman alphabet: The alphabet of the ancient Romans from which the alphabets of most modern western European languages, including English, are derived.

Roman Catholic Church: The designation of the church of which the pope or Bishop of Rome is the head, and that holds him as the successor of St. Peter and heir of his spiritual authority, privileges, and gifts.

Roman Empire: A Mediterranean Empire, centered in the Italian peninsula, that was the most powerful state in the region in the first four centuries A.D. The empire helped spread Greek culture throughout its territory. After the fourth century, the empire served as a Christianizing influence. Although the western half of the area fell to barbarian invasions in the fifth century, the eastern half, based in Constantinople, continued until 1453.

romance language: The group of languages derived from Latin: French, Spanish, Italian, Portuguese, and other related languages.

runoff election: A deciding election put to the voters in case of a tie between candidates.

Russian Orthodox: The arm of the Orthodox Eastern Church that was the official church of Russia under the czars.

Sahelian zone: Eight countries make up this dry desert zone in Africa: Burkina Faso, Chad, Gambia, Mali, Mauritania, Niger, Senegal, and the Cape Verde Islands. (*Also see* Club du Sahel.)

savanna: A treeless or near treeless plain of a tropical or subtropical region dominated by drought-resistant grasses.

secession: The act of withdrawal, such as a state withdrawing from the Union in the Civil War in the United States.

sect: A religious denomination or group, often a dissenting one with extreme views.

segregation: The enforced separation of a racial or religious group from other groups, compelling them to live and go to school separately from the rest of society.

self-sufficient: Able to function alone without help.

separatism: The policy of dissenters withdrawing from a larger political or religious group.

serfdom: In the feudal system of the Middle Ages, the condition of being attached to the land owned by a lord and being transferable to a new owner.

Seventh-day Adventist: One who believes in the second coming of Christ to establish a personal reign upon the earth.

shamanism: A religion of some Asians and Amerindians in which shamans, who are priests or medicine men, are believed to influence good and evil spirits.

Shia Muslims: Members of one of two great sects of Islam. Shia Muslims believe that Ali and the Imams are the rightful successors of Mohammed (also commonly spelled Muhammad). They also believe that the last recognized Imam will return as a messiah. Also known as Shiites. (*Also see* Sunni Muslims.)

Shiites *see* Shia Muslims

Shintoism: The system of nature- and hero-worship which forms the indigenous religion of Japan.

Sikh: A member of a politico-religious community of India, founded as a sect around 1500 and based on the principles of monotheism (belief in one god) and human brotherhood.

Sino-Tibetan language family: The family of languages spoken in eastern Asia, including China, Thailand, Tibet, and Burma.

slash-and-burn agriculture: A hasty and sometimes temporary way of clearing land to make it available for agriculture by cutting down trees and burning them.

slave trade: The transportation of black Africans beginning in the 1700s to other countries to be sold as slaves—people owned as property and compelled to work for their owners at no pay.

Slavic languages: A major subgroup of the Indo-European language family. It is further subdivided into West Slavic (including Polish, Czech, Slovak and Serbian), South Slavic (including Bulgarian, Serbo-Croatian, Slovene, and Old Church Slavonic), and East Slavic (including Russian Ukrainian and Byelorussian).

socialism: An economic system in which ownership of land and other property is distributed among the community as a whole, and every member of the community shares in the work and products of the work.

socialist: A person who advocates socialism.

Southeast Asia: The region in Asia that consists of the Malay Archipelago, the Malay Peninsula, and Indochina.

state: The politically organized body of people living under one government or one of the territorial units that make up a federal government, such as in the United States.

subcontinent: A land mass of great size, but smaller than any of the continents; a large subdivision of a continent.

Sudanic language group: A related group of languages spoken in various areas of northern Africa, including Yoruba, Mandingo, and Tshi.

suffrage: The right to vote.

Sufi: A Muslim mystic who believes that God alone exists, there can be no real difference between good and evil, that the soul exists within the body as in a cage, so death should be the chief object of desire, and sufism is the only true philosophy.

sultan: A king of a Muslim state.

Sunni Muslims: Members of one of two major sects of the religion of Islam. Sunni Muslims adhere to strict orthodox traditions, and believe that the four caliphs are the rightful successors to Mohammed, founder of Islam. (Mohammed is commonly spelled Muhammad, especially by Islamic people.) (*Also see* Shia Muslims.)

Taoism: The doctrine of Lao-Tzu, an ancient Chinese philosopher (about 500 B.C.) as laid down by him in the *Tao-te-ching*.

tariff: A tax assessed by a government on goods as they enter (or leave) a country. May be imposed to protect domestic industries from imported goods and/or to generate revenue.

terrorism: Systematic acts of violence designed to frighten or intimidate.

Third World: A term used to describe less developed countries; as of the mid-1990s, it is being replaced by the United Nations designation Less Developed Countries, or LDC.

topography: The physical or natural features of the land.

totalitarian party: The single political party in complete authoritarian control of a government or state.

trade unionism: Labor union activity for workers who practice a specific trade, such as carpentry.

treaty: A negotiated agreement between two governments.

tribal system: A social community in which people are organized into groups or clans descended from common ancestors and sharing customs and languages.

undeveloped countries *see* least developed countries

unemployment rate: The overall unemployment rate is the percentage of the work force (both employed and unemployed) who claim to be unemployed.

UNICEF: An international fund set-up for children's emergency relief: United Nations Children's Fund (formerly United Nations International Children's Emergency Fund).

untouchables: In India, members of the lowest caste in the caste system, a hereditary social class system. They were considered unworthy to touch members of higher castes.

Warsaw Pact: Agreement made May 14, 1955 (and dissolved July 1, 1991) to promote mutual defense between Albania, Bulgaria, Czechoslovakia, East Germany, Hungary, Poland, Romania, and the USSR.

Western nations: Blanket term used to describe mostly democratic, capitalist countries, including the United States, Canada, and western European countries.

workers' compensation: A series of regular payments by an employer to a person injured on the job.

World Bank: The World Bank is a group of international institutions which provides financial and technical assistance to developing countries.

world oil crisis: The severe shortage of oil in the 1970s precipitated by the Arab oil embargo.

Zoroastrianism: The system of religious doctrine taught by Zoroaster and his followers in the Avesta; the religion prevalent in Persia until its overthrow by the Muslims in the seventh century.

Bibliography

Africa

Algeria

Lorcin, Patricia M. E. *Imperial Identities: Stereotyping, Prejudice, and Race in Colonial Algeria.* London: I. B. Tauris Publishers, 1995.

MacMaster, Neil. *Colonial Migrants and Racism: Algerians in France, 1900–62.* New York: St. Martin's Press Inc., 1997.

Metz, Helen Chapin. *Algeria: A Country Study. Area Handbook Studies.* Washington, D.C.: Federal Research Division, Library of Congress. U.S. Government, Department of the Army, 1994.

Sahnouni, Mohamed. *The Lower Paleolithic of the Maghreb: Excavations and Analyses at Ain Hanech, Algeria. Cambridge Monographs in African Archaeology 42.* BAR International Series 689. Oxford: Hadrian Books Ltd., 1998.

Angola

Bollig, M. *When War Came the Cattle Slept: Himba Oral Traditions.* Koln: R. Koppe, 1997.

Ciment, J. *Conflict and Crisis in the Post-Cold War World: Angola and Mozambique: Postcolonial Wars in Southern Africa.* New York: Facts on File, 1997.

Etienne Dostert , P., ed. "The Republic of Angola." *Africa.* Harpers Ferry, WV: Stryker-Post Publications, 1997.

Kaplan, I. *Angola: A Country Study.* Washington, D.C.: American University, Foreign Area Studies, 1979.

Oliver, R. and B. Fagan. *Africa in the Iron Age, c. 500 BC to AD 1400.* New York: Cambridge University Press, 1975.

Benin

Africa on File. New York: Facts on File, Inc., 1995.

Decalo, Samuel. *Historical Dictionary of Benin.* 3rd ed. Lanham, Md.: Scarecrow Press, 1995.

Miller, Susan Katz. "Sermon on the Farm." *International Wildlife.* March/April 1992: 4951.

Botswana

Country Profile of Botswana. McLean, Va.: SAIC, 1998.

Jackson, A. *Botswana, 1939–1945: An African Country at War.* Oxford: Clarendon Press, 1999.

Morton, R. F., and J. Ramsay, eds. *Birth of Botswana: The History of the Bechuanaland Protectorate, 1910–1966.* Gaborone: Longman Botswana, 1988.

Pickford, P. and Pickford, B. *Okavango: The Miracle Rivers.* London: New Holland, 1999.

Ramsay, J.; Morton, B. and Morton, F. *Historical Dictionary of Botswana.* 3rd ed. Lanham, Md.: Scarecrow Press, 1996.

Burkina Faso

McFarland, Daniel Miles, and Lawrence A. Rupley. *Historical Dictionary of Burkina Faso.* Second Ed. *African Historical Dictionaries,* No. 74. Lanham, Md., and London: The Scarecrow Press, 1998.

Sankara, Thomas. *Women's Liberation and the African Freedom Struggle.* London: Pathfinder, 1990.

Wilks, Ivor. *The Mossi and Akan States.* In *History of West Africa.* Third Ed. Ed. J.F.A. Ajayi and Michael Crowder. Harlow, U.K.: Longman, 1985.

Burundi

Lemarchand, René. *Burundi: Ethnic Conflict and Genocide.* Woodrow Wilson Center Press, Cambridge University Press, 1997.

Ramsay, F. Jeffress. *Burundi in Global Studies: Africa.* Connecticut: Dushkin Publishing Group/Brow, Benchmark Publishers, 1995.

Weinstein, Warren, Robert Schire. *Political Conflict and Ethnic Strategies: A Case Study of Burundi.* New York: Maxwell School of Citizenship and Public Affairs, 1976.

Cameroon

Bjornson, Richard. *The African Quest for Freedom and Identity: Cameroonian Writing and the National Experience.* Bloomington: Indiana University Press, 1991.

DeLancey, Mark. *Historical Dictionary of the Republic of Cameroon.* 2nd ed. Metuchen, N.J.: Scarecrow Press, 1990.

"Odd Man in: Cameroon. (African nation to become the 52nd member of the Commonwealth of Nations)." *The Economist (US),* October 7, 1995, vol. 337, no. 7935, p. 51.

Takougang, Joseph. *African State and Society in the 1990s: Cameroon's Political Crossroads.* Boulder, Colo.: Westview Press, 1998.

Cape Verde

Carreira, Antonio. *The People of the Cape Verde Islands: Exploitation and Emigration.* Hamden, Conn.: Archon Books, 1982

Chilcote, Ronald H. *Amilcar Cabral's Revolutionary Theory and Practice: A Critical Guide.* Boulder, Colo.: Lynne Rienner Publishers, 1991.

Foy, Colm. *Cape Verde: Politics, Economics and Society.* London: Pinter Publishers, 1988.

Lobban, Jr. Richard A. *Cape Verde: Crioulo Colony to Independent Nation.* Boulder, Colo.: Westview Press, 1995.

Russell-Wood, A.J.R. *A World on the Move: The Portuguese in Africa, Asia, and America, 1415–1808.* New York: St. Martin's Press, 1993.

Central African Republic

Decalo, Samuel. *Psychoses of Power: African Personal Dictatorships.* Boulder, Colo.: Westview Press, 1989.

Kalck, Pierre. *Central African Republic: A Failure in De-colonisation.* Trans. Barbara Thomson. New York: Praeger, 1971.

———. *Central African Republic.* Santa Barbara, Calif.: Clio Press, 1993.

———. *Historical Dictionary of the Central African Republic.* 2nd ed. Metuchen, N.J.: Scarecrow Press, 1992.

O'Toole, Thomas. *The Central African Republic: The Continent's Hidden Heart.* Boulder, Colo.: Westview Press, 1986.

Titley, Brian. *Dark Age: The Political Odyssey of Emperor Bokassa.* Montreal: McGill-Queen's University Press, 1997.

Chad

Azevedo, Mario Joaquim. *Chad : A Nation in Search of its Future.* Boulder, Colo.: Westview Press, 1998.

Collelo, Thomas, ed. *Chad: A Country Study,* 2nd ed. Washington, DC: Government Printing Office, 1990.

Decalo, Samuel. *Historical Dictionary of Chad.* 2nd ed. Metuchen, NJ: Scarecrow Press, 1987.

Nolutshungu, Sam C. *Limits of Anarchy: Intervention and State Formation in Chad.* Charlottesville: University Press of Virginia, 1996.

Wright, John L. *Libya, Chad, and the Central Sahara.* Totowa, NJ: Barnes & Noble Books, 1989.

Comoros

"Under the Volcano." *Time International,* March 23, 1998, vol. 150, no. 30, p. 35.

The Comoros: Current Economic Situation and Prospects. Washington, D.C.: World Bank, 1983.

Newitt, Malyn. *The Comoro Islands: Struggle Against Dependency in the Indian Ocean.* Boulder, Colo: Westview, 1984.

Ottenheimer, Martin. *Historical Dictionary of the Comoro Islands.* Metuchen, N.J: Scarecrow Press, 1994.

Weinberg, Samantha. *Last of the Pirates: The Search for Bob Denard.* New York: Pantheon Books, 1994.

Congo, Democratic Republic of the

Bobb, F. Scott. *Historical Dictionary of Zaire.* Metuchen, N.J.: Scarecrow Press, 1988.

Kanza, Thomas. *The Rise and Fall of Patrice Lumumba.* Cambridge, Mass.: Schenkman, 1979.

Leslie, Winsome J. *Zaire: Continuity and Political Change in an Oppressive State.* Boulder, Colo.: Westview Press, 1993.

Lumumba, Patrice. *Congo, My Country.* With a foreword and notes by Colin Legum. Transl. by Graham Heath. London: Pall Mall Press, 1962.

Mokoli, Mondonga M. *State Against Development: The Experience of Post-1965 Zaire.* Westport, Conn.: Greenwood Press, 1992.

Congo, Republic of the

Clark, John F. "Elections, Leadership, and Democracy in Congo." *Africa Today* 41, no. 3 (1994): 41–60.

Decalo, Samuel, Virginia Thompson, and Richard Adloff. *Historical Dictionary of Congo.* Lanham, Md.: The Scarecrow Press, Inc., 1996.

Fegley, Randall. *The Congo.* World Bibliographical Series, vol. 162, Oxford: Clio Press, 1993.

Hilton, Anne. *The Kingdom of the Kongo.* Oxford: Clarendon Press, 1985.

Vansina, Jan. *The Tio Kingdom of the Middle Congo: 1880–1892.* London: Oxford University Press, 1973.

Cote d'Ivoire

Ajayi, J.F. Ade, and Michael Crowder, eds. *History of West Africa, 2.* New York: Colombia University Press, 1974.

Clark, John F. and David E. Gardinier, eds. *Political Reform in Francophone Africa.* Boulder, Colo.: Westview Press, 1997.

EIU Country Reports. London: Economist Intelligence Unit, April 1999.

Handloff, Robert E. *Côte d'Ivoire: A Country Study. Area Handbook Series.* Third Edition. Washington, DC: Federal Research Division, Library of Congress, 1991.

Mundt, Robert. *Historical Dictionary of the Ivory Coast.* Metuchen, N.J.: Scarecrow Press, 1995.

Djibouti

Darch, Colin. *A Soviet View of Africa: An Annotated Bibliography on Ethiopia, Somalia, and Djibouti.* Boston: G. K. Hall, 1980.

Koburger, Charles W. *Naval Strategy East of Suez: The Role of Djibouti.* New York: Praeger, 1992.

Schrader, Peter J. *Djibouti.* Santa Barbara, Calif.: Clio Press, 1991.

Tholomier, Robert. *Djibouti: Pawn of the Horn of Africa.* Metuchen, N.J.: Scarecrow, 1981.

Woodward, Peter. *The Horn of Africa: State Politics and International Relations.* London: I. B. Tauris, 1996.

Egypt

Daly, M.W., ed. *The Cambridge History of Egypt*. New York: Cambridge University Press, 1998.

Metz, Helen Chapin. *Egypt, a Country Study,* 5th ed. Washington, D.C.: Library of Congress, 1991.

Rubin, Barry M. *Islamic Fundamentalism in Egyptian Politics*. New York: St. Martin's Press, 1990.

Shamir, Shimon, ed. *Egypt from Monarchy to Republic: A Reassessment of Revolution and Change*. Boulder, Colo.: Westview Press, 1995.

Equatorial Guinea

Fegley, Randall. *Equatorial Guinea: An African Tragedy*. New York: P. Lang, 1989.

Klitgaard, Robert E. *Tropical Gangsters*. New York: Basic Books, 1990.

Liniger-Goumaz, Max. *Historical Dictionary of Equatorial Guinea*. 2nd ed. Metuchen, NJ: Scarecrow Press, 1988.

———. *Small is Not Always Beautiful: The Story of Equatorial Guinea*. Translated from the French by John Wood. Totowa, N.J.: Barnes & Noble Books, 1989.

Sundiata, I.K. *Equatorial Guinea: Colonialism, State Terror, and the Search for Stability*. Boulder, Colo.: Westview Press, 1990.

Eritrea

Connell, Dan. *Against All Odds*. Trenton, N.J.: Red Sea Press, 1993.

Doombos, Martin, et al., eds. *Beyond the Conflict in the Horn*. Trenton, NJ: Red Sea Press, 1992.

Gebremedhin, Tesfa G. *Beyond Survival: The Economic Challenges of Agriculture and Development in Post-Independence Eritrea*. Trenton, NJ: Red Sea Press, 1997.

Okbazghi, Yohannes. *Eritrea: A Pawn in World Politics*. Gainesville: University of Florida Press, 1991.

Ethiopia

Bahru, Z. *A History of Modern Ethiopia, 1855–1974*. Athens: Ohio University Press, 1991.

Hassen, M. *The Oromo of Ethiopia: A History 1570–1860*. Cambridge: Cambridge University Press, 1990.

Kaplan, S. *The Beta Israel (Falasha) in Ethiopia: From Earliest Times to the Twentieth Century*. New York: New York University Press, 1992.

Marcus, H. G. *A History of Ethiopia*. Berkeley: University of California Press, 1994.

Prouty, C. and Rosenfeld, E. *Historical Dictionary of Ethiopia*. Metuchen, NJ: The Scarecrow Press, Inc., 1994.

Gabon

Alexander, Caroline. *One Dry Season*. New York: Alfred A. Knopf, 1989.

Gall, Timothy L., ed. *Worldmark Encyclopedia of the Nations*. 9th ed. Detroit: Gale Research, 1998.

Iliffe, John. *Africans: The History of a Continent*. Cambridge University Press, 1995.

Ungar, Sanford J. *Africa: The People and Politics of an Emerging Continent*. New York: Simon & Schuster, 1989.

The Gambia

Else, David. *The Gambia and Senegal*. Oakland, Calif.: Lonely Planet Publications, 1999.

Gailey, Harry A. *Historical Dictionary of The Gambia*. 2nd ed. African Historical Dictionaries, No. 4. Metuchen, N.J.: Scarecrow Press, 1987.

Gall, Timothy L., ed. *Worldmark Encyclopedia of Cultures and Daily Life*. vol. 1: Africa 1998.

Vollmer, Jurgen. *Black Genesis, African Roots: A voyage from Juffure, The Gambia, to Mandingo country to the slave port of Dakar, Senegal*. New York: St. Martin's Press, 1980.

Wright, Donald R. "The world and a very small place in Africa (history of Niumi)." *Sources and Studies in World History*. New York: M.E. Sharpe, Armonk, 1997.

Ghana

Ardayfio-Schandorf, Elizabeth and Kate Kwafo-Akoto. *Women in Ghana: An Annotated Bibliography*. Accra: Woeli Publishing Services, 1990.

Berry, LaVerle. *Ghana: A Country Study. Area Handbook Series*. Third Edition. Washington, DC: Federal Research Division, Library of Congress, 1995.

Davidson, Basil. *Black Star: A View of the Life and Times of Kwame Nkrumah*. Boulder: Westview Press, 1989.

Glickman, Harvey, ed. *Political Leaders of Contemporary Africa South of the Sahara: A Biographical Dictionary*. New York: Greenwood Press, 1992.

Guinea

Africa on File. New York: Facts on File, 1995.

Gall, Timothy L., ed. *Worldmark Encyclopedia of Cultures and Daily Life*. vol. 1: Africa. Detroit: Gale Research, 1998.

Nelson, Harold D. et al, eds. *Area Handbook for Guinea. Foreign Area Studies*. Washington, DC.: American University, 1975.

O'Toole, Thomas. *Historical Dictionary of Guinea*. Third Edition. Lanham, MD: London: The Scarecrow Press, 1994.

Guinea-Bissau

Africa on File. New York: Facts on File, 1997.

EIU Country Reports. London: Economist Intelligence Unit, Ltd., April 8, 1999.

Forrest, Joshua. *Guinea-Bissau: Power, Conflict, and Renewal in a West African Nation*. Boulder: Westview Press, 1992.

Lopes, Carlos. *Guinea-Bissau: From Liberation Struggle to Independent Statehood*. Boulder, Colo.: Westview Press, 1987.

Pedlar, Frederick. *Main Currents of West African History, 19401978*. London: The Macmillan Press, Ltd., 1979.

Kenya

Cohen, W. David, and E. S. *Atieno Odhiambo. Burying SM: The Politics of Knowledge and the Sociology of Power in Africa*. Portsmouth, NH: Heinemann, 1992.

Miller, Norman, and Roger Yeager. *Kenya: The Quest for Prosperity* (2nd edition). Boulder, Colo.: Westview Press, 1994.

Mwaniki, Nyaga. "The Consequences of Land Subdivision in Northern Embu, Kenya." *The Journal of African Policy Studies,* 2(1), 1996.

Ogot, A. Bethwell. *Historical Dictionary of Kenya*. Metuchen, NJ: The Scarecrow Press, Inc., 1981.

Lesotho

Eldredge, E. A. *A South African Kingdom: The Pursuit of Security in Nineteenth-Century Lesotho*. Cambridge: Cambridge University Press.

Haliburton, G. *Historical Dictionary of Lesotho*. African Historical Dictionaries, No. 10, Metuchen, N.J.: The Scarecrow Press, Inc., 1977.

Khaketla, B. M. *Lesotho 1970: An African Coup Under the Microscope*. London: Hurst, 1971.

Liberia

Africa on File. New York: 1995 Facts on File, Inc., 1995.

Africa South of the Sahara. London: Europa Publishers, 1998.

Dunn, D. Elwood, and Svend E. Holsoe. *Historical Dictionary of Liberia*. African Historical Dictionaries, No. 38. Metuchen, N.J., and London: The Scarecrow Press, Inc., 1985.

Dunn, D. Elwood and S. Byron Tarr. *Liberia: A National Polity in Transition*. Metuchen, N.J., and London: The Scarecrow Press, Inc., 1988.

Nelson, Harold D., ed. *Liberia: A Country Study*. Third Edition. Washington, DC: The American University, 1985.

Libya

Gall, Timothy, L., ed. *Worldmark Encyclopedia of Cultures and Daily Life*. vol. 1: Africa. Detroit: Gale Research, 1998.

Haley, P. Edward. *Qadhafi and the United States Since 1969*. New York: Praeger, 1984.

Simonis, Damien, *et al. North Africa*. Hawthorn, Aus.: Lonely Planet Publications, 1995.

Simons, Geoff. *Libya: The Struggle for Survival*. New York: St. Martin's Press, 1996.

Vanderwalle, Dirk ed. *Qadhafi's Libya: 19691994*. New York: St. Martin's Press, 1995.

Madagascar

Allen, Philip M. *Madagascar: Conflicts of Authority in the Great Island*. Boulder, Colo.: Westview Press, 1994.

Covell, Maureen. *Historical Dictionary of Madagascar*. Lanham, Md.: Scarecrow Press, 1995.

Kent, Raymond K. *Early Kingdoms in Madagascar 1500–1700*. New York: Holt, Rinehart and Winston, 1970.

Stratton, Arthur. *The Great Red Island*. New York: Charles Scribner's Sons, 1964.

Verin, Pierre. *The History of Civilisation in North Madagascar*. Rotterdam: A.A. Balkema, 1986.

Malawi

Baker, Colin. *State of Emergency: Crisis in Central Africa, Nyasaland 1959–1960*. London: I. B. Tauris Publishers, 1997.

Crosby, Cynthia A. *Historical Dictionary of Malawi*. Metuchen, N.J.: The Scarecrow Press, 2nd edition, 1993.

Phiri, D. D. *From Nguni to Ngoni: A History of the Ngoni Exodus from Zululand and Swaziland to Malawi, Tanzania and Zambia*. Limbe, Malawi: Popular Publications, 1982.

Mali

Economist Intelligence Unit. *Country Profile: Mali, 1998*. London: The Economist, 1998.

Historical Dictionary of Mali. Second edition. Metuchen, New Jersey: Scarecrow Press, 1986.

Mann, Kenny. *Ghana, Mali, Songhay: The Western Sudan*. Parsippany, NJ: Dillon Press, 1996.

McIntosh, Roderick J. *The Peoples of the Middle Niger: The Island of Gold*. Malden, MA: Blackwell Publishers, 1998.

Mauritania

Africa. Hawthorn, Aus.: Lonely Planet Publications, 1995.

Goodsmith, Lauren. *The Children of Mauritania*. Minneapolis: Carolrhoda Books, 1993.

McLachlan, Anne and Keith. *Morocco Handbook*. Bath. Eng.: Footprint Handbooks. 1993.

Pazzanita, Anthony G. *Historical Dictionary of Mauritania*. Lanham, Md.: Scarecrow Press, Inc., 1966.

Thompson, Virginia, and Adloff, Richard. *The Western Saharans*. Totowa, N.J.: Barnes and Noble Books, 1980.

Mauritius

Bunge, Frederica M., ed. *Indian Ocean. Five Island Countries*. Washington, D.C.: Dept of the Army, 1983.

Bunwaree, Sheila S. *Mauritian Education in a Global Economy*. Stanley, Rose Hill, Mauritius: Editions de l'Océan Indien, 1994.

Butlin, Ron (ed). *Mauritian Voices. New Writing in English*. Newcastle Upon Tyne: Flambard Press, 1997.

Selvon, Sydney. *Historical Dictionary of Mauritius*. Metuchen, N.J.: Scarecrow Press, 1991.

Morocco

Cook, Weston, F. *The Hundred Year War for Morocco: Gunpowder and the Military Revolution in the Early Modern Muslim World*. Boulder, Colo.: Westview Press, 1985.

Jereb, James F. *Arts and Crafts of Morocco*. New York: Chronicle Books, 1996.

Hermes, Jules M. *The Children of Morocco*. Minneapolis: Carolrhoda Books, 1995.

Hoisington, William A., Jr. *Lyautey and the French Conquest of Morocco.* New York: St. Martins Press, 1995.

Simonis, Damien, et al. *North Africa.* Hawthorn, Aus.: Lonely Planet Publications, 1995.

Mozambique

Azevedo, Mario Joaquim. *Historical Dictionary of Mozambique.* Metuchen, N.J.: Scarecrow Press, 1991.

Davidson, Basil. *Africa in History.* New York: Simon and Schuster, 1991.

Magnin, Andre; and Jacques Soulillou, eds., *Contemporary Art of Africa.* New York: Harry N. Abrams, Inc. Publishers, 1996.

Newitt, M. D. D. *A History of Mozambique.* Bloomington: Indiana University Press, 1995.

Slater, Mike. *Mozambique.* London: New Holland Publishers Ltd, 1997.

Namibia

Breytenbach, Cloete. *Namibia: Birth of a Nation.* South Africa: LUGA Publishers, 1989.

Cliffe, Lionel. *The Transition to Independence in Namibia.* Boulder, Colo.: Lynne Rienner Publishers, 1994.

Namibia: A Nation Is Born. Washington, D.C.: U.S. Dept. of State, 1990.

Swaney, Deanna. *Zimbabwe, Botswana, and Namibia: A Lonely Planet Travel Guide.* Hawthorn, Australia, Lonely Planet Publications, 1995.

Niger

Beckwith, Carol. *Nomads of Niger.* New York: Henry N. Abrams, 1983.

Charlick, Robert. *Niger. Personal Rule and Survival in the Sahel.* Boulder, Colo.: Westview, 1991.

Cooper, Barbara. *Marriage in Maradi: Gender and Culture in a Hausa Society in Niger 1900-1989.* Portsmouth, N.H.: Heinemann, 1997.

Decalo, Samuel. *Historical Dictionary of Niger.* Metuchen, N.J.: Scarecrow Press, 1979.

Gall, Timothy L., ed. *Worldmark Encyclopedia of Cultures and Daily Life.* vol. 1: Africa. Detroit: Gale Research, 1998.

Nigeria

Diamond, Larry, Anthony Kirk-Greene, and Oyeleye Oyediran, eds. *Transition Without End: Nigerian Politics and Civil Society under Babangida.* Boulder, Colo.: Lynne Rienner Publishers, 1997.

Graf, William D. *The Nigerian State: Political Economy, State Class, and Political System in the Post-Colonial Era.* Portsmouth, N.H.: Heineman, 1988.

Ihonvbere, Julius O., and Timothy Shaw. *Illusions of Power: Nigeria in Transition.* Trenton, NJ: Africa World Press, 1998.

Osaghae, Eghosa E. *Crippled Giant: Nigeria Since Independence.* Bloomington: Indiana University Press, 1998.

Wesler, Kit W. *Historical Archaeology in Nigeria.* Trenton, N.J.: Africa World Press, 1998.

Rwanda

Hodd, M. *East African Handbook,* Chicago: Passport Books, 1994.

Nyrop, Richard., et al. *Rwanda: A Country Study.* Washington, D.C.: U.S. Government Printing Office, 1984.

Pierce, Julian R. *Speak Rwanda.* New York: Picador USA, 1999.

Webster, J.B., B.A. Ogot, and J.P. Chretien. "The Great Lakes Region: 1500–1800." In *The General History of Africa,* Volume V, Calif.: Heinemann, UNESCO, 1998

Sao Tome and Principe

Carreira, Antonio. *The People of the Cape Verde Islands: Exploitation and Emigration.* Connecticut: Archon Books, 1982.

Davidson, Basil. *Black Mother: The Years of the African Slave Trade.* Boston: Little, Brown, and Co., 1961.

Denny, L.M. and Donald I. Ray. *São Tomé and Príncipe: Economics, Politics and Society.* London and New York: Pinter Publishers, 1989.

Garfield, Robert. *A History of São Tomé Island, 1470–1655: The Key to Guinea.* San Francisco: Mellen Research University Press, 1992.

Hodges, Tony and Malyn Hewitt. *São Tomé and Príncipe: From Plantation Colony to Microstate.* Boulder and London: Westview Press, 1988.

Thomas, Hugh. *The Slave Trade.* New York: Touchstone Books-Simon and Schuster, 1997.

Senegal

Clark, Andrew Francis, ed. *Historical Dictionary of Senegal.* 2nd ed. African Historical Dictionaries, No. 23. Metuchen, N.J.: The Scarecrow Press, Inc., 1994.

Else, David. *The Gambia and Senegal.* Oakland, Calif.: Lonely Planet Publications, 1999.

Gellar, Sheldon. *Senegal: An African Nation between Islam and the West.* Second Edition. Boulder, Colo.: Westview Press, 1995.

Sharp, Robin. *Senegal: A State of Change.* Oxford, UK: Oxfam, 1994.

Seychelles

Bennett, George. *Seychelles.* Oxford, England; Santa Barbara, CA: Clio Press, 1993.

Franda, Marcus. *The Seychelles, Unquiet Islands.* Boulder, Colo.: Westview 1982.

McAteer, William. *Rivals in Eden: A History of the French Settlement and British Conquest of the Seychelles Islands, 1724–1818.* Sussex, Eng.: Book Guild, 1990.

Vine, Peter. *Seychelles.* London, Eng.: Immel Pub. Co., 1989.

Sierra Leone

Africa South of the Sahara. "Sierra Leone." London: Europa Publishers, 1997.

Alie, Joe. *A New History of Sierra Leone.* New York: St. Martin's Press, 1990.

Foray, Cyril P. *Historical Dictionary of Sierra Leone.* African Historical Dictionaries, No. 12. Metuchen, N.J.: The Scarecrow Press, 1977.

Gall, Timothy L., ed. *Worldmark Encyclopedia of Cultures and Daily Life.* vol. 1: Africa. Detroit: Gale Research, 1998.

Somalia

Barnes, Virginia Lee. *Aman: The Story of a Somali Girl.* New York: Pantheon, 1994.

Clarke, Walter S., and Jeffrey Ira Herbst. *Learning from Somalia: The Lessons of Armed Humanitarian Intervention.* Boulder, Colo.: Westview Press, 1997.

DeLancey, Mark. *Blood and Bone: The Call of Kinship in Somali Society.* Lawrenceville, N.J.: Red Sea, 1994.

———, et al, eds. *Somalia.* Santa Barbara, Calif.: Clio, 1988.

DeLancey, Mark. *Blood and Bone: The Call of Kinship in Somali Society.* Lawrenceville, N.J.: Red Sea, 1994.

Metz, Helen Chapin, ed. *Somalia: A Country Study.* 4th ed. Washington, DC: Library of Congress, 1993.

South Africa

Brynes, Rita M. ed. *South Africa: A Country Study.* Washington D.C.: U.S. Government Printing Office, 1997.

Mostert, N. *Frontiers: The Epic of South Africa's Creation and the Tragedy of the Xhosa People.* New York: Knopf, 1992.

Riley, E. *Major Political Events in South Africa, 1948–1990.* New York: Facts on File, 1991.

Thompson, L. M. *A History of South Africa.* London: Yale University Press, 1990.

Sudan

Alier, Abel. *Southern Sudan.* Reading:Ithaca Press,1990.

Bovill, Edward William. *The Golden Trade of the Moors.*Princeton, New Jersey: Marcus Wiener Publishers,1995.

Metz, Helen Chapin, ed. *Sudan: A Country Study.* Federal Research Division/Library of Congress,1991.

Stewart, Judy. *A Family in Sudan.* Minneapolis: Lerner Publishing Co., 1988.

Voll, John Obert and Sarah Potts Voll. *The Sudan: Unity and Diversity in a Multicultural State.* Boulder, Col.: Westview Press, 1985.

Swaziland

Bonner, P. M. "Swati II, c. 1826–1865." In *Black Leaders in Southern African History.* Edited by Christopher Saunders. London: Heinemann, 1979: 61–74.

Davies, R. H.; O'Meara, D. and Dlamini, S. *The Kingdom of Swaziland: A Profile.* London: Zed Books, 1985.

Grotpeter, J. J. *Historical Dictionary of Swaziland.* Metuchen, N.J.: The Scarecrow Press, 1975.

Kuper, H. *Sobhuza II: Ngwenyama and King of Swaziland.* London: Duckworth, 1978.

Williams, G. and B. Hackland. *The Dictionary of Contemporary Politics of Southern Africa.* New York: Macmillan, 1988.

Tanzania

Hodd, M. *East African Handbook.* Chicago: Passport Books, 1994.

Hughes, A. J. *East Africa.* Baltimore: Penguin Books, 1969.

Iliffe, J. *A Modern History of Tanganyika.* Cambridge, Eng.: Cambridge University Press, 1979.

Maddox, Gregory, et al., eds. *Custodians of the Land: Ecology and Culture in the History of Tanzania.* Athens: Ohio University Press, 1996.

Togo

Ajayi, J.F.A. and Michael Crowder, eds. *History of West Africa,* Vol. 1. London,Eng.: Congman Group Limited, 1971.

———. *History of West Africa,* Vol. 2. Londond, Eng.: Congman Group Limited, 1971.

Decalo, Samuel. *Historical Dictionary of Togo.* 3rd ed. Metuchen, NJ: Scarecrow Press, 1996.

Packer, George, *The Village of Waiting.* New York: Vintage Books, 1988.

Tunisia

Ali, Wijdan. *Modern Islamic Art: Development and Continuity.* Gainesville, Fla.: University Press of Florida, 1997.

Lancel, Serge. Nevill, Antonia (translator). *Hannibal.* Oxford, Eng.: Blackwell Publications, 1998.

Ling, Dwight L. *Morocco and Tunisia: A Comparative History.* Washington DC: University Press of America, 1979.

Memmi, Albert. *Pillar of Salt.* Boston: Beacon Press, 1992.

Uganda

Byrnes, Rita. ed. *Uganda: A Country Study.* Washington, D.C.: U.S. Government Printing Office, 1992.

Gall, Timothy L., ed. *Worldmark Encyclopedia of Cultures and Daily Life.* vol. 1: Africa. Detroit: Gale Research, 1998.

Hodd, M. *East African Handbook.* Chicago: Passport Books, 1994.

Jorgenson, Jan Jelmett. *Uganda: A Modern History.* New York: St. Martin's Press, 1981.

Nzita, Richard. *Peoples and Cultures of Uganda.* Kampala: Fountain Publishers, 1993.

Zambia

Burdette, M. M. *Zambia Between Two Worlds.* Boulder, Colo.: Westview Press, 1988.

Dresang, E. *The Land and People of Zambia.* Philadelphia: Lippincott, 1975.

Grotpeter, J. J.; Siegel, B. V. and Pletcher, J. R. *Historical Dictionary of Zambia.* Lanham, Maryland: The Scarecrow Press, Inc., 1998.

Zimbabwe

Dewey, William Joseph. *Legacies of Stone: Zimbabwe Past and Present.* Tervuren: Royal Museum for Central Africa, 1997.

Nelson, H. D. *Zimbabwe: A Country Study.* Washington, D. C.: U.S. Government Printing Office, 1983.

Rasmussen, R. Kent and Rubert, Steven, C. *Historical Dictionary of Zimbabwe.* 2nd ed. Metuchen, N.J.: The Scarecrow Press, 1990.

Sheehan, Sean. *Zimbabwe.* New York: M. Cavendish, 1996.

Americas

Antigua and Barbuda

Coram, Robert. *Caribbean Time Bomb: the United States' Complicity in the Corruption of Antigua.* New York: William Morrow and Company, Inc., 1993.

Dyde, Brian. *Antigua and Barbuda: the Heart of the Caribbean.* London: Macmillan Publishers, 1990.

Kurlansky, Mark. *A Continent of Islands: Searching for the Caribbean Destiny.* Reading, Mass.: Addison-Wesley Publishing Co., 1992.

Argentina

American University. *Argentina: A Country Study,* 3rd ed. Washington, DC: Government Printing Office, 1985.

Sarlo Sabajanes, Beatriz. *Jorge Luis Borges: A Writer on the Edge.* New York: Verso, 1993.

Timerman, Jacobo. *Prisoner Without a Name, Cell Without a Number.* New York: Knopf, 1981.

Tulchin, Joseph S. *Argentina: The Challenges of Modernization.* Wilmington, Del.: Scholarly Resources, 1998.

Worldmark Press, Ltd. *Worldmark Encyclopedia of Cultures and Daily Life.* vol. 2: Americas 1998.

Bahamas

Hamshere, Cyril. *The British in the Caribbean.* Cambridge, Mass.: Harvard University Press, 1972.

Kurlansky, Mark. *A Continent of Islands: Searching for the Caribbean Destiny.* Reading, Mass.: Addison-Wesley Publishing Co., 1992.

Marx, Jenifer. *Pirates and Privateers of the Caribbean.* Malabar, Fla.: Krieger Publishing Company, 1992.

Barbados

Beckles, Hilary. *A History of Barbados: From Amerindian Settlement to Nation-State.* New York: Cambridge University Press, 1990.

Pariser, Harry S. *Adventure Guide to Barbados.* Edison, N.J.: Hunter Publishing, 1995.

Wilder, Rachel, ed. *Barbados.* Boston: Houghton Mifflin Co., 1993

Belize

Bolland, O. Nigel. *Belize: A New Nation in Central America.* Boulder, Colo.: Westview, 1986.

Edgell, Zee. *Beka Lamb.* London: Heinemann, 1982.

Fernandez, Julio A. *Belize: Case Study for Democracy in Central America.* Brookfield, Vt.: Avebury, 1989.

Gall, Timothy L., ed. *Worldmark Encyclopedia of Cultures and Daily Life.* vol. 2: Americas. Detroit: Gale Research, 1998.

Mallan, Chicki. *Belize Handbook.* Chico, Calif.: Moon Publications, 1991.

Bolivia

Blair, David Nelson. *The Land and People of Bolivia.* New York: J.B. Lippincott, 1990.

Hudson, Rex A. and Dennis M. Hanratty. *Bolivia, a Country Study.* 3rd ed. Washington, DC: Government Printing Office, 1991.

Morales, Waltrand Q. *Bolivia: Land of Struggle.* Boulder, Colo.: Westview Press, 1992.

Lindert, P. van. *Bolivia : A Guide to the People, Politics and Culture.* New York: Monthly Review Press, 1994.

Parker, Edward. *Ecuador, Peru, Bolivia. Country fact files.* Austin, TX: Raintree Steck-Vaughn, 1998.

Brazil

Carpenter, Mark L. *Brazil, an Awakening Giant.* Minneapolis, MN: Dillon Press, 1987.

Levine, Robert M. *Historical Dictionary of Brazil.* Metuchen, N.J.: Scarecrow Press, 1979.

——— and John J. Crocitti. *The Brazil Reader: History, Culture, Politics.* Durham, N.C.: Duke University Press, 1999.

Poppino, Rollie E. *Brazil: The Land and People.* New York: Oxford University Press, 1973.

Roop, Peter, and Connie Roop. *Brazil.* Des Plaines, Ill.: Heinemann Interactive Library, 1998.

Canada

Bothwell, Robert. *A Short History of Ontario.* Edmonton: Hurtig Publishers Ltd., 1986.

Dickinson, John A. and Brian Young. *A Short History of Quebec.* Toronto: Copp Clark Pitman Ltd., 1993.

McNaught, Kenneth. *The Penguin History of Canada.* London: Penguin, 1988.

Morton, Desmond. *A Short History of Canada.* 2nd revised ed. Toronto: McClelland & Stewart Inc., 1994.

Woodcock, George. *A Social History of Canada.* Markham, Ont.: Penguin Books Canada, 1989.

Chile

Arriagada Herrera, Genaro. Trans. Nancy Morris. *Pinochet: The Politics of Power.* Boston : Allen & Unwin, 1988.

Blakemore, Harold. *Chile.* Santa Barbara, Calif.: Clio Press, 1988.

Collier, Simon. *A History of Chile.* Cambridge: Cambridge University Press, 1996.

Falcoff, Mark. *Modern Chile, 1970–89: A Critical History.* New Jersey: Transaction Books, 1989.

Hudson, Rex A., ed. *Chile, a Country Study.* 3rd ed. Federal Research Division, Library of Congress. Washington, D.C., 1994.

Colombia

Bushnell, David. The Making of Modern Colombia: A Nation in Spite of Itself. Berkeley: University of California Press, 1993.

Davis, Robert H. *Colombia.* Santa Barbara, Calif.: Clio Press, 1990.

———. *Historical Dictionary of Colombia.* Metuchen, N.J.: Scarecrow Press, 1977.

Hanratty, Dennis M., and Sandra W. Meditz. *Colombia: A Country Study.* 4th ed. Washington, DC: Federal Research Division, Library of Congress, 1990.

Pearce, Jenny. *Colombia: The Drugs War.* New York: Gloucester Press, 1990.

Posada Carbs, Eduardo. *The Colombian Caribbean: A Regional History, 1870–1950.* New York: Clarendon Press, 1996.

Costa Rica

Biesanz, Richard; Biesanz, Karen Zubris; Biezanz, Mavis Hiltunen. *The Costa Ricans.* New Jersey: Prentice Hall, 1982.

Gall, Timothy L., ed. *Worldmark Encyclopedia of Cultures and Daily Life.* vol. 2: Americas. Detroit: Gale Research, 1998.

Stone, Doris. *Pre-Columbian Man in Costa Rica.* Cambridge: Peabody Museum Press, 1977.

Todorov, Tzvetan. *The Conquest of America.* New York: Harper & Row, 1984.

Cuba

Balfour, Sebastian. *Castro.* New York: Longman, 1990.

Brune, Lester H. *The Cuba-Caribbean Missile Crisis of October 1962.* Claremont, Calif.: Regina Books, 1996.

Gall, Timothy L., ed. *Worldmark Encyclopedia of Cultures and Daily Life.* vol. 2: Americas. Detroit: Gale Research, 1998.

Rudolph, James D., ed. *Cuba: A Country Study,* 3rd ed. Washington, D.C.: U.S. Government Printing Office, 1985.

Wyden, Peter. *Bay of Pigs: The Untold Story.* New York: Simon & Schuster, 1979.

Dominica

Baker, Patrick L. *Centering the Periphery: Chaos, Order, and the Ethnohistory of Dominica.* Montreal: McGill-Queen's University Press, 1994.

Philpott, Don. *Caribbean Sunseekers: Dominica.* Lincolnwood, Ill.: Passport Books, 1996.

Whitford, Gwenith. "Mining on 'Nature Island': the Dominican Government's Resource Extraction Plans Anger Conservationists." *Alternatives Journal,* Winter 1998, vol. 24, no. 1, p. 9+.

Dominican Republic

Cambeira, Alan. *Quisqueya la bella: the Dominican Republic in Historical and Cultural Pperspective.* Armonk, N.Y.: M.E. Sharpe, 1997.

Horowitz, Michael M. *Peoples and Cultures of the Caribbean: An Anthropological Reader.* New York: Natural History Press, 1971.

Logan, Rayford W. *Haiti and the Dominican Republic,* New York: Oxford University Press, 1968.

Plant, Roger. *Sugar and Modern Slavery: Haitian Migrant Labor and the Dominican Republic.* Totowa, N.J.: Biblio Dist., 1986.

Moya Pons, Frank. *The Dominican Republic: A National History.* New Rochelle, N.Y.: Hispaniola Books, 1995.

Ecuador

Bork, Albert William. *Historical Dictionary of Ecuador.* Metuchen, N.J.: Scarecrow Press, 1973.

Hemming, John. *The Conquest of the Incas.* San Diego: Harcourt Brace Jovanovich, 1970.

Rathbone, John Paul. *Ecuador, the Galápagos, and Colombia.* London: Cadogan Books, 1991.

Roos, Wilma, and Omer van Renterghem. *Ecuador in Focus: a Guide to the People, Politics and Culture.* New York: Interlink Books, 1997.

El Salvador

Browning, David, *El Salvador: Landscape and Society.* London: Clarendon Press, 1971.

Flemion, Philip, *Historical Dictionary of El Salvador* Metuchen, N.J.: The Scarecrow Press, 1972.

Haggerty, Richard, *El Salvador: A Country Study.* Washington, D.C.: Department of the Army, 1990.

Grenada

Gall, Timothy L., ed. *Worldmark Encyclopedia of the Nations.* 9th ed. Detroit: Gale Research, 1998.

Gunson, Phil, et. al., eds. *The Dictionary of Contemporary Politics of Central America and the Caribbean.* New York: Simon and Schuster, 1991.

Rouse, Irving. *The Tainos: Rise and Decline of the People Who Greeted Columbus.* New Haven: Yale University Press, 1992.

Weeks, John, and Ferbel, Peter. *Ancient Caribbean.* New York: Garland Publishing, 1994.

Guatemala

Gall, Timothy L., ed. *Worldmark Encyclopedia of the Nations.* 9th ed. Detroit: Gale Research, 1998.

Handy, Jim. *Gift of the Devil: A History of Guatemala.* Toronto: Between the Lines Press, 1984.

Nyrop, Richard, (ed.). *Guatemala: A Country Study.* (Washington DC: Department of the Army, 1983.

Schele, Linda. *A Forest of Kings: the Untold Story of the Ancient Maya.* New York: Morrow, 1990.

South America, *Central America and the Caribbean,* 6th ed. London: Europa Publication, 1997.

Guyana

Daly, Vere T. *A Short History of the Guyanese People*. London: Macmillan Education, 1975.

Gall, Timothy L., *Worldmark Encyclopedia of the Nations*. 9th ed. Detroit: Gale Research, 1998.

Mecklenburg, Kurt K. *Guyana Gold*. Carlton Press, 1990.

Singh, Chaitram. Guyana: *Politics in a Plantation Society*. New York: Praeger Publishers, 1988.

Haiti

Abbott, Elizabeth. *Haiti: The Duvaliers and their Legacy*. New York: Simon and Schuster. 1991.

Aristide, Jean-Bertrand. *Dignity*. Charlottesville and London: University Press of Virginia, 1996.

Gall, Timothy L., ed. *Worldmark Encyclopedia of the Nations*. 9th ed. Detroit: Gale Research, 1998.

McFadyen, Deidre; LaRamee, Pierre (editors). *Haiti: Dangerous Crossroads*. Boston: South End Press, 1995.

Honduras

Gall, Timothy L., ed. *Worldmark Encyclopedia of the Nations*. 9th ed. Detroit: Gale Research, 1998.

Schulz, Donald E. and Deborah S. Schulz. *The United States, Honduras and the Crisis in Central America*. Boulder, Colo.: Westview Press, 1994.

Todorov, Tzvetan (translated by Richard Howard) *The Conquest of America*. New York: Harper & Row, 1984.

Jamaica

Bayer, Marcel. *Jamaica: A Guide to the People, Politics, and Culture*. Trans. John Smith. London, Eng.: Latin American Bureau, 1993.

Davis, Stephen. *Reggae Bloodlines: In Search of the Music and Culture of Jamaica*. New York: Da Capo Press, 1992.

Gall, Timothy L., ed. *Worldmark Encyclopedia of the Nations*. 9th ed. Detroit: Gale Research, 1998.

Sherlock, Philip, and Hazel Bennett. *The Story of the Jamaican People*. Princeton, NJ: Markus Wiener Publishers, 1998.

Stone, Carl. *Class, State, and Democracy in Jamaica*. New York: Praeger, 1986.

Mexico

Burke, Michael E. *Mexico: An Illustrated History*. New York: Hippocrene Books, 1999.

Briggs, Donald C. *The Historical Dictionary of Mexico*. Metuchen, N.J.: Scarecrow Press, 1981.

Gall, Timothy L., ed. *Worldmark Encyclopedia of the Nations*. 9th ed. Detroit: Gale Research, 1998.

National Geographic. 190:2 (Aug. 1996) "Emerging Mexico". Entire isssue devoted to Mexico.

Randall, Laura, ed. *Changing Structure of Mexico*. Armonk, N.Y.: M.E. Sharpe, 1966.

Williamson, Edwin. *The Penguin History of Latin America*. New York: Penguin Books, 1992.

Nicaragua

Gall, Timothy L, ed. *Worldmark Encyclopedia of the Nations*. 9th ed. Detroit: Gale Research, 1998.

Kagan, Robert. *A Twilight Struggle: American Power and Nicaragua. 19771990*. New York: Free Press, 1996.

Rudolph, James D., ed. *Nicaragua: A Country Study*. 2nd ed. Washington, D.C.: Government Printing Office, 1994.

Walker, Thomas W., ed. *Revolution & Counterrevolution in Nicaragua*. Boulder, Colo.: Westview Press, 1991.

Panama

Flanagan, E. M. *Battle for Panama: Inside Operation Just Cause*. Washington, D.C.: Brassey's, Inc., 1993.

Guevara Mann, Carlos. *Panamanian Militarism: A Historical Interpretation*. Athens, Oh.: Ohio University Center for International Studies, 1996.

Hedrick, Basil C. and Anne K. *Historical Dictionary of Panama*. Metuchen, N.J.: Scarecrow Press, 1970.

Major, John. *Prize possession: The United States and the Panama Canal, 1903–1979*. New York: Cambridge University Press, 1993.

Noriega, Manuel Antonio. *America's Prisoner: The Memoirs of Manuel Noriega*. 1st ed. New York : Random House, 1997.

Pearcy, Thomas L. *We Answer Only to God: Politics and the Military in Panama, 1903–1947*. Albuquerque: University of New Mexico Press, 1998.

Paraguay

Gall, Timothy L., ed. *Worldmark Encyclopedia of the Nations*. 9th ed. Detroit: Gale Research, 1998.

Paraguay: A Country Study. Washington: U.S. Government Printing Office, 1990.

Wiarda, Howard J. and Harvey F. Kline, eds. *Latin American Politics and Development*. Boulder: Westview Press, 1996.

Peru

Gall, Timothy L., ed. *Worldmark Encyclopedia of the Nations*. 9th ed. Detroit: Gale Research, 1998.

Hemming, John. *The Conquest of the Incas*. New York: Harcourt Brace Jovanovich, 1970.

Holligan de Dmaz-Lmmaco, Jane. Peru: A Guide to the People, Politics and Culture. New York: Interlink Books, 1998.

Hudson, Rex A., ed. *Peru in Pictures*. Minneapolis: Lerner, 1987.

———. *Peru: A Country Study*. 4th ed. Washington, D.C.: Library of Congress, Federal Research Division, 1993.

Strong, Simon. *Shining Path : Terror and Revolution in Peru*. New York : Times Books, 1992.

Puerto Rico

Fernandez, Ronald. *The Disenchanted Island: Puerto Rico and the United States in the Twentieth Century*. 2d ed. Westport, Conn.: Praeger, 1996.

Figueroa, Loida. *History of Puerto Rico.* New York: Anaya, 1974.

Gall, Timothy L., ed. *Worldmark Encyclopedia of Cultures and Daily Life.* vol. 2: Americas. Detroit: Gale Research, 1998.

Morales, Carrion, Arturo. *Puerto Rico: A Political and Cultural History.* New York: Norton, 1983.

Morris, Nancy. *Puerto Rico: Culture, Politics, Indentity.* Westport, Conn.: Praeger, 1995.

St. Kitts and Nevis

Gall, Timothy L., ed. *Worldmark Encyclopedia of the Nations.* 9th ed. Detroit: Gale Research, 1998.

Hamshere, Cyril. *The British in the Caribbean.* Cambridge, MA: Harvard University Press, 1972.

Moll, V.P. *St. Kitts-Nevis.* Santa Barbara, CA: Clio, 1994.

Olwig, karen Fog. *Global Culture, Island Identity: Continuity and Change in the Afro-Caribbean Community of Nevis.* Philadelphia: Harwood, 1993.

St. Lucia

Claypole, William, and Robottom, John, *Caribbean Story, Book Two: The Inheritors.* Essex: Longman, 1986.

Craton, Michael, *Testing the Chains: Resistance to Slavery in the British West Indies.* Ithaca: Cornell University Press, 1982.

Gall, Timothy L., ed. *Worldmark Encyclopedia of the Nations.* 9th ed. Detroit: Gale Research, 1998.

Gunson, Phil, et. al., (eds.), *The Dictionary of Contemporary Politics of Central America and the Caribbean.* New York: Simon and Schuster, 1991.

Weeks, John, and Ferbel, Peter, *Ancient Caribbean.* New York: Garland Publishing, 1994.

St. Vincent and the Grenadines

Bobrow, Jill and Dana Jinkins. *St. Vincent and the Grenadines: Gems of the Caribbean.* Waitsfield, Vt.: Concepts Publishing Inc., 1993.

Hamshere, Cyril. *The British in the Caribbean.* Cambridge, Mass.: Harvard University Press, 1972.

Philpott, Don. *Caribbean Sunseekers: St. Vincent & Grenadines.* Lincolnwood, Ill.: Passport Books, 1996.

Suriname

Chin, Henk E. *Surinam: Politics, Economics, and Society.* New York: F. Pinter, 1987.

Cohen, Robert. *Jews in Another Environment: Surinam in the Second Half of the Eighteenth Century.* New York: E.J. Brill, 1991.

Dew, Edward M. *The Trouble in Suriname, 1975–1993.* Westport, CT: Praeger, 1994.

Hoogbergen, Wim S. M. *The Boni Maroon Wars in Suriname.* New York: Brill, 1990.

Sedoc-Dahlberg, Betty, ed. *The Dutch Caribbean: Prospects for Democracy.* New York: Bordon and Breach, 1990.

Trinidad and Tobago

Bereton, Bridget, *A History of Modern Trinidad* (Portsmouth: Heinemann, 1981).

Black, Jan, et. al., A*rea Handbook for Trinidad and Tobago* (Washington D.C.: U.S. Government Printing Office, 1976).

Gunson, Phil, et. al., (eds.), *The Dictionary of Contemporary Politics of Central America and the Caribbean* (NY: Simon and Schuster, 1991).

Reddock, *Women Labour and Politics in Trinidad and Tobago: A History* (London: Zed Books 1994).

Weeks, John, and Ferbel, Peter, *Ancient Caribbean* (NY: Garland Publishing, 1994).

United States

Ayres, Stephen M. *Health Care in the Unites States: The Facts and the Choices.* Chicago: American Library Association, 1996.

Bacchi, Carol Lee. *The Politics of Affirmative Action: 'Women', Equality, and Category Politics.* London: Sage, 1996.

Barone, Michael. *The Almanac of American Politics.* Washington, D.C.: National Journal, 1992.

Bennett, Lerone. *Before the Mayflower: A History of Black America.* 6th ed. New York: Penguin, 1993.

Brinkley, Alan. *American History: A Survey.* 9th ed. New York: McGraw-Hill, 1995.

Brinkley, Douglas. *American Heritage History of the United States.* New York: Viking, 1998.

Carnes, Mark C., ed. *A History of American Life.* New York: Scribner, 1996.

Commager, Henry Steele (ed.). *Documents of American History.* Englewood Cliffs, N.J.: Prentice-Hall, 1988.

Davidson, James West. *Nation of Nations: A Narrative History of the American Republic.* 3rd ed. Boston, MA: McGraw-Hill, 1998.

Davies, Philip John. (ed.) *An American Quarter Century: US Politics from Vietnam to Clinton.* New York: Manchester University Press, 1995.

Donaldson, Gary. *America at War since 1945: Politics and Diplomacy in Korea, Vietnam, and the Gulf War.* Westport, Conn.: Praeger, 1996.

Foner, Eric. *The Story of American Freedom.* New York: Norton, 1998.

Garraty, John Arthur. *The American Nation: A History of the United States.* 9th ed. New York: Longman, 1998.

Goldfield, David, ed. *The American Journey: A History of the United States.* Upper Saddle River, NJ: Prentice Hall, 1998.

Hart, James David, ed.. *Oxford Companion to American Literature.* 6th ed. New York: Oxford University Press, 1995.

Hummel, Jeffrey Rogers. *Emancipating Slaves, Enslaving Free Men: A History of the Civil War.* Chicago: Open Court, 1996.

Jenkins, Philip. *A History of the United States.* New York: St. Martin's Press, 1997.

Kaplan, Edward S. *American Trade Policy, 1923–1995.* Westport, Conn.: Praeger, 1996.

Magill, Frank N., ed. *Great Events from History: North American Series*. rev. ed. Pasadena, CA: Salem Press, 1997.

Martis, Kenneth C. *The Historical Atlas of Political Parties in the United States Congress 1789–1989*. New York: Macmillan, 1989.

McNickle, D'Arcy. *Native American Tribalism: Indian Survivals and Renewals*. New York: Oxford University Press, 1993.

Mudd, Roger. *American Heritage Great Minds of History*. New York: Wiley, 1999.

Nash, Gary B. *The American People: Creating a Nation and a Society*. 4th ed., New York : Longman, 1998.

People Who Shaped the Century. Alexandria, VA: Time-Life Books, 1999.

Robinson, Cedric J. *Black Movements in America*. New York: Routledge, 1997.

Tindall, George Brown. *America: A Narrative History*. 5th ed. New York: Norton, 1999.

Virga, Vincent. *Eyes of the Nation: A Visual History of the United States*. New York: Knopf, 1997.

Woodward, C. Vann, ed. *The Comparative Approach to American History*. New York: Oxford University Press, 1997.

Uruguay

Gall, Timothy L., ed. *Worldmark Encyclopedia of the Nations*. 9th ed. Detroit: Gale Research, 1998.

Hudson, Rex A., and Sandra Meditz. *Uruguay: A Country Study*. Washington, DC : Federal Research Division, Library of Congress, 1992.

Weinstein, Martin. *Uruguay, Democracy at the Crossroads. Nations of Contemporary Latin America*. Boulder, Colo.: Westview Press, 1988.

Zlotchew, Clark M. Paul David Seldis, ed. *Voices of the River Plate : Interviews with Writers of Argentina and Uruguay*. San Bernadino, CA: Borgo Press, 1993.

Venezuela

Fox, Geoffrey. *The Land of People of Venezuela*. New York: HarperCollins, 1991.

Gall, Timothy L., ed. *Worldmark Encyclopedia of Cultures and Daily Life*. vol. 2: Americas. Detroit: Gale Research, 1998.

Haggerty, Richard A., ed. *Venezuela: A Country Study,* 4th ed. Washington, D.C.: Library of Congress, 1993.

Hellinger, Daniel. *Venezuela: Tarnished Democracy*. Boulder, Colo.: Westview Press, 1991.

Asia

Afghanistan

Adamec, Ludwig W. *Historical Dictionary of Afghanistan*. Metuchen, NJ: Scarecrow Press, 1991.

Gall, Timothy L., ed. *Worldmark Encyclopedia of the Nations*. 9th ed. Detroit: Gale Research, 1998.

Giradet, Edward. *Afghanistan: The Soviet War.* London: Croom Helm, 1985

Nyrop, Richard F. and Donald M. Seekins, eds. *Afghanistan: A Country Study*. 5th ed. Washington, DC: U.S. Government Printing Office, 1986.

Australia

Bassett, Jan. *The Oxford Illustrated Dictionary of Australian History*. New York: Oxford University Press, 1993.

Bolton, Geoffrey, ed. *The Oxford History of Australia*. New York: Oxford University Press, 1986–90.

Gunther, John. *John Gunther's Inside Australia*. New York: Harper & Row, 1972.

Heathcote, R.L. *Australia*. London: Longman, Scientific & Technical, 1994.

Rickard, John. *Australia, A Cultural History*. London: Longman, 1996.

Azerbaijan

Altstadt, Audrey. *The Azerbaijani Turks*. Stanford, Calif.: Hoover Institution Press, 1992.

Gall, Timothy L., ed. *Worldmark Encyclopedia of the Nations*. 9th ed. Detroit: Gale Research, 1998.

Nichol, James. "Azerbaijan." in *Armenia, Azerbaijan and Georgia*. Area Handbook Series, Washington, DC: Government Printing Office, 1995, pp. 81148.

Bahrain

Crawford, Harriet. *Dilmun and its Gulf Neighbors*. Cambridge: Cambridge University Press, 1998.

Gall, Timothy L., ed. *Worldmark Encyclopedia of the Nations*. 9th ed. Detroit: Gale Research, 1998.

Nugent, Jeffrey B., and Theodore Thomas, eds. *Bahrain and the Gulf: Past Perspectives and Alternate Futures*. New York: St. Martin's Press, 1985.

Robison, Gordon. *Arab Gulf States*. Hawthorn, Aus.: Lonely Planet Publications, 1996.

Bangladesh

Baxter, Craig. *Bangladesh: A New Nation in an Old Setting*. Boulder and London: Westview Press, 1984.

Bigelow, Elaine. *Bangladesh: the Guide*. Dhaka: AB Publishers, 1995.

O'Donnell, Charles Peter. *Bangladesh: Biography of a Muslim Nation*. Boulder and London: Westview Press, 1984.

Newton, Alex. *Bangladesh: a Lonely Planet Travel Survival Kit*. Hawthorne, Victoria, Australia: Lonely Planet, 1996.

Republic of Bangladesh. U.S. Department of State Background Notes, November 1997.

Bhutan

Aris, Michael. *Bhutan: The Early History of a Himalayan Kingdom*. Warminster, England: Aris & Phillips, 1979.

Crossette, Barbara. *So Close to Heaven: The Vanishing Buddhist Kingdoms of the Himalayas*. New York: A. A. Knopf, 1995.

Matles, Andrea, ed. *Nepal and Bhutan: Country Studies.* Washington, DC: Federal Research Division, Library of Congress, 1993.

Pommaret-Imaeda, Françoise. Lincolnwood, Ill.: Passport Books, 1991.

Strydonck, Guy van. *Bhutan: A Kingdom in the Eastern Himalayas.* Boston: Shambala, 1985.

Brunei Darussalam

Bartholomew, James. *The Richest Man in the World: The Sultan of Brunei.* London: Penguin Group, 1990.

Major, John S. *The Land and People of Malaysia & Brunei.* New York: HarperCollins 1991.

Pigafetta, Antonio. *Magellan's Voyage.* Trans. and ed. by R. A. Skelton. New Haven: Yale University Press, 1969.

Singh, D.S. Ranjit. *Brunei 1839–1983: The Problems of Poltical Survival.* London: Oxford University Press, 1984.

Vreeland, N. et al. *Malaysia: A Country Study.* Area Handbook Series. Fourth edition. Washington, D.C.: Department of the Army, 1984.

Cambodia

Barron, John and Anthony Paul. *Murder of a Gentle Land. The Untold Story of Communist Genocide in Cambodia.* New York: Reader's Digest Press. 1977.

Becker, Elizabeth. *When the War Was Over: The Voices of Cambodia's Revolution and Its People.* New York: Simon and Schuster. 1986.

Chandler, David P. *A History of Cambodia.* Boulder, Colo.: Westview Press. 1983.

Gall, Timothy L., ed. *Worldmark Encyclopedia of Cultures and Daily Life.* vol. 3: Asia. Detroit: Gale Research, 1998.

Ponchaud, Francois. *Cambodia Year Zero.* London: Allen Lane. 1978.

China

Bailey, Paul. *China in the Twentieth Century.* New York: B. Blackwell, 1988.

Cotterell, Arthur. *China: A Cultural History.* New York: New American Library, 1988.

Ebrey, Patricia B. *The Cambridge Illustrated History of China.* New York: Cambridge University Press, 1996.

Hsu, Immanuel Chung-yueh. *The Rise of Modern China.* 5th ed. New York: Oxford University Press, 1995.

Worden, Robert L., Andrea Matles Savada, and Ronald E. Dolan, eds. *China, a Country Study.* 4th ed. Washington, D.C.: Library of Congress, 1988.

Cyprus (Greek and Turkish zones)

Crawshaw, Nancy. *The Cyprus Revolt: The Origins, Development, and Aftermath of an International Dispute.* Winchester, Mass.: Allen & Unwin, 1978.

Gall, Timothy L., ed. *Worldmark Encyclopedia of Cultures and Daily Life.* vol. 3: Asia. Detroit: Gale Research, 1998.

Salem, Norma, ed. *Cyprus: A Regional Conflict and its Resolution.* New York: St. Martin's Press, 1992.

Solsten, Eric, ed. *Cyprus, a Country Study.* 4th ed. Washington, D.C. Government Printing Office, 1993.

Tatton-Brown, Veronica. *Ancient Cyprus.* Cambridge, Mass.: Harvard University Press, 1988.

Federated States of Micronesia

Denoon, Donald, ed. *The Cambridge History of the Pacific Islanders.* Cambridge: Cambridge University Press, 1997.

Gall, Timothy L., ed. *Worldmark Encyclopedia of the Nations.* 9th ed. Detroit: Gale Research, 1998.

Levesque, Rodrigu, ed. *History of Micronesia.* Honolulu: University of Hawaii Press, 1994.

Spate, Oskar. *The Pacific Since Magellan, vol. 1: The Spanish Lake.* Canberra: Australian National University Press, 1979.

Fiji

Gall, Timothy L., ed. *Worldmark Encyclopedia of the Nations.* 9th ed. Detroit: Gale Research, 1998.

Ogden, Michael R. "Republic of Fiji," *World Encyclopedia of Political Systems,* 3rd edition, New York: Facts on File, 1999.

Scarr, Deryck. *Fiji: A Short History.* Honolulu: The Institute for Polynesian Studies, 1984.

"Vaughn, Roger. "The Two Worlds of Fiji," *National Geographic,* October 1995, vol. 88, no. 4, p. 114.

India

Heitzen, James and Robert L. Worden, eds. *India: A Country Study.* Washington, DC: Federal Research Division, Library of Congress, 1996.

Robinson, Francis, ed. *The Cambridge Encyclopedia of India, Pakistan, Bangladesh, Sri Lanka, Nepal, Bhutan and the Maldives.* Cambridge: Cambridge University Press. 1989.

Schwartzberg, Joseph E., ed. *A Historical Atlas of South Asia.* 2nd impression. New York and Oxford: Oxford University Press, 1992.

Wolpert, Stanley. *India.* Berkeley: University of California Press, 1991.

Indonesia

Bellwood, Peter S. *Prehistory of the Indo-Malaysian Archipelago.* Honolulu, Hawaii: University of Hawaii Press, 1997.

Broughton, Simon, ed. *World Music: The Rough Guide.* London: The Rough Guides Ltd., 1994.

Cribb, Robert. *Historical Dictionary of Indonesia.* Metuchen, N.J.: Scarecrow Press, 1992.

Frederick, William H., ed. *Indonesia: A Country Study.* Washington, D.C.: Library of Congress, 1993.

Schwarz, Adam. *A Nation in Waiting: Indonesia in the 1990s.* Boulder, Colo.: Westview Press, 1994.

Iran

Albert, David H. *Tell the American People: Perspectives on the Iranian Revolution.* Philadelphia: Movement for a New Society, 1980.

Bacharach, Jere L. *A Near East Studies Handbook, 570–1974.* Seattle: University of Washington Press, 1974.

Famighetti, Robert (ed.). *The World Almanac and Book of Facts.* New York: St. Martin's Press, 1998.

Nyrop, Richard F. (ed.). *Area Handbook of Iran: A Country Study.* Washington, D.C.: The American University, 1978.

Salinger, Pierre. *America Held Hostage: The Secret Negotiations.* Garden City, N.Y.: Doubleday & Company, 1981.

Iraq

Bulloch, John. *Saddam's War: The Origins of the Kuwait Conflict and the International Response.* Boston: Faber and Faber, 1991.

Chaliand, Gerard, ed. *A People without a Country: The Kurds and Kurdistan.* New York: Olive Branch Press, 1993.

Mansfield, Peter. *A History of the Middle East.* New York: Viking, 1991.

Metz, Helen Chapin, ed. *Iraq: A Country Study.* 4th ed. Washington, D.C.: Library of Congress, Federal Research Division, 1990.

Simons, Geoff L. *Iraq: From Sumer to Saddam.* New York: St. Martin's Press, 1994.

Israel

Blumberg, Arnold. *The History of Israel.* Westport, Conn.: Greenwood Press, 1998.

———. *Zion Before Zionism.* Syracuse, N.Y.: Syracuse University Press, 1985.

Grant, Michael. *The History of Ancient Israel.* New York: Charles Scribner's Sons, 1984.

Metz, Helen Chapin. *Israel: A Country Study.* Washington, DC : Library of Congress, 1990

Sachar, Abram Leon. *A History of the Jews.* New York: Alfred A. Knopf, 1955.

Shanks, Hershel. *Ancient Israel.* Englewood Cliffs, N.J.: Prentice-Hall, 1988.

Japan

Demente, Boye Lafayette. *Japan Encyclopedia.* Lincolnwood, Ill.: NTC Publishing, 1995.

Dolan, Ronald E., and Robert L. Dolan, eds. *Japan, a Country Study.* 5th ed. Washington, D.C.: Library of Congress, 1992.

Perkins, Dorothy. *Encyclopedia of Japan: Japanese History and Culture, from Abacus to Zori.* New York: Facts on File, 1991.

Richardson, Bradley M. *Japanese Democracy: Power, Coordination, and Performance.* New Haven, Conn.: Yale University Press, 1997.

Thomas, J. E. *Modern Japan: A Social History Since 1868.* New York: Longman, 1996.

Jordan

Ali, Wijdan. *Modern Islamic Art: Development and Continuity.* University Press of Florida, Gainesville. Fla. 1997.

Contreras, Joseph and Christopher Dickey. "The Day After: King Hussein's Second Bout of Cancer Raises Questions About Jordan's Political Future." *Newsweek,* Vol. 132, No. 6, p. 38, August 10, 1998.

Shahin, Mariam. "Cracks Become Chasms," *The Middle East.* December 1997, No. 273, p. 13-14.

Kazakstan

Conflict in the Soviet Union: The Untold Story of the Clashes in Kazakhstan. New York: Human Rights Watch, 1990.

Edwards-Jones, Imogen. *The Taming of Eagles: Exploring the New Russia.* London: Weidenfeld & Nicolson, 1992.

Kalyuzhnova, Yelena. *Kazakstani Economy: Independence and Transition.* New York: St. Martin's Press, 1998.

Olcott, Martha Brill. *The Kazakhs.* Stanford, CA: Hoover Institution Press, Stanford University, 1987.

World Bank. *Kazakhstan: The Transition to a Market Economy.* Washington, DC: World Bank, 1993.

Kiribati

"Bones Found in '40 May Have Been Hers," *Honolulu Star-Bulletin,* December 3, 1998.

Krauss, Bob. "Heroes Heed Call From Sea," *Honolulu Advertiser,* March 10, 1999.

MacDonald, Barrie. *Cinderellas of the Empire: Towards a History of Kiribati and Tuvalu.* Canberra, Australia: Australian National University Press, 1982.

"New Meaning to the Term Down Under. The Tiny Islands of Kiribati, Tuvalu and Nauru Are Pressuring Australia to Reduce Greenhouse Gas Emissions.)" *The Economist (US),* September 27, 1997, vol. 344, no. 8036, p. 41.

"South Pacific Group Cuts French Ties," *Honolulu Star-Bulletin,* October 3, 1995.

Thompson, Rod. "Hilo Plans Gift for Christmas Isle Neighbors, " *Honolulu Star-Bulletin,* December 18, 1998.

Korea, Democratic People's Republic of

Gills, Barry K. *Korea Versus Korea: A Case of Contested Legitimacy.* New York: Routledge, Inc., 1996.

Oliver, Robert Tarbell. *A History of the Korean People in Modern Times: 1800 to the Present.* Newark,: University of Delaware Press, 1993.

Smith, Hazel. *North Korea in the New World Order.* New York: St. Martin's Press, 1996.

Soh, Chung Hee. *Women in Korean Politics.* 2nd ed. Boulder, Colo.: Westview Press, 1993.

Tennant, Roger. *A History of Korea.* London: Kegan Paul International, 1996.

Korea, Republic of (ROK)

Gills, Barry K. *Korea Versus Korea: A Case of Contested Legitimacy.* New York: Routledge, Inc., 1996.

Lone, Stewart. *Korea Since 1850.* New York: St. Martin's Press, 1993.

Oberdorfer, Don. *The Two Koreas: A Contemporary History.* Reading, Mass.: Addison-Wesley Pub. Co., 1997.

Oliver, Robert Tarbell. *A History of the Korean People in Modern Times: 1800 to the Present.* Newark: University of Delaware Press, 1993.

Tennant, Roger. *A History of Korea*. London: Kegan Paul International, 1996.

Kuwait

Ali, Wijdan. *Modern Islamic Art: Development and Continuity*. Gainesville, FL:

Anscombe, Frederick F. *The Ottoman Gulf: The Creation of Kuwait, Saudi Arabia and Qatar*. New York: Columbia University Press, 1997.

Cordesman, Anthony H. *Kuwait: Recovery and Security after the Gulf War*. Boulder, Colo.: Westview Press, 1997.

Crystal, Jill. *Oil and Politics in the Gulf: Rulers and Merchants in Kuwait and Qatar*. New York: Cambridge University Press, 1990.

Khadduri, Majid. *War in the Gulf, 1990–1991: The Iraq-Kuwait Conflict and its Implications*. New York: Oxford University Press, 1997.

Robison, Gordon. *Arab Gulf States*. Hawthorn, Aus.: Lonely Planet Publications, 1996.

Kyrgyzstan

Attokurov, S. *Kïrgïz Sanjïrasï*. Bishkek: Kyrgyzstan, 1995.

Bennigsen, Alexandre & S. Enders Wimbush. *Muslims of the Soviet Empire: A Guide*. Bloomington & Indianapolis: Indiana Universiity Press, 1986.

Huskey, Eugene. "Kyrgyzstan: The politics of demographic and economic frustration." in: *New States, New Politics: Building the Post-Soviet Nations*, edited by Ian Bremmer and Ray Taras, Cambridge & New York: Cambridge University Press, 1997, 655-680.

Krader, Lawrence. *The Peoples of Central Asia*. Bloomington, Indiana: Uralic and Altaic Series, vol. 26, 1966.

Olcott, Martha B. "Kyrgyzstan." in: *Kazakstan, Kyrgyzstan, Tajikistan, Turkmenistan and Uzbekistan* Area Handbook Series, Washington D.C.: Government Printing Office, 1997, 99-193.

Laos

Cummings, Joe. *A Golden Souvenir of Laos*. New York: Asia Books. 1996.

Kremmer, Chistopher. *Stalking the Elephant King. In Search of Laos*. Honolulu, University of Hawai'i Press. 1997.

Scott, Joanna C. *Indochina's Refugees: Oral Histories from Laos, Cambodia, and Vietnam*. Jeferson, NC: McFarland. 1989.

Stieglitz, P. *In a Little Kingdom*. New York: M. E. Sharpe. 1990.

Stuart-Fox, Martin. *A History of Laos*. New York: Cambridge University Press. 1997.

Lebanon

Abul-Husn, Latif. *The Lebanese Conflict: Looking Inward*. Boulder, Colo.: Lynne Rienner Publishers, 1998.

Bleaney. C.H. *Lebanon: Revised Edition*. World Bibliographical Series, Volume 2, Oxford, Eng.: Clio Press, Ltd., 1991.

King-Irani, Laurie. "War Gods Roar Again, Appear Unstoppable: Jet Streaks So Close She Could See Pilot," *National Catholic Reporter*. Vol. 34, No. 17, p. 9. February 27, 1998.

Norton, Augustus Richard. "Hizballah: From Radicalism to Pragmatism?" *Middle East Policy*. Vol. 5, No. 4, pp. 174-186. January 1998.

Tuttle, Robert. "Americans return to AUB," *The Middle East*. No. 272, p. 42. November 1997.

Malaysia

Bellwood, Peter S. *Prehistory of the Indo-Malaysian Archipelago*. Honolulu,: University of Hawaii Press, 1997.

Bevis, William W. *Borneo Log: The Struggle for Sarawak's Forests*. Seattle,: University of Washington Press, 1995.

Broughton, Simon, ed. *World Music: The Rough Guide*. London: The Rough Guides Ltd., 1994.

Kahn, Joel S. and Loh Kok Wah, Francis, eds. *Fragmented Vision: Culture and Politics in Contemporary Malaysia*. Honolulu,: University of Hawaii Press, 1992.

Kaur, Amarjit. *Historical Dictionary of Malaysia*. Metuchen, NJ: Scarecrow Press, 1993.

Maldives

Ellis, Kirsten. *The Maldives*. Hong Kong: Odyssey, 1993.

Republic of Maldives, Office for Women's Affairs. *Status of Women, Maldives*. Bangkok: UNESCO Principal Regional Office for Asia and the Pacific, 1989.

Marshall Islands

Johnson, Giff. *Collision Course at Kwajalein: Marshall Islanders in the Shadow of the Bomb*. Honolulu: Pacific Concerns Resource Center, 1984.

Langley, Jonathan, and Wanda Langley. "The Marshall Islands." *Skipping Stones,* Winter 1995, vol. 7, no. 5, p. 17+.

Levesque, Rodrigu, ed. *History of Micronesia*. Honolulu: University of Hawaii Press, 1994.

Scarr, Deryck. *History of the Pacific Islands: Kingdom of the Reefs*. Sydney: Macmillan Australia, 1990.

Mongolia

Bergholz, Fred W. *The Partition of the Steppe: The Struggle of the Russians, Manchus, and the Zunghar Mongols for Empire in Central Asia, 1619–1758*. New York: Peter Lang Publishing, Inc., 1993.

Bruun, Ole and Ole Odgaard, eds. *Mongolia in Transition*. Richmond, England: Curzon Press Ltd., 1996.

de Hartog, Leo. *Russia and the Mongol Yoke: The History of the Russian Principalities and the Golden Horde, 1221–1502*. London: British Academic Press, 1996.

Greenway, Paul. *Mongolia*. Hawthorn, Australia: Lonely Planet Publications, 1997.

Sanders, Alan J. K. *Historical Dictionary of Mongolia*. Lanham, Md: Scarecrow Press, Inc., 1996.

Myanmar (Burma)

Diran, Richard K. *The Vanishing Tribes of Burma*. New York: Amphoto Art, 1997.

Parenteau, John. *Prisoner for Peace: Aung San Suu Kyi and Burma's Struggle for Democracy*. Greensboro, N.C. : Morgan Reynolds, 1994.

Renard, Ronald D. *The Burmese Connection: Illegal Drugs and the Making of the Golden Triangle*. Boulder, Colo.: L. Rienner Publishers, 1996.

Rotberg, Robert I., ed. *Burma: Prospects for a Democratic Future*. Brookings Institute Press, 1998.

Silverstein, Josef. *The Political Legacy of Aung San*. Ithaca, N.Y.: Cornell University Press, 1993.

Nauru

Bunge, Frederica and Melinda Cooke, eds. *Oceania: a regional study*. Washington, D.C.: U.S. Government Printing Office, 1984.

Hanlon, David. *Remaking Micronesia*. Honolulu: University of Hawai'i Press, 1998.

McKnight, Tom. *Oceania: the geography of Australia, New Zealand, and the Pacific Islands*. Englewood Cliffs: Prentice Hall, 1995.

"New Meaning to the Term Down Under. The Tiny Islands of Kiribati, Tuvalu and Nauru Are Pressuring Australia to Reduce Greenhouse Gas Emissions.)" *The Economist (US)*, September 27, 1997, vol. 344, no. 8036, p. 41.

Viviani, Nancy. *Nauru: phosphate and political progress*. Honolulu: University of Hawaii Press, 1970.

Nepal

Bista, Dor Bahadur. *People of Nepal*. Kathmandu: Ratna Pustak Bhandar, 1987.

Chauhan, R. S. *Society and State Building in Nepal: From Ancient Times to Mid-Twentieth Century*. New Delhi: Sterling, 1989.

Matles, Andrea, ed. *Nepal and Bhutan: Country Studies*. Washington, D.C.: Federal Research Division, Library of Congress, 1993.

Sanday, John. *The Kathmandu Valley: Jewel of the Kingdom of Nepal*. Lincolnwood, IL.: Passport Books, 1995.

Seddon, David. *Nepal, a State of Poverty*. New Delhi: Vikas, 1987.

New Zealand

Belich, James. *Making Peoples: A History of the New Zealanders from Polynesian Settlement to the End of the Nineteenth Century*. Honolulu: University of Hawaii Press, 1996.

Mascarenhas, R.C. *Government and the Economy in Australia and New Zealand: The Politics of Economic Policy Making*. San Francisco: Austin & Winfield, 1996.

McKinnon, Malcolm. *Independence and Foreign Policy: New Zealand in the World since 1935*. Auckland: Oxford University Press, 1993.

Oddie, Graham, and Roy W. Perrett, ed. *Justice, Ethics, and New Zealand Society*. New York: Oxford University Press, 1992.

Rice, Geoffrey W., ed. *The Oxford History of New Zealand*. 2nd ed. New York: Oxford University Press, 1992.

Oman

Arab Gulf States: A Travel Survival Kit. 2d edition. Melbourne Aus.: Lonely Planet Publications, 1996.

Miller, Judith. "Creating Modern Oman: An Interview With Sultan Qabus." *Foreign Affairs*, May–June 1997, Vol. 76, No. 3, pp. 1318.

Molavi, Afshin. "Oman's Economy: Back on Track." *Middle East Policy*, January 1998, Vol. 5, No. 4, pp. 110.

Osborne, Christine. "Omani Forts win Heritage Award." *The Middle East*, November 1995, No. 250, pp. 4344.

Riphenburg, Carol. J. *Oman: Political Development in a Changing World*. Westport, CT: Praeger Publishers, 1998.

Pakistan

Blood, Peter R., ed. *Pakistan, a Country Study*. 6th ed. Washington, D.C.: Federal Research Division, Library of Congress, 1995.

Burki, Shahid Javed. *Historical Dictionary of Pakistan*. Metuchen, NJ: Scarecrow Press, 1991.

Mahmud, S. F. *A Concise History of Indo-Pakistan*. Karachi: Oxford University Press, 1988.

Schwartzberg, Joseph E., ed. *A Historical Atlas of South Asia*. 2nd impression. Oxford and New York: Oxford University Press, 1992.

Taylor, David (revised by Asad Sayeed). "Pakistan: Economy," in *The Far East and Australasia 1997*. London: Europa Publications, 1996, pp. 873-79.

Palau

Hanlon, David. *Remaking Micronesia*. Honolulu: University of Hawai'i Press, 1996.

Hijikata, Hisakatsu. *Society and Life in Palau*. Tokyo: Sasakawa Peace Foundation, 1993.

Liebowitz, Arnold. *Embattled Island: Palau's Struggle for Independence*. Westport, CT: Praeger, 1996.

Morgan, William. *Prehistoric Architecture in Micronesia*. Austin: The University of Texas Press, 1988.

Roff, Sue Rabbitt. *Overreaching in Paradise: United States Policy in Palau Since 1945*. Juneau, Alaska: Denali Press, 1991.

Papua New Guinea

Campbell, I.C. *A History of the Pacific Islands*. Berkeley: University of California Press, 1989.

"Death of a Peacemaker." *The Economist*, October 19, 1996, vol. 341, no. 7988, p. 41.

Grattan, C. Hartley. *The Southwest Pacific since 1900*. Ann Arbor: University of Michigan Press, 1963.

Sinclair, James. *Papua New Guinea: the First 100 Years*. Bathurst: Robert Brown and Associates, 1985.

Spriggs, Matthew. *The Island Melanesians.* Cambridge, Mass.: Blackwell Publishers, 1997.

Philippines

Brands, H.W. *Bound to Empire: The United States and the Philippines.* New York: Oxford Univ. Press, 1992.

Dolan, Ronald E., ed. *Philippines: A Country Study.* 4th ed. Washington, D.C.: Library of Congress, 1993.

Karnow, Stanley. *In Our Image: America's Empire in the Philippines.* New York: Random House, 1989.

Steinberg, David Joel. *The Philippines, a Singular and a Plural Place.* 3rd ed. Boulder, Colo.: Westview, 1994.

Thompson, W. Scott. *The Philippines in Crisis: Development and Security in the Aquino Era 198692.* New York: St. Martin's 1992.

Qatar

Abu Saud, Abeer. *Qatari Women Past and Present.* Essex, Eng.: Longman Group Limited, 1984.

Anscombe, Frederick F. *The Ottoman Gulf: The Creation of Kuwait, Sa'udi Arabia, and Qatar.* New York: Columbia Press, 1997.

Crystal, Jill. *Oil and Politics in the Gulf: Rulers and Merchants in Kuwait and Qatar.* New York: Cambridge University Press, 1995.

Samoa

Fialka, John J. "From Dots in the Pacific, Envoys Bring Fear, Fury to Global-Warming Talks," *Wall Street Journal,* September 31, 1997, A24.

Hallowell, Christopher. "Rainforest Pharmacist," *Audubon* 101:1 (January 1999), 28.

Holmes, Lowell D., and Ellen Rhoads Holmes. *Samoan Village Then and Now.* Orlando, Fla.: Holt, Rinehart and Winston Inc., 1974, repr. 1992.

Meleisea, Malama. *Change and Adaptation in Western Samoa.* Christchurch, New Zealand: MacMillan Brown Centre for Pacific Studies, 1992.

Samoa: A Travel Survival Kit. Sydney, Aus.: Lonely Planet Publications, 1996.

Sa'udi Arabia

Abir, Mordechai. *Saudi Arabia: Government, Society, and the Gulf Crisis.* London and New York: Routledge, 1993.

Caesar, Judith. *Crossing Borders: An American Woman in the Middle East.* Syracuse, NY: Syracuse University Press, 1997.

Long, David E. *The Kingdom of Saudi Arabia.* Gainesville, Fla.: University Press of Florida, 1997.

Peterson, J. J. *Historical Dictionary of Saudi Arabia.* Metuchen, N.J.: Scarecrow Press, 1993.

Wilson, Peter W. and Douglas F. Graham. *Saudi Arabia: The Coming Storm.* Armonk, N.Y.: M. E. Sharpe, 1994.

Singapore

Chew, Ernest and Edwin Chew, ed. *A History of Singapore.* New York: Oxford University Press, 1991.

Chiu, Stephen Wing-Kai. *City States in the Global Economy: Industrial Restructuring in Hong Kong and Singapore.* Boulder, Colo.: Westview Press, 1997.

Lee, W.O. *Social Change and Educational Problems in Japan, Singapore, and Hong Kong.* New York: St. Martin's Press, 1991.

LePoer, Barbara Leitch, ed. *Singapore: A Country Study.* 2nd ed. Washington, D.C.: Library of Congress, 1991.

Trocki, Carl A. *Opium and Empire: Chinese Society in Colonial Singapore.* Ithaca, N.Y.: Cornell University Press, 1990.

Solomon Islands

Bennett, Judith. *Wealth of the Solomons: A History of a Pacific Archipelago, 1800–1978.* Honolulu: University of Hawai'i Press, 1987.

Denoon, Donald, ed. *The Cambridge History of the Pacific Islanders.* Cambridge: Cambridge University Press, 1997.

White, Geoffrey and Lindstrom, Lamont, eds. *The Pacific Theater: island representations of World War II.* Honolulu: University of Hawai'i Press, 1989.

Sri Lanka

Anderson, John Gottberg, and Ravindral Anthonis, editors. *Sri Lanka.* Hong Kong: Apa Productions, 1993.

Baker, Victoria J. *A Sinhalese Village in Sri Lanka: Coping with Uncertainty.* Ft. Worth: Harcourt Brace College Publishers, 1998.

Robinson, Francis, editor. *The Cambridge Encyclopedia of Pakistan, Bangladesh, Sri Lanka, Nepal, Bhutan and the Maldives.* Cambridge: Cambridge University Press., 1989.

Ross, Russell R., et al. *Sri Lanka: A Country Study* (Area Handbook Series). Washington, D.C.: Library of Congress, 1990.

Vesilind, Pritt J. "Sri Lanka." *National Geographic,* January 1997, vol. 191, no. 1, pp. 110+.

Syria

Dourian, Kate. "City of Apamea, Once Lost in Sand, Partially Restored." *Washington Times.* September 3, 1994.

Katler, Johannes. *The Arts and Crafts of Syria.* Thames and Hudson: London, 1992.

Kayal, Michele. "Ruins to Riches." *Washington Post.* February 13, 1994.

LaFranchl, Howard. "Ancient Syria's History Rivals That of Egypt, Mesopotamia." *The Christian Science Monitor.* February 17, 1994.

Parmelee, Jennifer. "Tracking Agatha Christie." *The Washington Post.* August 25, 1991.

"Syria." *Background Notes.* Washington, D.C.: Central Intelligence Agency, November 1994.

Syria: A Country Report. Washinton, D.C.: Library of Congress Research Division, 1999.

Taiwan

Hood, Steven J. *The Kuomintang and the Democratization of Taiwan*. Boulder, Colo.: Westview, 1997.

Long, Simon. *Taiwan: China's Last Frontier*. New York: St. Martin's Press, 1990.

Marsh, Robert. *The Great Transformation: Social Change in Taipei, Taiwan Since the 1960s*. Armonk, N.Y.: M.E. Sharpe, 1996.

Shepherd, John Robert. *Statecraft and Political Economy on the Taiwan Frontier, 1600–1800*. Stanford, Calif.: Stanford University Press, 1993.

Tajikistan

Rakhimov, Rashid, et al. *Republic of Tajikistan: Human Development Report 1995,* Istanbul, 1995.

Rashid, Ahmed. *The Resurgence of Central Asia or Nationalism?,* Zed Books, 1995.

Thailand

Dixon, C. J. *The Thai Economy: Uneven Development and Internationalisation*. London: Routledge, 1999.

Pattison, Gavin and John Villiers. *Thailand*. New York: Norton, 1997.

West, Richard. *Thailand, the Last Domino: Cultural and Political Travels*. London: Michael Joseph, 1991.

Tonga

Aswani, Shankae and Michael Graves. "The Tongan Maritime Expansion." *Asian -Perspectives* 1998. 37: 135201.

———. *Island Kingdom: Tonga Ancient and Modern*. Christchurch, NZ: Canterbury University Press, 1992.

Denoon, Donald, ed. *The Cambridge History of the Pacific Islanders*. Cambridge: Cambridge University Press, 1997.

Perminow, Arne. *The Long Way Home: Dilemmas of Everyday Life in a Tongan Village*. Oslo: Scandinavian University Press, 1993.

Turkmenistan

Edwards-Jones, Imogen. *The Taming of Eagles: Exploring the New Russia*. London: Weidengeld and Nicolson, 1993.

Hunter, Shireen T. *Central Asia Since Independence*. Westport, CN: Praeger, 1996.

Kazakhstan, Kyrgyzstan, Tajikistan, Turkmenistan, and Uzbekistan, country studies. Washington: US Government Printing Office, 1997.

Maslow, Jonathan Evan. *Sacred Horses: The Memoirs of a Turkmen Cowboy*. New York: Random House, 1994.

Olcott, Martha Brill. *Central Asia's New States: Independence, Foreign Policy, and Regional Security*. Washington: United States Institute of Peace Press, 1996.

Tuvalu

Adams, Wanda. "'Double Ghosts' remembers early traders," *Honolulu Advertiser*, March 29, 1998.

Chappell, David. *Double Ghosts: Oceanic Voyagers on Euroamerican Ships*. Armonk, N.Y.: M.E. Sharpe Press, 1997.

Friendship and Territorial Sovereignty: Treaty Between the United States of America and Tuvalu, signed at Funafuti February 7, 1979. Washington, D.C.: U.S. Government Printing Office, 1985.

"New Meaning to the Term Down Under. The Tiny Islands of Kiribati, Tuvalu and Nauru Are Pressuring Australia to Reduce Greenhouse Gas Emissions.)" *The Economist (US)*, September 27, 1997, vol. 344, no. 8036, p. 41.

Kristoff, Nicholas D. "In Pacific, Growing Fear of Paradise Engulfed," *New York Times*, March 2, 1997.

United Arab Emirates

Ali Rashid, Noor. *The UAE Visions of Change*. Dubai: Motivate Publishing, 1997.

Forman, Werner. *Phoenix Rising: the United Arab Emirates, Past, Present & Future*. London: Harvill Press, 1996.

Peck, Malcolm C. *The United Arab Emirates: A Venture in Unity*. Boulder, Colorado: Westview Press, 1986.

Taryam, Abdullah Omran. *The Establishment of the United Arab Emirates 1950–85*. London: Croom-Helm, 1987.

Zahlan, Rosemarie Said. *The Making of the Modern Gulf States*. London: Unwin Hyman Ltd., 1989.

Uzbekistan

Alworth, Edward A. *The Modern Uzbeks: From the Fourteenth Century to the Present*. Stanford, Calif.: Hoover Institution Press, 1990.

Alworth, Edward A., ed. *Central Asia: 130 Years of Russian Dominance, A Historical Overview*. Durham, N.C.: Duke University Press, 1994.

MacLeod, Calum and Bradley Mayhew. *Uzbekistan*. Lincolnwood, Ill.: Passport Books, 1997.

Rashid, Ahmed. *The Resurgence of Central Asia: Islam or Nationalism?* Atlantic Highlands, N.J.: Zed Books, 1994.

Undeland, Charles and Nicholas Platt. *The Central Asian Republics: Fragments of Empire*. New York: The Asia Society, 1994.

Vanuatu

Allen, Michael, ed. *Vanuatu: Politics, Economics, and Ritual in Island Melanesia*. Sydney, Aus.: Academic Press, 1981.

Jennings, Jesse, ed. *The Prehistory of Polynesia*. Cambridge, MA: Harvard University Press, 1979.

Speiser, F. *Ethnology of Vanuatu*. translated by D. Stephenson. Hawaii: University of Hawaii Press, 1996 [1923].

Stanley, D. *South Pacific Handbook,* 3d ed. Chico, Calif.: Moon Publications, 1986.

Viet Nam

Hickey, Gerald Cannon. *Free in the Forest: Ethnohistory of the Vietnamese Central Highlands, 1954–1976*. New Haven: Yale University Press, 1982.

Kahin, George McT. *Intervention: How America Became Involved in Vietnam.* Garden City, N.Y.: Anchor Books. 1987.

Karnow, Stanley. *Vietnam. A History.* New York: The Viking Press. 1983.

Sheehan, Neil. *A Bright Shining Lie.* New York: Random House. 1988.

Thuy, Vuong G. *Getting to the Know the Vietnamese and Their Culture.* New York: Frederick Ungar Publishing. 1975.

Yemen

Al-Suwaidi, Jamal. *The Yemeni War of 1994: Causes and Consequences.* London: Saqi Books, 1995.

Carapico, Sheila. *Civil Society in Yemen : The Political Economy of Activism in Modern Arabia.* New York: Cambridge University Press, 1998.

Chaudhry, Kiren Aziz. *The Price of Wealth: Economies and Institutions in the Middle East.* Ithaca, NY: Cornell University Press, 1997.

Crouch, Michael. *An Element of Luck: To South Arabia and Beyond.* New York: Radcliffe Press, 1993.

Halliday, Fred. *Revolution and Foreign Policy: The Case of South Yemen, 1967–87.* New York: Cambridge University Press, 1990.

Europe

Albania

Battiata, Mary. "Albania's Post-Communist Anarchy," *Washington Post,* March 21, 1998, A1–A18.

Biberaj, Ekiz. *Albania: A Sicuakust Maverick.* Boulder, Colo.: Westview Press, 1990.

Durham, M.E. *High Albania.* Boston: Beacon Press, l985.

Pipa, Arshi. *The Politics of Language in Socialist Albania.* New York: Columbia University Press for Eastern European Monographs, l989.

Andorra

Carter, Youngman. *On to Andorra.* New York: W. W. Norton, 1964.

Deane, Shirley. *The Road to Andorra.* New York: William Morrow, 1961.

Duursma, Jorri. *Fragmentation and the International Relations of Micro-States: Self-determination and Statehood.* Cambridge: Cambridge University Press, 1996.

Armenia

Batalden, Stephen K. and Sandra L. Batalden. *The Newly Independent States of Eurasia: Handbook of Former Soviet Republics.* Phoenix, AZ: The Oryx Press, 1993.

Croissant, Michael P. *The Armenia-Azerbaijani Conflict: Causes and Implications.* Westport, Conn: Praeger, 1998.

Curtis, Glenn E., ed. *Armenia, Azerbaijan, and Georgia: Country Studies.* Washington, D.C.: Federal Research Division, Library of Congress, 1995.

Economist Intelligence Unit, The. *Country Profile: Georgia, Armenia, 1998–1999.* London: The Economist Intelligence Unit, 1999.

Goldberg, Suzanne. *Pride of Small Nations: The Caucasus and Post-Soviet Disorder.* Atlantic Heights, N.J.: Zed, 1994.

Kaeger, Walter Emil. *Byzantium and the Early Islamic Conquests.* Cambridge: University Press, 1992.

Lang, David Marshall. *Armenia: Cradle of Civilization.* London, Boston: Allen and Unwin, 1978.

McEcedy, Colin. *The New Penguin Atlas of Medieval History.* London: Penguin, 1992.

Austria

Barkey, Karen and Mark von Hagen. *After Empire: Multiethnic Societies and Nation-Building: the Soviet Union and Russian, Ottoman, and Habsburg Empires.* Boulder, Colo.: Westview Press, 1997.

Brook-Shepherd, Gordon. *The Austrians: A Thousand-Year Odyssey.* London: Harper Collins, 1996.

Johnson, Lonnie. *Introducing Austria: A Short History.* Riverside, Calif.: Ariadne, 1989.

Steininger, Rolf, and Michael Gehler, eds. *Österreich im 20. Jahrhundert.* 2 vols. Vienna: Böhlau, 1997.

Belarus

Gross, Jan Tomasz. *Revolution from Abroad: The Soviet Conquest of Poland's Western Ukraine and Western Belorussia.* Princeton, N.J: Princeton University Press, 1988.

Marples, David R. *Belarus: From Soviet Rule to Nuclear Catastrophe.* London: Macmillan, 1996.

Sword, Keith, ed. *The Soviet Takeover of the Polish Eastern Provinces, 1939–41.* New York: St. Martin's Press, 1991.

Zaprudnik, I.A. *Belarus: At a Crossroads in History.* Boulder, Colo.: Westview Press, 1993.

Belgium

Fitzmaurice, John. *The Politics of Belgium: A Unique Feudalism.* London: Hurst, 1996.

Files, Yvonne. *The Quest for Freedom: The Life of a Belgian Resistance Fighter.* Santa Barbara, Calif.: Fithian Press, 1991.

Hilden, Patricia. *Women, Work, and Politics: Belgium 1830–1914.* Oxford: Clarendon Press, 1993.

Hooghe, Liesbet. *A Leap in the Dark: Nationalist Conflict ad Federal Reform in Belgium.* Ithaca: Cornell University Press, 1991.

Warmbrunn, Werner. *The German Occupation of Belgium: 1940–1944.* New York: P. Lang, 1993.

Wee, Herman van der. *The Low Countries in Early Modern Times.* Brookfield, Vt.: Variorum, 1993.

Bosnia and Herzegovina

Burg, Steven L., and Paul S. Shoup. *The War in Bosnia-Herzegovina: Ethnic Conflict and International Intervention.* Armonk, N.Y.: M.E. Sharpe, 1999.

Filipovic, Zlata. *Zlata's Diary: A Child's Life in Sarajevo.* New York: Viking, 1994.

Glenny, Misha. *The Fall of Yugoslavia: The Third Balkan War.* New York: Penguin, 1996.

Lampe, John R. *Yugoslavia as History: Twice There Was a Country.* Cambridge: Cambridge University Press, 1996.

Pinson, Mark, ed. *The Muslims of Bosnia-Herzegovina: Their Historic Development from the Middle Ages to the Dissolution of Yugoslavia.* 2nd ed. Cambridge, Mass.: Harvard University Press, 1996.

Prstojevic, Miroslav. *Sarajevo Survival Guide.* Trans. Aleksandra Wagner with Ellen Elias-Bursac. New York: Workman Publishing, 1993.

Rogel, Carole. *The Breakup of Yugoslavia and the War in Bosnia.* Westport, Conn.: Greenwood, 1998.

West, Rebecca. *Black Lamb and Grey Falcon.* Reprint. New York: Penguin Books, 1982

Bulgaria

Crampton, R.J. *A Concise History of Bulgaria.* Cambridge, New York: Cambridge University Press, 1997.

Curtis, Glenn E., ed. *Bulgaria, a Country Study.* 2nd ed. Federal Research Division, Library of Congress. Washington, D.C., 1993.

Melone, Albert P. *Creating Parliamentary Government: The Transition to Democracy in Bulgaria.* Columbus: Ohio State University Press, 1998.

Minaeva, Oksana. *From Paganism to Christianity: Formation of Medieval Bulgarian Art (681–972).* Frankfurt am Main, New York: P. Lang, 1996.

Paskaleva, Krassira, ed. *Bulgaria in Transition: Environmental Consequences of Political and Economic Transformation.* Brookfield, VT: Ashgate Publications, 1998.

Sedlar, Jean W. *East Central Europe in the Middle Ages, 1000–1500. A History of East Central Europe,* 3. Seattle and London: University of Washington Press, 1994.

Croatia

Cuvalo, Ante. *The Croatian National Movement, 1966–72.* New York: Columbia University Press, 1990.

Glenny, Michael. *The Fall of Yugoslavia: The Third Balkan War.* New York: Penguin, 1992.

Irvine, Jill A. *The Croat Question: Partisan Politics in the Formation of the Yugoslav Socialist State.* Boulder, Colo.: Westview Press, 1993.

Tanner, Marcus. *Croatia: A Nation Forged in War.* New Haven, Conn.: Yale University Press, 1997.

Czech Republic

Bradley, J. F. N. *Czechoslovakia's Velvet Revolution: A Political Analysis.* New York: Columbia University Press, 1992.

Kalvoda, Josef. *The Genesis of Czechoslovakia.* New York: Columbia University Press, 1986.

Kriseova, Eda. *Vaclav Havel: The Authorized Biography.* New York: St. Martin's Press, 1978.

Leff, Carol Skalnik. *The Czech and Slovak Republics: Nation Versus State.* Boulder, Colo.: Westview Press, 1997.

Denmark

Kjrgaard, Thorkild. *The Danish Revolution, 1500–1800: An Ecohistorical Interpretation.* Cambridge: Cambridge University Press, 1994.

Monrad, Kasper. *The Golden Age of Danish Painting.* New York: Hudson Hills Press, 1993.

Pundik, Herbert. *In Denmark It Could Not Happen: The Flight of the Jews to Sweden in 1943.* New York: Gefen Publishing House, 1998.

Estonia

Gerner, Kristian, and Stefan Hedlund. *The Baltic States and the End of the Soviet Empire.* New York: Routledge, 1993.

Hiden, John and Patrick Salmon. *The Baltic Nations and Europe.* London and New York: Longman, 1994.

Iwaskiw, Walter R. *Estonia, Latvia, and Lithuania: Country Studies.* Washington, DC: Federal Research Division, Library of Congress, 1996.

Lieven, Anatol. *The Baltic Revolution.* New Haven and London: Yale University Press, 1993.

Taagepera, Rein. *Estonia: Return to Independence.* Boulder, Colo.: Westview Press, 1993.

Finland

Lander, Patricia Slade. *The Land and People of Finland.* New York: HarperCollins, 1990.

Maude, George. Historical Dictionary of Finland. Metuchen, N.J.: Scarecrow Press, 1994.

.Rajanen, Aini. *Of Finnish Ways.* New York: Barnes & Noble Books, 1981.

Schoolfield, George C. *Helsinki of the Czars: Finland's Capital, 1808–1918.* Columbia, SC: Camden House, 1996.

Singleton, Fred. *A Short History of Finland,* 2nd ed. Cambridge, Eng.: Cambridge University Press, 1998.

France

Agulhon, Maurice. *The French Republic, 1879–1992.* Cambridge, Mass.: B. Blackwell, 1993.

Corbett, James. *Through French Windows: An Introduction to France in the Nineties.* Ann Arbor, Mich.: University of Michigan Press, 1994.

Gildea, Robert. *France Since 1945.* Oxford: Oxford University Press, 1996.

Gough, Hugh, and John Horne. *De Gaulle and Twentieth-Century France.* New York: Edward Arnold, 1994.

Hollifield, James F., and George Ross, eds. *Searching for the New France.* New York: Routledge, 1991.

Noiriel, Gérard. *The French Melting Pot: Immigration, Citizenship, and National Identity.* Minneapolis, Minn.: University of Minnesota Press, 1996.

Northcutt, Wayne. *The Regions of France: A Reference Guide to History and Culture.* Westport, Conn.: Greenwood Press, 1996.

Young, Robert J. *France and the Origins of the Second World War.* Basingstoke, England: Macmillan, 1996.

Georgia

Braund, David. *Georgia in Antiquity: A History of Colchis and Transcaucasian Iberia, 550 BC–AD 562*. New York: Oxford University Press, 1994.

Goldstein, Darra. *The Georgian Feast: The Vibrant Culture and Savory Food of the Republic of Georgia*. New York : HarperCollins, 1993.

Schwartz, Donald V., and Razmik Panossian. *Nationalism and History: The Politics of Nation Building in Post-Soviet Armenia, Azerbaijan and Georgia*. Toronto, Canada : University of Toronto Centre for Russian and East European Studies, 1994.

Suny, Ronald Grigor. *The Making of the Georgian Nation*. 2nd ed. Bloomington, Ind.: Indiana University Press, 1994.

———, ed. *Transcaucasia, Nationalism, and Social Change: Essays in the History of Armenia, Azerbaijan, and Georgia*. Ann Arbor: University of Michigan Press, 1996.

Germany

Davies, Norman. *A History of Europe*. New York: Oxford University Press, 1996.

Dülffer, Jost. *Nazi Germany 1933–1945: Faith and Annihilation*. London: Arnold, 1996.

Eley, Geoff, ed. *Society, Culture, and the State in Germany: 1870–1930*. Ann Arbor: The University of Michigan Press, 1996.

Friedländer, Saul. *Nazi Germany and the Jews*. Volume I - "The Years of Persecution, 1933–1939." New York: HarperCollins, 1997.

Fulbrook, Mary. *A Concise History of Germany*. Updated edition. New York: Cambridge University Press, 1994.

Gies, Frances and Joseph. *Cathedral, Forge, and Waterwheel: Technology and Invention in the Middle Ages*. New York: HarperCollins, 1994.

Kramer, Jane. *The Politics of Memory: Looking for Germany in the New Germany*. New York: Random House, 1996.

Greece

Costas, Dimitris, ed. *The Greek-Turkish Conflict in the 1990s*. New York: St. Martin's Press, 1991.

Jouganatos, George A. *The Development of the Greek Economy, 1950–1991*. Westport, Conn.: Greenwood Press, 1992.

Laisné, Claude. *Art of Ancient Greece: Sculpture, Painting, Architecture*. Paris: Terrail, 1995.

Lawrence, A.W. *Greek Architecture*. New Haven, Conn.: Yale University Press, 1996.

Legg, Kenneth R. *Modern Greece: A Civilization on the Periphery*. Boulder, Colo.: Westview Press, 1997.

Pettifer, James. *The Greeks: The Land and People Since the War*. New York: Viking, 1993.

Hungary

Bartlett, David L. *The Political Economy of Dual Transformations: Market Reform and Democratization in Hungary*. Ann Arbor, MI: University of Michigan Press, 1997.

Corrin, Chris. *Magyar Women: Hungarian Women's Lives, 1960s–1990s*. New York: St. Martin's, 1994.

Hoensch, Jorg K. *A History of Modern Hungary, 1867–1994*. 2nd ed. New York: Longman, 1996.

Litvan, Gyorgy, ed. *The Hungarian Revolution of 1956: Reform, Revolt, and Repression, 1953–1956*. Trans. Janos M. Bak and Lyman H. Legters. New York: Longman, 1996.

Szekely, Istvan P. and David M.G. Newberry, eds. *Hungary: An Economy in Transition*. Cambridge: Cambridge University Press, 1993.

Iceland

Durrenberger, E. Paul. *The Dynamics of Medieval Iceland: Political Economy and Literature*. Iowa City: University of Iowa Press, 1992.

Jochens, Jenny. *Women in Old Norse Society*. Ithaca: Cornell University Press, 1995.

Lacy, Terry G. *Ring of Seasons: Iceland: Its Culture and History*. Ann Arbor: University of Michigan Press, 1998.

Roberts. David. *Iceland*. New York: H.N. Abrams, 1990.

Ireland

Breen, Richard. *Understanding Contemporary Ireland: State, Class, and Development in the Republic of Ireland*. New York: St. Martin's Press, 1990.

Daly, Mary E. *Industrial Development and Irish National Identity, 1922–1939*. Syracuse, N.Y.: Syracuse University Press, 1992.

Hachey, Thomas E. *The Irish Experience: A Concise History*. Armonk, N.Y.: M. E. Sharpe, 1996.

Harkness, D. W. *Ireland in the Twentieth Century: Divided Island*. Hampshire, England: Macmillan Press, 1996.

MacDonagh, Oliver, et al. *Irish Culture and Nationalism, 1750–1950*. New York: St. Martin's Press, 1984.

Sawyer, Roger. *"We Are But Women": Women in Ireland's History*. New York: Routledge, 1993.

Italy

Baranski, Zygmunt G., and Robert Lumley, ed. *Culture and Conflict in Postwar Italy: Essays on Mass and Popular Culture*. New York: St. Martin Press, 1990.

Duggan, Christopher. *A Concise History of Italy*. New York: Cambridge University Press, 1994.

Furlong, Paul. *Modern Italy: Representation and Reform*. New York: Routledge, 1994.

Ginsberg, Paul. *A History of Contemporary Italy: Society and Politics, 1943–1988*. London: Penguin, 1990.

Hearder, Harry. *Italy: A Short History*. New York: Cambridge University Press, 1990.

Holmes, George. *The Oxford History of Italy*. New York: Oxford University Press, 1997.

Latvia

Gerner, Kristian, and Stefan Hedlund. *The Baltic States and the End of the Soviet Empire*. London and New York: Routledge, 1993.

Iwaskiw, Walter R. *Estonia, Latvia, and Lithuania: Country Studies.* Washington, D.C.: Federal Research Division, Library of Congress, 1996.

Lieven, Anatol. *The Baltic Revolution.* New Haven and London: Yale University Press, 1993.

Plakans, Andrejs. *The Latvians: A Short History.* Stanford, Calif: Hoover Institution Press, Stanford University, 1995.

Liechtenstein

Background Notes: Liechtenstein. Washington, D.C.: U.S. Department of State, Bureau of Public Affairs, Office of Public Communication, Editorial Division, USGPO, 1989.

Duursma, Jorri C. *Fragmentation and the International Relations of Micro-States: Self-determination and Statehood.* Cambridge: Cambridge University Press, 1996.

The Principality of Liechtenstein: A Documentary Handbook. Vaduz: Press and Information Office of the Government of the Principality of Liechtenstein, 1967.

Raton, Pierre. *Liechtenstein: History and Institutions of the Principality.* Vaduz: Liechtenstein-Verlag AG, 1970.

Lithuania

Gerner, Kristian and Stefan Hedlund. *The Baltic States and the End of the Soviet Empire.* London/New York: Routledge, 1993.

Hiden, John and Patrick Salmon. *The Baltic Nations and Europe.* London and New York: Longman, 1994.

Hiden, John. *The Baltic States and Weimar Ostpolitik.* Cambridge: Cambridge University Press, 1987.

———. *The Baltic States: Years of Dependence, 1940–1980.* Berkeley: University of California Press, 1983.

Vardys, Vytas Stanley. *Lithuania: The Rebel Nation.* Boulder, Colo.: Westview Press, 1997.

Luxembourg

Barteau, Harry C. *Historical Dictionary of Luxembourg.* Lanham, Md.: Scarecrow Press, 1996.

Clark, Peter. *Luxembourg.* New York: Routledge, 1994.

Dolibois, John. *Pattern of Circles: An Ambassador's Story.* Kent, Oh.: Kent State University Press, 1989.

Hury, Carlo. *Luxembourg.* Oxford, England: Clio Press, 1981.

Newcomer, James. *The Grand Duchy of Luxembourg: The Evolution of Nationhood.* Luxembourg: Editions Emile Borschette, 1995.

Macedonia

Billows, Richard A. *Kings and Colonists: Aspects of Macedonian Imperialism.* New York: E.J. Brill, 1995.

Danforth, Loring M. *The Macedonian Conflict: Ethnic Nationalism in a Transnational World.* Princeton, NJ: Princeton University Press, 1995.

Kofos, Evangelos. *Nationalism and Communism in Macedonia: Civil Conflict, Politics of Mutation, National Identity.* New Rochelle, N.Y.: A.D. Caratzas, 1993.

Poulton, Hugh. *Who Are the Macedonians?* Bloomington: Indiana University Press, 1995.

Shea, John. *Macedonia and Greece: The Struggle to Define a New Balkan Nation.* Jefferson, NC: McFarland, 1997.

Malta

Berg, Warren G., *Historical Dictionary of Malta.* Lanham, Md.: Scarecrow, 1995.

Caruana, Carmen M., *Education's Role in the Socioeconomic Development of Malta.* Westport Connecticut: Praeger, 1992.

Europa World Yearbook, Vol. 2, 39th Edition, 1998.

Evans, J.D., *The Prehistoric Antiquities of the Maltese Islands.* New York: Oxford University Press, 1971.

Moldova

Belarus and Moldova: Country studies. Washington: Department of the Army, 1996.

Bruchis, Michael. *The Republic of Moldavia: From the Collapse of the Soviet Empire to the Restoration of the Russian Empire.* Transl. by Laura Treptow. Boulder, Col.: East European Monographs, 1996.

Hitchins, Keith. *Rumania, 1866–1947.* Oxford: Clarendon Press, 1994.

Papacostea, Serban. *Stephen the Great: Prince of Moldavia, 1457–1504.* Transl. by Seriu Celac. Bucharest: Editura Enciclopedica, 1996.

Treptow, Kurt W. *Historical Dictionary of Romania.* Lanham, Md.: Scarecrow, 1996.

Monaco

Duursma, Jorri. *Self-determination, Statehood, and International Relations of Micro-states: The Cases of Liechtenstein, San Marino, Monaco, Andorra, and the Vatican City.* New York: Cambridge University Press, 1996.

Edwards, Anne. *The Grimaldis of Monaco.* New York: William Morrow, 1992.

Sakol, Jeannie and Caroline Latham. *About Grace: An Intimate Notebook.* Chicago: Contemporary Books, 1993.

Netherlands

Andeweg, R.B. *Dutch Government and Politics.* New York: St. Martin's, 1993.

Fuykschot, Cornelia. *Hunger in Holland: Life During the Nazi Occupation.* Amherst, N.Y.: Prometheus Books, 1995.

Israel, Jonathan Irvine. *Dutch Primacy in World Trade, 1585–1740.* New York: Oxford University Press, 1990.

Schilling, Heinz. *Religion, Political Culture, and the Emergence of Early Modern Society: Essays in German and Dutch History.* New York: E.J. Brill, 1992.

Slive, Seymour. *Dutch Painting 1600–1800.* New Haven, Conn.: Yale University Press, 1995.

Norway

Berdal, Mats R. *The United States, Norway and the Cold War 1954–60.* New York: St. Martin's, 1997.

Heide, Sigrid. *In the Hands of My Enemy: A Woman's Personal Story of World War II.* Trans. Norma Johansen. Middletown, Conn.: Southfarm Press, 1996.

Jochens, Jenny. *Women in Old Norse Society.* Ithaca, N.Y.: Cornell University Press, 1995.

Poland

Blazyca, George and Ryszard Rapacki, eds. *Poland into the 1990s: Economy and Society in Transition.* New York: St. Martin's, 1991.

Engel, David. *Facing a Holocaust: The Polish Government-in-Exile and the Jews, 1943–1945.* Chapel Hill: University of North Carolina Press, 1993.

Staar, Richard F., ed. *Transition to Democracy in Poland.* New York: St. Martin's, 1993.

Steinlauf, Michael. *Bondage to the Dead: Poland and the Memory of the Holocaust.* Syracuse, N.Y.: Syracuse University Press, 1997.

Tworzecki, Hubert. *Parties and Politics in Post-1989 Poland.* Boulder, Colo.: Westview, 1996.

Portugal

Birmingham, David, *A Concise History of Portugal*, Cambridge: Cambridge University Press, 1993.

Hermano Saraiva, Jóse, *Portugal: A Companion History.* Manchester: Carcanet Press, 1995.

Herr, Richard ed. *The New Portugal: Democracy and Europe.* Berkeley: University of California at Berkeley, 1992.

Russell-Wood, A.J.R. *A World on the Move: The Portuguese in Africa, Asia, and America, 1415–1808*, New York: St. Martin's Press, 1993.

Winius, George D. ed., *Portugal, The Pathfinder: Journeys from the Medieval toward the Modern World, 1300–ca.1600*, Madison: The Hispanic Seminary of Medieval Studies, 1995.

Romania

Bachman, Ronald D., ed. *Romania: A Country Study.* 2d ed. Washington, D.C.: Library of Congress, 1991.

Commission on Security and Cooperation in Europe. *Human Rights and Democratization in Romania.* Washington, D.C.: Commission on Security and Cooperation in Europe, 1994.

Economist Intelligence Unit. *Country Report: Romania.* London: The Economist Intelligence Unit, 1999.

Treptow, Kurt W. and Marcel Popa, eds. *Historical Dictionary of Romania.* Lanham, Md.: Scarecrow, 1996.

Russian Federation

Barner-Barry, Carol, and Cynthia A. Hody. *The Politics of Change: The Transformation of the Former Soviet Union.* New York: St. Martin's Press, 1995.

Channon, John, and Robert Hudson. *The Penguin Historical Atlas of Russia.* London: Viking, 1995.

Curtis, Glenn E., ed. *Russia: A Country Study.* Washington: Library of Congress, 1998.

Daniels, Robert V., ed. *The Stalin Revolution: Foundations of the Totalitarian Era.* 4th ed. Boston: Houghton Mifflin, 1997.

Fitzpatrick, Sheila. *The Russian Revolution.* New York: Oxford University Press, 1994.

MacKenzie, David, and Michael W. Curran. *A History of Russia, the Soviet Union, and Beyond.* 5th ed. Belmont, CA: West/Wadsworth, 1999.

Raymond, Boris, and Paul Duffy. *Historical Dictionary of Russia.* Lanham, Md.: Scarecrow, 1998.

San Marino

Catling, Christopher. *Umbria, the Marches and San Marino.* Lincolnwood, Ill.: Passport Books, 1994.

Duursma, Jorri C. *Fragmentation and the International Relations of Micro-States.* Cambridge: Cambridge University Press, 1996.

United States Bureau of Public Affairs; Background Notes: *San Marino.* Washington

World Reference Atlas: *San Marino.* London: Darling Kindersley Publishing, Inc., 1998.

Slovakia

Goldman, Minton F. *Slovakia Since Independence: A Struggle for Democracy.* Westport, Conn.: Praeger, 1999.

Jelinek, Yeshayahu A. *The Parish Republic: Hlinka's Slovak People's Party: 1939–1945.* New York: Columbia University Press, 1976.

Kirshbaum, Stanislav J. *A History of Slovakia: The Struggle for Survival.* New York: St. Martin's Press, 1995.

Leff, Carol Skalnik. *The Czech and Slovak Republics: Nation Versus State.* Boulder, Colo.:Westview Press, 1997.

Slovenia

Cohen, Lenard. Broken Bonds: *The Disintegration of Yugoslavia.* Boulder, Colo.: Westview Press, 1993.

Fink-Hafner, Danica, and John R. Robbins, eds. *Making a New Nation: The Formation of Slovenia.* Brookfield, Vt. Aldershot: Dartmouth Publishing, 1997.

Owen, David. *Balkan Odyssey.* New York: Harcourt Brace, 1995.

Rogel, Carole. *The Breakup of Yugoslavia and the War in Bosnia.* Westport, Conn.: Greenwood Press, 1998.

Silber, Laura, and Allan Little. *Yugoslavia: Death of a Nation.* New York: TV Books, 1996.

Zimmerman, Warren. *Origins of a Catastrophe: Yugoslavia and its destroyers—America's last ambassador tells what happened and why.* New York: Times Books, 1996.

Spain

Cantarino, Vicente. *Civilización y Cultura de España,* 3rd. ed., Englewood Cliffs, N.J.: Prentice Hall, l995.

Hopper, John. *The New Spaniards.* Suffolk, Eng.: Penguin, l995.

Jordan, Barry. *Writings and Politics in the Franco's Spain.* London, Eng.: Routledge, 1990.

Leahy, Philippa. *Discovering Spain.* New York: Crestwood House, 1993.

Wernick, Robert. "For Whom the Bell Tolled. (the Spanish Civil War)." *Smithsonian.* April 1998, vol. 28, no. 1, pp. 110+.

Sweden

Palmer, Alan. *Bernadotte: Napoleon's Marshal, Sweden's King.* London: Murray, 1990.

Roberts, Michael. *From Oxenstierna to Charles XII: Four Studies.* New York: Cambridge University Press, 1991.

Rothstein, Bo. *The Social Democratic State: The Swedish Model and the Bureaucratic Problem of Social Reforms.* Pittsburgh: University of Pittsburgh, 1996.

Scobbie, Irene. *Historical Dictionary of Sweden.* Metuchen, N.J.: Scarecrow Press, 1995.

Switzerland

Bacchetta, Philippe, and Walter Wasserfallen, ed. *Economic Policy in Switzerland.* New York: St. Martin's Press, 1997.

Eu-Wong, Shirley. *Culture Shock!: Switzerland.* Portland, Or.: Graphic Arts Center Pub. Co., 1996.

Hilowitz, Janet Eve, ed. *Switzerland in Perspective.* New York: Greenwood Press, 1990.

New, Mitya. *Switzerland Unwrapped: Exposing the Myths.* New York: I.B. Tauris, 1997.

Steinberg, Jonathan. *Why Switzerland?* 2nd ed. Cambridge: Cambridge University Press, 1996.

Turkey

Fodor's Turkey. New York: Fodor's Travel Publications, 1999.

Mitchell, Stephen. *Anatolia: Land, Men, and Gods in Asia Minor.* New York: Oxford University Press, 1993.

Stoneman, Richard. *A Traveller's History of Turkey.* New York: Interlink Books, 1998.

Time-Life Books. *Anatolia: Cauldron of Cultures.* Alexandria, Va.: Time-Life Books, 1995.

Ukraine

Hosking, Geoffrey A., ed. *Church, Nation and State in Russia and Ukraine.* New York: St. Martin's Press, 1991.

Kuzio, Taras. *Ukraine under Kuchma: Political Reform, Economic Transformation and Security Policy in Independent Ukraine.* New York: St. Martin's Press, 1997.

Shen, Raphael. *Ukraine's Economic Reform: Obstacles, Errors, Lessons.* Westport, Conn: Praeger, 1996.

Subtelny, Orest. *Ukraine: A History.* Toronto: University of Toronto Press, 1994.

Wilson, Andrew. *Ukrainian Nationalism in the 1990s: A Minority Faith.* New York: Cambridge University Press, 1997.

United Kingdom

Abrams, M. H., ed. *The Norton Anthology of English Literature.* 2 vols. 6th ed. New York : Norton, 1993.

Cook, Chris. *The Longman Handbook of Modern British History, 1714–1995.* 3rd ed. New York: Longman, 1996.

Delderfield, Eric F. *Kings & Queens of England & Great Britain.* New York: Facts on File, 1990.

Figes, Kate. *Because of Her Sex: The Myth of Equality for Women in Britain.* London: Macmillan, 1994.

Foster, R.F., ed. *The Oxford History of Ireland.* New York: Oxford University Press, 1992.

Glynn, Sean. *Modern Britain: An Economic and Social History.* New York: Routledge, 1996.

Jenkins, Philip. *A History of Modern Wales, 1536–1990.* New York: Longman, 1992.

Judd, Denis. *Empire: The British Imperial Experience from 1765 to the Present.* London : HarperCollins, 1996.

The Oxford History of Britain. New York: Oxford University Press, 1992.

Powell, David. *British Politics and the Labour Question, 1868–1990.* New York: St. Martin's, 1992.

Vatican

Accattoli, Luigi. *Life in the Vatican with John Paul II.* Trans. Marguerite Shore. New York : Universe, 1998.

Reese, Thomas J. *Inside the Vatican: The Politics and Organization of the Catholic Church.* Cambridge, Mass.: Harvard University Press, 1996.

Rosa, Peter de. *Vicars of Christ: The Dark Side of the Papacy.* New York: Crown Publishers, 1988.

Volpini, Valerio. *Vatican City: Art, Architecture, and History.* New York, Portland House: Distributed by Crown Publishers, 1986.

Yugoslavia

Cohen, Lenard J. *Broken Bonds: Yugoslavia's Disintegration and Balkan Politics in Transition.* Boulder, CO: Westview, 1995.

Denitch, Bogdan Denis. *Ethnic Nationalism: The Tragic Death of Yugoslavia.* Minneapolis: University of Minnesota Press, 1996.

Dyker, David A., and Ivan Vejvoda, ed. *Yugoslavia and After: A Study in Fragmentation, Despair and Rebirth.* New York: Longman, 1996.

Lampe, John R. *Yugoslavia as History: Twice There Was a Country.* New York: Cambridge University Press, 1996.

Pavkovic, Aleksandr. *The Fragmentation of Yugoslavia: Nationalism in a Multinational State.* New York: St. Martin's Press, 1997.

Ramet, Sbrina P. *Balkan Babel: The Disintegration of Yugoslavia from the Death of Tito to Ethnic War.* Boulder, Colo.: Westview, 1996.

Index

H

S

REFERENCE